BIOLOGY
HOW LIFE WORKS

Vol. 2: From Organisms to the Environment, and Evolution

James Morris
BRANDEIS UNIVERSITY

Daniel Hartl
HARVARD UNIVERSITY

Andrew Knoll
HARVARD UNIVERSITY

Robert Lue
HARVARD UNIVERSITY

ANDREW BERRY, ANDREW BIEWENER, BRIAN FARRELL, N. MICHELE HOLBROOK, NAOMI PIERCE, ALAIN VIEL
HARVARD UNIVERSITY

W.H. Freeman and Company

A Macmillan Higher Education Company

PUBLISHER Susan Winslow
LEAD DEVELOPMENTAL EDITOR Lisa Samols
SENIOR DEVELOPMENTAL EDITOR Susan Moran
DEVELOPMENTAL EDITOR Erica Pantages Frost
EDITORIAL ASSISTANTS Yassamine Ebadat, Jane Taylor
REVIEW COORDINATOR Donna Brodman
PROJECT MANAGER Karen Misler
ART MANAGER Carolyn Deacy
EDITORIAL RESEARCH AND DEVELOPMENT Shannon Howard
MARKET DEVELOPMENT MANAGER Lindsey Veautour
ASSOCIATE DIRECTOR OF MARKETING Debbie Clare
LEAD ASSESSMENT AUTHOR Melissa Michael
ASSESSMENT AUTHORS/TEAM LEADERS Mark Hens, John Merrill, Randall Phillis, Debra Pires

ART AND MEDIA DIRECTOR Robert Lue, Harvard University
MANAGER OF DIGITAL DEVELOPMENT Amanda Dunning
SENIOR DEVELOPMENT EDITOR FOR TEACHING & LEARNING STRATEGIES Elaine Palucki
SENIOR MEDIA PRODUCER Keri Fowler
PROJECT EDITOR Robert Errera
MANUSCRIPT EDITOR Nancy Brooks
DESIGN MGMT. design
SENIOR ILLUSTRATION COORDINATOR Bill Page
ILLUSTRATIONS Imagineering
CREATIVE DIRECTOR Mark Mykytiuk, Imagineering
ART DIRECTOR Diana Blume
LAYOUT ARTIST Tom Carling, Carling Design Inc.
PHOTO EDITOR Christine Buese
PHOTO RESEARCHERS Jacquelin Wong and Deborah Anderson
PRODUCTION MANAGER Paul Rohloff
COMPOSITION MPS Limited
PRINTING AND BINDING Quad Graphics–Versailles

Library of Congress Control Number: 2012951531
ISBN-13: 978-1-4641-0428-2
ISBN-10: 1-4641-0428-X

© 2013 by W. H. Freeman and Company. All rights Reserved.

Printed in the United States of America

First printing

Macmillan
W. H. Freeman and Company
41 Madison Avenue
New York, NY 10010
Houndmills, Basingstoke RG21 6XS, England
www.whfreeman.com

*To all who are curious
about life and how it works.*

ABOUT THE AUTHORS

James R. Morris is Associate Professor in the Biology Department at Brandeis University. He teaches a wide variety of courses for majors and non-majors in evolution, genetics, genomics, anatomy, and health sciences. In addition, he teaches a first-year seminar focusing on Darwin's *On the Origin of Species*. He is the recipient of numerous teaching awards. His research focuses on the rapidly growing field of epigenetics, using the fruit fly *Drosophila melanogaster* as a model organism. He currently pursues this research with undergraduates in order to give them the opportunity to do genuine, laboratory-based research early in their scientific careers. Dr. Morris received a PhD in genetics from Harvard University and an MD from Harvard Medical School. He was a Junior Fellow in the Society of Fellows at Harvard University, has given talks to the public on current science at the Museum of Science in Boston, and works on promoting public understanding of personal genetics and genomics.

Daniel L. Hartl is Higgins Professor of Biology in the Department of Organismic and Evolutionary Biology at Harvard University. He has taught highly popular courses in genetics and evolution at both the introductory and advanced levels. His lab studies molecular evolutionary genetics and population genetics and genomics. Dr. Hartl is the recipient of the Samuel Weiner Outstanding Scholar Award and the Medal of the Stazione Zoologica Anton Dohrn, Naples. He is a member of the National Academy of Sciences and the American Academy of Arts and Sciences. He has served as President of the Genetics Society of America and President of the Society for Molecular Biology and Evolution. Dr. Hartl's PhD was awarded by the University of Wisconsin, and he did postdoctoral studies at the University of California, Berkeley. Before joining the Harvard faculty, he served on the faculties of the University of Minnesota, Purdue University, and Washington University Medical School. In addition to publishing more than 350 scientific articles, Dr. Hartl has authored or coauthored 30 books.

Andrew H. Knoll is Fisher Professor of Natural History in the Department of Organismic and Evolutionary Biology at Harvard University. He is also Professor of Earth and Planetary Sciences. Dr. Knoll teaches introductory courses in both departments. His research focuses on the early evolution of life, Precambrian environmental history, and the interconnections between the two. He has also worked extensively on the early evolution of animals, mass extinction, and plant evolution. He currently serves on the science team for NASA's mission to Mars. Dr. Knoll received the Phi Beta Kappa Book Award in Science for *Life on a Young Planet*. Other honors include the Paleontological Society Medal and Wollaston Medal of the Geological Society, London. He is a member of the National Academy of Sciences, the American Academy of Arts and Sciences, and the American Philosophical Society. He received his PhD from Harvard University and then taught at Oberlin College before returning to Harvard.

Robert A. Lue is Professor in the Department of Molecular and Cellular Biology and Director of Life Science Education at Harvard University. He regularly teaches courses in Harvard's first-year Life Sciences program and upper-level courses in cell biology. He has a long-standing commitment to interdisciplinary teaching and research, and chaired the faculty committee that developed an integrated science course to serve science majors and premedical students. Dr. Lue has also developed award-winning multimedia, including the animation "The Inner Life of the Cell." He has coauthored undergraduate biology textbooks and chaired education conferences on college biology for the National Academies and the National Science Foundation and on diversity in science for the Howard Hughes Medical Institute and the National Institutes of Health. He also founded and directs a Harvard Life Sciences outreach program that serves over 50 high schools. He received his PhD from Harvard University.

Andrew Berry is Lecturer in the Department of Organismic and Evolutionary Biology and an undergraduate advisor in the Life Sciences at Harvard University. He teaches courses in Harvard's first-year Life Sciences program, as well as courses on evolution and Darwin. His research interests are in evolutionary biology and the history of science. He has coauthored two books: *Infinite Tropics*, a collection of the writings of Alfred Russel Wallace, and *DNA: The Secret of Life*, which is part history, part exploration of the controversies swirling around DNA-based technology.

Andrew A. Biewener is Charles P. Lyman Professor of Biology in the Department of Organismic and Evolutionary Biology at Harvard University and Director of the Concord Field Station. He teaches both introductory and advanced courses in anatomy, physiology, and biomechanics. His research focuses on the comparative biomechanics and neuromuscular control of mammalian and avian locomotion, with relevance to biorobotics. He is currently Deputy Editor-in-Chief for the *Journal of Experimental Biology*. He also served as President of the American Society of Biomechanics.

Brian D. Farrell is Professor of Biology in the Department of Organismic and Evolutionary Biology at Harvard University and Curator of Entomology in the Museum of Comparative Zoology. He has collaborated with Los Niños de Leonardo y Meredith in the Dominican Republic to teach children about native insects, and participates in an All Taxa Biodiversity Inventory of the Boston Harbor Islands National Recreation Area. His research focuses on the interplay of adaptation and historical contingency in species diversification, particularly of beetles. In 2011–2012, Dr. Farrell was a Fulbright Scholar to the Universidad Autonoma de Santo Domingo.

N. Michele Holbrook is Charles Bullard Professor of Forestry in the Department of Organismic and Evolutionary Biology at Harvard University. She teaches an introductory course on biodiversity as well as advanced courses in plant biology. She studies the physics and physiology of vascular transport in plants with the goal of understanding how constraints on the movement of water and solutes between soil and leaves influences ecological and evolutionary processes.

Naomi E. Pierce is Hessel Professor of Biology in the Department of Organismic and Evolutionary Biology at Harvard University and Curator of Lepidoptera in the Museum of Comparative Zoology. She studies and teaches animal behavior and behavioral ecology. Her lab focuses on the ecology of species interactions, such as insect–host plant associations, and on the life-history evolution and systematics of Lepidoptera. She has also been involved in reconstructing the evolutionary tree of life of insects such as ants, bees, and butterflies.

Alain Viel is Director of Undergraduate Research and Senior Lecturer in the Department of Molecular and Cellular Biology at Harvard University. He teaches research-based courses as well as courses in molecular biology and biochemistry. He is a founding member of BioVisions, a collaboration between scientists, teaching faculty, students, and multimedia professionals that focuses on science visualization. Dr. Viel worked with his colleague and *Biology: How Life Works* coauthor Robert Lue on the animation "The Inner Life of the Cell."

Biology: How Life Works Assessment Authors

Melissa Michael, Lead Assessment Author, is Director for Core Curriculum and Assistant Director for Undergraduate Instruction for the School of Molecular and Cellular Biology at the University of Illinois at Urbana-Champaign. A cell biologist, she primarily focuses on the continuing development of the School's undergraduate curricula. She is currently engaged in several projects aimed at improving instruction and assessment at the course and program levels. She continues to work in several different arenas to improve undergraduate biology education.

Mark Hens is Associate Professor of Biology at the University of North Carolina at Greensboro. He has taught introductory biology at this institution since 1996. He is the director of his department's Introductory Biology Program and is chair of the university's General Education Council. In these administrative capacities, he leads efforts on his campus to establish learning objectives and develop assessment tools. Dr. Hens is a National Academies Education Mentor at the National Academies/HHMI Summer Institute for Undergraduate Education in Biology. He also serves on the advisory board of an NSF-funded project focused on the assessment of student learning in college science curricula.

John Merrill is Director of the Biological Sciences Program in the College of Natural Science at Michigan State University. This program administers the core biology course sequence required for all science majors. In recent years, Dr. Merrill has focused his research on teaching and learning. With the support of several NSF grants, he is exploring innovative classroom interventions coupled to enhanced assessment. A particularly active area is the use of computers to analyze student's written responses to conceptual assessment questions, with the goal of making it easier to use open-response questions in large-enrollment classes.

Randall Phillis is Associate Professor of Biology at the University of Massachusetts Amherst. He has taught in the majors introductory biology course at this institution for 19 years and is a National Academies Education Mentor in the Life Sciences. With help from the PEW Center for Academic Transformation (1999), he has been instrumental in transforming the introductory biology course to an active learning format that makes use of classroom communication systems. He also participates in an NSF-funded project to design model-based reasoning assessment tools for use in class and on exams. These tools are being designed to develop and evaluate student scientific reasoning skills, with a focus on topics in introductory biology.

Debra Pires is an Academic Administrator at the University of California, Los Angeles. She teaches the introductory courses in the Life Sciences Core Curriculum. Her research focuses on creating assessment tools to evaluate how well students understand concepts taught in the introductory courses and how well they retain those concepts during their time at UCLA. Student Learning Outcome (SLO)-centered assessments have become a major component of the introductory curriculum, and workshops with faculty in two departments have begun to help instructors develop rubrics and assessment strategies that are aligned with the goals of the long-term study.

BRIEF CONTENTS

PART 1   FROM CELLS TO ORGANISMS

CHAPTER 1	**LIFE** Chemical, Cellular, and Evolutionary Foundations	1-1
? **CASE 1**	*The First Cell: Life's Origins*	C1-1
CHAPTER 2	**THE MOLECULES OF LIFE**	2-1
CHAPTER 3	**NUCLEIC ACIDS AND THE ENCODING OF BIOLOGICAL INFORMATION**	3-1
CHAPTER 4	**TRANSLATION AND PROTEIN STRUCTURE**	4-1
CHAPTER 5	**ORGANIZING PRINCIPLES** Lipids, Membranes, and Cell Compartments	5-1
CHAPTER 6	**MAKING LIFE WORK** Capturing and Using Energy	6-1
CHAPTER 7	**CELLULAR RESPIRATION** Harvesting Energy from Carbohydrates and Other Fuel Molecules	7-1
CHAPTER 8	**PHOTOSYNTHESIS** Using Sunlight to Build Carbohydrates	8-1
? **CASE 2**	*Cancer: When Good Cells Go Bad*	C2-2
CHAPTER 9	**CELL COMMUNICATION**	9-1
CHAPTER 10	**CELL FORM AND FUNCTION** Cytoskeleton, Cellular Junctions, and Extracellular Matrix	10-1
CHAPTER 11	**CELL DIVISION** Variations, Regulation, and Cancer	11-1
? **CASE 3**	*You, From A to T: Your Personal Genome*	C3-2
CHAPTER 12	**DNA REPLICATION AND MANIPULATION**	12-1
CHAPTER 13	**GENOMES**	13-1
CHAPTER 14	**MUTATION AND DNA REPAIR**	14-1
CHAPTER 15	**GENETIC VARIATION**	15-1
CHAPTER 16	**MENDELIAN INHERITANCE**	16-1
CHAPTER 17	**BEYOND MENDEL** Sex Chromosomes, Linkage, and Organelles	17-1
CHAPTER 18	**THE GENETIC AND ENVIRONMENTAL BASIS OF COMPLEX TRAITS**	18-1
CHAPTER 19	**GENETIC AND EPIGENETIC REGULATION**	19-1
CHAPTER 20	**GENES AND DEVELOPMENT**	20-1
? **CASE 4**	*Malaria: Co-Evolution of Humans and a Parasite*	C4-2
CHAPTER 21	**EVOLUTION** How Genotypes and Phenotypes Change over Time	21-1
CHAPTER 22	**SPECIES AND SPECIATION**	22-1
CHAPTER 23	**EVOLUTIONARY PATTERNS** Phylogeny and Fossils	23-1
CHAPTER 24	**HUMAN ORIGINS AND EVOLUTION**	24-1

PART 2 FROM ORGANISMS TO THE ENVIRONMENT

CHAPTER 25	**CYCLING CARBON**	25-1
? CASE 5	*The Human Microbiome: Diversity Within*	C5-1
CHAPTER 26	**BACTERIA AND ARCHAEA**	26-1
CHAPTER 27	**EUKARYOTIC CELLS** Origins and Diversity	27-1
CHAPTER 28	**BEING MULTICELLULAR**	28-1
? CASE 6	*Agriculture: Feeding a Growing Population*	C6-2
CHAPTER 29	**PLANT STRUCTURE AND FUNCTION** Moving Photosynthesis onto Land	29-1
CHAPTER 30	**PLANT REPRODUCTION** Finding Mates and Dispersing Young	30-1
CHAPTER 31	**PLANT GROWTH AND DEVELOPMENT** Building the Plant Body	31-1
CHAPTER 32	**PLANT DEFENSE** Keeping the World Green	32-1
CHAPTER 33	**PLANT DIVERSITY**	33-1
CHAPTER 34	**FUNGI** Structure, Function, and Diversity	34-1
? CASE 7	*Predator–Prey: A Game of Life and Death*	C7-1
CHAPTER 35	**ANIMAL NERVOUS SYSTEMS**	35-1
CHAPTER 36	**ANIMAL SENSORY SYSTEMS AND BRAIN FUNCTION**	36-1
CHAPTER 37	**ANIMAL MOVEMENT** Muscles and Skeletons	37-1
CHAPTER 38	**ANIMAL ENDOCRINE SYSTEMS**	38-1
CHAPTER 39	**ANIMAL CARDIOVASCULAR AND RESPIRATORY SYSTEMS**	39-1
CHAPTER 40	**ANIMAL METABOLISM, NUTRITION, AND DIGESTION**	40-1
CHAPTER 41	**ANIMAL RENAL SYSTEMS** Water and Waste	41-1
CHAPTER 42	**ANIMAL REPRODUCTION AND DEVELOPMENT**	42-1
CHAPTER 43	**ANIMAL IMMUNE SYSTEMS**	43-1
? CASE 8	*Biodiversity Hotspots: Rain Forests and Coral Reefs*	C8-1
CHAPTER 44	**ANIMAL DIVERSITY**	44-1
CHAPTER 45	**ANIMAL BEHAVIOR**	45-1
CHAPTER 46	**POPULATION ECOLOGY**	46-1
CHAPTER 47	**SPECIES INTERACTIONS, COMMUNITIES, AND ECOSYSTEMS**	47-1
CHAPTER 48	**THE ANTHROPOCENE** Humans as a Planetary Force	48-1
	QUICK CHECK ANSWERS	Q-1
	GLOSSARY	G-1
	CREDITS/SOURCES	CS-1
	INDEX	I-1

VISION AND STORY OF

Dear Students and Instructors,

We wrote this book in recognition of recent and exciting changes in biology, education, and technology. There was a time when introductory biology could cover all of biology over the course of a single year. This is no longer possible. The amount of scientific information has grown exponentially, necessitating that we, as teachers, rethink the role of introductory biology and the resources that support it. One goal remains paramount: to help students think like biologists. To think like a biologist means understanding key concepts that span all of biology. It means being able to communicate in the shared language of biologists. It means recognizing the powerful ability of evolution to explain both the unity and diversity of life. It means thinking about how biological research can help solve some of the world's most pressing issues, from cancer to infectious diseases to biodiversity loss to climate change.

We have also noticed a change in the way biological problems are approached. We now have a "parts list" of genes and proteins for how life works, and many scientists today are focused on how the parts work together. As a result, we can no longer divide information into discrete topics. To prepare students for science as it is currently practiced, we must integrate concepts from different areas of biology as well as from other scientific disciplines.

What is particularly exciting for us as teachers is that the remarkable changes in the science of biology are paralleled by a new appreciation for and understanding of how students learn. There is now good evidence that teaching students only by lecturing does not lead to mastery of core concepts. Lecturing alone does not help develop the scientific skills and habits of mind that students need to become successful scientists and health-care workers or thoughtful, scientifically informed citizens. Students learn most effectively when they are actively involved in their learning and construct their own knowledge through a combination of lectures, problem solving, hands-on experiences, and collaborative work.

At the same time, technology is transforming how and where students access information. The Internet provides all kinds of information at a click. There is no need for a modern textbook to be a reference book. What, then, is the role of a textbook? A textbook needs to be selective, help students see connections between seemingly disparate topics, and make the material engaging and relevant. Technology is also making possible new and unprecedented ways to visualize biological processes and provide interactive ways for students to learn.

To support 21st-century student learning and instructor teaching, we feel that it is time to rethink what takes place both in and out of the introductory biology classroom and to reimagine the resources that can best support these efforts. *Biology: How Life Works* provides an integrated set of resources to engage students, encourage critical thinking, help students make connections, and provide a framework for further studies.

Sincerely,
The *Biology: How Life Works* author team

BIOLOGY: HOW LIFE WORKS

Rethinking Biology

For the *Biology: How Life Works* team, it has been an exciting experience to reimagine what a resource for today's students should look like. We began with the question, "What do we want students to understand and apply at the end of the course?" Once we decided where we wanted students to end up, we started asking other questions: What content should be covered? How should it be organized? How should it be delivered—in the narrative, a figure, an animation? What questions should we ask students to gauge their understanding? What questions should we ask to help students work toward that understanding? Through this process of questioning and answering, one point became clear—we'd have to start from scratch and build *Biology: How Life Works* from the ground up.

Rethinking biology means rethinking the text, visual program, and assessment.

Rethinking biology required rethinking the way a textbook is created. Ordinarily, textbooks are developed by first writing chapters, then making decisions about art and images, and finally, once the book is complete, assembling a test bank and ancillary media. This process dramatically limits integration across resources and reduces art, media, and assessments to supplements rather than essential resources for student learning.

Biology: How Life Works is the first project to develop three pillars of learning—the text, visual program, and assessment—at the same time. These three pillars are all tied to the same set of core concepts, share a common language, and use the same visual palette. In this way, the visual program and assessments are integral parts of student learning, rather than accessories to the text. In addition, every concept is conveyed and explored in multiple ways, allowing for authentic learning.

BIOLOGY: HOW LIFE WORKS

> I think the best selling point is that the text focuses on helping students make connections between the subfields of biology.
> — Cindee Giffen, Instructor
> University of Wisconsin, Madison

> I like its simplicity and focus on key concepts using relevant examples. Linkage to previous chapters was one of the strongest parts.
> — David Hicks, Instructor
> The University of Texas at Brownsville

> I really like the streamlined approach and emphasis on ideas and concepts rather than details and facts.
> — Scott Solomon, Instructor
> Rice University

> I like the evolution emphasis throughout. If evolution really is the central organizing principle in biology, it's about time somebody wrote an introductory textbook that reflects that.
> — David Lampe, Instructor
> Duquesne University

> I think it is more readable than the text we currently use and it does a better job of integrating the theme of evolution.
> — Kathryn Craven, Instructor
> Armstrong Atlantic University

Rethinking biology means rethinking THE TEXT

Biology: How Life Works includes a text that is uniquely *integrated*, *selective*, and *thematic*.

INTEGRATED

Textbooks commonly present biology as a series of minimally related chapters. This style of presentation lends itself to memorization and offers students little guidance on how concepts connect to one another and to the bigger picture. *Biology: How Life Works* moves away from this model and toward an integrated approach. Across the book, we present chemistry in context, and cover structure and function together. We introduce the flow of information in a cell where it makes the most conceptual sense and use cases as a framework for connecting and assimilating information.

SELECTIVE

With the ever-increasing scope of biology, it is unrealistic to expect the majors course or a textbook to cover everything. From the start, we envisioned *Biology: How Life Works* not as a reference book for all of biology, but as a resource focused on foundational concepts, terms, and experiments. We explain fundamental topics carefully, with an appropriate amount of supporting detail. This allows students to more easily identify, understand, and apply critical concepts. In this way, students will leave an introductory biology class with a framework on which to build.

THEMATIC

We wrote *Biology: How Life Works* with six themes in mind. Deciding on these themes in advance helped us make decisions about which concepts to include and how to organize them. Introduced in Chapter 1 and revisited throughout the text, the themes provide a framework that helps students see biology as a set of connected concepts. In particular, we emphasize the theme of evolution for its ability to explain and predict so many patterns in biology.

1. We learn how life works by applying the scientific method, which involves making observations, generating hypotheses, and testing hypotheses through experiment and observation.
2. Life works according to fundamental principles of chemistry and physics. All organisms share a limited number of molecules and chemical processes.
3. All of the chemical and physical functions of life are packaged within cells. Multicellular organisms function by the differentiation and coordinated operation of many cells.
4. Both the features that organisms share and those that set them apart are explained by evolution. Variation exists within as well as between species.
5. Organisms interact in nature, with basic features of anatomy, physiology, and behavior shaping the ecological systems that sustain life.
6. Humans have emerged as major agents in ecology and evolution. Our future welfare depends, in part, on improving our knowledge of how life works.

BIOLOGY: HOW LIFE WORKS

> I think the figures are much simpler for beginning students to understand. I found all of the figures in my chapter to be clear, concise, and distilled into the most relevant facts necessary to illustrate the concepts.
> — Dale Casamatta, Instructor
> University of North Florida

> The figures in this chapter [Chapter 22: Species and Speciation] are particularly good. These are not the only great figures, but they are among my favorites for how they demonstrate a concept so easily that might otherwise require a couple of paragraphs of text.
> — Matt Brewer, Instructor
> Georgia State University

> My initial reaction was "Wow." It [animation on gene expression] helped me visualise the spatial relationships associated with information flow at the cellular level, and I think it is thus likely to really help undergraduates.
> — Dave Kubien, Instructor
> University of New Brunswick

Rethinking biology means rethinking
THE VISUAL PROGRAM

The *Biology: How Life Works* visual program—all art and visual media—is an *integrated, engaging, visual framework* for understanding and connecting concepts.

INTEGRATED

Just as we decided on a consistent vocabulary to explain concepts and processes, we also created a consistent "visual language." This means that across *Biology: How Life Works*—whether students are looking at a figure in the book, watching an animation, or interacting with a simulation—they see a consistent use of color, shapes, and design. Having a coherent, integrated visual program allows students to recognize concepts they have already encountered, and assimilate new information.

ENGAGING

We have a shared goal with instructors—we want students to *want* to learn more. Cognitive science tells us visualization is tremendously effective at triggering student interest and passion. With that in mind, the *Biology: How Life Works* team committed to designing and developing a visual program with the same attention and care that goes into text development. Every image—still and in motion—engages students by being vibrant, clear, and approachable. The result is a visual environment that pulls students in, deepens their interest, and helps them see a world of biological processes.

VISUAL FRAMEWORK

Scientists often build a contextual picture, or visual framework, in their mind upon which they hang facts and connect ideas. To help students think like biologists, our visual program deliberately provides this type of framework. Individual figures present foundational concepts; Visual Synthesis figures tie together multiple concepts across multiple chapters; animations bring these figures to life, allowing students to explore concepts in space and time; and simulations have students interact with the concepts. Collectively, this visual framework gives students a way to contextualize information—to move seamlessly back and forth between the big picture and the details, from the basic to the complex.

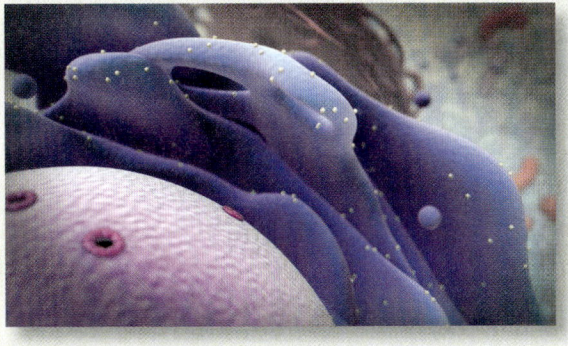

xi

BIOLOGY: HOW LIFE WORKS

> The pre-class questions are basic and yet not simple repeat-from-reading, and the other questions fit comfortably at the level where I think they'll challenge my students but not be so difficult they're unreachable for them. I particularly like that many of these questions involve interpreting data, or analyzing a case story.
> — ANNE CASPER
> EASTERN MICHIGAN UNIVERSITY

> These questions are great in that they really require a greater depth of understanding and ability to apply concepts than I've seen in any other assessment packages. There's a clear connection between the pedagogical approach of the textbook and the assessment materials.
> — SONJA PYOTT, INSTRUCTOR
> UNIVERSITY OF NORTH CAROLINA AT WILMINGTON

> The questions in these assessments seem to be superior, since they require higher-order thinking and application of knowledge, rather than just simple factual recall.
> — PEGGY ROLFSEN, INSTRUCTOR
> CINCINNATI STATE TECHNICAL AND COMMUNITY COLLEGE

> There are more suggestions about how to use various types of assessments—my current text gives examples of questions but it is left to me to imagine how I might use them.
> — LISA ELFRING
> UNIVERSITY OF ARIZONA

Rethinking biology means rethinking ASSESSMENT

Biology: How Life Works represents a groundbreaking departure from traditional assessment materials. Written by leaders in science education in collaboration with the authors, our assessments are fully *integrated* with the text and visual program, span the *full range of Bloom's Taxonomy,* and are properly *aligned*.

INTEGRATED

If assessment is important, it cannot be ancillary—it must be integral to the learning process. Each time an instructor asks a student to engage with *Biology: How Life Works*—whether it is reading a chapter, watching an animation, or working through an experiment—the opportunity to assess that experience exists. This unprecedented level of integration results from developing assessments alongside the text and visual program. This parallel development also allows for consistency of language and focus, helping students connect what they have read with the questions they are being asked to answer.

RANGE

For the first time, instructors in majors biology have access to a set of thoughtfully developed, peer-reviewed formative and summative assessments. Most test banks consist entirely of low-order, recall questions. Consequently, instructors wanting to teach and test at a range of levels are left with the time-consuming task of creating their own higher-order questions. The *Biology: How Life Works* assessment team has done the "heavy lifting" of question writing for instructors—providing assessments that span everything from recall to synthesis. They are designed to be used in a range of settings (pre-class, in-class, post-class, and exam) and come in a variety of formats (multiple choice, multiple true/false, free response). In addition to questions, our assessments include in-class activities and interactive, online exercises.

ALIGNED

Traditionally, assessment questions are provided as test banks, organized by chapter but not as a series designed to work together. We believe questions aren't just for testing—they are for teaching. To help instructors assess students in more meaningful ways, we wrote and organized sets of questions we call *progressions*. Progressions include formative and summative assessment—reading comprehension questions, in-class activities, post-class assignments, and exam questions—all properly aligned with the text's core concepts. Used in sequence, questions within a progression provide a connected learning path for students and a suggested teaching path for instructors.

BIOLOGY: HOW LIFE WORKS

AUTHORING *BIOLOGY: HOW LIFE WORKS*

Biology: How Life Works is authored by a team of nationally recognized scientists and educators.

James Morris, Lead Author, Brandeis University
Daniel Hartl, Lead Author, Harvard University
Andrew Knoll, Lead Author, Harvard University
Robert Lue, Lead Author and Visual Program Director, Harvard University
Andrew Berry, Author, Harvard University
Andrew Biewener, Author, Harvard University
Brian Farrell, Author, Harvard University
N. Michele Holbrook, Author, Harvard University
Naomi Pierce, Author, Harvard University
Alain Viel, Author, Harvard University

Melissa Michael, Lead Assessment Author, University of Illinois at Urbana-Champaign
Mark Hens, Assessment Author, University of North Carolina at Greensboro
John Merrill, Assessment Author, Michigan State University
Randall Phillis, Assessment Author, University of Massachusetts Amherst
Debra Pires, Assessment Author, University of California, Los Angeles

The mission of *Biology: How Life Works* is not to cover all of biology, but to provide a proper introduction. This selective approach requires a large team of educators who are experts in their respective fields. Because the authors possess an intimate knowledge of their fields, they were able to step back and consider where their fields have been, what questions are currently being asked, and where they are heading. It is only with a *collective* perspective that *Biology: How Life Works* is able to present contemporary biology in a complete but selective manner.

The multi-author approach was also necessary for the amount of writing, design, development, and collaboration this project required. With the text, visual program, and assessments all being developed in parallel, *Biology: How Life Works* needed more than one author. Additionally, the level of integration, creativity, and new ideas present in the three pillars of *Biology: How Life Works* was largely achieved by having a community of authors who were able to dialogue, discuss, and debate.

At the same time, *Biology: How Life Works* has a single voice. This is possible because every chapter, image, animation, and assessment flowed through one author, James Morris. In this way, it benefits from multi-author expertise while telling a single, cohesive story.

In sum, the size of the author team is purposeful and directly supports the goals of the project.

> I think it's time for an innovative approach to a biology text for majors. Your team of authors has the ideal combination of research and teaching experience to be successful in developing a superior text that reflects the contemporary biological sciences.
> — ERIK SKULLY
> TOWSON UNIVERSITY

> It has a really fresh narrative voice. . . . It kept the details to a minimum, but presented the information clearly enough that I think students will get the big picture without getting bogged down in the details. Where details were given, they were mostly based on exciting, new findings in the field.
> — LAURA BERMINGHAM
> UNIVERSITY OF VERMONT

> I've reviewed several chapters now, and I like the writing style. It seems more engaging and easy to read, and feels less like reading an encyclopedia than other textbooks do.
> — MARYJO WITZ
> MONROE COMMUNITY COLLEGE

KEY FEATURES SUPPORTING THE VISION

TOPIC COVERAGE AND ORDER:

The goal of a single narrative of biology is supported by the Table of Contents. a select number of important and purposeful changes. Below is an annotated table of

TABLE OF CONTENTS | BIOLOGY: HOW LIFE WORKS

PART 1: FROM CELLS TO ORGANISMS

CHAPTER 1	**LIFE** CHEMICAL, CELLULAR, AND EVOLUTIONARY FOUNDATIONS

CASE 1 *The First Cell: Life's Origins*

CHAPTER 2	**THE MOLECULES OF LIFE**
CHAPTER 3	**NUCLEIC ACIDS AND THE ENCODING OF BIOLOGICAL INFORMATION**
CHAPTER 4	**TRANSLATION AND PROTEIN STRUCTURE**
CHAPTER 5	**ORGANIZING PRINCIPLES** LIPIDS, MEMBRANES, AND CELL COMPARTMENTS
CHAPTER 6	**MAKING LIFE WORK** CAPTURING AND USING ENERGY
CHAPTER 7	**CELLULAR RESPIRATION** HARVESTING ENERGY FROM CARBOHYDRATES AND OTHER FUEL MOLECULES
CHAPTER 8	**PHOTOSYNTHESIS** USING SUNLIGHT TO BUILD CARBOHYDRATES

CASE 2 *Cancer: When Good Cells Go Bad*

CHAPTER 9	**CELL COMMUNICATION**
CHAPTER 10	**CELL FORM AND FUNCTION** CYTOSKELETON, CELLULAR JUNCTIONS, AND EXTRACELLULAR MATRIX
CHAPTER 11	**CELL DIVISION** VARIATIONS, REGULATION, AND CANCER

CASE 3 *You, From A to T: Your Personal Genome*

CHAPTER 12	**DNA REPLICATION AND MANIPULATION**
CHAPTER 13	**GENOMES**
CHAPTER 14	**MUTATION AND DNA REPAIR**
CHAPTER 15	**GENETIC VARIATION**
CHAPTER 16	**MENDELIAN INHERITANCE**
CHAPTER 17	**BEYOND MENDEL** SEX CHROMOSOMES, LINKAGE, AND ORGANELLES
CHAPTER 18	**THE GENETIC AND ENVIRONMENTAL BASIS OF COMPLEX TRAITS**
CHAPTER 19	**GENETIC AND EPIGENETIC REGULATION**
CHAPTER 20	**GENES AND DEVELOPMENT**

CASE 4 *Malaria: Co-evolution of Humans and a Parasite*

CHAPTER 21	**EVOLUTION** HOW GENOTYPES AND PHENOTYPES CHANGE OVER TIME
CHAPTER 22	**SPECIES AND SPECIATION**
CHAPTER 23	**EVOLUTIONARY PATTERNS** PHYLOGENY AND FOSSILS
CHAPTER 24	**HUMAN ORIGINS AND EVOLUTION**

Chapter 1 introduces evolution as a major theme of the book before discussing microevolution in Chapter 4 as a foundation for later discussions of conservation of metabolic pathways and enzyme structure (Chapters 6–8) and genetic and phenotypic variation (Chapters 14 and 15). After the chapters on the mechanisms of evolution (Chapters 21–24), we discuss diversity of all organisms in terms of adaptations and comparative features, culminating in ecology as the ultimate illustration of evolution in action.

The first set of chapters emphasizes three key aspects of a cell—information, homeostasis, and energy. Chemistry is taught in the context of biological processes, highlighting the principle that structure determines function.

I found [Chapter 2: The Molecules of Life] quite readable and it makes the chemistry seem much more relevant to organisms from the beginning.
UDO SAVALLI, INSTRUCTOR
ARIZONA STATE UNIVERSITY WEST

The genetics chapters start with genomes, move to mutation and genetic variation, and then consider inheritance to provide a modern, molecular look at genetic variation and how traits are transmitted.

AND STORY OF BIOLOGY: HOW LIFE WORKS

The book's chapters and sections are arranged in a familiar way to be used in a range of introductory biology courses, with contents, highlighting these changes and the reasons behind them.

PART 2: FROM ORGANISMS TO THE ENVIRONMENT

- **CHAPTER 25**   **CYCLING CARBON**

CASE 5   *The Human Microbiome: Diversity Within*

CHAPTER 26   **BACTERIA AND ARCHAEA**

CHAPTER 27   **EUKARYOTIC CELLS**
ORIGINS AND DIVERSITY

- **CHAPTER 28**   **BEING MULTICELLULAR**

CASE 6   *Agriculture: Feeding a Growing Population*

CHAPTER 29   **PLANT STRUCTURE AND FUNCTION**
MOVING PHOTOSYNTHESIS ONTO LAND

CHAPTER 30   **PLANT REPRODUCTION**
FINDING MATES AND DISPERSING YOUNG

CHAPTER 31   **PLANT GROWTH AND DEVELOPMENT**
BUILDING THE PLANT BODY

- **CHAPTER 32**   **PLANT DEFENSE**
KEEPING THE WORLD GREEN

CHAPTER 33   **PLANT DIVERSITY**

CHAPTER 34   **FUNGI**
STRUCTURE, FUNCTION, AND DIVERSITY

CASE 7   *Predator–Prey: A Game of Life and Death*

CHAPTER 35   **ANIMAL NERVOUS SYSTEMS**

CHAPTER 36   **ANIMAL SENSORY SYSTEMS AND BRAIN FUNCTION**

CHAPTER 37   **ANIMAL MOVEMENT**
MUSCLES AND SKELETONS

CHAPTER 38   **ANIMAL ENDOCRINE SYSTEMS**

CHAPTER 39   **ANIMAL CARDIOVASCULAR AND RESPIRATORY SYSTEMS**

CHAPTER 40   **ANIMAL METABOLISM, NUTRITION, AND DIGESTION**

CHAPTER 41   **ANIMAL RENAL SYSTEMS**
WATER AND WASTE

CHAPTER 42   **ANIMAL REPRODUCTION AND DEVELOPMENT**

CHAPTER 43   **ANIMAL IMMUNE SYSTEMS**

CASE 8   *Biodiversity Hotspots: Rain Forests and Coral Reefs*

CHAPTER 44   **ANIMAL DIVERSITY**

CHAPTER 45   **ANIMAL BEHAVIOR**

CHAPTER 46   **POPULATION ECOLOGY**

CHAPTER 47   **SPECIES INTERACTIONS, COMMUNITIES, AND ECOSYSTEMS**

- **CHAPTER 48**   **THE ANTHROPOCENE**
HUMANS AS A PLANETARY FORCE

[Chapter 25: Cycling Carbon] does a great job connecting the "halves"; providing examples in the text and asking questions along the way forces the student to give some thought to the topic.... Well done.
STEPHEN TRUMBLE
BAYLOR UNIVERSITY

We present the carbon cycle as a bridge between the molecular and organismal parts of the book, integrating ecology and diversity.

The plant chapters combine content in novel ways to better integrate content and provide context for biological processes, such as plant reproduction with the timing of reproductive events.

Diversity follows physiology in order to provide a basis for understanding the groupings of organisms.

- *Biology: How Life Works* includes chapters that don't traditionally appear in introductory biology texts, one in almost every major subject area. These novel chapters represent shifts toward a more modern conception of certain topics in biology:
  - The basis of complex traits (Chapter 18)
  - Human evolution (Chapter 24)
  - Cycling carbon (Chapter 25)
  - Multicellularity (Chapter 28)
  - Plant defenses (Chapter 32)
  - Human impact on the environment (Chapter 48)

BIOLOGY: HOW LIFE WORKS

CONNECTED LEARNING PATH

The authors of *Biology: How Life Works* rethought the standard pedagogical tools like chapter summaries and end-of-chapter questions. Each element purposefully relates to the same set of core concepts and functions as a connected learning path for students.

Core Concepts listed at the beginning of each chapter map to the chapter's numbered sections and give students a preview of what ideas they should know by the time they finish reading. We built each chapter around core concepts in order to focus on the most important and overarching ideas in any one topic.

Throughout the chapter, Quick Checks ask students to pause and make sure they're ready to move on to the next idea. These brief questions pointedly ask the student to demonstrate a full understanding of a tough topic before reading on.

Students revisit the core concepts in the Core Concepts Summary. Here, they can remind themselves of the big-picture ideas from the beginning of the chapter and see them in the context of key supporting ideas.

Students test their understanding of core concepts with Self-Assessments. These assessments, along with the online answer guides, allow students to see for themselves whether or not they have understood the major ideas of the chapter.

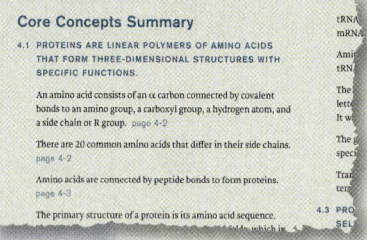

Learning doesn't end when the printed chapter ends. A wealth of media organized by core concept, including dynamic, zoomable Visual Synthesis figures, animations, and simulations, reinforce what students have read. Assessments give students opportunities to apply what they have learned and hone problem-solving skills.

Instructors can use and assign material from a set of questions we call progressions. Progressions offer questions designed as a sequence to use before, during, and after lecture. Progressions help instructors align the questions they ask on exams to the in-class activities and homework questions they have students engage in before an exam.

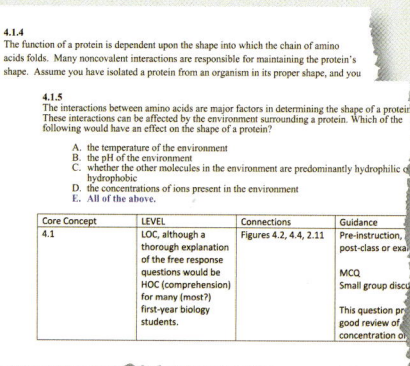

BIOLOGY: HOW LIFE WORKS

VISUAL SYNTHESIS FIGURES

Just as we integrate topics in biology to create a single narrative, it's our goal to extend integration to the media by creating a seamless path from the book to all online materials. Visual Synthesis figures represent one way that we integrate the text and media. In the book, a Visual Synthesis figure is a two-page spread and serves as a visual summary integrating concepts across multiple chapters. These figures help students look beyond chapter divisions and see how concepts combine to tell a single story. Students continue their exploration online, where they can interact with a dynamic Visual Synthesis figure. Zooming in and out, students can explore both the big picture and the details, building a framework for how concepts connect and relate. The Visual Synthesis figure also functions as a launch pad to other resources, like animations, simulations, and assessment questions.

> The course I teach strives to integrate concepts across multiple levels of biological complexity and Visual Synthesis would help with this goal.
> BRETTON KENT, INSTRUCTOR
> UNIVERSITY OF MARYLAND

> Visual Synthesis does a great job of summarizing the content of [Chapter 42: Reproduction] in terms of human development. It appears to be very comprehensive including the concepts of fertilization, cleavage (including both maternal and fetal transcription controls) all the way through cell differentiation and tissue development.
> CYNTHIA LITTLEJOHN, INSTRUCTOR
> UNIVERSITY OF SOUTHERN MISSISSIPPI

> The zoomable Visual Synthesis image is an incredibly powerful visual aid for students, allowing instructors to bridge the gap between the details and the larger concept.
> ROBERT MAXWELL, INSTRUCTOR
> GEORGIA STATE UNIVERSITY

> The zoomable digital media is an excellent tool for students and instructors! The interactive capabilities and ability to view content on different levels of complexity are unlike anything I have seen from other texts.
> JEANELLE MORGAN, INSTRUCTOR
> GAINESVILLE STATE COLLEGE

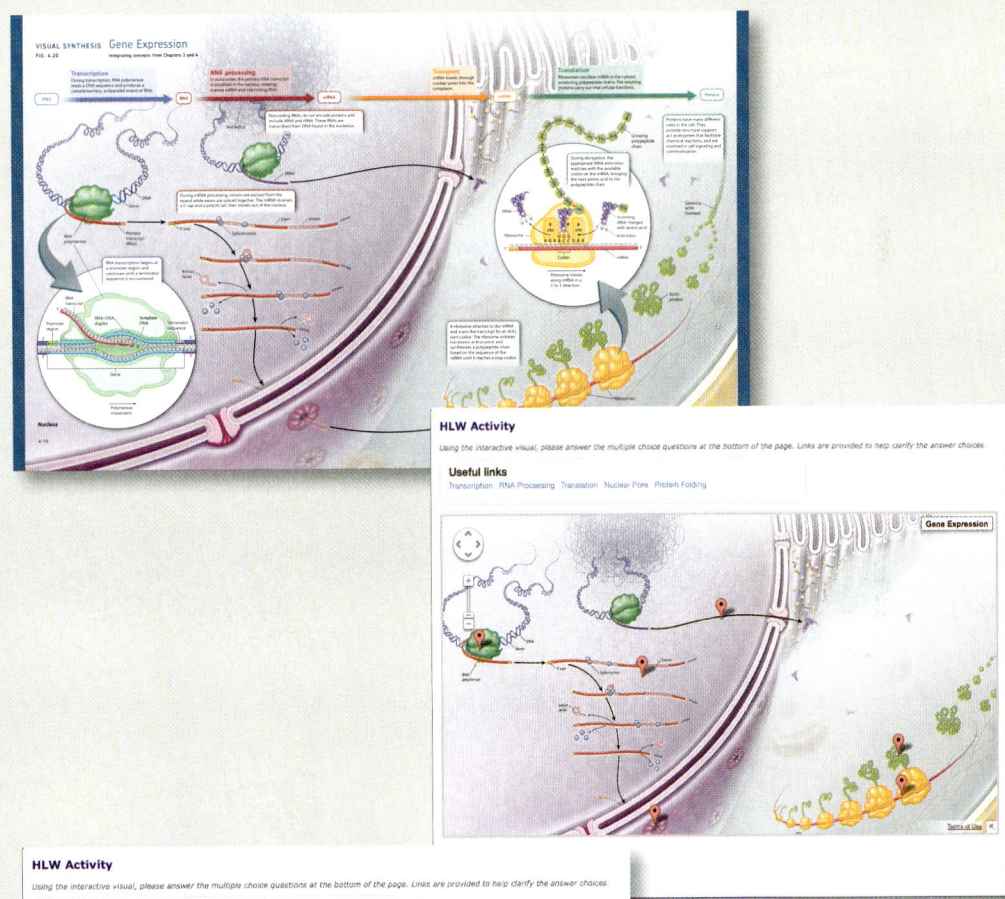

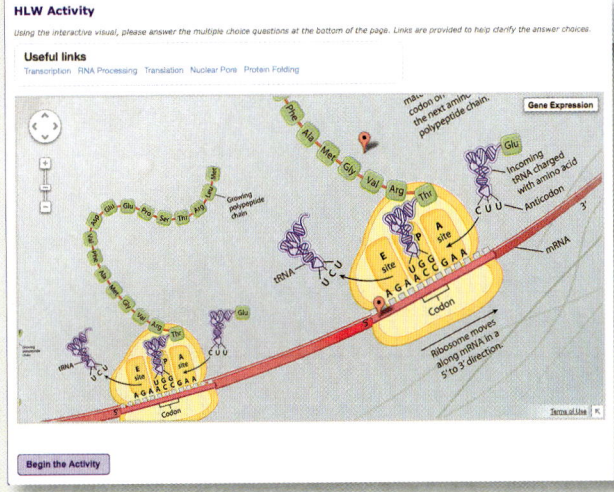

xvii

BIOLOGY: HOW LIFE WORKS

I would like to incorporate more animations, virtual labs, and real-life scenarios relating to the concepts we are learning in lecture. It is incredibly helpful when these resources are created by the textbook publisher, so they correspond directly to the material I am covering.
—Ashley Spring
Brevard Community College

Students have trouble visualizing cellular activities, so we constantly look for tools to enhance their understanding.
—Margaret Ott
Tyler Junior College

ANIMATIONS AND SIMULATIONS

Biological processes are not static—they are dynamic, always in motion. We use **animations** as an engaging and revealing way to bring biological processes to life, place them in context, and give students an intuitive sense of how these processes work.

Some concepts are best learned by doing, rather than by reading or watching. *Biology: How Life Works* **simulations** allow students to explore biological processes directly and problem solve by doing. Each simulation asks students to work within or break a system, and assessment questions lead students through the system so they can work through a core concept.

The animations and simulations are available independently but are also integrated into the online, zoomable *Visual Synthesis* figures. In exploring the *Visual Synthesis* figure, students can watch animations, answer assessment questions, and interact with simulations to guide them through the illustration.

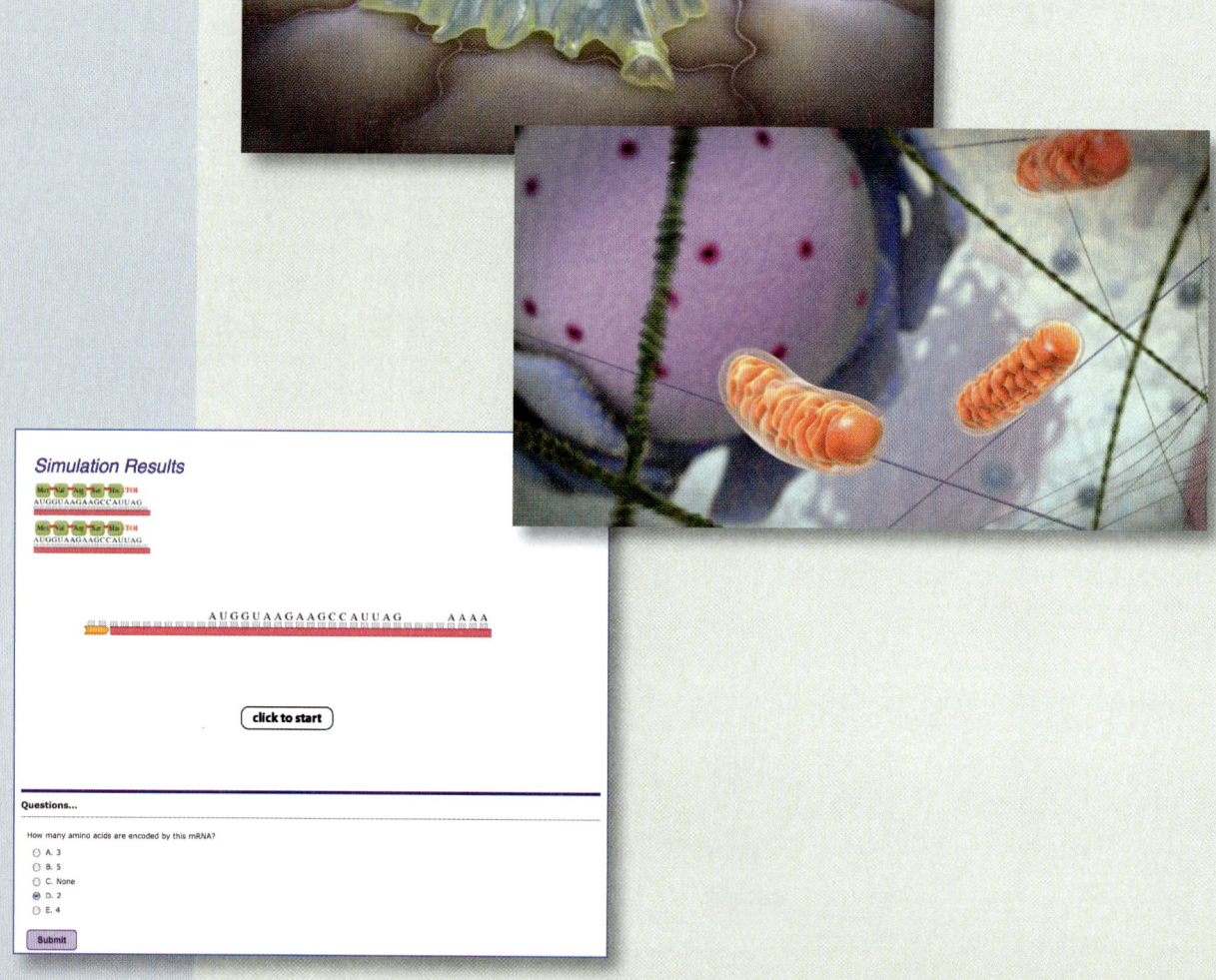

BIOLOGY: HOW LIFE WORKS

CASES AND CASE ACTIVITIES

Biology is best understood when presented using real and engaging examples as a framework for putting information together. One of the ways we provide this framework is through **cases**. Each case begins with a two-page essay, providing background and raising questions about the case's subject. Those questions are answered in subsequent chapters, reinforcing information students have already learned and providing a framework for learning new information. For example, our case about malaria is introduced before the set of chapters on evolution and is revisited in each of these chapters where it serves to reinforce important evolutionary concepts.

The *Case Activities* tie chapter core concepts and cases together, showing students again how what they are learning is relevant to their lives and to a larger picture of biology. We created several activities, varying in length and complexity, for each case. Activities can be assigned as homework or used in class, giving instructors the ability to choose what kind of activity best fits their course needs.

> The connection of the case study to the two sections in the chapter is a great way to maintain reader engagement and show the connection of material being presented to real-life problems and solutions.
> STEPHEN JURIS, INSTRUCTOR
> CENTRAL MICHIGAN UNIVERSITY

> This [Cancer: When Good Cells Go Bad] case study is perfect for this group of chapters because it covers cell division, uncontrolled cell growth, cell communication, genetic control, and transformation. In addition, this topic is relevant to the health of the students taking this course.
> SHANNON MCQUAIG, INSTRUCTOR
> ST. PETERSBURG COLLEGE

> I really like this [Cancer: When Good Cells Go Bad] case, including the integration of vaccinations and some public policy questions. This will help students see the real-world application of what they are learning.
> JOHN KAUWE, INSTRUCTOR
> BRIGHAM YOUNG UNIVERSITY

xix

BIOLOGY: HOW LIFE WORKS

> I liked the way that experiments were shown in *How Life Works*, particularly Fig 8.7. It summarized three experiments in a way that students could understand them.
> DAVID BYRES
> FLORIDA STATE COLLEGE JACKSONVILLE

> This was a great figure [Fig 13.1]. It achieves the goal, easy to immediately grasp, an appropriate experiment to highlight because it explains how the technique is done conceptually without any of the annoying details. It is perfect.
> VICTORIA CORBIN, INSTRUCTOR
> UNIVERSITY OF KANSAS

> I like the background portion of those figures in . . . *How Life Works*; they help understand the rationale behind the experiment or the technique that was used.
> PETER KOURTEV, INSTRUCTOR
> CENTRAL MICHIGAN UNIVERSITY

HOW DO WE KNOW? FIGURES AND ACTIVITIES

Biology is not just about what we know, but also *how we know what we know*. An important skill that introductory students must develop is asking questions and forming hypotheses like scientists. Therefore, we introduce key experiments in the narrative flow of the text and then visually unpack the steps of scientific inquiry in *How Do We Know?* figures. These figures emphasize that science is a process of asking and answering questions and convey an authentic sense of scientific inquiry.

The exploration of *How Do We Know?* continues with online activities that emphasize using the scientific method. A *How Do We Know?* activity begins with a tutorial that reviews the experimental design of the text figure, followed by questions that ask students to apply what they learned to new experiments, as well as predict how different variables shape the outcome of the experiment. Our goal is to ensure that students understand the concepts covered in the *How Do We Know?* figures, and that they are able to apply the concepts to new situations or experiments.

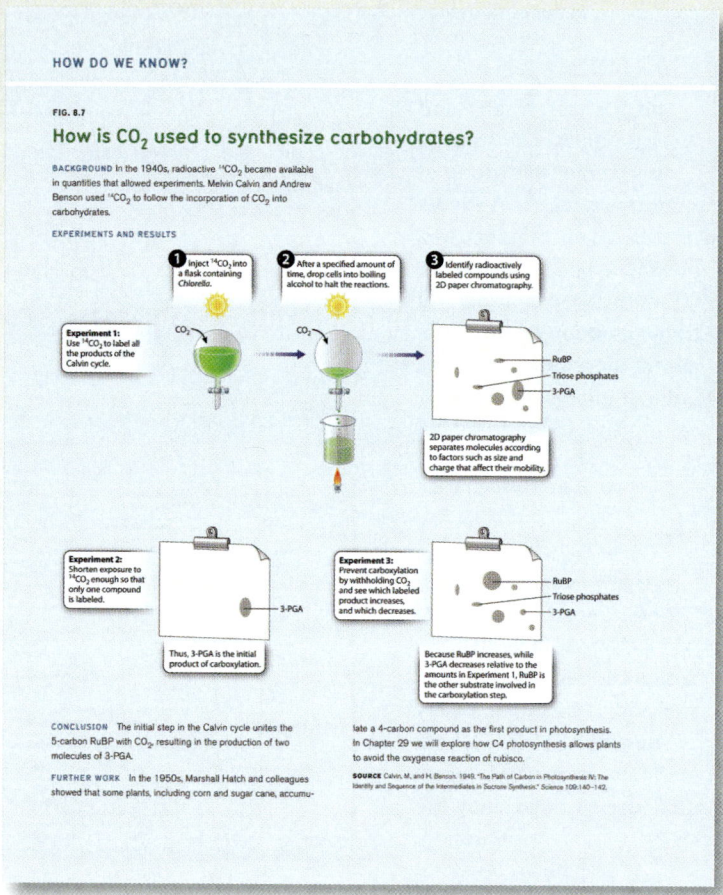

In keeping with our modern approach, we are taking full advantage of the online environment and making **Biology: How Life Works** dynamic. We will continually update and add to our collection of online assessments, visual media, and activities based on feedback, our continued interaction with the teaching community, and the needs of students and instructors.

ACKNOWLEDGMENTS

A book like this one is a team effort, and we have had the good fortune of working with a remarkable, caring, and dedicated team. First and foremost, we would like to thank the thousands of students we have collectively taught. Their curiosity, intelligence, and enthusiasm have been sources of motivation for all of us. Our teachers and mentors have likewise provided us with models of patience, persistence, and inquisitiveness that we bring into our own teaching and research. They encourage us to be lifelong learners, teachers, and scholars.

We all feel very lucky to be a partner with W. H. Freeman and Company. From the start, they have embraced our project, giving us the space and room to carve out something unique, while at the same time providing guidance, support, and input from the broader community of instructors and students.

Susan Winslow, our publisher, deserves special thanks for her ability to expertly juggle all aspects of this first edition, keeping a watchful eye on important trends in education and science, carefully listening to what we wanted to do, and providing a gentle nudge from time to time.

This book represents a genuine, rewarding, and unique partnership with a team of assessment leaders and authors. We especially thank and acknowledge the contributions of lead assessment author Melissa Michael and assessment team leaders Mark Hens, John Merrill, Randall Phillis, and Debra Pires. We all learned a great deal about pedagogy and how to use questions as tools for teaching, not just testing.

We also especially thank Mark Hens of the University of North Carolina at Greensboro for his contribution to several chapters, and Kirsten Weir, an award-winning science writer and editor, who helped us to set the tone, write, and put together the case introductions.

Lead developmental editor Lisa Samols has just the right touch—she has the ability to listen as well as offer intelligent suggestions; she is serious with a touch of humor; she is quiet but persistent. Senior developmental editor Susan Moran has an eye for detail and the uncanny ability to read the manuscript like a student. Developmental editor Erica Pantages Frost brought intelligence and thoughtfulness to her edits.

Karen Misler kept us all on schedule in a clear and firm, but always understanding and compassionate, way. Thanks also to Dusty Friedman, who helped us start this project on schedule. Carolyn Deacy expertly managed the art program, keeping all the various components coordinated and offering intelligent advice, and Bill Page coordinated our initial relationship with Imagineering.

Shannon Howard, who focused on editorial research and development, and Lindsey Veautour, our market development manager, are remarkable for their energy and enthusiasm, and their creativity in ways to reach out to instructors and students.

We thank Robert Errera for coordinating the move from manuscript to the page. We also thank Diana Blume, our art director, and Tom Carling, the layout artist. Together, they managed the look and feel of the book, coming up with creative solutions for page layout.

Imagineering under the patient and intelligent guidance of Mark Mykytiuk provided creative, insightful art to complement, support, and reinforce the text. Christine Buese, our photo editor, and Jacquelin Wong and Deborah Anderson, our photo researchers, provided us with a steady stream of stunning photos, and never gave up on hard-to-find shots.

On the digital media front, we thank Amanda Dunning for her editorial insight into making pedagogically useful media tools, and Keri Fowler for managing and coordinating the media and websites. They both took on this project with dedication, persistence, enthusiasm, and attention to detail that we deeply appreciate.

We are very grateful to Debbie Clare for her work in marketing, Elaine Palucki for her insight into teaching and learning strategies, Donna Brodman for coordinating the many reviewers, and Yassamine Ebadat and Jane Taylor for their consistent and tireless support.

We are also extremely grateful for all of the hard work and expertise of the sales representatives, regional managers, and regional sales specialists. We have enjoyed meeting and working with this dedicated sales staff, who are the ones that ultimately put the book in the hands of instructors for the first time.

We would also like to acknowledge Charles Smith and Neil Patterson for their early support of the author team and helping to foster the idea that would eventually become *How Life Works*, and special thanks to Charles for many spectacular photos that appear throughout the book.

Countless reviewers made invaluable contributions to this book and deserve special thanks. From catching mistakes to suggesting new and innovative ways to organize the content, they provided substantial input to the book. They brought their collective years of teaching to the project, and their suggestions are tangible in every chapter.

And none of this would have been possible without the support, inspiration, and encouragement of our families.

Contributors

Thank you to all the instructors who worked in collaboration with the authors and assessment authors to write Biology: How Life Works *assessments, activities, and exercises.*

Allison Alvarado, University of California, Los Angeles
Peter Armbruster, Georgetown University
Zane Barlow-Coleman, formerly of University of Massachusetts Amherst
James Bottesch, Brevard Community College
Jessamina Blum, Yale University
Jere Boudell, Clayton State University
David Bos, Purdue University
Laura Ciaccia West, Yale University*
Laura DiCaprio, Ohio University
Tod Duncan, University of Colorado Denver
Cindy Giffen, University of Wisconsin Madison
Paul Greenwood, Colby College
Stanley Guffey, University of Tennessee, Knoxville
Alison Hill, Duke University
Meg Horton, University of North Carolina at Greensboro
Kerry Kilburn, Old Dominion University
Jo Kurdziel, University of Michigan
David Lampe, Duquesne University
Brenda Leady, University of Toledo
Sara Marlatt, Yale University*
Kelly McLaughlin, Tufts University
Brad Mehrtens, University of Illinois at Urbana-Champaign
Nancy Morvillo, Florida Southern College
Jennifer Nauen, University of Delaware
Kavita Oommen, Georgia State University
Patricia Phelps, Austin Community College
Melissa Reedy, University of Illinois at Urbana-Champaign
Lindsay Rush, Yale University*
Sukanya Subramanian, Collin College
Michelle Withers, West Virginia University

*Graduate student, Yale University Scientific Teaching Fellow

Reviewers, Class Testers, and Focus Group Participants

Thank you to all the instructors who reviewed and/or class tested chapters, art, assessment questions, and other Biology: How Life Works *materials.*

Thomas Abbott, University of Connecticut
Tamarah Adair, Baylor University
Sandra Adams, Montclair State University
Jonathon Akin, University of Connecticut
Eddie Alford, Arizona State University
Chris Allen, College of the Mainland
Sylvester Allred, Northern Arizona University
Shivanthi Anandan, Drexel University
Andrew Andres, University of Nevada, Las Vegas
Michael Angilletta, Arizona State University
Jonathan Armbruster, Auburn University
Jessica Armenta, Lone Star College System
Brian Ashburner, University of Toledo
Andrea Aspbury, Texas State University
Nevin Aspinwall, Saint Louis University
Felicitas Avendano, Grand View University
Yael Avissar, Rhode Island College
Ricardo Azpiroz, Richland College
Jessica Baack, Southwestern Illinois College
Charles Baer, University of Florida
Brian Bagatto, University of Akron
Alan L. Baker, University of New Hampshire
Ellen Baker, Santa Monica College
Mitchell Balish, Miami University
Teri Balser, University of Florida
Paul Bates, University of Minnesota, Duluth
Michel Baudry, University of Southern California
Jerome Baudry, University of Tennessee, Knoxville
Mike Beach, Southern Polytechnic State University
Andrew Beall, University of North Florida
Gregory Beaulieu, University of Victoria
John Bell, Brigham Young University
Michael Bell, Richland College
Rebecca Bellone, University of Tampa
Anne Bergey, Truman State University
Laura Bermingham, University of Vermont
Aimee Bernard, University of Colorado, Denver
Annalisa Berta, San Diego State University
Joydeep Bhattacharjee, University of Louisiana, Monroe
Arlene Billock, University of Louisiana, Lafayette
Daniel Blackburn, Trinity College
Mark Blackmore, Valdosta State University
Justin Blau, New York University
Andrew Blaustein, Oregon State University
Mary Bober, Santa Monica College
Robert Bohanan, University of Wisconsin, Madison
Jim Bonacum, University of Illinois at Springfield
Laurie Bonneau, Trinity College
David Bos, Purdue University
James Bottesch, Brevard Community College
Jere Boudell, Clayton State University
Nancy Boury, Iowa State University
Matthew Brewer, Georgia State University
Mirjana Brockett, Georgia Institute of Technology
Andrew Brower, Middle Tennessee State University
Heather Bruns, Ball State University
Jill Buettner, Richland College

Stephen Burnett, Clayton State University
Steve Bush, Coastal Carolina University
David Byres, Florida State College at Jacksonville
James Campanella, Montclair State University
Darlene Campbell, Cornell University
Jennifer Campbell, North Carolina State University
John Campbell, Northwest College
David Canning, Murray State University
Richard Cardullo, University of California, Riverside
Sara Carlson, University of Akron
Jeff Carmichael, University of North Dakota
Dale Casamatta, University of North Florida
Anne Casper, Eastern Michigan University
David Champlin, University of Southern Maine
Rebekah Chapman, Georgia State University
Samantha Chapman, Villanova University
Mark Chappell, University of California, Riverside
P. Bryant Chase, Florida State University
Young Cho, Eastern New Mexico University
Tim Christensen, East Carolina University
Steven Clark, University of Michigan
Ethan Clotfelter, Amherst College
Catharina Coenen, Allegheny College
Mary Colavito, Santa Monica College
Craig Coleman, Brigham Young University
Alex Collier, Armstrong Atlantic State University
Sharon Collinge, University of Colorado, Boulder
Jay Comeaux, McNeese State University
Reid Compton, University of Maryland
Ronald Cooper, University of California, Los Angeles
Victoria Corbin, University of Kansas
Asaph Cousins, Washington State University
Will Crampton, University of Central Florida
Kathryn Craven, Armstrong Atlantic State University
Scott Crousillac, Louisiana State University
Kelly Cude, College of the Canyons
Stanley Cunningham, Arizona State University
Karen Curto, University of Pittsburgh
Bruce Cushing, The University of Akron
Rebekka Darner, University of Florida
James Dawson, Pittsburg State University
Elizabeth De Stasio, Lawrence University
Jennifer Dechaine, Central Washington University
James Demastes, University of Northern Iowa
D. Michael Denbow, Virginia Polytechnic Institute and State University Joseph Dent, McGill University
Terry Derting, Murray State University
Jean DeSaix, University of North Carolina at Chapel Hill
Donald Deters, Bowling Green State University
Hudson DeYoe, The University of Texas, Pan American
Leif Deyrup, University of the Cumberlands
Laura DiCaprio, Ohio University
Jesse Dillon, California State University, Long Beach
Frank Dirrigl, The University of Texas, Pan American
Kevin Dixon, Florida State University
Elaine Dodge Lynch, Memorial University of Newfoundland
Hartmut Doebel, George Washington University
Jennifer Doll, Loyola University, Chicago
Logan Donaldson, York University
Blaise Dondji, Central Washington University
Christine Donmoyer, Allegheny College
James Dooley, Adelphi University
Jennifer Doudna, University of California, Berkeley
John DuBois, Middle Tennessee State University
Richard Duhrkopf, Baylor University
Kamal Dulai, University of California, Merced
Arthur Dunham, University of Pennsylvania
Mary Durant, Lone Star College System
Roland Dute, Auburn University
Andy Dyer, University of South Carolina, Aiken
William Edwards, Niagara University
John Elder, Valdosta State University
William Eldred, Boston University
David Eldridge, Baylor University
Inge Eley, Hudson Valley Community College
Lisa Elfring, University of Arizona
Richard Elinson, Duquesne University
Kurt Elliott, Northwest Vista College
Miles Engell, North Carolina State University
Susan Erster, Stony Brook University
Joseph Esdin, University of California, Los Angeles
Jean Everett, College of Charleston
Brent Ewers, University of Wyoming
Melanie Fierro, Florida State College at Jacksonville
Michael Fine, Virginia Commonwealth University
Jonathan Fingerut, St. Joseph's University
Ryan Fisher, Salem State University
David Fitch, New York University
Paul Fitzgerald, Northern Virginia Community College
Jason Flores, University of North Carolina at Charlotte
Matthias Foellmer, Adelphi University
Barbara Frase, Bradley University
Caitlin Gabor, Texas State University
Michael Gaines, University of Miami
Jane Gallagher, The City College of New York, The City University of New York
Kathryn Gardner, Boston University
J. Yvette Gardner, Clayton State University
Gillian Gass, Dalhousie University
Jason Gee, East Carolina University
Topher Gee, University of North Carolina at Charlotte
Vaughn Gehle, Southwest Minnesota State University
Tom Gehring, Central Michigan University
John Geiser, Western Michigan University
Alex Georgakilas, East Carolina University
Peter Germroth, Hillsborough Community College
Arundhati Ghosh, University of Pittsburgh
Carol Gibbons Kroeker, University of Calgary
Phil Gibson, University of Oklahoma
Cindee Giffen, University of Wisconsin, Madison
Matthew Gilg, University of North Florida
Sharon Gillies, University of the Fraser Valley
Leonard Ginsberg, Western Michigan University
Florence Gleason, University of Minnesota
Russ Goddard, Valdosta State University
Miriam Golbert, College of the Canyons
Jessica Goldstein, Barnard College, Columbia University
Steven Gorsich, Central Michigan University
Sandra Grebe, Lone Star College System
Robert Greene, Niagara University
Ann Grens, Indiana University, South Bend
Theresa Grove, Valdosta State University
Stan Guffey, The University of Tennessee
Nancy Guild, University of Colorado, Boulder
Lonnie Guralnick, Roger Williams University
Laura Hake, Boston College
Kimberly Hammond, University of California, Riverside

Paul Hapeman, University of Florida
Luke Harmon, University of Idaho
Sally Harmych, University of Toledo
Jacob Harney, University of Hartford
Sherry Harrel, Eastern Kentucky University
Dale Harrington, Caldwell Community College and Technical Institute
J. Scott Harrison, Georgia Southern University
Diane Hartman, Baylor University
Mary Haskins, Rockhurst University
Bernard Hauser, University of Florida
David Haymer, University of Hawaii
David Hearn, Towson University
Marshal Hedin, San Diego State University
Paul Heideman, College of William and Mary
Gary Heisermann, Salem State University
Brian Helmuth, University of South Carolina
Christopher Herlihy, Middle Tennessee State University
Albert Herrera, University of Southern California
Brad Hersh, Allegheny College
David Hicks, The University of Texas at Brownsville
Karen Hicks, Kenyon College
Alison Hill, Duke University
Kendra Hill, South Dakota State University
Jay Hodgson, Armstrong Atlantic State University
John Hoffman, Arcadia University
Jill Holliday, University of Florida
Sara Hoot, University of Wisconsin, Milwaukee
Margaret Horton, University of North Carolina at Greensboro
Lynne Houck, Oregon State University
Kelly Howe, University of New Mexico
William Huddleston, University of Calgary
Jodi Huggenvik, Southern Illinois University
Melissa Hughes, College of Charleston
Randy Hunt, Indiana University Southeast
Tony Huntley, Saddleback College
Brian Hyatt, Bethel College
Jeba Inbarasu, Metropolitan Community College
Colin Jackson, The University of Mississippi
Eric Jellen, Brigham Young University
Dianne Jennings, Virginia Commonwealth University
Scott Johnson, Wake Technical Community College
Mark Johnston, Dalhousie University
Susan Jorstad, University of Arizona

Stephen Juris, Central Michigan University
Julie Kang, University of Northern Iowa
Jonghoon Kang, Valdosta State University
George Karleskint, St. Louis Community College at Meramec
David Karowe, Western Michigan University
Judy Kaufman, Monroe Community College
Nancy Kaufmann, University of Pittsburgh
John Kauwe, Brigham Young University
Elena Keeling, California Polytechnic State University
Jill Keeney, Juniata College
Tamara Kelly, York University
Chris Kennedy, Simon Fraser University
Bretton Kent, University of Maryland
Jake Kerby, University of South Dakota
Jeffrey Kiggins, Monroe Community College
Scott Kight, Montclair State University
Stephen Kilpatrick, University of Pittsburgh, Johnstown
Kelly Kissane, University of Nevada, Reno
David Kittlesen, University of Virginia
Jennifer Kneafsey, Tulsa Community College
Jennifer Knight, University of Colorado, Boulder
Ross Koning, Eastern Connecticut State University
David Kooyman, Brigham Young University
Olga Kopp, Utah Valley University
Anna Koshy, Houston Community College
Todd Kostman, University of Wisconsin, Oshkosh
Peter Kourtev, Central Michigan University
William Kroll, Loyola University, Chicago
Dave Kubien, University of New Brunswick
Allen Kurta, Eastern Michigan University
Ellen Lamb, University of North Carolina at Greensboro
Troy Ladine, East Texas Baptist University
David Lampe, Duquesne University
Evan Lampert, Gainesville State College
James Langeland, Kalamazoo College
John Latto, University of California, Santa Barbara
Brenda Leady, University of Toledo
Jennifer Leavey, Georgia Institute of Technology
Hugh Lefcort, Gonzaga University
Brenda Leicht, University of Iowa
Craig Lending, The College at Brockport, The State University of New York
Nathan Lents, John Jay College of Criminal Justice, The City University of New York
Michael Leonardo, Coe College
Army Lester, Kennesaw State University

Cynthia Littlejohn, University of Southern Mississippi
Zhiming Liu, Eastern New Mexico University
Jonathan Lochamy, Georgia Perimeter College
Suzanne Long, Monroe Community College
Julia Loreth, University of North Carolina at Greensboro
Jennifer Louten, Southern Polytechnic State University
Janet Loxterman, Idaho State University
Ford Lux, Metropolitan State College of Denver
Jose-Luis Machado, Swarthmore College
C. Smoot Major, University of South Alabama
Charles Mallery, University of Miami
Mark Maloney, Spelman College
Carroll Mann, Florida State College at Jacksonville
Carol Mapes, Kutztown University of Pennsylvania
Nilo Marin, Broward College
Diane Marshall, University of New Mexico
Heather Masonjones, University of Tampa
Scott Mateer, Armstrong Atlantic State University
Luciano Matzkin, The University of Alabama in Huntsville
Robert Maxwell, Georgia State University
Meghan May, Towson University
Douglas Meikle, Miami University
Michael McGinnis, Spelman College
Kathleen McGuire, San Diego State University
Maureen McHale, Truman State University
Shannon McQuaig, St. Petersburg College
Susan McRae, East Carolina University
Lori McRae, University of Tampa
Mark Meade, Jacksonville State University
Brad Mehrtens, University of Illinois at Urbana-Champaign
Michael Meighan, University of California, Berkeley
Richard Merritt, Houston Community College
Jennifer Metzler, Ball State University
James Mickle, North Carolina State University
Brian Miller, Middle Tennessee State University
Allison Miller, Saint Louis University
Yuko Miyamoto, Elon University
Ivona Mladenovic, Simon Fraser University
Marcie Moehnke, Baylor University
Chad Montgomery, Truman State University
Jennifer Mook, Gainesville State College
Daniel Moon, University of North Florida

Jamie Moon, University of North Florida
Jeanelle Morgan, Gainesville State College
David Morgan, University of West Georgia
Julie Morris, Armstrong Atlantic State University
Becky Morrow, Duquesne University
Mark Mort, University of Kansas
Nancy Morvillo, Florida Southern College
Anthony Moss, Auburn University
Mario Mota, University of Central Florida
Alexander Motten, Duke University
Tim Mulkey, Indiana State University
John Mull, Weber State University
Michael Muller, University of Illinois at Chicago
Beth Mullin, University of Tennessee, Knoxville
Paul Narguizian, California State University, Los Angeles
Jennifer Nauen, University of Delaware
Paul Nealen, Indiana University of Pennsylvania
Diana Nemergut, University of Colorado, Boulder
Kathryn Nette, Cuyamaca College
Jacalyn Newman, University of Pittsburgh
James Nienow, Valdosta State University
Alexey Nikitin, Grand Valley State University
Tanya Noel, York University
Fran Norflus, Clayton State University
Celia Norman, Arapahoe Community College
Eric Norstrom, DePaul University
Jorge Obeso, Miami Dade College
Kavita Oommen, Georgia State University
David Oppenheimer, University of Florida
Joseph Orkwiszewski, Villanova University
Rebecca Orr, Collin College
Don Padgett, Bridgewater State College
Joanna Padolina, Virginia Commonwealth University
One Pagan, West Chester University
Kathleen Page, Bucknell University
Daniel Papaj, University of Arizona
Pamela Pape-Lindstrom, Everett Community College
Bruce Patterson, University of Arizona, Tucson
Shelley Penrod, Lone Star College System
Roger Persell, Hunter College, The City University of New York
John Peters, College of Charleston
Chris Petrie, Brevard Community College
Patricia Phelps, Austin Community College

Steven Phelps, The University of Texas at Austin
Kristin Picardo, St. John Fisher College
Aaron Pierce, Nicholls State University
Debra Pires, University of California, Los Angeles
Thomas Pitzer, Florida International University
Nicola Plowes, Arizona State University
Crima Pogge, City College of San Francisco
Darren Pollock, Eastern New Mexico University
Kenneth Pruitt, The University of Texas at Brownsville
Sonja Pyott, University of North Carolina at Wilmington
Rajinder Ranu, Colorado State University
Philip Rea, University of Pennsylvania
Amy Reber, Georgia State University
Ahnya Redman, West Virginia University
Melissa Reedy, University of Illinois at Urbana-Champaign
Brian Ring, Valdosta State University
David Rintoul, Kansas State University
Michael Rischbieter, Presbyterian College
Laurel Roberts, University of Pittsburgh
George Robinson, The University at Albany, The State University of New York
Peggy Rolfsen, Cincinnati State Technical and Community College
Mike Rosenzweig, Virginia Polytechnic Institute and State University
Doug Rouse, University of Wisconsin, Madison
Yelena Rudayeva, Palm Beach State College
Ann Rushing, Baylor University
Shereen Sabet, La Sierra University
Rebecca Safran, University of Colorado
Peter Sakaris, Southern Polytechnic State University
Thomas Sasek, University of Louisiana, Monroe
Udo Savalli, Arizona State University
H. Jochen Schenk, California State University, Fullerton
Gregory Schmaltz, University of the Fraser Valley
Jean Schmidt, University of Pittsburgh
Andrew Schnabel, Indiana University, South Bend
Roxann Schroeder, Humboldt State University
David Schultz, University of Missouri, Columbia
Andrea Schwarzbach, The University of Texas at Brownsville

Erik Scully, Towson University
Robert Seagull, Hofstra University
Pramila Sen, Houston Community College
Alice Sessions, Austin Community College
Vijay Setaluri, University of Wisconsin
Jyotsna Sharma, The University of Texas at San Antonio
Elizabeth Sharpe-Aparicio, Blinn College
Patty Shields, University Of Maryland
Cara Shillington, Eastern Michigan University
James Shinkle, Trinity University
Rebecca Shipe, University of California, Los Angeles
Marcia Shofner, University of Maryland
Laurie Shornick, Saint Louis University
Jill Sible, Virginia Polytechnic Institute and State University
Allison Silveus, Tarrant County College
Kristin Simokat, University of Idaho
Sue Simon-Westendorf, Ohio University
Sedonia Sipes, Southern Illinois University, Carbondale
John Skillman, California State University, San Bernardino
Marek Sliwinski, University of Northern Iowa
Felisa Smith, University of New Mexico
John Sollinger, Southern Oregon University
Scott Solomon, Rice University
Morvarid Soltani-Bejnood, The University of Tennessee
Vladimir Spiegelman, University of Wisconsin, Madison
Chrissy Spencer, Georgia Institute of Technology
Kathryn Spilios, Boston University
Ashley Spring, Brevard Community College
Bruce Stallsmith, The University of Alabama in Huntsville
Jennifer Stanford, Drexel University
Barbara Stegenga, University of North Carolina, Chapel Hill
Patricia Steinke, San Jacinto College, Central Campus
Asha Stephens, College of the Mainland
Robert Steven, University of Toledo
Eric Strauss, University of Wisconsin, La Crosse
Sukanya Subramanian, Collin College
Mark Sugalski, Southern Polytechnic State University
Brad Swanson, Central Michigan University
Ken Sweat, Arizona State University
David Tam, University of North Texas
Ignatius Tan, New York University

William Taylor, University of Toledo
Christine Terry, Lynchburg College
Sharon Thoma, University of Wisconsin, Madison
Pamela Thomas, University of Central Florida
Carol Thornber, University of Rhode Island
Patrick Thorpe, Grand Valley State University
Briana Timmerman, University of South Carolina
Chris Todd, University of Saskatchewan
Gail Tompkins, Wake Technical Community College
Martin Tracey, Florida International University
Randall Tracy, Worcester State University
James Traniello, Boston University
Bibit Traut, City College of San Francisco
Terry Trier, Grand Valley State University
Stephen Trumble, Baylor University
Jan Trybula, The State University of New York at Potsdam
Alexa Tullis, University of Puget Sound
Marsha Turell, Houston Community College
Mary Tyler, University of Maine
Marcel van Tuinen, University of North Carolina at Wilmington
Dirk Vanderklein, Montclair State University
Jorge Vasquez-Kool, Wake Technical Community College
William Velhagen, New York University
Dennis Venema, Trinity Western University
Laura Vogel, North Carolina State University
Jyoti Wagle, Houston Community College
Jeff Walker, University of Southern Maine
Gary Walker, Appalachian State University
Andrea Ward, Adelphi University
Fred Wasserman, Boston University
Elizabeth Waters, San Diego State University
Douglas Watson, The University of Alabama at Birmingham
Matthew Weand, Southern Polytechnic State University
Michael Weber, Carleton University
Cindy Wedig, The University of Texas, Pan American
Brad Wetherbee, University of Rhode Island
Debbie Wheeler, University of the Fraser Valley
Clay White, Lone Star College System
Lisa Whitenack, Allegheny College
Maggie Whitson, Northern Kentucky University
Stacey Wild East, Tennessee State University
Herbert Wildey, Arizona State University and Phoenix College
David Wilkes, Indiana University, South Bend
Lisa Williams, Northern Virginia Community College
Elizabeth Willott, University of Arizona
Mark Wilson, Humboldt State University
Ken Wilson, University of Saskatchewan
Bob Winning, Eastern Michigan University
Candace Winstead, California Polytechnic State University
Robert Wise, University of Wisconsin, Oshkosh
D. Reid Wiseman, College of Charleston
MaryJo Witz, Monroe Community College
David Wolfe, American River College
Kevin Woo, University of Central Florida
Denise Woodward, Penn State
Shawn Wright, Central New Mexico Community College
Grace Wyngaard, James Madison University
Aimee Wyrick, Pacific Union College
Joanna Wysocka-Diller, Auburn University
Ken Yasukawa, Beloit College
John Yoder, The University of Alabama
Kelly Young, California State University, Long Beach
James Yount, Brevard Community College
Min Zhong, Auburn University

CONTENTS

About the Authors — iv
Brief Contents — vi
Preface — viii
Acknowledgments — xxi

? CASE 4 Malaria: Co-evolution of Humans and a Parasite — C4-2

CHAPTER 21 EVOLUTION
How Genotypes and Phenotypes Change over Time — 21-1

21.1 Genetic Variation — 21-1
- Population genetics is the study of patterns of genetic variation. — 21-1
- Mutation and recombination are the two sources of genetic variation. — 21-2
- Mutations can be harmful, neutral, or beneficial. — 21-2

21.2 Measuring Genetic Variation — 21-3
- To understand patterns of genetic variation, we require information about allele frequencies. — 21-3
- Early population geneticists relied on observable traits to measure variation. — 21-4
- Gel electrophoresis facilitates the detection of genetic variation. — 21-4
- DNA sequencing is the gold standard for measuring genetic variation. — 21-4
- HOW DO WE KNOW? How is genetic variation measured? — 21-5

21.3 Evolution and the Hardy–Weinberg Equilibrium — 21-6
- Evolution is a change in allele or genotype frequency over time. — 21-6
- The Hardy–Weinberg equilibrium describes situations in which allele and genotype frequencies do not change. — 21-6
- The Hardy–Weinberg equilibrium translates allele frequencies into genotype frequencies. — 21-7
- The Hardy–Weinberg equilibrium is the starting point for population genetic analysis. — 21-8

21.4 Natural Selection — 21-8
- Natural selection brings about adaptations. — 21-8
- The Modern Synthesis is a marriage between Mendelian genetics and Darwinian evolution. — 21-9
- Natural selection increases the frequency of advantageous mutations and decreases the frequency of deleterious mutations. — 21-10
- **?** What genetic differences have made some individuals more and some less susceptible to malaria? — 21-10
- Natural selection can be stabilizing, directional, or disruptive. — 21-10
- HOW DO WE KNOW? How far can artificial selection be taken? — 21-12
- Sexual selection increases an individual's reproductive success. — 21-13

21.5 Migration, Mutation, and Genetic Drift — 21-13
- Migration reduces genetic variation between populations. — 21-13
- Mutation increases genetic variation. — 21-13
- Genetic drift is particularly important in small populations. — 21-13

21.6 Molecular Evolution — 21-14
- The extent of sequence difference between species is a function of the time since the species diverged. — 21-15
- The rate of the molecular clock varies. — 21-15

CHAPTER 22 SPECIES AND SPECIATION — 22-1

22.1 The Biological Species Concept — 22-1
- The species is the fundamental evolutionary unit. — 22-1
- Reproductive isolation is the key to the biological species concept (BSC). — 22-2
- The BSC is more useful in theory than in practice. — 22-2
- The BSC does not apply to asexual or extinct organisms. — 22-3
- Ring species and hybridization complicate the BSC. — 22-4
- Ecology and evolution can extend the BSC. — 22-4

22.2 Reproductive Isolation — 22-5
- Pre-zygotic isolating factors occur before egg fertilization. — 22-5
- Post-zygotic isolating factors occur after egg fertilization. — 22-6

22.3 Speciation — 22-6
- Speciation is a by-product of the genetic divergence of separated populations. — 22-6
- Allopatric speciation is speciation that results from the geographical separation of populations. — 22-6

HOW DO WE KNOW? Can a vicariance event cause speciation?	22-8
Co-speciation is speciation that occurs in response to speciation in another species.	22-11
? How did malaria come to infect humans?	22-11
Can sympatric populations—those not geographically separated—undergo speciation?	22-12
Speciation can occur instantaneously.	22-13
22.4 Speciation and Selection	**22-15**
Speciation can occur with or without natural selection.	22-15
Natural selection can enhance reproductive isolation.	22-15
VISUAL SYNTHESIS Speciation	22-16

CHAPTER 23 EVOLUTIONARY PATTERNS
Phylogeny and Fossils   23-1

23.1 Reading a Phylogenetic Tree	**23-1**
Phylogenetic trees provide hypotheses of evolutionary relationships.	23-2
The search for sister groups lies at the heart of phylogenetics.	23-3
A monophyletic group consists of a common ancestor and all its descendants.	23-4
Taxonomic classifications are information storage and retrieval systems.	23-4
23.2 Building a Phylogenetic Tree	**23-5**
Homology is similarity by common descent.	23-5
Shared derived characters enable biologists to reconstruct evolutionary history.	23-6
The simplest tree is often favored among multiple possible trees.	23-6
Molecular data complement comparative morphology in reconstructing phylogenetic history.	23-8
HOW DO WE KNOW? Did an HIV-positive dentist spread the AIDS virus to his patients?	23-10
Phylogenetic trees can help solve practical problems.	23-10
23.3 The Fossil Record	**23-11**
Fossils provide unique information.	23-11
Fossils provide a selective record of past life.	23-12
Geological data indicate the age and environmental setting of fossils.	23-14
Fossils can contain unique combinations of characters.	23-16
HOW DO WE KNOW? Can fossils bridge the evolutionary gap between fish and tetrapod vertebrates?	23-18
Rare mass extinctions have altered the course of evolution.	23-18
23.4 Comparing Evolution's Two Great Patterns	**23-19**
Phylogeny and fossils complement each other.	23-19
Agreement between phylogenies and the fossil record provides strong evidence of evolution.	23-19

CHAPTER 24 HUMAN ORIGINS AND EVOLUTION   24-1

24.1 The Great Apes	**24-1**
Comparative anatomy shows that the human lineage branches off the great apes tree.	24-1
Molecular analysis reveals that our lineage split from the chimpanzee lineage about 5–7 million years ago.	24-3
HOW DO WE KNOW? How closely related are humans and chimpanzees?	24-3
The fossil record gives us direct information about our evolutionary history.	24-4
24.2 African Origins	**24-6**
Studies of mitochondrial DNA reveal that modern humans evolved in Africa.	24-6
HOW DO WE KNOW? When and where did the most recent common ancestor of all living humans live?	24-6
Neanderthals disappear from the fossil record as modern humans appear but have contributed to the modern human gene pool.	24-8
24.3 Distinct Features of Our Species	**24-9**
Bipedalism was a key innovation.	24-9
Adult humans share many features with juvenile chimpanzees.	24-10
Humans have large brains relative to body size.	24-11
The human and chimpanzee genomes help us identify genes that make us human.	24-12

24.4 Human Genetic Variation	24-12
The prehistory of our species has had an impact on the distribution of genetic variation.	24-13
The recent spread of modern humans means that there are few genetic differences between groups.	24-14
Some human differences have likely arisen by natural selection.	24-14
? What human genes are under selection for resistance to malaria?	24-15
24.5 Culture, Language, and Consciousness	24-15
Culture changes rapidly.	24-15
Is culture uniquely human?	24-16
Is language uniquely human?	24-17
Is consciousness uniquely human?	24-18

PART 2   FROM ORGANISMS TO THE ENVIRONMENT

CHAPTER 25   CYCLING CARBON — 25-1

25.1 The Short-Term Carbon Cycle	25-1
Photosynthesis and respiration are key processes in short-term carbon cycling.	25-2
The regular oscillation of CO_2 reflects the seasonality of photosynthesis in the Northern Hemisphere.	25-2
Human activities play an important role in the modern carbon cycle.	25-3
HOW DO WE KNOW? How much CO_2 was in the atmosphere 1000 years ago?	25-3
Carbon isotopes show that much of the CO_2 added to air over the past half century comes from burning fossil fuels.	25-4
HOW DO WE KNOW? What is the major source of the carbon dioxide that has accumulated in Earth's atmosphere over the last two centuries?	25-4
25.2 The Long-Term Carbon Cycle	25-6
Reservoirs and fluxes are key in long-term carbon cycling.	25-6
Physical processes add and remove CO_2 from the atmosphere.	25-7
Records of atmospheric composition over 400,000 years show periodic shifts in CO_2 content.	25-9
Variations in atmospheric CO_2 over hundreds of millions of years reflect plate tectonics and evolution.	25-11
25.3 The Carbon Cycle, Ecology, and Evolution	25-12
Food webs trace the movement of carbon through communities.	25-12
Trophic pyramids trace the flow of energy through communities.	25-13
The diversity of photosynthetic organisms reflects adaptation to a wide range of environments.	25-13
The diversity of respiring organisms reflects many sources of food.	25-14
The carbon cycle provides a framework for understanding life's evolutionary history.	25-14

? CASE 5   *The Human Microbiome: Diversity Within*   C5-1

CHAPTER 26   BACTERIA AND ARCHAEA — 26-1

26.1 Two Prokaryotic Domains	26-1
The bacterial cell is small but powerful.	26-1
Diffusion limits cell size in bacteria.	26-2
Horizontal gene transfer promotes genetic diversity in bacteria.	26-4
The Archaea form a second prokaryotic domain.	26-5
26.2 An Expanded Carbon Cycle	26-6
Many photosynthetic bacteria do not produce oxygen.	26-7
Many bacteria respire without oxygen.	26-8
Photoheterotrophs obtain energy from light but obtain carbon from preformed organic molecules.	26-8
Chemoautotrophy is a uniquely prokaryotic metabolism.	26-9
26.3 Other Biogeochemical Cycles	26-10
Bacteria and archaeons dominate Earth's sulfur cycle.	26-10
The nitrogen cycle is also driven by bacteria and archaeons.	26-10
26.4 The Diversity of Bacteria	26-12
Bacterial phylogeny is a work in progress.	26-12
HOW DO WE KNOW? How many kinds of bacterium live in the oceans?	26-13

What, if anything, is a bacterial species?	26-14
Proteobacteria are the most diverse bacteria.	26-15
The gram-positive bacteria include organisms that cause and cure disease.	26-15
Photosynthesis is widely distributed on the bacterial tree.	26-15

26.5 The Diversity of Archaea — 26-16

The archaeal tree has anaerobic, hyperthermophilic organisms near its base.	26-17
The Crenarchaeota and Euryarchaeota both include acid-loving microorganisms.	26-18
Euryarchaeote archaeons include heat-loving, methane-producing, and salt-loving microorganisms.	26-18
Thaumarchaeota may be the most abundant cells in the oceans.	26-18
HOW DO WE KNOW? How abundant are archaeons in the oceans?	26-19

26.6 The Evolutionary History of Prokaryotes — 26-20

Life originated early in our planet's history.	26-20
Prokaryotes have coevolved with eukaryotes.	26-21
? How do intestinal bacteria influence human health?	26-22

CHAPTER 27 EUKARYOTIC CELLS
Origins and Diversity — 27-1

27.1 The Eukaryotic Cell: A Review — 27-1

Internal protein scaffolding and dynamic membranes organize the eukaryotic cell.	27-1
In eukaryotic cells, energy metabolism is localized in mitochondria and chloroplasts.	27-2
The organization of the eukaryotic genome also helps explain eukaryotic diversity.	27-2
Sex promotes genetic diversity in eukaryotes and gives rise to distinctive life cycles.	27-3

27.2 Eukaryotic Origins — 27-4

? What role did symbiosis play in the origin of chloroplasts?	27-4
HOW DO WE KNOW? What is the evolutionary origin of chloroplasts?	27-5
? What role did symbiosis play in the origin of mitochondria?	27-6
? How did the eukaryotic cell originate?	27-7
In the oceans, many single-celled eukaryotes harbor symbiotic bacteria.	27-9

27.3 Eukaryotic Diversity — 27-9

Our own group, the opisthokonts, is the most diverse eukaryotic superkingdom.	27-10
Amoebozoans include slime molds that produce multicellular structures.	27-12
Archaeplastids are photosynthetic organisms, including land plants.	27-13
Other photosynthetic organisms occur in the stramenopiles and alveolates.	27-15
Photosynthesis spread through eukaryotes by repeated endosymbioses involving eukaryotic algae.	27-16
HOW DO WE KNOW? How did photosynthesis spread through the Eukarya?	27-18
The first branch of the eukaryotic tree may separate animals and slime molds from plants and diatoms.	27-19

27.4 The Fossil Record of Protists — 27-20

Fossils show that eukaryotes existed at least 1800 million years ago.	27-20
Protists have continued to diversify during the age of animals.	27-21

CHAPTER 28 BEING MULTICELLULAR — 28-1

28.1 The Phylogenetic Distribution of Multicellular Organisms — 28-1

Simple multicellularity is widespread among eukaryotes.	28-1
Complex multicellularity evolved several times.	28-3

28.2 Diffusion vs. Bulk Transport — 28-4

Diffusion is effective only over short distances.	28-4
Animals achieve large size by circumventing limits imposed by diffusion.	28-4
Complex multicellular organisms have structures specialized for bulk transport.	28-5

28.3 How to Build a Multicellular Organism — 28-6

Complex multicellularity requires adhesion between cells.	28-6
HOW DO WE KNOW? How do bacteria influence the life cycles of choanoflagellates?	28-6

How did animal cell adhesion originate? 28-7

Complex multicellularity requires communication between cells. 28-7

Complex multicellularity requires a genetic program for coordinated growth and cell differentiation. 28-8

28.4 Variations on a Theme: Plants vs. Animals 28-10

Cell walls shape patterns of growth and development in plants. 28-10

Animal cells can move relative to one another. 28-11

28.5 The Evolution of Complex Multicellularity 28-12

Complex multicellularity appeared in the oceans 575 to 555 million years ago. 28-12

Oxygen is necessary for complex multicellular life. 28-13

Land plants evolved from green algae that could carry out photosynthesis on land. 28-14

Regulatory genes have played an important role in the evolution of complex multicellular organisms. 28-15

HOW DO WE KNOW?
What controls color pattern in butterfly wings? 28-15

? CASE 6 *Agriculture: Feeding a Growing Population* C6-2

CHAPTER 29 PLANT STRUCTURE AND FUNCTION
Moving Photosynthesis onto Land 29-1

29.1 Plant Structure and Function: An Evolutionary Perspective 29-1

29.2 The Leaf: Acquiring CO_2 While Avoiding Dessication 29-2

CO_2 uptake results in water loss. 29-3

The cuticle restricts water loss from leaves but inhibits the uptake of CO_2. 29-4

Stomata allow leaves to regulate water loss and carbon gain. 29-5

CAM plants use nocturnal CO_2 storage to avoid water loss during the day. 29-6

C_4 plants suppress photorespiration. 29-7

HOW DO WE KNOW?
How do we know that C_4 photosynthesis suppresses photorespiration? 29-8

29.3 The Stem: Transport of Water Through Xylem 29-8

Xylem provides a low-resistance pathway for the movement of water. 29-9

Water is pulled through xylem by an evaporative pump. 29-10

HOW DO WE KNOW?
How large are the forces that allow leaves to pull water from the soil? 29-10

The structure of xylem conduits reduces the risks of collapse and cavitation. 29-11

29.4 The Stem: Transport of Carbohydrates Through Phloem 29-12

Carbohydrates are pushed through phloem by an osmotic pump. 29-13

Phloem feeds both the plant and the rhizosphere. 29-14

29.5 The Root: Uptake of Water and Nutrients from the Soil 29-15

Nutrient uptake by roots is highly selective. 29-15

Nutrient uptake requires energy. 29-16

Mycorrhizae enhance the uptake of phosphorus. 29-17

Symbiotic nitrogen-fixing bacteria supply nitrogen to both plants and ecosystems. 29-17

? How has nitrogen availability influenced agricultural productivity? 29-18

CHAPTER 30 PLANT REPRODUCTION
Finding Mates and Dispersing Young 30-1

30.1 The Plant Life Cycle and Evolution of Pollen and Seeds 30-1

The algal sister groups of land plants have one multicellular generation in their life cycle. 30-2

Bryophytes illustrate how the alternation of generations allows the dispersal of spores in the air. 30-2

Vascular plants evolved a large photosynthetic sporophyte generation. 30-4

In seed plants, the transport of pollen in air allows fertilization to occur in the absence of external sources of water. 30-6

Seeds enhance dispersal and establishment of the next sporophyte generation. 30-8

30.2 Flowering Plants 30-9

Flowers are reproductive shoots specialized for the production, transfer, and receipt of pollen. 30-9

The diversity of floral morphology is related to modes of pollination. 30-10

Angiosperms have mechanisms to increase outcrossing. 30-12

HOW DO WE KNOW?
Can pollinator shifts enhance rates of species formation? ... 30-13

Angiosperms delay provisioning their ovules until after fertilization. ... 30-15

Fruits enhance the dispersal of seeds. ... 30-15

30.3 Timing of Reproductive Events ... 30-17

Flowering time is affected by day length. ... 30-17

Photoreceptors enable plants to measure day length. ... 30-17

Vernalization prevents plants from flowering until winter has passed. ... 30-18

Dormant seeds can delay germination if they detect the presence of plants overhead. ... 30-18

HOW DO WE KNOW?
How do plants measure day length? ... 30-19

HOW DO WE KNOW?
How do seeds detect the presence of plants growing overhead? ... 30-20

? What is the basis for the spectacular increases in the yield of cereal grains during the Green Revolution? ... 30-21

30.4 Vegetative Reproduction ... 30-22

CHAPTER 31 **PLANT GROWTH AND DEVELOPMENT**
Building the Plant Body ... 31-1

31.1 Shoot Growth and Development ... 31-2

Shoots grow by adding new cells at their tips. ... 31-2

Stem elongation occurs primarily in a zone just below the apical meristem where new cells elongate. ... 31-3

The shoot apical meristem controls the production and arrangement of leaves. ... 31-4

The development of new apical meristems allows stems to branch. ... 31-5

Flowers grow from and consume shoot meristems. ... 31-6

31.2 Plant Hormones ... 31-6

Hormones affect the growth and differentiation of plant cells. ... 31-7

Auxin transport guides the development of vascular connections between leaves and stems. ... 31-8

? What is the developmental basis for the shorter stems of high-yielding rice and wheat? ... 31-9

Branching is affected by multiple hormones. ... 31-10

31.3 Secondary Growth ... 31-10

Shoots produce two types of lateral meristem. ... 31-10

The vascular cambium produces secondary xylem and phloem. ... 31-11

The cork cambium produces an outer protective layer. ... 31-12

Wood has both mechanical and transport functions. ... 31-12

31.4 Root Growth and Development ... 31-13

Roots grow by producing new cells at their tips. ... 31-14

The formation of new root apical meristems allows roots to branch. ... 31-15

The structures and functions of root systems are diverse. ... 31-15

31.5 The Environmental Context of Growth and Development ... 31-16

Plants orient the growth of their stems and roots by light and gravity. ... 31-17

HOW DO WE KNOW?
How do plants grow toward light? ... 31-17

Plants grow taller and branch less when light levels are low. ... 31-19

Roots elongate more and branch less when water is scarce. ... 31-19

Exposure to wind results in shorter and stronger stems. ... 31-20

Plants use day length as a cue to prepare for winter. ... 31-20

CHAPTER 32 **PLANT DEFENSE**
Keeping the World Green ... 32-1

32.1 Protection Against Pathogens ... 32-1

Plant pathogens infect and exploit host plants by a variety of mechanisms. ... 32-2

An innate immune system allows plants to detect and respond to pathogens. ... 32-3

Plants respond to infections by isolating infected regions. ... 32-4

Mobile signals trigger defenses in uninfected tissues. ... 32-5

HOW DO WE KNOW?
Can plants develop immunity to specific pathogens? ... 32-6

Plants defend against viral infections by producing siRNA. ... 32-6

A pathogenic bacterium provides a way to modify plant genomes. ... 32-7 xxxiii

32.2 Defense Against Herbivores — 32-8

Plants use mechanical and chemical defenses to avoid being eaten. — 32-8

Diverse chemical compounds deter herbivores. — 32-9

Some plants provide food and shelter for ants, which actively defend them. — 32-11

Grasses can regrow quickly following grazing by mammals. — 32-12

32.3 Allocating Resources to Defense — 32-13

Plants can sense and respond to herbivores. — 32-13

Plants produce volatile signals that attract insects that prey upon herbivores. — 32-14

Nutrient-rich environments select for plants that allocate more resources to growth than to defense. — 32-14

HOW DO WE KNOW?
Can plants communicate? — 32-14

Exposure to multiple threats can lead to trade-offs. — 32-16

32.4 Defense and Plant Diversity — 32-16

Pathogens, herbivores, and seed predators can increase plant biodiversity. — 32-16

The evolution of new defenses may allow plants to diversify. — 32-17

? Can modifying plants genetically protect crops from herbivores and pathogens? — 32-17

CHAPTER 33 PLANT DIVERSITY — 33-1

33.1 Plant Diversity: An Evolutionary Overview — 33-1

33.2 Bryophytes — 33-2

Bryophytes are small, simple, and tough. — 33-3

Bryophytes exhibit several cases of convergent evolution with the vascular plants. — 33-4

Sphagnum moss plays an important role in the global carbon cycle. — 33-5

33.3 Spore-Dispersing Vascular Plants — 33-5

Rhynie cherts provide a window into the early evolution of vascular plants. — 33-5

Lycophytes are the sister group of all other vascular plants. — 33-6

Ancient lycophytes included giant trees that dominated coal swamps about 320 million years ago. — 33-7

HOW DO WE KNOW?
Did woody plants evolve more than once? — 33-8

Ferns and horsetails are morphologically and ecologically diverse. — 33-9

An aquatic fern contributes to rice production. — 33-10

33.4 Gymnosperms — 33-11

Cycads and ginkgos are the earliest diverging groups of living gymnosperms. — 33-11

Conifers are forest giants that thrive in dry and cold climates. — 33-13

Gnetophytes are gymnosperms that have independently evolved xylem vessels and double fertilization. — 33-14

33.5 Angiosperms — 33-14

Angiosperm diversity remains a puzzle. — 33-15

Early diverging angiosperms have low diversity. — 33-15

Monocots develop according to a novel body plan. — 33-16

HOW DO WE KNOW?
When did grasslands expand over the land surface? — 33-18

Eudicots are the most diverse group of angiosperms. — 33-19

? What can be done to protect the genetic diversity of crop species? — 33-20

VISUAL SYNTHESIS Angiosperms: Structure and Function — 33-22

CHAPTER 34 FUNGI
Structure, Function, and Diversity — 34-1

34.1 Growth and Nutrition — 34-1

Hyphae permit fungi to explore their environment for food resources. — 34-2

Fungi transport materials within their hyphae. — 34-2

Not all fungi produce hyphae. — 34-2

Fungi are principal decomposers of plant tissues. — 34-3

Fungi are important plant and animal pathogens. — 34-4

Many fungi form symbiotic associations with plants and animals. — 34-5

Lichens are symbioses between a fungus and a green alga or a cyanobacterium. — 34-6

34.2 Reproduction — 34-7

Fungi proliferate and disperse using spores. — 34-7

Multicellular fruiting bodies facilitate the dispersal of sexually produced spores.	34-8
HOW DO WE KNOW? What determines the shape of fungal spores that are ejected into the air?	34-9
The fungal life cycle often includes a stage in which haploid cells fuse, but nuclei do not.	34-10
Genetically distinct mating types promote outcrossing.	34-11
Fungi that lack sexual reproduction have other means of generating genetic diversity.	34-11
34.3 Diversity	**34-12**
Fungi are highly diverse.	34-12
Fungi evolved from aquatic, unicellular, and flagellated ancestors.	34-12
Zygomycetes produce hyphae undivided by septa.	34-13
Glomeromycetes form endomycorrhizae.	34-14
The Dikarya produce regular septa during mitosis.	34-14
Ascomycetes are the most diverse group of fungi.	34-14
HOW DO WE KNOW? Can a fungus influence the behavior of an ant?	34-16
Basidiomycetes include smuts, rusts, and mushrooms.	34-17
? How do fungi threaten global wheat production?	34-19

? **CASE 7** *Predator–Prey: A Game of Life and Death*	**C7-1**

CHAPTER 35 ANIMAL NERVOUS SYSTEMS	**35-1**
35.1 Nervous System Function and Evolution	**35-1**
Animal nervous systems have three types of nerve cell.	35-2
Nervous systems range from simple to complex.	35-2
? What body features arose as adaptations for successful predation?	35-4
35.2 Neuron Structure	**35-4**
Neurons share a common organization.	35-5
Neurons differ in size and shape.	35-5
Neurons are supported by other types of cell.	35-6
35.3 Neuron Function	**35-6**
The resting membrane potential is negative and results in part from the movement of potassium ions.	35-6
Neurons are excitable cells that transmit information by action potentials.	35-8
Neurons propagate action potentials along their axons by sequentially opening and closing adjacent Na^+ and K^+ ion channels.	35-8
HOW DO WE KNOW? What is the resting membrane potential and what changes in electrical activity occur during an action potential?	35-11
Nerve cells communicate at synapses.	35-12
Signals between neurons can be excitatory or inhibitory.	35-13
35.4 Nervous System Organization	**35-15**
Nervous systems are organized into peripheral and central components.	35-15
Nervous systems have voluntary and involuntary components.	35-16
The nervous system helps to maintain homeostasis.	35-17
Simple reflex circuits provide rapid responses to stimuli.	35-18

CHAPTER 36 ANIMAL SENSORY SYSTEMS AND BRAIN FUNCTION	**36-1**
36.1 Animal Sensory Systems	**36-1**
Specialized sensory receptors detect diverse stimuli.	36-2
Stimuli are transmitted by changes in the firing rate of action potentials.	36-4
36.2 Smell and Taste	**36-5**
Smell and taste depend on chemoreception of molecules carried in the environment and in food.	36-5
36.3 Sensing Gravity, Movement, and Sound	**36-6**
Hair cells sense gravity and motion.	36-7
Hair cells detect the physical vibrations of sound.	36-8
? How have sensory systems evolved in predators and prey?	36-10
36.4 Vision	**36-10**
All animals use a similar photosensitive protein called opsin to detect light.	36-11
Animals see the world through different types of eyes.	36-11
The structure and function of the vertebrate eye underlie image processing.	36-13

Color vision detects different wavelengths of light.	36-14
Local sensory processing of light determines basic features of shape and movement.	36-15
HOW DO WE KNOW? How does the retina process visual information?	36-16

36.5 Brain Organization and Function — 36-17

The brain processes and integrates information received from different sensory systems.	36-17
The brain is divided into lobes with specialized functions.	36-18
Information is topographically mapped into the vertebrate cerebral cortex.	36-19

36.6 Memory and Cognition — 36-20

The brain serves an important role in memory and learning.	36-20
Cognition involves brain information processing and decision making.	36-21

CHAPTER 37 ANIMAL MOVEMENT
Muscles and Skeletons — 37-1

37.1 Muscles: Biological Motors That Generate Force and Produce Movement — 37-1

Muscles can be striated or smooth.	37-1
Skeletal muscle fibers are organized into repeating contractile units called sarcomeres.	37-2
Muscles contract by the sliding of myosin and actin filaments.	37-4
Calcium regulates actin–myosin interaction through excitation–contraction coupling.	37-6
Calmodulin regulates Ca^{2+} activation and relaxation of smooth muscle.	37-7

37.2 Muscle Contractile Properties — 37-8

Muscle length affects actin–myosin overlap and generation of force.	37-8
HOW DO WE KNOW? How does filament overlap affect force generation in muscles?	37-8
Muscle force and shortening velocity are inversely related.	37-9
Antagonist pairs of muscles produce reciprocal motions at a joint.	37-10
Muscle force is summed by an increase in stimulation frequency and the recruitment of motor units.	37-10
Skeletal muscles have slow-twitch and fast-twitch fibers.	37-11
? How do different types of muscle fiber affect the speed of predators and prey?	37-12

37.3 Animal Skeletons — 37-13

Hydrostatic skeletons support animals by muscles that act on a fluid-filled cavity.	37-13
Exoskeletons provide hard external support and protection.	37-14
The rigid bones of vertebrate endoskeletons are jointed for motion and can be repaired if damaged.	37-15

37.4 Vertebrate Skeletons — 37-16

Vertebrate bones form by intramembranous and endochondral ossification.	37-16
Joint shape determines range of motion and skeletal muscle organization.	37-17
Muscles exert forces by skeletal levers to produce joint motion.	37-18

CHAPTER 38 ANIMAL ENDOCRINE SYSTEMS — 38-1

38.1 An Overview of Endocrine Function — 38-1

The endocrine system helps to regulate an organism's response to its environment.	38-1
The endocrine system is involved in growth and development.	38-2
HOW DO WE KNOW? How are growth and development controlled in insects?	38-3
The endocrine system underlies homeostasis.	38-5

38.2 Properties of Hormones — 38-7

Three main classes of hormone are peptide, amine, and steroid hormones.	38-7
Hormonal signals are typically amplified.	38-8
Hormones act specifically on cells with receptors that bind the hormone.	38-11
Hormones are evolutionarily conserved molecules with diverse functions.	38-11

38.3 The Vertebrate Endocrine System — 38-12

The pituitary gland integrates diverse bodily functions by secreting hormones in response to signals from the hypothalamus.	38-13
Many targets of pituitary hormones are endocrine tissues that also release hormones.	38-14

Other endocrine organs have diverse functions.	38-15
? **How does the endocrine system influence predators and prey?**	**38-15**
38.4 Other Forms of Chemical Communication	**38-16**
Local chemical signals regulate neighboring target cells.	38-16
Pheromones are chemical compounds released into the environment to signal behavioral cues to other species members.	38-17

CHAPTER 39 ANIMAL CARDIOVASCULAR AND RESPIRATORY SYSTEMS 39-1

39.1 Delivery of Oxygen and Elimination of Carbon Dioxide	**39-1**
Diffusion governs gas exchange over short distances.	39-2
Bulk flow moves fluid over long distances.	39-2
39.2 Respiratory Gas Exchange	**39-3**
Many aquatic animals breathe through gills.	39-4
Insects breathe air through tracheae.	39-5
Terrestrial vertebrates breathe by tidal ventilation of internal lungs.	39-6
Mammalian lungs are well adapted for gas exchange.	39-7
The structure of bird lungs allows unidirectional airflow for increased oxygen uptake.	39-8
Voluntary and involuntary mechanisms control breathing.	39-9
39.3 Oxygen Transport by Hemoglobin	**39-10**
Hemoglobin is an ancient molecule with diverse roles related to oxygen binding and transport.	39-10
HOW DO WE KNOW? What is the molecular structure of hemoglobin and myoglobin?	**39-10**
Hemoglobin reversibly binds oxygen.	39-11
Myoglobin stores oxygen, enhancing delivery to muscle mitochondria.	39-12
Many factors affect hemoglobin–oxygen binding.	39-12
39.4 Circulatory Systems	**39-13**
Circulatory systems have vessels of different sizes to facilitate bulk flow and diffusion.	39-15
Arteries are muscular, elastic vessels that carry blood away from the heart under high pressure.	39-15
Veins are thin-walled vessels that return blood to the heart under low pressure.	39-16
Compounds and fluid move across capillary walls by diffusion, filtration, and osmosis.	39-16
? **How do hormones and nerves provide homeostatic regulation of blood flow as well as allow an animal to respond to stress?**	**39-17**
39.5 The Evolution, Structure, and Function of the Heart	**39-17**
Fish have two-chambered hearts and a single circulatory system.	39-18
Amphibians and reptiles have more three-chambered hearts and partially divided circulations.	39-18
Mammals and birds have four-chambered hearts and fully divided pulmonary and systemic circulations.	39-19
Cardiac muscle cells are electrically connected to contract in synchrony.	39-20
Cardiac output is regulated by the autonomic nervous system.	39-21

CHAPTER 40 ANIMAL METABOLISM, NUTRITION, AND DIGESTION 40-1

40.1 Patterns of Animal Metabolism	**40-1**
Animals rely on anaerobic and aerobic metabolism.	40-2
Metabolic rate varies with activity level.	40-3
? **Does body temperature limit activity level in predators and prey?**	**40-5**
Metabolic rate is affected by body size.	40-5
HOW DO WE KNOW? How is metabolic rate affected by running speed and body size?	**40-6**
Metabolic rate is linked to body temperature.	40-7
40.2 Animal Nutrition and Diet	**40-7**
Energy balance is a form of homeostasis.	40-7
VISUAL SYNTHESIS Homeostasis and Thermoregulation	40-8
An animal's diet must supply nutrients that it cannot synthesize.	40-10
40.3 Adaptations for Feeding	**40-12**
Suspension filter feeding is common in many aquatic animals.	40-12
Large aquatic animals apprehend prey by suction feeding and active swimming.	40-12

Jaws and teeth provide specialized food capture and mechanical breakdown of food. 40-13

40.4 Digestion and Absorption of Food 40-14

The digestive tract has regional specializations. 40-14

Digestion begins in the mouth. 40-15

The stomach is an initial storage and digestive chamber. 40-16

The small intestine is specialized for nutrient absorption. 40-17

The large intestine reabsorbs water and stores waste. 40-20

The lining of the digestive tract is composed of distinct layers. 40-21

Plant-eating animals have specialized digestive tracts that reflect their diets. 40-21

CHAPTER 41 ANIMAL RENAL SYSTEMS
Water and Waste 41-1

41.1 Water and Electrolyte Balance 41-1

Osmosis governs the movement of water across cell membranes. 41-2

Osmoregulation is the control of osmotic pressure inside cells and organisms. 41-3

Osmoconformers match their internal solute concentration to that of the environment. 41-4

Osmoregulators have internal solute concentrations that differ from that of their environment. 41-5

? Can the loss of water and electrolytes in exercise be exploited as a strategy to hunt prey? 41-6

41.2 Excretion of Wastes in Relation to Electrolyte Balance 41-7

The excretion of nitrogenous wastes is linked to an animal's habitat and evolutionary history. 41-7

Excretory organs work by filtration, reabsorption, and secretion. 41-8

Animals have diverse excretory organs. 41-9

Vertebrates filter blood under pressure through paired kidneys. 41-10

41.3 Structure and Function of the Mammalian Kidney 41-12

The mammalian kidney has an outer cortex and inner medulla. 41-12

Glomerular filtration isolates wastes carried by the blood along with water and small solutes. 41-12

The proximal convoluted tubule reabsorbs solutes by active transport. 41-14

The loop of Henle acts as a countercurrent multiplier to create a concentration gradient from the cortex to the medulla. 41-14

HOW DO WE KNOW?
How does the mammalian kidney produce concentrated urine? 41-16

The distal convoluted tubule secretes additional wastes. 41-17

The final concentration of urine is determined in the collecting ducts and is under hormonal control. 41-17

The kidneys help regulate blood pressure and blood volume. 41-19

CHAPTER 42 ANIMAL REPRODUCTION AND DEVELOPMENT 42-1

42.1 The Evolutionary History of Reproduction 42-1

Asexual reproduction produces clones. 42-2

Sexual reproduction involves the formation and fusion of gametes. 42-3

Many species reproduce both sexually and asexually. 42-4

Exclusive asexuality is often an evolutionary dead end. 42-4

HOW DO WE KNOW?
Do bdelloid rotifers only reproduce asexually? 42-6

42.2 Movement onto Land and Reproductive Adaptations 42-7

Fertilization can take place externally or internally. 42-7

r-strategists and *K*-strategists differ in number of offspring and parental care. 42-8

Oviparous animals lay eggs, and viviparous animals give birth to live young. 42-8

42.3 Human Reproductive Anatomy and Physiology 42-9

The male reproductive system is specialized for the production and delivery of sperm. 42-9

The female reproductive system produces eggs and supports the developing embryo. 42-11

Hormones regulate the human reproductive system. 42-13

42.4 Gamete Formation to Birth in Humans — 42-15

Male and female gametogenesis have shared and distinct features. — 42-15

Fertilization occurs when a sperm fuses with an oocyte. — 42-16

The first trimester includes cleavage, gastrulation, and organogenesis. — 42-17

The second and third trimesters are characterized by fetal growth. — 42-19

VISUAL SYNTHESIS Reproduction and Development — 42-20

Childbirth is initiated by hormonal changes. — 42-22

CHAPTER 43 ANIMAL IMMUNE SYSTEMS — 43-1

43.1 Innate Immunity — 43-1

The skin and mucous membranes provide the first line of defense against infection. — 43-2

Some cells act broadly against diverse pathogens. — 43-3

Phagocytes recognize foreign molecules and send signals to other cells. — 43-4

Inflammation is a coordinated response to tissue injury. — 43-5

The complement system participates in the innate and adaptive immune systems. — 43-6

43.2 Adaptive Immunity: B cells, Antibodies, and Humoral Immunity — 43-7

B cells produce antibodies. — 43-7

Mammals produce five classes of antibody with different biological functions. — 43-8

Clonal selection is the basis for antibody specificity. — 43-9

Clonal selection also explains immunological memory. — 43-10

Genomic rearrangement creates antibody diversity. — 43-10

HOW DO WE KNOW?
How is antibody diversity generated? — 43-11

43.3 Adaptive Immunity: T cells and Cell-Mediated Immunity — 43-13

T cells include helper and cytotoxic cells. — 43-13

T cells have T cell receptors on their surface. — 43-14

T cell activation requires the presence of antigen in association with MHC proteins. — 43-14

The ability to distinguish between self and non-self is acquired during T cell maturation. — 43-16

43.4 Three Infections: A Virus, Bacterium, and Eukaryote — 43-16

The flu virus evades the immune system through antigenic drift and shift. — 43-17

Tuberculosis is caused by a slow-growing, intracellular bacterium. — 43-18

The malaria parasite uses antigenic variation to change surface molecules. — 43-18

? CASE 8 *Biodiversity Hotspots: Rain Forests and Coral Reefs* — C8-1

CHAPTER 44 ANIMAL DIVERSITY — 44-1

44.1 A Tree of Life for More than a Million Animal Species — 44-1

Phylogenetic trees propose an evolutionary history of animals. — 44-1

Nineteenth-century biologists grouped animals by anatomical and embryological features. — 44-2

Molecular sequence comparisons have confirmed some relationships and raised new questions. — 44-4

44.2 The Simplest Animals: Sponges, Cnidarians, Ctenophores, and Placozoans — 44-5

Sponges are simple and widespread in the oceans. — 44-5

Cnidarians are the architects of life's largest constructions: coral reefs. — 44-7

Ctenophores and placozoans represent the extremes of body organization among early branching animals. — 44-8

44.3 Bilaterian Animals — 44-11

Lophotrochozoans make up nearly half of all animal phyla, including the diverse and ecologically important annelids and mollusks. — 44-11

Ecdysozoans include arthropods, the most diverse animals. — 44-14

HOW DO WE KNOW?
How did the diverse feeding appendages of arthropods arise? — 44-16

Deuterostomes include humans and other chordates, but also acorn worms and sea stars. — 44-18

Chordates include vertebrates, cephalochordates, and tunicates.		44-19
44.4 Vertebrate Diversity		44-21
Fish are the earliest-branching and most diverse vertebrate animals.		44-22
The common ancestor of tetrapods had four limbs.		44-24
Amniotes evolved terrestrial eggs.		44-25
44.5 The Evolutionary History of Animals		44-27
Fossils and phylogeny show that animal forms were initially simple but rapidly evolved complexity.		44-27
The animal body plans we see today emerged during the Cambrian Period.		44-27
Animals began to colonize the land 420 million years ago.		44-28
? How have coral reefs changed through time?		44-29
VISUAL SYNTHESIS Diversity through Time		44-30

CHAPTER 45 ANIMAL BEHAVIOR 45-1

45.1 Tinbergen's Questions	45-1
45.2 Genes and Behavior	45-2
The fixed action pattern is a stereotyped behavior.	45-2
The nervous system processes stimuli and evokes behaviors.	45-3
Hormones can trigger certain behaviors.	45-4
Breeding experiments can help determine the degree to which a behavior is genetic.	45-5
Molecular techniques provide new ways of testing the role of genes in behavior.	45-6
HOW DO WE KNOW? Can genes influence behavior?	45-7
45.3 Learning	45-8
Non-associative learning occurs without linking two events.	45-8
Associative learning occurs when two events are linked.	45-9
Learning takes many forms.	45-9
HOW DO WE KNOW? To what extent are insects capable of learning?	45-10
45.4 Orientation, Navigation, and Biological Clocks	45-11
Orientation involves a directed response to a stimulus.	45-11

Navigation is illustrated by the remarkable ability of homing in birds.	45-11
Biological clocks provide important time cues for many behaviors.	45-12
HOW DO WE KNOW? Does a biological clock play a role in birds' ability to orient?	45-12
45.5 Communication	45-13
Communication is the transfer of information between a sender and receiver.	45-14
Some forms of communication are complex and learned during a sensitive period.	45-15
Other forms of communication convey specific information.	45-15
45.6 Social Behavior	45-16
Group selection is a weak explanation of altruistic behavior.	45-16
Reciprocal altruism is one way that altruism can evolve.	45-17
Kin selection is based on the idea that it is possible to contribute genetically to future generations by helping close relatives.	45-18
45.7 Behavior and Sexual Selection	45-19
Patterns of sexual selection are governed by differences between the sexes in their investment in offspring.	45-20
Sexual selection can be intrasexual or intersexual.	45-20

CHAPTER 46 POPULATION ECOLOGY 46-1

46.1 Populations and Their Properties	46-1
Three key features of a population are its size, range, and density.	46-2
Population size can increase or decrease over time.	46-3
Carrying capacity is the maximum number of individuals a habitat can support.	46-5
Factors that influence population growth can be dependent on or independent of its density.	46-6
Ecologists estimate population size and density by sampling.	46-6
HOW DO WE KNOW? How many butterflies are there in a given population?	46-7

46.2 Age-Structured Population Growth — 46-8

Birth and death rates vary with age and environment. — 46-8

Survivorship curves record changes in survival probability over an organism's life-span. — 46-9

Patterns of survivorship vary among organisms. — 46-10

Reproductive patterns reflect the predictability of a species' environment. — 46-10

The life history of an organism shows trade-offs among physiological functions. — 46-11

46.3 Metapopulation Dynamics — 46-12

A metapopulation is a group of populations linked by immigrants. — 46-12

Island biogeography explains species diversity on habitat islands. — 46-14

? **How do islands promote species diversification?** — 46-15

Species coexistence depends on habitat diversity. — 46-16

HOW DO WE KNOW?
Can predators and prey coexist stably in certain environments? — 46-17

CHAPTER 47 SPECIES INTERACTIONS, COMMUNITIES, AND ECOSYSTEMS — 47-1

47.1. The Niche — 47-1

The niche is the ecological role played by a species in its community. — 47-2

The realized niche of a species is more restricted than its fundamental niche. — 47-3

47.2 Antagonistic Interactions Between Species — 47-3

Limited resources foster competition. — 47-4

Competition promotes niche divergence. — 47-4

Species compete for resources other than food. — 47-5

Predators and parasites can limit prey population size, minimizing competition. — 47-5

47.3 Mutualistic Interactions Between Species — 47-5

Mutualisms are interactions between species that benefit both participants. — 47-6

Mutualisms may evolve increasing interdependence. — 47-6

Mutualisms may be obligate or facultative. — 47-6

HOW DO WE KNOW?
Have aphids and their symbiotic bacteria coevolved? — 47-7

The costs and benefits of species interactions can change over time. — 47-8

47.4 Ecological Communities — 47-9

Species that live in the same place make up communities. — 47-9

A single herbivore species can affect other herbivores and their predators. — 47-9

Keystone species have disproportionate effects on communities. — 47-10

Disturbance can modify community composition. — 47-11

Succession describes the community response to new habitats or disturbance. — 47-12

47.5 Ecosystems — 47-13

Species interactions result in food webs that cycle carbon and other elements through ecosystems. — 47-13

Species interactions form trophic pyramids that transfer energy through ecosystems. — 47-15

Light, water, nutrients, and diversity all influence rates of primary production. — 47-15

HOW DO WE KNOW?
Does species diversity promote primary productivity? — 47-16

47.6 Biomes and Diversity Gradients — 47-16

Biomes reflect the interaction of Earth and life. — 47-19

Tropical biomes usually have more species than temperate biomes. — 47-20

? **Why are tropical species so diverse?** — 47-21

Evolutionary and ecological history underpins diversity. — 47-22

CHAPTER 48 THE ANTHROPOCENE
Humans as a Planetary Force — 48-1

48.1 The Anthropocene Period — 48-1

Humans are a major force on the planet. — 48-1

48.2 Human Influence on the Carbon Cycle — 48-3

As atmospheric carbon dioxide levels have increased, so has mean surface temperature. — 48-3

Changing environments affect species distribution and community composition. — 48-5

?	**How has climate change affected coral reefs around the world?**	48-7
	HOW DO WE KNOW? What is the effect of increased atmospheric CO_2 and reduced ocean pH on skeleton formation in marine algae?	48-9
	What can be done?	48-10
48.3	**Human Influence on the Nitrogen and Phosphorus Cycles**	48-11
	Nitrogen fertilizer transported to lakes and the sea causes eutrophication.	48-11
	VISUAL SYNTHESIS Succession: Ecology in Microcosm	48-12
	Phosphate fertilizer is also used in agriculture, but has finite sources.	48-14
	What can be done?	48-14
48.4	**Human Influence on Evolution**	48-15
?	**How has human activity affected biological diversity?**	48-15
	Humans play an important role in the dispersal of species.	48-17
	Humans have altered the selective landscape for many pathogens.	48-18
	Are amphibians ecology's "canary in the coal mine"?	48-19
48.5	**Scientists and Citizens in the 21st Century**	48-19

QUICK CHECK ANSWERS	**Q-1**
GLOSSARY	**G-1**
CREDITS/SOURCES	**CS-1**
INDEX	**I-1**

CASE 4

Malaria

CO-EVOLUTION OF HUMANS AND A PARASITE

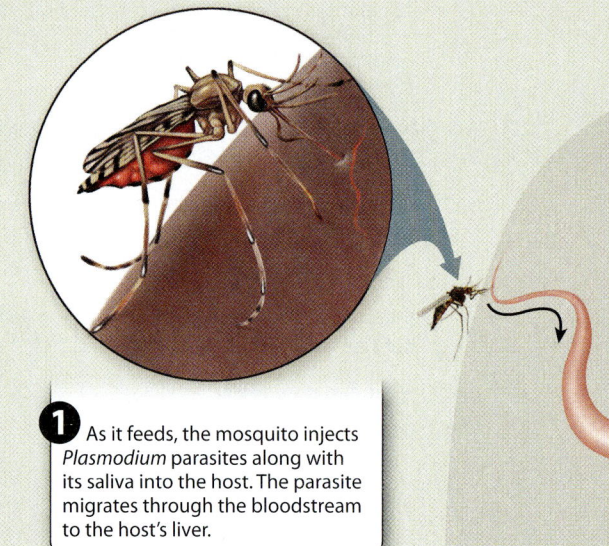

1 As it feeds, the mosquito injects *Plasmodium* parasites along with its saliva into the host. The parasite migrates through the bloodstream to the host's liver.

Most people would say the world has more than enough mosquitoes, but in 2010 scientists at the University of Arizona conjured up a new variety. The high-tech bloodsuckers were genetically engineered to resist *Plasmodium*, the single-celled eukaryote that causes malaria.

Normally, the parasite grows in the mosquito's gut and is spread to humans by the insect's bite. By altering a single gene in the mosquito's genome, the researchers had made the insects immune to the malaria parasite. The accomplishment is a noteworthy advance, the latest in a long line of efforts to stop *Plasmodium* in its tracks.

Malaria is one of the most devastating diseases on the planet. The World Health Organization estimates that 500 million people contract malaria annually, primarily in tropical regions. The disease is thought to claim about a million lives each year. Of those deaths, 85% to 90% occur in sub-Saharan Africa, mostly among children under 5.

As humans have tried a succession of weapons to defeat *P. falciparum*, the parasite has evolved, thwarting their efforts.

Four species of *Plasmodium* can cause malaria in humans. One of them, *P. falciparum*, is particularly dangerous, accounting for the vast majority of malaria fatalities. Over thousands of years, this parasite and its human host have played a deadly game of tug-of-war. As humans have tried a succession of weapons to defeat *P. falciparum*, the parasite has evolved, thwarting their efforts. And in turn, the tiny organism has helped to shape human evolution.

Plasmodium is a wily and complicated parasite, requiring both humans and mosquitoes to complete its life cycle. The part of the cycle in humans begins with a single bite from an infected mosquito. As the insect draws blood, it releases *Plasmodium*-laden saliva into the bloodstream. Once inside their human host, the *Plasmodium* parasites invade the liver cells. There they undergo cell division for several days, their numbers increasing. Eventually, they infect red blood cells, where they continue to grow and multiply. The mature parasites burst from the red blood cells at regular intervals, triggering malaria's telltale cycle of fever and chills.

Some of the freed *Plasmodium* parasites go on to infect new red blood cells; others divide to form gametocytes that travel through the victim's blood vessels. When another mosquito bites the infected individual, it takes up the gametocytes with its blood meal. Inside the insect, the parasite completes its life cycle. The gametocytes fuse to form zygotes. Those zygotes bore into the mosquito's stomach, where they form oocysts that give rise to a new generation of parasites. When the infected mosquito sets out to feed, the cycle begins again.

The battle between malaria and humankind has raged through the ages. Scientists have recovered *Plasmodium* DNA from the bodies of 3500-year-old Egyptian mummies —evidence that those ancient humans were infected with the malaria parasite. The close connection between humans, mosquitoes, and the malaria parasite almost certainly extends back much further.

In fact, people who hail from regions where malaria is endemic are more likely than others to have certain genetic signatures that offer some degree of protection from the parasite. That indicates that *Plasmodium* has been exerting evolutionary pressure on humankind for quite some time.

Meanwhile, we've done our best to fight back. For centuries, humans fought the infection with quinine, a chemical found in the bark of the South American cinchona tree. In the 1940s, scientists developed a more sophisticated drug based on the cinchona compound. That drug, chloroquine, was effective, inexpensive, easy to administer, and caused few side effects. As a result, it was widely used, and in the late 1950s, *P. falciparum* began showing signs of resistance to chloroquine. Within 20 years, resistance had spread to Africa, and today most strains of *P. falciparum* have evolved resistance to the once-potent medication.

Just as bacteria develop resistance to antibiotics, *Plasmodium* evolves resistance to the antiparasitic drugs designed to fight it. In poor, rural areas where malaria is prevalent, people often can't follow the recommended protocols for antimalarial treatment. Sick individuals may be able to afford only a few pills rather than the

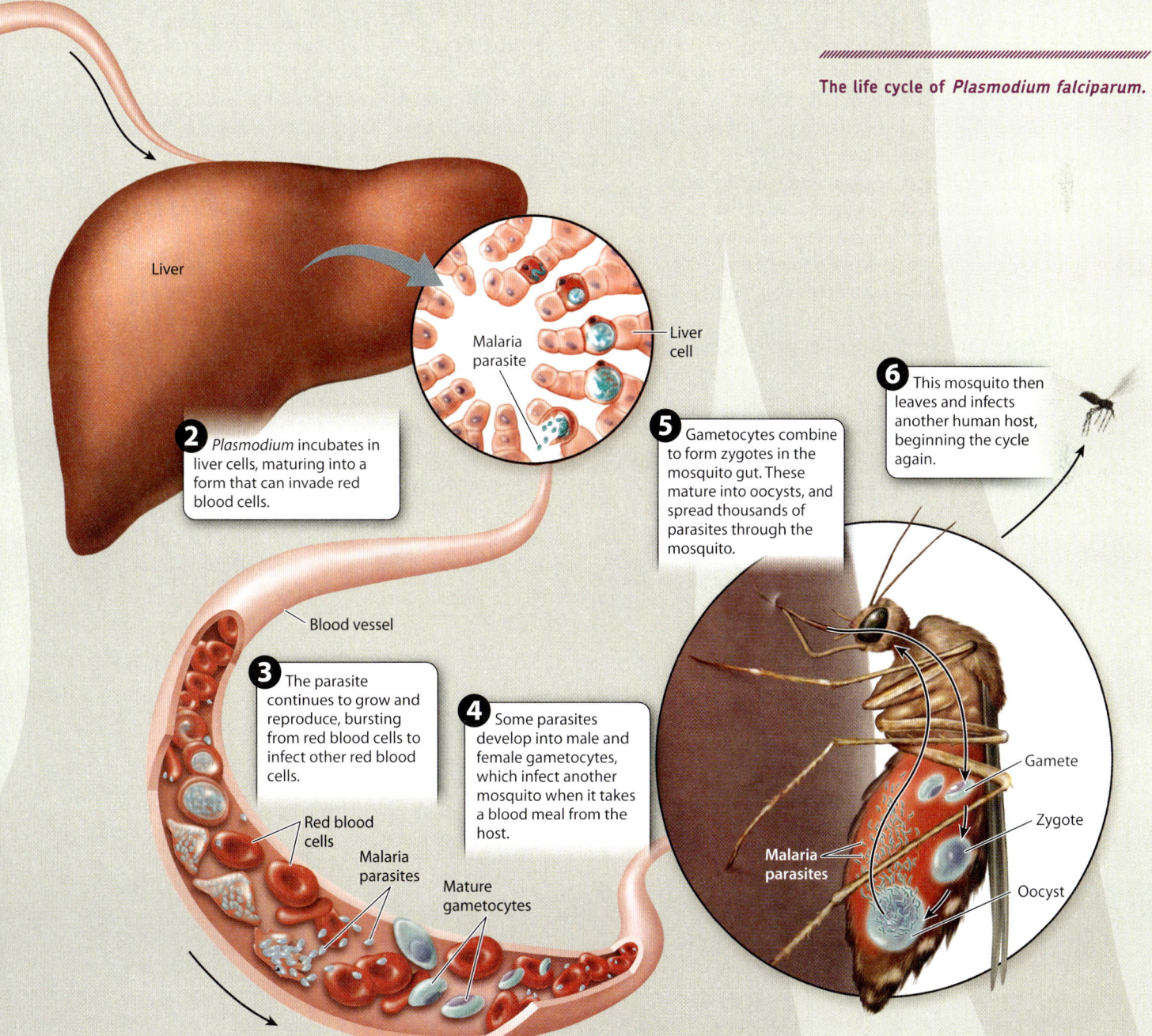

The life cycle of *Plasmodium falciparum*.

full recommended dose. The strength and quality of those pills may be questionable, and the drugs are often taken without oversight from a medical professional.

Unfortunately, inadequate use of the drugs fuels resistance. When the pills are altered or the course of treatment is abbreviated, not all *Plasmodium* parasites are wiped out. Those that survive in the presence of the drug are likely to evolve resistance to the drug. Because the resistant parasites have a survival advantage, the genes for drug resistance spread quickly through the population.

Since chloroquine resistance emerged, pharmaceutical researchers have developed a variety of new medications to prevent or treat malaria. Most, however, are far too expensive for people in the poverty-stricken regions where malaria is rampant. And just as was the case with chloroquine, almost as quickly as new drugs are developed, *Plasmodium* begins evolving resistance.

One of the more recent weapons added to the drug arsenal is artemisinin, a compound derived from the Chinese wormwood tree. It has turned out to be an effective and relatively inexpensive way to treat malaria infections. However, pockets of artemisinin resistance have already been uncovered in Southeast Asia. Public health workers now recommend that artemisinin be given in combination with other drugs. Treating infected patients with multiple drugs is more likely to wipe out *Plasmodium* in their bodies, reducing the chances that more drug-resistant strains will emerge.

Given the challenges of developing practical drugs to prevent or treat malaria, some scientists have turned their attention to other approaches. One goal is to produce a malaria vaccine. The parasite's complex life cycle makes that a complicated endeavor, though. Several vaccines are now in various stages of testing, and some show promise.

But researchers expect it will be years before a safe, effective vaccine for malaria could be available.

Other researchers are focusing their efforts on the mosquitoes that carry the parasite, rather than on *Plasmodium* itself. The genetically engineered insects created by the team in Arizona are a promising step in that direction.

The Arizona researchers set out to alter a cellular signaling gene that plays a role in the mosquito's life cycle. Mosquitoes normally live 2 to 3 weeks, and *Plasmodium* takes about 2 weeks to mature in the mosquito's gut. The researchers hoped to create mosquitoes that would die prematurely, before the parasite is mature. The genetic modification worked as planned. The engineered mosquitoes' life-spans were shortened by 18% to 20%. The genetic tweak also had a surprising side effect. The altered gene completely blocked the development of *Plasmodium* in the mosquitoes' guts. The engineered mosquitoes are incapable of spreading malaria to humans, regardless of how long they live.

While the finding was a laboratory success, it will be much harder to translate the results to the real world. To create malaria-free mosquitoes in the wild, scientists would have to release the genetically modified mosquitoes and hope that their altered gene spreads through the wild mosquito population. But that gene would be passed on only if it gave the insects a distinct evolutionary advantage. The engineered mosquitoes may be malaria free, but so far they're no fitter than their wild counterparts.

It's clear that slashing malaria rates will not be an easy task. Despite decades of research, insecticide-treated bed nets are still the best method for preventing the disease. For millennia, the malaria parasite has managed to withstand our efforts to squelch it, yet science continues to push the boundaries. Who will emerge the victor? Stay tuned.

? CASE 4 QUESTIONS

Answers to Case 4 questions can be found in Chapters 21–24.

1. What genetic differences have made some individuals more and some less susceptible to malaria? *See page 21-10.*
2. How did malaria come to infect humans? *See page 22-11.*
3. What human genes are under selection for resistance to malaria? *See page 24-15.*

CHAPTER 21

EVOLUTION

How Genotypes and Phenotypes Change over Time

Core Concepts

21.1 Genetic variation is the result of differences in DNA sequences.

21.2 Information about allele frequencies is key to understanding patterns of genetic variation.

21.3 Evolution is a change in the frequency of alleles or genotypes over time.

21.4 Natural selection leads to adaptations, which enhance the fit between an organism and its environment.

21.5 Migration, mutation, and genetic drift are non-adaptive mechanisms of evolution.

21.6 Molecular evolution looks at changes in DNA or amino acid sequences.

Variation is a fact of nature. A walk down any street reveals how variable our species is: Skin color and hair color, for example, vary from person to person. Until the publication in 1859 of Charles Darwin's *On the Origin of Species*, scientists tended to view all the variation we see in humans and other species as biologically unimportant. According to the traditional view at the time, not only were species individually created by God in their modern forms, but, because the Creator had a specific design in mind for each one, they were fixed and unchanging. Departures or variations from this divinely ordained type were therefore ignored.

Since Darwin, however, we have appreciated that a species does not conform to a type. Rather, a species consists of a range of variants. In our own species, people may be tall, short, dark-skinned, fair-skinned, and so on. Furthermore, variation is an essential ingredient of Darwin's theory because **natural selection** depends on the differential success—in terms of surviving and reproducing—of variants. Darwin changed how we view variation. Before Darwin, variation was irrelevant, something to be ignored, but after him it was recognized as the key to the evolutionary process.

21.1 GENETIC VARIATION

Variation is a major feature of the natural world. We humans are particularly good at noticing phenotypic variation among individuals of our own species. As we discussed in Chapter 16, a phenotype is an observable trait, such as human height or wing color in butterflies. Two factors contribute to phenotype: an individual's genotype, the set of alleles possessed by the individual at relevant genetic loci, and the environment in which the individual lives. We can take the environment out of the equation by looking directly at genotypic differences through sequencing DNA regions in multiple individuals. We now explore genetic variation directly, in terms of differences at the DNA sequence level.

Population genetics is the study of patterns of genetic variation.
Remarkably, in spite of a high degree of phenotypic variation, humans actually rank low in terms of overall genetic variation compared to other species. Specifically, if we ask how many DNA bases differ from one individual to another within the same species, we find that the fruit fly *Drosophila melanogaster* is about 10 times more genetically variable than we are. Even one of the most seemingly uniform species on

FIG. 21.1 Genetic diversity in Adelie penguins. Adelie penguins are uniform in appearance but are actually more genetically diverse than humans.

the planet, the Adelie penguin seen in **Fig. 21.1,** is two to three times more genetically variable than we are.

As we discuss in Chapter 22, a **species** consists of individuals that can exchange genetic material through interbreeding. From a genetic perspective, a species is therefore a group of individuals capable, through reproduction, of sharing alleles with one another. Individuals represent different combinations of alleles drawn from a single **gene pool,** that is, all the alleles present in all individuals in a species. The human gene pool includes alleles that cause differences in skin color, hair type, eye color, and so on. Each one of us has a different set of those alleles—alleles that cause brown hair and brown eyes, for example, or black hair and blue eyes—drawn from that gene pool.

Population genetics is the study of genetic variation in natural **populations,** which are interbreeding groups of organisms of the same species living in the same geographical area. What factors determine levels of variation in a population and in a species? Why are humans genetically less variable than penguins? What factors affect the distribution of particular variations? Population genetics addresses detailed questions about patterns of variation. And small differences, given enough time, can lead to the major differences we see among organisms today.

Mutation and recombination are the two sources of genetic variation.

Genetic variation has two sources: mutation and recombination. Mutation generates new variation and recombination shuffles mutations to create new combinations of mutations. In both cases, new alleles are formed, as shown in **Fig. 21.2.**

Mutations can be **somatic,** occurring in the body's tissues, or **germ-line,** occurring in the reproductive cells and therefore passed on to the next generation. From an evolutionary viewpoint, we are primarily interested in germ-line mutations. A somatic mutation affects only the cells descended from the one cell in which the mutation originally arose, and thus affects only that one individual. However, a germ-line mutation appears in every cell of an individual derived from the fertilization involving the mutation-bearing gamete, and thus in its descendants.

Mutations can be harmful, neutral, or beneficial.

Organisms, even the very simplest, are exquisitely complicated. In a large multicellular species such as humans, the level of organizational complexity is staggering. Consider the extraordinary process of development whereby, starting from a single cell at

CHAPTER 21 EVOLUTION: HOW GENOTYPES AND PHENOTYPES CHANGE OVER TIME

FIG. 21.2 The formation of new alleles through mutation and recombination.

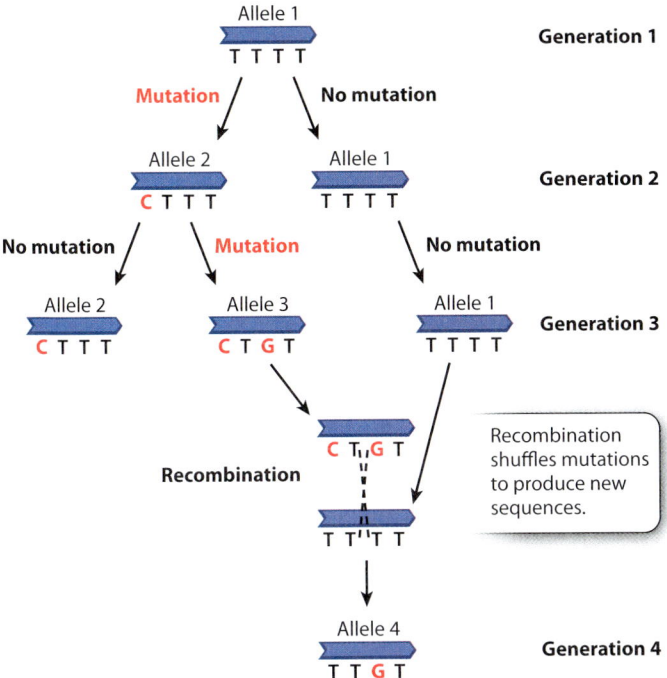

conception, an adult human being develops that consists of a few hundred trillion interacting, cooperating cells (Chapter 20). What is the likely impact of a random change—a mutation—on a complex piece of biological machinery? Random changes to the working parts of complex, organized systems are almost always **deleterious** (harmful). Imagine making a random change to the engine of a car, a relatively simple machine compared to complex biological machines. Pull out a hose here or put in a steel rod there, and you will almost certainly have a problem with your car.

Many mutations, however, have little or no effect on the organism. These mutations are neither deleterious nor advantageous because they occur in regions of the genome that are not functionally important. Such mutations are termed **neutral**.

Occasionally, a mutation occurs that is actually beneficial in terms of survival or reproduction. For example, a gene encoding hemoglobin might by chance acquire a mutation that allows it to take up and deliver oxygen to tissues more easily. Similarly, a gene encoding a particular enzyme might be mutated in such a way that it is more efficient at catalyzing a chemical reaction. Mutations like these are considered **advantageous** if they improve their carriers' chances of survival or reproduction. Advantageous mutations, as we will see, can increase in frequency in a population until eventually they are carried by every member of a species. These mutations are the ones that result in a species that is **adapted** to its environment—better able to survive and reproduce in that environment.

21.2 MEASURING GENETIC VARIATION

Mutations, whether harmful, neutral, or advantageous, are sources of genetic variation. The goal of population genetics is to make inferences about the evolutionary process from patterns of genetic variation in nature. The raw information for this comes from the rates of occurrence of alleles in populations, or **allele frequencies**.

To understand patterns of genetic variation, we require information about allele frequencies.

The allele frequency of an allele x is simply the number of x's present in the population divided by the total number of alleles. Consider, for example, pea color in Mendel's pea plants. In Chapter 16, we discussed how pea color (yellow or green) results from variation at a single gene. Two alleles of this gene are the dominant A (yellow) allele and the recessive a (green) allele. AA homozygotes and Aa heterozygotes produce yellow peas, and aa homozygotes produce green peas. Imagine that in a population every pea plant produces green peas, meaning that only one allele, a, is present: The allele frequency of a is 100%, whereas the allele frequency of A is 0%. When a population exhibits only one allele at a particular gene, we say that the population is **fixed** for that allele.

Now consider another population of 100 pea plants with genotype frequencies of 50% aa, 25% Aa, and 25% AA. (A genotype frequency is the proportion in a population of each genotype at a particular gene or set of genes.) These genotype frequencies give us 50 green-pea pea plants (aa), 25 yellow-pea heterozygotes (Aa), and 25 yellow-pea homozygotes (AA). What is the allele frequency of a in this population? Each of the 50 aa homozygotes has two a alleles and each of the 25 heterozygotes has one a allele. Of course, there are no a alleles in AA homozygotes. The total number of a alleles is thus $(2 \times 50) + 25 = 125$. To determine the allele frequency of a, we divide the number of a alleles by the total number of alleles in the population, 200 (because each pea plant is diploid, meaning that it has two alleles): $\frac{125}{200} = 62.5\%$. Because we are dealing with only two alleles in this example, the allele frequency of A is $100\% - 62.5\% = 37.5\%$

Thus, the allele frequencies of A and a provide a measure of genetic variation at one gene in a given population. In this example, we were given the genotype frequencies, and from this information we determined the allele frequencies. But how are genotype and allele frequencies measured? We consider three ways to measure genotype and allele frequencies in populations: observable traits, gel electrophoresis, and DNA sequencing.

→ **Quick Check 1** In the example of pea color in Mendel's pea plants, we worked with numbers of plants of each genotype to calculate allele frequencies. Can you figure out how to determine allele frequencies directly from genotype frequencies, even when we have no information on the number of individuals being studied?

Early population geneticists relied on observable traits to measure variation.

It would be a simple matter to measure genetic variation in a population if we could use observable traits. Then we could simply count the individuals displaying variant forms of a trait and have a measure of the variation of that trait's gene. However, as we saw in Chapter 18, this approach can work only rarely for two important reasons. First, many traits are encoded by a large number of genes. In these cases, it is difficult, if not impossible, to make direct inferences from a phenotype to the underlying genotype. Even apparently straightforward traits often prove to have a complicated genetic basis. For instance, human skin color is determined by at least six different genes. Second, the phenotype is a product of both the genotype and the environment.

Until the 1960s, there was only one workable solution: to limit population genetics to the study of phenotypes that are encoded by a single gene. As these are few, the number of genes that population geneticists could study was extremely small. Human blood groups, including the ABO system, provided an early example of a trait encoded by a single gene with multiple alleles. At this gene, there are three alleles in the population—*A*, *B*, and *O*—and therefore six possible genotypes, which result in four different phenotypes (**Table 21.1**).

Other instances in which phenotypic variation can be readily correlated with genotype include certain markings in invertebrates. For example, the coloring of the two-spot ladybug *Adalia bipunctata* is controlled by a single gene (**Fig. 21.3**). However, the genetic basis of most traits is not so simple.

Gel electrophoresis facilitates the detection of genetic variation.

Single-gene variation became much easier to detect in the 1960s with the application of gel electrophoresis. In Chapter 12, we saw how gel electrophoresis separates segments of DNA according to their size. Before DNA technologies were developed, the same basic process was applied to proteins to separate them according to their electrical charge and their size. In gel electrophoresis, the proteins being studied migrate through a gel when an electrical charge is applied, creating an electrical field. The rate at which the proteins move from one end of the gel to the other is determined by their charge and their size. Proteins with more negatively charged amino acids migrate more rapidly toward the positively charged end, and vice versa.

Early studies of protein electrophoresis focused on enzymes that catalyze reactions that can be induced to produce a dye when the substrate for the enzyme is added. If we add some of the substrate, we can see the locations of the proteins in the gel. **Fig. 21.4** shows this sort of experiment. Material from different individuals is loaded in each lane of the gel. One individual, a homozygote for a particular protein sequence, produces a single band on the gel. A heterozygote for two differently charged alleles produces two bands. So the bands in the gel can provide a visual picture of genetic variation.

DNA sequencing is the gold standard for measuring genetic variation.

Protein gel electrophoresis was a significant leap forward in our ability to detect genetic variation, but even this technique had significant limitations. We could only study enzymes because we needed to be able to stain specifically for enzyme activity, and we could only detect mutations that resulted in amino acid substitutions that changed a protein's mobility in the gel. Only with DNA sequencing did we finally have an unambiguous means of detecting all genetic variation in a stretch of DNA, whether in a coding region or not. The variations studied by modern population geneticists are differences in DNA sequence, such as a *T* rather than a *G* at a specified nucleotide position in a particular gene.

Calculating allele frequencies, then, simply involves collecting a population sample and counting the number of occurrences of a given mutation. Take the *A*-or-*G* mutation at a specific nucleotide position in

TABLE 21.1 The ABO blood system.

PHENOTYPE	GENOTYPE
A	*AA* or *AO*
B	*BB* or *BO*
AB	*AB*
O	*OO*

FIG. 21.3 A genetic difference in color in the two-spot ladybug, *Adalia bipunctata*, that results from variation in a single gene.

HOW DO WE KNOW?

FIG. 21.4

How is genetic variation measured?

BACKGROUND The introduction of protein gel electrophoresis in 1966 gave researchers the opportunity to identify differences in amino acid sequence in proteins both among individuals and, in the case of heterozygotes, within individuals. Proteins with different amino acid sequences run at different rates through a gel in an electric field. Often, a single amino acid difference is enough to affect the mobility of a protein in a gel.

METHOD Starting with crude tissue—the whole body of a fruit fly, or a blood sample from a human—we load the material on a gel, and turn on the current. The rate at which a protein migrates depends on its size and charge, both of which may be affected by its amino acid sequence. To visualize the protein at the end of the gel run, we use a biochemical indicator that produces a stain when the protein of interest is active. The result is a series of bands on the gel.

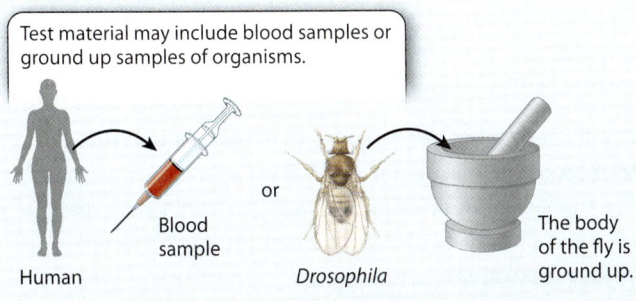

Test material may include blood samples or ground up samples of organisms.

Human — Blood sample — or — Drosophila — The body of the fly is ground up.

The samples are loaded into separate wells on the gel, and the proteins in each sample migrate toward the positive electrode according to their charge and size.

RESULTS The genotypes of eight individuals for a gene with two alleles are analyzed. Four are allele 1 homozygotes; two are allele 2 homozygotes; and two are heterozygotes. Note that the heterozygotes do not stain as strongly on the gel because each band has half the intensity of the single band in the homozygote. We can measure the allele frequencies simply by counting the alleles. Each homozygote has two of the same allele, and each heterozygote has one of each.

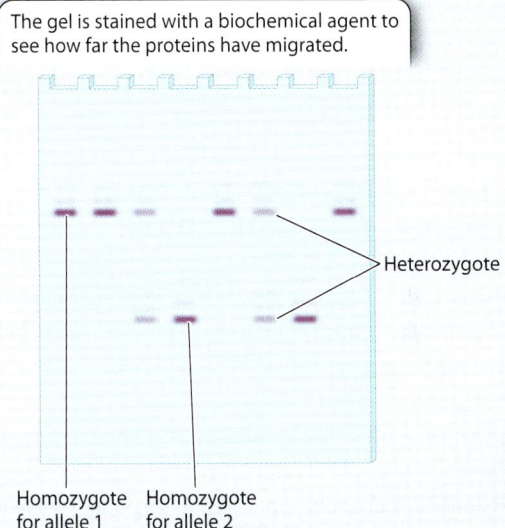

The gel is stained with a biochemical agent to see how far the proteins have migrated.

Homozygote for allele 1 — Homozygote for allele 2 — Heterozygote

Total number of alleles in the population = 8 × 2 = 16

Number of allele 1 in the population = 2 × (number of allele 1 homozygotes) + (number of heterozygotes) = 8 + 2 = 10

Frequency of allele 1 = $\frac{10}{16} = \frac{5}{8}$

Number of allele 2 in the population = 2 × (number of allele 2 homozygotes) + (number of heterozygotes) = 4 + 2 = 6

Frequency of allele 2 = $\frac{6}{16} = \frac{3}{8}$

Note that the two allele frequencies add to 1.

CONCLUSION We now have a profile of genetic variation at this gene for these individuals. Population genetics involves comparing data such as these with data collected from other populations to determine the forces shaping patterns of genetic variation.

FOLLOW-UP WORK This technique is seldom used these days because it is easy now to recover much more detailed genetic information about genetic variation from DNA sequencing.

SOURCE Lewontin, R. C., and J. L. Hubby. 1966. "A Molecular Approach to the Study of Genic heterozygosity in natural populations. II. Amount of variation and degree of heterozygosity in natural populations of *Drosophila pseudoobscura*." *Genetics* 54:595–609.

the fruit fly *Alcohol dehydrogenase (Adh)* gene. We can sequence the gene from 50 individual flies. We will then have 100 gene sequences from these diploid individuals. We find 70 sequences have an *A* and 30 have a *G* at a given position. Therefore, the allele frequency of *A* is $\frac{70}{100}$ = 0.7 and the allele frequency of *G* is 0.3. In general, in a sample of *n* diploid individuals, the allele frequency is the number of occurrences of that allele divided by twice the number of individuals.

→ **Quick Check 2** Population genetics data have become ever more resolved over time, from phenotypes that are determined by a single gene, to gel electrophoresis that looks at variation among genes that encode for enzymes, to analysis of the DNA sequence. What is the next logical step?

21.3 EVOLUTION AND THE HARDY–WEINBERG EQUILIBRIUM

Determining allele frequencies gives us information about genetic variation. Following that variation over time is key to understanding the genetic basis of evolution.

Evolution is a change in allele or genotype frequency over time.

At the genetic level, evolution is simply a change in the frequency of an allele or a genotype from one generation to the next. For example, if there are 200 copies of an allele that causes blue eye color in a population in generation 1 and there are 300 copies of that allele in a population of the same size in generation 2, evolution has occurred. In principle, evolution may occur without allele frequencies changing. For instance, even if, in our fruit fly example, the *A/G* allele frequencies stay the same from one generation to the next, the frequencies of the different genotypes (that is, of *AA*, *AG*, and *GG*) may change. This would be evolution *without* allele frequency change.

Evolution is therefore a change in the genetic makeup of a population over time. Note an important and often misunderstood aspect of this definition: *populations* evolve, not *individuals*. Note, too, that this definition does not specify a mechanism for this change. As we will see, many mechanisms can cause allele or genotype frequencies to change. Regardless of which mechanisms are involved, any change in allele frequencies, genotype frequencies, or both constitutes evolution.

The Hardy–Weinberg equilibrium describes situations in which allele and genotype frequencies do not change.

Allele and genotype frequencies change over time only if specific mechanisms act on the population. This principle was demonstrated independently in 1908 by the English mathematician G. H. Hardy and the German physician Wilhelm Weinberg, and has become known as the **Hardy–Weinberg equilibrium.** In essence, the Hardy–Weinberg equilibrium is the situation in which evolution does not occur.

The Hardy–Weinberg equilibrium specifies the relationship between allele frequencies and genotype frequencies when a number of key conditions are met. When these conditions are met—and the Hardy–Weinberg equilibrium holds—we can conclude that evolutionary mechanisms are not acting on the gene in the population we are studying. In many ways, then, the Hardy–Weinberg equilibrium is most interesting when we find instances in which allele or genotype frequencies *depart* from expectations. This finding implies that one or more of the conditions are *not* met and that evolutionary mechanisms are at work.

These are the required conditions for the Hardy–Weinberg equilibrium:

1. **There can be no differences in the survival and reproductive success of individuals**. Given two alleles, *A* and *a*, imagine an extreme case in which *a*, a recessive mutation, is lethal. All *aa* individuals die. Therefore, in every generation, there is a selective elimination of *a* alleles, meaning that the frequency of *a* will gradually decline (and the frequency of *A* correspondingly increase) over the generations. As we discuss below, we call this differential success of alleles **selection.**

2. **Populations must not be added to or subtracted from by migration**. Imagine a second population adjacent to the one we used in the preceding example in which all the alleles are *A* and all individuals have the genotype *AA*. Recall that when a population has only one allele of a particular gene, it is fixed for that allele. Now imagine that there is a sudden influx of individuals from the first population into the second. The frequency of *A* in the second population will change in proportion to the number of immigrants.

3. **There can be no mutation.** If *A* alleles mutate into *a* alleles, and vice versa, then again we will see changes in the allele frequencies over the generations. In general, because mutation is so rare, it is not a serious consideration on the timescales studied by population geneticists. However, as we have seen, mutation is the ultimate source of genetic variation, so it is a key input on which other evolutionary mechanisms act.

4. **The population must be sufficiently large to prevent sampling errors**. Chance events are more likely in small samples than in large ones. Campus-wide, a college's sex ratio may be close to 50:50, but in a small class of 8 individuals it is not improbable that we would have 6 women and 2 men (a 75:25 ratio). Sample size, in the form of population size, also affects the Hardy–Weinberg equilibrium such that it technically holds only for infinitely large populations. A change in the frequency of an allele due to the random effects of small population size is called **genetic drift.**

5. **Individuals must mate at random.** For the Hardy–Weinberg equilibrium to hold, mate choice must be made without regard to genotype. For example, an *AA* homozygote when offered a choice of mate from among *AA*, *Aa*, or *aa* individuals should choose at random. In contrast, **non-random mating** occurs when individuals do not mate randomly. For example, *AA* homozygotes might preferentially mate with other *AA* homozygotes. Non-random mating affects genotype frequencies from generation to generation, but does not affect allele frequencies.

The Hardy–Weinberg equilibrium translates allele frequencies into genotype frequencies.

Now that we have established the conditions required for a population to be in Hardy–Weinberg equilibrium, let us explore the idea in detail. In the example we looked at earlier, we know the frequency of the two alleles, one with *A* and the other with *G* in the *Adh* gene in *Drosophila*. What are the genotype frequencies? That is, how many *AA* homozygotes, *AG* heterozygotes, and *GG* homozygotes do we see? The Hardy–Weinberg equilibrium predicts genotype frequencies from allele frequencies.

The logic is simple. Random mating is the equivalent of putting all the population's gametes into a single pot and drawing out pairs of them at random to form a zygote. We therefore put in our 70 *A* alleles and 30 *G* alleles and pick pairs at random. What is the probability of picking an *AA* homozygote (that is, what is the probability of picking an *A* allele followed by another *A* allele)? The probability of picking an *A* allele is its frequency in the population, so the probability of picking the first *A* is 0.7. What is the probability of picking the second *A*? Also 0.7. What then is the probability of picking an *A* followed by another *A*? It is the product of the two probabilities: $0.7 \times 0.7 = 0.49$. Thus, the frequency of an *AA* genotype is 0.49. We take an identical approach to determine the genotype frequency for the *GG* genotype: Its frequency is 0.3×0.3, or 0.09.

What about the frequency of the heterozygote, *AG*? This is the probability of drawing *G* followed by *A*, or *A* followed by *G*. There are thus two separate ways in which we can generate the heterozygote. Its frequency is therefore $(0.7 \times 0.3) + (0.3 \times 0.7) = 0.42$.

We can generalize these calculations algebraically by substituting letters for the numbers we have computed to derive the Hardy–Weinberg equilibrium. If the allele frequency of one allele, *A*, is *p*, and the other, *G*, is *q*, then

$p + q = 1$ (because there are no other alleles at this site).

Genotypes	AA	AG	GG
Frequencies	p^2	$2pq$	q^2

Not only do these relationships predict genotype frequency from allele frequencies, but they work in reverse, too. **Fig. 21.5** shows how Hardy–Weinberg values are calculated from allele frequencies and, in the graph on the right, how allele and genotype frequencies are related. Note that we can use the graph to determine both allele frequencies for a given genotype frequency (to take a simple case, a 0.5 frequency of heterozygotes, *Aa*, implies that both *p* and *q* are 0.5) and genotype frequencies for a specified allele frequency.

We can do this mathematically as well. Knowing the genotype frequency of *AA*, for example, permits us to calculate allele frequencies: if, as in our *Adh* example, p^2 is 0.49 (that is, 49% of the population has genotype *AA*), then *p*, the allele frequency of *A*, is $\sqrt{0.49} = 0.7$. Because $p + q = 1$, then *q*, the allele frequency of *a*, is $1 - 0.7 = 0.3$.

Note that these relationships hold only if the Hardy–Weinberg conditions are met. If not, then allele frequencies can only be deduced by tallying alleles from genotype frequencies as described earlier in section 21.2.

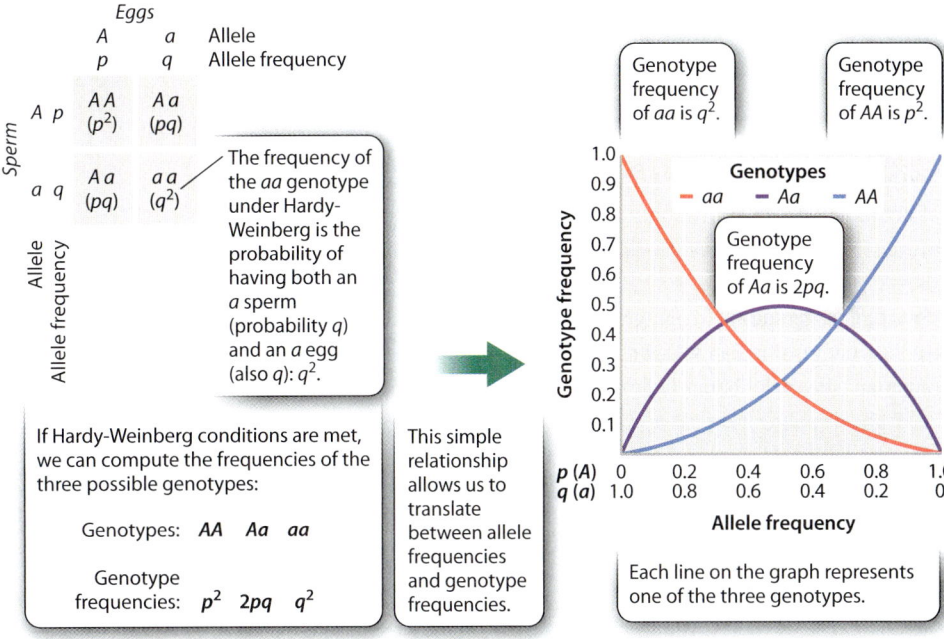

FIG. 21.5 Hardy-Weinberg Relation. The Hardy-Weinberg relation predicts genotype frequencies from allele frequencies, and vice versa.

The Hardy–Weinberg equilibrium is the starting point for population genetic analysis.

Recall our definition of evolution: a change in allele or genotype frequency from one generation to the next. Given this definition, it might seem odd to be discussing factors necessary for allele frequencies to stay the same. The Hardy–Weinberg equilibrium not only provides a means of converting between allele and genotype frequencies, but, critically, it also serves as an indicator that something interesting is happening in a population when it is not upheld.

If we find a population whose allele or genotype frequencies are not in Hardy–Weinberg equilibrium, we can infer that evolution has occurred. We can then consider, for the gene under study, whether the population is subject to selection, migration, mutation (unlikely because of its rarity), genetic drift, or non-random mating. These are the primary mechanisms of evolution. The Hardy–Weinberg equilibrium gives us a baseline from which to explore the evolutionary processes affecting populations. We will start by considering one of the most important evolutionary mechanisms: natural selection.

→ **Quick Check 3.** When we find a population whose allele frequencies are *not* in Hardy–Weinberg equilibrium, what can and can't we conclude about that population?

21.4 NATURAL SELECTION

Natural selection results in allele frequencies changing from generation to generation according to the allele's impact on the survival and reproduction of individuals. New mutations that are deleterious and eliminated by natural selection have no long-term evolutionary impact; ones that are beneficial, however, can result in adaptation to the environment over time.

Natural selection brings about adaptations.

The adaptations we see in the natural world—the exquisite fit of organisms to their environment—were typically taken by pre-Darwinian biologists as evidence of God's existence. Each species, they argued, was so well adapted—the desert plant so physiologically adept at coping with minimal levels of rainfall and the fast-swimming fish so hydrodynamically streamlined—that it must have been deliberately designed by a divine Creator.

With the publication of *On the Origin of Species* in 1859, Darwin, pictured in **Fig. 21.6**, overturned the biological convention of his day on two fronts. First, he showed that species are not unchanging; they have evolved over time. Second, he suggested a mechanism, natural selection, that brings about adaptation. Natural selection was a brilliant solution to the central problem of biology: how organisms come to fit so well in their environments. From where does the woodpecker get its powerful chisel of a bill? And the hummingbird its long delicate bill for probing the nectar stores in flowers? Darwin showed how a simple mechanism, without foresight or intentionality, could result in the extraordinary range of adaptations that all of life is testimony to.

For 20 years after originally conceiving the essence of his theory, Darwin collected supporting evidence. In 1858, however, he was spurred to begin writing *On the Origin of Species* by a letter from a little-known naturalist collecting specimens in what is today Indonesia. By a remarkable coincidence, Alfred Russel Wallace, shown in **Fig. 21.7**, had also developed the theory of evolution by natural selection. Aware that Darwin was interested in the problem but having no idea that Darwin was working on the same theory, Wallace wrote to Darwin in 1858 to see what he thought of his idea.

Suddenly, Darwin was confronted with the prospect of losing his claim on the theory that he had been quietly nurturing for the previous 20 years. But all was not lost. Darwin's colleagues arranged for the publication of a joint paper by Wallace and Darwin in 1858. This was done without consulting Wallace, who nonetheless never resented Darwin and afterward was careful to insist that the idea rightly belonged to the older man. It was Darwin's publication of *On the Origin of Species* in 1859 that brought both evolution and its underlying mechanism of adaptation, natural selection, to public attention. Wallace is only fleetingly mentioned in Darwin's great work, and Darwin is now the name, not Wallace, associated with the discovery.

Both Darwin and Wallace recognized their debt to the writings of a British clergyman, Thomas Malthus. In his *Essay on the Principle of Population,* first published in 1798, Malthus pointed out that natural populations have the potential to increase

FIG. 21.6 Charles Darwin, photographed at about the time he was writing *On the Origin of Species.*

FIG. 21.7 Alfred Russel Wallace, photographed in Singapore during his expedition to Southeast Asia.

geometrically, meaning that there is potentially an accelerating increase in population size. Imagine that human couples can have just four children (two males and two females), meaning that the population doubles every generation. Starting with a single couple, by the twentieth generation, the population will have grown to over a million—1,048,576 to be precise. Note that this growth assumes a rather modest rate of increase (twofold). Consider how much faster population size would ramp up in an insect that lays hundreds of eggs in its lifetime.

However, this geometric expansion of populations does not occur. In fact, population sizes are typically reasonably stable from generation to generation. This is because the resources upon which populations are dependent—food, water, places to live—are limited. This imbalance suggests that in each generation many fail to survive or reproduce; there simply is not enough food and other resources to go around. This implies in turn that individuals within a population must compete for these resources.

Which individuals will win the competition? Darwin and Wallace suggested that those that are best adapted would most likely survive and leave more offspring. Genetic variation among individuals results in some being more likely to survive and reproduce. As a result, the next generation will be enriched for these same advantageous alleles. Darwin used the term "natural selection" for the filtering process that acts against deleterious alleles and in favor of advantageous ones.

Competitive advantage is a function of how well an organism is adapted to its environment. A desert plant that is more efficient at minimizing water loss than another plant is better adapted to the desert environment. An organism that is better adapted to its environment is more fit. By fit, we do not mean physically fit, or even optimally adapted. **Fitness,** in this context, is a measure of the extent to which the individual's genotype is represented in the next generation. We say that the first plant's fitness is higher than the second's if it leaves more surviving offspring. Natural selection then acts over generations to increase the overall fitness of a population. A plant population newly arrived in a desert may be poorly adapted to its environment, but, over time, alleles that minimize water loss will increase under natural selection, resulting in a population that is better adapted to the desert.

Such changes in populations take time. Borrowing from the geologists of his day, Darwin recognized that time is a critical ingredient of his theory. Geologists had put forward a view of Earth's history that argued that large geological changes—like the carving out of the Grand Canyon—can be explained by simple day-to-day processes—in this case, erosion—operating over vast timescales. Darwin applied this worldview to biology. He recognized that small changes, like subtle shifts in the frequencies of alleles, could add up to major changes given long enough time periods. What might seem to us to be a trivial change over the short term can, over the long term, result in substantial differences among populations.

The Modern Synthesis is a marriage between Mendelian genetics and Darwinian evolution.

Darwinian evolution involves the change over time of the genetic composition of populations and is thus a genetic theory. Although Mendel published his genetic studies of pea plants in 1866, not long after *The Origin*, Darwin never saw them, so a key component of the theory was missing. However, instead of clarifying the role of genetic processes underlying natural selection, the rediscovery of Mendel's work in 1900 unexpectedly provoked a major controversy among evolutionary biologists. Some argued that Mendel's discoveries did not apply to most genetic variation because the traits studied by Mendel were **discrete,** meaning that they had clear alternative states, such as peas that were either yellow or green. Most of the variation we see in natural populations, in contrast, is **continuous,** meaning that variation occurs across a spectrum. Human height, for example, does not come in discrete classes. People are not either 5 feet tall or 6 feet tall and no height in between. Instead, they may be any height within a certain range.

How could Mendel's discrete traits account for the continuous variation seen in natural populations? This question was answered by the English theoretician Ronald Fisher, who realized that, instead of a single gene contributing to a trait like human height, there could be several genes that contribute to the trait. He argued that extending Mendel's theory to include multiple genes per trait could account for patterns of continuous variation that we see all around us.

Fisher's insight formed the basis of a synthesis between Darwin's theory of natural selection and Mendelian genetics that was forged during the middle part of the twentieth century. The product of this **Modern Synthesis** was our current theory of evolution.

Natural selection increases the frequency of advantageous mutations and decreases the frequency of deleterious mutations.

Natural selection increases the frequency of beneficial alleles, resulting in adaptation. In some cases, it can promote the **fixation** of beneficial alleles, meaning the allele has a frequency of 1. To start with, a new beneficial allele will exist as a single copy in a single individual (that is, as a heterozygote), but, under the influence of natural selection, the beneficial allele can eventually replace all the other alleles in the population. Natural selection that increases the frequency of a favorable allele is called **positive selection.**

As we have seen, the majority of mutations to functional genes are deleterious. In extreme cases, they are lethal to the individuals carrying them and are thus instantly eliminated from the population. Sometimes, however, natural selection is inefficient in getting rid of a deleterious allele. Consider a recessive lethal mutation, b (that is, one that is lethal only as a homozygote, bb, and has no effect as a heterozygote, Bb). When it first arises, all the other alleles in the population are B, which means that the first b allele that appears in the population must be paired with a B allele, resulting in a Bb heterozygote. Because natural selection does not act against heterozygotes in this case, the b allele may increase in frequency by chance alone (we discuss how this happens below). Only when two b alleles come together to form a bb homozygote does natural selection act to rid the population of the allele. Natural selection that decreases the frequency of a harmful allele is called **negative selection.**

Many human genetic diseases show this pattern: The deleterious allele is rare and recessive. Because it is rare, homozygotes are formed only infrequently. Remember that the expected frequency of homozygotes under the Hardy–Weinberg equilibrium is the square of the frequency of the allele in the population. Therefore, if the allele frequency is 0.01, we expect 0.01 × 0.01, or 1 in every 10,000 individuals, to be homozygous for it. Thus the genetic disease occurs rarely, and the allele remains in the population because it is recessive and not expressed as a heterozygote.

❓ CASE 4 Malaria: Co-Evolution of Humans and a Parasite
What genetic differences have made some individuals more and some less susceptible to malaria?

In addition to allowing alleles to be either eliminated or fixed, natural selection can also maintain an allele at some intermediate frequency between 0% and 100%. This form of natural selection is called **balancing selection,** and it acts to maintain two or more alleles in a population. A simple case is members of a species that face different conditions depending upon where they live. One allele might be favored by natural selection in a dry area, but a different one favored in a wet area. Taking the species as a whole, these alleles are maintained by natural selection at intermediate frequencies.

Another example of balancing selection occurs when the heterozygote's fitness is higher than that of either of the homozygotes, resulting in selection that ensures that both alleles remain in the population at intermediate frequencies. This form of balancing selection is called **heterozygote advantage,** and it is exemplified by human populations in Africa, where malaria has been a long-standing threat. Because the malaria parasite spends part of its life cycle in human red blood cells, mutations in the hemoglobin molecule that affect the structure of the red blood cells have a negative impact on the parasite and can reduce the severity of malarial attacks.

Two alleles of the gene for hemoglobin are A and S. The A allele codes for normal hemoglobin, resulting in fully functional, round red blood cells. The S allele causes the red blood cell to assume a sickle shape (**Fig. 21.8**). The S allele encodes a polypeptide that differs from the A allele's product in just a single amino acid, which is enough to cause the molecules to aggregate end to end, so the red blood cell is distorted into a sickle (Chapter 15).

In regions of the world with malaria, heterozygous individuals (SA) have an advantage over homozygous individuals (SS and AA). SS homozygotes are protected against malaria, but they are burdened with severe sickling disease. Sickle-shaped red blood cells can block capillaries, and therefore people with two copies of S are prone to debilitating, painful, and sometimes fatal episodes resulting from capillary blockage. AA homozygotes lack sickling disease but are vulnerable to malaria. SA heterozygotes, however, do not have severe sickling disease and have some protection from malaria. As a result, natural selection maintains both the S and A alleles in the population at intermediate frequencies.

In areas where there is no malaria, this balance is shifted. Many African-Americans, descended from Africans upon whom natural selection operated in favor of the heterozygote, still carry the S allele, even though the allele is no longer useful to them in their malaria-free environment. If natural selection were to run its course among African-Americans, the S allele would gradually be eliminated. The problem is that this is a slow process, and many more people will continue to suffer from sickle cell anemia before it is complete.

Natural selection can be stabilizing, directional, or disruptive.

Up to this point, we have followed the fate of individual mutations, which can increase, decrease, or be maintained at an intermediate frequency under the influence of natural selection. We can also look at the consequence of natural selection from a

FIG. 21.8 **The effect of a single base-pair mutation.** (a) Normal red blood cells look different from (b) sickled ones, whose shape has been distorted by hemoglobin molecules with the sickle variant.

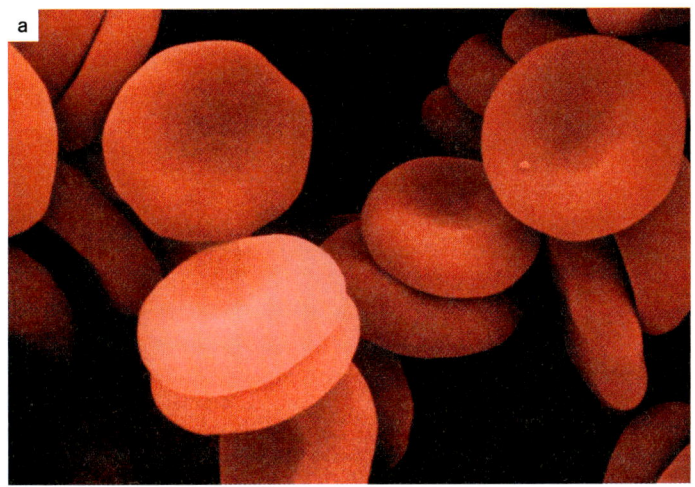
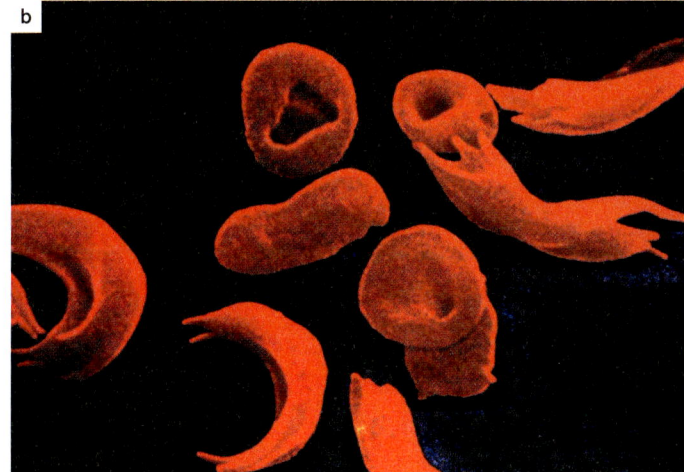

different perspective. Instead of following individual mutations, we can look at changes over time of a particular trait of an organism. For example we might track the evolution of height in a population, despite not knowing the specifics of the genetic basis of height differences. When we look at natural selection from this perspective, we see three types of patterns: **stabilizing, directional,** and **disruptive,** illustrated in **Fig. 21.9.**

Stabilizing selection maintains the status quo and acts against extremes. A good example is provided by human birth weight, a trait affected by many factors, including many fetal

FIG. 21.9 **Stabilizing, directional, and disruptive selection.** Selection can change the distribution of phenotypes (and therefore genotypes) in three different ways.

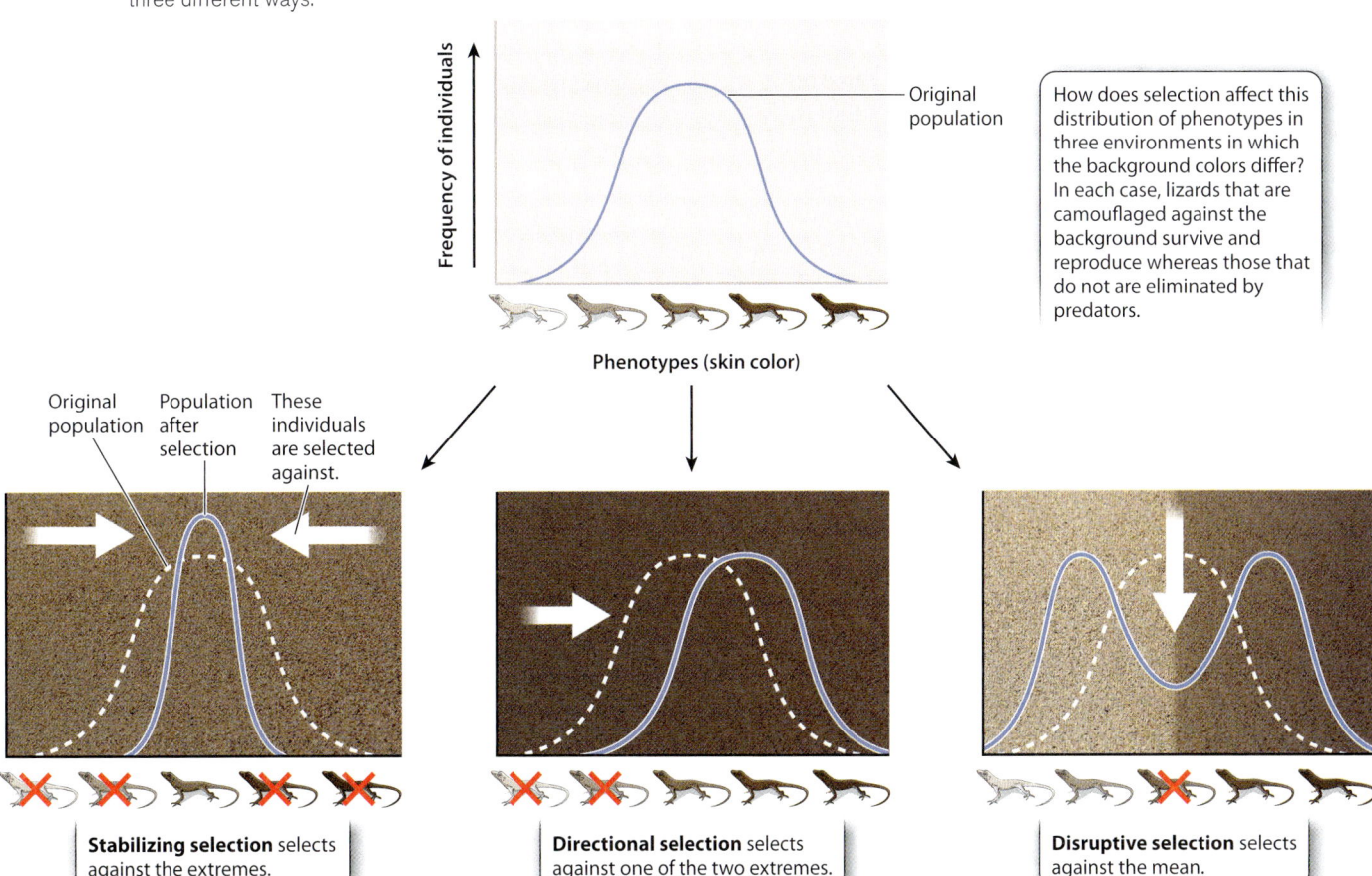

FIG. 21.10 Stabilizing selection. Stabilizing selection on human birth weight results in selection against babies that are either too small or too large.

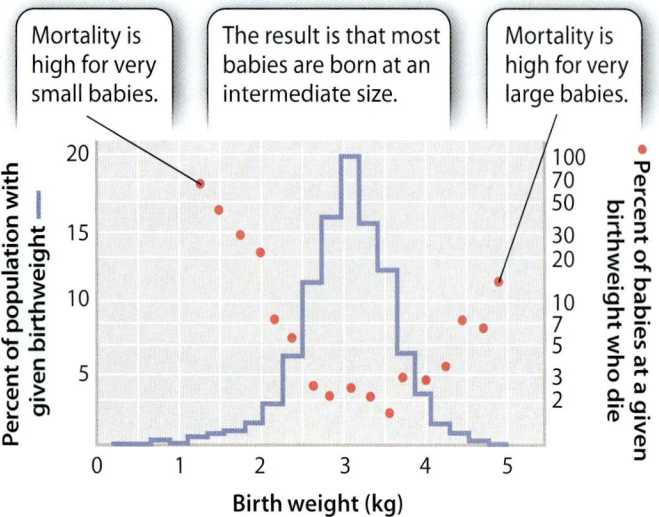

genes (**Fig. 21.10**). If a baby is too small, then its chances of survival after birth are low. However, if it is too big, there may be complications that endanger both mother and baby during delivery. Thus, the optimum birth weight is between these two extremes. In this case, natural selection acts against the extremes. The vast majority of natural selection is of this kind as deleterious mutations that cause a departure from the optimal phenotype are selected against.

Whereas stabilizing selection keeps a trait the same over time, **directional selection** leads to a change in a trait over time. A timely and concerning example is provided by drug resistance in many viruses, bacteria, and the malaria parasite. In the case of malaria, widespread use of the drug chloroquine has led to changes in the population of the malaria parasite in favor of parasites that are resistant to this drug. From the point of view of drug resistance, the population has moved in one direction since the 1960s toward increased resistance.

Artificial selection, which has been practiced by humans since at least the dawn of agriculture, is a form of directional selection. Artificial selection is analogous to natural selection, but the competitive element is removed. Successful genotypes are selected by the breeder, not through competition. Because it can be carefully controlled by the breeder, artificial selection is astonishingly efficient at generating genetic change. Dogs are basically domesticated wolves that, over just a few millennia of careful breeding, have produced morphologies ranging from the tiny Pekingese to the vast Great Dane and behaviors ranging from the herding abilities of Border Collies to the retrieval skills of a Labrador.

Practiced over many generations, artificial selection can create a population in which the selected phenotype is far removed from that of the starting population. **Fig. 21.11** shows

HOW DO WE KNOW?

FIG. 21.11

How far can artificial selection be taken?

BACKGROUND Begun in the 1890s and continuing to this day at the University of Illinois, this experiment is one of the longest-running biological experiments in history.

HYPOTHESIS Researchers hypothesized that there would be a limit to the extent to which the population could respond to continued directional selection, that there would be a point at which there was no longer a response to selection.

EXPERIMENT Corn was artificially selected for either high oil content or low oil content. Every generation, kernels showed a range of oil levels, but only the 12 kernels with the highest or the lowest oil content were used for the next generation.

RESULTS In the line selected for high oil content, the percentage of oil more than quadrupled, from about 5% to more than 20%. In the line selected for low oil content, the oil content fell so close to zero that it could no longer be measured accurately, and the selection was terminated. Both selected lines are completely outside the range of any phenotype observed at the beginning of the experiment.

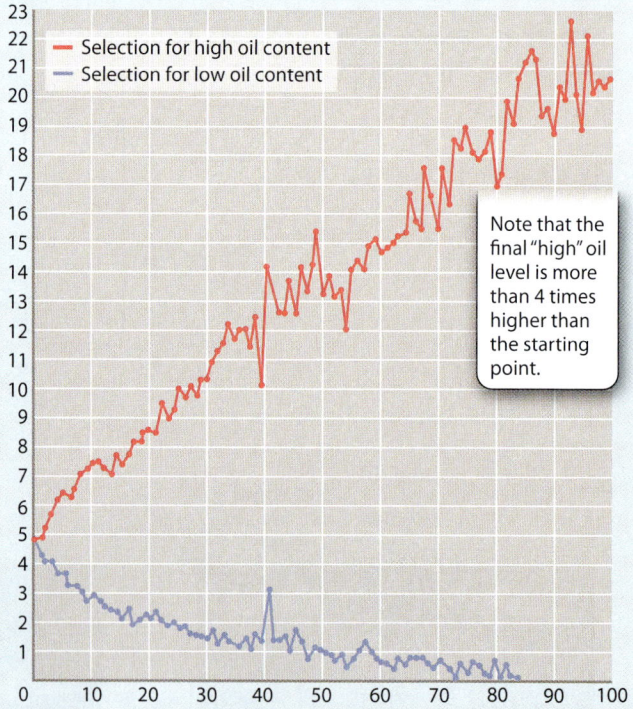

FOLLOW-UP WORK Genetic analysis of the selected lines indicates that the differences in oil content are due to the effects of at least 50 genes.

SOURCE Moose, S. P., J. W. Dudley, and T. R. Rocheford. 2004. "Maize Selection Passes the Century Mark: A Unique Resource for 21st Century Genomics." *Trends in Plant Science* 9:358–364.

the result of long-continued artificial selection for the oil content in kernels of corn.

A third mode of selection, known as **disruptive selection,** operates in favor of extremes and against intermediate forms. For example, we may see disruptive selection operating on the size of seed-eating birds' bills, with small (for delicate seeds) and large (for chunky seeds) bills favored over intermediate sizes that perform poorly with both delicate and chunky seeds. We explore this mechanism in more detail in the next chapter, as it can lead to the evolution of new species.

Sexual selection increases an individual's reproductive success.

Initially, Darwin was puzzled by features of organisms that seemed to reduce an individual's chances of survival. In a letter a few months after the publication of *The Origin*, he wrote, "The sight of a feather in a peacock's tail, whenever I gaze at it, makes me sick!" The tail is metabolically expensive to produce; it is an advertisement to potential predators; and it is a massive encumbrance in any attempt to escape a predator. How could such a feature evolve under natural selection?

In his 1871 book, *The Descent of Man,* Darwin introduced a solution to this problem. Natural selection is indeed acting to reduce the showiness and size of the peacock's tail, but another form of selection, **sexual selection,** is acting in the opposite direction. Sexual selection promotes traits that increase an individual's access to reproductive opportunities. Peahens are thought to select their mates on the basis of the showiness and size of a male's tail. In the absence of sexual selection, natural selection would act to minimize the size of the peacock's tail. Presumably, the peacocks' tails we see are a compromise, a trade-off between these conflicting demands of reproduction and survival (Chapter 45).

21.5 MIGRATION, MUTATION, AND GENETIC DRIFT

Selection is evolution's major driving force, enriching each new generation for the mutations that best fit organisms to their environments. However, as we have seen from the discussion of the Hardy–Weinberg equilibrium, it is not the only evolutionary mechanism. There are other forces that can cause allele frequencies to change. These are migration or gene flow, mutation, and the random effects of finite population size (that is, genetic drift). Like natural selection, these mechanisms can cause allele frequencies to change. Unlike natural selection, they do not lead to adaptations. Therefore, they are often considered non-adaptive evolutionary mechanisms.

Migration reduces genetic variation between populations.

Migration is the movement of individuals from one population to another, resulting in **gene flow,** the movement of alleles from one population to another. It is relatively simple to see how movements of individuals and alleles can lead to changes in allele frequencies. Consider two isolated island populations of rabbits, one white, the other black. Now imagine that the isolation breaks down—a bridge is built between the islands—and migration occurs. Over time, black alleles enter the white population and vice versa, and the allele frequencies of the two populations gradually become the same.

The consequence of migration is therefore the homogenizing of populations, making them more similar to each other and reducing genetic differences between them. Because populations are often adapted to their particular local conditions (think of dark-skinned humans in regions of high sunlight versus fair-skinned humans in regions of low sunlight), migration may be worse than merely non-adaptive—it may be maladaptive, in that it causes a decrease in a population's average fitness. Fair-skinned people arriving in an equatorial region are, for example, at risk of sunburn and skin cancer.

Mutation increases genetic variation.

As we saw earlier in this chapter, mutation is a rare event. This means that it is generally not important as an evolutionary mechanism that leads allele frequencies to change. However, as we have also seen, it is the source of new alleles and the raw material on which the other forces act. Without mutation, there would be no genetic variation, and therefore no evolution.

Genetic drift is particularly important in small populations.

Genetic drift is the random change in allele frequencies from generation to generation. By "random" change, we mean that frequencies can either go up or down simply by chance. An extreme case is a population **bottleneck,** which occurs when a population falls to just a few individuals. Consider a rare allele, A, with frequency of $\frac{1}{1000}$. Now imagine that habitat destruction reduces the population to just one pair of individuals, one of which is carrying A. The frequency of A in this new population is $\frac{1}{4}$ because each individual has two alleles, giving a total of four alleles. In other words, the bottleneck resulted in a massive change in allele frequencies. Another kind of bottleneck, called a **founder event,** may be important in the establishment of new populations, for example when just a few individuals arrive to colonize an island.

Earlier, we considered the fate of beneficial and harmful mutations under the influence of natural selection. What about neutral mutations? Natural selection, by definition, does not

FIG. 21.12 Genetic drift. The fate of neutral mutations is governed by genetic drift, the effect of which is more extreme in small populations than in large populations.

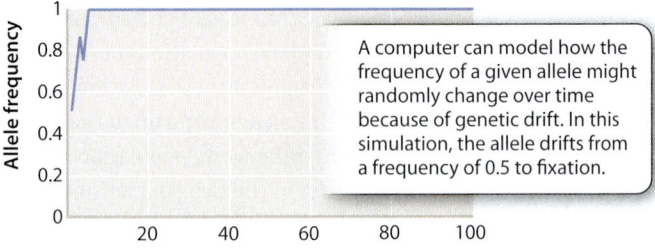

a. Population size = 4

A computer can model how the frequency of a given allele might randomly change over time because of genetic drift. In this simulation, the allele drifts from a frequency of 0.5 to fixation.

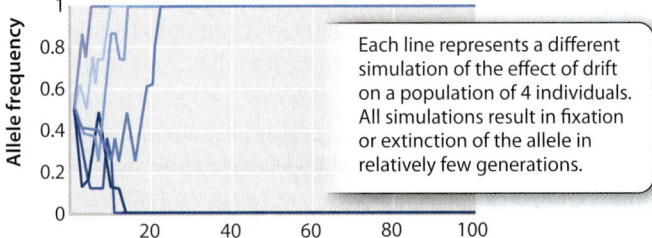

b. Population size = 4

Each line represents a different simulation of the effect of drift on a population of 4 individuals. All simulations result in fixation or extinction of the allele in relatively few generations.

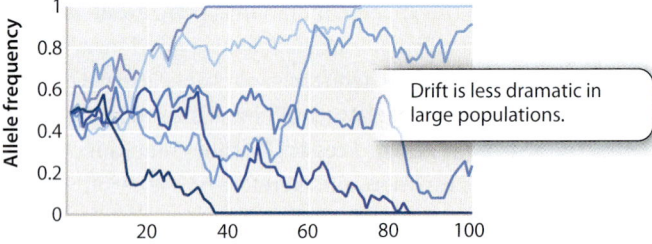

c. Population size = 40

Drift is less dramatic in large populations.

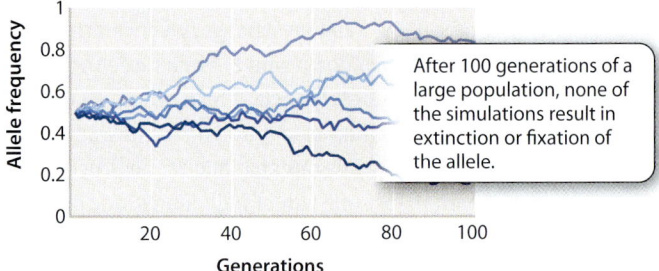

d. Population size = 400

After 100 generations of a large population, none of the simulations result in extinction or fixation of the allele.

govern the fate of neutral mutations. So what happens to them? Consider a neutral mutation, *m*, which has no effect on the fitness of its carrier. At first, it is in just a single heterozygous individual. What happens if that individual fails to reproduce (for reasons unrelated to *m*)? In this case, *m* will be lost from the population, but not by natural selection (which does not discriminate against *m*). Alternatively, the *m*-bearing individual might by chance leave many offspring (again for reasons unrelated to *m*), in which case the frequency of *m* will increase. In principle, it is possible over a long period of time for *m* to take over the population. At the end of the process, every member of the population is homozygous *mm*.

Like natural selection, genetic drift leads to allele frequency changes and therefore to evolution. Unlike natural selection, however, it does not lead to adaptations, since the alleles whose frequencies are changing do not affect an individual's ability to survive or reproduce.

The impact of genetic drift depends on population size (**Fig. 21.12**). If *m* arises in a very small population, its frequency will change rapidly, as shown in Figs. 21.12a and 21.12b. Imagine *m* arising in a population of just six individuals (or three pairs). Its initial frequency is 1 in 12, or about 8% (there are a total of 12 alleles because each individual is diploid). If, by chance, one pair fails to breed and the other two (including the one who is an *Mm* heterozygote) each produce three offspring, and all three of the *Mm* individual's offspring happen to inherit the *m* allele, then the frequency of *m* will increase to 3 in 12 (25%) in a single generation. In effect, genetic drift is equivalent to a sampling process. In a small sample, extreme departures from the expected outcome are common. Toss a coin 5 times, and you might well end up with zero heads.

On the other hand, if the population is large, as in Figs. 21.12c and 21.12d, then shifts in allele frequency from generation to generation are much smaller, typically less than 1%. A large population is analogous to a large sample size, in which we tend not to see marked departures from expectation. Toss a coin 1000 times, and you will end up with approximately 500 heads. In a small sample of coin tosses, we are much more likely to see marked departures from our 50:50 expectation than in a large sample. The same is true of genetic drift. It is likely to be much more significant in small populations than in large ones.

→ **Quick Check 4:** Why, of all the evolutionary forces, is selection the only one that can result in adaptation?

21.6 MOLECULAR EVOLUTION

How do DNA sequence differences arise among species? Imagine starting with two pairs of identical twins, one pair male and the other female. Now we place one member of each pair together on either side of a mountain range (**Fig. 21.13**). Let's assume the mountain range completely isolates the couples on each side. What, in genetic terms, will happen over time? The original pairs will found populations on each side of the mountain range. The genetic starting point is, in each case,

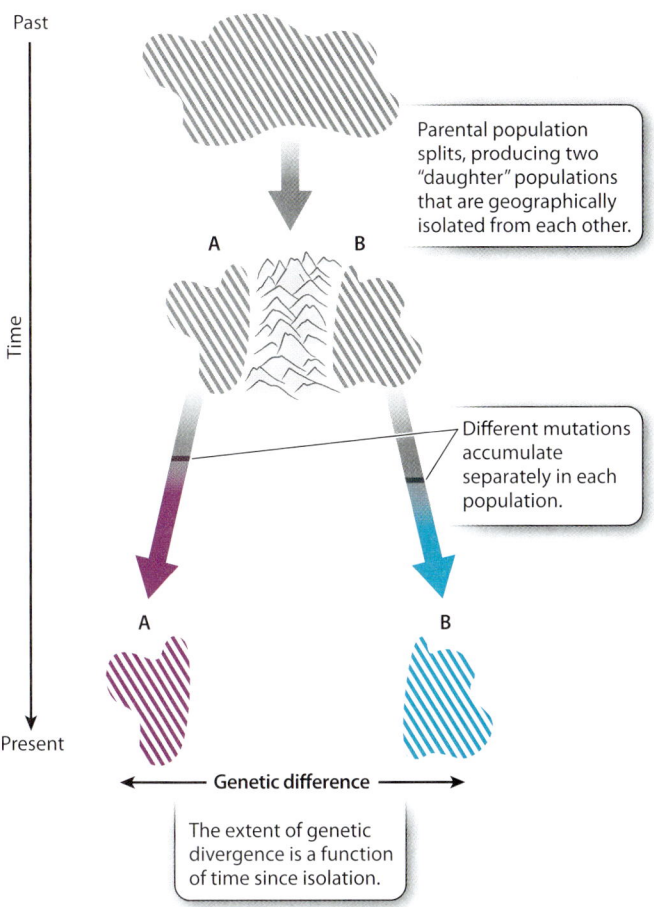

FIG. 21.13 Genetic divergence in isolated populations.

exactly the same, but, as the generations tick by, differences will accumulate between the two populations. Over time, mutations will occur in one population that will not have arisen in the other population, and vice versa. Because mutations occur at a low rate and are scattered randomly throughout the genome, the chance that both populations acquire the same mutations is practically zero.

A mutation in either population ultimately suffers one of three fates: It goes to fixation (either through genetic drift or through positive selection); it is maintained at intermediate frequencies (by balancing selection); or it is eliminated (either through natural selection or genetic drift). Different mutations will be fixed in each population. When we come back thousands of generations later and sequence the DNA of our original identical individuals' descendants, we will find that many differences have accumulated. The populations have diverged genetically. What we are seeing is evidence of **molecular evolution.**

Species are the biological equivalents of islands because they, too, are isolated. They are *genetically* isolated because, by definition, members of one species cannot exchange genetic material with members of another (Chapter 22). The amount of time that two species have been isolated from each other is the time since their most recent common ancestor. Thus, humans and chimpanzees, whose most recent common ancestor lived about 6-7 million years ago, have been isolated from each other for about 6-7 million years. Mutations arose and were fixed in the human lineage over that period; mutations, usually different ones, also arose and were fixed in the chimpanzee lineage over the same period. The result is the genetic difference between humans and chimpanzees. If we sequence a particular gene for both species, we find a number of sequence differences.

The extent of sequence difference between species is a function of the time since the species diverged.

The extent of genetic difference, or genetic divergence, between two species is a function of the time they have been genetically isolated from each other. The longer they have been apart, the greater the opportunity for mutation and fixation to occur in each population. This correlation between the time two species have been evolutionarily separated and the amount of genetic divergence between them is known as the **molecular clock.**

For a clock to function properly, it not only needs to keep time, but it also needs to be set. We set the clock using dates from the fossil record. For example, in a 1967 study, Vince Sarich and Allan Wilson deduced from fossils that the lineages that gave rise to the Old and New World monkeys separated about 30 million years ago. Finding that the amount of genetic divergence between humans and chimpanzees was about one-fifth of that between Old and New World monkeys, they concluded that humans and chimpanzees had been separated one-fifth as long, or about 6 million years. Although this is the generally accepted number today, Sarich and Wilson's result was revolutionary at the time, when it was thought that the two species were separated for as long as 25 million years.

The rate of the molecular clock varies.

Molecular clocks can be useful for dating evolutionary events like the separation of humans and chimpanzees. However, because the rates of molecular clocks vary from gene to gene, clock data should be interpreted cautiously. These rate differences can be attributed largely to differences in intensity of negative selection (which results in the elimination of harmful mutations) among different genes. The slowest molecular clock on record belongs to the histone genes, which encode the proteins around which DNA is wrapped to form chromatin (Chapter 3). These proteins are exceptionally similar in all organisms; only 2 amino acids (in a chain of about 100) distinguish plant and animal histones. Plants and animals last shared a common ancestor more than 1 billion years ago, which means, because each evolutionary

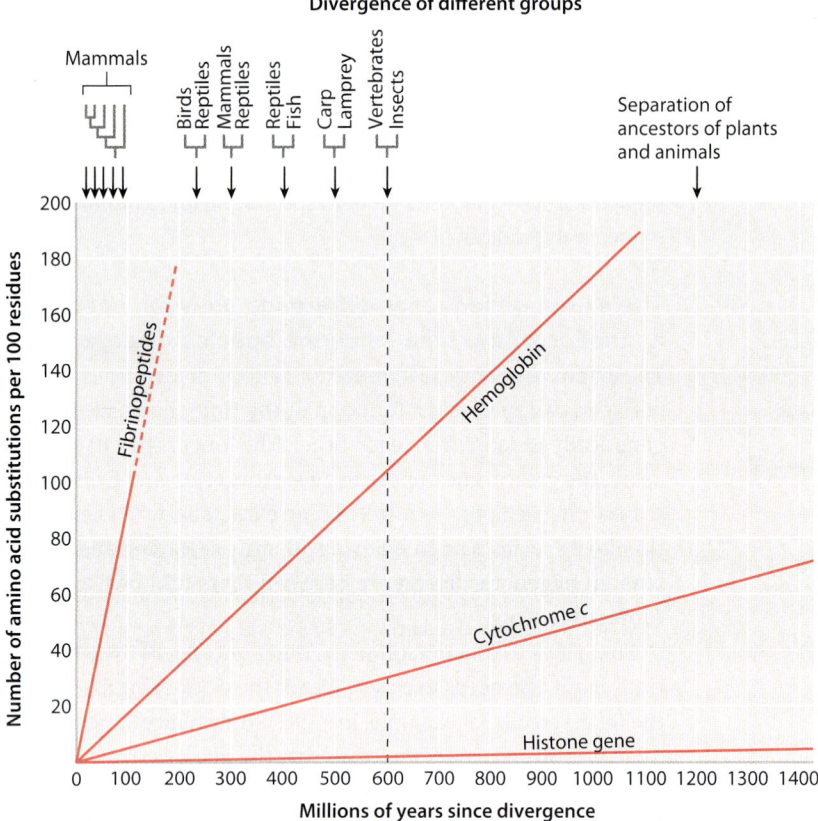

FIG. 21.14 The molecular clock. Different genes evolve at different rates because of differences in the intensity of negative selection.

Since vertebrates and insects diverged 600 mya, the amino acid sequences of cytochrome *c* have acquired 25 differences. Amino acid sequences of hemoglobin have acquired many more differences in the same time period.

lineage is separate, that there have been at least 2 billion years of evolution since they were in genetic contact. And yet, in all that time, the histones have hardly changed at all. Almost any amino acid change fatally disrupts the histone protein, preventing it from carrying out its proper function. Negative selection has thus been extremely effective in eliminating just about every amino acid-changing histone mutation over 2 billion years of evolution. The histone molecular clock is breathtakingly slow.

Other proteins are less subject to such rigorous nagative selection. Occasional mutations may therefore become fixed, either through drift (if they are neutral) or selection (if beneficial). The extreme case of a fast molecular clock is that derived from a **pseudogene,** a gene that is no longer functional. Because *all* mutations in a pseudogene are by definition neutral—there is no function for a mutation to disrupt, so a mutation is neither deleterious nor beneficial—we expect to see a pseudogene's molecular clock tick at a very fast rate. In the histone genes, virtually all mutations are selected against, constraining the rate of evolution; in pseudogenes, none are. **Fig. 21.14** vividly shows the varying rates of the molecular clock for different genes.

In the next chapter, we will see how genetic divergence between populations can lead to the evolution of new species.

Core Concepts Summary

21.1 GENETIC VARIATION IS THE RESULT OF DIFFERENCES IN DNA SEQUENCES.

Visible differences among members of a species (phenotypic variation) are the result of differences at the DNA level (genetic variation) as well as the influence of the environment. page 21-1

Mutation, a change in DNA sequence, is the ultimate source of genetic variation. This variation is reconfigured by recombination. page 21-2

Mutations can be somatic (in body tissues) or germ-line (in gametes), but germ-line mutations are the only ones that can be passed on to the next generation. page 21-2

When a mutation occurs in a gene, it creates a new allele. page 21-2

21.2 INFORMATION ON ALLELE FREQUENCIES IS KEY TO UNDERSTANDING PATTERNS OF GENETIC VARIATION.

An allele frequency is the number of occurrences of a particular allele divided by the total number of occurrences of all alleles of that gene in a population. page 21-3

In the past, population geneticists relied on observable traits determined by a single gene to measure genetic variation, but this method is not generally useful as most traits are influenced by many genes and by the environment. page 21-4

Allele frequencies can be measured by protein gel electrophoresis. page 21-4

DNA sequencing is now the standard technique for measuring allele frequencies. page 21-4

21.3 EVOLUTION IS A CHANGE IN THE FREQUENCY OF ALLELES OR GENOTYPES OVER TIME.

Allele frequencies change in response to selection, migration, mutation, and genetic drift. These are the primary mechanisms of evolution. page 21-6

The Hardy–Weinberg equilibrium describes situations in which allele frequencies do not change. In this way, it provides a null hypothesis against which to test whether or not evolution is occurring in a population. page 21-6

The Hardy–Weinberg equilibrium makes five assumptions. These assumptions are no selection, no migration, no mutation, no sampling error due to small population size, and random mating. page 21-6

The Hardy–Weinberg equilibrium allows allele frequencies and genotype frequencies to be calculated from each other. page 21-7

21.4 NATURAL SELECTION LEADS TO ADAPTATION, WHICH ENHANCES THE FIT BETWEEN AN ORGANISM AND ITS ENVIRONMENT.

Independently conceived by Charles Darwin and Alfred Russel Wallace, natural selection is the differential reproductive success of genetic variants. page 21-8

Under natural selection, a harmful allele is eliminated over time, and a beneficial one increases in frequency. Natural selection does not affect the frequency of neutral mutations. page 21-10

Natural selection can maintain alleles at intermediate frequencies by balancing selection. page 21-10

Changes in phenotype show that natural selection can be stabilizing, directional, or disruptive. page 21-11

In artificial selection, a form of directional selection, a breeder governs the selection process. page 21-12

Sexual selection involves the evolution of traits that increase an individual's access to members of the opposite sex. page 21-13

21.5 MIGRATION, MUTATION, AND GENETIC DRIFT ARE NON-ADAPTIVE MECHANISMS OF EVOLUTION.

Migration involves the movement of alleles between populations (gene flow) and tends to have a homogenizing effect. page 21-13

Mutation is the ultimate source of variation, but it also can change allele frequencies on its own. page 21-13

Genetic drift is a kind of sampling error, which is more marked in small populations. page 21-13

21.6 MOLECULAR EVOLUTION LOOKS AT CHANGES AT THE LEVEL OF DNA OR AMINO ACID SEQUENCES.

The extent of sequence difference between two species is a function of the time they have been genetically isolated from each other. page 21-15

Correlation between sequence differences among species and time since common ancestry of those species is known as the molecular clock. page 21-15

The rate of the molecular clock varies among genes because some genes are more selectively constrained than others. page 21-15

Self-Assessment

1. Differentiate between a phenotype and a genotype.
2. Name three types of mutations in terms of their effect on an organism.
3. Define genetic variation.
4. Describe three ways to measure genetic variation, and state which one is used today.
5. Given a set of genotype frequencies, calculate allele frequencies.
6. Define evolution.
7. Describe what happens to allele and genotype frequencies under the Hardy–Weinberg equilibrium.
8. Name and describe the five assumptions of the Hardy–Weinberg equilibrium.
9. Given a set of allele frequencies, calculate genotype frequencies if the population is in Hardy–Weinberg equilibrium.
10. Name and describe the five mechanisms of evolution.
11. Define natural selection and indicate how it is different from other mechanisms of evolution.
12. Explain how a molecular clock can be used to determine the time of divergence of two species.

Do you understand the chapter's Core Concepts? Log in to check your answers to the Self-Assessment questions, then practice what you've learned and reinforce this chapter's concepts by working through the problems and multimedia tutorials provided there.

www.biologyhowlifeworks.com

CHAPTER 22

SPECIES AND SPECIATION

Core Concepts

22.1 Species are reproductively isolated from one another.

22.2 Reproductive isolation is caused by barriers to reproduction before or after egg fertilization.

22.3 Speciation underlies the diversity of life on Earth.

22.4 Speciation can occur with or without natural selection.

Imagine for a moment a world without **speciation,** the process that produces new and distinct forms of life. Life would have originated and natural selection would have done its job of winnowing advantageous mutations from disadvantageous ones, but the planet would be inhabited by a single kind of generally adapted organism. Instead of the staggering biological diversity we see around us—current estimates are that between 10 and 100 million species call Earth home—there would be just a single life-form. From a biological perspective, the planet would be a decidedly dull place. Evolution is as much about adaptation, the result of natural selection, as it is about speciation, the engine that generates that breathtaking biodiversity.

22.1 THE BIOLOGICAL SPECIES CONCEPT

The definition of **species** has been a long-standing problem in biology. Many biologists respond to the problem in the same way that Darwin himself did. In *On the Origin of Species,* Darwin wrote, "No one definition has as yet satisfied all naturalists; yet every naturalist knows vaguely what he means when he speaks of a species." The difficulty of defining species has come to be called the species problem.

Here is the problem in a nutshell: The species, as an evolutionary unit, must by definition be fluid and capable of changing, giving rise through evolution to new species. The whole point of the Darwinian revolution is that species are not fixed. How, then, can we define something that changes over time, and, by the process of speciation, even leads to the origin of two species from one?

The species is the fundamental evolutionary unit.

We can plainly see biodiversity, but are what we call "species" real biological entities? Or was the term coined by biologists to simplify their description of the natural world? If so, species would then be convenient categories that derive not from nature but from our desire to categorize what we see around us. To test whether or not species are real, we can examine the natural world, measure some characteristic of different living organisms we see, and then plot these measurements on a graph. **Fig. 22.1** shows such a plot, graphing antenna length and wing length of three different types of butterfly. Note that the dots, representing individual organisms, fall into non-overlapping clusters. Each cluster is a species, and the fact that the clusters are distinct implies that species are biologically real. The distances we see between the dots within a cluster reflect variation from one individual to the next within a species (Chapter 21). Humans are highly variable, but overall we are more similar to one another than to our most

22-1

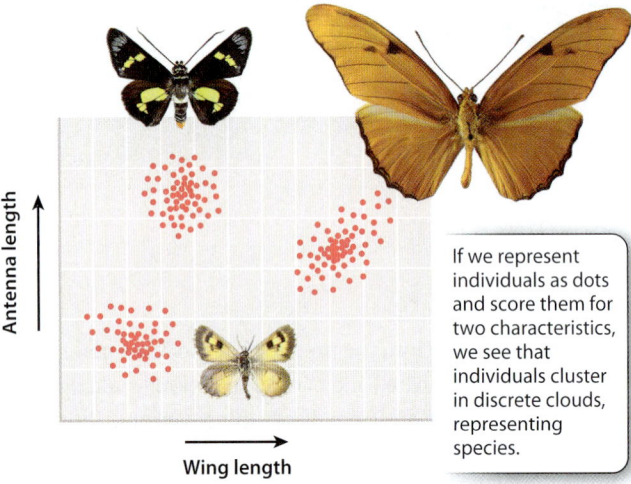

FIG. 22.1 Species clusters on the basis of two characteristics.

If we represent individuals as dots and score them for two characteristics, we see that individuals cluster in discrete clouds, representing species.

humanlike relative, the chimpanzee. We form a messy cluster, but that cluster does not overlap with the chimpanzee cluster.

The species, then, is a fundamental biological unit. Taxonomists group similar species into genera (singular, genus), and similar genera into families, and so forth on the basis of shared ancestry. The lines between these higher taxonomic categories, however, are arbitrary; their composition depends on the judgment of taxonomists. If, in their view, two species are enough alike, they are placed in the same genus. If they are sufficiently different, they are placed in separate genera.

Whether or not two individuals are members of the same species, however, is not a matter of judgment, but rather a reflection of their ability (or inability) to exchange genetic material by producing fertile offspring. Furthermore, the species is the ultimate evolutionary unit. Think back to Chapter 21 and consider a new advantageous mutation that appears initially in a single individual. That individual and its offspring inheriting the mutation will have a competitive advantage over other members of the population, and the mutation will increase in frequency until it reaches 100%. Migration among populations causes the mutation to spread further until all individuals within the species have it. The mutation spreads *within* the species but cannot spread beyond it. Therefore, the species is the key evolutionary unit. It is species that become extinct, and it is species that, through genetic divergence, give rise to new species.

Reproductive isolation is the key to the biological species concept (BSC).

Although biologists took until the middle part of the twentieth century to reach any kind of agreement on the definition of species, Alfred Russel Wallace, who discovered natural selection independently of Darwin, published a definition in 1865 that is almost identical to the one we use today:

> Species are merely those strongly marked races or local forms which when in contact do not intermix, and when inhabiting distinct areas are generally believed to have had a separate origin, and to be incapable of producing a fertile hybrid offspring.

Wallace's definition anticipates the key elements of the most widely used and generally accepted definition of a species, the **biological species concept (BSC)**. The BSC was described by the great evolutionary biologist Ernst Mayr (1904–2005) as follows:

> Species are groups of actually or potentially interbreeding populations that are reproductively isolated from other such groups.

Let us look at this definition closely. At the heart of it is the idea of reproductive compatibility. Members of the same species are capable of producing offspring together, whereas members of different species are incapable of producing offspring together. In other words, members of different species are reproductively isolated from one another. As Wallace and Mayr both realized, however, reproductive compatibility in fact entails more than just the ability to produce offspring. The offspring must be fertile and therefore capable of passing their genes on to their own offspring. For example, although a horse and a donkey, two different species, can mate to produce a mule, the mule is infertile. If, like a mule, a hybrid offspring is infertile, then it is a genetic dead end.

Note, too, the "actual or potential" part of Mayr's definition. An Asian elephant living on the island of Sri Lanka and one living in nearby India are considered members of the same species, *Elephas maximus*, even though Sri Lankan and Indian Asian elephants never have a chance to mate with each other in nature because they are geographically separated (**Fig. 22.2**).

The BSC is more useful in theory than in practice.

Species, then, are real biological entities and the BSC is the most useful way to define them. However, the BSC has shortcomings, the most important of which is that it can be difficult to apply. Imagine you are on a field expedition in a rainforest and you find two insects that look reasonably alike. Are they members of the same species or not? To use the BSC, you would need to test whether or not they are capable of producing fertile offspring. However, in practice, you probably will not have the time and resources to perform such a test. And even if you do, it is possible that members of the same species will not reproduce when you place them together in a petri dish in your field camp.

FIG. 22.2 Sri Lankan and Indian Asian elephants.

The conditions may be too unnatural or too threatening for the insects to behave in a normal way. And even if they do mate, you may not have time to determine whether or not the offspring are fertile. Wallace recognized this practical limitation of a reproduction-based definition of species, pointing out that in practice, "the test of hybridity cannot be applied in more than one case in ten thousand."

Thus, we should consider the BSC as a valid framework for thinking about species, but one that is difficult to test. On a day-to-day basis, therefore, biologists often use a rule of thumb called the **morphospecies concept.** Stated simply, the morphospecies concept holds that members of the same species usually look alike, which is why natural history guidebooks are useful. For example, the shape, size, and coloration of the bald eagle make it easy to determine whether or not a bird observed in the wild is a member of the bald eagle species. Here we are returning to those biological clusters of Fig. 22.1 in which members of a species fall close to one another within a single, discrete group.

Although the morphospecies concept is a useful and generally applicable rule of thumb, it is not an infallible way to identify species. Members of a species may not always look alike, but instead show different phenotypes called polymorphisms (literally, "many forms")—for example, color difference in some species of birds. In Chapter 21, we saw how minor genetic differences can result in two individuals within a species having very different appearances. Sometimes, males of a species may look different from females—think of the showy peacock, which differs dramatically from the relatively drab peahen. Young can be very different from old. Caterpillars that mature into butterflies are a striking example.

Whereas members of a species may look quite different from one another, the opposite can also be true: Members of different species can look quite similar. Some species of butterfly, for example, appear identical but have chromosomal differences that can be observed only with the aid of a microscope (**Fig. 22.3**). These chromosomal differences prevent the formation of interspecies hybrids, so those butterflies meet the BSC criterion—they are different species—but they do not meet the morphospecies criterion because they look so much alike.

The BSC does not apply to asexual or extinct organisms.

In addition to the difficulties of putting the BSC into practice, a second problem is that it overlooks some organisms. For example, because it is based on the sexual exchange of genetic information, the BSC cannot apply to species that reproduce asexually, as do many bacteria. Although some asexual species do occasionally exchange genetic information in a process

FIG. 22.3 Three species of *Agrodiaetus* butterflies. These similar-appearing butterflies are identifiable only by differences in their chromosome numbers.

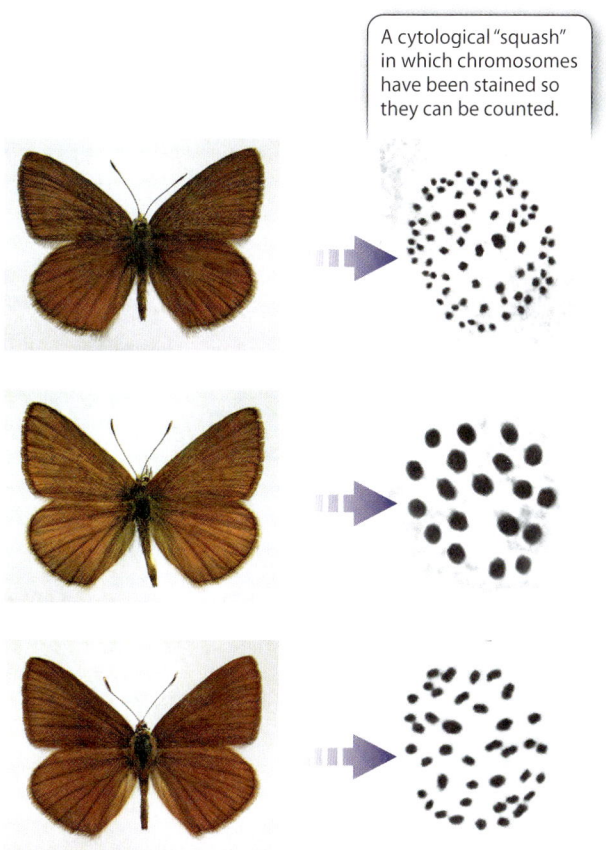

known as conjugation (Chapter 26), true asexual organisms do not fit the BSC.

In addition, because it depends on reproduction, the BSC obviously cannot be applied to species that have become extinct and are known only through the fossil record, like different "species" of trilobites and dinosaurs.

Ring species and hybridization complicate the BSC.

An unusual but interesting geographic pattern shown by **ring species** highlights another shortcoming of the BSC. Here we find that some populations within a species are reproductively isolated from each other, but others are not.

For example, populations of the greenish warbler, *Phylloscopus trochiloides,* are distributed in a large geographic loop around the Himalayas, as shown in **Fig. 22.4**. In Russia to the north, members of two neighboring populations do not interbreed. Therefore, according to the BSC, they are different species. However, the more western of the two Russian populations is capable of reproducing with the population to the south of it, and that population is capable of reproducing with the population to the east of it, and so on. Eventually, the loop of genetically exchanging populations comes all the way round to the more eastern of the two Russian populations. Thus, though members of the two Russian populations cannot exchange genes directly, they can do so indirectly, with the genetic material passing through many intermediate populations. This situation is more complicated than anything predicted by the BSC. The two Russian populations are reproductively isolated from each other but they are not genetically isolated from each other because of gene flow around the ring.

We also see a complicated situation in some groups of closely related species of plants. Despite apparently being good morphospecies that can be distinguished by appearance alone, many different species of willows (*Salix*), oaks (*Quercus*), and dandelions (*Taraxacum*) are still capable of exchanging genes with other species in their genera through **hybridization,** or interbreeding, between species. By the BSC, these different forms should be considered one large species because they are able to reproduce and produce fertile offspring. However, because they maintain their distinct appearances, natural selection must work against the hybrid offspring. This unusual phenomenon seems to occur mainly in plants.

Ecology and evolution can extend the BSC.

Because of these problems, much effort has been put into modifying and improving the BSC and many alternative definitions of species have been suggested. In general, these efforts highlight how difficult it is to make all species fit easily into one definition. The natural world truly defies neat categorization!

Several useful ideas, however, have come out of this literature. First is the notion that a species can sometimes be characterized by its **ecological niche,** which, as we will discuss further in Chapter 47, is a complete description of the role the species plays in its environment—its habitat requirements, its nutritional and water needs, and the like. It

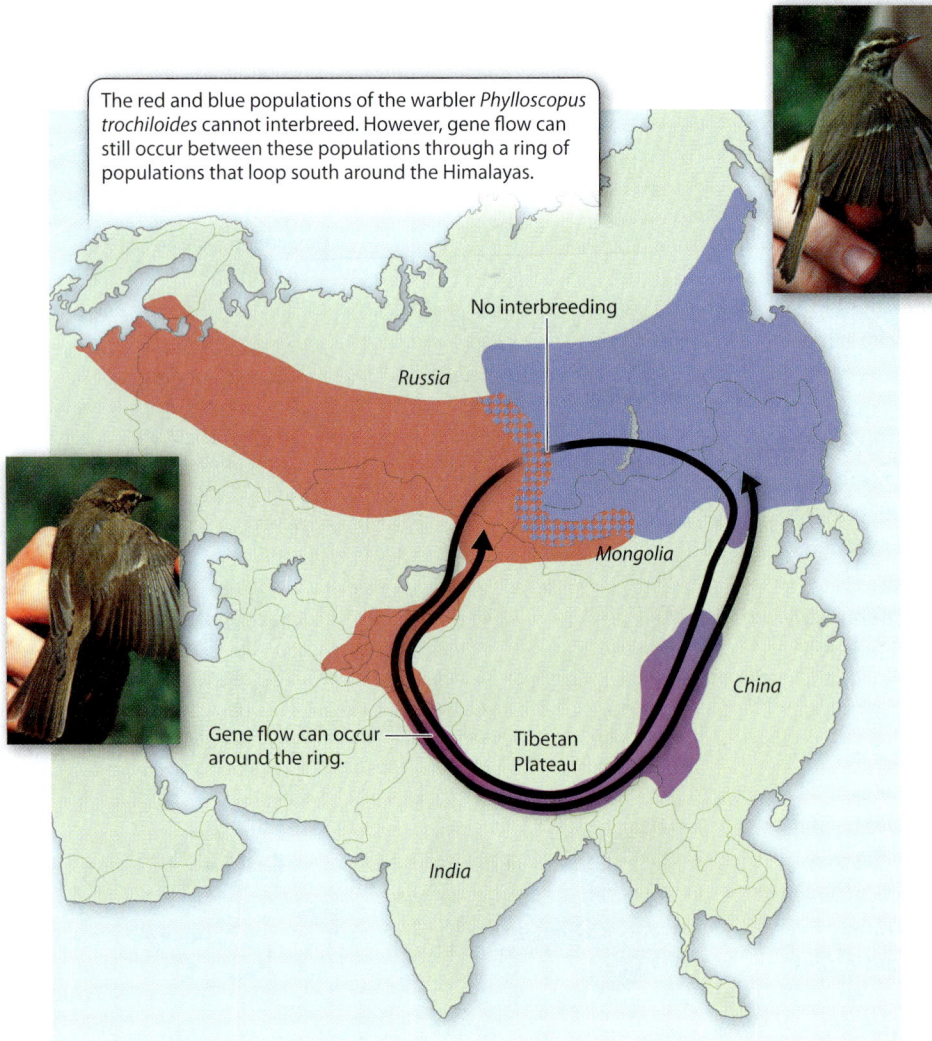

FIG. 22.4   Ring species. Ring species contain populations that are reproductively isolated from each other but can still exchange genetic material via other linking populations.

turns out that it is impossible for two species to coexist in the same location if their niches are too similar because competition between them for resources will inevitably lead to the extinction of one of them. This observation has given rise to the **ecological species concept (ESC),** the idea that there is a one-to-one correspondence between a species and its niche. Thus, we can determine whether or not asexual bacterial lineages are distinct species on the basis of differences or similarities in their ecological requirements. If two lineages have very different nutritional needs, for example, we can infer on ecological grounds that they are separate species.

Second is the **evolutionary species concept (EvSC),** which is the idea that members of a species all share a common ancestry and a common fate. It is, after all, species rather than individuals that become extinct. The EvSC requires that all members of a species are descended from a single common ancestor. It does not specify, however, on what scale this idea should be applied. All mammals derive from a single common ancestor that lived about 200 million years ago, but there are thousands of what we recognize as species of mammals that have evolved since that long-ago common ancestor. But, under a strict application of the EvSC, would we consider all mammals to be a single species? Similarly, siblings and cousins are all descended from a common ancestor, a grandmother, but that is surely not sufficient grounds to classify more distant relatives as a distinct species. The EvSC can be useful when thinking about asexual species, but, given the arbitrariness of the decisions involved in assessing whether or not the descendants of a single ancestor warrant the term "species," its utility is distinctly limited.

It is worth bearing these ecological and evolutionary considerations in mind when discussing and thinking about species. Although the BSC remains our most useful definition of species, the ESC and EvSC broaden and generalize the concept. For example, in the case of a normally asexual species that is difficult to study because conjugation is so infrequent, we can jointly apply the ESC and EvSC. We can use the ESC to loosely define the species in terms of its ecological characteristics (for example, its nutritional requirements), and we can refine that definition by using genetic analyses to determine whether the group is indeed a species by the EvSC's standard (that is, that all its members derive from a single common ancestor).

Despite the shortcomings of the BSC and the usefulness of alternative ideas, we stress that the BSC is the most constructive way to think about species. In particular, by focusing on reproductive isolation—the inability to produce viable, fertile offspring—the BSC gives us a means of studying and understanding speciation, the process by which two populations, originally members of the same species, become distinct.

→ **Quick Check 1** Why haven't we been able to come up with a single, comprehensive, and agreed-upon species concept?

22.2 REPRODUCTIVE ISOLATION

Factors that cause reproductive isolation are generally divided into two categories, depending on when they act. **Pre-zygotic** isolating factors act before the fertilization of an egg, and **post-zygotic** factors come into play after fertilization. In other words, pre-zygotic factors prevent fertilization from taking place, whereas post-zygotic factors result in the failure of the fertilized egg to develop into a fertile individual.

Pre-zygotic isolating factors occur before egg fertilization.

Most species are reproductively isolated by pre-zygotic isolating factors. Among animals, species are often **behaviorally isolated,** meaning that individuals only mate with other individuals based on specific courtship rituals, songs, and other kinds of behaviors. Chimpanzees may be our closest relative—and therefore the species we are most likely to confuse with our own—but a chimpanzee of the appropriate sex, however attractive to a chimpanzee of the opposite sex, fails to provoke even the faintest reproductive impulse in a human. In this case, the pre-zygotic reproductive isolation of humans and chimpanzees is behavioral.

Behavior does not play a role in plants, but pre-zygotic factors can still be important in reproductive isolation. Pre-zygotic isolation in plants can take the form of incompatibility between the incoming pollen and the receiving flower so fertilization fails to take place. We see similar forms of isolation between members of marine species, such as abalone, which simply discharge their gametes into the water. In these cases, membrane-associated proteins on the surface of sperm interact specifically with membrane-associated proteins on the surface of eggs of the same species but not with those of different species. These specific interactions ensure that a sperm from one abalone species, *Haliotis rufescens,* fertilizes only an egg of its own species and not an egg from *H. corrugata,* a closely related species.

In some animals, especially insects, incompatibility arises earlier in the reproductive process. The genitalia of males of the fruit fly *Drosophila melanogaster* are configured in such a way that they fit only with the genitalia of females of the same species. Attempts by males of *D. melanogaster* to copulate with females of another species of fruit fly, *D. virilis,* are thwarted by mechanical incompatibility.

In all three of these cases—plants, marine invertebrates, and insects—mating with members of one species (and reproductive isolation from other species) is promoted by **lock and key** systems that require both components, whether physical or biochemical, to match for a successful interaction to take place.

Both plants and animals may also be pre-zygotically isolated in time (**temporal separation**) and space (**ecological separation**). For example, closely related plant species may flower at different times, so there is no chance that the pollen of one will come into contact with the flowers of the other. Similarly, members of a nocturnal animal species simply will not encounter members of a closely related species that are active only during the day.

Closely related species that are separated by vast distances or physical features of the landscape obviously have no opportunity to exchange gametes, but spatial separation may also be more subtle. For example, the two Japanese species of ladybug beetle shown in **Fig. 22.5** can be found living side by side in the same field, but they feed on different plants. Because their life cycles are so intimately associated with their host plants (adults even mate on their host plants), these two species never breed with each other. This ecological separation is what causes their pre-zygotic isolation.

Post-zygotic isolating factors occur after egg fertilization.

Post-zygotic isolating factors involve mechanisms that come into play after fertilization of the egg. Typically, they involve some kind of **genetic incompatibility.** One example, which we saw earlier in Fig. 22.3 and will explore later in the chapter, is the case of two organisms with different numbers of chromosomes.

In some instances, the effect can be extreme. For example, the zygote may fail to develop after fertilization because the two parental genomes are sufficiently different to prevent normal development. In others, the effect is less obvious. Some matings between different species produce perfectly viable adults, as in the case of the horse–donkey hybrid, the mule. As we have seen, though, all is not well with the mule from an evolutionary perspective. The horse and donkey genomes are different enough to cause the mule to be sterile. As a general rule, the more closely related—and therefore genetically similar (Chapter 21)—a pair of species, the less extreme the genetic incompatibility between their genomes.

22.3 SPECIATION

Recognizing that species are groups of individuals that are reproductively isolated from other such groups, we are now in a position to recast the key question: How does speciation occur? Instead of asking, "How do new species arise?" we can ask, "How does reproductive isolation arise between populations?"

Speciation is a by-product of the genetic divergence of separated populations.

The key to speciation is the fundamental evolutionary process of genetic divergence between genetically separated populations. As we saw in Chapter 21, if a single population is split into two populations that are unable to interbreed, different mutations will appear by chance in the two populations. Like all mutations, these will be subject to genetic drift or natural selection, resulting over time in the genetic divergence of the two populations. Two separate populations that are initially identical will, over long periods of time, gradually become distinct as different mutations are introduced and propagated in each population. At some stage in the course of divergence, changes occur in one population that lead to its members being reproductively isolated from members of the other population (**Fig. 22.6**). It is this process that results in speciation.

Speciation—the development of reproductive isolation between populations—is, therefore, just a by-product of the genetic divergence of separated populations.

As Fig. 22.6 shows, speciation is typically a gradual process. If we try to cross members of two populations that have genetically diverged but not diverged far enough for full reproductive isolation (that is, speciation) to have arisen, we may find that the populations are **partially reproductively isolated.** They are not yet separate species, but the genetic differences between them are extensive enough that the hybrid offspring they produce have reduced fertility or viability compared to offspring produced by crosses between individuals within each population.

Allopatric speciation is speciation that results from the geographical separation of populations.

Reproductive isolation between two populations ultimately results in two species. Because this process requires genetic isolation between the diverging populations and because geography is

FIG. 22.5 **Ecological separation.** The ladybugs (a) *Henosepilachna yasutomii* and (b) *H. niponica* are reproductively separated from each other because they feed and mate on different host plants.

a.

b.

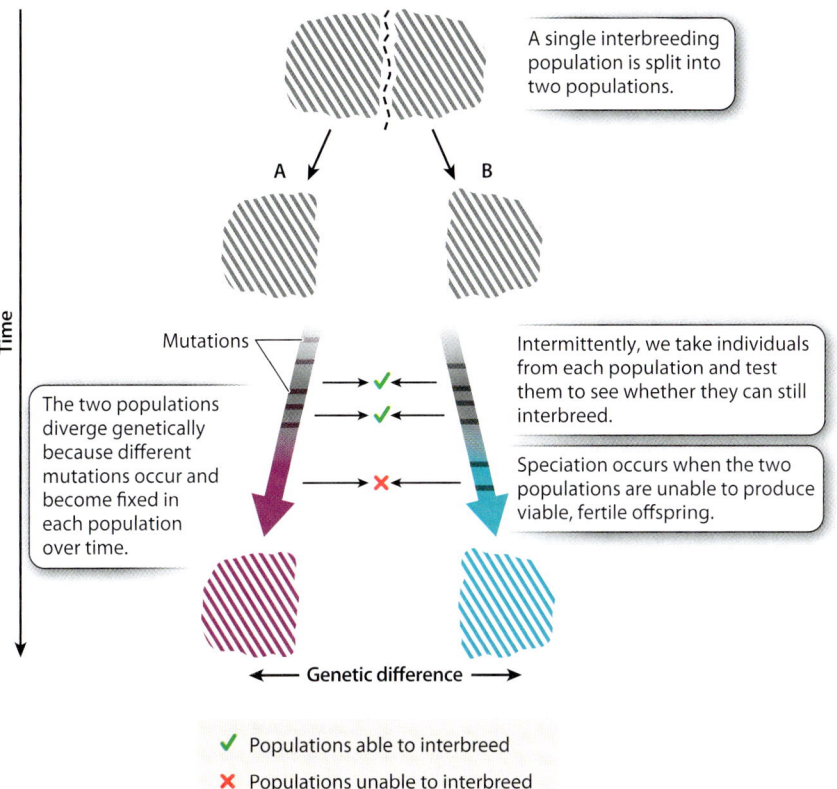

FIG. 22.6 Speciation. Speciation occurs when two populations that are genetically diverging become reproductively isolated from each other.

the easiest way to ensure genetic isolation, models of speciation focus on geography.

The process usually begins with the creation of **allopatric** (literally, "different place") populations, populations that are geographically separated from each other. Clearly, physically separating two populations is the simplest way to ensure that they become genetically isolated from each other. It is therefore not surprising that most speciation is thought to be allopatric. Note, however, that the size and ecology of a species must be considered in deciding whether or not two populations are truly allopatric. Two populations of salamanders separated by a large highway may be allopatric, as salamanders may not be able to cross the highway, whereas such a barrier would obviously not suffice to separate populations of birds. To assess, then, whether two populations are truly allopatric, it is necessary to take more than just location into account.

Because genetic divergence is typically gradual, we often find allopatric populations that have yet to evolve even partial reproductive isolation but which have accumulated a few population-specific traits. This genetic distinctness is sometimes recognized by taxonomists, who deem each geographic form a **subspecies** by adding a further designation after its species name. For example, Sri Lankan Asian elephants, subspecies *Elephas maximus maximus*, are generally larger and darker than Indian ones, subspecies *Elephas maximus indicus* (see Fig. 22.2).

How do populations become allopatric? There are two basic ways. The first is by **dispersal,** in which some individuals colonize a distant place, such as an island, far from the main source population. The second is by **vicariance,** in which a geographic barrier arises within a single population, separating it into two or more isolated populations. For example, when sea levels rose at the end of the most recent ice age, new islands formed along the coastline as the low-lying land around them was flooded. The populations on those new islands suddenly found themselves isolated from other populations of their species. This kind of island formation is a vicariance event.

Regardless of how the allopatric populations came about—whether through dispersal or vicariance—the outcome is the same. The two separated populations will diverge genetically until speciation occurs.

Often, vicariance-derived speciation events are the easiest to study because we can date the time at which the populations were separated if we know when the vicariance event occurred. One such event whose history is well known is the formation of the Isthmus of Panama between Central and South America, shown in **Fig. 22.7**. This event took place about 3.5 million years ago. As a result, populations of marine organisms in the western Caribbean and eastern Pacific that had formerly been able to interbreed freely were separated from each other. After a period of time, the result was the formation of many distinct species, each of whose closest relatives are on the other side of the isthmus.

Dispersal is important in a specific kind of allopatric speciation known as **peripatric speciation.** In this model, a few individuals from a **mainland population** (the central population of a species) disperse to a new location remote from the original population and evolve separately. This may be an intentional act of dispersal, or it could be an accident brought about by, for example, an unusual storm that blows migrating birds off their normal route. The result is a distant, isolated **island population.** "Island" in this case may refer to a true island—like Hawaii—or may simply refer to a patch of habitat on the mainland that is appropriate for the species but is geographically remote from the initial mainland population's habitat area. For a species adapted to life on mountaintops, a new island might be another previously

HOW DO WE KNOW?

FIG. 22.7

Can a vicariance event cause speciation?

BACKGROUND 3.5 million years ago, the Isthmus of Panama was not completely formed. Several marine corridors remained open, allowing interbreeding between marine populations in the Caribbean and the eastern Pacific. Subsequently, the gaps in the isthmus were plugged, separating the Caribbean and eastern Pacific populations and preventing gene flow between them.

HYPOTHESIS Patterns of speciation will reflect the impact of the vicariance event resulting from the closing of the direct marine connections between the Pacific and the Caribbean. Specifically, researchers predicted that each ancestor species (from the time before the formation of the isthmus) split into two "daughter" species, one in the Caribbean and the other in the Pacific. The closest relative of each current Pacific species, then, is predicted to be a Caribbean species (and vice versa).

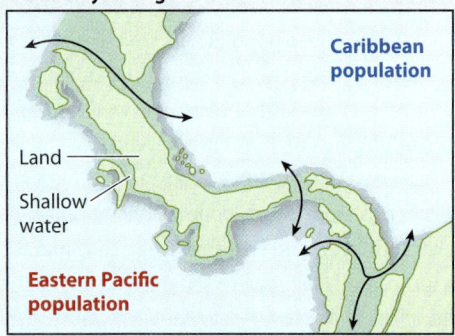

Interbreeding between Eastern Pacific and Caribbean populations of *Alpheus* was possible by the corridors that existed before the final formation of the Isthmus of Panama.

Interbreeding between Eastern Pacific and Caribbean populations is no longer possible because of the geographic barrier.

EXPERIMENT This study focused on 17 species of snapping shrimp in the genus *Alpheus,* a group that is distributed on both sides of the isthmus. The first step was to sequence the same segment of DNA from each species. The next step was to compare those sequences in order to reconstruct the phylogenetic relationships among the species.

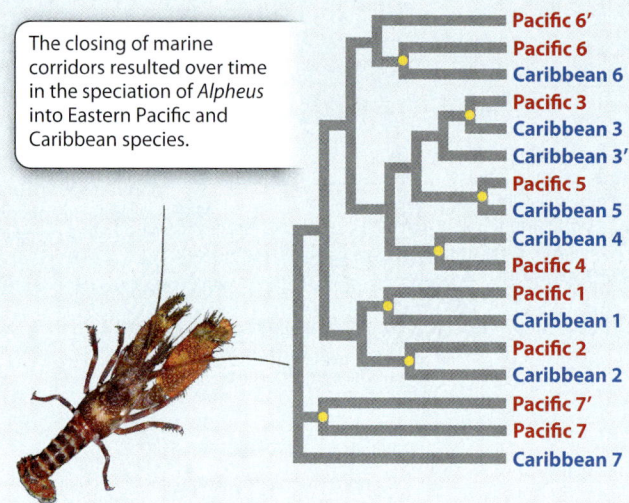

The closing of marine corridors resulted over time in the speciation of *Alpheus* into Eastern Pacific and Caribbean species.

RESULTS The phylogeny reveals that species show a distinctly paired pattern of relatedness: The closest relative of each species is one from the other side of the isthmus.

CONCLUSION That we see these consistent Pacific/Caribbean sister species pairings strongly supports the hypothesis that the vicariant event caused by the formation of the isthmus has driven speciation in *Alpheus*. Each Pacific/Caribbean pairing is derived from a single ancestral species (indicated by a red dot) whose continuous distribution between the Caribbean and eastern Pacific was disrupted by the formation of the isthmus. Here we see striking evidence of the role of vicariance in speciation.

FOLLOW-UP WORK Speciation is about more than just genetic differences between isolated populations, as shown here. Knowlton and colleagues also tested the different species for reproductive isolation and found that there were high levels of isolation between Caribbean/Pacific pairs. Given that we know that each species pair has been separated for at least 3 million years, we can then calibrate the rate at which new species are produced.

SOURCE Knowlton, N., L. A. Weigt, L. A. Solórzano, D. K. Mills, and E. Bermingham. 1993. "Divergence in Proteins, Mitochondrial DNA, and Reproductive Compatibility Across the Isthmus of Panama." *Science* 260 (5114):1629–1632.

uninhabited mountaintop. For a rainforest tree species, that new island might be a patch of lowland forest on the far side of a range of mountains that separates it from its mainland forest population.

The island population is classically small and often in an environment that is slightly different from that of the mainland population. The peripatric speciation model suggests that change accumulates faster in these peripheral isolates than in the large mainland populations, both because genetic drift is more pronounced in smaller populations than in larger ones and because the environment may differ between the mainland and island in a way that results in natural selection driving differences between the two populations. These mechanisms cause genetic divergence of the island population from the mainland one, ultimately leading to speciation.

It is possible to glimpse peripatric differentiation in action (**Fig. 22.8**). Studies of a kingfisher, *Tanysiptera galatea*, in New Guinea and nearby islands show it under way. There are eight recognized subspecies of *T. galatea,* three on mainland New Guinea (where they exist in large populations separated by mountain ranges) and five on nearby islands (where, because the islands are small, the populations are correspondingly small). The mainland subspecies are still quite similar to one another, but the island subspecies are much more distinct, suggesting that genetic divergence is occurring faster in the small island populations. If we wait long enough, these subspecies will probably diverge into new species.

Because dating such dispersal events is tricky—newcomers on an island tend not to leave a record of when they arrived—we can some times use vicariance information to study the timing of peripatric speciation. For instance, we know that the oldest of the Galápagos Islands were formed 4-5 million years ago by volcanic action. Some time early in the history of the Galápagos, individuals of a small South American finch species arrived there. Conditions on the Galápagos are very different from those on the South American continent, where the mainland population of this ancestral finch lived, and so the isolated island population evolved to become distinct from its mainland ancestor and eventually became a new species. The finches' subsequent dispersal among the other islands of the Galápagos has promoted further peripatric speciation (**Fig. 22.9**). The result was the evolution of 13 different species of finches, known today as Darwin's finches.

The Galápagos finches and their frenzy of speciation illustrate the important evolutionary idea of **adaptive radiation,** a bout of unusually rapid evolutionary diversification in which natural selection accelerates the rates of both speciation and adaptation. Adaptive radiation occurs when there are many

---

FIG. 22.8   Peripatric speciation in action among populations of New Guinea kingfishers.

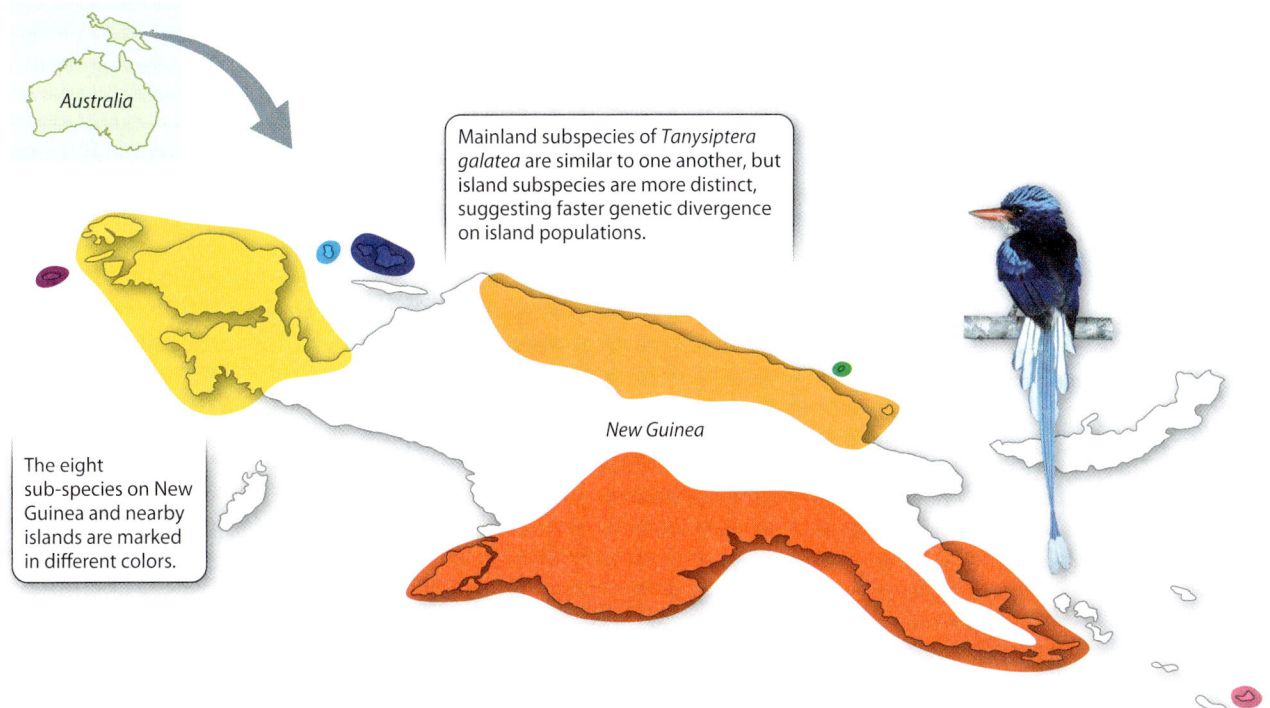

FIG. 22.9 Adaptive radiation in Darwin's finches. Today, there are 13 different species of finch on the Galápagos Islands.

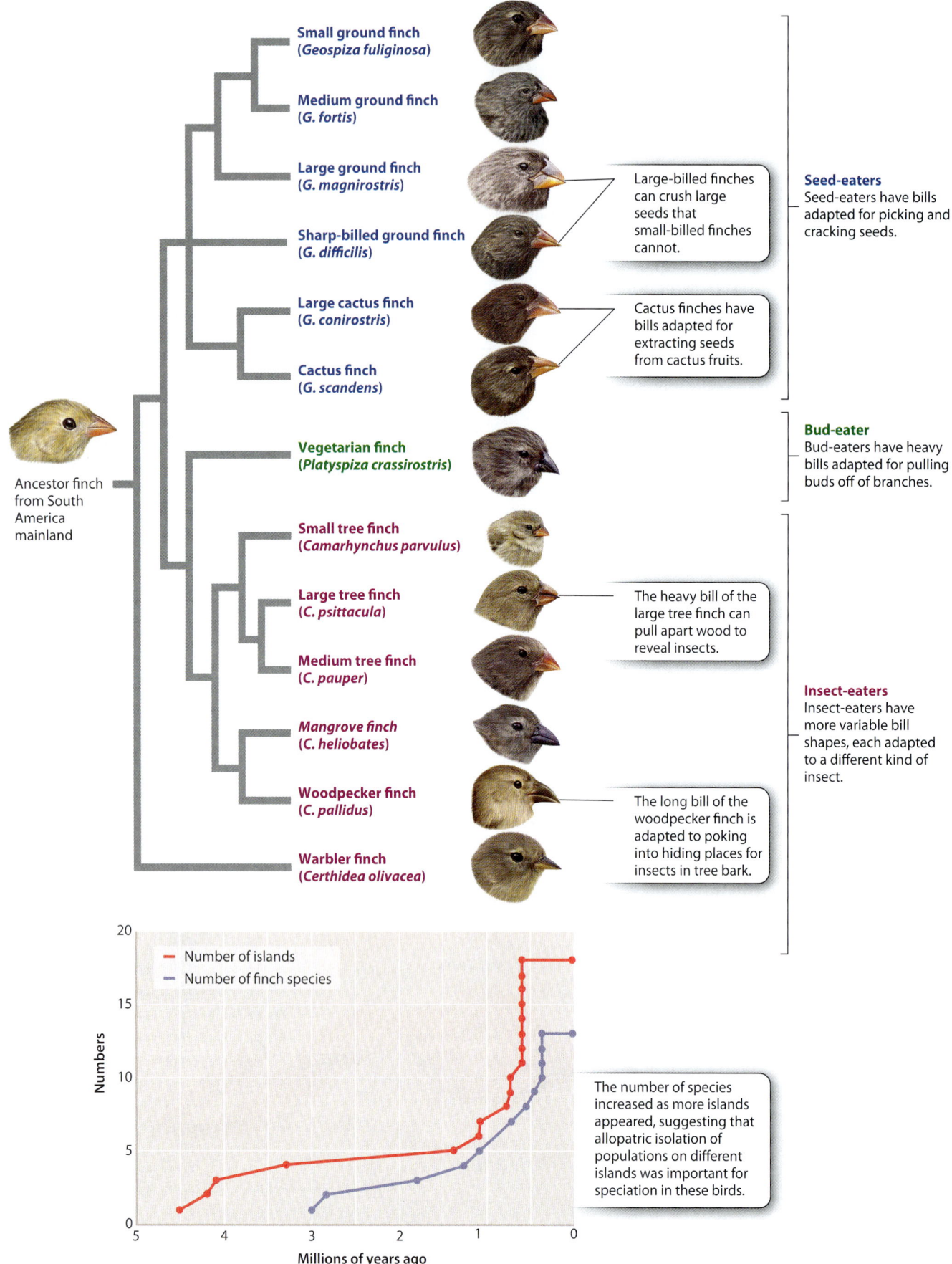

ecological opportunities available for exploitation. Consider the ancestral finch immigrants arriving on the Galápagos. A wealth of ecological opportunities was open and available. Until the arrival of the colonizing finches, there were no birds on the islands to eat the plant seeds, or to eat the insects on the plants, and so on. Suppose that the ancestral finches fed specifically on medium-sized seeds on the South American mainland (that is, they were medium-seed specialists, with bills that are the right size for handling medium seeds). On the mainland, they were constrained to that size of seed because any attempt to eat larger or smaller ones brought them into conflict with other species—a large-seed specialist and a small-seed specialist—that already used these resources. In effect, stabilizing selection (Chapter 21) was operating on the mainland to eliminate the extremes of the bill-size spectrum in the medium-seed specialists. A somewhat smaller bill, which is good for eating slightly smaller seeds, was selected against because its possessors competed poorly against a small-billed species that is a small-seed specialist. A slightly larger bill was similarly disadvantageous because its possessors ended up competing for large seeds with a species that is a large seed specialist.

When the medium-billed immigrant finches first arrived on the Galápagos, however, no such competition existed. Therefore, natural selection promoted the formation of new species of small- and large-seed specialists from the original medium-seed-eating ancestral stock. With the elimination of the competition-mediated stabilizing selection that kept the medium-billed finches medium billed on the mainland, selection actually operated in the opposite direction, favoring the large and small extremes of the bill-size spectrum because these individuals could take advantage of the abundance of unused resources, small and large seeds. It is this combination of emptiness—the availability of ecological opportunity—and the potential for allopatric speciation that results in adaptive radiation.

→   **Quick Check 2.** Why do we see so many wonderful examples of adaptive radiation on mid-ocean volcanic archipelagos like the Galápagos?

Co-speciation is speciation that occurs in response to speciation in another species.

As we have seen, physical separation is often a critical ingredient in speciation. Two populations that are not fully separated from each other—that is, there is gene flow between them—will typically not diverge from each other genetically, because genetic exchange homogenizes them. This is why most speciation is allopatric. Allopatric speciation brings to mind populations separated from each other by stretches of ocean or deserts or mountain ranges. However, separation can be just as complete even in the absence of geographic barriers. Consider an organism that parasitizes a single host species. Suppose that the host undergoes speciation, producing two daughter species. The original parasite population will also be split into two populations, one for each host species. Thus, the two new parasite populations are physically separated from each other and will diverge genetically, ultimately undergoing speciation. This divergence results in a pattern of coordinated host–parasite speciation called **co-speciation,** a process in which two groups of organisms speciate in response to each other and at the same time.

Phylogenetic analysis of lineages of parasites and their hosts that undergo co-speciation reveals trees that are similar for each group. Each time a branching event—that is, speciation—has occurred in one lineage, a corresponding branching event has occurred in the other (**Fig. 22.10**).

? CASE 4 Malaria: Co-Evolution of Humans and a Parasite
How did malaria come to infect humans?

Now let's look at a human parasite, *Plasmodium falciparum*, the causative agent of malaria. It had been suggested that *P. falciparum*'s closest relative is another *Plasmodium* species, *P. reichenowi,* found in chimpanzees, our closest living relative. Maybe *P. falciparum* and *P. reichenowi* were the products of co-speciation. When the ancestral population split millions of years ago to give rise to human and chimpanzee lineages, that population's parasitic *Plasmodium* population could also have

---

FIG. 22.10 Co-speciation. Parasites and their hosts often evolve together, and the result is similar phylogenies.

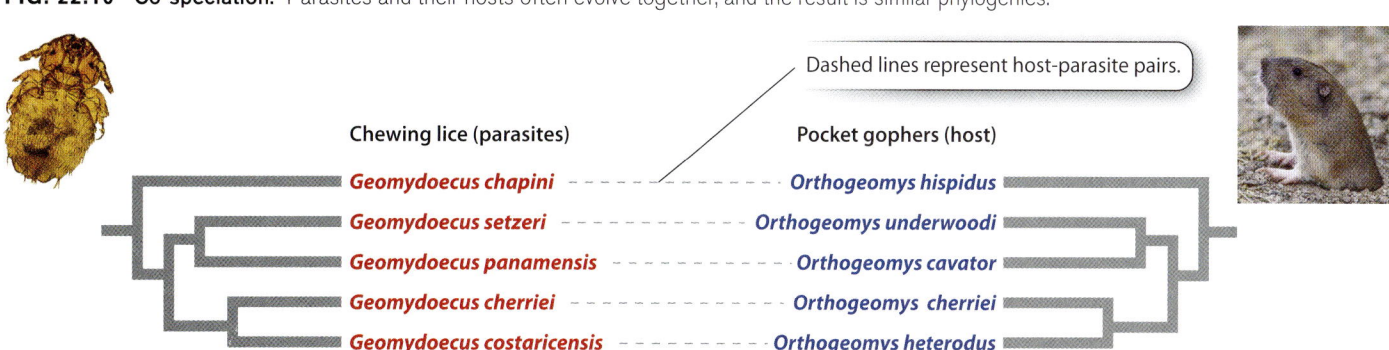

been split, ultimately yielding *P. falciparum* and *P. reichenowi*.

Recent studies, however, have disproved this hypothesis. We now know that *P. falciparum* was introduced to humans relatively recently from gorillas. Why doesn't the evolutionary history of *Plasmodium* follow the classical host–parasite co-speciation pattern? We know that this history is complex and is still being unraveled. However, we also know that the mosquito-borne phase of its life cycle facilitates transfer to new hosts. Malaria parasites are thus not as inextricably tied to their hosts as the pocket gopher lice in Fig 22.10. Presumably, it was just such a mosquito-mediated event that resulted in the introduction of *P. falciparum* into an ancestral human population in Africa from gorillas.

Can sympatric populations—those not geographically separated—undergo speciation?

Can speciation occur *without* complete physical separation of populations? Evolutionary biologists are still exploring this question. Recall how separated populations inevitably diverge genetically over time (see Fig. 22.6). If a mutation arises in population A after it has separated from population B, that mutation is present only in population A and may eventually become fixed (100% frequency) in that population, either through natural selection, if it is advantageous, or through drift, if it is neutral. Once the mutation is fixed in population A, it represents a genetic difference between populations A and B. Repeated independent fixations of different mutations in the two populations result over time in the genetic divergence of separated populations.

Now imagine that populations A and B are not completely separated and there is some gene flow between them. The mutation that arose in population A can, in principle, appear in population B. The new mutation may indeed have become fixed in population A, but a migrant from population A to population B may introduce the mutation to population B as well. Gene flow effectively negates the genetic divergence of populations. If there is gene flow, a pair of populations may change over time, but they do so together. How, then, can speciation occur if gene flow exists? The term we use to describe populations that are in the same geographic location is **sympatric** (literally, "same place"). So we can rephrase the question in technical terms: Can speciation occur sympatrically?

One school of thought insists that sympatric speciation cannot occur and that all speciation is allopatric. If speciation can and does occur sympatrically, one point is clear: Natural selection must act strongly to counteract the homogenizing effect of gene flow. Consider two sympatric populations of finch-like birds, represented in the graph in **Fig. 22.11**. One population begins to specialize on small seeds and the other on large seeds. If the two populations freely interbreed, no genetic differences between the two will occur and speciation will not take place. Now suppose that the offspring produced by the pairing of a big-seed specialist with a small-seed specialist is an individual best adapted to eat medium seeds, and there are no medium-sized seeds available in the environment. Natural selection will act against the hybrids, which will starve to death because there are no medium-sized seeds for them to eat and they are not well adapted to compete with the big- or small-seed specialists. Natural selection would then, in effect, eliminate the products of gene flow. So, although gene flow is occurring, it does not affect the divergence of the two populations because the hybrid individuals do not survive to reproduce. As discussed in Chapter 21, this form of natural selection, which operates against the middle of a spectrum of variation, is called disruptive selection.

Is there evidence for sympatric speciation? It turns out to be difficult to find evidence of sympatric speciation in nature, though it may not be especially rare in plants, as we will see. We might find two very closely related species in the same

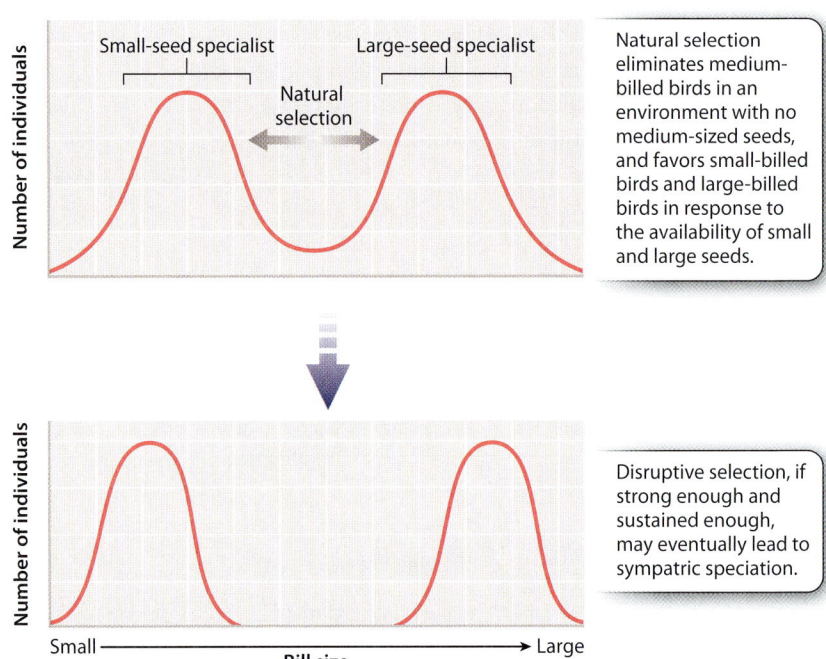

FIG. 22.11 Sympatric speciation through disruptive selection. Natural selection eliminates individuals in the middle of the spectrum.

location and argue that they must have arisen through sympatric speciation. However, there is an alternative explanation. One species could have arisen elsewhere by, for example, peripatric speciation and subsequently moved into the environment of the other one. In other words, the speciation occurred in the past in allopatry, and the two species are only currently sympatric because of migration after speciation occurred.

However, recent studies of fish in isolated lakes have provided strong, if not definite, evidence of sympatric speciation. In the volcanic mountains of Cameroon in West Africa, there are several small crater lakes inhabited by a diverse array of fish. In each case, all the fish species in a single lake can be shown to be one another's closest relatives. Presumably, all the species in a lake are descended from a single common ancestor, the fish species that first arrived in the lake. Each lake is small, so fish populations within them share a single habitat, implying that the speciation events that must have occurred to generate the current species diversity were sympatric.

This demonstration that sympatric speciation has in all probability occurred shows, at the least, that sympatric speciation is possible. We must recognize that we do not yet know just how much of all speciation is sympatric and how much is allopatric. **Fig. 22.12** summarizes modes of speciation based on geography.

→ **Quick Check 3** There are hundreds of species of cichlid fish in Lake Victoria in Africa. Some scientists argue that they evolved sympatrically, but recent studies of the lake suggest that it periodically dried out, leaving a series of small ponds. Why is this observation relevant to evaluating the hypothesis that these species arose by sympatric speciation?

Speciation can occur instantaneously.

Although speciation is typically a lengthy process, it can occasionally occur in a single generation, making it sympatric by definition. Typically cases of such **instantaneous speciation** are caused by hybridization between two species in which the offspring are reproductively isolated from both parents.

For example, hybridization in the past between two sunflower species, *Helianthus annuus* and *H. petiolaris* (the ancestor of the cultivated sunflower) has apparently given rise to three new sunflower species, *H. anomalus*, *H. paradoxus*, and *H. deserticola*

FIG. 22.12 Modes of speciation.

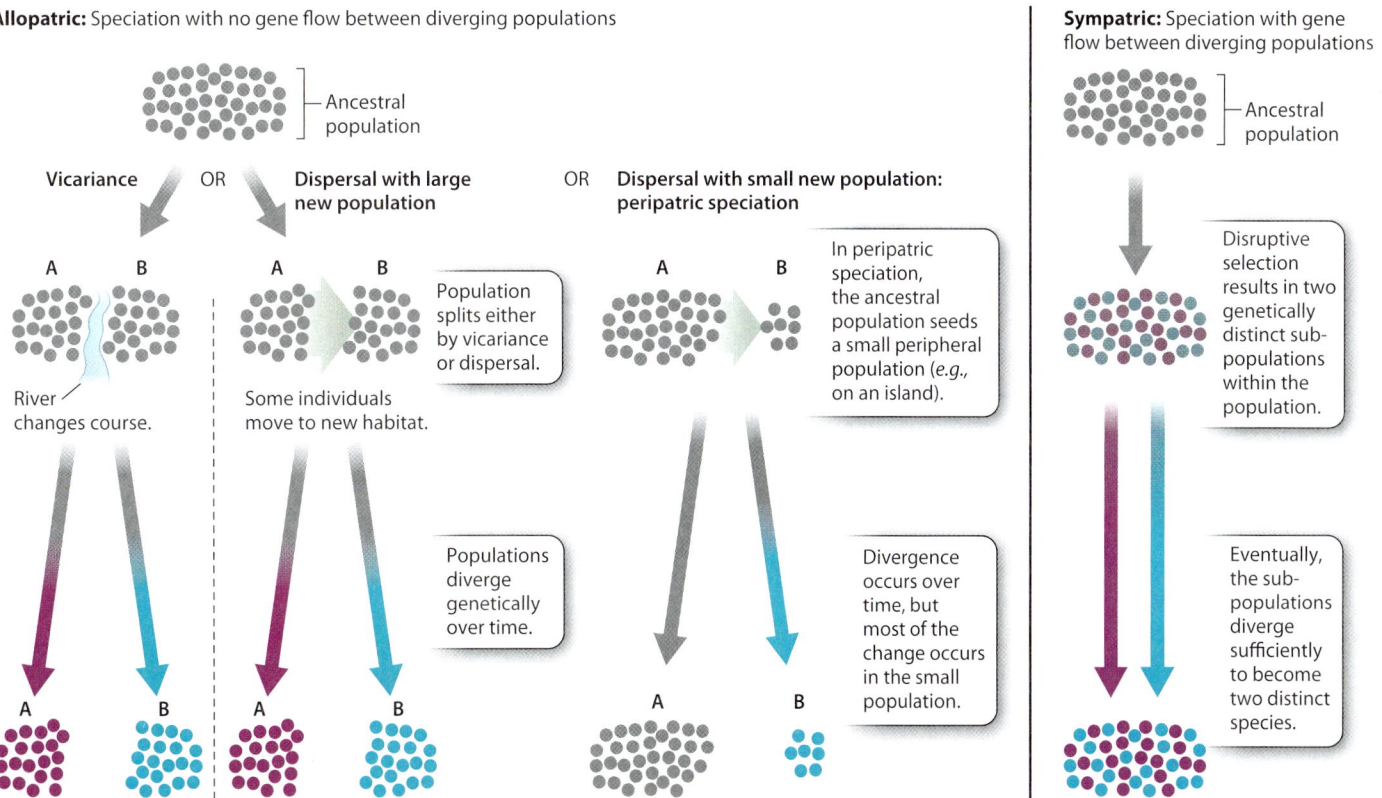

FIG. 22.13 Speciation through hybridization. In sunflowers, *Helianthus anomalus* (bottom panel) is the product of natural hybridization between *H. annuus* (top left) and *H. petiolaris* (top right).

(Fig. 22.13). *H. petiolaris* and *H. annuus* have probably formed innumerable hybrids in nature, virtually all of them inviable. However, a few—the ones with a workable genetic complement—are the ones that survived to yield these three daughter species. In this case, each one of these new species has acquired a different mix of parental chromosomes. It is this species-specific chromosome complement that makes all three distinct and reproductively isolated from the parent species and from one another.

In many cases of hybridization, chromosome numbers may change. Two diploid parent species with 5 pairs of chromosomes, for a total of 10 chromosomes each, may produce a hybrid with double the number of chromosomes (that is, the hybrid inherits a full paired set of chromosomes from each parental species)—a total of 20. In this case, the hybrid has four genomes rather than the diploid number of two. We call such a double diploid a **tetraploid.** In general, animals cannot sustain this kind of expansion in chromosome complement, but plants are more likely to do so. As a result, the formation of new species through **polyploidy**—multiple chromosome sets (Chapter 13)—has been relatively common in plants.

Polyploids may be allopolyploids, meaning that they are produced from hybridization of two different species. For example, related species of *Chrysanthemum* appear, on the basis of their chromosome numbers, to be allopolyploids. Alternatively, polyploids may be autopolyploids, meaning that they are derived from an unusual reproductive event between members of a single species. In this case, through an error of meiosis in one or both parents, the chromosome number is again increased. However, in contrast to the situation with allopolyploids, all the chromosomes derive from the single parental species. For example, *Anemone rivularis*, a plant in the buttercup family, has 16 chromosomes, and its close relative *A. quinquefolia* has 32. *A. quinquefolia* appears to be an autopolyploid derived from the joining of *A. rivularis* gametes with two full sets of chromosomes each.

So rampant is speciation by polyploidy in plants that it affects the pattern of chromosome numbers across all plants. In **Fig. 22.14,** we see the haploid chromosome numbers

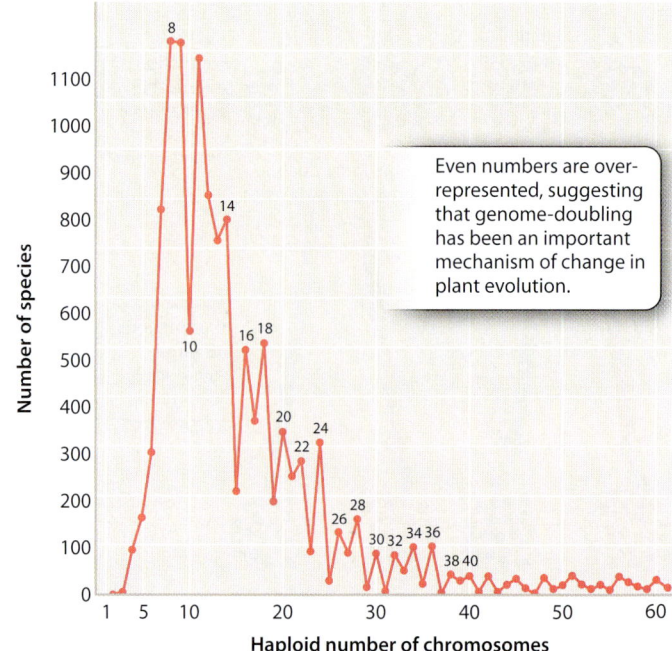

FIG. 22.14 Plant chromosome numbers. Plant chromosome numbers suggest that polyploidy has played an important role in plant evolution.

Even numbers are over-represented, suggesting that genome-doubling has been an important mechanism of change in plant evolution.

of thousands of plant species plotted on a graph. Note that, as the numbers get higher, even numbers tend to predominate, suggesting that the doubling of total chromosome complement (a form of polyploidy that will always result in an even number of chromosomes) is an important factor in plant evolution.

22.4 SPECIATION AND SELECTION

The association of both the origin of species and natural selection with Charles Darwin may make it seem that one cannot occur without the other. However, speciation can occur in the presence or absence of natural selection, and natural selection does not always lead to speciation.

Speciation can occur with or without natural selection.

Natural selection may or may not play a role in speciation. The genetic divergence of two populations can be entirely due to genetic drift, for example, with no role for natural selection.

We have also seen two ways in which natural selection can be involved in speciation. First, sympatric speciation requires some form of disruptive natural selection, as when hybrid offspring are competitively inferior. Second, allopatric speciation (and adaptive radiation) may be facilitated by natural selection. For example, when a peripheral population is in a new environment, natural selection will act to promote its adaptation to the new conditions, accelerating in the process the rate of genetic divergence between it and its parent population.

Natural selection can enhance reproductive isolation.

There is a third way in which natural selection may play an important role in speciation. Recall the two hypothetical bird populations, one with large bills and the other with small bills, whose medium-billed hybrid offspring are at a disadvantage because of a lack of medium-sized seeds in their environment. If a large-billed individual cannot distinguish between large- and small-billed individuals as potential mates (that is, there is a lack of pre-zygotic isolation between them), it will frequently make the "wrong" mate choice, picking a short-billed individual and paying a considerable evolutionary cost of producing poorly adapted hybrid offspring.

Now imagine a new mutation in the large-billed population that permits individuals carrying it to distinguish between the two groups of birds and to mate only with other large-billed individuals. Such a mutation would spread under natural selection because it would prevent the wasted reproductive effort of producing disadvantaged hybrid offspring. This is an example of **reinforcement of reproductive isolation,** or **reinforcement** for short. Reinforcement is the process by which diverging populations undergo natural selection in favor of enhanced pre-zygotic isolation to prevent the production of less fit hybrid offspring. In this case, enhanced pre-zygotic isolation takes the form of mating discrimination, that is, an increased ability to recognize and mate with members of one's own population.

The best evidence in support of reinforcement comes from a study of related fruit-fly species living either in allopatry (geographically separated) or sympatry (in the same place, without geographical barriers). Sympatric species evolve pre-zygotic isolating mechanisms more rapidly than allopatric species. Why would this be? When the two populations are geographically separated, the formation of less fit hybrids is impossible since the two populations do not interbreed, so a mutation that increases mating discrimination between the two groups would not be favored by natural selection. In sympatry, however, where the production of less fit hybrids is a day-to-day problem, such a mutation provides a fitness benefit, so natural selection will favor its spread in the population and reinforce reproductive isolation between the two groups.

Fig. 22.15 summarizes the evolutionary mechanisms that lead to speciation.

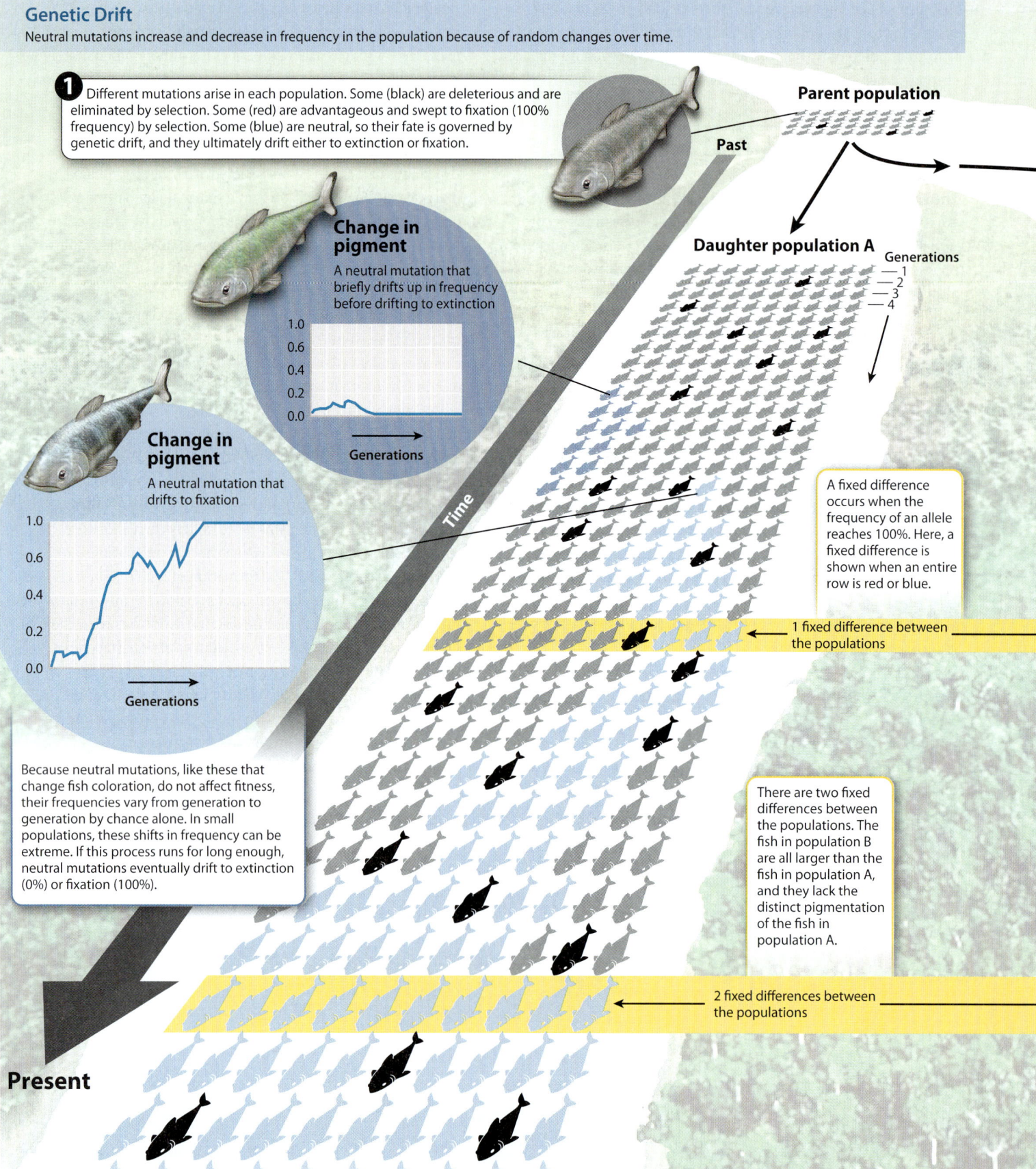

VISUAL SYNTHESIS: Speciation
FIG. 22.15 — Integrating concepts from Chapters 21 and 22

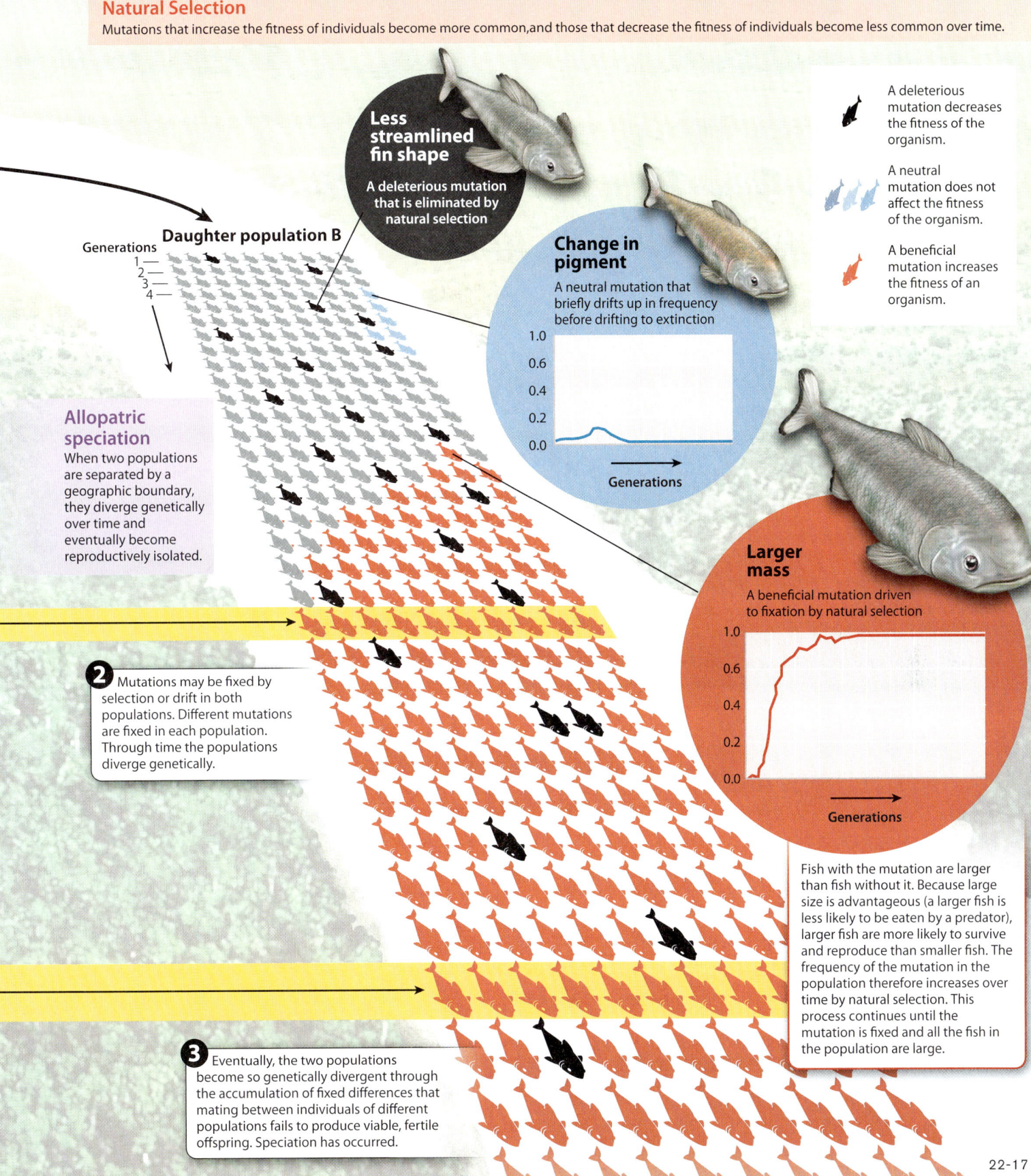

Core Concepts Summary

22.1 SPECIES ARE REPRODUCTIVELY ISOLATED FROM ONE ANOTHER.

The best working definition of a species is the biological species concept (BSC). page 22-2

The BSC states that species are groups of actually or potentially interbreeding populations that are reproductively isolated from other such groups. page 22-2

We cannot apply the reproduction-based BSC to asexual or extinct organisms. page 22-3

Ring species and hybridization further demonstrate that the BSC is not a comprehensive or watertight definition of species. page 22-4

The BSC is nevertheless especially useful because it emphasizes reproductive isolation. page 22-4

22.2 REPRODUCTIVE ISOLATION IS CAUSED BY BARRIERS TO REPRODUCTION BEFORE OR AFTER EGG FERTILIZATION.

Reproductive barriers can be pre-zygotic, occurring before egg fertilization. page 22-5

Pre-zygotic isolation may be behavioral, mediated by differences in the timing of reproduction or by ecological differences, or brought about by mechanical or biochemical reproductive incompatibilities. page 22-5

Reproductive barriers can be post-zygotic, occurring after egg fertilization. page 22-6

In post-zygotic isolation, mating occurs but genetic incompatibilities prevent the development of a viable, fertile offspring. page 22-6

22.3 SPECIATION UNDERLIES THE DIVERSITY OF LIFE ON EARTH.

Speciation is typically a by-product of genetic divergence that occurs as a result of the fixation of different mutations in two populations that are not regularly exchanging genes. page 22-6

If divergence continues long enough, chance differences will arise that result in reproductive barriers between the two populations. page 22-6

Most speciation is thought to be allopatric, involving two geographically separated populations. page 22-6

Geographic separation may be caused by dispersal, resulting in the establishment of a new and distant population, or by vicariance, in which the range of a species is split by a change in the environment. page 22-7

A special case of allopatric speciation by dispersal is peripatric speciation, in which the new population is small and outside the species' original range. page 22-7

Adaptive radiation, in which speciation occurs rapidly to generate a variety of ecologically diverse forms, is best documented on oceanic islands after the arrival of a single ancestral species. Darwin's finches on the Galápagos are a good example of adaptive radiation. page 22-9

Co-speciation occurs when one species undergoes speciation in response to speciation in another. In parasites and their hosts, for example, co-speciation can result in host and parasite phylogenies that have the same branching patterns. page 22-11

Speciation may be sympatric, meaning that there is no geographic separation between the diverging populations. For this type of speciation to occur, natural selection for two or more different types within the population must act so strongly that it overcomes the homogenizing effect of gene flow. page 22-12

22.4 SPECIATION CAN OCCUR WITH OR WITHOUT NATURAL SELECTION.

Separated populations can diverge as a result of genetic drift, with no role for natural selection. page 22-15

Natural selection may act on mutations that allow individuals to identify and mate with individuals that are more like themselves. This process is called reinforcement of reproductive isolation. page 22-15

Self-Assessment

1. Define the term "species."
2. Given a group of organisms, describe how you would test whether they all belong to one species or whether they belong to two separate species.
3. Name two types of organisms that do not fit easily into the biological species concept.
4. Explain how ecological and evolutionary considerations can help inform whether or not a group of organisms represents a single species.
5. Name four reproductive barriers and indicate whether each is pre- or post-zygotic.
6. Describe how genetic divergence and reproductive isolation are related to each other.
7. Differentiate between allopatric and sympatric speciation, and state which is thought to be more common.
8. Differentiate between allopatric speciation by dispersal and by vicariance, and give one example of each.
9. Describe how genetic drift can result in speciation.
10. Describe how natural selection can result in speciation.

Do you understand the chapter's Core Concepts? Log in to check your answers to the Self-Assessment questions, then practice what you've learned and reinforce this chapter's concepts by working through the problems and multimedia tutorials provided there.

www.biologyhowlifeworks.com

CHAPTER 23

EVOLUTIONARY PATTERNS

Phylogeny and Fossils

Core Concepts

23.1 A phylogenetic tree is a reasoned hypothesis of the evolutionary relationships of organisms.

23.2 A phylogenetic tree is built on the basis of shared derived characters.

23.3 The fossil record provides a direct glimpse of evolutionary history.

23.4 Phylogeny and fossils provide independent and corroborating evidence of evolution.

Nature displays, all around us, nested patterns of similarity among species. For example, as noted in Chapter 1, humans are more similar to chimpanzees than either humans or chimpanzees are to monkeys. Humans, chimpanzees, and monkeys, in turn, are more similar to one another than any of them is to a mouse. And humans, chimpanzees, monkeys, and mice are more similar to one another than any of them is to a catfish. This pattern of nested similarity was recognized more than 200 years ago and used by the great Swedish naturalist Carolus Linnaeus to classify biological diversity. A century later, Charles Darwin recognized this pattern as the expected outcome of a process of "descent with modification," or evolution.

Evolution produces two distinct but related patterns, both evident in nature. First is the nested pattern of similarities found among species on present-day Earth. The second is the historical pattern of evolution recorded by fossils. Life, in its simplest form, originated more than 3.5 billion years ago. Today, an estimated 10 million species inhabit the planet. Short of inventing a time machine, how can we reconstruct those 3.5 billion years of evolutionary history in order to understand the extraordinary events that have ultimately resulted in the biological diversity we see around us today? These two great patterns provide the answer.

Darwin recognized that the species he observed were the modified descendants of earlier ones. Distinct populations of an ancestral species separate and diverge through time, again and again, giving rise to multiple descendant species. The result is the pattern of nested similarities observed in nature (see Fig. 1.17). Just as a tree sprouts new branches on old ones and adds rings to a thickening trunk, the pattern of nested similarities among species strongly indicates a process of descent with modification and the accumulation of change. This history of descent with branching is called **phylogeny,** and is much like the genealogy that records our own family histories.

The evolutionary changes inferred from the patterns of relatedness among present-day species make predictions about the historical pattern of evolution we should see in the fossil record. For example, groups with features that we infer to have evolved earlier than others should appear earlier in time as fossils. Paleontological research reveals that the history of life is indeed laid out in the chronological order predicted on the basis of comparative biology. How do we reconstruct life's two great evolutionary patterns, and how do they compare?

23.1 READING A PHYLOGENETIC TREE

Chapter 22 introduced the concept of speciation, the set of processes by which physically, physiologically, or ecologically isolated subsets of a population diverge from one another to the point where they can no longer produce fertile offspring. As

FIG. 23.1 The relationship between speciation and a phylogenetic tree. The phylogenetic tree on the right depicts the evolutionary relationships that result from the two successive speciation events diagrammed on the left.

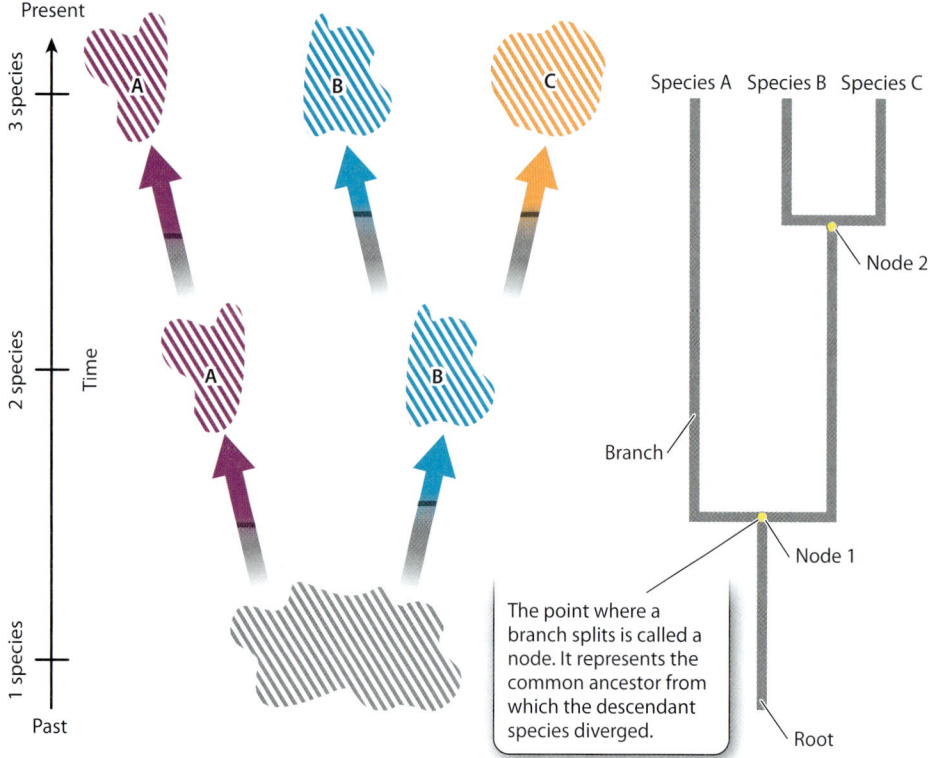

The point where a branch splits is called a node. It represents the common ancestor from which the descendant species diverged.

illustrated in **Fig. 23.1,** speciation can be thought of as a process of branching. Now consider what happens as this process occurs over and over in a lineage through time. As species proliferate, their evolutionary relationships to one another unfold in a treelike pattern, with present-day species as the tips of branches and their last common ancestors indicated by the point (called a **node**) from which they branched off.

Phylogenetic trees provide hypotheses of evolutionary relationships.

Phylogenetics is one of two related disciplines in systematics, the study of evolutionary and genetic relationships among organisms. The other is taxonomy, the classification of organisms.

The aim of taxonomy is to recognize and name groups of individuals as species, and, subsequently, to group closely related species into the more inclusive taxonomic group of the genus, and so on up through the taxonomic ranks—species, genus, order, class, phylum, kingdom, domain. Taxonomy, then, provides us with a hierarchical classification of species in more and more inclusive groups, giving us a convenient way to communicate information about the features each group possesses. So, if we want to tell someone about a small animal we have seen with fur, mammary glands, and extended finger bones that permit it to fly, we can give them this long description, or we can just say we saw a bat (or, a member of the Order Chiroptera. All the rest is understood (or can be looked up in a reference).

Phylogenetics, on the other hand, aims to discover the pattern of evolutionary relatedness among groups of species or other groups by comparing their anatomical or molecular features, and to depict these relationships as a **phylogenetic tree.** A phylogenetic tree is a hypothesis about the evolutionary history, or phylogeny, of the species. Phylogenetic trees are hypotheses because they represent the best model, or explanation, of the relatedness of organisms on the basis of all the existing data. However, as with any model or hypothesis, new data may require adjustments to the pattern of the tree. Many phylogenetic trees explore the relatedness of particular groups of individuals, populations, or species. We may, for example, want to understand how wheat is related to other, non-commercial grasses, or how pathogenic populations of *Escherichia coli* relate to more benign strains of the bacterium. At a much larger scale, universal similarities of molecular biology indicate that all living organisms are descended from a single common ancestor. This means that we can attempt to build a phylogenetic tree for all species, commonly referred to as the tree of life. In Part 2 of this book, we will make use of the tree of life and many smaller-scale phylogenetic trees to understand our planet's biological diversity.

Fig. 23.2 shows a phylogenetic tree for vertebrate animals. The informal name at the end of each branch represents a group of organisms, many of them familiar. We will sometimes find it useful to refer to groups of species this way (for example, "frogs," or the "Class Anura") rather than name all the individual species or list the characteristics they have in common. It is important, however, to remember that such named groups represent a number of member species. If, for example, we were able to zoom in on the branch labeled "Frogs," we would see that it consists of many smaller branches, each representing a distinct species of frog.

This tree provides information about evolutionary relationships among vertebrates. For example, it proposes that the closest living relatives of birds are crocodiles and alligators. The tree also proposes that the closest relatives of all tetrapod (four-legged) vertebrates are lungfish. Phylogenetic trees are built from careful analyses of the morphological and molecular attributes of the species or other groups under study.

FIG. 23.2 **A phylogeny of vertebrate animals.** The branching order constitutes a hypothesis of evolutionary relationships within the group.

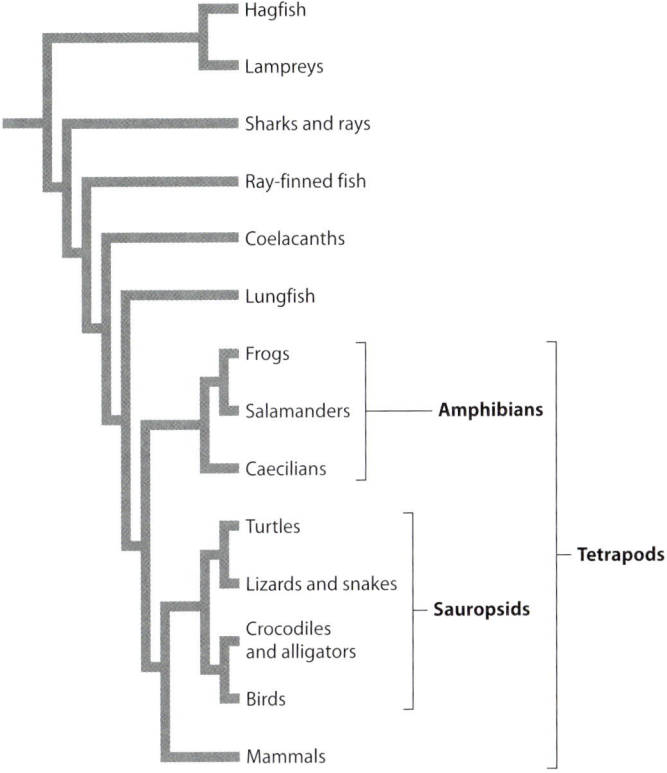

The search for sister groups lies at the heart of phylogenetics.

Two species, or groups of species, are considered to be closest relatives if they share a common ancestor not shared by any other species or group. In Fig. 23.2, for example, we can see that frogs are more closely related to salamanders than to any other group of organisms because frogs and salamanders share a common ancestor not shared by any other group. Similarly, lungfish are more closely related to tetrapods than to any other group. A lungfish may look more like a fish than it does a frog or salamander, but lungfish are more closely related to frogs and salamanders than they are to other fish because lungfish share a common ancestor with frogs and salamanders that was more recent than their common ancestor with other fish (Fig. 23.2).

Groups that are more closely related to each other than either of them is to any other group, like lungfish and tetrapods, are called **sister groups**. Simply put, phylogenetic hypotheses amount to determining sister-group relationships because the simplest phylogenetic question we can ask is which two of any three species (or other groups) are more closely related to each other than either is to the third. In this light, we can see that a phylogenetic tree is simply a set of sister-group relationships in which each addition of a species beyond the original three entails finding its sister group in the tree.

Closeness of relationship is then determined by looking to see how recently two groups share a common ancestor. Shared ancestry is indicated by a node, or branch point, on a phylogenetic tree. An important aspect of a node is that it can be rotated without changing the evolutionary relationships of the groups. **Fig. 23.3,** for example, shows four phylogenetic trees depicting evolutionary relationships among birds, crocodiles and alligators, and lizards and snakes. In all four trees, birds are a sister group to crocodiles and alligators because birds, crocodiles, and alligators share a common ancestor not shared by lizards and snakes. Information about evolutionary relationships therefore lies in the order of nodes over time, not the order of groups along the tips.

→ **Quick Check 1** Does either of these two phylogenetic trees indicate that humans are more closely related to lizards than to mice?

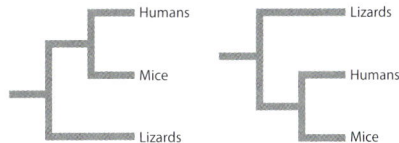

FIG. 23.3 **Depicting sister groups.** The four trees illustrate the same set of sister-group relationships.

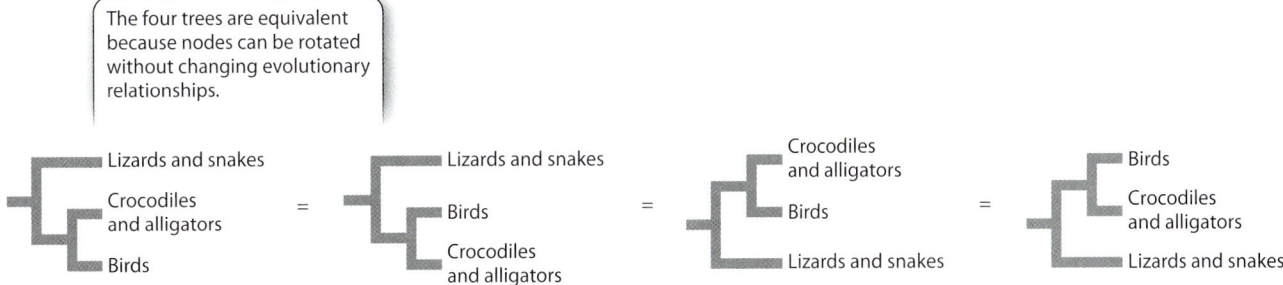

A monophyletic group consists of a common ancestor and all its descendants.

So far in our discussion, we have used the word "group" as a kind of summary word to mean all the species in some taxonomic entity under discussion. A more technical word is **taxon** (plural, taxa), with taxonomy providing a formal means of naming groups. In recent decades, biologists have worked to integrate evolutionary history with taxonomic classification. The resulting classification emphasizes groups that are **monophyletic,** meaning that all members share a single common ancestor not shared with any other species or group of species. In Fig. 23.2, the tetrapods are monophyletic because they all share a common ancestor not shared by any other taxa. Similarly, amphibians are monophyletic.

In contrast, consider the group of animals traditionally recognized as reptiles, which includes turtles, snakes, lizards, crocodiles, and alligators (see Fig. 23.2). The group "reptiles" excludes birds, although they share a common ancestor with the included animals. Such a group is considered **paraphyletic.** A paraphyletic group includes some, but not all, of the descendants of a common ancestor. Early zoologists separated birds from reptiles because they are so distinctive, but feathers and other unique features of birds do not tell us about their relationship to other vertebrates. In fact, many features of skeletal anatomy and DNA sequence strongly support the placement of birds as a sister group to the crocodiles and alligators.

There is a simple way to distinguish between monophyletic and paraphyletic groups, illustrated in **Fig. 23.4**. If in order to separate a group from the rest of the phylogenetic tree you need only to make one cut, the group is monophyletic. If you need at least two cuts to separate the group from the tree, it is paraphyletic.

Groupings that do not include the last common ancestor of all members are called **polyphyletic.** For example, clustering bats and birds together as flying tetrapods results in a polyphyletic group (Fig. 23.4).

Determining what sets of organisms are considered monophyletic groups is a main goal of phylogenetics because monophyletic groups include *all* descendants of a common ancestor and *only* the descendants of that common ancestor. This means that monophyletic groups alone show the evolutionary path a given group has taken since its origin. Omitting some members of a group, as in the case of reptiles and other paraphyletic groups, can provide a misleading sense of evolutionary history. By using monophyletic groups in taxonomic classification, we effectively convey our knowledge of their evolutionary history.

→ **Quick Check 2** Look at Fig. 23.4. Are fish a monophyletic group?

Taxonomic classifications are information storage and retrieval systems.

The nested pattern of similarities among species has been recognized by naturalists for centuries. As we saw in Chapter 22, closely related species group together on a single branch of the tree of life. In the vocabulary of formal classification, closely related species are grouped into a **genus** (plural, genera). Closely related genera, in turn, belong to a larger, more inclusive branch of the tree—they are classified as a **family.** Closely related families, in turn, form an **order,** orders form a **class,** classes form a **phylum** (plural, phyla), and phyla form a **kingdom,** each more inclusive taxonomic level occupying a successively larger limb on the tree (**Fig. 23.5**). Biologists today commonly refer to the three largest limbs of the entire tree of life as **domains** (the Eukarya, or eukaryotes; Bacteria; and Archaea, or archaeons, introduced in Chapter 1 and discussed more fully in Chapters 26 and 27).

The ranks of classification form a nested hierarchy, but the boundaries of ranks above the species level are arbitrary in that there

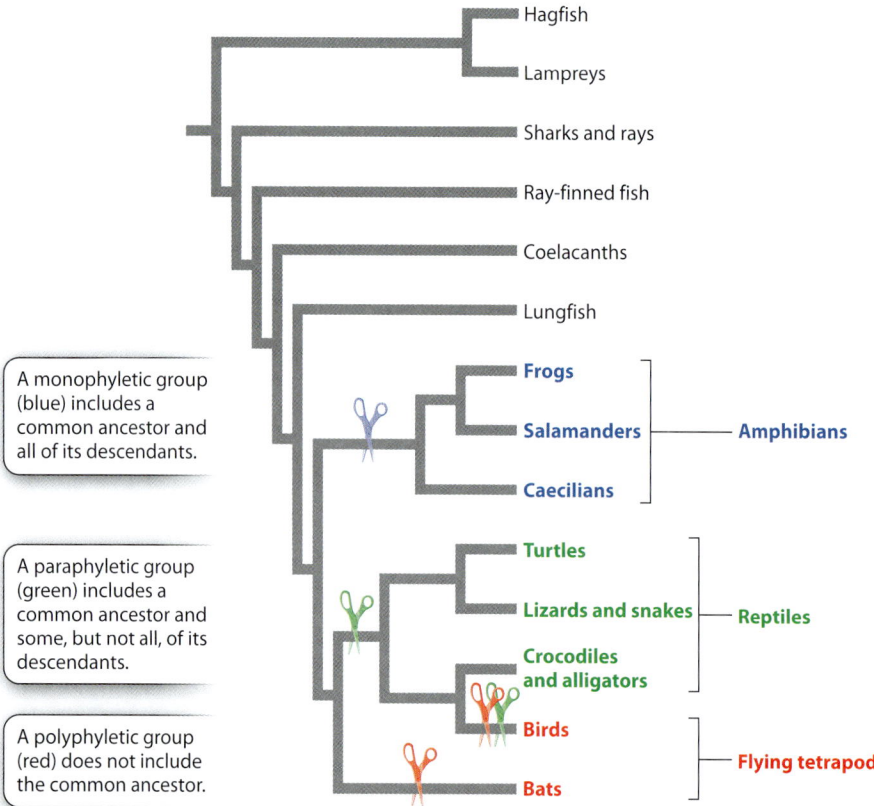

FIG. 23.4 **Monophyletic, paraphyletic, and polyphyletic groups.** Only monophyletic groups reflect evolutionary relationships because only they include all the descendants of a common ancestor.

CHAPTER 23 EVOLUTIONARY PATTERNS: PHYLOGENY AND FOSSILS

FIG. 23.5 Classification. Classification reflects our understanding of phylogenetic relationships, and the taxonomic hierarchy reflects the order of branching.

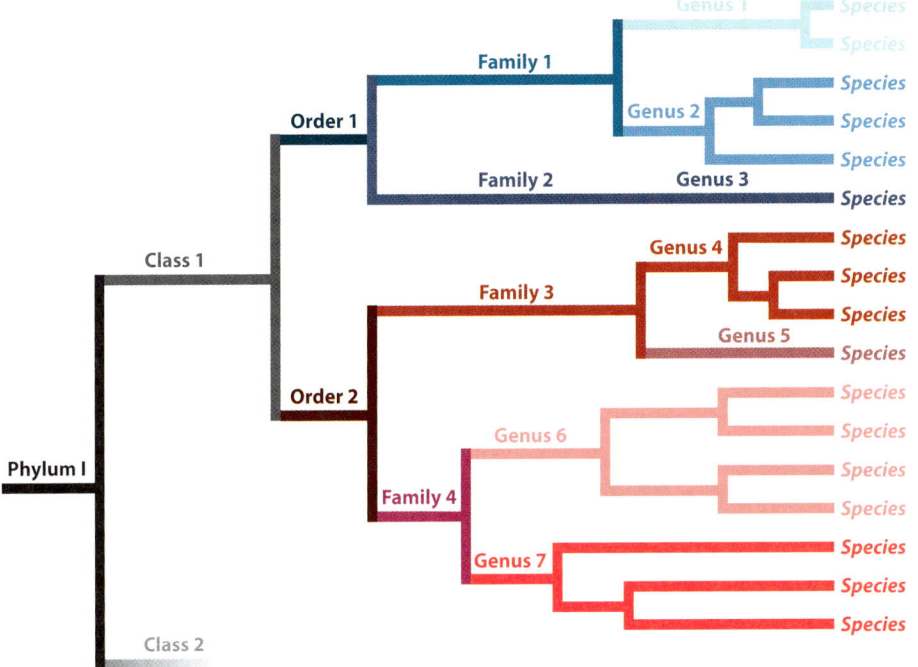

is nothing particular about a group that makes it, for example, a class rather than an order. For this reason, it is not necessarily true that orders or classes of, for example, birds and ferns are equivalent in any meaningful way. In contrast, sister groups are equivalent to one another in several ways—notably, in that they diverged from a single ancestor at a single point in time. Therefore, if one branch is a sister group that contains 500 species and the other has 6, the branches have experienced different rates of speciation, extinction, or both since they diverged. In Fig 23.5, Families 1 and 2 are sister groups but Family 1 has five species while Family 2 has just one.

23.2 BUILDING A PHYLOGENETIC TREE

Up to this point, we have focused on how to interpret a phylogenetic tree, a diagram that depicts the evolutionary history of organisms. But how do we infer their evolutionary history from a group of organisms? That is, how do we actually construct a phylogenetic tree? Biologists use characteristics of organisms to figure out their relationships. Similarities among organisms are particularly important in that similarities sometimes suggest shared ancestry. However, a key principle of constructing trees is that only *some* similarities are actually useful. Others can in fact be misleading.

Homology is similarity by common descent.

Phylogenetic trees are inferred by comparison of character states shared among different groups of organisms. **Characters** are the anatomical, physiological, or molecular features that make up organisms. To be useful for phylogenetic reconstruction, they must vary among but not within species and have a genetic basis. In general, characters have several observed conditions, called **character states.** In the simplest case, a character can be present or absent—lungs are present in tetrapods and lungfish, but absent in other vertebrate animals. Commonly, however, multiple character states are apparent—petals are a character of flowers, for example, and their arrangement can be considered a character state. Flowers can have many petals arranged in a helical pattern, many petals arranged in a whorl, few petals arranged in a whorl, or few petals fused into a tube. All species contain some character states that are shared with other members of their group, some that are shared with members of other groups, and some that are unique.

Character states can be similar for one of two reasons: The character state was present in the common ancestor of the two groups and retained over time (common ancestry), or the character state independently evolved in the two groups as an adaptation to similar environments (convergent evolution). Consider two examples. In the vertebrate phylogeny shown in **Fig. 23.6**, mammals and birds both

FIG. 23.6 Homology and analogy. A homology is a similarity that results from shared ancestry, while an analogy is a similarity that results from convergent evolution.

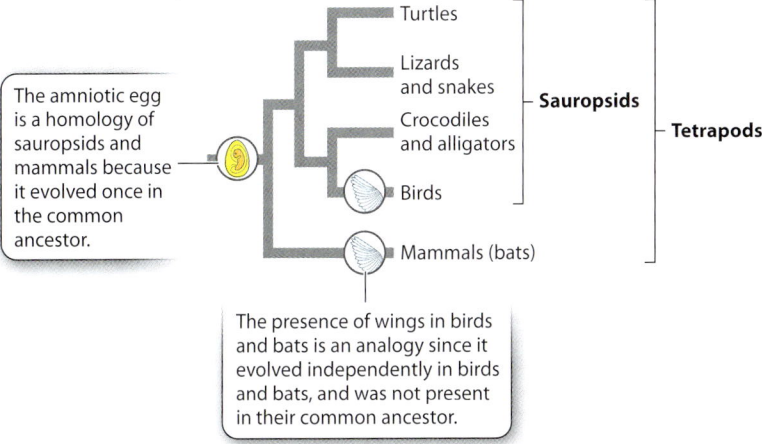

produce amniotic eggs. Because amniotic eggs occur only in groups descended from the common ancestor at the node connecting the mammal and sauropsid branches of the tree, we reason that birds and mammals each inherited this character from a common ancestor in which the trait first evolved. Characters that are similar because of descent from a common ancestor are said to be **homologous.**

Not all similarities arise in this way, however. Think of wings, a character exhibited by both birds and bats (which are mammals). Much evidence supports the view that wings in these two groups do not reflect descent from a common, winged ancestor but rather evolved independently in the two groups. Similarities due to independent adaptation by different species are said to be **analogous.** They are the result of convergent evolution.

Innumerable convergences less dramatic than wings are known. In some, we even understand the genetic basis of the convergence. For example, echolocation has evolved in bats and in dolphins, which are also mammals. Prestin is a protein in the hair cells of ears that is involved in hearing ultrasonic frequencies. Both bats and dolphins have evolved parallel changes in the *Prestin* genes, apparently convergent adaptations for echolocation. Similarly, unrelated fish that live in freezing water at the poles, Arctic and Antarctic, have evolved similar glycoproteins that act as molecular "antifreeze," preventing the formation of ice in their tissues.

→ **Quick Check 3** Fish and dolphins have many traits in common, including a streamlined body and fins. Are these traits homologous or analogous?

Shared derived characters enable biologists to reconstruct evolutionary history.

Because homologies result from shared ancestry, only they, and not analogies, are useful in constructing phylogenetic trees. However, it turns out that only some homologies are useful. For example, character states that are unique to a given species or other monophyletic group can't tell us anything about its sister group. They evolved after the divergence of the group from its sister group and so can be used to characterize a group but not to relate it to other groups. Similarly, homologies formed in the common ancestor of the entire group and therefore present in all its descendants do not help to identify sister-group relationships among the groups under consideration.

What we need to develop hypotheses of evolutionary relationship are homologies shared by some, but not all, of the members of the group under consideration. These are shared derived characters and are called **synapomorphies.** A derived character state is an evolutionary innovation (for example, the change from five toes to a single toe - the hoof - in the ancestor of horses and donkeys). When such a novelty arises in the common ancestor of two taxa, it is shared by both (thus the hoof is a synapomorphy defining horses and donkeys as sister groups).

In **Fig. 23.7**, we indicate the major synapomorphies that have helped us construct the phylogeny of vertebrates. For example, the lung is a character present in lungfish and tetrapods, but absent in other vertebrates. Thus, the presence of lungs provides one piece of evidence that lungfish are the sister group of tetrapods. Phylogenetic reconstruction on the basis of synapomorphies is called **cladistics.**

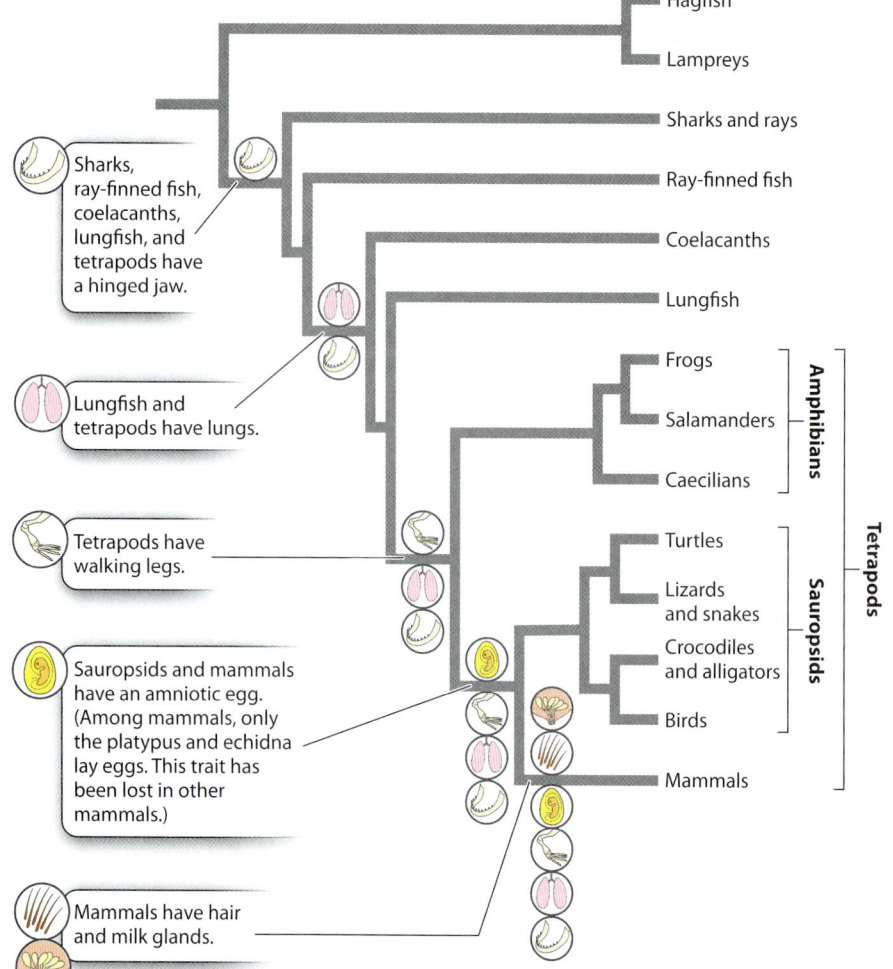

FIG. 23.7 Synapomorphies, or shared derived characters. Homologies that are present in some, but not all, members of a group help us to construct phylogenetic trees.

The simplest tree is often favored among multiple possible trees.

To show how synapomorphies help us chart out evolutionary relationships, let's consider

FIG. 23.8 Constructing a phylogenetic tree from shared derived traits. The strongest hypothesis of evolutionary relationships overall is the tree with the fewest number of changes because it minimizes the total number of independent origins of character states.

a. Character states

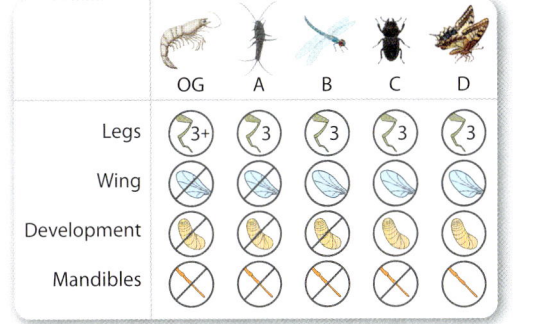

b. Possible phylogenetic trees

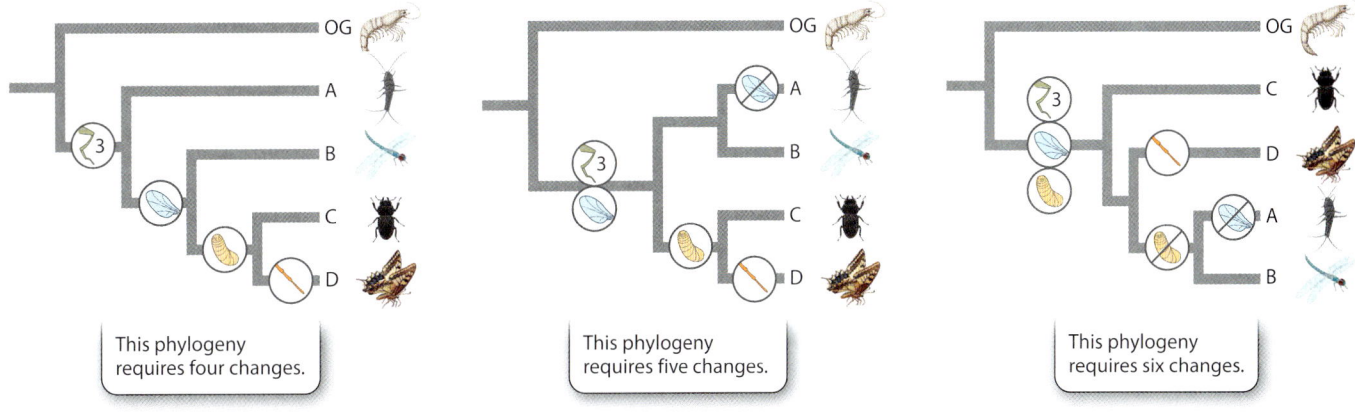

c. Phylogeny of Hexapoda (insects)

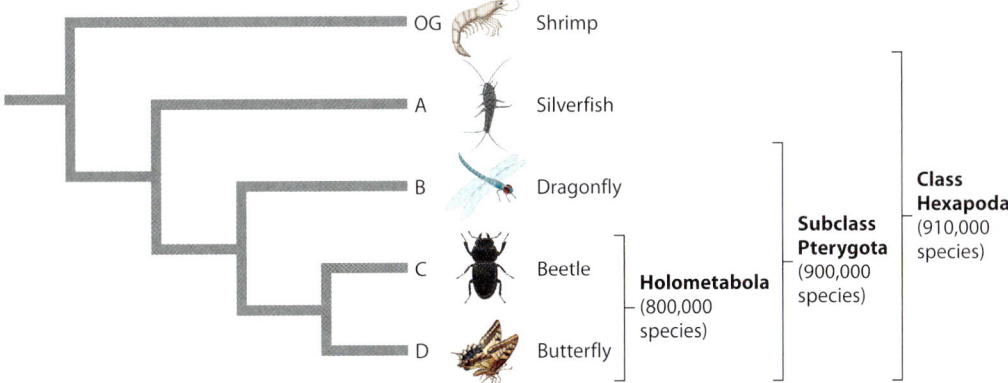

the simple example in **Fig. 23.8**. We begin with four species of animals in a group we wish to study, labeled "A" through "D," that we will call our ingroup; we believe them to be closely related to each other. For comparison, we have a species that we believe is outside this ingroup—that is, it falls on an earlier branch of the tree—and so is called an outgroup (labeled "OG" in Fig. 23.8). Each species in the ingroup and outgroup has a different combination of characters, such as leg number, presence of wings, and whether development of young to adult is direct or goes through a pupal stage (Fig. 23.8a).

We are interested in the relationships among species A–D and so focus on potential synapomorphies—character states shared by some but not all species within the group. For example, only C and D have pupae. This character suggests that C and D are more closely related to each other than either is to the other species. The alternatives are that C and D each evolved pupae independently or that pupae were present in the common ancestor of A–D but were lost in A and B.

How do we choose among the alternatives? Comparison with the outgroup shows that it does not form pupae, supporting the hypothesis that pupal development evolved within the ingroup. In practice, biologists examine multiple characters and choose the phylogenetic hypothesis that best fits all of the data.

How do we determine "best fit"? The phylogenetic tree shown on the left in Fig. 23.8b considers four characters and their various character states and reflects the sister-group relationship between C and D proposed earlier. This tree requires exactly four character-state changes during the evolution of these species: reduction of leg pairs to three in the group ABCD, wings in the group BCD, and pupae in group CD, plus a change of form of the mandible, or jaw, in species D.

Now consider the middle tree in Fig. 23.8b. It groups A and B together, and so differs from the tree on the left in requiring either a loss of wings in A, or an additional origin of wings in B, independent of that in the ancestor of CD—five changes in all. The tree on the right groups A and B together, and requires two extra steps for a total of six steps. No tree that we can construct from species A–D requires fewer than four evolutionary changes, so the left-hand version in Fig. 23.8b is the best available hypothesis of evolutionary relatedness. In fact, this is the phylogeny for a sample of species from the largest group of animals on Earth, the Hexapoda—or insects (Fig. 23.8c).

In general, trees with fewer character changes are preferred to ones that require more because they provide the simplest explanation of the data. This approach is an example of **parsimony,** that is, choosing the simpler of two or more hypotheses to account for a given set of observations. In systematics, parsimony suggests counting character changes on a phylogenetic tree to find the simplest tree for the data (the one with the fewest number of changes). Each change corresponds to a mutation (or mutations) in an ancestral species, and the more changes or steps we propose, the more independent mutations we must also hypothesize.

Note also that it isn't necessary to make decisions in advance about which characters are homologies and which are analogies. We can construct all possible trees and then choose the one requiring the fewest evolutionary changes. This is a simple matter for the example in Fig. 23.8, because four species can be arranged into only 15 possible different trees. As the number of groups increases, however, the number of possible trees connecting them increases as well, and dramatically so. There are 105 trees for 5 groups and 945 trees for 6 groups, and there are nearly 2 million possible trees for 10. For 50 groups the possibilities balloon to 3×10^{76}! Clearly, computers are required to sort through all the possibilities.

As is true for all hypotheses, phylogenetic hypotheses can be strongly supported or weakly supported. Biologists use statistical methods to evaluate a given phylogenetic hypothesis. Available character data may not strongly favor any hypothesis. When support for a specific branching pattern is weak, biologists commonly depict the relationships as unresolved and show multiple groups diverging from one node, rather than just two, as we have seen. Such branching patterns are not meant to suggest that multiple species diverged simultaneously, but instead to indicate that we lack the data to choose unequivocally among several different hypotheses of relationships. In Part 2 of this book, we will encounter unresolved branches in a number of groups. These shouldn't be read as admissions of defeat, but instead as problems awaiting resolution and opportunities for future research.

Molecular data complement comparative morphology in reconstructing phylogenetic history.

Trees can be built using anatomical features as characters, but increasingly, tree construction relies on molecular data. The amino acids at particular positions in the primary structure of a protein can be used as characters, as can the nucleotides at specific positions along a strand of DNA.

From genealogy to phylogeny, tracing mutations in DNA or RNA sequences has revolutionized the reconstruction of historical genetic connections. Whether we are tracing the paternity of the children of Sally Hemings, mistress of Thomas Jefferson, identifying the origin of a recent cholera epidemic in Haiti, or placing baleen whales near the Hippopotamus family in a phylogenetic tree of mammals, molecular data give insights through their highly detailed information.

Much of what we know already about molecular evolution comes from phylogenetic comparisons previously established by comparative morphology, the study of structures and forms in different taxa. It is important to recognize that there is nothing about molecular data that provides a better record of history than does comparative morphology; molecular data simply provide more details. Indeed, for microbes and viruses, there is very little morphology available, so molecular information is critical for phylogenetic reconstruction in these microscopic taxa. Once a gene or other stretch of DNA or RNA with suitable levels of variation is identified in two or more species, sequences are obtained and aligned to identify homologous nucleotide sites. Analyses of this kind have typically involved comparisons of sequences of about 1000 nucleotides from one or more genes. Increasingly, though, the availability of whole-genome sequences is changing the way we do molecular phylogenetics: Rather than comparing the sequences of a few genes, we compare the sequences of entire genomes.

The process of using molecular data is conceptually similar to the process described earlier for morphological data. Through comparison to an outgroup, we can identify derived and ancestral molecular characters (whether DNA nucleotides or amino acids) and generate the phylogeny on the basis of synapomorphies as before.

FIG. 23.9 Phylogenetic trees of DNA sequences based on (b) synapomorphies and (c) distance.

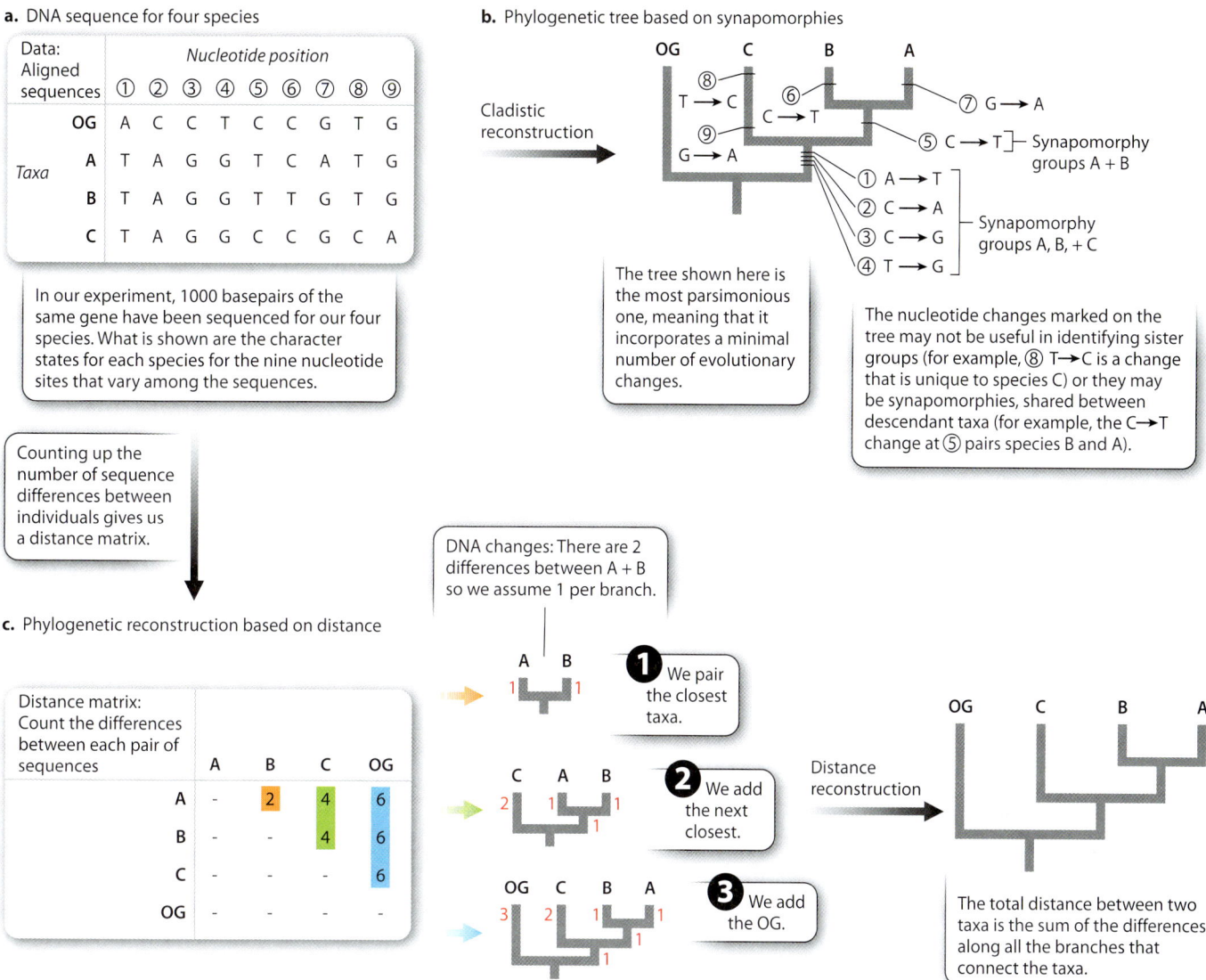

An alternative method of reconstruction is based on distance rather than synapomorphies. Here the premise is simple: The descendants of a recent common ancestor will have had relatively little time to evolve differences, whereas the descendants of an ancient common ancestor have had a lot of time to evolve differences. Thus similarity (or low distance) indicates the recency of common ancestry.

Underpinning this approach is the assumption that the rate of evolution is constant (otherwise, a pair of taxa with a recent common ancestor could be more different than expected because of an unusually fast rate of evolution). This rate-constancy assumption is less likely to be violated when we are using molecular data than when we are using morphological data. Recall from Chapter 21 that the molecular clock is based on the observation of constant accumulation of genetic divergence through time. **Fig. 23.9** shows a simple DNA sequence dataset which we can analyze either on the basis of synapomorphies (Fig. 23.9b) or on the basis of distance (Fig. 23.9c). Note that both give the same result in this case.

Molecular data are often combined with morphological data in analyses, and each can also serve as an independent assessment of the other. Not surprisingly, results from analyses of each kind of data are commonly compatible, at least for plants and animals rich in morphological characters.

Today, the single largest library of taxonomic information is the National Institutes of Health's genetic data storage facility, called GenBank. As of this writing, GenBank gives users access to more than 100 billion observations (mostly nucleotides) collected under more than 400,000 taxonomic names. A growing web resource is Encyclopedia of Life, which is gathering additional

HOW DO WE KNOW?

FIG. 23.10

Did an HIV-positive dentist spread the AIDS virus to his patients?

BACKGROUND In the late 1980s, several patients of a Florida dentist contracted AIDS. Molecular analysis showed that the doctor was HIV-positive.

HYPOTHESIS It was hypothesized that the patients acquired HIV during dental procedures carried out by the infected dentist.

METHOD Two HIV samples (denoted 1 and 2 in the figure) were obtained from several people, including the dentist (Dentist 1 and Dentist 2), several of his patients (Patients A through G), and other HIV-positive individuals chosen at random from the local population (LP). In addition, a strain of HIV from Africa (HIVELI) was included in the analysis.

RESULTS Biologists constructed a phylogeny based on the nucleotide sequence of a rapidly evolving gene in the genome of HIV. Because the gene evolves so quickly, its mutations preserve a record of evolutionary relatedness on a very fine scale. HIV in some of the infected patients — patients A, B, C, E, and G — were more similar to the dentist's HIV than they were to samples from other infected individuals. Some patients' sequences, however, did not cluster with the dentist's, suggesting that these patients, D and F, had acquired their HIV infections from other sources.

CONCLUSION HIV phylogeny makes it highly likely that the dentist infected several of his patients. The details of how the patients were infected remain unknown, but rigidly observed safety practices make it unlikely that such a tragedy could occur again.

FOLLOW-UP WORK Phylogenies based on molecular sequence characters are now routinely used to study the origin and spread of infectious diseases, such as swine flu.

SOURCE Hillis, D. M., J. P. Huelsenbeck, and C. W. Cunningham. 1994. "Application and Accuracy of Molecular Phylogenies." *Science* 264:671–677.

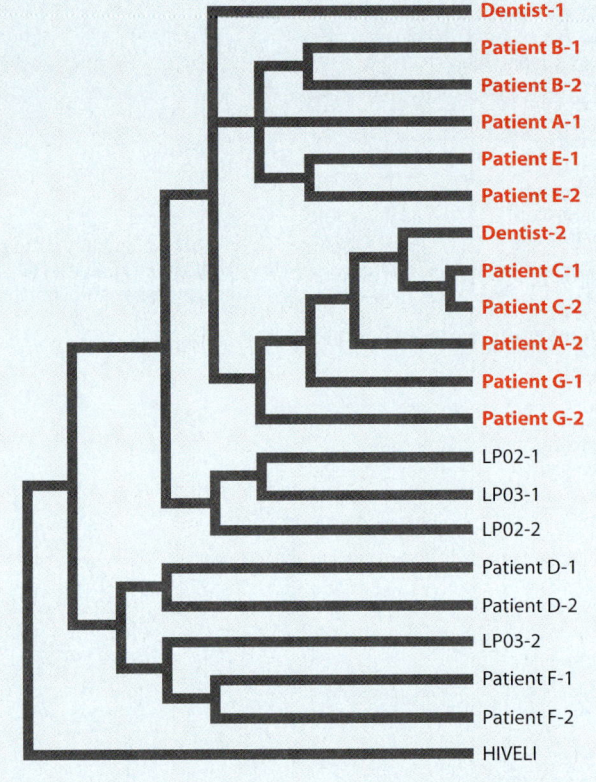

biological information about species, including ecology, geographic distributions, photographs, and sounds in pages for individual species that are easy to navigate. Another web-based resource, the Tree of Life, provides information on phylogenetic trees for many groups of organisms.

Phylogenetic trees can help solve practical problems.

The sequence of changes on a tree from its base to its tips documents evolutionary changes that have accumulated through time. Trees suggest which lineages are older than others, and which traits came first and which followed later. Proper phylogenetic placement thus reveals a great deal about evolutionary history, and it can have practical consequences as well. For example, oomycetes, microorganisms responsible for potato blight and other important diseases of food crops, were long thought to be fungi because they look a lot like some fungal species. The discovery, using molecular characters, that oomycetes belong to a very different group of eukaryotic organisms, has opened up new possibilities for understanding and controlling these plant pathogens. Similarly, in 2006, researchers used DNA sequences to identify the Malaysian parent population of the lime swallowtails that had become an invasive species in the Dominican Republic.

Phylogenetics solved a famous case in which an HIV-positive dentist in Florida was accused of infecting his patients (**Fig. 23.10**). Nucleotide sequence in the genome of HIV evolves so rapidly that biologists can build phylogenetic trees that trace the spread of specific strains from one individual to the next. Phylogenetic study of HIV present in samples from several infected patients, the dentist, and other individuals provided evidence that the dentist had, indeed, infected his patients.

Similarly, phylogenetic studies of influenza virus strains show their origins and subsequent movements among geographic regions and individual patients. Today, there is a growing effort to use particular DNA sequences as a kind of fingerprint or barcode for tracking biological material. Such information could quickly identify samples of shipments as meat from endangered species, or track newly emerging pests. The Consortium for the Barcode of Life has already accumulated on its website species-specific DNA barcodes for more than 100,000 species. Phylogenetic evidence provides a powerful tool for evolutionary analysis and is useful across timescales ranging from months to the entire history of life, from the rise of epidemics to the origins of metabolic diversity. Few other breakthroughs in science have had as broad an impact as that of phylogenetic systematics and the new sources of molecular data that have carried it so far.

23.3 THE FOSSIL RECORD

Phylogenies based on living organisms provide hypotheses about evolutionary pattern. Branches toward the base of the tree occurred earlier than those near the top, and characters change and accumulate along the path from the root to the tips. Fossils provide direct documentation of ancient life, and so, in combination, fossils and phylogenies provide strong complementary insights into evolutionary history.

Fossils provide unique information.

Fossils can and do provide evidence for phylogenetic hypotheses, showing, for example, that organisms that branch early in phylogenies appear early in the geologic record. But the fossil record does more than this. First, fossils enable us to calibrate phylogenies in terms of time. It is one thing to infer that mammals diverged from the common ancestor of birds, crocodiles, lizards, and turtles before crocodiles and birds diverged from a common ancestor (see Fig. 23.7), but another matter to state that birds and crocodiles diverged about 220 million years ago, whereas the lineage represented today by mammals branched from other vertebrates about 100 million years earlier. As we saw in Chapter 21, estimates of divergence time can be made using molecular sequence data, but all such estimates must be calibrated using fossils.

The evolutionary relationship between birds and crocodiles highlights a second kind of information provided by fossils. Not only do fossils record past life, they also provide our only record of extinct species. The phylogeny in Fig. 23.7 contains a great deal of information, but it is silent about dinosaurs. Fossils demonstrate that dinosaurs once roamed Earth, and details of skeletal structure place birds among the dinosaurs in the vertebrate tree. Indeed, some remarkable fossils from China show that the dinosaurs most closely related to birds had feathers (**Fig. 23.11**).

FIG. 23.11 *Microraptor gui*, a remarkable fossil discovered in approximately 125-million-year-old rocks from China. The structure of its skeleton identifies *M. gui* as a dinosaur, yet it had feathers on its arms, tail, and legs.

FIG. 23.12 The Grand Canyon. Erosion has exposed layers of sedimentary rock that record Earth history.

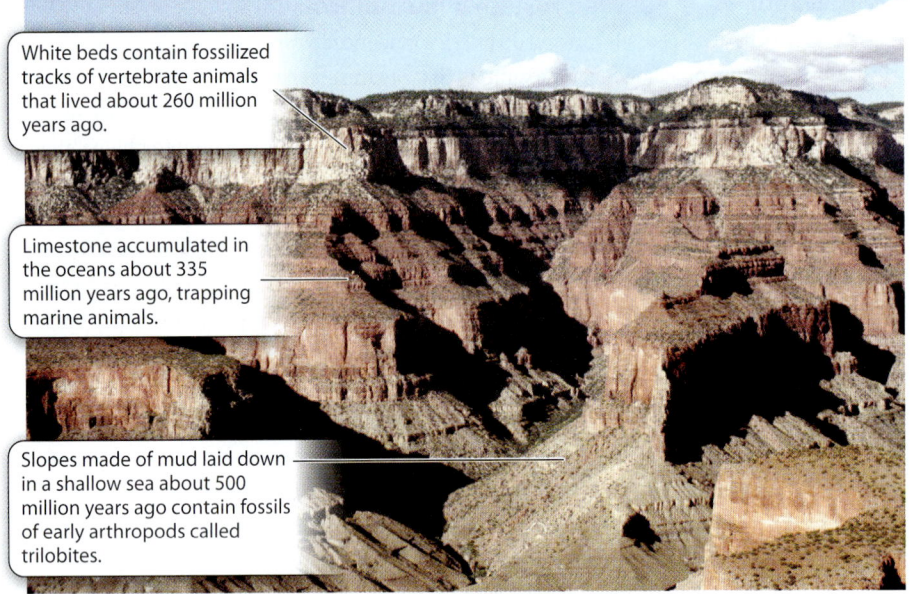

White beds contain fossilized tracks of vertebrate animals that lived about 260 million years ago.

Limestone accumulated in the oceans about 335 million years ago, trapping marine animals.

Slopes made of mud laid down in a shallow sea about 500 million years ago contain fossils of early arthropods called trilobites.

A third, and also unique, contribution of fossils is that they place evolutionary events in the context of Earth's dynamic environmental history. Again, dinosaurs illustrate the point. As discussed in Chapter 1, geologic evidence from several continents suggests that a large meteorite triggered drastic changes in the global environment 65 million years ago, leading to the extinction of most of the dinosaurs. In fact, at half a dozen moments in the past, large environmental disturbances sharply decreased Earth's biological diversity. These events, called mass extinctions, have played a major role in shaping the course of evolution.

Fossils provide a selective record of past life.

Fossils are the remains of once-living organisms, preserved through time in sedimentary rocks. The feathered dinosaur shown in Fig. 23.11 is an example of one kind of fossil. If we wish to use fossils to complement phylogenies based on modern organisms, we must understand how fossils form and how the processes of formation govern what is and is not preserved.

For all its merits, the fossil record should not be thought of as a complete dictionary of everything that ever walked, crawled, or swam across our planet's surface. Fossilization requires burial, as when a clam dies on the seafloor and is quickly covered by sand, or a leaf falls to the forest floor and ensuing floods cover it in mud. Through time, accumulating sediments harden into sedimentary rocks such as those exposed so dramatically in the walls of the Grand Canyon (**Fig. 23.12**). Without burial, the remains of organisms are eventually recycled by biological and physical processes, and no fossil forms. In general, the fossil record of marine life is more complete than that for land-dwelling creatures because marine habitats are more likely than those on land to be places where sediments accumulate and become rock. Thus, trees and elks living high in the Rocky Mountains have a low probability of fossilization, whereas clams and corals on the shallow seafloor are commonly buried and become fossils.

Biological factors contribute to the incompleteness of the fossil record. Most fossils preserve the hard parts of organisms, those features that resist decay after death. For animals, this usually means mineralized skeletons. Clams and snails that secrete shells of calcium carbonate have excellent fossil records. More than 80% of the clam species found today along California's coast also occur as fossils in sediments deposited during the past million years. In contrast, nematodes, tiny worms that may be the most abundant animals on Earth, have no mineralized skeletons and almost no fossil record. The wood and pollen of plants, which are made in part of decay-resistant organic compounds, enter the fossil record far more commonly than

FIG. 23.13 Trace fossils. These footprints in 150-million-year-old rocks record both the structure and behavior of the dinosaurs that made them.

FIG. 23.14 A fossil of *Opabinia regalis*, an early relative of the arthropods, from the 505-million-year-old Burgess Shale.

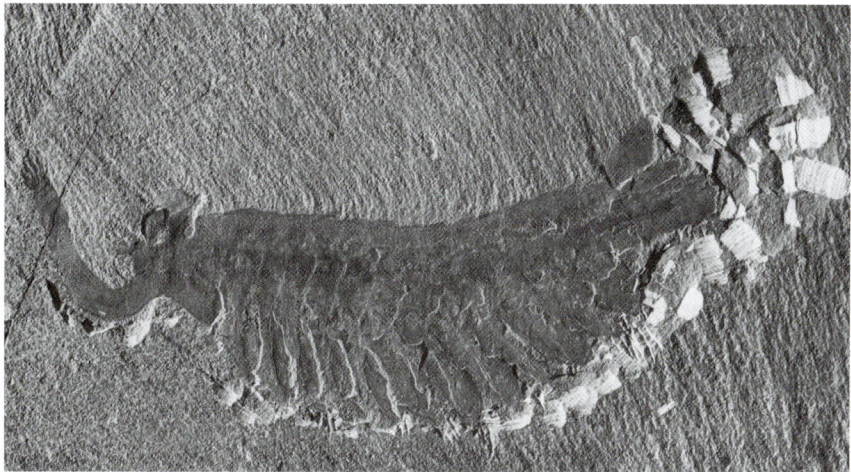

do flowers. And, among unicellular organisms, the skeleton-forming diatoms, radiolarians, and foraminiferans have exceptionally good fossil records, whereas most amoebas are unrepresented.

Together, then, the properties of organisms (do they make skeletons or other features that resist decay after death?) and environment (did the organisms live in a place where burial was likely?) determine the probability that an ancient species will be represented in the fossil record. Fortunately, for species that make mineralized skeletons and live on the shallow seafloor, the fossil record is very good, preserving a detailed history of evolution through time.

Organisms that lack hard parts can leave a fossil record in two other, distinctive ways. Many animals leave tracks and trails as they move about or burrow into sediments. These **trace fossils,** from dinosaur tracks to the feeding trails of snails and trilobites, preserve a record of both anatomy and behavior (**Fig. 23.13**).

Organisms can also contribute **molecular fossils** to the rocks. Most biomolecules decay quickly after death. Proteins and DNA, for example, generally break down before they can be preserved, although, remarkably, a sizable fraction of the Neanderthal genome has been pieced together from DNA in 40,000-year-old bones. Other molecules, especially lipids like cholesterol, are more resistant to decomposition. Sterols, bacterial lipids, and some pigment molecules can accumulate in sedimentary rocks, documenting organisms that rarely form conventional fossils, especially bacteria and single-celled eukaryotes.

Rarely, unusual conditions preserve fossils of unexpected quality, including animals without shells, delicate flowers or mushrooms, fragile seaweeds or bacteria, even the embryos of plants and animals. For example, 505 million years ago, during the Cambrian Period, a sedimentary rock formation called the **Burgess Shale** accumulated on a relatively deep seafloor covering what is now British Columbia. Waters just above the basin floor contained little or no oxygen, so that when mud swept into the basin, entombed animals were sealed off from scavengers, disruptive burrowing activity, and even bacterial decay. For this reason, Burgess rocks preserve a remarkable sampling of marine life during the initial diversification of animals (**Fig. 23.14**).

The **Messel Shale** formed more recently—about 50 million years ago—in a lake in what is now Germany. Release of toxic gases from deep within the Messel Lake suffocated local animals, and their carcasses settled into oxygen-poor muds on the lakefloor. Fish, birds, mammals, and reptiles are preserved as complete and articulated skeletons, and mammals retain impressions of fur and color patterning (**Fig. 23.15**). Plants were also beautifully preserved, as were insects, some with a striking iridescence still intact. Messel rocks provide a truly outstanding snapshot of life on land as the age of mammals began.

FIG. 23.15 The Messel Shale, Germany. Even the furry tail of this extinct squirrel-like mammal is preserved in this remarkable 50-million-year-old fossil.

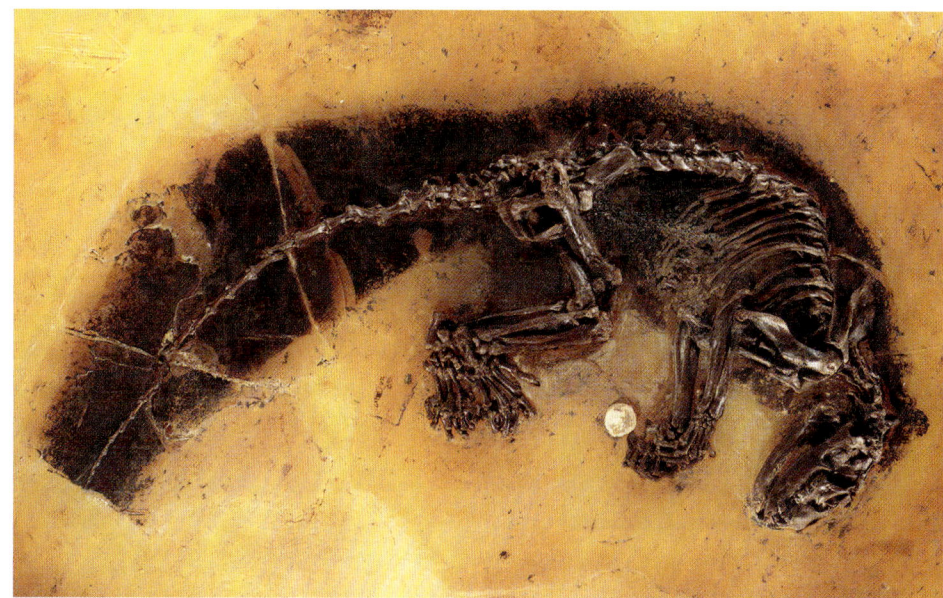

In general, the fossil record preserves some aspects of biological history well and others poorly. Fossils provide a good sense of how the forms, functions, and diversity of skeletonized animals have changed over the past 500 million years. The same is true for land plants and unicellular organisms that form mineralized skeletons. These fossils shed light on major patterns of morphological evolution and diversity change through time; their geographic distributions record the movements of continents over millions of years; and the radiations and extinctions they document show how life responds to environmental change, both gradual and catastrophic.

Geological data indicate the age and environmental setting of fossils.

How do we know the ages of fossils? Beginning in the nineteenth century, geologists recognized that groups of

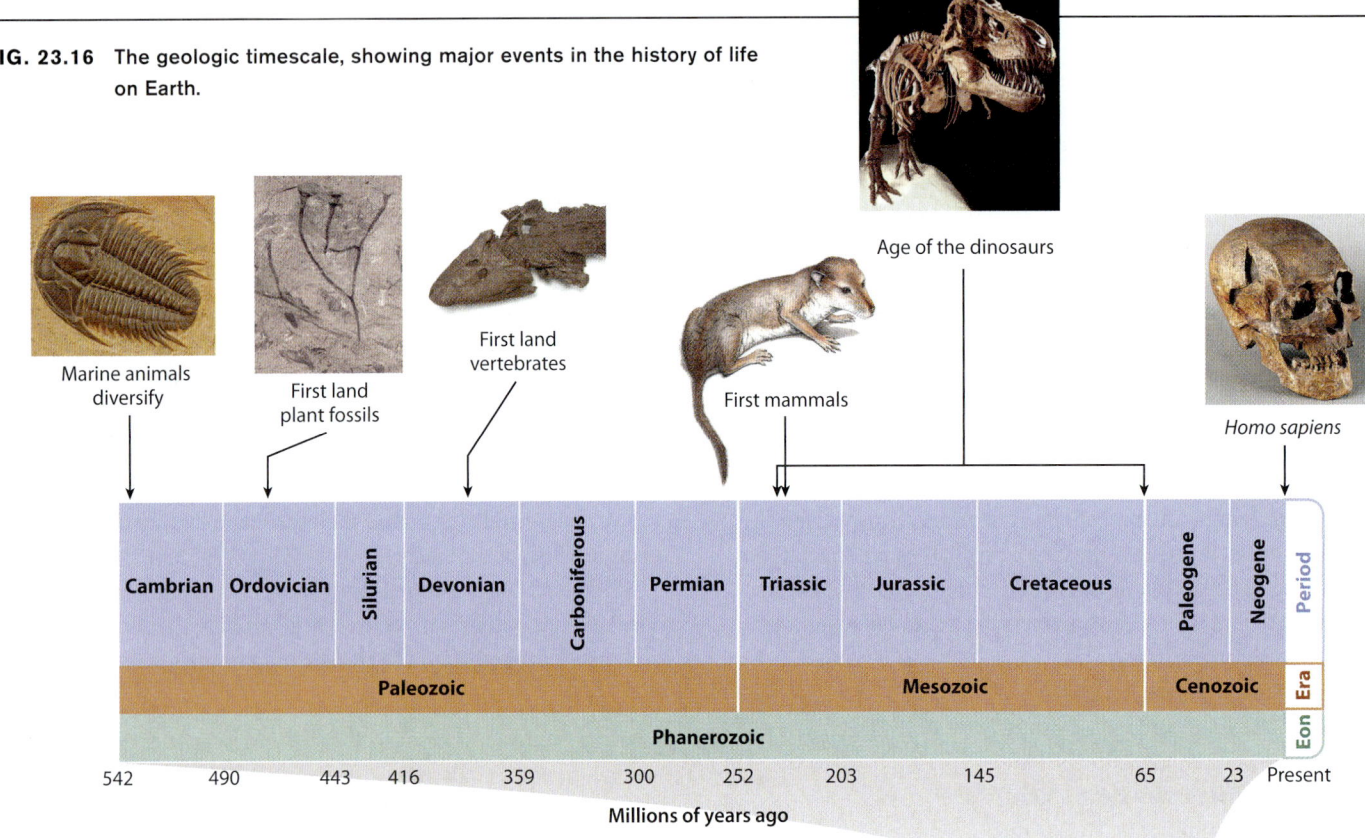

FIG. 23.16 The geologic timescale, showing major events in the history of life on Earth.

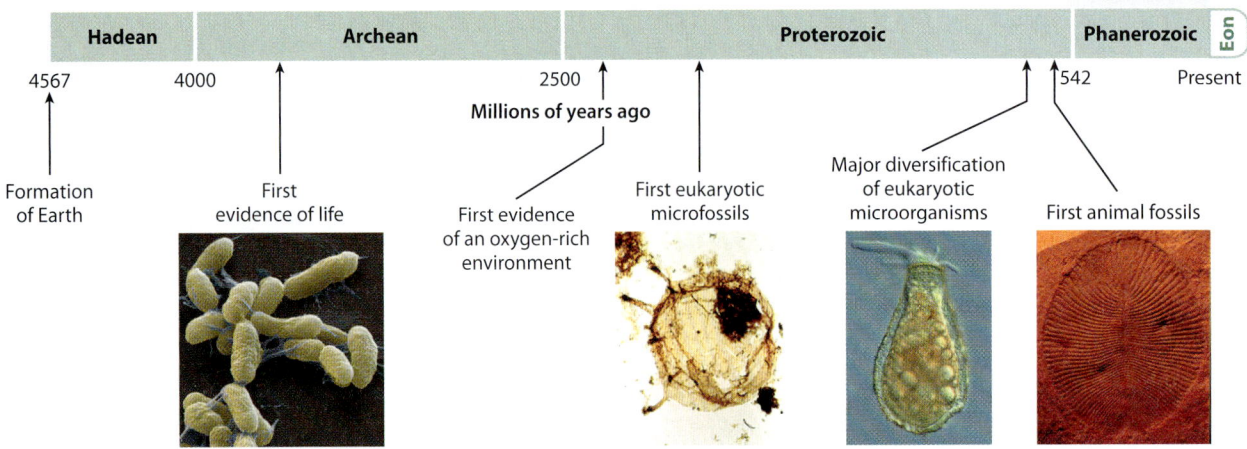

fossils change systematically from the bottom of a sedimentary rock formation to its top. As more of Earth's surface was mapped and studied, it became clear that certain fossils always occur in layers that lie beneath (and so are older than) layers that contain other species. From these patterns, geologists concluded that fossils mark time in Earth history. At first, geologists didn't know why fossils changed from one bed to the next, but after Darwin the reason became apparent: Fossils record the evolution of life on Earth. They eventually mapped out the **geologic timescale**, the series of time divisions that mark Earth's long history (**Fig. 23.16**).

The layers of fossils in sedimentary rocks can tell us that some rocks are older than others, but they cannot by themselves provide an absolute age. Calibration of the timescale became possible with the discovery of radioactive decay. In Chapter 2, we discussed isotopes, variants of an element that differ from one another in the number of neutrons they contain. Many isotopes are unstable and spontaneously break down to form other, more stable isotopes. In the laboratory, scientists can measure how fast unstable isotopes decay. Then, by measuring the amounts of the unstable isotope and its stable daughter inside a mineral, they can determine when the mineral formed.

Archaeologists commonly use the radioactive decay of the isotope carbon-14, or ^{14}C, to date wood and bone. As shown in **Fig. 23.17**, cosmic rays continually generate ^{14}C in the atmosphere. Through photosynthesis, carbon dioxide that contains ^{14}C is incorporated into wood, and animals also incorporate small amounts of ^{14}C into their tissues when they eat plant material. After death, the unstable ^{14}C in these materials begins to break down, losing an electron to form ^{14}N, a stable isotope of nitrogen. Laboratory measurements indicate that half of the ^{14}C in a given sample will decay to nitrogen in 5730 years, a period called its **half-life** (Fig. 23.17). Armed with this information, scientists can measure the amount of ^{14}C in an archaeological sample and, by comparing it to the amount of ^{14}C in a sample of known age—annual rings in trees, for example, or yearly growth coral skeletons—determine the age of the sample.

Because its half-life is so short (by geological standards), ^{14}C is useful only in dating materials younger than 50,000 to 60,000 years. Beyond that, there is too little ^{14}C left to measure accurately. Older geological materials are commonly dated using the radioactive decay of uranium (U) to lead (Pb): ^{238}U breaks down to ^{206}Pb with a half-life of 4.47 billion years, and ^{235}U decays to ^{207}Pb with a half-life of 704 million years. The calibration of the geologic timescale is based mostly on radiometric ages of volcanic ash beds found within sedimentary rock that contain key fossils, as well as volcanic rocks that intrude into (and so are younger than) layers of rock containing fossils. In turn, the age of fossils provide calibration points for phylogenies.

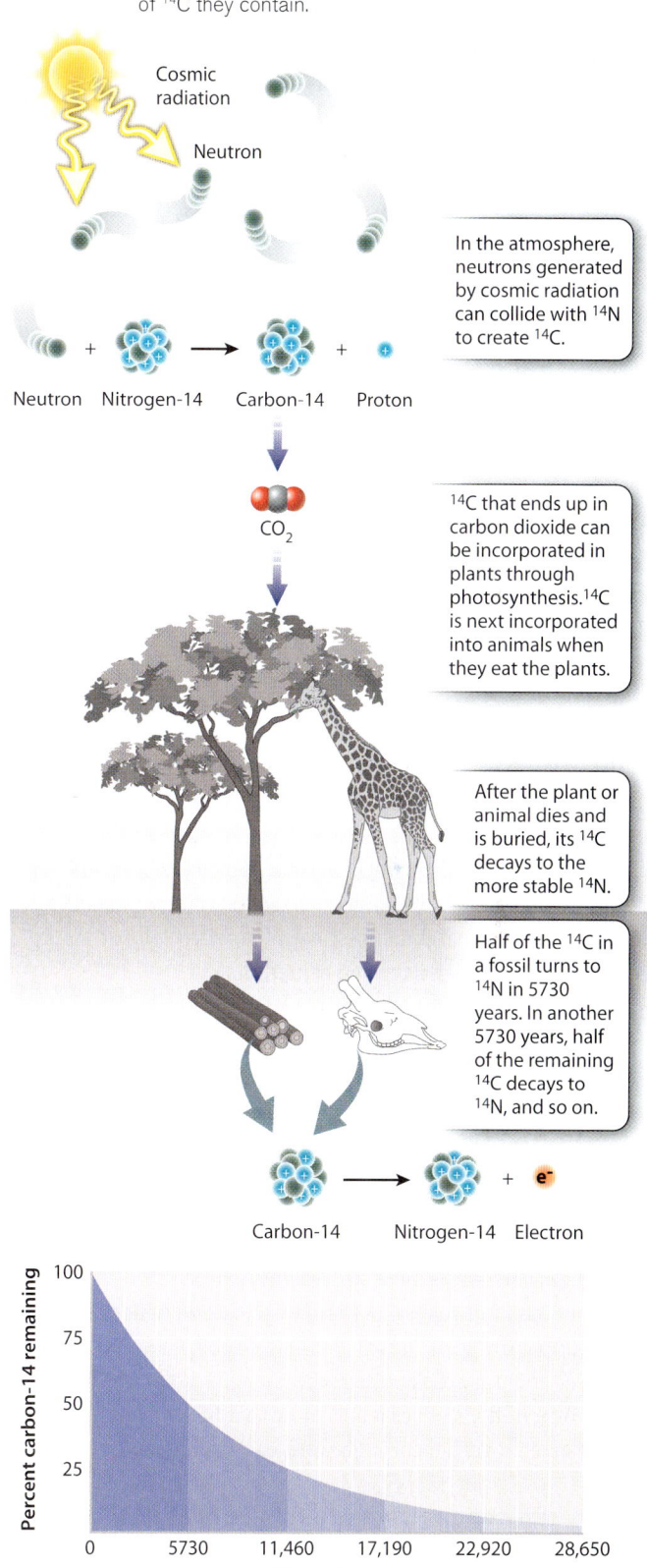

FIG. 23.17 ^{14}C **decay.** Scientists can determine the age of relatively young materials such as wood and bone from the amount of ^{14}C they contain.

In the atmosphere, neutrons generated by cosmic radiation can collide with ^{14}N to create ^{14}C.

^{14}C that ends up in carbon dioxide can be incorporated in plants through photosynthesis. ^{14}C is next incorporated into animals when they eat the plants.

After the plant or animal dies and is buried, its ^{14}C decays to the more stable ^{14}N.

Half of the ^{14}C in a fossil turns to ^{14}N in 5730 years. In another 5730 years, half of the remaining ^{14}C decays to ^{14}N, and so on.

FIG. 23.18 Pangaea, 290 million years ago. Plate tectonics has shaped and reshaped Earth's geography through time. The white near the South Pole is glacial ice.

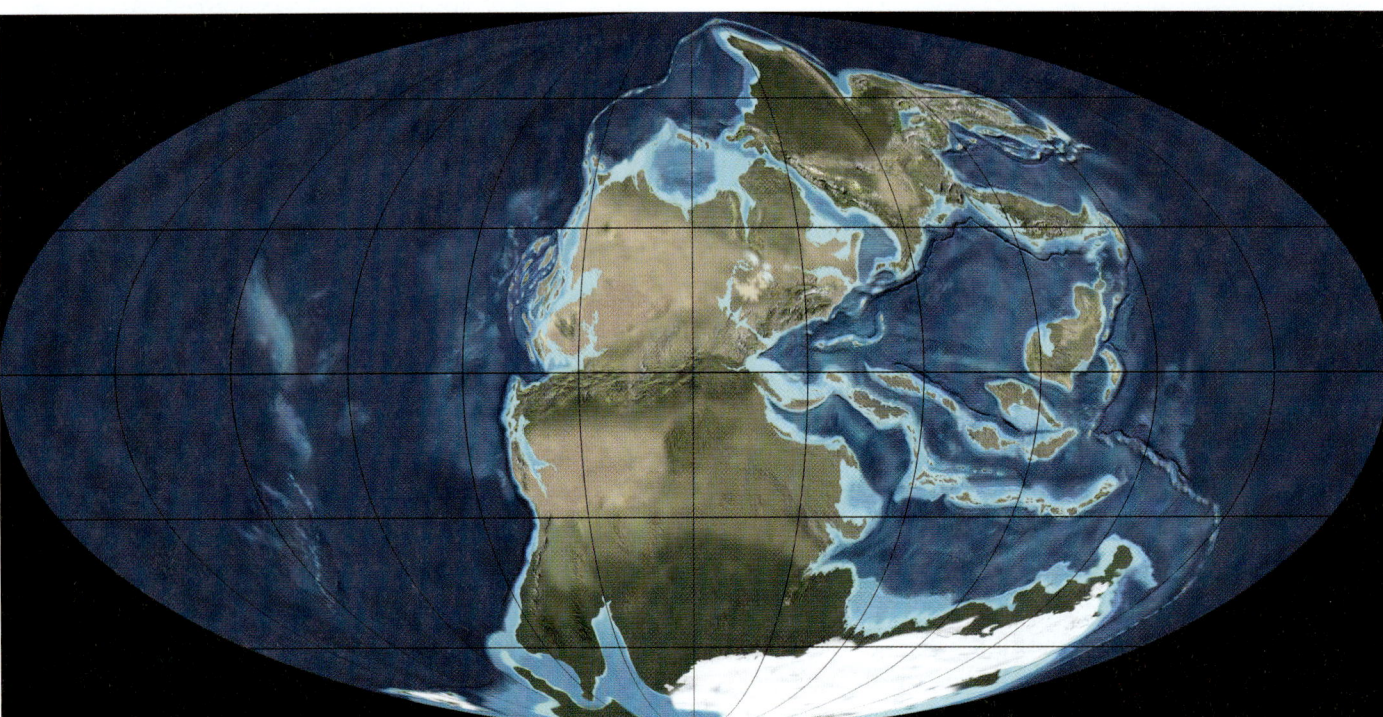

The sedimentary rocks that contain fossils also preserve, encrypted in their physical features and chemical composition, information about the environment in which they formed. Sandstone beds, for example, may have rippled surfaces, like the ripples produced by currents that we see today in the sand of a seashore or a lake margin. Enrichment in pyrite (FeS_2) or other iron minerals may signal oxygen depletion.

We might think our moment in geologic time is representative of the Earth as it has always existed, but nothing could be further from the truth. In the location and sizes of its continents, ocean chemistry, and atmospheric composition, the Earth we experience is unlike any previous state of the planet. Today, for example, the continents are distributed widely over the planet's surface, but 290 million years ago they were clustered in a supercontinent called Pangaea (**Fig. 23.18**). Oxygen gas permeates most surface environments of Earth today, but 3 billion years ago, there was no O_2 anywhere. And, just 20,000 years ago, 2 km of glacial ice stood where Boston lies today. Sedimentary rocks record the changing state of Earth's surface over billions of years and show that life and environment have changed together through time, each influencing the other.

Fossils can contain unique combinations of characters.

Phylogenies hypothesize impressive morphological and physiological shifts through time—amphibians from fish, for example, or land plants from green algae. Do fossils capture a record of these transitions as they took place?

Let's begin with an example introduced earlier in this chapter. Phylogenies based on living organisms generally place birds as the sister group to crocodiles and alligators, but birds and crocodiles are decidedly different from each other in structure—birds have wings, feathers, toothless bills, and a number of other skeletal features distinct from those of crocodiles. In 1861, just 2 years after publication of *On the Origin of Species*, German quarry workers discovered a remarkable fossil that remains paleontology's most famous example of a transitional form. *Archaeopteryx lithographica*, now known from 11 specimens splayed for all time in fine-grained limestone, lived 150 million years ago. Its skeleton shares many characters with dromaeosaurs, a group of small, agile dinosaurs, but several features—its pelvis, its braincase, and, especially, its winglike forearms—are distinctly birdlike. Spectacularly, the fossils preserve evidence of feathers. *Archaeopteryx* clearly suggests a close relationship between birds and dinosaurs, and phylogenetic reconstructions that include information from fossils show that many of the characters found today in birds accumulated through time in their dinosaur ancestors. And, as noted earlier, even feathers first evolved in dinosaurs (**Fig. 23.19**).

Tiktaalik roseae and other skeletons in rocks deposited 375 to 362 million years ago record an earlier but equally fundamental transition: the colonization of land by vertebrates. Phylogenies

CHAPTER 23 EVOLUTIONARY PATTERNS: PHYLOGENY AND FOSSILS

FIG. 23.19 Dinosaurs and birds. A number of dinosaur fossils link birds phylogenetically to their closest living relatives, the crocodiles. The fossil at the bottom is *Archaeopteryx*.

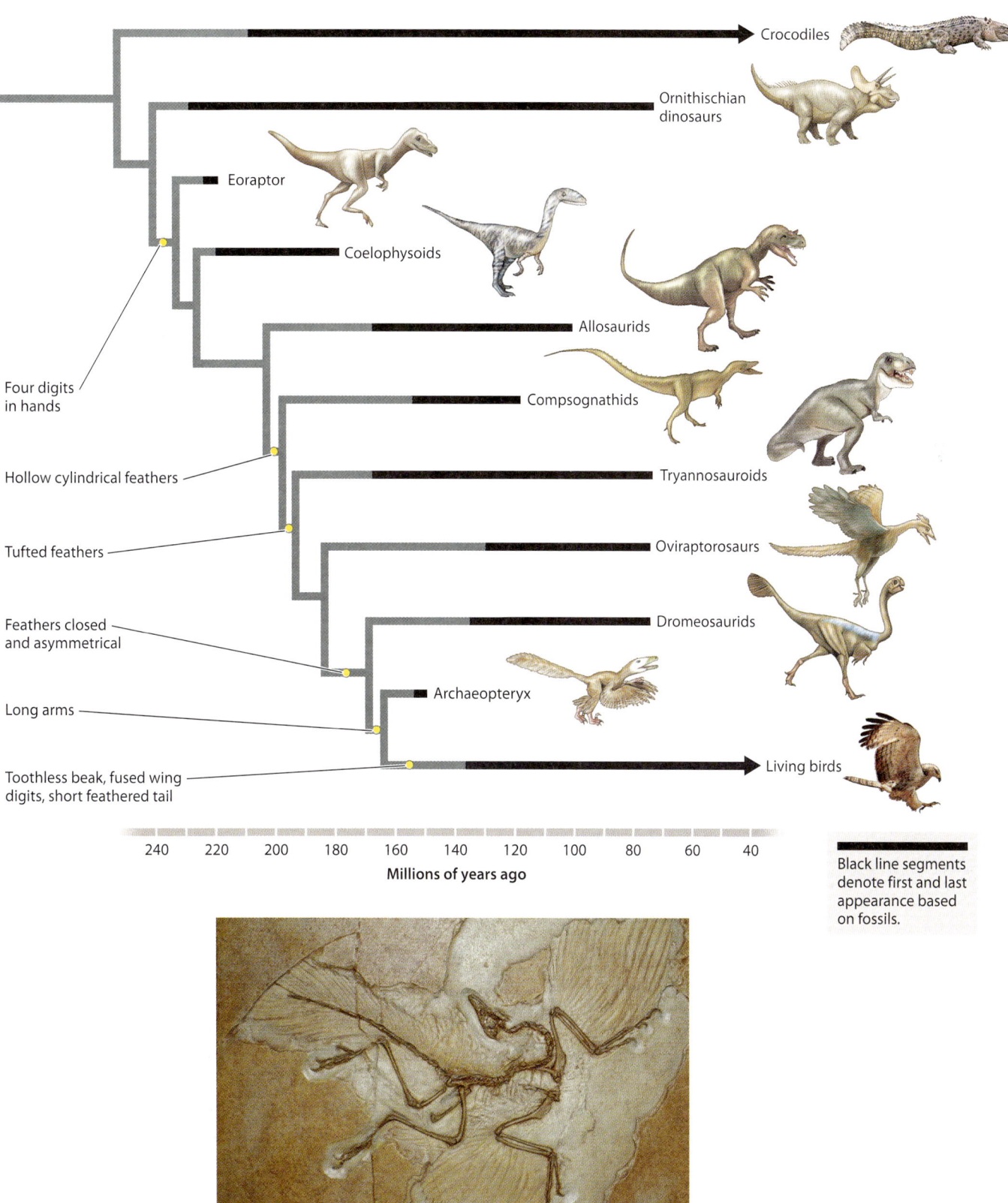

HOW DO WE KNOW?

FIG. 23.20

Can fossils bridge the evolutionary gap between fish and tetrapod vertebrates?

BACKGROUND Phylogenies based on both morphological and molecular characters indicate that fish are the closest relatives of four-legged land vertebrates.

HYPOTHESIS Land vertebrates evolved from fish by modifications of the skeleton and internal organs that made it possible for them to live on land.

OBSERVATION Fossil skeletons 390 to 360 million years old show a mix of features seen in living fish and amphibians. Older fossils have fins, fishlike heads, and gills, and younger fossils have weight-bearing legs, skulls with jaws able to grab prey, and ribs that help ventilate lungs. Paleontologists predicted that key intermediate fossils would be preserved in 380–370-million-year-old rocks. In 2004, Edward Daeschler, Neil Shubin, and Farish Jenkins discovered *Tiktaalik*, a remarkable fossil that has fins, scales, and gills like fish, but wrist bones and fingers, an amphibian-like skull, and a true neck (which fish lack).

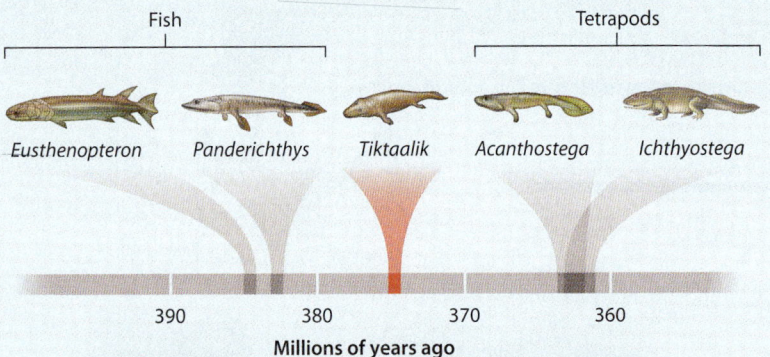

CONCLUSION Fossils confirm the phylogenetic prediction that tetrapod vertebrates evolved from fish by the developmental modification of limbs, skulls, and other features.

FOLLOW-UP WORK Research into the genetics of vertebrate development shows that the limbs of fish and amphibians are shaped by similar patterns of gene expression, providing further support for the phylogenetic connection between the two groups.

SOURCE Daeschler, E. B., N. H. Shubin, and F. A. Jenkins, Jr. 2006. "A Devonian Tetrapod-like Fish and the Evolution of the Tetrapod Body Plan." *Nature* 440:757–763.

23-18

show that all land vertebrates, from amphibians to mammals, are descended from fish. As seen in **Fig. 23.20**, *Tiktaalik* had fins, gills, and scales like other fish of its day, but its skull was flattened, more like that of a crocodile than a fish, and it had a functional neck and ribs that could support its body—features today found only in tetrapods. Along with other fossils, *Tiktaalik* captures key moments in the evolutionary transition from water to land, confirming the predictions of phylogeny.

→ **Quick Check 4** You have just found a novel vertebrate skeleton in 200-million-year-old rocks. How would you integrate this new fossil species into the phylogenetic tree depicted in Fig. 23.7?

Rare mass extinctions have altered the course of evolution.

Fossils show that life originated more than 3.5 billion years ago, but animals appeared much later, only about 600 million years ago. **Fig. 23.21** graphs the number of fossil genera found for each period of the Phanerozoic (literally, the age of visible animals) Eon. Recorded diversity was low at the beginning of the Phanerozoic and, in general, has increased through time. Nonetheless, the graph clearly shows that at several moments in the past 542 million years, recorded diversity dropped catastrophically. These **mass extinctions** spelled the end of many previously important groups of species, but they opened up new possibilities for evolution.

The best-known mass extinction occurred 65 million years ago, at the end of the Cretaceous Period (Fig. 23.21). On land, dinosaurs disappeared abruptly, following more than 150 million years of dominance in terrestrial ecosystems. In the oceans, ammonites, cephalopod mollusks that had long been abundant predators, also became extinct, and most skeleton-forming microorganisms in the oceans disappeared as well. As discussed in Chapter 1, a large body of geologic evidence supports the hypothesis that this biological catastrophe was caused by the impact of a giant meteorite. The mass extinction at the end of the Cretaceous removed much of the biological diversity that had built up through genetic changes over millions of years, but it had another, equally important evolutionary consequence. Survivors of the extinction could proliferate with little competition from other populations. Mammals, our own branch of the animal tree, diversified across continents once dinosaurs had disappeared from the landscape.

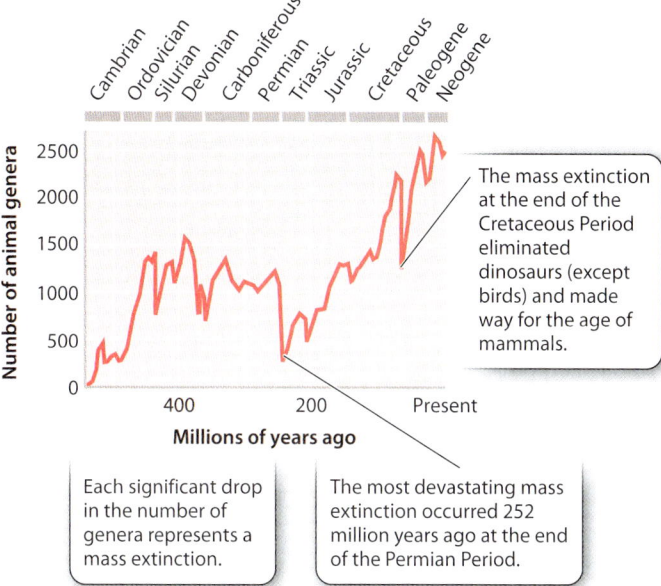

FIG. 23.21 Mass extinctions documented by compilations of fossil diversity through time. Though they eliminate much of the life found on Earth at the time, mass extinctions allow new groups to proliferate and diversify.

Each significant drop in the number of genera represents a mass extinction.

The most devastating mass extinction occurred 252 million years ago at the end of the Permian Period.

The mass extinction at the end of the Cretaceous Period eliminated dinosaurs (except birds) and made way for the age of mammals.

An even larger mass extinction took place much earlier at the end of the Permian Period, 252 million years ago (Fig. 23.21). More than 90% of all genera recorded in late Permian oceans disappeared, especially animals like corals and sponges with heavy skeletons. There is no compelling evidence for meteorite impact, so geologists have hypothesized that this mass extinction resulted from the catastrophic effects of massive volcanic eruptions.

At the end of the Permian Period, most continents were gathered into the supercontinent Pangaea (see Fig. 23.18), and so a huge ocean covered more than half the Earth. Levels of oxygen in the deep waters of this ocean fell, the result of sluggish circulation and warm seawater temperatures. Then, a massive outpouring of ash and lava erupted across what is now Siberia. Enormous emissions of carbon dioxide and methane from the volcanoes caused global warming (so even less oxygen reached deep oceans) and ocean acidification (so it was difficult for animals and algae to secrete calcium carbonate skeletons).

The three-way insult of lack of oxygen, ocean acidification, and global warming doomed many species on land and in the seas. Seascapes dominated by corals and shelled invertebrates called brachiopods disappeared. After the extinction of these organisms, new populations of mollusks, new forms of coral, and other animals we see today rose to prominence. Again, major environmental disruption removed diversity that had slowly accumulated through many millions of years but also opened up new evolutionary possibilities for the survivors.

Five great mass extinctions occurred over the past 500 million years, but only the two at the end of the Permian and Cretaceous Periods so thoroughly changed the course of evolution. At present, we live at a time of accelerating ocean acidification and global warming. Will future geologists recognize our evolutionary moment as a time of mass extinction? We discuss this possibility in Chapter 48.

23.4 COMPARING EVOLUTION'S TWO GREAT PATTERNS

The diversity of life we see today is the result of evolutionary processes playing out over geologic time. Evolutionary process can be studied by experiment, both in the field and in the laboratory, but evolutionary history is another matter. There is no experiment we can do to determine why the dinosaurs became extinct—we cannot rerun the events of 65 million years ago, this time without the meteorite impact. The history of life must be reconstructed from evolution's two great patterns: the nested similarity observed in the forms and macromolecular sequences of living organisms, and the direct historical archive of the fossil record.

Phylogeny and fossils complement each other.

The great advantage of reconstructing evolutionary history from living organisms is that we can use a full range of features—skeletal morphology, cell structure, DNA sequence—to generate phylogenetic hypotheses. The disadvantage of using comparative biology is that we lack evidence of extinct species, the time dimension, and the environmental context. This, of course, is where the fossil record comes into play. Fossil evidence has strengths and limitations that complement the evolutionary information in the living organisms.

We can use phylogenetic methods based on DNA sequences to infer that birds and crocodiles are closely related, but only fossils can show that the evolutionary link between birds and crocodiles runs through dinosaurs. And only the geologic record can show that mass extinction removed the dinosaurs, paving the way for the emergence of modern mammals. Paleontologists and biologists work together to understand evolutionary history. Biology provides a functional and phylogenetic framework for the interpretation of fossils, and fossils provide a record of life's history in the context of continual planetary change.

Agreement between phylogenies and the fossil record provides strong evidence of evolution.

Phylogenies based on morphological or molecular comparisons of living organisms make hypotheses about the timing of evolutionary changes through Earth history. We humans are a case in point. As discussed in more detail in Chapter 24, comparisons of DNA sequences suggest that chimpanzees are our closest living relatives. This is hardly surprising, as simple observation shows that chimpanzees and humans share many features. However, among other differences, chimpanzees have smaller stature, long arms that facilitate knuckle-walking (the arms help support the body as the chimpanzee moves forward), a more prominent snout, and larger teeth. Fossils painstakingly unearthed over the past century show that the morphological features that mark us as human accumulated through a series of speciation events. Six to seven million years ago, when the human and chimpanzee lineages first split, the two newly diverged taxa looked much more alike than humans and chimpanzees do today.

The agreement of comparative biology and the fossil record can be seen at all scales of observation. Humans form one tip of a larger branch that contains all members of the primate family. Primates, in turn, are nested within a larger branch occupied by mammals, and mammals nest within a still larger branch containing all vertebrate animals, which include fish. This arrangement predicts that the earliest fossil fish should be older than the earliest fossil mammals, the earliest fossil mammals should be older than the earliest primates, and the earliest primates older than the earliest humans. This is precisely what the fossil record shows (**Fig. 23.22**).

The agreement between fossils and phylogenies drawn from living organisms can be seen again and again when we examine different branches of the tree of life or, for that matter, the tree as a whole. All phylogenies indicate that microorganisms diverged early in evolutionary history, and mammals, flowering plants, and other large complex organisms diverged more recently. The tree's shape implies that diversity has accumulated through time, beginning with simple organisms and later adding complex macroscopic forms.

The geologic record shows the same pattern. For nearly 3 billion years of Earth history, microorganisms dominate the fossil record, with the earliest animals appearing about 600 million years ago, the earliest vertebrate animals 520 million years ago, the earliest tetrapod vertebrates about 360 million years ago, the first mammals

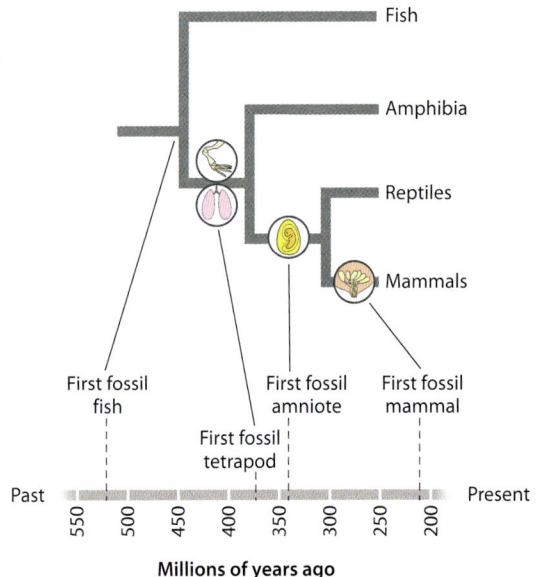

FIG. 23.22 Phylogeny and fossils of vertebrates. The branching order of the phylogeny corresponds to the order of appearance of each group in the fossil record.

210 million years ago, the first primates perhaps 55 million years ago, the oldest fossils of our own species a mere 200,000 years ago. Similarly, if we focus on photosynthetic organisms, we find a record of photosynthetic bacteria beginning at least 3500 million years ago, algae 1200 million years ago, simple land plants 470 million years ago, seed plants 370 million years ago, flowering plants about 140 million years ago, and the earliest grasses 70 million years ago.

In Part 2, we explore the evolutionary history of life in some detail. Here, it is sufficient to draw the key general conclusion: The fact that comparative biology and fossils, two complementary but independent approaches to reconstructing the evolutionary past, yield the same history is powerful evidence of evolution.

Core Concepts Summary

23.1 A PHYLOGENETIC TREE IS A REASONED HYPOTHESIS OF THE EVOLUTIONARY RELATIONSHIPS OF ORGANISMS.

The nested pattern of similarities seen among organisms is a result of descent with modification and can be represented as a phylogenetic tree. page 23-1

The order of branches on a phylogenetic tree indicates the sequence of events in time. page 23-2

Sister groups are more closely related to one another than they are to any other group. page 23-3

A node is a branching point on a tree, and it can be rotated without changing evolutionary relationships. page 23-3

A monophyletic group includes all the descendants of a common ancestor, and it is considered a natural grouping of organisms based on shared ancestry. page 23-4

A paraphyletic group includes some, but not all, of the descendants of a common ancestor. page 23-4

A polyphyletic group includes organisms from distinct lineages based on shared characters, but it does not include a common ancestor. page 23-4

Organisms are classified into domain, kingdom, phylum, class, order, family, genus, and species. page 23-4

23.2 A PHYLOGENETIC TREE IS BUILT ON THE BASIS OF SHARED DERIVED CHARACTERS.

Characters, or traits, existing in different states are used to build phylogenetic trees. page 23-5

Homologies are similarities based on shared ancestry, while analogies are similarities based on independent adaptations. page 23-6

Homologies can be ancestral, unique to a particular group, or present in some, but not all, of the descendants of a common ancestor (shared derived characters). page 23-6

Only shared derived characters, or synapomorphies, are useful in constructing a phylogenetic tree. page 23-6

Molecular data provide a wealth of characters that complement other types of characters in building phylogenetic trees. page 23-8

Phylogenetic trees can be used to understand evolutionary relationships of organisms and solve practical problems, such as how viruses evolve over time. page 23-10

23.3 THE FOSSIL RECORD PROVIDES A DIRECT GLIMPSE OF EVOLUTIONARY HISTORY.

Fossils are the remains of organisms preserved in sedimentary rocks. page 23-11

The fossil record is imperfect because fossilization requires burial in sediment, sediments accumulate episodically and discontinuously, and fossils typically preserve only the hard parts of organisms. page 23-12

Radioactive decay provides a means of dating rocks. page 23-14

Archaeopteryx and *Tiktaalik* are two fossil organisms that document, respectively, the bird–dinosaur transition and the fish–tetrapod transition. page 23-16

The history of life is characterized by rare mass extinctions, including the extinction at the end of the Cretaceous Period 65 million years ago in which the dinosaurs (other than birds) became extinct, and the extinction at the end of the Permian Period 252 million years ago, the largest documented mass extinction in the history of Earth. page 23-18

23.4 PHYLOGENY AND FOSSILS PROVIDE INDEPENDENT AND CORROBORATING EVIDENCE OF EVOLUTION.

Phylogeny makes use of living organisms, and the fossil record supplies a record of species that no longer exist, absolute dates, and environmental context. page 23-19

Data from phylogeny and fossils are often in agreement, providing strong evidence for evolution. page 23-19

Self-Assessment

1. Give two examples of a nested pattern of similarity among organisms.

2. With reference to a phylogenetic tree, show how a nested pattern of similarity is the necessary result of evolution.

3. Distinguish among monophyletic, paraphyletic, and polyphyletic groups, and give an example of each.

4. List the levels of classification, from the least inclusive (species) to the most inclusive (domain).

5. Describe two traits that are homologous and two that are analogous.

6. Name a type of homology that is useful in building phylogenetic trees and explain why this kind of homology, and not others, is useful.

7. Name three types of fossil.

8. Explain why there are gaps in the fossil record.

9. Explain how the fossil record can be used to determine both the relative and the absolute timescales of past events.

10. Describe the significance of *Archeopteryx* and *Tiktaalik*.

11. Describe how mass extinctions have shaped the ecological landscape.

Do you understand the chapter's Core Concepts? Log in to check your answers to the Self-Assessment questions, then practice what you've learned and reinforce this chapter's concepts by working through the problems and multimedia tutorials provided there.

www.biologyhowlifeworks.com

CHAPTER 24

HUMAN ORIGINS AND EVOLUTION

Core Concepts

24.1 Anatomical, molecular, and fossil evidence shows that the human lineage branches off the great apes tree.

24.2 Phylogenetic analysis of mitochondrial DNA and the *Y* chromosome shows that our species arose in Africa.

24.3 During the 5–7 million years since the most recent common ancestor of humans and chimpanzees, our lineage acquired a number of distinctive features.

24.4 Human history has had an important impact on patterns of genetic variation in our species.

24.5 Culture is a potent force for change in modern humans.

Charles Darwin carefully avoided discussing the evolution of our own species in *On the Origin of Species*. Instead, he wrote only that he saw "open fields for far more important researches," and that "Light will be thrown on the origin of man and his history." Darwin, an instinctively cautious man, realized that the ideas presented in *On the Origin of Species* were controversial enough without his adding humans to the mix. He presented his ideas on human evolution to the public only when, 12 years later, he published *The Descent of Man* in 1871.

As it turned out, Darwin's delicate sidestepping of human origins had little effect. The initial print run of *The Origin* sold out on the day of publication, and the public was perfectly capable of reading between the lines. The Victorians found themselves wrestling with the book's revolutionary message: that humans are a species of ape. The mild-mannered author, living in virtual seclusion at his country house in rural Kent, had initiated what is arguably the greatest of all intellectual revolutions.

Darwin's conclusions remain controversial to this day among the general public, but they are not controversial among scientists. The evidence that humans are descended from a line of apes whose modern-day representatives include gorillas and chimpanzees is compelling. We know now that about 5–7 million years ago the family tree of the great apes split, one branch ultimately giving rise to chimpanzees and the other to our species. It is those 5–7 million years that hold the key to our humanity. It was over this period—brief by evolutionary standards—that the attributes that make our species so remarkable arose. This chapter discusses what happened over those 5–7 million years and how we came to be the way we are.

24.1 THE GREAT APES

We can approach the question of our place in the tree of life in three different ways: through comparative anatomy, molecular analysis, and the fossil record. In this section, we use data from all three sources as we apply the standard methods of phylogenetic reconstruction (Chapter 23) to figure out the evolutionary relationships between humans and other mammals.

Comparative anatomy shows that the human lineage branches off the great apes tree.

There are about 400 species of **primates,** which include prosimians (lemurs, bushbabies, monkeys, and apes (**Fig. 24.1**). All primates share a number of general features that distinguish them from other mammals, including nails rather than claws and eyes on the front of the face instead of the side, allowing stereoscopic (that is, three-dimensional) vision. Most primates also have some form of opposable thumb (a thumb that can touch the finger tips of the same hand; contrast the motion of your thumb with that of your non-opposable big toe). Prosimians are thought to represent a separate primate lineage

FIG. 24.1 The primate family tree, showing the evolutionary relationships of prosimians, monkeys, and apes.

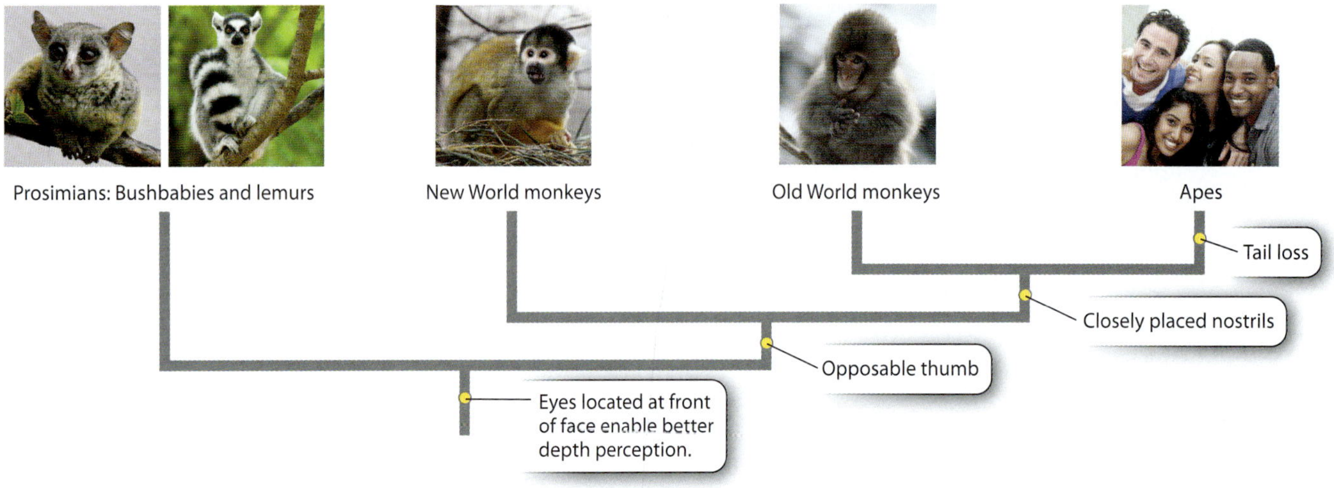

from the one that gave rise to humans. Lemurs, which today are confined to the island of Madagascar, are thus only distantly related to humans.

Monkeys underwent independent bouts of evolutionary change in the Americas and in Africa and Eurasia, so the family tree is split along geographic lines into New and Old World monkeys. Though both groups have evolved similar habits, there are basic distinctions. There are differences between the teeth of the two groups, and in New World monkeys the nostrils tend to be widely spaced, whereas in Old World species they are closer together.

One line of Old World monkeys gave rise to the apes, which lack a tail and show more sophisticated behaviors than other monkeys. The apes are split into two groups, the lesser and the great apes (**Fig. 24.2**). Lesser apes include the fourteen species of gibbon,

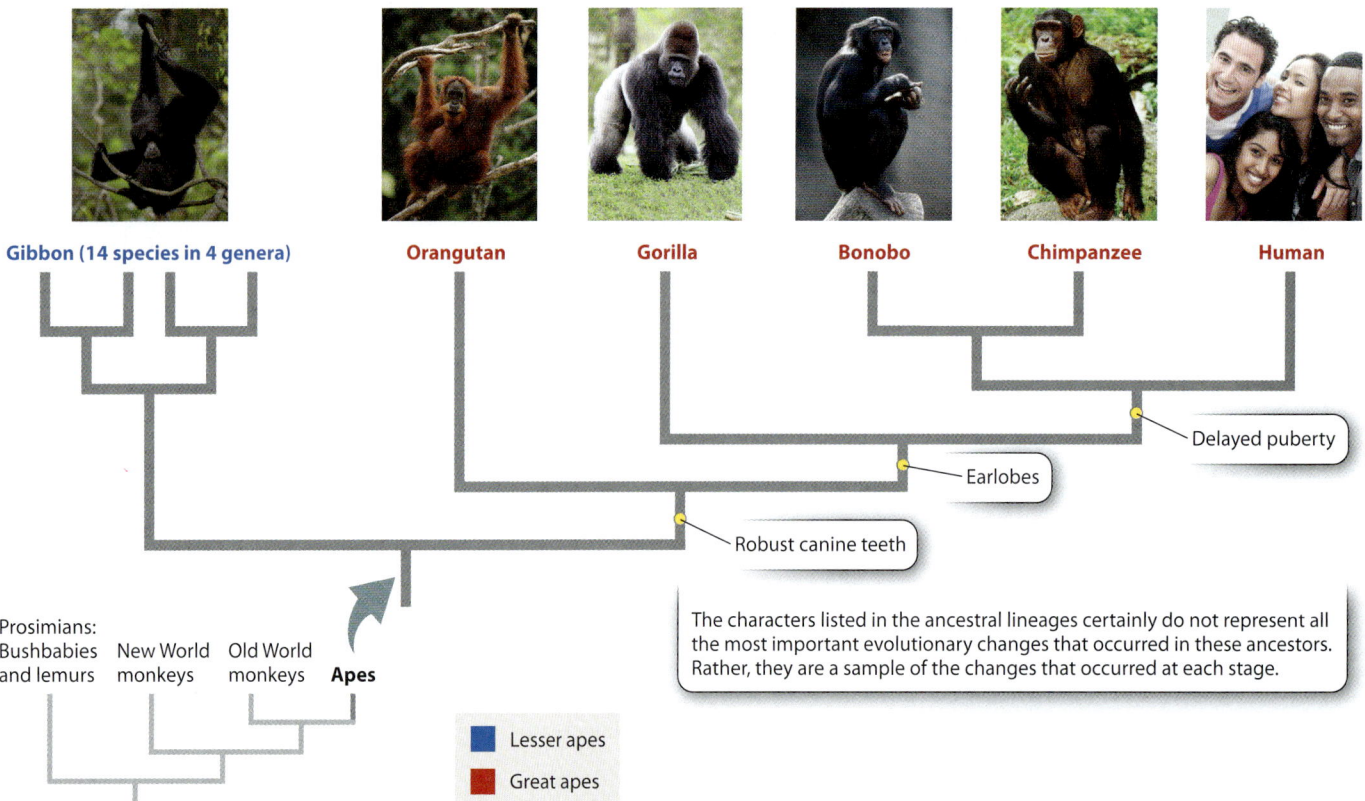

FIG. 24.2 The family tree of the lesser and great apes. The apes consist of two major groups, the lesser apes and great apes. The great apes group includes humans.

The characters listed in the ancestral lineages certainly do not represent all the most important evolutionary changes that occurred in these ancestors. Rather, they are a sample of the changes that occurred at each stage.

CHAPTER 24 HUMAN ORIGINS AND EVOLUTION

all of which are found in Southeast Asia. The great apes include the orangutan, gorilla, chimpanzee and, strictly speaking, humans. Taxonomists classify all the descendants of a specified common ancestor as belonging to a monophyletic group (Chapter 23). Thus, humans, blue whales, and hedgehogs are all mammals because all three are descended from the first mammal, the original common ancestor of all mammals. Because we trace our ancestry to the common ancestor of orangutans, gorillas, and chimpanzees, we, too, are a member of the monophyletic great ape group.

Molecular analysis reveals that our lineage split from the chimpanzee lineage about 5-7 million years ago.

Which great ape is most closely related to humans? That is, which is our sister group? Traditional approaches of reconstructing evolutionary history by comparing anatomical features failed to distinguish between two candidates, gorillas and chimpanzees. The implication is that humans, gorillas, and chimpanzees split into separate lineages at about the same time. It was only with the introduction of molecular methods of assessing evolutionary relationships—through the comparison of DNA and protein amino acid sequences from the different species—that we had the answer. Our closest relative is the chimpanzee (Fig. 24.2)—or, more accurately, the chimpanzees, plural, because there are two closely related chimpanzee species, the smaller of which is often called the bonobo.

Just how closely are humans and chimpanzees related? To answer this question, we need to know the timing of the evolutionary split that led along one fork to chimpanzees and along the other to us. As we saw in Chapter 21, DNA sequence differences accumulate between isolated populations or species, and they do so at a more or less constant rate. As a result, the extent of the sequence difference between two species is a good indication of the amount of time they have been separate lineages, that is, the amount of time since their last common ancestor.

The first thorough comparison of DNA molecules between humans and chimpanzees was carried out before the advent of DNA sequencing methods by Mary-Claire King and Allan Wilson at the University of California at Berkeley (**Fig. 24.3**). One of their methods to measure molecular differences between species relied on DNA–DNA hybridization (Chapter 12). Two complementary strands of DNA in a double helix can be separated by heating the sample.

HOW DO WE KNOW?

FIG. 24.3

How closely related are humans and chimpanzees?

BACKGROUND Phylogenetic analysis based on anatomical characteristics had established that chimpanzees are closely related to humans. Mary-Claire King and Allan Wilson used molecular techniques to determine *how* closely related the two species are.

HYPOTHESIS Despite marked anatomical and behavioral differences between the two species, the genetic distance between the two is small, implying a relatively recent common ancestor.

METHOD When two complementary strands of DNA are heated, the hydrogen bonds pairing the two helices are broken at around 95°C and the double helix denatures, or separates (Chapter 12). Two complementary strands with a few mismatches separate at a temperature slightly lower than 95°C because fewer hydrogen bonds hold the helix together. Many more mismatches between the two sequences results in an even lower denaturation temperature. Using hybrid DNA double helices with one strand contributed by each species—humans and chimpanzees, in this case—and determining their denaturation temperature, King and Wilson could infer the genetic distance (the extent of genetic divergence) between the two species.

Perfect complementarity: 95°C denaturation

Some mismatch: 93°C denaturation

More mismatch: 91°C denaturation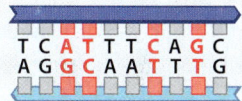

RESULTS King and Wilson found that human–chimpanzee DNA molecules separated at a temperature approximately 1°C less than the temperature at which human–human DNA molecules separate. This difference could be calibrated on the basis of studies of other species whose genetic distances were known from other methods. The DNA of humans differs from that of chimpanzees by about 1%.

CONCLUSION AND INTERPRETATION King and Wilson noted the discrepancy between the extent of genetic divergence (small) and the extent of anatomical and behavioral divergence (large) between humans and chimpanzees. They suggested that one way in which relatively little genetic change could produce extensive phenotypic change is through differences in gene regulation (Chapter 19). A small genetic change in a control region responsible for switching a gene on and off might have major consequences for the organism.

FOLLOW-UP WORK The sequences of the chimpanzee and human genomes allow us to compare the two sequences directly, and these data confirm King and Wilson's observations.

SOURCE King, M. C., and A. C. Wilson. 1975. "Evolution at Two Levels in Humans and Chimpanzees." *Science* 188:107–116.

If the two strands are not perfectly complementary, as is the case if there is a basepair mismatch (for example, a G paired with a T rather than a C), less heat is required to separate the strands.

King and Wilson used this fact to examine the differences between a human strand and the corresponding chimpanzee strand. They inferred the extent of DNA sequence divergence from the melting temperature and made a striking discovery: Human and chimpanzee DNA differ in sequence by just 1%.

King and Wilson's conclusions have been confirmed by subsequent research. Now that we have both the human and chimpanzee genome sequences in hand, we can literally count the differences between the two sequences: The genomes differ by about 1%. The overall amount of difference between human and chimpanzee genomes goes up somewhat if you take into account segments of the genomes that are absent in one or other of the species (that is, sequences where DNA has been inserted or deleted in one lineage or the other).

Because the amount of sequence difference is correlated with the length of time the two species have been isolated, we can convert the divergence results into an estimate of the timing of the split between the human and chimpanzee lineages. That split occurred about 5–7 million years ago. All the extraordinary characteristics that set our species apart from the rest of the natural world—those attributes that are ours and ours alone—arose in just 5–7 million years.

→ **Quick Check 1** Did humans evolve from chimpanzees? Explain.

The fossil record gives us direct information about our evolutionary history.

Molecular analysis is a powerful tool for comparing species and populations within species. It allows us to compare humans and chimpanzees and look at differences among groups of humans or groups of chimpanzees. But we are limited in what we can study because we need samples from living individuals. We can study the end products of evolution but not the intermediate stages, which have long since disappeared and cannot provide us with DNA samples. For a full picture of human evolution, we must turn to fossils (Chapter 23).

For the first several million years of human evolution, all the fossils from the human lineage are found in Africa. This is not surprising. Charles Darwin himself noted that it was likely that the human lineage originated there, as humans' two closest relatives, chimpanzees and gorillas, live only in Africa. The fossil material varies in quality, and a great deal of ingenuity is often required to reconstruct the appearance and attributes of an individual from fragmentary fossil material. It's hard to determine whether two fossil specimens with slight differences belong to the same or different species. And, of course, interbreeding—the criterion that defines a species (Chapter 22)—cannot be applied to fossils. As a result, experts disagree over the details of the human fossil record. However, several robust general conclusions may be drawn.

The members of all the different species in the lineage leading to humans are called **hominins.** Let's start with the earliest known member of the hominin lineage, *Sahelanthropus tchadensis* (**Fig. 24.4**). Discovered in Chad in 2002, the skull of *S. tchadensis* combines both modern (human) and ancestral features. *S. tchadensis*, which has been dated to about 7 million years ago and has a chimpanzee-sized brain but hominin-type brow ridges, probably lived shortly after the split between the hominin and chimpanzee lineages, although this conclusion is controversial.

An important early hominin, dating from about 4.4 million years ago, is a specimen of *Ardipithecus ramidus* from

FIG. 24.4 **The skull of *Sahelanthropus tchadensis*.** This skull, which is about 7 million years old and was found in Chad, has both human and chimpanzee features.

FIG. 24.5 **Lucy, a specimen of *Australopithecus afarensis*.** Lucy was fully bipedal, establishing that the ability to walk upright evolved at least 3.2 million years ago.

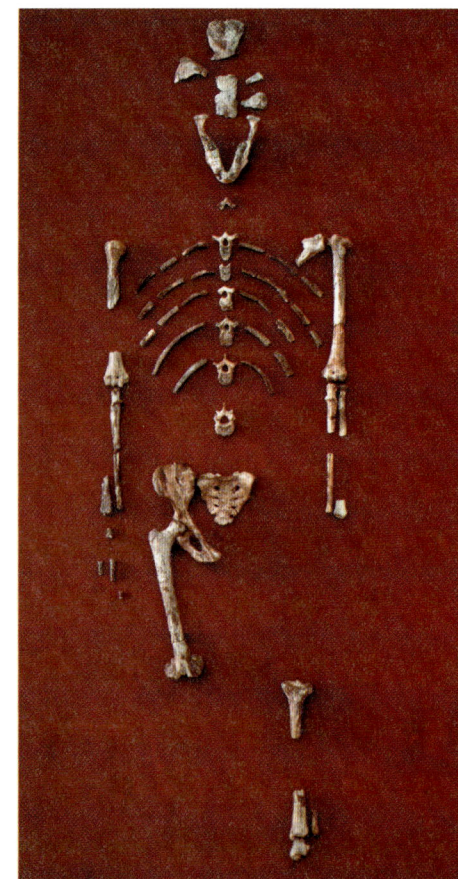

Ethiopia. This individual, known as **Ardi,** was capable of walking upright, using two legs on the ground but all four limbs in the trees. An unusually complete early hominin fossil, **Lucy,** found in 1974 at Hadar, Ethiopia, represents the next step in the evolution of hominin gait: She was fully **bipedal**, habitually walking upright (**Fig. 24.5**). Her name comes from the Beatles' song "Lucy in the Sky with Diamonds," which was playing in the paleontologists' field camp when she was unearthed. This fossil dates from around 3.2 million years ago. Lucy was a member of the species *Australopithecus afarensis* and was much smaller than modern humans, less than 4 feet tall, and had a considerably smaller brain (even when the difference in body size is taken into consideration). In many ways, then, Lucy was similar to the common ancestor of the human and chimpanzee lineages, except that she was bipedal.

Remarkably, the hominin lineage produced many different species in Africa, so that there were at times as many as three species living at the same time. **Fig. 24.6** shows the main hominin species and the longevity of each in the fossil record, along with suggested evolutionary relationships among the species. Note that all hominins have a common ancestor, but not all of these lineages lead to modern humans, producing instead other branches of the hominin tree that ultimately went extinct.

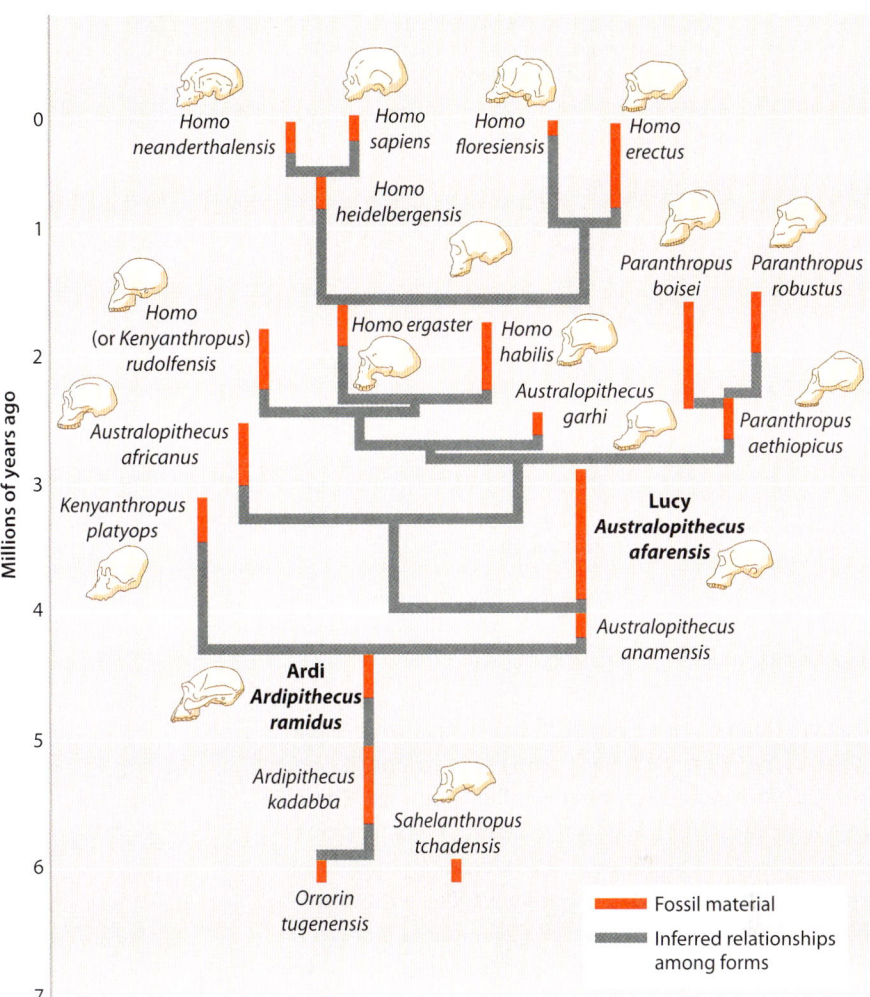

FIG. 24.6 **Hominin lineages.** Three main lineages, *Ardipithecus, Australopithecus,* and *Homo* existed at various and sometimes overlapping times in history. The *Homo* lineage led to modern humans, *Homo sapiens.*

A number of trends can be seen when we look over the entire record. Body size increased, and most striking is the increase in size of the cranium and therefore, by inference, of the brain, as shown in **Fig. 24.7**.

The fossil record indicates that about 2 million years ago, the hominin lineage ventured out of Africa. The first hominin that left Africa is sometimes called *Homo ergaster* and sometimes *Homo erectus.* The confusion stems from the controversies that surround the naming of fossil species. Some researchers contend that *H. ergaster* is merely an early form of *H. erectus.* We designate this first hominin *Homo ergaster* and a later, descendant species, *H. erectus* (see Fig. 24.6). The naming details, however, are relatively unimportant. What matters is that some hominins first left Africa about 2 million years ago. Fossils have been found throughout Eurasia, though not in Australia or the Americas. It was from relatives of *Homo ergaster* that our species, *Homo sapiens,* derived.

Another species closely related to *Homo sapiens* was *Homo neanderthalensis,* whose fossils appear in Europe and the Middle East. Thicker boned than us, and with flatter heads that contained brains about the same size as, or slightly larger than, ours, **Neanderthals** first appeared in the fossil record around 600,000 years ago and disappeared around 30,000 years ago. As we will see, genetic analysis suggests that this disappearance was perhaps not as complete as the fossil record suggests.

Another species that probably ultimately derived from the original *H. ergaster* emigration from Africa became extinct only about 12,000 years ago. This was *H. floresiensis,* known popularly as the Hobbit. *H. floresiensis* is peculiar: Limited to the Indonesian island of Flores, adults were only just over 3 feet tall. Some have suggested that *H. floresiensis* is not a genuinely distinct species, but, rather, is an aberrant *H. sapiens.* Plenty of evidence, however, suggests that *H. floresiensis* is a distinct species derived from an archaic *Homo* species, probably *H. erectus.* Mammals often evolve small body size on islands because of the limited availability of food.

FIG. 24.7 Increase in brain size (as inferred from fossil cranium volume) over 3 million years.

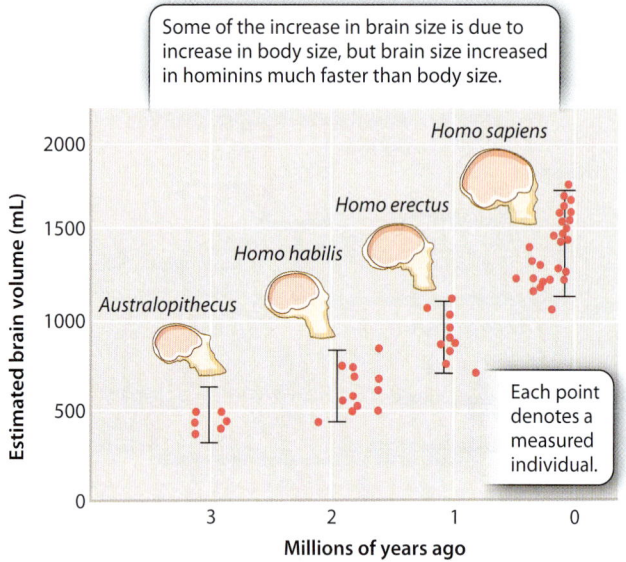

24.2 AFRICAN ORIGINS

For a long time, it was argued that modern humans, *Homo sapiens*, derive from the *Homo ergaster* populations that spread around the world starting about 2 million years ago. This idea is called the **multiregional hypothesis** of human origins because it implies that different *Homo ergaster* populations throughout the Old World evolved in parallel, with some limited gene flow among them, to each produce modern *H. sapiens* populations. In short, modern human traits evolved convergently in multiple populations (Chapter 23). This idea was overturned in 1987 by another study from Allan Wilson's laboratory, which instead suggested that modern humans arose much more recently from *Homo ergaster* descendants (sometimes called *Homo heidelbergensis*; see Fig 24.6) in Africa, about 200,000 years ago. This newer idea is the **out-of-Africa hypothesis** of human origins.

→ **Quick Check 2** What do the multiregional and out-of-Africa hypotheses predict about the age of the common ancestor of all humans living today?

Studies of mitochondrial DNA reveal that modern humans evolved in Africa.

To test these two hypotheses about human origins, Rebecca Cann, a student in Allan Wilson's laboratory, chose to analyze DNA sequences to reconstruct the human family tree. Specifically, she studied sequences of a segment of mitochondrial DNA from people living around the world (**Fig. 24.8**).

HOW DO WE KNOW?

FIG. 24.8

When and where did the most recent common ancestor of all living humans live?

BACKGROUND With the availability of DNA sequence analysis tools, it became possible to investigate human prehistory by comparing the DNA of living people from different populations.

HYPOTHESIS The multiregional hypothesis of human origins suggests that our most recent common ancestor was living at the time *Homo ergaster* populations first left Africa, about 2 million years ago.

METHOD Rebecca Cann compared mitochondrial DNA (mtDNA) sequences from 147 people from around the world. This approach required a substantial amount of mtDNA from each individual, which she acquired by collecting placentas from women after childbirth. Instead of sequencing the mtDNA, Cann inferred differences among sequences by digesting the mtDNA with different restriction enzymes, each of which cuts DNA at a specific sequence (Chapter 12). If the sequence is present, the enzyme cuts. If any base in the recognition site of the restriction enzyme has changed in an individual, the enzyme does not cut. The resulting fragments were then separated using gel electrophoresis. By using 12 different enzymes, Cann was able to assay a reasonable proportion of all the mtDNA sequence variation present in the sample.

Here, we see a sample dataset for four people and a chimpanzee. There are seven varying restriction sites:

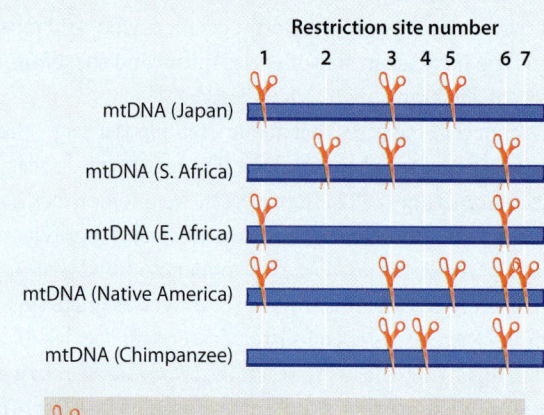

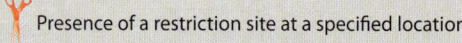

ANALYSIS The data were converted into a family tree by using shared derived characters, described in Chapter 23. By mapping the pattern of changes in this way, the phylogenetic relationships of the mtDNA sequences can be reconstructed. For example, the derived "cut" state at restriction site 1 shared by the East African, Japanese, and Native American sequences implies that the three groups had a common ancestor in which the mutation that created the new restriction site occurred.

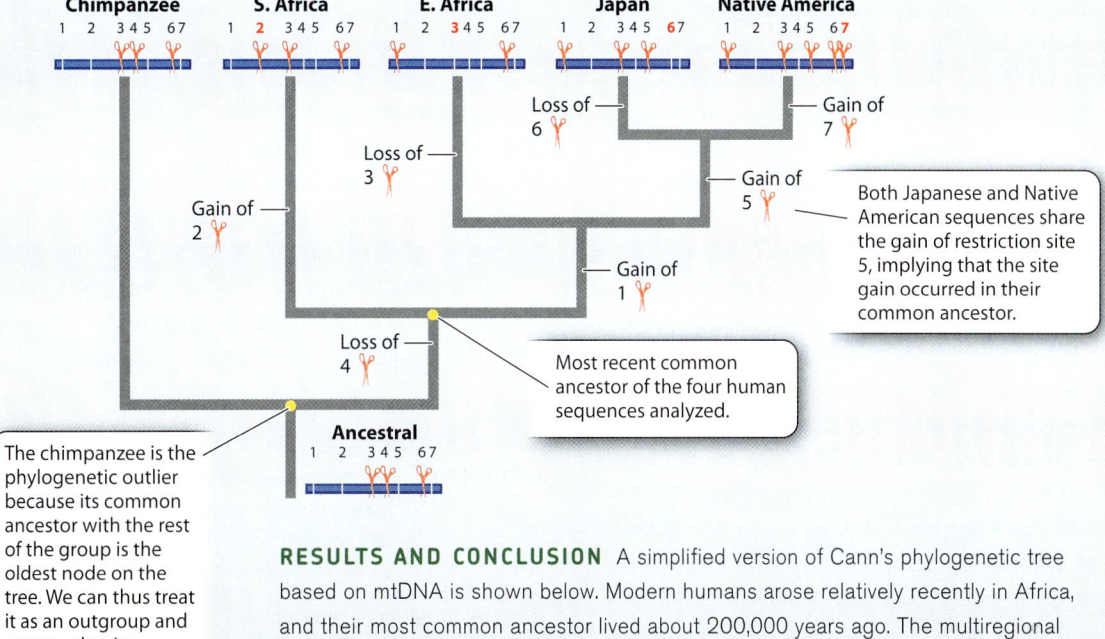

RESULTS AND CONCLUSION A simplified version of Cann's phylogenetic tree based on mtDNA is shown below. Modern humans arose relatively recently in Africa, and their most common ancestor lived about 200,000 years ago. The multiregional hypothesis is rejected. Also, the data revealed that all non-Africans derive from within the African family tree, implying that *H. sapiens* left Africa relatively recently. The Cann study gave rise to the out-of-Africa theory of human origins.

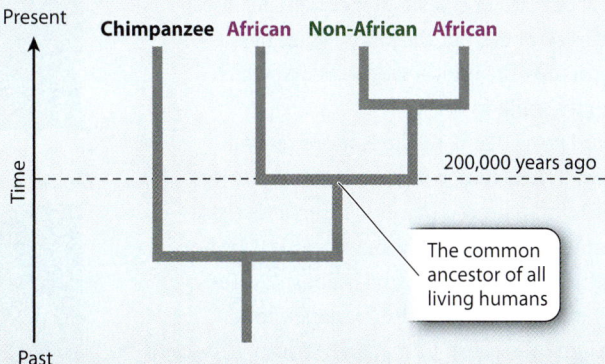

FOLLOW-UP WORK This study set the stage for an explosion in genetic studies of human prehistory that used the same approach of comparing sequences in order to identify migration patterns. Today, with molecular tools vastly more powerful than those available to Cann, studies of the distribution of genetic variation are giving us a detailed pattern, even on relatively local scales, of demographic events.

SOURCE Cann, R. L., M. Stoneking, and A. C. Wilson. 1987. "Mitochondrial DNA and Human Evolution." *Nature* 325:31–36.

Mitochondrial DNA (mtDNA) is a small circle of DNA, about 17,000 base pairs long, found in every mitochondrion (Chapter 17). Cann chose to study mtDNA for several reasons. Although a typical cell contains a single nucleus with just two copies of nuclear DNA, each cell has many mitochondria, each carrying multiple copies of mtDNA. Thus, mtDNA is much more abundant than nuclear DNA and therefore easier to extract. Most important, however, is its mode of inheritance. All your mtDNA is inherited from your mother in the egg she produces because sperm do not contribute mitochondria to the zygote. This means that there is no opportunity for genetic recombination between different mtDNA molecules, so the only way in which sequence variation can arise is through mutation. In nuclear DNA, by contrast, differences between two sequences can be introduced through both mutation and recombination. Recombination obscures genealogical relationships because it mixes segments of DNA with different evolutionary histories.

Mitochondrial DNA sequences offered Cann a clear advantage. With sequence information from the mtDNA of 147 people from around the world, Cann reconstructed the human family tree (Fig. 24.8). The tree contained two major surprises. First, the two deepest branches—the ones that come off the tree earliest in time—are African. All non-Africans are branches off the African tree. The implication is that *Homo sapiens* evolved in Africa and only afterward did populations migrate out of Africa and become established elsewhere. This finding contradicted the expectations of the multiregional theory, which predicted that *Homo sapiens* evolved independently in different locations throughout the Old World.

Second, the tree is remarkably shallow, that is, even the most distantly related modern humans have a relatively recent common ancestor. By calibrating the rate at which mutations occur in mtDNA, Cann was able to estimate the time back to the common ancestor of all modern humans as about 200,000 years. Subsequent analyses have somewhat refined this estimate, but the key message has not changed: The data contradict the multiregional theory, which predicted a number closer to 2 million years.

Two hundred thousand years ago is 10 times more recent than 2 million years ago, so this change represents a revolution in our understanding of human prehistory. But surely it is risky to make such major claims on the basis of a single study. Maybe there's something peculiar about mtDNA, and it doesn't offer an accurate picture of the human past. To find independent evidence for a recent origin of modern humans in Africa, another dataset was required. One such dataset comes from studies of the Y chromosome, another segment of human DNA that does not undergo recombination (Chapter 17). When an approach similar to Cann's was used to reconstruct the human family tree using Y chromosome DNA sequences, the result was completely in agreement with the mtDNA result: The human family is young, and it arose in Africa.

Neanderthals disappear from the fossil record as modern humans appear but have contributed to the modern human gene pool.

If we exclude *H. floresiensis*, the last of the nonmodern humans were the Neanderthals (**Fig. 24.9**), who lived in Europe and Western Asia until about 30,000 years ago. What happened? Were they eliminated by the first population of *Homo sapiens* to arrive in Europe (a group known as **Cro-Magnon** for the site in France from which specimens were first described)? Or did they interbreed with the Cro-Magnons, and, if so, does that mean that Neanderthals should be included among our direct ancestors? The controversy surrounding this question flares up from time to time when paleontologists find specimens that have anatomical features intermediate between the heavyset Neanderthal form and the lighter human one.

In 1997, we thought we had a clear answer. Matthias Krings working in Svante Pääbo's lab in Germany extracted intact stretches of mtDNA from 30,000-year-old Neanderthal material. If Neanderthals had interbred with our ancestors, we would expect to see evidence of their genetic input in the form of Neanderthal mtDNA sequences in the modern human gene pool. However, the

---

FIG. 24.9 (a) *Homo sapiens* and (b) *Homo neanderthalensis*. Neanderthals disappeared from Europe about 30,000 years ago when *H. sapiens*, called Cro-Magnon after the site at which they were discovered, first appeared in Europe.

a. Cro-Magnon (modern)

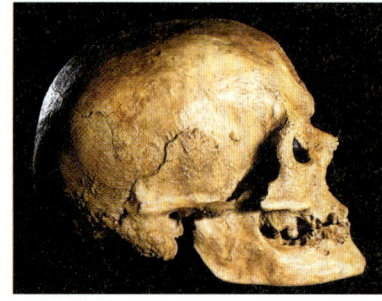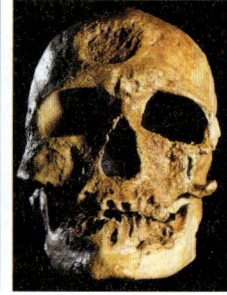

b. Neanderthal

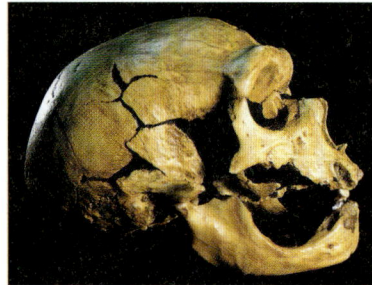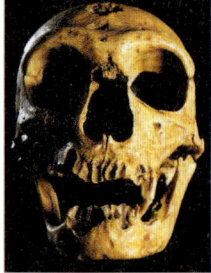

Neanderthal brains were slightly larger than ours, but their skulls were shaped differently. The Neanderthal forehead was much less pronounced than ours, and the skull in general lower.

Neanderthal sequence was strikingly different from that of modern humans. That we see nothing even close to the Neanderthal mtDNA sequence in modern humans strongly suggested that Neanderthals did not interbreed with our ancestors (**Fig. 24.10a**).

This conclusion, however, has been reversed following remarkable technological advances that permitted Pääbo and others to sequence the entire Neanderthal genome (rather than just its mtDNA). Careful population genetic analysis revealed that our ancestors did interbreed with Neanderthals, and that 1% to 4% of the genome of every non-African is Neanderthal-derived. A scenario emerged: As populations of our ancestors headed from Africa into the Middle East, they encountered Neanderthals, and it was then, before modern humans had spread out across the world, that interbreeding took place (**Fig. 24.10b**).

How can we reconcile these two apparently contradictory results? The mtDNA study indicates there was no genetic input from Neanderthals in modern humans, but the whole-genome study suggests that in fact there was. One possible solution lies in the difference in patterns of transmission between mtDNA and genomic material (Chapter 17). Recall that mtDNA is maternally inherited because sperm do not contribute mitochondrial material. Your mtDNA, whether you are male or female, is derived solely from your mother. One possibility is that the Neanderthal mtDNA lineage has been lost through genetic drift (Chapter 21). An alternative is that the discrepancy in the ancient DNA stems from a sex-based difference in interbreeding. If Neanderthal females did not interbreed with our ancestors, Neanderthal mtDNA would not have entered the modern human gene pool. If only male Neanderthals interbred with our ancestors, we would expect exactly the pattern we observe: no Neanderthal mtDNA in modern humans, but Neanderthal genomic DNA in the modern human gene pool.

It is likely that our interpretation of these events will always be speculative. The basic result, however, forces us to reimagine the ancestry of a large portion of the human population. Every non-African on the planet is part Neanderthal, albeit a very small part.

→ **Quick Check 3** Explain how genetic data suggest that only male Neanderthals interbred with our ancestors.

FIG. 24.10 Evidence for and against early interbreeding between Neanderthals and early humans. Studies using only mtDNA (a) did not support the idea of interbreeding, but later studies using whole genomes (b) did.

a. mtDNA phylogeny for Neanderthals and modern humans

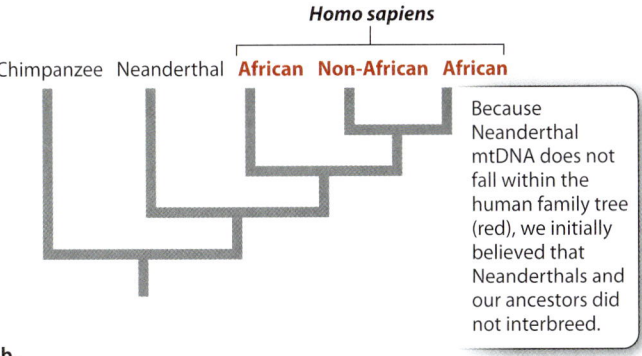

Because Neanderthal mtDNA does not fall within the human family tree (red), we initially believed that Neanderthals and our ancestors did not interbreed.

b.

However, studies of the full Neanderthal genome revealed that there was interbreeding, probably in the Middle East as *H. sapiens* first emerged from Africa about 60,000 years ago.

— Initial *H. sapiens* out-of-Africa migration
— Expansion of *H. sapiens* populations after interbreeding with Neanderthals

24.3 DISTINCT FEATURES OF OUR SPECIES

Many extraordinary changes in anatomy and behavior occurred in the 5–7 million years since our lineage split from the lineage that gave rise to the chimpanzees. Fossils tell us a great deal about those changes, especially when high-quality material such as Lucy or Ardi is available, but in general this is an area in which fossils are hard to come by and there is a lot of speculation. Speculation is especially common when we try to explain the reasons behind the evolution of a particular trait. Why, for example, did language evolve? It is easy enough to think of a scenario in which natural selection favors some ability to communicate—maybe language arose to facilitate group hunting. There are plenty of plausible ideas on the subject, but, in most cases, no evidence, so it is impossible to distinguish among competing hypotheses. We can, however, be confident that the events that produced language occurred in Africa, and, through paleontological studies of past environments, we can conclude that humans evolved in an environment similar in many ways to today's East African savanna.

Bipedalism was a key innovation.

The shift from walking on four legs to walking on two was probably one of the first changes in our lineage. Many primates are partially bipedal—chimpanzees, for example, may be observed knuckle-walking—but ours is the only species that is wholly bipedal. As we've seen, Lucy, dating from about 3.2 million years ago, was already bipedal, and 4.4 million years ago, Ardi was partially bipedal. We can further refine our estimate of when full bipedalism arose from the evidence of a trace fossil. A set of 3.5-million-year-old fossil footprints discovered in Laetoli, Tanzania, reveal a truly upright posture.

Becoming bipedal is not simply a matter of standing up on hind legs. The change required substantial shifts in a number of basic anatomical characteristics, described in **Fig. 24.11**.

Why did hominins become bipedal? Maybe it gave them access to berries and nuts located high on a bush. Maybe it allowed them to scan the vicinity for predators. Maybe it made long-distance travel easier. And, very important, bipedalism freed up our ancestors' hands. No longer did they need their hands for locomotion, so for the first time there arose the possibility that specialized hand function could evolve. Most primates have some kind of opposable thumb, but the human version is much more refined. The human thumb has three muscles that are not present in the thumb of chimpanzees, and these allow much finer motor control of the thumb. Tool use, present but crude in chimpanzees, can be much more subtle and sophisticated with a human hand.

Bipedalism also made it possible to carry material over long distances. Hominins could then set up complex foraging strategies whereby some individuals supply others with resources. Exactly when sophisticated tool use arose in our ancestors is controversial, but it is indisputable that bipedalism contributed. Similarly, the ability to manipulate food with the hands, and to carry material using the hands rather than the mouth, likely permitted the evolution of the human jaw—indeed, the entire facial structure—in such a way that language became a possibility.

Adult humans share many features with juvenile chimpanzees.

King and Wilson's 1975 discovery that the DNA of humans and chimpanzees are 99% identical has recently been confirmed by DNA sequence comparison of the human and chimpanzee genomes. It turns out that the two genomes are extraordinarily similar: All our genes are also present in the chimpanzee and their sequences are extremely similar, suggesting that the functions of the proteins they code for are the same. If we are so similar, how can we account for the extensive differences between the two species?

In their original paper, King and Wilson suggested that much of the most significant evolution along the hominin lineage came about through changes in the regulation of genes (Chapter 19). A small change—one that causes a gene to be transcribed at a different stage of development, for example—can have a major effect. In other words, King and Wilson introduced a model whereby small changes in the software could have a major impact even though the basic hardware is the same.

One of the gene-regulatory pathways that changed may be responsible for human **neoteny**, the long-term evolutionary process in which the timing of development is altered so that a sexually mature organism still retains the physical characteristics of the juvenile form. As early as 1836, French naturalist Étienne Geoffroy Saint-Hilaire noted that a young orangutan on exhibit in

FIG. 24.11 The shift from four to two legs. Anatomical changes were required for the shift, especially in the structure of the skull, spine, legs, and feet.

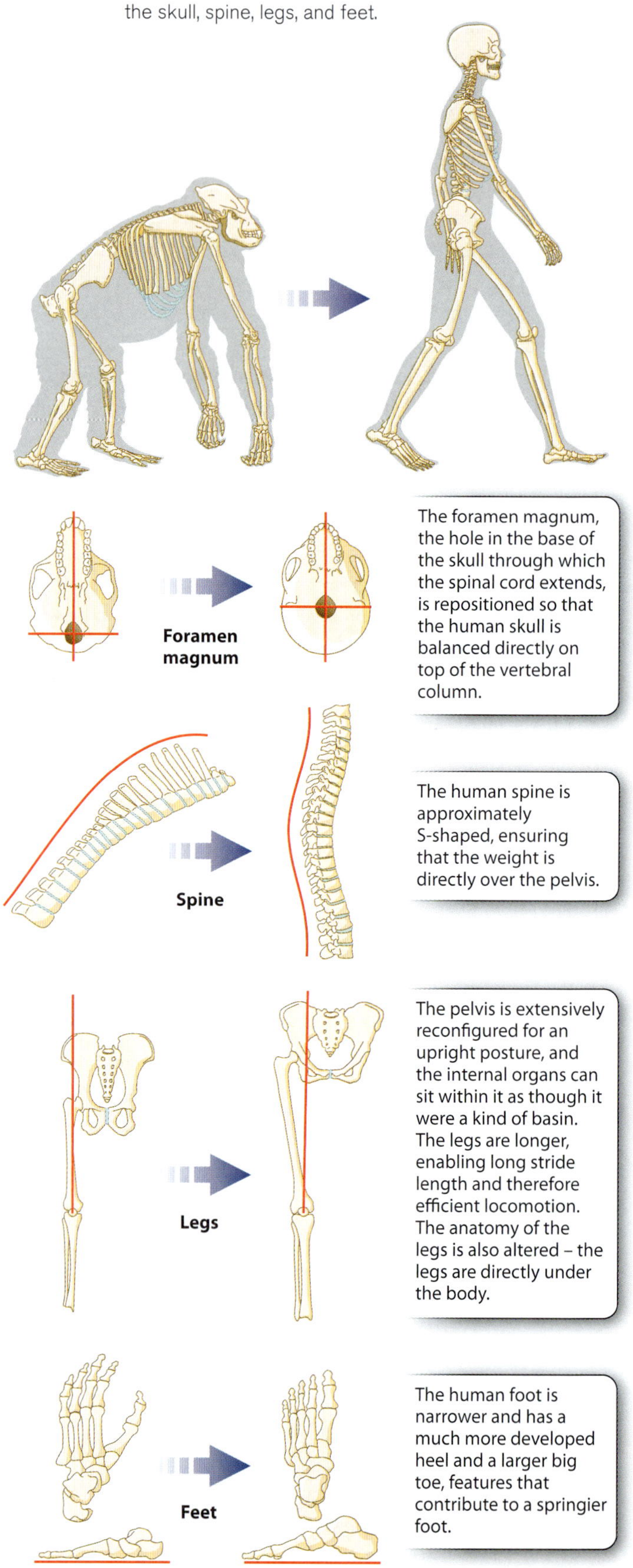

Foramen magnum: The foramen magnum, the hole in the base of the skull through which the spinal cord extends, is repositioned so that the human skull is balanced directly on top of the vertebral column.

Spine: The human spine is approximately S-shaped, ensuring that the weight is directly over the pelvis.

Legs: The pelvis is extensively reconfigured for an upright posture, and the internal organs can sit within it as though it were a kind of basin. The legs are longer, enabling long stride length and therefore efficient locomotion. The anatomy of the legs is also altered – the legs are directly under the body.

Feet: The human foot is narrower and has a much more developed heel and a larger big toe, features that contribute to a springier foot.

FIG. 24.12 A juvenile and an adult orangutan. Humans look more like juvenile orangutans than adult orangutans, suggesting that humans may be neotenous great apes.

Paris looked considerably more like a human than the adult of its own species (**Fig. 24.12**).

Several human attributes support this model, including our large heads (and correspondingly large brains), a feature of juvenile great apes. A second human attribute is our lack of hair. The juveniles of other great apes are not as hairless as humans, but they're considerably less hairy than adult apes. A third attribute is the position of the foramen magnum at the base of the skull. In primate development, the foramen magnum starts off in the position it occupies in the adult human and then, in nonhuman great apes, migrates toward the back of the skull. Adult humans have retained the juvenile great ape foramen magnum position. Finally, it has been suggested that our mentality, with its questioning and playfulness, is equivalent in many ways to that of a juvenile ape rather than to that of the comparatively inflexible adult ape.

In keeping with King and Wilson's idea, this shift in development could conceivably be achieved with relatively little genetic change. All that it might take would be a few changes to the regulatory switches that control the timing of development.

Humans have large brains relative to body size.

In mammals, brain size is typically correlated with body size. Humans are relatively large-bodied mammals, but our brains are large for even our body size (**Fig. 24.13**). It is our large brains that have allowed our species' success, extraordinary technological achievements, and at times destructive dominion over the planet.

What factors promoted the evolution of the large human brain? Again, speculation is common, and it is unlikely we will ever have a definitive answer. Because a large brain is metabolically expensive to produce and to maintain, we can conclude that natural selection must have acted in favor of large brains. What are the selective factors? Here are some possibilities:

- *Tool use.* Bipedalism permitted the evolution of manual dexterity, which in turn requires a complex nervous organization if delicate hands are to be useful.

- *Social living.* Groups require coordination, and coordination requires some form of communication and the means of integrating and acting upon the information conveyed. One scenario, for example, sees group hunting as critical in the evolution of the brain: Natural selection favored those individuals who cooperated best as they pursued large prey.

- *Language.* Did the evolution of language drive the evolution of large brains? Or did language arise as a result of having large brains? Again, we will probably never know, but it is tempting to speculate that the brain and our extraordinary powers of communication evolved in concert.

Probably, as with bipedalism, there was no single factor but a mix of elements that worked together to result in the evolution of a large brain size. We tend to focus on brain size because it is a convenient stand-in for mental power and because we can measure it in the fossil record by making the reasonable assumption that the volume of a fossil's cranium reflects the size of its brain. However,

FIG. 24.13 Brain size plotted against body size for different species. Humans have large brains for their body size.

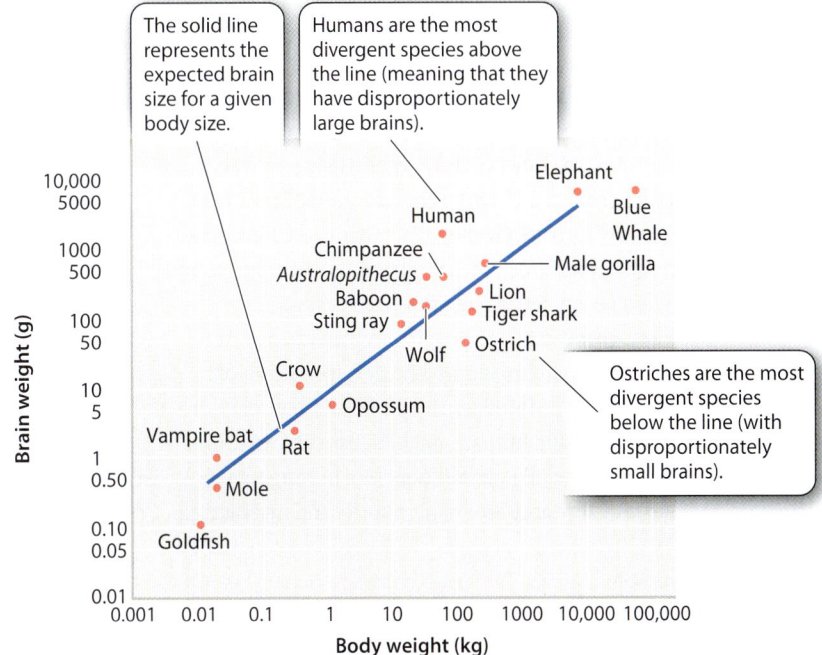

it should be emphasized that our minds are not the products solely of larger brains. Rather, the reorganization of existing structures and pathways is the more important product of the evolution of the brain. Not only was the brain expanding over those 7 million years of human evolution, it was also being rewired.

The human brain evolved through natural selection. What we have today is a learning machine capable of generating many more skills and abilities than just those that directly enhance evolutionary fitness. For example, let's assume that the brain evolved to make group hunting more efficient. The result is that the brain does indeed enhance our ability to hunt together, but there are numerous by-products of this brain that take its abilities beyond just hunting. Playing the piano, for instance, has nothing to do with group hunting, and yet the brain allows us to do it. A brain evolved for a mundane task like group hunting lies at the heart of much that is wonderful about humanity.

The human and chimpanzee genomes help us identify genes that make us human.

The key genetic differences between humans and chimpanzees must lie in the approximately 1% of DNA sequence that differs between the two genomes. What do we see when we compare human and chimpanzee genomes?

A gene that has attracted a lot of interest is *FOXP2*, a member of a large family of evolutionarily conserved genes that encode transcription factors that play important roles in development. Individuals with mutations in *FOXP2* often have difficulty with speech and language. Interestingly, other animals whose *FOXP2* has been knocked out also have communication impairments. Songbirds are less capable of learning new songs, and the ultrasonic "songs" of mice are disrupted. In addition to its effects on the brain, *FOXP2* plays a role in the development of many tissues.

Studies of the gene's amino acid sequence reveal a pattern of extreme conservation. The sequences in mice and chimpanzees, whose most recent common ancestor lived about 75 million years ago, differ by a single amino acid (**Fig. 24.14**). However, two amino acids are present in humans but absent in chimpanzees. Studies of Neanderthal DNA have shown that Neanderthals possessed the modern human version of *FOXP2*. The presence of the same version of the gene in both species suggests that the differences arose in the hominin line before the split between our species and Neanderthals, which occurred at least 600,000 years ago.

FOXP2 gives us a glimpse of the genetic architecture of the traits that are likely to be important in the determination of "humanness." Those two differences in amino acid sequence are intriguing, but they do not a human make. Substitute the human *FOXP2* gene into a mouse and you will not get a talking mouse. The critical genetic differences are many, subtle, and interacting.

→ **Quick Check 4** The *FOXP2* gene is sometimes called the "language gene." Why is this name inaccurate?

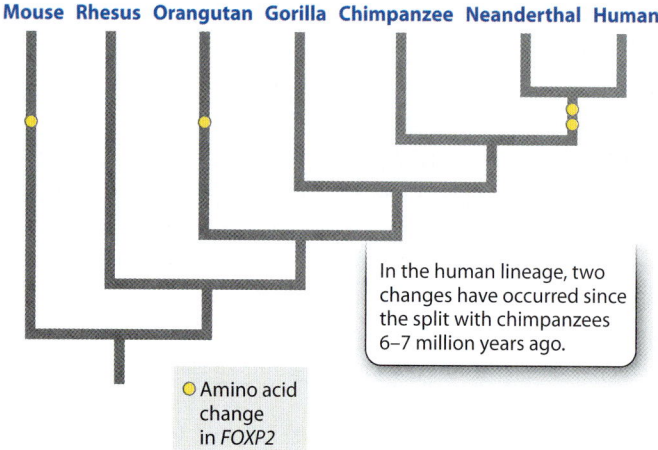

FIG. 24.14 A family tree of mammals showing amino acid changes in *FOXP2*. The gene is highly conserved, but two changes are seen in the Neanderthals and modern humans.

24.4 HUMAN GENETIC VARIATION

So far we have treated humans as all alike, and in many ways, of course, we are. But there are also many differences from one person to the next. Those differences ultimately have two sources: genetic variation and differences in environment (Chapter 18). A person may be born with dark skin, or a person born with pale skin may acquire darker skin—a tan—in response to exposure to sun.

The differences we see from one person to the next are deceptive. Despite appearances, ours is not the most genetically variable species on Earth. While it is certainly true that everyone alive today (except for identical twins) is genetically unique, our species is actually rather low in overall amounts of genetic variation. Modern estimates based on comparisons of many human DNA sequences indicate that, on average, about 1 in every 1000 base pairs differs among individuals (that is, our level of DNA variation is 0.1%). That's about 10 times less genetic variation than in fruit flies (which nevertheless all look the same to us) and about two to three times less than in Adélie penguins, which look strikingly similar to one another (see Fig. 21.1).

Why, then, are we all so phenotypically different if there is so little genetic variation in our species? Given the large size of our genome, a level of variation of 0.1% translates into a great many genetic differences. Our genome consists of approximately 3 billion base pairs, so 0.1% variation means that 3 million bp differ between any two people chosen at random. Many of those differences are in noncoding DNA, but some fall in regions of DNA that encode proteins and therefore influence the phenotype, so there is a fairly large reservoir of genetic variation in humans. When those mutations are reshuffled by recombination, we get the vast array of genetic combinations present in the human population.

The prehistory of our species has had an impact on the distribution of genetic variation.

The reasons for our species' relative lack of genetic variation compared to other species lie in prehistory, and factors affecting the geographical distribution of that variation also lie in the past. Studies of the human family tree initiated by Rebecca Cann's original mtDNA analysis are giving a detailed picture of how our ancestors colonized the planet and how that process affected the distribution of genetic variation across populations today. Detailed analyses of different populations, often using mtDNA or Y chromosomes, allow us to reconstruct the history of human population movements.

As we have seen, *Homo sapiens* arose in Africa. Perhaps 60,000 years ago, populations started to venture out through the Horn of Africa and into the Middle East (**Fig. 24.15**). The first phase of colonization took our ancestors through Asia and into Australia by about 50,000 years ago. Not until about 15,000 years ago did the first modern humans cross from Siberia to North America to populate the New World.

Other colonizations were even later. Despite its closeness to the African mainland, the first humans arrived in Madagascar only about 2000 years ago, and the colonists came from Southeast Asia, not Africa. Madagascar populations to this day bear the genetic imprint of this surprising Asian input. The Pacific Islands were among the last habitable places on Earth to be colonized during the Polynesians' extraordinary seaborne odyssey from Samoa, which began about 2000 years ago. Hawaii was colonized about 1500 years ago, and New Zealand only 1000 years ago.

By evolutionary standards, the beginning of the spread of modern humans out of Africa about 60,000 years ago is very recent. There has therefore been relatively little time for differences to accumulate among regional populations, and most of the variation in human populations today arose in ancestral populations before any humans left Africa. When we compare levels of variation in a contemporary African population to that in a non-African population, like Europeans or Asians, we find there is more variation in the African population. This is because the individuals that left Africa 60,000 years ago to found populations in the Middle East and beyond were a relatively small sample of

FIG. 24.15 Human migratory routes. Tracking the spread of mitochondrial DNA mutations around the globe allows us to reconstruct the colonization history of our species.

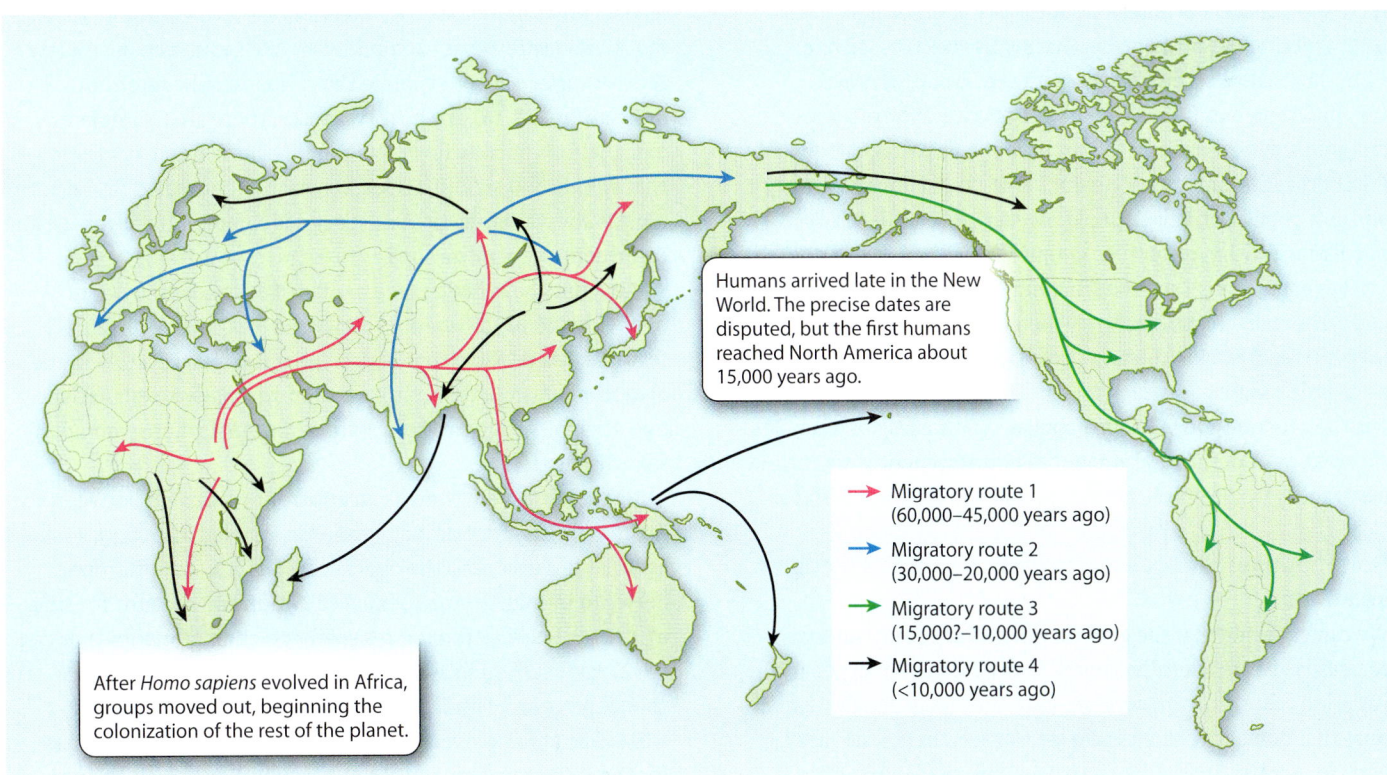

the total amount of genetic variation then present in the human population. Non-African populations therefore began with less genetic variation, and that initial lower variation is reflected in their genetic profiles today.

The recent spread of modern humans means that there are few genetic differences between groups.
Because the out-of-Africa migration was so recent, the genetic differences we see among geographical groups—sometimes called races—are minor. This fact is highly counterintuitive. We see many superficial differences between an African and a Caucasian, such as skin color, facial form, and hair type, and assume that these superficial differences must reflect extensive genetic differences. This assumption made sense when the standard theory about the origin of modern humans was the multiregional one. If European and African populations really had been geographically isolated from each other for as long as 2 million years, then we would expect significant genetic differences between the populations.

We expect isolated populations to diverge genetically over time as different mutations occur in each population. The longer two populations have been isolated from each other, the more genetic differences between them we expect to see (Chapter 21). Isolation lasting 2 million years implies that the differences are extensive, but isolation of 60,000 years suggests they are relatively few. Patterns of genetic variation among different human populations support the hypothesis that human populations dispersed as recently as 60,000 years ago. What we see when we look at genetic markers—variable A's, T's, G's, and C's in human DNA—is that there is indeed very little genetic differentiation by what sometimes is called race.

In short, there's a disconnect: Different groups may *look* very different, but, from a genetic perspective, they're not very different at all. Statistical analyses have shown that approximately 85% of the total amount of genetic variation in humans occurs within a population (for example, Swedes); 8% occurs between populations within races (for example, in Caucasians, between Swedes and Italians); and the remaining 7% occurs between races. The characteristics we use when we assess an individual's ethnicity, such as skin color, eye type, and hair form, are encoded by genetic variants that lie in that 7%. If Earth were threatened with destruction and only one population—Swedes, for example—survived, 85% of the total amount of human genetic variation that exists today would still be present in that Swedish population.

Some human differences have likely arisen by natural selection.
We can conclude that the genetic variants affecting traits we can easily see are overrepresented in the 7% of human genetic variation that sorts by race. If we look at other genetic variants, ones that don't affect traits that we can see, there is no racial pattern. An African is as likely to have a particular base-pair mutation in a gene as a European. So why are visible traits so markedly different among races and others are not? Given the short amount of time (by evolutionary standards) since all *Homo sapiens* were in Africa, it is likely that the differences we see between groups are the product of selection.

People with dark skin tend to live in lower latitudes with high levels of solar radiation, and people with light skin tend to live in higher latitudes with low levels of solar radiation. It is likely that natural selection is responsible for the physical differences between these populations. Assuming that the ancestors of non-African populations were relatively dark-skinned, what selective factors can account for the loss of pigmentation?

A likely factor is an essential vitamin, vitamin D, which is particularly important in childhood because it is needed for the production of bone. A deficiency of vitamin D can result in the skeletal malformation known as rickets. The body can synthesize vitamin D, but the process requires ultraviolet radiation. Heavily pigmented skin limits the entry of UV radiation into cells and so limits the production of vitamin D. This does not present a difficulty in parts of the world where there is plenty of sunlight, but it can be problematical in regions of low sunlight. Presumably, natural selection favored lighter skin in the ancestors of Eurasian populations because lighter skin favored the production of the vitamin.

Some aspects of body shape and size may also have been influenced by natural selection. In hot climates, where dissipating body heat is a priority, a tall and skinny body form has evolved. Exemplified by East African Masai, this body type maximizes the ratio of surface area to volume and thus aids heat loss. In colder climates, by contrast, selection has favored a more robust, stockier body form, as exemplified by the Inuit, who have a low ratio of surface area to volume that promotes the retention of heat (**Fig. 24.16**). In these two cases, these are plausible explanations of body form. We should bear in mind, however, that simple one-size-fits-all explanations of human difference are almost always simplistic. Our species is complex and diverse—and often defies generalizations.

Attempts have been made to identify the adaptive value of obvious visual differences between races, such as facial features. It's possible that natural selection played a role in the evolution of these differences, but an alternative explanation, one originally suggested by Charles Darwin, is more compelling: sexual selection.

As we have seen, there is an apparent mismatch between the extent of difference among groups in visible characters, such as facial features, and the overall level of genetic difference between human groups. Sexual selection can account for this mismatch because it operates solely on characteristics that can readily be seen, such as the peacock's tail. As we learn more about the genetic underpinnings of the traits in question, we will be able to investigate directly the factors responsible for the differences we see among groups. Sexual selection is discussed in Chapter 45.

FIG. 24.16 Evolutionary responses of body shape to climate. (a) Heat-adapted Masai in Kenya and (b) cold-adapted Inuit in Greenland.

❓ CASE 4 Malaria: Co-Evolution of Humans and a Parasite

What human genes are under selection for resistance to malaria?

We see evidence of regional genetic variation in response to local challenges, especially those posed by disease. Malaria, for example, is largely limited to warm climates because it is transmitted by a species of mosquito that can survive only in these regions. Historically, the disease has been devastating in Africa and the Mediterranean. As we saw in Chapter 21, the sickle allele of hemoglobin, *S*, has evolved to be present at high frequencies in these regions because in heterozygotes it confers some protection against the disease. But in homozygotes, the *S* allele is highly detrimental because it causes sickle-cell anemia. Homozygotes for the allele encoding normal hemoglobin are also at a disadvantage because they are entirely unprotected from the parasite.

The *S* allele is beneficial only in the presence of malaria. If there is no malaria in an area, the *S* allele is disadvantageous, so natural selection presumably acted rapidly to eliminate it in the ancestors of Europeans when they arrived in malaria-free regions. The continued high frequency of the *S* allele in Africans, some Mediterranean populations, and in populations descended recently from Africans (such as African-Americans) is, however, a reflection of the response of natural selection to a regional disease.

The hemoglobin genes are not the only genes that are under selection for resistance to malaria. *Glucose-6-phosphate dehydrogenase* (*G6PD*), a gene involved in glucose metabolism, is one of several other genes implicated. People who are heterozygotes for a mutation in the *G6PD* gene—and therefore have a G6PD enzyme deficiency—can develop severe anemia when they eat certain foods (most notably fava beans; hence, the condition is called favism). People who are heterozygotes for a mutation in the *G6PD* gene, however, also have increased resistance to malaria, apparently because they are better at clearing infected red blood cells from their bloodstream. In areas where malaria is common, the advantage of malaria resistance offsets the disadvantage of favism.

Detailed evolutionary analysis of mutations in *G6PD* shows that favism has arisen multiple times, each time selectively favored because of its role in the body's response to the malaria parasite. As expected, favism, like sickle-cell anemia, is mainly a feature of populations in malarial areas or of populations whose evolutionary roots lie in these areas.

24.5 CULTURE, LANGUAGE, AND CONSCIOUSNESS

Culture is generally defined as a body of learned behavior that is socially transmitted among individuals and passed down from one generation to the next. Culture has permitted us to transcend our biological limits. To take a simple example, clothing and ingeniously constructed shelters have enabled us to live in extraordinarily inhospitable parts of the planet, like the Arctic (**Fig. 24.17**). The capacity to innovate coupled to the ability to transmit culture is the key to the success of humans.

Culture changes rapidly.

Culture, of course, changes over time. In many ways, cultural change is responsible for our species' extraordinary achievements. Genetic evolution is slow because it involves mutation followed by changes in allele frequencies that take place over many generations. Cultural change, on the other hand, can occur much more rapidly. Ten years ago, nobody had heard of smartphones, but today, millions of people own them. Or think of the speed at which a change in clothing style—a shift from flared to straight leg jeans, for example—spreads through a population. Today's human population as a whole is genetically almost identical to the population when your grandparents were young. But think of the cultural changes that have occurred in the 50 years or so between your grandparents' youth and your own.

Despite this clear contrast between biological evolution and cultural change, we should not necessarily think of the two processes as independent of each other. Sometimes cultural change drives biological evolution.

A good example of the interaction between cultural change and biological evolution is the evolution of lactose tolerance in populations for which domesticated animals became an important source of dairy product. Most humans are lactose intolerant. Lactose, a sugar, is a major component of mammalian milk, including human breast milk. We have an enzyme, lactase, that breaks down lactose in the gut, but, typically, the enzyme is produced only in the first years of life, when we are breast-feeding. Once a child is weaned, lactase production is turned off. Lactase, however, is clearly a useful enzyme to have if there is a major dairy component to your diet.

Archaeological and genetic analyses indicate that cattle were domesticated probably three separate times in the past 10,000 years: in the Middle East, in East Africa, and in the Indus Valley. In at least two of these cases, there has been subsequent human biological evolution in favor of lactose tolerance, that is, continued lactase production throughout life. Analysis of the gene region involved in switching lactase production on and off has revealed mutations in European lactose-tolerant people that are different from those in African lactose-tolerant people, implying independent, convergent evolution of this trait in the two populations. Furthermore, we see evidence that these changes have evolved very recently, implying that they arose as a response to the domestication of cattle. Here, we see the interaction between cultural change and biological evolution. Biological evolution of continued lactase production has resulted from the change in a cultural practice, namely cattle domestication.

Is culture uniquely human?

Even nonprimates are known to be capable of learning from another member of their own species—culture is not uniquely human. Bird songs, for instance, may vary regionally because juveniles learn the song from local adults. In a famous case, shown in **Fig. 24.18a**, small birds called blue tits (*Parus caeruleus*) in Britain learned from each other how to peck through the aluminum caps of milk bottles left on doorsteps by milkmen to reach the rich cream at the top of the milk. Presumably, one individual discovered accidentally how to peck through a milk bottle cap and the others then imitated it.

Nor is teaching, one way in which culture is transmitted, limited to humans. In teaching, one individual tailors the information available to another in order to facilitate learning. Teaching and imitation together make learning highly efficient. Adult meerkats teach young meerkats how to handle prey by giving them the opportunity to interact with live prey (**Fig. 24.18b**).

The most sophisticated example of nonhuman culture is shown by chimpanzees. Detailed studies of several geographically isolated populations have revealed 39 culturally transmitted behaviors, such as ways of using tools to catch insects, that are specific to a particular population (**Fig. 24.18c**). West African Chimpanzees in Guinea and the Ivory Coast both use tools to help them harvest the same species of army ants for food (and avoid the painful defensive bites of the ants), but each population has its own distinct way of doing this. Chimpanzees, then, are like us in having regional variation in culture.

Despite the cultural achievements of other species, it is clear that culture

FIG. 24.17 The power of culture. Inventions (such as clothing) have allowed our species to expand its geographic range.

FIG. 24.18 Nonhuman culture. (a) English blue tits steal cream from a milk bottle on the doorstep. (b) Adult meerkats teach their young how to handle their prey. (c) Chimpanzees in different populations have devised different ways of using tools to hunt insects.

in *Homo sapiens* is very far removed from anything seen in the natural world. We are amazingly adept at imitation, learning, and acquiring culture. Teaching, learning, and cultural transmission all benefit from another extraordinary human attribute, language.

Is language uniquely human?

The short answer is again no, language is not unique to humans. The waggle dance of a worker bee on its return to the hive communicates information on the direction, distance, and nature of a food resource (Chapter 45). Vervet monkeys use warning vocalizations to specify the identity of a potential predator: They have one call for "leopard," for example, and another for "snake." However, most animal languages are limited. Even if a chimpanzee wished to communicate conversationally as we do, it would not be able to because its larynx is not capable of such subtle vocalizations.

Attempts have therefore been made to explore the linguistic capabilities of chimpanzees by teaching individual chimpanzees sign language or other forms of visual communication (**Fig. 24.19**). In a number of famous experiments, chimpanzees were able to acquire extensive vocabularies of signs and to construct elementary sentences. It is difficult, however, to evaluate the meaning of this research because the animals' achievements are entirely the product of the training program. Perhaps the limitation is most clearly seen when we try to teach an animal to use metaphor. Time, for example, is often described using a spatial metaphor. An event in a time line on the left-hand side of a page is understood to come before an event on the right in time. Not even the best-trained chimpanzees can grasp this concept.

Grammar provides a set of rules that allow the combination of words into a virtually infinite array of meanings. Noam Chomsky, father of modern linguistics, has pointed out that all human languages are basically similar from a grammatical viewpoint. Humans have what he has called a "universal grammar" that would lead a visiting linguist from another planet to conclude that all Earth's languages are dialects of the same basic language. That universal grammar is in some way hard wired into the human

FIG. 24.19 Dr. Susan Savage Rumbaugh with Panbanisha, a female bonobo who learned to communicate using sign language. Chimpanzees and bonobos are able to learn and use sign language to express words and simple sentences.

brain in such a way that every human infant spontaneously strives to acquire language. The specific attributes of the language depend on the baby's environment—a baby in France will learn French, and one in Japan will learn Japanese—but the basic process is similar in every case. Chimpanzees, and the entire natural world, lack the drive toward the acquisition of a grammatical language.

Is consciousness uniquely human?

Descartes famously wrote, "Cogito, ergo sum"—"I think, therefore I am." Can we legitimately rewrite his statement to declare, "Animals think, therefore they are"? With the growth of the animal rights movement, particularly in reference to the treatment of animals in factory farms, this question is of more than academic interest. We now have many examples of animal thinking from a range of species, including, not surprisingly, chimpanzees and gorillas.

Perhaps more remarkable are the examples that come from animal species that are not closely related to us. In experiments carried out by Alex Kacelnik in Oxford, England, a pair of New Caledonian Crows was presented with two pieces of wire, one straight and the other hooked, and offered a food reward that could be obtained only by using the hooked wire. One member of the pair, the male, disregarded the experiment and flew off with the hooked wire. The female, however, having discovered that she could not get the food reward with the straight wire, went to some considerable trouble to bend a hook into the straight wire. She succeeded in getting the food. It is difficult to deny that the crow thought about the problem and was able to solve it, perhaps in the same way as we would. Definitions of consciousness are contested, but, as with language and culture, it seems clear that other species are capable of some form of conscious thought.

The evolutionary biologist Theodosius Dobzhansky once said, "All species are unique, but humans are uniquest." Our "uniquest" status is not derived from having attributes absent in other species, but from the extent to which those attributes are developed in us. Human language, culture, and consciousness are extraordinary products of our extraordinary brains. Nevertheless, as Darwin taught us and as we should never forget, we are fully a part of the natural world.

Core Concepts Summary

24.1 ANATOMICAL, MOLECULAR, AND FOSSIL EVIDENCE SHOWS THAT THE HUMAN LINEAGE BRANCHES OFF THE GREAT APES TREE.

Anatomical features indicate that primates are a monophyletic group that includes prosimians, monkeys, and apes. The apes in turn include the lesser and great apes. The great apes include orangutans, gorillas, chimpanzees, and humans. page 24-1

Analysis of sequence differences between humans and our closest relatives, chimpanzees, indicate that our lineage split from chimpanzees 5-7 million years ago. page 24-3

Lucy, an unusually complete specimen of *Australopithecus afarensis,* demonstrates that our ancestors were bipedal by about 3.2 million years ago. page 24-4

Human lineage fossils occur only in Africa until about 2 million years ago, when *Homo ergaster* migrated out of Africa to colonize the Old World. page 24-5

24.2 PHYLOGENETIC ANALYSIS OF MITOCHONDRIAL DNA AND THE Y CHROMOSOME SHOWS THAT OUR SPECIES AROSE IN AFRICA.

Studies of mitochondrial DNA (mtDNA) suggest that the time back to the common ancestor of modern humans is about 200,000 years, implying that modern humans (*Homo sapiens*) arose in Africa (the out-of-Africa theory). page 24-6

The mtDNA out-of-Africa pattern is supported by Y-chromosome analysis, which also shows a recent African origin of modern humans. page 24-8

Analysis of Neanderthal DNA from 30,000-year-old material indicates that, as the ancestors of non-African humans emigrated from Africa, they interbred with the Neanderthals. page 24-8

Our species originated in Africa and subsequently colonized the rest of the planet, starting about 60,000 years ago. page 24-9

24.3 DURING THE 5–7 MILLION YEARS SINCE THE MOST RECENT COMMON ANCESTOR OF HUMANS AND CHIMPANZEES, OUR LINEAGE ACQUIRED A NUMBER OF DISTINCTIVE FEATURES.

The development of bipedalism involved a wholesale restructuring of anatomy. page 24-9

Neoteny is the acquisition of sexual maturity in an otherwise juvenile state; humans are neotenous, exhibiting many traits as adults that chimpanzees exhibit as juveniles. page 24-10

There are many possible selective factors that explain the evolution of our large brain, including tool use, social living, and language. page 24-11

FOXP2, a transcription factor involved in brain development, may be important in language, as mutations in the gene that encodes FOXP2 are implicated in speech pathologies. page 24-12

24.4 HUMAN HISTORY HAS HAD AN IMPORTANT IMPACT ON PATTERNS OF GENETIC VARIATION IN OUR SPECIES.

Because our ancestors left Africa very recently in evolutionary terms, there has been little chance for genetic differences to accumulate among geographically separated populations. page 24-13

Humans have very little genetic variation, with only about 1 in every 1000 base pairs varying among individuals. page 24-14

Most of the variation segregates within populations. As much as 85% of the total amount of genetic variation in humans can be found within a single population. Only about 7% of human genetic variation segregates between races. page 24-14

Some racial differences, such as skin color and resistance to malaria, have probably arisen via natural selection. page 24-14

Other racial differences have probably arisen via sexual selection. page 24-15

24.5 CULTURE IS A POTENT FORCE FOR CHANGE IN MODERN HUMANS.

Cultural evolution and biological evolution may interact, as in the case of the evolution of lactose tolerance in regions where cattle were domesticated. page 24-16

Other animals possess simple versions of culture, language, and even consciousness, but the capabilities of our species in all three are truly exceptional. page 24-16

Self-Assessment

1. Describe what evidence suggests that chimpanzees are the closest living relatives of humans.

2. Explain the out-of-Africa theory of human origins and how studies of mitochondrial DNA and the Y chromosome support it.

3. List three anatomical differences between chimpanzees and humans, and explain how these changes facilitated walking upright.

4. Given the high genetic similarity of humans and chimpanzees, how can we account for the differences we see between the two species?

5. Describe three possible selective factors underlying the evolution of large brains in our ancestors.

6. Explain how differences among different human populations arose by natural and sexual selection.

7. Provide arguments for and against the idea that culture, language, and consciousness are uniquely human.

Do you understand the chapter's Core Concepts? Log in to check your answers to the Self-Assessment questions, then practice what you've learned and reinforce this chapter's concepts by working through the problems and multimedia tutorials provided there.

www.biologyhowlifeworks.com

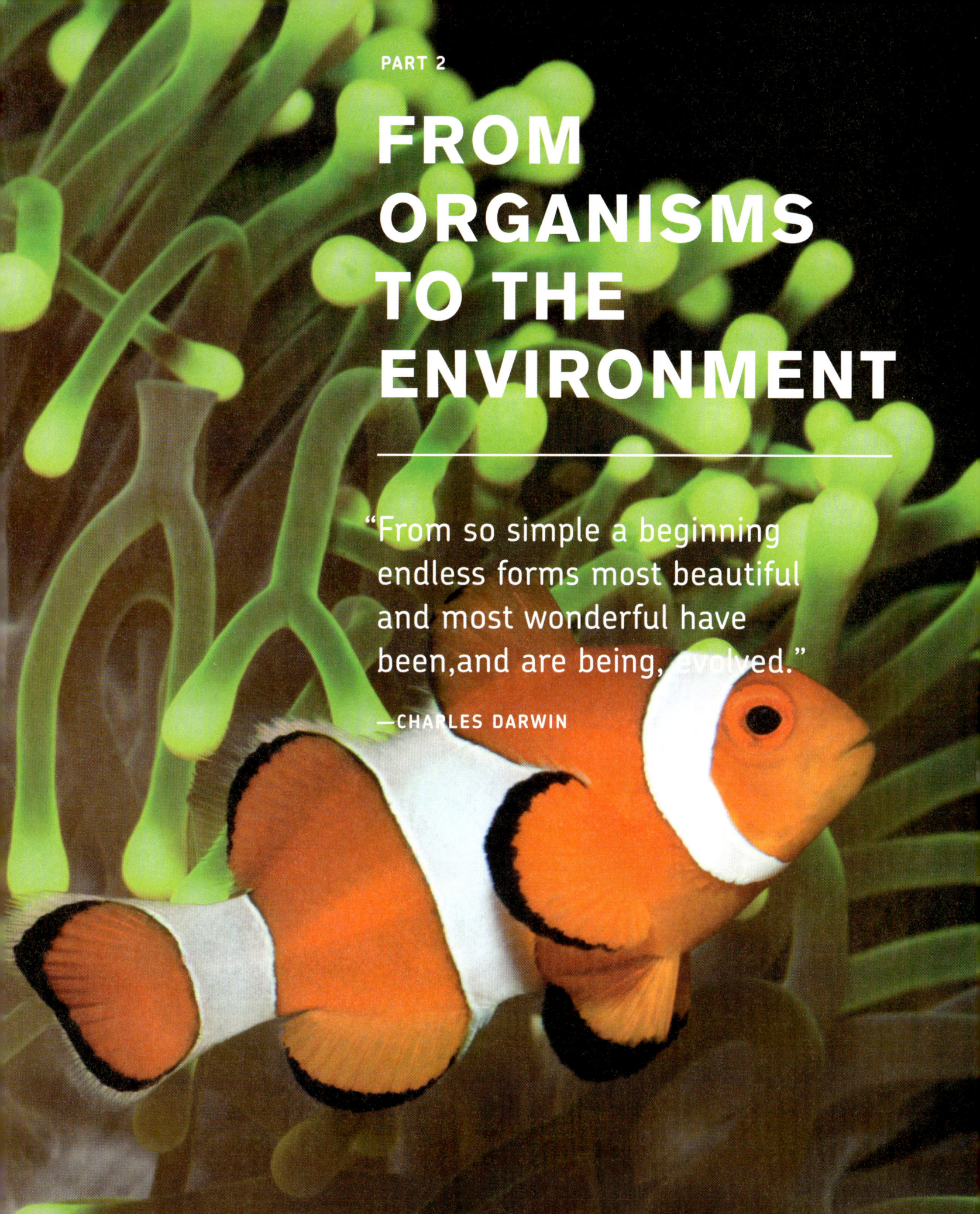

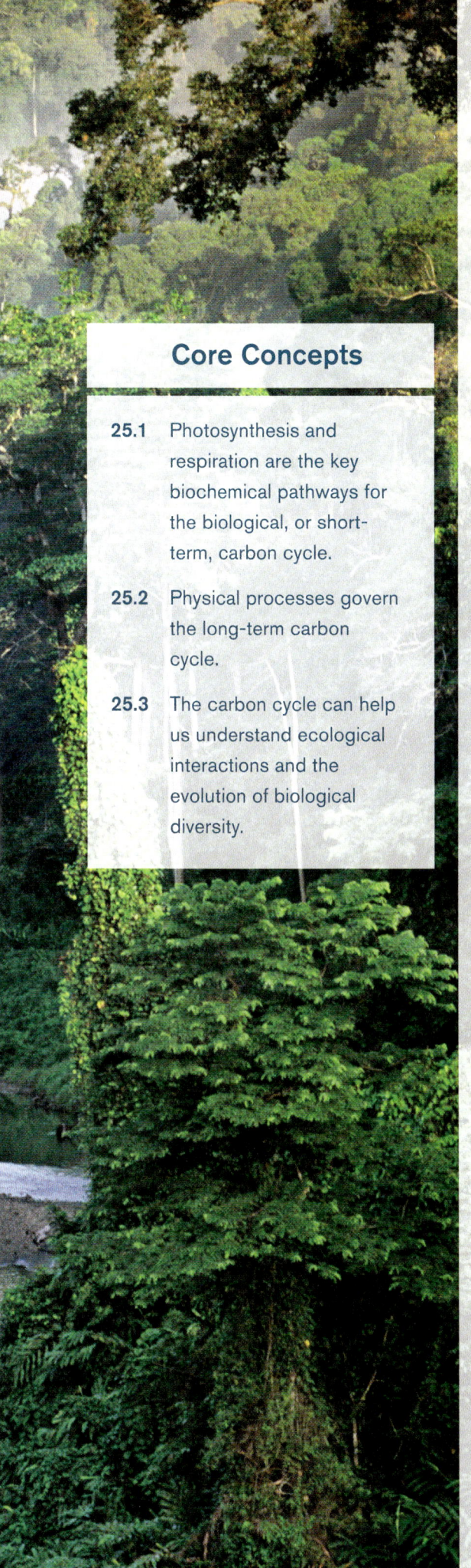

CHAPTER 25

CYCLING CARBON

Core Concepts

25.1 Photosynthesis and respiration are the key biochemical pathways for the biological, or short-term, carbon cycle.

25.2 Physical processes govern the long-term carbon cycle.

25.3 The carbon cycle can help us understand ecological interactions and the evolution of biological diversity.

In 1958, American chemist Charles Keeling began a novel program to monitor the atmosphere. At the time he began his study, scientists had only a vague notion of how carbon dioxide (CO_2) behaves in air. Many thought that CO_2 might vary unpredictably from time to time, and from place to place. Keeling decided to find out if in fact it did. From five towers set high on Mauna Loa in Hawaii, 3400 m above sea level, he sampled the atmosphere every hour and measured its composition with an infrared gas analyzer. Within the first few years of his project, a pattern of seasonal oscillation became apparent: CO_2 concentration in the air reached its annual high point in spring and then declined by about six parts per million (ppm, by volume) to a minimum in early fall (**Fig. 25.1**). Such regular variation was unexpected: What could be causing it?

Perhaps the pattern was local. Airplane traffic to Hawaii peaked in winter, and maybe that was affecting the air around Mauna Loa. As measurements continued through a number of years, however, the pattern persisted, even as trends in air traffic changed. Moreover, monitoring stations in many other parts of the globe gave similar results. Knowing that the Mauna Loa measurements are representative of the atmosphere as a whole allows us to appreciate what an annual variation of 6 ppm really means. With calendar-like regularity, approximately 47 billion metric tons of CO_2 were entering and leaving the atmosphere annually—that's 13 billion metric tons of the element carbon (the rest of the mass is oxygen). To explain such a pattern, we must consider processes that affect the planet as a whole.

A second pattern is evident in Fig. 25.1. Summertime removal of CO_2 from the atmosphere does not quite balance winter increase, with the result that the amount of CO_2 in the atmosphere on any given date is greater than it was a year earlier. Atmospheric CO_2 levels have therefore increased steadily through the period of monitoring. For example, mean annual CO_2 levels, about 315 ppm in 1958, reached 397 ppm by May 2012. This is an increase of more than 25%, and there is no indication that the rise will slow or stop any time soon. What causes this increase? What might be its consequences?

Recognizing the processes reflected in the Keeling curve provides a crucial first step toward understanding the **carbon cycle,** the intricately linked network of biological and physical processes that shuttles carbon among rocks, soil, oceans, air, and organisms. The carbon cycle, in turn, provides a glimpse of the interactions that underpin ecology and promote biological diversity. It focuses our attention on the ways that physical and biological processes interact to determine the properties of environments. And it provides a basis for assessing the role that humans play in our environmental present and future.

25.1 THE SHORT-TERM CARBON CYCLE

What causes atmospheric carbon dioxide to vary through the year and over a timescale of decades? To answer this question, we need to ask what processes introduce CO_2 into the atmosphere and what processes remove it. CO_2 is added to the atmosphere by (1) geologic inputs, mainly from volcanoes and mid-ocean ridges; (2) biological inputs, especially respiration; and (3) human activities, including deforestation and the burning of fossil fuels. Processes that remove CO_2 from the atmosphere include (1) geologic removal, especially by the chemical weathering of rocks, and (2) biological

FIG. 25.1 The Keeling curve. The Keeling curve provides a record of atmospheric carbon dioxide concentrations over half a century, measured from an observatory at Mauna Loa, Hawaii.

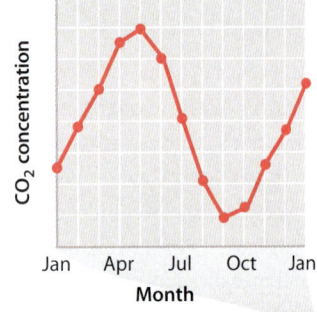

CO₂ concentrations regularly cycle up and down over the course of a year.

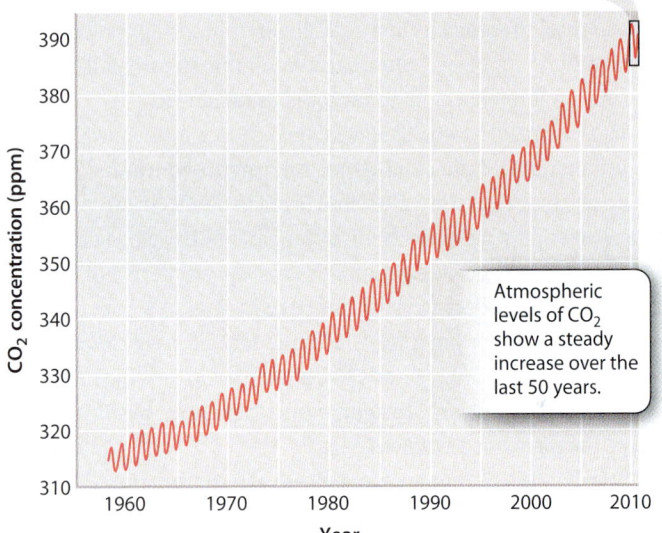

Atmospheric levels of CO₂ show a steady increase over the last 50 years.

removal, mainly through photosynthesis. As we will see, geologic processes play key roles in the carbon cycle on timescales of centuries and longer. Only biology can account for the annual rhythm of the Keeling curve.

Photosynthesis and respiration are key processes in short-term carbon cycling.

To grow, all organisms require both carbon and energy. As we saw in Chapter 8, photosynthetic organisms convert energy from the sun into ATP, which they use to reduce CO_2 to sugar. As photosynthesis pulls CO_2 out of the atmosphere (or water, for aquatic organisms), the carbon is transferred to carbohydrates, and oxygen is given off as a by-product. The summary chemical reaction is shown here:

$$6CO_2 + 6H_2O \rightarrow C_6H_{12}O_6 + 6O_2$$

It is estimated that photosynthetic organisms remove 190 billion metric tons of carbon (in nearly 700 billion metric tons of CO_2) from the atmosphere each year, about 60% by land plants and the rest by phytoplankton and seaweeds in the oceans. This annual photosynthetic removal of CO_2 equals about 25% of the total CO_2 in the atmosphere. In the absence of processes that restore CO_2 to the atmosphere, high rates of photosynthesis would use up the atmospheric reservoir of carbon dioxide in just a few years. But the Keeling curve tells us that atmospheric CO_2 is *not* decreasing rapidly—it is rising. Moreover, the geologic record indicates that photosynthesis has fueled ecosystems for billions of years without causing CO_2 depletion.

What processes return CO_2 to the atmosphere, balancing photosynthetic removal? In Chapter 7, we discussed aerobic respiration. Humans and many other organisms gain both the energy and carbon needed for growth from organic molecules (that is, by consuming food). As explained in Chapter 7, aerobic respiration uses oxygen to oxidize organic molecules to carbon dioxide, converting chemical energy in the organic compounds to ATP for use in cellular processes. Aerobic respiration can be summarized by this chemical equation:

$$C_6H_{12}O_6 + 6O_2 \rightarrow 6CO_2 + 6H_2O$$

Plants respire, too, converting the chemical energy in carbohydrates into useful work. On the present-day Earth, the amount of CO_2 returned annually to the atmosphere by aerobic respiration and related processes is about equal to the amount that plants and algae remove by photosynthesis. That is why photosynthesis does not deplete the CO_2 in air.

As we discussed in Chapter 8, the equations for oxygenic photosynthesis and aerobic respiration mirror each other. Photosynthesis continually converts CO_2 to organic molecules, using energy from the sun, and respiration harvests the chemical energy stored in organic molecules, releasing CO_2 and water in the process. Photosynthesis and aerobic respiration are complementary metabolic pathways: Each uses the products of the other and generates a new supply of substrates for its complement. The result is a cycle (**Fig. 25.2**).

The regular oscillation of CO₂ reflects the seasonality of photosynthesis in the Northern Hemisphere.

Now that we have identified the major processes in the short-term carbon cycle, we can examine what makes atmospheric CO_2 level oscillate seasonally. Except in a narrow band around the equator, photosynthesis is seasonal, with higher rates in the summer and lower rates in the winter. Respiration, in contrast, remains more or less constant through the year.

FIG. 25.2 The short-term carbon cycle. The complementary metabolic processes of photosynthesis and respiration drive the short-term cycling of carbon through the biosphere.

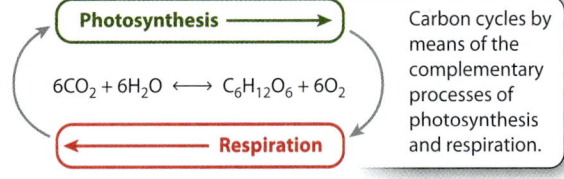

Carbon cycles by means of the complementary processes of photosynthesis and respiration.

Even cursory examination shows that land on Earth is distributed asymmetrically, with more land—and hence more plants—in the Northern Hemisphere than in the Southern Hemisphere. For this reason, global atmospheric CO_2 declines through the northern summer, when the ratio of photosynthesis to respiration is highest, and then increases through fall and winter, when the ratio is reversed. The result is the seasonal oscillation of atmospheric CO_2 levels documented by Keeling.

Human activities play an important role in the modern carbon cycle.

Now let us ask about the second major pattern evident in the Keeling curve. Although CO_2 levels oscillate on an annual basis, the overall pattern is one of sustained increase. During the 1960s, the observed increase was less than 1 ppm each year. In the first decade of the new millennium, it has been closer to 2 ppm. The current level of atmospheric CO_2 is 25% higher than it was 50 years ago. Why has CO_2 input into the atmosphere outstripped removal for the past half century? Should we consider this pattern unusual?

Answering the second question helps us to address the first. Conclusions about whether it is unusual for atmospheric CO_2 to increase 25% in 50 years can be answered only if we can compare the Keeling curve to longer records of atmospheric CO_2 levels. Before Keeling, no one systematically collected air samples. Fortunately, however, nature did it for us. When glacial ice forms in Antarctica and Greenland, it traps tiny bubbles of air. Each year, snowfall gives rise to a new layer of ice, and as layers accumulate through time their bubbles preserve a history of the atmosphere. **Fig. 25.3** shows the amount of CO_2 in air bubbles trapped in

---

HOW DO WE KNOW?

FIG. 25.3

How much CO_2 was in the atmosphere 1000 years ago?

BACKGROUND Direct measurements of the atmosphere show that CO_2 has increased by 25% over the past 50 years. To know whether or not such a change is unusual, we need to know CO_2 levels over a longer time interval.

OBSERVATION Scientists recognized that the snow that accumulates on glaciers to form new layers of ice is initially full of small spaces that are in contact with the air. Through time, as snow changes to ice, these spaces become sealed off from the surrounding environment and form bubbles, preserving a minute sample of air at the time of ice formation.

MEASUREMENTS Physical layering and chemical indicators that vary on an annual basis enabled scientists to assign ages to ice samples taken from cores drilled through the Law Dome, a large ice dome in Antarctica. The youngest ice samples overlap in age with the first part of the Keeling data, providing a check that air samples trapped in ice have CO_2 concentrations similar to those in air measured directly over the same interval.

Ice core drilled from a glacier in Antarctica

Viewed under the microscope, glacial ice contains small bubbles that trap samples of air.

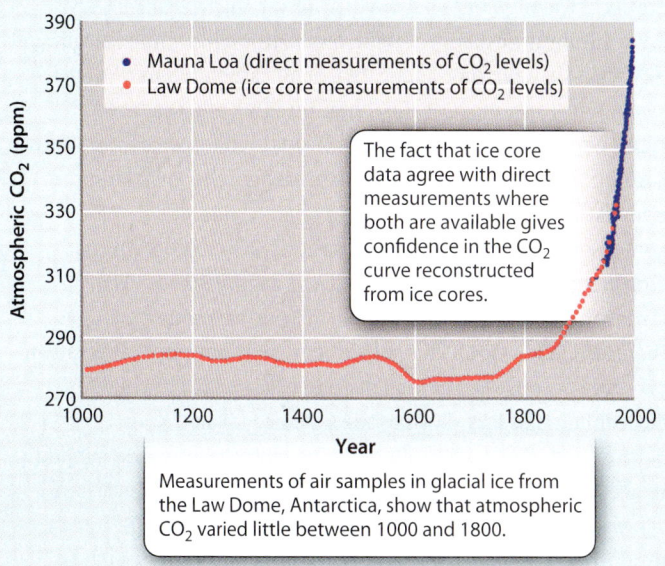

Measurements of air samples in glacial ice from the Law Dome, Antarctica, show that atmospheric CO_2 varied little between 1000 and 1800.

The fact that ice core data agree with direct measurements where both are available gives confidence in the CO_2 curve reconstructed from ice cores.

CONCLUSIONS Before the Industrial Revolution, atmospheric CO_2 levels had varied little over 1000 years, generally falling between 270 and 280 ppm. From this, scientists have concluded that current changes in atmospheric CO_2 are unusual on the timescale of the past millennium.

SOURCE Etheridge, D. M., L. P. Steele, R. L. Langenfelds, R. J. Francey, J.-M. Barnola, and V. I. Morgan. 1996. "Natural and Anthropogenic Changes in Atmospheric CO_2 over the Last 1000 Years from Air in Antarctic Ice and Firn." *Journal of Geophysical Research* 101:4115–4128.

Antarctic ice that has accumulated over the past 1000 years. Ice samples show that CO_2 levels actually began to increase slowly in the 1800s, the time of major transformations in mining, manufacturing, and transportation we call the Industrial Revolution. Before that, however, atmospheric CO_2 levels had varied little since the Middle Ages, staying at 270–280 ppm for centuries on end. On a 1000-year timescale, then, big changes in CO_2 abundance happened only once, over the last 200 years. Atmospheric CO_2 inputs and outputs were approximately in balance until the Industrial Revolution.

Fig. 25.3 shows that there is a **correlation** between the increasing CO_2 content of the atmosphere and human activities. A correlation simply indicates that two events or processes occur together. Can we actually demonstrate that humans have played a role in recent atmospheric change? By itself, correlation does not establish **causation**, a relationship in which one event leads to another. We might propose instead that recent CO_2 increase reflects natural processes and only matches the period of industrialization by coincidence.

We can test the hypothesis that human activities have contributed to the increases in atmospheric CO_2 measured at Mauna Loa and in ice cores by making careful measurements of a chemical detail: the isotopic composition of atmospheric CO_2.

→ **Quick Check 1** How does the graph in Fig. 25.3 suggest that human activities have influenced carbon dioxide levels in the atmosphere? What might be a plausible alternative hypothesis?

Carbon isotopes show that much of the CO_2 added to air over the past half century comes from burning fossil fuels.

In Chapter 2, we saw that many elements have several isotopes, atoms of the element that vary in atomic mass because they have different numbers of neutrons. Carbon has three isotopes: ^{12}C (with six neutrons, about 99% of all carbon atoms); ^{13}C (with seven neutrons, most of the remaining 1%); and the rare ^{14}C (with eight neutrons, about one part per trillion of atmospheric carbon). Beginning more than 40 years ago, the Austrian-born chemist Hans Suess measured the relative abundances of the three isotopes of carbon in atmospheric CO_2, a program that continues today. He observed that the proportion of ^{13}C in atmospheric CO_2 has declined as the total amount of CO_2 has increased (**Fig. 25.4**). This subtle change in isotopic composition occurred because the carbon dioxide being added to the atmosphere has less ^{13}C than the CO_2 already in the air.

How does this result influence the hypotheses we developed to account for increasing atmospheric CO_2 levels? The various candidate sources of increasing carbon dioxide have different proportions of ^{13}C and ^{12}C. Potential CO_2 contributors such as volcanic gases and inorganic carbon dissolved in the oceans do

HOW DO WE KNOW?

FIG. 25.4

What is the major source of the carbon dioxide that has accumulated in Earth's atmosphere over the past two centuries?

BACKGROUND Chemical analyses show that CO_2 levels in the current atmosphere are about 40% higher than they were at the time of the American Revolution. This rise coincides with major advances in manufacturing and transportation, which are driven by the burning of fossil fuels.

HYPOTHESIS The burning of fossil fuels, central to the emergence of modern industrial societies, has been the principal source of the measured increase of CO_2 in the atmosphere.

EXPERIMENT AND RESULTS Hans Suess measured the abundance of the two stable isotopes of carbon, ^{13}C and ^{12}C, in air samples and demonstrated that the ratio of ^{13}C to ^{12}C in atmospheric CO_2 has decreased over the second half of the twentieth century. Additional measurements of coral skeletons and wood show that the ratio of ^{13}C to ^{12}C in air has been decreasing over the entire period during which atmospheric CO_2 levels have been increasing.

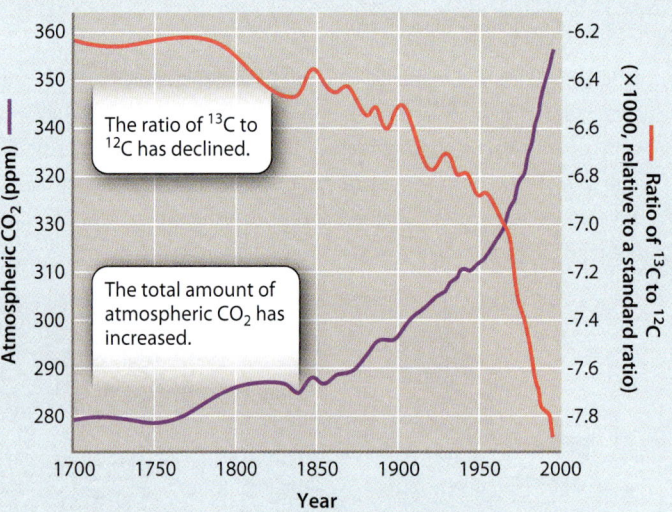

DISCUSSION The ratio of ^{13}C to ^{12}C in CO_2 added to the atmosphere over the past 200 years is lower than that in CO_2 already in the air. The ^{13}C:^{12}C ratio of CO_2 emitted by volcanoes is too high to account for the data, as is that of CO_2 released from the oceans. In contrast, organic matter formed by photosynthesis has just the right isotopic composition to account for the isotopic measurements.

FOLLOW-UP WORK Further studies, using ^{14}C, the third, unstable isotope of C, also show a strong decrease in the proportional abundance of ^{14}C in atmospheric CO_2 over time. While the burning of vegetation to clear land for agriculture results in CO_2 with the right ratio of ^{13}C to ^{12}C, modern vegetation has far too much ^{14}C to account for observed changes in the ^{14}C of air. Only fossil fuels have the right ratios of all three carbon isotopes—^{12}C, ^{13}C, and ^{14}C—to account for the pattern of isotopic change in atmospheric CO_2 measured by Suess and others. In fact, ^{14}C data from the high Rocky Mountains indicate that if fossil fuels were the principal source of added CO_2, carbon dioxide levels should have increased by a little less than 2 ppm per year between 2004 and 2008—which is very close to the increase actually measured.

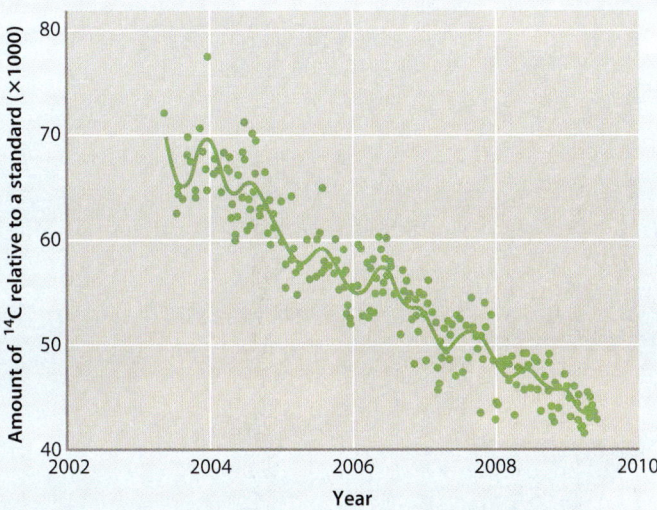

CONCLUSION Fossil fuel burning by industrialized societies has been, and continues to be, a principal source of carbon dioxide build-up in Earth's atmosphere.

SOURCES Revelle, R., and H. E. Suess. 1957. "Carbon dioxide exchange between atmosphere and ocean and the question of an increase of atmospheric CO_2 during the past decades." *Tellus* 9:18–27; Turnbull, J. C., S. J. Lehman, J. B. Miller, R. J. Sparks, J. R. Southon, P. P. Tans. 2007. "A new high precision $^{14}CO_2$ time series for North American continental air." *Journal of Geophysical Research-Atmospheres* 112, no. D11. Article Number: D11310. doi: 10.1029/2006JD008184.

not have the right isotopic composition to explain the observed change. Organic carbon in living organisms and in soil has the right ^{13}C composition. Therefore, processes that cause a net conversion of organic matter to carbon dioxide must be adding CO_2 to the atmosphere.

Here's where the other isotope of carbon, ^{14}C, comes in. Measurements also show that the sources of added carbon dioxide are depleted in ^{14}C relative to CO_2 already in the air. Modern organic matter contains too much ^{14}C to account for the observed pattern, but ancient organic matter—the coal, petroleum, and natural gas burned as fossil fuels—is isotopically just right. Chemical analyses, therefore, support the hypothesis that human activities cycle carbon in amounts high enough to affect the chemical composition of the atmosphere.

That atmospheric CO_2 is increasing is not a hypothesis; it is a measurement. That the burning of fossil fuels plays an important role in this increase is also unambiguously documented by chemical analyses. Current debate focuses not on these observations, but instead on the more difficult problem of understanding how increasing CO_2 may affect climate. We discuss this debate in Chapter 48. For now, however, we can use some relevant numbers to reveal another important aspect of the carbon cycle.

As shown in the top graph in **Fig. 25.5**, humans add about 8 billion metric tons of carbon to the atmosphere each year, nearly

FIG. 25.5 Carbon dioxide sources and sinks. Much of the carbon dioxide introduced by humans into the atmosphere has not stayed there. About half of it has accumulated as dissolved inorganic carbon in the oceans or as new biomass on the continents.

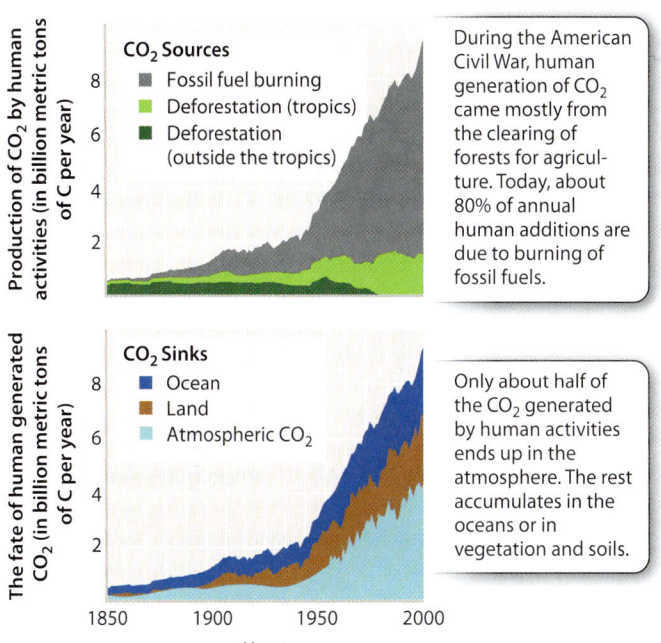

all of it as carbon dioxide. Much of this material comes from power plants, airplanes, ships, and automobiles, which oxidize ancient organic matter to CO_2 on a grand scale. Further important inputs come from the conversion of forests to agriculture or pastureland. Clearing and burning forest trees also oxidizes organic carbon to CO_2.

Interestingly, if we compare our best estimates of nonrespiratory human CO_2 generation since 1958 with the measured increase in atmospheric CO_2, we find that about half of the CO_2 released by human activities *did not* end up in the air. Where did it go? This has been a major question for years, and it now appears that much of the "missing carbon" has been stored as inorganic carbon in the form of CO_2, bicarbonate (HCO_3^-), and carbonate (CO_3^{2-}) ions dissolved in the oceans. The bottom graph in Fig. 25.5 shows an estimate of the various places where human-produced CO_2 ends up. These observations indicate that Earth's carbon cycle is more complicated than a simple exchange of carbon between photosynthesis and respiration. There are additional storehouses of carbon and additional processes that move carbon from one store to another, topics we discuss next.

25.2 THE LONG-TERM CARBON CYCLE

Up to this point, we have considered only the short-term carbon cycle: exchanges over days, years, and decades driven by the biological processes of photosynthesis and respiration, and altered in recent times by human activities. Geologic evidence, however, tells us that on longer timescales, carbon dioxide levels in air have changed dramatically. These longer-term changes mean that we must consider additional contributions to the carbon cycle: physical processes, including volcanism and climate change. Indeed, the complete carbon cycle links Earth's physical and biological processes, providing a foundation for understanding the interconnected histories of life and environment through our planet's long history.

Reservoirs and fluxes are key in long-term carbon cycling.

To understand how biological and physical processes interact to govern carbon dioxide levels in the air, we must first look at how carbon is distributed among its various **reservoirs,** the places where carbon resides in the Earth. Reservoirs of carbon include organisms, the atmosphere, soil, the oceans, and sedimentary

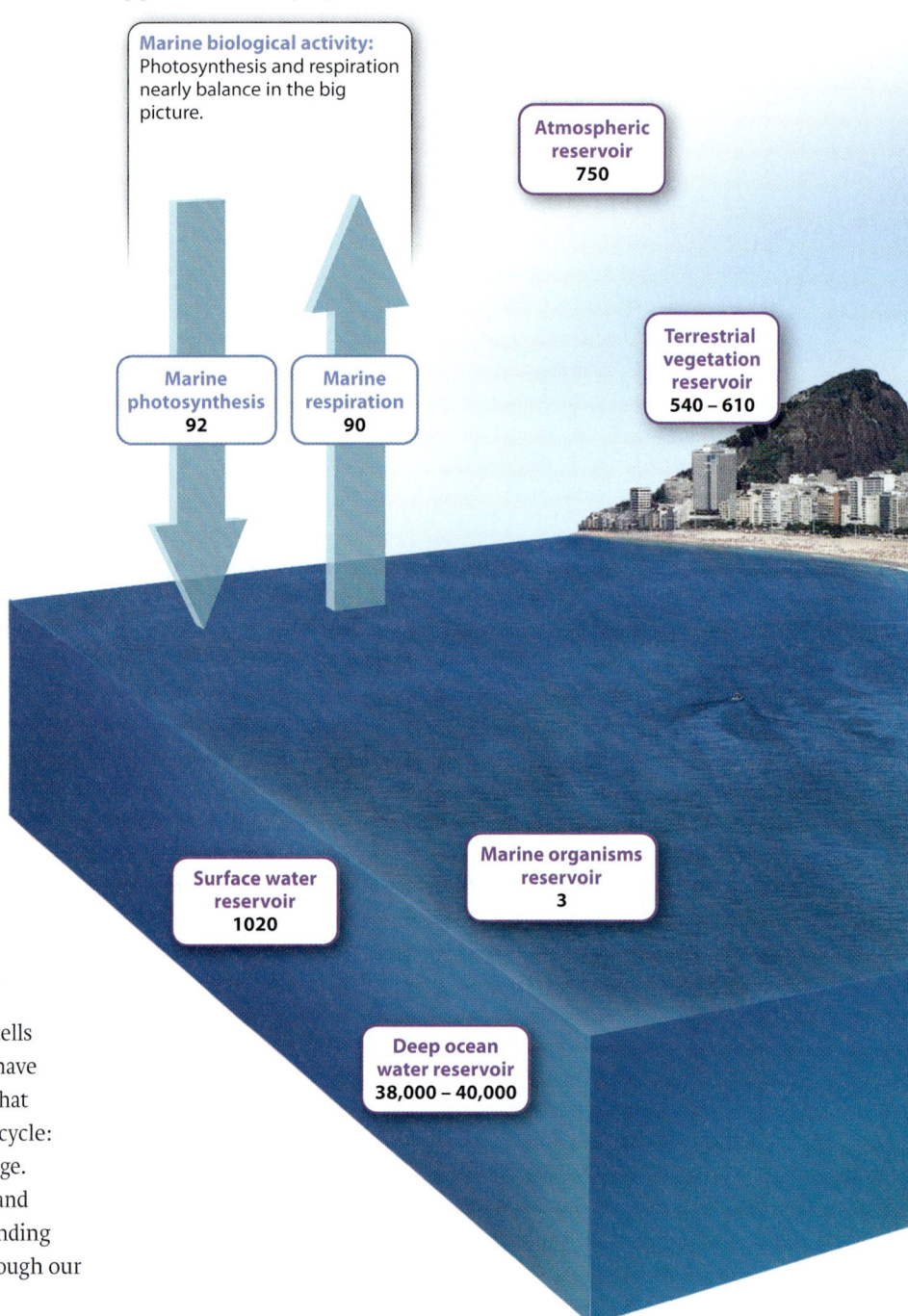

FIG. 25.6 Movement of carbon between reservoirs. The numbers for each reservoir are gigatons (1000 million metric tons) of carbon, and those for fluxes are gigatons of carbon per year.

rocks, as shown in the purple boxes in **Fig. 25.6.** How much carbon is stored in each of these reservoirs?

If we add up all the carbon contained in the total mass of organisms living on land, that is, in their **biomass,** that amount of carbon is just a bit smaller than the amount of carbon stored as CO_2 in the atmosphere. Soil, by comparison, stores as much carbon as do land organisms and the

The biggest carbon reservoir of all, however, lies beneath our feet. Calcium carbonate minerals (CaCO₃), which form limestone, and organic matter preserved in sedimentary rocks dwarf all other carbon reservoirs combined by three orders of magnitude. Coal, petroleum, and natural gas make up only a small part of the organic matter stored in sedimentary rocks, but as we have seen, they play an important role in the modern carbon cycle.

Fluxes are the rates at which carbon flows from one reservoir to another. The sensitivity of different reservoirs to change depends on the relative sizes of the reservoir and of the movement of material into and out of it. When fluxes are large relative to the size of the reservoir, reservoir size can change rapidly. As mentioned earlier, the amount of carbon stored as CO_2 in air is not that much larger than the annual fluxes into and out of the atmosphere. For this reason, atmospheric CO_2 abundance can be influenced by a number of processes at work in the carbon cycle.

Physical processes add and remove CO_2 from the atmosphere.

Physical processes, like biological ones, are capable of adding and removing CO_2 from the atmosphere. A key process is volcanism: Volcanoes and mid-ocean ridges release an estimated 0.1 billion metric tons of carbon (as CO_2) into the atmosphere each year. More CO_2, about 0.05 billion metric tons of carbon per year, is released by the slow oxidation of coal, oil, and other ancient organic material in sedimentary rocks exposed at the Earth's surface. In nature, bacteria and fungi accomplish most of this oxidation by respiring old organic molecules. Burning fossil fuel accelerates this process dramatically, increasing a hundredfold the rate at which sedimentary organic carbon is oxidized.

Carbon dioxide is removed from the atmosphere by chemical reactions between air and exposed rocks, a process called chemical weathering. How can air and rock interact? To understand how weathering fits into the carbon cycle, we must first understand that carbon dioxide reacts with rainwater to form carbonic acid (H_2CO_3). This acid slowly reacts with rock-forming minerals, generating (among other substances) calcium and bicarbonate (HCO_3^-) ions that are transported by rivers to the oceans. Within the oceans, the calcium and bicarbonate ions react with each other to form calcium carbonate (CaCO₃) minerals that accumulate on the seafloor, forming limestone. In essence, carbon moves from

atmosphere combined, mostly as slowly decaying organic compounds. In the oceans, the amount of carbon contained in living organisms is actually very small. Much more resides as inorganic carbon dissolved in the water—most of it in the deep oceans as CO_2, bicarbonate, and carbonate ions. Perhaps fortunately for us, the oceans have sopped up some of the CO_2 generated by human activities.

FIG. 25.7 Clams and corals with skeletons made up of calcium carbonate.

the atmosphere to the sediments, where it can be stored for millions of years. The two linked processes of chemical weathering and mineral precipitation in the oceans can be summarized by a simple chemical reaction (here, CaSiO₃ stands in for rock-forming minerals as a whole):

$$CaSiO_3 + CO_2 \rightarrow CaCO_3 + SiO_2$$

In present-day oceans, the carbonate minerals that accumulate on the seafloor are precipitated predominantly as skeletons and shells. Clams, corals, and many other organisms form beautiful and functional shells of CaCO₃ (**Fig. 25.7**). Silica (SiO₂) also precipitates from the ocean water largely as skeletons, especially as tiny but exquisite shells fashioned by single-celled algae called diatoms (**Fig. 25.8**). Skeleton formation is an example of **biomineralization**, the precipitation of minerals by organisms. Biomineralization provides yet another link between Earth and life, and one that has varied through evolutionary history.

The annual carbon fluxes associated with geologic processes are small relative to the yearly inputs and outputs from photosynthesis and respiration. As we see in Fig. 25.6, however, photosynthesis and respiration more or less cancel each other out. In the long-term carbon cycle, it is the net result of these rapid biological processes—photosynthesis input minus respiratory output—that matters. Is CO₂ produced or consumed when all biological processes are considered together?

Today, the amount of carbon incorporated into organic matter by photosynthesis is actually a bit larger than the amount returned to the atmosphere by respiration. The difference amounts to about 0.05 billion metric tons per year. This number, you will note, is on the same order as volcanic and weathering fluxes. The "excess" organic carbon produced by photosynthesis is deposited in sediments as they accumulate on the seafloor, sustaining a small but constant leak of material from the short-term biological components of the carbon cycle to long-term geologic reservoirs.

One final set of geological processes completes the long-term carbon cycle. The Earth's crust is constantly in motion, propelled by heat within the underlying mantle. **Plate tectonics** is the name given to this dynamic movement of our planet's outer layer. New crust forms at spreading centers, the places where molten rock ascends from the

FIG. 25.8 Silica (SiO₂) skeletons of diatoms captured by an electron microscope. Most SiO₂ leaves the ocean as diatom skeletons.

FIG. 25.9 Geological processes that drive the long-term carbon cycle. Numbers are estimates, in gigatons, of annual fluxes of carbon.

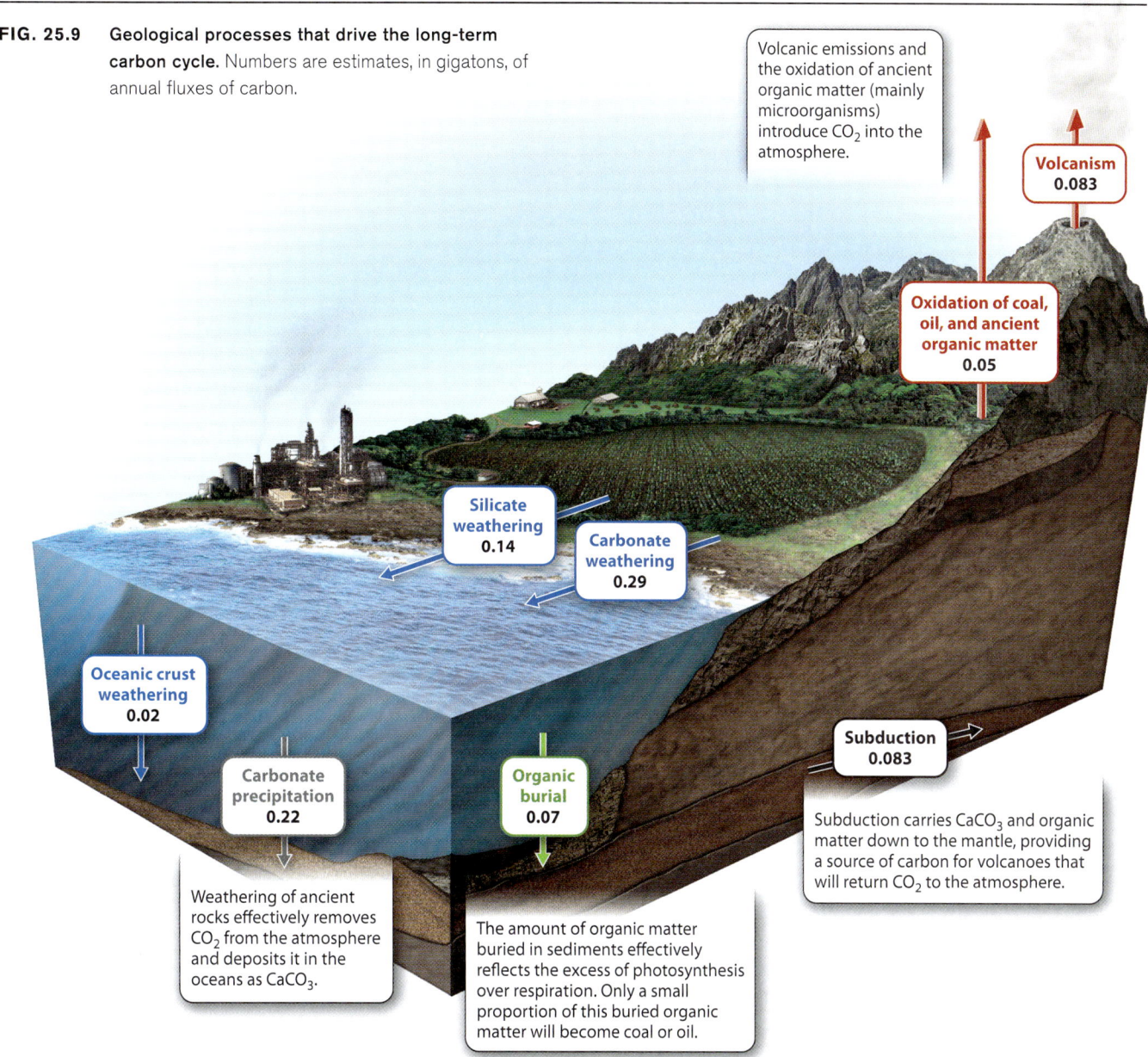

mantle. Old crust is destroyed in subduction zones, where one slab of crust slides beneath another, returning material to the mantle. Plate tectonics provides geology's best explanation for the formation of mountains and ocean basins, and it plays an important role in the long-term carbon cycle. The crust that descends into subduction zones carries with it sediments, including carbonate minerals and sedimentary organic matter. Subduction removes carbon from Earth's surface, but this carbon will be recycled to the surface as carbon dioxide emitted from volcanoes and mid-ocean ridges. **Fig. 25.9** shows the physical processes at work in Earth's long-term carbon cycle.

→ **Quick Check 2** If plate tectonics form a chain of high mountains, would you expect atmospheric CO_2 to increase or decrease?

Records of atmospheric composition over 400,000 years show periodic shifts in CO_2 content.

Earlier in this chapter, we discussed evidence that before the Industrial Revolution, atmospheric CO_2 levels had not changed appreciably for 1000 years or more. Longer-term records, however, show that the carbon dioxide levels in air can change substantially through time.

FIG. 25.10 Atmospheric CO_2 content for the past 400,000 years. These measurements were recorded from air bubbles trapped in glacial ice at Vostok, Antarctica.

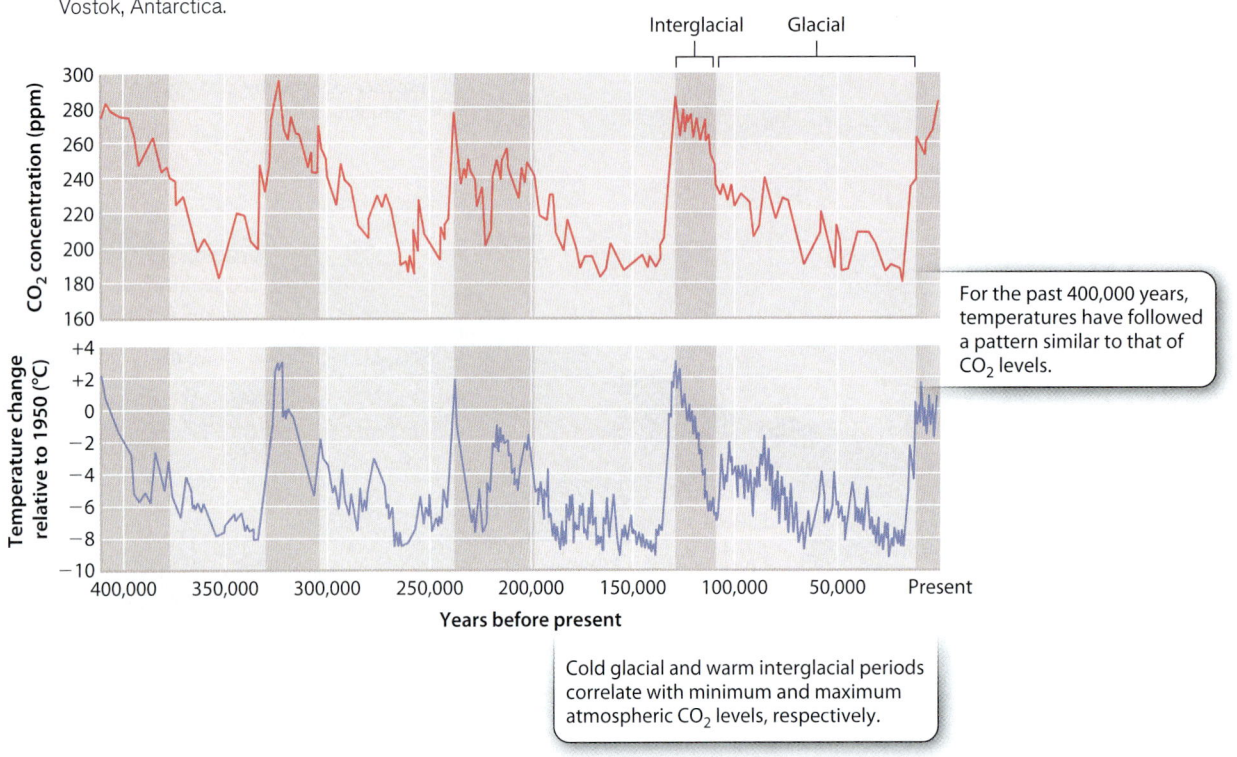

For the past 400,000 years, temperatures have followed a pattern similar to that of CO_2 levels.

Cold glacial and warm interglacial periods correlate with minimum and maximum atmospheric CO_2 levels, respectively.

At Vostok, high on the Antarctic ice sheet, glacial ice records more than 400,000 years of environmental history (**Fig. 25.10**). As shown in the top graph of Fig. 25.10, the youngest samples show about 285 ppm CO_2 in the atmosphere, consistent with direct measurements of air, including the first years of the Keeling curve. Notice, however, that 20,000 years ago, CO_2 levels were much lower—about 180 ppm. In fact, the Vostok ice core in its entirety shows that atmospheric CO_2 has oscillated between 285 ppm and 180 ppm for at least 400,000 years. On long timescales, therefore, the natural variations in the carbon cycle can be large.

The bottom graph in Fig. 25.10 shows an estimate of surface temperature obtained by chemical analysis of oxygen isotopes in ice from the same glacier. Over the last 400,000 years, Antarctic temperature has oscillated between peaks of a few degrees warmer than today's temperature and temperatures as much as 6° to 8° C colder than present. Interestingly, the temperature and carbon dioxide curves closely parallel each other. Carbon dioxide is known to be an effective **greenhouse gas,** meaning that it allows incoming solar radiation to reach Earth's surface but traps heat that is reemitted from land and sea. Therefore, it is not surprising that temperature and atmospheric CO_2 levels show the parallel history documented in the figure.

The two curves correlate closely with one further phenomenon, the periodic growth and decay of continental ice sheets. Large glaciers expanded in the Northern and Southern hemispheres a few million years ago, ushering in an ice age. Today, we live in an interglacial interval, when climate is relatively mild, but 20,000 years ago thick sheets of ice extended far enough away from the poles to cover the present site of Boston (**Fig. 25.11**). The repeated climatic shifts recorded in ice cores reflect periodic variations in the amount and distribution of solar radiation on Earth's surface, which are caused by oscillating changes in Earth's orbit around the sun.

The temperature and CO_2 increases recorded by Vostok ice between 20,000 and 10,000 years ago coincide with the last great retreat of continental ice sheets. What processes might explain how atmospheric CO_2 could increase by 100 ppm in just a few thousand years, as glaciers began to retreat? Can the short-term carbon cycle processes of photosynthesis and respiration account for this much carbon? Certainly, the amount of forests on the Earth's surface has varied through the past 500,000 years as ice sheets grew and decayed, but forests expand as glaciers shrink, so changes in forests cannot account for a pattern of *increasing* atmospheric CO_2 with the retreat of glaciers. Volcanism and weathering also fail to account for the observed pattern. There is no evidence that volcanic activity has waxed and waned in a pattern that could explain observed CO_2 variations. And rates of weathering, which remove CO_2 from the atmosphere, should increase as temperature rises, but carbon dioxide levels have actually increased. Something else must be going on.

Scientists continue to debate why atmospheric CO_2 oscillates in parallel with glacial expansion and retreat. Proposed

FIG. 25.11 Earth's most recent ice age. Twenty thousand years ago, glacial ice covered much of North America and Europe, as well as mountain ranges such as the Andes and Himalayas. Sea ice expanded markedly in both northern and southern oceans.

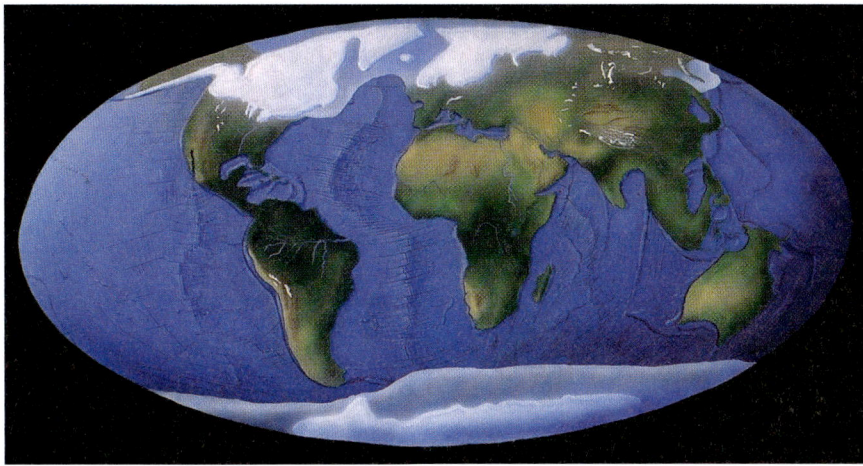

mechanisms suggest interactions involving the ocean and its large reservoir of inorganic carbon. For example, it has been hypothesized that during glacial advances, the circulation of carbon-rich deep-ocean waters back to the sea surface slows, causing more inorganic carbon to accumulate in the deep sea. With glacial retreat, the oceans circulate more vigorously, returning CO_2 to the surface and then to the atmosphere. Whatever the explanation, the historical record of the past 400,000 years shows that climate can and does change without any input from humans, something we must take into account when considering our climatic future (Chapter 48).

Variations in atmospheric CO_2 over hundreds of millions of years reflect plate tectonics and evolution.

The atmospheric record trapped in glacial ice extends backward less than 1,000,000 years. To estimate atmospheric CO_2 levels for earlier intervals of Earth history, we must rely on computer models and measurements of chemical or paleontological features of ancient rocks that are thought to reflect atmospheric CO_2 at their time of formation. One substitute for direct measurements of ancient carbon dioxide levels is the anatomy of fossil leaves: Experiments show that stomata, the small pores on leaf surfaces (Chapter 29), decrease in density as atmospheric CO_2 levels increase. The chemistry of ancient soils is also thought to reflect atmospheric history. Such indirect observations and analyses come with a large degree of uncertainty, but most Earth scientists accept at least the broad pattern of atmospheric history shown in **Fig. 25.12**.

For the past 30 million years or so, atmospheric CO_2 levels have probably not exceeded those recorded today at Mauna Loa (Fig. 25.12). Before that, higher levels of CO_2—perhaps 4 to 6 times the 1958 level—characterized the Mesozoic Era (252 to 65 million years ago), the age of dinosaurs. Levels more like today's levels existed earlier, during much of the late Paleozoic Era (350–252 million years ago), another time of extensive glaciation. And even before that, about 500 million years ago, early in the Paleozoic Era, atmospheric CO_2 may have reached values as high as 15 to 20 times the present-day level.

Not even variations in ocean circulation can account for the scale of these variations. On timescales of millions of years, atmospheric CO_2 and, hence, climate are determined by changes in the rate of organic carbon burial in sediments, continental weathering of rocks uplifted into mountains, and volcanic gas release. All these processes reflect the action of Earth's great physical engine: plate tectonics. Seafloor formation and destruction together influence rates of volcanic gas release, and mountain formation strongly influences rates of weathering and, therefore, the fluxes of sediments that bury organic carbon beneath the seafloor.

But life plays a role, even here. For example, the large drop in atmospheric CO_2 suggested for the mid-Paleozoic Era, 400–350 million years ago, is thought to reflect the evolution of a new player in the carbon cycle: woody plants. The evolution of trees

FIG. 25.12 Atmospheric CO_2 content for the past 542 million years. Models and indirect estimates from geological measurements suggest that CO_2 levels have varied substantially through Earth's history.

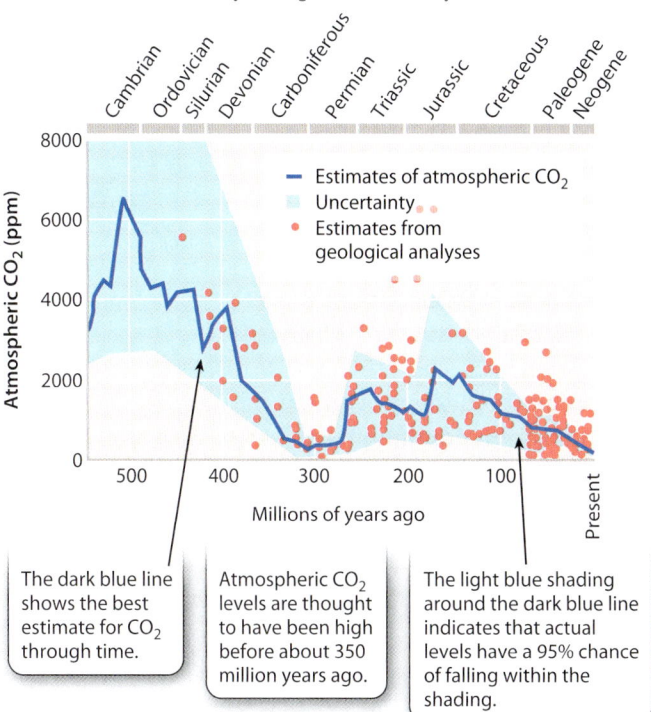

The dark blue line shows the best estimate for CO_2 through time.

Atmospheric CO_2 levels are thought to have been high before about 350 million years ago.

The light blue shading around the dark blue line indicates that actual levels have a 95% chance of falling within the shading.

increased the size of the carbon reservoir on land and ushered in an important new mechanism for removing carbon from the air and ultimately to sedimentary rocks. That mechanism was the burial of plant material on land, forming peat and, eventually, coal.

25.3 THE CARBON CYCLE, ECOLOGY, AND EVOLUTION

As you sit quietly by a woodland pond, the carbon cycle is at work all around you. In the forest, plants generate carbohydrates as they photosynthesize, and algae and photosynthetic bacteria are doing the same in the pond. These organisms are **primary producers,** generating organic compounds that will provide food for other organisms in the local environment.

An insect grazes on leaves, and ducks feed on both insects and algae. Ducks and insects obtain the carbon they need for growth and reproduction from the foods they eat, and they also gain energy by respiring food molecules. Such organisms are **consumers,** and they include all the animals in both pond and forest. Beneath our feet, similar interactions play out on a microscopic scale, especially among the fungi, protists, and bacteria that recycle the organic carbon in soil back to CO_2.

The complementary metabolic processes of photosynthesis and respiration cycle carbon through forest and pond communities. Furthermore, as we will see in the next chapter, complementary metabolic processes also cycle nitrogen, sulfur, and other elements required for life. By continually recycling materials, biogeochemical cycles, which involve both biological and physical processes, sustain life over long intervals. In their absence, life could hardly have persisted for 4 billion years.

Food webs trace the movement of carbon through communities.

In biological communities across the planet, carbon and energy are packaged by photosynthetic organisms and transferred to heterotrophs (Chapter 6) that graze on plants or photosynthetic microorganisms. Carnivorous organisms may eat these grazers, and they, in turn, may provide food for still other carnivores. Eventually, having passed from one consumer to another, the carbon originally fixed by photosynthesis is returned to the atmosphere by the respiration of fungi and other **decomposers** that break down dead tissues.

The linear transfer of carbon from one organism to another is called a **food chain.** Because most heterotrophs within a community can consume or be consumed by a number of other species, biologists often prefer to speak of **food webs,** a term that provides a better sense of the complexity of interactions within the carbon cycle (**Fig. 25.13**). Food webs define the interactions among organisms in ponds, forests, and many other habitats. Simply put, the carbon cycle underpins the ecological structure of biological communities (Chapter 47).

FIG. 25.13 A food web. Food webs trace the cycling of carbon through communities.

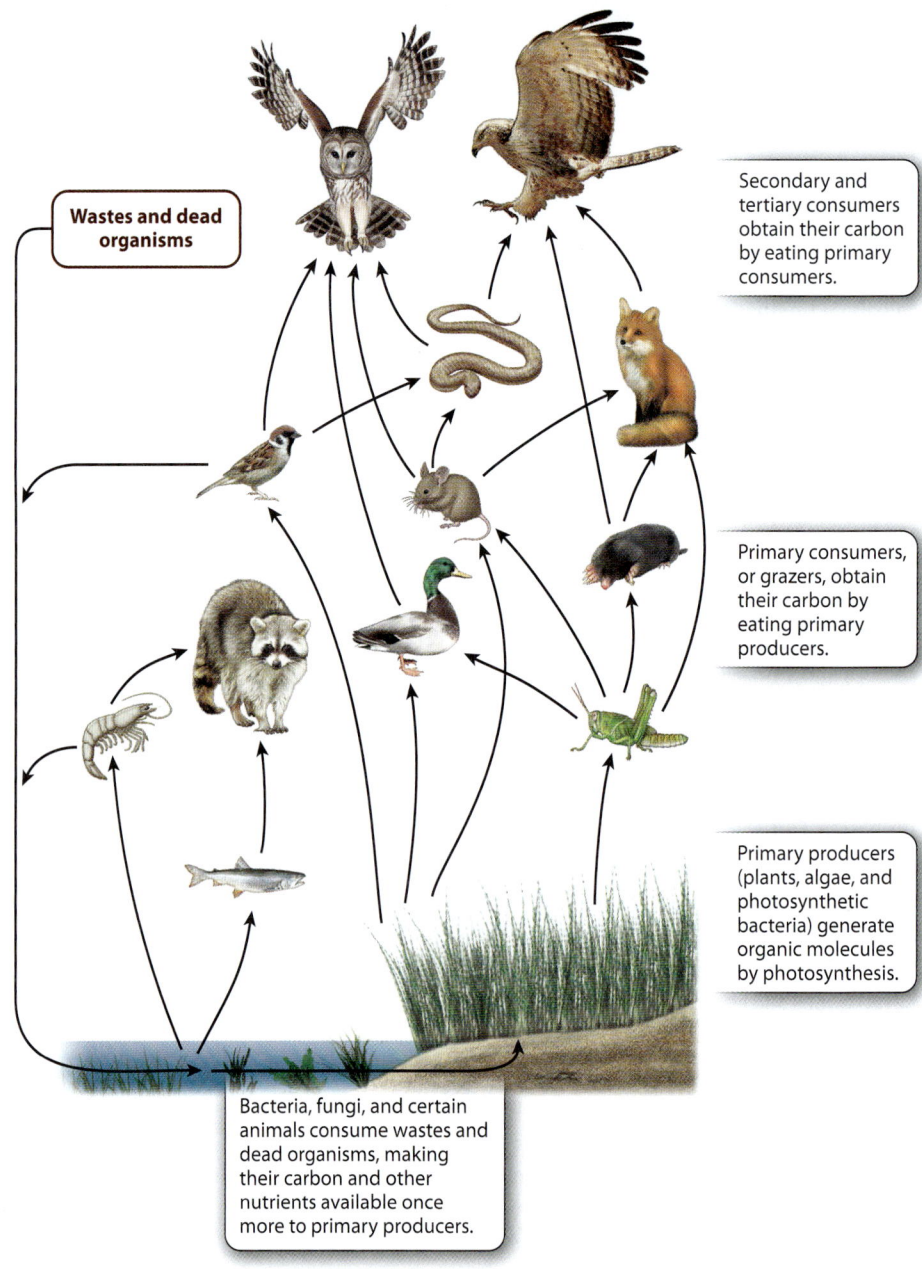

Secondary and tertiary consumers obtain their carbon by eating primary consumers.

Primary consumers, or grazers, obtain their carbon by eating primary producers.

Primary producers (plants, algae, and photosynthetic bacteria) generate organic molecules by photosynthesis.

Bacteria, fungi, and certain animals consume wastes and dead organisms, making their carbon and other nutrients available once more to primary producers.

Trophic pyramids trace the flow of energy through communities.

Throughout this chapter and earlier ones, we have emphasized that photosynthesis and respiration transfer energy as they cycle carbon. Carbon is removed from the environment by photosynthetic organisms and passed from one consumer to another, eventually returning to the environment by respiration carried out by consumers and decomposers. At the same time as they remove carbon from the atmosphere, photosynthetic organisms also harvest energy from sunlight and convert it into chemical energy that is stored in the bonds of organic molecules. Consumers store some of the energy in their food as fat, starch, or other organic molecules, but they use much of their food energy to do work or generate heat. The energy used for work or dissipated as heat cannot be captured again by photosynthetic organisms. Whereas carbon cycles through communities, energy transfer is directional.

Within a biological community, the various producers and consumers are commonly grouped into functionally similar sets called trophic groups, from *trophé*, the Greek word for "nourishment." Diagrams showing the amount of energy available at each level to feed the next are called **trophic pyramids** (**Fig. 25.14**). These diagrams have a pyramidal shape because the biomass of primary producers generally much larger than the biomass of primary consumers, and the biomass of primary consumers is, in turn, much greater than that of secondary consumers, and so on. The reason for this is that energy transfer from level to level is inefficient. Because of wastes, work, and heat dissipation, only about 10% to 15% of the energy available in biomass at one level gets incorporated into biomass at the next level. Like food webs, trophic pyramids relate community structure to the fundamental biological processes at work in the carbon cycle.

→ **Quick Check 3** Why are antelopes more abundant than lions in the African savanna?

The diversity of photosynthetic organisms reflects adaptation to a wide range of environments.

An estimated 500,000 species of photosynthetic organisms fuel the carbon cycle in all but a few deep-sea and subterranean environments. In terms of carbon and energy metabolism, all these species do pretty much the same thing. Why, then, is there such a diversity of photosynthetic organisms? Why don't just a few species dominate the world's photosynthesis?

The example of the pond and forest helps to make the basis of photosynthetic diversity clear. In the forest community, the leaves of several different tree species form a photosynthetic canopy above the forest floor. Below, shrubs grow, making use of light not absorbed by the leaves above them. And below the shrubs are grasses, herbs, ferns, and mosses that can grow in the reduced light levels of the forest floor. In the adjoining pond, a few species of aquatic plants line the water's edge, but beyond that algae and photosynthetic bacteria dominate photosynthesis, with some species anchored to the pond bottom and others floating in the water column.

On the largest scale, then, the reason for photosynthetic diversity is obvious—cyanobacteria, algae, and land plants all perform comparable metabolic functions in different habitats. This kind of specialization for different environments also applies on a finer scale, although it is complicated by grazing and environmental disturbances such as fire and landslides (Chapter 47). In the forests of New England, the tree species that thrive in wet bottomlands differ from those found on well-drained hillsides. And seasonally dry woodlands in southern California support yet another set of plant species. In general, plants of varying size, shape, and physiology inhabit physically and biologically distinct environments, and the same is true of photosynthetic organisms in water.

Thus, the immense diversity of photosynthetic organisms found today reflects not so much evolutionary variations in the biochemistry of photosynthesis (although some of that occurs; see Chapter 29) as structural and physiological adaptations. These

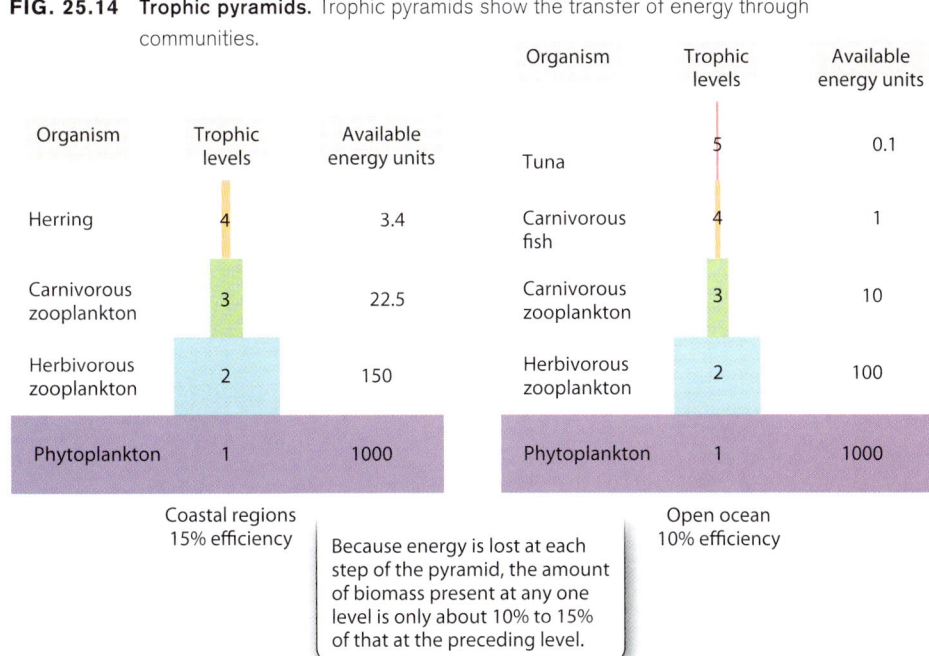

FIG. 25.14 Trophic pyramids. Trophic pyramids show the transfer of energy through communities.

Because energy is lost at each step of the pyramid, the amount of biomass present at any one level is only about 10% to 15% of that at the preceding level.

FIG. 25.15 Microbial communities in hot springs in New Zealand. This environment and others that few eukaryotes tolerate provide a glimpse of how the carbon cycle worked before algae, amoebas, plants, and animals evolved.

The carbon cycle provides a framework for understanding life's evolutionary history.

The world did not always support the biological diversity we see today. Indeed, 2 billion years ago, the carbon cycle was anchored by photosynthetic bacteria and microbial heterotrophs because there were no plants or animals (**Fig. 25.15**). Subsequently, algae gained an ecological foothold as nutrients such as nitrate became more widely available, and single-celled eukaryotic heterotrophs expanded their reach by gathering food in ways not possible for bacteria—that is, by ingesting cells and other types of particulate food. With the evolution of multicellularity, animals and seaweeds added further complexity to carbon cycling in the sea. Eventually, some aquatic green algae evolved the capacity to live on land, and animals soon followed, building unprecedented levels of diversity on our planet.

The history of life is one of accumulating variety and complexity, based mostly on the great biological components of the carbon cycle: photosynthesis and respiration. Indeed, the carbon cycle did more than provide a framework for accumulating adaptations allow the effective gathering of light, nutrients, and—critical to life on land—water, in widely varying local environments. Natural selection, acting on local populations, links the diversity of photosynthetic organisms to the carbon cycle.

The diversity of respiring organisms reflects many sources of food.

About 10 million species help to cycle carbon through respiration. These include the plants, algae, and bacteria that generate carbohydrates by photosynthesis, as well as legions of animals, fungi, and microorganisms that obtain both carbon and energy from preexisting organic compounds.

Heterotrophic bacteria, amoebas, and humans may use essentially the same biochemical pathway to respire organic molecules, but they differ markedly in how they feed and, therefore, what they can eat. Bacteria (and also fungi) absorb molecules from their environment, but amoebas and many other eukaryotic microorganisms can capture and ingest cells—they are capable of predation. Animals capture prey, as well, but commonly feed on organisms far too large for an amoeba to eat. As photosynthetic organisms have adapted structurally and physiologically to local environments across the globe, heterotrophs have adapted by means of locomotion, mouth and limb specialization, perception, and behavior to obtain their food.

FIG. 25.16 Oxygen levels in Earth's atmosphere during the history of life, based on analysis of oxygen-sensitive minerals in sedimentary rocks.

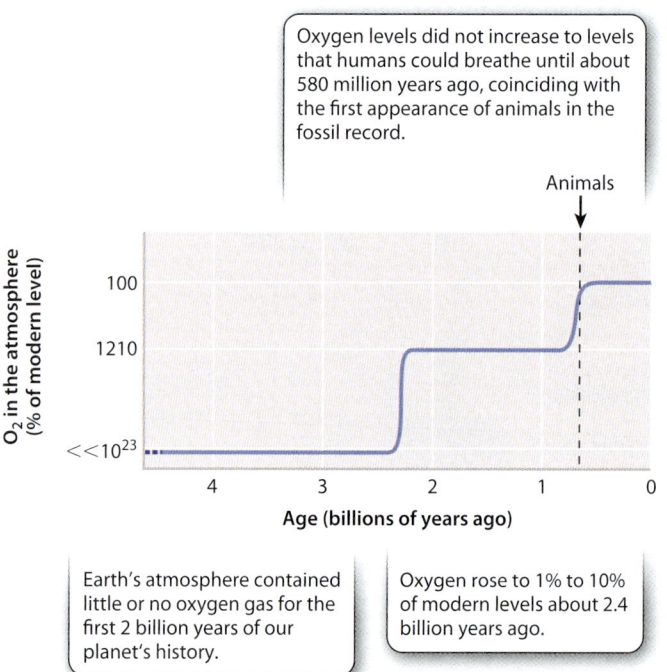

Oxygen levels did not increase to levels that humans could breathe until about 580 million years ago, coinciding with the first appearance of animals in the fossil record.

Earth's atmosphere contained little or no oxygen gas for the first 2 billion years of our planet's history.

Oxygen rose to 1% to 10% of modern levels about 2.4 billion years ago.

diversity; it changed the very nature of Earth surface environments in ways that enabled new types of organism to evolve.

When life began, the atmosphere and oceans contained little or no oxygen gas. How did our present environment come to be? The key lies in the formulas for photosynthesis and respiration, shown in Fig. 25.2. Photosynthesis and respiration not only cycle carbon, they cycle oxygen (and water) as well. When photosynthesis and respiration are in balance, the oxygen generated by photosynthesis is completely consumed by respiration. But when some of the organic carbon generated by photosynthesis is buried in sediments, oxygen production can exceed oxygen consumption. In other words, because sedimentary organic matter burial can break the tight coupling depicted in Fig. 25.2, it can facilitate an increase in O_2 levels that may accumulate in the atmosphere and oceans. The geologic history of oxygen shown in **Fig. 25.16** results from the interactions through time between the carbon (and sulfur) cycle with plate tectonics. Our oxygen-rich world is the result.

In the chapters that follow, we examine life's diversity in terms of function, phylogeny, and ecology, but don't be fooled into thinking that these are discrete topics: They are inextricably intertwined. The remarkable diversity of life on Earth reflects the ecology and evolution of populations that interact to cycle carbon through the biosphere.

Core Concepts Summary

25.1 PHOTOSYNTHESIS AND RESPIRATION ARE THE KEY BIOCHEMICAL PATHWAYS FOR THE BIOLOGICAL, OR SHORT-TERM, CARBON CYCLE.

Keeling began a program to monitor carbon dioxide levels in the atmosphere in 1958. Two patterns emerged: (1) carbon dioxide levels oscillate on a yearly basis, and (2) carbon dioxide levels have increased steadily to the present day since measurements began. page 25-1

The first pattern, seasonal oscillation, can be explained by the imbalance between photosynthesis and respiration in the summer and the winter months. page 25-2

The second pattern, of steady increase, can be extended further back in time by measuring the composition of air trapped in arctic gas bubbles. These measurements indicate that carbon dioxide levels in the atmosphere remained steady for more than a thousand years, but then started to increase in the mid-1800s, the time of the Industrial Revolution. page 25-3

The ratio of different isotopes of carbon in the atmosphere indicates that most of the carbon added to the atmosphere in recent decades comes from human activities, particularly the burning of fossil fuels. page 25-4

25.2 PHYSICAL PROCESSES GOVERN THE LONG-TERM CARBON CYCLE.

Physical processes, which act at slower rates and over longer time scales, are the major drivers of the long-term carbon cycle. page 25-6

In following carbon around its cycle, it is important to consider where carbon is stored (reservoirs) and rates at which it is added or removed from these stores (fluxes). page 25-6

Important reservoirs of carbon include organisms, the atmosphere, soils, the oceans, and sedimentary rocks. Sedimentary rocks are by far the largest carbon reservoir. page 25-7

Physical processes that remove CO_2 from the atmosphere include the weathering of rocks, and those that add CO_2 to the atmosphere include volcanoes and the oxidation of ancient organic matter in sedimentary rocks. page 25-7

The short- and long-term carbon cycles are linked by the slow leakage of organic matter from biological communities to sediments accumulating on the seafloor. page 25-8

Over the last 400,000 years, CO_2 levels in the atmosphere have gone through periodic shifts that reflect repeated cycles of glaciation. page 25-10

CO_2 levels in the atmosphere 500 million years ago may have been as much as 15 to 20 times higher than present-day levels. Over timespans of hundreds of millions of years, movements of Earth's plates and the evolution of new life-forms are major players in the carbon cycle. page 25-11

25.3 THE CARBON CYCLE CAN HELP US UNDERSTAND ECOLOGICAL INTERACTIONS AND THE EVOLUTION OF BIOLOGICAL DIVERSITY.

Photosynthetic and other autotrophic organisms are primary producers, converting carbon dioxide into organic molecules. page 25-12

Heterotrophic organisms that eat primary producers are called primary consumers. Other heterotrophs that prey on these grazers are called secondary consumers. page 25-12

Interacting networks of species, linked by predator–prey interactions, are called food webs, and they are an important feature of the carbon cycle. page 25-12

Communities contain a much greater biomass of producers than consumers because the movement of energy through communities is inefficient. page 25-13

Photosynthetic organisms are well adapted to obtaining sunlight and nutrients in different environments. page 25-13

Respiring organisms are adapted to obtain different sources and sizes of food. page 25-14

The evolution of life on Earth can be considered a long history during which organisms have evolved different ways of obtaining energy and carbon from their environment. page 25-14

The interactions between biological and geologic processes at work in the carbon cycle have not only produced our current biological diversity. They are also responsible for Earth's oxygen-rich atmosphere and oceans. page 25-15

Self-Assessment

1. Name two processes that drive the short-term carbon cycle.
2. Draw and explain the curve representing changing atmospheric levels of CO_2 over the course of a year.
3. Draw and explain the curve representing changing atmospheric levels of CO_2 over the past 150 years.
4. Explain one way by which the source of the increase in CO_2 in the atmosphere can be determined.
5. Name processes that influence the long-term carbon cycle.
6. Draw and explain the curve representing changing atmospheric levels of CO_2 over the last 400,000 years.
7. Trace the flow of carbon and energy through biological communities.

Do you understand the chapter's Core Concepts? Log in to check your answers to the Self-Assessment questions, then practice what you've learned and reinforce this chapter's concepts by working through the problems and multimedia tutorials provided there.

www.biologyhowlifeworks.com

CASE 5

The Human Microbiome

DIVERSITY WITHIN

Clostridium difficile doesn't grab many headlines. Perhaps it should. This bacterium is thought to be the most common cause of hospital-acquired infections in the United States. It sickens about 165,000 U.S. hospital patients each year, causing fevers, chills, diarrhea, and severe intestinal pain—and adds more than $1 billion annually to the nation's health care costs. Altogether, the infection kills about 28,000 people every year, according to the Centers for Disease Control and Prevention.

Now a promising (if stomach-turning) new treatment is gaining traction: fecal transplants. The vast majority of *C. difficile* cases occur after patients have taken antibiotics for other infections. The drugs kill the beneficial bacteria normally found in a healthy gut, allowing *C. difficile* to proliferate. Now some pioneering researchers have begun to replace the missing microbes by transplanting fecal material from healthy donors into patients sick with *C. difficile*. Some doctors have reported that the procedure alleviates symptoms in more than 80% of patients.

While transplanting fecal bacteria may not yet be mainstream medicine, there is no question that our gut, as well as our skin, mouth, nose, and urogenital tract, is home to a diversity of bacterial species. No one has yet identified the many microbes living on and within our bodies, but estimates suggest that each of us contains 10 times as many bacterial cells as human cells. The relationship between "us and them" is a type of symbiosis, in which both humans and bacteria benefit from the close association. Indeed, the microbiome, as our assemblage of bacterial houseguests is known, affects human health in many important ways.

Some of the most mysterious, and most intriguing, members of our microbiome live in our gut. Bacteria living in our intestines help us break down food and synthesize vitamins. On the other hand, abnormal communities of microbes in the gut have been linked to Crohn's disease, colon cancer, and nonalcoholic fatty liver disease.

Your own personal gut flora is a unique collection of species—even identical twins can have surprisingly different microbiomes. This variability has made it difficult for scientists to decipher exactly what role individual species play in health and in disease. The sheer number of bacteria in your gut only compounds the problem. One international team of scientists recently reported that each human gut contains at least 160 different bacterial species.

Despite these challenges, researchers have made strides toward understanding the unique bacterial communities within our intestinal tract. In 2011, scientists compared all the DNA sequences from gut microbes of people living in Europe, America, and Japan. They found that each individual fell into one of three enterotypes, or "gut types." Most people had a bacterial mix dominated by organisms from the genus *Ruminococcus*. But others had gut types dominated by the genus *Bacteroides* or the genus *Prevotella*. The three different gut compositions appear to produce different vitamins and may even make a person more, or less, susceptible to certain diseases.

No one has yet identified the many microbes living on and within our bodies, but estimates suggest that each of us contains 10 times as many bacterial cells as human cells.

Surprisingly, the researchers reported that enterotype was unrelated to gender, body mass index, or nationality. So what determines your gut type? A follow-up study suggested that enterotypes were associated with long-term dietary patterns. People who eat a diet high in animal protein and saturated fats are more likely to contain a gut community dominated by *Bacteroides*. People who stick to

The human microbiome. Each part of the body has different groups of bacteria.

a high-carbohydrate diet rich in plant-based nutrition are more likely to host an abundance of *Prevotella*.

The researchers don't know yet what a person's enterotype means for his or her health. But scientists are eager to explore the link between microbiome and physical well-being. Meanwhile, various researchers are exploring the microbiome of other organisms—specifically, cows.

As it happens, there are good reasons to peer into a cow's gut. The cow's rumen (one of the four stomach chambers) contains microbes that break down the cellulose in plant material. Without them, cattle wouldn't be able to extract adequate nutrition from the grass they graze. Those microbes are of particular interest to scientists at the Department of Energy (DOE), who are interested in converting material containing cellulose, such as switchgrass, into biofuels.

As researchers look for more sustainable replacements for fossil fuels, many have pinned their hopes on biofuels. But the enzymes currently used to break down cellulose aren't efficient or cost effective enough to produce biofuels on an industrial scale. That's where the cows come in. DOE scientists recently analyzed the microbial genes present in the cow rumen. They discovered nearly 28,000 genes for proteins that metabolize carbohydrates, and at least 50 new proteins that may help break down cellulose in ways that will produce biofuels more efficiently and more cheaply.

Mammals such as cows and humans are hardly the only organisms that have symbiotic relationships with bacteria. All living animals have their own microbiomes. Even tiny single-celled eukaryotes harbor still-tinier bacterial guests.

Symbiotic bacteria provide a variety of benefits to their hosts. Consider the bobtail squid, *Euprymna scolopes*, which possesses a specialized organ to house bacteria. The bacteria, *Vibrio fischeri*, are luminescent, that is, capable of generating light. The squid use their *Vibrio*-filled light organs like spotlights, projecting light downward to match light from the surface as a way to camouflage themselves from would-be predators below.

In many cases, animals and their bacterial symbionts are so intimately connected that it's hard to separate one from the other. Many human gut bacteria are poorly studied because scientists haven't been able to grow them outside the body. But some species have taken their intimate bacterial partnership a step further.

Aphids are small insects that feed on plant sap. They rely on the bacteria *Buchnera aphidicola* to synthesize certain amino acids for them that aren't available from their plant-based diet. The partnership between aphids and *Buchnera* dates back as far as 250 million to 150 million years. Over time, the organisms have become completely dependent on one another for survival. The *Buchnera* bacteria actually live inside the aphid's cells. In some ways, the bacteria resemble organelles rather than independent organisms. Intracellular symbionts may not be so unusual. In fact, mitochondria and chloroplasts—key organelles of eukaryotic cells—are thought to have originated from symbiotic bacteria. Over time, the progenitor bacteria became so intertwined with their host cells that they developed into an integral part of the cells themselves. Studying simple sapsucking aphids, it turns out, may offer insights into the evolutionary history of all eukaryotic cells.

Clearly, there are many good reasons to learn more about our bacterial comrades. From evolution to energy to human health, our microbiomes hold great promise—and great mystery. We've evolved hand in hand with these bacteria for millions of years. Let's hope it takes less time to uncover their secrets.

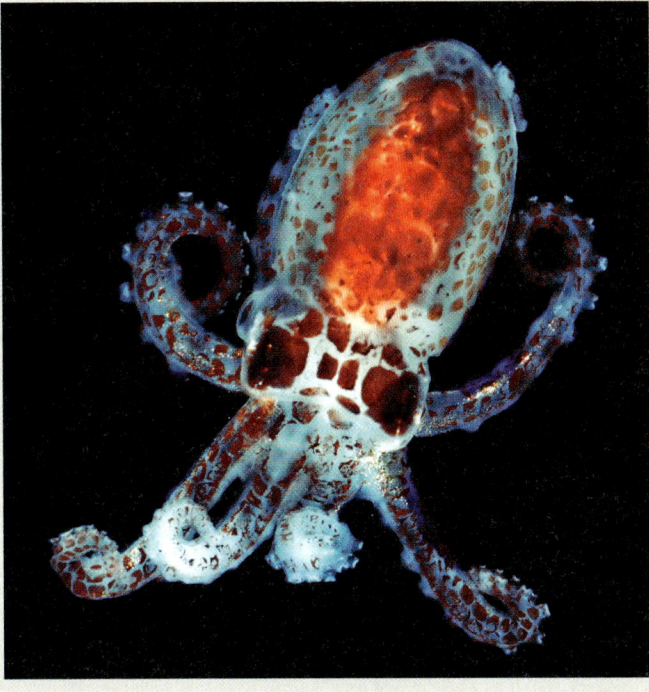

The bobtail squid and its luminescent bacteria. This squid harbors bacteria that emit light in specialized organs.

? CASE 5 QUESTIONS

Answers to Case 5 questions can be found in Chapters 26 and 27.

1. **How do intestinal bacteria influence human health?** *See page 26-22.*
2. **What role did symbiosis play in the origin of chloroplasts?** *See page 27-4.*
3. **What role did symbiosis play in the origin of mitochondria?** *See page 27-6.*
4. **How did the eukaryotic cell originate?** *See page 27-7.*

CHAPTER 26

BACTERIA AND ARCHAEA

Core Concepts

26.1 The tree of life has three main branches, called domains: Eukarya, Bacteria, and Archaea.

26.2 Bacteria and Archaea are notable for their metabolic diversity.

26.3 In addition to their key roles in the carbon cycle, Bacteria and Archaea are critical to the biological cycling of sulfur and nitrogen.

26.4 The extent of bacterial diversity was recognized only when sequencing technologies could be applied to non-culturable bacteria.

26.5 The diversity of Archaea has only recently been recognized.

26.6 The earliest forms of life on Earth were Bacteria and Archaea.

In Chapter 25, we outlined the carbon cycle as it operates in the oxygen-rich environments of our daily existence. But can carbon cycle through the deep waters of the Black Sea, within black muds beneath swamps and marshlands, or in other habitats where oxygen is limited or absent? Can biological communities survive without oxygen? The answer, emphatically, is yes. Indeed, life existed on Earth for more than a billion years before oxygen-rich habitats first appeared on our planet.

Exploration of seemingly alien oxygen-poor environments on today's Earth reveals further aspects of the carbon cycle. These habitats are populated by organisms that neither produce nor consume oxygen, yet are able to cycle carbon. The organisms responsible for this expanded cycle share one fundamental feature: They have prokaryotic cell organization. Prokaryotes, which include bacteria and archaeons, lack a nucleus (Chapter 5). We'll look at the carbon cycle again later in this chapter, but first we discuss **Bacteria** and **Archaea,** the prokaryotic domains of microscopic organisms so critical to the cycling of carbon and other elements essential to life. We examine their diversity and how they have evolved through time.

26.1 TWO PROKARYOTIC DOMAINS

Chapter 5 outlined the two distinct ways that cells are organized internally. Eukaryotic cells, like those that make up our bodies, have a membrane-bound nucleus and organelles that form separate compartments for many cell functions. Prokaryotic cells have a simpler organization. No membrane surrounds the prokaryotic cell's DNA, and there is little in the way of cell compartments. Prokaryotic cell organization is an ancestral character for life as a whole—that is, it is a feature that was present in the last common ancestor of all organisms alive today. The group defined traditionally as the prokaryotes, however, is paraphyletic in that it excludes some descendants of the last common ancestor of all living things, namely eukaryotic organisms (Chapter 23).

Bacteria and Archaea are the two great domains characterized by prokaryotic cell structure. These organisms are present almost everywhere on Earth's surface. What they lack in complexity of cell structure, these tiny cells more than make up for in their dazzling metabolic diversity. As will become clear over the course of this chapter, Bacteria and Archaea underpin the efficient operation of ecosystems on our planet.

The bacterial cell is small but powerful.

Because of their small size and deceptively simple cell organization, bacteria were long dismissed as primitive organisms, distinguished mostly by the eukaryotic features they lack: They have no membrane-bounded nuclei, no energy-producing organelles, no sex. This point of view turns out to be more than a little misleading. Bacteria are the diverse and remarkably successful products of nearly 4 billion years of evolution. On present-day Earth, bacterial cells outnumber eukaryotic cells by several orders of magnitude.

FIG. 26.1 A bacterial cell. The cells of Bacteria (and Archaea) do not have a membrane-bound nucleus or other organelles.

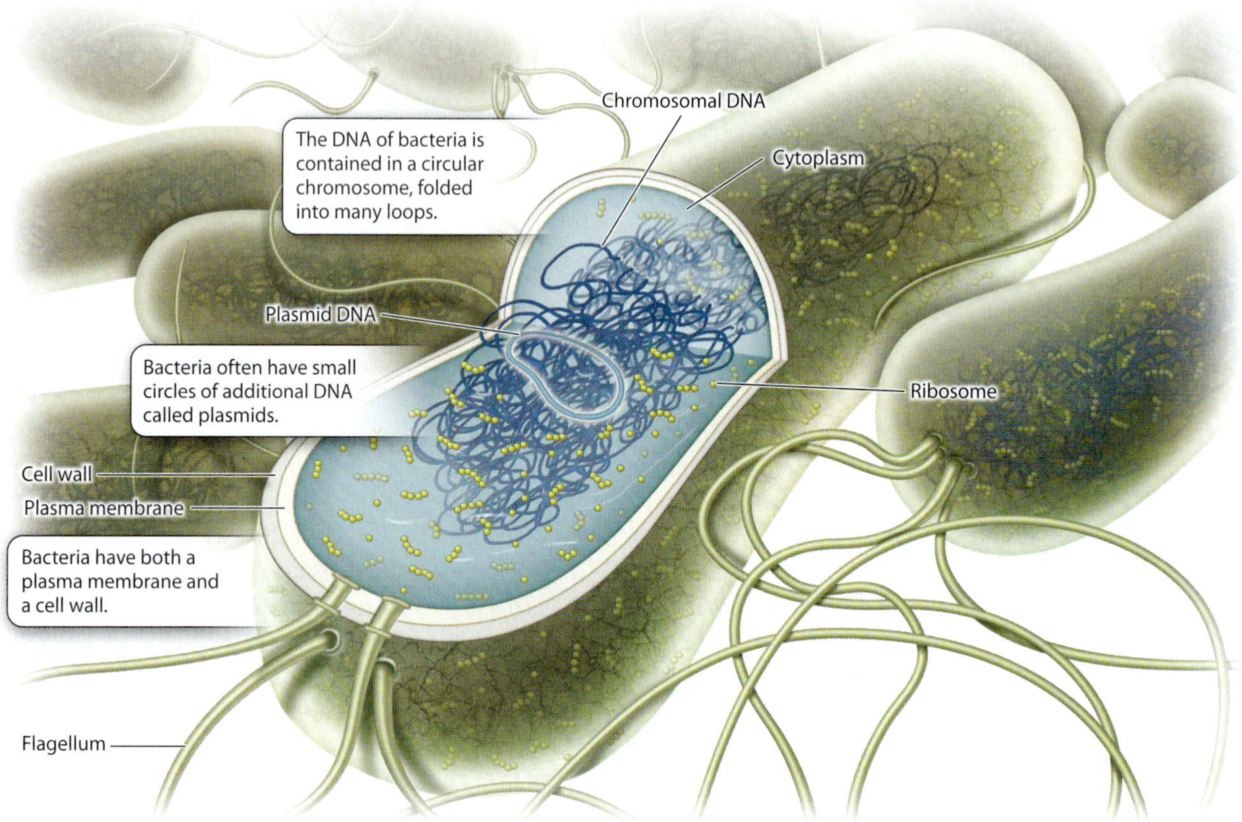

Labels on figure:
- Chromosomal DNA
- Cytoplasm
- The DNA of bacteria is contained in a circular chromosome, folded into many loops.
- Plasmid DNA
- Bacteria often have small circles of additional DNA called plasmids.
- Ribosome
- Cell wall
- Plasma membrane
- Bacteria have both a plasma membrane and a cell wall.
- Flagellum

Fig. 26.1 illustrates the bacterial cell, which was briefly introduced in Chapter 5. The cell's DNA is present in a single circular chromosome, unlike the multiple linear chromosomes characteristic of eukaryotic cells. Many bacteria carry additional DNA in the form of **plasmids,** small circles of DNA that replicate independently of the cell's circular chromosome. In general, plasmid DNA is not essential for the cell's survival, but it may contain genes that have adaptive value under specific environmental conditions. No nuclear membrane separates DNA from the surrounding cytoplasm, and so transcribed mRNA is immediately translated into proteins by ribosomes.

Bacteria lack the membrane-bounded organelles found in eukaryotic cells. Instead, cell processes such as metabolism are carried out by proteins that float freely in the cytoplasm or are embedded in the plasma membrane. A few bacteria, notably the photosynthetic bacteria, contain internal membranes. The light reactions of photosynthetic bacteria take place in association with membranes distributed within the cytoplasm.

Structural support is provided by a **cell wall** made of **peptidoglycan,** a complex polymer of sugars and amino acids (Chapter 10). Some bacteria have thick walls made up of multiple peptidoglycan layers, while others have thin walls surrounded by an outer layer of lipids. For many years, it was believed that bacteria lacked the cytoskeletal framework that organizes cytoplasm in eukaryotic cells. However, careful studies now show that bacteria do have an internal scaffolding of proteins that plays an important role in determining the shape, polarity, and other spatial properties of bacterial cells.

Diffusion limits cell size in bacteria.

Most bacterial cells are tiny: The smallest are only 200–300 nanometers (nm) in diameter, and relatively few are more than 1–2 micrometers (μm) long. Why are bacteria so small? The answer has to do with **diffusion,** the random motion of molecules, a critical process introduced in Chapter 5. When there is a concentration difference, diffusion results in net movement of molecules from a region of higher concentration to a region of lower concentration. If you could watch the movement of any particular molecule, you would see it is random, sometimes moving in one direction and sometimes in another. On average, however, more molecules move from a region with a higher concentration of the molecule to a region with a lower concentration of the molecule than in the other direction. Net movement stops only when the two regions achieve equal concentration, but diffusion continues. Photosynthetic bacteria gain the carbon dioxide they need by the diffusion of CO_2 from

CHAPTER 26

BACTERIA AND ARCHAEA

Core Concepts

26.1 The tree of life has three main branches, called domains: Eukarya, Bacteria, and Archaea.

26.2 Bacteria and Archaea are notable for their metabolic diversity.

26.3 In addition to their key roles in the carbon cycle, Bacteria and Archaea are critical to the biological cycling of sulfur and nitrogen.

26.4 The extent of bacterial diversity was recognized only when sequencing technologies could be applied to non-culturable bacteria.

26.5 The diversity of Archaea has only recently been recognized.

26.6 The earliest forms of life on Earth were Bacteria and Archaea.

In Chapter 25, we outlined the carbon cycle as it operates in the oxygen-rich environments of our daily existence. But can carbon cycle through the deep waters of the Black Sea, within black muds beneath swamps and marshlands, or in other habitats where oxygen is limited or absent? Can biological communities survive without oxygen? The answer, emphatically, is yes. Indeed, life existed on Earth for more than a billion years before oxygen-rich habitats first appeared on our planet.

Exploration of seemingly alien oxygen-poor environments on today's Earth reveals further aspects of the carbon cycle. These habitats are populated by organisms that neither produce nor consume oxygen, yet are able to cycle carbon. The organisms responsible for this expanded cycle share one fundamental feature: They have prokaryotic cell organization. Prokaryotes, which include bacteria and archaeons, lack a nucleus (Chapter 5). We'll look at the carbon cycle again later in this chapter, but first we discuss **Bacteria** and **Archaea**, the prokaryotic domains of microscopic organisms so critical to the cycling of carbon and other elements essential to life. We examine their diversity and how they have evolved through time.

26.1 TWO PROKARYOTIC DOMAINS

Chapter 5 outlined the two distinct ways that cells are organized internally. Eukaryotic cells, like those that make up our bodies, have a membrane-bound nucleus and organelles that form separate compartments for many cell functions. Prokaryotic cells have a simpler organization. No membrane surrounds the prokaryotic cell's DNA, and there is little in the way of cell compartments. Prokaryotic cell organization is an ancestral character for life as a whole—that is, it is a feature that was present in the last common ancestor of all organisms alive today. The group defined traditionally as the prokaryotes, however, is paraphyletic in that it excludes some descendants of the last common ancestor of all living things, namely eukaryotic organisms (Chapter 23).

Bacteria and Archaea are the two great domains characterized by prokaryotic cell structure. These organisms are present almost everywhere on Earth's surface. What they lack in complexity of cell structure, these tiny cells more than make up for in their dazzling metabolic diversity. As will become clear over the course of this chapter, Bacteria and Archaea underpin the efficient operation of ecosystems on our planet.

The bacterial cell is small but powerful.

Because of their small size and deceptively simple cell organization, bacteria were long dismissed as primitive organisms, distinguished mostly by the eukaryotic features they lack: They have no membrane-bounded nuclei, no energy-producing organelles, no sex. This point of view turns out to be more than a little misleading. Bacteria are the diverse and remarkably successful products of nearly 4 billion years of evolution. On present-day Earth, bacterial cells outnumber eukaryotic cells by several orders of magnitude.

FIG. 26.1 **A bacterial cell.** The cells of Bacteria (and Archaea) do not have a membrane-bound nucleus or other organelles.

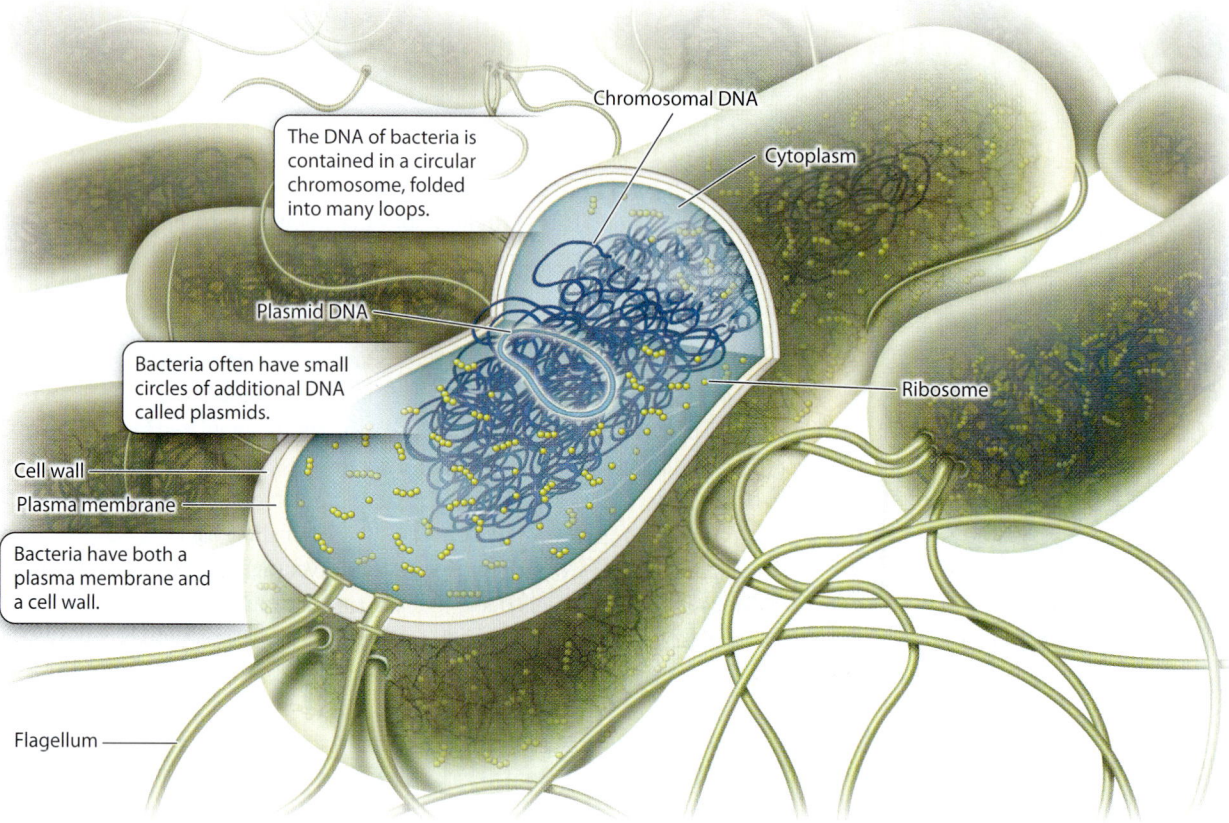

Fig. 26.1 illustrates the bacterial cell, which was briefly introduced in Chapter 5. The cell's DNA is present in a single circular chromosome, unlike the multiple linear chromosomes characteristic of eukaryotic cells. Many bacteria carry additional DNA in the form of **plasmids,** small circles of DNA that replicate independently of the cell's circular chromosome. In general, plasmid DNA is not essential for the cell's survival, but it may contain genes that have adaptive value under specific environmental conditions. No nuclear membrane separates DNA from the surrounding cytoplasm, and so transcribed mRNA is immediately translated into proteins by ribosomes.

Bacteria lack the membrane-bounded organelles found in eukaryotic cells. Instead, cell processes such as metabolism are carried out by proteins that float freely in the cytoplasm or are embedded in the plasma membrane. A few bacteria, notably the photosynthetic bacteria, contain internal membranes. The light reactions of photosynthetic bacteria take place in association with membranes distributed within the cytoplasm.

Structural support is provided by a **cell wall** made of **peptidoglycan,** a complex polymer of sugars and amino acids (Chapter 10). Some bacteria have thick walls made up of multiple peptidoglycan layers, while others have thin walls surrounded by an outer layer of lipids. For many years, it was believed that bacteria lacked the cytoskeletal framework that organizes cytoplasm in eukaryotic cells. However, careful studies now show that bacteria do have an internal scaffolding of proteins that plays an important role in determining the shape, polarity, and other spatial properties of bacterial cells.

Diffusion limits cell size in bacteria.

Most bacterial cells are tiny: The smallest are only 200–300 nanometers (nm) in diameter, and relatively few are more than 1–2 micrometers (μm) long. Why are bacteria so small? The answer has to do with **diffusion,** the random motion of molecules, a critical process introduced in Chapter 5. When there is a concentration difference, diffusion results in net movement of molecules from a region of higher concentration to a region of lower concentration. If you could watch the movement of any particular molecule, you would see it is random, sometimes moving in one direction and sometimes in another. On average, however, more molecules move from a region with a higher concentration of the molecule to a region with a lower concentration of the molecule than in the other direction. Net movement stops only when the two regions achieve equal concentration, but diffusion continues. Photosynthetic bacteria gain the carbon dioxide they need by the diffusion of CO_2 from

CHAPTER 26 BACTERIA AND ARCHAEA

FIG. 26.2 Cell shape and size in Bacteria and Archaea.

a. *Streptococcus*, strings of spheroidal or coccoidal bacteria

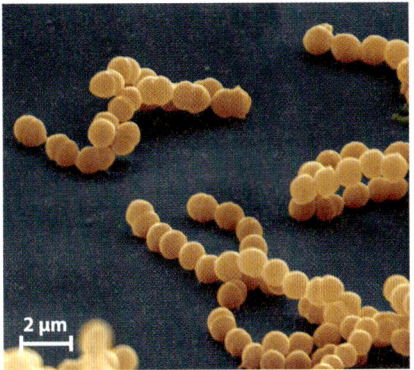

b. *E. coli*, bacterial rods

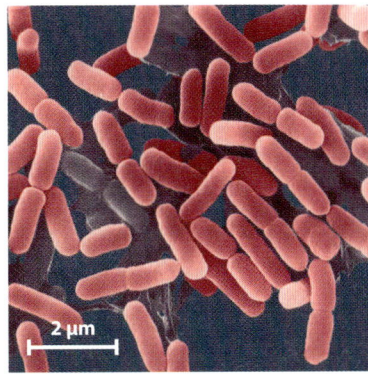

c. *Haloquadratum walsbyi*, a square archaeon that lives in salt ponds

d. *Streptomyces*, helical bacteria that produce antibiotics

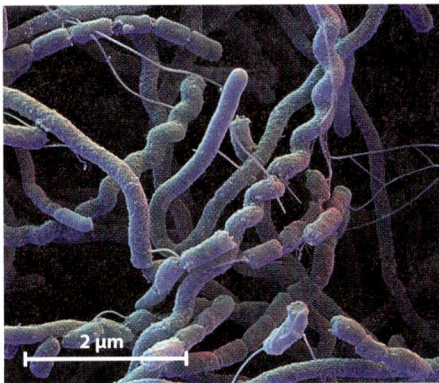

e. A myxobacterium, a bacterium in which cells aggregate to form fruiting bodies

the environment into the cell, and that is also how respiring bacteria take in small organic molecules and oxygen.

Diffusion explains why bacterial cells tend to be small. A small cell has more surface area in proportion to its volume, and so the interior parts of a small cell are closer to the surrounding environment than those of a larger cell. As a consequence, slowly diffusing molecules do not have to travel far to reach every part of a small cell's interior. The surface area of a spherical cell—the area available for taking up molecules from the environment—increases as the square of the radius. However, the cell's volume—the amount of cytoplasm that is supported by diffusion—increases as the cube of the radius. Therefore, as cell size increases, it becomes harder to supply the cell with the materials needed for growth. For this reason, most bacterial cells are small spheres, rods, spirals, or filaments that facilitate diffusion of molecules into cell interiors (**Fig. 26.2**).

A few exceptional bacteria exceed 100 μm in maximum dimension. For example, *Thiomargarita namibiensis*, a bacterium that lives in oxygen-poor sediments off the coast of southwestern Africa, has a total volume about 100,000,000 times larger than that of *Escherichia coli* (**Fig. 26.3**). But, in one sense, *T. namibiensis* cheats: 98% of its volume is taken up by a large vacuole, so the metabolically active cytoplasm is restricted to a thin film around the cell's periphery. Thus, the distance through which nutrients move by diffusion is only a few miocrometers, as in many other bacteria.

Some bacteria are multicellular, forming simple filaments or sheets of cells. More unusual are myxobacteria, which aggregate to form multicellular reproductive structures that are composed of several distinct cell types (see Fig. 26.2e).

FIG. 26.3 *Thiomargarita namibiensis*, the largest known bacterial cell. The volume of the cell is mostly taken up by a vacuole, so the active cytoplasm is only a few micrometers thick.

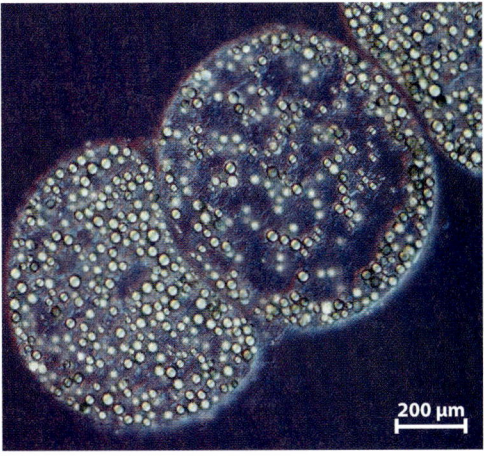

Horizontal gene transfer promotes genetic diversity in bacteria.

Bacterial genomes are generally smaller than those of eukaryotes, in part because bacteria lack the large stretches of noncoding DNA characteristic of eukaryotic chromosomes (Chapter 13). The streamlining of the bacterial genome confers certain benefits. For example, bacteria can reproduce rapidly when nutrients required for growth are available in the local environment. Bacteria replicate their DNA from only one or a small number of initiation sites, so genome size can influence rate of reproduction.

Bacteria do not undergo meiotic cell division and cell fusion. That is, they lack the sexual processes characteristic of eukaryotic organisms (Chapter 11). Despite this, bacterial populations display remarkable genetic diversity. For example, different strains of *Pseudomonas aeruginosa*, a common disease-causing organism, may differ in genome size by nearly a factor of 2 (3.7 million base pairs compared to 7.1 million base pairs), yet these strains are quite similar in function.

How does this variation arise? In eukaryotic organisms, genes generally pass from parent to offspring. Bacteria also inherit most of their genes from parental cells. However, bacterial cells can also obtain new genes from distant relatives. Called **horizontal gene transfer,** this process is a major source of genetic diversity in bacteria.

How does DNA move from one bacterial cell into another independently of cell division? Some bacteria synthesize thin strands of membrane-bound cytoplasm called pili that connect them to other cells (**Fig. 26.4a**). A pilus provides a migration route for the direct cell-to-cell transfer of DNA. This process, called **conjugation,** commonly transfers plasmids from one cell to another, spreading novel genes throughout a population. Genes that confer resistance to antibiotics are a well-studied example of horizontal gene transfer by conjugation. Biologists interested in genetic engineering use conjugation in the laboratory to introduce genes into cells. But conjugation is just the beginning of genetic exchange in bacteria.

Genes can be transferred from one cell to another without any direct bridge between

FIG. 26.4 Horizontal gene transfer in bacteria. DNA shown in red originates from the donor cell. DNA shown in blue is that of the recipient cell.

a. DNA transfer by conjugation

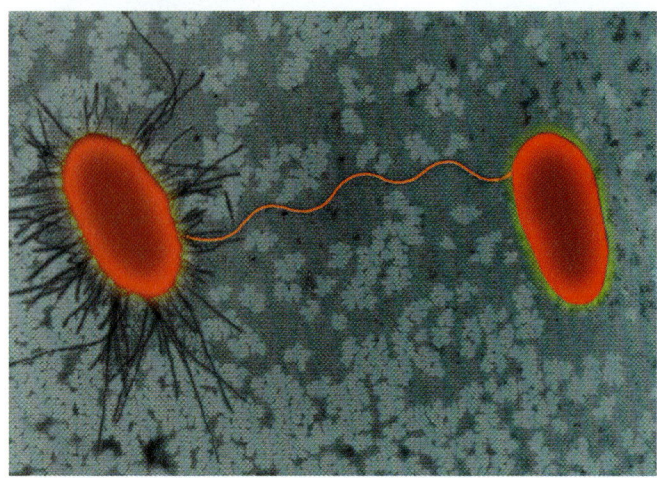

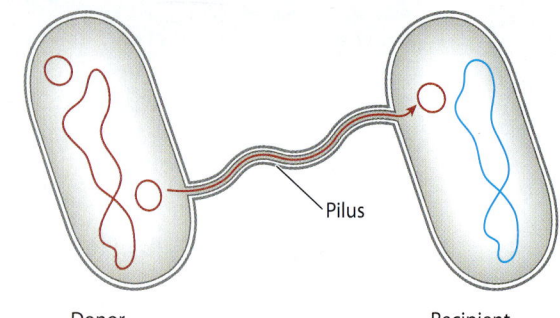

In conjugation, DNA (usually a plasmid) from a donor cell is transferred through a pilus into the recipient cell.

b. DNA transfer by transformation

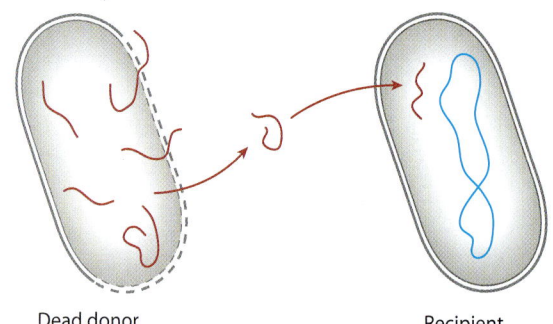

In transformation, DNA released into the environment by dead cells is taken up by a recipient cell.

c. DNA transfer by transduction

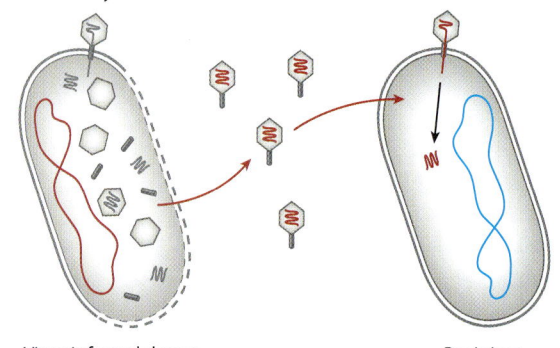

In transduction, DNA is transferred from a donor to a recipient cell by a virus.

the cells. DNA released to the environment by cell breakdown can be taken up by other cells, a process called **transformation** (**Fig. 26.4b**). Transformation was revealed when experiments showed that harmless strains of the bacteria causing pneumonia could be transformed into virulent strains by exposure to media containing dead cells of disease-causing strains (Chapter 3). Scientists reasoned that the transformation occurred because some substance was being taken up by the living bacteria. The "transforming substance" was later shown to be DNA.

Viruses provide a third mechanism of horizontal gene transfer. Recall from Chapter 19 that viruses in bacterial cells sometimes integrate their DNA into the host bacterial DNA and persist within the cells as they grow and divide. Before leaving the cell to infect others, the viral DNA removes itself from the bacterial genome. This excision is not always precise, and sometimes additional genetic material from the bacterial host is incorporated into the virus. Viruses released from their host cell go on to infect others, bringing host-derived genes with them. Horizontal gene transfer by means of viruses is called **transduction** (**Fig. 26.4c**). It is common in nature and is also widely used in the laboratory to to introduce novel genes into bacteria for medical research.

Horizontal gene transfer allows bacterial cells to gain beneficial genes from organisms distributed throughout the bacterial domain and beyond. Indeed, bacteria everywhere are constantly reshaping their genomes. Bacteria, therefore, are not the poor cousins of eukaryotes, unable to generate genetic diversity by sexual recombination. Instead, bacteria have highly efficient mechanisms for adding and subtracting genes that permit them to evolve and adapt rapidly to local conditions. Perhaps the most widely discussed and worrisome manifestation of horizontal gene transfer is the rapid spread of antibiotic resistance among bacteria (Chapter 48).

→ **Quick Check 1** In eukaryotes, sexual reproduction is the main process by which new gene combinations are generated. How do bacteria generate new gene combinations in the absence of sexual reproduction?

The Archaea form a second prokaryotic domain.

For many years, all prokaryotic cells were classified as bacteria. However, in 1977, George Fox and Carl Woese published a revolutionary hypothesis. From comparisons of RNA molecules from the small subunits of ribosomes (and later, comparisons of the genes for these RNAs), Fox

FIG. 26.5 **A three-branched tree of life, based on sequence comparisons of the genes for the small subunit of ribosomal RNA.** Archaea and Bacteria form distinct branches, although they inherited prokaryotic cell organization from a common ancestor.

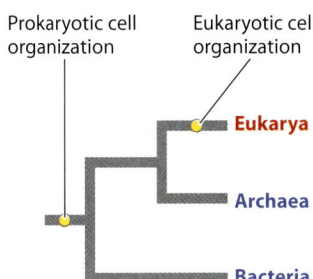

and Woese argued that prokaryotic organisms actually fall into two distinct groups, as different from each other as either is from eukaryotes. If this is true, the tree of life must have three great branches: the Bacteria, the Eukarya, and the limb recognized by Fox and Woese, called the Archaea (**Fig. 26.5**). Research since the late 1970s confirms this hypothesis. Like bacterial cells, archaeal cells are prokaryotic—they have no membrane-bound nucleus and their genes are arrayed along a single circular chromosome. Moreover, cell size in Archaea is limited by diffusion, and genetic diversity is promoted by horizontal gene transfer, much as in Bacteria.

In detail, however, the Archaea are quite distinctive (**Table 26.1**). Their membranes are made from lipids different from the fatty acids found in bacterial and eukaryotic membranes.

TABLE 26.1 Principal Differences among Archaea, Bacteria, and Eukarya

CHARACTERISTIC	ARCHAEA	BACTERIA	EUKARYA
Cell contains a nucleus and other membrane-bound organelles	No	No	Yes
DNA occurs in a circular form*	Yes	Yes	No
Ribosome size	70S	70S	80S
Membrane lipids ester-linked[†]	No	Yes	Yes
Photosynthesis with chlorophyll	No	Yes	Yes
Capable of growth at temperatures greater than 80°C	Yes	Yes	No
Histone proteins present in cell	Yes	No	Yes
Operons present in DNA	Yes	Yes	No
Introns present in most genes	No	No	Yes
Capable of methanogenesis	Yes	No	No
Sensitive to the antibiotics chloramphenicol, kanamycin, and streptomycin	No	Yes	No
Capable of nitrogen fixation	Yes	Yes	No
Capable of chemoautotrophy	Yes	Yes	No

*Eukaryote DNA is linear. [†]Archaea membrane lipids are ether-linked.

Archaea also show a diversity of molecules in their cell walls, but none has the peptidoglycan characteristic of Bacteria or the cellulose and chitin found in most eukaryotic cell walls.

Archaea differ from Bacteria in another intriguing way. DNA transcription in archaeons employs RNA polymerase and ribosomes more similar to those of eukaryotes than to bacteria (Chapter 3). Moreover, many of the antibiotics that target protein synthesis in bacteria are ineffective against archaeons, suggesting fundamental differences in translation as well.

Many of the microorganisms first identified as Archaea have unusual physiological properties or inhabit extreme environments. For example, some archaeons live in acid mine water at pH 1 (or less!). Others live in water salty enough to precipitate NaCl (**Fig. 26.6**), or in deep-sea hydrothermal vents where temperatures can exceed 100°C. We now know, however, that many archaeons live under less extreme conditions in soils, lakes, and the sea. Indeed, they may be the most abundant organisms throughout much of the ocean.

→ **Quick Check 2** Why were Archaea originally thought to be simply unusual forms of Bacteria? What lines of evidence showed this domain to form a distinct branch on the tree of life?

26.2 AN EXPANDED CARBON CYCLE

In our discussion of energy metabolism in Chapters 7 and 8 and of the carbon cycle in Chapter 25, we emphasized oxidation–reduction (redox) chemistry. In redox reactions, a pair of molecules reacts, one molecule becoming more oxidized and the other becoming more reduced. If we follow the transfer of electrons in a redox reaction, we see that the molecule that is reduced gains electrons, whereas the molecule that is oxidized loses electrons. Reduction therefore requires a source of electrons, or an electron donor, and oxidation requires a sink for electrons, or an electron acceptor.

In photosynthesis, as we discussed in Chapter 8, carbon dioxide (CO_2) is reduced to carbohydrates, and water is oxidized to oxygen gas (O_2). Water is the electron donor needed to reduce CO_2, generating O_2 as a by-product. In respiration, as described in Chapter 7, organic molecules are oxidized to CO_2, and O_2 is reduced to water. In this case, O_2 serves as the electron acceptor needed to oxidize organic molecules.

In all known photosynthetic eukaryotes, the photosynthetic reaction is **oxygenic**, or oxygen producing. There is a good reason for this. Water occurs nearly everywhere and, in the oxygen-rich environments where most eukaryotic organisms thrive, no other molecule is available to donate electrons. Similarly, essentially all respiration in eukaryotic cells is **aerobic**, or oxygen utilizing. Again, this makes good chemical sense because oxidation of carbohydrates using oxygen generates far more energy than can be obtained using other oxidants. Nonetheless, where oxygen is limited or absent, other electron donors are available for photosynthesis, and other electron acceptors can be used for respiration. The organisms that can use these alternative electron donors and electron acceptors are Bacteria and Archaea.

We've already noted that Bacteria and Archaea differ from Eukarya in cell structure, biochemistry, and genome organization, but it is the distinctive ways in which Bacteria and Archaea gather carbon and harvest energy that make them indispensable parts of nature. Prokaryotic organisms are thus the real foundation of the carbon cycle, capable of building and dismantling organic molecules throughout the full breadth of habitats on Earth.

FIG. 26.6 **Archaeons in San Francisco Bay.** The ponds are flooded and then evaporated to precipitate table salt. Salt-loving archaeons, made conspicuous by their purple pigment, thrive here.

FIG. 26.7 Microbial mats. Mats, like the example in (a) from Western Australia, cover tidal flats and lagoons where salinity or other environmental factors inhibit animals and seaweeds. As shown by the example from Cape Cod in (b), microbial mats have sharp vertical gradients of light, oxygen, and other chemical compounds, and so support nearly all the energy metabolisms found in the expanded carbon cycle (c).

a.

b.

The blue-green pigment identifies cyanobacteria, the oxygenic photosynthetic bacteria that dominate the well-lit, oxygen-rich mat surface.

The purple pigment marks purple bacteria, anoxygenic photosynthetic bacteria that live where there is light but no oxygen gas.

In the black subsurface layer, anaerobic respiration and fermentation support microbial populations.

c.

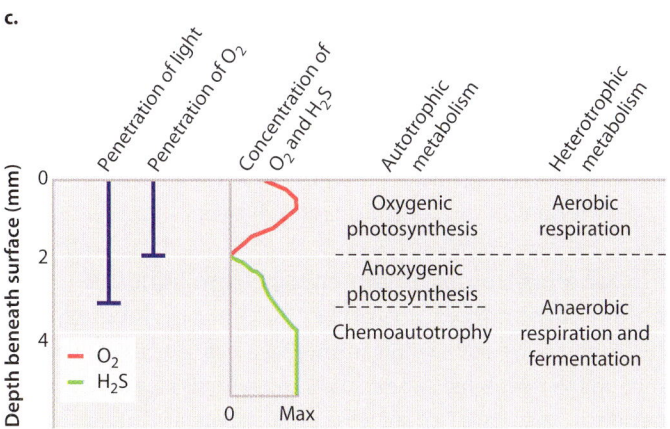

Many photosynthetic bacteria do not produce oxygen.

In saline bays along the west coast of Australia and many other low-lying tropical coastlines, carpets of deep blue-green cover tidal flats along restricted lagoons (**Fig. 26.7a**). Called microbial mats, these carpets are densely packed communities of (mostly) bacteria and archaeons that thrive where animals and seaweeds cannot grow. The top surface of a microbial mat is dominated by cyanobacteria, bacteria that photosynthesize like plants and algae, using water as an electron donor and releasing oxygen gas (**Fig. 26.7b**). It is the photosynthetic pigments in the cyanobacteria that impart the blue-green color to microbial mats. The cyanobacteria (which synthesize their own organic molecules) and other, heterotrophic bacteria (which get their carbon from preformed organic molecules) respire aerobically, using O_2 to oxidize organic molecules to CO_2. Thus, at the mat surface, prokaryotic microorganisms carry out a biological carbon cycle much like that for plants and animals described in Chapter 25.

What happens beneath the mat surface is a different story. Mats generate a great deal of organic matter, and aerobic respiration of this material consumes O_2 rapidly. In the mat interior below the level at which cyanobacteria supply oxygen during the day, O_2 is supplied only slowly by diffusion, with the result that O_2 becomes depleted within a few millimeters of the mat surface. Light is also greatly reduced beneath the mat surface, but in many mats, light actually penetrates a few millimeters deeper than oxygen (**Fig. 26.7c**). What organisms live where light is present and oxygen absent?

If we label CO_2 with the radioisotope ^{14}C, we find that carbon dioxide continues to be reduced to carbohydrates within this subsurface zone of the mat. In fact, the newly generated carbohydrates, identified by the ^{14}C they have incorporated, occur in cells with bright purple or green pigments (**Fig. 26.7b**). These are **anoxygenic** photosynthetic bacteria, photosynthetic microorganisms that harvest light energy to drive the synthesis of carbohydrates, but which do not gain electrons from water and so do not generate oxygen gas in the process.

Anoxygenic photosynthetic bacteria absorb sunlight using **bacteriochlorophyll,** a light-harvesting pigment closely related to the chlorophyll found in plants, algae, and cyanobacteria. **Fig. 26.8a** shows the chemical structure of one form of bacteriochlorophyll, called bacteriochlorophyll *a*, and one form of chlorophyll, called chlorophyll *a*. Note the similarity in their overall structure. The two molecules absorb light of similar wavelengths as well, though small differences in their chemical structures account for small differences in which wavelengths they absorb more strongly (**Fig. 26.8b**).

Anoxygenic photosynthesis differs from oxygenic photosynthesis in another important way: It employs only a single photosystem. As discussed in Chapter 8, the amount of energy that can be captured this way is not sufficient to pull electrons from (that is, to oxidize) water and then use those electrons to reduce CO_2, and so oxygenic photosynthesis has two photosystems in series. With only a single

FIG. 26.8 Structures and absorption spectra of bacteriochlorophyll a and chlorophyll a. Note that bacteria absorb wavelengths of light that are outside the visible spectrum.

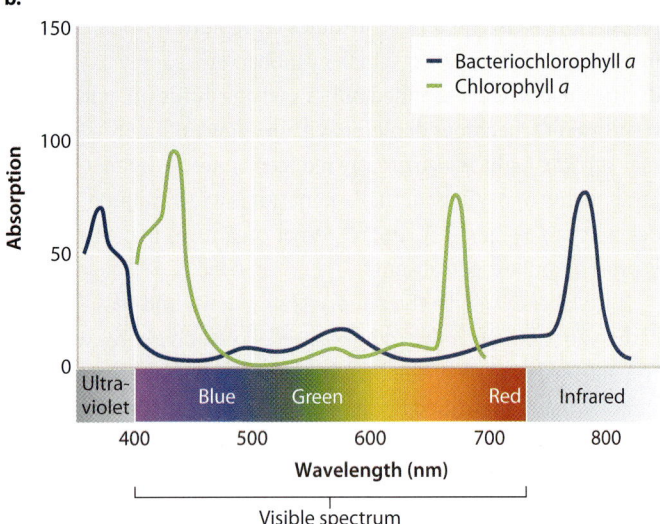

The differences in structure (shown in red) account for differences in wavelengths absorbed by the two molecules.

photosystem, bacteria that carry out anoxygenic photosynthesis use electron donors that donate their electrons more easily than water. As a result, these microorganisms do not release O_2 as a by-product. Electron donors in anoxygenic photosynthesis include hydrogen sulfide (H_2S), hydrogen gas (H_2), ferrous iron (Fe^{2+}), and even the arsenic compound arsenite (AsO_3^{3-}). Where O_2 is present, these compounds themselves become oxidized to other molecules and so are not available for photosynthesis. Therefore, microorganisms that conduct anoxygenic photosynthesis are restricted to sunlit habitats where O_2 is absent.

Many bacteria respire without oxygen.

In microbial mat communities, heterotrophic microorganisms in the upper, oxygen-rich portion of the mat respire aerobically, just as we do, gaining carbon by taking in organic molecules from their surroundings and energy by using oxygen as an electron acceptor for the oxidation of those molecules. Does heterotrophy simply cease below the level where O_2 disappears?

Not at all. Careful measurements show that organic molecules can be oxidized in the absence of O_2. Oxygen is not unique in its capacity to accept electrons in redox reactions involving organic molecules, but where O_2 is present it is used preferentially for respiration. This is because more energy is released by the oxidation of organic compounds by oxygen gas than by any alternative oxidant. In habitats where O_2 is absent, oxidized forms of nitrogen (NO_3^-), sulfur (SO_4^{2-}), manganese (Mn^{4+}), iron (Fe^{3+}), and even arsenic (AsO_4^{3-}) are used as electron acceptors in cellular respiration.

Fermentation provides an alternative to cellular respiration as a way of extracting energy from organic molecules. Whereas cellular respiration is the full oxidation of carbon compounds to CO_2, fermentation is the partial oxidation of carbon compounds to molecules that are less oxidized than CO_2 (Chapter 7). Fermentation has one advantage over cellular respiration in that it does not require an external electron acceptor, such as O_2. On the other hand, fermentation yields only a modest amount of energy.

Fermentation plays an important role in oxygen-poor environments that are rich in organic matter, such as landfills and the digestive tracts of animals. Here, the breakdown of organic molecules typically requires more than one organism, each able to metabolize different intermediates. These cooperating groups of fermenters enable the breakdown of substances that could not be metabolized by any one organism alone. Groups of fermenters thus form what is essentially an ecological solution to the metabolically difficult problem of gaining energy from complex organic molecules. Fermentation is widespread in Bacteria and Archaea, but except for yeasts, which are world-class fermenters, fermentation is of only minor importance in the energy metabolism of most Eukarya.

Therefore, where O_2 is present, many different kinds of organisms participate in the carbon cycle. Photosynthetic plants, algae, and cyanobacteria all transform CO_2 into organic molecules, and animals, fungi, single-celled eukaryotes, bacteria, and archaeons return CO_2 to the environment by aerobic respiration. In Chapter 25, we noted that this biological carbon cycle leaks a bit. In other words, some of the organic carbon produced by photosynthesis escapes aerobic respiration and accumulates in oxygen-depleted waters or sediments. Only prokaryotic heterotrophs can complete the recycling of organic carbon in oxygen-poor environments. Thus, in carrying out anaerobic respiration or a diversity of fermentation reactions, prokaryotes play a major role in the carbon cycle. The key point is that prokaryotic organisms are required to sustain Earth's carbon cycle, while eukaryotic organisms are optional.

Photoheterotrophs obtain energy from light but obtain carbon from preformed organic molecules.

In Chapter 6, we saw that organisms have two sources of energy: the sun (phototrophs) and chemical compounds (chemotrophs). Organisms also have two sources of the carbon needed for growth and other functions: inorganic molecules like CO_2 (autotrophs) and organic molecules like glucose (heterotrophs). Taken together, this means that there are four ways that organisms acquire the energy and carbon they need (see Fig. 6.1).

Up to this point, we have considered only two of these ways. Plants, algae, and cyanobacteria are photoautotrophs, gaining energy from the sunlight and carbon from CO_2, whereas animals, fungi, and many prokaryotes are chemoheterotrophs, gaining energy and carbon from organic molecules taken up from the environment. Among the Bacteria and Archaea, additional metabolisms are possible. Some microorganisms use the energy from sunlight to make ATP, just as plants do, but rather than reducing CO_2 to make their own organic molecules, they rely on organic molecules obtained from the environment as the source of carbon for growth and other vital functions. These organisms are known as **photoheterotrophs.**

Photoheterotrophy can be advantageous in environments rich in dissolved organic compounds. It allows organisms to use all their absorbed light energy to make ATP while directing all absorbed organic molecules toward growth and reproduction. Examples of organisms capable of photoheterotrophy include heliobacteria and most green nonsulfur bacteria.

Chemoautotrophy is a uniquely prokaryotic metabolism.

On the deep seafloor, animals are relatively uncommon, their abundance limited by the slow descent of organic matter from surface oceans. Locally, however, where hydrothermal springs punctuate the deep seafloor, animal populations can be remarkably dense (**Fig. 26.9**). Why are animals so abundant around hydrothermal vents?

A clue comes from microbial mats. In mats where oxygen penetrates more deeply than light, CO_2 continues to get reduced into organic molecules along the base of the zone containing oxygen. What is happening within the mat?

The organisms that incorporate CO_2 into organic molecules deep within microbial mats and the organisms that provide food for abundant animals around deep-sea hydrothermal vents employ similar strategies for obtaining energy and carbon. Like plants and other autotrophs, they gain carbon by reducing CO_2 to form carbohydrates. However, they obtain the energy to fuel this process not from sunlight but from chemical reactions. Hence, these microorganisms are called **chemoautotrophs.**

Our discussions of respiration emphasized that energy released by the oxidation of organic matter is used to generate ATP (Chapter 7).

Chemoautotrophic microorganisms also use chemical reactions to generate ATP, but they use inorganic molecules present in their local environment. Chemoautotrophic prokaryotes use the oxidation of molecules such as H_2, H_2S, and even Fe^{2+} to generate the ATP and reducing power required to incorporate CO_2 into organic molecules. In chemoautotrophic metabolism, the electron acceptor is most commonly O_2 or nitrate (NO_3^-).

Because chemoautotrophy requires access to both oxidized and reduced molecules, chemoautotrophs tend to live along the interface between oxygen-rich and oxygen-poor environments. Microbial mats commonly have a sharp boundary between oxygen-rich and oxygen-poor layers, and at mid-ocean hydrothermal ridges, chemoautotophic prokaryotes (and the animals that harvest them) thrive where reduced gases released from Earth's interior meet oxygen-rich seawater. The ability to use inorganic sources of chemical energy is widespread among Bacteria and Archaea, but entirely absent in Eukarya.

→ **Quick Check 3** How could the biological carbon cycle have worked on the primitive Earth, where oxygen gas was essentially absent from the atmosphere and oceans?

FIG. 26.9 Deep-sea hydrothermal vents. Chemoautotrophs fuel the carbon cycle in these environments.

26.3 OTHER BIOGEOCHEMICAL CYCLES

Let's think further about prokaryotic metabolism. Anoxygenic photosynthetic bacteria commonly use reduced sulfur compounds as electron donors for photosynthesis (instead of water), and bacterial heterotrophs often use oxidized sulfur compounds as electron acceptors in anaerobic respiration (instead of oxygen gas). By reducing oxidized sulfur compounds in the course of anaerobic respiration, microorganisms generate the reduced sulfur compounds required for anoxygenic photosynthesis. The reverse is also true: Oxidation of sulfur compounds during anoxygenic photosynthesis generates the oxidized sulfur molecules used in anaerobic respiration.

Once again, we see complementary metabolic pathways. The expanded carbon cycle of Bacteria and Archaea also promotes the cycling of sulfur, nitrogen, and other biologically important elements.

Bacteria and archaeons dominate Earth's sulfur cycle.

Our bodies are mostly carbon, oxygen, and hydrogen. We also contain about 0.2% sulfur by weight. Sulfur is a component of the amino acids cysteine and methionine and therefore is present in many proteins (Chapter 2). Sulfur is also present in iron–sulfur clusters that are key components of enzymes fundamental to electron transfer. It occurs, as well, in some membranes and is found in a number of vitamins and other cofactors, including coenzyme A, thiamine (vitamin B_1), and biotin. Where do we get the sulfur we need? From the food we eat—protein in the steak or fish on our dinner plate provides a rich source of sulfur. Where do cows and fish get the sulfur they need? Like us, from the food they eat. Food chains don't go on forever, so somewhere in the system, some organisms must take up inorganic sulfur from the environment. **Primary producers,** organisms that reduce CO_2 to form carbohydrates, fill this role.

On land, plants take up sulfate (SO_4^{2-}) ions from the soil and reduce them within their cells to hydrogen sulfide (H_2S) that can be incorporated into cysteine and other biomolecules (**Fig. 26.10**). This process is called **assimilation.** Algae and photosynthetic bacteria do much the same thing in lakes, rivers, and oceans. Why don't primary producers take up H_2S directly? There are two answers to this question. First, H_2S is rapidly oxidized in the presence of oxygen and so does not occur in environments where oxygenic photosynthesis is common. Second, H_2S is generally toxic to eukaryotic organisms, so plants and algae do not thrive where it is abundant. (The H_2S produced within eukaryotic cells has a short lifetime and is restricted to intracellular sites distant from those that are vulnerable to H_2S.)

We've now seen half the sulfur cycle, the conversion of sulfate to H_2S within cells. How did sulfate molecules get into the soil in the first place, and what happens to biological sulfur

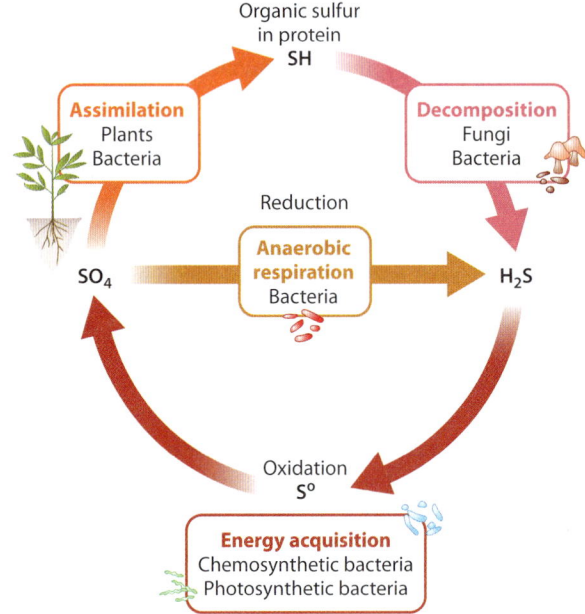

FIG. 26.10 The sulfur cycle. Plants take up sulfur as sulfate ions for incorporation into proteins and other compounds, and fungi release sulfur during decomposition. However, the major role in cycling sulfur through the biosphere is played by microbes.

compounds when cells die? After death, fungi and bacteria decompose cells, returning carbon, sulfur, and other compounds to the environment. Reduced sulfur compounds released from decomposing cells are then oxidized by bacteria and archaeons, completing the cycle (Fig. 26.10).

How do those reduced sulfur compounds get oxidized? There are two pathways, both of which involve microorganisms (Fig. 26.10). Chemoautotrophic bacteria react H_2S with O_2 to obtain energy, producing sulfate in the process. And in anoxygenic photosynthesis, bacteria use electrons from H_2S to form carbohydrates from CO_2, again generating sulfate as a by-product. The sulfate produced by these processes is consumed by heterotrophic bacteria living in oxygen-free environments. Sulfate rather than oxygen is used as the final electron acceptor in respiration and is reduced to H_2S. In fact, most of the sulfur cycled biologically is used to drive energy metabolism in oxygen-poor environments, not to build proteins.

Note that eukaryotes use neither H_2S for photo- (or chemo-) synthesis nor sulfate in respiration. Thus, the biological sulfur cycle is completed by bacteria and archaeons alone.

The nitrogen cycle is also driven by bacteria and archaeons.

The model provided by the sulfur cycle can be extended to other biologically important elements. Nitrogen is the fourth most abundant element in the human body, contributing 4% by weight

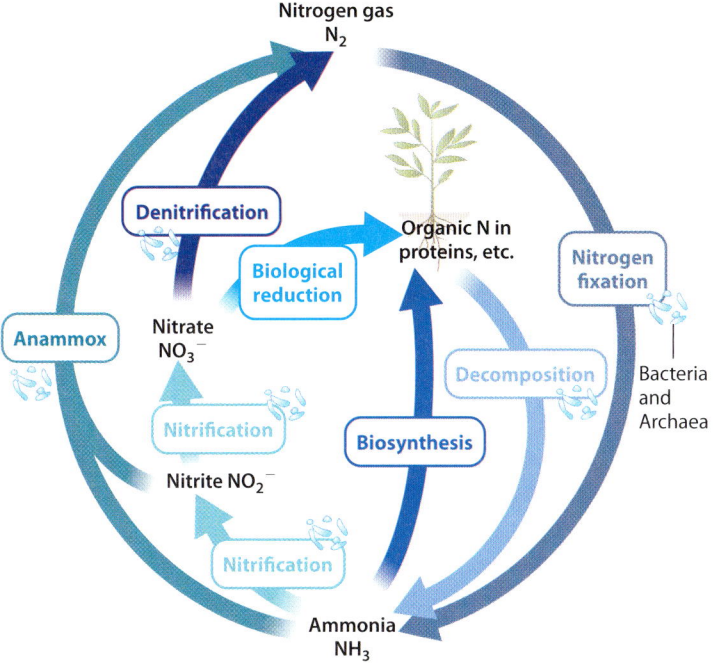

FIG. 26.11 **The nitrogen cycle.** Bacteria and archaeons make possible nitrogen fixation, nitrification and denitrification, as well as anammox.

by the decomposition of dead cells can be taken up by primary producers, but commonly it is oxidized to nitrate (NO_3^-) by chemoautotrophic bacteria, a process called **nitrification** (see Fig. 26.11). This process works efficiently in both soil and water, and so nitrate is the most common form of nitrogen available to life in both environments.

Anaerobic respiration completes the cycle by returning N_2 to the atmosphere. Some bacteria can use nitrate as an electron acceptor in respiration, a process known as **denitrification** (see Fig. 26.11). Because denitrification is so efficient at returning nitrogen to the atmosphere as nitrogen gas, nitrogen fixation must continually make nitrogen available for organisms. Without nitrogen fixation, the nitrogen cycle would slowly wind down, and the productivity of Earth's biota—that is, of all its organisms—would wind down with it. This is why nitrogen fixation is such an important process.

Until the 1990s, the nitrogen cycle was described—completely, it was thought—in terms of the metabolic reactions just discussed. Then, a new and unexpected form of nitrogen metabolism was discovered in the oceans. A number of prokaryotic microorganisms function as chemoautotrophs by reacting ammonium ion (NH_4^+) with nitrite (NO_2^-) to gain energy. The reaction is shown here:

$$NH_4^+ + NO_2^- \rightarrow N_2 + 2H_2O$$

as a component of proteins, nucleic acids, and other compounds. As was true of sulfur, primary producers incorporate inorganic forms of nitrogen into biomolecules, and we get the nitrogen we need from our food. As we'll see, however, there is a twist.

Let's walk through the biological nitrogen cycle (**Fig. 26.11**). Carbon is a minor component of the atmosphere and sulfur only a trace constituent, but the air we breathe consists mostly (79%) of nitrogen gas (N_2). Indeed, most of the nitrogen at Earth's surface resides in the atmosphere. Although nitrogen is plentiful, the nitrogen available to primary producers in many environments is limited. Why should this be? The simple answer is that plants cannot make use of nitrogen gas and neither can algae. Only certain bacteria and archaeons can reduce N_2 to ammonia (NH_3), a form of nitrogen that can be incorporated into biomolecules.

The process of converting N_2 into a biologically useful form such as ammonia is called **nitrogen fixation,** one of the most important reactions ever evolved by organisms. Farmers plant soybeans to regenerate nitrogen nutrients in soils, but it is not the soybean that fixes nitrogen. Rather, soybean roots have nodules that harbor nitrogen-fixing bacteria (**Fig. 26.12**). That is, plants like soybean have entered into an intimate partnership with nitrogen-fixing bacteria to obtain the biologically useful forms of nitrogen needed for growth.

The nitrogen cycle, then, begins with nitrogen fixation, and it continues through many additional metabolisms, most of them confined to bacteria and archaeons. Ammonia released

FIG. 26.12 **Roots of a soybean plant.** The nodules contain bacteria capable of reducing nitrogen gas to ammonia.

Note that this is a redox reaction, like other reactions involved in energy metabolism. In this case, the ammonium ion is oxidized to nitrogen gas, and nitrite is reduced to nitrogen gas, with water produced as a by-product. The reaction, called **anammox** (**an**aerobic **amm**onia **ox**idation), provides only a small amount of energy, but it supports many bacteria and archaeons in oxygen-depleted parts of the ocean, providing a major route by which fixed nitrogen returns to the atmosphere as N_2 (see Fig. 26.11).

The biological nitrogen cycle, then, begins and ends with prokaryotic metabolisms. Bacteria and archaeons fix N_2 into a biologically available form, nitrifying bacteria oxidize ammonia to nitrate, and still other prokaryotes gain energy from denitrification and anammox, returning N_2 to the air. Unlike the carbon and sulfur cycles, the nitrogen cycle does not rely at any point on elements stored as minerals in rocks and sediments. Most of Earth's nitrogen was released to the atmosphere as gas early in our planet's history, and so movement into sediments and out of volcanoes contributes relatively little to the cycle. Humans, however, contribute in important ways, most notably through the industrial synthesis of nitrate fertilizer (Chapter 48).

→ **Quick Check 4** How would the nitrogen cycle operate in the absence of bacteria and archaeons?

26.4 THE DIVERSITY OF BACTERIA

Traditionally, bacterial taxonomy was a laborious process. Microbiologists grew bacteria in beakers or on petri dishes, carefully separating and enriching individual populations until only one type of bacterium—that is, a pure culture—was present. Species were identified on the basis of details of cell structure and division observed under the microscope, metabolic capabilities established through experiments and reactions to chemical stains or antibiotics. By the 1980s, a few thousand species had been described in this way.

With molecular sequencing technologies, however, bacterial species are now characterized by DNA sequences, especially the genes for the small subunit of ribosomal RNA (rRNA). A novel application of sequencing technology paved the way for modern studies of bacterial diversity and ecology. Microbiologists reasoned that the same procedures used to extract, amplify, and sequence genes from pure cultures in the lab could be used to study bacteria in nature. That is, extracting and reading gene sequences directly from soil or seawater could reveal the diversity of the microbial communities in these environments.

The results were stunning. Most bacteria in nature—perhaps 99% or more—are species that have not been cultured in the lab. Indeed, species well characterized by studies of pure cultures commonly represent only minor components of natural communities. To give just one illustration, a bacterium called SAR11 was originally identified on the basis of gene sequences amplified from seawater. SAR11 caught the attention of microbiologists because it makes up about a third of all sequences identified in the samples from the surface ocean. How could we be ignorant of a bacterium that must be among the most abundant organisms on Earth? The answer is simple: No one had cultured it. Recently, SAR11 has been cultured. Formally described as *Pelagibacter ubique* (*ubique* is Latin for "everywhere"), this bacterium is now known to be a heterotroph that plays an important role in recycling organic molecules dissolved in the sea.

More recently, the application of shotgun sequencing techniques (Chapter 13) to the environmental sampling of genomes has allowed microbiologists to begin matching uncultured populations with specific metabolisms. That is, we need no longer be content to know from molecular sequences what is living in a given environment. Now we can begin to understand what the different microorganisms are doing there. For this reason, another revolution in our understanding of the microbial world lies just around the corner (**Fig. 26.13**).

Bacterial phylogeny is a work in progress.

In Chapter 23, we introduced the concept of molecular phylogeny, in which gene sequences are compared to draw conclusions about evolutionary relatedness among populations in nature. Gene sequencing made it possible to test traditional bacterial groupings based on form and physiology. Some traditionally defined groups—for example, the cyanobacteria—passed the test and stand today as a well-defined bacterial group. Many, however, did not.

Molecular sequence comparisons bring enormous new possibilities to the study of prokaryotic diversity and evolution, but they introduce new problems as well. One set of problems arises because major groups of Bacteria and Archaea separated from one another billions of years ago. Inherited sequence similarities can be masked by continuing molecular evolution within groups. That is, specific nucleotides in gene sequences may change not once but multiple times through a long evolutionary history, erasing molecular features that were once the same because of inheritance from a common ancestor. Also, in some groups rates of sequence evolution appear to be higher than in others. As a result, these groups have especially divergent sequences and so may misleadingly fall toward the bottoms of phylogenetic trees.

Another problem arises because of horizontal gene transfer. Phylogenies may falsely group distantly related bacteria because they contain genes passed on by conjugation, transformation, or transduction. Statistical analyses of complete bacterial genomes suggest that at least 85% of all genes in bacterial genomes have been transferred horizontally at least once. Such apparently rampant gene exchange might well doom attempts to build a bacterial tree of life, and some biologists argue that bacterial

HOW DO WE KNOW?

FIG. 26.13

How many kinds of bacterium live in the oceans?

BACKGROUND The oceans are full of bacteria—by some estimates, the sea harbors more than 10^{29} bacterial cells. How diverse are these bacterial communities?

METHOD 1 Different types of bacterium can be recognized by the unique nucleotide sequences of their genes. In fact, individual types of bacterium can be identified on the basis of nucleotide sequence in a relatively small region of the gene for the small subunit of ribosomal RNA (rRNA). This molecular "barcode" can be sequenced quickly and accurately for samples of DNA drawn from the environment.

Using this strategy, scientists associated with the International Census of Marine Life obtained samples of seawater and seafloor sediment from deep-ocean environments. In the laboratory, the rRNA sequences were amplified by PCR, separated by gel electrophoresis, and sequenced, thus providing a library of tags that document bacterial diversity.

METHOD 2 In another set of experiments, scientists collected samples of seawater from the Sargasso Sea, and from these they amplified whole genomes—more than a billion nucleotides' worth. They used small-subunit rRNA and other genes to characterize species richness, and also characterized a large assortment of additional sequence data to understand genetic and physiological diversity.

RESULTS Both surveys found that the bacterial diversity of marine environments is much higher than had been thought. The barcode survey (method 1) found as many as 20,000 distinct types of bacteria in a single liter of seawater and estimated that as many as 5–10 million different microbes may live in the world's oceans. Some of these bacteria are abundant and widespread, but most are rare. The genomic survey (method 2) of the Sargasso Sea found more than 1800 distinct bacteria in sea-surface samples and, remarkably, identified more than 1.2 million previously unknown genes within these genomes.

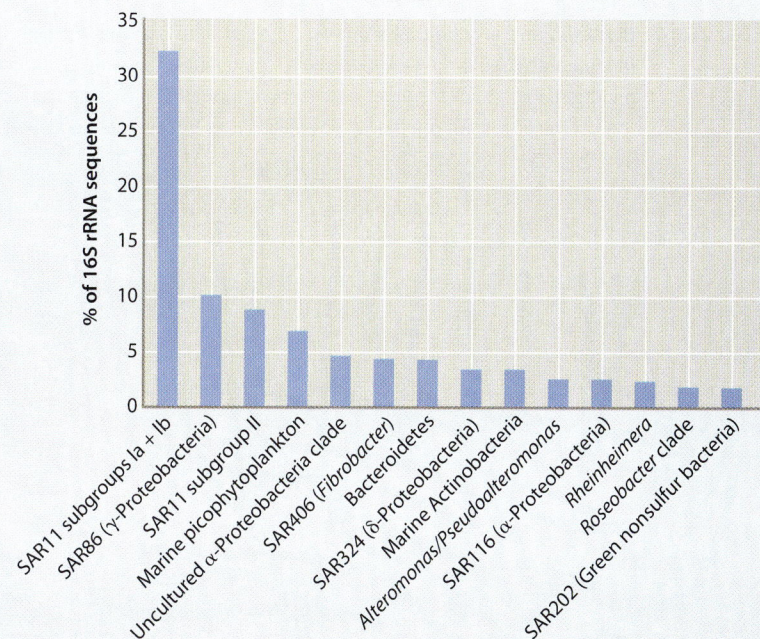

Relative abundance of different bacterial groups in samples from the Sargasso Sea, based on percent representation of small subunit rRNA sequences. Note the abundance of SAR11, only recently characterized.

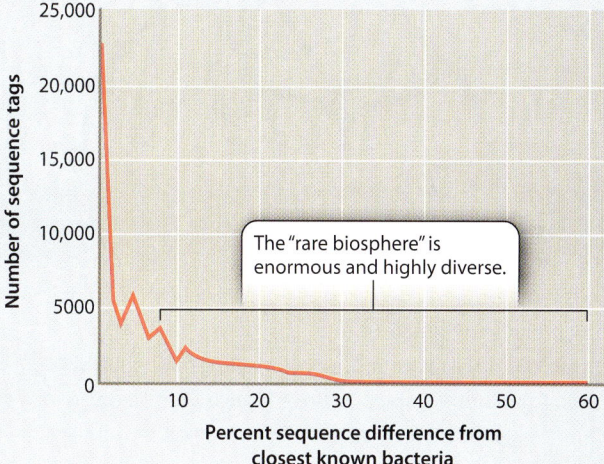

Computer-aided comparison of sequences from the world's oceans shows that about 75% of all sequences collected are similar to other sequences in the library of sequence tags. However, 25% are distinctive, forming a rare but enormously diverse pool of bacterial types.

CONCLUSION The bacterial diversity of the oceans is huge and still largely unexplored. Both barcode and genomic surveys are continuing, and promise a more nearly complete accounting of marine diversity. Together, these surveys will help us understand how marine bacteria function and how they sort into communities.

SOURCES Venter, J. C., et al. 2004. "Environmental Genome Shotgun Sequencing of the Sargasso Sea." *Science* 304:66–74; Sogin, M. L., et al. 2006. "Microbial Diversity in the Deep Sea and the Underexplored 'Rare Biosphere.'" *Proceedings of the National Academy of Sciences USA* 103:12115–12120.

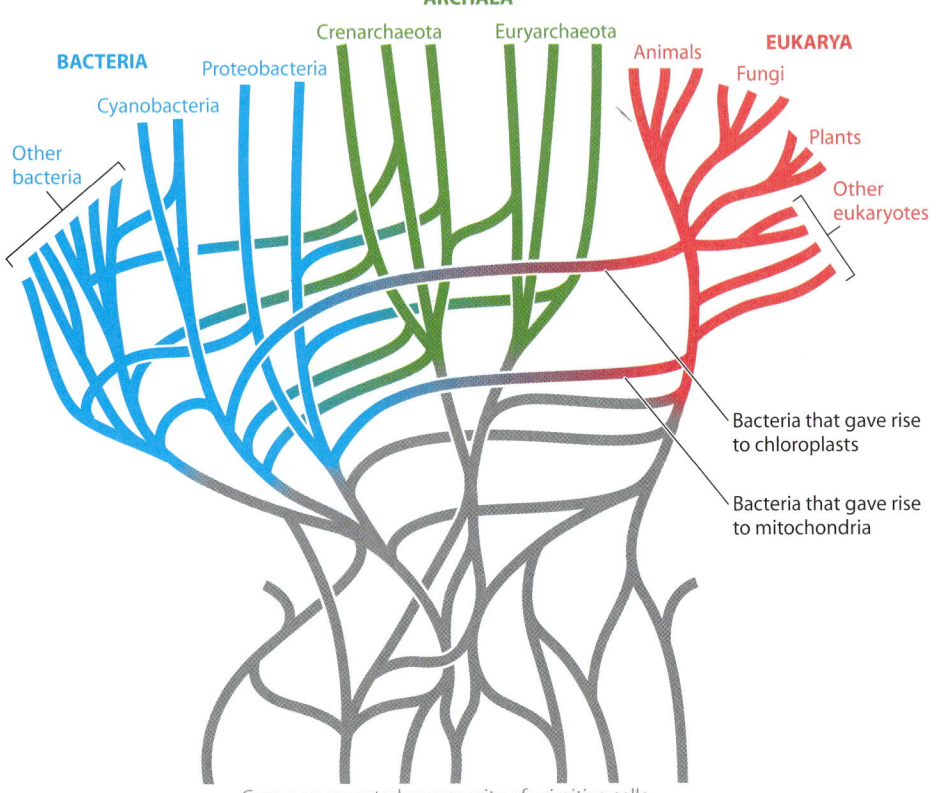

FIG. 26.14 Bacterial and archaeal phylogeny. Some scientists argue that the evolution of prokaryotes should not be viewed as a tree but as a series of branches that diverge and then come back together, representing genes that diverge from one another but then come to reside together in new organisms because of horizontal gene transfer.

evolution is more properly depicted as a complex bush with many connecting branches (**Fig. 26.14**).

Other microbiologists are more optimistic. Small-subunit rRNA genes, for example, show little evidence of horizontal transfer and so may faithfully record the evolutionary history of the organisms in which they occur. Nonetheless, we need to be cautious when using bacterial trees to study the evolution of specific metabolic or physiological characteristics. Nitrogen fixation, photosystems, drug tolerance—such features can and do depend on genes known to jump from branch to branch on the tree.

At present, molecular signatures have been extremely useful in identifying major groups of bacteria, but less clear in establishing branching order among these groups. About 50 major groups of bacteria (each one very roughly equivalent to a eukaryotic phylum) have been recognized. Nearly half of these groups are not represented in pure cultures. Rather, they have been identified only by the cloning of genes from environmental samples. To support our discussion of bacterial diversity, we present a phylogeny based on comparisons of whole genomes in **Fig. 26.15**. As you examine this figure, remember that microbiologists continue to debate branching order on the bacterial tree.

What, if anything, is a bacterial species?

What do we mean when we talk about bacterial species? As we saw in Chapter 22, Ernst Mayr defined species as "groups of interbreeding natural populations that are reproductively isolated from other such groups." This widely accepted definition is called the biological species concept (Chapter 22), but a moment's reflection will show that it can't be applied to the most abundant organisms on Earth—Bacteria and Archaea. In the absence of sexual reproduction, how do we think about species in the prokaryotic domains?

Some microbiologists argue that we should abandon the concept of species when talking about bacteria, especially since distantly related bacteria can exchange genes. Nonetheless, nature is full of microbial populations with well-defined features of form and function.

In the age of molecular sequence comparisons, bacterial species have sometimes been defined operationally: Two populations belong to the same species if the gene sequences of their small-subunit rRNA are more than 97% identical. It would be useful, however, to conceive of bacterial species in terms of population-level processes, much as Ernst Mayr did for eukaryotic organisms. In plants, animals, and single-celled eukaryotes, interbreeding ensures that populations share the same pool of genes, providing a counterforce to mutations and local selection pressures that promote divergence.

A similar process exists in bacteria. In early research on the population genetics of bacteria, biologists observed that genetic diversity in laboratory cultures gradually increases through time and then rapidly decreases with the emergence of a successful variant that outcompetes the rest. The episodic loss of diversity is called **periodic selection.** Some biologists argue that this process provides a means of recognizing bacterial species: Populations subject to the same episodes of periodic selection belong to a single species. This is a useful way to define bacterial species, but other definitions are possible, and as yet there is no consensus among microbiologists.

Proteobacteria are the most diverse bacteria.

Proteobacteria are far and away the most diverse of all bacterial groups (Fig. 26.15). Defined largely by similarities in gene sequences of rRNA, Proteobacteria include many of the organisms that populate our expanded carbon cycle and other biogeochemical cycles. For example, this phylum includes anoxygenic photosynthetic bacteria and chemoautotrophs that oxidize NH_3, H_2S, and Fe^{2+}, as well as bacteria able to respire using SO_4^{2-}, NO_3^-, or Fe^{3+}.

Many Proteobacteria have evolved intimate ecological relationships with eukaryotic organisms. Some of these are mutually beneficial, such as the nitrogen-fixing bacteria in soybean root nodules. Other Proteobacteria, however, are our worst pathogens: the rickettsias that cause typhus and spotted fever, vibrios that trigger cholera, salmonellas responsible for typhoid fever and food poisoning, and *Pseudomonas aeruginosa* strains that infect burn victims and others whose resistance is reduced. *Escherichia coli*, famous as a laboratory model organism (see Fig. 26.4b), is usually a harmless presence in our intestinal tract, but pathogenic strains can cause serious diarrhea and even death.

The gram-positive bacteria include organisms that cause and cure disease.

In 1884, the Danish scientist Hans Christian Gram developed a diagnostic dye that, after washing, is retained by bacteria with thick peptidoglycan walls, but not by those with thin walls. Molecular sequence comparisons confirm that most **gram-positive bacteria** (those that retain the dye) form a well-defined branch of the bacterial tree (Fig. 26.15). One species in this group, *Bacillus subtilis*, is well known to biologists because it has been the focus of many studies about basic molecular function in bacteria. All of us harbor *Streptococcus* species in our mouths, and many of us have suffered from strep throat, produced by streptococcal pathogens. More serious infections, including some forms of meningitis, are also associated with streptococci. *Staphylococcus* bacteria are commonly found on human skin, but only occasionally do we suffer from maladies such as staph infection, toxic shock syndrome, or food poisoning induced by staphylococcal bacteria.

In contrast to these pathogens, gram-positive bacteria called streptomycetes have proved invaluable to medicine because of a remarkable property. More than half of the species in this group secrete compounds that kill other bacteria and fungi in their vicinity. Streptomycetes have provided us with tetracycline, streptomycin, erythromycin, and numerous other antibiotics that combat infectious disease.

Photosynthesis is widely distributed on the bacterial tree.

As noted earlier, molecular studies confirm that all bacteria capable of oxygenic photosynthesis form a single branch, the **cyanobacteria**. Cyanobacteria can be found in

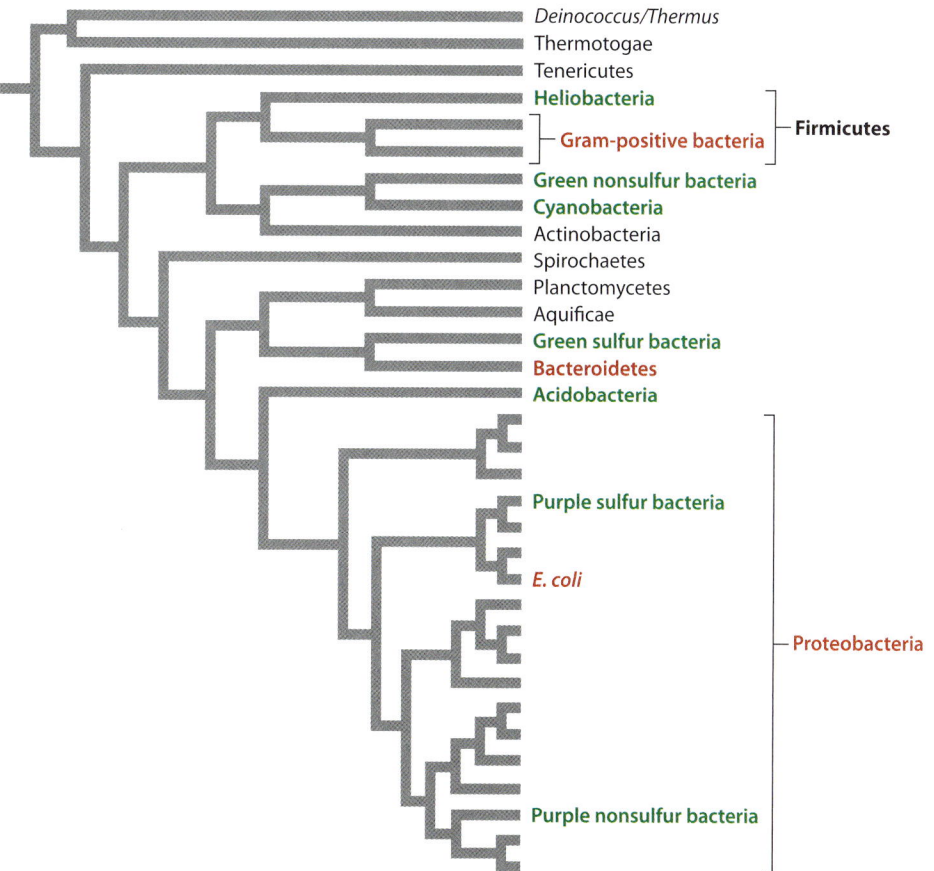

FIG. 26.15 Evolutionary relationships among bacteria, inferred from data from complete genomes. Photosynthetic groups are shown in green (the Acidobacteria have one known photoheterotrophic member). Other groups discussed in the text are shown in red.

environments that range from deserts to the open ocean. Their diverse species include unicellular rods and spheroidal cells, as well as multicellular balls and filaments (**Fig. 26.16**). Some filamentous cyanobacteria can even form several different cell types, including specialized cells for nitrogen fixation and resting cells that provide protection when the local environment does not favor growth.

In contrast to the monophyly of cyanobacteria, molecular sequence comparisons show that *anoxygenic* photosynthesis is distributed widely on the bacterial tree (see Fig. 26.15). Think of the purple layer beneath the surface of many microbial mats (see Fig. 26.7). The vivid color comes from the light-sensitive pigments of the aptly named purple bacteria, found within the Proteobacteria. Purple bacteria are capable of photoautotrophic growth using bacteriochlorophyll and a single photosystem. Most, however, show evidence of metabolic diversity and can grow heterotrophically in the absence of light or appropriate electron donors. The green layer beneath the surface of some microbial mats reflects the pigments of another group of photosynthetic bacteria, called green sulfur bacteria because they commonly gain electrons from hydrogen sulfide and deposit elemental sulfur on their walls. Unlike most groups of photosynthetic bacteria, the green sulfur bacteria are highly intolerant of oxygen gas. Oddly, a species of green sulfur bacteria has been found in hydrothermal rift vents 2 km beneath the surface of the ocean. Light doesn't penetrate to these depths, so if these organisms harvest light, it must come from incandescent lavas as they erupt.

Light-harvesting bacteria occur on several other branches of the bacterial tree—the photoheterotrophic heliobacteria, for example, and the green nonsulfur bacteria often found in freshwater hot springs. Despite a century of research, our knowledge of photosynthesis within the bacterial tree remains incomplete, as illustrated by the discovery in 2007 of unique light-harvesting bacteria in Yellowstone National Park. Genomic surveys of hot springs turned up evidence of a previously unknown bacteriochlorophyll-containing microorganism, and gene sequencing established that this organism belongs to the phylum Acidobacteria. There is no evidence for carbon fixation, however, suggesting that the Yellowstone bacterium is a photoheterotroph. To date, these unusual microorganisms have not been grown in pure culture, so as yet they are difficult to study.

FIG. 26.16 Diversity of form among cyanobacteria.

a. *Aphanothece*, small single cells

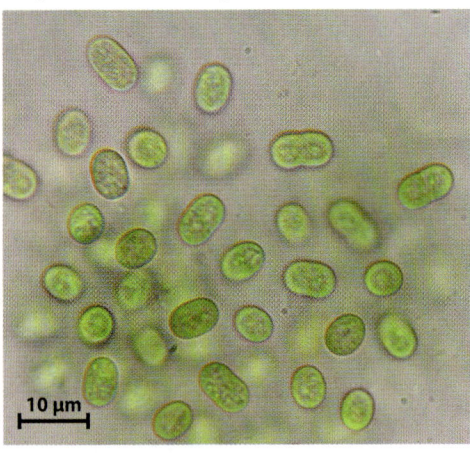

b. *Chroococcus*, larger cells bound into a colony by a common extracellular envelope made of polysaccharides

c. *Oscillatoria*, filaments with no cell differentiation

d. *Anabaena*, strings of undifferentiated cells at either end of the filament, flanking two elongated resting cells, and in the middle, a heterocyst specialized for nitrogen fixation

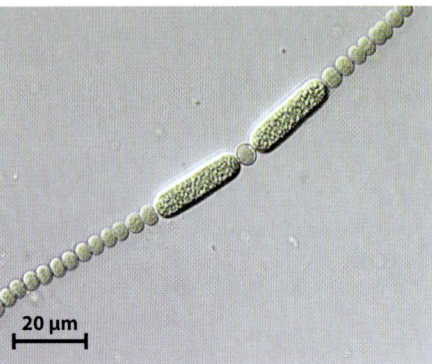

26.5 THE DIVERSITY OF ARCHAEA

We have already noted that Archaea and Bacteria are both prokaryotic, but that they differ in their membrane and wall composition, molecular mechanisms for transcription and translation, and many other biological features (see Table 26.1). As a result, the two groups tend to thrive in very different environments. Many archaeons tolerate environmental extremes such as heat and acidity. In fact, for many environmental conditions, archaeons define the known limits of life. What do these disparate environments have in common? One proposal is that

FIG. 26.17 Phylogeny of Archaea, showing the evolution of temperature preference, salt tolerance, acid tolerance, and production of methane within the domain.

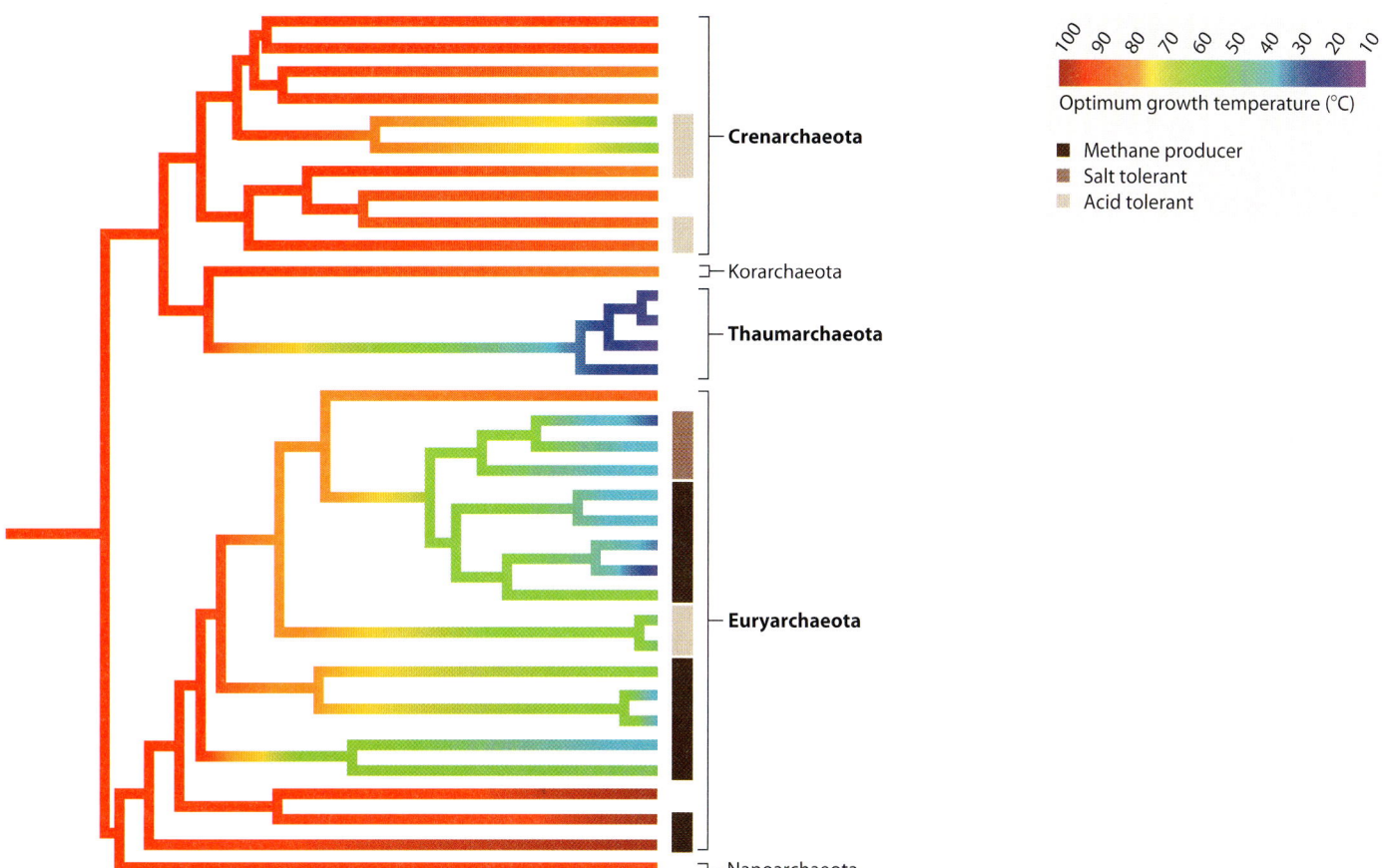

archaeons thrive in environments where the energy available to fuel cell activities is limited—where there is no light, or a limited abundance of fuels and oxidants for chemoautotrophy or respiration, or both. That is, archaeons commonly live where energy resources are too low to support bacteria and eukaryotes.

Bacteria and Archaea diverged from their last common ancestor more than 3 billion years ago. Just as Bacteria have evolved a phylogenetic diversity that defies easy categorization, Archaea also include many distinct types of organism. Many microbiologists recognize three major divisions of Archaea, called the **Crenarchaeota**, **Euryarchaeota**, and **Thaumarchaeota**, but a few poorly known groups may indicate further major branches within the domain (for example, the Korarchaeota and Nanoarchaeota noted in **Fig. 26.17**).

The archaeal tree has anaerobic, hyperthermophilic organisms near its base.

Archaeons found near the base of the Crenarchaeota and Euryarchaeota branches exhibit a remarkable ability to grow and reproduce at temperatures of 80°C or more (Fig. 26.17).

Called **hyperthermophiles,** these organisms do not simply tolerate high temperature, they require it. At present, the world record for high temperature growth is held by *Methanopyrus kandleri,* which has been shown to grow at 122°C among hydrogen- and CO_2-rich hydrothermal vents on the seafloor beneath the Gulf of California. Most of these heat-loving archaeons are anaerobic, living as anaerobic respirers or as chemoautotrophs that obtain energy from the oxidation of hydrogen gas.

Why do anaerobic hyperthermophiles sit on the lowest branches of the archaeal tree? Many biologists believe that these features characterized the first Archaea and therefore imply that the Archaea evolved in hot environments with no oxygen. As we discussed in Chapter 23, geologists agree that oxygen gas was not present on the early Earth, consistent with the phylogenetic inference that early Archaea were anaerobic. A hot environment for early Archaea can be understood in two ways. Either the entire ocean was hot—for which there is little evidence—or Archaea first evolved in hot springs where local supplies of hydrogen and metals enabled them to live as chemoautotrophs. Because some versions of the bacterial tree also show hyperthermophilic

FIG. 26.18 Headwaters of the Rio Tinto, southwestern Spain. Archaeons thrive in these highly acidic waters, pH 1–2.

The red color comes from iron sulfate ions that form in highly acidic water.

species on their lowest branches, it has been suggested that the last common ancestor of *all* living organisms lived in a hot environment.

→ **Quick Check 5** If early branches on the bacterial and archaeal trees are dominated by hyperthermophilic microorganisms, does this mean that the early oceans were very hot?

The Crenarchaeota and Euryarchaeota both include acid-loving microorganisms.

In many parts of the globe where humans mine coal or metal ores, abandoned mines fill up with very acidic water. Called acid mine drainage, this water has been acidified through the oxidation of pyrite (FeS_2) and other sulfide minerals to produce sulfuric acid (H_2SO_4). The pH of these waters is commonly around 1 to 2, and can actually fall below 0. To us, acid mine drainage is an environmental catastrophe, but a number of archaeons thrive in it (**Fig. 26.18**). Modified protein and membrane structures enable them to tolerate their harsh surroundings, making it possible to use sulfur compounds in the environment for both chemoautrophic and heterotrophic growth. What is the limit to growth at low pH? *Picrophilus torridus*, a euryarchaeote archaeon, grows optimally at 60°C and in environments with a pH of 0.7—the pH of battery acid. It is capable of growth in environments down to pH 0—

a record, as far as we know. In fact, *Picrophilus torridus* requires an acidic environment. It will not grow in acid rainwater because the pH is too high.

Euryarchaeote archaeons include heat-loving, methane-producing, and salt-loving microorganisms.

Some euryarchaeote archaeons, like *Picrophilus torridus*, are thermophilic—heat-loving. Many other euryarchaeotes, uniquely among organisms, generate natural gas, or methane (CH_4), as a by-product of fermentation or chemoautotrophic metabolism. Methanogenic archaeons play an important role in the carbon cycle, recycling organic molecules back to simple gases in environments where the supplies of electron acceptors for respiration (oxygen gas, sulfate ion, ferric iron) are limited. Methanogenic archaeons are prominent in peats and lake bottoms where sulfate levels are low and bacterial sulfate reduction is therefore limited. Most methanogenic organisms live in soil or sediments, but some live in the rumens (a specialized chamber in the gut) of cows, helping to maximize the energy yield from ingested grass. Methane doesn't last long in the atmosphere—it is quickly oxidized to carbon dioxide. However, the steady supply of methane generated by archaeons contributes significantly to the greenhouse properties of the atmosphere. For this reason, biologists are interested in how methanogenic archaeons will respond to 21st-century climate change.

A related group of euryarchaeote archaeons illustrates the limits of tolerance to another environmental condition—salinity, or saltiness. Halophilic (salt-loving) archaeons are photoheterotrophs that use the protein bacteriorhodopsin to absorb energy from sunlight (see Fig. 26.6). They live in waters that are salty enough to precipitate table salt (NaCl). As is correspondingly true of extreme acidophiles, these extreme halophiles require high salt conditions and cannot live in dilute water.

Thaumarchaeota may be the most abundant cells in the oceans.

Another surprise of environmental sequencing has been the discovery that the immense volumes of ocean beneath the surface waters contain vast numbers of archaeons that have not yet been cultured in the lab. In the deep ocean, these organisms may be nearly as abundant as the combined number of all bacteria on and below the surface. Recently segregated as the Thaumarchaeota, these archaeons are chemoautotrophs, deriving energy from the oxidation of ammonia (**Fig. 26.19**). Thus, like the discovery of SAR11, the discovery of marine thaumarchaeotes makes it clear that before the still-emerging age of environmental genetics and genomics, we simply didn't have a very good idea of which microorganisms lived where and did what in the oceans. Fortunately, that situation is changing rapidly.

HOW DO WE KNOW?

FIG. 26.19

How abundant are archaeons in the oceans?

BACKGROUND Early exploration of microbial diversity in the oceans revealed that archaeons are tremendously abundant. Just how abundant are they, and what metabolisms do they employ to obtain energy and carbon?

METHOD Scientists sampled seawater throughout the depth of the ocean at a test site in the northern Pacific Ocean. To the samples, they added molecular tags bound to fluorescent molecules that are visible under the microscope. The tags were RNA sequences that bind to small-subunit ribosomal RNA genes known to be useful in identifying different types of bacteria and archaeons. In this way, separate fluorescent markers were attached to thaumarchaeote archaeons, euryarchaeote archaeons, and bacteria.

RESULTS By counting the cells marked by different fluorescent tags in seawater samples, the biologists showed that bacteria dominate microbial communities in near-surface seawater, but that thaumarchaeotes are as numerous as bacteria in deeper waters (Fig. 26.19a).

CONCLUSION Marine thaumarchaeota, unknown in 1990, are now known to be among the most abundant organisms in the oceans.

FOLLOW-UP WORK Archaea are among the most abundant of all cells in the world's oceans. How do these cells live? Microbial samples from water known to be sites of nitrification (that is, the conversion of ammonia to nitrite or nitrate) contained an abundance of thaumarchaeote cells. The high numbers of thaumarchaeotes from these communities supported the hypothesis that they are nitrifiers—the higher the abundance of thaumarchaeote cells (Fig. 26.19b, dark blue line), the higher the amount of nitrite (Fig. 26.19b, light blue line). Thaumarchaeotes were grown in pure culture and shown to grow by consuming ammonia (Fig. 26.19b, red line). In other words, they oxidize ammonia to provide the ATP and reducing power needed to incorporate CO_2 into organic molecules. Therefore, they play a major role in the marine nitrogen cycle, thriving where sources of carbon and energy for other types of metabolism are scarce.

a.

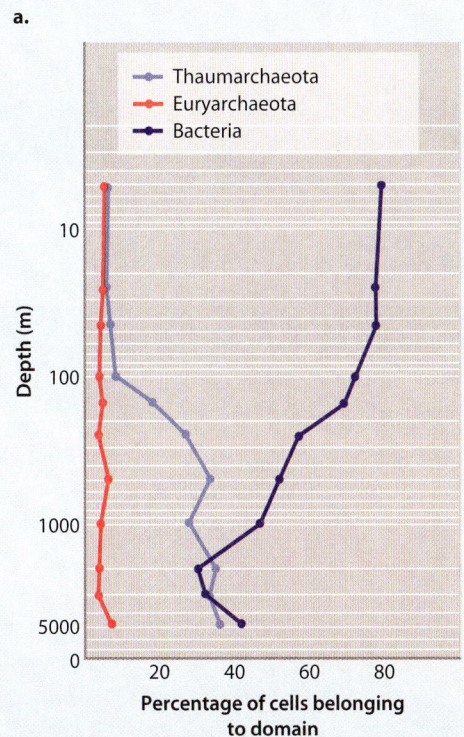

b.

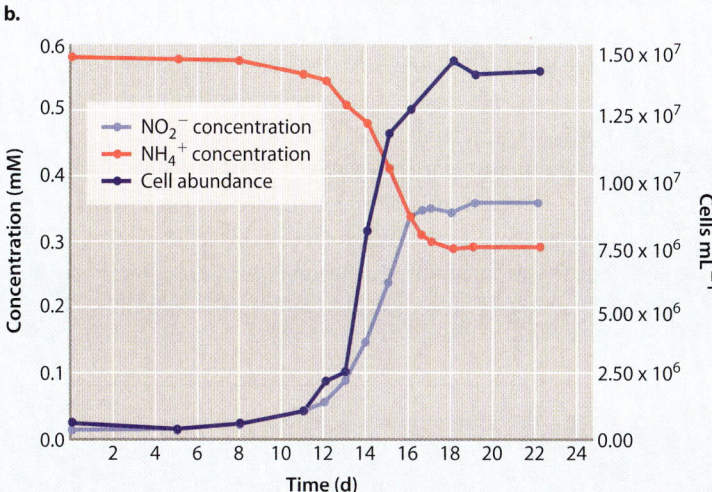

Growth of a marine thaumarchaeote in a medium containing ammonium chloride and bicarbonate as the only sources of energy and carbon, respectively. As cell number increased, ammonia (the ammonium ion NH_4^+) was increasingly converted to nitrite (NO_2^-), supporting the hypothesis that these cells are ammonia-oxidizing chemoautotrophs.

SOURCES Karner, M. B., E. F. DeLong, and D. M. Karl. 2001. "Archaeal Dominance in the Mesopelagic Zone of the Pacific Ocean." *Nature* 409:507–510; Könneke, K., A. E. Bernhard, J. R. de la Torres, C. B. Walter, J. B. Waterbury, and D. A. Stahl. 2005. "Isolation of an Autotrophic Ammonia-Oxidizing Marine Archaeon." *Nature* 437:543–546.

Surveys of cell abundance through a depth profile of the Pacific Ocean show that bacteria dominate cell numbers near the surface, but thaumarchaeotes make up about 40% of all cells in deeper waters.

26.6 THE EVOLUTIONARY HISTORY OF PROKARYOTES

On the tree of life, the plants and animals of our familiar existence populate only the most recent branches. For most of Earth history, life was entirely microbial. Perhaps surprisingly given their small size, Bacteria and Archaea have left a fossil record that allows paleontologists to reconstruct aspects of life's early evolution. Sedimentary rocks preserve microscopic fossils of ancient bacteria, including the photosynthetic cyanobacteria (**Fig. 26.20**). Limestones and related rocks preserve **stromatolites,** layered structures that record sediment accumulation by microbial communities (**Fig. 26.21**). Some especially well preserved ancient rocks also contain molecular fossils—biomolecules, usually lipids, that resist decay and so preserve a chemical signature of ancient life.

Life originated early in our planet's history.

The Earth is about 4.5 billion years old, but few sedimentary rocks have survived destruction by geologic processes and remain to document our planet's earliest history. Sedimentary rocks deposited nearly 3.5 billion years ago in Western Australia and South Africa provide some of our earliest records of Earth's history. Remarkably, they preserve a scarce but discernible signature of life. Tiny organic structures preserved in these rocks have been interpreted as microfossils, although this conclusion remains controversial. More persuasive evidence of life in early oceans is provided by stromatolites. Stromatolites in ancient rocks are strikingly similar to modern structures formed by microbial communities, and so provide evidence of life that goes back about 3.5 billion years (Fig. 26.21). Also, because purely physical processes cannot explain the isotopic composition of carbon in very old limestones and organic matter, we can conclude that a biological carbon cycle existed 3.5 billion years ago (and, from the evidence of 3.8-billion-year-old rocks in Greenland, possibly earlier).

The chemistry of Earth's oldest sedimentary rocks also shows that the early atmosphere and ocean contained little or no free oxygen. What would the carbon cycle look like on an oxygen-free planet? We know the answer from our discussion of prokaryotic metabolism. Photosynthesis can proceed in oxygen-poor waters,

FIG. 26.20 Bacterial microfossils. (a) Bacteria that probably metabolized iron in 1.9-billion-year-old rock from the Gunflint Formation in northwest Ontario, Canada. (b) Cyanobacteria preserved in 800-million-year-old silicified marine carbonates from the Draken Formation in Spitsbergen, Norway. (c) A short cyanobacterial filament from silicified tidal flat carbonates of the Bilyakh Group in northern Siberia.

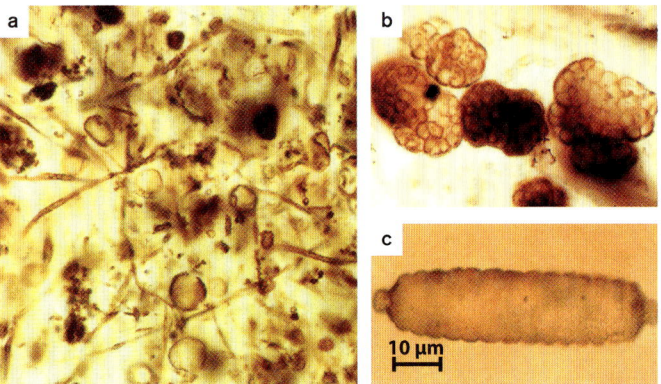

FIG. 26.21 Stromatolites. Modern day stromatolites at Shark Bay, Western Australia (top) show microbial communities building the domed structures exhibited by stromatolites in ancient rocks (bottom). Stromatolites are among the earliest records of life on Earth.

using H_2S, H_2, or Fe^{2+} as electron donors, and fermenters and anaerobic respirers can recycle carbon using electron acceptors such as SO_4^{2-} and Fe^{3+}. Geologic evidence suggests that iron played a particularly important role in cycling carbon for the first billion years of evolutionary history.

The evolution of cyanobacteria, with their capacity for oxygenic photosynthesis, made possible the accumulation of oxygen in the atmosphere and oceans. As discussed in Chapter 25, oxygen began to accumulate in the atmosphere about 2.4 billion years ago, making possible aerobic respiration and, eventually, our present-day carbon cycle. Both microfossils and molecular fossils indicate that cyanobacteria and other photosynthetic bacteria continued to dominate primary production until about 800 million years ago. Thus, our modern carbon cycle, powered overwhelmingly by oxygenic photosynthesis and with major participation by eukaryotic organisms, may have existed for only the last 20% of Earth's history.

Prokaryotes have coevolved with eukaryotes.

Even after the rise of eukaryotic cells and, later, plants and animals, Bacteria and Archaea have remained essential to the functioning of biogeochemical cycles and hence to the maintenance of habitable environments on Earth. Prokaryotic organisms have also radiated into the many novel environments made possible by eukaryotes. We noted earlier that bacteria fix nitrogen in nodules formed on soybean roots. Soybeans and nitrogen-fixing bacteria evolved together, a linkage called **coevolution** (Chapter 47).

Coevolutionary relationships between prokaryotic microorganisms and eukaryotes are common in nature. As noted earlier, cows metabolize grass with the help of methanogenic archaeons in the rumen, a specialized digestive organ. The microbes enable the cow to gain nutrition from plant materials that could otherwise not be digested, and the benefit to the microorganisms is a steady supply of food. Similarly, clams that live near methane vents in the Gulf of Mexico have evolved specialized organs that house methane-oxidizing bacteria, providing them with a reliable source of nutrition.

Bacteria can affect animal behavior as well as nutrition. For example, some squid species attract mates by emitting light from specialized organs full of bioluminescent bacteria. The bacteria provide the squid with a mechanism for communication in the dark deep-sea environment, and the squid feeds the bacteria. Interestingly, many different kinds of bacteria live in the waters that surround the squid, but only the bioluminescent species accumulates in the squid's light organ. The squid apparently sends molecular signals that attract just those bacteria.

Bacteria can also influence animal reproduction. Earth supports more than a million species of insects, and perhaps two-thirds of these harbor the proteobacterial parasite *Wolbachia* within their tissues. A heterotroph that gains nutrition

FIG. 26.22 A large population of *Wolbachia* bacteria (fluorescing green) in the embryo of a fruit fly. *Wolbachia* can influence the development and reproductive biology of their animal hosts.

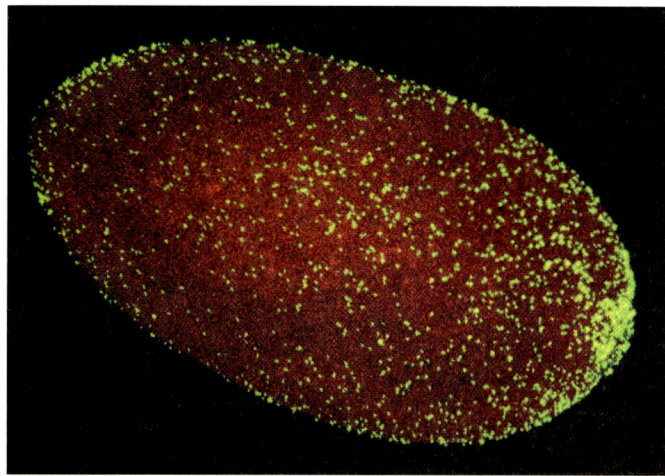

by metabolizing organic compounds supplied by its host, *Wolbachia* infects its host's reproductive tissues and passes from one generation to the next in the host's eggs (**Fig. 26.22**). Remarkably, *Wolbachia* strains ensure the integrity of their biological homes by manipulating the reproductive biology of the host. Many *Wolbachia* strains alter host reproductive cells so that the insect host can successfully reproduce only with partners infected by the same *Wolbachia* strain. Because *Wolbachia* cells are transferred by eggs and not sperm, they have little use for males. Consequently, some cause male host embryos to die, thus increasing the proportion of females. Others cause genetically male embryos to develop as females, or enable eggs to develop into female embryos without fertilization.

Wolbachia have another effect on their insect hosts that is highly relevant to human health: They commonly inhibit infection by viruses and the eukaryotic parasites that cause malaria. Recent research shows that *Wolbachia* populations in some fly species protect their hosts against infection by RNA viruses. Scientists introduced these bacteria into mosquitos that transmit dengue fever, a tropical disease that infects 50–100 million people each year. The *Wolbachia*-infected mosquitos were strongly resistant to the Dengue virus, which causes dengue fever. Moreover, when introduced into the wild, these virus-resistant forms rapidly expanded to dominate regional mosquito populations. *Wolbachia* parasites present a promising path to controlling dengue fever, malaria, and other insect-borne diseases.

All around us, plants, animals, and microorganisms live in intimate association—sometimes to our detriment, but commonly in mutually beneficial relationships that influence how we grow, develop, eat, and behave. Humans are no exception.

? CASE 5 The Human Microbiome: Diversity Within
How do intestinal bacteria influence human health?

It has been estimated that the bacteria (and, to a much lesser extent, archaeons) in and on your body outnumber your own cells 10 to 1. Estimates of the number of microbial species that inhabit our bodies vary, but some 750 types have been identified in the mouth alone, and the list remains incomplete (**Fig. 26.23**). An equal number resides in the colon, and still more live on the skin and elsewhere. Some are transients, entering and leaving the body within a few bacterial generations. Others are specifically adapted for life within humans, forming complex communities of interacting species.

At present, much research is focused on the bacteria within our intestinal tracts. We commonly think of gut bacteria as harmful, but this perception arises from only a small—albeit devastating—subset of our microbial guests. Illnesses known to be caused by bacteria within our bodies include cholera, dysentery, and tuberculosis. Once the leading causes of death in New York and London, they remain major killers in Africa. Other diseases, not previously thought to be of microbial origin, are now known to be induced by bacteria—ulcers, for example, and stomach cancer, both mediated by the acid-tolerant bacterium *Helicobacter pylori*. More often than not, however, intestinal bacteria have a beneficial effect on our well-being. They help to break down food in our digestive system and secrete vitamins and other biomolecules into the colon for absorption into our tissues. Molecular signals from bacteria guide the proper development of cells that line the interior of our intestines.

A fertilized human egg has no bacteria attached to it, so our microscopic passengers arrive by colonization—by infection, if you will. Research on children from Europe and rural Africa clearly shows the importance of diet in establishing our gut microbiota (**Fig. 26.24**). The African children, raised on a diet low in fat and animal protein but rich in plant matter, have gut microbiotas enriched in species of the phylum Bacteroidetes that are known to help digest cellulose. European children raised on a typical Western diet rich in sugar and animal fat lack these bacteria, harboring instead diverse members of the phylum Firmicutes. We do not understand the full ramifications of these differences, but active research suggests that they may help to explain the relative prominence in Western societies of allergies and other disorders involving the immune system.

FIG. 26.24 Diet and the intestinal microbiota. Studies of gut bacteria in children from (a) rural Africa and (b) Western Europe show markedly different communities.

a.

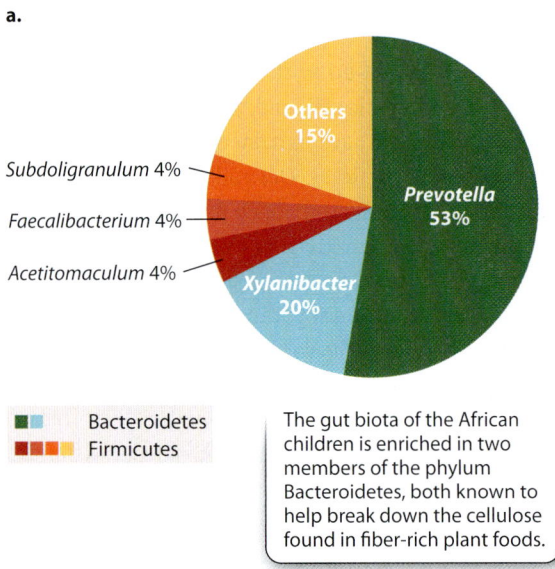

The gut biota of the African children is enriched in two members of the phylum Bacteroidetes, both known to help break down the cellulose found in fiber-rich plant foods.

FIG. 26.23 The human microbiome. This chart provides a rough guide to the distribution of the 3000 or more species of bacteria found on and within a healthy human adult, showing the locations of bacteria that had been sequenced as of 2010.

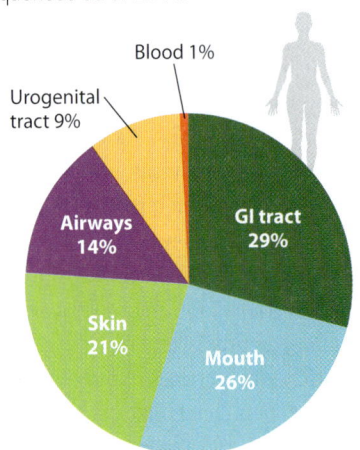

b.

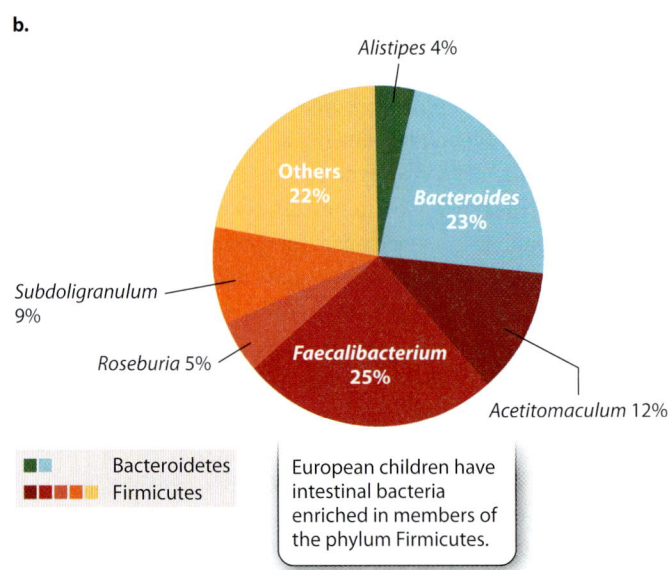

European children have intestinal bacteria enriched in members of the phylum Firmicutes.

We also influence our gut biota by ingesting antibiotics. Although prescribed for the control of pathogens, most antibiotics kill a wide range of bacteria and so can change the balance of our intestinal microbiota. Clinical studies show that the incidence of inflammatory bowel disease, a disabling inflammation of the colon, increases in humans who have been treated with antibiotics.

To understand the relationships between the human microbiome and human health, we need to know what constitutes a healthy gut biota, and this means studying the metabolism, ecology, and population genetics of the bacteria in our bodies. We need to know which bacteria have evolved to take advantage of the environments provided by the human digestive system, and which are "just passing through." Future visits to the doctor may include routine genetic fingerprinting of our bacterial biota, as well as treatments designed to keep our microbiome—and, so, ourselves—healthy.

Core Concepts Summary

26.1 THE TREE OF LIFE HAS THREE MAIN BRANCHES, CALLED DOMAINS: EUKARYA, BACTERIA, AND ARCHAEA.

Prokaryotic cells are cells that lack a nucleus; they include Bacteria and Archaea. page 26-1

Bacteria are small and lack membrane-bound organelles. page 26-1

The bacterial genome is circular. Some bacteria also carry smaller circles of DNA called plasmids. page 26-2

Diffusion limits size in bacterial cells. page 26-2

Bacteria can obtain DNA by horizontal gene transfer from organisms that may be distantly related. page 26-4

Like Bacteria, Archaea lack a nucleus, but form a second prokaryotic domain distinct from Bacteria. page 26-5

Some archaeons are extremophiles, living in extreme environments characterized by low pH, high salt, or high temperatures, but others live in less extreme environments like the upper ocean or soil. page 26-6

26.2 BACTERIA AND ARCHAEA ARE NOTABLE FOR THEIR METABOLIC DIVERSITY.

Bacteria are capable of oxygenic photosynthesis, using water as a source of electrons and producing oxygen as a by-product, and of anoxygenic photosynthesis, using electron donors other than water, such as H_2S, H_2, and Fe^{2+}. page 26-7

Some bacteria and archaeons are capable of anaerobic respiration, in which NO_3^-, SO_4^{2-}, Mn^{4+}, and Fe^{3+} serve as the electron acceptor instead of oxygen gas. page 26-8

Many bacteria and archaeons obtain energy from fermentation, which involves the partial oxidation of organic molecules and the production of ATP by substrate-level phosphorylation. page 26-8

Many photosynthetic bacteria are photoheterotrophs, obtaining energy from sunlight and using preformed organic compounds as a source of carbon instead of CO_2. page 26-9

Chemoautotrophy, in which chemical energy is used to convert CO_2 to organic molecules, is unique to Bacteria and Archaea. page 26-9

26.3 IN ADDITION TO THEIR KEY ROLES IN THE CARBON CYCLE, BACTERIA AND ARCHAEA ARE CRITICAL TO THE BIOLOGICAL CYCLING OF SULFUR AND NITROGEN.

Plants and algae can take up sulfur and incorporate it into proteins, but bacteria and archaeons dominate the sulfur cycle by means of oxidation and reduction reactions that are linked to the carbon cycle. page 26-10

Bacteria and archaeons can to reduce nitrogen gas to ammonia in a process called nitrogen fixation. page 26-11

The nitrogen cycle also involves oxidation and reduction reactions that are linked to the carbon cycle. page 26-11

26.4 THE EXTENT OF BACTERIAL DIVERSITY WAS RECOGNIZED ONLY WHEN SEQUENCING TECHNOLOGIES COULD BE APPLIED TO NON-CULTURABLE BACTERIA.

Traditionally, bacterial groups were recognized by morphology, physiology, and the ability to take up specific stains in culture. page 26-12

Direct sequencing of ribosomal RNA genes from organisms in soil and seawater samples revealed new groups of bacteria. page 26-12

Proteobacteria are the most diverse group of bacteria and are involved in many of the biogeochemical cycles that are linked to the carbon cycle. page 26-15

Gram-positive bacteria include important disease-causing strains as well as species that are principal sources of antibiotics. page 26-15

Photosynthetic bacteria are not limited to a single branch of the bacterial tree. page 26-15

26.5 THE DIVERSITY OF ARCHAEA HAS ONLY RECENTLY BEEN RECOGNIZED.

Archaeons tend to thrive where energy available for growth is limited. page 26-16

Archaea are divided into three major groups, the Crenarchaeota, Thaumarchaeota, and Euryarchaeota. page 26-17

Archaeons at the base of the Crenarchaeota and Euryarchaeota are hyperthermophiles, meaning that they grow at high temperatures. page 26-17

A number of archaeons grow in highly acidic waters, such as those associated with acid mine drainage. page 26-18

Some archaeons generate methane, and others tolerate high-salt conditions. page 26-18

Thaumarchaeotes may be the most abundant cells in the oceans. page 26-18

26.6 THE EARLIEST FORMS OF LIFE ON EARTH WERE BACTERIA AND ARCHAEA.

Evidence for the early history of life on Earth comes from microfossils, fossilized structures called stromatolites, and the isotopic composition of rocks and organic matter. page 26-20

Fossils indicate that life on Earth originated more than 3.5 billion years ago. page 26-20

The early atmosphere and ocean contained little or no free oxygen. page 26-20

Oxygen began to accumulate in the atmosphere and oceans about 2.4 billion years ago as a result of the success of cyanobacteria utilizing oxygenic photosynthesis. page 26-21

Prokaryotic metabolisms were not only essential in the early history of the Earth, but are also vital today, as many forms of life depend on biogeochemical cycles and metabolisms unique to Bacteria and Archaea. page 26-21

Most animals, including humans, live in intimate association with bacteria, which in turn affect our health. page 26-22

Self-Assessment

1. Name and describe the three domains of life.
2. Describe shared and contrasting features of bacterial and archaeal cells.
3. Explain how prokaryotic cells obtain nutrients and how this process puts constraints on their size.
4. Describe how surface area and volume change with size.
5. Explain how photosynthesis can occur without the production of oxygen, and how respiration can occur without requiring oxygen.
6. Describe the roles of bacteria and archaeons in the sulfur and nitrogen cycles.
7. Explain how horizontal gene transfer complicates our understanding of evolutionary relationships among bacteria and archaeons.
8. Name and describe three major groups of Bacteria.
9. Name and describe three major groups of Archaea.
10. State the age of the Earth and the time when life is thought to have first originated.

Do you understand the chapter's Core Concepts? Log in to check your answers to the Self-Assessment questions, then practice what you've learned and reinforce this chapter's concepts by working through the problems and multimedia tutorials provided there.

www.biologyhowlifeworks.com

CHAPTER 27

EUKARYOTIC CELLS

Origins and Diversity

Core Concepts

27.1 Eukaryotic cells are defined by the presence of a nucleus, but other features, particularly a dynamic cytoskeleton and membranes, explain their success in diversifying.

27.2 The endosymbiotic hypothesis proposes that the chloroplasts and mitochondria of eukaryotic cells were originally free-living bacteria that were incorporated into a host cell.

27.3 Eukaryotes have historically been divided into four kingdoms—animals, plants, fungi, and protists—but are now divided into at least seven superkingdoms.

27.4 The fossil record extends our understanding of eukaryotic diversity by providing perspectives on the timing and environmental context of eukaryotic evolution.

Eukaryotic cells have a nucleus—that's their defining characteristic. As we will see, however, it is quite another set of features that distinguishes eukaryotes in terms of function and the capacity for evolutionary innovation. Eukaryotic diversity—the animals, plants, fungi, and protists we are familiar with—does not reflect exceptional metabolic versatility. In fact, compared to Bacteria and Archaea, the Eukarya display only a limited range of metabolisms. For example, oxygenic photosynthesis is widespread among eukaryotes, but anoxygenic photosynthesis is unknown. Furthermore, aerobic respiration is present almost everywhere, but anaerobic respiration has been reported in only a handful of species. Chemoautotrophy has not been documented in eukaryotes at all. Many eukaryotes can ferment organic substrates—yeasts are particularly important in this regard—but most use fermentation only as a supplementary metabolism and lack the diversity of reactions known among prokaryotic organisms. The reasons for eukaryotic success lie elsewhere, in their dazzling diversity of shapes and sizes.

27.1 THE EUKARYOTIC CELL: A REVIEW

Chapter 5 introduced the fundamental features of eukaryotic cells. Here, we revisit those attributes (**Fig. 27.1**) and consider their consequences for function and diversity. How do eukaryotic cells differ from bacteria and archaeons, and how do these differences explain the roles that eukaryotes play in modern ecosystems?

Internal protein scaffolding and dynamic membranes organize the eukaryotic cell.

All cells require a mechanism to maintain spatial order in the cytoplasm. As we saw in Chapter 26, bacteria and archaeons rely primarily on walls that support the cell from the outside, along with a framework of proteins within the cytoplasm. Eukaryotes also use an internal scaffolding of proteins, mostly filaments of actin and microtubules, to organize the cell. This **cytoskeleton** (Chapter 10), found in all eukaryotic cells, differs in one key property from the rigid protein framework of bacteria: It can be remodeled quickly, enabling cells to change shape.

This characteristic provides some eukaryotes with new possibilities for movement, and it permits them to engulf particles, including other cells, something that prokaryotic cells simply cannot do. Packaged in membrane-lined vesicles, the particles can be transported and processed inside the cell by endocytosis. The same process also works in reverse, and it is then called exocytosis. In exocytosis, vacuoles package newly formed molecules or cytoplasmic waste in vesicles and move them to the cell surface for excretion (Chapter 5). These vesicles and the molecules they carry are

FIG. 27.1 A eukaryotic cell. Eukaryotic cells have a nucleus and extensive internal compartmentalization.

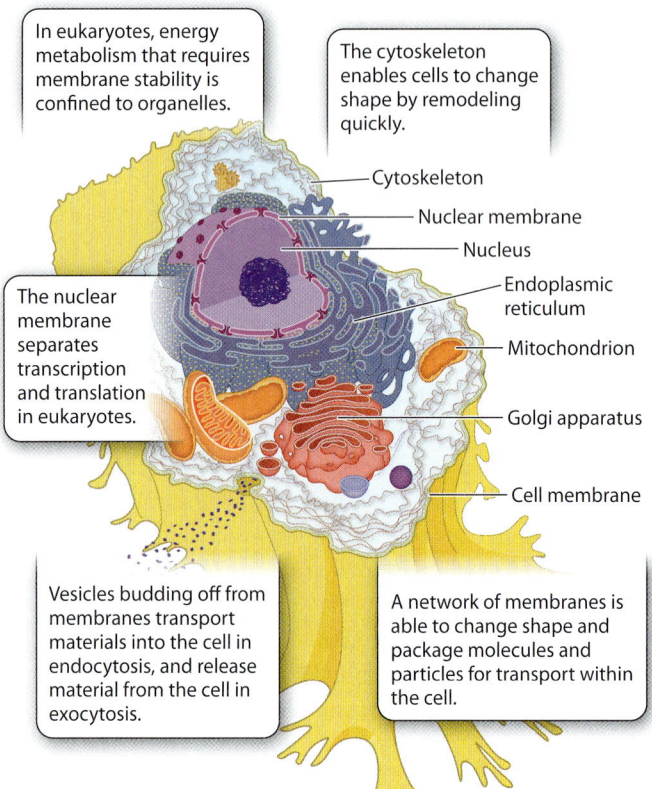

transported through the cytoplasm by means of molecular motors associated with the cytoskeleton. In this way, molecules move through the cell at speeds much greater than diffusion would allow. A major consequence is that eukaryotic cells can be much larger than most bacteria.

Dynamic cytoskeletons require dynamic membranes, and eukaryotes maintain within their cells a remarkable network of membranes called the endomembrane system (Chapter 5). This network includes the nuclear envelope, an assembly of membranes that runs through the cytoplasm called the endoplasmic reticulum (ER) and Golgi apparatus, and a cell or plasma membrane that surrounds the cytoplasm. Moreover, small pieces of membrane can bud off to form tiny vesicles that package molecular signals, nutrients, and wastes within the cell.

All membranes of the endomembrane system are interconnected, either directly or by the movement of vesicles. Many are also capable of changing shape rapidly. In fact, the different membranes are interchangeable in the sense that material originally added to the endoplasmic reticulum may in time be transferred to the cell membrane or nuclear membrane. Biologists like to say that the membranes of eukaryotic cells are in dynamic continuity. Membrane stability, required for energy metabolism, is confined to the mitochondria and chloroplasts.

In eukaryotic cells, energy metabolism is localized in mitochondria and chloroplasts.

We have already noted that, relative to prokaryotic organisms, eukaryotes are fairly limited in the ways they obtain carbon and energy. Moreover, the metabolic processes that power eukaryotic cells take place only in specific organelles—aerobic respiration in the mitochondrion (Chapter 7) and photosynthesis in the chloroplast (Chapter 8). Only limited anaerobic processing of food molecules takes place within the cytoplasm.

Because energy metabolism is confined to these organelles, eukaryotes free the rest of the cell for other, complementary functions. Unlike bacteria, which absorb individual molecules for respiration, single-celled eukaryotic heterotrophs can engulf particulate food, including other cells (**Fig. 27.2**). The cells engulf food particles and package them inside a membrane vesicle, which is then transported into the cytoplasm in a process called phagocytosis. Within the cytoplasm, enzymes break down the particles into molecules that can be processed by the mitochondria. Animals can ingest larger foodstuffs, including other animals and plants. In consequence, eukaryotes can exploit sources of food not readily available to bacterial heterotrophs, which feed on individual molecules. This ability opens up a great new ecological possibility—predation—increasing the complexity of interactions among organisms.

The functional flexibility of eukaryotic cells also allows photosynthetic eukaryotes to interact with their environment in ways that photosynthetic bacteria cannot. Unicellular algae (which are eukaryotes) can move effectively through the water column vertically as well as horizontally and therefore can seek and exploit local patches of nutrients. Diatoms, discussed shortly, go one step further. Large internal vacuoles allow them to store nutrients when they are plentiful for later use. Plants, of course, can capture sunlight many meters above the ground, and so have a tremendous advantage on land, as long as their leaves can obtain water and nutrients from the soil in which the plants are rooted.

Taken together, the dynamic cytoskeleton and membrane system, and the related compartmentalization of basic energy metabolism, go a long way toward explaining why eukaryotic cells have so many shapes and prokaryotes have so few.

The organization of the eukaryotic genome also helps explain eukaryotic diversity.

The prokaryotic genome has evolved in such a way that it can replicate quickly. Bacteria and Archaea absorb available nutrients quickly, so both rapid deployment of metabolic proteins and reproduction are key to ecological and evolutionary success. As the majority of a prokaryote's DNA is arrayed in a single circular chromosome, speed of replication allows for speed of reproduction. As a result, selection favors those strains of Bacteria and Archaea that retain only the genetic material vital to the organism.

FIG. 27.2 **Phagocytosis.** The flexibility of the eukaryotic cytoskeleton and membrane system enables eukaryotic cells to engulf food particles, including other cells. This sequence shows an amoeba engulfing a yeast cell.

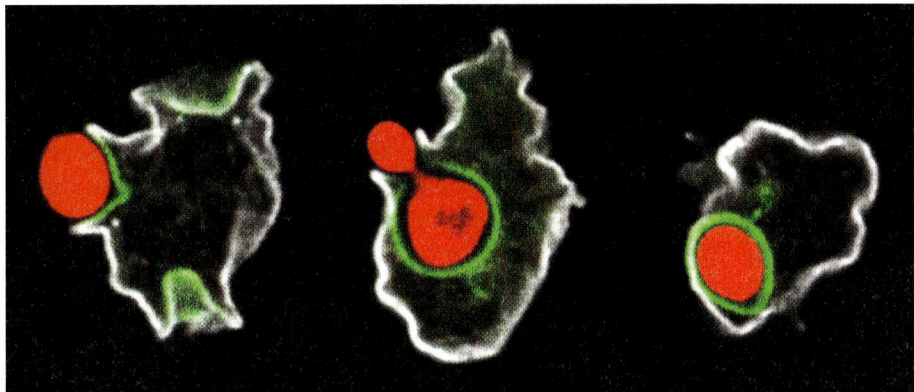

Eukaryotes have multiple linear chromosomes and can begin replication from many sites on each one, enabling them to replicate multiple strands of DNA simultaneously and rapidly (Chapter 12). This ability relieves the evolutionary pressure for streamlining, allowing eukaryotic genomes to build up large amounts of DNA that do not code for proteins. Most of this additional DNA was originally considered to have no function and, indeed, was called "junk DNA." That view, however, has lost ground in recent years, as the complete nature of the genome has come to be better understood (Chapter 13). At least some of the DNA that does not code for proteins has functions in gene regulation (Chapter 19). This regulatory DNA gives eukaryotes the fine control of gene expression required for both multicellular development and complex life cycles, two major features of eukaryotic diversity.

The evolutionary success of eukaryotes really rests on a combination of features. The innovations of dynamic cytoskeletal and membrane systems gave eukaryotes the structure required for larger cells with complex shapes and the functional ability to ingest other cells. Thus, early unicellular eukaryotes did not gain a foothold in microbial ecosystems by outcompeting bacteria and archaeons. Instead, they succeeded by evolving novel functions. Along with the capacity to remodel cell shape, eukaryotes evolved complex patterns of gene regulation, which in turn enabled unicellular eukaryotes to evolve complex life cycles and multicellular eukaryotes to generate multiple, interacting cell types during growth and development. These abilities opened up still more possibilities for novel functions, which we explore in this and later chapters.

→ **Quick Check 1** How did the evolutionary expansion of eukaryotic organisms change the way carbon is cycled through biological communities?

Sex promotes genetic diversity in eukaryotes and gives rise to distinctive life cycles.

In Chapter 26, we saw how Bacteria and Archaea generate genetic diversity by means of horizontal gene transfer. Horizontal gene transfer has been documented in eukaryotic species, but it is relatively uncommon. How then do eukaryotes build and maintain genetic diversity within populations? The answer is sex.

Sexual reproduction involves the formation of **gametes** by meiosis and their fusion by fertilization (Chapters 11 and 42). Sex promotes genetic variation in two simple ways. First, meiotic cell division results in daughter cells (gametes or spores) that are genetically unique—that is, the combinations of alleles in each gamete are different from each other and from the parental cell—as a result of recombination and independent assortment. Second, in fertilization, new combinations of genes are brought together by the fusion of gametes. Interestingly, a few eukaryotic groups have lost the capacity for sexual reproduction. In the best-studied of these eukaryotes, tiny animals called bdelloid rotifers, genetic diversity is actually high, maintained by high rates of horizontal gene transfer. Apparently, horizontal gene transfer and sexual reproduction are alternative and, to a large extent, mutually exclusive ways of maintaining genetic diversity within populations.

As we saw in Chapter 11, meiotic cell division results in cells with one set of chromosomes. Such cells are **haploid.** Sexual fusion brings two haploid ($1n$) cells together to produce a **diploid** ($2n$) cell that has two sets of chromosomes. The life cycle of sexually reproducing eukaryotes, then, necessarily alternates between haploid and diploid states.

Many single-celled eukaryotes normally exist in the haploid stage and reproduce asexually by mitotic cell division (**Fig. 27.3a**). An example is provided by the green alga *Chlamydomonas*. Typically, starvation or other environmental stress can cause two cells to fuse, forming a diploid cell, or **zygote.** The zygote formed by these single-celled eukaryotes commonly functions as a resting cell. It covers itself with a protective wall and then lies dormant until environmental conditions improve. In time, further signals from the environment induce meiotic cell division, resulting in four genetically distinct haploid cells that emerge from their protective coating to complete the life cycle.

As shown in **Fig. 27.3b**, some single-celled eukaryotes normally exist as diploid cells. For example, in diatoms, single-celled eukaryotes commonly found in lakes, soils, and the oceans, cells are mostly diploid and reproduce asexually by mitotic cell division to make more diploid cells. Because of the constraints of their mineralized skeletons, diatoms become smaller with each asexual division. Once a critical size is reached, meiotic cell division is triggered, producing haploid gametes that fuse to regenerate the diploid state as a round, thick-walled cell.

FIG. 27.3 **Eukaryotic life cycles.** (a and b) The life cycles of single-celled eukaryotes differ in the proportion of time spent as haploid (1n) versus diploid (2n) cells. (c) The life cycles of animals have many mitotic divisions between formation of the zygote and meiosis. (d) Vascular plants have two multicellular phases.

a. Unicellular eukaryote with prominent haploid phase

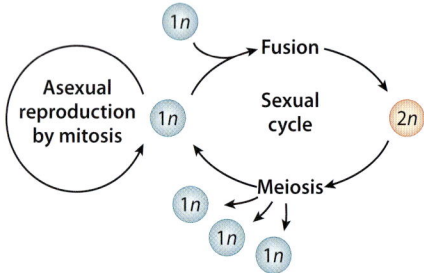

b. Unicellular eukaryote with prominent diploid phase

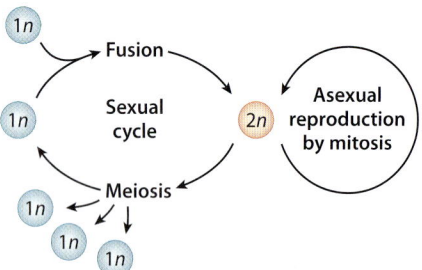

c. Animal

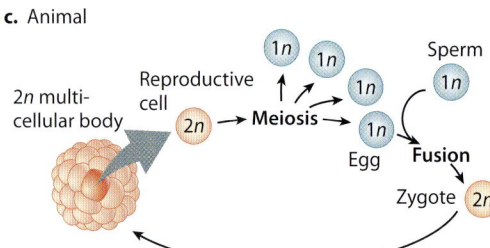

d. Vascular plant

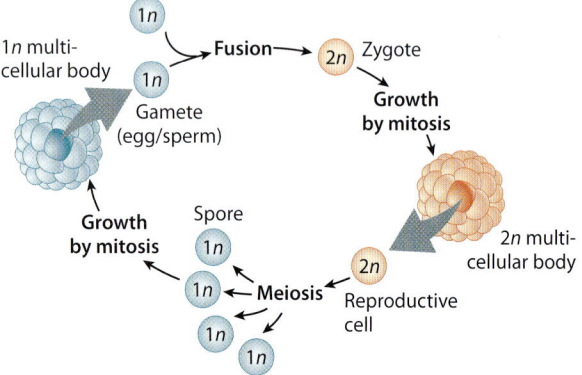

This cell grows and then germinates to form actively growing, skeletonized cells. In diatoms, then, short-lived gametes constitute the only haploid phase of the life cycle.

The two life cycles just introduced are similar in many ways. Both involve haploid cells that fuse to form diploid cells, and diploid cells that undergo meiotic cell division to generate haploid cells. Both life cycles also commonly include cells capable of persisting in a protected form when the environment becomes stressful. Why some single-celled eukaryotes usually occur as haploid cells and others generally occur as diploid cells remains unknown—a good question for continuing research.

Sexual reproduction has never been observed in some eukaryotes, but most species appear to be capable of sex, even if they reproduce asexually most of the time. Variations on the eukaryotic life cycle can be complex, especially in parasitic microorganisms that have multiple animal hosts. All, however, have the same fundamental components as the two life cycles described here.

In succeeding chapters, we discuss multicellular organisms in detail. Here we note only that animals, plants, and other complex multicellular organisms have life cycles with the same features as those just discussed. The big difference is that in animals, the zygote divides many times to form a multicellular diploid body before a small subset of cells within the body undergoes meiotic cell division to form haploid gametes (eggs and sperm). During fertilization, the egg and sperm combine sexually to form a zygote (**Fig. 27.3c**). In animals, as in diatoms, the only haploid phase of the life cycle is the gamete. As we will see in Chapter 32, plants have two multicellular phases in their life cycle, one haploid and one diploid (**Fig. 27.3d**). Like single-celled eukaryotes, many plants and animals can reproduce asexually. Humans and other mammals are unusual in that we have no capacity for asexual reproduction.

27.2 EUKARYOTIC ORIGINS

In the preceding section, we reviewed the basic features of eukaryotic cells. The distinctive cellular organization of Eukarya differs markedly from the simpler cells of Bacteria and Archaea. How did eukaryotes come to form such a remarkably diverse branch of the tree of life? In Case 5: The Human Microbiome, we saw that symbioses are common, even in humans. Is the relationship between humans and bacteria unusual, or are symbioses so fundamental that they played key roles in the evolution of the eukaryotic cell?

? CASE 5 The Human Microbiome: Diversity Within
What role did symbiosis play in the origin of chloroplasts?

The chloroplasts found in plant cells closely resemble certain photosynthetic bacteria, specifically cyanobacteria. The molecular workings of photosynthesis are nearly identical in the two, and the way that internal membranes organize the photosynthetic machinery of cyanobacteria closely resembles the way that the thylakoid membranes organize the photosynthetic machinery of chloroplasts (Chapter 8). The Russian botanist Konstantin Sergeevich Merezhkovsky recognized this similarity more than a century ago. He also understood that corals and some other organisms harbor algae within their tissues as symbionts that aid the growth of their host. A **symbiont** is an organism that lives in closely evolved association with another species. This association, called **symbiosis,** is discussed in Chapter 47.

Putting these two observations together, Merezhkovsky came up with a radical hypothesis. Chloroplasts, he argued, originated as symbiotic cyanobacteria that through time became permanently incorporated into their hosts. Merezhkovsky's hypothesis of chloroplast origin by **endosymbiosis** (a symbiosis in which one partner lives within the other) was difficult to test with the tools available in the early twentieth century, and his idea was dismissed, more neglected than disproved, by most biologists.

In 1967, American biologist Lynn Margulis resurrected the endosymbiotic hypothesis, supporting her arguments with new types of data made possible by the then-emerging techniques of cell and molecular biology (**Fig. 27.4**). Transmission electron microscopy showed that structural similarities between chloroplasts and cyanobacteria extend to the submicrometer level, for example in

HOW DO WE KNOW?

FIG. 27.4

What is the evolutionary origin of chloroplasts?

HYPOTHESIS Chloroplasts evolved from cyanobacteria living as endosymbionts within a eukaryotic cell.

OBSERVATION Chloroplasts and cyanobacteria have closely similar internal membranes that organize the light reactions of photosynthesis (Fig. 27.4a).

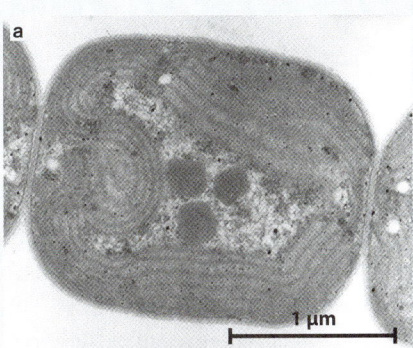
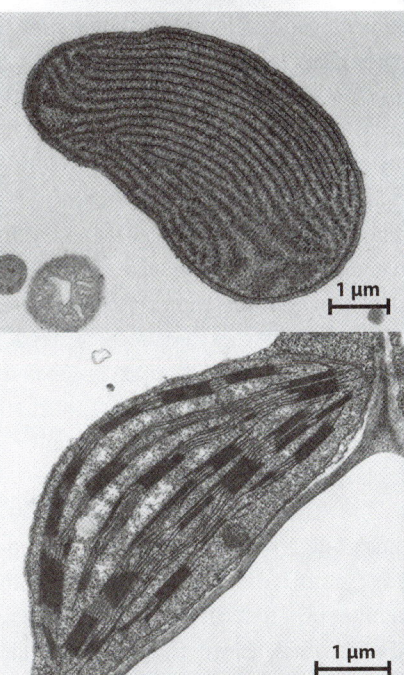

EXPERIMENT Like cyanobacteria, chloroplasts have DNA organized in a single circular chromosome. Phylogenies based on molecular sequence comparisons place chloroplasts among the cyanobacteria (Fig. 27.4b).

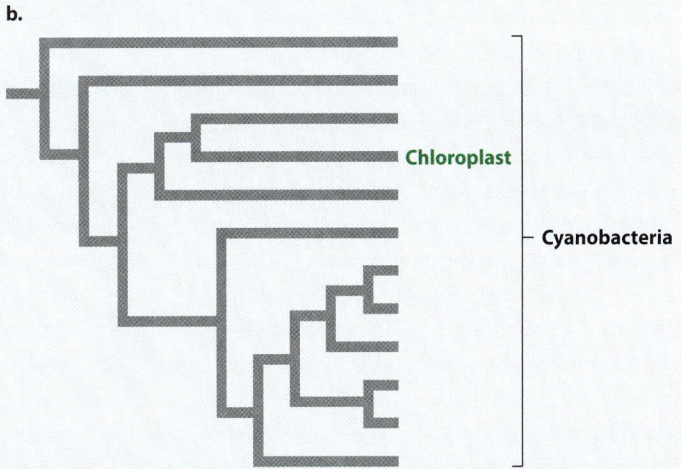

CONCLUSION Molecular and electron microscope data support the hypothesis that chloroplasts originated as endosymbiotic cyanobacteria.

Transmission electron microscopy shows that cyanobacteria (top), red algal chloroplasts (middle), and plant chloroplasts (bottom) have very similar internal membranes that organize the light reactions of photosynthesis.

SOURCE Giovannoni, S. J., S. Turner, G. J. Olsen, S. Barns, D. J. Lane, and N. R. Pace. 1988. "Evolutionary Relationships Among Cyanobacteria and Green Chloroplasts." *Journal of Bacteriology* 170:3584–3592.

the organization of the photosynthetic membranes (Fig. 27.4a). It also became clear that the chloroplasts in red and green algae (and in land plants) are separated from the cytoplasm that surrounds them by two membranes. This is expected if a cyanobacterial cell had been engulfed by a eukaryotic cell. The inner membrane corresponds to the cell membrane of the cyanobacterium, and the outer one is part of the engulfing cell's plasma membrane. In addition, the biochemistry of photosynthesis was found to be essentially the same in cyanobacteria and chloroplasts. Both use two linked photosystems, a common mechanism for extracting electrons from water, and the same reactions to reduce carbon dioxide (CO_2) into organic matter (Chapter 8).

Such observations kindled renewed interest in the endosymbiotic hypothesis, but the decisive tests were made possible by another, and unexpected, discovery. Chloroplasts, it turns out, have their own DNA, organized into a single circular chromosome, like that of bacteria. Armed with this knowledge, investigators could use the tools of molecular sequence comparison (Chapter 23). The sequences of nucleotides in chloroplast genes closely match those of cyanobacterial genes but are strikingly different from sequences of genes in the nuclei of photosynthetic eukaryotes (Fig. 27.4b). This finding provides strong support for the hypothesis of Merezkhovsky and Margulis. Chloroplasts are indeed the descendants of symbiotic cyanobacteria that lived within eukaryotic cells.

Eukaryotes were able to acquire photosynthesis because they could engulf and retain cyanobacterial cells. But engulfing another microorganism was just the beginning. The cyanobacteria most closely related to chloroplasts have 2000 to 3000 genes. In contrast, among photosynthetic eukaryotes, chloroplast gene numbers vary from just 60 to 200. Where did the rest of the cyanobacterial genome go? Some genes may simply have been lost through evolution; unneeded chloroplast genes may have been lost if similar nuclear genes could supply chloroplast requirements. Many, however, were transported to the nucleus when chloroplasts broke or by hitching a ride with viruses. The nuclear genome of the flowering plant *Arabidopsis* contains several thousand genes of cyanobacterial origin. Some code for proteins destined for use within the chloroplast.

Although foreign to mammals and other vertebrate animals, symbiosis between a heterotrophic host and photosynthetic partner is, in fact, common throughout the eukaryotic domain. Reef corals, for example, harbor photosynthetic cells that live symbiotically within their tissues. Some corals have even lost the capacity to capture food from surrounding waters. *Tridacna*, the giant clam of tropical Pacific waters, obtains its nutrition from symbiotic algae that live within its tissues.

Until recently, most biologists agreed that chloroplasts originated from endosymbiotic cyanobacteria only once, in a common ancestor of the green algae and red algae. Now,

FIG. 27.5 A second example of chloroplast endosymbiosis. The photosynthetic amoeba *Paulinella chromatophora* acquired photosynthesis by endosymbiosis independently of and more recently than the endosymbiosis that gave rise to chloroplasts in green plants and other eukaryotes.

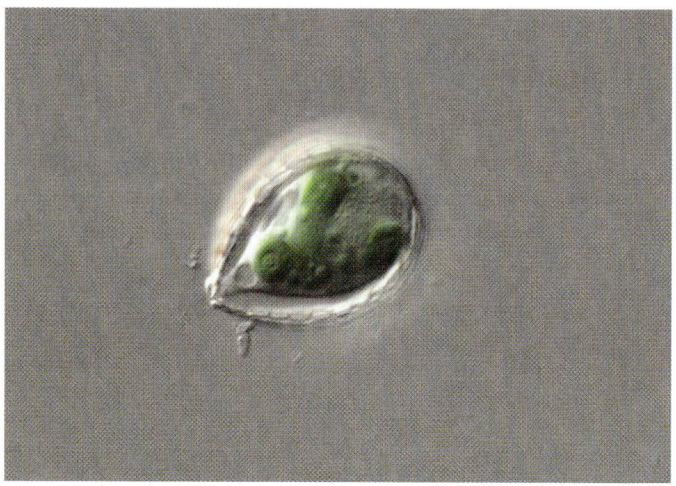

remarkably, a second case of cyanobacterial endosymbiosis has come to light. *Paulinella chromatophora* is a photosynthetic amoeba (**Fig. 27.5**). Its chloroplast genes show that its chloroplast originated from a branch of the cyanobacteria different from and arising long after the one that gave rise to chloroplasts in other photosynthetic eukaryotes. About a third to a half of the ancestral genome remains in *P. chromatophora* chloroplasts, suggesting that this organism reflects chloroplast evolution still in progress.

CASE 5 The Human Microbiome: Diversity Within
What role did symbiosis play in the origin of mitochondria?

Like chloroplasts, mitochondria closely resemble free-living bacteria in organization and biochemistry. And like chloroplasts, mitochondria contain DNA that confirms their close phylogenetic relationship to a form of bacteria—in this case proteobacteria (Chapter 26). Like chloroplasts, then, mitochondria originated as endosymbiotic bacteria.

Mitochondria are also like chloroplasts in having a small genome. Indeed, the mitochondrial genome is dramatically reduced compared to the ancestral proteobacterial genome, in most eukaryotes containing only a handful of functioning genes. Human mitochondria, for example, code for just 13 proteins and 24 RNAs. Once again, many genes from the original bacterial endosymbiont migrated to the nucleus, where they still reside.

Most eukaryotic cells contain mitochondria, but a few single-celled eukaryotic organisms found in oxygen-free environments do not. Biologists earlier hypothesized that these eukaryotes evolved

FIG. 27.6 Transmission electron microscope images comparing the mitochondrion within the aerobic eukaryotic microorganism *Euplotes* (left) with a hydrogen-producing mitochondrion adapted for anaerobic metabolism in the closely related genus *Nictotherus* (right). The organelle in *Nictotherus* is halfway between a normal mitochondrion and a hydrogenosome, supporting the view that these two organelles are descended from a common ancestor.

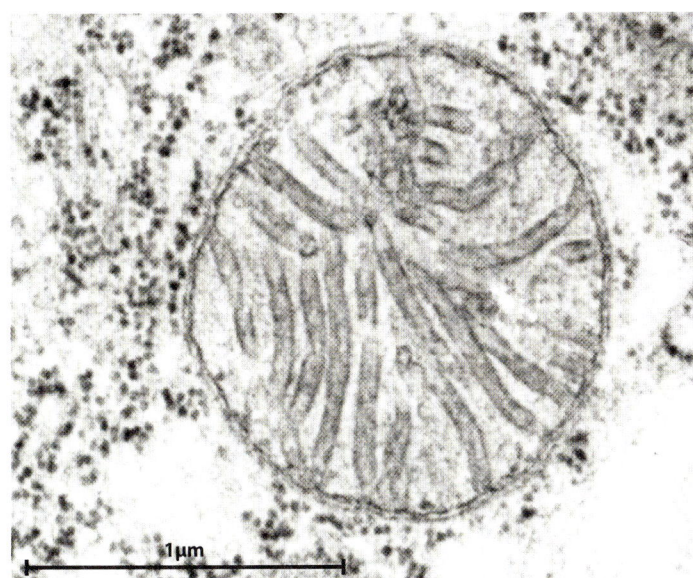
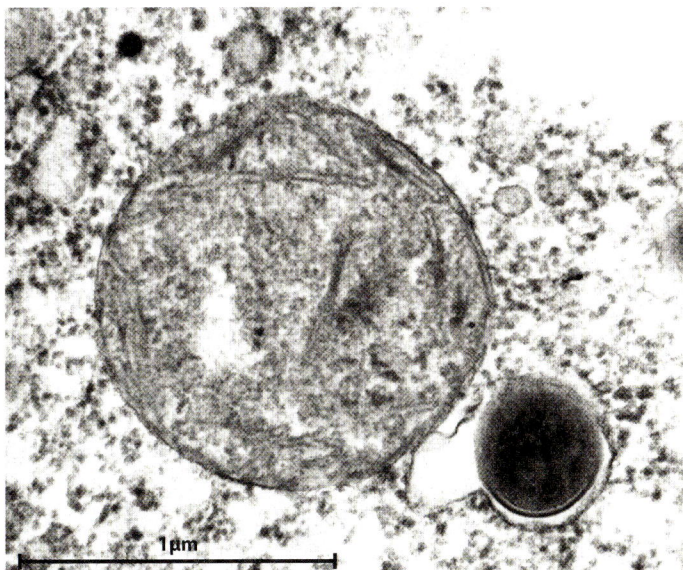

before the endosymbiotic event that established mitochondria in cells having a nucleus, but that proposal turns out to be wrong. We know this because of the propensity for genes to migrate from the endosymbiont to the host's nucleus. Every mitochondria-free eukaryote examined to date has relic mitochondrial genes in its nuclear genome, providing support for a second hypothesis, that eukaryotic cells without mitochondria had them once but have lost them. In fact, many eukaryotes that lack mitochondria contain small organelles called hydrogenosomes that generate ATP by anaerobic processes (**Fig. 27.6**). These organelles have little or no DNA, but genes of mitochondrial origin in the cells' nuclei code for proteins that function in the hydrogenosome. Thus, hydrogenosomes appear to be highly altered mitochondria adapted to life in oxygen-poor environments. Other anaerobic eukaryotes contain still smaller organelles called mitosomes that also appear to be remnant mitochondria.

? CASE 5 The Human Microbiome: Diversity Within
How did the eukaryotic cell originate?

If mitochondria originated as proteobacteria and chloroplasts are descended from cyanobacteria, where does the rest of the eukaryotic cell come from? Analysis of the nuclear genome alone provides no clear picture because genes of bacterial, archaeal, and purely eukaryotic origin are all present. As discussed, many nuclear genes originated with the mitochondria and chloroplasts acquired from specific bacteria. However, genes from other groups of bacteria also reside in the eukaryotic nucleus,

recording multiple episodes of horizontal gene transfer through evolutionary history. In contrast, some genes are present only in eukaryotes and apparently evolved after the domain originated. Still others are clearly related to the genes of Archaea, including the genes that govern DNA transcription and translation.

Two starkly different hypotheses have been proposed to explain this mix of genes. Some biologists believe that the host for mitochondrion-producing endosymbiosis was itself a true eukaryotic cell, with nucleus, cytoskeleton, and endomembrane system, but only limited ability to derive energy from organic molecules (**Fig. 27.7**). In this view, nuclear genes in Eukarya resemble those of Archaea because the primordial eukaryotic host cell was closely related to Archaea. Others, however, argue that no eukaryotic cell existed before there were mitochondria. Instead, they propose that the eukaryotic cell as a whole began as a symbiotic association between a proteobacterium and an archaeon (Fig. 27.7). The proteobacterium became the mitochondrion and provided many genes to the nuclear genome. The archaeon provided other genes, including those used to transcribe DNA and translate it into proteins.

Biologists continue to debate these alternatives. Both hypotheses address the hybrid nature of the eukaryotic genome, but neither explains the origins and evolution of the nucleus, linear chromosomes, the eukaryotic cytoskeleton, or a cytoplasm subdivided by ever-changing membranes. There is no consensus on the question of eukaryotic origins—it is one of biology's deepest unanswered questions, awaiting novel observations by a

FIG. 27.7 Two hypotheses for the origin of the eukaryotic cell. Both hypotheses lead to the evolution of a mitochondrion-bearing protist.

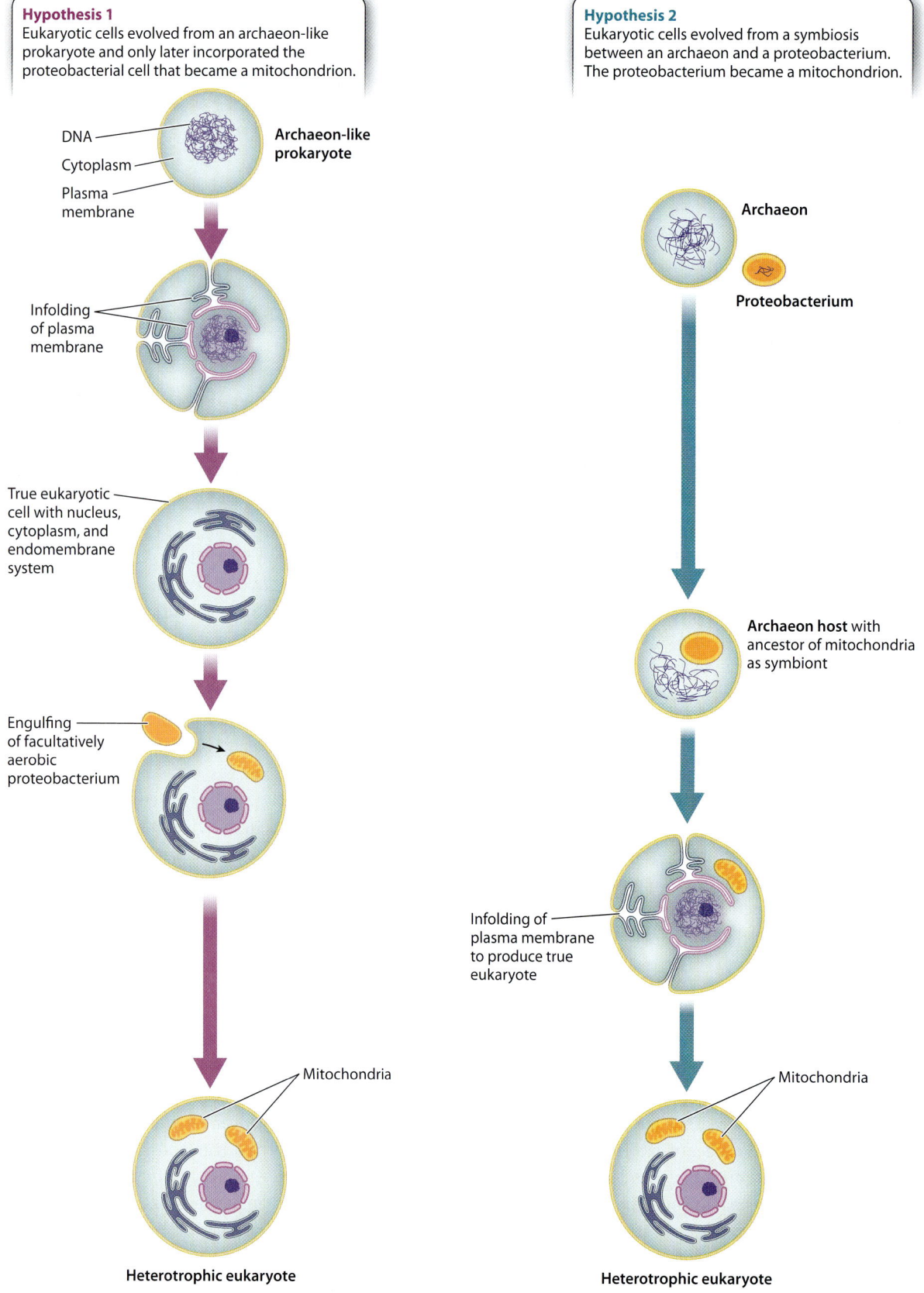

FIG. 27.8 Scanning electron microscope image of bacterial cells on the surface of a single-celled eukaryote found in oxygen-depleted sediments in the Santa Barbara Basin, off the coast of California. The bacteria are hypothesized to metabolize the H_2S in this environment, thereby protecting the eukaryote that they enclose.

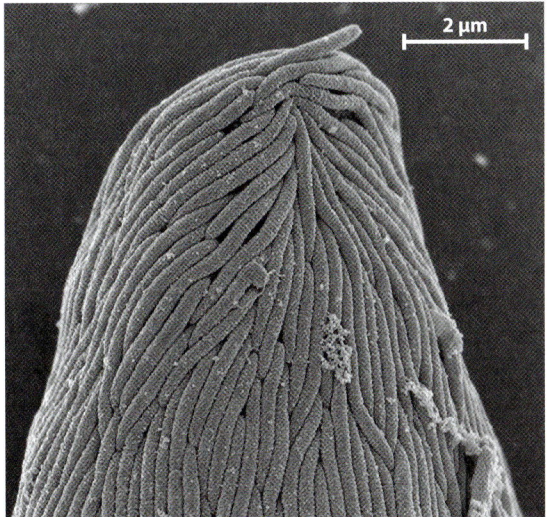

new generation of biologists. Once a dynamic cytoskeleton became coupled to a flexible membrane system, however, the evolutionary possibilities of eukaryotic form were established.

In the oceans, many single-celled eukaryotes harbor symbiotic bacteria.

The evolution of the eukaryotic cell is one of intimate associations between formerly free-living organisms. An unexpected symbiosis was recently discovered in the Santa Barbara Basin, a local depression in the seafloor off the coast of southern California. Sediments accumulating in this basin contain little oxygen but large amounts of hydrogen sulfide (H_2S) generated by anaerobic bacteria within the sediments. We might predict that eukaryotic cells would be uncommon in these sediments. Not only is oxygen scarce, but sulfide actively inhibits respiration in mitochondria. It came as a surprise, then, that samples of sediment from the Santa Barbara Basin contain large populations of single-celled eukaryotes.

Some of these cells have hydrogenosomes, mitochondria altered by evolution to generate energy where oxygen is absent. Others, however, appear to thrive by supporting populations of symbiotic bacteria on or within their cells. In one case, in which rod-shaped bacteria cover the surface of their eukaryotic host (**Fig. 27.8**), research has shown that the bacteria metabolize sulfide in the local environment, thereby protecting their host from this toxic substance. Scientists hypothesize that the bacteria benefit from this association as well because they get a free ride through the sediments, enabling them to maximize chemoautotrophic growth by remaining near the boundary between waters that contain oxygen and those rich in sulfide.

The diversity of these eukaryotic–bacterial symbioses is remarkable—and poorly studied. What has been learned to date, however, shows that single-celled eukaryotes have evolved numerous symbiotic relationships with chemoautotrophic bacteria, associations that feed and protect the eukaryotes, enabling them to colonize habitats where most eukaryotes cannot live. Clearly, then, the types of symbioses that led to mitochondria and chloroplasts on the early Earth were not rare events, but basic associations between cells that continue to evolve today.

27.3 EUKARYOTIC DIVERSITY

Historically, Domain **Eukarya** was divided into four kingdoms: plants, animals, fungi, and protists. Plants, animals, and fungi received special consideration for the obvious reason that they include large and relatively easily studied species. All remaining eukaryotes were grouped together as **protists,** which were defined as organisms having a nucleus but lacking other features specific to plants, animals, or fungi. The term "protist" is frowned on by some biologists because it doesn't refer to a monophyletic grouping of species—that is, an ancestral form and all its descendants. Nonetheless, as an informal term that draws attention to a group of organisms sharing particular characteristics, "protist" usefully describes the diverse world of microscopic eukaryotes and seaweeds.

Two other terms scorned by systematists but embraced by ecologists are "algae" and "protozoa." **Algae** are photosynthetic protists. They may be microscopic single-celled organisms or the highly visible, multicelled organisms we call seaweed. **Protozoa** are heterotrophic protists. These are almost exclusively single-celled organisms. "Algae" and "protozoa" in fact are simple and useful terms that have little phylogenetic meaning but which convey a great deal of information about the structure and function of these organisms.

Protist cells exhibit remarkable diversity. Some have cell walls, while others do not. Some make skeletons of silica (SiO_2) or calcium carbonate ($CaCO_3$), and others are naked or live within **tests,** or "houses," constructed exclusively of organic molecules. Some are photosynthetic, while others are heterotrophic. Some move by beating flagella, and others, like *Amoeba*, can extend fingers of cytoplasm, called pseudopodia, to move and capture food. Most are aerobic, but there are anaerobic protists as well. Surprisingly, most of these distinctive features have evolved multiple times and so do not define phylogenetically coherent groups. As a result, our understanding of evolutionary pattern in Eukarya had to wait for the molecular age.

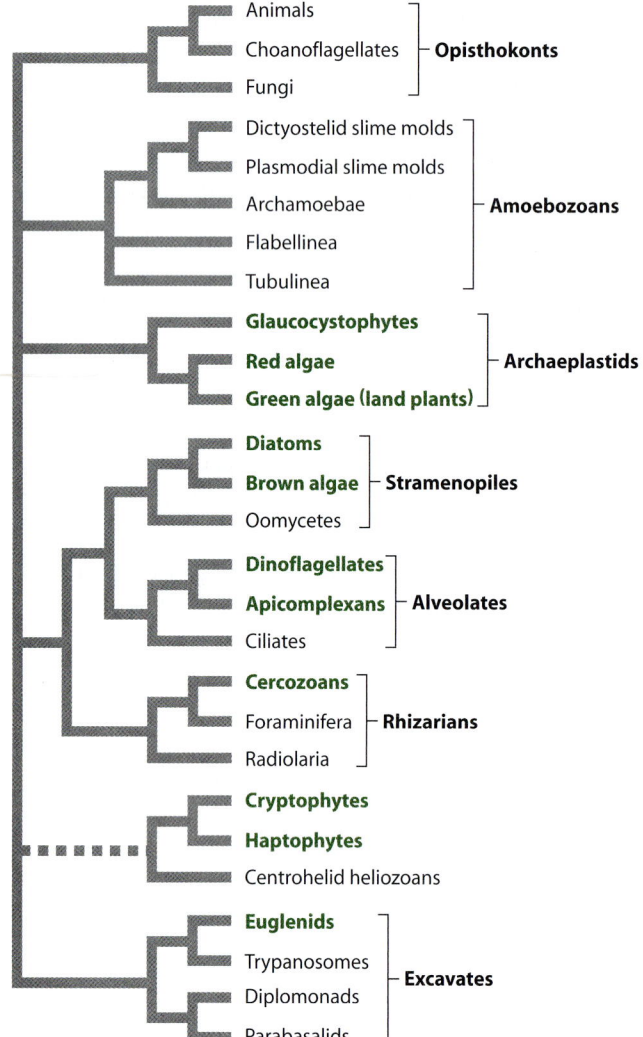

FIG. 27.9 The eukaryotic tree of life showing major groups. Branches with photosynthetic species (shown in green) are distributed widely. Additional branches, including animals, foraminiferans, radiolarians, and ciliates, contain species that harbor photosynthetic symbionts.

Fig. 27.9 shows a phylogenetic tree of eukaryotes as biologists currently understand it. As in the case of Bacteria, it is easier to identify major groups of Eukarya than to establish their evolutionary relationships to one another. Molecular sequence comparisons consistently recognize seven major groups, commonly called **superkingdoms,** noted in the figure. Animals fall within one of these superkingdoms, the Opisthonkonta, whereas plants fall within another, the Archaeplastida. Most, but not all, of the divisions into these major eukaryotic branches are strongly supported by data from genes and genomes. A few groups, notably the cryptophytes, haptophytes, and centrohelid heliozoans, seem to wander from position to position from one phylogenetic analysis to the next, sometimes moving together, sometimes separately. Whether the well-resolved superkingdoms constitute the only major limbs on the eukaryotic tree is less certain. Many microscopic protists remain unstudied, and we may yet be surprised by what they ultimately show.

In the following sections, we introduce the major features of eukaryotic diversity. The species numbers we present indicate known taxonomic diversity, and we emphasize that most protistan species remain unknown. Many biologists estimate that only about 1 in 10 protist species has been described so far. As this is also the case for animals and fungi (but arguably not for land plants because they are all visible to the naked eye), the relative numbers of species within different superkingdoms may reflect something like the actual distribution of biodiversity among eukaryotes.

Our own group, the opisthokonts, is the most diverse eukaryotic superkingdom.

Over the past 250 years, biologists have described about 1.8 million species. Of these, 75% or so fall within the Opisthokonta (**Fig. 27.10**), a name derived from Greek words meaning "posterior pole," calling attention to the fact that cell movement within this group is propelled by a single flagellum attached to the posterior end of the cell. Not all cells in this superkingdom move around. In humans, for example, the flagellum is limited mostly to sperm. Opisthokonts are heterotrophic, although some species harbor photosynthetic symbionts. Animals are the most diverse and conspicuous opisthokonts. More than 1.3 million animal species, mostly insects and their relatives, have been described. Fungi are also diverse, with more than 75,000 described species, including the visually arresting mushrooms (Fig. 27.10a). Here, however, we focus on opisthokont protists, somewhat poorly known microorganisms that hold clues to the origins of complex multicellularity in this superkingdom (Chapter 28).

It isn't easy to find morphological characters that unite all opisthokonts. We face this difficulty partly because animals and fungi have diverged so strikingly from what must have been the ancestral condition of the group. However, as molecular sequence comparisons began to reshape our understanding of eukaryotic phylogeny, it soon became clear that fungi and animals are closely related.

Even more closely related to the animals are **choanoflagellates,** a group of mostly unicellular protists characterized by a ring of microvilli, fingerlike projections that form a collar around the cell's single flagellum (Fig. 27.10b). About

FIG. 27.10 Opisthokonts. This superkingdom includes animals and fungi, as well as several protistan groups, including choanoflagellates, the closest protistan relatives of animals.

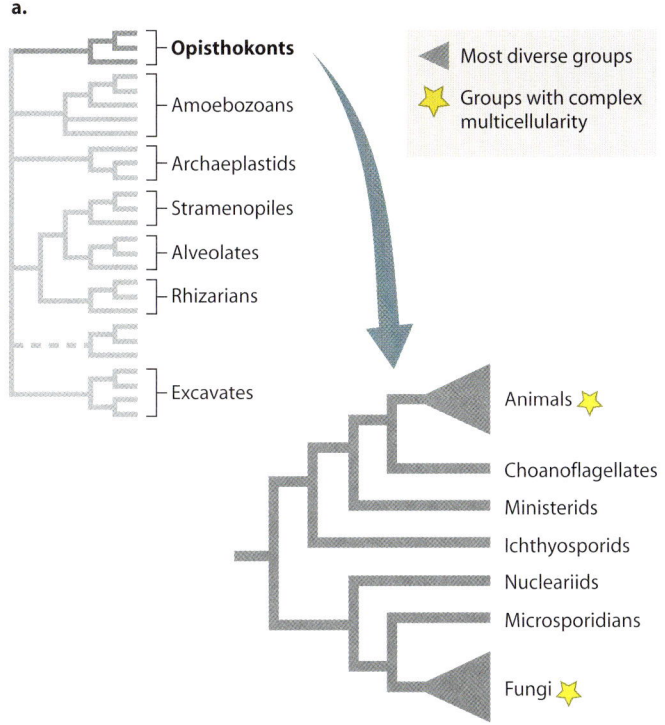

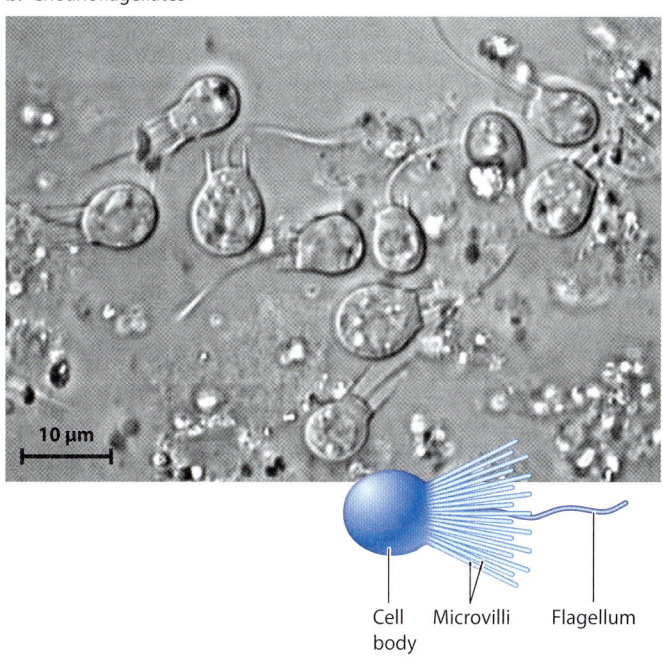

b. Choanoflagellates

150 choanoflagellate species have been described from marine and freshwater environments, where they prey on bacteria. As early as 1841, the close similarity between choanoflagellates and the collared feeding cells of sponges suggested that these minute organisms might be our closest protistan relatives. This view gained widespread popularity with the discovery in 1880 of *Proterospongia haeckeli*, a choanoflagellate that lives in colonies of cells joined by adhesive proteins.

Molecular sequence comparisons now confirm the close relationship between animals and choanoflagellates. For example, the complete genome of the choanoflagellate *Monosiga brevocollis* shows that a number of signaling molecules known to play a role in animal development are present in our choanoflagellate relatives, although their function in these single-celled organisms remains largely unknown. More generally, many genes once thought to be unique features of animals have now been identified in choanoflagellates, underscoring the deep evolutionary roots of animal biology.

Microsporidia form another group of single-celled opisthokonts. Microsporidia are parasites that live inside animal cells. Only their spores survive in the external environment, where they await the opportunity to infect a host and complete their life cycle. Microsporidia infect all animal phyla, so the approximately 1000 known species probably represent only the tip of the iceberg. More than a dozen species have been isolated from human intestinal tissues. Microsporidia infections can be particularly devastating in AIDS patients. Beyond their importance in human health, microsporidians have attracted the attention of biologists because of features they *lack*: Microsporidian cells have no aerobically respiring mitochondria, no Golgi apparatus, and no flagella. Furthermore, they have a highly reduced metabolism and among the smallest genomes of any known eukaryote. The cellular simplicity of microsporidians does not mean that they are early-evolved organisms, however. Molecular sequencing studies show that microsporidians are the descendants of more complex organisms. Their simplicity then is an adaptation for life as an intracellular parasite. It is now widely accepted that microsporidians are closely related to the fungi.

Other protistan opisthokonts have been identified in recent years. These are mostly parasites of aquatic animals, but like the choanoflagellates described earlier, their biology is beginning to illuminate the evolutionary path to complex multicellularity in animals and fungi.

FIG. 27.11 Amoebozoans. This group includes protists with an amoeboid stage in their life cycle.

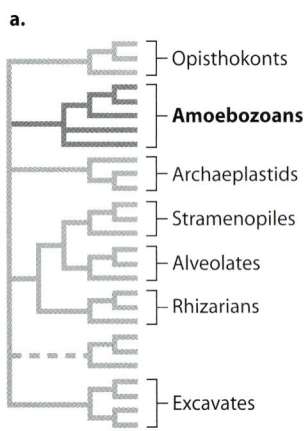

a.

b. *Amoeba proteus*, with clearly visible pseudopodia

c. *Entamoeba histolytica*, the infectious agent in amoebic dysentery

Amoebozoans include slime molds that produce multicellular structures.

As their name implies, the superkingdom **Amoebozoa** (**Fig. 27.11**) is a group of eukaryotes with amoeba-like cells that move and gather food by means of pseudopodia (illustrated by *Amoeba proteus* in Fig. 27.11b). More than 1000 amoebozoan species have been described. In some amoebozoan species, large numbers of cells aggregate to form multicellular structures as part of their life cycle.

Although there are amoeba-like cells that are not members of the Amoebozoa superkingdom, all members of this group are basically amoeboid in cellular organization, at least at some stage of their life cycle. Amoebozoans play an important role in soils as predators on other microorganisms. Some amoebozoans also influence human health. *Entamoeba histolytica* (Fig. 27.11c), an anaerobic protist that causes amoebic dysentery, is responsible for 50,000–100,000 deaths every year.

Beyond considerations of human health, the amoebozoans of greatest biological interest are the slime molds. In plasmodial slime molds (**Fig. 27.12**), haploid cells fuse to form zygotes that subsequently undergo repeated rounds of mitosis but not cell division to form colorful, often lacy structures visible to the naked eye (Fig. 27.12a). These structures, called plasmodia, are **coenocytic,** which means they contain many nuclei within one giant cell. The plasmodia can move and so seek out and feed on the bacteria and small fungi commonly found on bark or plant litter on the forest floor. Triggered by poorly understood environmental signals, plasmodia eventually differentiate to form stalked structures called sporangia that can be 1 to 2 mm high (Fig. 27.12b). Within the mature sporangium, cell walls form around the many nuclei, producing discrete cells that undergo meiosis, generating haploid spores that disperse into the environment. Germination of these spores begins the life cycle anew.

Cellular slime molds (**Fig. 27.13**) spend most of their life cycle as solitary amoeboid cells feeding on bacteria in the soil.

FIG. 27.12 Plasmodial slime molds. (a) Plasmodia, such as this pretzel slime mold, are coenocytic structures containing many nuclei. (b) Plasmodia generate sporangia, stalked structures that produce spores for dispersal.

Starvation, however, causes cells to produce the chemical signal cyclic AMP (Chapter 9), which induces as many as 100,000 cells to aggregate into a large multicellular slug-like form (Fig. 27.13a). This "slug" can migrate by a coordinated movement of its cells governed by actin and myosin, the same proteins that control muscle movement in animals. Migrating "slugs" can forage for food, leaving behind a trail of mucilage as they move—that's why they're called slime molds. Like plasmodial slime molds, the "slugs" of cellular slime molds eventually differentiate to form stalk-borne sporangia that produce the spores responsible for dispersal of the organism and renewal of the life cycle (Fig. 27.13b).

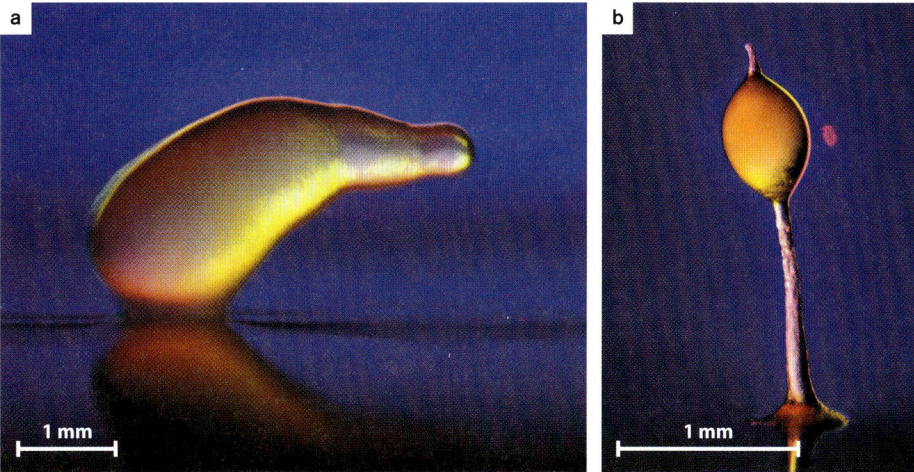

FIG. 27.13 Cellular slime molds. (a) Starvation causes amoeboid feeding cells of cellular slime molds to aggregate into a multicellular "slug" that can migrate along surfaces, leaving behind a trail of slime. (b) "Slugs" differentiate to form stalked sporangia that produce spores.

Slime molds are not true intermediates in the evolution of multicellularity—that is, they are not closely related to animals, fungi, or other groups in which complex multicellularity evolved. Nonetheless, the cellular slime mold *Dictyostelium* has been extensively studied by biologists as a model organism because it has much to teach us about fundamental eukaryotic processes of cell signaling and differentiation, cellular motility, and mitosis.

Archaeplastids are photosynthetic organisms, including land plants.

Next to the opisthokonts, the most conspicuous and diverse eukaryotes fall within the superkingdom **Archaeplastida,** another major branch of the eukaryotic tree (**Fig. 27.14a**). This is where we find the land plants, whose 300,000 described species dominate eukaryotic biomass on this planet (Chapter 33). The archaeplastids contain three major phylogenetic groups, and with the exception of a few parasitic and saprophytic (living on dead matter) species, all are photosynthetic. This indicates that the last common ancestor of living archaeplastids was photosynthetic. In fact, archaeplastids are the direct descendants of the protist that first evolved chloroplasts from endosymbiotic cyanobacteria. Some molecular analyses also place the group that includes cryptophyte and haptophyte algae within this branch, but this remains a topic of continuing research.

The least conspicuous group of archaeplastids are the glaucocystophytes (**Fig. 27.14b**), a small group of single-celled algae found in freshwater ponds and lakes. We might skip over them entirely, except for one illuminating observation: Glaucocystophyte chloroplasts appear to retain more features of the ancestral cyanobacterial endosymbiont than any other algae. Their chloroplasts have walls of peptidoglycan, the same molecule found in cyanobacterial walls, and their photosynthetic pigments include biliproteins, also found in cyanobacteria.

More abundant and diverse is the second major archaeplastid group, the red algae (**Fig. 27.14c**). About 5000 species are known, mostly from marine environments. Most are multicellular and some have differentiated tissues and complex morphology. Like other archaeplastids, most also have walls made of cellulose. The principal photosynthetic pigments in red algae are a form of chlorophyll called chlorophyll *a* and the biliproteins also found in glaucocystophytes and most cyanobacteria.

Red algae can be conspicuous as seaweeds on the shallow seafloor. One subgroup of red algae, known as the coralline algae, secretes skeletons of calcium carbonate ($CaCO_3$) that deter predators and help the algae withstand the energy of waves in shallow marine environments. If you examine the margin of coral reefs, where waves crash into the reef, you will often see that it is coralline algae and not the corals themselves that resist breaking waves and so enable reefs to expand upward into wave-swept environments. While most common and diverse in near-shore environments, red algae can also grow where light penetrates most deeply in the ocean. Their photosynthetic pigments enable them to take advantage of the short (blue) wavelengths of light.

Red algae are important as food in certain cultures, for example, laver in the British Isles (particularly in Wales, in cakes similar to oatcakes) and nori in Japan. Without knowing it, you often ingest carageenan, a polysaccharide synthesized by red algae. Carageenan molecules stabilize mixtures of biomolecules

FIG. 27.14 Archaeplastids. (a) Archaeplastids are a photosynthetic group descended from the protist that acquired photosynthesis from an endosymbioic cyanobacterium. They include the (b) glaucocystophytes, (c) red algae, and (d) green algae.

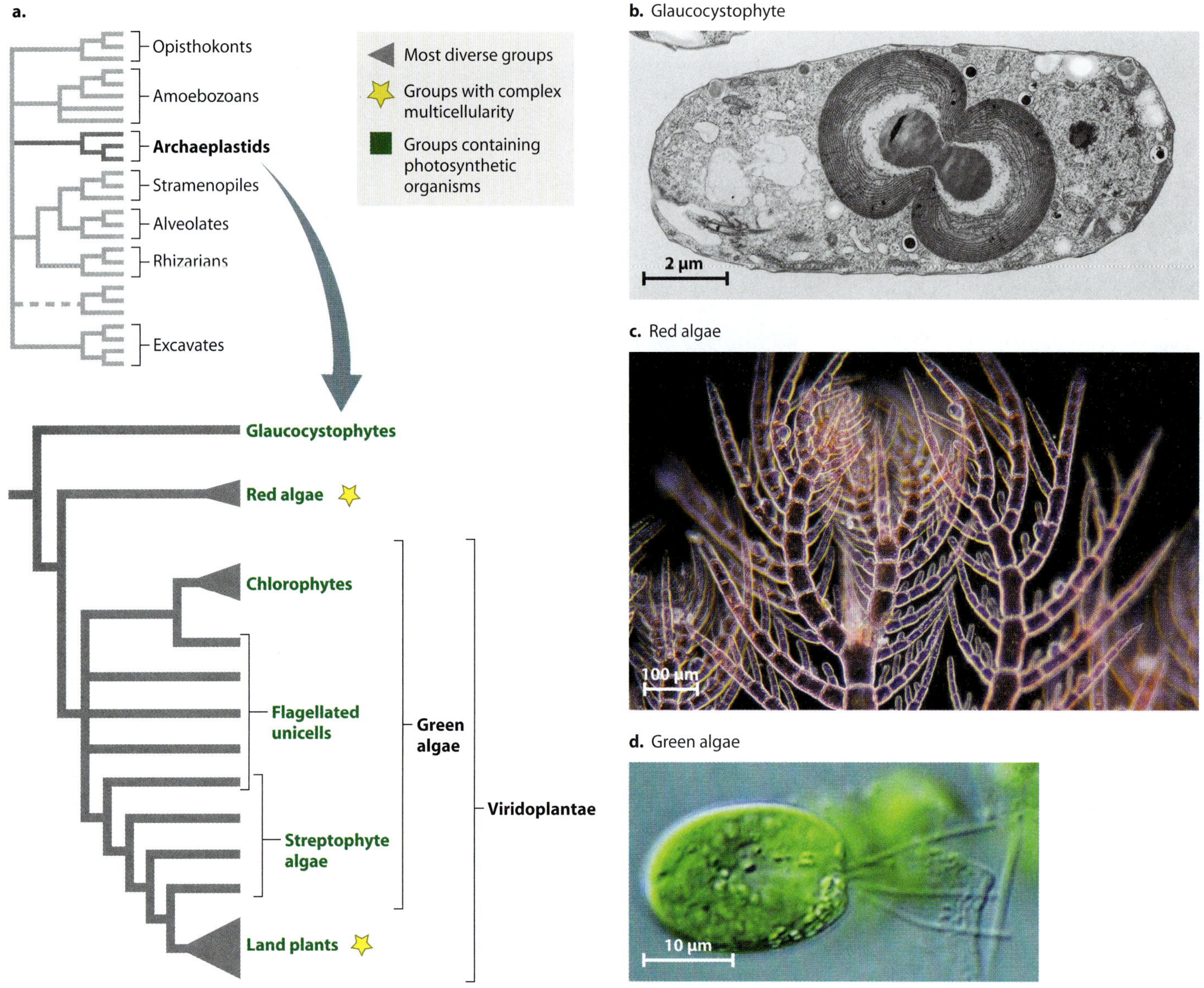

and promote gel formation, and so they are used in products ranging from ice cream to toothpaste. Agar, another red algal polysaccharide, is found universally in microbiology laboratories because the gel it forms at room temperature provides a nearly ideal substrate for bacterial growth.

The land plants and their algal relatives form Viridoplantae, the third branch of Archaeplastida. In this chapter, we are concerned with the protistan members of this group, the green algae (**Fig. 27.14d**). Green algae include a diverse assortment of species (about 10,000 have been described) that live in marine and, especially, freshwater environments. Green algae range from tiny single-celled flagellates to meter-scale seaweeds (**Fig. 27.15**). They all are united by two features: the presence of chlorophyll *a* and chlorophyll *b* in chloroplasts that have two membranes, and a unique attachment for flagella.

It is thought that green algae originated as small flagellated cells in seawater, and today most species that branch early on the green algal tree live as photosynthetic cells in the water. These species are called phytoplankton, and one of them is illustrated in Fig. 27.15a. The larger diversity of green algae, however, is found on two branches that diverged from these flagellated ancestors. One branch, the chlorophytes, radiated mostly in the sea and includes common seaweeds such as sea lettuce, which is seen globally on seashores as flotsam. The chlorophyte branch also includes relatively large and complex seaweeds that can form structures up to tens of centimeters long in tropical and temperate oceans (Fig. 27.15e). Moreover, chlorophyte green algae have played a major role in laboratory studies of photosynthesis (*Chlamydomonas*, Fig. 27.15a) and multicellularity (*Volvox*, Fig. 27.15d).

FIG. 27.15 Diversity of green algae. (a) *Chlamydomonas reinhardtii*, a tiny flagellated unicell widely used in laboratory research on photosynthesis; (b) *Spirogyra*, a freshwater green alga with distinctive helical chloroplasts; (c) a desmid, among the most diverse and widespread of all green algae in freshwater; (d) *Volvox*, a simple multicellular organism; and (e) *Acetabularia*, a macroscopic (but single-celled!) green alga found in coastal marine waters in tropical climates.

For humans, however, the more important branch of the green algal tree is the one that diversified on land. Called streptophytes, the species on this branch show a progression of form from unicells at the base through simple cell clusters and filaments on intermediate branches (see Figs. 27.15b and 21.15c), to complex multicellular algae that form the closest relatives of land plants (Chapter 33). As was true of choanoflagellates and animals, green algae display features of cell biology and genomics that tie them unequivocally to land plants.

→ **Quick Check 2** Do plants and animals have a common multicellular ancestor?

Other photosynthetic organisms occur in the stramenopiles and alveolates.

The superkingdom **Stramenopila** forms a diverse and distinctive branch of the eukaryotic tree (see Fig. 27.9). The group includes unicellular organisms and giant kelps, algae and protozoa, free-living cells and parasites. All have an unusual flagellum that bears two rows of stiff hairs (**Fig. 27.16**). Most also have a second, smooth flagellum.

FIG. 27.16 Stramenopiles. This electron micrograph of a unicellular stramenopile shows a flagellum with two rows of stiff hairs. Like this cell, most stramenopiles also have a second, smooth flagellum.

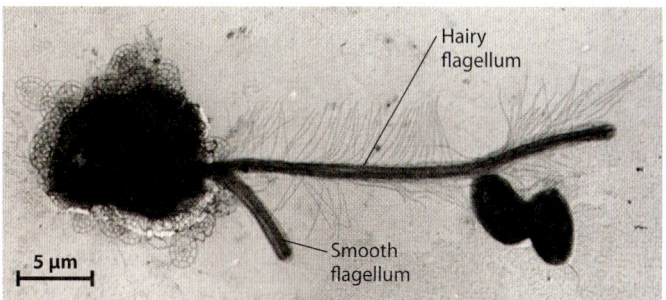

FIG. 27.17 Kelps, complex multicellular brown algae that can form "forests" tens of meters high in the ocean.

Despite the presence of heterotrophic species, most stramenopiles are photosynthetic. There are about a dozen groups of stramenopile algae, of which the diatoms and the brown algae deserve special mention. Brown seaweeds are common along rocky shorelines across the world, and in the Sargasso Sea, huge masses of ropy brown algae (aptly named *Sargassum*) float at the surface. Easily the most impressive brown algae are the kelps, giant seaweeds that form forests above the seafloor (**Fig. 27.17**). Kelps not only develop functionally distinct tissues, but have evolved distinct organs specialized for attachment, photosynthesis, flotation, and reproduction.

The most diverse stramenopiles, however, are the diatoms (**Fig. 27.18**). Recognized by their distinctive skeletons of silica, diatoms account for half of all primary production in the sea (and, so, nearly a quarter of all photosynthesis on Earth). The 10,000 known diatom species thrive in environments that range from wet soil to the open ocean.

Among heterotrophic stramenopiles, the oomycetes, or water molds, are particularly interesting. Originally classified as fungi, these multicellular pathogens cause a number of plant diseases, including the potato blight that devastated Ireland during the 1840s (Chapter 33). Other species infect animals, including humans.

Like stramenopiles, the superkingdom **Alveolata** is a diverse group. It includes dinoflagellates (**Fig. 27.19a**), less than half of which contain chloroplasts and are photosynthetic, and ciliates (**Fig. 27.19b**), heterotrophic protists that have two nuclei in each cell and numerous short flagella called cilia. The group is united by the presence of cortical alveoli, small vesicles packed beneath the cell surface that, in some species at least, store calcium ions for use by the cell (**Fig. 27.19c**). Dinoflagellates are important as photosynthesizers in many parts of the oceans, and they affect humans when they form red tides, large and toxic blooms in coastal waters. Red tides occur when nutrients are supplied to coastal waters in large amounts. This may occur for natural reasons, but it commonly reflects human activities such as sewage seeps or runoff from highly fertilized croplands.

Scientists have described nearly 4000 dinoflagellate and 8000 ciliate species, but it may be the third principal group of alveolates, the Apicomplexans, that attracts most scientific attention. The reason is simple: The apicomplexan *Plasmodium falciparum* and related species cause malaria, a lethal disease that sickened millions and killed more than 600,000 people in 2011 (Case 4: Malaria). Although *Plasmodium* cells are heterotrophic, they contain vestigial chloroplasts that suggest evolutionary loss of photosynthesis. Recently, this hypothesis has been confirmed by the discovery in Sydney Harbour, Australia, of photosynthetic cells closely related to parasitic *Plasmodium*.

Photosynthesis spread through eukaryotes by repeated endosymbioses involving eukaryotic algae.

A curious feature of the eukaryotic tree is that photosynthetic eukaryotes are distributed widely but discontinuously among its branches (see Fig. 27.9). How can we explain this pattern?

Two hypotheses have been suggested: Either photosynthesis was established early in eukaryotic evolution and was subsequently lost in some lineages, or eukaryotes acquired photosynthesis multiple times by repeated episodes of endosymbiosis. The second hypothesis, it turns out, is correct. The twist is that most of the symbionts involved in the spread of photosynthesis across eukaryotic phylogeny were photosynthetic eukaryotes, not cyanobacteria.

Once more, DNA sequence comparisons provide the decisive data. We can construct phylogenetic trees for photosynthetic

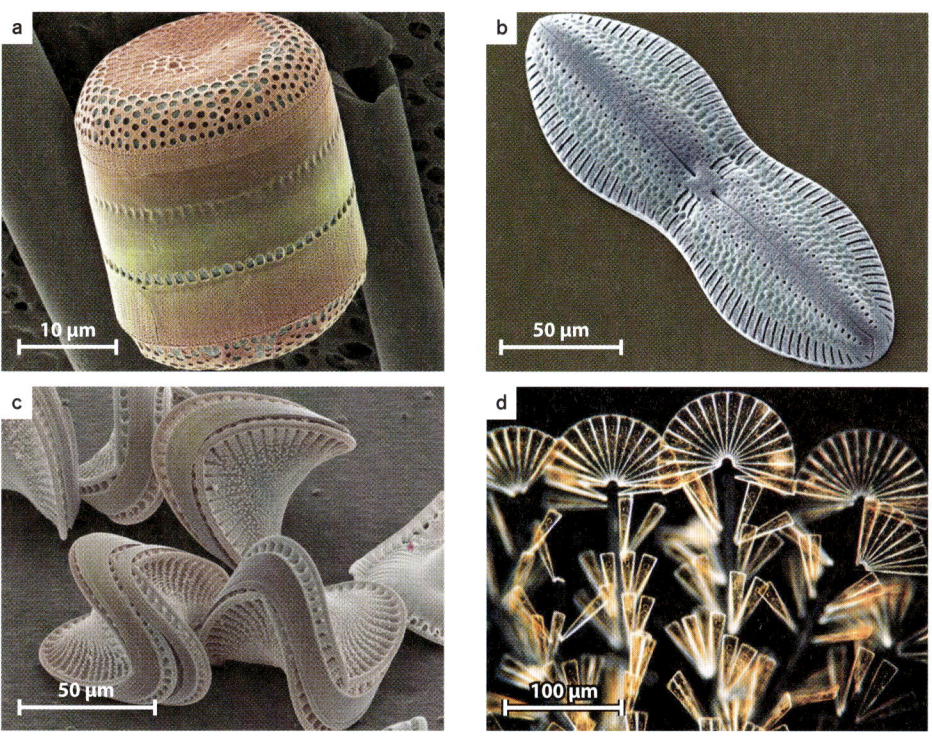

FIG. 27.18 Diatoms. Diatoms are the most diverse stramenopiles, and among the most diverse of all protists. Their shapes, outlined by their tiny skeletons made of silica (SiO_2) can be (a) like a pill-box, (b) elongated, or (c) twisted around the long axis of the cell. Some diatom species form colonies, like those shown in (d).

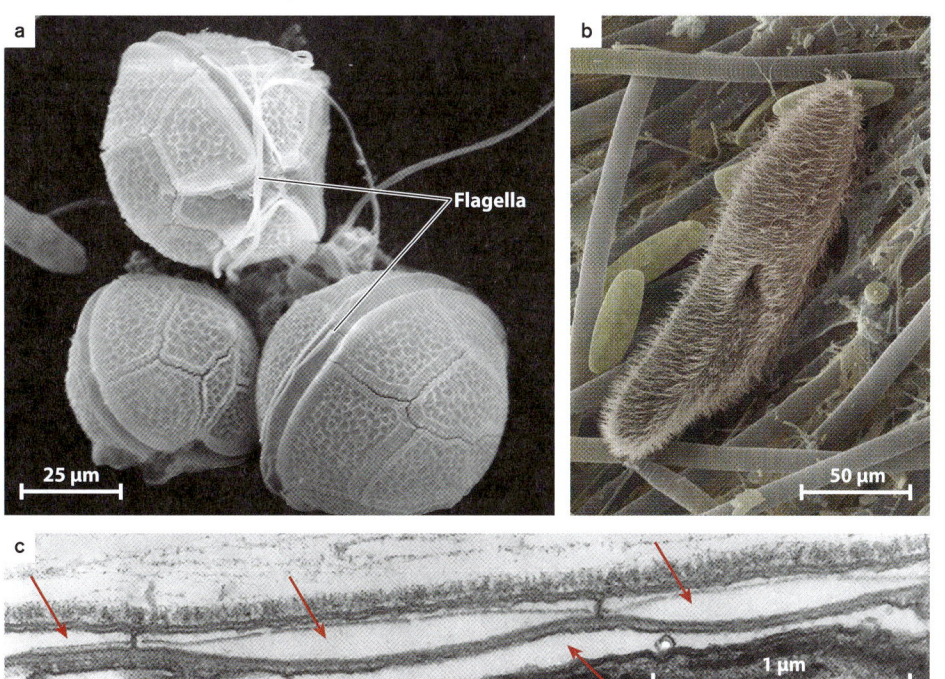

FIG. 27.19 Alveolates. (a) Dinoflagellates have cell walls made of interlocking plates of cellulose. (b) A ciliate has many short flagella, or cilia, that line the cell. (c) The arrows point to alveoli, flattened vesicles just beneath the surface of the cell that give alveolates their name.

eukaryotes using chloroplast genes and then repeat the exercise using nuclear genes. **Fig. 27.20a** shows a phylogenetic tree for chloroplasts. Notice that this tree differs substantially from the branching order in trees based on comparison of nucleotide sequences for nuclear genes, such as the tree shown in **Fig. 27.20b**. The differences between the chloroplast and nuclear trees favor the hypothesis that photosynthesis spread through the eukaryotes by means of repeated symbiotic events.

Let's look more closely at the chloroplast phylogeny in Fig. 27.20a. The lower part of the phylogeny shows that chloroplasts form a monophyletic grouping nested within the cyanobacteria. This is a principal line of evidence supporting the endosymbiotic origin of chloroplasts. Moving upward into the chloroplast branch, if we look only at the branches shown in black, we see the phylogeny indicated by nuclear genes for the archaeplastids—glaucocystophytes are sister to red algae and green algae. But notice the branches in green that sprout from within the green algae, the chlorarachniophytes and the euglenids. Both of these groups have chloroplasts that contain the same pigments as green algae (chlorophyll *a* and chlorophyll *b*), and molecular sequence comparison clearly indicates that their chloroplasts are closely related to the chloroplasts of green algae. Nuclear genes, however, place these groups far from the archaeplastids. Chlorarachniophytes are found within the superkingdom Rhizaria, and euglenids fall among the superkingdom Excavata. How do we explain this discordance? The answer is that chlorarachniophytes gained photosynthesis by establishing green algal cells (containing chloroplasts derived from an earlier symbiotic assimilation of cyanobacteria) as endosymbionts. Euglenids did the same thing, independently.

HOW DO WE KNOW?

FIG. 27.20

How did photosynthesis spread through the Eukarya?

BACKGROUND Many different branches on the eukaryotic tree include photosynthetic species, commonly interspersed with non-photosynthetic lineages. Did eukaryotes gain photosynthesis once, by the endosymbiotic incorporation of a cyanobacterium, and then lose this capacity multiple times? Or, is the history of photosynthetic endosymbioses more complicated, involving multiple events of symbiont capture and transformation?

a. Phylogenetic relationships based on chloroplast genes

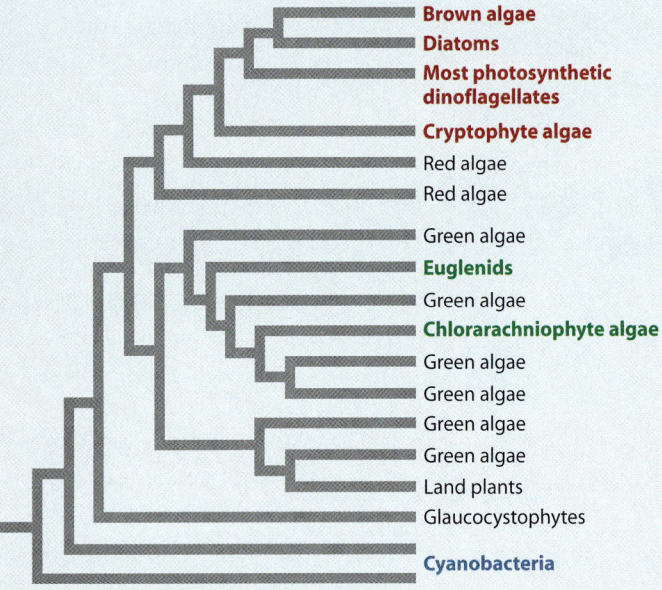

b. Phylogenetic relationships based on nuclear genes

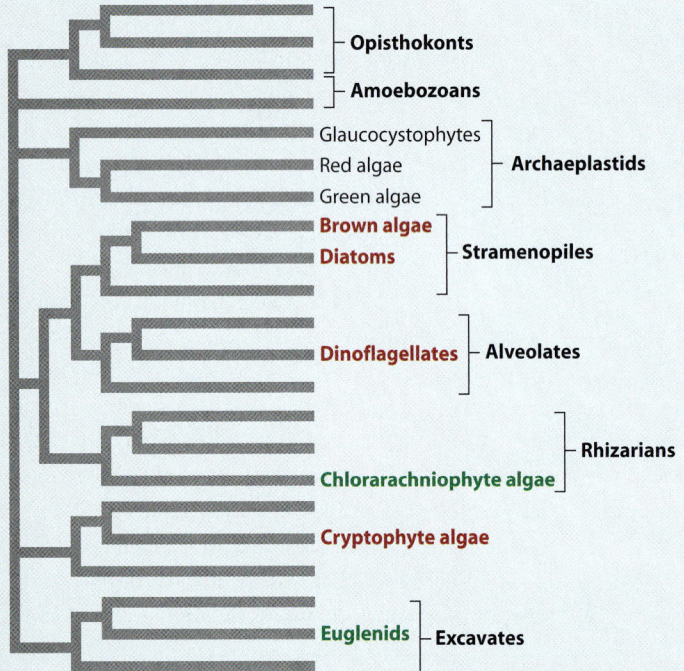

HYPOTHESIS 1 Photosynthesis was established early in eukaryotic evolution and was subsequently lost in some lineages.

HYPOTHESIS 2 Eukaryotes acquired photosynthesis multiple times by repeated episodes of endosymbiosis.

PREDICTIONS The two hypotheses make different predictions. If photosynthesis were acquired in a single common ancestor of all living eukaryotes and then lost in some lineages, we would expect chloroplast and nuclear gene phylogenies to show the same pattern of branching. If photosynthesis were acquired multiple independent times, we would expect different chloroplast and nuclear phylogenies to show different patterns.

EXPERIMENT Chloroplasts have DNA, so scientists developed a phylogenetic hypothesis based on molecular sequence comparison of chloroplast genes (Fig. 27.20a). This was then compared with a second phylogeny, for all eukaryotes, based on molecular sequence comparison of nuclear genes (Fig. 27.20b).

RESULTS Molecular sequence comparisons show that evolutionary relationships among chloroplasts do not mirror those based on nuclear genes. The chlorarachniophyte algae and photosynthetic euglenids (shown in green) fall in a monophyletic group with the green algae when chloroplast DNA is analyzed (Fig. 27.20a). However, these groups lie far from green algae in the phylogeny based on nuclear genes (Fig. 27.20b). Similarly, brown algae, diatoms, most photosynthetic dinoflagellates, and cryptophyte algae (shown in red) form a monophyletic group with the red algae in the phylogeny based on chloroplast DNA (Fig. 27.20a), but analysis of nuclear genes places them in different groups (Fig. 27.20b).

CONCLUSION The chloroplast and nuclear phylogenies show different patterns of branching, supporting the hypothesis that photosynthesis spread through the Eukarya by means of multiple eukaryotic endosymbionts.

SOURCE Hackett, J. D., H.S. Yoon, N. J. Butterfield, M. J. Sanderson, and D. Bhattacharya. 2007. "Plastid Endosymbiosis: Sources and Timing of the Major Events." In *Evolution of Primary Producers in the Sea*, edited by P. G. Falkowski and A. H. Knoll, 109–132. Boston: Elsevier Academic Press.

Now focus on the branch dominated by red algae. Once again, the branches shown in red indicate photosynthetic eukaryotes whose nuclear genes place them far away from red algae on phylogenetic trees. Thus, groups like the photosynthetic stramenopiles (diatoms, kelps, and others), photosynthetic dinoflagaellates, and a few others also appear to have arisen by endosymbiosis involving another photosynthetic endosymbiont that was eukaryotic—in this case, a red algal cell.

Several features of cell biology support this view. We noted earlier that chloroplasts in green and red algae are bounded by two membranes, but most of the algae shown in red and green in Fig. 27.20 have four membranes around their chloroplasts. The presence of four membranes is expected under the hypothesis of eukaryotic endosymbionts: The inner two membranes are those that bounded the chloroplast in the green or red algal endosymbiont, the third membrane is derived from the outer cell membrane of the symbiont, and the fourth, outer membrane is part of the cell membrane of the host (**Fig. 27.21**).

Chlorarachniophyte (and, on the red branch, cryptophyte) algae also retain another feature that supports the eukaryotic origin of their chloroplasts: a small organelle called a nucleomorph. The nucleomorph in chlorarachniophyte algae is, in fact, the remnant nucleus of a green algal symbiont. It retains only a few genes, but those show close phylogenetic affinity with green algae. Nucleomorph genes in cryptophyte algae similarly provide evidence for a red algal symbiont.

How many endosymbiotic events are required to account for the diversity of photosynthetic eukaryotes? There was the primary endosymbiosis that gave rise to green and red algae, two further events that established chloroplasts derived from green algae in euglenids and chlorarachniophytes, and at least one to three events that established chloroplasts derived from red algae in cryptophytes, haptophytes, and stramenopiles.

The dinoflagellates add a remarkable coda to this accounting. Like stramenopiles, most dinoflagellates have chloroplasts derived from red algae. Three dinoflagellate groups, however, have chloroplasts that are clearly derived from green, cryptophyte, and haptophyte algae. Include *Paulinella* and at least 8 to 10 endosymbiotic events are required to explain eukaryotic photosynthesis. Eukaryotes never evolved oxygenic photosynthesis—only the cyanobacteria accomplished that. Instead, eukaryotes appropriated the biochemistry of cyanobacteria and spread it throughout the domain by the horizontal transfer of entire cells.

→ **Quick Check 3** How do the molecular sequences of genes in chloroplasts show how photosynthesis spread throughout the eukaryotic domain?

The first branch of the eukaryotic tree may separate animals and slime molds from plants and diatoms.

Molecular sequence comparisons have succeeded in dividing eukaryotic species into the seven superkingdoms, but determining relationships among these major branches has proved difficult. This uncertainty is apparent in Fig. 27.9, which shows seven major branches of eukaryotic organisms but, for the most part, does not suggest how they relate to one another. Biologists call this an unrooted tree because we cannot go from the tips of the finest branches (present-day organisms) and trace backward through successive branch points to arrive at the "root"—the last common ancestor of the total group.

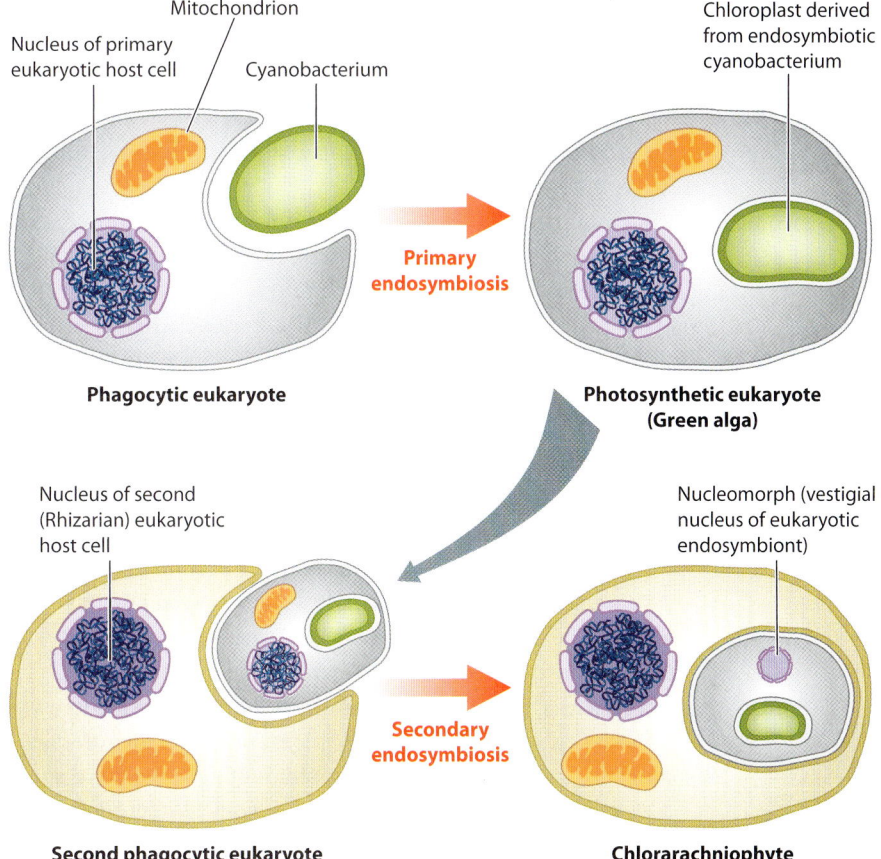

FIG. 27.21 Successive endosymbiotic events that established photosynthesis in chlorarachniophyte algae.

As discussed in Chapter 23, biologists use synapomorphies, features shared by some but not all members of a group through descent from a common ancestor, to recognize sister groups, or closest evolutionary relatives. Skull shape, for example, unites birds and crocodiles, while separating them from other tetrapod vertebrates. Can we find characters that will allow us to identify sister-group relationships among the eukaryotic superkingdoms? It isn't easy, because the groups diverged more than a billion years ago and have accumulated so many differences in their cellular features that synapomorphies among superkingdoms are hard to identify. This is true of molecular sequences as well as features of cell shape and internal structure.

A leading hypothesis for sorting the superkingdoms reflects a seemingly minor feature of gene organization, but one that has so far been successful in dividing known eukaryotes into two groups. In opisthokonts and amoebozoans, genes that code for the enzymes dihydrofolate reductase and thymidylate synthase do not lie next to each other on a single chromosome. In all other eukaryotes, the two genes are fused together along one chromosome. As the two genes are also separate in bacteria, this observation suggests that the gene fusion observed in many eukaryotes is a synapomorphy that unites most eukaryotes and separates them from the branch we share with fungi and amoebozoans. Other lines of evidence support this interpretation, particularly the observation that opisthokont and amoebozoan cells have only one flagellum, but other eukaryotes (if they have them at all) usually have two or more.

This being the case, we can draw the tree in two different ways, as shown in **Fig. 27.22**. The hypothesis that opisthokonts and amoebozoans are sisters and that these two groups together are sister to all other eukaryotes (the top tree) is consistent with the data on gene location and flagella number. But so is an alternative hypothesis—that the opisthokonts are sister to all other eukaryotes, with gene fusion occurring near the base of the limb that contains archaeplastids, stramenopiles, alveolates, rhizarians, and excavates (the bottom tree). At present, we lack data to choose between these hypotheses, but the resolution of branching pattern in eukaryotic phylogeny will help us understand which features were present in the last common ancestor of living eukaryotes and how the immense diversity of eukaryotic organisms evolved from those humble beginnings.

27.4 THE FOSSIL RECORD OF PROTISTS

Not all algae and protozoa fossilize easily, but those that do provide a striking record of evolution that both predates and parallels the better-known fossil record of animals. Fossilized protists do not contain DNA, nor do they preserve the features of cell biology that define eukaryotes, such as nuclear membranes and organelles. Thus, for a fossil to be identified as eukaryotic, it must preserve other features of morphology found today in Eukarya, but not in Bacteria or Archaea. For this reason, fossils that can be identified with certainty as eukaryotic are limited to those left by eukaryotes that synthesized distinctive and preservable cell walls at some stage of their life cycle.

Fossils show that eukaryotes existed at least 1800 million years ago.

Sedimentary rocks deposited in coastal marine environments 1800–1400 million years ago contain microscopic fossils (called **microfossils** for short) up to about 300 μm in diameter. Many of these have complicated wall structures that identify them as eukaryotic. Interlocking plates, long and branching arms, and complex internal layering have been imaged by both scanning and transmission electron microscopes (**Fig. 27.23a**). Comparison with living organisms suggests that such fossils could only be formed by organisms with a cytoskeleton and endomembrane system, the hallmarks of eukaryotic biology.

At present, it appears that the eukaryotic domain first appeared no later than 1800 million years ago and possibly much earlier, although the record is difficult to trace in older rocks. None of these earliest eukaryotic microfossils can be placed with confidence into one of the living superkingdoms. Conceivably, at least some of the oldest protistan microfossils could predate the last common ancestor of living eukaryotes.

In sedimentary rocks thought to be about 1200 million years old, we find the oldest fossils that can be linked clearly to a living group of eukaryotes (**Fig. 27.23b**). These fossils preserve features found today only in the red algae. We can therefore

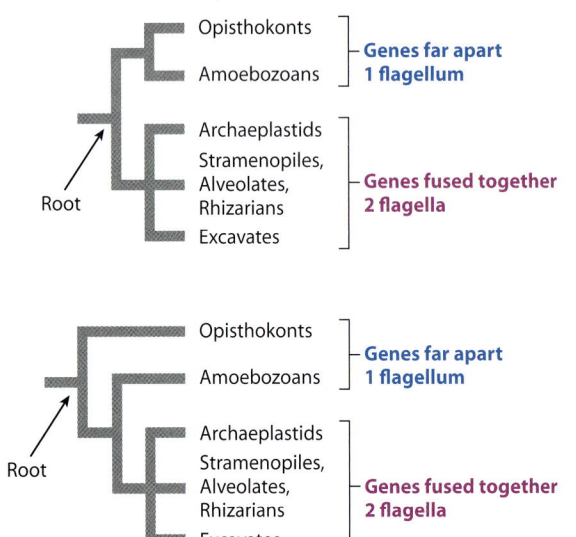

FIG. 27.22 Two possible phylogenies for eukaryotes based on number of flagella and whether the genes for dihydrofolate reductase and thymidylate synthase are close or far apart.

FIG. 27.23 **Eukaryotic fossils in Precambrian sedimentary rocks.** (a) Simple protist in 1500-million-year-old rocks from Australia. (b) Simple multicellular red algae in 1200-million-year-old rocks from Canada. (c) Amoebozoan test in 750-million-year-old rocks from the Grand Canyon, Arizona.

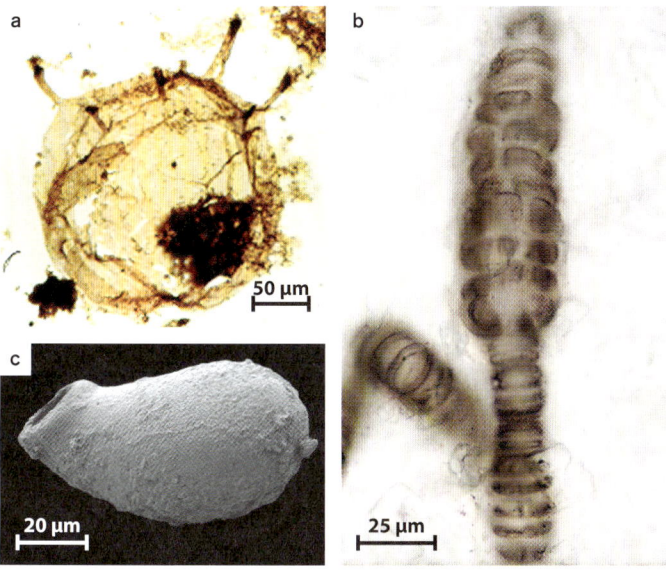

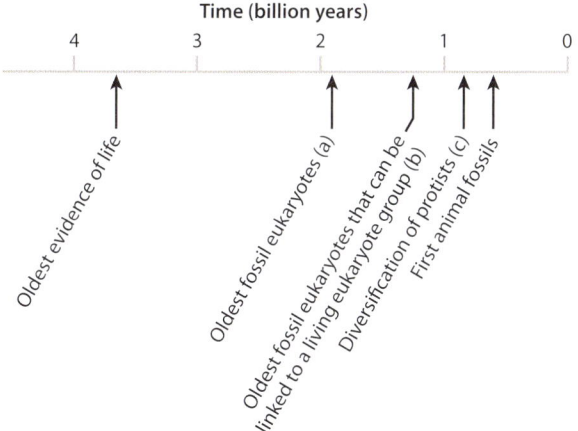

appears to have declined while oxygen increased. It may be, then, that eukaryotic organisms diversified as the oceans and atmosphere began to draw closer to their modern states.

Protists have continued to diversify during the age of animals.

Modern protistan diversity took shape alongside evolving animals during the Phanerozoic Eon, 542 million years ago to the present. For example, foraminiferans and radiolarians, marine protozoans with shells of calcium carbonate and silica, respectively, appear for the first time in rocks deposited near the beginning of the Phanerozoic Eon and have continued to diversify since that time (**Fig. 27.24**). Some algal seaweeds also evolved skeletons of calcium carbonate at this time, which diversified along with animals in Phanerozoic oceans. Why did algae and protozoans evolve mineralized skeletons? Presumably to keep from being eaten.

Marine protists underwent a new round of radiation during the Mesozoic Era, 252–65 million years ago. The most significant changes occurred among photosynthesizers. During the preceding Paleozoic Era, green algae and cyanobacteria were the primary photosynthesizers in the oceans, but as the Mesozoic Era began,

FIG. 27.24 (a) Foraminifera and (b) Radiolaria, showing skeletons of calcium carbonate and silica, respectively.

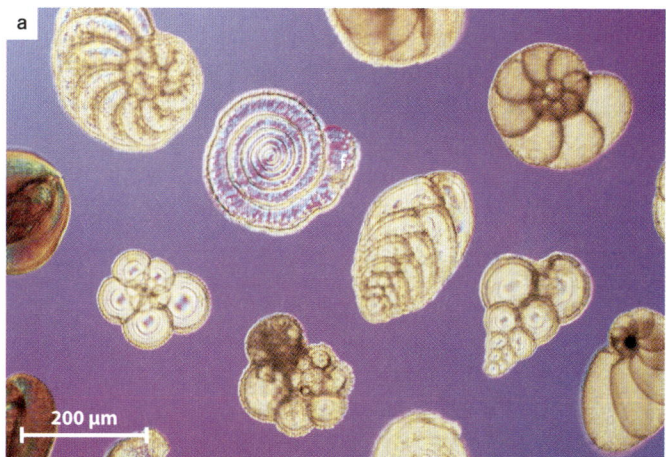

conclude that by 1200 million years ago Archaeplastida had already diverged from other eukaryotic superkingdoms, photosynthesis had become established in eukaryotes, and simple multicellularity had already evolved within the domain.

Much more eukaryotic diversity is recorded in rocks 800–700 million years old, including green algae and the remains of test-forming amoebozoans (**Fig. 27.23c**). Molecular fossils further indicate the presence of ciliates and dinoflagellates in the emerging eukaryotic world and show that algae were expanding to become major photosynthesizers in the oceans. Until about 800 million years ago, subsurface waters of the oceans were commonly poorly supplied with oxygen and were sometimes rich in sulfide, as well. After this time, sulfide

FIG. 27.25 Coccolithophorids. (a) Coccolithophorid algae, like *Emiliania huxleyi*, form tiny scales of calcium carbonate. (b) A satellite image captures a coccolithophorid bloom off the coast of Newfoundland, visible as light blue streamers of calcium carbonate in the otherwise dark blue ocean.

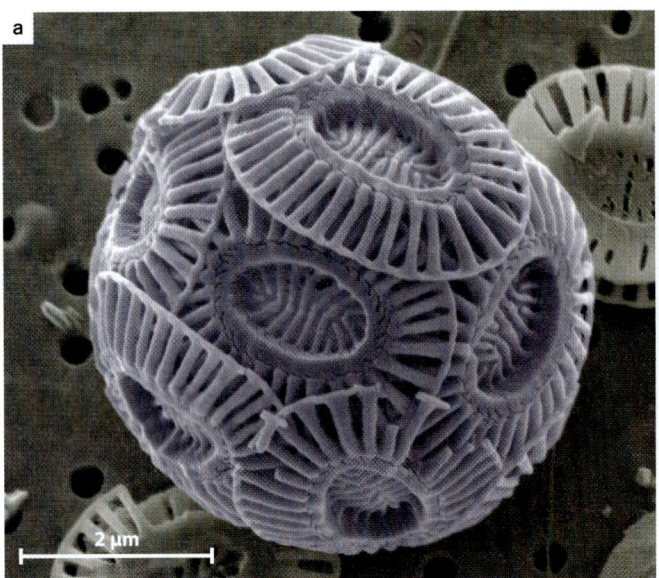

new groups of photosynthetic dinoflagellates and coccolithophorid algae appeared, followed by diatoms. Today, these three groups dominate photosynthesis in many parts of the sea.

Within the oceans, diatom and coccolithophorid evolution had another important consequence. The silica skeletons of diatoms are now the principal means by which SiO_2 is transferred from seawater to sediments. Similarly, the $CaCO_3$ scales made by coccolithophorids and the skeletons of planktonic foraminifera are a primary source of carbonate sediments on the seafloor (**Fig. 27.25**). In this way, the ongoing evolution of protists in the age of animals has continued to transform marine ecosystems.

Today, microscopic eukaryotes dominate photosynthetic production in coastal oceans and play a major role in biogeochemical cycles. And for thousands of protists, plants and animals are simply habitats, ripe for colonization—we need look no further than the many eukaryotic vectors of human disease. The key conclusion is that a remarkable, and still largely undocumented, diversity of eukaryotic organisms has evolved through time to form an integral part of ecosystems across our planet, shaping in ways both positive and negative the world in which we live.

Core Concepts Summary

27.1 EUKARYOTIC CELLS ARE DEFINED BY THE PRESENCE OF A NUCLEUS, BUT OTHER FEATURES, PARTICULARLY A DYNAMIC CYTOSKELETON AND MEMBRANES, EXPLAIN THEIR SUCCESS IN DIVERSIFYING.

Eukaryotic cells have a network of proteins inside the cell that allow them to change shape, move, and transfer substances in and out of the cell. page 27-1

Eukaryotic cells have dynamic membranes that facilitate movement and feeding. page 27-2

Eukaryotic cells compartmentalize their machinery for energy metabolism into mitochondria and chloroplasts, freeing the rest of the cell to interact with the environment in novel ways not available to prokaryotes. page 27-2

The eukaryotic genome is larger than that of prokaryotes, allowing new mechanisms of gene regulation. page 27-3

Eukaryotes reproduce sexually, which promotes genetic diversity, but cannot on its own help us understand their great diversity. page 27-3

27.2 THE ENDOSYMBIOTIC HYPOTHESIS PROPOSES THAT THE CHLOROPLASTS AND MITOCHONDRIA OF EUKARYOTIC CELLS WERE ORIGINALLY FREE-LIVING BACTERIA THAT WERE INCORPORATED INTO A HOST CELL.

The endosymbiotic hypothesis is based on physical, biochemical, and genetic similarities between chloroplasts and cyanobacteria, and between mitochondria and proteobacteria. page 27-5

Chloroplasts and mitochondria have their own genomes, but their genomes are small relative to free-living bacteria, to which

CHAPTER 27 EUKARYOTIC CELLS: ORIGINS AND DIVERSITY

they are closely related, mostly because of gene migration to the nuclear genome. page 27-6

Symbiosis between a heterotrophic host and photosynthetic partner is common throughout the eukaryotic domain; reef-forming corals are an example. page 27-6

Photosynthesis spread through the Eukarya by means of multiple independent events involving a protozoan host and a eukaryotic endosymbiont. page 27-6

Most eukaryotic cells have mitochondria, but a few do not. Evidence suggests that cells lacking mitochondria once had them but lost them. page 27-6

The eukaryotic nuclear genome contains genes unique to Eukarya, but also genes related to Bacteria and Archaea, suggesting either that the ancestor of the modern eukaryotic cell was a primitive eukaryote or an archaeon that engulfed a bacterium. page 27-7

27.3 EUKARYOTES HAVE HISTORICALLY BEEN DIVIDED INTO FOUR KINGDOMS—ANIMALS, PLANTS, FUNGI, AND PROTISTS—BUT ARE NOW DIVIDED INTO AT LEAST SEVEN SUPERKINGDOMS.

The terms "protist," "algae," and "protozoa" are useful, but do not describe monophyletic groups. page 27-9

Protists are eukaryotes that do not have features of animals, plants, and fungi; algae are photosynthetic protists; and protozoa are heterotrophic protists. page 27-9

The seven superkingdoms of eukaryotes are Opisthokonta, Amoebozoa, Archaeplastida, Stramenopila, Alveolata, Rhizaria, and Excavata. page 27-10

Opisthokonts are the most diverse eukaryotic superkingdom. They include animals and fungi, as well as choanoflagellates (our closest protistan relatives). page 27-10

Amoebozoans produce multicellular structures by the aggregation of amoeba-like cells, and include organisms that cause human disease and those important in biological research. page 27-12

The archaeplastids are photosynthetic, include land plants, and are divided into three major groups. page 27-13

Algae occur within the stramenopiles and several other groups. page 27-15

The earliest branch of the eukaryotic tree may separate Opisthokonta and Amoebozoa from the other superkingdoms, although this is still debated. page 27-19

27.4 THE FOSSIL RECORD EXTENDS OUR UNDERSTANDING OF EUKARYOTIC DIVERSITY BY PROVIDING PERSPECTIVES ON THE TIMING AND ENVIRONMENTAL CONTEXT OF EUKARYOTIC EVOLUTION.

Fossils in sedimentary rocks as old as 1800 million years have unmistakable signs of eukaryotic cells, including complicated wall structures. page 27-20

The earliest fossil eukaryotes that can be placed into one of the present-day superkingdoms are 1200 million years old and belong to the red algae. page 27-20

Eukaryotic fossils diversified around 800 million years ago, coinciding with an increase in oxygen and a decline in sulfide in the atmosphere and oceans. page 27-21

Protists continue to diversify to the present. Green algae and cyanobacteria dominated primary production in earlier oceans, but since about 200 million years ago dinoflagellates, coccolithophorids, and diatoms have become the primary photosynthetic organisms in the oceans. page 27-21

Protists have evolved to take advantage of the environments provided by animal and plants; these include many protists that cause disease in humans. page 27-22

Self-Assessment

1. List key features that distinguish a eukaryotic cell from a prokaryotic cell.

2. Describe the forms of energy metabolism found in eukaryotes.

3. Describe the origin and evolution of the chloroplast and mitochondrion.

4. Present two hypotheses for the origin of the eukaryotic cell.

5. Name the seven superkingdoms of eukaryotes and an organism in each one.

6. State when the eukaryotic cell is first thought to have evolved and the evidence that supports this date.

Do you understand the chapter's Core Concepts? Log in to check your answers to the Self-Assessment questions, then practice what you've learned and reinforce this chapter's concepts by working through the problems and multimedia tutorials provided there.

www.biologyhowlifeworks.com

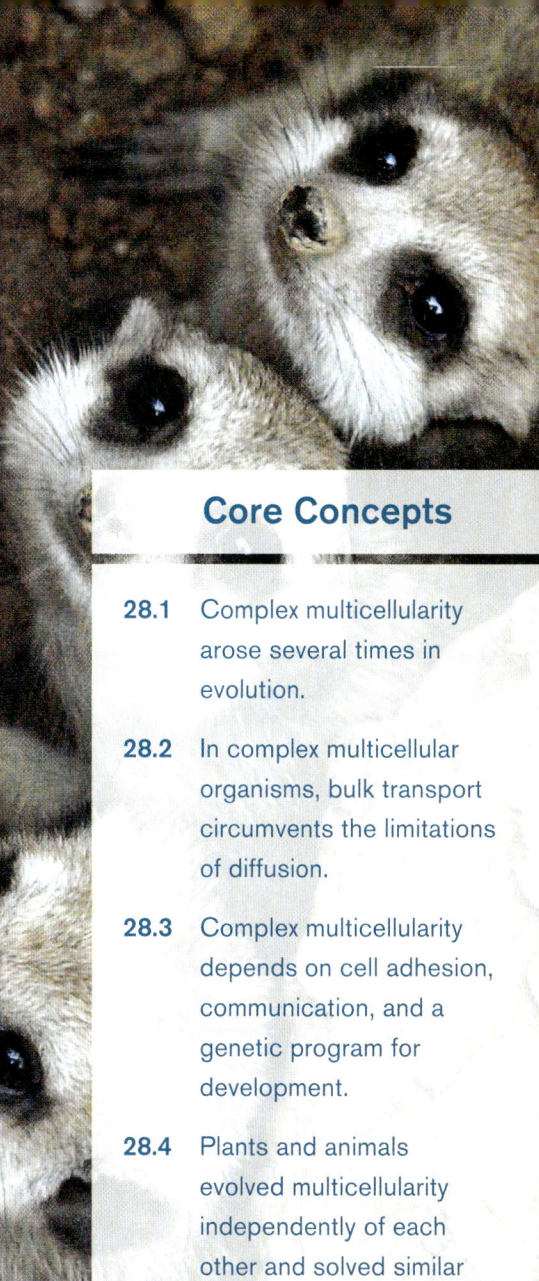

CHAPTER 28

BEING MULTICELLULAR

Core Concepts

28.1 Complex multicellularity arose several times in evolution.

28.2 In complex multicellular organisms, bulk transport circumvents the limitations of diffusion.

28.3 Complex multicellularity depends on cell adhesion, communication, and a genetic program for development.

28.4 Plants and animals evolved multicellularity independently of each other and solved similar problems with different sets of genes.

28.5 The evolution of large and complex multicellular organisms, which required abundant oxygen, is recorded in the fossil record.

In the human body, several trillion cells work in close coordination, enabling us to sense our surroundings and respond to them, move, eat, grow, and reproduce. Humans are an example of a **complex multicellular organism.** Our cells are specialized for specific functions, so our body as a whole can perform a broad range of tasks, but individual cells for the most part cannot. We require trillions of cells to accomplish the same broad range of biological activities that can be performed by single-celled protists and bacteria.

Complex multicellular organisms are relative latecomers in the history of life, arising more than 3 billion years after microorganisms first evolved. In subsequent chapters, we discuss the biology of three groups that have evolved complex multicellularity: animals, land plants, and fungi. First, however, we need to ask, how did complex multicellular organisms arise from simple precursors, and what new ways of life were made possible by their evolution? As we will see, complex multicellularity depended on several key evolutionary innovations. Each step built on the previous one, leading to the evolution of organisms with tissues and organs able to transport oxygen, signaling molecules, and nutrients rapidly over long distances.

28.1 THE PHYLOGENETIC DISTRIBUTION OF MULTICELLULAR ORGANISMS

As we discussed in Chapter 26, most prokaryotic organisms are composed of a single cell, although some form simple filaments or live in colonies. A few types of bacterium, notably some cyanobacteria, develop several different cell types. No bacteria, however, develop macroscopic bodies with distinct types of tissues. Only eukaryotes have evolved those.

Simple multicellularity is widespread among eukaryotes.

A recent survey of eukaryotic organisms recognized 119 major groups within the superkingdoms discussed in Chapter 27. Of these, 83 contain only single-celled organisms, predominantly cells that engulf other microorganisms or ingest small organic particles, photosynthetic cells that live suspended in the water column, or parasitic cells that live within other organisms. Each of the 36 remaining branches exhibits some cases of **simple multicellularity,** mostly in the form of filaments, hollow balls, or sheets of little-differentiated cells (**Fig. 28.1**).

Simple multicellular eukaryotes share several properties. In many, adhesive molecules cause adjacent cells to stick together, but there is relatively little communication or transfer of resources between cells and little differentiation of specialized cell types. Most or all of the cells in simple multicellular organisms retain a full range of functions, including reproduction, and so the penalty paid by the organism for individual cell death is usually small. Importantly, in simple multicellular organisms, every cell is in direct contact with the external environment, at least during phases of the life cycle when the cells must acquire nutrients.

FIG. 28.1 Simple multicellularity, in which cells show little differentiation and remain in close contact with the external environment. Many groups of eukaryotic organisms display simple multicellularity: (a) *Uroglena*, a stramenopile alga found commonly in lakes; (b) *Epistylis*, a stalked ciliate protozoan that lives attached to the surfaces of fish and crabs; and (c) *Prasiola*, a bladelike green alga only a single cell thick, found on tree trunks and rock surfaces.

Most simple multicellular species are forms of algae, although stalklike colonies of particle feeders have evolved at least three times, and simple filamentous fungi absorb organic molecules as sources of carbon and energy. In addition, four eukaryotic groups have achieved simple multicellularity by a different route, aggregating during just one stage of the life cycle. Slime molds are the best-known example (Chapter 27).

Six other groups (two algal and three protozoan, plus a group of fungi) have evolved what is called **coenocytic** organization. In these organisms, the nucleus divides multiple times, but the nuclei are not partitioned into individual cells. The result is large cells—sometimes even visible to the naked eye—with many nuclei. The green algae *Codium* and *Caulerpa* are common examples of coenocytic organisms found along the shorelines of temperate and tropical seas, respectively (**Fig. 28.2**). Coenocytic rhizarians on the deep seafloor can be 10 to 20 cm long. There is no evidence that any of these groups evolved from truly multicellular ancestors, nor have any given rise to complex multicellular descendants.

Why did simple multicellular organisms evolve? One selective advantage is that multicellularity helps organisms to avoid protozoan predators. In an illuminating experiment, single-celled green algae were grown in the presence of a protistan predator. Within 10 to 20 generations, most of the algae were living in eight-cell colonies that were essentially invulnerable to predator attack. Another advantage is that multicellular organisms may

FIG. 28.2 Coenocytic organization, in which nuclei divide repeatedly but cytokinesis does not occur. Two different groups of green algae have evolved large size as coenocytic organisms: (a) *Codium*, found abundantly along temperate coastlines; and (b) *Caulerpa*, common in shallow tropical seas and aquaria.

FIG. 28.3 **Complex multicellular organisms.** Complex multicellularity evolved independently in (a) red algae, (b) brown algae, (c) land plants, (d) animals, and (e) fungi (at least twice).

be able to maintain their position on a surface or in the water column better than their single-celled relatives. Seaweeds, for example, live anchored to the seafloor in places where light and nutrients support growth. And in colonial heterotrophs such as the stalked ciliate *Epistylis* (see Fig. 28.1b), the coordinated beating of flagella assists feeding by directing currents of food-laden water toward the cells. As we will see, complex multicellular organisms evolved from simple multicellular ancestors, but most taxonomic groups containing simple multicellular organisms never gave rise to complex descendants.

Complex multicellularity evolved several times.

Complex multicellular organisms mark the biological world of our daily existence. Plants, animals, and, if you look a little more closely, mushrooms and complex seaweeds define our perceptions of landscapes and the coastal ocean (**Fig. 28.3**). Plants and animals differ in many ways, but they have some fundamental features in common. Both have highly developed molecular mechanisms for adhesion between cells, and they have specialized structures that allow cells to communicate with one another. Plants and animals also both display complex patterns of cellular and tissue differentiation, guided by networks of regulatory genes. All complex multicellular organisms exhibit these properties. In general, in complex multicellular organisms only a small subset of all cells contributes to reproduction. Cell or tissue loss can be lethal for the entire organism. Notably, complex multicellular organisms have a three-dimensional organization, so only some cells are in direct contact with the environment.

The presence of both exterior and interior cells is an important characteristic of complex multicellular organisms. Because some cells are buried within tissues, relatively far from the exterior of the organism, they do not have direct access to nutrients. Therefore, interior cells cannot grow as fast as surface cells unless there is a mechanism to transfer resources from one cell to another. Also, interior cells do not receive signals directly from the environment, even though all cells must be able to respond to environmental signals if the organism is to grow, reproduce, and survive. Complex multicellular organisms, therefore, require mechanisms for transferring environmental signals received by cells at the body's surface to interior cells, where genes will be activated or repressed in response. That process has a familiar ring to it. Development in complex multicellular organisms can be

FIG. 28.4 The seven superkingdoms on the eukaryotic tree of life, showing that complex multicellular organisms (found on branches marked by blue) evolved at least six times.

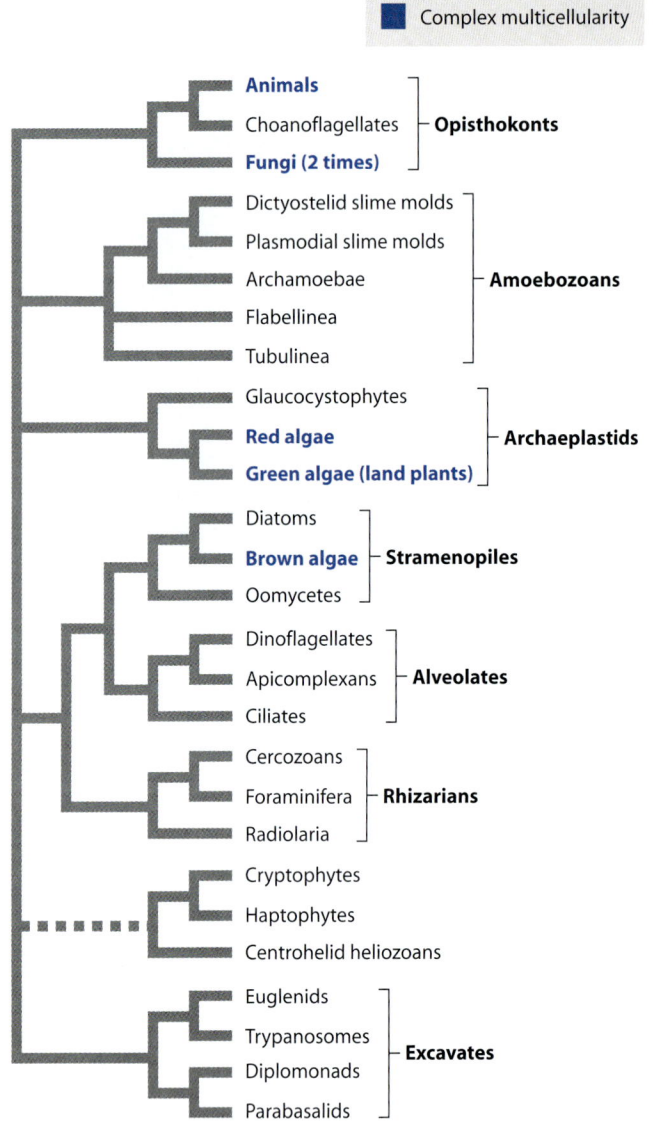

defined as increasing or decreasing gene expression in response to molecular signals from surrounding cells (Chapter 20).

Complex multicellularity evolved at least six separate times in different eukaryotic groups (**Fig. 28.4**). Obviously, complex multicellularity characterizes animals, but it also evolved at least twice in the fungi, once in the green algal lineage that gave rise to land plants, once in the red algae, and once in the brown algae, producing the giant kelps that form forests in the sea.

→ **Quick Check 1** How do simple multicellular organisms differ from complex multicellular organisms?

28.2 DIFFUSION VS. BULK TRANSPORT

How does oxygen get from the air in your lungs into your bloodstream? How does atmospheric carbon dioxide get into leaves? How does ammonia get from seawater into the cells of planktonic algae? The answer to all three questions is the same: by **diffusion.** But oxygen absorbed by your lungs doesn't reach your toes by diffusion alone—it is transported actively, and in bulk, by blood pumped through your circulatory system. Bulk transport of oxygen, nutrients, and signaling molecules, at rates and across distances far larger than can be achieved by diffusion alone, lies at the functional heart of complex multicellularity.

Diffusion is effective only over short distances.

Diffusion is the random motion of molecules, with net movement occurring when there are areas of higher and lower concentration of the molecules (Chapter 5). Because diffusion supplies key molecules for metabolism, it exerts a strong constraint on the size, shape, and function of cells. Diffusion is effective only over small distances and so it strictly limits the size and shape of bacterial cells, as discussed in Chapter 26. Diffusion also constrains how eukaryotic organisms function.

Oxygen provides a good example. Most eukaryotes require oxygen for respiration. If a cell or tissue must rely on diffusion for its oxygen supply, its thickness is limited by the concentration difference in oxygen between the cell or tissue and its environment. The concentration difference is dependent on the amount of oxygen in the environment and the rate at which oxygen is used for respiration inside the organism. In shallow water that is in direct contact with the atmosphere, diffusion permits animals to reach thicknesses of about 1 mm to 1 cm, depending on their metabolic rate. Of course, a quick swim along the seacoast will reveal many animals much larger than this. And, obviously, *you* are larger than this. How do you and other large animals get enough oxygen to all of your cells to enable you to survive?

Animals achieve large size by circumventing limits imposed by diffusion.

Sponges can reach overall dimensions of a meter or more, but they actually consist of only a few types of cell that line a dense network of pores and canals and so remain in close contact with circulating seawater (**Fig. 28.5a**). The large size of a sponge is therefore achieved without placing metabolically active cells at any great distance from their environment. Likewise, in jellyfish, active metabolism is confined to thin tissues that line the body. Essentially, a large flat surface is folded up to produce a three-dimensional structure (**Fig. 28.5b**). The jellyfish's bell-shaped body is often thicker than the metabolically active

FIG. 28.5 **Circumventing limits imposed by diffusion.** (a) Sponges can attain a large size because the many canals in their bodies ensure that all cells are in close proximity to the environment. (b) Jellyfish also have thin layers of metabolically active tissue, but their familiar bell can be relatively thick because it is packed with metabolically inert molecules (the mesoglea, or "jelly").

Complex multicellular organisms have structures specialized for bulk transport.

We have seen how humans circumvent the limitations of oxygen diffusion. Many other animals similarly rely on the active circulation of fluids to transport oxygen and other essential molecules, including food and molecular signals, across distances far larger than those that could be traversed by diffusion alone. Indeed, without a mechanism like bulk transport, animals could not have achieved the range of size, shape, and function familiar to us.

Bulk transport is the means by which molecules move through organisms at rates beyond those possible by diffusion across a concentration gradient (**Fig. 28.6**). In animals, the active pumping of blood through the circulatory system supplies oxygen to tissues that may be more than a meter distant from the lungs (Fig. 28.6a). Bulk transport carries the organic molecules required for respiration large distances from the intestinal cells that absorb these molecules from the

tissue, but its massive interior is filled by materials that are not metabolically active. This material constitutes the mesoglea, the jellyfish's "jelly." The mesoglea provides structural support but does not require much oxygen.

But what about us? How does the human body circumvent the constraints of diffusion? Our lungs gather the oxygen we need for respiration, but the lung is a prime example of diffusion in action, not a means of avoiding it. Because lung tissues have a very high ratio of surface area to volume, oxygen can diffuse efficiently from the air you breathe into lung tissue (Chapter 39). A great deal of oxygen can be taken in this way, but how does it get from the lungs to our brains or toes? The distances are far too large for diffusion to be effective. The answer is that oxygen binds to molecules of hemoglobin in red blood cells and is carried through the bloodstream to distant sites of respiration. We circumvent diffusion by actively transporting oxygen through our bodies. More generally, the circumvention of diffusion is what makes complex multicellularity possible.

FIG. 28.6 **Bulk transport.** (a) The circulatory system in animals and (b) the vascular system in plants allow these organisms to get around the size limits of diffusion.

a. Animal circulatory system

b. Plant vascular system

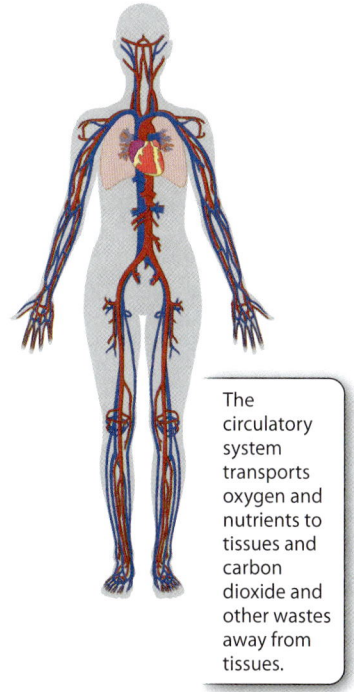
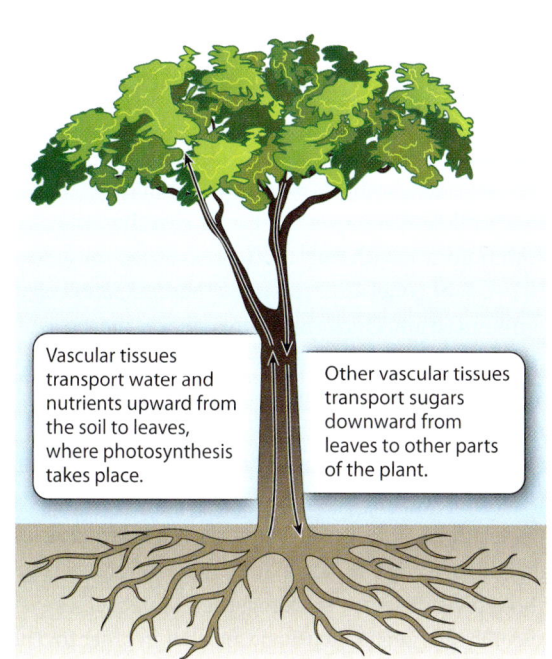

The circulatory system transports oxygen and nutrients to tissues and carbon dioxide and other wastes away from tissues.

Vascular tissues transport water and nutrients upward from the soil to leaves, where photosynthesis takes place.

Other vascular tissues transport sugars downward from leaves to other parts of the plant.

digestive tract. (Again, the molecules are transported actively in the bloodstream.) Endocrine signaling molecules such as hormones (Chapter 9) also move rapidly through the body by means of the blood and other fluids. Bulk transport is key to complex multicellularity in humans and most other animals.

Complex organisms other than animals also rely on bulk transport. A redwood tree must transport water upward from its roots to leaves that may be 100 meters above the soil. If plants relied on diffusion to transport water, they would be only a few millimeters tall. How, then, do they move water? Plants move water by bulk transport through a system of specialized tissues powered by the evaporation of water from leaf surfaces (Fig. 28.6b; Chapter 29). Vascular plants also have specialized tissues for the transport of nutrients and signaling molecules upward and downward through roots, stems, and leaves. Thus, like animals, plants rely on specialized tissues to transport essential materials over long distances at rates many times faster than could be accomplished by diffusion alone.

The circumvention of diffusion can be seen in other groups characterized by complex multicellularity. Fungi transport nutrients through networks of filaments that may be meters long, relying on osmosis to pump materials from sites of absorption to sites of metabolism (Chapter 34). The giant kelps have an internal network of tubular cells that transports molecules through a body that can be tens of meters long. In general, when some cells within an organism are buried within tissues, far from the external environment, bulk transport is required to supply those cells with molecules needed for metabolism.

→ **Quick Check 2** How do mechanisms for bulk transport enable organisms to achieve large size?

28.3 HOW TO BUILD A MULTICELLULAR ORGANISM

We have emphasized three general requirements for complex multicellular life: Cells must stick together; they must communicate with one another; and they must participate in a network of genetic interactions that regulates cell division and differentiation. Once these are in place, the stage is set for the formation of specialized tissues and organs observed in plants, animals, and other complex multicellular organisms.

Complex multicellularity requires adhesion between cells.

If a fertilized egg is to develop into a complex multicellular organism, it must divide many times, and the cells produced from those divisions must stick together. Moreover, they must retain a specific relationship with one another in order for the developing organism to function. Chapter 10 introduced the molecular mechanisms that hold multicellular organisms together. We begin by reviewing those mechanisms briefly before placing them in evolutionary context.

In animals, proliferating cells commonly become organized into sheets of cells called epithelia that line the inside and cover the outside of the body. Within epithelia, adjacent cells adhere to one another by means of transmembrane proteins, especially cadherins, that form molecular bonds between cells. Epithelial cells also secrete an extracellular matrix made of proteins and glycoproteins and attach themselves to this matrix by means of other transmembrane proteins called integrins. Cadherins, integrins, and additional transmembrane proteins thus provide the molecular mechanisms for adhesion in animals (Chapter 10).

Cadherins and integrins do not occur in other complex multicellular organisms, but all such organisms require some means of keeping their cells stuck together. How do they do it? Plants also synthesize adhesive molecules that bind cells into tissues, but in this case the molecules are polysaccharides called pectins (pectins are what make fruit jelly jell). Molecules that control adhesion between adjacent cells are less well known in other complex multicellular organisms, but the general picture is clear. Without adhesion there can be no complex multicellularity, and different groups of eukaryotes have employed distinct types of molecular "glue" for this purpose.

HOW DO WE KNOW?

FIG. 28.7

How do bacteria influence the life cycles of choanoflagellates?

BACKGROUND Choanoflagellates are the closest protistan relatives of animals, so an understanding of cell adhesion, cell differentiation, and multicellularity in these organisms stands to illuminate early events in the evolution of animal multicellularity. Nicole King and her colleagues cultured the choanoflagellate species *Salpingoeca rosetta* in the laboratory. High-speed videos revealed that these choanoflagellates differentiate into several distinct cell types in the course of their life cycle. Actively swimming cells can be fast swimmers or slow swimmers that differ in shape and ornamentation (Fig. 28.7a). When swimming cells make contact with a solid surface, they can differentiate a stalk that tethers them to the substrate (Fig. 28.7b). Simple multicellular colonies in the shape of rosettes (Fig. 28.7c) were also seen when the choanoflagellates were cultured in the presence of many bacteria, including *Algoriphagus machipongonesis*.

How did animal cell adhesion develop?

The genome of the choanoflagellate *Monosiga brevicollis* provides fascinating insight into the evolution of adhesion molecules. Choanoflagellates, the closest protistan relatives of animals, are unicellular microorganisms. Therefore, it came as a surprise that the genes of *M. brevicollis* code for many of the same protein families that promote cell adhesion in animals. The choanoflagellate genome contains genes for both cadherin and integrin proteins. Clearly, these proteins are not supporting epithelia in *M. brevicollis*, so what are they doing?

One approach to an answer came from the observation that cells stick not only to one another but also to their substrates. Proteins that originally evolved to promote adhesion to sand grains or a rock surface may have been co-opted by evolution for cell–cell adhesion. It is also possible that adhesion molecules originally facilitated the capture of bacterial cells, improving predation by causing prey to adhere to the predator.

An intriguing clue to this problem comes from careful laboratory studies of choanoflagellates (**Fig. 28.7**). Although choanoflagellates are fundamentally unicellular, simple multicellular structures can be induced in a number of species. Multicellularity is induced by molecular signals, and, unexpectedly, the source of these signals is a bacterium, the preferred prey of the choanoflagellates. When they detect their food bacteria, the choanoflagellates form a novel multicellular structure. The function of adhesion molecules in these choanoflagellates has not yet been determined, but these observations support the hypothesis that they facilitate predation.

To date, cadherins have been found only in choanoflagellates and animals, but proteins of the integrin complex extend even deeper into eukaryotic phylogeny, occurring in single-celled protists that branch near the base of the opisthokont superkingdom (Chapter 27). Such a phylogenetic distribution provides strong support for the general hypothesis that cell adhesion in animals resulted from the redeployment of protein families that evolved to perform other functions before animals diverged from their closest protistan relatives.

Complex multicellularity requires communication between cells.

In complex multicellular organisms, it is not sufficient for cells to stick together. They must also be able to communicate. As discussed in Chapter 9, cells communicate by molecular signals. A signaling molecule (generally a protein) synthesized by one cell binds with a receptor molecule (also a protein) on the surface of a second cell, essentially flipping a molecular switch that activates or represses gene expression in the receptor cell's nucleus. Plants

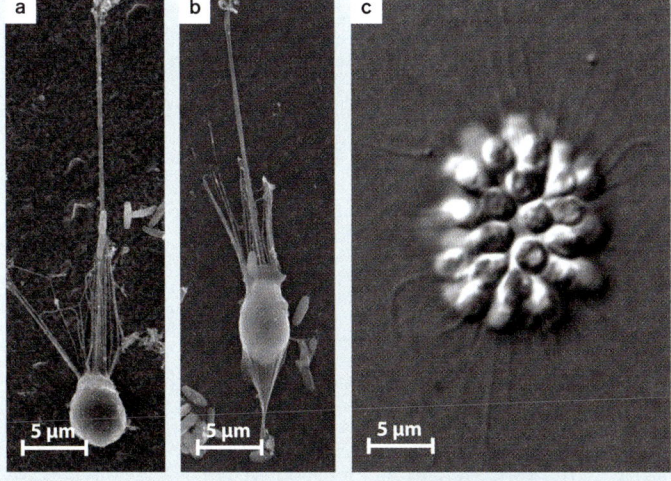

Salpingoeca rosetta choanoflagellates can exist as single cells that swim through their watery environment (a); they can be stalked cells that live attached to surfaces (b); or, in the presence of specific bacteria, they can form simple multicellular structures (c).

HYPOTHESIS The bacterium *Algoriphagus machipongonesis* induces the choanoflagellate to form rosette-like colonies.

EXPERIMENT The scientists grew the chonaoflagellates in a medium containing only *A. machipongonesis*.

RESULTS The multicellular state formed (Fig. 28.7c).

CONCLUSION Choanoflagellates differentiate several cell types upon cues from the environment. In particular, molecular signals from prey bacteria can induce a simple form of multicellularity. The direct ancestors of animals may have gained simple multicellularity by a similar route.

FOLLOW-UP WORK Research is continuing to investigate the molecular mechanisms of cell differentiation, cell adhesion, and signal reception in these distant relatives of ours.

SOURCE Dayel, M., R. A. Alegado, S. R. Fairclough, T. C. Levine, S. A. Nichols, K. McDonald, and N. King. 2011. "Cell Differentiation and Morphogenesis in the Colony-Forming Choano-flagellate *Salpingoeca rosetta*." *Developmental Biology* 357:73–82.

and animals both have signaling pathways based on receptor kinases, and, as in the case of adhesion molecules, these signaling pathways, or components of the pathways, have also been identified in their close protistan relatives.

Evidently, many of the signaling pathways used for communication between cells in complex multicellular organisms first evolved in single-celled eukaryotes. Again we can ask, what function did molecular signals and receptors have in the protistan ancestors of complex organisms?

All cells have transmembrane receptors that respond to signals from the environment. In some cases, the signal is supplied by food organisms (such as the bacteria that induce simple multicellularity in choanoflagellates), and in some cases the cells sense nutrients, temperature, or oxygen level. Single-celled eukaryotes also communicate with other cells within the same species, not least to ensure that two cells can find each other to fuse in sexual reproduction. Signaling between two cells within an animal's body can be seen as a variation on this more general theme of cellular responses to other cells and the physical environment.

While all eukaryotic cells have molecular mechanisms for communication between cells, complex multicellular organisms have distinct cellular pathways for the movement of molecules from one cell to another. For example, animals more complex than sponges have **gap junctions,** protein channels that allow ions and signaling molecules to move from one cell into another (**Fig. 28.8**). Gap junctions not only help cells to communicate with their neighbors, they allow targeted communication between a cell and only one or a few of the cells that surround it (Chapter 10).

Complex photosynthetic organisms, in contrast, have an intrinsic barrier to intercellular communication: the cell wall. Careful observation of plant cell walls using the electron microscope reveals that these walls have tiny holes through which thin strands of cytoplasm extend from one cell to the next (**Fig. 28.9**). Called **plasmodesmata,** these intercellular strands are lined by extensions of the cell membrane and contain tubules connecting the endomembrane systems of the two cells. Like gap junctions, plasmodesmata permit signaling molecules to pass between cells in a targeted fashion.

Other complex multicellular organisms also have channels that connect adjacent cells. As similar channels do not occur in most other eukaryotic organisms, they appear to represent an important step in the evolution of complex multicellularity.

Complex multicellularity requires a genetic program for coordinated growth and cell differentiation.

All of the cells in your body derive from a single fertilized egg, and most of those cells contain the same genes. Yet your body contains about 200 distinct cell types precisely arranged in a variety of tissues and organs. How can two cells with the same genes become different cell types? The answer lies in development, the system of gene regulation that guides growth from zygote to adult (Chapter 20).

Development is the product of molecular communication between cells. Cells have different fates depending on which genes are switched off or on by the molecular signals they receive. Different signals induce the reorganization of cytoskeletons and alter the production of proteins. As a result, a stem cell may become

FIG. 28.8 Gap junctions. Gap junctions facilitate the targeted transport of ions and molecules between adjacent cells in animals.

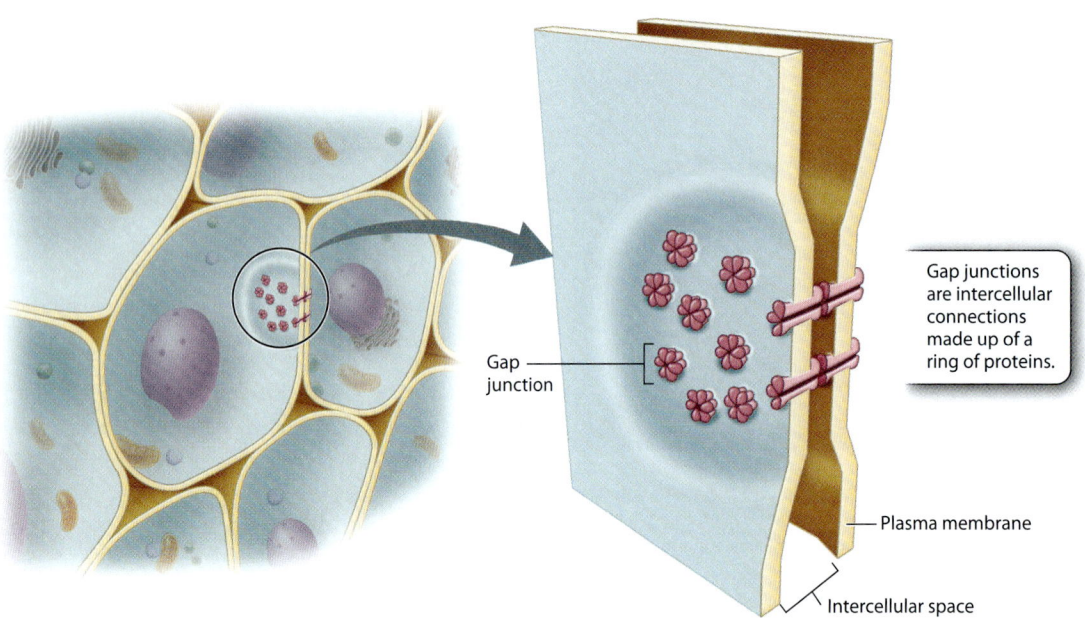

FIG. 28.9 Plasmodesmata. Plasmodesmata facilitate the movement of ions and molecules between cells in land plants and in complex red and brown algae.

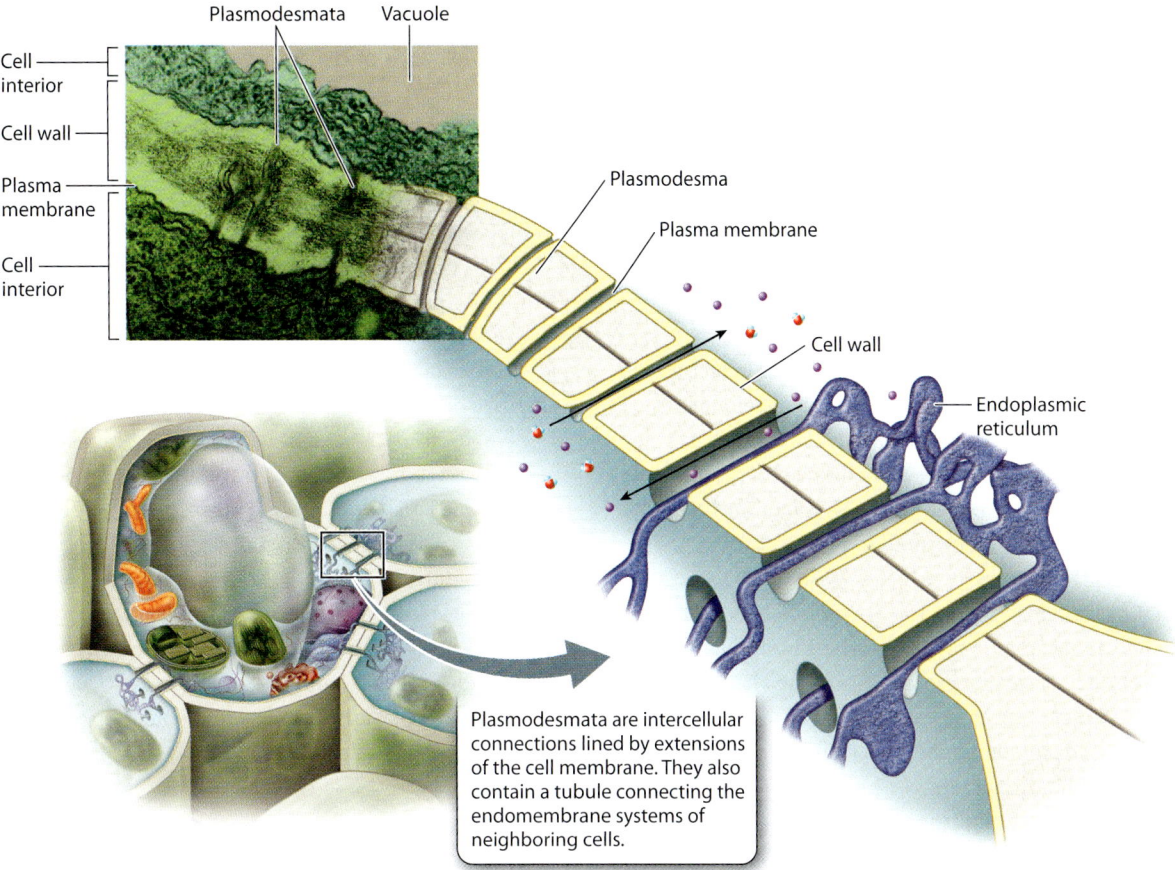

Plasmodesmata are intercellular connections lined by extensions of the cell membrane. They also contain a tubule connecting the endomembrane systems of neighboring cells.

an epithelial cell, or a muscle cell, or a neuron. This observation leads to another question: What causes the same gene to be turned on in one cell and off in another? The ultimate answer is that two cells in the same developing organism can be exposed to very different environments.

When we think about development as a process of programmed cell division and differentiation, the link to unicellular ancestors becomes clearer. Many biological innovations accompanied the evolution of complex multicellularity, but, as noted in Chapter 27, the differentiation of distinct cell types is not one of them. Many unicellular organisms have life cycles in which different cell types alternate in time, depending on environmental conditions. For example, if we experimentally starve dinoflagellate cells, two cells will undergo sexual fusion to form morphologically and physiologically distinct resting cells protected by thick walls. When food becomes available again, the cells undergo meiotic cell division to form new feeding cells. Many other single-celled eukaryotes differentiate resting cells in response to environmental cues, especially deprivation of nutrients or oxygen.

The innovation of complex multicellularity was to differentiate cells in *space* instead of time. In a three-dimensional multicellular organism, only surface cells are in direct contact with the outside environment. Interior cells are exposed to a different physical and chemical environment because the availability of nutrients, oxygen, and light decreases with increasing depth within tissues. In effect, there is a gradient of environmental signals within multicellular organisms. It is, therefore, possible that in the earliest organisms with three-dimensional multicellularity, a nutrient or oxygen gradient triggered differentiation of interior cells, causing cell differentiation of oxygen- or nutrient-starved interior cells much as these conditions induce the expression and repression of specific genes in their single-celled relatives. With increasing genetic control of cellular responses to signaling gradients, the seeds of complex development were sown.

A fascinating example supporting this hypothesis has been reported from research on green algae. The simple multicellular organism *Volvox* (see Fig. 27.15) has two types of cells, vegetative cells that photosynthesize and control movement of the organism, and reproductive cells. Cell differentiation in *Volvox* is regulated by a gene also involved in cell remodeling during the life cycle of *Volvox*'s single-celled relative *Chlamydomonas*, supporting the hypothesis that the spatial differentiation of cells in multicellular organisms began with the redeployment of genes that regulate cell differentiation in single-celled ancestors.

Bulk transport, discussed earlier with regard to the transport of nutrients and water within complex multicellular organisms, also affects developmental signaling. It increases the distance over which signals can travel, circumventing the limitations of diffusion and making possible spatially specific patterns of differentiation along the path of signal transport. For example, in animals the endocrine system releases hormones directly into the bloodstream, enabling them to affect cells far from those within which they formed (Chapter 38). Thus, the sex hormones estrogen and testosterone are synthesized in reproductive organs, but regulate development throughout the body, contributing to the differences between males and females.

The genome of the choanoflagellate *Monosiga brevicollis*, discussed earlier, has been a treasure trove of information on the antiquity of signaling molecules deployed in animal development. In addition to proteins that govern cell adhesion and epithelial cohesion in animals, *M. brevicollis* expresses a number of proteins active in animal cell differentiation. For example, signaling based on receptor kinases (Chapter 9), long thought to be restricted to animals, occurs in *M. brevicollis*. In other cases, individual domains of signaling proteins occur in the *M. brevicollis* genome, but not the complex multidomain proteins formed by animals. This is true, for example, of several protein complexes that are important in development. Similarly, molecules that play an important role in plant development are also being identified in the genomes of morphologically simple green algae. The key point is that genome sequences interpreted in light of eukaryotic phylogeny are now enabling biologists to piece together the patterns of gene evolution that accompanied the evolution of morphological complexity and diversity in plants and animals.

28.4 VARIATIONS ON A THEME: PLANTS VS. ANIMALS

Plants and animals are the best-known examples of complex multicellular organisms. The study of phylogenetic relationships makes it clear that complex multicellularity evolved independently in the two groups. In other words, they do not share a common ancestor that was multicellular. We can see the results of these separate evolutionary events by examining cell adhesion, communication, and development in both plants and animals.

Both plants and animals have evolved sophisticated systems that cause adjacent cells to adhere to each other and that promote the targeted movement of signaling molecules between cells. Plant and animal mechanisms, however, must differ, because plant cells have cell walls and animal cells do not. Likewise, plants and animals have evolved similar genetic logic to govern development, but use mostly distinct sets of genes. In both plants and animals, many proteins switch other genes on or off, so that the spatial organization of multicellular organisms arises from networks of interacting genes and their protein products. Ancestral plants and animals simply recruited distinct families of genes to populate regulatory networks.

Cell walls shape patterns of growth and development in plants.

The plant cell wall (Chapter 5), made of cellulose, imparts structural support to cells and, in fact, provides the mechanical support that allows plants to stand erect (Chapter 29). The presence of cell walls has largely determined the evolutionary fate of plants. For example, because their cells cannot engulf particles or absorb organic molecules, most plants gain carbon and energy only through photosynthesis. Moreover, because all plant cells except eggs and sperm are completely surrounded by cell walls, they have no pseudopodia and no flagella (in conifers and flowering plants, even sperm have lost flagella; see Chapter 30). This being the case, plant cells cannot move, either during development or to obtain nutrients, evade predators, or escape stressful conditions.

The inability of plant cells to move has major consequences for development. At the level of the cell, the entire program of growth and development involves cell division, cell expansion (commonly by developing large vacuoles in cell interiors), and cell differentiation. The mechanical consequence is that plant growth is confined to **meristems** (**Fig. 28.10**), populations of actively dividing cells at the tips of stems and roots (Chapter 31). More

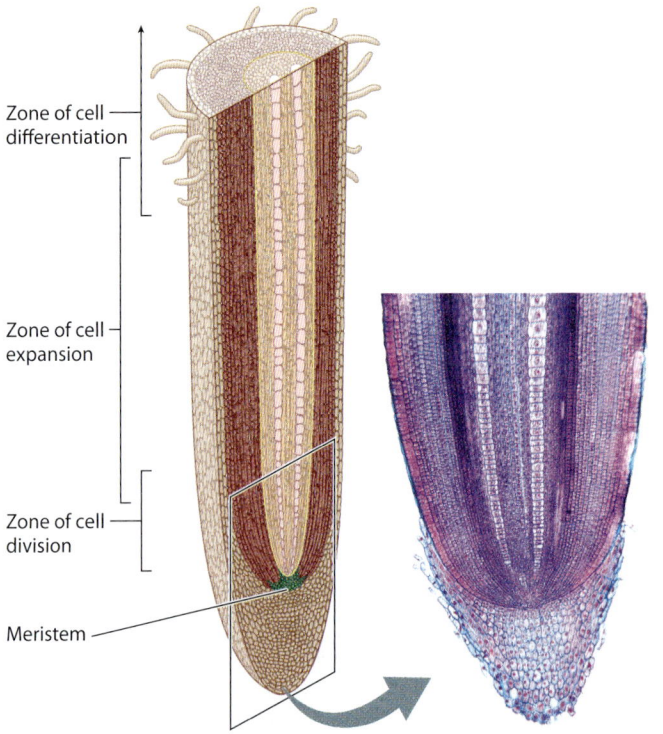

FIG. 28.10 A meristem. Because of their walls, plant cells can't move. For this reason, roots grow by cell division of the undifferentiated cells in the meristem region at the root tip.

or less permanently undifferentiated cells in meristem regions repeatedly undergo mitosis. A few millimeters from the region of active cell division in a stem or root meristem, cells stop dividing and begin to expand. Within another millimeter or so, this activity is curtailed as well. There, individual cells respond to signaling molecules that induce differentiation, forming the mature cells that will function in photosynthesis, storage, bulk transport, or mechanical support. In general, then, plant development involves cell division, adhesion among the cells that form from meristems, adsorption of water and nutrients supplied by adjacent cells, and differentiation into distinct cell types that govern the function of the plant as a whole.

Because cell walls render plants immobile, plants have evolved mechanisms to transport water and nutrients from the soil to leaves that may be tens of meters distant without the use of any moving parts or even the expenditure of ATP. And because plants are anchored in place, they are unable to move in response to unfavorable growth conditions. For this reason, plants have intricate systems that feed information from the environment to their meristems. Heat, drought, floods, and fire can all leave their mark in altered patterns of growth. In addition, plants can't flee predators, so they have evolved mechanical structures (hairs and spines) and poisons to keep from being eaten. Indeed, it has been suggested that, in terms of responding to the environment, growth plays a role in plants similar to that played by behavior in animals.

The details of plant growth, development, reproduction, and function are presented in Chapters 29–33. Here, the point to bear in mind is that the properties of a pine tree or a rice plant depend fundamentally on the basic features of cell adhesion, intercellular communication, and regulatory network to guide development, all carried out under the constraints imposed by cell walls.

Animal cells can move relative to one another.

Having outlined the growth and development of plants, we can appreciate anew the remarkable ballet that characterizes development in animal embryos (**Fig. 28.11;** Chapter 42). Fertilized eggs undergo several rounds of mitosis to form a ball of undifferentiated cells called a **blastula.** Then something happens that has no parallel in plant development: Unconstrained by cell walls, animal cells can move relative to one another. Blastula cells migrate, becoming reorganized into a hollow ball that folds inward at one location to form a layered structure called the **gastrula.**

Gastrula formation brings new populations of cells into direct contact with one another, inducing patterns of molecular signaling and gene regulation that begin the long process of growth and tissue specification. As cells proliferate, molecular signals are expressed in some cells and diffuse into others, generating what amounts to a three-dimensional molecular map of the organism that guides cell differentiation. Gradients in signaling molecules define top, bottom, front, back, left,

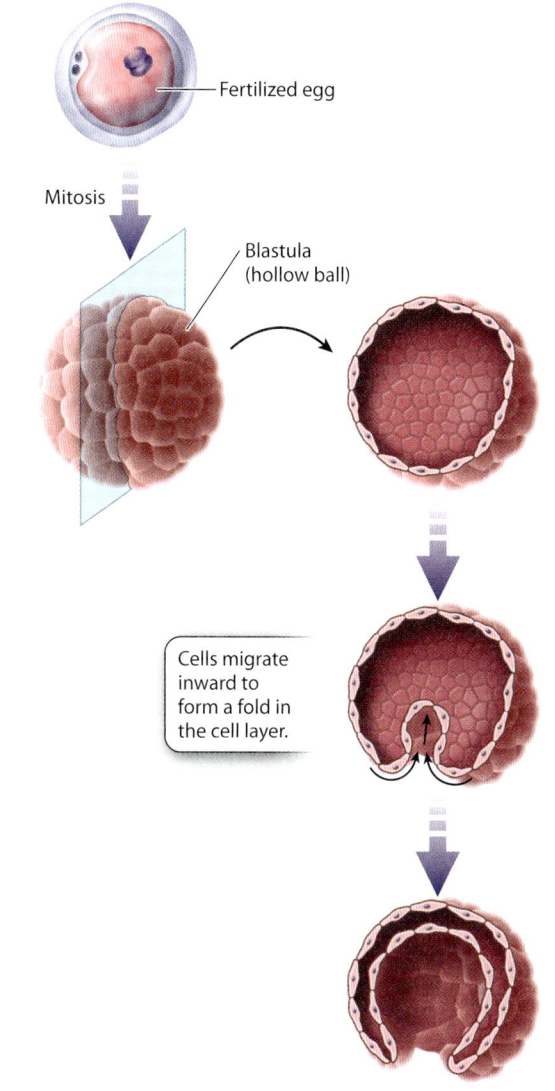

FIG. 28.11 Gastrulation, the movement of cells during embryogenesis that transforms a blastula to a gastrula. Unlike plant cells, animal cells move during development.

and right. Plants do much the same thing, but because animal cells can move during development, animal embryos are not restricted to growth only from localized regions like meristems. Cell division and tissue differentiation occur throughout the developing animal body.

Because they are not constrained by cell walls, animals can form organs with moving parts—muscles that power active transport of food and fluids and allow movement. Thus, animals have possibilities for function that are far different from those of plants. If the environment offers a challenge, like drought or predators, animals can respond by changing their behavior. For example, they can move to a new location when threatened or stressed, as is discussed more fully in Chapter 37.

Plants and animals, then, display contrasting patterns of development and function that reflect both their independent origins from different groups of protists and the constraints imposed by cell walls in plants.

→ **Quick Check 3** How do plants and animals differ in the ways their cells adhere, communicate, and differentiate during development?

28.5 THE EVOLUTION OF COMPLEX MULTICELLULARITY

The gulf between an amoeba and a lobster seems vast, but research over the past decade increasingly shows how this apparent gap was bridged through a series of evolutionary innovations. Throughout this chapter, we have emphasized that complex multicellular organisms require mechanisms for cell adhesion, modes of communication between cells, and a genetic program to guide growth and development. Not only were all three required for the evolution of complex multicellular organisms, they also had to be acquired in a specific order. If the products of cell division don't stick together in a pattern that has some usefulness in function, there can be no complex multicellularity. Adhesion, however, is not sufficient. It must be followed by mechanisms for communication between cells. Moreover, cells must be able to send molecular messages in a targeted or spatially specific pattern, facilitated in animals by gap junctions and in plants by plasmodesmata.

With these requirements in place, evolution would favor the increase and diversification of the genes that regulate growth and development, making possible more complex morphology and anatomy. The biological stage was finally set for the functional key to complex multicellularity: the differentiation of tissues and organs that govern the bulk transport of fluids, nutrients, signaling molecules, and oxygen through increasingly large and complex bodies. This differentiation freed organisms from the tight constraints imposed by diffusion. When all these features are placed onto phylogenies, they show both the predicted order of acquisition and the tremendous evolutionary consequences of complex multicellularity (**Fig. 28.12**).

Complex multicellularity appeared in the oceans 575 to 555 million years ago.

In Chapter 23, we noted that phylogenies based on living organisms make predictions about the fossil record. Characters and groups associated with lower branches in phylogenies should appear as earlier fossils than those associated with later branches. Although fossils don't preserve molecular features, the record supports the phylogenetic pattern shown in Fig. 28.12.

Single-celled protists are found in sedimentary rocks as old as 1800 million years, and a number of simple multicellular forms have been discovered that are nearly as old (see Fig. 27.23). Nonetheless, the oldest fossils that record complex multicellularity occur in rocks that were deposited only 575 to 555 million years ago. **Fig. 28.13a** shows one of the oldest known animal fossils, a frondlike form preserved in 575-million-year-old rocks from Newfoundland, near the eastern tip of Canada. This fossil and many others found in rocks of this age are enigmatic. There is complex morphology to be sure, but it is difficult

FIG. 28.12 The evolution of complex multicellularity. Phylogenetic relationships of complex (a) animals and (b) plants, with their close evolutionary relatives, show similar patterns of character accumulation and its consequences for biological diversity. Estimated species numbers are shown.

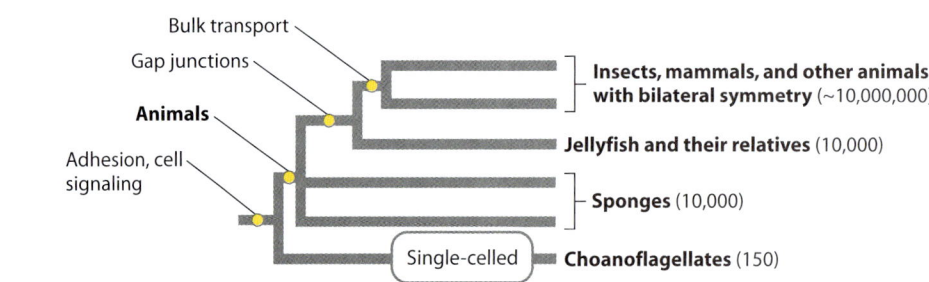

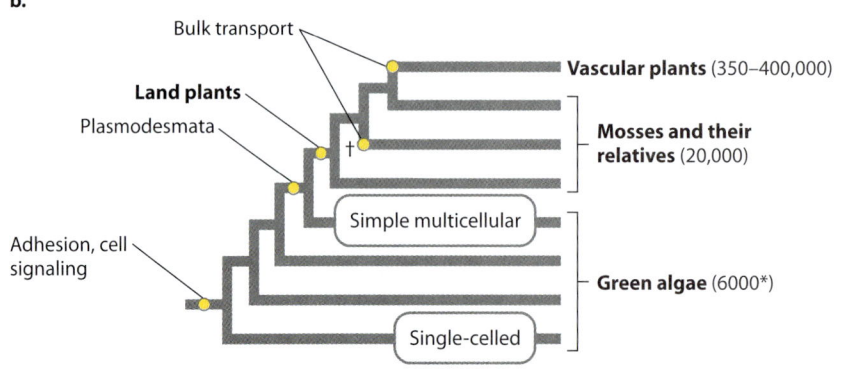

* Only species on the branch leading to land plants
† Limited development of transport tissues in some mosses

FIG. 28.13 The earliest fossils of animals. (a) The oldest known macroscopic animal fossil is a frondlike form made up of tubular structures, preserved in 575-million-year-old rocks from Newfoundland. (b) The oldest known animal fossil showing bilateral symmetry is a mollusk-like form preserved in 560–555-million-year-old rocks from northern Russia. (c) Some of the oldest tracks and trails of mobile animals are common in rocks 555 million years old and younger.

to understand these forms by comparing them to the body plans of living animals. The Newfoundland fossils show no evidence of head or tail, no limbs, and no opening that might have functioned as a mouth. They appear to be very simple organisms that gained both carbon and oxygen by diffusion. These fossils are long and wide, but they are not thick. In fact, it is likely that active metabolism occurred only along the surfaces of these structures.

Such animals lie near the base of the animal tree (Chapter 44), and probably would have developed under the guidance of regulatory genes similar to those present in modern sponges or, perhaps, jellyfish. By 560–555 million years ago, more complex animals, with distinct head and tail, top and bottom, left and right enter the record (**Fig. 28.13b**). At the same time, we begin to see tracks and trails made by animals whose muscles enabled them to move across and through surface sediments (**Fig. 28.13c**), an indirect but complementary record of increasing animal complexity. These, then, are the first known occurrences of the types of animal that dominate ecosystems today, both in the sea and on land. These animals must have possessed a diverse array of genes to guide development, something akin to those seen today in insects or snails.

Clearly, by 575–555 million years ago, the animal tree was beginning to branch. This, however, raises a new question: Why did complex multicellularity appear so much later than simple multicellular organisms?

Oxygen is necessary for complex multicellular life.

Earlier in the chapter, we discussed the key biological requirements for complex multicellularity. There is a critical environmental requirement as well: the presence of oxygen. Only the oxidation of organic molecules by O_2 provides sufficient energy to support biological communities that include large and active predators like wolves and lions. Not only do other potential electron acceptors for respiration, like sulfate and ferric iron, occur in much lower abundances, they are not gases and so do not accumulate in air. And only oxygen that is present in concentrations approaching those of the present day is able to diffuse into the interior cells of large, active organisms. On our planet—and probably, on all planets—no other oxidant is both sufficiently abundant and has the oxidizing power needed to support the biology of large complex organisms.

On the present-day Earth, lake or ocean bottoms with oxygen levels below about 10% of surface levels support few animals, and those that do tend to be tiny. Large, active, and diverse animals can live only in oxygen-rich environments. The clear restrictions on animal form and diversity observed along an environmental gradient of oxygen availability have important implications for the distribution of large, active, and diverse animals through time. We noted in Chapter 26 that oxygen first began to accumulate in the atmosphere and surface oceans about 2400 million years ago, but the chemistry of sedimentary rocks deposited after that time tells us that oxygen

FIG. 28.14 Oxygen and the evolution of complex multicellularity.

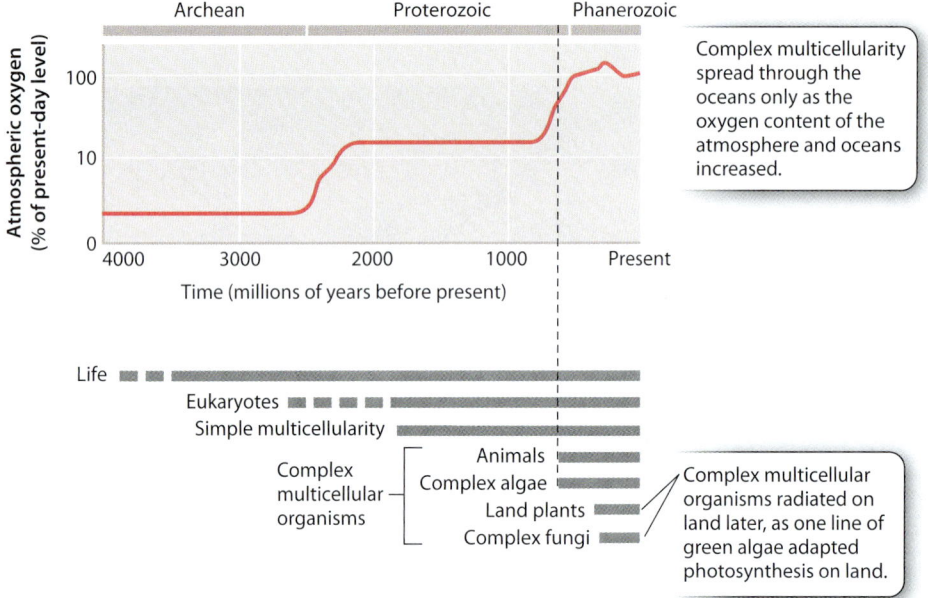

levels remained low for a very long time. Our modern world of abundant oxygen came to exist only 580–560 million years ago, about the time when fossils first record complex multicellular organisms (**Fig. 28.14**).

Scientists continue to debate the causes of this environmental transformation, but the correspondence in time between oxygen enrichment and the first appearance of complex animals (and algae) suggests that more oxygen permitted greater size. Greater size in turn created opportunities for tissue differentiation, leading to the bulk transport of nutrients and signaling molecules. The evolution of bulk transport in turn permitted still larger size, setting up a positive feedback that eventually resulted in the complex marine organisms we see today.

With morphological complexity came new functions, including predation on other animals. Protozoan predators capture other microorganisms, but do so one cell at a time. In contrast, animals that obtain food by filtering seawater gather cells by the thousands, and larger animals can eat smaller ones, opening up endless possibilities for evolutionary specialization and, therefore, diversification. Among photosynthetic organisms, multicellular red and green algae were able to establish populations in wave-swept coastlines and other environments where simpler organisms lacked the mechanical strength to hold their position along the shoreline.

The key point is that complex multicellular organisms didn't succeed by doing the same things as simpler eukaryotes. In an oxygen-rich world, complex multicellularity spread through the oceans because it opened up new and unprecedented evolutionary possibilities.

Land plants evolved from green algae that could carry out photosynthesis on land.

Complex multicellular land plants originated about 460 million years ago, well after marine animals and algae. In some ways, the pattern of evolution of land plants parallels that of animals. Ancestral green algae evolved molecular means for cell–cell adhesion and communication between cells, and the algae that form the sister group to land plants share with them the innovation of plasmodesmata. Then, as they developed larger and more complex three-dimensional bodies, plant ancestors evolved a succession of genes and their protein products to guide cell division, expansion, and differentiation. Land plants, however, had a different challenge from that of animals. The diversification of plants across the land surface requires that photosynthesis, evolved earlier by aquatic organisms, be carried out within tissues bathed by air. It also requires that nutrients and water be absorbed from the soil and transported throughout the plants rather than simply taken in by diffusion from the surrounding water.

FIG. 28.15 Evolution of complex fungi. (a) This treelike fossil, discovered in Saudi Arabia, is actually a giant fungus that towered over early land plants 375 million years ago. (b) This image shows the interior of the fungus, which consists of tubes that transported nutrients through the large body.

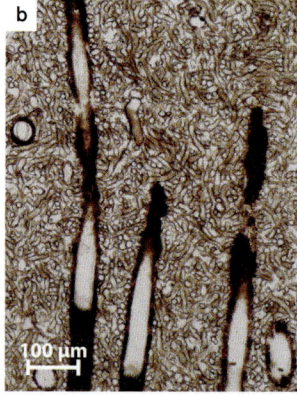

CHAPTER 28   BEING MULTICELLULAR

Fossils indicate that by 400 million years ago, plants with specialized tissues for bulk transport of water and nutrients had begun to spread across the continents. As plant biomass built up on land, it created opportunities for additional types of complex multicellular organism. In particular, the fungi, which decompose organic matter, radiated by taking advantage of this resource (**Fig. 28.15**), evolving complex multicellularity in two distinct lineages (Chapter 34).

Regulatory genes played an important role in the evolution of complex multicellular organisms.

Complex multicellular organisms account for a large proportion of all eukaryotic species, and the evolutionary consequences of complex multicellularity are seen clearly when we compare the diversity of complex multicellular groups with their simpler phylogenetic sisters. Choanoflagellates, for example, number about 150 species, simple animals such as sponges and jellyfish about 20,000, and complex animals with circulatory systems that circumvent diffusion perhaps as many as 10 million. Plants and their relatives show a comparable pattern. There are about 6000 species of green algae on the branch that includes land plants, but about 400,000 species of anatomically complex land plants capable of bulk transport. Similarly, complex fungi and red algae include more species by an order of magnitude than their simpler relatives.

How did this immense diversity arise? At one level, the answer is functional and ecological. Complex multicellular organisms can perform a range of functions that simpler organisms cannot, and so they have evolved many specific types of interaction with other organisms and the physical environment. This explanation, however, prompts another question: What is the genetic basis for the bewildering range of sizes and shapes displayed by complex multicellular organisms?

The answer has to do with the network of genes that guide development. Since the 1950s, biologists have learned a remarkable amount about this genetic network and its host of interacting genes. Many of the genes involved in development are what we might consider middle managers: A molecular signal induces the expression of a gene, and the protein product, in turn, prompts the expression or repression of another gene. It is the complex interplay of these genetic switches—turning specific genes on in one cell and off in another, depending on where those cells occur in the developing body—that results in crabs with large pincers or butterflies with patterned wings (**Fig. 28.16**).

It makes sense that if regulatory genes guide development of the body in each complex multicellular

HOW DO WE KNOW?

FIG. 28.16

What controls color pattern in butterfly wings?

BACKGROUND Butterfly wings commonly show a striking pattern of color, including circular features known as eyespots. Eyespots have adaptive value, for example in deterring predation by birds.

HYPOTHESIS The expression of regulatory genes during wing development governs eyespot formation on wing surfaces.

EXPERIMENT Paul Brakefield and his colleagues mapped the expression of a regulatory gene called *Distalless* in developing butterfly wings. *Distalless* genes were fused to genes for green fluorescent protein (GFP), a protein that glows vivid green when illuminated under blue light. GFP would then be expressed along with *Distalless*, allowing the spatial pattern of *Distalless* expression in developing wings to be followed by GFP visualization.

RESULTS The expression pattern of *Distalless* in developing wings closely resembles the pattern of eyespots on the wing. Moreover, mutations that change the spatial pattern of *Distalless* expression result in different eyespot patterns.

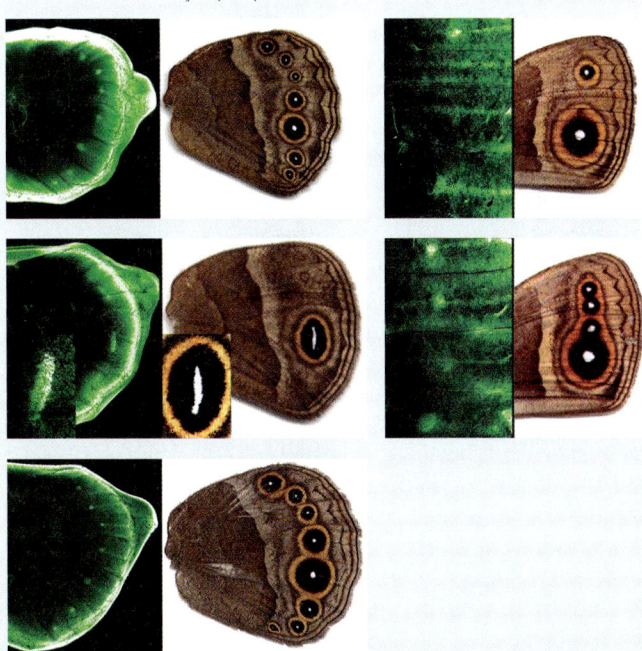

The five pairs show the pattern of *Distalless* gene expression and eyespot development in wild-type (top left) and mutant butterfly wings. High levels of *Distalless* expression (green dots and ovals) define the centers of eyespots.

CONCLUSION Regulatory genes play an important role in butterfly wing coloration, and mutations in these genes can account for differences in wing coloration among species.

SOURCE Brakefield, P. M., J. Gates, D. Keys, F. Kesbeke, P. J. Wijngaarden, A. Monteiro, V. French, and S. B. Carroll. 1996. "Development, Plasticity and Evolution of Butterfly Eyespot Patterns." *Nature* 384: 236-242.

species, then mutations in regulatory genes may account for many of the differences we observe among different species. Our growing understanding of how developmental genes underpin evolutionary change has given rise to a whole new field of biology called evolutionary-developmental biology, or **evo-devo** for short.

In evo-devo research, scientists compare the genetic programs for growth and development in species found on different branches of phylogenetic trees. In addition to wanting to discover the molecular mechanisms that guide development in individual species, these scientists aim to understand the genetic differences associated with differences in form and function among species. Evo-devo is an exciting and rapidly expanding field of research because it is helping to illuminate the long-suspected relationship between the development of individuals and patterns of evolutionary relatedness among species. Continuing research promises to shed new light on the similarities and differences among plants, animals, and other complex multicellular organisms.

Core Concepts Summary

28.1 COMPLEX MULTICELLULARITY AROSE SEVERAL TIMES IN EVOLUTION.

Bacteria are unicellular or form simple multicellular structures. page 28-1

Eukaryotes are unicellular or multicellular and exhibit both simple and complex multicellularity. page 28-1

Simple multicellularity involves the adhesion of cells with little cell differentiation. Complex multicellularity involves cell adhesion, cell signaling, and differentiation and specialization among cells. page 28-2

Complex multicellular organisms have evolved independently at least six times: in animals, vascular plants, red algae, brown algae (the kelps), and at least twice in the fungi. page 28-3

28.2 IN COMPLEX MULTICELLULAR ORGANISMS, BULK TRANSPORT OF MOLECULES CIRCUMVENTS THE LIMITATIONS OF DIFFUSION.

Diffusion is the random motion of molecules, with net movement occurring from regions of higher to regions of lower concentration. It generally acts over small distances, placing limits on the size of multicellular organisms. page 28-4

Bulk transport is an active process that allows multicellular organisms to nourish cells located far from the external environment, thereby circumventing the constraints imposed by diffusion. page 28-5

28.3 COMPLEX MULTICELLULARITY DEPENDS ON CELL ADHESION, COMMUNICATION, AND A GENETIC PROGRAM FOR DEVELOPMENT.

Animal and plant cells are organized into tissues characterized by specific molecular attachments between cells. page 28-6

Choanoflagellates express some of the same proteins that permit cell adhesion in animals, even though choanoflagellates are unicellular. Experiments suggest that choanoflagellates use these proteins to capture bacteria, not for cell adhesion. page 28-7

Gap junctions in animals and plasmodesmata in plants allow cells to communicate with each other in a targeted fashion. page 28-7

Complex multicellularity involves the genetic programming of cells so that they differentiate in space. page 28-8

A number of gene families known to play developmental roles in multicellular organisms also occur in their unicellular relatives, where at least some of them play a role in life cycle differentiation. page 28-9

28.4 PLANTS AND ANIMALS EVOLVED MULTICELLULARITY INDEPENDENTLY OF EACH OTHER AND SOLVED SIMILAR PROBLEMS WITH DIFFERENT SETS OF GENES.

The cell wall characteristic of plant cells provides structural and mechanical support, but does not allow plant cells to move. page 28-10

The cell wall of plants has led to distinct solutions to the problems of cell adhesion, cell communication, and development. For example, plants grow by the activity of meristems, populations of actively dividing cells at the tips of stems and roots. page 28-10

Animal cells do not have cell walls, allowing cell movement that is not possible in plants. For example, during animal development, embryos undergo gastrulation, a process in which cells migrate inward to form a layered structure called a gastrula. page 28-11

28.5 THE EVOLUTION OF LARGE AND COMPLEX MULTICELLULAR ORGANISMS, WHICH REQUIRED ABUNDANT OXYGEN, IS RECORDED IN THE FOSSIL RECORD.

Oxygen may be required for the evolution of complex multicellularity because of its chemical properties and its abundance. page 28-13

The evolution of complex multicellularity in the fossil record correlates with increases in atmospheric oxygen that occurred 580-560 million years ago. page 28-14

Complex multicellular organisms evolved later on land, as ancestral plants evolved the capacity to photosynthesize surrounded by air rather than water. page 28-14

Evolutionary-developmental biology, or evo-devo, is a field of research that looks at both individual development and evolutionary patterns in an attempt to understand the developmental changes that allowed organisms to diversify and adapt to changing environments. page 28-15

Self-Assessment

1. Describe differences between simple and complex multicellularity.
2. Describe the phylogenetic distribution of complex multicellularity.
3. Explain how diffusion limits the size of organisms.
4. Explain how multicellular organisms get around the size limits imposed by diffusion.
5. Describe a key difference between multicellular plants and animals.
6. Describe environmental changes recorded by the sedimentary rocks that contain the oldest fossils of large active animals.

Do you understand the chapter's Core Concepts? Log in to check your answers to the Self-Assessment questions, then practice what you've learned and reinforce this chapter's concepts by working through the problems and multimedia tutorials provided there.

www.biologyhowlifeworks.com

CASE 6

Agriculture

FEEDING A GROWING POPULATION

You may have noticed that while your grocery store or supermarket has several types of apple for sale, there is usually only one type of banana. In fact, it's possible that every banana you have ever eaten has been the same variety. Now the popular yellow fruit may be in trouble.

Years ago, the Western world favored a banana variety known as the Gros Michel. In the first part of the twentieth century, a devastating fungal infection called Panama disease wiped out Gros Michel plantations around the world. In the 1950s, banana growers turned instead to the Cavendish, a variety that displayed some natural resistance to Panama disease. Half a century later, the Cavendish remains the top-selling banana in supermarkets in North America and beyond.

Yet the reign of the Cavendish may be coming to an end. For years, banana growers have battled black sigatoka, a fungal infection that can cause losses of 50% or more of the banana yield in infected regions. Making matters worse, Panama disease is once again posing a threat. A strain of the disease has emerged against which the Cavendish plants have no resistance. The disease has spread across Asia and Australia, and experts fear it's only a matter of time before it reaches prime banana-growing regions in Latin America.

Agriculture has transformed our planet and our species; without it, modern civilization would not be possible.

For those who enjoy bananas sliced into their breakfast cereal, a loss of bananas would be annoying. For the millions of people in tropical countries who depend on bananas and related plantains for daily sustenance, the destruction of those crops would be much more serious. Agriculture has transformed our planet and our species; without it, modern civilization would not be possible. But as the case of the banana shows, there are challenges to overcome if we're to continue feeding a growing population.

Cultivated bananas are vulnerable to infection in a way that wild bananas are not. The bananas that we eat are sterile; they do not produce seeds. This means they must be propagated vegetatively by replanting cuttings taken from parent plants. As a result, banana plantations contain virtually no genetic diversity. As Panama disease and black sigatoka are proving, that lack of diversity can have disastrous effects.

As sterile clones, bananas are particularly vulnerable to disease outbreaks. Other agricultural crops face similar threats. For example, a new, virulent strain of yellow wheat rust fungus emerged in the Middle East in 2010. In its first year, the disease wiped out as much as half of Syria's wheat crop and has since spread to several other countries in the region. Meanwhile, crop scientists have warned that Ug99, a new, virulent strain of an even more damaging fungal wheat pathogen that first surfaced in eastern Africa, could devastate global wheat crops as it spreads.

Wheat was one of the first crops that our ancestors cultivated when agriculture emerged 10,000 years ago. The early farmers learned to choose wheat plants that were easiest to harvest and whose seeds germinated without delay. Over time, wheat evolved by this process of artificial selection. Those earliest farmers set in motion a long chain of genetic modifications that ultimately led to the high-yield wheat varieties grown today.

In the wild, natural selection favored wheat whose ears were protected by tough barbs that broke apart

Domestication of wheat. Wild wheat (top and bottom left) and modern domesticated wheat (top and bottom right). The ears of wild wheat are protected by tough barbs (the long spikes), and the grains fall easily from the plant, enhancing dispersal. The ears of domesticated wheat remain intact, making the grain easier to harvest.

into individual units for efficient dispersal across the landscape. Farmers, on the other hand, preferred wheat plants whose ears didn't separate, and so were easier to harvest. Such features would interfere with the plant's ability to self-seed in the wild. By selecting for traits that would be disadvantageous in nature, humans created a plant dependent on humans for its survival. The co-dependency between domestic plants and people is a hallmark of agriculture.

In addition to selecting desirable traits in their crop plants, early agriculturalists altered the environment to be favorable for those plants. They provided water, chose planting sites to ensure optimal sunlight, and removed nearby plants that would compete with the crops for vital resources such as water and nutrients. By changing the community structure and the availability of resources to support growth and reproduction, humans allowed domesticated crops to thrive.

By the 1940s, scientists had developed industrially fixed nitrogenous fertilizers that significantly boosted plant growth. They had also learned enough about genetics and DNA to begin applying that knowledge to plant breeding. These advances led to an amazingly productive period during the 1960s and 1970s now known as the Green Revolution. During that time, plant yields ballooned. Food production actually outstripped population growth through the second half of the twentieth century.

Today, we are entering a new phase of agriculture, with genetic engineering being added to the traditional approaches of crossbreeding and artificial selection. Most of the corn and soy products now consumed in the United States come from plants that have been genetically engineered to resist certain pests and herbicides. Some of the genes employed come from different species altogether. For example, corn and cotton plants are often engineered to contain a gene from the bacterium *Bacillus thuringiensis* (Bt) that confers pest resistance.

At a glance, it would seem that we've successfully taken domesticated species under our control, and indeed, we've made great strides in increasing yields to feed an ever-growing population. But evolution never takes a break. Crop pests and pathogens continue to evolve ways to evade our savviest plant breeders.

The nature of modern agriculture only compounds the problem. Most modern farms consist largely of monocultures, single types of crop each grown over a large area. Growing only a single crop at a time makes it much easier to mechanize planting and harvesting. In a given field of corn or wheat, therefore, the individual plants are often genetically quite similar to one another. With so little genetic variation, each plant is vulnerable to pathogens in exactly the same way. This means that a pathogen that can overcome one plant's natural defenses has the potential to wipe out the entire field.

Lately, many consumers are taking a hard look at these problems and at the source of the food on their plates. Critics of modern agriculture argue that monocultures, with their heavy reliance on fungicides and pesticides, are unsustainable and unhealthy. Proponents of genetic modification and high-tech agriculture argue that we will need every tool to meet food-production demands for the expanding human population. Sometime in 2011, the human population surpassed 7 billion people. The population continues to grow, expanding at its fastest rate at any time in human history. By 2100, experts predict, our planet will be home to more than 9 billion people. One point that both sides agree on is that unless agricultural productivity continues to increase, more and more land will be needed to produce food.

In many parts of the world, agricultural productivity is limited by the availability of water or nutrients or both. Global climate change is predicted to cause regional droughts and flooding that may further weaken the global food supply. There will be no easy solutions as we move forward into civilization's next phase of agriculture. One thing that is certain, however, is that understanding how plants grow, reproduce, and protect themselves from pests will help us survive in an increasingly crowded world.

? CASE 6 QUESTIONS

Answers to Case 6 questions can be found in Chapters 29–34.

1. How has nitrogen availability influenced agricultural productivity? *See page 29-18.*
2. What is the basis for the spectacular increases in the yield of cereal grains during the Green Revolution? *See page 30-21.*
3. What is the developmental basis for the shorter stems of high-yielding rice and wheat? *See page 31-9.*
4. Can modifying plants genetically protect crops from herbivores and pathogens? *See page 32-17.*
5. What can be done to protect the genetic diversity of crop species? *See page 33-20.*
6. How do fungi threaten global wheat production? *See page 34-19.*

CHAPTER 29

PLANT STRUCTURE AND FUNCTION

Moving Photosynthesis onto Land

Core Concepts

29.1 The evolution of land plants from aquatic ancestors introduced a major challenge for photosynthesis: acquiring CO_2 without losing excessive amounts of water.

29.2 Leaves have a waxy cuticle that retards water loss but inhibits the diffusion of CO_2, and pores, called stomata, that regulate CO_2 gain and water loss.

29.3 Xylem allows vascular plants to replace water evaporated from leaves with water pulled from the soil.

29.4 Phloem transports carbohydrates that support the growth and respiration of non-photosynthetic organs such as stems and roots.

29.5 Roots expend energy to obtain nutrients from the soil.

In the ocean, where life first evolved, organisms are bathed in seawater that mirrors the osmotic concentration of their cells. In contrast, on land, organisms are surrounded by air and are at risk of drying out. Excessive water loss, or **desiccation,** is a constant challenge for life on land, especially for photosynthetic organisms, which expose large surface areas to the air in order to obtain sunlight and carbon dioxide (CO_2). In Chapter 27, we saw that land plants evolved from aquatic green algae. What structural and functional innovations enabled plants to colonize the land so successfully?

The story begins about 470 million years ago, when a lineage of green algae began the transition from water to land. There are few fossils to show us what the first land plants looked like. We know they were small, at most only a few centimeters tall. These early colonists resembled their algal ancestors; they had no means of obtaining water from the soil and, at best, only a limited capacity to restrict water loss from cells. Therefore, their water content and their ability to carry out photosynthesis would have fluctuated wildly.

Today, descendants of those first land plants dominate terrestrial habitats, having evolved the ability to draw water from the soil and limit water loss from their leaves. These are the **vascular plants,** and their evolution transformed the physical and biological environment on land. Vascular plants can carry out photosynthesis even when the soil surface is dry and can sustain the hydration of photosynthetic leaves elevated as much as 100 meters into the air. In many ways, the capacity to control the gain and loss of water made possible the extraordinary evolutionary success of vascular plants.

29.1 PLANT STRUCTURE AND FUNCTION: AN EVOLUTIONARY PERSPECTIVE

Phylogeny provides a road map of land plant evolution. We will reserve detailed discussion of plant diversity until we have discussed how plants function, and we begin now with a quick sketch of evolutionary relationships among land plants. **Fig. 29.1** shows that land plants form a monophyletic group descended from green algae. The land surface, in turn, is covered by two groups of plants: the **bryophytes** (which include the mosses, along with the less-familiar liverworts and hornworts) and the vascular plants. For many years, botanists treated both bryophytes and vascular plants as monophyletic, but phylogenetic analysis based on molecular sequences makes it clear that bryophytes are a paraphyletic grouping, including some of but not all the descendants of a common ancestor. Note in Fig. 29.1 that the descendants of the last common ancestor of bryophytes also include vascular plants.

Vascular plants make up more than 95% of all land plant species found today, and they similarly dominate the photosynthetic output of terrestrial environments. The phylogenetic tree shows that vascular plants form a monophyletic group, with four main subgroups, three of which are probably familiar to you. **Ferns and horsetails** form a monophyletic group, as do plants that reproduce by seeds. Molecular sequence comparisons suggest that living seed plants can be subdivided into two monophyletic groups, gymnosperms and angiosperms. The **gymnosperms** include pine trees and other conifers. The **angiosperms,** or flowering plants, include oak trees, grasses, and sunflowers.

The fourth major vascular plant group, the **lycophytes,** may be less familiar. This group is particularly interesting to scientists because it forms the sister group to all other vascular plants. Keep Fig. 29.1 in mind as you read the next several chapters because, while the major groups of vascular plants share many features, they have each evolved distinctive variations in growth, anatomy, and reproduction that help to explain their present-day diversity and distribution.

The evolution of vascular plants was a game-changing event in the history of life. To appreciate why this is so, we have to understand the "game" played by the lineages that preceded the vascular plants. In particular, how are the bryophytes (the three groups of nonvascular land plants) able to survive on land? Bryophytes do not produce roots and so are totally dependent upon surface water to keep their photosynthetic cells hydrated. As the environment dries, so do they.

All cells require a high degree of hydration to function. As bryophytes dry out, they lose the ability to carry out photosynthesis. However, many bryophytes exhibit **desiccation tolerance,** a suite of biochemical traits that allows their cells to survive extreme dehydration without damage to membranes or macromolecules (**Fig. 29.2**). Bryophytes can then resume photosynthesis when surface water is once again available. In

FIG. 29.2 Desiccation tolerance in bryophytes. *Syntrichia princeps*, a moss, can withstand complete desiccation (right) but regains photosynthetic capacity rapidly when rewetted (left).

contrast to bryophytes, vascular plants maintain the hydration of their photosynthetic cells by drawing upon water in the soil. As a result, vascular plants can photosynthesize across a wider range of environments and conditions than can bryophytes.

Bryophytes provide insights into how the first plants may have dealt with the challenge of carrying out photosynthesis on land. The ability to survive desiccation is likely to have been critical. But we must avoid the temptation to think of living bryophytes as representative of the ancestral state. Bryophyte diversity has been shaped by their long coexistence with vascular plants. Indeed, bryophytes thrive today only in places where roots would not provide a significant advantage—on rocks and the trunks of trees, for example, and in regions too dry or too cold for roots to penetrate the soil.

We will return to bryophytes when we examine plant reproduction (Chapter 30) and diversity (Chapter 33). For now, however, let's focus on the vascular plants.

29.2 THE LEAF: ACQUIRING CO$_2$ WHILE AVOIDING DESICCATION

Fig. 29.3 shows the general features of a vascular plant. Aboveground, we see three major types of organ—leaves, stems, and reproductive organs—which collectively form the **shoot**. **Roots,** which are generally hidden from view in the ground, make

FIG. 29.1 The phylogenetic tree of plants. Plants can be divided into two groups: the vascular plants, which actively control their hydration, and the bryophytes, which do not.

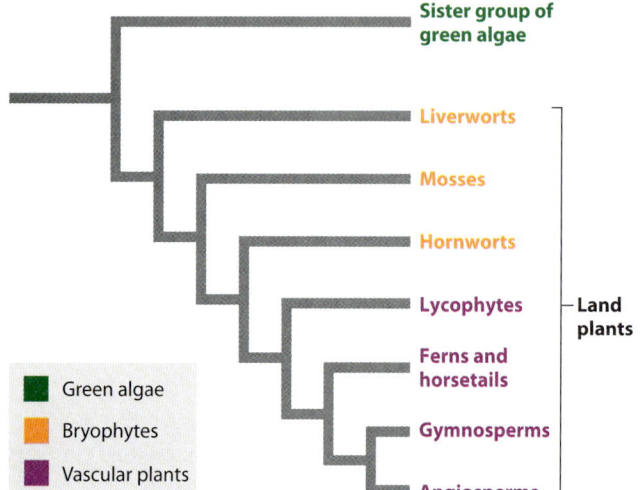

CHAPTER 29 PLANT STRUCTURE AND FUNCTION: MOVING PHOTOSYNTHESIS ONTO LAND

FIG. 29.3 Major organs of a vascular plant.

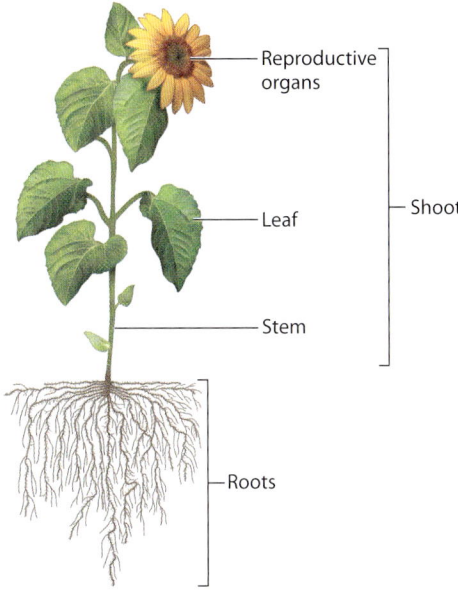

up the fourth major organ system. We can answer the question of how vascular plants photosynthesize on land by examining how leaves, stems, and roots work together.

The **leaf** is the principal site of photosynthesis in vascular plants. A cross section shows the leaf's three major tissues: sheets of cells called the **epidermis** that line the leaf's upper and lower surfaces; loosely packed photosynthetic cells that make up the **mesophyll** (literally, "middle leaf"); and **veins,** the system of vascular conduits that connects the leaf to the rest of the plant (**Fig. 29.4**). We discuss how water and carbohydrates are transported through the vascular system in sections 29.3 and 29.4. In this section, we explore how the structure of leaves allows vascular plants to photosynthesize in air without drying out.

CO_2 uptake results in water loss.

Water is the resource that most often limits a plant's ability to grow and function. Water availability is the single most important factor limiting agricultural yields, and rainfall patterns have a major impact on the structure and productivity of natural ecosystems (Chapter 47). Water has such a significant effect on the growth and functioning of plants because plants use large amounts of water. Why do plants require so much water?

The answer is *not* because water is consumed in supplying electrons for photosynthesis (Chapter 8). That process accounts for less than 1% of the water required by vascular plants. Instead, most of a plant's need for water arises as a consequence of CO_2 uptake in air.

As we saw in Chapter 25, CO_2 is a minor constituent of air, about 400 parts per million at present. This low concentration limits the rate at which CO_2 can diffuse into the leaf. Therefore, how fast a plant can take up CO_2 depends in large part on the degree to which leaves can expose their photosynthetic cells to the surrounding air. If we look inside a leaf, we see that each mesophyll cell is largely

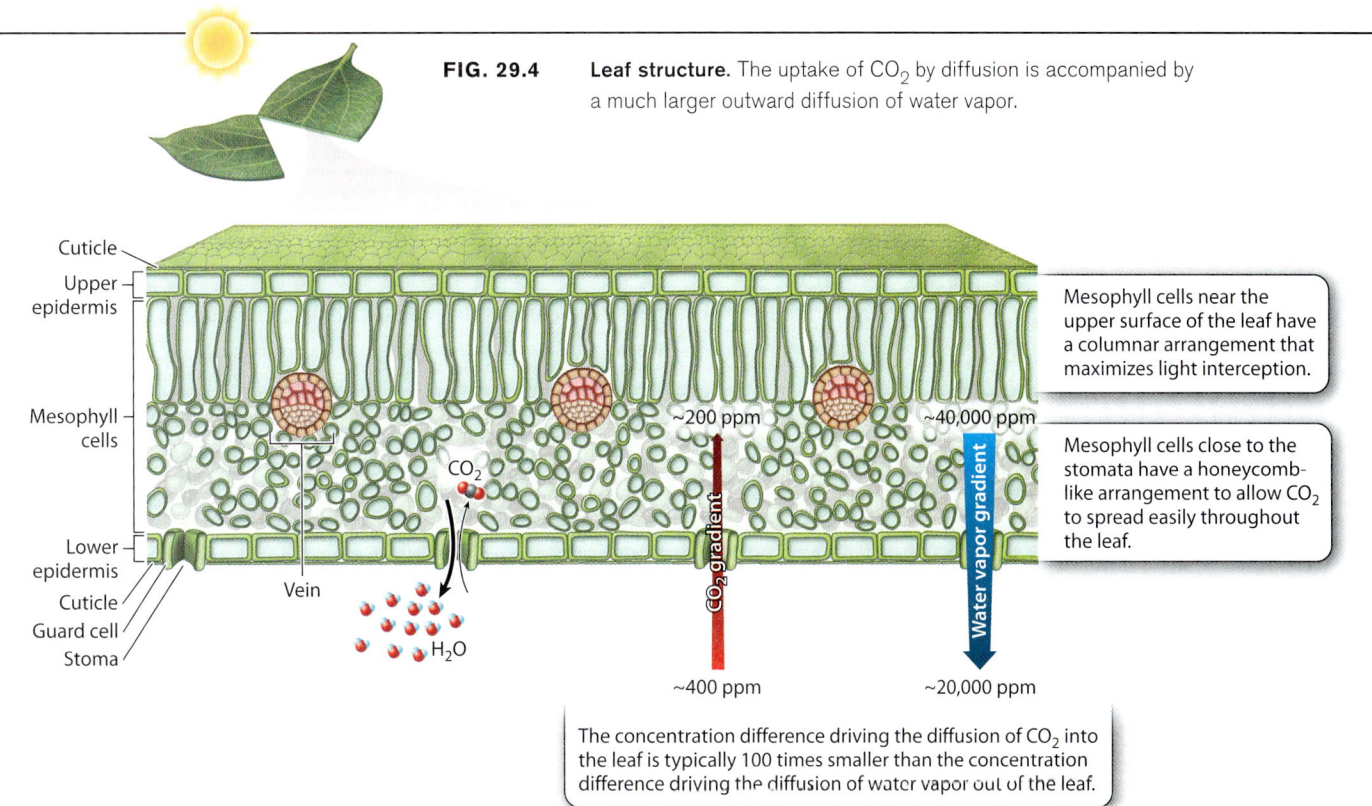

FIG. 29.4 **Leaf structure.** The uptake of CO_2 by diffusion is accompanied by a much larger outward diffusion of water vapor.

surrounded by air. The mesophyll cells obtain the CO_2 that they need for photosynthesis directly from these air spaces. If the air spaces within the leaf were completely sealed off, photosynthesis would quickly run out of CO_2 and come to a halt. But they're not sealed off—the leaf's air spaces are connected to the air surrounding the leaf by pores in the epidermis. As photosynthesis draws down the concentration of CO_2 molecules within the leaf's air spaces relative to the concentration of CO_2 in the outside air, the difference in concentration between the inside and the outside of the leaf causes CO_2 to diffuse into the leaf, replenishing the supply of CO_2 for photosynthesis.

The ability of leaves to draw in CO_2 comes at a price, however: If CO_2 can diffuse into the leaf, water vapor can diffuse out (Fig. 29.4). Furthermore, the rate at which water vapor diffuses out of a leaf is much greater than the rate at which CO_2 diffuses inward. Molecules diffuse from regions of higher concentration to regions of lower concentration, and the rate of diffusion is proportional to the difference in concentration. On a sunny summer day, the difference in water vapor concentration between the air spaces within a leaf and the air outside can be more than 100 times larger than the difference in concentration of CO_2. Add the fact that water is lighter than CO_2, and so diffuses 1.6 times faster *for the same concentration gradient*, and it becomes clear why, on a sunny summer day, several hundred water molecules are lost for every molecule of CO_2 acquired for photosynthesis.

Because of this difference, the water costs associated with photosynthesis in air are large. A sunflower leaf, for example, can lose an amount equal to its total water content in as little as 20 minutes. To put this rate of water loss in perspective, you would have to drink about 2 liters per minute to survive a similar rate of loss. Vascular plants can sustain such high rates of water loss because they can access the only consistently available source of water on land: the soil. The evaporative loss of water vapor from leaves, commonly called **transpiration,** is the end point of the movement of water from the soil, through the bodies of vascular plants, and then, as water vapor, into the atmosphere. Therefore, the challenge of keeping photosynthetic cells hydrated is met partly by the continual supply of water from the soil and partly by limiting the rate at which water already in leaf tissues is lost to the atmosphere.

→ **Quick Check 1** Why do plants transpire?

The cuticle restricts water loss from leaves but inhibits the uptake of CO_2.

Epidermal cells secrete a waxy **cuticle** on their outer surface that limits water loss. Without a cuticle, the humidity within the internal air spaces would drop and the photosynthetic mesophyll cells would dry out. Interestingly, it is the evaporation of water from the exterior face of epidermal cells that brings waxes secreted by the cell's plasma membrane up to the surface of the cell wall. There the wax molecules self-organize into a protective layer that builds up until water loss essentially ceases. This process of cuticle

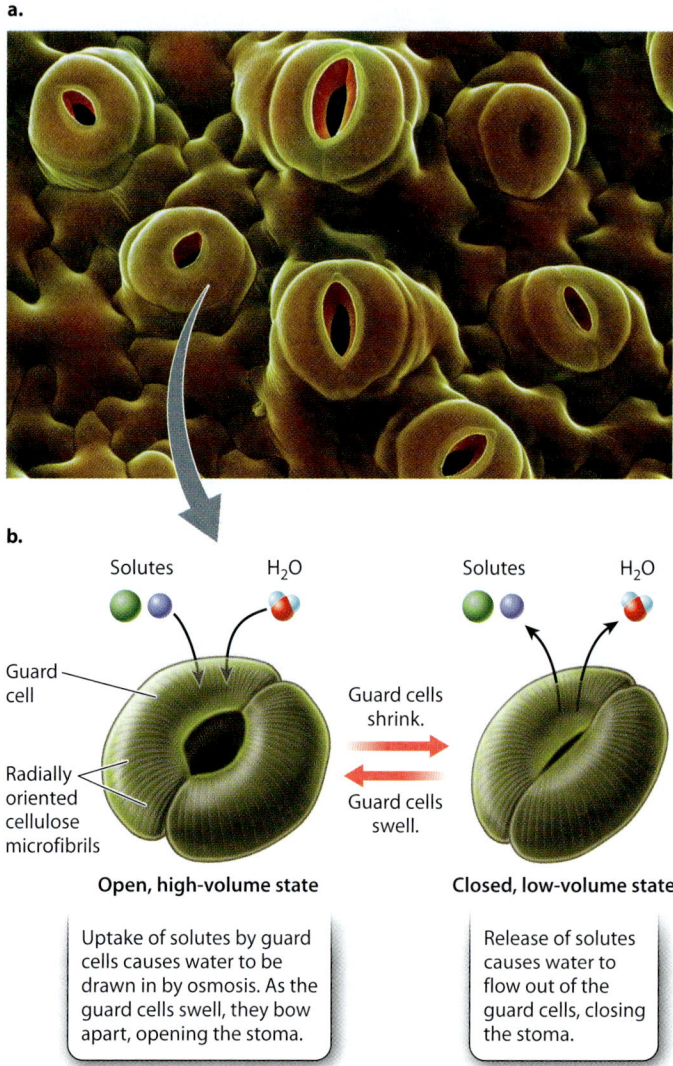

FIG. 29.5 **Stomata.** (a) A scanning electron microscope (SEM) image of stomata on a leaf surface; (b) guard cells of a stoma in open and closed positions.

formation ensures the rapid repair of any damage to the cuticle that would allow water to evaporate. The mechanism is far from perfect, however, because the cuticle prevents CO_2 from diffusing into the leaf even as it restricts water vapor from diffusing out of it.

As noted above, small pores in the epidermis allow CO_2 to diffuse into the leaf (**Fig. 29.5a**). These pores, called **stomata** (singular, stoma), regulate the diffusion of gases between the interior of the leaf and the atmosphere. Stomata can be numerous, up to 1000 per mm^2. Yet because each one is small, less than 1% to 2% of the leaf surface is actually covered by these pores. Thus, even with stomata, the epidermis is a significant barrier to the diffusion of CO_2 and water vapor.

The epidermis with its stomatal pores represents a compromise between the challenges of providing food (CO_2 uptake) while preventing thirst (water loss). However, because both the dryness of the air and the wetness of the soil are variable, leaves must be able to alter the porosity of the leaf epidermis to maintain a balance between the rates of water loss to the atmosphere and water delivery from the soil.

Stomata allow leaves to regulate water loss and carbon gain.

Stomata are more than just holes in the leaf epidermis; they are hydromechanical valves that can open and close. Each stoma consists of two **guard cells** surrounding a central pore. The guard cells can shrink or swell, changing the size of the pore between the guard cells (**Fig. 29.5b**). How does this valve system work?

Cellulose microfibrils in the guard cell walls are oriented radially, that is, wrapped around the cell. The radially oriented microfibrils make it easier for swelling guard cells to expand in length rather than in girth. Because the guard cells are firmly connected at their ends, this increase in length causes them to bow apart, opening the stoma. Conversely, a decrease in cell volume causes the pore to close.

Guard cells control their volume by altering the concentration of solutes, for example the ions K^+ and Cl^-, in their cytoplasm. An increase in solute concentration causes water to flow into the cell by osmosis, whereas a decrease in solutes causes water to flow out of the cell. Recall from Chapter 5 that osmosis is the diffusion of water across a semipermeable membrane, such as the cell's plasma membrane, from a region of higher water concentration to a region of lower water concentration. An increase in the concentration of solutes in guard cells is accompanied by a corresponding decrease in the concentration of water molecules. Thus, as guard cells add solutes, water diffuses into the cells, causing them to swell.

Stomata close by the same process, only in reverse. A decrease in solute concentration results in a higher concentration of water molecules inside a guard cell relative to the solution in the surrounding cell wall space, and water diffuses out of the cell, causing it to shrink.

The accumulation of solutes inside guard cells requires an expenditure of energy in the form of ATP, which drives the uptake of solutes across the plasma membrane (Chapter 5). Closing, in contrast, does not require direct energy expenditure. Thus, the default or resting state of guard cells is "closed"; this is the state that conserves the hydration of the leaf. In angiosperms (the flowering plants), K^+ and Cl^- ions are transferred between guard cells and an adjacent epidermal cell, so an increase in the volume of one cell type is accommodated by a decrease in the other. This reciprocal change in cell size allows the stomatal pore to open much wider than it could if the guard cells operated on their own (**Fig. 29.6**). It takes approximately 10 minutes for angiosperm stomata to go from fully closed to fully open. During this time, the volume of each guard cell can increase as much as threefold.

To function effectively, stomata must open and close in response to both CO_2 demand and water loss. Stomata are thus key sites for the processing of physiological information. Light stimulates stomatal opening, while high levels of CO_2 inside the leaf (a signal that CO_2 is being supplied faster than photosynthesis can take it up) cause stomata to close. When the soil can't supply enough water to keep up with evaporation from the leaf, guard cells become dehydrated and shrink in volume, closing the stomata. Guard cell volume also changes in response to signaling molecules. For example, abscisic acid, a hormone produced during drought, causes stomata to close. This signal allows leaves to close their stomata early enough to prevent dehydration, rather than responding only when water has already been lost. Clearly, then, stomata are a key innovation in the evolution of land plants. Not surprisingly, fossils indicate that stomata evolved early in the colonization of land.

Before leaving the topic of how stomata affect rates of water loss from leaves, we pause to consider the following question: Is there any evidence that plants control the temperature of their leaves by regulating transpiration? The answer, perhaps surprisingly, is no. Although our own experience leads us to think of evaporation (for example, sweating) as a means of preventing overheating, plants transpire when their leaves are too hot *or* too cold. When leaves are above their optimal temperature for photosynthesis, cooling through transpiration is beneficial. But when temperatures are

---

FIG. 29.6 **Opening and closing of guard cells in a fern and an angiosperm.** (a) Guard cells in ferns shrink and swell independently of other cells, limiting the size of the stomatal pore; (b) in angiosperms, changes in guard cell volume are matched by reciprocal changes in the adjacent epidermal cells.

a. *Nephrolepis exaltata* (fern)

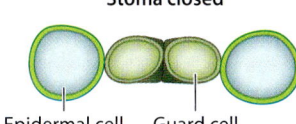

b. *Tradescantia virginiana* (angiosperm)

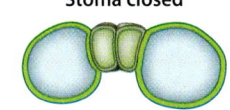

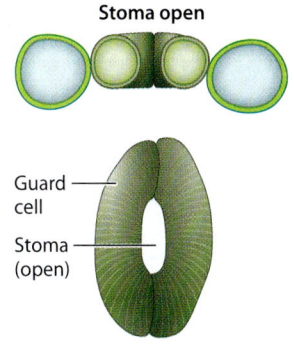

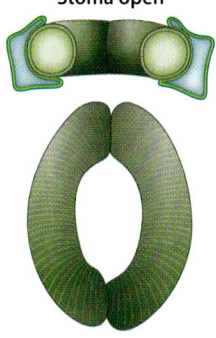

below the optimum, transpiration results in leaf temperatures that are even less favorable. In addition, the hot environments in which evaporative cooling would be beneficial often coincide with low availability of water in the soil. For this reason, plants in arid regions tend to rely on mechanisms other than evaporative cooling, such as reflective hairs and waxes and small leaf size, to prevent their leaves from overheating.

CAM plants use nocturnal CO$_2$ storage to avoid water loss during the day.

In most plants, water loss and photosynthesis both peak at the same time, when the sun is high overhead. What if a plant could open its stomata to capture CO$_2$ at night, when cool air limits rates of evaporation?

In fact, a number of plants have evolved precisely this mechanism, called **crassulacean acid metabolism,** or **CAM,** to help balance CO$_2$ gain and water loss (**Fig. 29.7**). CAM (named after the Crassulaceae family of plants, which uses this pathway), provides a system for overnight storage, converting CO$_2$ into a form that will not diffuse away. The storage form of CO$_2$ is produced by the activity of the enzyme PEP carboxylase, which combines a dissolved form of CO$_2$ (bicarbonate ion, HCO$_3^-$) with a 3-carbon compound called phosphoenol pyruvate (PEP). The resulting product is a 4-carbon organic acid that is stored in the cell's vacuole.

When the sun comes up the next morning, the stomata close, conserving water. At the same time, the 4-carbon organic acids are retrieved from the vacuole and decarboxylated—that is, their CO$_2$ is released into the cytoplasm. Because the stomata are now closed, this newly released CO$_2$ does not escape from the leaf. Instead, it diffuses into the chloroplast, where it can become incorporated into carbohydrates by the Calvin cycle. As we saw in Chapter 8, the Calvin cycle requires a continual supply of energy in the form of ATP and NADPH and so can operate only in the light.

The buildup of organic acid during the night results in a marked decrease in the pH of vacuoles. In 1815, the British naturalist Benjamin Heyne wrote that some of the plants in his garden in India tasted "as acid as sorrel" early in the morning, but by afternoon the acid taste had disappeared.

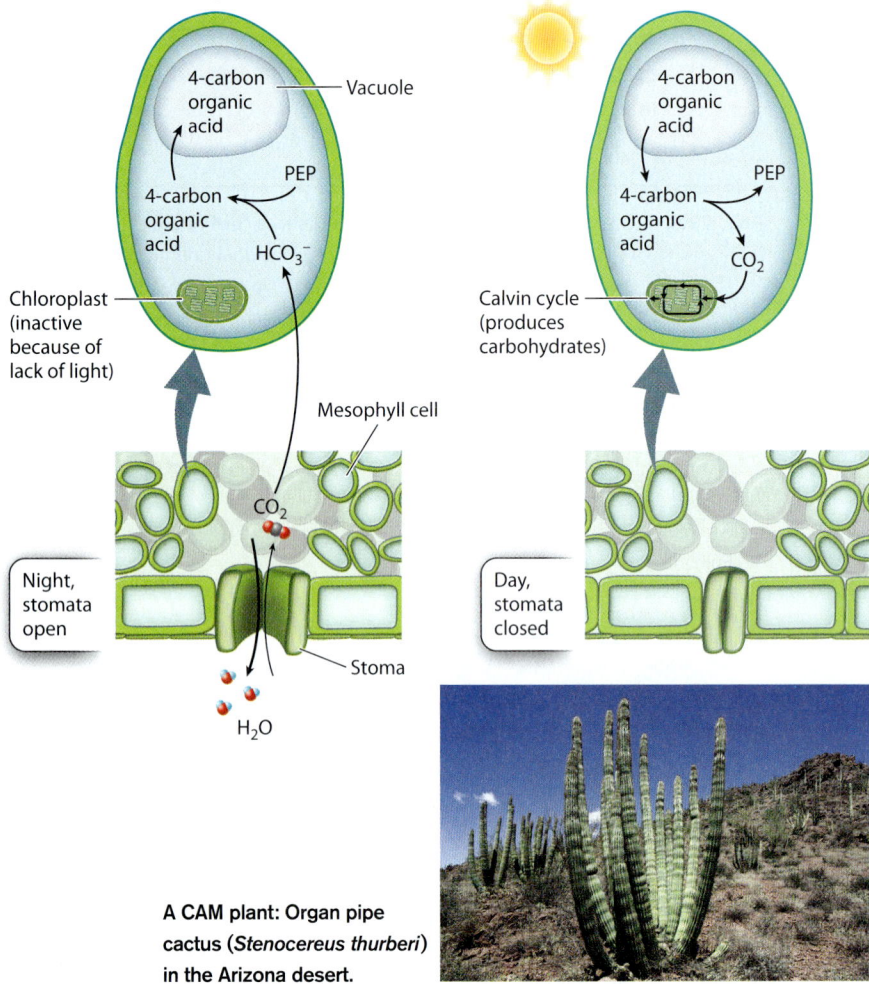

FIG. 29.7 CAM photosynthesis. CAM plants store CO$_2$ at night in the vacuole; leaves can then photosynthesize during the day without opening their stomata.

A CAM plant: Organ pipe cactus (*Stenocereus thurberi*) in the Arizona desert.

The nocturnal opening of stomata exhibited by CAM plants greatly increases the amount of CO$_2$ gain per unit of water loss. Indeed, the CO$_2$: H$_2$O exchange ratio for CAM plants is in the range of 1 : 50, nearly 10 times higher than that of plants that open their stomata during the day. However, CAM has a drawback. The rates of photosynthetic carbohydrate production by CAM plants tend to be low because ATP is needed to drive the uptake of organic acids into the vacuole and because only so much organic acid can accumulate in the vacuole. CAM is therefore most common in habitats such as deserts, where water conservation is crucial, and among **epiphytes,** plants that grow high in the canopy of other plants, without contact with the soil.

Because CAM uses enzymes (for example, PEP carboxylase) and compartments (for example, vacuoles) that exist in all plant cells, it is not surprising that it has evolved multiple times. CAM occurs in all four groups of vascular plants shown in Fig. 29.1. It is most widespread in the angiosperms, occurring in 5% to 10% of species. Vanilla orchids and Spanish moss, two well-known epiphytes, as well as many cacti, exhibit CAM.

C₄ plants suppress photorespiration.

Photorespiration adds another wrinkle to the challenge of acquiring CO_2 while minimizing water loss. Recall from Chapter 8 that either CO_2 or O_2 can be a substrate for rubisco, the key enzyme in the Calvin cycle. CO_2 as a substrate leads to the production of carbohydrates through photosynthesis; O_2 as a substrate results in a net loss of energy and the release of CO_2, a process called photorespiration. (Photorespiration is similar to aerobic respiration only in the sense that it uses O_2 and releases CO_2. However, plants do not gain energy; they lose it.)

Photorespiration presents a significant challenge for land plants because air contains approximately 21% O_2 but less than 0.04% CO_2. Although rubisco reacts more readily with CO_2 than with O_2, the sheer abundance of O_2 means that O_2 is a substrate for rubisco some of the time. At moderate leaf temperatures, O_2 is the substrate instead of CO_2 as often as 1 time out of 4. At higher temperatures, O_2 is even more likely to be the substrate for rubisco, and the rates of photorespiration are yet higher. The reason is that higher temperatures decrease both rubisco's affinity for CO_2 relative to O_2 and the solubility of CO_2 relative to O_2 in water.

Some plants have evolved a way to reduce the energy and carbon losses associated with photorespiration. These are the **C₄ plants**, which suppress photorespiration by increasing the concentration of CO_2 in the immediate vicinity of rubisco. C₄ plants take their name from the fact that they, like CAM plants, use PEP carboxylase to produce 4-carbon organic acids that subsequently supply the Calvin cycle with CO_2. The Calvin cycle produces 3-carbon compounds (Chapter 8), and plants that do not use 4-carbon organic acids to supply the Calvin cycle with CO_2 are thus referred to as **C₃ plants**.

Both CAM and C₄ plants produce 4-carbon organic acids as the entry point for photosynthesis. However, in CAM plants, CO_2 capture and the Calvin cycle take place *at different times*; in C₄ plants, they take place *in different cells*. It is the spatial separation between CO_2 capture and the Calvin cycle that allows C₄ plants to suppress photorespiration.

C₄ plants intitially capture CO_2 in mesophyll cells by means of PEP carboxylase, which combines a dissolved form of CO_2 (bicarbonate ion, HCO_3^-) with the 3-carbon compound PEP. This produces 4-carbon organic acids that diffuse through plasmodesmata into the **bundle sheath** (**Fig. 29.8**), a cylinder of cells that surrounds each vein. Once inside bundle-sheath

FIG. 29.8 C₄ photosynthesis. C₄ plants suppress photorespiration by concentrating CO_2 in bundle-sheath cells.

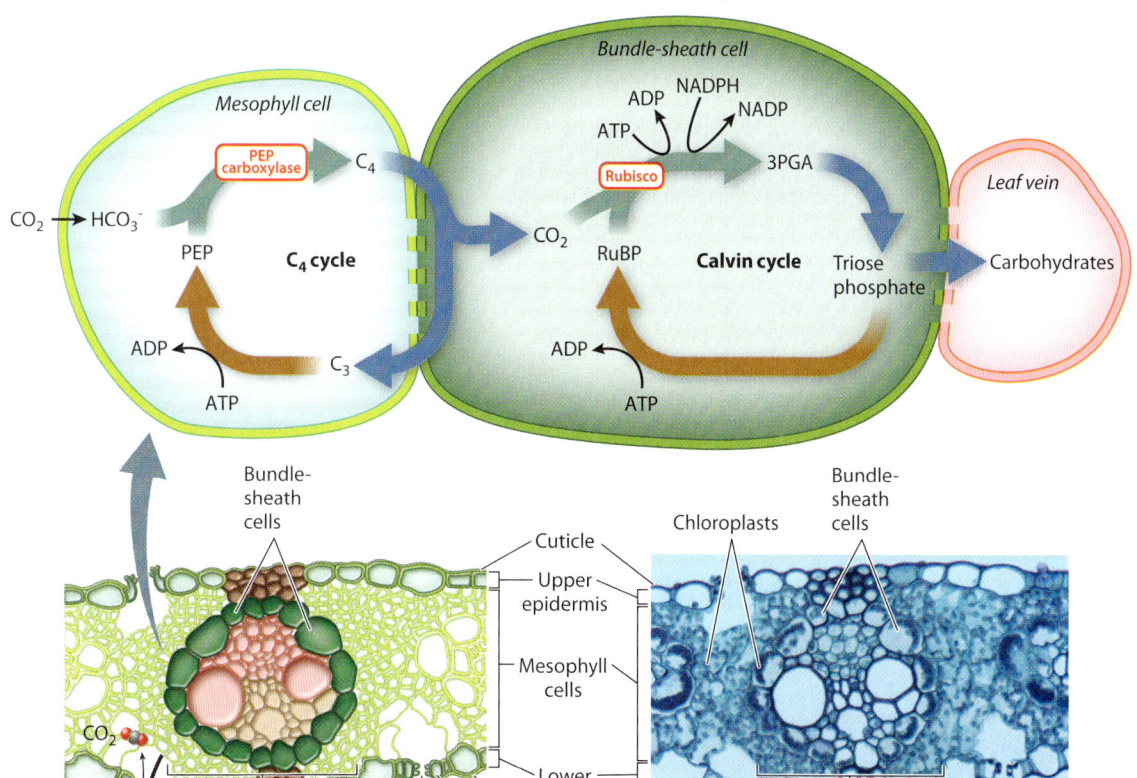

cells, the 4-carbon compounds are decarboxylated, releasing CO_2 that is then incorporated into carbohydrates through the Calvin cycle (Chapter 8). The C_4 cycle is completed as the 3-carbon molecules generated during decarboxylation diffuse back to the mesophyll, where ATP is used to re-form PEP. Because the C_4 cycle operates much faster than the Calvin cycle, the concentration of CO_2 within bundle-sheath cells builds up, reaching levels as much as five times higher than in the air surrounding the leaf. The increase in the concentration of CO_2 in bundle-sheath cells makes it much less likely that rubisco will use O_2 as a substrate.

C_4 plants have high rates of photosynthesis because they do not suffer the losses in energy and reduced carbon associated with photorespiration (**Fig. 29.9**). In addition, C_4 plants often have a more favorable $CO_2 : H_2O$ exchange ratio than plants that lack the ability to concentrate CO_2. However, C_4 photosynthesis has a greater energy requirement than conventional (C_3) photosynthesis, as ATP must be used to regenerate PEP in the C_4 cycle. Thus, C_4 photosynthesis confers an advantage in hot, sunny environments where rates of photorespiration would otherwise be high. C_4 photosynthesis has evolved as many as 20 times, but is most common among tropical grasses and plants of open habitats with warm temperatures. C_4 plants include a number of important crops, especially maize (corn), sugarcane, and sorghum.

→ **Quick Check 2** Why does the formation of 4-carbon organic acids increase the efficiency of water use in CAM plants but suppress photorespiration in C_4 plants?

HOW DO WE KNOW?

FIG. 29.9

How do we know that C_4 photosynthesis suppresses photorespiration?

BACKGROUND Studies using radioactively labeled CO_2 showed that some species initially incorporate CO_2 into 4-carbon compounds instead of the 3-carbon compounds that are the first products in the Calvin cycle. These C_4 plants also have high rates of photosynthesis. Is this a new, more efficient photosynthetic pathway? Or do C_4 plants have high rates of photosynthesis because they are able to avoid the carbon and energy losses associated with photorespiration?

HYPOTHESIS C_4 plants do not exhibit photorespiration.

METHOD "Air tests," as these experiments were first called, compared rates of photosynthesis in normal air (21% O_2) and in an experimental gas mixture in which the concentration of O_2 is only 1%. When the concentration of O_2 is low, rubisco has a low probability of using O_2 (instead of CO_2) as a substrate, and thus photorespiration does not occur.

RESULTS

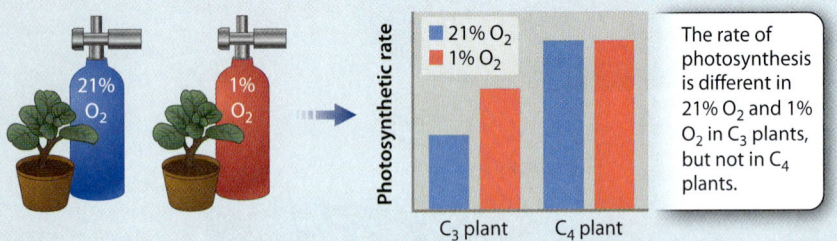

The rate of photosynthesis is different in 21% O_2 and 1% O_2 in C_3 plants, but not in C_4 plants.

CONCLUSION Photosynthesis in C_4 plants is not affected by differences in O_2 concentration, indicating that significant photorespiration is not occurring in these plants. In contrast, the photosynthetic rate of the C_3 plants increased in the low O_2 environment, indicating that photorespiration depresses rates of photosynthesis in 21% O_2.

FOLLOW-UP WORK The higher photosynthetic efficiency of C_4 photosynthesis has prompted efforts, so far unsuccessful, to incorporate this pathway into C_3 crops such as rice.

SOURCE Bjorkman, O., and J. Berry. 1973. "High-Efficiency Photosynthesis." *Scientific American* 229:80–93.

29.3 THE STEM: TRANSPORT OF WATER THROUGH XYLEM

On a summer day, a tree can transport many hundreds of liters of water from the soil to its leaves. This is an impressive feat given that it is accomplished without any moving parts. Even more remarkable, trees and other plants transport water without any direct expenditure of energy. The upward transport of water is possible because the structure of vascular plants allows them to use the evaporation of water from leaves to pull water from the soil.

Fig. 29.10 shows the cross section of a sunflower stem. Like a leaf, it has a surface layer of epidermal cells. This layer encloses thin-walled **parenchyma** cells in the interior. Notice that the stem also contains differentiated tissues that lie in a ring near the outside of the stem. These are the vascular tissues, which form a continuous pathway that extends from near the tips of the roots, through the stem, and into the network of veins within leaves. The outer tissue, called **phloem,** transports carbohydrates from leaves to the rest of the plant body. The inner tissue, called **xylem,** transports water and nutrients from the roots to the leaves.

Xylem provides a low-resistance pathway for the movement of water.

Water travels with relative ease through xylem because of the structure of the water-transporting cells within this tissue. As they develop, these cells become greatly elongated. When they complete their growth, they secrete an additional wall layer that is very thick and which contains lignin, a chemical compound that increases mechanical strength. Most remarkable is the final stage of development, when the nucleus and cytoplasm are lost, leaving behind only the cell walls.

These thick walls form a hollow conduit through which water can flow (**Fig. 29.11**). Water enters and exits xylem conduits through **pits,** circular or ovoid regions in the walls where the lignified cell wall layer is not produced. Instead, pits contain only the thin, water-permeable cell wall that surrounded each cell as it grew. As we will see, pits play an important role in xylem because they allow the passage of water, but not air, from one conduit to another.

Xylem conduits can be formed from a single cell or multiple cells stacked to form a hollow tube. Unicellular conduits are called **tracheids** (Fig. 29.11a), and multicellular conduits are called **vessels** (Fig. 29.11b). Because tracheids are the product of a single cell, they are typically 4 to 40 μm in diameter and no more than 2 to 3 cm long. Vessels, which are made up of individual cells called **vessel elements,** can be much wider and longer. Vessel diameters range from 5 to 500 μm, and lengths can be up to several meters.

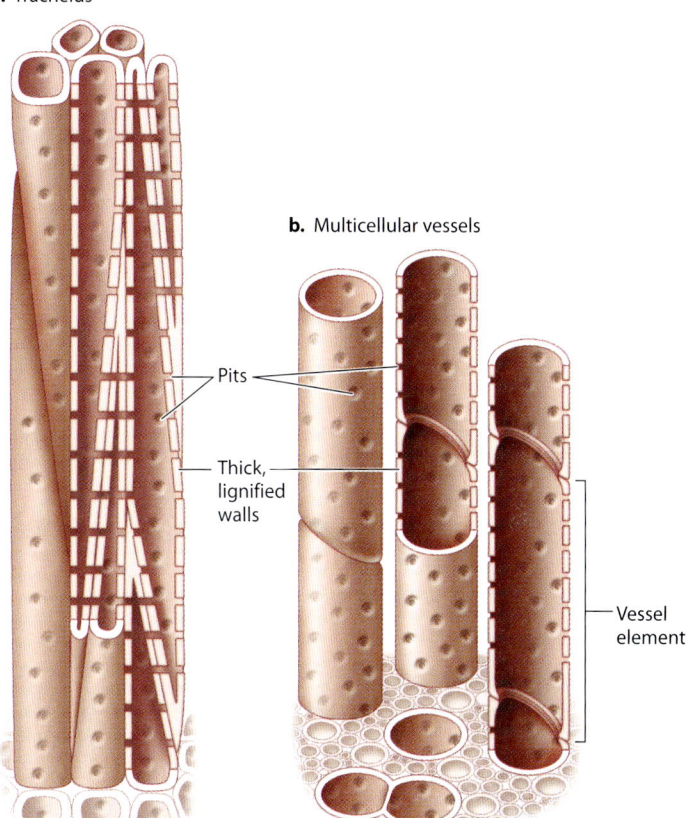

FIG. 29.11 **Xylem structure.** Xylem conduits are formed by hollow cells with stiff walls and no cell content or membrane. (a) Single-celled tracheids and (b) multicellular vessels are interconnected by pits.

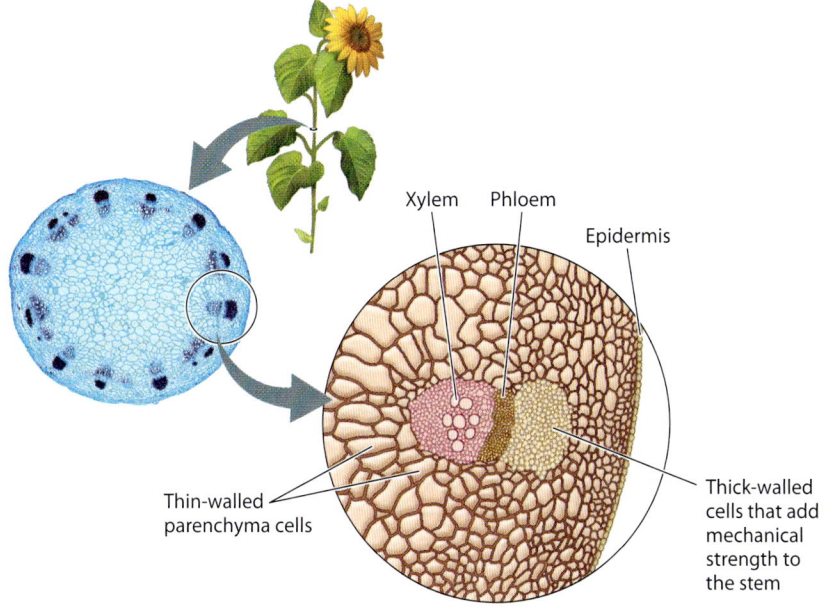

FIG. 29.10 A cross section of the stem of a sunflower showing the two vascular tissues, xylem and phloem.

Water enters a tracheid through pits, travels upward through the conduit interior, and then flows outward through other pits into an adjacent, partially overlapping, tracheid. Water also enters and exits a vessel through pits. In contrast to tracheids, however, once the water is inside the vessel, little or nothing blocks the flow of water from one cell to the next. That is because during the development of a xylem vessel, the end walls of the vessel elements are digested away, allowing water to flow along the entire length of the vessel without having to cross any pits. At the end of a vessel, however, the water must flow through pits if it is to enter an adjacent vessel and thereby continue its journey from the soil to the leaves.

The rate at which water moves through xylem is a function of both the number of conduits and their size. Conduit length determines how often water must flow across pits, which exert a significant resistance to flow. Conduit width also has a strong effect on the rate of water transport. Like the flow through pipes, water flow through xylem conduits is greater when the conduits are wider. The flow is proportional to the radius of the conduit raised to the fourth power, so doubling the radius increases flow sixteenfold. Because vessels are both longer and wider than tracheids, plants with vessels achieve greater rates of water transport than is possible through tracheids alone. Tracheids are the most common conduits in lycophytes, ferns and horsetails, and gymnosperms, whereas vessels are the principal conduit in angiosperms.

Water is pulled through xylem by an evaporative pump.

If you cut a plant's roots off under water, the leaves continue to transpire for some time. The persistence of transpiration when roots are absent demonstrates that the driving force for water transport is not generated in the roots, but instead comes from the leaves. In essence, water is pulled through xylem from above rather than being pushed from below.

The forces pulling water upward through the plant are large. Not only must these forces be able to lift water against gravity, they must also be able to pull water from the soil, which becomes increasingly difficult as the soil dries. In addition, they must be able to overcome the physical resistance associated with moving water through the plant itself. To replace water lost by transpiration with water pulled from the soil, the leaves must exert forces that are many times greater than the suctions that we can generate with a vacuum pump (**Fig. 29.12**). How do leaves exert this force?

HOW DO WE KNOW?

FIG. 29.12

How large are the forces that allow leaves to pull water from the soil?

BACKGROUND The idea that water is pulled through the plant by forces generated in leaves was first suggested in 1895. However, without a way to measure these forces, there was no way to know how large they were.

HYPOTHESIS In 1912, German physiologist Otto Renner hypothesized that leaves are able to exert stronger suctions than could be generated with a vacuum pump.

EXPERIMENT Renner measured the rate at which water flowed from a reservoir into the cut tip of a branch of a transpiring plant. He next cut off about 10 cm of the branch and attached a vacuum pump in place of the transpiring plant. He then compared the flow rate through the branch when attached to the transpiring plant with the flow rate generated by the vacuum pump.

RESULTS Renner found that the flow rates generated by transpiring plants were two to nine times greater than the flow rates generated by the vacuum pump.

CONCLUSION A transpiring plant pulls water through a branch faster than a vacuum pump. Therefore, the pulling force generated by a transpiring plant must be greater than that of a vacuum pump. Because vacuum pumps create suctions by reducing the pressure in the air, the maximum suction that a vacuum pump can generate is 1 atmosphere. Thus, the maximum height that one can lift water using a vacuum pump is 10 m (33 feet). The forces that pull water through plants have no comparable limit because they are generated within the hydrated cell wall. This is why plants can pull water from dry soils and why they can grow to more than 100 m in height.

FOLLOW-UP WORK In 1965, Per Fredrick Scholander developed a pressurization technique that uses reverse osmosis to measure the magnitude of the suctions exerted by leaves of intact plants. He used this technique to show that the leaves at the top of a Douglas fir tree 100 m tall exert greater pulling forces than do leaves closer to the ground.

SOURCES Renner, O. 1925. "Zum Nachweis negativer Drucke im Gefässwasser bewurzeler Holzgewachse." *Flora* 118/119:402–408.; Scholander, P. F., H. T. Hammel, E. D. Bradstreet, and E. A. Hemmingson. 1965. "Sap Pressure in Vascular Plants." *Science* 148:339–346.

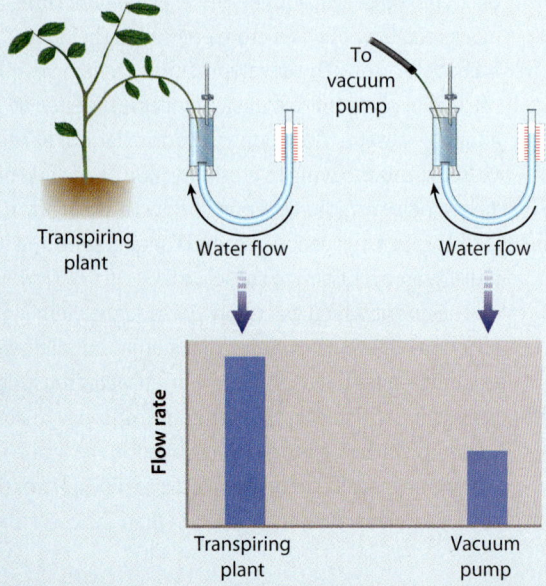

FIG. 29.13 Xylem transport from roots to leaves. Water is pulled through the plant by forces generated in the leaves, preventing them from drying out.

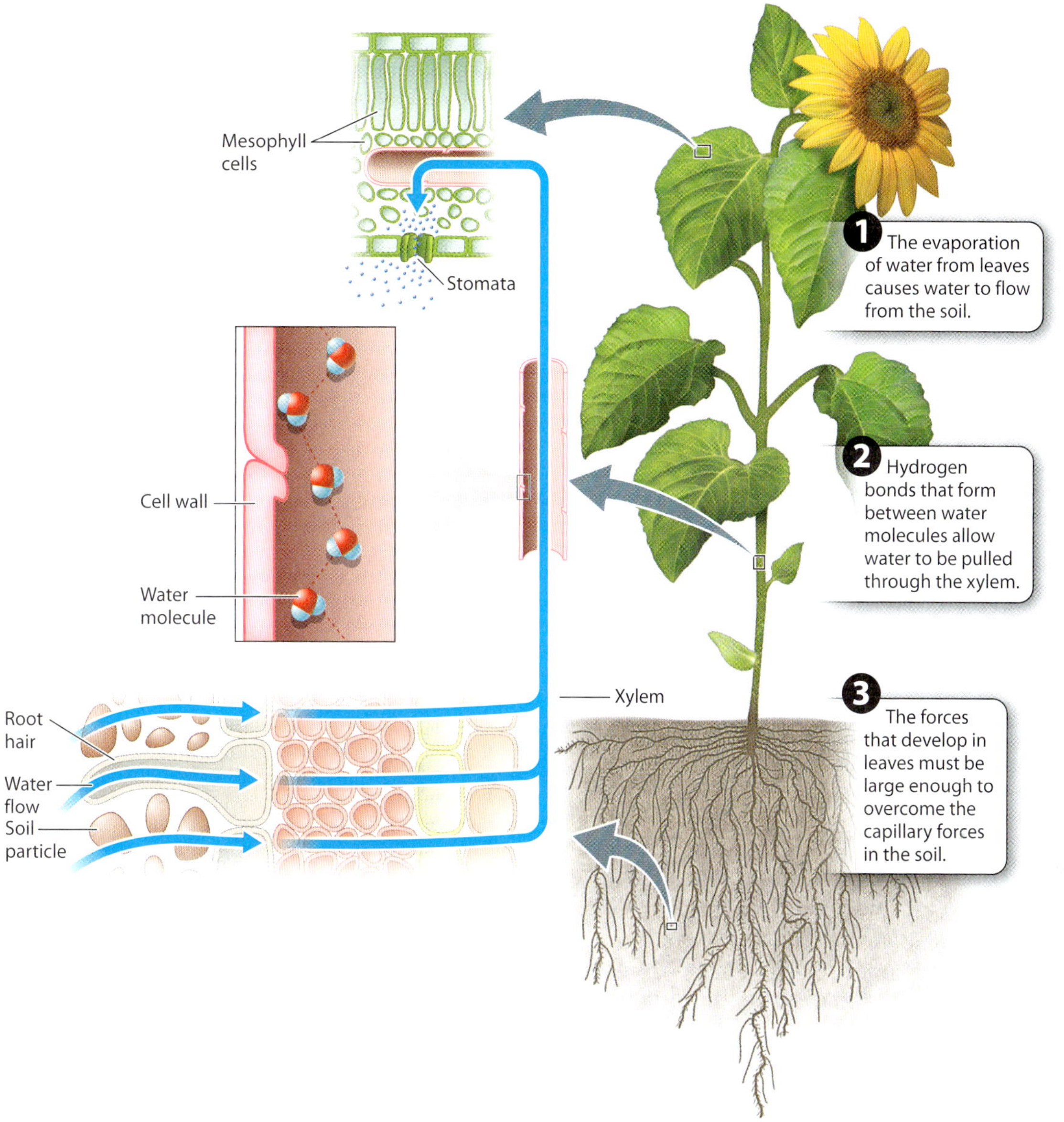

1 The evaporation of water from leaves causes water to flow from the soil.

2 Hydrogen bonds that form between water molecules allow water to be pulled through the xylem.

3 The forces that develop in leaves must be large enough to overcome the capillary forces in the soil.

When stomata are open, water evaporates from the walls of cells lining the intercellular air spaces of leaves. The partial dehydration of the cell walls creates a force that pulls water towards the sites of evaporation. One hypothesis for how this force is generated is that water is pulled by capillary action into spaces between the cellulose microfibrils in the cell wall, the same reason that water is drawn into a sponge. A second hypothesis is that the pectin gel in which the cellulose microfibrils are embedded causes water to flow into the partially dehydrated cell walls by osmosis. Osmotic forces can be generated in cell walls because the negatively charged pectin network restricts the diffusion of positively charged ions, much as the plasma membrane maintains a high concentration of solutes in the cytoplasm.

Once generated by the evaporation of water from leaves, this force is transmitted through the xylem, beginning in the leaf veins, then down through the stem, and out through the roots to the soil (**Fig. 29.13**). Water can be pulled through xylem because of the strong hydrogen bonds that form between water molecules. This mechanism of water transport only works, however, if there is a continuous column of water in the xylem that extends from the roots to the leaves.

The structure of xylem conduits reduces the risks of collapse and cavitation.

The fact that water is pulled from the soil means that xylem conduits must be structured in such a way as to minimize two distinct risks (**Fig. 29.14**). The first is the danger of collapse (Fig. 29.14a). If you suck too hard on a drinking straw, it will collapse inward, blocking flow. Much the same thing can happen in xylem. Although in metabolic terms, lignin is more costly to produce than cellulose, lignin makes conduit walls rigid, reducing the risk of collapse.

The second danger associated with pulling water through the xylem is **cavitation**, which occurs when the water in a conduit is abruptly replaced by water vapor. Because cavitation disrupts the continuity of the water column, cavitated conduits can no longer transport water from the soil. Cavitation in xylem results from the presence of microscopic gas bubbles in the water that are sufficiently large that they expand under the tensile, or pulling, force exerted by the leaves. The microscopic bubbles that cause cavitation can be formed in either of two ways. The first is if an air bubble is pulled through a pit because of lower pressure in the water compared to the air (Fig. 29.14b). The larger the tensile forces exerted by leaves, the more likely it is that air will be pulled across a pit. Thus, higher rates of transpiration increase the risk of cavitation.

The second way that gas bubbles can form within xylem is if gases come out of solution during freezing (Fig. 29.14c). Gases are much less soluble in ice than in water, so as a conduit freezes, dissolved gas molecules aggregate into tiny bubbles that can cause cavitation when the conduit thaws. Wide vessels are especially vulnerable to cavitation at freezing temperatures. The susceptibility of wide vessels to cavitation partly explains why boreal (that is, subarctic) forests contain few angiosperms, which commonly have large vessel diameters.

Once cavitation has occurred, the liquid phase will not re-form as long as tension persists within the xylem. Thus, xylem is organized in ways that minimize the likelihood and impact of cavitation. For example, xylem consists of many conduits in parallel, so the loss of any one conduit to cavitation does not result in a major loss of transport capacity. Similarly, as water flows from the soil to the leaves, it passes from one conduit to another, each one of finite length. The likelihood that cavitation will spread is thereby reduced because air can be pulled through pits only when the tensile forces in the xylem are large.

→ **Quick Check 3** How is the statement that water is transported through the xylem without requiring any input in energy by the plant both correct and incorrect?

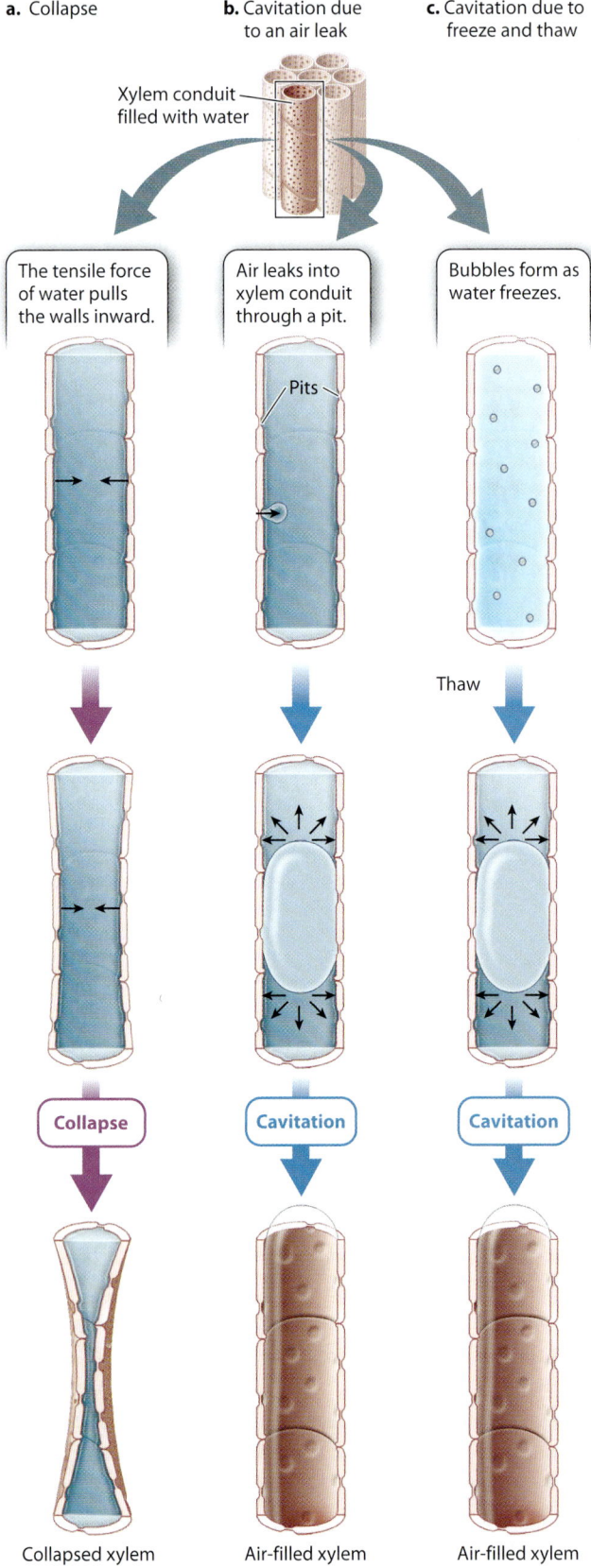

FIG. 29.14 Collapse and cavitation. The forces exerted by transpiring leaves can cause xylem conduits to collapse inward or to cavitate, in both cases blocking flow.

a. Collapse
b. Cavitation due to an air leak
c. Cavitation due to freeze and thaw

29.4 THE STEM: TRANSPORT OF CARBOHYDRATES THROUGH PHLOEM

Much of the body of a vascular plant is devoted to the uptake and transport of the water required by leaves rather than to photosynthesis. In contrast, all cells of the algal ancestors of plants would have been capable of photosynthesis. Roots and stems contribute indirectly to photosynthesis but produce little or no carbohydrate themselves. Thus, although vascular plants are photosynthetic, a large part of their body must be supplied internally with food.

Carbohydrates are pushed through phloem by an osmotic pump.

Phloem transport takes place through multicellular **sieve tubes,** which are composed of highly modified cells called **sieve elements** that are connected end to end (**Fig. 29.15**). During development, sieve elements lose much of their intracellular structure, including the nucleus and the vacuole. At maturity, sieve elements retain an intact plasma membrane that encloses a modified cytoplasm containing only smooth endoplasmic reticulum and a small number of organelles, including mitochondria. Cellular functions such as protein synthesis are carried out by an adjacent **companion cell,** to which the sieve element is connected by numerous plasmodesmata. Sieve elements are linked by **sieve plates,** which are modified end walls with large (1 to 1.5 μm diameter) pores. The plasma membrane of adjacent sieve elements is continuous through each of these pores, so each multicellular sieve tube can be considered a single cytoplasm-filled compartment. **Phloem sap** is the sugar-rich solution that flows through both the lumen of the sieve tubes and the sieve plate pores.

Phloem transports carbohydrates as sucrose (glucose plus fructose) or larger sugars, assembled from monosaccharides in the cytoplasm. Phloem also transports amino acids, inorganic forms of nitrogen, and ions including K+, which are present in much lower concentrations. Finally, phloem transports informational molecules such as hormones, protein signals, and even RNA. Thus, phloem forms a multicellular highway for the movement of raw materials and signaling molecules across the entire length of the plant.

Phloem transports its molecular cargo from **source** to **sink** (**Fig. 29.16**). In plants, sources are the regions that produce

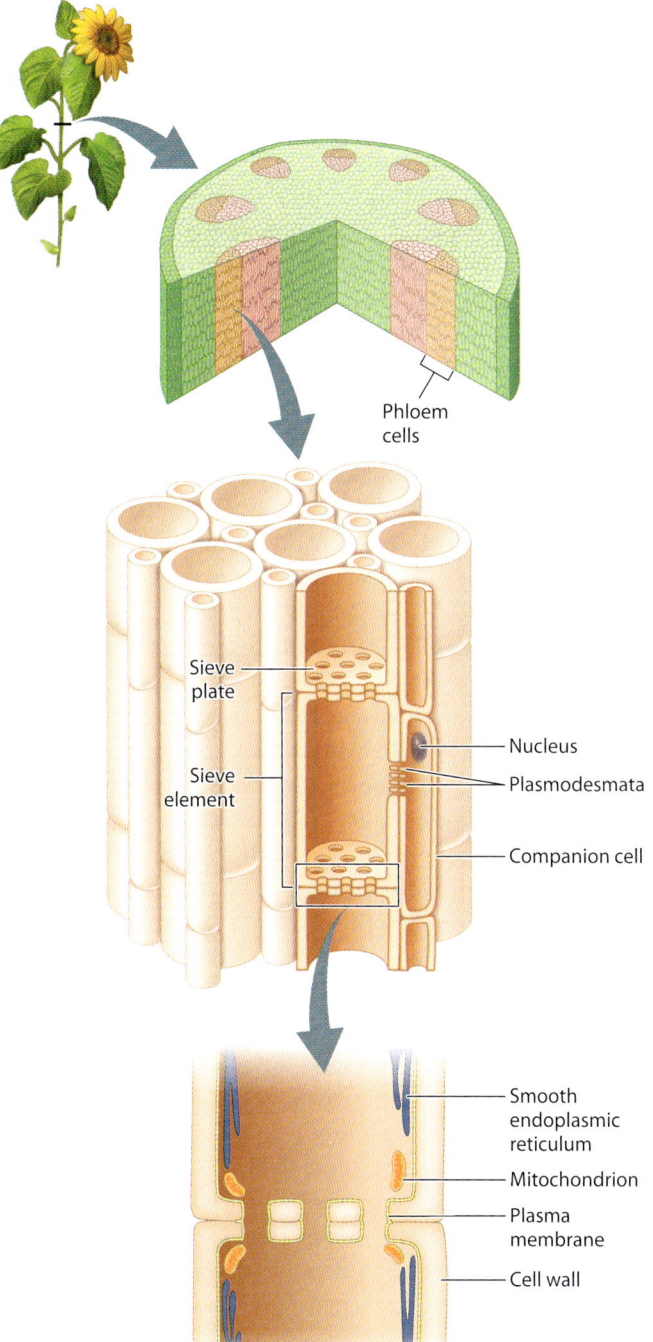

FIG. 29.15 Sieve tubes and associated companion cells.

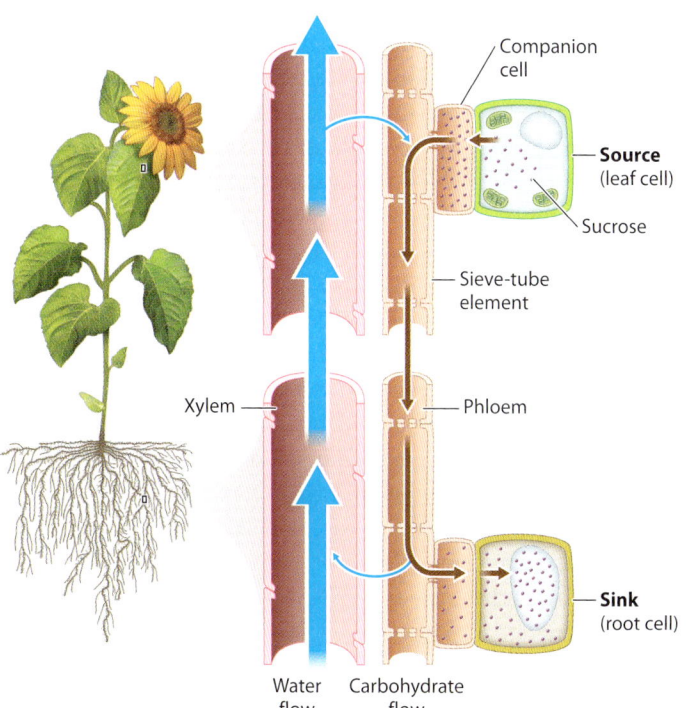

FIG. 29.16 **Phloem transport from source to sink.** Sources produce carbohydrates or supply them from storage, and sinks require carbohydrates for growth and respiration.

or store carbohydrates. For example, leaves are sources because they produce carbohydrates by photosynthesis, and potato tubers (the part we eat) are sources because they can supply stored carbohydrates to the rest of the plant body. Sinks are any portion of the plant that needs carbohydrates to fuel growth and respiration—examples are roots, young leaves, and developing fruits. Whereas the direction of xylem transport is always up toward the leaves, the direction of phloem transport can be either up or down, depending on where the source and sink are relative to each other.

How does phloem transport sugars from source to sink? In some plants, active transporters powered by ATP move sucrose into phloem. The buildup of sugar concentration causes water to be drawn into the phloem by osmosis. Because the cell walls of the sieve tube resist being stretched outward to accommodate the influx of water, they press back (inward) against the cytoplasm. Their resistance to stretching increases the turgor pressure (Chapter 5) at the source end of a sieve tube. At sinks, sugars are transported out of the phloem into surrounding cells. This withdrawal of sugars causes water to leave the sieve tube, again by osmosis, reducing turgor pressure at the sink end. It is the difference in turgor pressure that drives the movement of phloem sap from source to sink.

The water that exits the phloem can be used locally to support the enlargement of sink cells or it can be carried back to the leaves in the xylem. Thus, some of the water in the phloem sap is recirculated in the xylem (Fig. 29.16). The volume of water that moves through the phloem, however, is tiny compared to the amount that must be transported through the xylem to replace water lost by transpiration. Therefore, the number and size of xylem conduits greatly exceeds the number and size of sieve tubes.

Phloem transport is sometimes referred to as "translocation," a term that pairs well with "transpiration." Yet in almost every way, phloem and xylem are a study in contrasts. In phloem, the plant generates the gradient that drives transport, whereas water moving through xylem is driven by the difference in hydration between soil and air. Furthermore, water is pulled upward through xylem conduits, while transport through phloem is more of a push.

These fundamental differences in function explain the cytological differences between xylem and phloem: Phloem conduits retain an intact plasma membrane and modified cytoplasm, whereas xylem conduits retain only their cell walls. What xylem and phloem have in common is that both are essential in enabling vascular plants to carry out photosynthesis on land. Moreover, like xylem, phloem is subject to risks that arise from the way flow through sieve tubes is generated.

Damaged sieve tubes are at risk of having their contents leak out, pushed out by high turgor pressures in the phloem. Damage is an ever-present danger because the sugar content of phloem makes it an attractive target for insects. Cell damage activates sealing mechanisms that repair breaks in sieve tubes. In some respects, these mechanisms are comparable to the formation of blood clots in humans, except that phloem can seal itself much more rapidly, typically in less than a second.

→ **Quick Check 4** How is phloem able to transport carbohydrates from the shoot to the roots, as well as from the roots to the shoot (although not at the same time)?

Phloem feeds both the plant and the rhizosphere.

All the cells in a plant's body contain mitochondria since all cells need a constant supply of ATP. Typically, about 50% of the carbohydrates produced by photosynthesis in one day are converted back to CO_2 by respiration within 24 hours.

Carbohydrates that are not immediately consumed in respiration can be used as raw materials for growth, or they can be stored for later use. Carbohydrates stored within roots and stems, or in tubers (specialized storage organs such as potatoes), can support new growth in the spring or following a period of drought. Stored reserves can also be used to repair mechanical damage or replace leaves consumed by insects or grazing mammals. Like their green algal ancestors, vascular plants store carbohydrates primarily as starch, a large molecule that is not soluble and so does not affect the osmotic balance of cells.

What determines how carbohydrates become distributed within the plant? Where phloem sap ends up is determined by the sinks. Phloem transport to reproductive organs appears to have priority over movement to stems and leaves, and these have priority over movement to roots. In Chapter 31, we discuss the role that hormones play in controlling the growth and development of plants and how these hormones may influence the ability of different sinks to compete successfully for resources delivered by the phloem.

Phloem also supplies carbohydrates to organisms outside the plant. A fraction of the carbohydrates transported to the roots spills out into the **rhizosphere,** the soil layer that surrounds actively growing roots. This supply of carbohydrates stimulates the growth of soil microbes. As a result, the densities of microbial organisms near roots are much greater than in the rest of the soil. These soil bacteria are important decomposers of soil organic matter, which is rich in nutrients such as nitrogen and phosphorus. Thus, the release of carbohydrates by roots into the soil is thought to enhance the ability of roots to acquire nutrients from the soil.

29.5 THE ROOT: UPTAKE OF WATER AND NUTRIENTS FROM THE SOIL

As anyone who has dug a hole in a forest or a field can attest, plants make a lot of roots (**Fig. 29.17**). Laid end to end, the roots of a single corn plant would extend over 600 km. A major reason for such extensive investment belowground is that, with the important exception of CO_2, everything that a vascular plant needs to build and sustain its body enters through its roots. This includes, of course, water, as well as nutrients such as nitrogen, potassium, and phosphorus.

Roots are an evolutionary innovation of the vascular plants; there were no roots when the ancestors of plants first moved onto land. The first roots would have encountered a land surface much less hospitable than today's soils. To a large extent soils have a biological origin. The processes by which roots obtain nutrients from the soil accelerate rates of chemical weathering, stimulate the growth of microbes, and slow erosion, all of which contribute to the development of soils.

Nutrient uptake by roots is highly selective.

Xylem moves both water and nutrients (for example, nitrogen and phosphorus) from the root to the rest of the plant. Xylem, however, is never in direct contact with the soil. The living cells that surround the xylem allow roots to exert a high degree of selectivity over the movement of nutrients from the soil. By taking up only the nutrients they need, roots can concentrate certain nutrients; they can also prevent the entry of substances that could cause damage.

FIG. 29.17 Roots and root hairs. Extensive branching of roots (a) and the presence of root hairs (b) create a large surface area in contact with the soil.

FIG. 29.18 Water movement through endodermal cells. As water carrying nutrients flows from the soil to the xylem of roots, it must pass through the plasma membranes of endodermal cells, which act as a selective barrier.

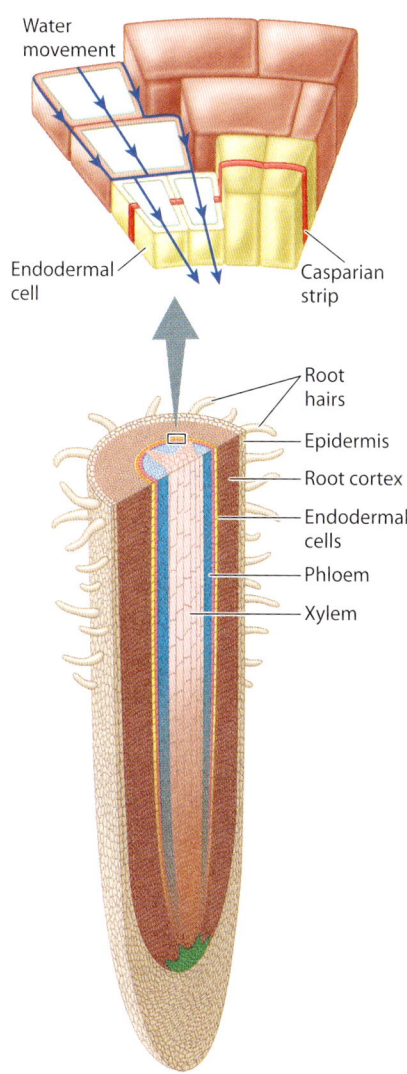

Like leaves and stems, roots have an outer cell layer of **epidermis** (**Fig. 29.18**). Epidermal cells in active areas of the root produce slender outgrowths known as **root hairs.** Root hairs require little material to build, but they greatly increase the surface area of the root available for absorbing nutrients. Inside the epidermis is the **cortex,** composed of parenchyma cells. Depending on species, the cortex can vary in thickness from a single cell to many cells. Xylem and phloem are located at the center of the root and are surrounded by the **endodermis,** a layer of cells that acts as a gatekeeper controlling the movement of nutrients into the xylem.

Water flowing across the epidermis and the root cortex can travel through the cells, crossing plasma membranes, and can also flow through the relatively porous walls that surround each cell. At the endodermis, flow through the cell wall is blocked by the **Casparian strip,** a thin band of hydrophobic material that encircles each cell. Water can gain access to the xylem only by passing through the cytoplasm of the endodermal cells.

The significance of the Casparian strip is that it allows the root to control which materials enter the xylem. How does it do that? In order for solutes to pass through the plasma membrane into the cytoplasm, they must be transported by ATP-powered proteins in the membranes of endodermal cells. These transporters move only certain compounds through the endodermis and actively exclude others. Thus, the solute composition of xylem sap—the fluid moving through the xylem—does not mirror the solute composition of water in the soil. By using the energy of ATP to control the movement of materials across the plasma membrane of endodermal cells, roots can concentrate nutrients that are needed in large amounts (for example, nitrogen) and exclude others that can damage plant cells (for example, sodium).

Nutrient uptake requires energy.

Roots expend a great deal of energy acquiring nutrients from the soil. We have already seen that energy is required to move nutrients across the cell membranes of the endodermis on their journey to the xylem. Energy is also required to make specific nutrients in the soil accessible to the root. And still more energy is required to support symbioses with soil microorganisms that take up nutrients and transfer them to plants. The mitochondria of root cells must, therefore, respire at high rates to provide energy needed for nutrient uptake. Their high respiration rates explain why plants are harmed by flooding: Waterlogged soils slow the diffusion of oxygen, limiting respiration.

The most abundant elements in the plant body—carbon, hydrogen, and oxygen—are obtained primarily from CO_2 and H_2O. All the other elements in the plant's body come from the soil. To understand how roots acquire nutrients, let's first look at how nutrients move within the soil.

Nutrients move through soil by diffusing through films of water that surround individual soil particles. In addition, when transpiration rates are high, the flow of water through the soil helps to convey nutrients toward the root. The uptake of nutrients by roots decreases their concentration at the root surface, creating a region of lower concentration toward which nutrients diffuse from the soil. However, because diffusion transports compounds efficiently only over short distances, each root is able to acquire nutrients only from a small volume of soil. This limitation explains why plants produce so many roots and so many root hairs: Both structures increase the volume of soil from which a root can obtain nutrients. It also explains why roots elongate more or less continuously, exploiting new regions of soil to obtain nutrients required for growth.

Nutrients vary dramatically in the ease and speed with which they move through the soil. For example, nitrate (NO_3^-) and sulfate (SO_4^-) are extremely mobile. In contrast, interactions with clay minerals keep zinc and inorganic phosphate (P_i) attached to mineral surfaces. One way roots gain access to highly immobile nutrients is by releasing protons or compounds that make the environment more acidic. Nutrients stuck to soil particles dissolve as water becomes acidic and diffuse to the root.

Soil microorganisms—both fungi and bacteria—are more adept than plants at obtaining nutrients. Fungi (Chapter 34) are particularly good at mobilizing phosphorus, and some bacteria can

FIG. 29.19 Mycorrhizae. These fungus–root associations allow plants to use carbohydrates to help gain nutrients from the soil.

Ectomycorrhizae

- Epidermis
- Root cortex
- Endoderm
- Phloem
- Xylem
- Fungal sheath

Fungal strands extending from the root

Fungal cells surround but do not penetrate root cells. Carbon and nutrients are exchanged through the plasma membrane.

Endomycorrhizae

- Fungal cells
- Arbuscule

Fungal cells penetrate inside root cells, enhancing carbon and nutrient exchange.

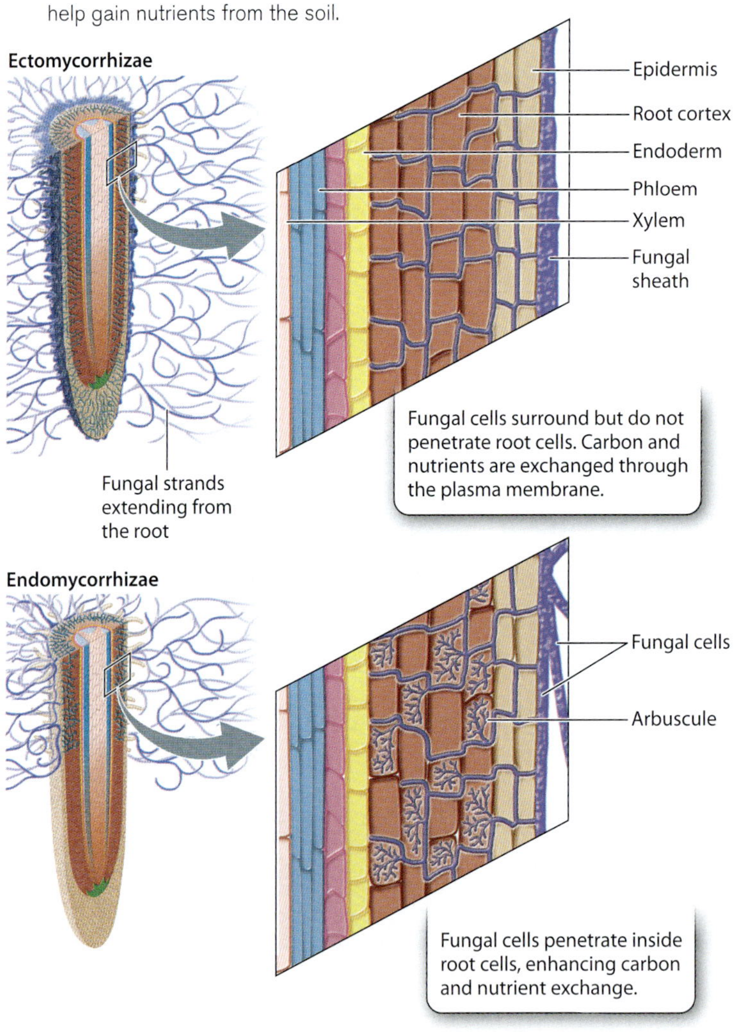

convert gaseous nitrogen (N_2) into a chemical form that can be incorporated into proteins and nucleic acids (Chapter 26). What plants have at their disposal is a renewable source of carbohydrates. Using these carbohydrates, plants feed symbiotic fungi and bacteria, in effect exchanging carbon for phosphorus or nitrogen.

Mycorrhizae enhance the uptake of phosphorus.

Plants are able to access soil nutrients that their roots can't reach by forming partnerships with fungi. **Mycorrhizae** are symbioses between roots and fungi that enhance nutrient uptake. Mycorrhizae are particularly important in enhancing the uptake of phosphorus, one of the most immobile nutrients in soil. The fungi absorb phosphate ions and transport some of them to root tissues; the fungi receive carbohydrates produced in leaves and transported to roots by the phloem.

Why are fungi so effective at obtaining phosphorus from the soil? Roots are thin and root hairs thinner, but the cells produced by fungi are thinner still. Elongate fungal cells can extend more than 10 cm from the root surface, greatly increasing the absorptive surface area in contact with the soil by acting more or less like "super root hairs." Mycorrhizae aid in nutrient uptake because solutes can be transported through the fungal cytoplasm much faster than they can diffuse through the soil. Furthermore, because fungal cells are so thin, they can penetrate small spaces between soil particles and thus access nutrients that would otherwise be unavailable. Finally, mycorrhizal fungi secrete enzymes that release phosphorus from soil organic matter and enhance the mobility of inorganic phosphorus in the soil.

Mycorrhizae are of two main types (**Fig. 29.19**). **Ectomycorrhizae** produce a thick sheath of fungal cells that surrounds the root tip. Some of these cells penetrate the walls of root cells, producing a dense network that surrounds individual root cells. The root and fungal cells are then able to exchange carbon and nutrients through their plasma membranes. **Endomycorrhizae** do not form structures that are visible on the outside of the root. Instead, they form highly branched structures, called arbuscules, that protrude into the interior of root cells. Despite this invasion, the plasma membranes of both cell types remain intact. The structure of the arbuscules increases the contact area between the two cell types, enhancing carbon and nutrient exchange.

Mycorrhizal associations are costly to the plant, consuming between 4% and 20% of its total carbohydrate production. When phosphorus is readily available from the soil, the percentage of mycorrhizal roots decreases, indicating some level of control by the plant. Nevertheless, mutualistic associations between roots and fungi occur in more than 85% of plant species. Mycorrhizal roots are thus the norm rather than an exception. Fossils from more than 400 million years ago suggest that interactions between plants and fungi may be ancient and perhaps helped plants gain a foothold on land.

Symbiotic nitrogen-fixing bacteria supply nitrogen to both plants and ecosystems.

Nitrogen is the nutrient that plants must acquire in greatest abundance from the soil, and in many ecosystems nitrogen is the nutrient that most limits growth. Nitrogen itself is not rare; 79% of Earth's atmosphere is nitrogen gas (N_2). However, as discussed in Chapter 26, plants and other eukaryotic organisms cannot use N_2 directly. Only certain bacteria and archaeons have the metabolic ability to convert N_2 into a biologically useful compound, a process known as nitrogen fixation. Nitrogen-fixing bacteria transform N_2 into forms such as ammonia (NH_3) that can be used to build proteins, nucleic acids, and other compounds. Some of these usable forms of nitrogen are released into the soil, making them available for use by plants and other organisms.

Although many nitrogen-fixing bacteria live freely in the soil, some live in plant roots. When the level of available soil nitrogen is low, some plants form symbiotic relationships with nitrogen-fixing bacteria (**Fig. 29.20**). These interactions are highly species

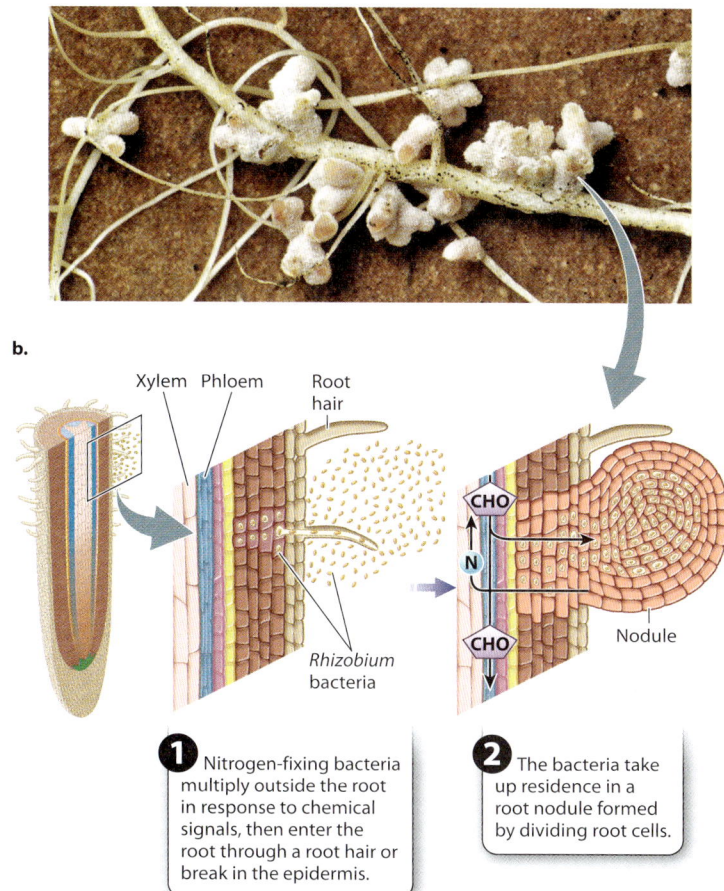

FIG. 29.20 Symbiosis between roots and nitrogen-fixing bacteria. (a) Symbiotic root nodules on alfalfa; (b) the formation of a root nodule.

a.

b. Xylem Phloem Root hair

Rhizobium bacteria

CHO N CHO Nodule

① Nitrogen-fixing bacteria multiply outside the root in response to chemical signals, then enter the root through a root hair or break in the epidermis.

② The bacteria take up residence in a root nodule formed by dividing root cells.

specific. In response to chemical signals transmitted between the root and bacteria, bacteria multiply within the rhizosphere. At the same time, root cells begin dividing, leading to the formation of a **root nodule.** Nitrogen-fixing bacteria enter the nodule either through root hairs or through breaks in the epidermis. Even though the bacteria appear to have taken up residence within the cytoplasm of nodule cells, in fact each one is contained within a membrane-bound vesicle produced by the host plant cell.

The plant supplies its nitrogen-fixing partners with carbohydrates transported by the phloem and carries away the products of nitrogen fixation in the xylem. As in mycorrhizal symbioses, the carbon cost to the plant is high: It is estimated that as much as 25% of these plants' total carbohydrate production goes to nitrogen-fixing bacteria.

Relatively few plant species have evolved symbiotic associations with nitrogen-fixing bacteria, but those that do—especially members of the legume (bean) family—provide an important way in which biologically available nitrogen enters into ecosystems. In addition, nitrogen fixation in legume roots helps to sustain fertility in many agricultural systems.

? CASE 6 Agriculture: Feeding a Growing Population
How has nitrogen availability influenced agricultural productivity?

Harvesting crops removes nutrients—including nitrogen—from an ecosystem. These nutrients must be returned if the land is to to be agriculturally productive. In the early days of agriculture, the practice of shifting cultivation solved the problem by leaving fields unplanted for long periods between crops. During these periods, nitrogen-fixing bacteria replenished soil nitrogen, either in symbiotic association with roots or as free-living bacteria in the soil.

As populations grew, fields could not be left unplanted for long periods. Nutrients removed during harvest had to be replaced or the fertility of fields would decline. Organic fertilizers such as crop residues, manure, and even human feces provide a way to recycle other nutrients removed during harvest, but unless they are brought in from off the farm, they add little or no new nitrogen to the system.

Crop rotation has been widely practiced throughout the history of agriculture to resupply the soil with nitrogen. Every few years, the field is planted with legumes harboring nitrogen-fixing bacteria. Typically, the legume rotation is not harvested, but instead is treated as "green manure" that is allowed to decay and replenish soil nitrogen. During the eighteenth and nineteenth centuries, legume rotation contributed to the higher crop yields needed to support increasing populations. Therefore, along with the mechanization of agriculture, nitrogen-fixing symbioses contributed to the social conditions that fueled the Industrial Revolution.

By the second half of the nineteenth century, more abundant and more concentrated supplies of nitrogen fertilizers were needed to promote crop growth to feed growing populations. Mining of guano and sodium nitrate mineral deposits, both found in arid regions such as the northern deserts of Chile, helped supply the demand for the nitrogen needed to increase crop production. With time, however, the reserves of these natural nitrogen sources declined and a mechanism was needed to obtain nitrogen fertilizer from the vast storehouse of nitrogen in the atmosphere.

An industrial method of fixing nitrogen was developed by German chemist Fritz Haber in the years 1908–1911 and was subsequently adapted for industrial production by Karl Bosch. Their method allows humans to achieve what, until then, had been an entirely prokaryotic activity. Like nitrogen-fixing bacteria, the Haber–Bosch process requires large inputs of energy, but unlike bacteria, it occurs at high temperatures and pressures. The prospect of an essentially unlimited supply of nitrogen in a form that plants could make use of has proved irresistible. Industrial fertilizers produced by the Haber–Bosch process now account for more than 99% of fertilizer use. Industrial fixation of nitrogen today is approximately equal to the entire rate of natural fixation of nitrogen, which is largely due to bacteria, although a small amount of nitrogen is fixed by lightning.

As we discuss in Chapter 48, human use of fertilizer is altering the nitrogen cycle on a massive scale, posing a threat to biodiversity and the stability of natural ecosystems. Yet abundant nitrogen fertilizers are a cornerstone of the Green Revolution. The high-yielding varieties of wheat and rice introduced in the mid-twentieth century achieve their high levels of growth and grain production only when abundantly supplied with nitrogen. The fact that plants such as legumes (including beans, soybeans, alfalfa, and clover) form symbiotic relationships with nitrogen-fixing bacteria suggests a possible way around this problem. If it were possible to engineer nitrogen fixation into crops such as wheat and rice, either by inducing them to form symbiotic relationships with nitrogen-fixing bacteria or by transferring the genes that would enable them to fix nitrogen on their own, would that eliminate the need for nitrogen fertilizers? The answer is yes—but with a caveat. As we have seen, nitrogen-fixation requires a large amount of energy, diverting resources that could otherwise be used to support growth and reproduction.

Core Concepts Summary

29.1 THE EVOLUTION OF LAND PLANTS FROM AQUATIC ANCESTORS INTRODUCED A MAJOR CHALLENGE FOR PHOTOSYNTHESIS: ACQUIRING CO_2 WITHOUT LOSING EXCESSIVE AMOUNTS OF WATER.

Early branching land plant lineages, commonly grouped as bryophytes, balance CO_2 gain and water loss passively, photosynthesizing in wet conditions and tolerating desiccation in dry ones. page 29-2

Vascular plants actively regulate the hydration of their photosynthetic cells. page 29-5

29.2 LEAVES HAVE A WAXY CUTICLE THAT RETARDS WATER LOSS BUT INHIBITS THE DIFFUSION OF CO_2, AND PORES, CALLED STOMATA, THAT REGULATE CO_2 GAIN AND WATER LOSS.

The low concentration of CO_2 in the atmosphere forces plants to expose their photosynthetic cells directly to the air. The outward diffusion of water vapor leads to a massive loss of water. page 29-3

The waxy cuticle on the outside of the epidermis slows rates of water loss from leaves but also slows the diffusion of CO_2 into leaves. page 29-4

Stomata are pores in the epidermis that open and close, allowing CO_2 to enter into the leaf but also allowing water vapor to diffuse out of the leaf. page 29-5

CAM plants capture CO_2 at night when evaporative rates are low. During the day, they close their stomata and use this stored CO_2 to supply the Calvin cycle during the day, resulting in increases in the exchange ratio of CO_2 and H_2O. page 29-6

C_4 plants suppress photorespiration by concentrating CO_2 in bundle-sheath cells. The buildup of CO_2 in bundle-sheath cells results from the rapid production of C_4 compounds in the mesophyll cells, which then diffuse into the bundle-sheath cells and release CO_2 that can be used in the Calvin cycle. page 29-7

29.3 XYLEM ALLOWS VASCULAR PLANTS TO REPLACE WATER EVAPORATED FROM LEAVES WITH WATER PULLED FROM THE SOIL.

Water flows through xylem conduits from the soil to the leaves. page 29-9

Xylem is formed from cells that lose all their cell contents as they mature. page 29-9

Xylem conduits have thick, lignified cell walls. Water flows into xylem conduits across small thin-walled regions called pits. page 29-9

Xylem conduits are of two types: tracheids and vessels. Tracheids are unicellular xylem conduits; vessels are formed from many cells. page 29-9

The driving force for water movement in xylem is generated by the evaporation of water from leaves. Because of the strong hydrogen bonds between water molecules, water can be pulled from the soil and transported through xylem. page 29-11

Pulling water through xylem creates the risk of mechanical failure, either by the collapse inward of conduit walls or by cavitation in which a gas bubble expands to fill the entire conduit. page 29-12

29.4 PHLOEM TRANSPORTS CARBOHYDRATES AND SUPPORTS THE GROWTH AND RESPIRATION OF NON-PHOTOSYNTHETIC ORGANS SUCH AS STEMS AND ROOTS.

Phloem transports carbohydrates to the non-photosynthetic portions of the plant for use in growth and respiration. page 29-13

Phloem transport occurs through sieve tubes, which are formed from elongate cells that are connected end to end by sieve plates. Sieve plates contain large pores that allow phloem sap to flow from one sieve tube cell to another. page 29-13

The active loading of solutes brings water into sieve tubes by osmosis, increasing the turgor pressure. At sites of use, the removal of solutes leads to an outflow of water and a drop in turgor pressure. page 29-14

The difference in turgor pressure drives the movement of phloem sap from source to sink. page 29-14

Carbohydrates exuded from roots provide carbon and energy sources for the rhizosphere microbial community. page 29-14

29.5 ROOTS EXPEND ENERGY TO OBTAIN NUTRIENTS FROM THE SOIL.

The endodermis allows roots to control which materials enter the xylem. The uptake of solutes by roots is therefore highly selective. page 29-16

Symbiotic associations with fungi allow plants to exchange carbohydrates for assistance in obtaining phosphorus from the soil. page 29-17

Plants supply carbohydrates to symbiotic nitrogen-fixing bacteria in exchange for nitrogen. page 29-17

Nitrogen fixation results in a net increase in ecosystem nitrogen. page 29-18

Agricultural productivity is closely linked to nitrogen supply. page 29-18

Self-Assessment

1. Explain why vascular plants are better able to sustain photosynthesis than are bryophytes.

2. Explain why transpiration is best explained as a consequence of acquiring CO_2 from the atmosphere, as opposed to temperature regulation, nutrient transport, or utilization of water as a substrate in photosynthesis.

3. Draw and label the features of a leaf that are necessary to maintain a well-hydrated interior while still allowing CO_2 uptake from the atmosphere.

4. Diagram how the movement of solutes results in the opening and closing of stomata.

5. Diagram how the enzyme PEP carboxylase allows CAM plants to reduce transpiration rates and C_4 plants to suppress photorespiration.

6. Contrast how transport is generated in xylem and in phloem.

7. Describe the structure of xylem conduits and sieve tubes in relation to their function.

8. Explain why phloem is necessary to produce roots.

9. List four adaptations of roots that enhance the uptake of nutrients from the soil.

Do you understand the chapter's Core Concepts? Log in to check your answers to the Self-Assessment questions, then practice what you've learned and reinforce this chapter's concepts by working through the problems and multimedia tutorials provided there.

www.biologyhowlifeworks.com

CHAPTER 30

PLANT REPRODUCTION

Finding Mates and Dispersing Young

As plants moved onto land, the challenges of carrying out photosynthesis in air were matched by the difficulties of completing their life cycle. The algal ancestors of land plants relied on water currents to carry sperm to egg and disperse their offspring. On land, the first plants were confronted with the challenge of moving gametes and offspring through air. Air is less buoyant than water, provides a poor buffer against changes in temperature and ultraviolet radiation, and increases the risk of drying out. As plants diversified on land, structures evolved that allow both gametes and offspring to survive being transported through air. For these structures to be useful, plant life cycles had to undergo radical change. The modification of land plant life cycles is a major theme in plant evolution.

The gametes and offspring of many land plants are carried passively by the movement of the air (or in a few cases, water—think coconuts). However, some plants evolved the capacity to harness animals as transport agents. In particular, the flowers and fruits of angiosperms influence animal behavior through their colors, scents, and food resources. Animals attracted by the food provided by flowers (or in a few cases, by the false promise of a mate) transport male gametes from one flower to another. Animals that eat fruits or inadvertently carry them attached to their fur are key agents of seed dispersal. As plants diversified on land, coevolution with animals emerged as a second major theme in plant evolution.

Core Concepts

30.1 The plant life cycle evolved in ways that enhance the ability to unite gametes and disperse offspring on land.

30.2 Angiosperms (flowering plants) attract and reward animal pollinators, and they provide resources for seeds only after fertilization.

30.3 Plants have sensory systems that control the timing of flowering and seed germination.

30.4 Many plants also reproduce asexually.

30.1 THE PLANT LIFE CYCLE AND EVOLUTION OF POLLEN AND SEEDS

Plants did not evolve the ability to reproduce on land all at once. Early diverging lineages evolved the capacity to disperse their offspring through the air, but they still required water for fertilization. How could these plants exist in both water and air? They couldn't—at least not at the same time. Instead, early land plants evolved a life cycle in which one generation, or phase of the life cycle, released sperm into a moist environment and the following generation dispersed offspring through the air. It was only with the evolution of pollen and seeds that plants no longer depended on external sources of water for fertilization. Yet the pattern of alternating generations remains a key feature of the plant life cycle.

30-1

FIG. 30.1 The phylogeny of land plants, showing the evolutionary trajectory of the environmental setting for fertilization and dispersal.

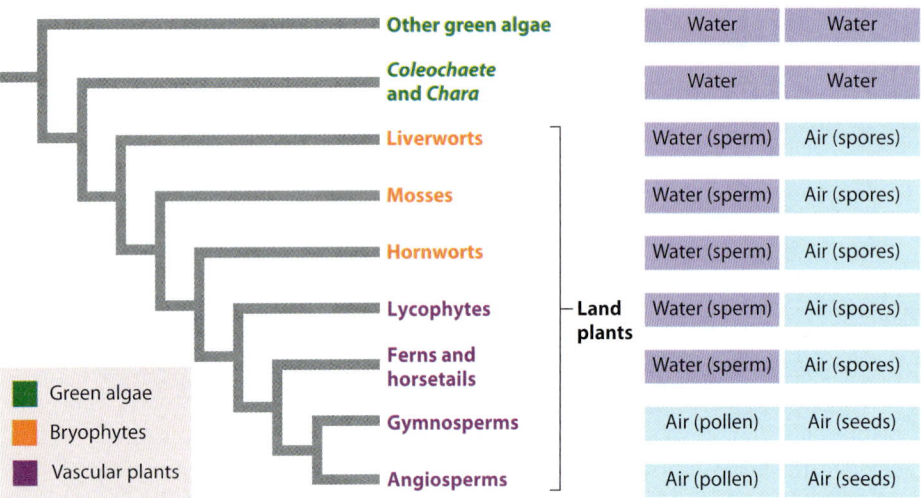

The algal sister groups of land plants have one multicellular generation in their life cycle.

Molecular sequence comparisons show that two groups of green algae are most closely related to land plants (**Fig. 30.1**): *Coleochaete*, an inconspicuous green pincushion that grows on aquatic plants or rocks in freshwater environments, and the erect stonewort *Chara* (and closely related genera) found along the margins of lakes and estuaries (**Fig. 30.2**). Like all sexually reproducing eukaryotes (Chapter 27), these algae alternate between a diploid ($2n$) phase and a haploid ($1n$) phase. The diploid phase is generated by the fusion of haploid gametes to form a zygote, and the haploid phase is generated from diploid cells by meiosis.

In animals, the multicellular body consists of diploid cells. *Coleochaete* and *Chara* have a multicellular body as well, shown in Fig. 30.2, but it consists entirely of haploid cells. Specialized cells of the haploid body produce haploid eggs and sperm by mitosis within multicellular reproductive organs. The egg is retained within the female reproductive organ and the sperm are released into the water. Fertilization results when one egg and one sperm fuse to form a diploid zygote. The zygote then undergoes meiosis to produce haploid cells that give rise to a new multicellular haploid generation (**Fig. 30.3**). In *Coleochaete*, the haploid products of meiosis disperse by swimming through the water, propelled by flagella. In *Chara*, the zygotes disperse, carried by water currents until they settle and undergo meiosis.

In *Chara* and *Coleochaete*, both fertilization and dispersal take place in water. Thus, these algae are as dependent on water as fish are. Living along the margins of lakes and streams, however, *Chara* and *Coleochaete* are occasionally left high and dry by drought. Their photosynthetic bodies shrivel quickly, but these algae survive because their zygotes form a protective wall that allows the cell inside to tolerate exposure to air. When water once again covers the resting zygotes, they undergo meiosis and the resulting haploid cells escape from their protective coat. This is the condition from which the more complicated life cycles of land plants evolved.

Bryophytes illustrate how the alternation of generations allows the dispersal of spores in the air.

As plants moved onto land, their life cycle pulled them in two directions. Fertilization requires that plants remain near the ground, where water is most likely to be found, but dispersal in air requires that they grow tall, away from that water. In Chapter 29, we saw that the bryophytes—mosses, hornworts, and liverworts—make up the earliest branches on the phylogenetic tree of land plants (see Fig. 30.1). Mosses, hornworts, and liverworts share many reproductive traits. Here, we use *Polytrichum commune*, a moss common in forests and the edges of fields, to illustrate how movement onto land was accompanied by the evolution of two multicellular generations: one specialized for fertilization and the other for dispersal.

FIG. 30.2 *Coleochaete* (left) and *Chara* (right), the two green algae most closely related to land plants.

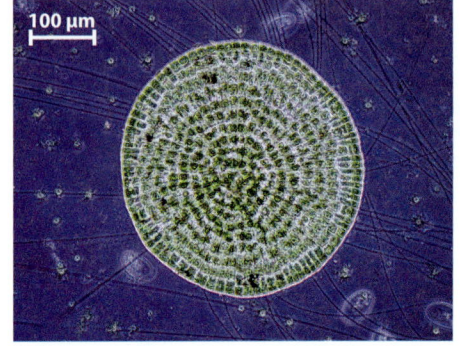 

FIG. 30.3 The life cycle of the green alga *Chara corallina*. The only diploid cell is the zygote.

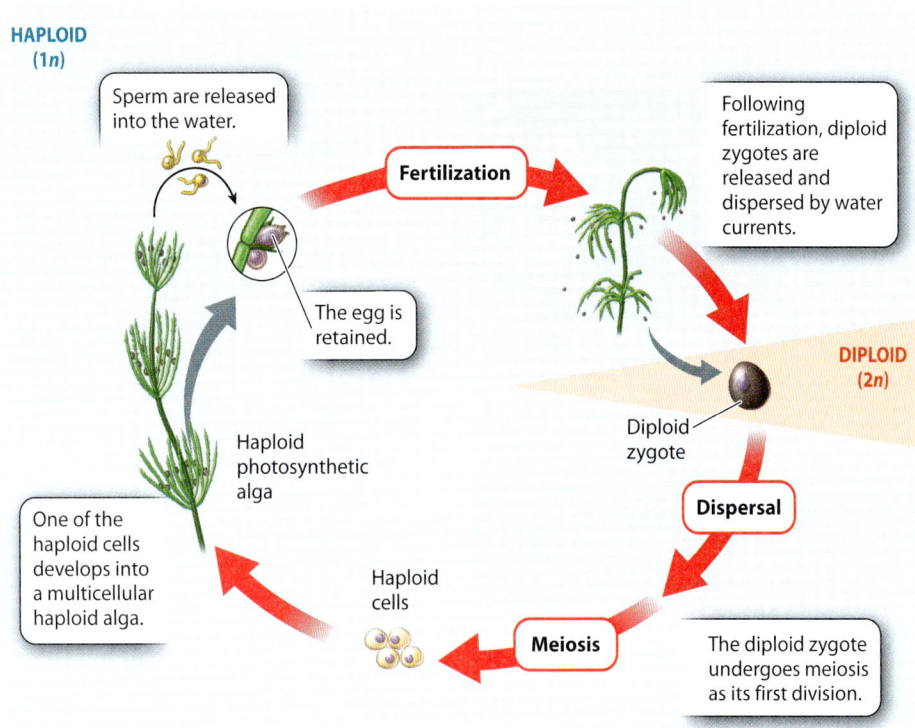

Like *Chara* and *Coleochaete*, *Polytrichum* has a photosynthetic body—the familiar green tufts observed on the forest floor—that is made of haploid cells and that forms gametes by mitotic division of specialized cells. Sperm travel to eggs retained within reproductive organs, and the fusion of egg and sperm gives rise to a diploid zygote.

At this point, the life cycles of the green algae and *Polytrichum* diverge. In the moss, the zygote does not undergo meiosis and it does not disperse. Instead, the zygote is retained within the female reproductive organ, where it divides repeatedly by mitosis to produce a new multicellular generation—this one made of diploid cells (**Fig. 30.4**). Some cells within this diploid body eventually undergo meiosis, giving rise to haploid **spores,** cells that disperse and give rise to a new haploid generation. Because the diploid multicellular generation gives rise to spores, it is called the **sporophyte**, and because the haploid multicellular generation gives rise to gametes, it is called the **gametophyte**. The resulting life cycle, in which a haploid gametophyte and a diploid sporophyte follow one after the other, is called **alternation of generations,** and it describes the basic life cycle of all land plants.

Because land plants are descended from green algae with only one haploid generation, we can infer that the diploid sporophyte of land plants is an evolutionary novelty on this branch of the phylogenetic tree. How might this new generation have made it possible for plants to colonize land?

To answer this question, we need to identify the sporophyte of *Polytrichum*. Late in the growing season, the green tufts of this moss sprout an extension—a small brown capsule at the end of a cylindrical stalk

FIG. 30.4 The life cycle of the moss *Polytrichum commune*. The multicellular sporophyte generation is a major innovation of land plants.

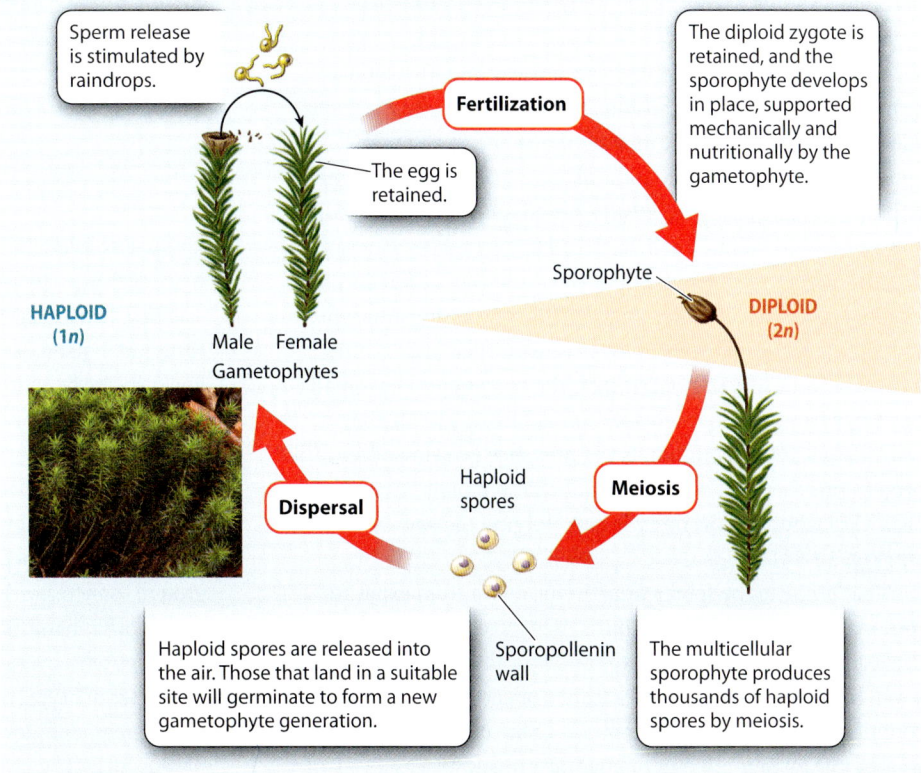

FIG. 30.5 Moss sporangia. (a) When spores of *Polytrichum commune* are mature, a cap falls off, allowing the salt shaker–like sporangium to release the spores. (b) The sporangia of *Ptychomitrium polyphyllum* have a ring of teethlike projections that curl back as they dry, flinging spores into the air.

a few centimeters high (**Fig. 30.5a**). This is the sporophyte, which originated from a fertilized egg. Because the fertilized egg was retained within the female reproductive organ, the sporophyte grows directly out of the gametophyte's body. Thus, the *Polytrichum* specimen pictured in Fig. 30.5a shows both the gametophyte (green) and the sporophyte (brown). The gametophyte is photosynthetically self-sufficient. In contrast, the sporophyte obtains water and nutrients needed for its growth from the gametophyte.

The multicellular sporophyte enhances the ability of plants to disperse on land. The capsule at the top of the *Polytrichum* sporophyte is a **sporangium**, a structure in which many thousands of cells undergo meiosis, producing large numbers of haploid spores. Spores are ideally suited for transport through the air because, being small, they can be carried for thousands of kilometers by the wind. But at the same time, their tiny size puts spores at risk of being washed from the air by raindrops. The sporangia of many bryophytes release their spores only when the air is dry. In *Polytrichum*, the sporangium has small openings from which it releases spores the way a salt shaker releases salt. More common is the presence of a ring of teethlike projections that curl over the top of the sporangia. These bend backward when air humidity is low, flinging the spores into the air (**Fig. 30.5b**).

What protects the spores from drying out or from exposure to damaging ultraviolet radiation as they travel through the air? Earlier, we noted that the zygotes of *Chara* and *Coleochaete* secrete a wall that protects the zygote within from desiccation. The wall contains **sporopollenin,** a complex mixture of polymers that is remarkably resistant to environmental stresses such as ultraviolet radiation and desiccation. In mosses and other land plants, it is the spore, not the zygote, that secretes sporopollenin. Thus, early land plants had a capacity present in their algal ancestors—the ability to synthesize sporopollenin—but the timing of gene expression changed in these plants. Coated in sporopollenin, the spores cannot swim like meiotic products of *Coleochaete*. Instead, they are dispersed by wind to new locations, where they germinate to form a new haploid generation. Thus, the evolution of the multicellular sporophyte generation permitted land plants to make many spores and disperse them across large distances on land.

What about fertilization, the other point in the life cycle where *Chara* and *Coleochaete* require water? Like their algal ancestors, *Polytrichum* and other bryophytes rely on water to transport sperm to egg. Sperm released into the environment swim through thin films of water in soil and on the surface of the egg-bearing gametophyte. In addition, bryophytes tend to be small, so the distance between male and female gametes does not exceed the distance over which the sperm can swim, and they produce their gametes near the ground, where the likelihood of encountering a continuous film of water is greatest. A reliance on external water for fertilization restricts the growth of bryophytes to habitats that are, at least intermittently, wet. Many bryophytes release their sperm only when agitated by raindrops. Raindrops signal the presence of surface water, and their impact can splash sperm farther than they could swim on their own.

→ **Quick Check 1** Give two reasons why spores are better suited for dispersal on land than are sperm.

Vascular plants evolved a large photosynthetic sporophyte generation.

The first two lineages of vascular plants—the lycophytes and the ferns and horsetails—depend on swimming sperm for fertilization and disperse by spores that are released into the air (see Fig. 30.1). In these ways, their life cycle is similar to that of the bryophytes. However, the evolution of vascular systems had a major impact on their life cycle because it allowed them to grow tall. Xylem and phloem occur only in the sporophyte generation. This makes sense because gametophytes must remain small and close to the ground to increase the chances of fertilization. Spore dispersal, on the other hand, is enhanced by height, and spore production increases with overall size. An important difference between the life cycles of bryophytes and the spore-dispersing vascular plants is that in bryophytes it is the gametophyte that is the conspicuous photosynthetic generation, while in vascular plants it is the sporophyte generation that dominates in both physical size and photosynthetic output.

To illustrate how the evolution of vascular plants affected plant reproduction, let us examine the life cycle of the bracken fern *Pteridium aquilinum*, beginning with the largest component, the diploid sporophyte generation (**Fig. 30.6**). Bracken sporophytes appear to consist entirely of leaves because the stem grows underground. If you examine a bracken leaf closely, you may find tiny brown packets along its lower margin. These are sporangia, and each contains diploid cells that undergo meiosis to generate haploid spores. As in mosses, the spores become covered by a thick wall containing sporopollenin.

The structure of the sporangium aids in spore dispersal. The sporangium wall has a distinct ridge of asymmetrically thickened cells that, upon drying, produce a motion like that of a slingshot that hurls the spores away from the leaf surface to be carried off by air currents. Most spores travel only short distances, but a few may end up far from their parent. Bracken ferns can be found on every continent except Antarctica, as well as on isolated oceanic islands, testimony to the ability of spores to travel long distances.

Spores germinate to produce the haploid gametophyte generation. You have probably never seen a fern gametophyte—or never noticed one. Typical of most ferns, the bracken gametophyte is less than 2 cm long and only one to a few cells thick (Fig. 30.6). The small size of the fern gametophyte has led many to assume that the haploid generation is a weak link in the fern life cycle. In fact, because fern gametophytes tolerate drying out, they often live longer and withstand stressful environmental conditions better than the much larger sporophyte generation.

Most fern gametophytes are capable of producing both male and female gametes, although typically they produce only one type. Chemical signals released by gametophytes producing female gametes (eggs) stimulate gametophytes growing nearby to produce male gametes (sperm) that swim to the egg through a film of water. However, if fertilization does not occur, many fern gametophytes eventually produce both male and female gametes. The union of a male and a female gamete forms a diploid zygote, which is supported by the gametophyte as it begins to grow. Eventually, it forms leaves and roots that allow it to become a physiologically independent, diploid sporophyte.

Ferns, as well as the other spore-dispersing plants (lycophytes, horsetails, and bryophytes), release swimming sperm and thus are able to reproduce only when conditions are wet. What if a plant could eliminate that requirement? Seed plants did just that.

→ **Quick Check 2** In what ways is fern reproduction similar to moss reproduction? In what ways is fern reproduction different from moss reproduction?

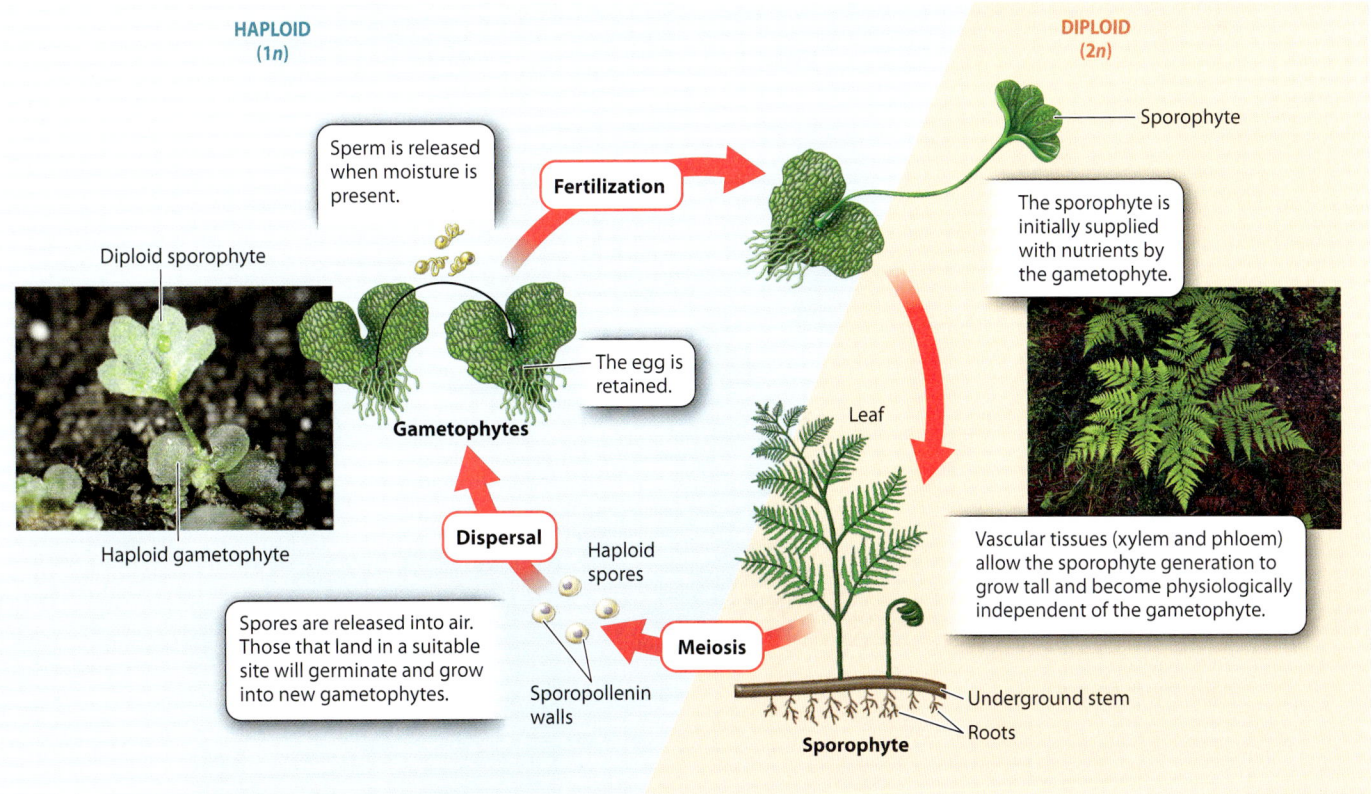

FIG. 30.6 **The life cycle of the bracken fern *Pteridium aquilinum*.** This life cycle illustrates the alternation between a small, free-living gametophyte generation and a taller, vascularized sporophyte generation.

In seed plants, the transport of pollen in air allows fertilization to occur in the absence of external sources of water.

What are the advantages of being able to disperse gametes and offspring? Where environments are variable, outcrossing (sexual fusion with a genetically different member of the population) is advantageous because it generates diverse genotypes. Thus, the greater the distance a sperm travels from its parent, the greater the likelihood that it will encounter a genetically distinct egg.

There are also advantages to dispersing offspring. Imagine what would happen if a plant's spores or seeds simply dropped straight down onto the ground below the parent plant. All the plant's offspring would be attempting to grow in a very small space that might contain enough nutrients and light for only a few to establish themselves. In contrast, the odds are greater that more of the offspring dispersed by wind or other means will settle in a less crowded spot. In addition, dispersal helps offspring avoid pathogens and parasites. Viruses and bacteria travel easily from individual to individual in a densely packed population. An individual that settles far from the parent plant is less likely to encounter a pathogen.

Let's return to the question of how seed plants are able to reproduce—to bring male and female gametes together—when conditions are dry. Standing beneath a pine tree on a spring day, you will see the answer to this question all around you: tiny yellow grains that cover the ground and everything else close to the tree. These are pollen grains, and they are produced by all seed plants. It is the evolution of pollen that liberated seed plants from the need to release swimming sperm into the environment. Pollen allows the sperm-producing gametophytes to be transported through

FIG. 30.7 The life cycle of the loblolly pine tree, *Pinus taeda*.

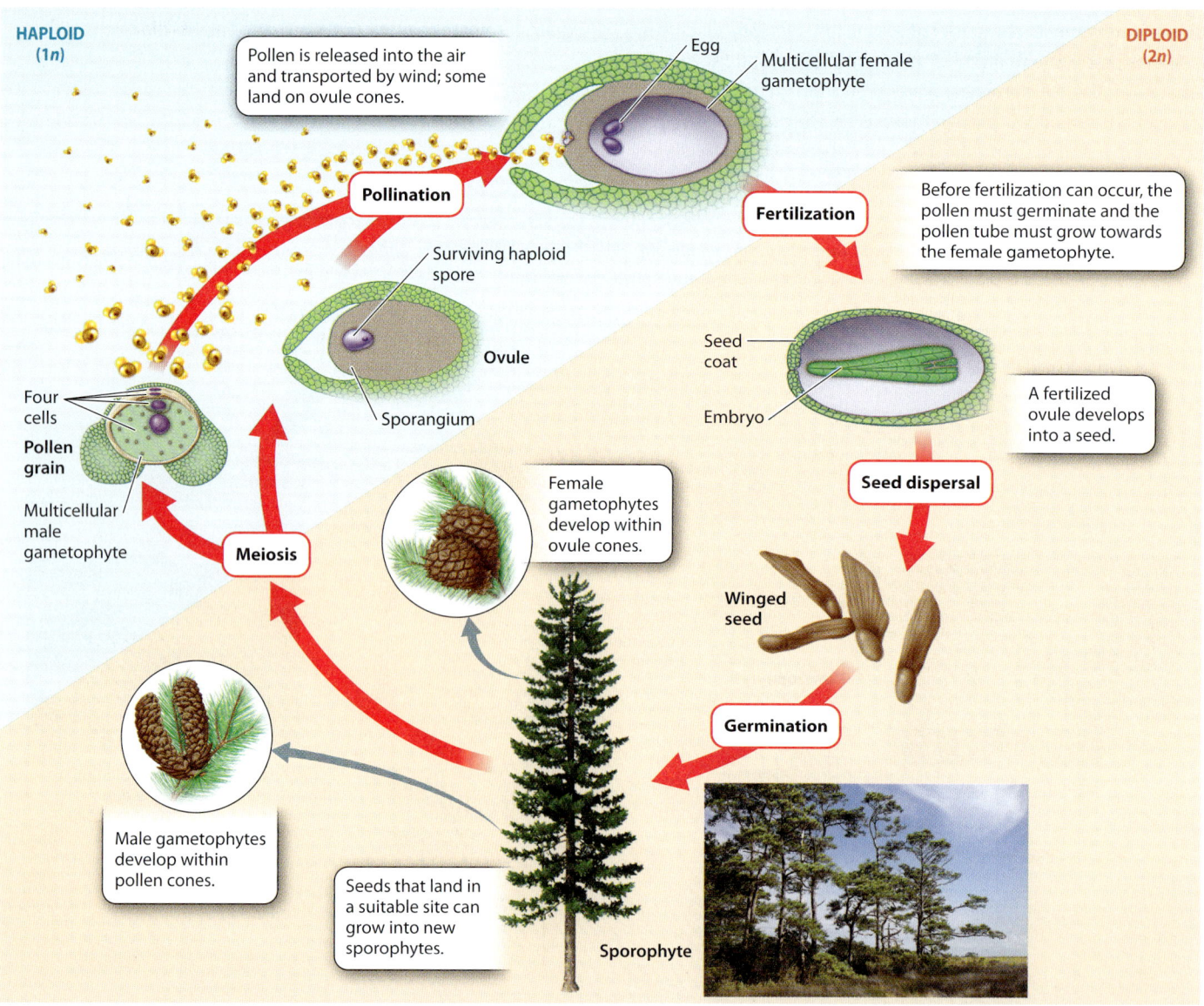

the air and thus allows fertilization to take place in the absence of external pools of water. To understand how this happens, let's look at the life cycle of a pine tree (**Fig. 30.7**).

In pines, as in all vascular plants, the conspicuous multicellular generation is the diploid sporophyte. As they mature, pines form reproductive structures called cones. In fact, pines develop two types of cones: **ovule cones,** which typically occur on upper branches and produce female gametophytes and female gametes, and **pollen cones,** commonly found in clusters near the tips of branches lower in the tree, which produce male gametophytes and male gametes.

If we examine the pollen cones closely, we see that the helically arranged structures that make up the cone are modified leaves that produce sporangia on their surface. This arrangement, again, is much like that of the bracken fern, whose sporophyte develops sporangia on leaf surfaces. As in ferns, the sporangia in the pollen cones contain cells that undergo meiosis, giving rise to haploid spores whose walls contain sporopollenin.

So far, the pine life cycle closely resembles that of the bracken fern. It is what happens next that sets pines—and all seed plants—apart from other vascular plants. In pines, each spore produced within a pollen cone divides mitotically to form a multicellular gametophyte *inside the spore wall.* These are male gametophytes, because they produce only male gametes (sperm). **Pollen** is the multicellular male gametophyte contained within the spore wall.

In pine, the male gametophyte consists of only four cells at the time the pollen is shed from the parent plant. Pine pollen is released into the air and transported by wind. For fertilization to occur, a pollen grain must land on an ovule cone, where the female gametes or eggs are produced.

Ovule cones contain sporangia, each one surrounded by a jacket of protective tissue. In these cones, each sporangium has only one cell that undergoes meiosis and thus produces only four spores. Three of these spores abort spontaneously, leaving a single functional spore inside the sporangium. This spore undergoes repeated mitoses to form a multicellular female gametophyte consisting of a few thousand haploid cells, one or more of which differentiate as eggs. The female gametophyte fills the sporangium, surrounded and protected by the tissues that envelop the sporangium wall. This entire structure is the **ovule.**

Pollination takes place when the pollen is carried by the wind to an ovule. To reach the egg, however, the pollen must germinate and the male gametophyte must produce a **pollen tube** that grows outward through an opening in the sporopollenin coat. In pines, pollination occurs as much as 15 months before fertilization. In fact, it is the arrival of the pollen that stimulates the development of the female gametophyte. When the female gametes are ready to be fertilized, the male gametophyte's pollen tube grows to the female gametophyte, attracted by chemical signals. The sperm travel down the pollen tube and one of them fuses with the egg to form a zygote.

Seed plant reproduction differs from that of spore-dispersing plants in three important ways. The first is that pollination allows fertilization to occur without the male gametes' ever being exposed to the external environment. The second is that the relationship between sporophyte and gametophyte has been reversed completely from that found in bryophytes. The gametophyte, so prominent in mosses, is reduced to a small number of cells that depend entirely on their parent sporophyte for nutrition (**Fig. 30.8**). The third difference is

---

FIG. 30.8 Life-cycle evolution in land plants. A major trend in the evolution of land plants is a decrease in the size and independence of the gametophyte generation and a corresponding increase in the prominence of the sporophyte generation.

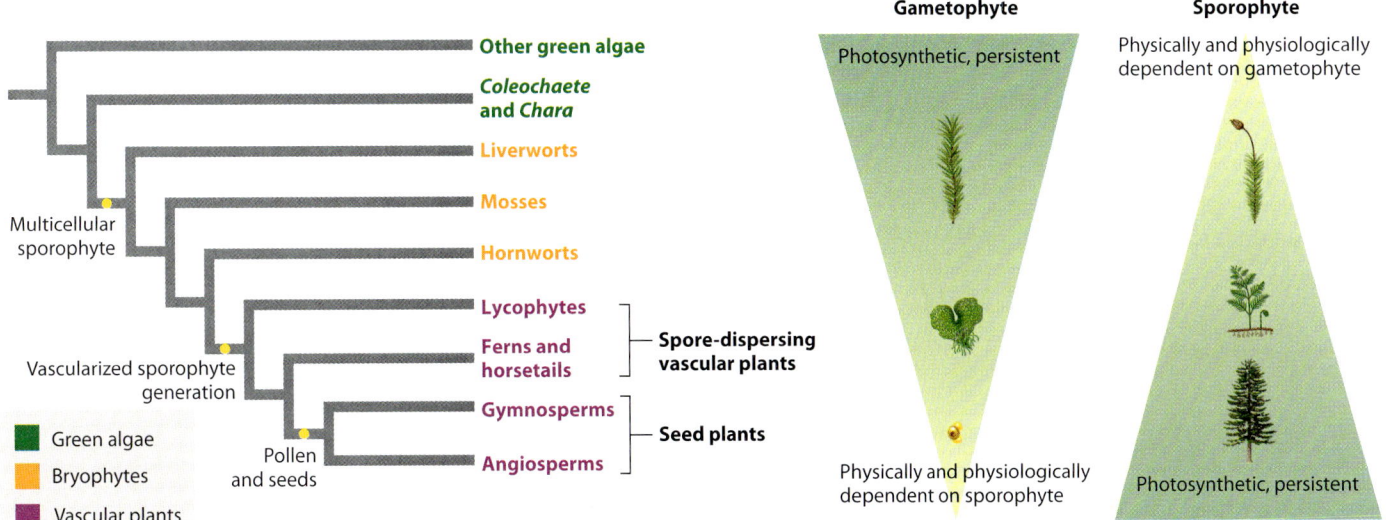

that an ovule, when fertilized, develops into a **seed.** Seeds are multicellular structures that allow offspring to disperse away from the parent plant.

→ **Quick Check 3** How do the male and female gametophyte generations differ in seed plants?

Seeds enhance dispersal and establishment of the next sporophyte generation.

A seed contains structures from three generations (**Fig. 30.9**). On the outside is the protective **seed coat,** which is formed from tissues that surround the sporangial wall and thus is a product of the diploid sporophyte. At the center is the embryo, which develops from the zygote and represents the next sporophyte generation. In gymnosperms such as pine trees, the embryo is surrounded by the haploid female gametophyte, which provides the raw materials that support the growth of the embryo.

Compared to a spore, a seed is more likely to form a new, photosynthetically self-supporting plant after dispersal. Because seeds are multicellular structures, they are able to store resources that the embryo can draw on during germination. As they mature, most seeds lose water. And as water content falls, the seeds' metabolic activity drops to extremely low levels. The combination of low metabolic activity and stored resources allows seeds to survive for long periods, in many cases one to several years, and sometimes much longer.

A final advantage of seeds over spores is that some seeds exhibit **dormancy.** A dormant seed can delay germination even when conditions for growth, notably temperature and moisture, are favorable. Dormancy prevents seeds from germinating at the wrong time, for example on a warm day at a time of year when the seedlings would be unable to survive. It also prevents seeds from germinating all at once, thus spreading the risk of making the transition from embryo to seedling over a larger time span and range of conditions.

Seeds span over 11 orders of magnitude in size (**Fig. 30.10**). The 100-nanogram (ng) seeds of orchids are so small (100 ng = 0.0001 mg) that they cannot germinate successfully without obtaining nutrients from a symbiotic fungus. At the other end of the spectrum are the 30-kg seeds of the coco-de-mer, a palm tree found on islands in the Indian Ocean. Large seeds, being better provisioned, are well adapted for the deep shade of a forest understory. However, large seeds typically disperse less far and have shorter life-spans, and they often have little or no dormancy. In contrast, small seeds can persist in the soil until the right combination of conditions triggers germination.

FIG. 30.9 **Structure of a pine seed.** A pine seed contains tissues from three generations: the seed coat formed from diploid sporophyte layers, the haploid female gametophyte, and the diploid embryo, which will grow into the next sporophyte generation.

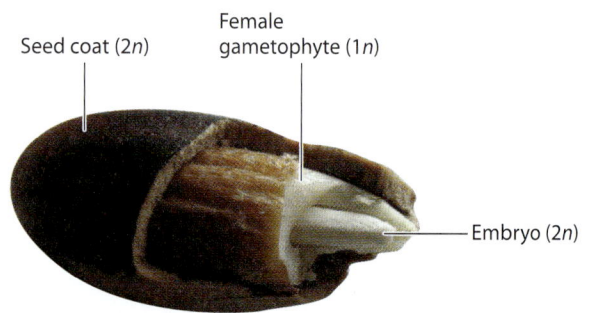

FIG. 30.10 **Seed diversity.** Seeds vary tremendously in size, from (top) the dustlike orchid seeds to (bottom) the large seeds of the coco-de-mer.

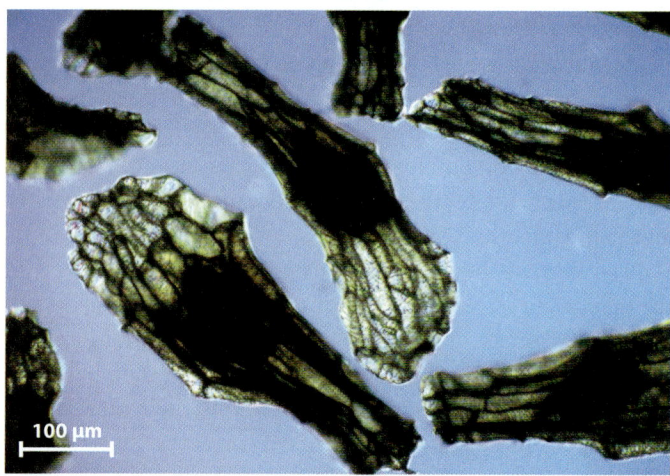

→ **Quick Check 4** Name three advantages of seeds over spores in terms of their ability to disperse.

30.2 FLOWERING PLANTS

Let's say that you want to send a postcard to a friend, but your only means of doing so is to throw the card into the air and hope the wind carries it to your friend's front door. How many cards would you have to loft in order to achieve a high probability of success? The answer, of course, is *many*, and this is precisely the reason that pines generate huge numbers of pollen grains each year. Suppose instead that you can give your card to the letter carrier whose route goes by your friend's house—your odds of success are now much greater. This analogy gives some idea of a key evolutionary innovation that appeared in the angiosperms. The "letter carriers" in angiosperm fertilization are animals, and the evolution of flowers is what allowed angiosperms to use animals to transport pollen.

Animal pollination increases the efficiency of pollination over what could be achieved by plants such as pines that depend entirely on the wind. That is, a higher proportion of pollen reaches an egg, so the plant can afford to produce less pollen. A second way in which angiosperm reproduction is efficient in its use of resources is that the female gametophyte in angiosperms is even smaller than the female gametophyte in gymnosperms. As a result, fewer resources are wasted when an ovule goes unfertilized. Here, we examine how angiosperms reproduce, beginning with flowers, which when fertilized develop into fruits.

Flowers are reproductive shoots specialized for the production, transfer, and receipt of pollen.

Flowers are spectacularly diverse in size, color, scent, and form (**Fig. 30.11**). Yet all flowers have the same basic organization: concentric whorls of floral organs (**Fig. 30.12;** Chapter 20). The center two whorls are made up of ovule-producing

FIG. 30.12 The four whorls of organs in a flower: carpels, stamens, petals, and sepals.

The outer two whorls, petals and sepals, serve to attract pollinators and to protect the flower as it develops.

The central two whorls, carpels and stamens, produce ovules and pollen, respectively.

FIG. 30.11 Flower diversity. Shown here are (a) a lady's-slipper orchid (*Cypripedium reginae*); (b) a magnolia (*Magnolia grandiflora*); (c) French lavender (*Lavandula stoechas*); and (d) a tropical tree (*Brownea grandiceps*).

carpels and pollen-producing **stamens.** Here, we see an important difference between angiosperms and gymnosperms. Most flowers produce both pollen and ovules, whereas in gymnosperms, pollen and ovules are produced in separate structures. The significance of this arrangement in angiosperms is that because the sites of pollen and ovule formation are in proximity, a pollinator can deliver pollen to one plant, and take up pollen to carry to another in a single visit. Approximately 12% of flowering plant species produce flowers that form only pollen or ovules, but phylogenetic analysis makes it clear that all these species descend from ancestors whose flowers had both stamens and carpels. Such "unisexual" flowers are often found in plants that have reverted to wind pollination.

Let's look at the floral organs in more detail, starting with the carpels at the center of the flower (Fig. 30.12). Each flower produces several carpels (typically 3 to 5, but sometimes more than 20), but because the carpels are often fused, it may seem as though there is only one. Each carpel has a hollow **ovary** at the base in which one to many ovules develop. The ovary protects the ovules from being eaten or damaged by animals. The fact that the ovules are completely enclosed within the carpel is what gives rise to the name "angiosperm," which is from Greek words meaning "vessel" and "seed." In gymnosperms, the ovules are not enclosed in a vessel but exposed, and the name, again from the Greek, literally means "naked seeds."

The ovary makes it impossible for pollen to land directly on the ovule surface. Instead, carpels commonly have a cylindrical stalk at the top, called the **style.** The surface at the top of the style is called the **stigma.** Pollen is captured on the sticky or feathery stigma, where it germinates and extends its pollen tube down through the style to reach the egg (**Fig. 30.13**). In some plants the style can be more than 10 cm long—corn silks, for instance, are styles.

Immediately surrounding the carpels are the stamens. A stamen can have a leaflike structure bearing sporangia on its surface, but more commonly it forms a filament that supports a structure known as the **anther,** which contains several sporangia in which pollen is produced. In most flowers, the anther splits open, exposing the pollen grains. Once exposed, the pollen can come into contact with the body of a visiting pollinator or, in the case of a wind-pollinated species, be carried off by the wind. However, in some plants, for example tomato, small holes open at the top of anthers. To extract the pollen, bees land on the flower and vibrate at just the right frequency to shake loose the pollen inside. Orchids and milkweeds disperse their pollen in a completely different way. Instead of releasing individual pollen grains, they disperse their pollen all together in a package with a sticky tag that attaches to a visiting pollinator.

The outer whorls of the flower produce neither pollen nor ovules but instead contribute to reproductive success in other ways. Most flowers have two outer whorls, of which

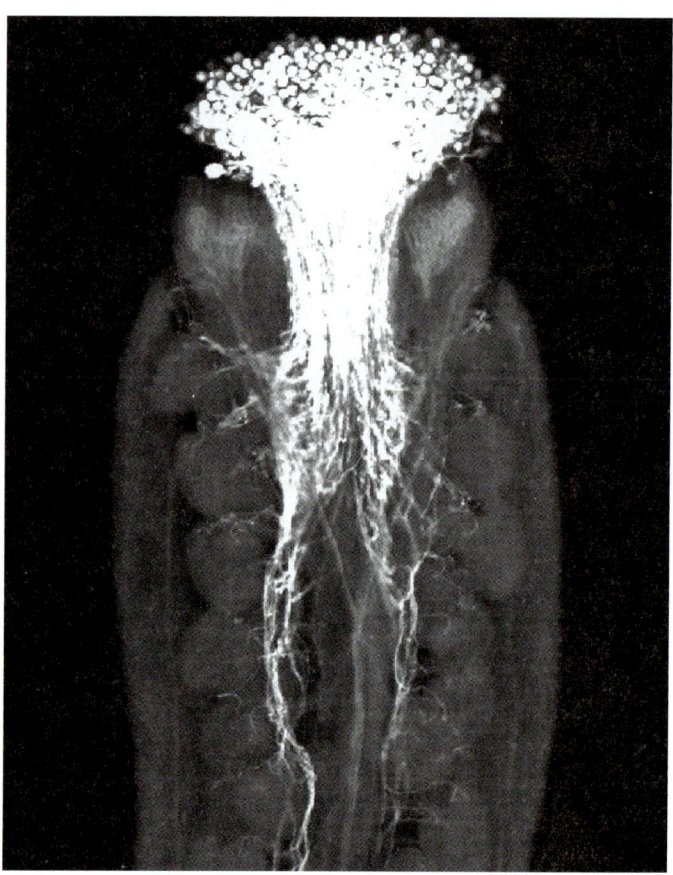

FIG. 30.13 Germination and growth of pollen tubes. Pollen tubes of *Arabidopsis thaliana* can be seen because they have been stained with a fluorescent dye.

the outermost is made up of **sepals.** Sepals, which are often green, encase and protect the flower during its development. In contrast, **petals** are frequently brightly colored and distinctively shaped. Their role is to attract and orient animal pollinators. In addition to serving as visual cues, petals in many flowers produce volatile oils. These are the source of the distinctive odors, some pleasant, some decidedly not, that many flowers use to advertise their presence to pollinators.

→ **Quick Check 5** What is the name and function of the structures in each whorl of a flower?

The diversity of floral morphology is related to modes of pollination.

The evolutionary histories of the angiosperms and their animal pollinators are closely intertwined. A rapid increase in angiosperm diversity occurred between 65 and 100 million years ago, about the same time as the diversification of bees (Hymenoptera) and butterflies (Lepidoptera). The evolution of flowers allowed

FIG. 30.14 Flowers providing food or other rewards for their pollinators. (a) A butterfly collecting pollen from flowers of a *Psiguria* vine; (b) a hummingbird visiting *Ravnia triflora* for nectar; and (c) a bat visiting *Agave palmeri* for nectar. (d) *Rhizanthes lowii* is pollinated by flies attracted to its smell of rotting flesh.

animals to specialize in new ways. At the same time, animal pollinators greatly facilitated the movement of pollen between plants within the same population.

So far, we have seen how flowers communicate their presence through both scent and color. For pollination to be reliable, however, it must be in the interest of the animal pollinator to move repeatedly (and with some constancy) between flowers of the same species. Flowers earn fidelity from their pollinators by providing rewards, frequently in the form of food (**Fig. 30.14**). Many insects consume some of the pollen itself, while many flowers also produce nectar, a sugar-rich solution attractive to bats and birds as well as to insects.

Some species provide rewards other than food. For example, many orchids secrete chemicals that male bees need to make their own sexual pheromones. Other flowers provide an enclosed and sometimes heated chamber in which insects can aggregate. But some flowers do not provide rewards at all: Instead, they "trick" their pollinators into visiting. For example, flowers that look and smell like rotting flesh attract flies, and orchids that look like a female bee even emit similar pheromones to attract

FIG. 30.15 *Ophrys ciliata*, an orchid from Sicily, that mimics the shape and color of female bees. These flowers induce males to attempt mating. Through repeated attempts on different flowers, the males transfer pollen.

male bees (**Fig. 30.15**). Male bees find these orchids irresistible and, as they attempt to copulate with the flower, deliver and receive pollen.

Producing rewards and attractants such as petals and scents requires resources. Thus, while animals can provide more efficient pollination than can the wind, the resources spent may be wasted if nonpollinating visitors "steal" pollen or nectar. For this reason, many flowers have mechanisms to protect their investments. For example, tubular cardinal flowers provide rewards to hummingbirds able to probe their depths, and snapdragons provide rewards only to insects that are strong enough to push apart their petals.

Many plants have evolved flowers that match rewards and attractants to the metabolic needs and sensory capabilities of their pollinators (see Fig. 30.14). For example, flowers pollinated by larger pollinators such as bats and birds produce copious amounts of nectar. Bat-pollinated flowers tend to be large, white or cream colored, and strongly scented, all traits appropriate to the size and nocturnal habits of their pollinators. The flowers are positioned among branches and leaves in such a way that they can be found by echolocation. In contrast, bird-pollinated flowers are often red, a color insects cannot see, and have no scent. Because interactions between plants and pollinators affect gene flow, they can affect rates of speciation. A recent radiation in columbine occurred as the plants shifted from bumble bees to hummingbirds to hawkmoths as their primary pollinator (**Fig. 30.16**).

Angiosperms are thought to have evolved in tropical forests. Pollen cannot move freely through the air in the high density of foliage that is present year round in these forests, favoring the evolution of animal pollination. However, as angiosperms diversified and spread around the globe, some evolved wind pollination once again. About 20% of present-day angiosperm species release their pollen into the wind. The reversal to wind pollination is associated with habitats where pollinators are not reliably abundant, such as dry regions.

Wind pollination is most effective in species that are locally abundant. For example, many grasses rely on wind for pollination. Wind pollination is also common in temperate-zone trees, which produce flowers in the early spring, before leaves expand. Wind-pollinated species tend to produce large anthers and stigmas, the latter often branched like a feather duster, well adapted to capture pollen from the air. The sepals and petals of wind-pollinated plants have no attractive role and so tend to be small. Wind-pollinated species produce large amounts of pollen. Hay-fever is an allergic reaction to pollen, virtually all of which comes from wind-pollinated species.

Angiosperms have mechanisms to increase outcrossing.

Plants do not have the elaborate courtship rituals that many animal species use to select an appropriate mate. Nevertheless, angiosperms have a wide range of mating systems. At one extreme are **self-compatible** species in which pollen and eggs produced by flowers on the same plant can unite and produce viable offspring. Self-compatible species are able to reproduce even when they are physically isolated from other individuals of the same species or when pollinators are rare. Many weedy species and the majority of crops are self-compatible.

At the other extreme are species in which pollen must be transferred between different plants. Approximately half of all of angiosperm species are **self-incompatible,** meaning that pollination by the same or a closely related individual does not lead to fertilization. Such self-recognition is based on the proteins produced from self-incompatibility genes, or S-genes. If the pollen's proteins match those of the carpel, the pollen either fails to germinate or germinates but the pollen tube grows slowly and eventually stops elongating. S-genes may have originated as a defense against fungal invaders, but subsequently evolved as a means of promoting outcrossing. Because there can be dozens of S-gene alleles in a population, only pollen transfers between closely related individuals are blocked. Apples are examples of self-incompatible plants, explaining why apple orchards always contain a mixture of different varieties.

HOW DO WE KNOW?

FIG. 30.16

Can pollinator shifts enhance rates of species formation?

BACKGROUND Columbines (*Aquilegia* species) are diverse and have visually striking flowers. Columbine flowers have nectar spurs, which are modified petals that form tubular outgrowths. Blue flowers with short nectar spurs are pollinated by bumblebees, red flowers with medium nectar spurs are pollinated by hummingbirds, and white or yellow flowers with long nectar spurs are pollinated by hawkmoths. The size of the nectar spur corresponds to the length of the tongue of the pollinator.

HYPOTHESIS The recent radiation of columbines corresponds with shifts from short-tongued pollinators (bumblebees) to ones with increasingly longer tongues (hummingbirds, hawkmoths).

METHOD Researchers mapped changes in flower color and nectar spur length onto a phylogenetic tree of columbines. To do this, they assigned each of 25 species examined to one of three groups based on length of tongue, flower color, and most common pollinator (bumblebees, hummingbirds, or hawkmoths). The group is indicated by the color circle at the tip of each branch in the phylogenetic tree in the figure. The researchers then did a statistical analysis that suggested the most likely pollinator type at each node representing an ancestor, also indicated by color circles in the figure. They used this information to infer the history of pollinator shifts during the radiation of this group.

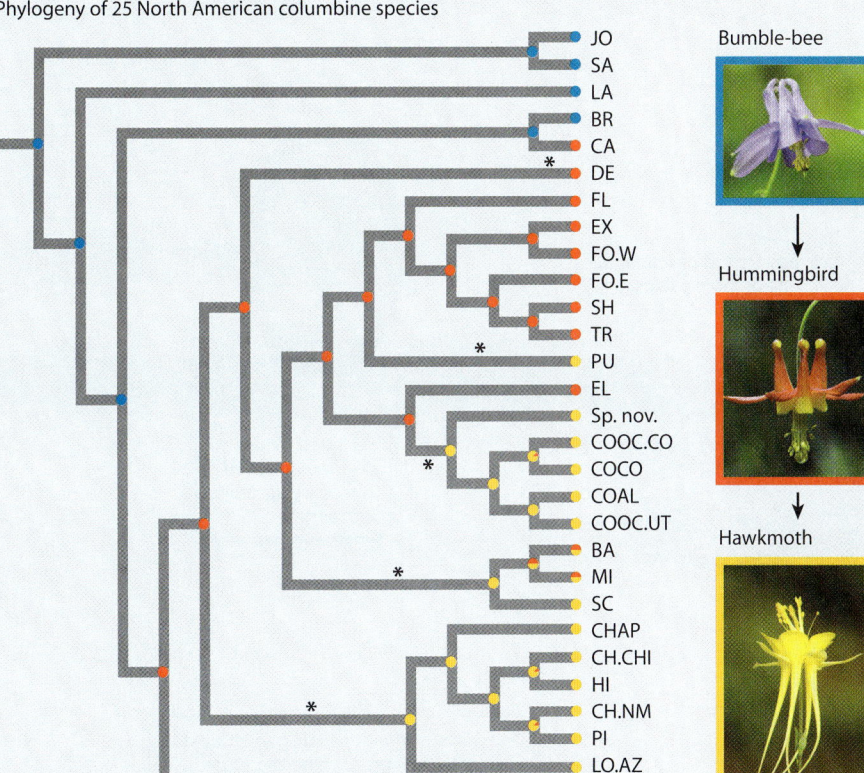

RESULTS Within the recent radiation of columbines, there are multiple instances of shifts from a shorter-tongued to a longer-tongued pollinator, and no reversals.

CONCLUSION The evolution of floral diversity and increasing nectar-spur length in columbines is associated with pollinator shifts. Large changes in nectar-spur length occur disproportionately at speciation events, indicating a key role in adaptive radiation.

FOLLOW-UP WORK Developmental studies show that the diversity of nectar spur length evolved as a result of changes in cell shape.

SOURCES Whittall, J. B. and S. A. Hodges. 2007. "Pollinator Shifts Drive Increasingly Long Nectar Spurs in Columbine Flowers." *Nature* 447:706–709; Puzey, J. R., S. J. Gerbode, S. A. Hodges, E. M. Kramer, and L. Mahadevan. 2011. "Evolution of Spur Length Diversity in *Aquilegia* Petals Is Achieved Solely Through Cell Shape Anisotropy." *Proceedings of the Royal Society, London, Series B*, 279:1640–1645.

- Short nectar spurs, blue petals, visited by bumble bees
- Medium-length nectar spurs, red petals, visited by hummingbirds
- Long nectar spurs, white or yellow petals, visited by hawkmoths

* indicates a likely evolutionary shift in pollinator

FIG. 30.17 Reproduction in angiosperms. (a) The angiosperm life cycle; (b) double fertilization, in which one sperm fuses with the egg to form a zygote, and another sperm fuses with the diploid gametophyte cell to form a triploid cell.

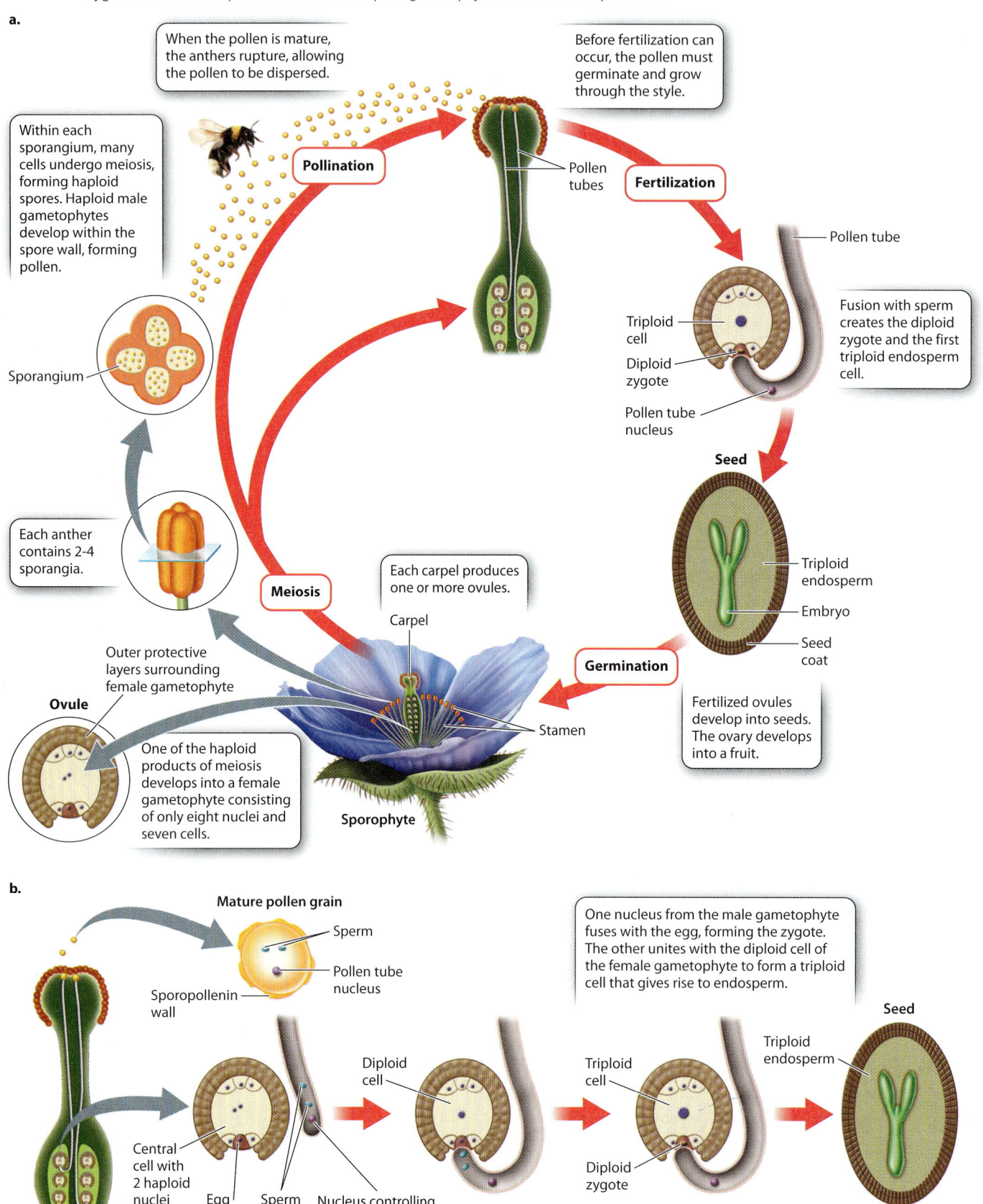

FIG. 30.18 Endosperm. Endosperm develops from the triploid cell produced by double fertilization.

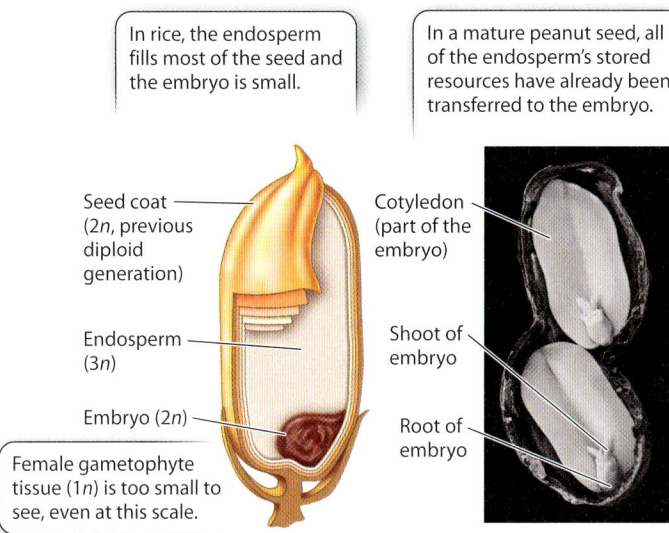

In rice, the endosperm fills most of the seed and the embryo is small.

In a mature peanut seed, all of the endosperm's stored resources have already been transferred to the embryo.

- Seed coat (2n, previous diploid generation)
- Endosperm (3n)
- Embryo (2n)
- Female gametophyte tissue (1n) is too small to see, even at this scale.
- Cotyledon (part of the embryo)
- Shoot of embryo
- Root of embryo

Angiosperms delay provisioning their ovules until after fertilization.

A major trend in the evolution of plants is the decreasing size and independence of the gametophyte generation and the increasing prominence of the sporophyte generation, as we have seen. In angiosperms, this trend is taken even further (**Fig. 30.17a**). The male gametophyte has only three cells, one of which controls the growth of the pollen tube, while the other two are male gametes or sperm. In most species, the female gametophyte contains only eight nuclei, arranged as six haploid cells plus a central cell containing two nuclei. One of the haploid cells gives rise to the egg, and the two nuclei within the central cell fuse either before or during fertilization to form a diploid cell. The importance of this cell will become apparent. The female gametophyte of angiosperms is too small to support the growth of the embryo, as it does in gymnosperms. How, then, do flowering plants provide nutrition for the embryo to grow?

This is where the diploid cell of the female gametophyte comes into play. During pollination, the two sperm travel down the pollen tube and enter the ovule (**Fig. 30.17b**). One of the sperm fuses with the egg to form a zygote, just as occurs in all plants. The other sperm, however, unites with the diploid gametophyte cell to form a triploid (3n) cell. This triploid cell undergoes many mitotic divisions, forming a new tissue called **endosperm**. In angiosperm seeds, it is the endosperm that supplies nutrition to the embryo. The process in which two sperm from a single pollen tube fuse with the egg and the diploid cell is called **double fertilization**.

What advantage is gained by double fertilization and endosperm formation? In pines and other gymnosperms, the female gametophyte provides food for the embryo. This provisioning occurs whether or not the egg is successfully fertilized, and so is potentially a significant waste of resources. In angiosperms, endosperm development occurs only when fertilization has occurred. In many flowering plants, including corn, wheat, and rice, the mature seed consists largely of endosperm (**Fig. 30.18**). The embryo itself is small. As the seeds germinate, the embryo draws resources from the endosperm to help it become established. It is no overstatement to suggest that the human population is sustained mostly by endosperm produced in the seeds of cereal crops.

In other angiosperms, growth of the embryo consumes all the endosperm before seed dispersal. For example, a peanut has virtually no endosperm left at the time it is dispersed. All the resources that were previously held within the endosperm have been transferred into two large embryonic leaves, called cotyledons, which form the two halves of the peanut we eat.

Fruits enhance the dispersal of seeds.

The transformation of a flower into a **fruit** is as remarkable as the metamorphosis of a caterpillar into a butterfly (**Fig. 30.19**). As the fertilized egg develops to form an embryo and the endosperm

FIG. 30.19 Developmental transition from flower to fruit. Fruits, in this case an apple, develop from ovaries and provide a means of dispersing seed.

FIG. 30.20 Fruits and seed dispersal. (a) The seeds of the milkweed *Asclepias syriaca* are dispersed by wind; (b) a squirrel carries a fruit it will bury for later use; (c) a robin is eating berries of mountain ash (*Sorbus americana*); (d) cocklebur (*Xanthium strumarium*) fruits are hitching a ride on the fur of a mule deer.

In a number of species, including apples, bananas, and tomatoes, a gaseous hormone called ethylene triggers fruit ripening. It is the production of ethylene that explains why placing a ripe apple and an unripe banana together hastens the ripening of the banana. Tomatoes and bananas are usually picked in an immature state and are subsequently ripened by exposure to ethylene. The fruits can therefore be transported while they are still hard and less prone to damage. Because ethylene synthesis requires oxygen, the unripe fruits are typically stored in a low-oxygen environment. Without this precaution, the ethylene produced by the ripening of even a single fruit could set off a chain reaction in which all the stored fruit would ripen at once.

The tremendous morphological diversity of fruits reflects the many ways in which angiosperm seeds are dispersed (**Fig. 30.20**). The winged fruit of an ash tree is carried away on the wind, and a coconut can float on ocean currents for many months, protected from exposure to salt water by its thick outer husk. The burdock fruit has small hooks that attach to fur (or clothing)—this small hitchhiker was the inspiration for Velcro. Other fruits are not dispersed, but instead open to expose their seeds. For example, as milkweed fruits mature and dry out, they split apart to release hundreds of seeds, each with a tuft of silky hairs that floats them away on even a light breeze. Still others, such as *Impatiens* (jewelweed), produce fruits that forcibly eject their seeds.

proliferates around it, the ovary wall develops into a fruit, stimulated by signaling molecules produced by the endosperm. In a tomato, for example, the fleshy fruit is a mature ovary that encloses the seeds. In some plants, other parts of the flower may also become incorporated into the fruit. For instance, the fleshy part of an apple is formed from the outer part of the ovary combined with the base of the petals and sepals. Other fruits, like pineapple, develop from multiple flowers.

Fruits serve two functions: They protect immature seeds from being preyed on by animals, and they enhance dispersal once the seeds are mature. In the case of fruits dispersed by animals, it is essential that the fruit not be consumed before the seeds are able to withstand a trip through their disperser's digestive tract. Thus, immature fleshy fruits are physically tough and their tissues highly astringent. As they ripen, the fruits are rapidly transformed in texture, palatability, and color. Ripening converts starches to sugars and loosens the connections between cell walls so that the fruit becomes softer.

Animals are probably the most important agents of seed transport, in many cases attracted by the nutritious flesh of the fruit. In fleshy fruits, the seeds are protected by a hard seed coat and pass unharmed through the animals' digestive tract. In some species, this passage actually enhances germination rates. In many cases, however, the seed is consumed, and successful dispersal occurs when the animals gather more seeds than they eat. Squirrels are a good example, storing many acorns that never get eaten. Humans represent an important dispersal agent, responsible for transporting seeds long distances, sometimes far outside their usual range. In some cases, accidentally introduced plant species can disrupt native communities.

→ **Quick Check 6** How does the fruit relate to structures found in the flower?

30.3 TIMING OF REPRODUCTIVE EVENTS

As plants moved onto land, they encountered environments with distinct seasonality. In temperate regions, temperatures vary from summer to winter; in most tropical habitats, wet seasons alternate with dry ones. Thus, in addition to overcoming the challenges of completing their life cycle on land, plants evolved mechanisms to control the timing of reproduction. This is true for all plants, which must produce gametes at a time when they can find suitable mates and which must disperse their young in a season suitable for their growth.

However, timing is particularly important for angiosperms. Not only must individuals of the same species flower at the same time, but they must do so when they can attract enough pollinators for efficient pollination. Furthermore, angiosperms are unique among plants in including species with life-spans of only one to two years. For these short-lived species, which are most commonly found in temperate regions, there is little margin for error in the timing of flower production and seed germination. In this section, we explore how plants gain information about their environment that enables them to regulate when flowers are produced and seeds germinate.

Flowering time is affected by day length.

In the 1910s, Wightman Garner and Henry Allard, scientists at the U.S. Department of Agriculture, set out to understand why Maryland Mammoth, a variety of tobacco, does not produce flowers during the summer like other tobacco plants. Instead, these plants continue to grow throughout the summer, reaching heights several times that of a typical tobacco plant and producing many more leaves. Only in autumn does Maryland Mammoth finally produce flowers, too late in the year for seeds to mature.

One possible explanation for the observed delay in flowering time is that Maryland Mammoth must achieve a greater total size before it produces flowers. However, when Garner and Allard grew Maryland Mammoth plants in a heated greenhouse during the winter, the plants flowered at the same time and the same size as other tobacco plants. Therefore, size does not determine when Maryland Mammoth flowers. And because the heated greenhouse was set to the same temperatures as those found during the summer, temperature could also be eliminated as a factor. Garner and Allard were left with only one more environmental variable that was different between the summer and the winter: the length of the day. They proposed that plants can measure and respond to day length.

In a series of experiments in which day length was artificially shortened during summer and extended during winter, Garner and Allen showed that in Maryland Mammoth, as well as in many other plants, flowering is controlled by day length. They coined the term **photoperiodism** to describe the effect of the "photo period," or day length, on flowering. **Short-day plants,** like Maryland Mammoth, flower only when the day length is less than a critical value. When the light period exceeds this threshold, a short-day plant continues to produce new leaves, but no flowers are formed. In contrast, **long-day plants,** which include radish, lettuce, and some varieties of wheat, flower only when the light period exceeds a certain length. Not all plants are photoperiodic. Flowering of **day-neutral plants** is independent of any change in day length.

Sensitivity to day length ensures that plants flower only when they are large enough to support the development of a large number of seeds *and* only when there is enough time for those seeds to mature before the onset of winter. A short-day plant that germinates in the spring will not flower until late summer, when the decreasing day length falls below the critical value for that species. This flowering strategy allows short-day plants to grow as large as possible and yet switch from producing leaves to producing flowers in time to complete seed development before winter. In contrast, many long-day plants germinate in the summer. Because they do not flower until a critical day length is exceeded, they do not flower in autumn, but instead during the following spring.

Understanding how photoperiod affects flowering has many commercial applications. For example, chrysanthemums are short-day plants and their natural flowering season is autumn. By artificially exposing plants to short day lengths, growers can induce chrysanthemums to flower at any time of the year. Photoperiodism can complicate the ability to grow crop species at latitudes different from the ones in which they evolved. For example, soybean is a short-day plant that normally germinates in the spring and flowers as the day length decreases in late summer. When planted in the tropics, where the day length is close to 12 hours throughout the year, soybean flowers when the plant is still quite small, reducing the number of seeds that can eventually be harvested. For this reason, crop species often become less sensitive to photoperiod as they are domesticated. For example, wild varieties of tobacco are photoperiodic, whereas cultivated tobacco, *Nicotiana tabaccum*, is day neutral.

Photoreceptors enable plants to measure day length.

The first requirement for measuring day length is that a plant be able to sense when it is day and when it is night. To accomplish this, many plants produce **photoreceptors,** molecules whose chemical properties are altered when they absorb light. We saw in Chapter 9 how a signal can be conveyed to a cell by a change in the shape of a receptor molecule within the cell or on its surface. Photoreceptors in plants work similarly. The absorption of light by a photoreceptor produces a signal that triggers changes in the cell's metabolism or alters patterns of gene expression. Your eye relies on thousands of photoreceptors that work on the same principle as those in plants.

Although photoreceptors provide information about the presence of light, they do not, by themselves, explain how leaves are able to measure day length. An early clue to this ability came when scientists discovered that interrupting the dark period with a brief exposure to light was sufficient to alter the flowering response. In these experiments, both short-day plants and long-day plants were grown under a photoperiod that would normally cause the short-day plants to flower and the long-day plants to produce only leaves. A brief exposure to light during the dark period reversed this outcome. The short-day plants no longer produced flowers, but the long-day plants did. The reverse experiment, in which plants were placed under light-tight cover for a short period during the daytime, did not have any effect.

If no further experiments had been done, the scientists would have concluded— incorrectly—that plants were measuring the length of the dark period. More extensive experiments demonstrated that a plant's sensitivity to a "night interruption" varies according to a circadian—or 24-hour—rhythm. We now know that photoperiodism results from an interaction between photoreceptors that are activated by light and the product of a gene whose expression is controlled by the plant's circadian clock (**Fig. 30.21**). Circadian clocks are biochemical mechanisms found in all organisms that oscillate with a 24-hour period and are coordinated with the day-night cycle.

Other experiments demonstrated that day length is sensed by the leaves and not by the buds where the flowers will form. When conditions are right for flowering, a signal is transported from the leaves to the growing points of a plant. A flowering signal (named "florigen") had been hypothesized to exist for more than 70 years, but its nature had not been known, although widely sought. Only with the development of techniques for genetic analysis were scientists able to investigate the genetic basis of photoperiodism and determine the identity of this important signal.

Vernalization prevents plants from flowering until winter has passed.

In a number of temperate-zone species, flowering can be induced only if the plant has experienced a prolonged period of cold temperatures, a process known as **vernalization.** Vernalization prevents plants from flowering prematurely in late summer or autumn. Plants are forced to wait until the following spring to flower, when the chances for successful pollination and seed development are greater.

An important feature of vernalization is that even parts of the plant that form after the cold stimulus is long past seem to "remember" that they received the required cold treatment. This "memory" of winter is important because many plants that overwinter grow rapidly in the spring. What is the basis of this "memory" of winter and how is it transmitted to new growth?

Plant species differ in the ways that vernalization affects the regulatory pathways that lead to flowering. These differences likely reflect the fact that the angiosperms evolved in tropical latitudes. As different groups migrated into temperate regions they independently evolved mechanisms for controlling flowering in highly seasonal environments.

One common mechanism acts through chromatin remodeling. The DNA of all eukaryotic organisms is bound with proteins to form chromatin. We saw in Chapter 19 that modifications to chromatin, such as DNA methylation, can change gene expression. By this means, a record of developmental events can be inscribed into a cell's DNA. In *Arabidopsis*, for example, vernalization results in chromatin remodeling that turns off a gene whose protein product represses flowering. Chromatin remodeling is stable through mitotic divisions, explaining why newly formed parts of a plant "remember" winter. However, the slate is wiped clean during meiosis, and the requirement for vernalization is reinstated with each generation.

Dormant seeds can delay germination if they detect the presence of plants overhead.

Seed germination is a one-way street. Once a seed begins to germinate, it quickly reaches a point of no return and it either becomes established as a photosynthetically competent seedling or it dies. Dormant seeds may delay germination even though environmental conditions, notably temperature and moisture, are favorable. Dormancy increases the probability that some offspring will survive to maturity by extending the time over which the seeds germinate. From a farmer's perspective, however, dormancy is a problem because it means that some fraction of the seeds put into the ground will not germinate in that year. Therefore, agriculturalists have paid close attention to what environmental factors induce germination.

In some species, dormancy is enforced by a hard, tough seed coat. In such species, the seed coat must be weakened mechanically before the seed will germinate, either by passing through an animal's digestive system or by physical abrasion. In other plants, dormancy is controlled by the embryo. Two environmental variables are particularly effective at inducing such seeds to germinate. One is exposure to a prolonged period of cold temperatures, which signals that winter has past and thus prevents seeds from germinating on warm days in autumn. The second is that some seeds, particularly very small ones, will germinate only if they have been exposed to light.

When scientists first began to study the effects of light on seed germination they found that red light is particularly good at stimulating germination, while far-red light inhibits germination. A simple "flip-flop" experiment by H. A.

HOW DO WE KNOW?

FIG. 30.21

How do plants measure day length?

BACKGROUND *Arabidopsis thaliana*, a model organism for plant biology, is a long-day plant, flowering only when the light period exceeds about 16 hours. Genetic studies have shown that the *CONSTANS (CO)* gene is essential in allowing *Arabidopsis* to flower in response to day length. To determine the mechanism underlying photoperiodism, researchers set out to determine how this gene is affected by both the plant's circadian clock and the presence or absence of light.

HYPOTHESIS The ability to respond to day length in *Arabidopsis* results from a circadian (daily) increase and decrease in *CO* expression, as well as interactions with light.

EXPERIMENT 1 The researchers determined how *CO* mRNA varies as a function of time of day and presence or absence of light.

RESULT 1 *CO* mRNA levels follow a circadian up-and-down pattern. This pattern occurred independent of whether the plant was exposed to long or short days or even kept continually in light. The timing of these oscillations is such that high levels of *CO* mRNA coincide with the light period only when day-lengths are long.

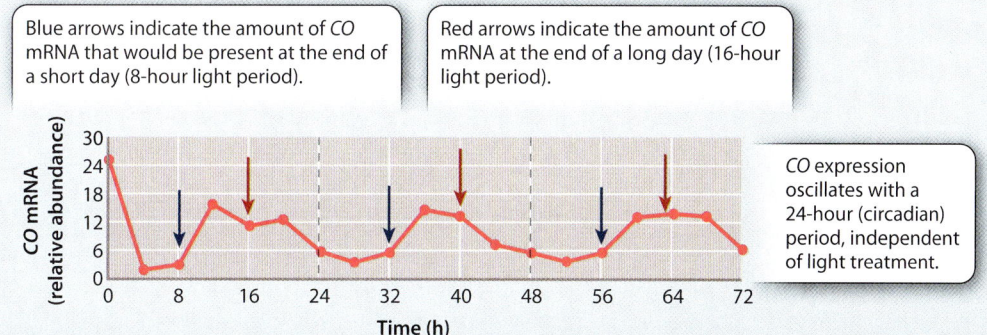

EXPERIMENT 2 Protein levels were measured following exposure to two days of light or dark. Protein extracts were first separated by gel electrophoresis and then probed with antibodies. Darker bands correspond to greater staining by protein-specific antibodies and thus provide visual information on the relative amount of the CO protein.

RESULT 2 The much darker band after exposure to two days of light indicates that CO protein accumulates only in the light. In the dark, the CO protein does not accumulate.

CONCLUSION Flowering in *Arabidopsis* is triggered by long days because there is overlap between high rates of CO expression and stabilization of the CO protein by light.

FOLLOW-UP WORK Studies with *Arabidopsis* mutants indicate a role for specific photoreceptors in stabilizing CO protein. Studies on other species indicate that photoperiod detection involves both the circadian clock and photoreceptors, but the way the CO protein is regulated by light varies.

SOURCES Suárez-López, P., K. Wheatley, F. Robson, H. Onouchi, F. Valverde, and G. Coupland. 2001. "*CONSTANS* Mediates Between the Circadian Clock and the Control of Flowering in *Arabidopsis*." *Nature* 410:1116–1120; Valverde, F., A. Mouradov, W. Soppel, D. Ravenscroft, A. Samach, and G. Coupland. 2004. "Photoreceptor Regulation of CONSTANS Protein in Photoperiodic Flowering." *Science* 303:1003–1006.

HOW DO WE KNOW?

FIG. 30.22

How do seeds detect the presence of plants growing overhead?

BACKGROUND To study the effect of light on seed germination, scientists exposed lettuce seeds that had been kept continually in the dark to different wavelengths of light and then counted what fraction of the seeds germinated. Red light had the greatest ability to stimulate germination, but surprisingly, far-red light inhibited germination such that fewer seeds germinated than in the control seeds, which were kept in darkness. In the rush to conduct more experiments, petri dishes with the light-treated seeds piled up by the sink until someone noticed that the seeds that had been experimentally treated with far-red light were now germinating.

HYPOTHESIS This observation suggested that the inhibitory effect of far-red light can be overcome by a subsequent exposure to red light.

EXPERIMENT The scientists exposed lettuce seeds to red and far-red light in an alternating pattern, ending with either red light or far-red light. They then placed the seeds in the dark for two days and afterward counted the number of seeds that had germinated.

RESULTS When the lettuce seeds were exposed to red light last, nearly 100% of the seeds germinated. By contrast, when the last exposure of the seeds was to far-red light, the percentage of seeds that germinated was dramatically reduced.

Lettuce Seeds Germinating After Exposure to Red and Far-red Light in Sequence

SEQUENCE OF LIGHT EXPOSURE	GERMINATION (%)
Dark (control)	8.5
Red	98
Red, Far-red	54
Red, Far-red, Red	100
Red, Far-red, Red, Far-red	43
Red, Far-red, Red, Far-red, Red	99
Red, Far-red, Red, Far-red, Red, Far-red	54
Red, Far-red, Red, Far-red, Red, Far-red, Red	98
TIME →	

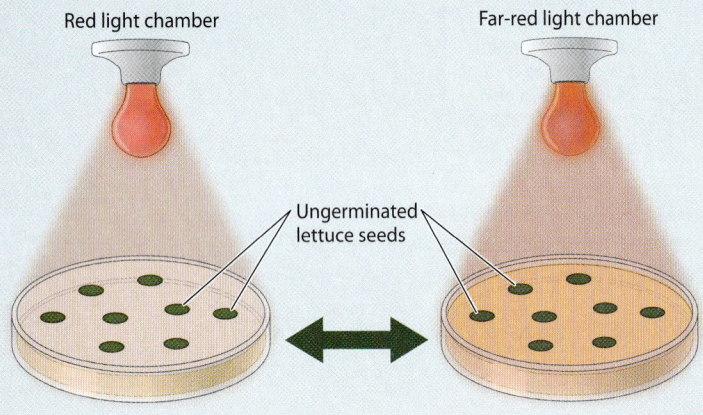

CONCLUSION Seed germination in lettuce is triggered by exposure to red light and is inhibited by exposure to far-red light in a reversible fashion. As a result, plants are able to track changes in the relative amount of red and far-red light, which provides information on the presence or absence of plants overhead (see Fig. 30.23).

FOLLOW-UP WORK Additional studies showed that this result was due to a single pigment in a photoreceptor, now known as phytochrome, that is converted into an active form by red light and reversibly converted into an inactive form by far-red light.

SOURCE Borthwick, H. A., S. B. Hendricks, M. W. Parker, E. H. Toole, and V. K. Toole. 1952. "A Reversible Photoreaction Controlling Seed Germination." *Proceedings of the National Academy of Sciences* 38:662–666.

Borthwick and colleagues showed that the inhibitory effect of far-red light could be overcome by a subsequent exposure to red light (**Fig. 30.22**). Ultimately it was shown that only a single photoreversible pigment was needed to detect the exposure to both far-red and red light.

That pigment is present in **phytochrome,** a photoreceptor that switches back and forth between two stable forms depending on its exposure to light (**Fig. 30.23a**). Red light causes phytochrome to change into a form that absorbs primarily far-red light (P_{fr}); far-red light causes phytochrome to change back into the red-light-absorbing form (P_r). Because red light stimulates seed germination, we know that P_{fr} is the active form of phytochrome.

In the dark, P_{fr} slowly converts back into the inactive, P_r, form. The light-independent conversion of P_{fr} to P_r allows a seed that had been exposed to red light, but failed to germinate, to have its phytochrome reset so that it can again respond to red light.

Phytochrome allows dormant seeds to respond to the presence of plants overhead. For a small seed that has few reserves, germinating in the shade of another plant could be fatal. How does phytochrome allow plants to detect the presence of other plants? As sunlight passes through leaves, the red wavelengths are absorbed by chlorophyll but the far-red wavelengths are not. The ratio of red to far-red light is thus much higher under an open sky than it is in the shade of another plant (**Fig. 30.23b**). As a result, the proportion of phytochrome that is in the active, P_{fr}, form is much greater in open environments than in the forest understory. Furthermore, the ratio of red to far-red light is independent of light intensity. Thus, the ratio is a reliable signal, even for seeds covered by a thin layer of soil, that leaves are overhead.

Phytochrome plays other important roles in plant development. As we will discuss in Chapter 31, phytochrome allows plants to alter their growth in the presence of neighboring plants to compete more effectively for sunlight. Phytochrome also maintains the accuracy of the plant's circadian clock by resetting it each day at dawn. Finally, phytochrome is one of the photoreceptors that plants use to determine changes in day length.

? **CASE 6 Agriculture: Feeding a Growing Population**
What is the basis for the spectacular increases in the yield of cereal grains during the Green Revolution?
Mechanisms to control the timing of flower production and seed germination have evolved to optimize reproductive success under natural conditions. Agricultural conditions, however, are quite different. When Norman Borlaug, an American agronomist, led a project in the mid-twentieth century to make Mexico self-sufficient in bread wheat, his group encountered problems of flowering time and seed germination. Borlaug's aim was to breed wheat for enhanced resistance to infection by a devastating class of fungal pathogens known as rusts. To do this, he and his group hand-pollinated plants that exhibited resistance to rusts with ones that grew well or had seeds that produced good flour.

Under natural conditions, wheat usually self-pollinates. To introduce new sources of pollen required that the anthers of each flower be removed by forceps before the flowers opened. Each flower had to be covered to prevent unwanted pollen from coming into contact with the flower. Finally, when the stigma was receptive, it had to be dusted with pollen collected from a plant with the desired characteristics. Under the hot sun, this is backbreaking work.

To speed the work, Borlaug proposed growing two crops of wheat each year. During summer, he conducted controlled crosses in the central highlands of Mexico, and then as winter approached he took his seeds north to the Yaqui Valley. There temperatures were warm enough for him to carry out another round of crosses during the winter.

FIG. 30.23 Detection of shading by phytochrome. Phytochrome detects the relative absence of red light under the canopy.

a.

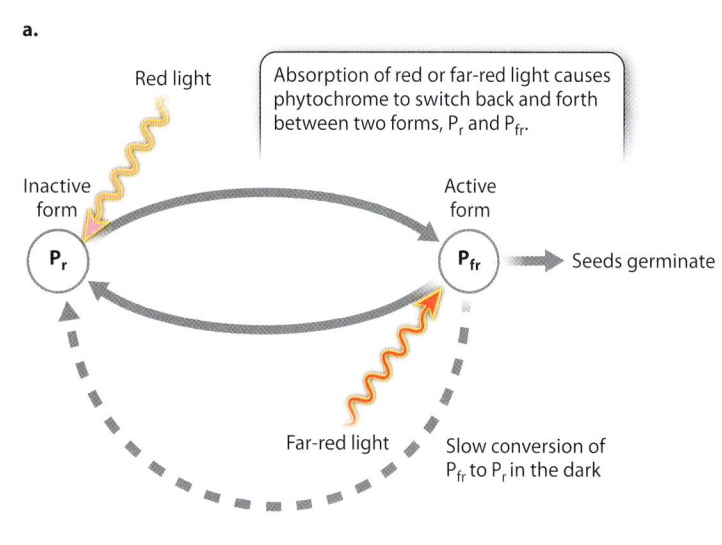

b.

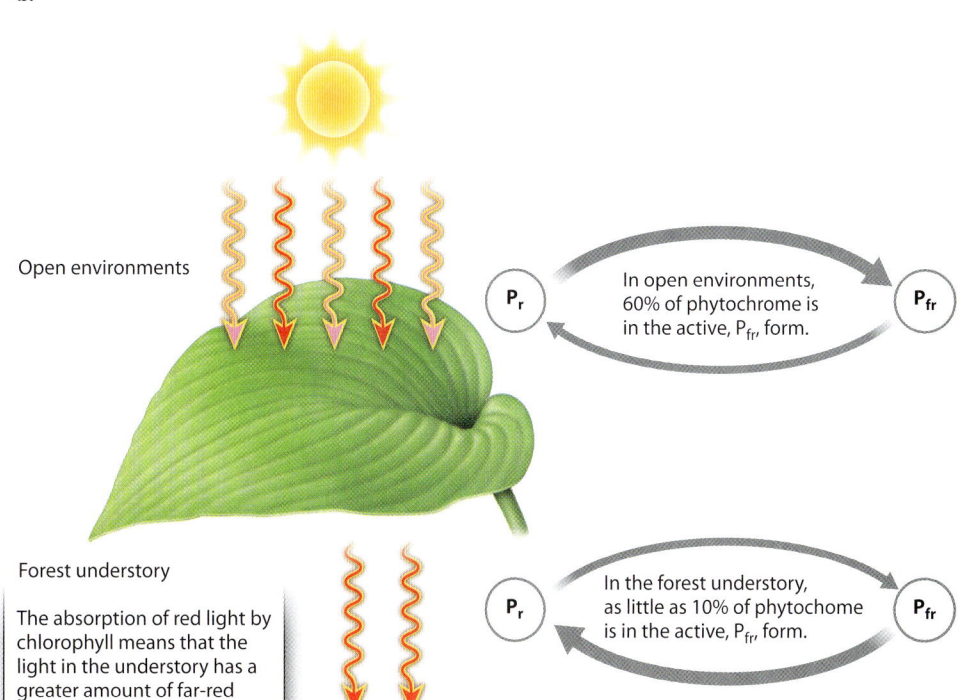

Wheat is a long-day plant that flowers in the long days of summer. By successively growing plants during the lengthening days of summer in the southern highlands and the shorter days of winter in the lowland north, Borlaug was unconsciously selecting for plants whose flowering was less sensitive to photoperiod. Although not the original goal of the project, the diminishing sensitivity to photoperiod had an important impact. The new varieties were day neutral, flowering when they achieved a certain size. As a result, Borlaug's varieties could be planted across a wide geographic range.

Before Borlaug's rust-resistant seeds would be ready to be grown throughout the world, however, one more problem had to be solved. Having succeeded in breeding wheat varieties that were disease resistant, Mexican farmers applied fertilizer in large amounts. Under these conditions, the new wheat varieties produced so many seeds that stalks became top heavy, falling over in the wind.

Borlaug knew that an answer to this problem was to cross his plants with wheat varieties that had shorter, sturdier stems. But given the selective advantage of height in natural populations, where was he to find such a plant? Japanese wheat breeders had identified and preserved a mutant dwarf plant that occurred spontaneously, and Borlaug obtained 80 seeds of this dwarf variety. However, in his haste to incorporate this new plant into his breeding program, he made a serious error. He forgot that the Japanese wheats, which were adapted to a more northerly climate, had to be vernalized before they would flower.

Because the Japanese wheats had not been exposed to a prolonged period of cold, they did not flower at the same time as the Mexican wheats. And when they did produce a few flowers, the blooms were of such low quality that no viable seeds were produced. As luck would have it, Borlaug had planted only 72 of the 80 seeds. Somehow 8 seeds had escaped his notice. Eventually, Borlaug found these at the bottom of the original envelope in which they had been sent. This time he took care to vernalize the seeds, and when the new plants were formed they flowered readily.

Much hard work and many thousands of hand pollinations lay ahead. But by 1963, more than 95% of Mexican wheat cultivation made use of Borlaug's high-yielding semidwarf wheat varieties, resulting in wheat yields six times greater than in 1944, the year he began work in Mexico. In 1965, the seeds were exported in large numbers, first to India and Pakistan and soon to the rest of the world. Used in combination with greater investments in fertilizer and irrigation, these varieties prevented the famines that had been predicted by many as a result of rising human populations. For his work in alleviating world hunger, Borlaug was awarded the Nobel Peace Prize in 1970.

30.4 VEGETATIVE REPRODUCTION

Eukaryotic organisms can reproduce asexually as well as sexually (Chapters 11 and 27). Asexual reproduction results in new individuals that are genetically identical to the parent. If the new individuals can disperse by walking, swimming, or floating away, asexual reproduction allows organisms to proliferate and spread.

In sexual reproduction, plants disperse by spores and seeds. What prevents asexually produced individuals from growing so close to the parent plant that they are in direct competition for resources? In mosses and liverworts, tiny plantlets form by mitosis at the base of shallow cups. Raindrops landing in this cup can dislodge the tiny plantlets, splashing some away from the parent plant.

Most plants that reproduce asexually do so by growing to a new location and only then producing a new plant. You need look no further than a front lawn to find evidence of such **vegetative reproduction.** Horizontal stems allow new upright grass shoots to be produced at a distance from the site where the parent plant originally germinated. Strawberries, bamboo, and spider plants (a common houseplant) also spread vegetatively by forming horizontal stems. In many cases, the connections that were needed initially to produce a new plant become severed. When

FIG. 30.24 Vegetative growth of aspen trees. Each distinct group of trees, distinguished by leaf color, is a group of genetically identical individuals produced by vegetative reproduction.

of another plant (**Fig. 30.23b**). As a result, the proportion of phytochrome that is in the active, P_{fr}, form is much greater in open environments than in the forest understory. Furthermore, the ratio of red to far-red light is independent of light intensity. Thus, the ratio is a reliable signal, even for seeds covered by a thin layer of soil, that leaves are overhead.

Phytochrome plays other important roles in plant development. As we will discuss in Chapter 31, phytochrome allows plants to alter their growth in the presence of neighboring plants to compete more effectively for sunlight. Phytochrome also maintains the accuracy of the plant's circadian clock by resetting it each day at dawn. Finally, phytochrome is one of the photoreceptors that plants use to determine changes in day length.

? **CASE 6 Agriculture: Feeding a Growing Population**
What is the basis for the spectacular increases in the yield of cereal grains during the Green Revolution?
Mechanisms to control the timing of flower production and seed germination have evolved to optimize reproductive success under natural conditions. Agricultural conditions, however, are quite different. When Norman Borlaug, an American agronomist, led a project in the mid-twentieth century to make Mexico self-sufficient in bread wheat, his group encountered problems of flowering time and seed germination. Borlaug's aim was to breed wheat for enhanced resistance to infection by a devastating class of fungal pathogens known as rusts. To do this, he and his group hand-pollinated plants that exhibited resistance to rusts with ones that grew well or had seeds that produced good flour.

Under natural conditions, wheat usually self-pollinates. To introduce new sources of pollen required that the anthers of each flower be removed by forceps before the flowers opened. Each flower had to be covered to prevent unwanted pollen from coming into contact with the flower. Finally, when the stigma was receptive, it had to be dusted with pollen collected from a plant with the desired characteristics. Under the hot sun, this is backbreaking work.

To speed the work, Borlaug proposed growing two crops of wheat each year. During summer, he conducted controlled crosses in the central highlands of Mexico, and then as winter approached he took his seeds north to the Yaqui Valley. There temperatures were warm enough for him to carry out another round of crosses during the winter.

FIG. 30.23 Detection of shading by phytochrome. Phytochrome detects the relative absence of red light under the canopy.

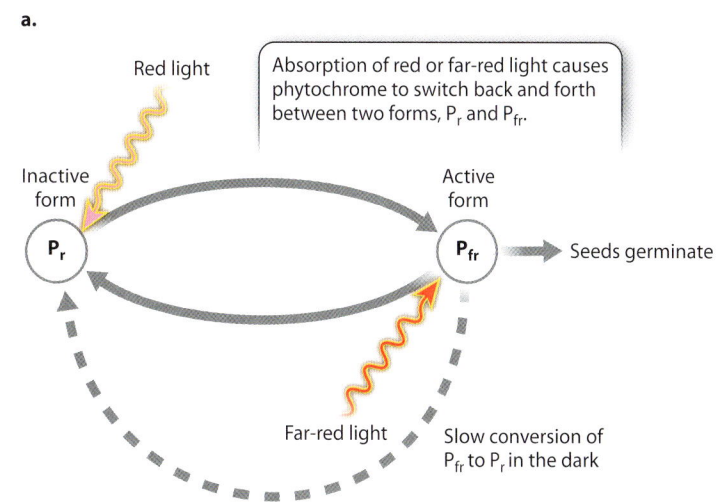

Wheat is a long-day plant that flowers in the long days of summer. By successively growing plants during the lengthening days of summer in the southern highlands and the shorter days of winter in the lowland north, Borlaug was unconsciously selecting for plants whose flowering was less sensitive to photoperiod. Although not the original goal of the project, the diminishing sensitivity to photoperiod had an important impact. The new varieties were day neutral, flowering when they achieved a certain size. As a result, Borlaug's varieties could be planted across a wide geographic range.

Before Borlaug's rust-resistant seeds would be ready to be grown throughout the world, however, one more problem had to be solved. Having succeeded in breeding wheat varieties that were disease resistant, Mexican farmers applied fertilizer in large amounts. Under these conditions, the new wheat varieties produced so many seeds that stalks became top heavy, falling over in the wind.

Borlaug knew that an answer to this problem was to cross his plants with wheat varieties that had shorter, sturdier stems. But given the selective advantage of height in natural populations, where was he to find such a plant? Japanese wheat breeders had identified and preserved a mutant dwarf plant that occurred spontaneously, and Borlaug obtained 80 seeds of this dwarf variety. However, in his haste to incorporate this new plant into his breeding program, he made a serious error. He forgot that the Japanese wheats, which were adapted to a more northerly climate, had to be vernalized before they would flower.

Because the Japanese wheats had not been exposed to a prolonged period of cold, they did not flower at the same time as the Mexican wheats. And when they did produce a few flowers, the blooms were of such low quality that no viable seeds were produced. As luck would have it, Borlaug had planted only 72 of the 80 seeds. Somehow 8 seeds had escaped his notice. Eventually, Borlaug found these at the bottom of the original envelope in which they had been sent. This time he took care to vernalize the seeds, and when the new plants were formed they flowered readily.

Much hard work and many thousands of hand pollinations lay ahead. But by 1963, more than 95% of Mexican wheat cultivation made use of Borlaug's high-yielding semidwarf wheat varieties, resulting in wheat yields six times greater than in 1944, the year he began work in Mexico. In 1965, the seeds were exported in large numbers, first to India and Pakistan and soon to the rest of the world. Used in combination with greater investments in fertilizer and irrigation, these varieties prevented the famines that had been predicted by many as a result of rising human populations. For his work in alleviating world hunger, Borlaug was awarded the Nobel Peace Prize in 1970.

30.4 VEGETATIVE REPRODUCTION

Eukaryotic organisms can reproduce asexually as well as sexually (Chapters 11 and 27). Asexual reproduction results in new individuals that are genetically identical to the parent. If the new individuals can disperse by walking, swimming, or floating away, asexual reproduction allows organisms to proliferate and spread.

In sexual reproduction, plants disperse by spores and seeds. What prevents asexually produced individuals from growing so close to the parent plant that they are in direct competition for resources? In mosses and liverworts, tiny plantlets form by mitosis at the base of shallow cups. Raindrops landing in this cup can dislodge the tiny plantlets, splashing some away from the parent plant.

Most plants that reproduce asexually do so by growing to a new location and only then producing a new plant. You need look no further than a front lawn to find evidence of such **vegetative reproduction.** Horizontal stems allow new upright grass shoots to be produced at a distance from the site where the parent plant originally germinated. Strawberries, bamboo, and spider plants (a common houseplant) also spread vegetatively by forming horizontal stems. In many cases, the connections that were needed initially to produce a new plant become severed. When

FIG. 30.24 Vegetative growth of aspen trees. Each distinct group of trees, distinguished by leaf color, is a group of genetically identical individuals produced by vegetative reproduction.

this happens, only genetic evidence can be used to determine if a plant is the result of sexual or asexual reproduction.

Vegetative reproduction also occurs in woody plants. For example, when a redwood tree falls over, new upright stems can form along the now-horizontal trunk. Perhaps the most impressive example of the ability to spread vegetatively is the quaking aspen (*Populus tremuloides*), a common tree of the Rocky Mountains that produces brilliant yellow foliage in autumn (**Fig. 30.24**). Aspens reproduce vegetatively by producing upright stems from a spreading root system. Although each shoot lives for less than 200 years, a single individual can persist for tens of thousands of years by continuing to produce new stems. The current record holder is an aspen clone in Utah, consisting of nearly 50,000 stems, that is estimated to be 80,000 years old.

Plants are able to reproduce vegetatively because of how they build their bodies. As we will see in the next chapter, growth and development occur throughout a plant's life, allowing plants to produce new upright "individuals" from roots or horizontal stems.

Core Concepts Summary

30.1 THE PLANT LIFE CYCLE EVOLVED IN WAYS THAT ENHANCE THE ABILITY TO UNITE GAMETES AND DISPERSE OFFSPRING ON LAND.

The life cycle of the algal sister groups of land plants has one multicellular haploid generation. page 30-2

The life cycle of land plants is characterized by an alternation of haploid and diploid generations, one generation specialized for fertilization and the other for dispersal. page 30-3

Gametes are haploid cells produced by mitosis from cells in the multicellular haploid generation, which is referred to as the gametophyte. page 30-3

Spores are haploid cells produced by meiosis from the multicellular diploid generation, which is referred to as the sporophyte generation. page 30-3

Bryophytes have a free-living haploid gametophyte that makes gametes and a dependent diploid sporophyte that makes spores that are released into the air. page 30-4

In spore-dispersing vascular plants, both the gametophyte generation and the sporophyte generation can survive on their own. page 30-4

Bryophytes and spore-dispersing vascular plants rely for fertilization on swimming sperm released into the environment. page 30-4

A trend in the evolution of plants is the progressive reduction in size and independence of the gametophyte and the increasing role of the sporophyte. page 30-7

Pollen is the male gametophyte that develops within the spore wall, and so can be transported through the air. page 30-7

The female gametophyte of seed plants is retained and nourished by the sporophyte. The female gametophyte, together with the surrounding layers of protective tissues, is called an ovule. Ovules, when fertilized, develop into seeds. page 30-7

Pollination is the transport of pollen to the ovule, where fertilization takes place. page 30-7

Seeds contain the diploid embryo, a seed coat, and, in gymnosperms, nourishment in the form of the female gametophyte. page 30-8

30.2 ANGIOSPERMS (FLOWERING PLANTS) ATTRACT AND REWARD ANIMAL POLLINATORS, AND THEY PROVIDE RESOURCES FOR SEEDS ONLY AFTER FERTILIZATION.

The pollen and ovules of gymnosperms are produced in separate structures, whereas the pollen and ovules of angiosperms often form in one structure, the flower. page 30-10

Flowers consist of four whorls of organs: ovule-bearing carpels, pollen-producing staments, petals, and sepals. page 30-10

Many flowers attract animals because they provide a reward, such as food, shelter, or chemicals. Animals, in turn, transfer pollen. page 30-11

Flowering plants have genetic and structural mechanisms to prevent self-fertilization. page 30-12

Double fertilization, which is unique to angiosperms, is the formation of a diploid zygote and triploid endosperm. The endosperm nourishes the embryo that develops from the zygote. page 30-15

The fertilization of an ovule triggers the development of the ovary wall into a fruit, a structure that enhances seed dispersal. page 30-15

30.3 PLANTS HAVE SENSORY SYSTEMS THAT CONTROL THE TIMING OF FLOWERING AND SEED GERMINATION.

Many plants are photoperiodic, meaning that their flowering is controlled by the amount of daylight. page 30-17

Plants measure day length through an interaction between photoreceptors and a gene regulated by an internal clock. page 30-18

Vernalization is a requirement that a plant be exposed to prolonged cold temperatures before flowering can be induced. page 30-18

Dormant seeds must be triggered to germinate by appropriate environmental signals. page 30-18

Some seeds will not germinate unless they have been previously exposed to a prolonged cold period. page 30-18

Phytochrome is a protein photoreceptor that allows plants to determine the amount of red versus far-red light. Because light filtered through leaves has lower levels of red wavelengths, phytochrome allows plants to detect the presence of plants overhead. page 30-20

Plant reproduction is central to agriculture because it allows plant breeders to create new genotypes. page 30-21

30.4 MANY PLANTS ALSO REPRODUCE ASEXUALLY.

Vegetative reproduction produces plants that are genetically identical to the parent plant. page 30-22

Vegetative reproduction can occur when a plant spreads horizontally and develops new upright shoots. page 30-22

Self-Assessment

1. Explain how the evolution of alternation of generations is an adaptation for reproduction on land.

2. Diagram the relationship between the sporophyte and gametophyte generations in bryophytes, ferns, gymnosperms, and angiosperms. Show the relative sizes and physical interactions (if any) of the two generations.

3. Explain how the organs in the different whorls of angiosperm flowers interact to promote fertilization.

4. Contrast the investments that angiosperms and gymnosperms make and the structures that they produce to enhance pollination.

5. Diagram the structure of a mature angiosperm and a mature gymnosperm seed, indicating the ploidy ($1n$, $2n$, $3n$) of each tissue and its role in seed development and germination.

6. Explain why a short-day plant that germinates in the spring will not flower until late summer and why a long-day plant that germinates at the end of summer will not flower until late the following spring.

7. Diagram how the amount of red versus far-red light allows plants to determine if they are in the open or in the forest understory.

8. Describe how vernalization can have an effect in cells that were not formed at the time of the cold treatment.

Do you understand the chapter's Core Concepts? Log in to check your answers to the Self-Assessment questions, then practice what you've learned and reinforce this chapter's concepts by working through the problems and multimedia tutorials provided there.

www.biologyhowlifeworks.com

CHAPTER 31

PLANT GROWTH AND DEVELOPMENT

Building the Plant Body

Core Concepts

31.1 In plants, upward growth by stems occurs at shoot apical meristems, populations of totipotent cells that produce new cells for the lifetime of the plant.

31.2 Hormones are chemical signals that influence the growth and differentiation of plant cells.

31.3 Lateral meristems allow plants to grow in diameter, increasing their mechanical stability and the transport capacity of their vascular system.

31.4 The root apical meristem produces new cells that allow roots to grow downward into the soil, enabling plants to obtain water and nutrients.

31.5 Plants respond to light, gravity, and wind through changes in internode elongation and the development of leaves and branches.

In the previous chapters, we have seen how two major innovations played key roles in enabling plants to thrive on land. The first is an internal vascular system capable of pulling water from the soil, allowing plants to elevate their leaves into the air to capture sunlight (Chapter 29). The second is a diploid sporophyte generation that allows plants to disperse by spores carried through the air (Chapter 30). These and other features of vascular plants depended on new ways of building the plant body.

The fossil record shows that the first vascular plants were little more than small photosynthetic stems with simple sporangia at their tips. Wood, leaves, and roots all came later. Today, most landscapes include trees with thick woody stems, extensive root systems, numerous leaves, and complex reproductive structures. To understand how these innovations evolved, we need to explore the ways that plants grow and develop.

To begin, let us remind ourselves briefly how our own bodies develop from fertilized eggs (Chapter 20). Humans and other mammals grow by the repeated division of cells throughout the body. Many cells migrate from one place to another, especially early in development. Embryonic stem cells, formed by the first divisions of the fertilized egg, have the potential to give rise to many different cell types. Cells with this ability are termed **totipotent**. As development proceeds, however, cells lose this totipotency and proceed down a genetically determined developmental pathway. Functionally distinct cell types differentiate as growth proceeds, forming tissues and organs. Once the body is mature, growth essentially stops.

In contrast, vascular plants grow according to entirely different principles, with cell division confined to discrete populations of totipotent cells called **meristems** and growth commonly lasting throughout the life of the plant. A bristlecone pine in western North America may be several thousand years old, but it continues to sprout new branches, new leaves, and new cones every year. The capacity for continued growth and development lies at the heart of how plants meet the challenges they face because they are rooted in one place.

Plants respond to the world around them not by moving about, but by modifying their size and shape. Plants acquire resources by growing to them, and they compete with neighboring plants by growing beyond them. Thus, growth fulfills many of the same roles in plants that behavior does in animals (Chapter 45). And although plants can't move out of the way of danger, they can rebuild after fire or frost damage, or after being struck by a falling tree. Even reproduction requires the continuing formation of new structures.

This chapter focuses on how vascular plants build their bodies. We begin by asking how the shoot system, consisting of leaves, stems, and reproductive structures, is formed. We then turn to the development of root systems, asking how plants are able to grow belowground. Finally, we consider how plants sense the environment and how they respond to what they sense by modifying their development.

FIG. 31.1 **Shoot organization.** A grape vine (a) and barrel cacti (b) look very different, but their shoot systems are built from the same repeating units of nodes and internodes (c).

Shoots are made up of repeating units of nodes and internodes; one or more leaves are attached at each node.

31.1 SHOOT GROWTH AND DEVELOPMENT

In tropical forests, vines weave their way upward through the branches of trees. Many of these vines have long, thin stems with widely spaced leaves (**Fig 31.1a**). In contrast, a barrel cactus, living in the desert, has thick stems and leaves modified to form sharp spines (**Fig. 31.1b**). Although they look nothing alike, these two plants are constructed in the same way. Plant shoots are modular, meaning that they are formed of repeating units. Each unit consists of a **node,** the point where one or more leaves are attached, and an **internode**, the segment between two nodes (**Fig. 31.1c**). In vines the internodes are long and the leaves large, whereas in cacti the internodes are short and the leaves thin and sharp. The modular nature of plants helps us understand how the capacity for continued growth gives rise to the tremendous variation in plant form that we see around us.

Shoots grow by adding new cells at their tips.

At the very tip of each branch, commonly hidden beneath an array of young leaves called **leaf primordia** (singular, primordium), lies a tiny dome of cells called the **shoot apical meristem** (**Fig. 31.2**). The shoot apical meristem is a group of totipotent cells that, like embryonic stem cells in animals, gives rise to new tissues. As we will see, plants have meristems in several parts of their bodies, but upward growth occurs exclusively in and just below the shoot apical meristem. This is where cell division occurs, generating all the new cells that serve

FIG. 31.2 **The shoot apical meristem.** The shoot apical meristem consists of undifferentiated cells that divide rapidly, giving rise to all the cells in stems and leaves. (a) The seedling is rowan (*Sorbus acuparia*). (b) This scanning electron micrograph shows the shoot apical meristem of tomato (*Solanum lycopersicum*).

FIG. 31.3 The zones of cell division, elongation, and maturation. The photo to the right shows a longitudinal section of the shoot apical meristem of coleus (*Solenostemon scutellariodes*).

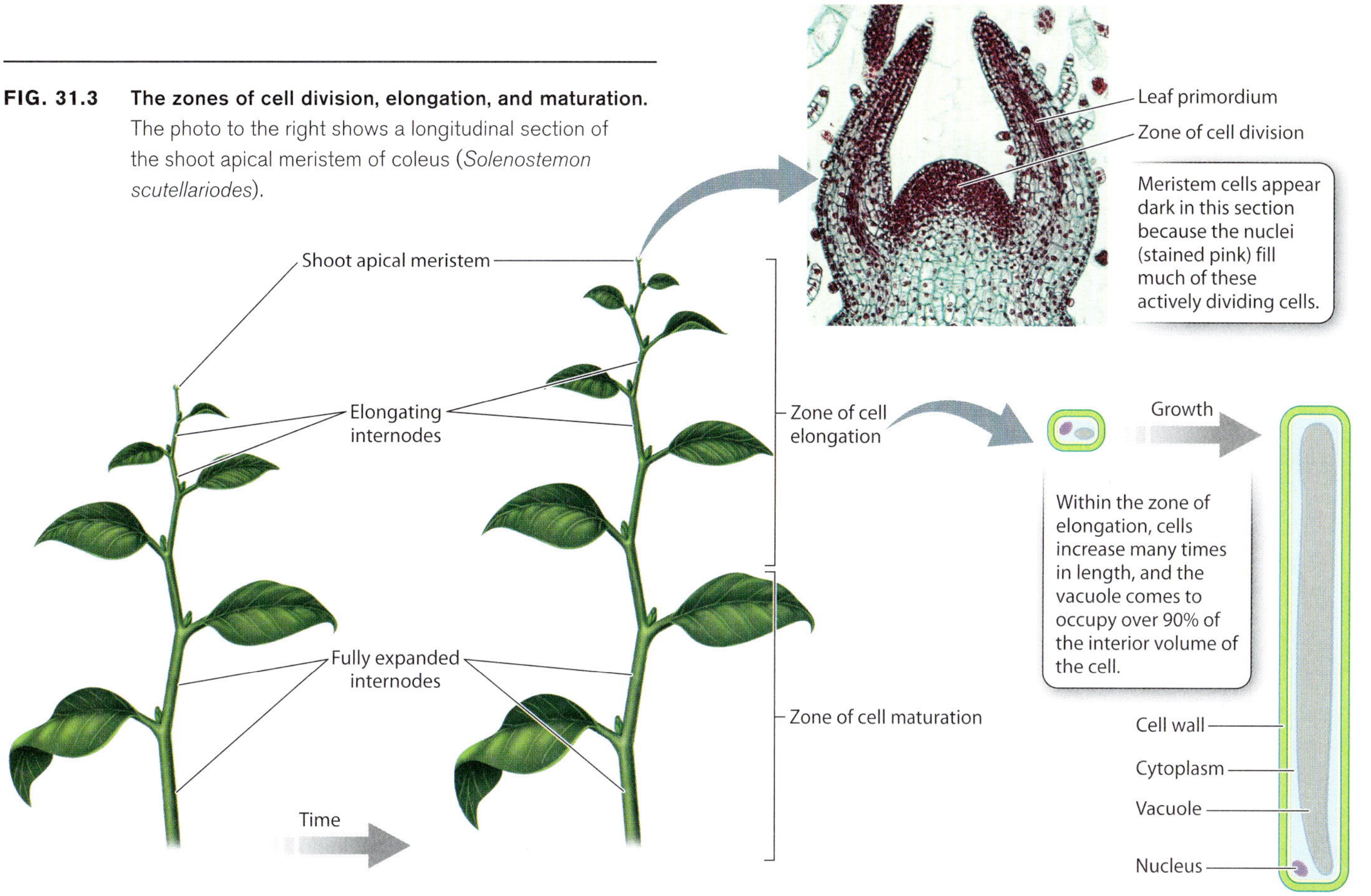

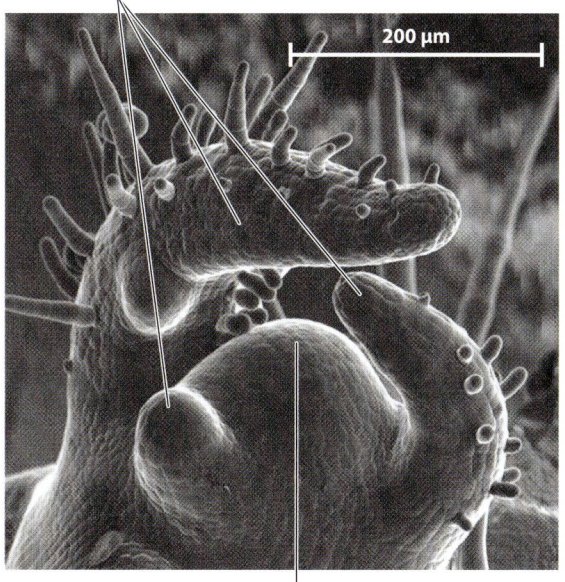

b. Developing leaves (leaf primorida)

Shoot apical meristem

to elongate stems. The shoot apical meristem also initiates leaves and produces new meristems, which allow plants to branch.

An important characteristic of the shoot meristem is that it maintains a constant size even though it is a region of active cell division. As new cells are added near the shoot tip, cells that are farther from the shoot tip cease to divide. This means that the cells of the meristem are constantly being lost and replaced. Cells near the shoot tip maintain their totipotency by the expression of **meristem identity genes,** which contribute to meristem stability and function. The expression of meristem identity genes is controlled by a network of chemical signals produced by cells at the very tip of the stem. Because these signals can be transmitted only over a limited distance, only the cells close to the shoot tip express meristem identity genes.

→ **Quick Check 1** What are three functions of the shoot apical meristem?

Stem elongation occurs primarily in a zone just below the apical meristem where new cells elongate.

In animals, cell division and cell enlargement typically go hand in hand. In plants, most of the increase in cell size occurs after mitotic cell division is complete. Cells that are too far from the shoot tip to express meristem identity genes cease to divide, yet they continue to grow. This results in a zone of cell elongation located just beneath the shoot apical meristem (**Fig. 31.3**). It is here that most of the elongation of stems occurs. If you imagine all the cells in the stem of a 100-m redwood tree at the size they were when they left the zone of cell division—the tree would be less than 5 m tall.

FIG. 31.4 Diversity of leaf arrangements.

Alternate   Opposite   Whorled

In the zone of cell elongation, each cell grows many times more in length than in width. The reason is that the strong cellulose microfibrils wrapped around the cell make it difficult for the cell to expand in girth. The large central vacuole that characterizes mature plant cells forms when the cell is in the zone of cell elongation. In fact, most of the increase in cell volume is due to the uptake of water and solutes that fill the vacuole. This explains in part why plant growth is markedly reduced during periods of drought.

As we move even farther from the shoot tip, cells reach their final size and complete their differentiation into the mature cell types in leaves and stems. This organization into successive zones of cell division, cell elongation, and cell maturation allows stems to grow without any predetermined limit to their length. It also means that we can follow the time course of development by moving along the stem from the tip toward the base.

Once cells have matured, they no longer expand. Thus, during the upward growth of the shoot, it is the production and elongation of cells at the shoot tip that lift the meristem ever higher into the air. Imagine that, as a 10-year-old, you carved your initials into a tree. Twenty years later, your initials will be exactly the same distance from the ground as they were on the day you inscribed them. The letters, however, will be wider—a phenomenon we discuss later in this chapter.

The shoot apical meristem controls the production and arrangement of leaves.

In most plants, leaves are the principal sites of photosynthesis. Because light is needed for photosynthesis, the arrangement of the leaves along a stem has a major impact on their function. Each species has a characteristic number of leaves attached at each node along the stem. Some species have only a single leaf at each node, whereas others have two or more. How the leaves are positioned around the stem varies in a predictable fashion (**Fig. 31.4**). The regular placement of leaves reduces the shading of one leaf by another and thus enhances the ability of plants to obtain sunlight.

Leaves begin as small bumps, the leaf primordia, which form on the sides of the shoot apical meristem (see Fig. 31.2). The regular arrangement of leaves around the stem is controlled by the fact that each successive leaf primordium is located as far away as possible from all previously formed primordia. One hypothesis proposed to explain this developmental pattern is that the diffusion of chemical signals from developing leaf primordia creates regions where the growth of new primordia is inhibited. The result is an arrangement that prevents leaves from being produced one on top of another.

As noted previously, the earliest vascular plants were simple branching stems, and photosynthesis took place along the length of the shoot. As evolution proceeded, however, some branch systems became flattened, their axes growing in a plane that facilitated the capture of light. These planar branches lost the capacity for continued growth. By about 380 million years ago, the planar branches became modified into structures recognizable as leaves (**Fig. 31.5**).

The evolution of leaves required three developmental changes. First, the genetic program to produce a three-dimensional stem became modified to form flattened organs. Second, apical meristem identity genes were down-regulated and leaf identity genes up-regulated, resulting in a specialized organ incapable of continuing growth. And third, new meristems evolved, enabling leaves to expand into flattened photosynthetic structures that capture sunlight. In fern leaves, these meristems are located along the leaf margin. In pine needles, they occur at the base of each needle. In flowering plants, meristematic cells can be distributed throughout the developing leaf, making possible a diversity of leaf shape. In contrast to cells in the shoot apical meristems which can divide continuously throughout the lifetime of a plant, leaf meristematic cells divide only for a relatively short period of time. This explains why leaves grow to a final size.

When we think of leaves, it is the green photosynthetic ones that first come to mind. However, many plants produce leaves that are specialized for functions other than photosynthesis, including climbing, trapping insects, and attracting pollinators (**Fig. 31.6**). Plants that overwinter produce **bud scales** that protect shoot apical meristems from desiccation and damage

FIG. 31.5 **The evolution of leaves.** Fossils show that leaves evolved from branches that (1) lost the ability for continued growth, (2) developed in a plane, and (3) expanded the photosynthetic surface to fill the area between veins. The photo shows *Archaeopteris*, one of the earliest plants with flattened leaves.

FIG. 31.6 Non-photosynthetic functions of leaves: (a) protection (bud scales); (b) climbing (tendrils); (c) trapping insects (spines); and (d) attracting pollinators (color).

Early vascular plants lacked leaves. Their shoots consisted of photosynthetic stems.

In some of these early vascular plants, side branches lost their apical meristems and became flattened into a single plane.

The development of new meristems allowed tissues to form between these side branches, leading to the development of leaves.

due to cold. Bud scales may not look like leaves, but they form from leaf primordia and are arranged in the same way around the stem as the green leaves produced in spring.

The development of new apical meristems allows stems to branch.

Vascular plants evolved the ability to branch even before they evolved either roots or leaves. Branching was important to these first plants because it allowed them to produce more sporangia. Branching allows present-day plants to support greater numbers of both reproductive structures and leaves.

In lycophytes and in ferns and horsetails, branching occurs when the shoot apical meristem divides in two, giving rise to two stems. In seed plants, branches grow out from **axillary buds** (also

FIG. 31.7 **Axillary buds.** In seed plants, branching results from axillary buds that are produced at nodes.

called lateral buds), which are meristems that form at the base of each leaf (**Fig. 31.7**). Axillary buds have the same structure and developmental potential as the apical meristem and express the same meristem identity genes. However, the axillary buds remain dormant until triggered to grow, remaining attached to the stem even after leaves are shed. Thus, axillary buds provide seed plants with many points along their stem where new branches can form.

Flowers grow from and consume shoot meristems.

Just as leaves and branches grow from meristems, so, too, do flowers. Flowers can be produced at the tip of a plant or shoot as a result of the conversion of the apical meristem into a floral meristem, or at the base of leaves as the products of axillary buds. As with leaves, floral meristems lose their capacity for continued growth. All the cells differentiate entirely during flower development.

As we saw in Chapter 30, flowering is triggered by florigen, a protein produced in leaves and transported through the phloem to apical meristems and axillary buds. Florigen triggers the transition to floral meristems by initiating the down-regulation of meristem identity genes and the up-regulation of genes that govern floral identity. The arrangement of primordia is altered to produce four whorls that give rise to the sepals, petals, stamens, and carpels. Once the meristem is launched along the trajectory for flower development, the identity of each whorl is controlled by the expression of homeotic genes (Chapter 20). Three classes of genes, referred to as A, B, and C, are expressed in overlapping rings around the meristem and serve as master controllers for the development of specific floral organs (**Fig. 31.8**).

FIG. 31.8 **Flower development.** The identity of floral organs is specified by the spatial expression of three classes of homeotic genes called A, B, and C.

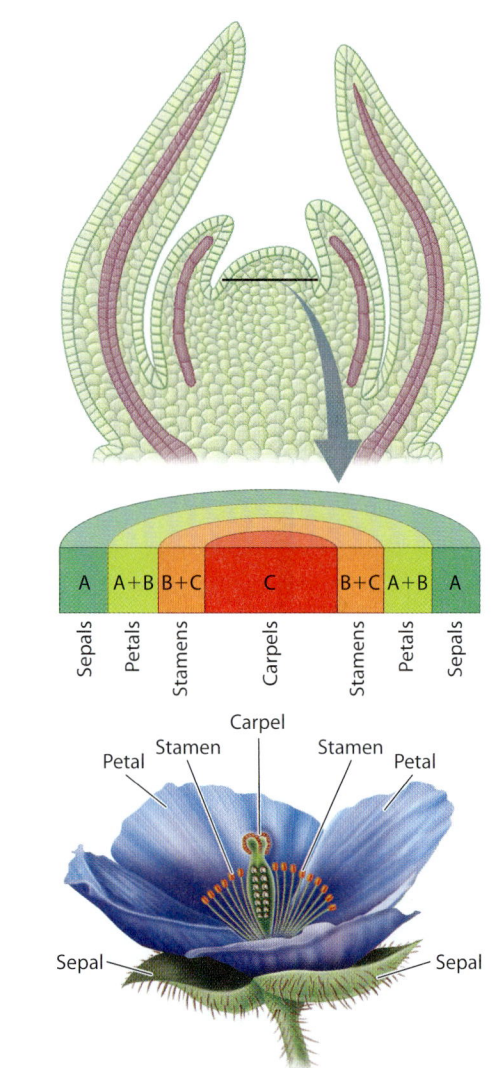

31.2 PLANT HORMONES

A fundamental question in the development of multicellular organisms is what controls the differentiation of individual cells. In considering the formation of plant stems, we can ask: (1) What controls the arrangement of leaf primordia and the elongation of internodes? (2) What ensures that the differentiation of xylem and phloem cells produces a continuous vascular system extending from the roots to the leaves? (3) What determines whether an axillary bud remains dormant or develops into a new stem?

The answer, in each case, is that chemical signals guide the growth and differentiation of plant cells. Earlier, we discussed basic principles of cell signaling and communication (Chapter 9). Chemical signals play a central role in plant development because cell fate is determined by position. Some of the chemical signals important in plant development have highly localized effects. For example, signals that cause cells to express meristem identity genes are restricted to the region close to the shoot tip. Other signaling molecules, or their chemical precursors, enter the vascular system and are transported across the entire length of the plant. Because plants grow and develop in many places at once, signaling molecules that can move from one part of the plant to another play an important role in coordinating growth and development at the whole plant level.

Hormones affect the growth and differentiation of plant cells.

Hormones are chemical signals that influence both physiology and development. They play a central role in establishing and maintaining basic patterns of plant development, as well as coordinating growth in different parts of the plant in response to both internal and external environmental factors. A key feature of hormones is that they are effective in extremely low concentrations. In plants, hormones can be synthesized throughout the plant body, although they are most often produced in meristems. In addition, plant hormones can affect both the cells in which they are produced as well as cells located in distant regions of the plant.

Plant hormones influence cell growth and differentiation by altering patterns of gene expression as well as rates of cell division and cell expansion. How a cell responds to a particular hormone depends both upon its developmental stage and its position within the plant. In addition, the effect of any one hormone is often affected by the presence of one or more other hormones. This observation explains how a small number of hormones is able to shape the ongoing development and growth of plants.

TABLE 31.1 THE FIVE MAJOR HORMONES

HORMONE	SYNTHESIS AND TRANSPORT	EFFECTS
Auxin Indole-3-acetic acid	Synthesized primarily in shoot apical meristems and young leaves. Transported by polar transport and through phloem.	Induces cell wall extensibility. Inhibits outgrowth of axillary meristems. Stimulates formation of root meristems. Stimulates differentiation of procambial cells. Involved in phototropism and gravitropism. Stimulates ethylene synthesis.
Gibberellic acid	Synthesized in the growing regions of both roots and shoots.	Stimulates stem elongation and cell division. Mobilizes seed resources for developing embryo. Stimulates cytokinin synthesis.
Cytokinins Zeatin	Synthesized in both root and shoot meristems. Transported in both xylem and phloem.	Stimulates cell division. Delays leaf senescence. Acts synergistically with auxin.
Ethylene	Synthesized in elongating regions of roots and stems, as well as in developing fruits. Gaseous; precursors can be transported in xylem.	Reduces cell elongation. Triggers fruit ripening. Formation of air spaces in roots.
Abscisic acid	Synthesized in root cap, developing seeds, and leaves. Transported from roots to leaves in xylem.	Maintains seed dormancy. Stimulates root elongation. Triggers closing of stomata. Antagonistic with ethylene.

Traditionally, five major plant hormones have been recognized: auxin, gibberellic acid, cytokinins, ethylene, and abscisic acid (**Table 31.1**). One characteristic of these hormones is that each triggers a wide range of developmental effects. For example, ethylene can lead to fruit ripening (Chapter 30) and also slows the elongation of roots and stems. Several new chemicals with hormone-like properties have been recently discovered. Some of these have broad effects on plant development (e.g., strigolactone), whereas others appear to be more specific (e.g., florigen).

To illustrate how hormones influence plant growth and development, we consider three aspects of shoot development: the differentiation of vascular tissues, the elongation of internodes, and the formation of lateral branches.

Auxin transport guides the development of vascular connections between leaves and stems.

During the initial growth of leaf primordia, the resources needed can be met by diffusion. As the young leaves increase in size, their demand for water and carbohydrates quickly exceeds what can be supplied by diffusion. If they are to grow to even a fraction of their full size, leaves must establish vascular connections with the xylem and phloem in the stem.

The plant hormone **auxin** was first identified by its ability to cause shoots to elongate. However, auxin has an even more remarkable property: It is transported preferentially from young leaves to the differentiated vascular tissues in the stem. In this way, the movement of auxin creates a template that guides vascular differentiation.

Auxin is synthesized and secreted by cells in young, expanding leaves. In the cell wall, auxin carries no net charge and thus can enter surrounding cells by diffusing across the plasma membrane. However, because the pH of the cytoplasm is higher than the pH of the cell wall, auxin loses a proton. As a result, auxin has a net negative charge when inside the cell, preventing it from moving easily across the plasma membrane. For auxin to leave the cell, a specific plasma membrane transport protein known as PIN is required. By restricting the placement of PIN proteins to only one face of the cell, plants can control the direction of auxin movement. Each cell in a plant's body has an apical end and a basal end that are maintained independent of gravity or orientation. PIN proteins typically are located only on the basal side of cells, the side farthest away from the shoot tip. As a result, the prevailing direction of auxin movement is in the apical-to-basal (top-to-bottom) direction. The coordinated movement of auxin across many cells is referred to as **polar transport** (**Fig. 31.9**).

FIG. 31.9 Polar transport of auxin. The polar transport of auxin results in the movement of this hormone from the tip of each developing leaf to the base of the plant.

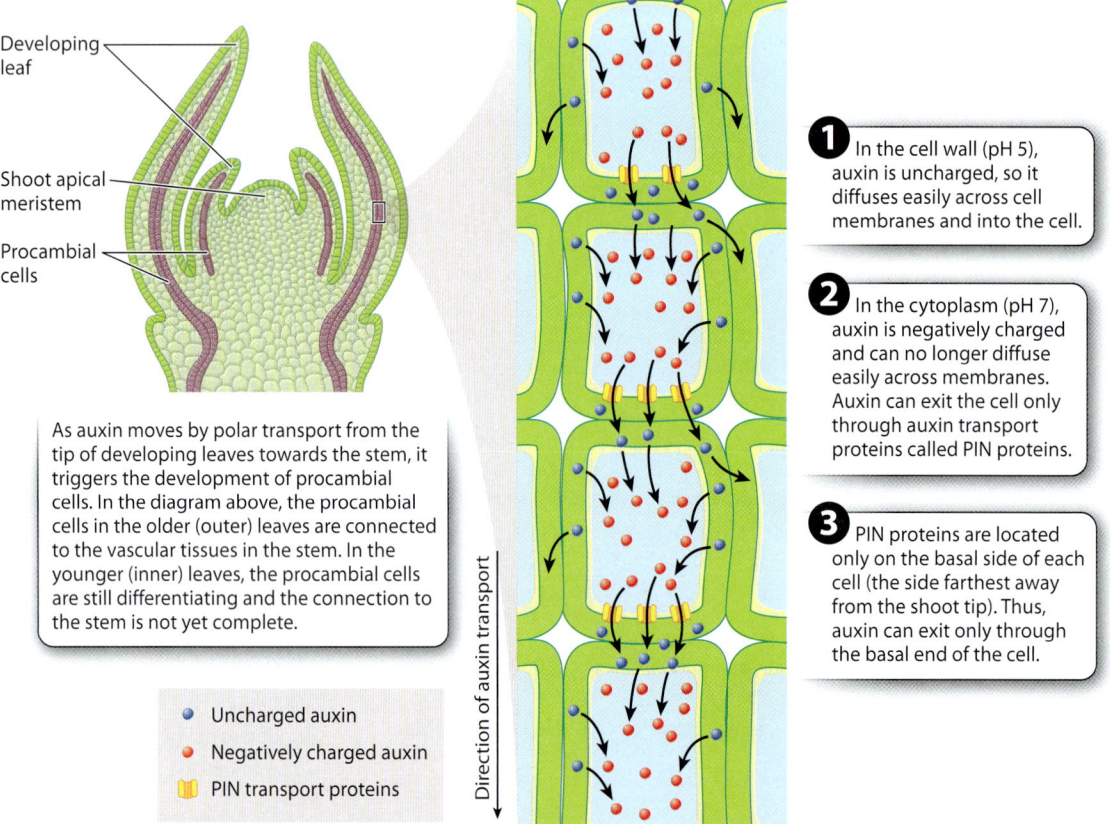

Auxin affects cells both by stimulating their elongation and increasing their production of PIN proteins. Thus, auxin transport results in linear rows of elongated cells as outflow from the basal side of one cell leads to enhanced auxin concentration in the cell immediately below. The elongated cells that carry auxin from the developing leaves to the stem differentiate to become **procambial cells.** Procambial cells retain the capacity for cell division and ultimately give rise to both xylem and phloem. The common origin of xylem and phloem from procambial cells explains why these two tissues are always found in close association throughout the length of the plant.

As auxin moves by polar transport out of developing leaves it is drawn towards the vascular tissues in the stem. The reason this happens is that auxin moves *between* cells by diffusion. Recall from Chapter 29 that molecules diffuse from regions of higher concentration to regions of lower concentration. The vascular tissues within the stem create a region of low auxin concentration because they transport auxin towards the roots faster than it can move through undifferentiated cells. In this way, the polar transport of auxin that originates in developing leaf primordia is responsible for establishing the vascular connections with the stem that allow leaves to grow and function.

The vascular system of an elongating stem is built from the successive addition of procambial strands that initiate in the developing leaves. The simplest arrangement, which is characteristic of the lycophytes, is a single cylinder of vascular tissues, with xylem at the center and phloem on the outside. The procambial strands that connect each leaf to the stem merge with this central cylinder.

In seed plants, the most common configuration is a ring of **vascular bundles** near the outside of the stem. The bundles have xylem conduits toward the center of the stem and phloem nearer the outside of the stem (**Fig. 31.10**). Typically, each leaf develops vascular connections with several of the stem's vascular bundles. The region between the epidermis and the vascular bundles is the **cortex,** and the region inside the ring of vascular bundles is the **pith.** The placement of the vascular bundles, which contain the stiff xylem conduits, near the outside of the stem increases the stem's ability to support its leaves without bending over.

? CASE 6 Agriculture: Feeding a Growing Population
What is the developmental basis for the shorter stems of high-yielding rice and wheat?

In natural ecosystems, competition for sunlight creates a strong selective advantage for growing tall. Compare the slender vine in Fig. 31.1, which grows in a dense forest, with the compact barrel cactus that grows where there is little danger of being shaded by a neighboring plant. Whereas the ancestors of today's crops competed for sunlight by growing taller, a key element of the high-yielding crop varieties developed during the Green Revolution is their shorter stems. These shorter-stemmed plants can support the much-larger seed mass induced by high rates of

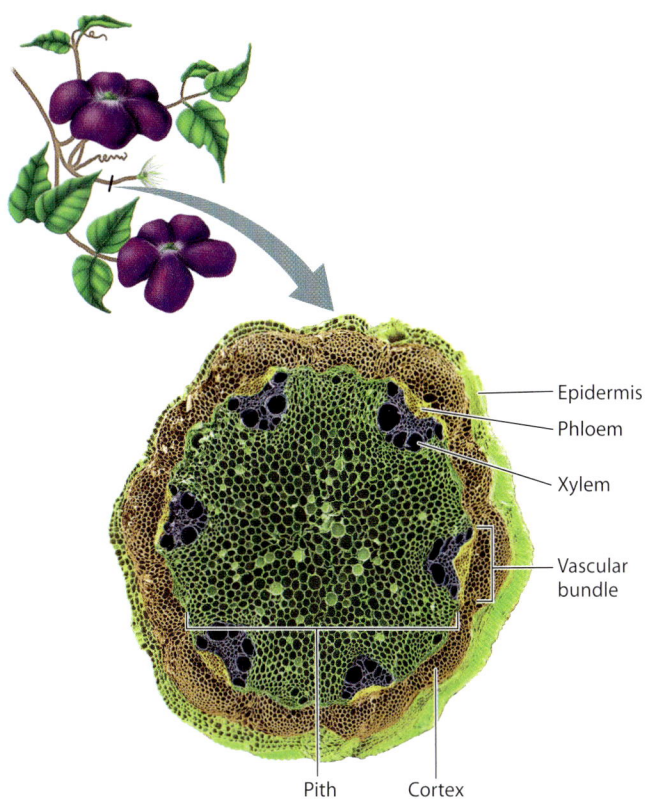

FIG. 31.10 Arrangement of vascular bundles. In most seed plants, the vascular bundles are arranged in a ring near the outside of the stem.

fertilizer application (Chapter 29). Moreover, by investing less in stems, these plants can allocate a greater percentage of their resources to reproduction.

At the time that the semidwarf varieties were developed, the underlying basis for their reduced stem growth was not known. What happens inside plants of these high-yielding crop varieties to make the stems shorter? Understanding how plants control intermodal elongation provides some insight.

The extent to which stems elongate can vary even within a single plant. For example, a tree in the temperate zone produces widely spaced green leaves during spring, but as autumn approaches the internodes cease elongating. As a result, bud scales become tightly packed together. This dramatic change in internode elongation is controlled by hormones.

Gibberellic acid is a plant hormone that stimulates the elongation of stems. It was first identified as a chemical produced by fungal pathogens that caused the stems of young rice plants to grow to many times their normal length. Giberellic acid increases internode elongation by reducing the force needed to cause cell walls to expand.

The genes responsible for reduced stem growth have now been identified in both wheat and rice. The semidwarf wheats have reduced sensitivity to gibberellic acid, and the shorter rice varieties are deficient in the biosynthetic enzymes needed to produce this hormone. In both cases, the result is shorter but sturdier plants.

Branching is affected by multiple hormones.

Branching allows plants to support more leaves, but producing branches too close together results in leaves that overlap and shade one another. What causes axillary buds to break dormancy and begin to grow? The answer to this question demonstrates how the interplay of different hormones can affect growth. In this case, the interplay of hormones produced in distinct parts of the plant communicates to axillary buds when conditions are right for the growth of new branches.

Removal of the shoot tip stimulates the outgrowth of axillary buds along that branch or stem (**Fig. 31.11**), suggesting that the shoot apical meristem somehow suppresses the growth of axillary buds. This suppression is referred to as **apical dominance.** If the shoot tip is cut off and a paste containing auxin is applied to the cut end of the stem, the axillary buds remain dormant. If instead the shoot tip is treated with a substance that inhibits auxin transport, branching is induced. Thus, branching is suppressed by the movement of auxin from the shoot tip toward the root. Apical dominance prevents branches from being formed too close to the shoot tip and also allows new branches to form should the shoot meristem be damaged.

Auxin suppresses the growth of axillary buds by inhibiting the synthesis of **cytokinins,** hormones that stimulate the outgrowth of axillary buds by increasing the rate of cell division (Table 31.1). When the apical meristem is removed, cytokinin synthesis increases in the stem nodes, triggering the growth of axillary buds. Recent studies indicate that this top-down control is complemented by the action of another hormone, **strigolactone,** which is made in roots and transported upward in the xylem. Plants that are unable to synthesize strigolactone produce many more branches than do wild-type plants. This indicates that strigolactone inhibits the outgrowth of axillary buds.

Clearly, the development of axillary buds is affected by multiple hormones, although the details of how they interact and under what conditions they are synthesized in greatest amounts are not yet known. One hypothesis is that strigalactones are transported to the shoot when the root system is unable to supply the water and nutrients required by new branches.

31.3 SECONDARY GROWTH

Shoot apical meristems enable plants to grow in length and to branch. But a plant unable to grow in diameter would not become very tall before buckling under its own weight. And a plant unable to add vascular tissue could not supply its ever-increasing number of leaves with the water needed for photosynthesis. Growth in diameter thus serves two functions: It strengthens the stem, and it increases the capacity of the vascular system. Both are essential for supporting and supplying large numbers of leaves.

Primary growth is the increase in length made possible by apical meristems. An increase in diameter is referred to as **secondary growth** because it results from a new type of meristem, one that forms only after elongation below the growing tip is complete. The fossil record shows that the ability to grow in diameter evolved millions of years after primary growth, and it did so independently in several lineages, testimony to the benefits of being able to compete for sunlight by growing tall. Today, secondary growth occurs almost exclusively in seed plants, although this was not always the case in Earth history.

Shoots produce two types of lateral meristem.

Lateral meristems are the source of the new cells that allow plants to grow in diameter. Lateral meristems are similar to shoot apical meristems in their ability to produce new cells that grow and differentiate. Lateral meristems, however, differ from their apical counterparts in several key ways. First, unlike shoot apical meristems, which are located at the shoot tip, lateral meristems form cylinders or, more properly, cones of cells that extend the entire length of the stem. Second, because lateral meristems form only after elongation is complete, the new cells they produce grow in diameter but not in length. Finally, as a stem becomes thicker, the number of meristem cells needed to encircle the stem also increases. Thus, lateral meristems become larger over time.

Plants produce two distinct lateral meristems that together result in secondary growth. One of these, the **vascular cambium** (plural, cambia), is the source of new xylem and

FIG. 31.11 Apical dominance.

Shoot apical meristem intact — The growth of axillary buds is suppressed when the shoot apical meristem is intact.

Shoot apical meristem cut off — When the shoot apical meristem is removed, growth of axillary buds along that stem is stimulated.

phloem. Secondary xylem is familiar to you: It's wood. Plants with secondary growth are often described as woody, whereas plants that lack secondary growth and so are produced entirely by apical meristems are described as herbaceous. In a large tree, most of the plant's mass consists of wood, generated from the vascular cambium.

As the activities of the vascular cambium increase the plant's diameter, epidermis formed during primary growth eventually ruptures. Thus, plants with secondary growth have a second lateral meristem, the **cork cambium,** which renews and maintains an outer layer that protects the stem against herbivores, mechanical damage, desiccation, and, in some species, fire. These tissues (along with secondary phloem) are commonly known as bark. The word "cambium" derives from the Latin word *cambiare*, meaning "to change." As we will see, the development of lateral meristems fundamentally changes the character of stems.

The vascular cambium produces secondary xylem and phloem.

The vascular cambium forms at the interface between xylem and phloem. In stems, the vascular cambium derives from procambial cells within each vascular bundle and parenchyma cells in between the bundles. In forming vascular cambium, both types of cell take on a new identity, becoming meristem cells.

The vascular cambium produces new cells that differentiate on both its sides. Those to the inside of the vascular cambium become **secondary xylem,** and those on the outside develop as **secondary phloem** (**Fig. 31.12**). When you peel the bark off a stem, it separates from the wood at the vascular cambium. This is because the undifferentiated and actively dividing cells of the vascular cambium are weak and easily pulled apart. The vascular cambium forms a continuous layer, one to several cells thick, which surrounds the xylem and runs nearly the entire length of the plant.

The high rates of water loss by leaves associated with their uptake of CO_2 for photosynthesis (Chapter 29) means that plants must produce more secondary xylem than secondary phloem. As the vascular cambium forms new secondary xylem cells along its inner face, the newly formed cells push the vascular cambium outward, much the way the elongation of newly formed cells pushes the shoot apical meristem upward. The vascular cambium accommodates this outward movement by adding new cells that increase its circumference. In contrast, the secondary phloem, once formed, is unable to produce any additional cells. As the production of secondary xylem pushes the vascular cambium outward, the phloem becomes stretched and, eventually, damaged. Consequently, new secondary phloem must be continually produced.

Over time, xylem conduits lose their ability to transport water. Thus, in long-lived trees, the functional xylem is located in a **sapwood** layer adjacent to the vascular cambium, while

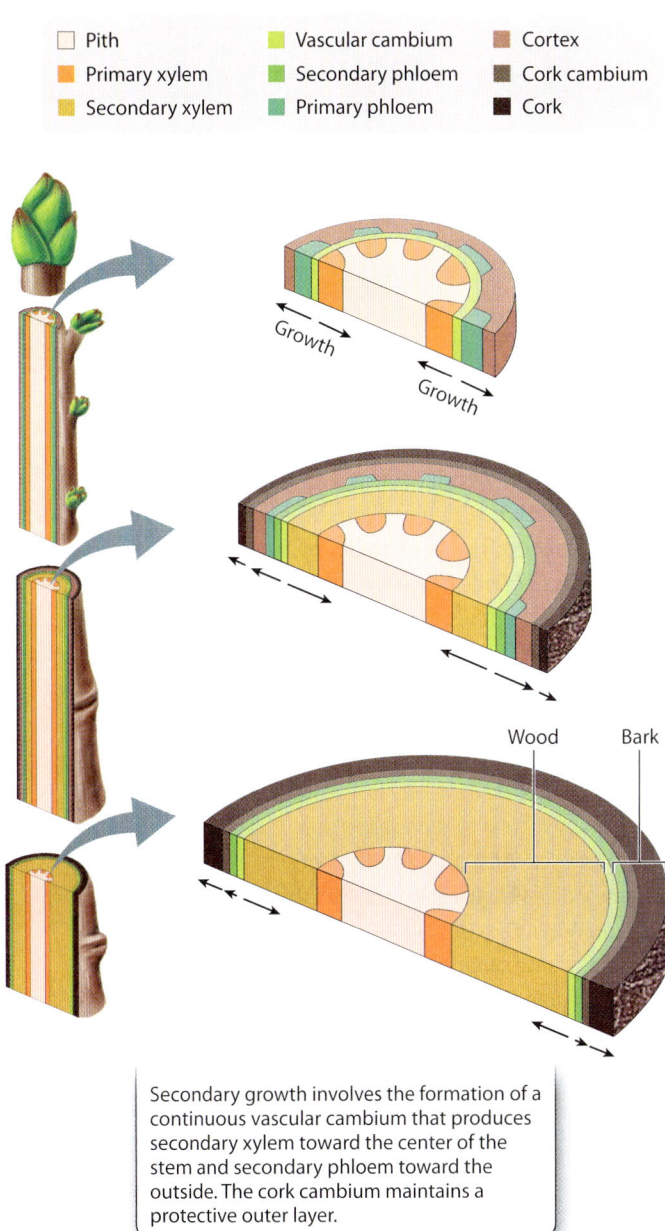

FIG. 31.12 Lateral meristems. The vascular cambium is the source of new xylem and phloem, while the cork cambium maintains an outer layer of bark.

Secondary growth involves the formation of a continuous vascular cambium that produces secondary xylem toward the center of the stem and secondary phloem toward the outside. The cork cambium maintains a protective outer layer.

the **heartwood,** in the center of the stem, does not conduct water. Heartwood is often darker in color than sapwood because it contains large amounts of chemicals such as resins. For this reason, heartwood is typically more resistant to decay than sapwood.

In many plants in seasonal climates, decreases in the size of secondary xylem cells at the end of the growing season result in **growth rings,** which make it possible to determine a tree's age. Growth rings also provide a window into the past because by measuring the width of each ring, it is possible to determine

FIG. 31.13 Annual growth. In seasonal climates, tree rings provide a record of the annual accumulation of wood. The tree rings shown are from red pine (*Pinus resinosa*).

- Wide growth rings indicate higher growth rates and thus suggest that the tree experienced more favorable conditions in recent years.
- Narrow growth rings indicate that this tree grew slowly when it was young, perhaps because it was shaded by another tree.

how much wood was laid down in a given year (**Fig. 31.13**). Wood production is higher when the environmental conditions that enable photosynthesis are more favorable. Thus, annual growth rings provide a record of past climatic conditions such as temperature and moisture, as well as events such as fire.

The cork cambium produces an outer protective layer.

The role of the cork cambium is to maintain a protective layer around a stem that is actively increasing in diameter (see Fig. 31.12). The cork cambium forms initially from cortex cells that dedifferentiate—that is, they regress to their earlier state—to become meristem cells. However, with time, this layer of actively dividing cells becomes increasingly cut off from its source of carbohydrates—the phloem. When the cork cambium becomes sufficiently cut off from its carbohydrate supply, a new cork cambium forms within the secondary phloem. Thus, each cork cambium is relatively short lived, and in many cases, new cork cambia form in patches rather than as continuous layers. As a result, the bark of many trees has a patchy and fissured appearance (**Fig. 31.14**).

The cells produced by the cork cambium are called cork. As cork cells mature, they become coated with **suberin,** a waxy compound that protects against mechanical damage and the entry of pathogens, while also forming a barrier to water loss. However, this layer of waxy, nonliving cells impedes the inward diffusion of oxygen needed to meet the respiratory demands of living cells. The outer bark cells are less tightly packed in regions called **lenticels,** allowing oxygen to diffuse into the stem (**Fig. 31.15**).

→ **Quick Check 2** Why do plants have two types of lateral meristem?

Wood has both mechanical and transport functions.

We have seen that as a tree grows, its secondary xylem fulfills the two roles of secondary growth. The first is to provide trees with the strength and stability to stand upright and support large crowns. The second is to transport water and nutrients from the roots to the leaves. In most gymnosperms, a single cell type fulfills both functions. Cells on the xylem side of the vascular cambium differentiate to form tracheids (**Fig. 31.16a**). As discussed in Chapter 29, tracheids are single-celled xylem conduits that have thick lignified cell walls and are nonliving and empty at maturity. The wood of a pine tree, for example, is relatively homogeneous in structure, consisting almost entirely of tracheids. This type of wood does not vary dramatically from species to species.

Angiosperms, by contrast, exhibit a great diversity in wood structure, as the cells mature into several distinct types that separate the functions of water conduction and mechanical support (**Fig. 31.16b**). **Fibers** are narrow cells with extremely thick walls and almost no lumen; their primary function is support. In contrast, **vessel elements** can be extremely wide cells; their primary function is water transport (Chapter 29). The largest vessel elements are close to 0.5 mm in diameter and thus visible to the naked eye. Vessel elements can be much wider than tracheids because the fibers provide support. And because vessel elements develop open connections that link them end to end

---

FIG. 31.14 Outside (a) and inside (b and c) of the outer bark of a pine tree, showing how successive cork cambia are produced. The growth of many, discontinuous cork cambia results in the patchy bark of pine trees.

a. Outside of tree, ponderosa pine (*Pinus ponderosa*)

b. Cross-section of bark, white pine (*Pinus strobus*)

c.

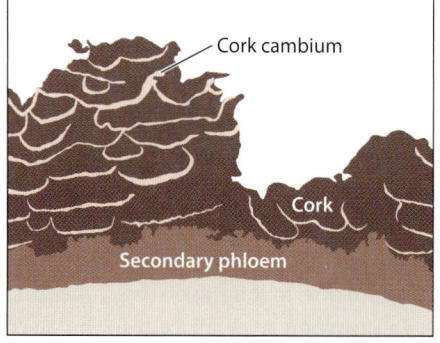

Cork cambium
Cork
Secondary phloem

FIG. 31.15 Lenticels in (a) water birch (*Betula occidentalis*) and (b) basswood (*Tilia americana*). Lenticels allow oxygen to diffuse through the outer bark to supply the respiratory needs of the cells within the stem.

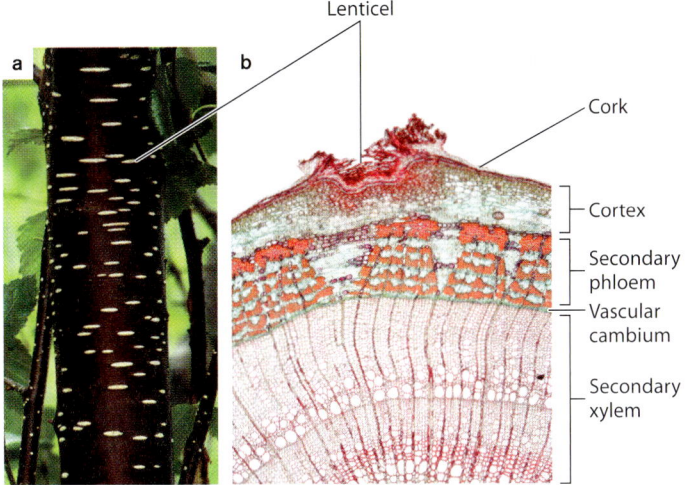

with other vessel elements, creating multicellular xylem **vessels,** they allow angiosperms to produce wood that is more efficient at transporting water than the wood of gymnosperms.

The ability to form xylem vessels has a number of important consequences. One of these is that angiosperms can have higher rates of water flow through their stems than is possible with wood that contains only tracheids. As a result, angiosperms sustain higher rates of CO_2 uptake and thus higher growth rates.

In addition, because vessels can be much more efficient than tracheids at transporting water, angiosperms can devote a smaller fraction of their wood to water conduction, and more of their stem volume to other cell types. The wood of some angiosperms contains many fibers, in extreme cases resulting in wood that is so dense that it sinks. Such dense woods are extremely strong, allowing trees to create spreading crowns with outstretched branches. Such "hardwoods" are useful as flooring materials and furniture. The wood of other angiosperms contains a high percentage of living cells instead of fibers, and this wood has a very low density. Balsa trees, whose light wood is used in building model airplanes, are a good example.

31.4 ROOT GROWTH AND DEVELOPMENT

The roots that evolved in vascular plants enabled them to obtain water and nutrients from the soil and provided a mechanical anchor for plants capable of growing tall (Chapter 29). As in the case of leaves and wood, the fossil record shows that the first vascular plants lacked roots. These plants would have been limited to extremely wet environments in which they could absorb water through horizontal, and possibly submerged, stems. The evolution of roots, then, allowed vascular plants to expand into drier habitats and ensured the stability of taller stems.

The fossil record also indicates that roots evolved independently in the lycophytes and the lineage leading to the other vascular plants. Nevertheless, the roots of all vascular plants share many features. This suggests that the physical challenges of

FIG. 31.16 Wood structure in (a) gymnosperms and (b) angiosperms. These scanning electron micrographs show (a) Douglas fir (*Pseudotsuga menziesii*) and (b) English elm (*Ulmus procera*).

a. Gymnosperms

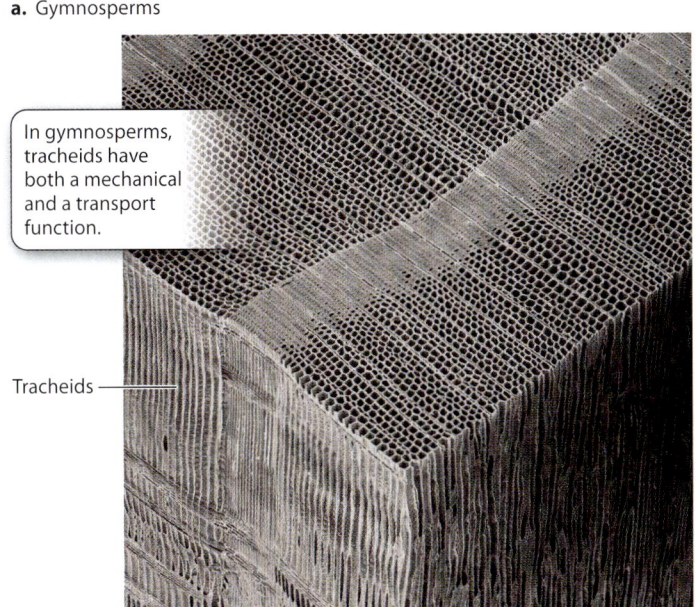

b. Angiosperms

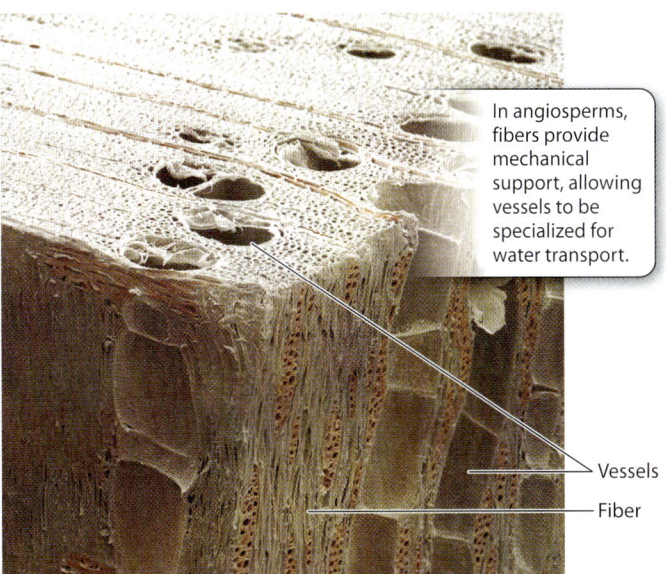

growing through a dense medium such as soil and the demands of obtaining water and nutrients from that medium place strong constraints on root structure. Here, we explore the growth and development of the plant body belowground, emphasizing how the elongation and branching of roots enables vascular plants to obtain the water and nutrients they need from the soil.

Roots grow by producing new cells at their tips.

In Chapter 29, we saw that plants produce large numbers of thin roots to obtain sufficient nutrients from the soil. And to obtain water during periods without rain, plants produce roots that penetrate deep into the soil. Growing roots into the soil presents challenges different from those associated with elongating stems into the air. Although the dangers of drying out are much reduced belowground, deeper soil layers can have densities approaching that of concrete. How are the thin and permeable roots needed for water and nutrient uptake able to penetrate and grow through the soil?

At one level, roots grow in the same way as shoots, with new cells produced at their tip. The **root apical meristem** serves as the source of new cells. After cell division, these cells undergo elongation and complete their differentiation only after elongation has ceased (**Fig. 31.17**). Thus, as we move back from the elongating root tip, we encounter successive regions of cell division, elongation, and maturation, just as we saw in stems.

Roots, however, differ from stems in several important ways. One important difference is that roots are much thinner than stems. Plants with thin roots can more efficiently produce the large surface area needed for optimal uptake of water and nutrients from the soil. Thin roots can also more easily bend around soil particles and make use of channels created by invertebrate animals that burrow through the soil. A second major difference is that the root apical meristem is covered by a **root cap** that protects the meristem as it grows through the soil. As the root elongates through the soil, root cap cells are rubbed off or damaged. Thus, the root cap is continually supplied with new cells produced by the root apical meristem. A third difference between roots and stems, and one that can be seen without a microscope, is that the root apical meristem does not produce any lateral organs, whereas the shoot apical meristem makes leaves. Root hairs, the single-celled outgrowths of root epidermal cells that greatly increase the surface area for uptake (Chapter 29), are formed only after elongation has ceased.

Let's now look at how roots develop the specialized cells and tissues that they need to function. A cross section taken through the zone of cell maturation shows a series of concentric tissue layers, starting with the epidermis on the outside, followed by the root cortex, and then the endodermis (Fig. 31.17). Recall that the endodermis plays a key role in the selective uptake of nutrients (Chapter 29). The Casparian strip, a waxy band surrounding each endodermal cell, forces all materials that will enter the xylem to cross the plasma membranes of these cells.

At the center of the root lies a single vascular bundle in which the xylem is arranged as a fluted cylinder, with phloem nestled in each groove. In cross section, the xylem is arranged in a star shape, the number of points of the star differing in different plant species.

As in stems, the hormone auxin plays a central role in specifying which cells differentiate into xylem and phloem. Auxin produced by developing leaves is transported downward through the mature stem and into the roots by the phloem. At the end of the differentiated phloem, auxin moves by polar transport from cell to cell until it reaches the root apical meristem and the root cap. As auxin moves through the elongation zone, it triggers the formation of procambial cells that subsequently differentiate into phloem and xylem.

In this way, the production of auxin by developing leaves and the

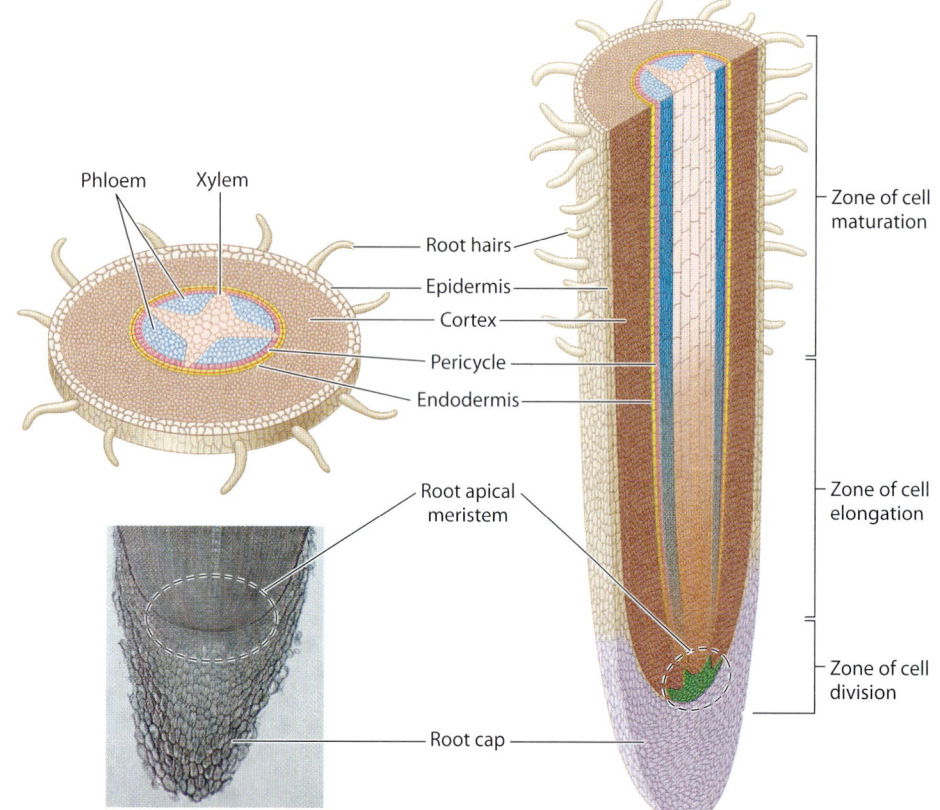

FIG. 31.17 The structure of a typical root, including the root apical meristem and root cap.

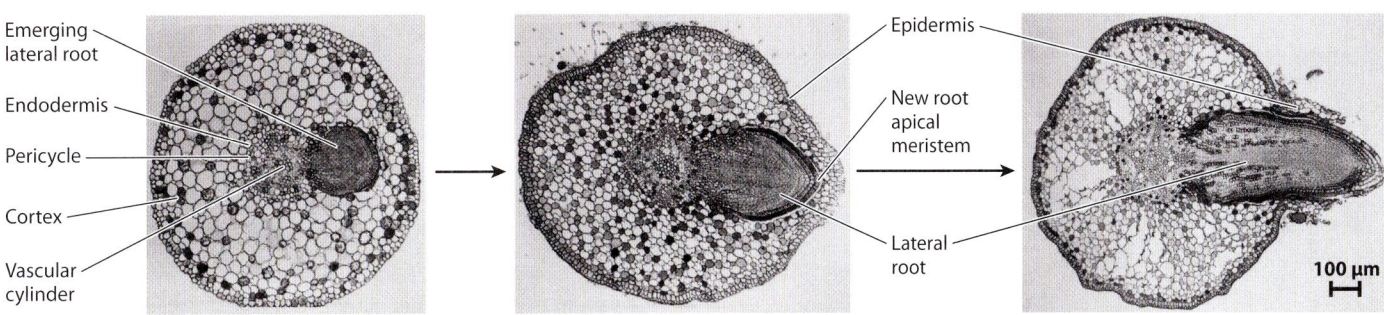

FIG. 31.18 Root pericycle. New root meristems develop from the pericycle and the developing lateral root grows through the root cortex before reaching the soil. The light micrographs show willow (*Salix*) roots.

polar transport of auxin from cell to cell ensures that vascular continuity is established between roots and leaves. Phloem extends closer to the root tip than do fully lignified xylem conduits. This reflects the fact that the apical meristem can obtain water and nutrients directly from the soil but depends entirely on the leaves for carbohydrates needed for growth.

So far, we have described only the primary growth of roots. However, roots can undergo secondary growth as well. This allows roots to increase in diameter and to add new xylem and phloem. As in stems, the vascular cambium forms at the interface between the xylem and the phloem. Indeed, the vascular cambium must be continuous from the shoot to the roots to ensure that the secondary xylem and phloem provide an unbroken connection between the roots and the leaves. In older roots with extensive secondary growth, a cork cambium is also formed. As in stems, the cork cambium renews the outer layers of cells that protect the living tissues from damage and predation.

The formation of new root apical meristems allows roots to branch.

Compared to stems, roots branch like crazy. Even a plant that has only a single stem aboveground will have many tens of thousands of elongating root tips. A major reason that roots branch so much is that the uptake of water and nutrients is concentrated in a zone near the root tip where root hairs are most abundant. Thus, branching is necessary to create enough root surface area to supply the water and nutrient needs of the leaves.

For a root to branch, a new root apical meristem must be formed. In roots, new meristems develop from the **pericycle,** a single layer of cells just to the inside of the endodermis (**Fig. 31.18**). The pericycle is immediately adjacent to the vascular bundle, so the new root is connected to the vascular system right from the beginning. But because the new root meristem develops internally, the new root must grow through the endodermis, cortex, and epidermis before reaching the soil.

One consequence of initiating new root meristems from the pericycle is that new roots can form anywhere along the root. As a result, plants have tremendous flexibility in the development of their root systems and roots can proliferate locally in response to nutrient abundance. For example, nitrate (NO_3^-), the most common form of nitrogen in soil, triggers the formation of lateral roots by causing auxin to accumulate locally within the root tissues nearest to the nitrate-rich soil. Only a relatively small number of a plant's roots, however, continue to elongate and branch indefinitely. Most new roots are active for only a short time, after which they die and are shed. This turnover of fine roots allows the root system to respond dynamically to changes in the availability of water and nutrients but represents a major expenditure of carbon and energy for the plant.

Up to this point, we have described how new roots form from existing roots. New roots can also be produced by stems. This is important in plants that lack secondary growth, many of which produce horizontal stems that grow on or beneath the soil surface, for example many lycophytes and ferns. In these plants, new roots form at each successive node, so the uptake capacity of the root system is directly coupled to the elongation of the stem.

When the stem of a plant is severed—and if the shoot tip is not allowed to dehydrate—new roots often form at the cut end. The new root meristems are stimulated by auxin, which is produced by the young leaves and accumulates, by polar transport, at the cut end. The ability of cut stems to form new roots allows plants to survive and in some cases proliferate following mechanical damage. It is also the means by which many commercially important plants are propagated.

→ **Quick Check 3** How is root development similar to and how is it different from stem development?

The structures and functions of root systems are diverse.

In the preceding sections, we focused on the growth and branching of individual roots. How a plant's roots are distributed in the soil has a tremendous impact on the ability of the root system to supply the shoot with water. Some plants produce very deep roots that provide them with access to a groundwater supply that is independent of the variability of rainfall. This can be advantageous, but it comes at the cost of producing and maintaining these

FIG. 31.19 Root diversity. (a) Climbing roots of poison ivy; (b) prop roots of a *Pandanus* tree; (c) storage roots of yams; and (d) breathing roots of black mangrove (shown at low tide).

non-photosynthetic structures. Roots can penetrate into mine shafts and caves more than 50 m below the soil surface.

At the other extreme are plants that produce shallow and spreading root systems. For example, barrel cacti produce roots that penetrate less than 10 cm into the soil but extend out several meters from the base of the plant. These roots cast a wide but shallow net to capture as much water as possible from intermittent rain showers.

Some plants produce distinctive roots whose principal role is other than the absorption of water and nutrients from the soil (**Fig. 31.19**). For example, climbing plants such as poison ivy produce roots along their stems that allow them to adhere to the sides of trees (Fig. 31.19a). The conspicuous prop roots produced by some tropical trees provide mechanical stability that allows these plants to grow tall despite their slender stems (Fig. 31.19b).

Many plants produce swollen roots in which resources such as starch can be stored (Fig. 31.19c). Belowground storage allows plants to produce new leaves and shoots quickly following either damage or a period in which drought or cold has caused the shoots to die. Cassava and sweet potato are examples of economically important species with storage roots.

Perhaps the most unusual rooting structures are the "breathing" roots produced by trees that grow in water. For example, black mangroves grow along the seashore in tropical regions and produce pencil-sized roots that extend vertically upward out of the sandy soil (Fig. 31.19d). Breathing roots do not actually breathe; instead, they contain internal air spaces that provide an easier pathway for oxygen to diffuse into the roots than the waterlogged soil. For this reason, breathing roots must extend above the surface of the water. Thus, the breathing roots of black mangrove provide an estimate of the water level at high tide.

31.5 THE ENVIRONMENTAL CONTEXT OF GROWTH AND DEVELOPMENT

We take it for granted that shoots grow up and roots grow down. But if you lay a plant on its side, its stems will turn upward and its roots will bend downward. Similarly, if you place a houseplant near a window, the plant will bend over and grow toward the light. Plants are able to modify their growth in response to both gravity and light. They can sense and respond to the world around them in other ways as well. For example, trees planted close together produce stems that are taller and thinner than those of trees planted far apart. Is this simply what happens when crowded plants compete for the resources that they need to grow? No. Experiments have shown that the taller, thinner stems result from individual plants' sensing the presence of their neighbors.

Plants gain information about the world around them in three principal ways. Photoreceptors, discussed in Chapter 30, provide information about the availability of light needed to drive photosynthesis. Receptors triggered by mechanical forces provide plants with information about physical influences such as gravity and wind. Chemical receptors allow plants to detect and respond to the presence of specific chemicals, as well as chemical gradients. When one of these sensory receptors is triggered, it sets in motion a chain of events that leads to a change in the way the plant develops and grows. Hormones play a key role in translating information gained by the plant's sensory receptors into an appropriate developmental response.

Plants orient the growth of their stems and roots by light and gravity.

How do stems and roots control their orientation so that they grow in the right direction? In most cases, light provides the most useful signal to guide the growth of stems. Roots and seeds that germinate underground use several cues, including gravity. **Tropism** is the bending or turning of an organism in response to an external signal such as light or gravity. Plant stems are positively **phototropic,** meaning they bend toward the light. Plant stems are also negatively **gravitropic,** meaning they grow upward against the force of gravity. In contrast, plant roots grow down and away from the light. Thus, roots are positively gravitropic and negatively phototropic.

In 1880, Charles Darwin, working with his son Francis, carried out some of the earliest research on phototropism. Observing that grass seedlings bend toward light, Darwin conducted experiments to determine what part of the plant detects light. When he covered just the tip of the young grass plants, they no longer grew toward the light, but when he buried them in fine black sand so that only their tips were exposed, they did. This experiment established the shoot tip as the site where the light was perceived but left open the question of what kind of signal caused the plants to bend (**Fig. 31.20**).

In Chapter 30, we saw that plants produce photoreceptors that allow them to detect and respond to light. If a plant is exposed to

HOW DO WE KNOW?

FIG. 31.20

How do plants grow toward light?

BACKGROUND In 1926, Fritz Went, a Dutch graduate student, followed up on the experiments of Charles and Francis Darwin. In particular, Went sought to understand how the perception of light by the tip of the plant led to a growth response farther down the stem. He reasoned that if the signal from the shoot tip was a chemical, he could isolate it by cutting the tip off of a growing plant and putting it on a block of gelatin. The gelatin would act like a trap, capturing any chemical that moved out of the shoot tip.

HYPOTHESIS A chemical signal links the perception of light by the shoot tip with the growth of the stem toward light.

EXPERIMENT AND RESULTS Went cut off the tips of young oat plants (*Avena sativa*) and placed them on gelatin blocks for a period of 1 to 4 hours. Removing the tip of the plant caused the plants to stop elongating. When the gelatin block was placed on top of the cut plant (Experiment 1), growth resumed. This experiment demonstrated that a growth-inducing chemical was exported from the tip of the young oat plants.

Interestingly, Went found that if the block of gelatin was not centered on the tip but instead was a bit off center, the plant bent away from the side of the gelatin block (Experiment 2). In this case, the side of the plant exposed to the gelatin block grew faster, causing bending, because it received more of the chemical signal. This experiment demonstrates that increasing the amount of auxin on one side of a stem results in bending.

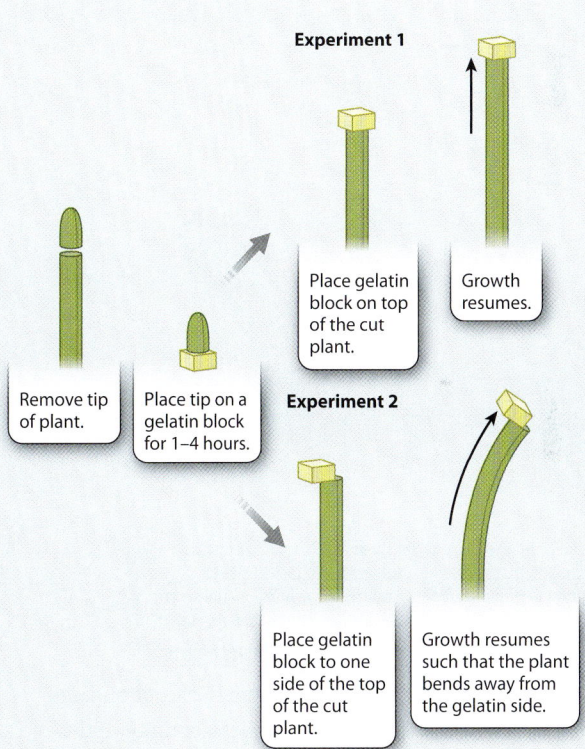

CONCLUSION A signal is produced in the tip of the plant and travels downward through the stem, toward the roots. This signal stimulates growth. Light causes the signal to move to the opposite side, stimulating growth on that side, which in turn causes the plant to bend toward the light. This signal is now known to be a plant hormone, called auxin from the Latin word *augere*, "to increase."

FOLLOW-UP WORK Went followed up his experiments by determining how the level of auxin relates to the degree of curvature. This assay became known as the Avena curvature test, after the experimental species. Other experiments showed that auxin moves in a directed fashion by polar transport. Later, auxin was shown to be indole-3-acetic acid (IAA).

SOURCES Darwin, C., and F. Darwin. 1880. *The Power of Movement in Plants*. London: John Murray; Went, F. W. 1926. "On Growth-Accelerating Substances in the Coleoptile of *Avena sativa*." *Proceedings of the Royal Academy of Amsterdam* 30:10–19.

FIG. 31.21 Phototropism in shoots, mediated by the hormone auxin.

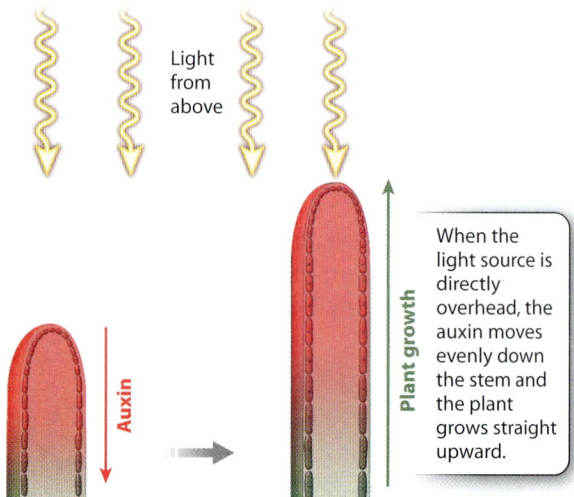

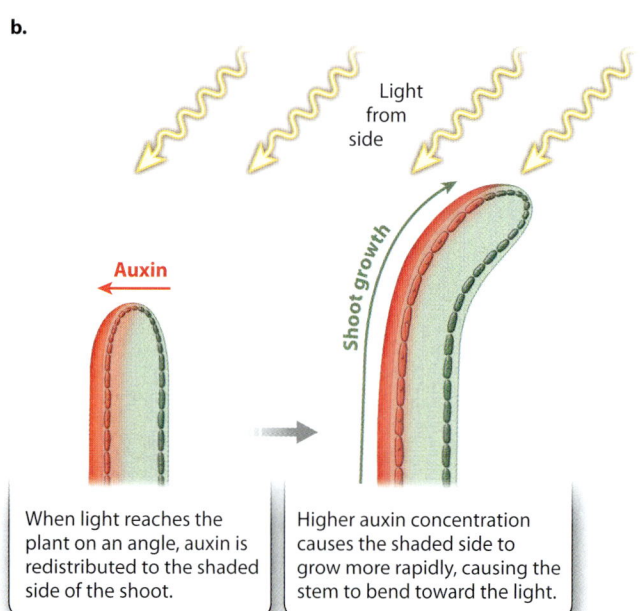

light from only one side, only the photoreceptors on the illuminated side are activated. Phototropism in plants involves a photoreceptor that absorbs blue wavelengths of light. The absorption of blue light by the photoreceptor causes auxin to be transported preferentially to the shaded side of the stem. This results in high concentrations of auxin on the shaded side and lower auxin levels on the illuminated side. Because auxin stimulates cell expansion in stems, the shoot grows faster on the shaded side than it does on the illuminated side, causing the plant to bend toward the light (**Fig. 31.21**).

Phototropism in roots is similar, in that illuminating a root from the side causes auxin to be redistributed toward the shaded side. In this case, auxin is transported to the shaded side within the root cap, then moves upward into the zone of cell elongation. However, higher auxin concentrations *decrease* the rate of cell elongation in roots, causing roots to bend *away* from light. The reason that auxin redistribution leads to such a different outcome in roots compared to shoots is that in roots high concentrations of auxin trigger the production of the hormone ethylene, which reduces cell elongation.

A similar mechanism orients plants with respect to gravity. Moving a plant to a horizontal position results in the transport of auxin to the lower side. The accumulation of auxin on the lower surface causes stems to bend up and roots to turn down. How is the plant able to detect which way is up, which way is down? Recall that plants store carbohydrates as starch. Starch is a large molecule that has a density greater than water. Starch-filled organelles thus sink to the bottom of the cell. The weight of these organelles pressing on the cytoskeleton or membranes is thought to provide plants with information regarding their orientation relative to gravity. In roots, for which gravity is a critical source of information regarding orientation, specialized gravity-sensing cells in the root cap contain large starch-filled organelles known as **statoliths** (**Fig. 31.22**). In stems, starch-filled organelles in bundle-sheath cells are thought to play a similar role.

FIG. 31.22 Root orientation with respect to gravity. Starch-filled statoliths settle to the bottom of cells, triggering a redistribution of auxin to the lower side of the root, where it slows elongation, thus causing the root to bend downward.

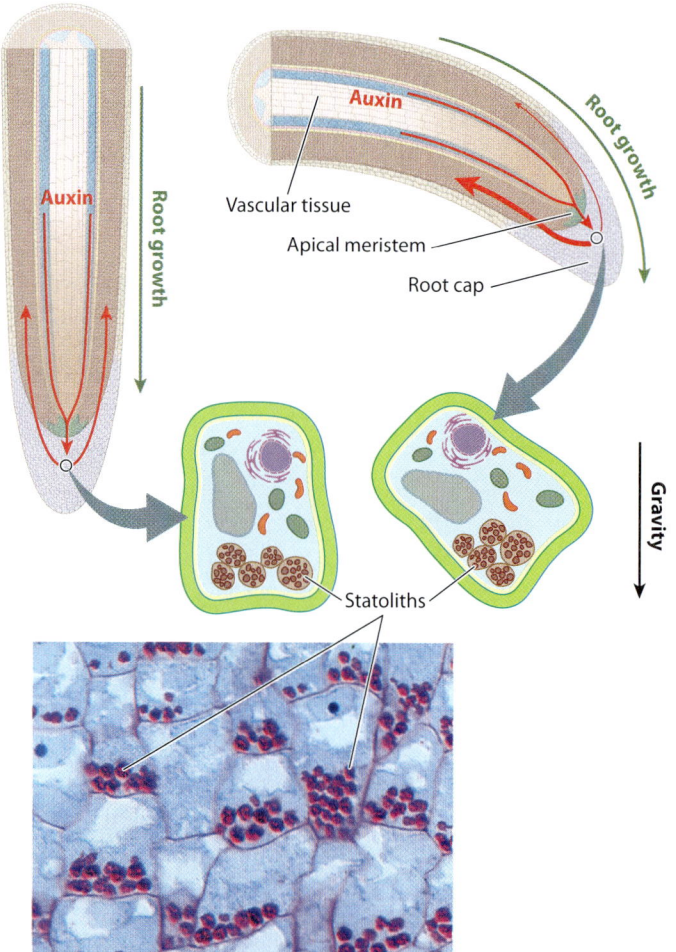

FIG. 31.23 Seedlings grown in the light and the dark (left) or in full sun and shade (right).

Plants grow taller and branch less when light levels are low.

A newly germinated seedling must reach the soil surface before its stored resources are used up or it will die. Seeds that germinate in the dark produce completely white seedlings with thin, highly elongated internodes and suppressed leaf development (**Fig. 31.23**). By putting all their resources into elongation, seedlings grown in the dark maximize their chances of reaching the light. If and when they do, internodal elongation slows markedly and green leaves expand. These dramatic changes are triggered by photoreceptors, which signal the presence of light. The transformation from a dark-grown to a light-exposed seedling involves several photoreceptors, including one that absorbs blue wavelengths and one, phytochrome, that is activated by the absorption of far-red wavelengths.

Even after a plant emerges above the soil, the ability to sense and respond to light can enhance its chances for survival. Some species are adapted for shade, producing thin leaves that efficiently use low levels of light; others grow best in direct sunlight. When such sun-adapted species are grown in the shade of another plant, they direct their resources primarily into growing tall. These plant species produce stems that have long, thin internodes and fewer branches, and they invest less in root growth underground.

Interestingly, the effect is much less if the shade is produced by anything other than another plant. The reason for this is that chlorophyll absorbs red wavelengths, but not far-red wavelengths of light. Thus, as light passes through leaves it becomes depleted in red wavelengths. As we saw in Chapter 30, phytochrome is a photoreversible photoreceptor that shuttles between a red-light and a far-red-light absorbing form. It allows plants to detect when they are underneath another plant. It also allows them to sense when they are surrounded by other plants that could grow to shade them in the future because light reflected off their neighbors' leaves is also depleted in red wavelengths of light. Thus, phytochrome provides an early-warning system of the presence of nearby competitors.

Roots elongate more and branch less when water is scarce.

The developmental sensitivity of plant shoots is mirrored underground. When a plant experiences drought, it produces more roots, and these roots penetrate farther into the soil, in part because they produce fewer lateral branches. By producing deeper roots, a plant increases its chances of reaching moister soil.

The root cap appears to play an important role in allowing roots to sense and respond to the amount of water in the soil. Because the root cap is not well connected to the plant's vascular system, the water status of its cells provides a good indication of the moisture content of the soil. As soils dry, root cap cells produce a hormone called **abscisic acid,** commonly abbreviated as ABA (Table 31.1). Abscisic acid stimulates root elongation by suppressing ethylene synthesis. Ethylene affects the growth of cells by influencing the orientation of their cellulose microfibrils. Cells treated with ethylene exhibit an increase in diameter growth and lower rates of elongation. Thus, by inhibiting the production of ethylene, the production of ABA in drought-influenced roots leads to roots that penetrate deeper into the soil.

Exposure to wind results in shorter and stronger stems.

In the 1980s, scientists studied the genetic model plant *Arabidopsis thaliana* to learn which genes are expressed in response to different hormones. The experimental treatment involved spraying *A. thaliana* plants with different solutions containing hormones; control plants were sprayed with water. The treatments were successful in that several genes were strongly up-regulated. However, the same genes were also turned on in the plants sprayed with water. Further experiments showed that what the plants were responding to was neither the hormones nor the water, but the fact that their stems were bent back and forth when the plants were sprayed. This led to the discovery of touch-sensitive genes in plants, which are activated by mechanical perturbation.

This discovery confirmed what foresters and horticulturists had long known. Plants exposed to wind produce stems that are shorter and wider than ones grown in more protected sites. In fact, commercial greenhouses often install fans, in part so that their plants will produce stems robust enough that the plants can thrive when moved outdoors. Flexing a stem back and forth triggers an increase in the synthesis of ethylene. Cells exposed to ethylene expand more in diameter and less in length, resulting in shorter and thicker stems.

Plants use day length as a cue to prepare for winter.

In Chapter 30, we saw how plants can use day length to control when flowers are produced. Not surprisingly, day length also triggers other developmental events, particularly ones that allow plants that persist for more than a year to prepare for winter. One such developmental change is the formation of storage organs, for example in roots. Carbohydrates that accumulate in these structures can support the growth of new leaves and stems the following spring.

A second developmental change in response to day length is the formation of overwintering buds. As the days begin to shorten, plants stop producing photosynthetic leaves and begin forming bud scales (see Fig. 31.6a). Bud scales surround and protect the meristem from ice and water loss. The formation of bud scales accompanies a series of metabolic changes that allow meristems to remain in a dormant state throughout the winter. For example, plugs are produced that block the plasmodesmata between the meristem and the rest of the plant. These plugs prevent any growth-stimulating compounds from reaching the meristem.

The hormone abscisic acid contributes to the formation of overwintering buds, similarly to its role in promoting seed dormancy. Many temperate trees must experience sustained exposure to cold temperatures before they will break dormancy, thus preventing a warm period in the autumn from inappropriately triggering growth. In the spring, lengthening days and warmer temperatures reverse the metabolic changes that accompany dormancy and allow growth to resume.

This section has focused on the environmental context of plant growth and development—how plants alter their development to enhance the uptake of water, nutrients, and sunlight, and to cope with physical stresses such as winter. In the next chapter, we focus on the biological context of plant growth and development—how plants defend themselves from being eaten or parasitized.

Core Concepts Summary

31.1 IN PLANTS, UPWARD GROWTH BY STEMS OCCURS AT SHOOT APICAL MERISTEMS, POPULATIONS OF TOTIPOTENT CELLS THAT PRODUCE NEW CELLS FOR THE LIFETIME OF THE PLANT.

Stems are built from repeating modules consisting of nodes, where one or more leaves are attached, and internodes, which are the regions of stem between each node. page 31-2

The shoot apical meristem, located at the tip of each stem, is the source of new cells from which the stem and the leaves are formed. page 31-2

Stems contain zones of cell division, elongation, and maturation. Most of the increase in size occurs in the zone of cell elongation. page 31-3

Leaves develop from leaf primordia produced by the shoot apical meristem. page 31-4

Leaves differ from stems in that they differentiate fully and thus have no capacity for continued growth. page 31-4

In seed plants, branching occurs at axillary buds at the base of each leaf. page 31-5

Shoot apical meristems and axillary buds can be transformed into floral meristems, at which point they lose their capacity for continued growth. page 31-6

31.2 HORMONES ARE CHEMICAL SIGNALS THAT INFLUENCE THE GROWTH AND DIFFERENTIATION OF PLANT CELLS.

Hormones are chemical signals that affect both physiology and development and are effective in extremely low concentrations. Traditionally five plant hormones have been identified: auxin, gibberellic acid, cytokinins, ethylene, and abscisic acid. page 31-7

The polar transport of auxin through PIN proteins guides the development of vascular channels (xylem and phloem) that connect leaves and stems. page 31-8

Xylem and phloem differentiate from procambial cells. page 31-9

In most seed plants, the stem's vascular bundles are organized as a ring. page 31-9

Gibberellic acid causes internodal elongation of shoots. page 31-9

Apical dominance is the suppression of the outgrowth of axillary buds caused by hormones produced by roots (strigolactone) and shoots (auxin). page 31-10

31.3 LATERAL MERISTEMS ALLOW PLANTS TO GROW IN DIAMETER, INCREASING THEIR MECHANICAL STABILITY AND THE TRANSPORT CAPACITY OF THEIR VASCULAR SYSTEM.

The formation of two lateral meristems—the vascular cambium and the cork cambium—allow plants to grow in diameter. page 31-10

The vascular cambium produces secondary xylem toward the center and phloem toward the outside. page 31-11

The cork cambium produces a protective outer layer. page 31-12

Gymnosperm tracheids function in both mechanical support and water transport. Angiosperm xylem produces fibers that influence the strength of wood and vessel elements that function in water transport. page 31-12

31.4 THE ROOT APICAL MERISTEM PRODUCES NEW CELLS THAT ALLOW ROOTS TO GROW DOWNWARD INTO THE SOIL, ENABLING PLANTS TO OBTAIN WATER AND NUTRIENTS.

The root apical meristem is the source of new cells that allow roots to elongate. page 31-14

The root apical meristem has a cap that protects it from damage as the root grows through the soil. page 31-14

The root's vascular tissues are located in the center of the root, surrounded by the pericycle, the endodermis, the cortex, and the epidermis. page 31-14

Lateral roots develop from new root apical meristems that form in the pericycle. page 31-15

Roots alter their development in response to conditions in the soil. page 31-15

31.5 PLANTS RESPOND TO LIGHT, GRAVITY, AND WIND THROUGH CHANGES IN INTERNODE ELONGATION AND THE DEVELOPMENT OF LEAVES AND BRANCHES.

Plant stems grow toward the light and away from gravity as a result of redistribution of auxin. page 31-18

Starch-filled statoliths allow roots to sense and respond to gravity. The weight of statoliths alters the lateral distribution of auxin, causing roots to bend downward. page 31-18

Seedlings germinating in the dark produce elongated internodes and suppress leaf expansion to increase their ability to reach the soil surface before their stored resources are depleted. page 31-19

Some plants respond to the presence of neighboring plants by elongating more rapidly and suppressing branching, which allow them to maximize height growth. page 31-19

Mechanical stress, such as the bending of stems in the wind, results in stems that are short and wide. page 31-20

Decreases in photoperiod trigger plants to prepare their meristems for winter by producing bud scales and reducing internode elongation. page 31-20

Self-Assessment

1. Diagram the zones of cell division, elongation, and maturation, and explain why this organization allows stems to grow without a predetermined limit to their length.

2. Give two examples of how changes in internode elongation allow plants to alter the development of their stems.

3. Name one role of the plant hormone auxin and describe how auxin is transported within a plant.

4. Explain why a plant that has a vascular cambium also has a cork cambium.

5. Explain why the vascular cambium forms a continuous sheath that runs from near the tips of the branches to near the tips of the roots, whereas the cork cambium is discontinuous in both space and time.

6. List three structural differences between roots and shoots that allow roots to grow through the soil. Explain how you would tell whether an isolated piece of a plant came from the shoot or from the root.

7. Describe how lateral roots form and how lateral roots enhance the ability of plants to obtain nutrients and water from the soil.

8. Discuss the similarities and differences in the ways stems and roots respond to gravity and light.

9. Give an example of how a plant's ability to sense its environment improves the plant's chances for survival and reproduction.

Do you understand the chapter's Core Concepts? Log in to check your answers to the Self-Assessment questions, then practice what you've learned and reinforce this chapter's concepts by working through the problems and multimedia tutorials provided there.

www.biologyhowlifeworks.com

CHAPTER 32

PLANT DEFENSE

Keeping the World Green

Core Concepts

32.1 Plants have evolved mechanisms to protect themselves from infection by pathogens, which include viruses, bacteria, fungi, worms, and even parasitic plants.

32.2 Plants use chemical, mechanical, and ecological defenses to protect themselves from being eaten by herbivores.

32.3 The production of defenses is costly, resulting in trade-offs between protection and growth.

32.4 Interactions among plants, pathogens, and herbivores contribute to the origin and maintenance of plant diversity.

Plants have complicated interactions with animals. As we saw in Chapter 30, many plants rely on animals for pollination or seed dispersal. In return, plant seeds provide a rich source of food for animals, and leaves also furnish a vast, if lower-quality, food resource. The plant body also provides habitats for an array of viruses, bacteria, and protists, not to mention other plants, that grow on or within their tissues. How do plants defend themselves against the animals that would eat them? How do they defend against pathogens? The answers to these questions help us understand how plant populations are distributed in nature, and they help us solve practical problems as well. In the United States alone, crop damage by herbivores and pathogens costs farmers more than $2 billion every year.

The English poet Alfred, Lord Tennyson, created an indelible image: "Nature, red in tooth and claw." But that image does not recognize that some of the fiercest battles in nature are waged between plants and their predators.

Plants can't run away and hide, and so they lack a key defense employed by animals. Nonetheless, they can sense danger and are able to deploy a diverse and powerful defensive arsenal. Plant defenses include physical and chemical deterrents as well as aerial surveillance, armed guards, and sticky traps.

Of particular importance to humans, plants produce a wide array of chemicals that alter the metabolism and influence the behavior of animals that seek to consume them. These molecules also interact with our own biochemistry, and so compounds derived from plants have been used since antiquity as sources of medicine. Our early ancestors turned to the leaves and bark of willow trees in much the same way that we rely on aspirin—and for good reason. Willows produce a compound that deters herbivores and in humans relieves aches and pain; this compound in fact provides the chemical basis of aspirin. Compounds in rosemary improve the taste of food, and the compound quinine, derived from the tropical cinchona tree, is used in the treatment of malaria. Belladonna, a relative of the tomato, produces toxins that the ancient Romans used to eliminate rivals, but in very low doses it relieves gastrointestinal disorders. In fact, nearly 30% of all medicines are based on chemicals that plants produce to defend themselves.

32.1 PROTECTION AGAINST PATHOGENS

In September 1845, potato plants across Ireland began to die. Their leaves turned black and the entire plant began to rot. When the potatoes were dug up, those that were not already rotten soon shriveled and decayed into foul-smelling and inedible mush. The biological agent responsible for the failure of the potato crop looked like a fungus, but in fact was an oomycete called *Phytophthora infestans*, a type of protist (Chapter 27).

32-1

FIG. 32.1 A potato plant infected with *Phytophthora infestans*, an oomycete that causes blight.

P. infestans is one example of the diverse pathogens that infect plants. Plant pathogens include viruses, bacteria, fungi, nematode worms, and even other plants, but what they all have in common is the ability to grow on and in the tissues of plants, extracting resources to support their own growth and causing an array of disease symptoms. The outcome for the plant is often death (**Fig. 32.1**). The fact that most plants appear disease free is testimony to their ability to defend themselves. Plants have an innate immune system that allows them to recognize and respond to pathogens that would otherwise exploit them. In the Irish potato famine, which lasted from 1845–52, these defenses were breached, allowing *Phytophthora infestans* to spread and grow with devastating effects on the potato crop and, in consequence, the people of Ireland.

Plant pathogens infect and exploit host plants by a variety of mechanisms.

The epidermis, with its thick walls and waxy cuticle, is a plant's first line of defense against pathogens. Viruses and bacteria cannot penetrate the cuticle and so commonly enter through wounds or on the piercing mouthparts of insects and nematodes. Stomata form a natural entry point that enables bacteria, oomycetes, and fungi to gain access to leaves (**Fig. 32.2**). Some bacteria and fungi even secrete chemicals that prevent stomata from closing, improving their ability to infect plant tissues. Other pathogens, notably parasitic plants and some species of fungi, secrete enzymes that weaken epidermal cell walls. These invaders then force their way through the weakened epidermis and gain access to the plant interior. Mechanical cultivation can damage crop plants, accelerating the spread of pathogens.

Once inside the plant, fungi can spread by growing through plant cell walls. Viruses move from one plant cell to another through plasmodesmata. Because the genomes of most plant viruses are single strands of RNA, they can travel by the plant's natural mechanism for moving messenger RNA between cells. The plant vascular system provides a highway that allows pathogens to move over longer distances. Viruses are transported within the phloem, whereas bacteria and fungi move through the xylem. During the 1950s, commercial banana production in Mesoamerica was nearly wiped out by *Fusarium oxysporum*, a fungal pathogen whose spores are carried in the xylem. More recently, the California wine industry has been threatened by Pierce's disease, which results from infection by *Xylella fastidiosa*, a bacterium that also moves through the xylem.

Biotrophic pathogens obtain resources from living cells, whereas **necrotrophic pathogens** kill cells before colonizing them. Defenses effective against one type are not necessarily effective against the other. Thus, plants respond differently

FIG. 32.2 **Entry points for pathogens.** A scanning electron micrograph shows *Phytophthora infestans* gaining access to the leaf interior by growing through stomata.

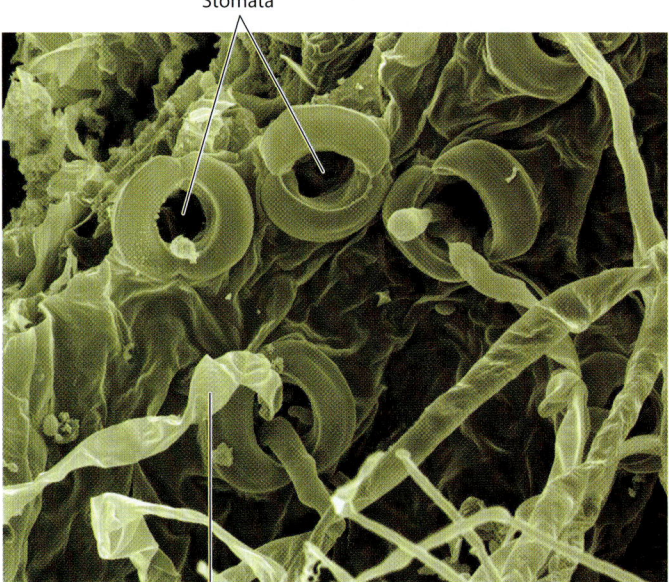

FIG. 32.3 **Parasitic plants.** (a) *Cuscuta*, or dodder, parasitizes the stems of many plants. It does not photosynthesize and instead receives everything it needs from its host. (b) *Rafflesia*, which produces the world's largest flower, also obtains everything it needs from its host. (c) Mistletoe carries out photosynthesis, but relies on its host plant for water and nutrients.

to these two classes of pathogens. Viruses are biotrophic pathogens; they must hijack a living cell's machinery to reproduce. Bacteria and fungal pathogens are present in both groups. For example, wheat leaf rust (*Puccinia* species) is a biotrophic fungus that significantly reduces wheat yields (Chapter 34). Rice blast (*Magnaporthe grisea*) is a necrotrophic fungus that each year reduces rice production by an amount that would feed 60 million people.

Parasitic plants obtain resources by infecting other plants (**Fig. 32.3**). Parasitic plants produce structures that penetrate into the stems or roots of other plants, eventually tapping into the host plant's vascular system. Some, like some mistletoes, are xylem parasites: They obtain water and nutrients from the xylem of the host plant without expending the costs of producing roots themselves. Other plant pathogens target both the xylem and the phloem. In the most extreme cases, the parasite receives all its carbohydrates from its host and carries out no photosynthesis of its own. *Rafflesia*, famous for producing the largest flower in the world, grows entirely within the tissues of its host, emerging only when it produces its giant flowers. The ability to parasitize other plants has evolved independently multiple times. Currently, more than 4000 species of parasitic plants have been identified.

How do parasitic plants find an appropriate host? Mistletoe fruits, which are eaten and dispersed by birds, have a sticky layer that adheres to the stems and branches of trees that the plant will parasitize. One of the most destructive plant pathogens is witchweed (*Striga* species), which infects the roots of a number of crop species, including corn, sorghum, and sugarcane. *Striga* seeds remain dormant in the soil until they detect chemicals released from the roots of suitable host species. The chemicals that trigger the germination of *Striga* seeds are strigolactones (Chapter 31), a plant hormone also involved in establishing symbiotic interactions between plant roots and mychorrhizal fungi (Chapter 29) and inhibiting the outgrowth of axillary buds. *Striga* infects more than 40% of all agricultural areas in Africa, driving down crop yields, in some cases to zero.

An innate immune system allows plants to detect and respond to pathogens.

Most plants are at risk of infection by only a small subset of the many plant pathogens known. **Host plants** of a particular pathogen are the species that can be infected by that pathogen. Not all infections have a significant effect on the host plant, however. **Virulent** pathogens are able to overcome the host plant's defenses and lead to disease. In contrast, **avirulent** pathogens damage only a small part of the plant because the host plant is able to contain the infection. Whether an interaction is virulent or avirulent depends on the genotypes of both the pathogen and its host plant.

In your own body, an innate immune system provides a generalized first line of defense against pathogens (Chapter 43). Plants, too, have an innate immune system that allows them to detect pathogens and mount an appropriate response. A key feature of the immune system is that plants, like animals, can recognize a cell or virus as foreign. How do they accomplish this?

Plant cells have protein receptors (Chapter 9) that bind to molecules produced by pathogens and recognize them as foreign to the plant body. Binding changes the conformation of the receptor, triggering a cascade of reactions that enhance the plant's ability to resist infection.

FIG. 32.4 **The plant immune system.** (a) In basal resistance, plasma membrane receptors recognize molecules produced by broad classes of pathogens. (b) Specific resistance depends on *R* genes that allow plant cells to identify and deactivate AVR proteins produced by specific pathogens.

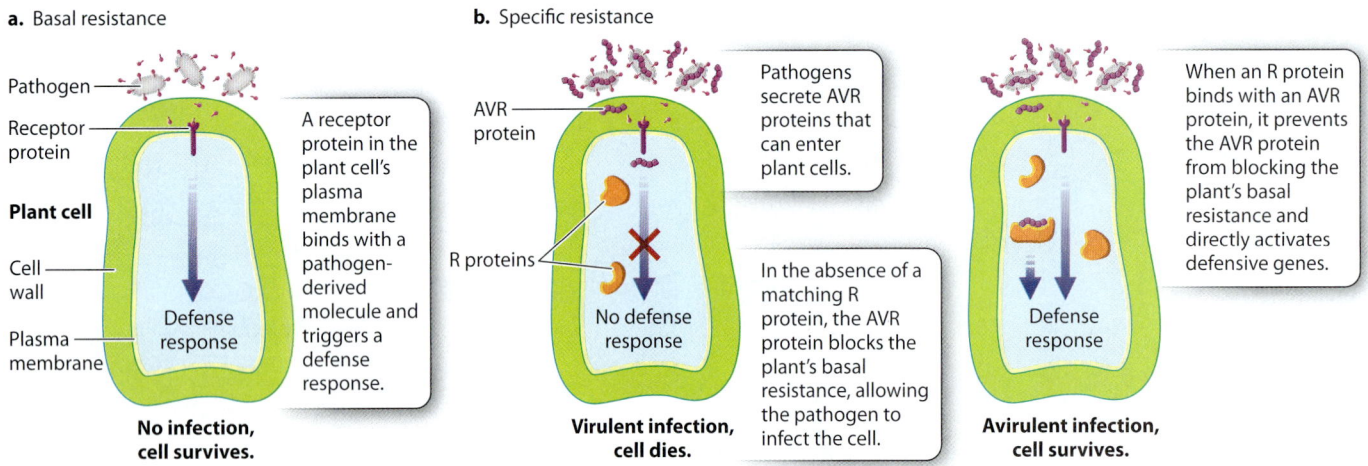

The plant immune system has two parts, one general, or basal, and the other specific (**Fig. 32.4**). The basal branch of the plant immune system consists of receptors located on the plasma membrane (Fig 32.4a). These receptors recognize highly conserved molecules generated by broad classes of pathogens. Examples of these molecules are flagellin, present in the flagella of bacteria, and chitin, a component of the fungal cell wall. When one of these molecules binds to the plant's receptor, an array of defense mechanisms are triggered that help protect the plant from infection.

The second branch of the plant immune system allows plants to resist specific pathogens (Fig. 32.4b). It also consists of receptors, in this case located inside the cell rather than on the plasma membrane. Plants harbor hundreds of these receptors, called **R proteins**, each expressed by a different gene —these genes are collectively called ***R* genes** (the "R" stands for "resistance"). Pathogens produce proteins called AVR proteins that enter into plant cells and facilitate infection ("AVR" stands for "avirulence"). Some AVR proteins block the basal branch of the immune system, while others cause the cell to secrete molecules that the pathogen needs to support its own growth. Each R protein recognizes a specific AVR protein. When an R protein binds with a pathogen-derived AVR protein, it both prevents the AVR protein from blocking the plant's defenses and activates additional defenses. Because this branch of the immune system depends on interactions between specific plant and pathogen genes, it is commonly called the gene-for-gene model of plant immunity.

The large number of *R* genes is the result of an evolutionary arms race that occurs as pathogens evolve ways to evade their host's detection systems. Pathogens improve their success rate when mutation results in an AVR protein that does not bind with any of the R proteins produced by its target. For example, *P. infestans*, the pathogen responsible for the Irish potato famine, produces many more AVR proteins than related *Phytophthora* species. The large number of AVR proteins found in *P. infestans* increases the probability that at least one of these proteins is not detected by the infected plant. Conversely, when pathogens are successful, natural selection favors plant populations that produce new R proteins capable of binding to the AVR proteins of the invader.

→ **Quick Check 1** Can the specific (*R*-gene-mediated) branch of the plant immune system defend against both biotrophic and necrotrophic pathogens?

Plants respond to infections by isolating infected regions.

Once a plant has detected a pathogen, how does it protect itself? One way is by reinforcing its natural barriers against pathogens. For example, plants commonly defend themselves by strengthening their cell walls, closing its stomata, and plugging their xylem. Or the plant can counterattack. Plants produce a variety of antimicrobial compounds, including ones that attack bacterial or fungal cell walls. A third type of defense is known as the **hypersensitive response.** In this case, uninfected cells surrounding the site of infection rapidly produce large numbers of reactive oxygen species, which trigger cell wall reinforcement and cause the cells to die (**Fig. 32.5**). The death of surrounding cells create a barrier of dead tissue that prevents the spread of biotrophic pathogens and slows the growth of necrotrophic ones.

Plants can defeat pathogens by isolating infected regions to a far greater degree than is possible in most animals. For example, plants respond to the presence of xylem-transported pathogens by plugging xylem conduits. If the infection is caught in time, the pathogen is prevented from spreading throughout the plant. But if containment fails and the pathogen spreads, the plant

FIG. 32.5 Hypersensitive response. A tobacco plant infected with tobacco mosaic virus responds by producing a region of dead tissue that surrounds the site of infection and prevents the virus from spreading to other parts of the plant.

Site of initial infection

plant were subsequently exposed to the virus, they suffered little or no damage (**Fig. 32.7**). Ross called this ability to resist future infections **systemic acquired resistance (SAR).** Initially reported for viral infections, SAR occurs in response to a wide range of pathogens, although only when infection results in necrosis due to either a hypersensitive response or a necrotrophic pathogen.

How do uninfected leaves acquire resistance to pathogens? One hypothesis is that a chemical signal is transported from the infected region through the phloem. This chemical signal then triggers the expression of genes encoding many of the same proteins that defend against the pathogen in infected cells. The identity of the mobile signal has proved elusive, but experiments show that salicylic acid—the chemical basis of aspirin—is required for SAR. In experiments in which salicylic acid accumulation was blocked, plants exposed to pathogens failed to develop SAR. Furthermore, the experimental exposure of plants to salicylic acid induces SAR. The methylated form of salicylic acid, known familiarly as oil of wintergreen, is also a potent inducer of defense responses. Methyl salicylic acid is volatile and vaporizes readily, giving rise to the characteristic wintergreen smell. It has been proposed that the transport of methyl salicylic acid vapor through the air may play a role in signaling the presence of pathogens to more distant regions of the plant.

may contribute to its own demise by blocking its entire vascular system. Virulent pathogens that move through the xylem cause a variety of disorders collectively referred to as vascular wilt diseases (**Fig. 32.6**). The widespread death of native elm trees in both Europe and North America is the result of a vascular wilt disease caused by a fungus introduced from Asia, but first identified in the Netherlands and therefore referred to as Dutch elm disease.

Mobile signals trigger defenses in uninfected tissues.

In 1961, American plant pathologist A. F. Ross reported that when individual tobacco leaves were infected with the tobacco mosaic virus (TMV), they developed necrotic patches characteristic of a hypersensitive response. However, when uninfected leaves of the same

FIG. 32.6 Vascular wilt diseases. Pathogens that are transmitted through the xylem can kill their host if plant defenses that prevent their spread also block water transport to the leaves. (a) Panama disease, which affects bananas, is caused by a fungal pathogen. (b) Pierce's disease, which affects grapes, is caused by a pathogenic bacterium.

HOW DO WE KNOW?

FIG. 32.7

Can plants develop immunity to specific pathogens?

BACKGROUND Tobacco is susceptible to an infectious disease that causes the leaves to turn a mottled yellow, hence the name of the disease—tobacco mosaic disease. In the 1890s, experiments with filters showed that the infectious agent was smaller than a bacterium, a finding that led to the discovery of viruses. Some plants succumb to the disease, but others do not. Are the plants that survive the initial infection immune from further attacks?

HYPOTHESIS Plants that survive infection with tobacco mosaic virus (TMV) have acquired immunity and therefore resist further attack.

EXPERIMENT American plant pathologist A. F. Ross infected one leaf on a tobacco plant with TMV. One week later, he exposed another leaf on the same plant to the virus.

RESULTS Plants that had been previously exposed to TMV showed no signs of infection when exposed to the virus a second time.

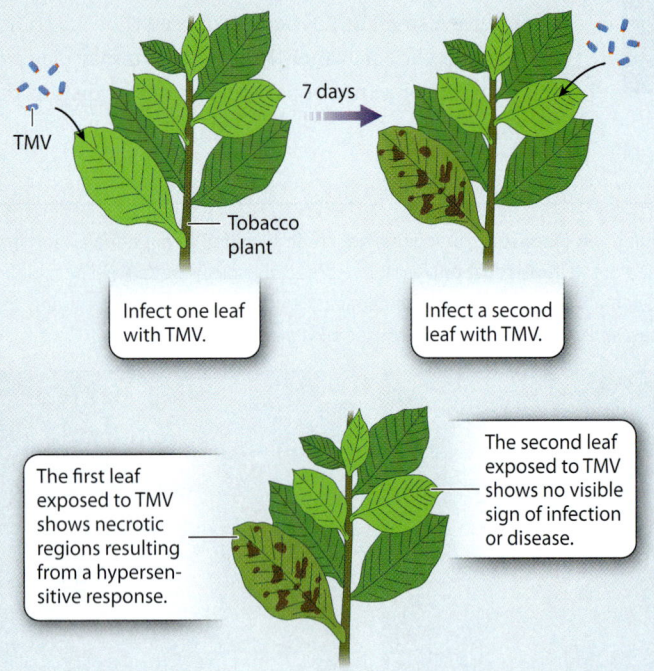

CONCLUSION Tobacco leaves that have never been exposed to TMV can acquire resistance to the pathogen if another leaf on the same plant has been previously exposed. This result indicates that a signal has been transmitted from the originally infected leaf to the undamaged parts of the plant, and that the transmitted signal subsequently triggers the development of an immune response that protects the plant from further infection.

FOLLOW-UP WORK Plants are now known to acquire immunity to a wide range of pathogens in addition to viruses. The genetic basis for the immune response is a system in which proteins encoded by plant *R* genes recognize specific AVR proteins produced by pathogens.

SOURCE Ross, A. F. 1961. "Systemic Acquired Resistance Induced by Localized Virus Infections in Plants." *Virology* 14:340–358.

Systemic acquired resistance increases the ability of uninfected tissues to resist infection. In general, these responses are not highly specific. Rather, SAR activates defenses effective against broad classes of pathogens. Only in the case of viral infections, discussed next, is the response targeted to an individual pathogen.

→ **Quick Check 2** How does the plant hypersensitive response differ from systemic acquired resistance (SAR)?

Plants defend against viral infections by producing siRNA.

Viruses can be potent infectious agents, but plants have evolved responses to viral infection—including a hypersensitive response—that are much like those mounted against other pathogens. In addition, plants have a form of defense that targets the virus itself (**Fig. 32.8**). Most plant viruses have genomes made of single-stranded RNA (ssRNA). During the replication of the viral genome, double-stranded RNA molecules (dsRNA) are formed. The production of double-stranded viral RNA opens the viral genome to counterattack by its host plant. Because plant cells do not normally make double stranded RNA, the replicating viral genomes are identified as foreign. Enzymes produced by the plant cell cleave the double-stranded RNA molecules into small pieces of 21 to 24 nucleotides, forming fragments called **small interfering RNA, or siRNA** (Chapter 3). These fragments play a role in targeting and destroying single-stranded RNA molecules that have a complementary sequence—that is, the viral genome (Chapter 19). The virus is unable to replicate and thus cannot spread. In this way, the plant cell uses RNA chemistry to eliminate viral infections.

When a cell is attacked by a virus, the siRNA molecules that are produced in response to the initial infection can also spread, allowing the plant to acquire immunity against specific viruses. The siRNA molecules move through plasmodesmata, enter the phloem, and from there spread throughout the plant.

FIG. 32.8 Plant defense by small interfering RNAs (siRNAs). siRNAs target complementary single-stranded RNA for destruction.

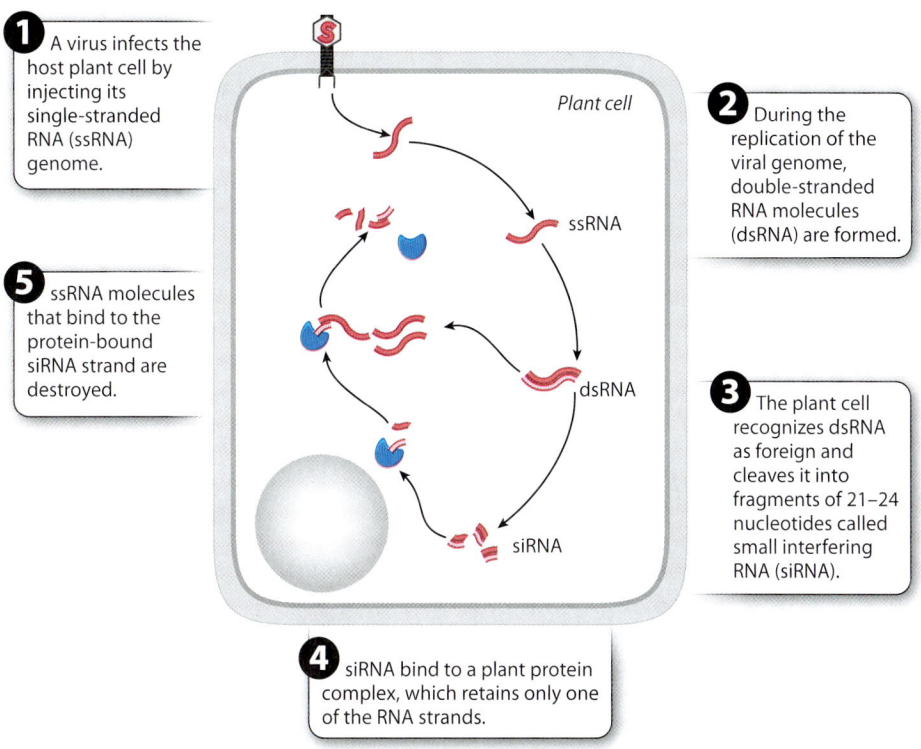

Thus, the systemic response to viral infection involves both general defense responses mediated by salicylic acid and the transport of highly specific molecules that can target and destroy viral genomes throughout the plant.

A pathogenic bacterium provides a way to modify plant genomes.

Some pathogenic bacteria alter their host plant's biology by inserting their own genes into the host's genome. One such pathogen, *Rhizobium radiobacter* (formerly known as *Agrobacterium tumifaciens*), is beneficial to humans, despite the fact that it can cause widespread damage to crops such as apples and grapes. The reason? This bacterium's naturally evolved ability to incorporate its DNA into the genome of its host plant provides a surprisingly easy way for biotechnologists to engineer plants.

R. radiobacter is a soil bacterium with virulent strains that infect the roots and stems of many plants. Virulent bacteria enter plants through wounds and move through the plant's cell walls, propelled by flagella. A tumor, or gall, forms at the point of infection, often where the stem emerges from the soil, and for this reason *R. radiobacter* infection is called crown gall disease. *R. radiobacter* alters the growth and metabolism of infected cells by inserting some of its own genes into the chromosomal DNA of the host plant (**Fig. 32.9**).

FIG. 32.9 The formation of a crown gall tumor. *Rhizobium radiobacter*, a soil bacterium, can insert genes into the plant genome, altering the growth and metabolism of the plant in ways that enhance the growth of *R. radiobacter* bacteria.

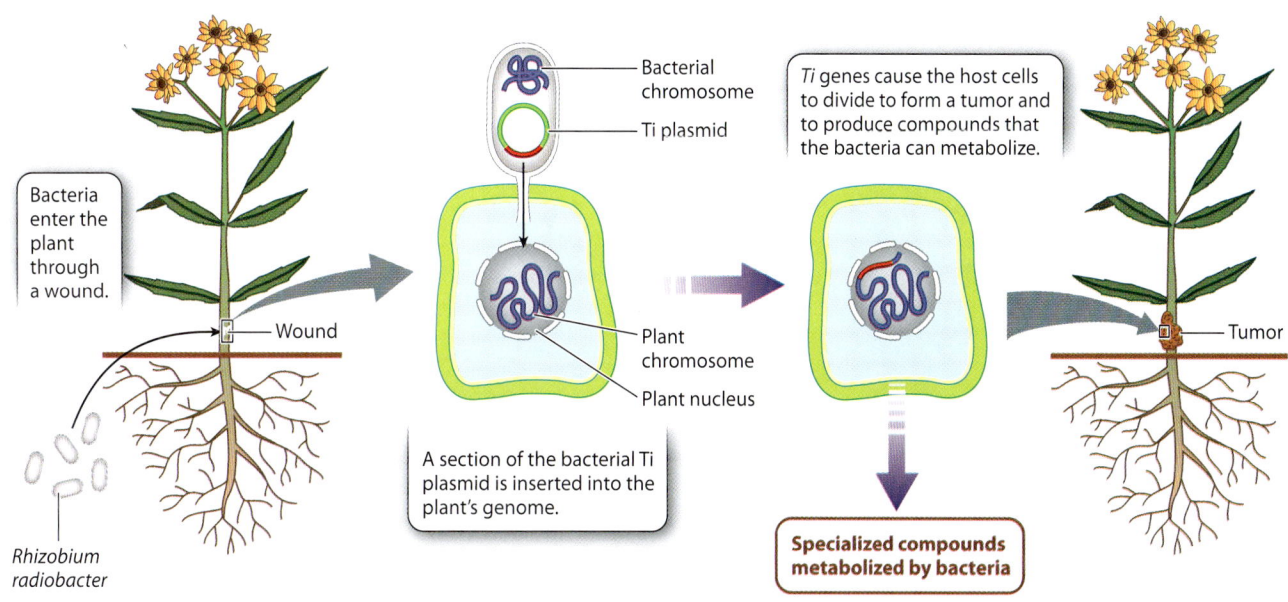

Virulent strains of *R. radiobacter* contain a plasmid, called the **Ti plasmid**, which is a small (~200 kb) circular DNA molecule containing the genes that will be integrated into the host cell's genome, as well as all the genes needed to make this transfer. Some of the transferred genes—including ones that result in the synthesis of the two plant hormones auxin and cytokinin (Chapter 31)—cause infected cells to proliferate and form a gall. Other transferred genes induce cells in the gall to produce specialized compounds that the bacteria use as sources of carbon and nitrogen for growth. Note that because plant cells do not move (in contrast to animal cells), there is no possibility that the tumor-forming cells will spread to other parts of the plant.

Biologists interested in improving crops and other plants make use of *R. radiobacter*'s ability to insert genes into the host genome. The first step is to replace the genes in the Ti plasmid that will be inserted into the plant genome. Genes for gall formation are removed, and genes of interest are inserted. Genes of interest include those whose products increase the plant's nutritional value or confer resistance against disease. Researchers need some way to identify which cells have taken up the genes, and so a marker gene, for example one conferring antibiotic resistance, is typically included. In this way, cells that survive an application of the antibiotic are known to have successfully incorporated the Ti plasmid into their genome.

32.2 DEFENSE AGAINST HERBIVORES

The first land plants were followed closely in time by the ancestors of spiders and scorpions. Fossils show that these earliest land animals soon began to feed on plant fluids and tissues. As plants diversified, so did herbivores, and the pressure on plants to defend themselves against organisms from caterpillars to cows only increased. Given that they are unable to run away, plants seem like they would be the ideal food. But plants have evolved a diversity of mechanisms that deter would-be consumers. Because these mechanisms use resources that could otherwise be directed toward growth and reproduction, plants have also evolved means of deploying their defenses in a cost-effective manner. Here, we explore the mechanical, chemical, developmental, and even ecological means by which plants protect themselves in a world teeming with hungry herbivores.

Plants use mechanical and chemical defenses to avoid being eaten.

Milkweeds (*Asclepias* species) are commonly found in open fields and roadsides across North America. They illustrate well how plants protect themselves from herbivorous animals.

FIG. 32.10 How monarch caterpillars feed on milkweed. Young monarch caterpillars disarm milkweed defenses by digging trenches, while older, larger caterpillars sever major veins. Both methods prevent the sticky latex from flowing into the feeding area.

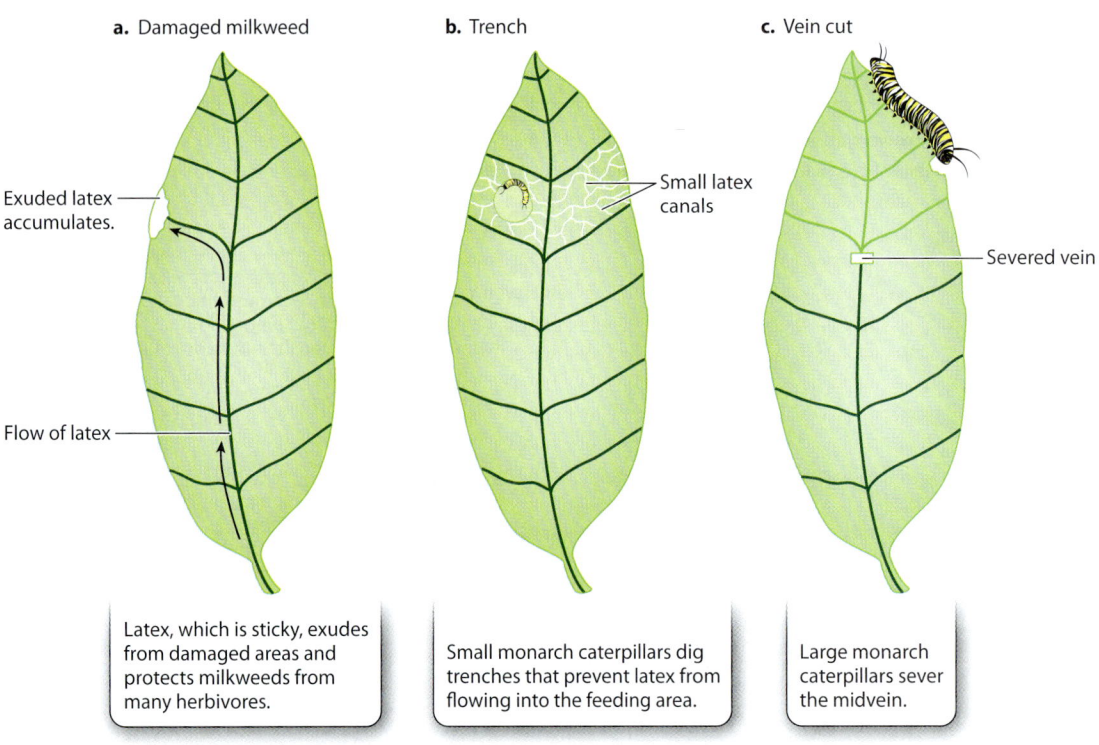

In fact, most generalist herbivores, which eat a diversity of plants, do not use milkweeds as a food source. Only specialists, such as caterpillars of the monarch butterfly (*Danaus plexippus*), can feed on these well-defended plants.

In part, milkweed's defenses are mechanical. The leaves of many milkweed species are covered with dense hairs. An insect crawling on a leaf must make its way through as many as 3000 hairs per square centimeter, a journey likely more daunting to an insect than wading into a blackberry thicket is to a human. Monarch caterpillars may spend as much as an hour mowing off these hairs before they begin to feed. This long handling time represents a significant cost. To understand why monarch caterpillars make this investment, let's look at milkweed's other defenses.

A network of extracellular canals filled with a white sticky liquid called **latex** runs through the milkweed leaf, both within the major veins and extending into the regions between veins (**Fig. 32.10**). Latex canals are under pressure in the intact leaf. Thus, when the leaf is damaged, the latex flows out, sticking to everything it comes into contact with and rapidly congealing on exposure to air. A feeding insect is in danger of having its tiny mouthparts glued together, with potentially fatal consequences. Small monarch caterpillars cut trenches into the leaf that prevent latex from flowing onto the region where they are feeding. Larger caterpillars sever the midvein, relieving the pressure within the latex canals downstream of the cut. The caterpillars are then able to feed on the part of the leaf that lies beyond the severed midvein without risk.

Milkweed latex is not only sticky but also toxic. In particular, milkweed latex contains high concentrations of cardenolides, steroid compounds that cause heart arrest in animals. (In small doses, these compounds are used to treat irregular heart rhythms in humans.) Monarch caterpillars do not metabolize the cardenolides that they consume, but instead sequester them within specific regions of their bodies. Storing these compounds provides the monarchs with a chemical defense against their own predators. Because of the presence of these cardenolides, monarch caterpillars and even the adult butterflies are highly toxic to most birds and other animals that would otherwise eat them.

FIG. 32.11 Mechanical defenses. Spines on the beach palm (a) make it hard for animals to climb or penetrate its stem, and hairs on the leaves of nettle (b) and *Arabidopsis* (c) deter much smaller herbivores.

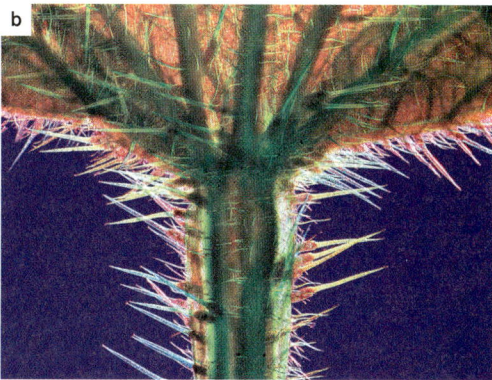
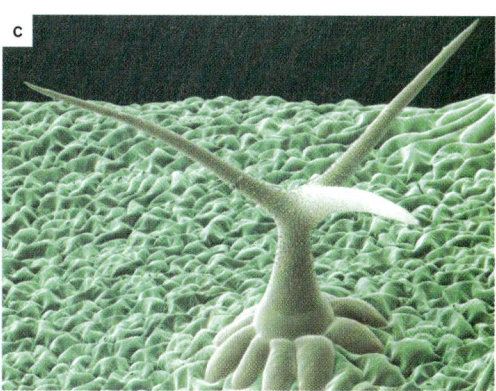

Most plant species have mechanical or chemical defenses against herbivores, or even both. Hairs are common on leaves, and, as anyone who has ever rubbed against a stinging nettle knows, these hairs are sometimes armed with chemical irritants. Grasses and some other plants have a hard, mineral defense consisting of silica (SiO_2) plates formed within epidermal cells. Silica functions as a mechanical defense because it wears down insect mouthparts, so the insects feed less efficiently and grow more slowly. On a larger scale, some tropical trees have prickles, spines, or thorns (**Fig. 32.11**). Even something as simple as how tough a leaf is can have a large impact on the probability of the leaf's being eaten. In fact, leaves are most likely to be eaten while they are still growing and their tissues not fully hardened. Nonetheless, the most spectacular diversity in plant defenses lies in the realm of chemistry.

Diverse chemical compounds deter herbivores.

The cardenolides produced by milkweeds are only one example of the vast chemical arsenal available to plants for protection.

Some plants produce **alkaloids,** nitrogen-bearing compounds that damage the nervous system of animals. Commonly bitter tasting, alkaloids include such well-known compounds as nicotine, caffeine, morphine, theobromine (found in chocolate), quinine (a treatment for malaria), strychnine, and atropine. Alkaloids are a costly defense because they are rich in nitrogen, an essential and often limiting element that plants need to build proteins for photosynthesis. However, they are specific in their action and therefore very small concentrations are effective. One of the first chemotherapy drugs (**Table 32.1**), vincristine, is an alkaloid extracted from the Madagascar periwinkle. It inhibits microtubule polymerization and thus prevents cell division (Chapter 11). Vincristine does not harm the plant's own cells because it is stored within an extracellular system of latex-containing canals.

A second group of defensive compounds, the **terpenes,** do not contain nitrogen. As a result, plants can produce them for defense without having to make use of nitrogen that could otherwise be used in protein synthesis. Small terpenes are volatile, vaporizing easily, and so make up many of the essential oils associated with plants. The distinctive smells of lemon peel, mint, sage, menthol, pine resins, and geranium leaves are all due to terpenes. While we find these odors pleasant, they are feeding deterrents to mammals such as squirrels and moose. In addition, terpenes obstruct the growth and metabolism of both fungi and insects. As a result, pyrethrin, a terpene derivative extracted from chrysanthemums, is marketed commercially as an insecticide.

Other compounds derived from terpenes interfere with insect development. For example, phytoecdysteroids are plant-produced

TABLE 32.1   PLANT COMPOUNDS USED IN MEDICINE: A SAMPLING

CLASS	EXAMPLE	PLANT SOURCE	MEDICAL USE
Alkaloids	Atropine	Deadly nightshade	Dilate pupils; treat extremely low heart rate
	Vincristine	Madagascar periwinkle	Treat cancer
	Codeine	Opium poppy	Cough suppressant
	Morphine	Opium poppy	Painkiller
	Quinine	Cinchona tree	Treat malaria
	Ephedrine	*Ephedra* (Mormon tea)	Asthma relief
	Turocurare (curare)	*Strychnos toxifera*	Muscle relaxant
	Digitoxin	Foxglove	Treat congestive heart failure and arrhymthia
	Berberine	*Berberis* (barberry)	Antibiotic, anti-inflammatory
Terpenoids	Menthol	Peppermint	Topical analgesic; mouthwash
	Camphor	Camphor laurel	Anti-itch creams
	Eucalyptol	*Eucalyptus* species	Mouthwash; cough suppressant
	Taxol	Pacific yew	Treat cancer
	Artemisinin	*Artemisia annua*	Treat malaria
Phenols	Salicylic acid	Willow tree	Treat aches and pains

replicas of the hormones that induce molting in insects. Exposure to these compounds cause insects to molt prematurely and therefore keeps their populations in check. Taxadiene, precursor for the chemotherapy drug taxol, is a terpene that was first extracted from the bark of the Pacific yew tree.

Phenols form the third main class of defensive compounds, illustrated by the **tannins** found widely in plant tissues. Tannins bind with proteins, reducing their digestibility. Plants store tannins in cell vacuoles, and so these compounds come into contact with the protein-rich cytoplasm only when cells are damaged. Herbivores attempting to feed on tannin-producing plants obtain a poor reward for their efforts. For this reason, natural selection favors individuals that avoid tannin-rich plants. Many unripe fruits are high in tannins, and the unpleasant experience of biting into an unripe banana illustrates how tannins deter consumers. For thousands of years, humans have taken advantage of the protein-binding properties of tannins, using extracts from tree bark to process animal skins by "tanning" to produce leather.

Some of the chemical defenses used by plants are stored in a separate compartment from the enzymes that activate them. When you bite into a plant belonging to the cabbage family, the chemicals stored in the vacuole comingle with enzymes in the cytosol. The enzymes then catalyze the production of mustard oils from these compounds. Mustard oils give cabbage and its relatives their distinctive smell. They serve as feeding deterrents because they interfere with insect growth.

Cassava roots, an important food crop in Africa and South America, release the toxin hydrogen cyanide when their cells are damaged. Before cassava roots can be safely consumed, the roots must be ground and the cyanide-producing chemicals removed in running water. Although some cassava varieties, so-called sweet cassava, lack these chemicals and can be eaten without any special preparation, the bitter varieties are preferred crops in many cases because they often have higher yields.

A final chemical defense found in plants is protein based. Plants and animals use the same 20 amino acids to construct proteins, but some plants produce additional amino acids as well. Plants do not incorporate these additional amino acids into their proteins, but herbivores that ingest them do. The resulting proteins can no longer fulfill their function. Thus, insect herbivores that consume nonprotein amino acids grow slowly and often die early. A second protein-based defense is the production of antidigestive proteins called **protease inhibitors.** These proteins bind to the active site of enzymes that break down proteins in the herbivore's digestive system. This prevents proteins from being broken down into their individual amino acids and therefore reduces the nutritional value of the plant tissue. Insects that feed on plants that produce protease inhibitors have reduced growth rates.

Some plants provide food and shelter for ants, which actively defend them.

While many plants have mechanical defenses against herbivores and nearly all have at least one form of chemical defense, a smaller number of species have evolved ecological defenses against herbivory. These plants "employ" animals as bodyguards in exchange for shelter or nourishment, much the way that many plants provide food or other rewards to animals that transfer pollen and disperse seeds (Chapter 30).

Nectar is typically associated with flowers. However, many plants produce nectar in glands located on their leaves (**Fig. 32.12**). These extrafloral nectaries, or nectar-producing regions, attract ants, which move actively throughout the plant in search of nectar. When the ants encounter either the eggs or larvae of other insects, they consume these as well. The value of this relationship to the plant can be demonstrated easily: When ants are prevented experimentally from patrolling certain branches, those branches suffer higher rates of herbivore damage.

A small number of plants, several hundred in total, have evolved a much closer relationship with ants. These so-called ant-plants provide both food and shelter for an entire colony of ants, which then defend their host. The bullhorn acacia (*Acacia cornigera*), which grows in dry areas of Mexico and Mesoamerica, produces hollow spines at the base of each leaf, as well as protein- and lipid-rich food bodies on the tips of its expanding leaves. Symbiotic ants (*Pseudomyrmex ferruginea*)

FIG. 32.12 Ants attracted by an extrafloral nectary. These ants get a food reward and also consume any insect eggs or larvae they encounter on the plant.

FIG. 32.13 The symbiosis between ants and the bullhorn acacia. (a) Ants live in the plant's thorns, and (b) the plant provides food (the yellow food bodies). The ants vigorously defend their home against animals and plants alike.

live in the spines and feed on the food bodies (**Fig. 32.13**). To protect this source of food and shelter, the ants actively defend the plant. When the ants detect an invader, they produce an alarm pheromone that causes the entire colony to swarm over the plant, attacking any insects or vertebrates (including scientists) that they encounter. The ants will attack the growing tip of a vine that might climb up the host and even destroy plants growing in the soil beneath the host *Acacia*.

This system works well until it comes time to reproduce. Without some mechanism to control the hypervigilant ants, any pollinator that tried to visit one of the host plant's flowers would be subject to the same treatment as a truly unwanted guest. To prevent pollinators from being turned away, flowers of the bullhorn acacia release a chemical that repels the resident ants, allowing pollinators to visit the flowers unharmed.

The interactions between ant-plants and their ants are highly coevolved. The ant species are found only in association with their host plants. And while ant-plants can be grown without ants, an undefended plant would rapidly be consumed in nature because ant-plants lack the chemical defenses present in related species that do not host ants.

Grasses can regrow quickly following grazing by mammals.

By far, the most important vertebrate herbivores are the bison, zebra, antelope, and their relatives that populate the great plains of East Africa and North America. Studies of herbivore damage in an East African savanna indicate that vertebrate herbivores annually consume approximately the same amount of leaf biomass as do herbivorous insects.

Grasses are well suited to cope with grazing mammals (**Fig. 32.14**). The vertically oriented leaves grow from stems with limited internodal elongation. This allows the shoot apical meristem to remain close to the ground, where it is less likely to be damaged. Grasses have persistent zones of cell division and elongation at the base of each leaf. Because grass leaves elongate from their base, grasses can replace leaf tissues removed by a grazing mammal, fire, or a lawn mower without having to add to the length of the stem. Only when grasses produce their flowering stalks does the apical meristem emerge above the leaves.

Grasses are thought to have evolved in forest environments. In fact, grass-dominated ecosystems have become widespread only

FIG. 32.14 Adaptation for grazing. The ability of grass leaves to elongate from the base allows grasses to regrow following grazing.

Prior to flowering, the apical meristem of grasses is close to the ground, and there is a persistent zone of cell division and elongation at the base of each leaf blade.

Therefore, if the top is cut off...

...the apical meristem is not damaged and the leaf blades can grow back.

during the past 25 million years. The fossil record shows that as grasslands expanded, grazing mammals evolved along with them. Horses, for example, originated as small browsers that fed on the leaves of small trees and shrubs. As grasslands expanded, larger horse species with teeth adapted to eating grass evolved. Why would specialized teeth be advantageous? The small plates of silica in grass cells appear to contribute to mechanical support and may provide mechanical defense against herbivores. Horses and other grazers that feed on these grasses wear down their tooth enamel grinding against the silica, so teeth with a thick enamel layer are favored by natural selection. It is not clear whether grasses evolved in response to grazing or to disturbances such as fire, but without question, their distinctive pattern of meristem activity minimizes the damage imparted by grazing mammals.

32.3 ALLOCATING RESOURCES TO DEFENSE

By now, it should be clear that plants invest substantial resources in defense, resources that might otherwise be used for growth and reproduction. Given the considerable cost of plant defenses, their benefits must be correspondingly large. Obviously, if a commonly encountered pathogen or herbivore is lethal, the resources invested in defense provide reproductive advantage. But what if the pathogen or herbivore usually damages only half of a plant's leaves? Or 10%? Or encounters the plant only infrequently?

In nature, the balance between cost and benefit is achieved in several different ways. When a threat is constant or generalized, plants mount **constitutive defenses**—defenses that are always active. When a threat is uncommon, however, there is an advantage to mounting a defense only when the threat is encountered. A defense that is mounted only when the plant senses the threat is an **inducible defense.**

Plants can sense and respond to herbivores.

When a plant is attacked by herbivores, it begins to express a new set of genes. Many of these genes increase the production of chemical defenses, while others stimulate the mechanical reinforcement of cell walls. In addition, herbivore damage triggers the synthesis of **jasmonic acid,** a signal that is transmitted through the phloem. Exposure to jasmonic acid induces the transcription of defensive genes even in parts of the plant not directly affected by herbivores. Methyl-jasmonic acid is highly volatile and can move through the air to reach portions of the plant that are not actively importing carbohydrates through the phloem.

→ **Quick Check 3** What would happen to a plant in which jasmonic acid synthesis is blocked?

Another way in which plants can sense and respond to herbivores is illustrated by coyote tobacco (*Nicotiana attenuata*), a relative of tobacco native to dry open habitats of the American West (**Fig. 32.15**). For more than 15 years, biologists have intensively studied the mechanisms by which *N. attenuata* defends itself against herbivores, mechanisms that provide a fascinating window into how plants protect themselves under natural conditions.

FIG. 32.15 *Nicotiana attenuata*, the coyote tobacco. This species has been developed as a model organism for studying how plants protect themselves against herbivores. (a) *N. attenuata* in a Utah desert; (b) *Manduca sexta*, a specialist herbivore, feeding on *N. attenuata*.

Nicotiana attenuata is attacked by a wide variety of herbivores, both generalists and specialists. Which species cause the most damage, however, varies significantly from year to year. This means that *N. attenuata* must either produce costly defenses for herbivores that might not show up in any particular year or be able to identify and respond to specific threats.

A major component of *N. attenuata*'s defense against being eaten is the production of the alkaloid nicotine. This defense is metabolically expensive because it contains nitrogen. Nicotine is an effective defense against many generalist herbivores. However, *N. attenuata* is also attacked by a specialist herbivore, the tobacco hornworm caterpillar, *Manduca sexta*. These caterpillars, which can reach 7 cm in length, have evolved the ability to sequester and secrete nicotine. By this means, these caterpillars are able to consume leaves with high levels of nicotine. If nicotine is not an effective defense against this voracious herbivore, what can *N. attenuata* do to protect itself?

It turns out that *N. attenuata* is able to recognize specific chemicals in the saliva of *M. sexta*. When damaged by most herbivores—or a pair of scissors—*N. attenuata* responds by producing nicotine. But when attacked by *M. sexta*, *N. attenuata* does not waste time and resources synthesizing nicotine. Instead, it protects itself in an entirely different way, by attracting other insects that will attack its herbivores.

Plants produce volatile signals that attract insects that prey upon herbivores.

How does *N. attenuata* attract its insect allies? A hint comes from the smell of a new-mown lawn. This distinctive smell originates from chemicals released from the cut blades of grass. Some of the chemicals released when plants are damaged represent nothing more than the exposure of cell interiors to the air. Others, however, are produced specifically in response to damage.

Studies first conducted in the 1980s showed that undamaged plants increase their synthesis of chemical defenses when neighboring plants are attacked by herbivores (**Fig. 32.16**). Although these plants were described in the popular press as "talking trees," the plants under attack are not altruistically warning their neighbors. Instead, subsequent studies suggest that these plants are sending signals directed at other insects. When *N. attenuata* is attacked by *M. sexta* caterpillars, it produces volatile signals that attract insects that prey on the eggs and larvae of *M. sexta*. The adage "The enemy of my enemy is my friend" seems to hold true when it comes to protecting plants from herbivore damage. Neighboring plants sense these chemical signals and respond by ramping up their own defenses.

Nutrient-rich environments select for plants that allocate more resources to growth than to defense.

Plant species vary significantly in the extent to which they allocate resources to defense. Because defenses are costly to produce, you might think that plants growing in nutrient-rich habitats should be the best defended. However, consider the counterargument: Plants growing on nutrient-poor soils might invest heavily in defense because they cannot afford to replace tissues lost to herbivores. Such plants would grow relatively slowly because their resources were being used to build defenses rather than to produce new leaves and roots. By

---

HOW DO WE KNOW?

FIG. 32.16

Can plants communicate?

BACKGROUND Plants respond to herbivore damage by producing chemicals that make their tissues less palatable. Field experiments in which caterpillars were added to target plants suggested that undamaged neighboring plants increased their production of defensive chemicals, whereas plants farther away did not.

HYPOTHESIS Volatile chemicals released from plants that have been attacked by herbivores elicit a defensive response in undamaged plants.

EXPERIMENT One set of plants was placed downwind of other plants in which two leaves were torn, simulating herbivore damage, and a set of control plants was placed downwind of an empty chamber. The concentrations of defensive compounds in all sets of plants were measured and compared.

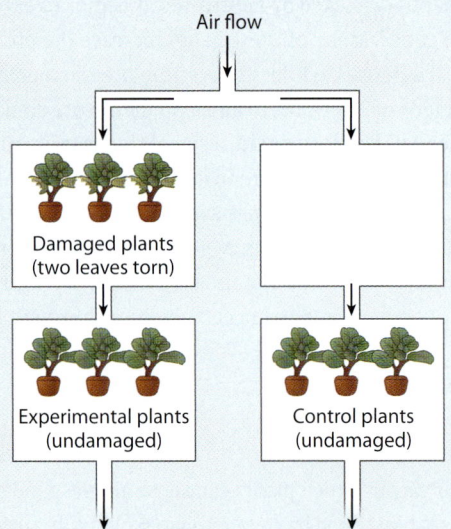

contrast, plants from nutrient-rich habitats would favor growth over defense because they could more readily replace lost or damaged tissues.

This situation is an example of a **trade-off**: Something is gained and, at the same time, something is lost. You can't have your cake and eat it, too—you have to make a choice. With plants, there can be a trade-off between growth and defense in which allocating resources for one means that those resources cannot be allocated for the other.

A recent study in the Amazon rain forest has focused on a possible trade-off between growth and defense. Researchers compared species found on nutrient-rich clay soils with closely related species found on nutrient-poor sandy soils. They found that species from the nutrient-rich clay soils invested less heavily in defenses than did species from the sandy sites. Because the clay-soil species divert fewer resources to defense, they grow faster in either soil type compared to sandy-soil species—as long as they are grown under a net that excludes herbivores (**Fig. 32.17**). However, when grown out in the open, each species grows best in the soil type on which it is naturally found.

The clay-soil species grow poorly on sandy soils when not protected by a net because, without ample soil nutrients

RESULTS When measured 52 hours after the initial damage, both the damaged plants and the undamaged experimental plants exhibited elevated levels of defensive compounds in their leaves compared to control plants.

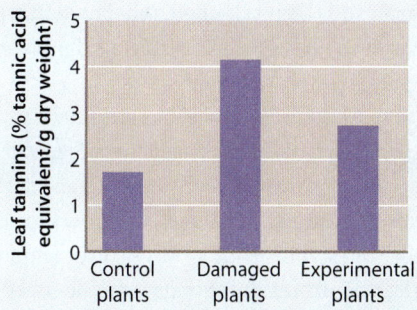

CONCLUSION Volatile chemicals released from damaged plant tissues can trigger the production of defensive chemicals in undamaged plants.

FOLLOW-UP WORK The volatile chemicals released from damaged plants have been identified and shown to attract predatory animals that feed on the herbivores or parasitize their eggs or both. Thus, when plants respond to damaged neighbors they are "eavesdropping" on signals that likely evolved to attract predatory insects, rather than to warn neighboring plants.

SOURCE Baldwin, I. T., and J. C. Schulze. 1983. "Rapid Changes in Tree Leaf Chemistry Induced by Damage: Evidence for Communication Between Plants." *Science* 221:277–279.

FIG. 32.17 A trade-off between growth and defense. Plants from nutrient-rich clay habitats can overcome herbivory by growing fast, whereas plants from low-nutrient sandy habitats grow slowly and invest more in defense. The photos show that, without defenses, clay-soil plants are prone to being eaten by herbivores.

Clay soil plants growing in sandy soils

Unprotected from herbivores

Protected from herbivores

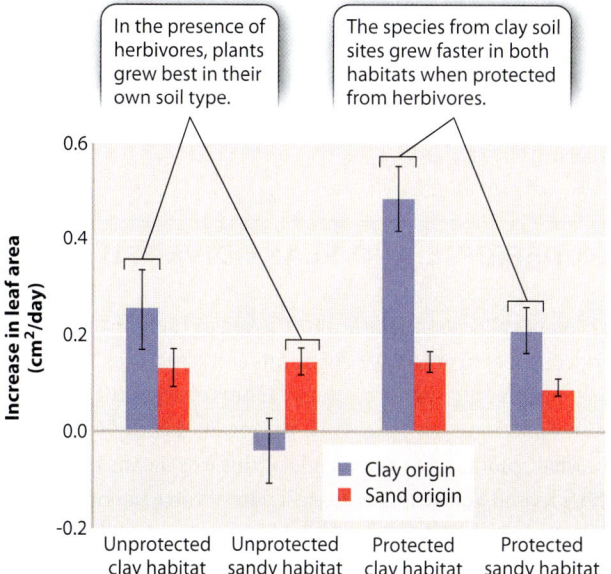

available, they are unable to replace tissues consumed by herbivores. Conversely, when grown on the nutrient-rich soils, species from the sandy habitats are poor competitors because they allocate resources that could have been used for growth to defensive compounds and structures. What we see is a classic example of a trade-off.

→ **Quick Check 4** Does the trade-off between growth and defense favor a single species that dominates in all soil types or different species specialized for each habitat?

Exposure to multiple threats can lead to trade-offs.

In the real world, plants are routinely confronted with more than one threat. A plant may be attacked simultaneously by both a pathogen and an herbivore, or it might be in danger of being shaded by a fast-growing neighbor at the same time that it must defend its leaves from being eaten. Given that the resources a plant can draw on are limited, how do plants respond to multiple threats?

Often, the response to multiple threats suggests a trade-off. For example, exposure to a pathogen can reduce the ability to respond to a subsequent herbivore attack. Tobacco plants in which systemic acquired resistance has been activated as a result of exposure to tobacco mosaic virus are more susceptible to the hornworm caterpillar, *Manduca sexta*. Studies indicate that cross talk between the signaling pathways for responding to pathogens versus herbivores is responsible for the reduced ability to respond subsequently to herbivores.

There can also be trade-offs between threats resulting from competition with neighboring plants and attack by herbivores. For example, when the phytochrome receptor in plants detects the presence of neighboring plants (Chapter 31), plants not only allocate more resources to growing tall but also produce fewer defensive chemicals in response to herbivore damage. Once again, interactions among the signaling molecules associated with each of these processes are responsible for mediating the trade-off.

32.4 DEFENSE AND PLANT DIVERSITY

Many plants allocate a significant fraction of their resource and energy budgets to defensive chemicals and structures. Despite this commitment, insects consume approximately 20% of all plant production each year, while pathogens destroy a substantial additional amount. The magnitude of this loss means that pathogens and herbivores must be a potent force in both ecology and evolution, shaping competitive interactions by allowing some species to prosper and limiting the success of others. In nature, plant fitness may be strongly influenced by the capacity to deter herbivores and resist pathogens.

Interactions between plants and their consumers are thought to have influenced patterns of plant diversification. Just as pollinating insects diversified along with angiosperms (the flowering plants), so, too, did plant-eating insects. The fact that angiosperms and plant-eating insects radiated at the same time is consistent with an important role for plant–insect coevolution in generating diversity. Further evidence that defense has influenced plant diversification comes from plant genes and secondary compounds. For example, the *R* genes that are instrumental in pathogen recognition form one of the largest gene families in plants, while chemical defenses against herbivores include about 6000 alkaloids and more than 10,000 terpenoid compounds. Such observations suggest that plants are locked into an evolutionary arms race with herbivores and pathogens. This arms race is very much in evidence in the efforts by farmers to limit crop losses by reducing the numbers of the pathogens and herbivores in fields and orchards.

Pathogens, herbivores, and seed predators can increase plant diversity.

The lowland rain forests of eastern Ecuador have the highest diversity of tree species recorded anywhere: More than 1000 species of trees were identified in a 500 m × 500 m plot. What allows so many tree species to coexist is one of the great questions of modern biology. In the 1970s, two ecologists, Daniel Janzen and Joseph Connell, independently proposed that interactions with host-specific pathogens, herbivores, or seed predators play an important role.

The Janzen–Connell hypothesis, as it is now known, is based on the observation that seeds or seedlings often suffer high rates of mortality when growing directly beneath an adult tree of the same species (**Fig. 32.18**). Seedlings that grow farther away have a better chance of escaping pathogens that are common in the soil underneath an adult tree. By a similar argument, seeds that disperse far from their maternal plant may escape notice by seed predators, which often gather in higher densities near trees with mature fruit. Such density-dependent mortality provides rare species with an advantage over more common species. In other words, the offspring of a rare species is more likely to survive because its seeds are more likely to be dispersed into a location where there are no neighboring adult plants of the same species. By favoring the survival of less-common species, density-dependent mortality prevents one species from outcompeting all others.

In the tropical rain forests of Ecuador and elsewhere, only a small number of tree species are common; the majority of tree species are rare. The ability to persist at low population density allows rare species to escape herbivore and pathogen damage, but requires reliable ways of getting pollen from one plant to another. Thus, animal pollination, herbivory, and infection by pathogens are thought to play complementary roles in promoting angiosperm diversity.

FIG. 32.18 Density-dependent mortality promotes tree diversity. Seedlings of rare species have a greater probability of surviving because they are more likely to escape infection by species-specific pathogens.

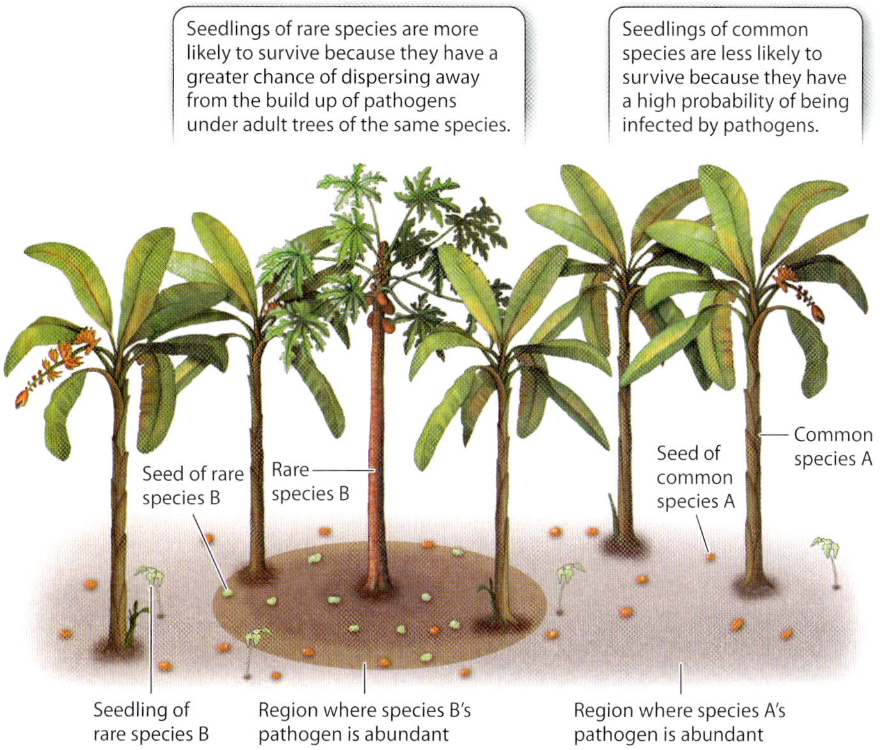

groups of plants. Reasoning that these features represent a novel defense, they asked whether the lineages that evolved latex or resin canals were more diverse than closely related groups that lacked this innovation. In 14 of the 16 groups examined, the lineages with the protective canals were significantly more diverse than their closest relatives that lacked this innovation. This pattern supports the hypothesis that the evolution of latex and resin canals provided plants with the freedom to expand into new habitats.

Escalation of defenses, however, is not the only possible evolutionary outcome. Phylogenetic research shows that, as milkweeds diversified, many of the newly evolved species produced *fewer* chemical defenses such as latex canals and cardenolides and instead replaced tissues by growing quickly following damage. The presence of specialized herbivores such as monarch caterpillars that can disarm milkweeds' defenses may help explain why these more recent plant species reduce investments in defense in favor of more rapid growth.

The evolution of new defenses may allow plants to diversify.

American scientists Paul Ehrlich and Peter Raven observed that closely related species of plants were fed upon by closely related species of butterflies. In 1964, they cited this observation as evidence that plants and their herbivores are caught up in a coevolutionary arms race: Plants evolve new forms of defense, and herbivores evolve mechanisms to overcome these defenses. In particular, Ehrlich and Raven hypothesized that the evolution of novel defenses has been an important force in the diversification of both plants and herbivorous insects.

An innovation in defense may allow a plant population to expand into new areas, increasing its potential to evolve into new species. Similarly, innovation in overcoming plant defenses may allow an insect or pathogen population access to new plant resources. Ehrlich and Raven's "escape and radiate" hypothesis thus predicts a burst of diversification following the evolution of a key innovation that allows plants to avoid being eaten. This hypothesis makes intuitive sense, but only with the advent of phylogenetic trees based on DNA sequence comparisons (Chapter 23) could it be tested. Biologists knew that latex or resin canals evolved independently in more than 40 different

? CASE 6 Agriculture: Feeding a Growing Population
Can modifying plants genetically protect crops from herbivores and pathogens?

The density-dependence of herbivore and pathogen damage is a particular challenge for agriculture as practiced in most countries. When acre after acre are covered by a single crop species, successful invaders will not be slowed by low target density. Large expanses of a single crop allow pest populations to build up to high levels, and the tendency to plant genetically uniform varieties increases the losses in yield. Not surprisingly, then, herbivores and pathogens exact a major toll on agricultural crops, and protecting crops from damage is a high priority for farmers.

The production of pesticides and herbicides increased dramatically in the second half of the twentieth century. Together with irrigation and fertilizers, these compounds are important contributors to the increased yields of the Green Revolution. However, large-scale application of pesticides is not without risk. In addition to the potential for toxic effects, widespread chemical treatments provide a strong selective force for the evolution of resistance. As pests become resistant to pesticides, one response is to apply more or stronger chemicals. The consequences of this approach, however, can spiral out of control as chemical applications indiscriminately kill off the natural predators of the pests themselves.

During the 1980s, farmers applied increasing amounts of pesticides to rice paddies in South Asia, yet crop losses continued to increase. This dangerous course was reversed by the introduction of a program first developed in the 1950s, in which farmers devote substantial time and effort determining what pests are actually present, whether or not they are at damaging levels, and whether or not there are natural predators of the pests. Only after understanding the ecology of the farm are pests removed, starting with mechanical means, then biological control, and, only as a last resort, industrial pesticides. This approach, called integrated pest management, allows farmers to minimize routine pesticide use and therefore prevent pest species from evolving resistance.

To further improve the ability of crops to defend themselves against herbivores and pathogens, crop breeders must emulate—on a faster timescale—the role played by evolution in selecting for resistance to newly evolving pests. For example, Norman Borlaug's original task was to develop wheat varieties for Mexico that were resistant to fungal pathogens. To do this, he obtained seeds of wheat varieties from around the world that showed evidence of pathogen resistance. He then crossed these disease-resistant varieties with ones adapted to the local conditions. Although it took thousands of hand-pollinations, Borlaug was successful at introducing resistance genes into wheat varieties that grew well in Mexico.

Crop breeding is a powerful way to alter the genetic makeup of cultivated species. However, because it relies on sexual reproduction, crop breeding is limited to the genetic diversity that exists within varieties that can interbreed. In contrast, *Rhizobium radiobacter* allows scientists to introduce specific genes found in distantly related organisms into cultivated species. One of the first commercial uses of this technology was to introduce chemical defenses from a bacterium into crop plants.

Bacillus thuringiensis (*Bt*), a soil-dwelling bacterium that produces proteins toxic to insects, has been used to control insect outbreaks in agriculture since the 1920s. At first, spores containing the toxins or the toxins themselves were applied directly to plants. In 1985, the genes that encode the toxins from *B. thuringiensis* were inserted into tobacco, where their expression conferred substantial protection against insects. Today, *Bt*-modified crops are some of the most widely planted genetically modified plants.

The introduction of *Bt* crops markedly reduced the application of pesticides, to the benefit of both farmworkers and the environment. However, the danger of planting *Bt*-expressing crops widely is that constant exposure to *Bt* toxins increases the probability that pest populations will evolve resistance. This possibility is of particular concern to organic farmers, who continue to use traditional application of *B. thuringiensis* spores to control insect outbreaks. To prevent the evolution of resistance, U.S. farmers are required by law to plant a fraction of their fields with non-*Bt*-expressing

FIG. 32.19 *Bt*-crops co-planted with non-*Bt* varieties to prevent the development of resistance.

plants (**Fig. 32.19**). The proximity of non-*Bt*-expressing plants allows pests that are not resistant to *Bt* toxins to survive and interbreed with individuals feeding on the *Bt*-expressing plants, thus slowing the evolution of *Bt*-resistance. The efficacy of this practice, however, is called into question by reports of *Bt*-resistant pests and the need to apply increasing amounts of industrial pesticides to *Bt* crops.

If there is one practical lesson to be learned from our experience with *Bt*, it is that pathogens and herbivores will continue to evolve new ways of circumventing the defenses of crop plants. Agriculture will continue to require new forms of crop protection. While the use of chemical pesticides will probably remain widespread, crop protection will increasingly be based on the application of genetics, through both crop breeding and biotechnology. In either case, success will depend upon the availability of existing genes honed by natural selection. Without access to genes involved in pathogen recognition (R genes) or the production of chemical and mechanical defenses, we will have the tools, but not the raw materials, to protect our crops.

In some cases, crop protection will rely on genes from distantly related organisms, as illustrated by *Bt*. Commonly, however, defense strategies for crops will rely on genes from other plants, often ones that are closely related. For example, the specific R genes used for pathogen recognition typically function only within a single species or closely related species. Thus, recent efforts to develop potatoes that are resistant to *Phytophthora infestans* focus on transferring R genes from a wild relative of potato. The continuation of such efforts requires a commitment to safeguarding the genetic diversity found in natural populations of crop species and their close relatives. Only by tapping into these natural genetic resources can breeders produce crops that stay one step ahead of the constantly evolving threats of pathogens and herbivores.

Core Concepts Summary

32.1 PLANTS HAVE EVOLVED MECHANISMS TO PROTECT THEMSELVES FROM INFECTION BY PATHOGENS, WHICH INCLUDE VIRUSES, BACTERIA, FUNGI, WORMS, AND EVEN PARASITIC PLANTS.

Pathogens enter plants through damaged tissue or stomata. page 32-2

Plant pathogens spread by growing or moving through the plant vascular system. Viruses move in the phloem, whereas bacteria and fungi move in the xylem. page 32-2

Biotrophic pathogens gain resources from living cells; necrotrophic pathogens kill cells before colonizing them. page 32-2

Plants have an innate immune system that allows them to detect and respond to pathogens. page 32-3

One part of the innate immune system acts generally and recognizes highly conserved molecules produced by broad classes of pathogens. page 32-4

A second part of the innate immune system acts specifically and consists of resistance or *R* genes that encode R proteins, which recognize pathogen-derived AVR proteins. page 32-4

Plants respond to pathogens by actively killing the cells surrounding the infection (the hypersensitive response) and by sending a signal to uninfected tissues so that they can mount a defense (systemic acquired resistance). page 32-5

The bacterium *Rhizobium radiobacter* infects plants by inserting some of its genes into the plant's genome, resulting in the formation of a tumor and providing a way to genetically engineer plants. page 32-7

32.2 PLANTS USE CHEMICAL, MECHANICAL, AND ECOLOGICAL DEFENSES TO PROTECT THEMSELVES FROM BEING EATEN BY HERBIVORES.

Plant defenses against herbivory include dense hairs, latex, and chemicals. page 32-9

Alkaloids are nitrogen-bearing compounds that damage the nervous system of animals. page 32-10

Terpenes are volatile compounds that give rise to many of the essential oils we associate with plants; they deter herbivores. page 32-10

Tannins bind with proteins, reducing their digestibility. page 32-11

Ant-plants, such as the bullhorn acacia, provide food and shelter for ants, which defend their host plant. page 32-11

32.3 THE PRODUCTION OF DEFENSES IS COSTLY, RESULTING IN TRADE-OFFS BETWEEN PROTECTION AND GROWTH.

Plants produce both constitutive defenses, which are produced whether or not a threat is present, as well as inducible defenses, which are triggered when a plant detects that it is being attacked. page 32-13

Plants produce volatile signals that attract insects that prey on herbivores. page 32-14

Experiments show that plants growing on nutrient-rich environments allocate more resources to growth than to defense. page 32-14

A trade-off is sometimes observed between plant growth and defense, in which plants favor one or the other, but cannot have both. page 32-15

32.4 INTERACTIONS AMONG PLANTS, PATHOGENS, AND HERBIVORES CONTRIBUTE TO THE ORIGIN AND MAINTENANCE OF PLANT DIVERSITY.

The Janzen–Connell hypothesis proposes that interactions with pathogens and herbivores increase plant diversity. page 32-16

The "escape and radiate" pattern of plant evolution suggests that plants undergo a burst of diversification following the evolution of a new form of defense against a pathogen or an herbivore. page 32-17

Herbivores and pathogens are a major concern for agriculture. page 32-17

Crop protection includes the use of chemical pesticides, integrated pest management, application of spores or toxins from *Bacillus thuringiensis* to plants, and inserting genes that encode for toxins from *B. thuringiensis* to make *Bt*-modified plants. page 32-18

Self-Assessment

1. Describe how pathogens enter and move within the plant body.
2. Distinguish between biotrophic and necrotrophic plant pathogens.
3. Name and describe two components of the plant innate immune system.
4. Describe and contrast the hypersensitive response and systemic acquired resistance.
5. Describe a feature of *Rhizobium radiobacter* that makes it a useful tool in biotechnology.

6. Describe three ways that plants protect themselves from being eaten by herbivores.

7. Explain why there are often trade-offs between plant growth and plant defense.

8. Explain how pathogens and herbivores can increase plant biodiversity.

9. Draw a phylogenetic tree that illustrates an "escape and radiate" pattern of diversification for plants that evolve novel defenses.

10. Describe one benefit and one disadvantage of herbicide and pesticide use in agriculture.

CHAPTER 33

PLANT DIVERSITY

Core Concepts

33.1 Plant diversity is dominated by angiosperms, which make up approximately 90% of all plant species found today.

33.2 The bryophytes diverged before the evolution of vascular plants, and they grow in environments where the ability to obtain water from the soil does not provide an advantage.

33.3 Spore-dispersing vascular plants are small, often epiphytic plants that usually grow in moist environments.

33.4 Gymnosperms produce seeds and woody stems and are most common in seasonally cool or dry regions.

33.5 Angiosperm diversity is partly explained by animal pollination.

In previous chapters, we explored how plants are built and how they function (Chapter 29); how these organisms, anchored to the ground, complete a sexual life cycle (Chapter 30); how the complex plant body develops from a fertilized egg (Chapter 31); and how plants defend themselves against predators and pathogens in terrestrial environments (Chapter 32). With this knowledge in hand, we are equipped to explore the diversity of land plants. How have evolutionary modifications of development and life cycle contributed to the patterns of plant diversity observed in forests, prairies, and tundra? Why are some groups of plants more diverse than others? How have changes in Earth's climate influenced the evolutionary history of plants? How did a small number of plant species come to support the nutritional needs of 7 billion humans?

33.1 PLANT DIVERSITY: AN EVOLUTIONARY OVERVIEW

In 1879, Charles Darwin wrote to botanist and long-time friend J. D. Hooker, "The rapid development as far as we can judge of all the higher plants within recent geological times is an abominable mystery." The mystery, according to Darwin, was the difficulty in explaining the relatively sudden and recent diversification of flowering plants, or angiosperms, in evolutionary history. If we tabulate the numbers of species within each major branch on the phylogenetic tree shown in **Fig. 33.1**, one fact stands out. Of the nearly 400,000 species of plants present today, approximately 90% are angiosperms. Thus, a central question in the study of plant evolution is: Why are angiosperms so much more diverse than any other group of plants?

Furthermore, angiosperms are relative newcomers. They appeared in the fossil record about 140 million years ago. The oldest known evidence of land plants is found in rocks approximately 465 million years old. Thus, for more than 300 million years, terrestrial vegetation was made up of plants other than angiosperms. During this period, lycophytes, ferns and horsetails, and gymnosperms, as well as many groups of now-extinct plants, dominated the land, forming forests and landscapes unlike those that surround us today.

Once the flowering plants gained an ecological foothold, however, the number of angiosperm species increased at an unprecedented rate. As the number of angiosperm species rose, the number of species of other groups fell. Yet the total number of plant species increased dramatically. Thus, angiosperms' hold on plant diversity is the result of both squeezing out other groups and increasing the number of species that can coexist (**Fig. 33.2**).

Angiosperms transformed the environments into which they diversified. With their multicellular xylem vessels, angiosperms can transport water through their stems at higher rates than plants that rely entirely on single-celled tracheids. Their uptake of CO_2 for photosynthesis increases as well because CO_2 uptake is directly linked to water lost to the atmosphere (Chapter 29). As a result, angiosperms can support higher rates of transpiration and photosynthesis than other groups of plants.

Recent climate simulations suggest that angiosperms may have been necessary for the formation of tropical rain forests as we know them today. Without the higher rates of

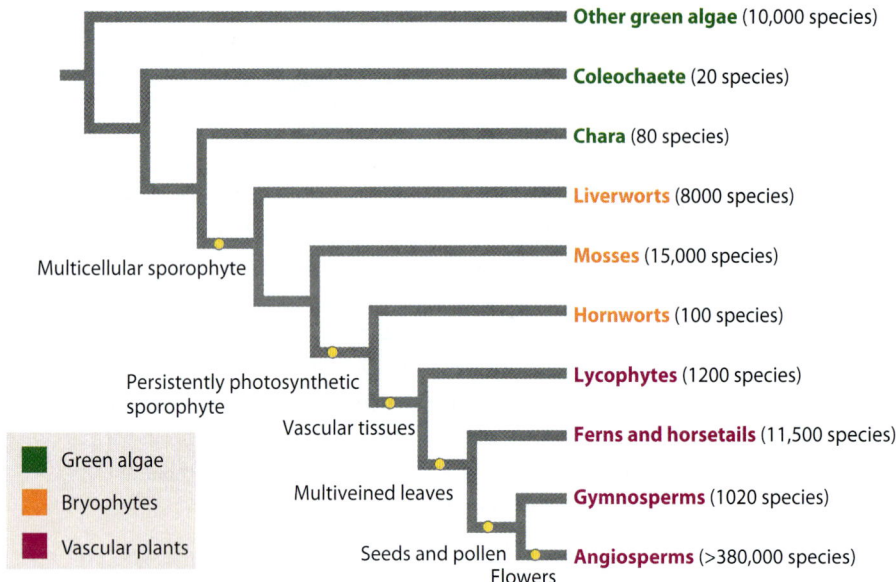

FIG. 33.1 Phylogenetic tree of land plants. The number of species is shown in parentheses. Note that 90% of all land plant species are angiosperms.

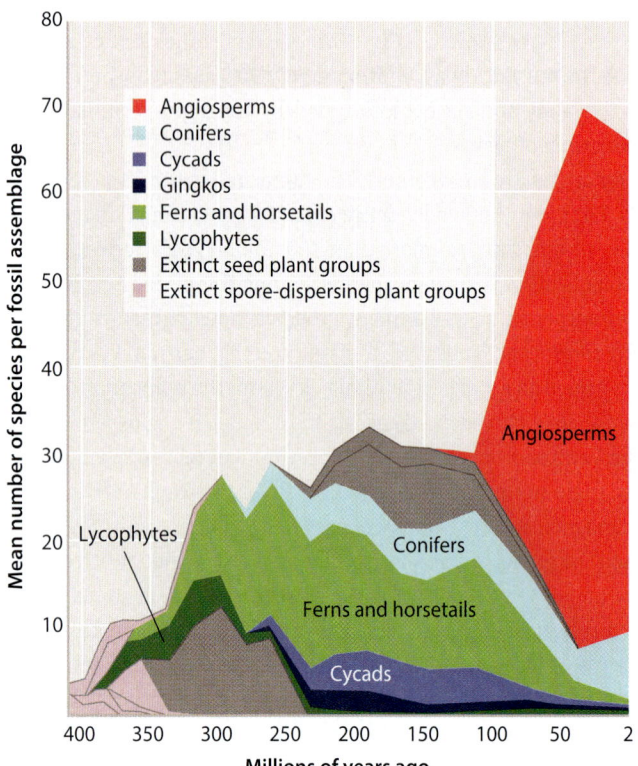

FIG. 33.2 The diversity of fossil plant assemblages through time. First appearing about 140 million years ago, angiosperms have come to dominate plant diversity.

transpiration exhibited by angiosperms, many tropical regions would have higher temperatures and lower rainfall. The warmer, drier conditions would have been less conducive to the luxuriant plant growth found in tropical rain forests. The new tropical forests, with their dense shade and humid understories, provided new habitats into which both angiosperms and non-angiosperms could evolve. Thus, although early angiosperms outcompeted many older lineages for space and light, their evolution also stimulated continued evolution in at least some of the groups that preceded them.

33.2 BRYOPHYTES

Liverworts, mosses, and hornworts are referred to as **bryophytes**. They do not form a monophyletic group, but instead a paraphyletic group. Recall from Chapter 23 that a monophyletic group includes a common ancestor and *all* of its descendants, whereas a paraphyletic group includes a common ancestor and *some* of its descendants. Nevertheless, liverworts, mosses, and hornworts are discussed as a group and merit their own name because they share so many features despite having evolved independently for hundreds of millions of years.

As the living representatives of the first plant lineages to diverge after plants moved onto land, bryophytes provide us with insights into how plants gained a foothold on the terrestrial environment. This knowledge is particularly useful because the first plants are poorly represented in the fossil record. Only tiny spores and fragments of a cuticle-like covering record these early events.

Fig. 33.1 shows that liverworts, mosses, and hornworts diverged before the evolution of lignified xylem, about 425 million years ago. Thus, bryophytes add to our understanding of the evolutionary history of plants by showing how plants that lack xylem and phloem—key innovations in the evolution of plants—are able to survive in terrestrial environments alongside vascular plants. Of course, we must not lose sight of the fact that bryophytes have continued to evolve as the conditions for life on land have changed over the past 400+ million years. Where bryophytes grow and what they look like have been shaped by their long coexistence with vascular plants.

→ **Quick Check 1** Look at Fig. 33.1 and name the paraphyletic groups indicated by the tree.

Bryophytes are small, simple, and tough.

Mosses are the most widely distributed of the bryophyte lineages and the most diverse, with about 15,000 species. You may have seen moss growing on a shady log or on top of rocks by a stream. However, mosses grow in all terrestrial environments, from deserts to tropical rain forests. Liverworts (about 8000 species) and hornworts (about 100 species) are less widespread and also less diverse. In considering the diversity of these three lineages, let's start by reviewing some of the features common to all bryophytes.

The most obvious feature that unites the bryophytes is that they are small, most of them only a couple of centimeters in height (**Fig. 33.3**). The major constraint on bryophyte size is thought to arise from their mode of fertilization. Bryophytes release their sperm into the environment. These sperm must swim through films of surface water or be transported by the splash of a raindrop to meet a female gamete. Sperm can travel in this manner only a relatively limited distance and only if water is present. A few mosses grow to be more than half a meter tall, but these giants of the bryophyte world are found only in the understory of very wet forests.

Bryophytes are not only small, they also have simple bodies. Some produce only a flattened photosynthetic structure called a **thallus.** Others consist of slender stalks and have a leafy appearance. However, these leaf-like structures are quite different from the leaves of vascular plants in that they are only one to several cells thick and lack internal air spaces or a water-conducting system. Both thalloid and leafy species are found in the liverworts, whereas all mosses are of the leafy type and all hornworts are of the thalloid type.

These conspicuous and persistently photosynthetic bryophyte bodies represent the haploid generation. Recall from Chapter 30 that a key innovation that appeared in the land plants is the alternation between a multicellular gamete-producing generation composed entirely of haploid cells and a multicellular spore-producing generation composed of diploid cells. The closest green algal relatives to the land plants have a life cycle in which there is only a single multicellular phase rather than two. The multicellular phase of the green algal relatives corresponds to the haploid, gamete-producing generation of land plants (the gametophyte generation) because both are formed entirely of haploid cells and produce gametes. Thus, the diploid, spore-producing generation (the sporophyte generation) represents a new component of the life cycle that evolved as plants moved onto land.

In bryophytes, the sporophyte remains physically attached to and, in varying degrees, nutritionally dependent upon the gametophyte. In a few liverworts, the sporophyte remains embedded within the thallus, releasing spores only when the gametophyte dies and its body decays. In this case, the multicellular sporophyte generation serves only to amplify the number of haploid spores. In most bryophytes, the sporophyte extends several centimeters above the gametophyte, greatly increasing the chances of spores being dispersed through the air. In a few cases, such as the liverwort *Marchantia* (Fig. 33.3b), tiny sporophytes grow from the top of upright structures produced by the gametophyte.

In mosses and liverworts, the sporophyte is short lived, drying out after the spores are dispersed. However, in hornworts, the sporophyte can live nearly as long as the gametophyte because it can produce new cells at its base, similar to the way grass blades elongate (Chapter 32). Thus, bryophytes illustrate an important trend in the evolution of

FIG. 33.3 **Bryophyte diversity.** Bryophytes include (a) mosses (*Polytricum commune* is shown); (b) liverworts (*Marchantia berteroanna*, a thalloid liverwort, is shown), and (c) hornworts (*Anthoceros* species).

plants, an increase in the persistence of the spore-producing generation.

Because bryophytes do not produce lignified xylem conduits, they cannot pull water from the soil. Instead, they absorb water and CO_2 through their surfaces, which have little or no waxy cuticle. Bryophytes can absorb enough water to remain metabolically active when the environment is wet, but they must be able to tolerate desiccation when the environment is dry. Thus, while the small size and lack of differentiated structures might suggest that bryophytes are delicate, nothing could be further from the truth. Bryophytes can be found from the equator to both the latitudinal and altitudinal limits of vegetation, and from swamps to deserts. They are the only plants that grow on the Antarctic continent. In fact, the entire flora of that continent consists of approximately 100 moss species and 25 species of liverworts.

Nevertheless, because bryophytes are so small, they are poor competitors for light and space. Instead, they thrive in local environments where roots do not provide an advantage—for example, on the surfaces of rocks. Many bryophytes live on the branches and trunks of trees rather than on the ground. Plants that grow on other plants are called **epiphytes** (from the Greek words *epi*, "on," and *phyton*, "plant"). Bryophytes are well suited to grow in this way because they are not dependent upon the soil as a source of water.

Bryophytes exhibit several cases of convergent evolution with the vascular plants.

Bryophytes are interesting partly because they are so different from the more familiar vascular plants. However, given that they have long evolved in parallel with vascular plants, it is not surprising that bryophytes have evolved similar solutions to environmental challenges, a process referred to as convergent evolution (Chapter 23). For example, some mosses depend on insects to transport their spores (**Fig. 33.4a**). These mosses produce brightly colored sporangia, which emit volatile chemicals that mimic compounds released by rotting flesh or herbivore dung. These chemicals attract insects, and the insect legs become covered with spores when they land on the sporangia. When the insect subsequently lands on a real pile of dung, some of the spores fall off into what is, for them, an ideal site for supporting the growth of a new gametophyte.

Another example of convergent evolution is the presence in some mosses and liverworts of cells specialized for the transport of water and carbohydrates. It is clear from the structure of these cells that they evolved independently from the xylem and phloem of vascular plants. In particular, the water-conducting cells in bryophytes do not have lignified cell walls and thus are not sufficiently rigid to pull water from the soil. Nevertheless, water and carbohydrates can move more efficiently from one part of the plant to the other through these elongated cells. Not surprisingly, these internal transport cells are found in the largest of the bryophytes (**Fig. 33.4b**).

As discussed in Chapter 29, stomata are pores in the epidermis of vascular plants that can open and close, providing an active means of controlling water loss. Stomata also occur on the sporophytes of some mosses and hornworts. It is unknown whether the presence of these stomata is a case of convergent evolution. Answering that question will require further study of the structure and development of bryophyte stomata.

→ **Quick Check 2** If the stomata of bryophytes and vascular plants are not an example of convergent evolution, what is an alternative explanation?

FIG. 33.4 Moss specializations. (a) The yellow moose-dung moss, *Splachnum luteum*, attracts insects that then transport its spores. (b) *Dawsonia superba* has internal transport tissues that allow it to grow more than 40 cm tall.

Sphagnum moss plays an important role in the global carbon cycle.

In most ecosystems, bryophytes make only a small contribution to the total biomass. The one exception is **peat bogs**, wetlands in which dead organic matter accumulates. A major component of peat bogs is sphagnum moss, any of the several hundred species of the genus *Sphagnum*. These mosses play a key role in creating wet and acidic conditions that slow rates of decomposition. They have specialized cells that hold onto water, much like a sponge, and they secrete protons that acidify the surrounding water (**Fig. 33.5**). Unlike many plants with roots, sphagnum moss thrives under wet, acidic conditions. In addition, sphagnum mosses produce phenols, organic compounds that slow decomposition under waterlogged conditions.

Peat bogs occupy only 2% to 3% of the total land surface, but they store large amounts of organic carbon—on the order of 65 times the amount released each year from the combustion of fossil fuels. Thus, any increase in the decomposition rate of these naturally formed peats has the potential to increase significantly the CO_2 content of Earth's atmosphere.

What might cause an increase in the decomposition rate? Peat bogs are vulnerable to changes in the water table. As water levels fall, the peat becomes exposed to the air, allowing it to be broken down by microbes. Most peat bogs are located in northern latitudes, where the effects of climate change are predicted to be greatest. If rising summer temperatures result in higher rates of transpiration from surrounding forests, the result could be a lowering of the water table and a relatively rapid release of CO_2 from this moss-dominated ecosystem. Thus, peat bogs have the potential to accelerate rates of climate change over the next century.

33.3 SPORE-DISPERSING VASCULAR PLANTS

As noted in Chapters 29 and 30, vascular plants differ from bryophytes in several key aspects of form and life cycles. In vascular plants the diploid sporophyte is larger than the gametophyte and becomes capable of supplying all of its own needs. The sporophyte achieves greater height and independence with the help of vascular tissues that pull water from the soil and distribute carbohydrates throughout the plant body.

Vascular plants can be divided into two groups according to how they complete their life cycle. Lycophytes, as well as ferns and horsetails, disperse by spores and rely on swimming sperm for fertilization, just as occurs in bryophytes. Gymnosperms and angiosperms are seed plants: They are dispersed through the movement of seeds and fertilized through the movement of pollen. In this section, we focus on the spore-dispersing vascular plants: lycophytes, and ferns and horsetails. In both groups, the sporophyte dominates in terms of physical size; the gametophyte is only a few centimeters, thalloid in structure, and typically tolerant of desiccation. Thus, our discussion emphasizes the sporophytes of these plants. However, it is important to remember that because their gametophytes release sperm into the environment, they, like the bryophytes, are dependent on surface moisture for fertilization.

Rhynie cherts provide a window into the early evolution of vascular plants.

Fossils recovered from a single site in Scotland document key stages in the evolution of vascular plants. Four hundred million years ago, near the present-day village of Rhynie, a wetland community thrived in a volcanic landscape. Mineral-laden springs similar to those found today in Yellowstone Park encased the newly deposited remains of local populations in chert, a mineral made of silica (SiO_2). These fossils preserve plants, fungi, protists, and small animals in remarkable detail. Rhynie cherts provide our

FIG. 33.5 **Sphagnum moss.** Peat bogs have low rates of organic decomposition due to the wet and acidic conditions created by sphagnum moss.

Peat bog

Water-holding cell

Sphagnum moss

Photosynthetic cell

FIG. 33.6 Rhynia, a fossil vascular plant from the 400-million-year-old Rhynie chert. All vascular plants are descended from simple plants such as this.

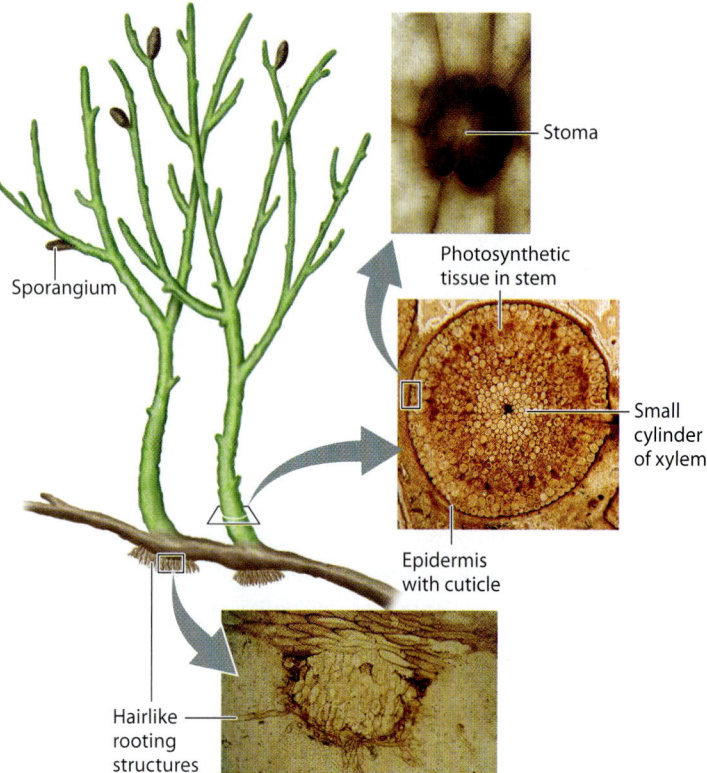

best view of early vascular plants, which were small (up to about 15 cm tall) photosynthetic structures that branched repeatedly, forming sporangia at the tips of short side branches (**Fig. 33.6**). These plants had no leaves, and their only rooting structures were small hairlike extensions on the lower parts of stems that ran along the ground. The stems had a cuticle with stomata, and a tiny cylinder of vascular tissue ran through the center. Many of the gametophytes also had erect cylindrical stems, not at all like the reduced gametophytes of living vascular plants.

Rhynie cherts preserve another important glimpse of early plant evolution: Some fossils have a branched sporophyte, but no evidence of vascular tissue. Evidently, plants evolved the upright stature we associate with living vascular plants before they evolved the differentiated tissues of the vascular system (xylem and phloem). Once vascular tissues had evolved, however, they played a key role in shaping subsequent evolution of the plant sporophyte.

Lycophytes are the sister group of all other vascular plants.

One of the fossil populations in Rhynie cherts is notably larger and more complex than the others. These fossils show leaf-like structures arranged on the stem in a spiral pattern. Through the stem runs a thick-lobed cylinder of xylem (**Fig. 33.7**). These are among the earliest fossils of lycophytes, an early-branching group of vascular plants that can still be found from tropical rain forests to Arctic tundra. Because lycophytes are the earliest branching group of vascular plants, they are the sister group to all other vascular plants (see Fig. 33.1).

Fossils indicate that the leaves and roots of lycophytes evolved independently from those of ferns, horsetails, and seed plants. A distinctive feature of lycophyte leaves is that they have only a single vein, in contrast to the more complex leaf venation found in other vascular plants. About 1200 species of lycophytes can be found today. Plants assigned to the genus *Lycopodium* and its close relatives (**Fig. 33.8a**) grow low to the ground and produce branching stems covered by small leaves arranged in a helical pattern. Roots are produced from stems growing along the ground, while spores form in sporangia borne on leaves, either those arranged along the stems or those clustered near stem tips.

A second group of lycophytes, called *Selaginella*, have leaves that may be helically arranged around the stem or occur in four distinct rows with the leaves held in a single plane, giving these branched stems a distinctive appearance (**Fig. 33.8b**). Many species of

FIG. 33.7 An early lycophyte preserved in the 400-million-year-old Rhynie chert. The fossil has helically arranged leaflike structures, adventitious roots (that is, roots that arise from the shoot and not from the root system), and a central strand of xylem, features much like those of living lycophytes.

FIG. 33.8 Lycophyte diversity. Living lycophytes include about 1200 species, in three major groups: (a) *Lycopodium annotinum*, showing yellow sporangia-bearing leaves at the shoot tips of this 5-to-30-cm tall plant; (b) *Selaginella wildenowii*, in a close-up showing the flattened arrangement of its 3-mm-long leaves; and (c) *Isoetes lacustris*, an aquatic lycophyte that grows from 8 to 20 cm tall.

Selaginella live as epiphytes in tropical forests. A few are vines and some, called resurrection plants, live in seasonally dry habitats where they tolerate desiccation, as more commonly do bryophytes.

The most distinctive living lycophytes are the quillworts (genus *Isoetes*), which are small plants that live along the margins of lakes and slow moving streams (**Fig. 33.8c**). A few species are desiccation tolerant and so can live in ponds that dry up in summer. Uniquely among living lycophytes, they have a vascular cambium. Seeing a quillwort today, you would never guess that its relatives once included enormous trees.

Ancient lycophytes included giant trees that dominated coal swamps about 320 million years ago.

Fossils show that ancient lycophytes evolved additional features convergently with seed plants, including a vascular cambium and cork cambium that enabled them to form trees up to 40 m tall and, in some cases, reproductive structures similar to seeds (**Fig. 33.9**). Indeed, swamps that formed widely about 320 million years ago were dominated by tree-sized lycophytes. Unlike the vascular cambium of seed plants, the vascular cambium of these

FIG. 33.9 Giant lycophytes. (a) Artist's vision of giant lycophytes. (b) A fossil grove of lycophyte trees showing rooting structures and thick stems. (c) Fossil showing the seedlike reproductive structures, about 1 to 2 cm long, evolved by these ancient trees.

lycophytes (**Fig. 33.10**) produced relatively little secondary xylem and no secondary phloem. Instead, the giant lycophytes relied on thick bark for mechanical support. Thus, tree-sized lycophytes were markedly different from the trees familiar to us today.

These forest giants were not outcompeted by seed plants, but persisted for millions of years in swampy environments alongside early seed plants. It was environmental change that spelled their doom. When changing climates dried out the swamps, the large

HOW DO WE KNOW?

FIG. 33.10

Did woody plants evolve more than once?

OBSERVATIONS Today, the vascular cambium occurs almost exclusively among seed plants, with only limited development of secondary xylem in some adder's tongue ferns and quillwort lycophytes (*Isoetes* species). The fossil record, however, shows that other plants, now extinct, also had a vascular cambium.

HYPOTHESIS The vascular cambium evolved convergently in several different groups of vascular plants.

EXPERIMENT AND RESULTS The vascular cambium is recorded in fossils by xylem cells in rows oriented radially in the stem or root. Thus, the giant tree lycophytes of Carboniferous coal swamps (the top photo at the lower right) had a vascular cambium, as did extinct tree-sized relatives of the horsetails (the bottom photo) and other extinct horsetail relatives. Phylogenetic trees generated from morphological features preserved in fossils unambiguously show that woody lycophytes, woody horsetail relatives, and the group of seed plants and progymnosperms did not share a common ancestor that had a vascular cambium. Moreover, anatomical research shows that the vascular cambium of extinct woody lycophytes and the giant horsetails *Archaecalamites* and *Calamites* generated secondary xylem but not secondary phloem, unlike the vascular cambium of seed plants. Living horsetails do not have a vascular cambium, but fossils show that they are descended from ancestors that did make secondary xylem and have lost this trait through evolution.

CONCLUSION Fossils and phylogeny support the hypothesis that the vascular cambium and, hence, wood evolved more than once, reflecting a strong and persistent selection for tall sporophytes among vascular plants.

SOURCE Taylor, T. N., E. L. Taylor, and M. Krings. 2009. *Paleobotany*. Amsterdam: Elsevier.

Phylogenetic tree tips:
- X *Rhynia*
- *Lycopodium*
- *Selaginella*
- *Isoetes*
- X *Lepidodendron*
- X *Psilophyton*
- Adder's tongue ferns
- Whisk ferns
- Extant horsetails
- X *Calamites*
- X *Archaecalamites*
- X *Sphenophyllum*
- Ferns
- X Progymnosperms
- Seed plants

Legend:
- ● Vascular cambium present (yellow)
- ● Vascular cambium lost (blue)
- X Extinct

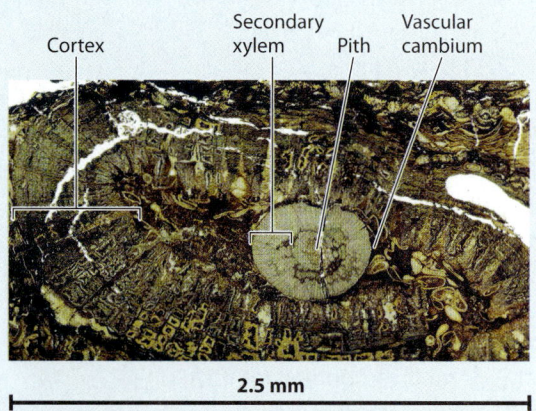

A small branch of *Lepidodendron*, an ancient giant lycophyte

Labels: Cortex, Secondary xylem, Pith, Vascular cambium

2.5 mm

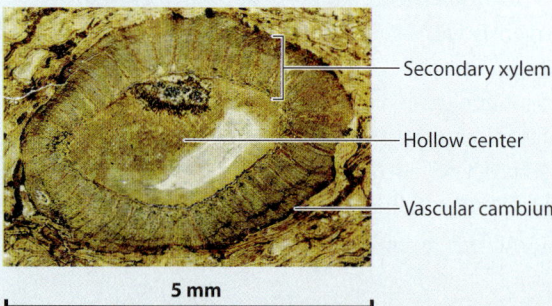

A small stem of *Calamites*, an ancient giant horsetail

Labels: Secondary xylem, Hollow center, Vascular cambium

5 mm

FIG. 33.11 Diversity of ferns and horsetails. (a) The adder's tongue fern *Ophioglossum*, (b) the whisk fern *Psilotum*, (c) the horsetail *Equisetum*, (d) a marattiod fern, (e) an aquatic fern (*Salvinia*), and (f) a polypod fern (in the middle is *Athyrium felix-femina*, the common lady-fern).

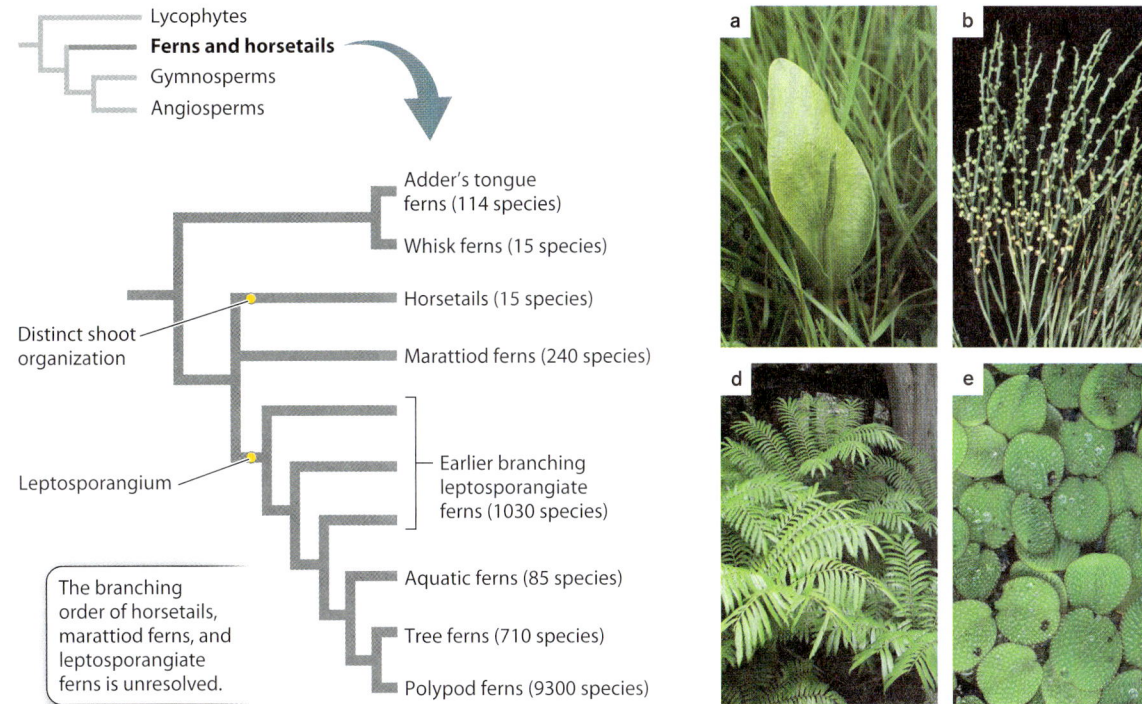

lycophytes disappeared, along with their habitat. Little *Isoetes* was left as the only living reminder of a once-dominant group of plants. However, the other legacy of these ancient forests is the coal deposits we mine today. As trees died and fell over into the swamp, their bodies decomposed slowly. Over time, as material accumulated and became buried, the combined action of high temperature and pressure converted the dead organic matter first into peat and then into the carbon-rich material we call coal. Thus, a major source of energy used today by humans derives originally from photosynthesis carried out by these lycophyte trees.

Ferns and horsetails are morphologically and ecologically diverse.

Ferns and horsetails form a monophyletic group that are the sister group to the seed plants (**Fig. 33.11**). The majority of species in this group are in fact ferns, which can be recognized by their distinctive leaves (Fig. 33.11) that uncoil during development from tightly wound "fiddleheads" (**Fig. 33.12**). This group also includes horsetails and whisk ferns, which traditionally were considered as distinct groups because of their unique body organizations. Whisk ferns have photosynthetic stems without leaves (see Fig. 33.11b) and they also do not form roots. They thus look similar to fossils of early vascular plants. Molecular sequence comparisons, however, support the hypothesis that whisk ferns are the simplified descendants of plants that produced both leaves and roots and that they are members of the same lineage as ferns.

The horsetails, represented by 15 living species of the genus *Equisetum*, produce tiny leaves arranged in whorls, giving the stem a jointed appearance (see Fig. 33.11c). Horsetail stems are hollow, and their cells accumulate high levels of silica, earning them the name "scouring rush." Today, horsetails are small plants,

FIG. 33.12 A fern leaf uncoiling from a "fiddlehead."

FIG. 33.13 A tree fern and an aquatic fern. (a) Silver tree ferns (*Cyathea dealbata*) grow 10 m tall or more. (b) *Azolla* ferns floating on the surface of still freshwater. Each fern is about 10 mm in diameter.

are by far the most diverse group of ferns (about 9300 species), and their distinctive sporangia, with their high potential for dispersal, may have played a role in their evolutionary success.

Although the fossil record shows that ferns originated more than 360 million years ago, the radiation of the polypod ferns occurred after the rise of the angiosperms. Thus, many of the fern species present today are likely to have evolved to occupy habitats newly created by the development of angiosperm forests. Approximately 40% of living ferns are epiphytes, many of them growing in tropical rain forests dominated by angiosperm trees.

most less than a meter tall. However, tree-sized versions grew side by side with the ancient lycophyte trees (see Fig. 33.9).

Fern leaves can be large—in some cases up to 5 m long, although typically the photosynthetic surfaces are divided into smaller units called **pinnae**. Fern stems frequently grow underground, and only the leaves emerge into the air. Internode elongation can be very short, so that all the leaves are bunched together. Some species, such as bracken ferns, produce horizontal stems with widely spaced leaves. This arrangement allows bracken to spread rapidly. Because bracken is toxic to cattle and other livestock, its tendency to invade pastures is a serious problem.

Vascular tissue allows fern sporophytes to grow large. However, because ferns do not exhibit secondary growth, there are limits to how tall they can grow. Tree ferns can grow to more than 10 m by producing thick roots that descend from the leaves to the ground, adding both vascular capacity and mechanical support (**Fig. 31.13a**). Other ferns grow tall by producing leaves that twine around the stems of other plants, using the other plant for support. At the other extreme are tiny aquatic ferns such as *Salvinia* (see Fig. 33.11e) and *Azolla* (**Fig. 33.13b**) that float on the surface of the water.

Horsetails, whisk ferns, and a few other fern groups make large sporangia similar to those produced by early vascular plants. Most ferns, however, have a distinctive sporangium (called a leptosporangium) whose wall is only a single cell thick. Polypod ferns (see Fig. 33.11f), in particular, have leptosporangia with a line of thick-walled cells that run along the sporangium surface. When the spores mature, the sporangium dries out and these cells contract, forcibly ejecting spores into the air. Polypod ferns

Epiphytic ferns are exposed to much more light than plants established on the ground. However, the canopy environment poses its own challenges, notably the danger of dislodgement and the risk of drying out. As a result, epiphytic ferns have adaptations that allow them to survive in the forest canopy. For example, staghorn ferns (**Fig. 33.14**) produce two leaf types: erect leaves that are photosynthetic and produce sporangia and basal leaves that grow close to the stem of the host tree, forming a pocket in which organic matter accumulates. The roots of staghorns can grow into the pocket of organic matter, where they are protected from damage and desiccation.

An aquatic fern contributes to rice production.

Azolla is a free-floating aquatic fern that has been used for centuries in Asia as a biofertilizer of rice paddies (see Fig. 33.13b). Its success as a fertilizer is due to its symbiotic association with nitrogen-fixing cyanobacteria. *Azolla* is common worldwide on still water such as ponds and quiet streams. Its leaves contain an internal cavity in which colonies of nitrogen-fixing cyanobacteria become established. The cyanobacteria provide *Azolla* with a ready source of nitrogen, while the fern provides shelter and carbohydrates. Rates of nitrogen fixation by the cyanobacteria are reported to be greater when the cyanobacteria are housed within an *Azolla* leaf than when these organisms live freely in the environment.

When added to newly planted rice fields, this fast-growing fern suppresses weed growth by forming a dense, floating layer. At the same time, it provides organic nitrogen that eventually

CHAPTER 33 PLANT DIVERSITY

FIG. 33.14 The staghorn fern. This epiphyte produces two distinct types of leaf that enable it to thrive in the canopies of angiosperm trees.

finds its way into the developing rice plants. The floating fertilizer factories provided by *Azolla* may have made possible the long history of rice cultivation in Asia.

33.4 GYMNOSPERMS

The seed—and equally important, pollen—are evolutionary innovations that played important roles in the success of seed plants. Seed plants do not require external water for fertilization. Instead, the male gametophyte is transported through the air in a pollen grain. Successful pollination leads to fertilization and the formation of a seed. Seeds are multicellular structures that enhance dispersal success by transporting the embryo together with resources that it can draw on during germination. Before these innovations, the ancestors of the seed plants evolved the ability to grow large woody stems through the formation of both a vascular and a cork cambium. Therefore, woody stems and roots are the ancestral condition in seed plants.

The gymnosperms are one of two major groups of seed plants. The living gymnosperms include four groups: cycads, ginkgos, conifers, and gnetophytes (**Fig. 33.15**). However, the fossil record shows that more than a dozen additional groups, now extinct, were once present. These include a number of early seed plants that branched off before any living gymnosperm groups evolved. The floodplains adjacent to the coal swamps in which giant lycophytes thrived were populated by early seed plants with large fern-like leaves that produced large fleshy seeds. Gymnosperms probably reached their greatest diversity during the time of the dinosaurs. At that time, groups like cycads and conifers lived alongside a larger number of now-extinct groups.

Cycads and ginkgos are the earliest diverging groups of living gymnosperms.

The unobservant eye often confuses cycads with palm trees because both have unbranched stems and large leaves

FIG. 33.15 Seed plant diversity, showing a hypothesis of phylogenetic relationships among the gymnospermous plants.

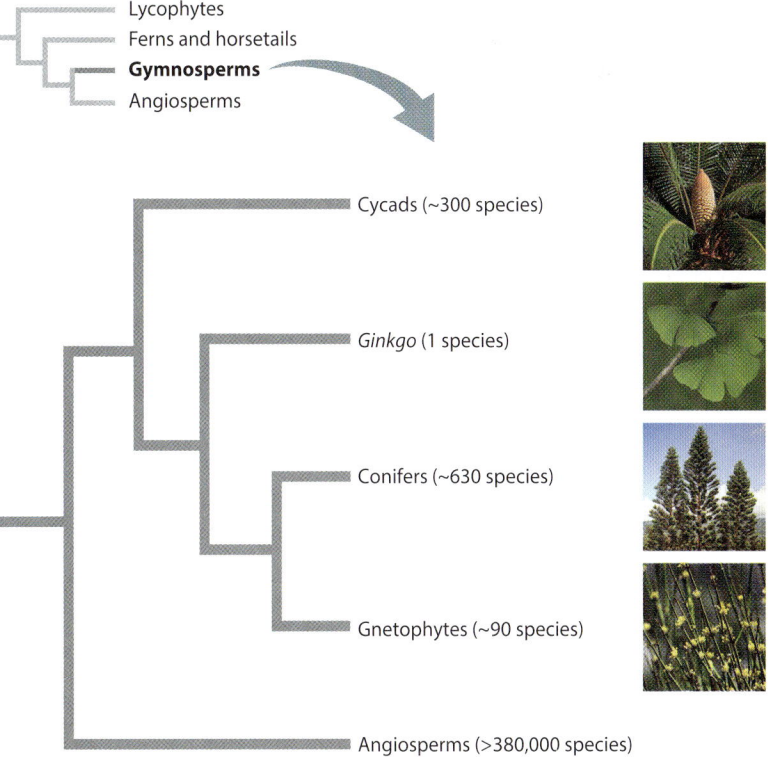

FIG. 33.16 Cycads. (a) Most living cycads have thick, unbranched stems that support large leaves and cones. Shown here is *Encephalartos transvenosus*. (b) Many cycads, such as *Encephalartos transvenosus*, have large cones. (c) Mature seeds are large (here showing those of *Cycas circinalis*), and contain a relatively large female gametophyte that provides nutrition for the embryo and germinating sporophyte.

(**Fig. 33.16**). However, the presence of large cones rather than flowers clearly marks cycads as gymnosperms. Cycads tend to have low photosynthetic rates and be slow growing. Although they have large stems, their vascular cambium produces little additional xylem, and most of their bulk is made up of a large pith and cortex. They form symbiotic associations with nitrogen-fixing cyanobacteria that allow them to grow in nutrient-poor environments. The apical meristem of cycads is protected by a dense layer of bud scales, enabling them to survive wildfires.

The approximately 300 species of cycads occur most commonly in tropical and subtropical regions. Most species have a limited and often fragmented distribution, suggesting that they were much more widespread in the past. Most cycad populations are small and often vulnerable to extinction. Cycads may owe their persistence in part to the fact that they rely on insects for pollination, as well as to their ability to live in nutrient-poor and fire-prone environments. Insects provide more efficient pollen transfer than wind, thus allowing species with small populations to reproduce successfully. Many cycads are pollinated by beetles, which are attracted by chemical signals produced by the cones. The insects use the large cones for shelter, and the pollen cones provide them with food. Following fertilization, brightly colored, fleshy seed coats attract a variety of birds and mammals that serve as dispersal agents. Because of their small numbers and fragmented distribution, two-thirds of all cycads are on the International Union for the Conservation of Nature's "red list" of threatened species, the highest percentage of any group of plants.

Cycads are an ancient lineage, with fossils at least 280 million years old. During the time of the dinosaurs, they were among the most common plants. Cycad diversity and abundance declined markedly during the Cretaceous Period, at the same time as the angiosperms rose to ecological prominence. A recent study shows, however, that all the present-day species within this group date from a burst of speciation that took place approximately 12 million years ago. At that time, changing continental positions and ocean circulation patterns resulted in increasingly cooler and more seasonal climates, providing new habitats for emerging species while dooming older species to extinction.

A second group of gymnosperms has only a single living representative: *Ginkgo biloba*. Like cycads, ginkgos date back about 270 million years, and during the time of the dinosaurs, they were common trees in temperate forests. Also like cycads, ginkgos declined in abundance, diversity, and geographic distribution over the past 100 million years, as a result of both changing climates and angiosperm diversification. *Ginkgo biloba* forms tall, branched trees, with fan-shaped leaves that turn a brilliant yellow in autumn (**Fig. 33.17**). Once again, like cycads, *Ginkgo* develops fleshy seeds, perhaps a feature inherited from their common seed plant ancestor.

Ginkgos have long been cultivated in China, especially near temples, and it is unclear whether any truly natural populations exist in the wild. Today, however, *Ginkgo* once again has a global

FIG. 33.17 *Ginkgo biloba*. (a) The fan-shaped leaves of the ginkgo turn yellow in autumn and are then shed. (b) Its fleshy seeds enclose a large female gametophyte, as in cycads.

distribution, as its ability to tolerate pollution and other stresses associated with urban environments has made it a tree of choice for street plantings. The only drawback is that the fleshy seeds produce a chemical that smells like rancid butter. For this reason, individuals that produce pollen, as opposed to seeds, are strongly preferred.

Conifers are forest giants that thrive in dry and cold climates.

The approximately 630 species of conifers are, for the most part, trees. Conifers include the tallest (over 100 m) and oldest (more than 5000 years) trees on Earth (**Fig. 33.18**). Well-known conifers include pines, junipers, and redwoods. Conifer xylem consists almost entirely of tracheids, resulting in wood that is strong for its weight and has relatively uniform mechanical properties. Conifers have traditionally served as masts for sailing ships, and today are used for telephone poles. Conifers supply much of the world's timber, as well as the raw material for producing paper.

Most conifers are evergreen, meaning that they retain their often needle-like leaves throughout the year. In some species, the leaves are remarkably long lived, remaining green and photosynthetic for decades. Many conifers produce resin canals in their wood, bark, and leaves that deter insects and fungi. Conifer resins are harvested and distilled to produce turpentine and other solvents. In addition, conifer resins are the source of a number of important chemicals, including the drug taxol, which is used in the treatment of cancer.

In Chapter 30, we examined the life cycle of pine. Pollen is produced in small cones, and ovules develop in larger cones that mature slowly as the fertilized ovules develop into seeds. Conifers are wind pollinated, and most species also rely on wind for seed dispersal. Some conifers, however, produce fleshy and often brightly colored tissues associated with their seeds that attract birds and other animals. In junipers, for example, the entire seed cone becomes fleshy. Juniper seeds are used as a seasoning; they are what gives gin its distinctive flavor.

Conifers dominate the vast boreal forests of Canada, Alaska, Siberia, and northern Europe. Conifers generally increase in abundance as elevation increases, and they are common in dry areas, such as the western parts of North America and much of Australia. Only a small number of conifer species are found in

FIG. 33.18 Conifers. Giant sequoia (*Sequoiadendron giganteum*) can reach 85 meters in height and have stems that are 8 meters in diameter.

tropical latitudes, most commonly at higher elevations. Before the rise of the angiosperms, however, conifers were much more evenly distributed across the globe, including in the lowland tropics. One of the great questions in the evolutionary history of plants is why conifers were displaced from tropical latitudes by the angiosperms but continued to thrive in cold and dry environments.

An important difference between angiosperms and conifers is that angiosperms transport water in the multicellular xylem conduits called vessels, whereas conifers transport water in single-celled tracheids. In wet and warm environments, the xylem vessels of angiosperms can be both wide and long, increasing the efficiency of water transport through their stems. Conifer tracheids, which form from single cells, are much smaller. Thus, one hypothesis is that the evolution of highly conductive xylem vessels gave angiosperms a competitive advantage in many environments: They could support higher rates of photosynthesis and grow larger leaves. Once conifer populations began to dwindle, they could have disappeared completely because wind pollination, to be effective, requires large populations. In colder or drier regions, the dangers of cavitation due to drought or freeze followed by thaw (Chapter 29) constrain the size of angiosperm vessels. This helps level the playing field in the competition between conifers and angiosperms and explains the greater coexistence of conifers and angiosperms in cold or dry climates.

Gnetophytes are gymnosperms that have independently evolved xylem vessels and double fertilization.

Gnetophytes are a small group of gymnosperms (about 90 species) made up of three morphologically different genera (**Fig. 33.19**). *Welwitschia mirabilis* is found only in the deserts of southwestern Africa. It produces only two straplike leaves that elongate continuously from their base. *Gnetum* occur in tropical rain forests, where they form woody vines and small trees with large broad leaves. *Ephedra*, which is native to arid regions, produces shrubby plants with photosynthetic stems and small leaves.

Gnetophytes have long attracted the attention of plant biologists because their members exhibit a number of traits typically associated with angiosperms, notably the formation of multicellular xylem vessels and double fertilization, although the double fertilization in gnetophytes does not result in the formation of endosperm. Indeed, for years many botanists viewed gnetophytes as the sister group of angiosperms. DNA analysis, however, indicates that gnetophytes are not closely related to the angiosperms but are, instead, closely related to conifers. Thus, both vessels and double fertilization evolved independently and convergently in this group.

→ **Quick Check 3** Double fertilization evolved independently in gnetophytes and flowering plants. Name two other features usually associated with angiosperms that evolved independently in gymnosperms.

FIG. 33.19 Gnetophytes. The three genera in this group include some of the most distinctive seed plants: (a) *Welwitschia mirabilis*; (b) *Gnetum* species; and (c) *Ephedra* species.

33.5 ANGIOSPERMS

The evolution of the angiosperms completely transformed life on land. Not only did the total number of plant species increase significantly, but the productivity of terrestrial ecosystems increased as well. As we have seen, the evolution of angiosperms affected the abundance and distribution of other plant groups. In addition, the higher productivity of angiosperm-dominated environments increased the food available for insects, birds, and mammals. All these animal groups diversified along with the flowering plants.

Angiosperm diversity remains a puzzle.

The question of why angiosperms are so diverse has occupied biologists since before Darwin's time. Despite their efforts, a definitive answer has so far not emerged. Diversity can result from high rates of species formation, and also from low rates of species loss. Interestingly, it may be that low rates of species loss have played an important role in the evolution of the angiosperms. Studies of the fossil record suggest that the rate at which new species appear is comparable for angiosperms and gymnosperms. However, angiosperm species disappear from the fossil record at a lower rate than gymnosperm species.

Flowers may contribute to angiosperm diversity by making it less likely that a rare species will become extinct. Wind-pollinated plants can reproduce only when their populations have a relatively high density. Since there is a high probability that no wind-carried pollen will fall on isolated individuals, species with low population density have a relatively high likelihood of extinction. But animal pollinators actively searching for rare species are much more likely to find them, and so animal-pollinated angiosperm populations can persist at low population densities.

A second feature of angiosperm reproduction that makes it likely that rare species will persist is the extreme reduction in the size of the female gametophyte (Chapter 30). Its small size is possible because the endosperm produced by double fertilization allows angiosperms to delay provisioning their offspring until after fertilization. The price in energy and resources that an angiosperm must pay to take a chance on being fertilized is much lower than in gymnosperms.

Tropical rain forests illustrate how the ability to reproduce at low population density contributes to angiosperm diversity. Rain forests may have several hundred different plant species growing on a single hectare of land, and most of these species are rare.

Early diverging angiosperms have low diversity.

Fig. 33.20 shows a phylogenetic tree of angiosperms. Note that while angiosperms as a whole are remarkably diverse, the early-branching groups of flowering plants are not. Early-branching angiosperms, however, provide important clues to the evolution of flowering plants.

The last common ancestor of angiosperms and living gymnosperms is thought to have lived more than 300 million years ago. In the absence of a closely related sister group, what tools can we use to understand the origin of the angiosperms? The first unambiguous fossils of angiosperms appear in rocks

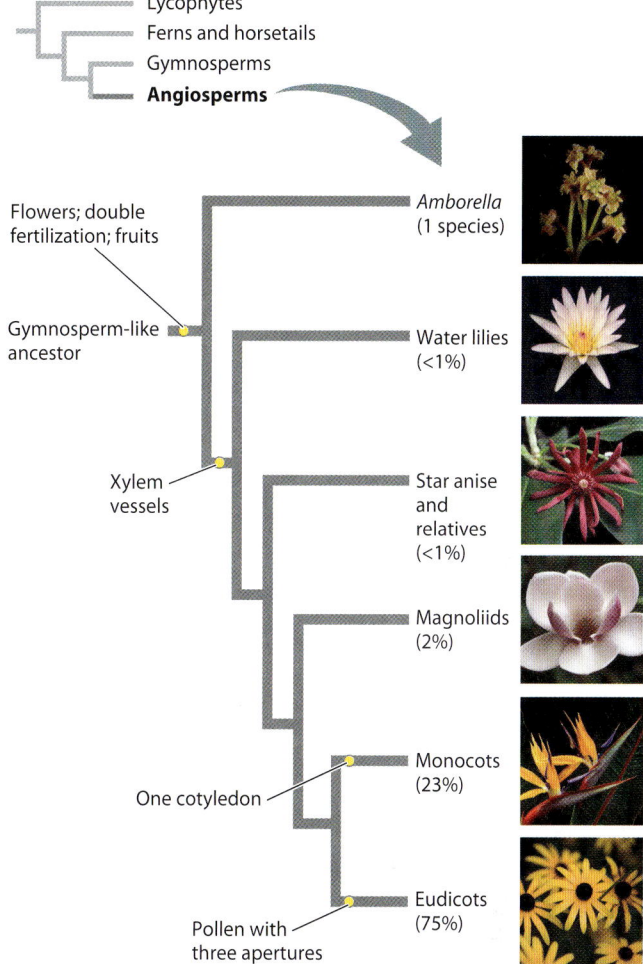

FIG. 33.20 Angiosperm phylogeny. Most angiosperm species belong to one of two subgroups, the monocots and the eudicots.

about 140 million years ago. These fossils belong to groups that branch early on the angiosperm phylogenetic tree, but they leave many questions unanswered regarding the ecological context of angiosperm evolution.

To gain further insights into the type of habitat in which the first angiosperms appeared, let's look at where living plants from the earliest-branching lineages of angiosperms are found today. Most occur in the understory of tropical rain forests; water lilies, which grow in shallow ponds, are one notable exception. This observation suggests that angiosperms may have evolved in wet and shady habitats. How could life in the understory of a tropical forest dominated by conifers have contributed to the evolution of flowers and xylem vessels?

Recall from Chapter 30 that flowers are a plant's way of enlisting animals to help them reproduce. Brightly colored petals and volatile scents advertise the presence of flowers, while rewards such as

FIG. 33.21 Star anise, an early-branching angiosperm. These flowers are pollinated by beetles.

nectar encourage pollinators to return repeatedly to individuals of the same species. Because many flowers produce pollen and ovules close to each other, the movement of animals from one flower to another provides an efficient mode of pollen transfer.

Insect pollination may have allowed the first angiosperms to exploit understory habitats of tropical rain forests. Wind provides an effective means of pollen transfer in open habitats or in plants that can produce reproductive structures at the top of the canopy. In the dense understory of a tropical forest, the likelihood that air movements alone would result in fertilization is close to zero. All of the living members of the earliest angiosperm lineages produce flowers that are visited by insects. **Fig. 33.21** shows the flower of star anise (*Illicium verum*), which is visited by beetles and whose fruit is the source of a popular seasoning used in Asian cooking.

Let's turn now to a second important feature of angiosperms, the formation of wood containing xylem vessels. As described in Chapter 31, most gymnosperms produce only tracheids, a single cell type that serves both the mechanical and the water transport roles of the stem. Angiosperms, in contrast, have two cell types, allowing them to separate the functions of mechanical support and water transport: Vessel elements stack to produce multicellular vessels through which water is transported, and thick-walled, elongate fibers provide mechanical support. This cellular specialization enables angiosperms to produce wood with larger diameter vessels and thus much higher water transport capacity than found in tracheid-only plants. Faster water transport in turn allows greater stomatal opening and thus higher rates of photosynthesis.

In the humid understory of a tropical forest, where transpiration rates are extremely low, there would seem to be little advantage in producing wood with a greater capacity for water transport. In fact, the sole existing species of the earliest-diverging angiosperm lineage—*Amborella trichopoda*—produces only tracheids. Water lilies also lack vessels, but since they do not produce wood it is less clear whether this is an ancestral or a derived feature. Furthermore, early-diverging angiosperm lineages that do produce vessels do not conduct water at higher rates than those that do not. Thus, xylem vessels in angiosperms appear not to have evolved, at least initially, as a means of enhancing water transport.

Instead, the formation of separate cell types for water transport (vessel elements) and mechanical support (fibers) may have given the early angiosperms greater flexibility in form, allowing them to grow toward small patches of sunlight that filtered through the forest canopy. In the heavily shaded tropical rain forest understory, the more opportunistic growth forms exhibited by early-diverging angiosperm groups, including vines and sprawling shrubs, may have provided a distinct advantage.

The fossil record provides evidence that for the first 30 to 40 million years of their evolution, angiosperms were neither diverse nor ecologically dominant. In fact, fossil pollen indicates that between about 140 and 100 million years ago, gnetophytes diversified just as much as angiosperms. Later in the Cretaceous Period, however, beginning about 100 million years ago, angiosperm diversity began to increase at a much higher rate. Trees belonging to the magnoliids, the branch of the angiosperm tree that today includes magnolias, black pepper, and avocado, emerged as ecologically important members of forest canopies. More important, however, was the divergence of the two groups that would come to dominate both angiosperm diversity and the ecology of many terrestrial environments: the monocots and the eudicots. Together, monocots and eudicots make up more than 95% of angiosperm species. What features may have contributed to their spectacular diversity?

Monocots develop according to a novel body plan.

Monocots or monocotyledons make up nearly one quarter of all angiosperms. Monocots come in all shapes and sizes and are found in virtually every terrestrial habitat on Earth

(**Fig. 33.22**). Coconut palms, hanging Spanish moss (a relative of pineapple), and tiny floating duckweeds, less than 2 mm across, are all monocots, as are the sea grasses of tropical lagoons and the lilies and daffodils in gardens. Some monocots grow in arid regions—including agave, which is used to make tequila. Others, such as orchids, grow primarily as epiphytes in tropical forests. A few monocots form vines—for example, the rattan palms, whose stems are used to produce rattan furniture. Monocots are the most important group of plants in terms of what we eat. Grasses, for example corn, rice, wheat, and sugar cane, are staples in the diets of most people. Other monocots that make their way to our dinner table include banana, yams, ginger, asparagus, pineapple, and vanilla.

Monocots take their name from the fact that they have one embryonic seed leaf, or cotyledon, whereas all other angiosperms have two. However, monocots are distinct in form in so many other ways that one rarely has to count the number of cotyledons to identify a member of this large, monophyletic group. Monocots represent a major evolutionary departure in the way plants build their bodies, in that a vascular cambium is never formed. How has this group been so successful despite the loss of this key innovation? To understand, we first explore in more detail the novel body plan of the monocots.

In monocot leaves, the major veins are typically in parallel and the base of the leaf surrounds the stem, forming a continuous sheath. This type of leaf base means that only one leaf can be attached at any node, consistent with the formation of a single cotyledon. Although monocots do not form a vascular cambium, they can still produce stems that are quite large. For example, both corn and coconut palms are monocots. In monocots, all of the increase in stem diameter occurs in a narrow zone immediately below the apical meristem. As a result of this lateral expansion, the vascular bundles become distributed throughout the stem instead of being arranged in a ring as in all other seed plants.

The lack of a vascular cambium has a profound impact on the way monocots form roots. Because individual roots cannot increase their vascular capacity, monocots continuously initiate new roots from their stems. Thus, the root systems of monocots are more similar to those found in ferns and lycophytes than in other seed plants. Finally, monocot flowers typically produce organs in multiples of 3 (for example, 3, 6 or 9 stamens), whereas eudicot flower organs are most commonly in multiples of 4 or 5.

Monocots have a relatively poor fossil record, in part because they do not produce wood. What factors might have led to such a radical revision in how they build their bodies? The question requires that we look at where and how living monocots grow. One hypothesis is

FIG. 33.22 Monocot diversity. This diverse group of angiosperms includes (a) *Agave shawii*, a desert succulent; (b) *Leucojum vernum*, a spring wildflower; (c) bamboo, a forest grass, and (d) *Costus* species, which grow in the rain forest understory.

that monocots evolved from ancestors that produced creeping, horizontal stems as they grew along the shores of lakes and other wetlands. Many monocots today, including the earliest diverging groups of monocots, grow in such habitats, and many of the features of the monocot body plan are well suited to environments with loose substrates, flowing water, and fluctuating water levels. For example, their leaf base provides a firm attachment that prevents leaves from being pulled off by flowing water. Furthermore, many monocots produce strap-shaped leaves that elongate from a persistent zone of cell division and expansion located at the base of the leaf blade. By continually elongating from the base, monocot leaves can extend above fluctuating water levels. Finally, it is easier to imagine evolutionary changes that affected the vascular system occurring within an environment that makes only modest demands for water transport.

The environmental context surrounding the evolution of monocots is something we may never know for sure. We thus turn our attention to one of the most diverse groups within the monocots—the grasses. Many grasses produce stems that grow

HOW DO WE KNOW?

FIG. 33.23

When did grasslands expand over the land surface?

BACKGROUND Today, prairies, steppes, and other grasslands occur widely in the interiors of continents. Grasses, however, do not fossilize readily. How, then, can we understand how grasslands developed through time?

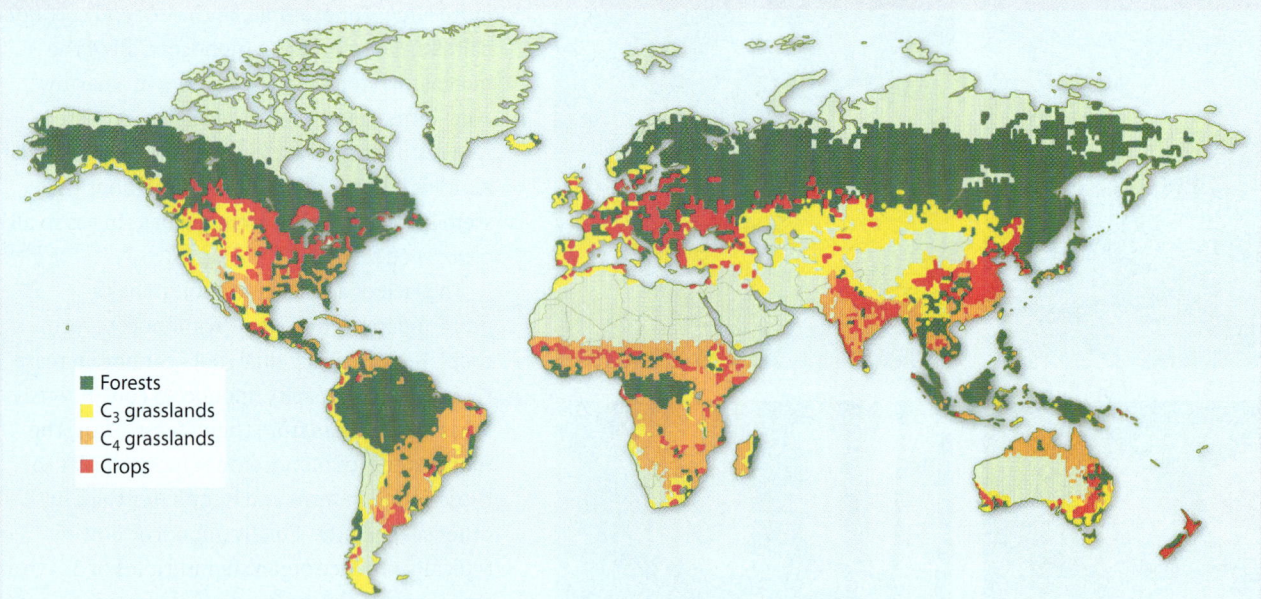

HYPOTHESIS Grasslands expanded as climate changed over the past 50 million years.

OBSERVATIONS AND EXPERIMENTS Grasses commonly make phytoliths, small structures of silica (SiO_2) in their cells. These preserve well, providing a direct record of grass expansion. Moreover, mammals that feed on grasses evolved high-crowned teeth, which also preserve well, giving us an indirect record of grassland history. Finally, C_4 grasses, which are adapted to hot sunny environments with limited rainfall, have a distinctive carbon isotopic composition imparted by the initial fixation of CO_2 by PEP carboxylase (Chapter 29). Measurements of ^{13}C and ^{12}C in mammal teeth, soil carbonate minerals, and more recently, tiny amounts of organic matter incorporated into phytoliths, allow scientists to track the C_4 grasslands through time.

horizontally and branch, allowing them to cover large areas. Grasses have linear leaves that elongate from the base, allowing them to survive grazing as well as fire and drought. Many grasses have evolved the ability to tolerate dry environments by producing roots that extend deep into the soil. In addition, C$_4$ photosynthesis has evolved within the grasses multiple times. As described in Chapter 29, C$_4$ photosynthesis allows plants to avoid photorespiration and thus photosynthesize with greater efficiency. Grasses are among the most successful group of plants, becoming widespread within the past 20 million years as climates changed (**Fig. 33.23**). Today, nearly 30% of terrestrial environments are grasslands.

→ **Quick Check 4** What are some of the distinctive features of monocots?

Eudicots are the most diverse group of angiosperms.

Eudicots first appear in the fossil record about 125 million years ago and, by 80 to 90 million years ago, most of the major groups present today can be distinguished. Today, there are estimated to be

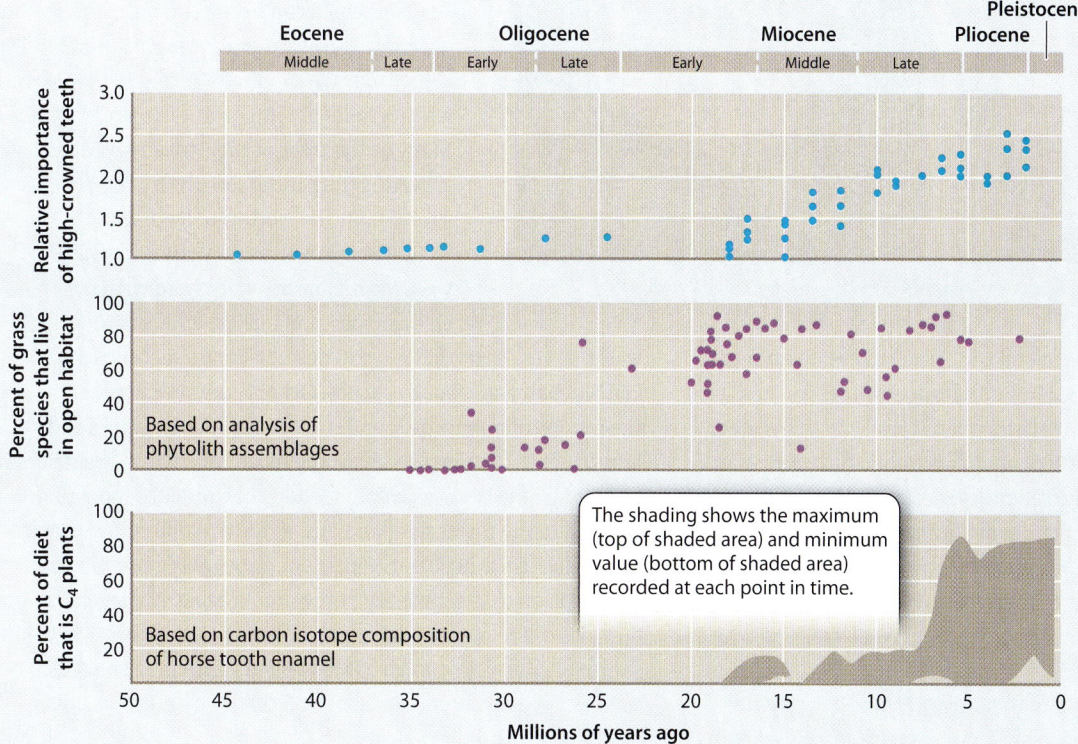

RESULTS Studies of phytoliths, mammal tooth structure, and carbon isotopic composition of teeth enamel from the North American midcontinent clearly show that grasslands expanded 15 to 20 million years ago, and C$_4$ grasslands expanded, later, about 6 to 8 million years ago.

CONCLUSION In North America, grasslands expanded as atmospheric CO$_2$ levels declined and climates became drier. Other continents show evidence of a similar linkage of grassland expansion to climate change.

FOLLOW-UP WORK At present, atmospheric CO$_2$ levels are increasing rapidly, which may affect the competitive abilities of C$_3$ and C$_4$ grasses. Scientists are working to understand how global change will influence grasslands and other vegetation.

SOURCES Stömberg, C. A. E. 2011. "Evolution of Grasses and Grassland Ecosystems." *Annual Review of Earth and Planetary Sciences* 39:517–544. Map is from Edwards, E. J., C. O. Osborne, C. A. E. Stromberg, S. A. Smith, and the C$_4$ Grasses Consortium. 2010. "The Origins of C$_4$ Grasslands: Integrating Evolutionary and Ecosystem Science." *Science* 328:587–591.

FIG. 33.24 Eudicot diversity. Eudicots include (a) red oak trees (*Quercus rubra*), (b) tropical passionflower vines (*Passiflora caerulea*), (c) *Banksia* shrubs, native to Australia, and (d) beavertail cactus (*Opuntia basilaris*).

approximately 160,000 species of eudicots, nearly three-quarters of all angiosperm species (**Fig. 33.24**). Eudicots are well represented in the fossil record, in part because their pollen is easily distinguished. Each eudicot pollen grain has three openings from which the pollen tube can grow, whereas pollen in all other seed plants has only a single opening. Eudicots take their name from the fact that they produce two cotyledons, whereas monocots produce only one. But because the magnoliids and early angiosperm lineages also have two cotyledons this largest of all angiosperm groups is referred to as the "eu-" or "true" dicots.

Many eudicots produce highly conductive xylem. High rates of water transport, and thus high rates of photosynthesis, may explain why eudicot trees were able to replace magnoliid trees as the most ecologically important members of forest canopies. Today, tropical rain forest trees are an important component of the diversity of eudicots. Important eudicot trees in temperate regions include oaks, willows, and eucalyptus.

At the other end of the size spectrum are the herbaceous eudicots, which represent a substantial fraction of eudicot diversity. Herbaceous plants do not form woody stems. Instead, the aboveground shoot dies back each year rather than withstand a period of drought or cold. At the extreme are annuals, herbaceous plants that complete their life cycle in less than a year, persisting during the unfavorable period as seeds. Annuals are unique to angiosperms and almost entirely occur within eudicots. Why might eudicots be successful as both herbs and trees? One factor is that the ability to produce highly conductive xylem may allow herbaceous eudicots to grow quickly and to produce inexpensive and thus easily replaced stems. Examples of herbaceous eudicots include violets, buttercups, and sunflowers.

The herbaceous growth form appears to have evolved many times within eudicots as tropical groups represented by woody plants expanded into temperate regions. For example, the common pea is an herbaceous plant that resides in many vegetable gardens; its relatives include many tropical rain forest trees. Peas are members of the legume, or bean, family, some of which can form symbiotic interactions with nitrogen-fixing bacteria (Chapter 29). The ability to trade carbon for nitrogen may explain why legumes are important in many habitats. Many of the trees of the African savannas, as well as trees in seasonally dry regions throughout the world, are legumes.

Eudicots are diverse in other ways as well. Most parasitic plants and virtually all carnivorous plants are eudicots, as are water-storing cacti that grow in deserts. Some, such as roses and blueberry, are woody shrubs. Others grow as epiphytes in the canopy of rain forests, and still others, such as grapes and honeysuckle, are vines. Perhaps the most unusual are the strangler fig trees. These begin as epiphytes and then produce roots that descend to the forest floor and fuse to form a solid cylinder that "strangles" their host tree. Finally, eudicots make a significant contribution to the diversity of the dinner table. Apples, carrots, pumpkins, and potatoes are all eudicots, as are coffee, cacao (the source of chocolate), and tea. Many plants with oil-rich seeds such as olives, walnuts, soybean, and canola are eudicots, as are buckwheat and quinoa.

The reproduction, growth, and physiology of angiosperms are summarized in **Fig. 33.26** on pages 33-22 and 33-23.

? CASE 6 Agriculture: Feeding a Growing Population
What can be done to protect the genetic diversity of crop species?

Of the more than 400,000 species of plants on Earth today, we eat surprisingly few. Of those we do eat, nearly all are angiosperms. The only major exceptions are the female gametophyte of a small

FIG. 33.25 "Centers of origin" where different crop species were domesticated.

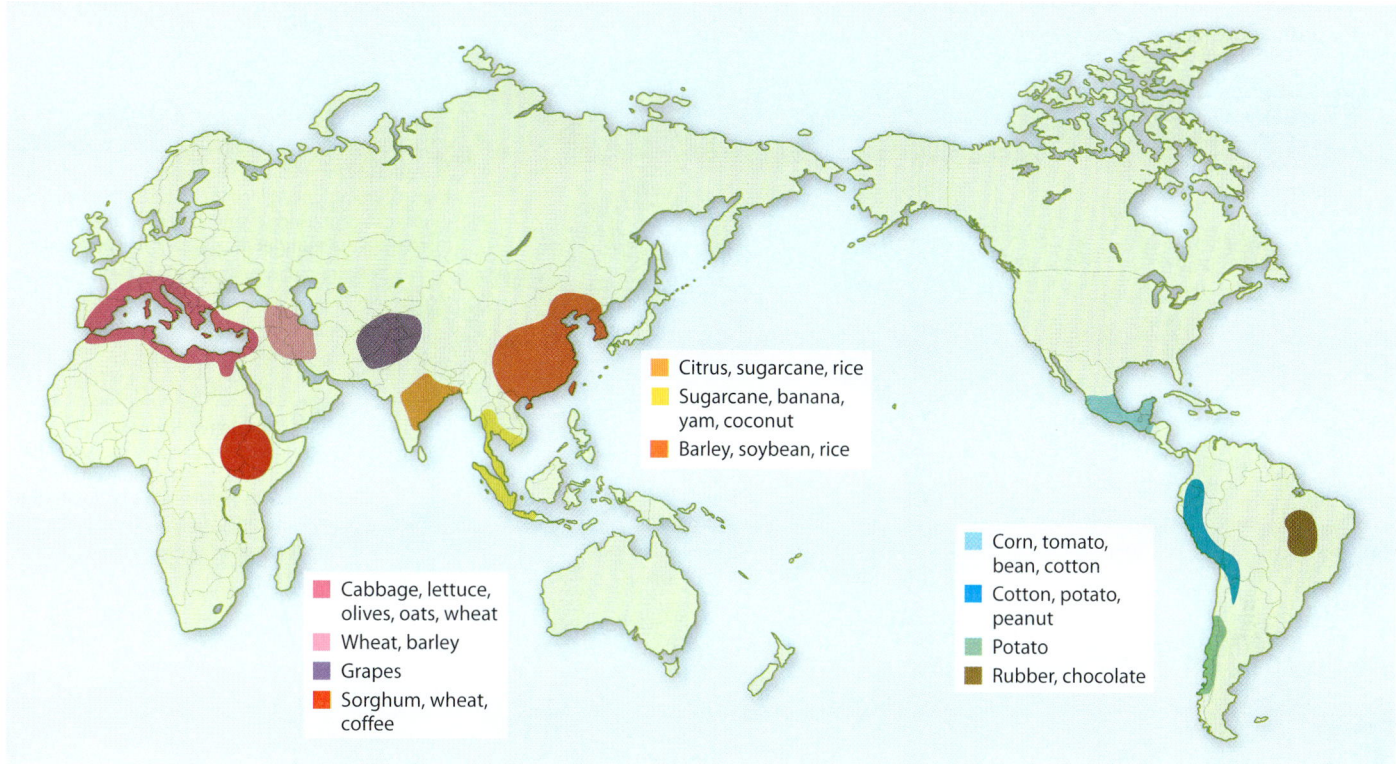

number of gymnosperms (pine nuts and ginkgos), as well as the young leaves of a few ferns. Most of the plants we eat come from species that are grown in cultivation, many of which have, as a consequence of artificial selection, lost the ability to survive and reproduce on their own. Of the approximately 200 such domesticated species, 12 account for over 80% of human caloric intake. Just three—wheat, rice, and corn—make up more than two-thirds.

Crop breeders select for varieties that grow well under cultivation and are easy to harvest, but a consequence of selective breeding is a decrease in genetic diversity. Thus, not only do we depend upon a small number of species for food, agriculture itself contributes to a narrowing of plant diversity. Because pathogens and pests continue to evolve, any loss in genetic diversity of crop species creates substantial risks. For example, in 1970, the fungal pathogen *Bipolaris maydis*, a disease-causing organism that had previously destroyed less than 1% of the U.S. corn crop annually, destroyed more than 15% of the corn crop in a single year. The severity of the epidemic was due to the genetic uniformity of the hybrid corn varieties planted at that time. In natural plant populations, genetic variation for pathogen resistance makes it highly unlikely that a newly evolved pathogen strain will be able to infect every plant.

Nicolai Vavilov, a Russian botanist and geneticist, was one of the first to recognize the importance of safeguarding the genetic diversity of both crop species and their wild relatives. In the early twentieth century, he mounted a series of expeditions to collect seeds from around the globe. Vavilov's observations of cultivated plants and their relatives led him to hypothesize that plants had been domesticated in specific locations (**Fig. 33.25**), from which they had subsequently expanded as a result of human migration and commerce. Vavilov called these regions "centers of origin," and he believed that they coincided with the centers of diversity for both the crop species and its wild relatives. In World War II, during the siege of Leningrad (1941–1944) in which more than 700,000 people perished, scientists at the Vavilov Institute sought to protect what was then the world's largest collection of seeds. Their dedication—at least one of the self-appointed caretakers died of starvation, despite being surrounded by vast quantities of edible seeds—illustrates the priceless nature of the genetic diversity on which our food supply rests.

Today, seed banks help preserve the genetic diversity of crop species and their wild relatives. However, seed banks can store only a fraction of the genetic diversity present in nature. To help make up the difference, some have suggested establishing protected areas that coincide with Vavilov's centers of origin. As we will see in the next chapter, new threats lie just over the horizon. Meeting these threats will require that every genetic resource be brought to bear.

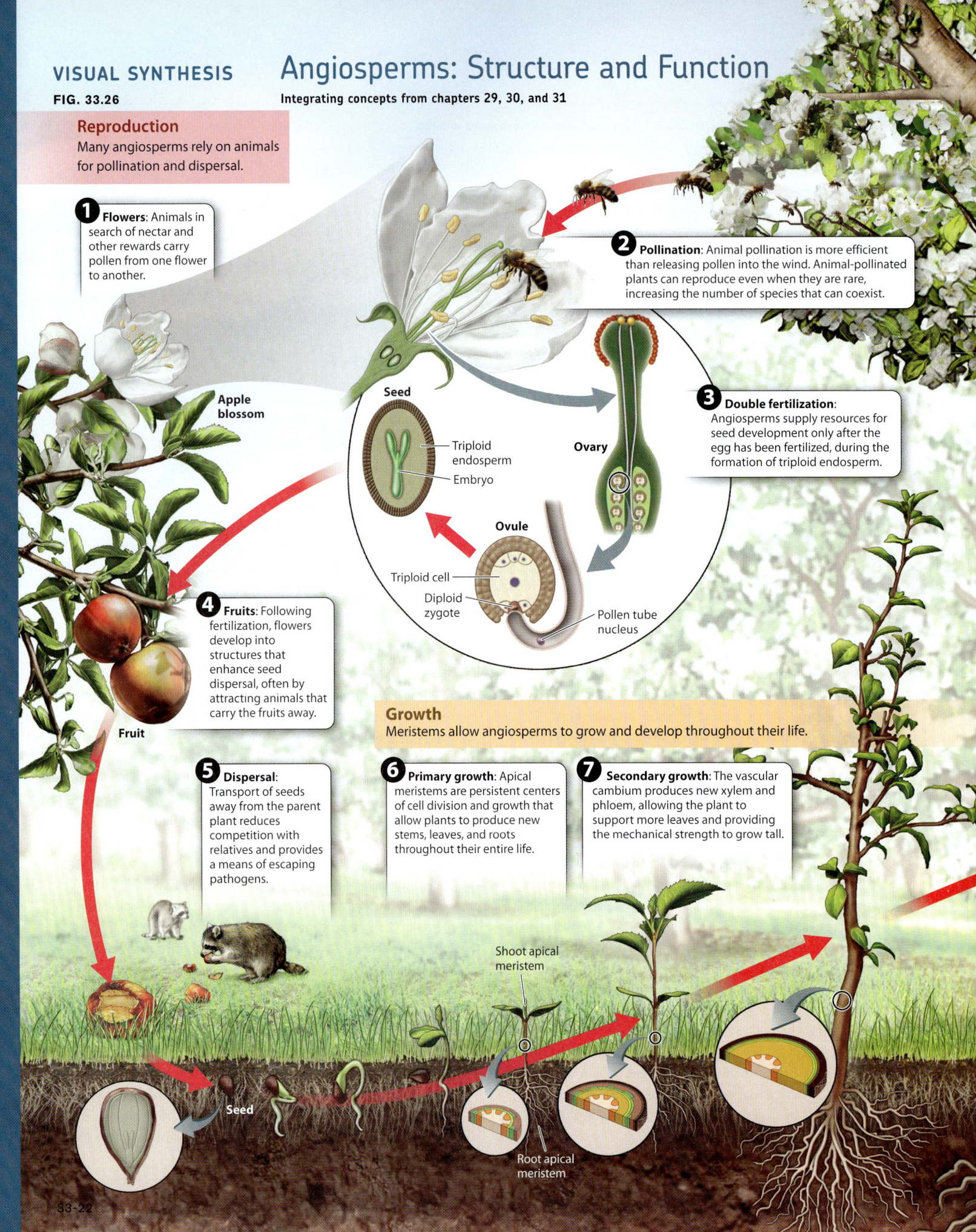

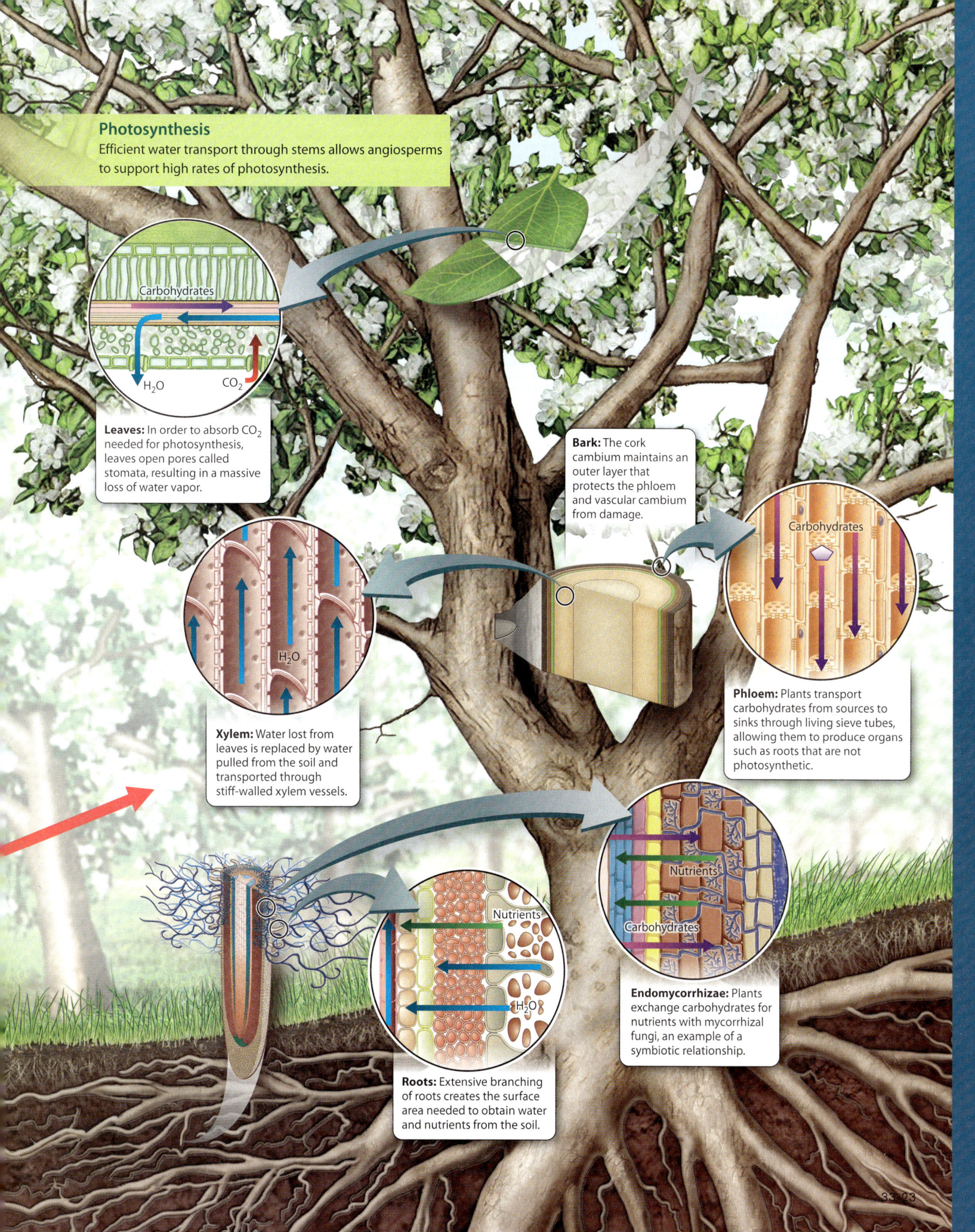

Core Concepts Summary

33.1 PLANT DIVERSITY IS DOMINATED BY ANGIOSPERMS, WHICH MAKE UP APPROXIMATELY 90% OF ALL PLANT SPECIES FOUND TODAY.

There are thought to be nearly 400,000 species of plants living today. Of these, approximately 90% are angiosperms. page 33-1

The other 10% of plant species is distributed among the other six major groups of plants. page 33-1

Angiosperms first appear in the fossil record about 140 million years ago, more than 300 million years after plants first moved onto land. page 33-1

The evolution of angiosperms resulted in a rapid and dramatic increase in total plant diversity. page 33-1

As angiosperm diversity increased, other plant groups, such as gymnosperms and ferns, declined in diversity. page 33-1

Moist tropical rain forests dominated by angiosperms provided new types of habitat into which other plants could evolve. page 33-2

33.2 BRYOPHYTES DIVERGED BEFORE THE EVOLUTION OF VASCULAR PLANTS, AND THEY GROW IN ENVIRONMENTS WHERE THE ABILITY TO OBTAIN WATER FROM THE SOIL DOES NOT PROVIDE AN ADVANTAGE.

Bryophytes constitute a paraphyletic group consisting of three groups of plants: mosses, liverworts, and hornworts. page 33-2

Bryophytes are small plants that produce one of two morphological types: a flattened thallus or an upright leafy type. They do not form roots, but instead absorb water through their surfaces. page 33-3

The persistent component of the life cycle is the gametophyte. Sporophytes range from tiny and non-photosynthetic in some liverworts to relatively long-lived and photosynthetic in hornworts. page 33-3

There are several examples of convergent evolution between bryophytes and vascular plants, including insect dispersal of spores in some mosses and the evolution of internal transport cells in some mosses and liverworts. page 33-4

Sphagnum moss is the dominant plant of peat bogs. page 33-5

Sphagnum moss produces water-holding cells that allow it to soak up water, and it acidifies the environment. Both characteristics help slow decomposition, so large amounts of organic carbon build up year after year. page 33-5

33.3 SPORE-DISPERSING VASCULAR PLANTS ARE SMALL, OFTEN EPIPHYTIC PLANTS THAT USUALLY GROW IN MOIST ENVIRONMENTS.

Only two groups of seedless vascular plants have living relatives today: lycophytes, and ferns and horsetails. page 33-5

Three hundred million years ago, lycophytes included large trees that dominated swamp forests. Today, lycophytes are small plants that grow either in the forest understory as epiphytes, or occur in shallow ponds. page 33-6

Ferns and horsetails are morphologically diverse. Ferns produce large leaves that uncoil as they grow; horsetails have tiny leaves and whisk ferns with no leaves at all. page 33-9

Although ferns have a long history, most present-day species are the result of a radiation that occurred after the rise of the angiosperms. page 33-10

33.4 GYMNOSPERMS PRODUCE SEEDS AND WOODY STEMS AND ARE MOST COMMON IN COLD OR DRY REGIONS.

Of the many groups of seed plants that have evolved, only two can be found today: the gymnosperms, with fewer than 1000 species, and the angiosperms, with more than 380,000 species. page 33-11

Gymnosperms are composed of four groups of woody plants: cycads, ginkgos, conifers, and the gnetophytes. page 33-11

Cycads produce large leaves on stout, unbranched stems. Although they once were widely distributed, they now occur in small, fragmented populations, primarily in the tropics and subtropics. Many cycads are insect pollinated, and all form symbiotic associations with nitrogen-fixing bacteria. page 33-12

Ginkgo is the single living species of a group that was distributed globally before the evolution of the angiosperms. *Ginkgo* is wind pollinated and produces tall, branched trees. page 33-12

Conifers include the tallest and longest-lived trees on Earth. Wind-pollinated and largely evergreen, they are found primarily in cool to cold environments. page 33-13

Before the angiosperms appeared, conifers were widespread. Their almost complete absence in the tropics and persistence in temperate regions may be due to their dependence upon wind pollination and xylem formed entirely of tracheids. page 33-14

The gnetophytes are a small group, containing only three genera and few species, that has independently evolved xylem vessels. page 33-14

33.5 ANGIOSPERM DIVERSITY IS PARTLY EXPLAINED BY ANIMAL POLLINATION AND XYLEM VESSELS.

Angiosperm diversity may result as much from low rates of extinction as from high rates of species formation. page 33-15

Plants having flowers can reproduce even if they are far apart, allowing rare species to persist and reproduce. page 33-15

Xylem vessels make it possible for angiosperms to have a diversity of form and to grow toward light. page 33-16

The angiosperm phylogeny indicates a major split between two diverse groups, monocots and eudicots. page 33-16

Monocots include grasses, coconut palms, bananas, ginger, and orchids. page 33-17

Monocots have a single cotyledon, or embryonic seed leaf, and they do not form a vascular cambium. page 33-17

Eudicots are diverse, including many familiar plants such as legumes, roses, cabbage, pumpkin, coffee, tea, and cacao, which yields chocolate. page 33-20

Eudicots are characterized by pollen grains with three openings through which the pollen tube can grow. page 33-20

Modern agriculture is based on just a few plant species with low genetic diversity, making it vulnerable to pests and pathogens. page 33-21

Self-Assessment

1. Using information in Fig. 33.2, draw a pie chart that depicts proportional plant diversity just before the appearance of the angiosperms.

2. Imagine a world in which mosses, liverworts, and hornworts formed a monophyletic group. How would your ability to infer what the first land plants looked like be affected?

3. Describe three environments that allow bryophytes to coexist with vascular plants.

4. Describe the habitats in which lycophytes are found today.

5. List three ways that ferns, which lack secondary growth, are able to elevate their leaves and thus access more sunlight.

6. Describe how fern diversity has been affected by the evolution of the angiosperms.

7. Contrast the ways in which the evolution of angiosperms has affected the distribution of cycads and of conifers.

8. Explain how xylem produced by conifers differs from that of angiosperms and how that difference may have influenced the present-day distribution of conifers.

9. Compare the movement of pollen in an animal-pollinated angiosperm and a wind-pollinated conifer, noting what features of angiosperm reproduction increase the efficiency (or lower the costs) of pollen transfer.

10. Name several features that might account for the diversity and success of angiosperms and discuss possible advantages of these features.

Do you understand the chapter's Core Concepts? Log in to check your answers to the Self-Assessment questions, then practice what you've learned and reinforce this chapter's concepts by working through the problems and multimedia tutorials provided there.

www.biologyhowlifeworks.com

CHAPTER 34

FUNGI

Structure, Function, and Diversity

Core Concepts

34.1 Fungi are heterotrophic eukaryotes that feed by absorption.

34.2 Fungi reproduce both sexually and asexually, and disperse by spores.

34.3 Next to the animals, fungi are the most diverse group of eukaryotic organisms.

In tropical rain forests, vascular plants dominate the biomass. Animals eat the plants, obtaining food for growth and reproduction, but they feed selectively: They consume leaves, fruits, and seeds, but largely avoid a much more abundant tissue—wood. In fact, most leaves also escape grazing and eventually fall, dead, onto the forest floor. Animals, too, may be eaten by predators, but some die of other causes and contribute their remains to the soil. Given this constant rain of biological materials, we might expect wood, leaves, and animal carcasses to accumulate in great piles, but a walk through the rain forest reveals that the forest floor is remarkably clean. What happened to all the dead plant and animal tissues?

On land, the organisms principally responsible for the decomposition of plant and animal tissues are **fungi,** one of the most abundant and diverse groups of eukaryotic organisms. Many of us have seen mushrooms and toadstools in a meadow or woodland, but most fungal biomass and most of the metabolic work that fungi do occurs within the soil, out of sight. Fungi play a critical role in cycling carbon because of their ability to locate and break down the complex molecules and bulky tissues in plant and animal bodies. Moreover, because they form intimate relationships with living plants and animals, fungi affect the growth and reproduction of many other organisms. For example, fungi associated with plant roots dramatically increase plant productivity by enhancing the uptake of mineral nutrients from the soil (Chapter 29). Other fungi are major agricultural pests (Chapter 32). Fungi enable us to turn wheat into bread, barley into beer, and milk into cheese, but still other fungi cause athlete's foot, yeast infections, and, especially in patients with compromised immune systems, overwhelming infections that can lead to death.

Whether we are interested in the fundamental workings of communities and ecosystems, human health, or agriculture and food production, no discussion of biological diversity is complete without consideration of the fungi.

34.1 GROWTH AND NUTRITION

Fungi are heterotrophs, meaning that they depend on pre-existing organic molecules for both carbon and energy. Unlike animals, fungi do not have organs that enable them to ingest food and break it down in a digestive cavity (Chapter 40). Instead, fungi absorb organic molecules directly through their cell walls. This mode of feeding presents two major problems. First, although simple molecules like amino acids and sugars pass readily across the cell wall, more complicated molecules do not. Fungi secrete a diversity of enzymes that break down complex organic molecules like starch, cellulose, or lignin into simpler compounds that can be absorbed. Fungi, then, digest their food first and take it into the body afterward.

34-1

A second challenge is to find food in the environment. Other heterotrophic organisms such as bacteria, protists, and animals commonly move through their habitat, actively searching for food. Fungi, however, have no means of locomotion, and so these organisms use the process of growth itself to find nourishment.

Hyphae permit fungi to explore their environment for food resources.

Most fungi are multicellular, consisting of highly branched filaments called **hyphae** (**Fig. 34.1**). The hyphae are slender, typically 10 to 50 times thinner than a human hair. The numerous long, thin hyphae provide fungi with a large surface area for absorbing nutrients.

Hyphae maintain their slender form by growing only at their tips. Elongating hyphae thus penetrate ever farther into their environment, encountering new food resources as they grow. Where resources are low, growth is slow or may stop entirely. On the other hand, when fungi encounter a rich food resource, they grow rapidly and branch repeatedly, forming a network of branching hyphae called a **mycelium.** Mycelia can grow to be quite large—the largest known individual of the fungus *Armillaria ostoyae* covers over 2000 acres in the Blue Mountains of Oregon and weighs many hundreds of tons. We may be impressed by blue whales and redwoods, but in fact some fungi are the largest living organisms on Earth.

A strong but flexible cell wall is key to hyphal growth and nutrient transport. In fungi, cell walls are made of **chitin**, the same compound as found in the exoskeletons of insects. Chitin is a modified polysaccharide that contains nitrogen. Fungal cell walls are thinner than plant cell walls because of the chemical makeup of chitin. The wall prevents cells from rupturing when exposed to dilute solutions such as those in freshwater environments and in the soil. Moreover, the walls keep cells from expanding as water flows into the cytoplasm by osmosis. This inflow of water generates positive turgor pressure (Chapters 5 and 29), providing the force that enables fungi to explore their surroundings as the hyphal tip is pushed ever deeper into the local environment.

Fungi transport materials within their hyphae.

Fungi can transport food and signaling molecules across long distances in mycelia, so they are able to grow between resource patches and to produce reproductive structures such as mushrooms that rise up above the ground. Molecules taken up actively from the environment drive water into the cell by osmosis and thus increase turgor pressure. At the same time, growth and respiration consume these molecules, resulting in a decrease in turgor pressure. Such differences in turgor pressure along hyphae drives bulk flow in a manner similar to phloem transport in plants (Chapter 29). Bulk flow carries materials obtained in a nutrient-rich location so they can fuel hyphal elongation across nutrient-poor locations. Transport by bulk flow also allows fungi to build relatively large reproductive structures aboveground. All the raw materials used to build a mushroom must be transported from hyphae in contact with the soil or a rotting log.

Cytoplasmic continuity is essential for the long-distance movement of materials within mycelia. In early-diverging groups, the hyphae have many nuclei but no cell walls to separate them. In later-diverging groups, nuclear divisions are accompanied by the formation of **septa,** walls that partially divide the cytoplasm into separate cells (**Fig. 34.2**). Each septum contains one or more pores that allow water and solutes to move freely between cells. Septa play an important role when hyphae are damaged. Injury activates sealing mechanisms that plug pores in the septa, preventing the loss of pressurized cytoplasm.

→ **Quick Check 1** Name two roles of hyphae.

Not all fungi produce hyphae.

Yeasts are single-celled fungi found in moist, nutrient-rich environments. In these circumstances, hyphae, which are used to search for food, are unnecessary. All yeasts are descended from ancestors that developed hyphae, but hyphal development has been lost independently several times. Thus, the lack of hyphae is an example of convergent evolution, in which the same trait develops independently in different lineages in response to similar selective pressures.

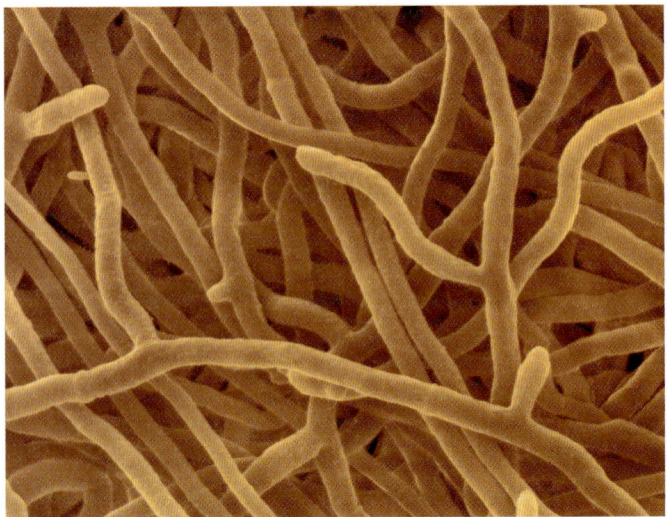

FIG. 34.1 Fungal hyphae, thin filaments that have enormous surface area to facilitate absorption of nutrients. Hyphae of *Trichophyton rubrum*, a fungus that infects the skin of humans, as seen under a scanning electron microscope.

FIG. 34.2 Septa separating individual cells in fungi. Septa, partitions between cells, have pores that allow transport of solutes along the hypha.

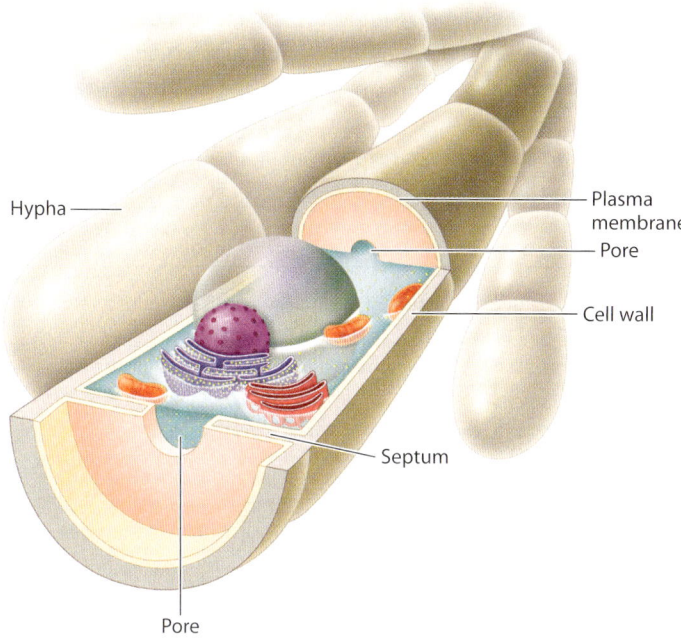

Most yeasts divide by budding (Chapter 42). A small outgrowth increases in size and eventually breaks off to form a new cell (**Fig. 34.3**). The localized outgrowth that results in budding is similar to growth by elongation at hyphal tips. In fact, some yeasts can form hypha-like structures under certain conditions. For example, in *Candida albicans*, a yeast that infects humans, hyphae are produced during the transition of the fungus from a harmless to a pathogenic form.

Yeasts are common on the surfaces of plants, and to a lesser extent on the surfaces and in the guts of animals. Humans have long used the yeast *Saccharomyces cerevisiae* to ferment plant carbohydrates to produce leavened bread and alcoholic beverages. *S. cerevisiae* is also an important model organism used in laboratory studies of genetics.

Fungi are principal decomposers of plant tissues.

Most fungi use dead organic matter as their source of energy and raw materials. As noted in Chapter 25, on land the organic matter in dead tissues on and within soils far exceeds the amount in the living biomass. Thus, the ground beneath our feet provides tremendous resources for heterotrophic growth.

Bacteria and some protists can use this resource, but for the most part it is the fungi that convert dead organic matter back to carbon dioxide and water. This helps to keep the biological carbon cycle in balance (Chapter 25) and returns nutrients in leaves to the soil where they will be available for new plant growth.

A brief consideration shows why fungi have the advantage on a forest floor or in soil. Like fungi, bacteria secrete enzymes that break down organic molecules in their immediate environment, but in the soil bacteria have only a limited capacity to move from place to place. Flagella can propel them a few micrometers through a thin film of water, but when local resources become depleted, bacteria must wait for new food to appear. In contrast, hyphal growth allows fungi to search actively through the soil for new food resources. Furthermore, because fungi can penetrate their food, they are particularly well suited for breaking down the bodies of multicellular organisms, such as the stems of trees.

Fungi can gain nutrition from dead animal, protozoan, or even bacterial cells, but the most abundant biomolecules on and within soils are cellulose and lignin, the principal components of plant cell walls (Chapter 31). Cellulose in particular is a rich source of carbon and energy. Cellulose is difficult to degrade because individual cellulose polymers bind tightly to one another. In addition, in wood and other lignified plant tissues, cellulose microfibrils occur in intimate association with lignin, making the cellulose even harder to get at. Lignin is even more difficult for enzymes to break apart, in part because it lacks a regular chemical structure. A few bacteria are known to

FIG. 34.3 The yeast *Saccharomyces cerevisiae*. Yeasts are unicellular fungi that often divide by budding smaller cells off of larger ones.

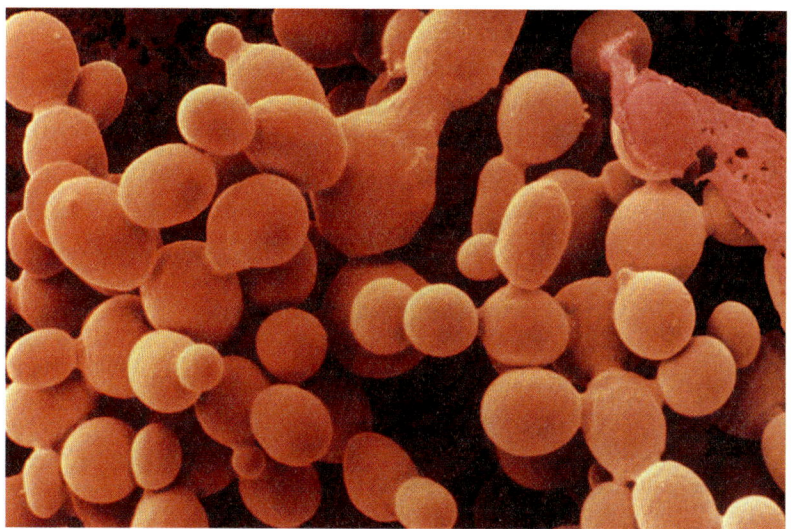

FIG. 34.4 Decomposition of wood by fungi. (a) A dead tree being decomposed by brown rot fungi. (b) These scanning electron microscope images show (left) normal cedar wood and (right) rotted cedar wood from the coffin in the tomb of King Midas in Gordion, Turkey.

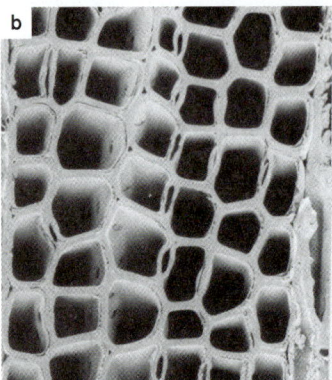 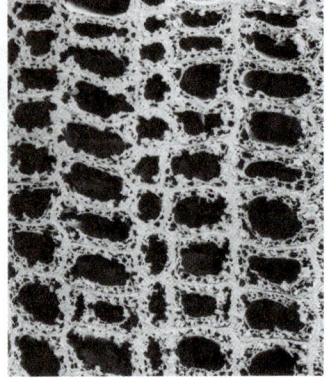

decompose lignin, but in nature, fungi account for most of the decomposition of wood (**Fig. 34.4**).

To break down lignified plant cell walls, fungi require oxygen. However, oxygen concentrations are typically low in water-logged environments, and in that case decay proceeds slowly if at all. For this reason, woody biomass, once submerged in water, tends to accumulate, leading to the buildup of partially decayed plant material as peat. Over geologically long time intervals, peat becomes coal, an energy staple of the industrial world that owes its existence to the inhibition of fungal decay.

Fungi are important plant and animal pathogens.

Fungi not only feed on dead plants and animals, but they are also well adapted to infect living tissues (**Fig. 34.5**). Plants, for example, are vulnerable to a diverse array of fungal pathogens—rusts, smuts, and molds—that cause huge losses in agricultural production, as we discuss in section 34.4.

Invasion is the key to success for plant pathogens (Chapter 32). Most plants have physical or chemical means of inhibiting fungal invasion, so successful pathogens must be able to get past these defenses. In many cases, fungi infect plants through wounds, which provide a route around a plant's outer defenses. Some fungi enter through stomata, while others penetrate epidermal cells directly, degrading the wall with enzymes and then pushing their hyphae into the plant interior by turgor pressure. The extent of infection usually depends on the ability of hyphae to expand into new tissues. Vascular wilt fungi (discussed in Chapter 32) send their hyphae through xylem vessels, enabling them to spread throughout the entire plant. Old photographs of towns in the eastern United States show streets shaded by

FIG. 34.5 Fungal infection of living tissue. (a) This stained section of heart tissue (pink) is infected with the fungus *Candida albicans* (purple threadlike structures). (b) A scanning electron micrograph shows hyphae of *Ceratocystis ulmi*, a vascular wilt fungus, growing in xylem vessels of an elm tree.

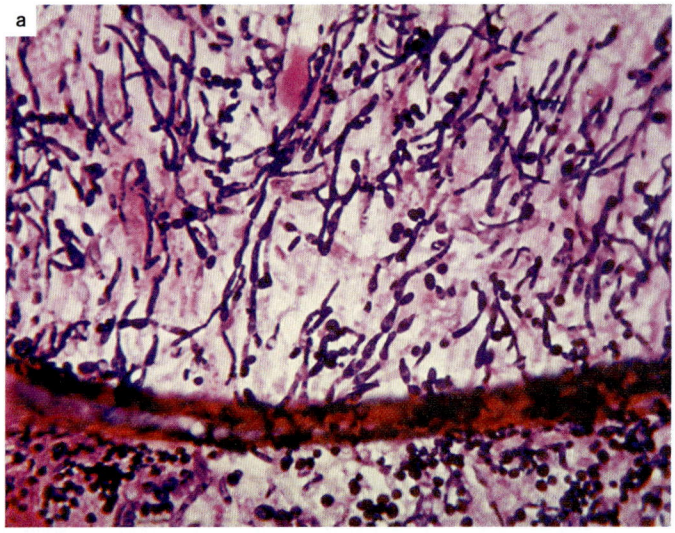

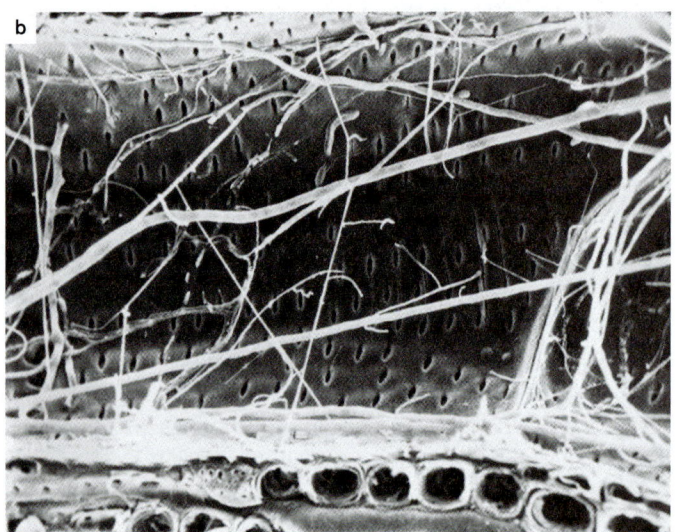

American elm trees, but today nearly all those trees are dead, victims of a vascular wilt fungus introduced into the country early in the twentieth century.

Aboveground, plant infections are usually transmitted by fungal spores, carried either by the wind or on the bodies of insects. For example, many fungi that attack fruits are transmitted by insects also seeking those fruits; yeasts that feed on nectar within flowers hitch rides on pollinators. Some fungi secrete sugar-rich droplets that attract insects, which then transport their spores to other sites.

Belowground, infection is typically transmitted by hyphae that penetrate the root. Hyphae can feed on organic matter in the soil until the root of a suitable host plant is encountered. The persistence of fungal hyphae in soils means that populations of pathogenic fungi can build up in the soil. This is why seedlings germinating close to an adult tree of the same species are at risk of infection, whereas seedlings of distantly related species may be spared. By favoring the establishment of seedlings of less-common species, fungal pathogens are thought to play an important role in maintaining high levels of species diversity in tropical rain forests (Chapter 32).

Insects and other invertebrates are also susceptible to fungal pathogens. Certain fungi exploit living animals in a unique way: by trapping them. Some fungal predators form sticky traps with their hyphae, whereas others actually lasso their prey (**Fig. 34.6**). The hyphae of these fungi form rings that can inflate within seconds, trapping anything within the ring. Nematodes, tiny worms common in soils, inadvertently pass through the rings, setting off hyphal inflation. The nematodes are then trapped, providing food for this fungus.

Fungal infection is relatively rare in vertebrates. Severe infections are more frequent among fish and amphibians than in mammals, perhaps because fungi grow poorly at mammalian body temperatures. An apparent exception is the fungal infection that has caused dramatic declines in North American bat populations, but the fungi infect the bats during their winter hibernation, when the bats have lowered body temperatures that conserve energy. In humans, most fungal infections are annoying rather than life threatening—athlete's foot and yeast infections are prime examples. However, in individuals with immune systems compromised by HIV or other illnesses, fungal infection can be fatal.

Many fungi form symbiotic associations with plants and animals.

While some interactions between species benefit the fungus at the expense of its host, others benefit both partners. In Chapter 29, we saw that mycorrhizal fungi supply plant roots with nutrients such as phosphorus from the soil and in return receive carbohydrates from their host (**Fig. 34.7**). There are two main types of mycorrhizae. The hyphae of **ectomycorrhizal** fungi surround, but do not penetrate, root cells. In contrast, the hyphae of **endomycorrhizal** fungi penetrate into root cells, where they produce highly branched structures that provide a large surface area for nutrient exchange. As we will see in section 34.3, the endomycorrhizal fungi constitute a monophyletic group dating back over 400 million years. In contrast, ectomycorrhizae are present in multiple groups and appear to have evolved more than once.

FIG. 34.7 Mutualistic relationship between fungi and plants. Mycorrhizal fungi (white) surround roots (brown) and branch extensively in the soil.

FIG. 34.6 A fungal lasso. Hyphae of *Drechslerella doedycoides* form inflated rings that capture nematode worms moving through the soil.

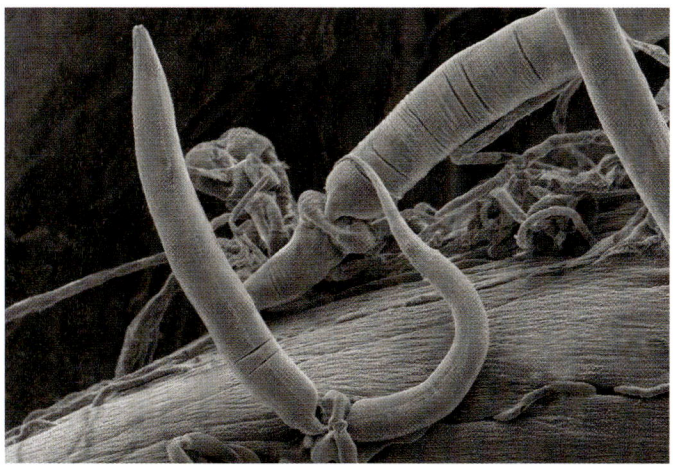

FIG. 34.8 Lichens. A lichen is a mutualistic association between a fungus and a photosynthetic microorganism.

Other fungi, called **endophytes,** live within leaves. While endophytes are much less studied than mycorrhizae, they may be as widespread. Endophytic hyphae grow within cell walls and in the spaces between cells. Long thought to be harmless, endophytes are now recognized as beneficial. They may help the host plant by producing chemicals that deter pathogens and herbivorous insects.

Mutually beneficial associations between fungi and animals are much less common, but a few examples are known, none more striking than insects that actually grow fungi for food. The fungi benefit because the insects provide shelter, food, and protection from predators and pathogens. Insect–fungal agriculture has evolved at least three times: in leaf-cutter ants that live in tropical forests (Chapter 48), in a group of African termites, and in some wood-boring beetles. These beetles maintain fungal gardens within tunnels that they excavate in damaged or dead trees.

Lichens are symbioses between a fungus and a green alga or a cyanobacterium.

Lichens are familiar sights in many environments, often forming colorful growths on rocks or tree trunks (**Fig. 34.8**). Lichens look, function, and even reproduce as single organisms, but they are actually stable associations between a fungus and a photosynthetic microorganism, usually a green alga but sometimes a cyanobacterium. Nearly 15% of all known fungal species grow as lichens.

The dual nature of lichens was first proposed by a Swiss botanist, Simon Schwendener, in 1867, but his hypothesis found little favor at the time. The existence of a composite "organism" challenged the idea that life could be divided into discrete categories such as animals and plants and raised the question of how an association of distinct species could function as an integrated whole. Given what we now understand about the widespread nature of mutualisms, the opposition to Schwendener's proposal may seem surprising. Yet it was only in 1939, when Eugen Thomas showed that lichens could be separated into their individual parts and then reassembled, that their dual nature became widely accepted.

Lichens consist mostly of fungal hyphae, with the photosynthetic algae or cyanobacteria forming a thin layer just under the surface (**Fig. 34.9**). The hyphae anchor the lichen to a rock or tree, aid in the uptake and retention of water and nutrients, and produce chemicals that protect against excess light and herbivorous animals. In turn, the photosynthetic partners provide a source of reduced carbon, such as carbohydrates. Cyanobacteria living in lichens also provide a source of fixed nitrogen, like ammonia. The two partners exchange nutrients through fungal hyphae that tightly encircle or even penetrate the walls of the photosynthetic cells. As a result, lichens are able to thrive at sites where neither partner could exist on its own.

It remains an open question whether either partner can exist independently in nature. In cases where the fungal partner can be grown in the laboratory in culture, it produces a relatively undifferentiated hyphal mass. Thus, while the bulk of the lichen structure comes from the fungus, chemical signals from the photosynthetic partner influence the form and shape of the fungus.

Surprisingly, these associations are not species specific. There are approximately 13,500 known lichens, but only about 100 participating photosynthetic species. This means that different lichens can have the same algae or cyanobacteria.

FIG. 34.9 **Lichen anatomy.** The fungus provides structure, and photosynthetic algae form a thin layer under the surface. The scanning electron micrograph shows a section through *Xanthoria flammea*.

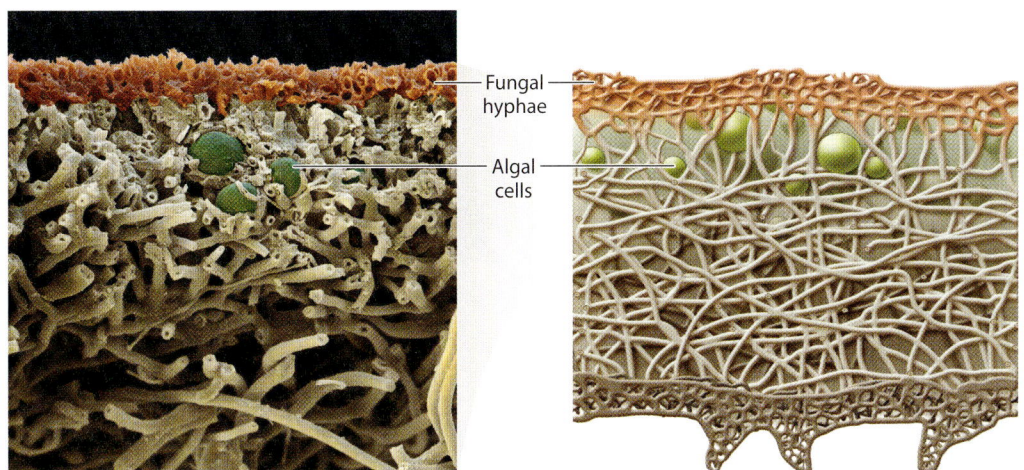

Lichens consist mostly of fungal hyphae that take up water and nutrients from the soil.

Photosynthetic algal cells form a thin layer just under the surface of the lichen.

Conversely, a given fungal partner may associate with more than one photosynthetic species, without affecting the lichen's outward form. Because of the intimate association between lichen morphology and fungal species, the lichen and fungus are assigned the same scientific name.

Lichens spread asexually by fragmentation or through the formation of dispersal units consisting of a single photosynthetic cell surrounded by hyphae. Sexual reproduction, at least by the fungi, is common, and the photosynthetic cells reproduce asexually by mitotic cell division. Whether sexual or asexual reproduction is more important in allowing lichens to establish in new habitats, however, is not known.

Lichens are remarkable for their ability to grow on the surfaces of rocks and tree trunks. They are among the first colonizers of lava flows and the barren land left after glacial retreat. In some habitats, for example Arctic tundra and semiarid landscapes, lichens make a measurable contribution to net productivity. How do lichens obtain the resources they need in these barren environments? They obtain nutrients from rainfall or by secreting organic acids that help release some nutrients from even rocky surfaces. Not surprisingly, lichens in these harsh environments grow slowly. Some lichens grow on soil, for example reindeer "moss," which is eaten by caribou. Reindeer moss is thought to compete successfully for space by secreting chemicals that hinder the growth of plants.

Given their small size and exposure to the environment, lichens must be able to tolerate drying out from time to time. All lichens have a high tolerance for dessication, and can tolerate wide fluctuations in temperature and light. In contrast, lichens are quite sensitive to air pollution, particularly sulfur dioxide (SO_2).

For this reason, lichen growth is sometimes used as an indicator of industrial pollution.

34.2 REPRODUCTION

Like plants, fungi face two challenges in completing their life cycles. First, to maintain genetic diversity within populations, they must find other individuals with which to mate. Second, they must be able to disperse from one place to another. The ways that fungi generate genetic diversity are unique. However, fungal adaptations for dispersal at least broadly resemble those of plants. Fungi rely on wind, water, or animals to carry spores through the environment. Recall from Chapter 30 that spores are specialized cells well adapted for dispersal and long-term survival.

We most often come into contact with fungi through their reproductive structures. Fungal spores commonly cause respiratory illness, and mushrooms attract our attention because they are delicious or poisonous. Yet many aspects of fungal reproduction are not easily observed, including life-cycle features that distinguish fungi from other eukaryotes. Fungi have a diverse set of reproductive strategies. In this section, we emphasize the most general principles and patterns.

Fungi proliferate and disperse using spores.

The tissues, living or dead, that support fungal growth are sometimes hard to come by and have a patchy distribution. For this reason, fungi must be able to disperse from one food source to another. The extensive networks of hyphae within soils or host organisms can spread locally but cannot disperse over great

FIG. 34.10 Fungal spores dispersed by wind.

a. Devil's snuffbox (*Lycoperdon perlatum*)

b. Black bread mold (*Rhizopus*)

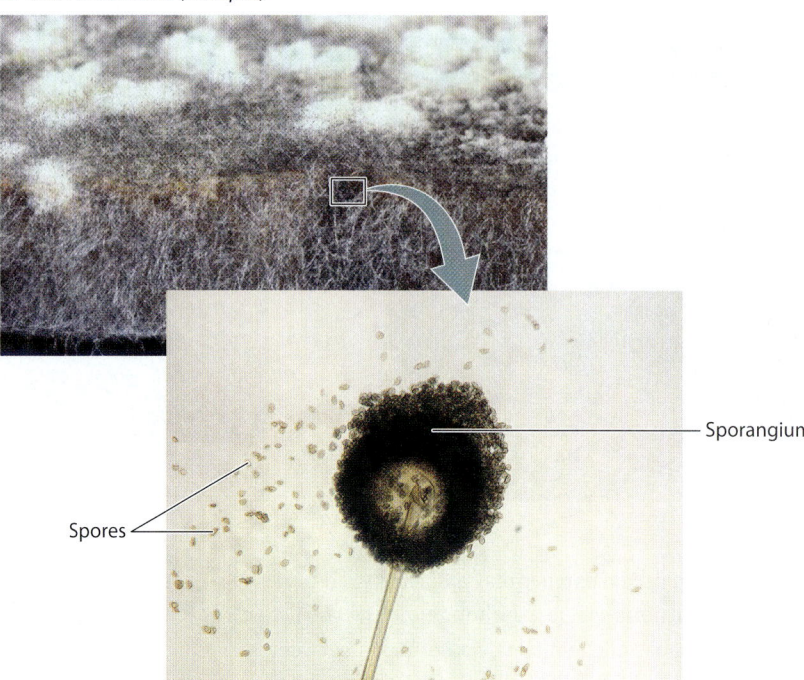

distances, so fungi produce spores that can be carried by the wind (**Fig. 34.10a**), in water, or attached to (or within) animals. In early-diverging fungi that still live in aquatic environments, spores have flagella that allow them to swim. The great majority of fungi, however, live on land, and their spores have no flagella. Instead, the spores are encased in a thick wall that protects them as they are dispersed over habitats unsuitable for growth.

The probability that any given spore will come to rest in a favorable habitat is low, so fungi produce huge numbers of spores. Fungal spores remain viable, able to grow if provided with an appropriate environment, for periods ranging from only a few hours in some species to many years in others. Thus, spores allow fungi to use resources that are patchy in time as well as in space. In fact, a shortage of resources is one of the cues that triggers spore formation.

Spores can form by meiotic cell division as part of sexual reproduction, and they can also form asexually. Asexual spores are formed by mitotic cell division and therefore are genetically identical to their parent. Asexual spores allow fungi to proliferate and disperse to new environments. In many species, asexual spores are produced within sporangia that form at the ends of erect hyphae, facilitating the release of the spores into the air. A close look at a moldy piece of bread reveals that the surface is covered with hyphae carrying sporangia containing asexual spores (**Fig. 34.10b**).

Fungal species vary in the extent to which they rely on asexual as opposed to sexual spore production. In some groups, such as the lineage that includes the most common mushrooms, sexual reproduction is the dominant means of spore production. On the other hand, in a small number of fungi that includes the endomycorrhizal species, spores produced by meiotic cell division have never been observed. Although sexual reproduction appears to be absent in some fungal species, asexual fungi have other mechanisms for producing genetic diversity, as discussed later in this section.

Multicellular fruiting bodies facilitate the dispersal of sexually produced spores.

Fungi employ an astonishing array of mechanisms to enhance spore dispersal. Particularly conspicuous are the multicellular **fruiting bodies** produced by some fungi, which facilitate the dispersal of sexually produced spores. (Fungal fruiting bodies should not be confused with the fruits of flowering plants, which are unrelated and very different structures.) Mushrooms, stinkhorns, puffballs, bracket fungi, truffles, and many other well-known structures are fungal fruiting bodies.

Fruiting bodies are highly ordered and compact structures compared to the mycelia from which they grow, yet they are constructed entirely of hyphae (**Fig. 34.11**). In fact, in many cases their mechanisms of spore dispersal demand a high degree of structural precision. Yet we know little of the developmental processes that allow tip-growing hyphae to produce such complex and regular structures.

FIG. 34.11 Fruiting bodies, complex multicellular structures built from hyphae. Shown are fruiting bodies of *Hygrophorus miniatus*, known commonly as the vermillion waxcap.

The fruiting bodies of many fungi rise above the ground or grow from the trunks of dead trees, so the sexually produced spores are released high above the ground. But elevation by itself is not enough to ensure dispersal. Thus, many fungi forcibly eject their spores, achieving velocities of more than 1 m/s that allow the tiny spores to penetrate and travel beyond the layer of stagnant air that surrounds the fruiting body. Studies have shown that spores move efficiently through air (**Fig. 34.12**). Other fungi rely on external agents such as raindrops or animals to move their spores around. In section 34.3 we will encounter examples of the diverse dispersal mechanisms found in the fungi.

HOW DO WE KNOW?

FIG. 34.12

What determines the shape of fungal spores that are ejected into the air?

BACKGROUND Many fungi disperse their spores by ejecting them forcibly into the air. For spores to be picked up by the wind, however, they must escape a layer of still air called the boundary layer that lies close to the ground. Escape from the boundary layer is easier if spores have a size and shape that minimize drag, the resistance of air to the movement of an object.

HYPOTHESIS In fungi that eject their spores into the atmosphere, natural selection favors spore shapes that minimize drag.

EXPERIMENT Working from computer models for minimizing drag on airplane wings, scientists modeled spore shapes that minimize drag. These models took into account spore size as well as the physical characteristics of the air through which spores travel. The scientists then measured spore shape for more than 100 species of spore-ejecting fungi.

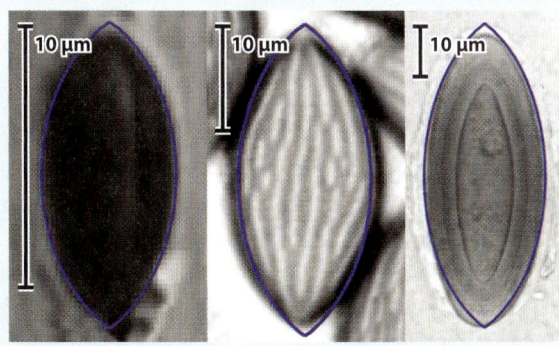

Fungal spores with outlines (blue) of computer-modeled shapes that minimize drag.

RESULTS Nearly three-quarters of the examined fungi had spore shapes that came within 1% of the shape calculated to minimize drag. Related species that do not eject spores forcibly were less likely to have drag-minimizing shapes.

CONCLUSION Natural selection has acted on fungi to facilitate spore dispersal by wind.

SOURCE Roper, M., R. E. Pepper, M. P. Brenner, and A. Pringle. 2008. "Explosively Launched Spores of Ascomycete Fungi Have Drag-Minimizing Shapes." *Proceedings of the National Academy of Sciences, USA* 105: 20583–20588.

The fungal life cycle often includes a stage in which haploid cells fuse, but nuclei do not.

Like other sexually reproducing eukaryotes, fungi have life cycles that include haploid ($1n$) and diploid ($2n$) stages. The nuclei in fungal hyphae are haploid, and the fungal life cycle is therefore similar to haploid-dominant life cycles found among eukaryotic organisms (see Fig. 27.3a). In haploid-dominant organisms, asexual reproduction involves the production of haploid spores by mitosis, while sexual reproduction involves the fusion of haploid cells (often differentiated as gametes) to form a diploid zygote, which undergoes meiosis as its first division. However, sexual reproduction in fungi differs from all other haploid-dominant organisms in one important respect: In fungi, the fusion of haploid cells is not immediately followed by the fusion of their nuclei.

In most fungi, the sexual phase of the life cycle involves the fusion of hyphal tips rather than specialized reproductive cells, or gametes. For mating to occur, two hyphae grow together and release enzymes that digest their cell walls at the point of contact. This allows the cell contents of the two hyphal cells to merge, forming a single cell with two haploid nuclei. In most sexually reproducing organisms, when two gametes merge, their nuclei fuse almost instantly to form a diploid zygote. In fungi, however, the cytoplasmic union of two cells (**plasmogamy**) is not always followed immediately by the fusion of their nuclei (**karyogamy**). Instead, the haploid nuclei retain their independent identities, resulting in what is referred to as a **heterokaryotic** ("different nuclei") stage (**Fig. 34.13**). In the heterokaryotic stage, cells have nuclei from two parental hyphae, but the nuclei remain distinct. The heterokaryotic stage ends with karyogamy, which leads to the formation of a diploid zygote. The zygote divides by meiotic cell division, giving rise to sexually produced haploid spores.

The separation between plasmogamy and karyogamy is unique to the fungi, and as will be discussed in the next section, the elaboration of this difference is a major theme in the evolution of fungi. In the earliest diverging fungal lineage, karyogamy follows fast on the heels of plasmogamy and there is no heterokaryotic stage. In some groups, the heterokaryotic stage consists of a single cell with many haploid nuclei. However, in other groups,

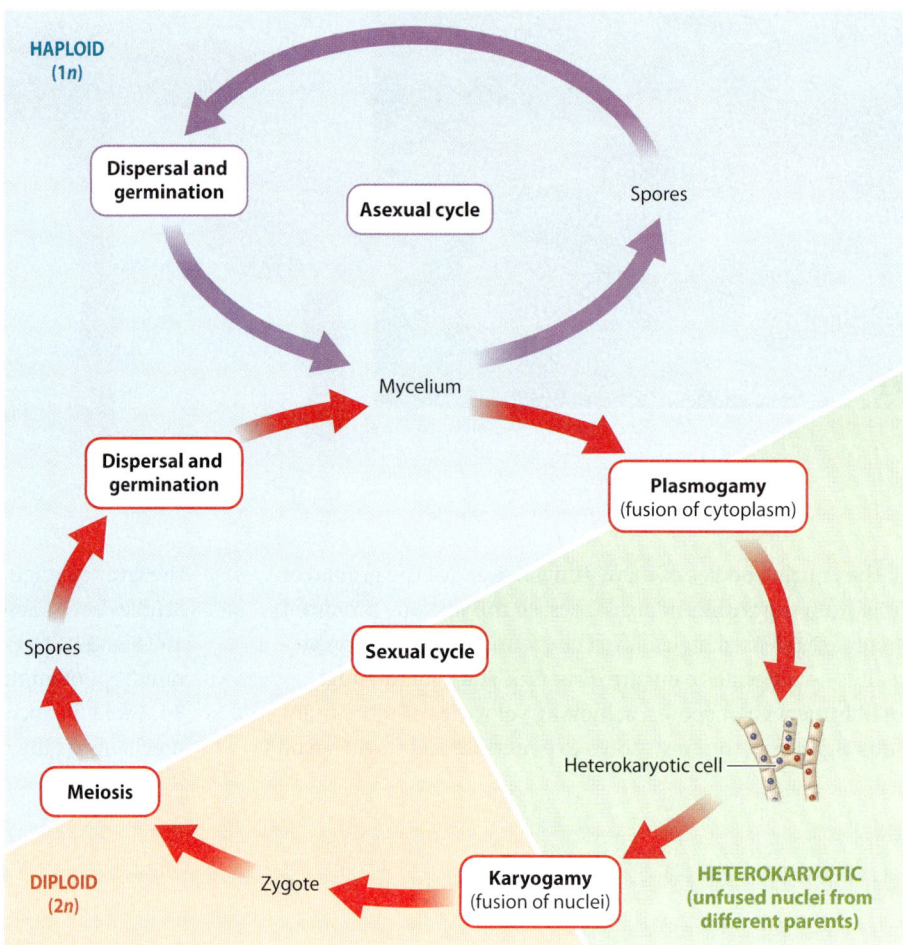

FIG. 34.13 Generalized fungal life cycle. In fungi, plasmogamy and karyogamy are separated by a heterokaryotic stage.

plasmogamy is followed by mitosis, which produces hyphae in which each cell contains two haploid nuclei, one from each parent. Heterokaryotic cells with just two genetically distinct haploid nuclei are referred to as **dikaryotic,** or **$n + n$,** cells. In fungi with dikaryotic cells, called dikaryotic fungi, there may be a small number of dikaryotic cells or extensive hyphal development made up of dikaryotic cells. For example, some edible mushrooms found on market shelves consist entirely of dikaryotic hyphae.

Dikaryotic fungi account for more than 98% of all known fungal species, suggesting that the separation of plasmogamy and karyogamy in time and space may confer an evolutionary advantage. One major advantage can be seen when we think about the environments where fungi live. Mating takes place principally within the soil or inside the trunks of rotting trees, sites where hyphae are most likely to come into contact. Dispersal, however, is most effective when spores can be released into the air. The separation of plasmogamy and karyogamy allows mating and spore production to occur where each is most effective. In some dikaryotic fungi, however, the separation of plasmogamy and karyogamy is neither distant in space nor long in time. In these fungi, the highly branched dikaryotic hyphae may serve primarily as a means of increasing the number of cells in which karyogamy will eventually take place, therefore leading to the production of more sexually produced spores.

At present, we really don't know why fungi proliferate $n + n$ cells rather than forming a multicellular diploid phase, and there may be no single answer to the question of why the dikaryotic fungi have this unique cell type.

→ **Quick Check 2** What is the difference between a diploid cell and a dikaryotic cell?

Genetically distinct mating types promote outcrossing.

Under most conditions found in nature, genetically diverse populations persist better than those lacking diversity. Indeed, sexual reproduction is widely viewed as a means of promoting genetic diversity for long-term ecological success in variable environments (Chapters 11 and 42). With the exception of early-diverging aquatic lineages, fungi do not produce male and female gametes. How, then, is sex accomplished, and what prevents an individual from mating with itself?

Many individual fungi of a species look similar but nevertheless have different **mating types** that are genetically determined and prevent self-fertilization. The mating type of an individual is determined by a mating-type gene. Fertilization can take place only between individuals that have different alleles at the mating type gene. In some species, there are only two mating-type alleles. In this case, mating patterns are identical to those for species with male and female sexes. If the two mating-type alleles are spread evenly throughout the population, an individual should be able to mate with 50% of the general population.

Some fungi have more than two mating-type alleles. These fungi have a greater likelihood of encountering a compatible genotype in the general population. That likelihood is even higher for fungi in which mating type is determined by two different mating-type genes that each have multiple alleles. Fungi in the group that includes the common mushrooms can have as many as 20,000 different mating-type alleles. In that case, the odds of finding a compatible mate are close to 100%.

Genetic compatibility is a prerequisite for mating. How do fungi go about finding a suitable mate? The answer is that they secrete chemical signals that attract fungi with mating types different from their own. The chemical nature of these signals differs among groups.

Fungi that lack sexual reproduction have other means of generating genetic diversity.

Asexual reproduction occurs in most groups of fungi. Indeed, about 20% of fungi appear to lack sexual reproduction altogether. Among these are such well-known groups as *Penicillium* (the source of the antibiotic penicillin), *Aspergillus* (the major industrial source of vitamin C), and all the endomycorrhizal fungi.

How are asexual species able to persist without a mechanism for generating genetic diversity? The short answer is, they don't. Fungal species that lack an observable sexual cycle have another mechanism of generating genetic diversity: the crossing over of DNA during mitosis. Such species are described as **parasexual.** The parasexual cycle (**Fig. 34.14**) is not in fact a coordinated cycle but rather a series of four events. First, two hyphae fuse, forming a heterokaryotic cell. Then, karyogamy produces a diploid nucleus from two genetically distinct haploid nuclei—so far, a familiar pattern. Next, however, during mitosis, crossing over may occur between the two sets of chromosomes. Crossing over is common

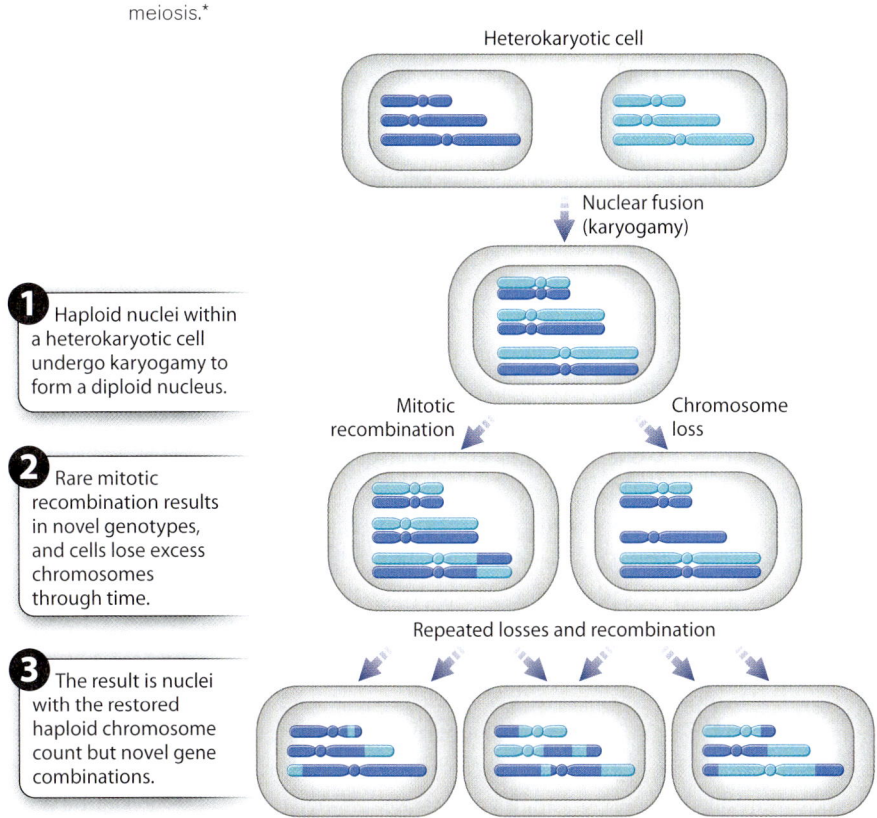

FIG. 34.14 The parasexual cycle in fungi. This cycle can generate genetic diversity without meiosis.*

① Haploid nuclei within a heterokaryotic cell undergo karyogamy to form a diploid nucleus.

② Rare mitotic recombination results in novel genotypes, and cells lose excess chromosomes through time.

③ The result is nuclei with the restored haploid chromosome count but novel gene combinations.

*Adapted from http://what-when-how.com/molecular-biology/parasexual-cycle-molecular-biology/

during meiosis (Chapter 11) but rare during mitosis. The haploid chromosome number is restored not by meiotic division but by the progressive loss of chromosomes. Like sexual reproduction, these processes produce novel gene combinations. The key difference between parasexuality and sexual reproduction is that parasexual species do not undergo meiosis.

Entirely asexual species of fungi posed a problem for early taxonomists because these species lack many of the morphological characters such as fruiting bodies or sporangium-bearing stalks previously used to define different groups. With the advent of DNA sequence data, it has become increasingly clear that many asexual species are closely related to sexual forms, indicating that their evolutionary history as asexual species is not long. A major exception are the Glomeromycota, for which fossils as old as 450 million years have been reported.

34.3 DIVERSITY

When Aristotle classified the living world into animals and plants, he grouped fungi with plants because of their lack of motility. As discussed in Chapter 27, molecular sequence comparisons now make it clear that fungi are actually related more closely to animals, such that it would not be amiss to refer to mushrooms and molds as our "cousins."

Fungi share a number of features with animals: Their motile cells, when present, have a single flagellum attached to their posterior ends; fungi and many animals synthesize chitin; and fungi store energy as glycogen, as do animals. Nevertheless, the last common ancestor of fungi and animals was a single-celled microorganism that lived in aquatic environments about 1 billion years ago. Multicellularity evolved independently in fungi and animals (Chapter 28), and so the fungi have many features that are unique in the biological world.

Fungi are highly diverse.

By any measure, fungi are diverse, although just how diverse is not known. About 75,000 species have been formally described, but estimates of true fungal diversity run as high as 5 million species. Among eukaryotes, the only organisms more diverse than fungi are the animals. The availability of DNA sequence data has greatly advanced our understanding of phylogenetic relationships within the fungi, which are shown in **Fig. 34.15**. The tree clearly shows how the characters present in familiar mushrooms accumulated through the course of evolution: first chitinous cell walls, then hyphae, then regularly placed septa, and finally the complex multicellular reproductive bodies we call mushrooms. Motile single cells are found only in the most ancient lineages, while dikaryotic cells and complexity of form are found only on the most recent branches.

The evolutionary history of fungi can be viewed as the transition from a flagellated ancestral form that lived in aquatic environments to soil-dwelling forms in which hyphae divided

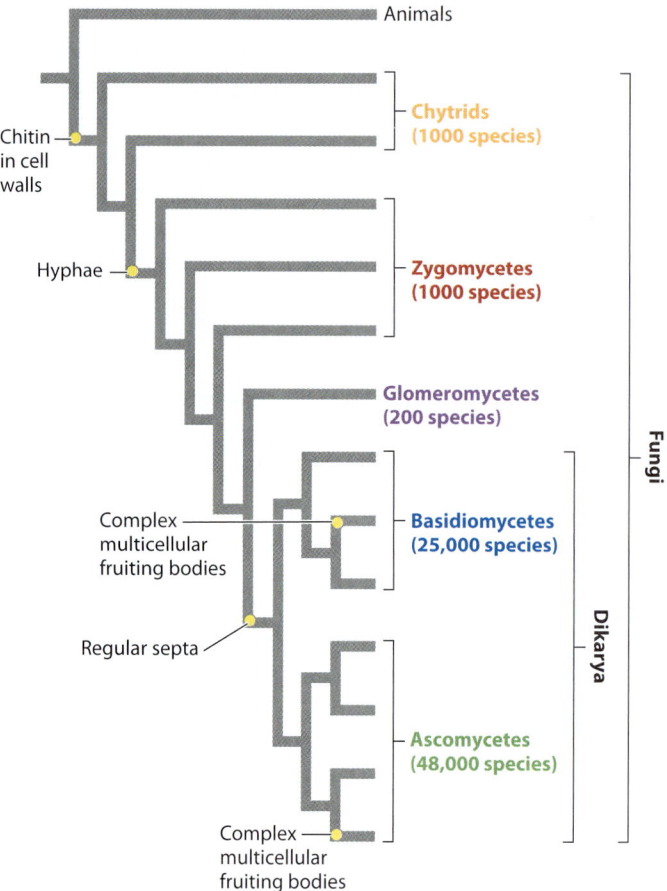

FIG. 34.15 A phylogeny of the fungi. Approximate numbers of described species are given in parentheses.

by septa developed complex morphologies visible to the naked eye. As we saw in the preceding section, another evolutionary trend is the progressive separation in time between plasmogamy and karyogamy in fungal life cycles. Many of the distinguishing features of fungi evolved as adaptations to the challenges faced by nonmotile, absorptive heterotrophs on land.

We also see that the numbers of species are not spread evenly across the phylogeny. Instead, more ancient lineages include less than 2% of known species, while the two dikaryotic groups include more than 98% of known fungi. Clearly, dikaryotic fungi are well adapted to many different habitats, including other organisms, both living and dead.

→ **Quick Check 3** From the phylogenetic tree shown in Fig. 34.15, would you say that complex multicellular fruiting bodies evolved once or twice (independently) in fungi?

Fungi evolved from aquatic, unicellular, and flagellated ancestors.

Fungi are opisthokonts, members of the eukaryotic superkingdom that includes animals (Chapter 27). Among fungi, the Chytridiomycota (commonly referred to as **chytrids**)

FIG. 34.16 Chytrids. Chytrids are aquatic fungi that attach to decomposing organic matter by rhizoids, seen in the lower half of the image. This is *Chytriomyces hyalinus*.

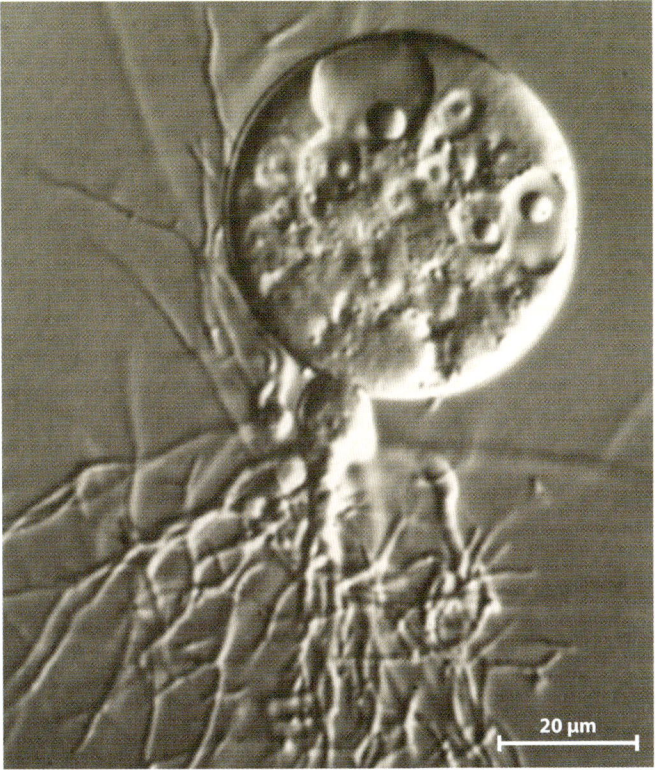

lie at the base of the phylogenetic tree. This is a small group of about 1000 species found in aquatic or moist environments (**Fig. 34.16**). Many chytrids are single cells with walls of chitin. They may also form short multinucleate structures, but they lack the well-defined hyphae characteristic of other fungi. Chytrids, therefore, do not form a true mycelium, although in some species elongated cellular outgrowths called rhizoids penetrate into organic substrates. Rhizoids anchor the organism in place and absorb food molecules. Chytrids generally disperse by flagellated spores. Sexual reproduction appears to be rare, but there is substantial life-cycle diversity within this group. Chytrids lack a heterokaryotic stage but form flagellated gametes that swim through their aqueous environment.

Most chytrids are decomposers, and a few of them are pathogenic. Notably, the chytrid *Batrachochytrium dendrobatidis* is associated with widespread mortality in amphibian populations (Chapter 48). Molecular clock analysis estimates that chytrids began to diversify 700 to 800 million years ago.

Zygomycetes produce hyphae undivided by septa.

Moving through the phylogenetic tree, we next encounter several groups traditionally united as the Zygomycota, or **zygomycetes** (see Fig. 34.15), which share a number of traits.

Collectively, zygomycetes make up less than 1% of known fungal diversity. Some are decomposers, specializing on dead leaves, animal feces, and food—there are probably zygomycetes in your kitchen. Others live on and in plants, animals, and even other fungi. Zygomycetes have many typical fungal traits, including the growth of a mycelium and the production of aerial spores. These traits may have evolved as adaptations for finding food and dispersing on land. The loss of flagellated spores may also reflect adaptation for life on land. Unlike more complex fungi, zygomycetes do not form regular septa along their hyphae, nor do they produce multicellular fruiting bodies.

Earlier, we introduced the black bread mold *Rhizopus*, which is a zygomycete (see Fig. 34.10b). *Rhizopus* and its relatives are specialists on substrates containing abundant, easy-to-digest carbon compounds, such as bread, ripe fruits, and the dung of herbivorous animals. These fungi consume their substrates rapidly, and once their meal is finished, they release large numbers of aerial spores to locate another food source.

Sexual reproduction occurs when two compatible hyphal tips fuse to form a thick-walled structure containing many nuclei of each mating type. Karyogamy and meiosis are followed by germination of the haploid cell to form an elevated stalk. Each stalk develops a sporangium that contains spores produced asexually by mitotic cell division. If you look closely at a bread mold, you can see the white filamentous hyphae and small black sporangia on slender stalks. Each sporangium can produce as many as 100,000 spores that are dispersed by the wind, explaining how these organisms seem to get everywhere.

Although zygomycete fungi do not form multicellular fruiting bodies, some have evolved truly spectacular means of dispersal. *Pilobilus*, which consumes the dung of herbivorous animals, has a life cycle similar to *Rhizopus* except that instead of releasing individual spores, it forcibly ejects the entire sporangium (**Fig. 34.17**). Turgor pressure generated in the supporting stalk propels the sporangia as far as 2 m. Light-sensitive pigments in the stalk's hyphae control the

FIG. 34.17 Spore dispersal in a zygomycete. In *Pilobilus*, light-sensing pigments orient the turgor-propelled dispersal of the sporangia.

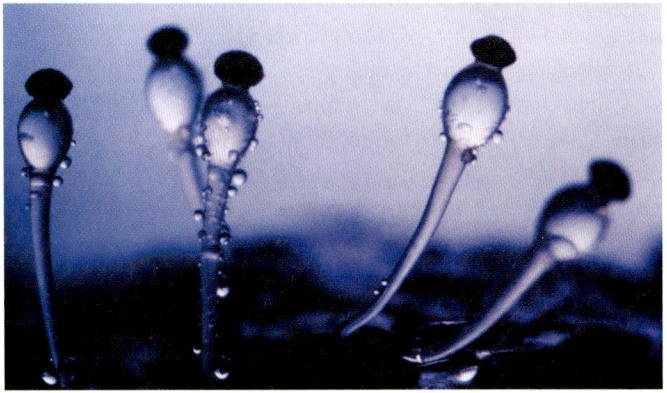

orientation of this water cannon, ensuring that the spores have the best chance of escaping the dung pile and landing on vegetation that is attractive to grazing herbivores. Feeding herbivores disperse the spores further, while supplying them with fresh dung for food. The remarkable dispersal of *Pilobilus* sporangia benefits other species, as well: Parasitic nematode worms hitch a ride on the sporangia to find a new host.

Glomeromycetes form endomycorrhizae.

The Glomeromycota, or **glomeromycetes**, are a monophyletic group of apparently low diversity (there are about 200 described species) but tremendous ecological importance. All known glomeromycetes occur in association with plant roots. Most famous are those that form endomycorrhizae. Molecular clock estimates suggest that glomeromycetes diverged from other fungi 500 to 600 million years ago, before the evolution of vascular plants. Establishment of glomerophyte–root symbiosis changed the course of evolution for both participants. Today, most vascular plant species harbor endomycorrhizae, dramatically increasing nutrient uptake from the soil.

Glomeromycetes are hard to study because they cannot be grown independently of their plant partners, and so many unanswered questions remain. For example, no evidence of sexual reproduction has ever been found in this group. Yet genetic studies of glomeromycete populations show some genetic diversity, which is thought to result from parasexual processes (see Fig. 34.14). Glomerocytes produce extremely large (0.1 to 0.5 mm diameter) multinucleate spores asexually. These large spores can persist in soils until they come into contact with an uninfected root.

The Dikarya produce regular septa during mitosis.

The final major node, or branching point, in the fungal phylogeny marks the origin of the **Dikarya,** a vast group that includes about 98% of all described fungal species. A key feature of the dikaryotic fungi is that every mitotic division is accompanied by the formation of a new septum. This innovation allows these fungi to control the number of nuclei within each cell and thus to proliferate dikaryotic cells. The Dikarya include all the edible mushrooms, the yeast species used in the production of beer, bread, and cheese, the major wood-rotting fungi, and pathogens of both crops and humans.

The Dikarya contain two monophyletic subgroups (see Fig. 34.15), each named for the shape of the cell in which the key reproductive processes of nuclear fusion (karyogamy) and meiosis take place. The Ascomycota, or **ascomycetes,** are sometimes called sac fungi because nuclear fusion and meiosis take place in an elongated saclike cell called an ascus (*askos* is Greek for "leather bag"). The Basidiomycota, or **basidiomycetes,** are popularly known as club fungi because the nuclear fusion and meiosis take place in a club-shaped cell called a basidium. ("Basidium" is derived from the Greek *basis,* which means "base" or "pedestal" in Greek; the reference is to the position of sexually produced spores at the tips of this club-shaped body.)

Although ascomycetes and basidiomycetes share the same basic life cycle, they diverged from each other as much as 600 to 500 million years ago and evolved complex multicellular structures independently of each other. The morel in the supermarket is an ascomycete, whereas the shiitake mushroom next to it is a basidiomycete (**Fig. 34.18**). In addition to asci and basidia, the two groups differ in several other features. For example, the septal pores that connect adjacent cells within hyphae are distinct in the two groups, as are the ways that cells can plug the pores to minimize damage to a mycelium. Also, when ascomycetes reproduce, mating cells form and their nuclei fuse in relatively close succession. Thus, the heterokaryotic stage is brief. In contrast, in the basidiomycetes, the nuclei of two fused cells may remain separate for a long time. In this case, hyphae containing both types of nuclei (dikaryotic hyphae) can continue to absorb nutrients from the substrate for a while before nuclear fusion occurs.

Ascomycetes are the most diverse group of fungi.

Ascomycetes make up 64% of all known fungal species. They include important wood-rotting fungi, many ectomychorrizal species, and significant pathogens of both animals and plants. Ascomycetes form the fungal partner in most lichens—approximately 40% of all ascomycete species occur in lichens. Ascomycetes loom large in human history, contributing the baker's and brewer's yeasts used to make bread and beer, the antibiotic penicillin, and model systems employed in laboratory investigations of eukaryotic cell biology and genetics. Ascomycetes are used to produce soy sauce, sake, rice vinegar, and

FIG. 34.18 Two groups of dikaryotic fungi. The morel (a) is an ascomycete, whereas the shiitake mushroom (b) is a basidiomycete.

miso, and to transform milk into Brie, Camembert, and Roquefort cheeses. Athlete's foot and other skin infections are caused by ascomycetes, as are many more serious fungal diseases.

Ascomycetes include both species that form multicellular fruiting bodies and ones that form unicellular yeasts. A remarkable feature of ascomycete fruiting bodies is that they contain both dikaryotic and haploid cells. The proliferation of dikaryotic cells leads to the production of many asci, while the growth of haploid cells contributes to the bulk of the fruiting body. Two ascomycete lineages consist largely of yeasts. The earliest branching members in these groups, however, produce hyphae. Thus, the unicellular nature of yeasts is a derived feature evolved through loss of hyphae and is not an ancestral condition.

Fig. 34.19 illustrates the life cycle of a common ascomycete, the brown cup fungus. In ascomycetes, meiosis is followed by a single round of mitosis resulting in asci that contain eight haploid spores. In many ascomycetes, fruiting bodies elevate the asci on one or more cup-shaped surfaces, from where they are easily caught and carried away by wind. When mature, the spores are ejected from the top of the asci, expelled by turgor pressure.

In the fruiting bodies of some ascomycetes, the asci are completely enclosed by a layer of tissue and thus must be dispersed by other organisms rather than by the wind. The truffles prized in cooking provide an example. The edible truffle is the fruiting body of an ascomycete that grows as an ectomycorrhizal fungus on tree roots. Not only do truffles encase their spores in protective tissues,

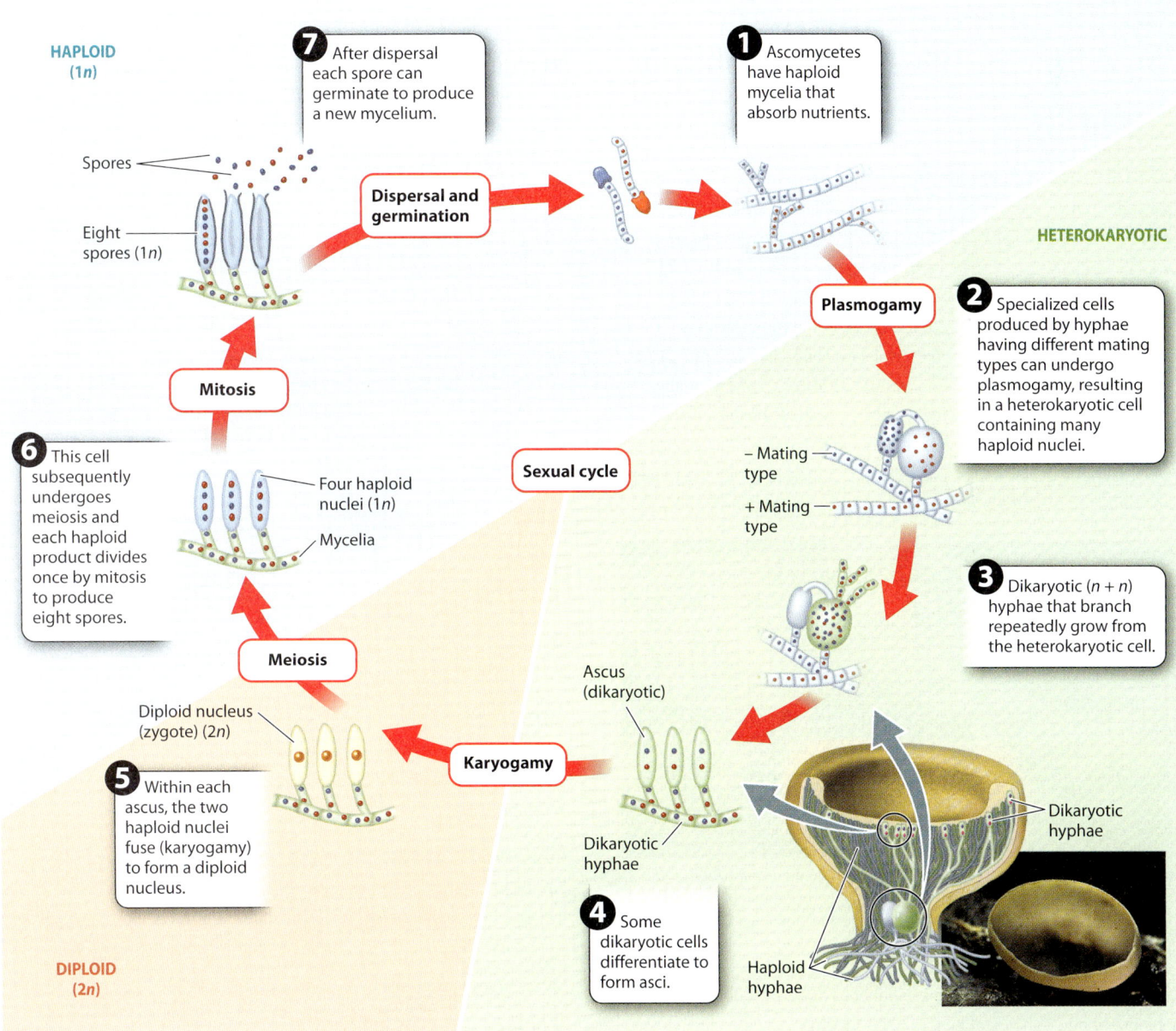

FIG. 34.19 The sexual life cycle of a common ascomycete, the brown cup fungus (*Peziza* species).

but they also develop underground. How, then, can their spores be dispersed? Developing truffles release androstenol, a hormone also produced by boars before mating. The hormone attracts female pigs, which unearth and consume the fruiting body. The spores pass through the pig's digestive tract without damage and are released into the environment in feces, thereby dispersing the spores.

Another ascomycete is thought to have played a role in the events that unfolded in Salem, Massachusetts, in 1692. That year, several girls came down with an unknown illness manifested by delirium, hallucinations, convulsions, and a crawling sensation on the skin. The girls were thought to be bewitched, and subsequent accusations led to the infamous Salem witch trials. They were executed for witchcraft, but today many scholars prefer a simpler diagnosis for the illness: poisoning by the ascomycete fungus ergot. Ergot is a common pathogen of rye and related grasses, and ingestion of contaminated plants can lead to the symptoms reported in Salem. Like many defensive compounds produced by plants, alkaloid molecules produced by ergot and their derivatives are used in medicine in low doses, such as in the treatment of migraines.

Ascomycetes are even known for producing so-called zombie ants, a topic explored in **Fig. 34.20**.

HOW DO WE KNOW?

FIG. 34.20

Can a fungus influence the behavior of an ant?

BACKGROUND A curious death ritual unfolds in the rain forest of Thailand. Spores of the ascomycete fungus *Ophiocordyceps* infect *Camponotus leonardi* ants, growing hyphae inside their bodies. The fungus eventually kills the ant, and fruiting bodies emerge from the dead ant's head to disperse spores that begin the life cycle anew. *C. leonardi* nests and forages for food high in the forest canopy. Infected ants, however, undergo convulsions that cause them to fall to the forest floor. The infected ants wander erratically and are unable to climb more than a few meters above the ground before convulsions make them fall again. In their final act, the ants bite into leaves and die. The fungus is then able to complete its life cycle within the humid forest understory. Is the ants' behavior induced by the fungi?

An infected ant attached to a leaf by having bitten into it in a "death grip."

HYPOTHESIS Parasitic fungi manipulate ant behavior to complete their own reproductive cycle.

OBSERVATIONS Infected and uninfected ants were observed for many hours and behavioral events were recorded.

RESULTS Infected ants have repeated convulsions, but uninfected ants show no such behavior. Transitions from erratic wandering to a "death grip," in which the ant bites into a leaf, occurred at about the same time. Dissections showed that the ants' death grip results from jaw-muscle wasting caused by the fungus.

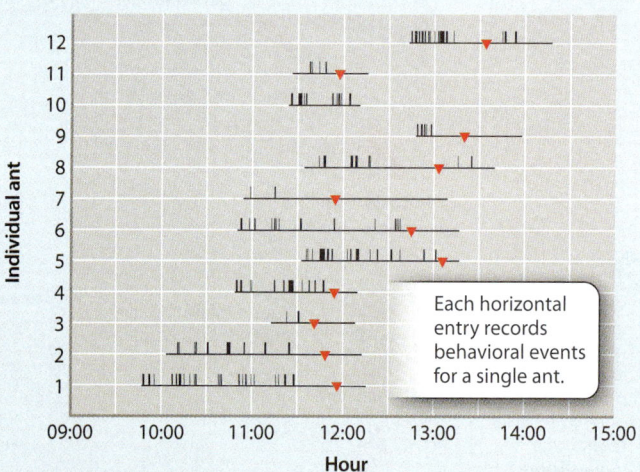

Each horizontal entry records behavioral events for a single ant.

Observations of numerous infected ants (vertical bars indicate convulsions; red triangles indicate the time of ant biting).

CONCLUSION The fungi change the behavior of infected ants, inducing a stereotypical behavior that facilitates completion of the fungus' life cycle. The molecular basis of this manipulation is not yet known.

FOLLOW-UP WORK The zombie-like behavior induced by the fungus is an example of what has been called the extended phenotype: the idea that a phenotype should include the effects that an organism has on its environment—in this case, its effects on host behavior.

SOURCE Hughes, D. P., S. B. Andersen, N. L. Hywel-Jones, W. Himaman, J. Billen, and J. J. Boomsma. 2011. "Behavioral Mechanisms and Morphological Symptoms of Zombie Ants Dying from Fungal Infections." *BMC Ecology* 11:13–22.

FIG. 34.21 Basidiomycetes. (a) Corn smut (*Ustilago maydis*); (b) pseudoflowers that form when plants are infected by the rust *Puccinia monoica*; (c) a toadstool fungus—the highly toxic *Amanita*.

Basidiomycetes include smuts, rusts, and mushrooms.

The basidiomycetes, which make up 34% of all described fungal species, include three major groups (**Fig. 34.21**). Two of these—the smuts and rusts—consist primarily of plant pathogens. Smut fungi, which infect the reproductive tissues of grasses and related plants, take their name from the black sooty spores that they produce. *Ustilago maydis*, the corn smut, turns developing corn kernels into soft gray masses that are a culinary delicacy in Mexico. Because smuts infect seeds, their spread through crops is magnified by the harvest, storage, and eventual sowing of seeds. Before the introduction of chemical seed treatments in the 1930s, infection by *Tilletia* (the stinking smut) commonly caused losses of up to 50% of the wheat harvest. Smuts remain a problem in parts of the world where untreated seeds are still planted.

Rusts can also have profound impacts on plant function and development. For example, the rust *Puccinia monoica* alters leaf development in its host plant *Arabis*, resulting in a pseudoflower that attracts insects by visual and olfactory cues as well as nectar rewards. These pseudopollinators transport spores, thus enhancing outcrossing as well as dispersal of the fungi but provide no service to the plant. We discuss the biology of rusts in greater detail in the next section when we examine Ug99, a highly virulent wheat rust first reported in Uganda in 1999.

The third basidiomycete group is characterized by the formation of multicellular fruiting bodies, including the iconic toadstools with their central stalk and umbrella-like cap. However, this group also includes the diverse shapes seen in stinkhorns, puffballs, and bracket fungi (**Fig. 34.22**).

FIG. 34.22 Diverse fruiting bodies of basidiomycetes. (a) Stinkhorn (*Dictyophora indusiata*); (b) puffball (*Calvatia gigantea*); (c) bracket fungi (*Laetiporus sulphureus*).

FIG. 34.23 The sexual life cycle of a basidiomycete that forms multicellular fruiting bodies.

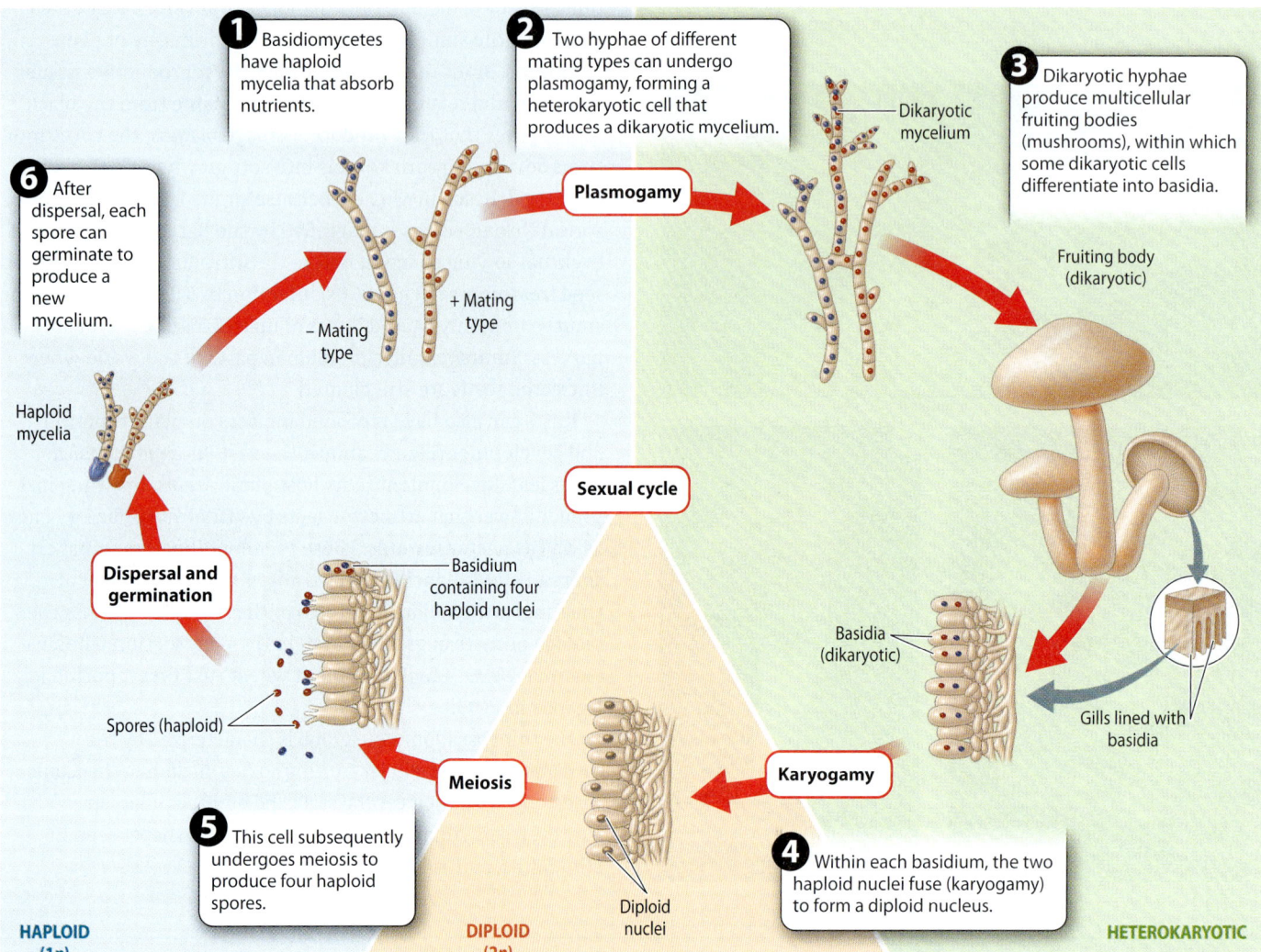

Fig. 34.23 shows the life cycle of a typical basidiomycete mushroom. The life cycle is similar to that of ascomycetes (see Fig. 34.19), except that the fruiting body is made up entirely of dikaryotic hyphae instead of a combination of dikaryotic and haploid hyphae as in ascomycetes. In addition, nuclear fusion (karyogamy) takes place in a specialized cell called a basidium rather than an ascus. Finally, the haploid products of meiosis (the spores) do not undergo mitosis, so four spores are produced from each basidium rather than eight spores from each ascus.

→ **Quick Check 4** When you eat a mushroom, what stage of its life cycle are you consuming?

Most species in this group live by forming ectomycorrhizal associations or by decomposing wood and other substrates. The fruiting bodies that we see elevated above a rotting log or the soil are connected to extensive mycelia that provide resources for fruiting-body development.

Elevation alone does not ensure dispersal because the spores must penetrate a boundary layer of air that surrounds the fruiting body. Many basidiomycetes depend on surface tension to catapult their spores through the air. The placement of their four spores on the ends of short stalks is the key to this mechanism. Both the spores and the supporting basidial cell actively secrete solutes that cause water to condense on their surfaces. At first, the droplets are independent, but as they grow they eventually come into contact with one another, and at this point the high surface tension of water causes them to merge into a single, smooth droplet. This change in shape shifts the center of mass of the water with such force that it catapults the spore into the air.

Basidiomycetes that form conspicuous fruiting bodies have diverse mechanisms for dispersing their spores. For example, puffballs function like bellows: The impact of raindrops forces the loose, dry spores out of a small hole at the tip of the ball (see Fig. 34.10a). Bird's nest fungi function like splash cups: In this case, raindrops displace groups of spores (the "eggs" in the nest) onto nearby vegetation. Herbivores consume the spores when they feed on the vegetation and transport them to a new dung pile. Stinkhorns produce spores in a stinky, sticky mass that attracts insects, which then carry off spores that stick to their legs and body.

? CASE 6 Agriculture: Feeding a Growing Population
How do fungi threaten global wheat production?

Throughout much of human history, crops have been vulnerable to infection by *Puccinia graminis*, the black stem rust of wheat. *P. graminis* infects wheat leaves and stems through their stomata and then extends within the plant's living tissues to fuel its own growth. Rust-colored pustules erupt along the stem and then release a vast number of spores, leaving the host plant to wither and die. Crop losses from *P. graminis* can be devastating. In the past, both the United States and the Soviet Union tried to develop black stem rust as a biological weapon. The U.S. version was cluster bombs (since destroyed) containing turkey feathers smeared with rust spores, produced with the aim of disrupting crop yields in enemy countries.

Many rusts have a life history in which they alternate between different host species. For example, to complete its life cycle, *P. graminis* must infect wheat first and then barberry (*Berberis* species). After a particularly severe outbreak of wheat rust during World War I, in which crop losses equaled one-third of total U.S. wheat consumption, a major program of barberry eradication was started to bring stem rusts under control. This program reduced the threat, but did not totally eliminate it.

The Green Revolution seemed to give growers a decisive advantage over stem rust. During World War II, Norman Borlaug, whose original training was in plant pathology, was sent to Mexico by the Rockefeller Foundation to help combat crop losses due to stem rust. Borlaug's breeding efforts led to resistant wheat varieties. Much of this resistance could be attributed to variation in a small number of genes. Stem rusts were apparently vanquished—that is, until the appearance of Ug99, a wheat stem rust first identified in Uganda in 1999 (hence the name). Ug99 is capable of defeating all major resistance genes. Given that wheat accounts for 20% of daily global calorie consumption by humans and that 90% of cultivated wheat has little or no resistance to Ug99, the appearance of this resistant fungus is cause for major concern. At present, Ug99 shows every sign of spreading (**Fig. 34.24**)—not surprising given that a single hectare of infected wheat produces upward of a billion spores. It currently

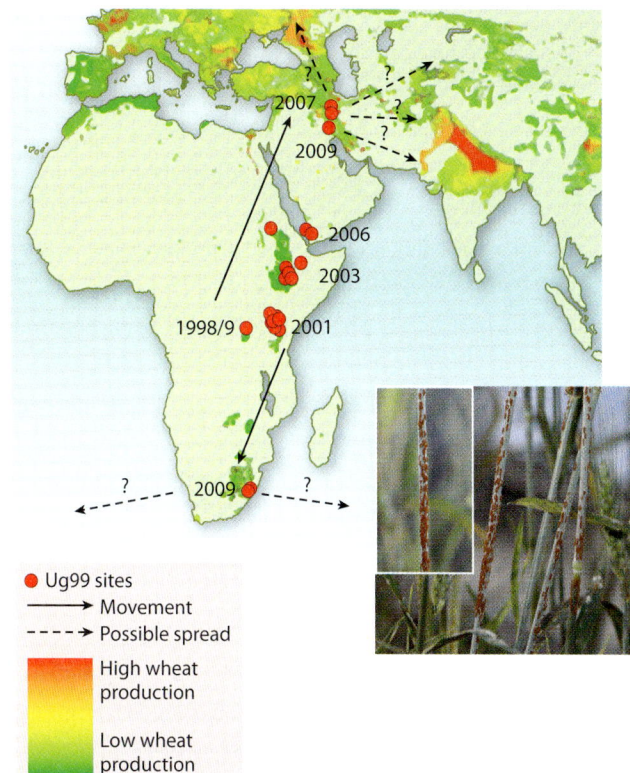

FIG. 34.24 **The spread of Ug99.** Ug99 is a highly resistant stem rust of wheat and a major threat to global food production.

devastates wheat production in Kenya and has been found in Yemen (2006), Iran (2007), and South Africa (2009). As you read this, Ug99 is poised to reach the major wheat growing regions of Turkey and South Asia. Even more rapid spread is possible if spores accidentally lodge on cargo or airline passengers.

Just before his death in 2009, Norman Borlaug urged the world to take this threat seriously. Today, plant breeders are actively searching ancestral wheat varieties for genetic sources of resistance to Ug99, while other scientists are trying to understand how plants defend themselves against rusts and what makes Ug99 so virulent. The resurgence of *P. graminis* as a serious threat to world food production underscores the importance of safeguarding the genetic diversity of agricultural species so that the inevitable evolutionary battles with pathogenic species may be waged successfully.

Core Concepts Summary

34.1 FUNGI ARE HETEROTROPHIC EUKARYOTES THAT FEED BY ABSORPTION.

Fungi are heterotrophs that depend on preformed organic molecules for both carbon and energy. page 34-1

Fungi break down their food and then absorb it. page 34-1

Fungi use the process of growth to find and obtain food in their environment. page 34-2

Most fungi have numerous branched filaments called hyphae, which they use to absorb nutrients. page 34-2

When fungi encounter a rich source of food, their hyphae form a network called a mycelium. page 34-2

Fungal cell walls are made of chitin, the same compound found in the exoskeletons of insects. page 34-2

In early fungal groups, the hyphae form a continuous compartment, with many nuclei but with no cell walls, but in later evolving groups, nuclear divisions are accompanied by the formation of septa that divide the cytoplasm into separate cells. page 34-2

Some fungi, such as yeasts, do not produce hyphae. page 34-2

Most fungi feed on dead organic matter. page 34-3

Fungi are critical elements of the carbon cycle, converting dead organic matter back into carbon dioxide and water. page 34-3

Many fungi feed on living animals and plants, causing diseases. page 34-4

Fungi have repeatedly evolved mutually beneficial relationships with plants and animals. page 34-5

Lichens are stable associations between a fungus and a photosynthetic microorganism that look, function, and even reproduce as single organisms. page 34-6

34.2 FUNGI REPRODUCE BOTH SEXUALLY AND ASEXUALLY, AND DISPERSE BY SPORES.

Fungi produce haploid spores that are dispersed by wind, water, or animals. page 34-8

Spores can be produced asexually (by mitosis) or sexually (by cell fusion and meiosis). page 34-8

Many fungi produce spores within multicellular fruiting bodies that enhance spore dispersal. page 34-8

The fungal life cycle is similar to haploid-dominant life cycles found in many eukaryotic organisms, except that plasmogamy (cell fusion) and karyogamy (nuclear fusion) are often separated, resulting in a stage in which there are genetically distinct haploid nuclei in one cell, called a heterokaryotic stage. page 34-10

Genetically determined mating types prevent self-fertilization in many fungi. page 34-11

Asexual fungal species have mechanisms of generating genetic diversity, including a parasexual cycle, which involves crossing over in mitosis and chromosome loss. page 34-11

34.3 NEXT TO THE ANIMALS, FUNGI ARE THE MOST DIVERSE GROUP OF EUKARYOTIC ORGANISMS.

About 75,000 fungal species have been formally described, but total diversity may be as high as 5 million species. page 34-12

The chytrids, a small group of about 1000 species, are at the base of the fungal tree. page 34-12

Zygomycetes produce hyphae without septa and make up less than 1% of known fungal diversity. page 34-13

The glomeromycetes are a group of fungi of low diversity but tremendous ecological importance due to their association with plant roots. page 34-14

Most fungi belong to the Dikarya (ascomycetes and basidiomycetes), which form septa along hyphae. During the heterokaryotic stage, these fungi produce cells that contain two haploid nuclei, one from each parent. page 34-14

Ascomycetes include wood-rotting fungi, many ectomychorrizal species, plant and animal pathogens, the fungal partner in most lichens, and baker's and brewer's yeasts. page 34-14

Basidiomycetes include familiar toadstools and puffballs, as well as plant pathogens. page 34-17

Self-Assessment

1. Describe two ways in which fungi differ from other heterotrophic organisms in how they obtain and digest their food.

2. Explain how hyphae and cell walls made of chitin allow fungi to obtain nutrients from their environment.

3. Describe how fungi contribute to the terrestrial (land-based) carbon cycle.

4. Name the two organisms that make up a lichen and describe how each partner benefits from the association.

5. Describe how fungi disperse.

6. Draw the life cycle of an ascomycete, and indicate the heterokaryotic stage.

7. Draw the life cycle of a basidiomycete, and indicate the heterokaryotic stage.

8. Name and describe several key innovations in the evolutionary history of fungi that allowed them to move from water to land.

9. Explain how the evolution of vascular plants has provided opportunities for fungal diversification.

Do you understand the chapter's Core Concepts? Log in to check your answers to the Self-Assessment questions, then practice what you've learned and reinforce this chapter's concepts by working through the problems and multimedia tutorials provided there.

www.biologyhowlifeworks.com

CASE 7

Predator–Prey

A GAME OF LIFE AND DEATH

In his 1850 poem *In Memoriam A.H.H.*, Alfred, Lord Tennyson wrote of "Nature, red in tooth and claw." The phrase has endured as a powerful description of the brutality of wild nature. Thanks to nature documentaries and our human fascination with wild animals, most of us have witnessed plenty of images of this red-toothed nature: a lion chasing down a zebra, a hawk scooping a mouse into its talons, a shark sinking its teeth into a sea lion.

Predator–prey interactions are important in every ecosystem on Earth. After all, every living animal can be classified as either "predator" or "prey"—and often, both labels apply. To a fly, a toad is a fearsome predator. To a snake, that same toad may be choice prey.

Very few species enjoy the luxury of having no natural predators; these are called top predators. For the vast majority of animals, the threat of predation is simply a fact of life. Not surprisingly, that threat has been a powerful evolutionary force. Predator–prey dynamics have influenced both the evolution of individual organisms and the shape of the ecosystems in which they live.

On Isle Royale, an island in Lake Superior, ecologists have been studying moose and wolves since 1958 in the longest-running study of a single predator–prey system in the world. The moose arrived on the island about 100 years ago, presumably by swimming about 15 miles from the nearest shoreline. They arrived to find a predator-free paradise, and the moose population quickly exploded. Then, about 1950, a pair of wolves crossed an ice bridge to Isle Royale. As the wolf population grew, the moose population declined, until a delicate balance was established.

As a predator population grows, the prey population shrinks—a seemingly logical relationship. But predator–prey interactions are complex and sometimes even counterintuitive. Removing a top predator can actually reduce biodiversity. Sea stars, for instance, prey on mussels in the rocky intertidal zone of the Pacific Northwest. When sea star numbers fall, mussel numbers increase. The mussels crowd out other species, such as barnacles and seaweed, that compete for space on the rocks. The result is a drop in the overall number of species in that habitat.

The same patterns have been found among birds of prey in the Italian Alps. These raptors—including the goshawk and four types of owl—are the top predators of their food chains. Spanish researchers found that locations where the birds were present had a greater diversity of trees, birds, and butterflies than did comparable control sites.

One reason that the predator and prey populations eventually achieve a delicate balance is that both are well adapted to their roles. Predation has exerted powerful evolutionary pressure that has influenced body plans over the long term. That pressure has shaped predators by giving them claws, teeth, venom, and powerful muscles for hunting prey.

Of course, predation pressure has also shaped prey. Some animals, including some insects and frogs, have evolved toxins to deter would-be predators. These toxic species often exhibit warning colors that tell the predators to steer clear. Other prey animals have evolved camouflage colors and

> **Predator–prey dynamics have influenced both the evolution of individual organisms and the shape of the ecosystems in which they live.**

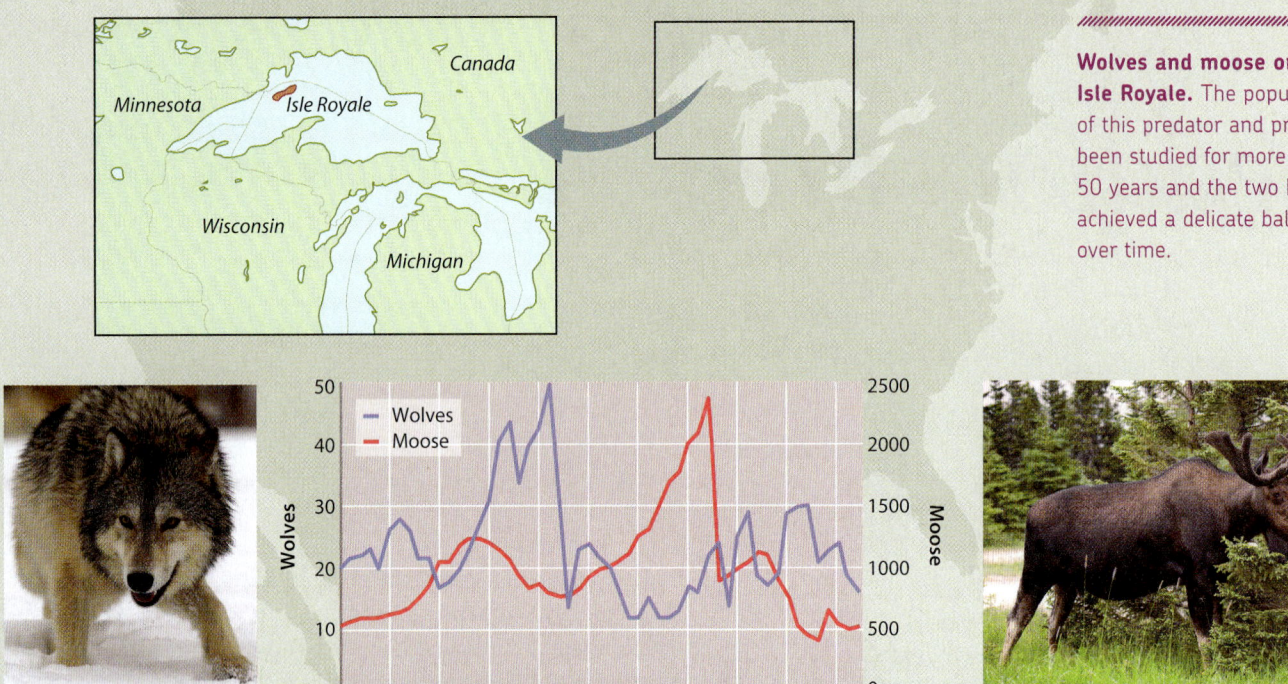

Wolves and moose on Isle Royale. The populations of this predator and prey have been studied for more than 50 years and the two have achieved a delicate balance over time.

forms to help them hide from hungry carnivores. Still others have developed protective behaviors, such as living together in herds for security against predation.

Often, predator and prey evolve in lockstep, driving each other's adaptations. Clams may have evolved thicker shells to protect themselves from hungry crabs. In turn, the crabs evolved larger, stronger claws for cracking clamshells. This type of coevolution is often described as an evolutionary arms race. And it has happened time and time again.

The interactions between eaters and eaten have also influenced the evolution of physiological systems. The skills that a predator needs to hunt its prey and that its prey needs to escape depend in large part on adaptations of their sensory systems, musculoskeletal systems, nervous systems, and even their circulatory and respiratory systems.

Consider sensory systems. Animals rely on visual, auditory, tactile, and chemical stimuli to warn them of approaching predators—or to guide them to suitable prey.

A bat relies on sonar to locate insects. A spider responds to the flutter of silk when an insect becomes ensnared in its web. Gazelles are always attentive, watching and listening for signs of a cheetah or lion in the grass.

When predators are nearby, prey species experience fear and anxiety. Stress hormones produced by prey animals can influence an animal's biology in a number of ways. In snowshoe hares, levels of the stress hormone cortisol increase when predators such as lynx and coyote are plentiful. The hormones trigger behaviors, such as alertness and fearfulness, that help the hares avoid becoming a lynx's lunch. But the behavioral benefit comes at a cost. Research has shown that stressed hares give birth to fewer and smaller offspring.

In short, predation has left a physical imprint on both predator and prey, from nose to tail—their body shapes, their behaviors, their physiology, even their muscle fibers have been influenced over time by predator–prey interactions.

In the modern world, our own species, *Homo sapiens*, occupies a unique position at the top of the food web. While humans are occasionally killed by animals such as bears or sharks, we are, for the most part, no longer constrained by the fear of predation by other species. But it wasn't always so. Some scientists have proposed that fear of predation among early humans led to the evolution of cooperative social behavior and large brains. Others have theorized that it was our hunting of other animals that led us to evolve big brains and the ability to work together. Perhaps both factors played a role. Either way, the importance of the predator–prey system can't be ignored. To understand our roots, it seems, we must take a good look at nature, red in tooth and claw.

❓ CASE 7 QUESTIONS

Answers to Case 7 questions can be found in Chapters 35–41.

1. **What body features arose as adaptations for successful predation?** *See page 35-4.*
2. **How have sensory systems evolved in predators and prey?** *See page 36-10.*
3. **How do different types of muscle fiber affect the speed of predators and prey?** *See page 37-12.*
4. **How does the endocrine system influence predators and prey?** *See page 38-15.*
5. **How do hormones and nerves provide homeostatic regulation of blood flow as well as allow an animal to respond to stress?** *See page 39-17.*
6. **Does body temperature limit activity level in predators and prey?** *See page 40-5.*
7. **Can the loss of water and electrolytes in exercise be exploited as a strategy to hunt prey?** *See page 41-6.*

CHAPTER 35

ANIMAL NERVOUS SYSTEMS

Core Concepts

35.1 Animal nervous systems allow organisms to sense and respond to the environment, coordinate movement, and regulate internal functions of the body.

35.2 The neuron is the basic functional unit of the nervous system. Neurons have dendrites that receive information and axons that transmit information.

35.3 The electrical properties of neurons allow them to communicate rapidly with one another.

35.4 Animal nervous systems can be organized into central and peripheral components.

Gazelle grazing on an African plain are alert to the threat of predators nearby. Members of the herd scan the distant landscape for visual cues that might reveal the stealthy approach of a cheetah (**Fig. 35.1**). Olfactory (smell) and auditory (sound) cues from the environment can also warn of the cheetah's approach. Herd members communicate the presence of the cheetah to one another by the directed attention of a dominant male or the pawing of a hoof. At what point is the threat real? When does each animal decide to stop grazing and turn to run? What direction should the escape take to keep the herd together as a group?

The cheetah scans the horizon, spotting the grazing gazelle with its keen vision. The cheetah's ability to recognize wind direction enables it to move downwind of the gazelle to minimize auditory and olfactory cues that might otherwise be carried by the air toward suspecting prey. At what point does the cheetah decide to sprint in pursuit of its quick-running prey?

The perceptions and reactions of these two animals depend on the activity of their nervous systems. In this and the following chapter, we explore the organization of the nervous system, including the brain and the nerve cells that compose it. The nervous system has the remarkable ability to process all kinds of information. It takes in information from the environment through the animal's sensory organs and processes that information through the activity of networks of nerve cells. The results of that processing guide an animal's motor responses. Such processes underlie the perceptions and interaction of cheetah and gazelle as predator and prey, as well as the complex emotional and cognitive responses triggered, for example, by seeing a friend or relative after a long separation.

35.1 NERVOUS SYSTEM FUNCTION AND EVOLUTION

Except for sponges, all multicellular animals possess a **nervous system,** a network of many interconnected nerve cells. This network allows an animal to sense and respond to the environment, coordinate the action of muscles, and control the internal function of its body. Nerve cells, or **neurons,** are the basic functional units of nervous systems.

Nervous systems first evolved in simple animals. These early nervous systems could sense basic features of the environment, such as light, temperature, chemical odors, and physical forces. Guided by these cues, early nervous systems could trigger movements useful for obtaining food, finding a mate, or choosing a suitable habitat. As animals evolved more complex bodies with increasingly specialized organ systems for respiration, circulation, digestion, and reproduction, their nervous systems also became more complex. These more complex nervous systems could regulate internal body functions in response to sensory cues received from the environment. They

FIG. 35.1 Predator and prey. Interactions of predator and prey, such as this cheetah and gazelle, are mediated by the nervous system.

also made possible more sophisticated behaviors that relied on improved decision making, learning, and memory. These abilities, possessed by many invertebrate and vertebrate animals, increased their fitness by improving their ability to survive, find and select a mate, and reproduce.

Animal nervous systems have three types of nerve cell.

All animal nervous systems are made up of three types of nerve cell: **sensory neurons, interneurons,** and **motor neurons.** Each type of neuron has a different function. Sensory neurons receive and transmit information about an animal's environment or its internal physiological state. These neurons respond to physical features such as temperature, light, and touch or to chemical signals such as odor and taste. Interneurons process the information received by sensory neurons and transmit it to different body regions, communicating with motor neurons at the end of the pathway to produce suitable responses. For example, a motor neuron may stimulate a muscle to contract to produce movement. Other motor neurons may adjust an animal's internal physiology, constricting blood vessels to adjust blood flow or causing wavelike contractions of the gut to aid digestion. As a result, nervous system function is fundamental to **homeostasis,** the ability of animals, organs, and cells to actively regulate and maintain a stable internal state.

Most nerve cells have fiberlike extensions that receive information and other extensions that transit information. The result is a network of interconnected nerve cells that form a circuit. These circuits allow information to be received, processed, and delivered. In most animals, interneurons form specialized circuits that may consist of a diverse array of nerve cells having different shapes, sizes, and chemical properties.

The ability to process information first evolved with the formation of **ganglia** (singular, ganglion). Ganglia are groups of nerve cell bodies that process sensory information received from a local, nearby region, resulting in a signal to motor neurons that control some physiological function of the animal. They can be thought of as relay stations or processing points in nerve cell circuits. Eventually, the centralized concentration of neurons that we call a **brain** evolved, and this organ was able to process increasingly complex sensory stimuli received from the environment or anywhere in the body. Neurons in the brain form complex circuits, allowing the brain to mediate a broad array of behaviors, as well as to learn and retain memories of past experiences.

Nervous systems range from simple to complex.

Even single-celled animals have cell-surface receptors that enable them to sense and respond to their environment. By contrast, most multicellular animals have a nervous system. In general, the organization and complexity of an animal's nervous system reflects its lifestyle. Sessile, or immobile, animals, like sea anemones or clams, have relatively simple sense organs and nervous systems. In contrast, active animals such as arthropods (which include insects, spiders, and crustaceans), squid, and vertebrates have more sophisticated sense organs linked to a brain and more complex nervous systems.

The sponges are the only multicellular organisms that lack a nervous system (**Fig. 35.2**). Their simple body plan means that a nervous system is not required to coordinate the functions of different body regions. Instead, groups of cells within sponges respond to local chemical and physical cues.

Among multicellular organisms, the simplest nervous systems are found in cnidarians, which are radially symmetric animals such as jellyfish and sea anemones (Fig. 35.2). The nervous system of

CHAPTER 35 ANIMAL NERVOUS SYSTEMS

FIG. 35.2 A simplified phylogenetic tree of multicellular animals. All multicellular organisms, except sponges, have a nervous system.

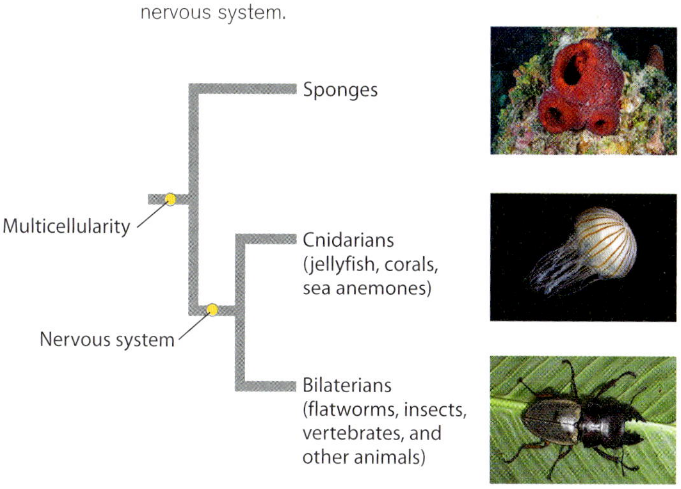

these animals has relatively few neurons and no ganglia or central brain (**Fig. 35.3a**). In these animals, the nerve cells are arranged like a net. The nerve net of a sea anemone is organized around its body cavity. At one end of its body is a mouth that opens into a central body cavity where food is digested. Sensory neurons sensitive to touch signal when one of the anemone's tentacles has captured food prey, then motor neurons react to coordinate the tentacle's movement toward the mouth for feeding and digestion. Sensory neurons responding to a noxious stimulus signal the sea anemone to retract or detach from its substrate.

More complex nerve connections develop in the head end of flatworms, segmented annelid worms, insects, and cephalopod mollusks, such as squid and octopus. These animals are all bilaterians, animals with symmetrical right and left sides and a distinct front and back end. Flatworms are representatives of one of the earliest branching groups of bilaterians, and their bodies are well adapted for forward locomotion. Sensory neurons are most numerous in the animal's front end, so the flatworm can perceive the environment ahead of it as it moves forward. The flatworm's nervous system can compare inputs from different sensory neurons and then send signals to the appropriate muscles to move toward a target. Paired ganglia (one on each side of the animal) in the head end of flatworms handle this processing task (**Fig. 35.3b**).

In animals with an organized nervous system, information is transmitted to different regions along the length of the animal's body by distinct **nerve cords** and **nerves** (Fig. 35.3). These are bundles of the long fiberlike extensions from multiple nerve cells.

In earthworms, paired ganglia, each made up of many interneurons, are located in each of the body's segments (**Fig. 35.3c**). These segmental ganglia help to regulate the muscles within each body segment that move the earthworm's body. Paired ganglia near the head regulate the activity of motor neurons that control movements of the mouth for feeding. Ganglia serve to regulate key processes in local regions and organs of the animal's body. For example, local ganglia assist in processing information from the eyes (Chapter 36) or control the digestive state of an animal's gut (Chapter 40).

Centralized collections of neurons forming a brain are present in annelid worms, insects, and mollusks (**Figs. 35.3c–35.3e**). Sense organs also developed in the head region, from eyespots in flatworms that respond to light to more sophisticated image-forming eyes in octopus and squid. The evolution of a brain in these invertebrate animals enabled them to learn and perform complex behaviors. Many of these nervous system capabilities are shared with vertebrate animals.

The evolution of a brain is also a hallmark of vertebrate animals (**Fig. 35.3f**). It enabled them to evolve complex behaviors that rely on learning and memory and, in certain vertebrates, the ability to reason. At the same time, animal brains became increasingly complex in the number of nerve cells they contain (fruit fly, about 0.25 million; cockroach, about 1 million; mouse, about 75 million; humans, about 100 billion), in the number of connections between neurons, and in the variety of nerve cell features.

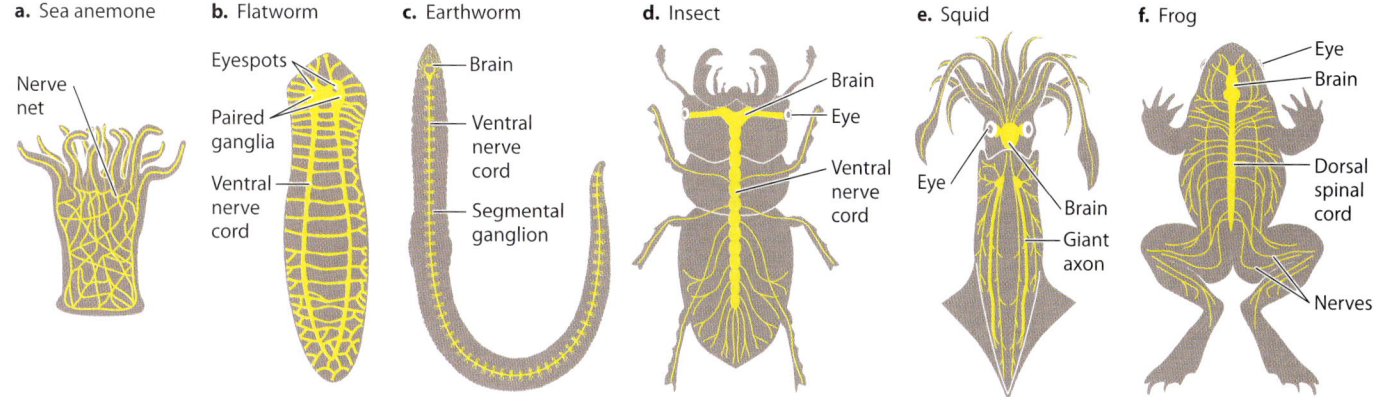

FIG. 35.3 Animal nervous systems. (a) Sea anemones have a nerve net; (b) flatworms have paired ganglia; (c) earthworms, (d) insects, (e) squid, and (f) frogs have a brain.

Although animal nervous systems differ in organization and complexity, their nerve cells have fundamentally similar molecular and cellular features and use fundamentally the same mechanisms to communicate with neighboring cells. These mechanisms have been retained over the course of evolution to code and reliably transmit information. The ability to receive sensory information from the environment, and transmit and process this information within a nervous system, is therefore shared across a wide diversity of animal life, contributing to the success of multicellular animals.

? CASE 7 Predator–Prey: A Game of Life and Death
What body features arose as adaptations for successful predation?

A notable feature of most animal nervous systems is that nervous system tissue, including specialized sense organs such as eyes, becomes concentrated at one end of the body. For example, the paired ganglia and eyespot of the flatworm are located at one end of its body, as are the brain and sense organs of the earthworm, squid, and insect, as shown in Fig. 35.3. The concentration of nervous system components at one end of the body, defined as the "front," is referred to as **cephalization.** Cephalization is a key feature of the body plan of most multicellular animals.

Cephalization is thought be an adaptation for forward locomotion because it allows animals to take in sensory information from the environment ahead of them as they move forward. As the quality and amount of sensory information increased, brain size and complexity increased. Cephalization is also considered to be an adaptation for predation, allowing animals to better detect and capture prey.

Cephalization evolved independently multiple times in different animal groups and is therefore thought to confer certain advantages. These advantages likely include the ability to sense environmental stimuli encountered by forward motion and to process this information quickly to enable a suitable behavioral response. Although cephalization is a feature of many animals, it has been particularly well studied in vertebrates. In vertebrates, the brain, many sense organs, and mouth are all located in the head region. Vertebrates also evolved several novel features, including a jaw, teeth, and tongue. These are all thought to be adaptations for successful predation, or more generally the acquisition and processing of food.

As a result of evolutionary selection for enhanced sensory perception and the ability to respond to important cues in the environment, the brain, sensory organs, and nervous system of many animals are complex in their organization. These are linked to more sophisticated decision-making abilities that allow for a broad range of behaviors, as well as to improved memory and learning. These abilities are critical to the success of both predators and prey and underlie the complex behavioral interactions that occur among members of a species when they mate, reproduce and disperse, and care for their young.

35.2 NEURON STRUCTURE

We have seen that nerve cells can be classified as sensory neurons, interneurons, or motor neurons. All share the same basic organization, adapted to their common function in receiving and transmitting information. Neurons have extensions that receive information and relay it to the cell body. Other extensions transmit information away from the cell body. Although all neurons share this basic plan, they differ remarkably in size and number of extensions, according to their specific function.

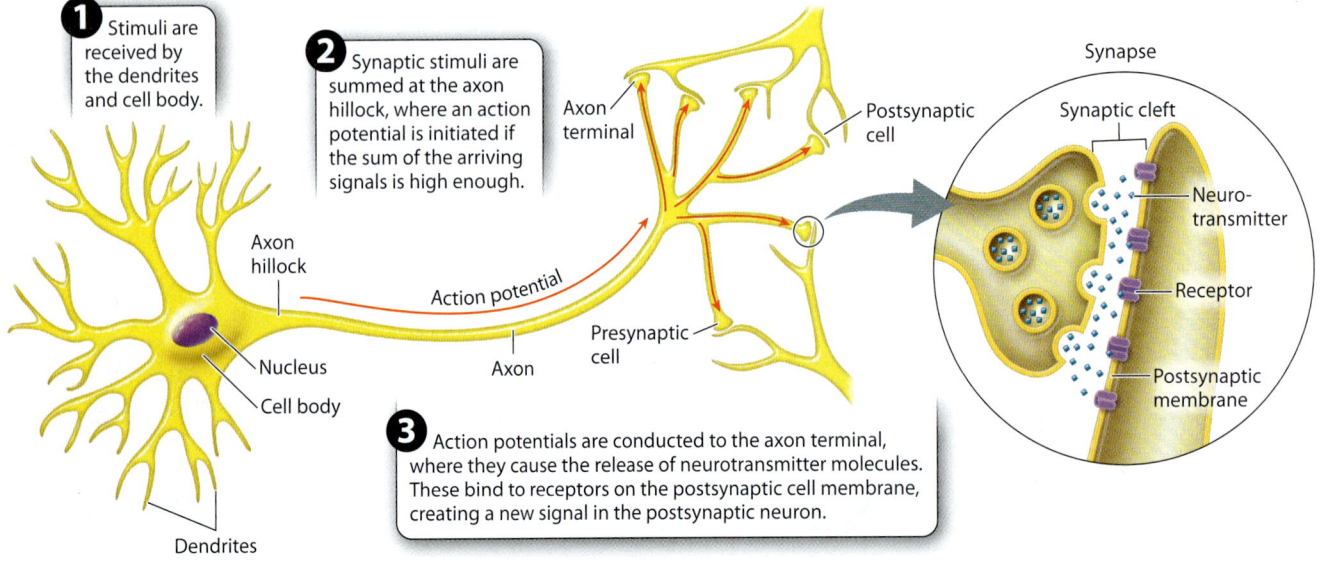

FIG. 35.4 **Neuron organization.** Nerve cells typically have dendrites that receive signals and axons that transmit signals.

CHAPTER 35 ANIMAL NERVOUS SYSTEMS

FIG. 35.5 **Variation in neuron shape.** Neurons display a diversity of shapes, reflecting their different functions.

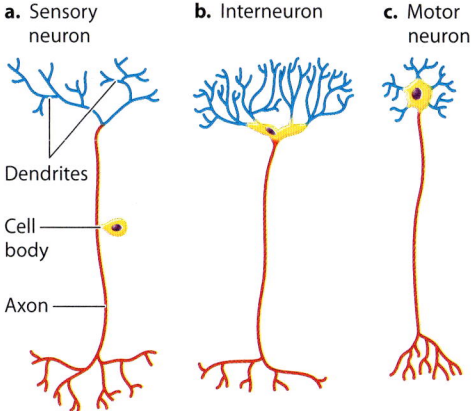

Neurons share a common organization.

Neurons all have some basic features in common. These include a large cell body from which emerge two kinds of fiberlike extension, the **dendrites** and **axons** (**Fig. 35.4**). These extensions are the input and output ends of the nerve cell. Both types of cellular extension can be highly branched, enabling neurons to communicate over large distances with many other cells. A neuron's dendrites receive signals from other nerve cells or, in the case of sensory nerves, from specialized sensory endings. These signals travel along the dendrites to the neuron's cell body. At the junction of the cell body and its axon, the **axon hillock,** the signals are summed. If the sum of the signals is high enough, the neuron fires an **action potential,** or nerve impulse, that travels down the axon. An action potential is a brief electrical signal transmitted from the cell body along one or more axon branches.

Axons generally transmit signals away from the nerve's cell body. The end of each axon forms a swelling called the axon terminal. An axon terminal communicates with a neighboring cell through a junction called a **synapse.** A space, the **synaptic cleft,** separates the end of the axon of the presynaptic cell and the neighboring postsynaptic cell. The synaptic cleft is commonly only about 10 to 20 nm wide.

How does a signal cross the synaptic cleft? Molecules called **neurotransmitters** convey the signal from the end of the axon to the post-synaptic target cell. The arrival of a nerve signal at the axon terminal triggers the release of neurotransmitter molecules from vesicles located in the terminal. The vesicles fuse with the axon's membrane, releasing neurotransmitter molecules into the synaptic cleft. The molecules diffuse across the synapse and bind to receptors on the plasma membrane of the target cell. The binding of neurotransmitters to these receptors causes a change in the electrical charge across the membrane of the receiving postsynaptic cell, continuing the signal in the second cell. Most neurons communicate by hundreds to thousands of synapses that they form with other cells. The massive number of interconnections enables the formation of complex nerve cell circuits and the transmission of information from one circuit to others in the nervous system.

Some neurons do not synapse with other neurons, but instead with other types of cell that produce some physiological response in the animal. Examples are muscle cells, which contract in response to nerve signals, and secretory cells in glands, which release hormones in response to nerve signals.

→ **Quick Check 1** How are signals transmitted from one end of a nerve cell to the other, and from one neuron to another?

Neurons differ in size and shape.

In spite of their common organization, neurons show a remarkable diversity in size and shape. Some neurons, such as interneurons, are very short, whereas others, like the motor neuron that extends from your spinal cord to your big toe, are several feet long. The position of the cell body can also vary in different neurons (**Fig. 35.5**). Some sensory neurons involved in smell, sight, and taste have just two extensions coming off the cell body (Fig. 35.5a), whereas other neurons have many more (Fig. 35.5b and 35.5c).

The cell bodies of some neurons have distinctive shapes. For example, pyramidal neurons, as their name suggests, have triangular-shaped cell bodies (**Fig. 35.6**). These cells, located

FIG. 35.6 **Pyramidal cell.** (a) These neurons, drawn by Santiago Ramón y Cajal (1852–1934), show the triangular cell body, highly branched dendrites, and single axon. (b) The triangular shape of the cell body is clear in the photomicrograph at right.

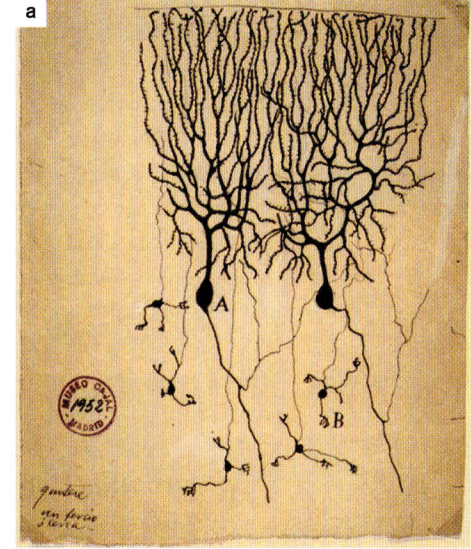
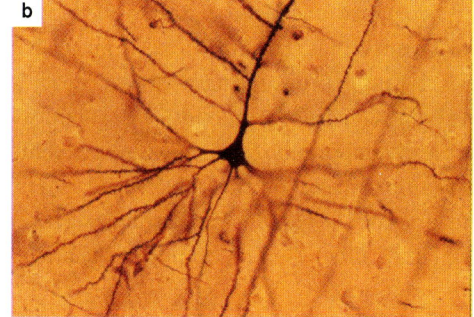

in the mammalian brain, were elegantly drawn by the Spanish neuroanatomist Santiago Ramón y Cajal, who shared the 1906 Nobel Prize in Medicine or Physiology with the Italian scientist Camillo Golgi "in recognition of their work on the structure of the nervous system."

The degree of branching of a neuron's extensions can vary from one neuron to the next. A neuron's dendrites form a finely branched field that receives inputs from many other nerve cells or receptors. A neuron with more dendritic branches receives inputs from more sources than one with fewer branches. Some neurons receive information only from nearby regions through a few dendrites that branch close to the cell body. Other neurons receive information from many branching dendrites and transmit their output along a single axon branch or only a few axon branches. Pyramidal cells provide an example (Fig. 35.6). They have many highly branched dendrites that converge on the cell body, but just a single axon. Pyramidal cells illustrate a general organizational feature of nervous systems: Information arriving from many sources often converges on a smaller subset of neurons.

Axons typically have fewer branches than dendrites. Signals from all the dendrites, converging on the neuron cell body, determine the strength, or frequency, and timing of signals carried by the neuron's axon. The process of bringing together information gathered from different sources is referred to as the integration of information. The degree of branching and the number of synapses that a neuron makes with neighboring nerve cells, therefore, reflect how information is processed and integrated by the cell as part of a larger circuit.

Neurons are supported by other types of cell.

Neurons in many body regions, and in particular within the brain, are supported by other types of cell that do not themselves transmit electrical signals. **Glial cells** are a major class of supporting cell. Indeed, the human brain has more glial cells than neurons. Glial cells surround neurons and provide them with nutrition and physical support. During development, glial cells help orient neurons as they develop their connections.

Star-shaped glial cells called **astrocytes** contribute to the blood–brain barrier, a set of structural adaptations of the blood vessels supplying the brain that prevent pathogens and toxic compounds in the blood from entering the brain. Astrocytes surround blood vessels in the brain, limiting the size of compounds that can diffuse from the blood into the brain. Because nerve cells have limited capacity to regenerate after being damaged, protection of the brain and spinal cord is critically important. Nevertheless, lipid-soluble compounds such as alcohol, certain anesthetics, and other drugs readily diffuse across the blood–brain barrier. As a result, these compounds affect the functioning of the brain and an animal's

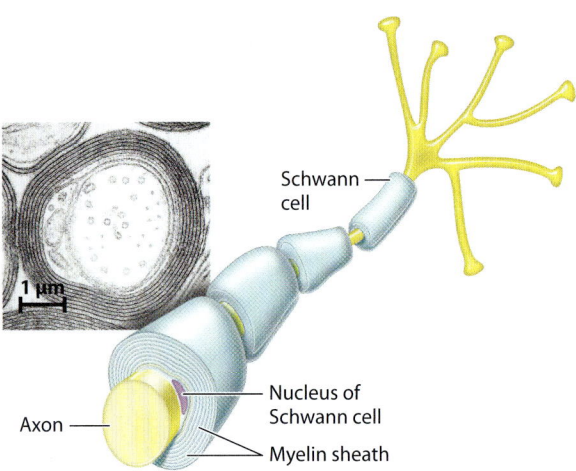

FIG. 35.7 **Schwann cell.** In vertebrates, Schwann cells insulate the axons of sensory and motor neurons.

mental state. They can also lead to damage and even destruction of neurons in the brain.

Glial cells also provide electrical insulation to vertebrate neurons. Glial cells form multiple lipid-rich layers or sheaths, called **myelin,** that wrap around the axons of neurons (**Fig. 35.7**). Myelin gives many nerves their glistening white appearance. Glial cells called **Schwann cells** insulate sensory and motor neurons, and glial cells called **oligodendrocytes** insulate cells in the brain and spinal cord. By insulating the axon, myelin greatly increases the speed of nerve signal transmission. As a result, the nervous system can respond rapidly to stimuli, as we will see after first discussing how nerve cells transmit electrical signals.

35.3 NEURON FUNCTION

In addition to sharing a common organization, neurons share another feature: They are electrically excitable cells. All neurons encode information by changes in their membrane voltage and transmit information in the form of electrical signals. Nerve signals travel at high speeds, ranging up to 200 m/s (450 mph). Although the speed of nerve signals is fast by biological standards, it is slow by comparison with the speed of electrical signals transmitted by a wire. The speed of signal transmission affects how nervous systems are organized and function.

The resting membrane potential is negative and results in part from the movement of potassium ions.

How are the electrical signals transmitted by animal nervous systems generated within each neuron? The key to producing an electrical signal in a nerve cell is the movement of positively and negatively charged ions across the cell's membrane. This movement creates a change in electrical charge across the membrane that constitutes the electrical signal.

There is a difference in electrical charge across the cell membrane because unequal numbers of charged ions are located outside and inside the cell. The difference in electrical charge is the voltage and is measured in volts. The charge difference between the inside and outside of a neuron due to differences in charged ions is the cell's **membrane potential**. Other cells also have membrane potentials, but only nerve cells and muscle cells (Chapter 37) respond to changes in their membrane potential. Therefore, these cells are considered electrically excitable. Changes in membrane potential provide the basis for transmitting signals between neurons and from neurons to muscle cells.

At rest, when no signal is being received or sent, the cell's membrane voltage is negative on its inside relative to its outside (**Fig. 35.8**). The resting membrane potential of the cell is said to be **polarized.** This means that there is a buildup of negatively charged ions on the inside surface of the cell's plasma membrane and positively charged ions on its outer surface. The negative voltage across the membrane at rest is referred to as the cell's **resting membrane potential.** The resting membrane potential ranges from -40 to -85 millivolts (mV) depending on the type of nerve cell and most commonly is about -65 to -70 mV. The voltage of the cell's interior can be measured with respect to its outside voltage by small glass electrodes on the inside and outside of the cell.

At rest, nerve (and muscle) cells have a greater concentration of sodium (Na^+) ions outside the cell and a greater concentration of potassium (K^+) ions inside the cell. This distribution of ions results in part from the action of the sodium-potassium pump, discussed in Chapter 5. The sodium-potassium pump uses the energy of ATP to move three Na^+ ions outside the cell for every two K^+ ions moved in (Fig. 35.8). The action of the pump makes the inside of the cell less positive, and therefore more negative, than the outside of the cell.

The exact value of the resting membrane potential, however, depends on the movement of K^+ ions back out of the cell by passive diffusion through potassium ion channels (Fig. 35.8). As discussed in Chapter 5, ion channels are protein pores embedded in the cell membrane's lipid bilayer. The most important ions that move across the cell membrane are sodium (Na^+), potassium (K^+), chloride (Cl^-), and calcium (Ca^{2+}). When a nerve cell is at rest, more K^+ ion channels are open, giving K^+ greater permeability compared with all other ions. As a result, K^+ ions move out of the cell. The movement of K^+ ions from the inside to the outside causes positive ions to build up on the outside of the cell, whereas negatively charged ions (largely Cl^- and proteins) remain on the inside of the cell, making the inside of the nerve cell more negative than its outside.

The relative proportion of ions does not by itself determine the resting membrane potential. This is because the number of charged ions that build up at the cell's surface is a tiny fraction of the total number of charged ions and proteins located inside and outside of the cell. It is the *movement* of K^+ ions relative to other ions, particularly Na^+ ions, that largely determines the resting membrane potential.

FIG. 35.8 **Neuron resting membrane potential.** The resting potential of a neuron is negative and results primarily from the movement of K^+ ions out of the cell.

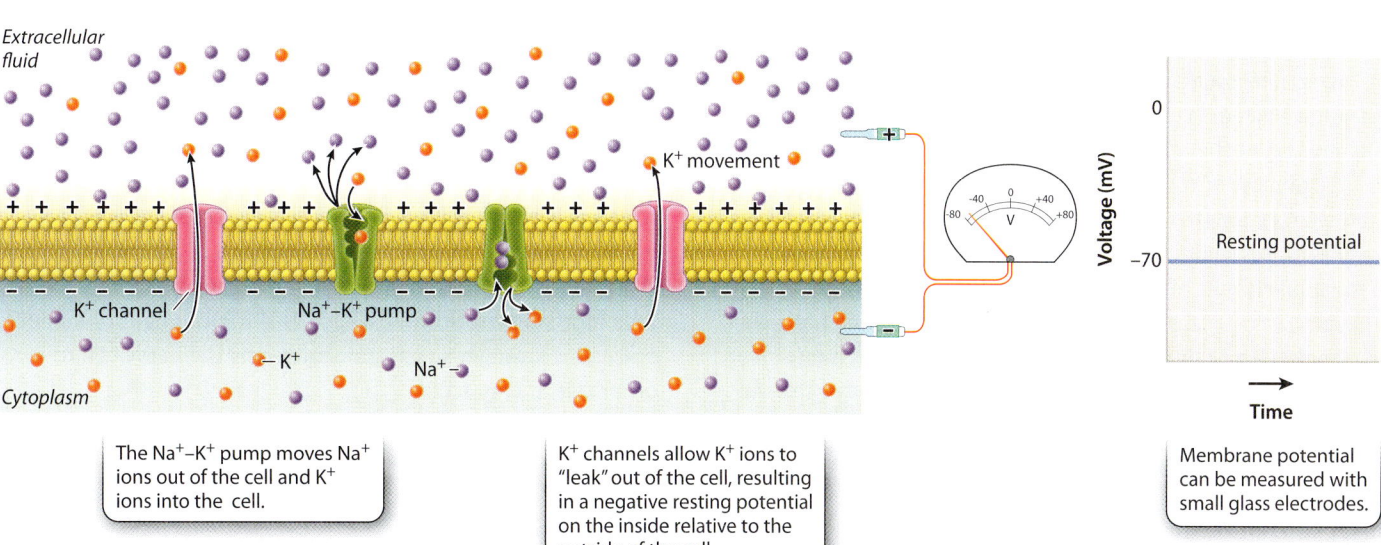

Neurons are excitable cells that transmit information by action potentials.

When a nerve cell is excited, its membrane potential becomes less negative—that is, the inside becomes less negative, or more positive, than the outside of the cell. The increase in membrane potential is therefore referred to as a **depolarization** of the membrane. At the axon hillock of the cell body, this depolarization causes voltage-gated sodium channels to open, allowing Na⁺ ions to enter the cell. Voltage-gated channels open and close in response to changes in membrane potential. Voltage-gated sodium channels are closed at the resting membrane potential but open rapidly in response to a depolarization of the cell membrane (**Fig. 35.9**).

The influx of Na⁺ causes a flow of positive charge, like a current, along the inside of the cell, toward more negatively charged nearby regions. This flow of charge is reduced at longer distances and quickly dissipates so that the membrane potential returns to a resting state unless additional excitatory stimuli further depolarize the membrane.

If the excitatory signal is strong enough to depolarize the membrane of the nerve cell body to a voltage of approximately 15 mV above the resting membrane potential (about −50 mV), the nerve fires an **action potential** at the axon hillock. An action potential is a rapid, short-lasting rise and fall in membrane potential, as shown in Fig. 35.9. The critical depolarization voltage of −50 mV required for an action potential is the cell's **threshold potential.** Nerve cells in different animals have slightly different resting and threshold potentials, but all operate similarly with respect to firing an action potential. When the threshold potential is exceeded, the nerve fires an action potential in an all-or-nothing fashion. This means that once the threshold potential has been exceeded, the magnitude of the action potential is always the same and is independent of the strength of the stimulating input. At this point, a rapid spike in voltage to approximately +40 mV occurs over a very brief time period, 1 to 2 msec.

An action potential results from dramatic changes in the state of voltage-gated Na⁺ and voltage-gated K⁺ channels in the axon membrane. As the cell membrane crosses its threshold potential, a large number of voltage-gated Na⁺ channels suddenly open, allowing Na⁺ ions to rush inside the cell. The rise in voltage that results causes additional voltage-gated Na⁺ channels to open. This is an example of **positive feedback,** in which a signal (depolarization) causes a response (open voltage-gated Na⁺ channels) that leads to an enhancement of the signal (more depolarization) that leads to an even larger response (more open voltage-gated Na⁺ channels). During the rising phase of the action potential, voltage-gated K⁺ channels remain closed, keeping K⁺ ions inside the cell.

As you can see from Fig. 35.9, the rising phase of the action potential (rapid depolarization) is followed by the falling phase (rapid repolarization). What causes this sudden reversal of the membrane potential? There are two important factors.

First, voltage-gated Na⁺ channels close. These gated Na⁺ channels do not close in response to the change in voltage, but instead close automatically after a brief period of time. Second, voltage-gated K⁺ channels open. These channels, like voltage-gated Na⁺ channels, open in response to the change in voltage, but they are slower to respond than the voltage-gated Na⁺ channels. As a result of the closing of voltage-gated Na⁺ channels and the opening of voltage-gated K⁺ channels, the membrane voltage peaks and then rapidly falls as K⁺ ions diffuse out of the axon through open voltage-gated K⁺ channels.

The voltage inside the axon does not return immediately to the resting membrane potential. Instead, it briefly falls below the resting potential (in what is known as hyperpolarization or undershoot), then returns to the resting potential after another few milliseconds as K⁺ channels close to restore the resting concentration of Na⁺ and K⁺ ions on either side of the cell membrane. The continuous action of the sodium-potassium pumps also helps to reestablish the resting membrane potential.

The period during which the inner membrane voltage falls below and then returns to the resting potential is the **refractory period** (Fig. 35.9). During the refractory period, a neuron cannot fire a second action potential. The refractory period results in part from the fact that when voltage-gated Na⁺ channels close, they require a certain amount of time before they will open again in response to a new wave of depolarization. In addition, open voltage-gated K⁺ channels make it difficult for the cell to reach the threshold potential. The duration of the refractory period varies for different kinds of nerve cells but limits a nerve cell's fastest firing frequency to less than 200 times per second. Most nerve cells fire at lower rates.

It might seem that many Na⁺ and K⁺ ions need to diffuse across the axon membrane to produce an action potential. However, only about 1 in 10 million available Na⁺ and K⁺ ions need to cross the membrane. Thus, enough ions are always available to generate repeated action potentials.

The nature of the action potential, which is both all-or-nothing and stereotyped, means that the action potential itself contains little information. Most neurons code information by changing the rate and temporal pattern of action potentials that they transmit. Generally, a higher firing frequency codes for a more intense stimulus, such as a brighter light or louder noise, or a stronger signal transmitted by the nerve cell to other cells it contacts.

Neurons propagate action potentials along their axons by sequentially opening and closing adjacent Na⁺ and K⁺ ion channels.

Once initiated, an action potential propagates along the axon. The local membrane depolarization initiated at the axon hillock triggers the opening of neighboring voltage-gated Na⁺ channels

FIG. 35.9 **An action potential.** An action potential results from the opening and closing of voltage-gated ion channels and allows the membrane potential briefly to be positively charged.

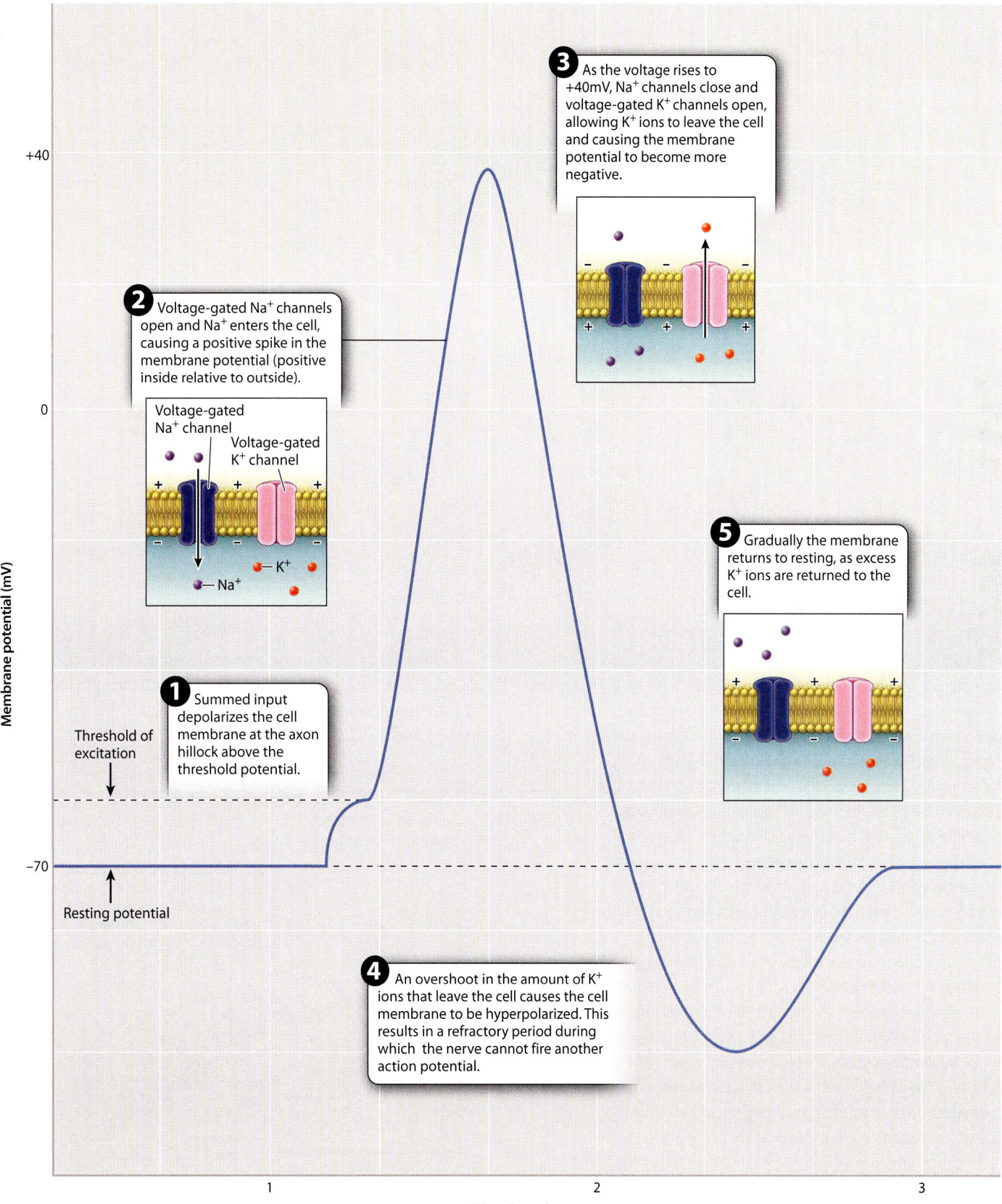

FIG. 35.10 Propagation of action potentials. (a) Local membrane potential depolarization triggers the opening of nearby voltage-gated Na⁺ channels, producing an action potential that spreads along the membrane. (b) The action potential moves in only one direction (shown here as left to right) because of the refractory period.

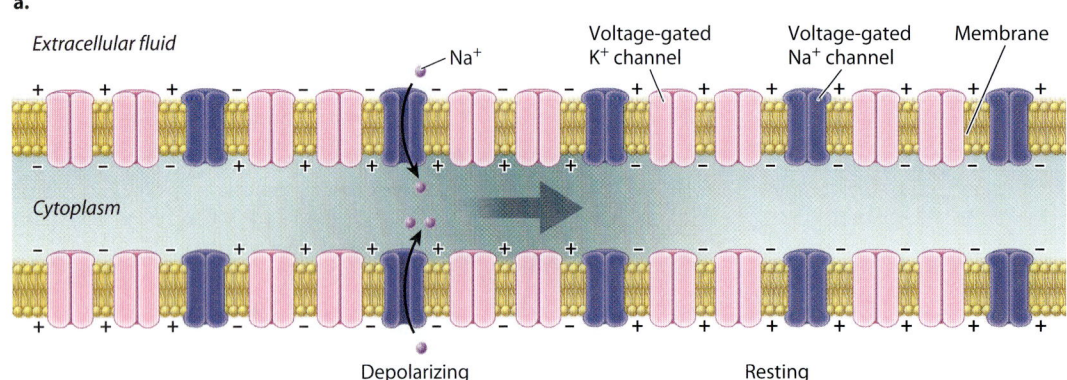

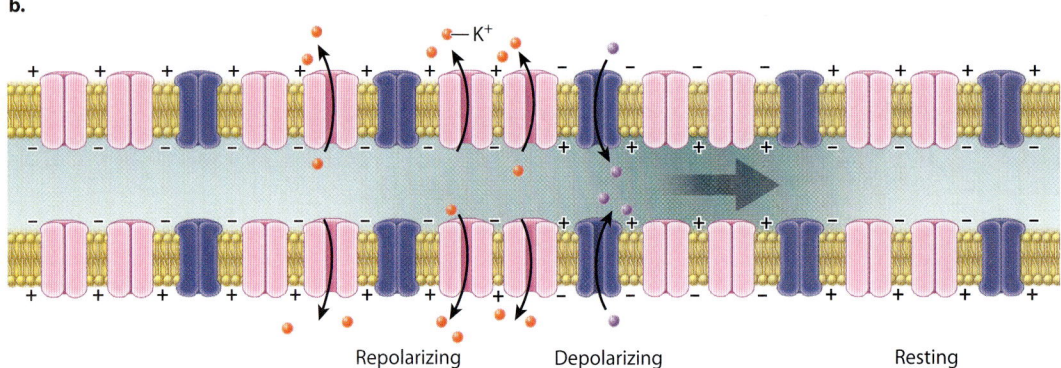

farther along the axon (**Fig. 35.10**). The inward sodium current depolarizes the membrane above threshold, triggering the opening of nearby voltage-gated Na⁺ channels still farther along the axon. By this means, the depolarization—the location of the action potential spike—spreads down the axon (Fig. 35.10a). Neighboring voltage-gated K⁺ channels subsequently also open and close to reestablish a resting membrane potential after an action potential has fired. Action potentials are thus **self-propagating**. Action potentials propagate only in one direction, normally from the cell body at the axon hillock to the axon terminal. The refractory period following an action potential prevents the membrane from reaching threshold and firing an action potential in the reverse direction (Fig. 35.10b).

The conduction speed of action potentials is limited by the neuron's membrane properties and the size of its axon. Yet fast transmission speeds are essential for enabling rapid responses to stimuli that may require communication over long distances within an animal's body. For example, when you mistakenly touch something very hot, your fingers sense and relay this information rapidly to ensure that you withdraw your hand quickly and avoid worse injury.

Vertebrate axons, as we have seen, are wrapped in myelin (see Fig. 35.7). The myelin sheath insulates the axon's membrane, spreading the charge from a local action potential over a much greater distance along the axon's length than would be possible in

FIG. 35.11 Saltatory conduction. Myelination insulates axons, enabling faster conduction of action potentials.

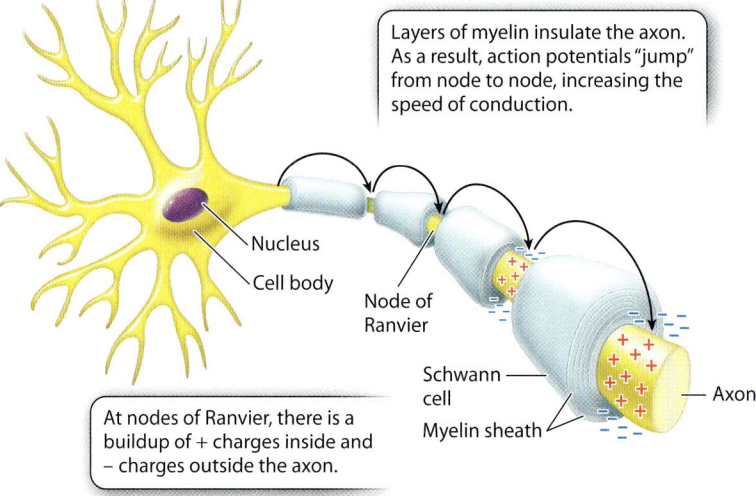

the absence of myelin. At regular intervals, the axon membrane is exposed at sites called **nodes of Ranvier** that lie between adjacent segments wrapped with myelin (**Fig. 35.11**). Voltage-gated Na⁺ and K⁺ channels are concentrated at these nodes. As a result, rather than being conducted in a continuous fashion, as is the case for unmyelinated axons, the action potentials in myelinated axons "jump" from node to node. This **saltatory propagation** (from the Latin *saltus*, "jump") of axon potentials by myelinated axons greatly increases the speed of signal transmission, enabling vertebrate animals to respond rapidly to stimuli in their environment.

→ **Quick Check 2** Multiple sclerosis is a disease in which the immune system attacks and destroys the myelin sheath surrounding nerve cells. What effect would you expect the loss of myelin to have on the speed of nerve impulses?

The British neurophysiologists Alan Hodgkin and Andrew Huxley first worked out the electrical properties of the axon plasma membrane by studying the giant axon in squid that stimulates contractions of their body muscles for swimming (**Fig. 35.12**). For this work, they shared the 1963 Nobel Prize in

HOW DO WE KNOW?

FIG. 35.12

What is the resting membrane potential and what changes in electrical activity occur during an action potential?

BACKGROUND In order to record the voltage of the inside of a nerve cell relative to the outside, a small electrical recording device, a microelectrode, was inserted into a neuron. The technique is more easily performed on large cells, such as the squid giant axon. This axon, as its name suggests, is quite large, measuring 0.5 mm in diameter (Fig. 35.12a). The large diameter of the axon allows electrical signals to be propagated to the muscles very quickly. Vertebrate neurons are much smaller; they rely on myelin sheaths rather than large size for rapid conduction of electrical signals.

EXPERIMENT In 1939, two British neurophysiologists, Alan Hodgkin and Andrew Huxley, inserted a microelectrode into a squid giant axon with a reference electrode on the outside (Fig. 35.12b). They then used a separate set of electrodes (not shown) to depolarize the cell to threshold, triggering an action potential.

RESULTS Fig. 35.12c is a trace from Hodgkin and Huxley's 1939 paper showing the resting membrane potential and the course of an action potential recorded by the electrode inside a giant squid axon. Note that the resting potential is negative on the inside of the axon relative to the outside, and that the action potential is a rapid spike in potential, with the inside of the cell quickly becoming positive, then negative again. The large size of the spike was a surprise.

FOLLOW-UP WORK This work was performed before the ion channels responsible for the changes in current across the membrane were identified. Subsequent work focused on identifying these channels and their roles in generating the action potential.

SOURCES Hodgkin, A. L., and A. F. Huxley. 1939. "Action Potentials Recorded from Inside a Nerve Fibre." *Nature* 144:710–711; R. D. Keynes. 1989. "The Role of Giant Axons in Studies of the Nerve Impulse." *Bioessays* 10:90–94.

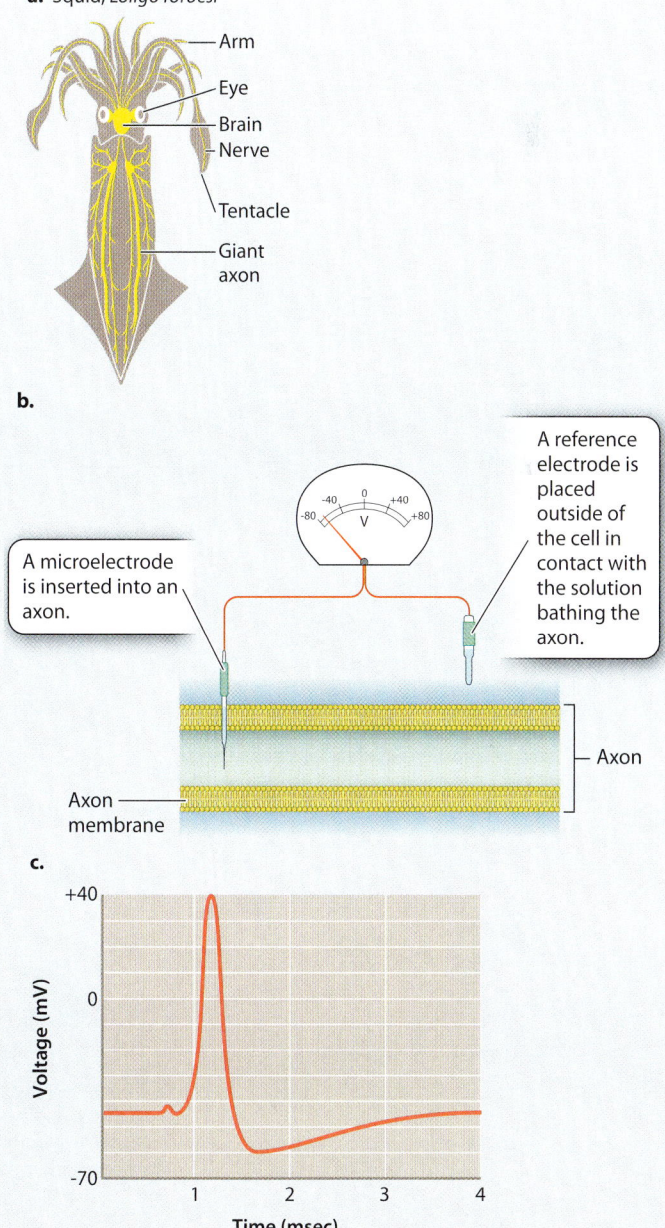

Physiology or Medicine along with John Eccles, who worked out the membrane properties of synapses, which we explore next.

Neurons communicate at synapses.

We have seen that nerve cells communicate at specialized junctions called synapses. There are two types of synapse, electrical and chemical. Electrical synapses provide direct electrical communication through gap junctions that form between neighboring cells (Chapter 10). They enable rapid communication but limit the ability to process and integrate information. Electrical synapses are found in the giant axons of squid, as well as in large motor nerves in fish for escape swimming. Electrical synapses can also be found in the mammalian brain. They likely help to speed up information processing, but their role in the brain has not been well studied.

Chemical synapses are by far the more common type of synapse in animal nervous systems. The signals conveyed at chemical synapses are chemicals called neurotransmitters, which are contained within small vesicles in the knoblike axon terminal. When an action potential reaches the end of an axon (**Fig. 35.13**), the resulting depolarization induces voltage-gated Ca^{2+} ion channels to open. These channels are found only in the axon terminal membrane. Because of their higher concentration outside the cell, Ca^{2+} ions diffuse through these channels into the axon terminal. In response to the rise in Ca^{2+} concentration, the vesicles fuse with the presynaptic membrane and release neurotransmitter molecules into the synaptic cleft by exocytosis. The neurotransmitters

FIG. 35.13 A chemical synapse. Neurons communicate with other neurons or with muscle cells by releasing neurotransmitters at a synapse.

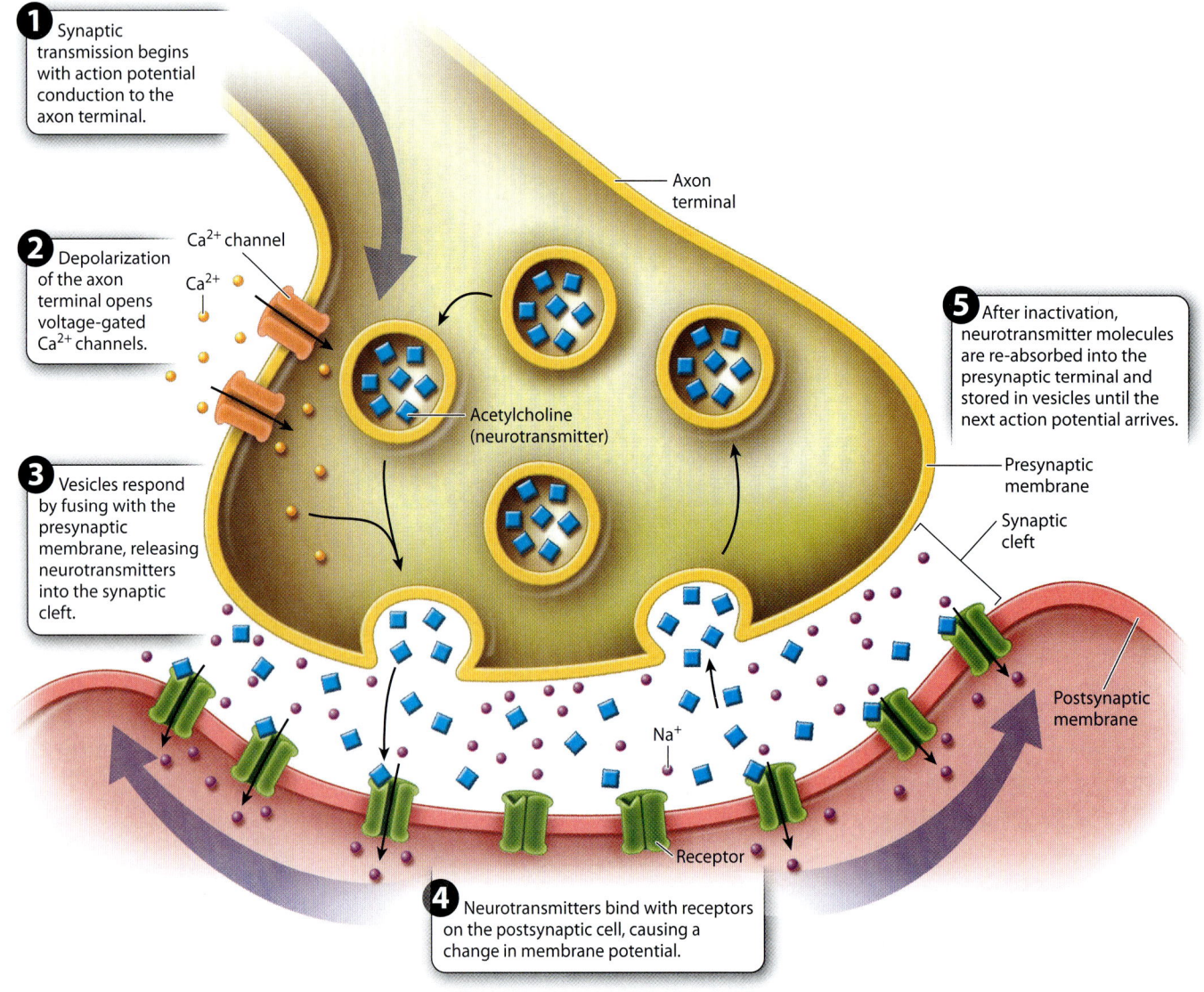

diffuse rapidly across the cleft and bind to postsynaptic membrane receptors of the neighboring cell. The binding of neurotransmitters causes a change in the postsynaptic cell membrane potential, allowing the signal to propagate along the next neuron.

The neurotransmitter's effect on the postsynaptic membrane ends shortly after binding. Neurotransmitters become unbound from the receptor, and neurotransmitters in the synaptic cleft are either broken down by a deactivating enzyme or taken up again by the presynaptic cell and reassembled in newly formed vesicles. The structural and chemical properties of the synapse ensure that communication proceeds in a single direction between nerve cells.

Signals between neurons can be excitatory or inhibitory.

Neurotransmitters bound to postsynaptic membrane receptors can either stimulate or inhibit the firing of action potentials in the postsynaptic neuron. The binding of some neurotransmitters depolarizes the postsynaptic membrane, making it more positive. The positive change in membrane potential is termed an **excitatory postsynaptic potential,** or **EPSP.** In contrast, the binding of other neurotransmitters can hyperpolarize the postsynaptic membrane, making it more negative. In this case, the negative change in membrane potential is called an **inhibitory postsynaptic potential,** or **IPSP.** Membrane depolarization makes a neuron more likely to respond and transmit an action potential, whereas membrane hyperpolarization inhibits the neuron from sending an action potential.

Excitatory synapses tend to transmit relevant information between nerve cells, and inhibitory synapses often serve to filter out unimportant information. For example, when a grazing gazelle senses a predator, its nervous system must transmit the key sensory information about the presence of a predator but filter out irrelevant information such as the flies buzzing over its back and ears. Inhibitory synapses are also important for controlling the timing of muscle activity required for coordinated movement, discussed in Chapter 37.

How does neurotransmitter binding change the postsynaptic membrane potential? A neurotransmitter binds to receptors that trigger the opening or closing of ligand-gated ion channels in the postsynaptic membrane. The binding of excitatory neurotransmitters causes Na$^+$ channels to open. The postsynaptic membrane becomes depolarized because the opening of Na$^+$ channels allows Na$^+$ ions to enter the cell. In contrast, the binding of inhibitory neurotransmitters causes Cl$^-$, or sometimes K$^+$, channels to open. Cl$^-$ ions diffuse into the cell or K$^+$ ions diffuse out of the cell through these channels, causing the membrane potential to become more negative, or hyperpolarized, than the outside.

Different types of nerve cell contain different neurotransmitters, but each type of nerve cell releases only one type of neurotransmitter. A particular type of nerve cell exerts either

FIG. 35.14 Integration of information. Multiple synapses—often hundreds of them—from communicating neurons enable a neuron to integrate diverse sources of information.

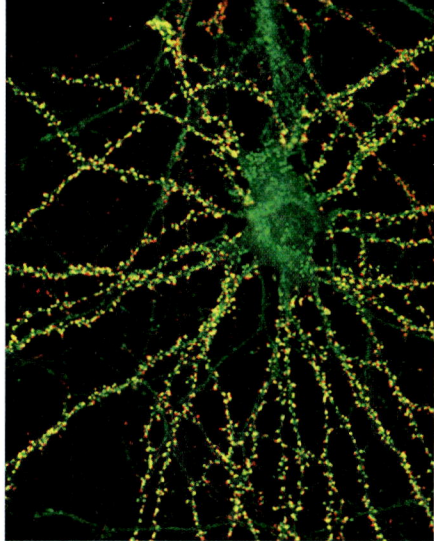

Each yellow-green dot represents an axon terminal from another neuron synapsing on the dendrites of this (green) postsynaptic cell.

excitatory or inhibitory effects on neighboring cells, but not both. In contrast, postsynaptic membranes typically have multiple types of membrane receptors that bind different kinds of neurotransmitters. Thus, postsynaptic cells can receive both excitatory and inhibitory signals. Commonly, the dendrites or cell body of a single postsynaptic nerve cell make connections to hundreds or even thousands of axons from other nerve cells (**Fig. 35.14**).

The postsynaptic nerve cell sums the excitatory and inhibitory synaptic inputs (the summed EPSPs and IPSPs) that it receives through its dendrites and cell body (**Fig. 35.15**). If the sum of synaptic stimulation results in a membrane potential that exceeds threshold at the axon hillock, the neuron fires an action potential, communicating an impulse to cells that it in turn contacts. If the sum of inputs does not exceed threshold, no action potential fires (Fig. 35.15a).

EPSPs and IPSPs can be summed over time and space. When summed over time, the frequency of synaptic stimuli determines whether the postsynaptic cell fires an action potential (Fig. 35.15b). This is referred to as **temporal summation.** When summed over space, the number of synaptic stimuli received from different regions of the postsynaptic cell's dendrites determines if the cell fires an action potential (Fig. 35.15c). This is referred to as **spatial summation.** Sometimes, excitatory and inhibitory signals may cancel each other out (Fig. 35.15d). Temporal and spatial summation of EPSPs and IPSPs are the fundamental forms of information processing carried out by the nervous system. They provide mechanisms for determining whether a particular sensory

FIG. 35.15 Summation of excitatory and inhibitory postsynaptic potentials. EPSPs and IPSPs can be summed in time (b) or space (c), or even cancel each other out (d). Widely spaced EPSPs (a) do not sum.

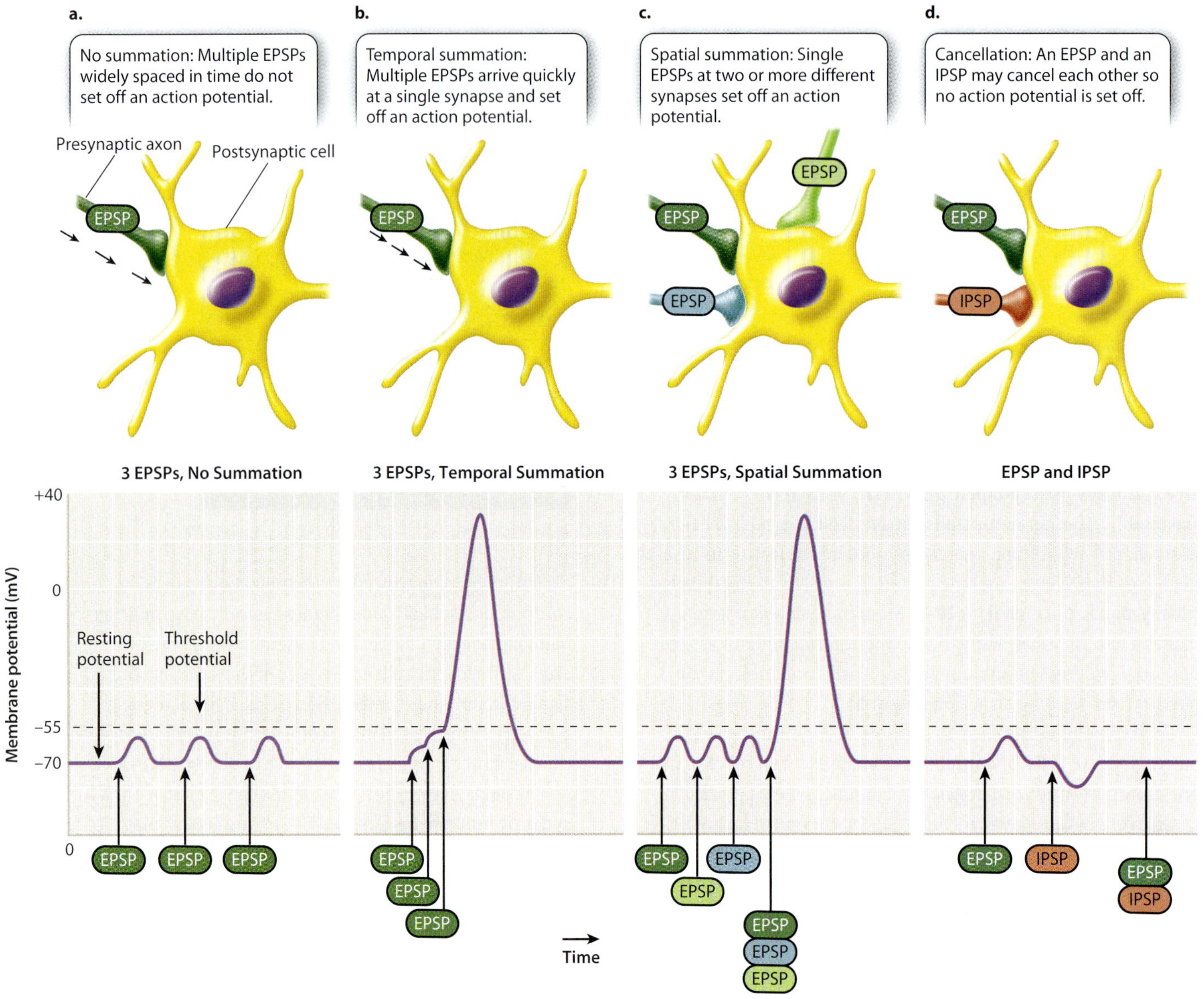

a. No summation: Multiple EPSPs widely spaced in time do not set off an action potential.

b. Temporal summation: Multiple EPSPs arrive quickly at a single synapse and set off an action potential.

c. Spatial summation: Single EPSPs at two or more different synapses set off an action potential.

d. Cancellation: An EPSP and an IPSP may cancel each other so no action potential is set off.

stimulus is responded to or ignored. The role of temporal and spatial summation is explored further in Chapter 36.

More than 25 neurotransmitters are now recognized, and more are likely to be discovered. The amino acids glutamate (excitatory), glycine (inhibitory), and GABA (inhibitory) are key neurotransmitters that operate at synapses in the brain. The related amino acid derivatives dopamine, norepinephrine, and serotonin are also neurotransmitters acting in the brain. Simple peptides serve as neurotransmitters for sensory neurons. More recently, two gases, nitrous oxide and carbon monoxide, have been discovered to function as messengers between nerve cells in various regions of the nervous system, even though they do not behave as standard neurotransmitters that bind to membrane receptors.

Acetylcholine is one of the key neurotransmitters produced in a variety of nerve cells. It is the excitatory neurotransmitter released by motor neurons to stimulate muscle fibers at a type of synapse called the motor endplate (**Fig. 35.16**). All vertebrate motor synapses rely on acetylcholine and are therefore excitatory only. The release of acetylcholine produces a depolarization of the muscle cell membrane, causing the muscle to contract.

→ **Quick Check 3** Acetylcholinesterase is the enzyme that breaks down and inactivates acetylcholine. Some nerve gases used as chemical weapons block acetylcholinesterase. What effect would such nerve gases have on muscle contraction?

FIG. 35.16 Vertebrate motor endplate.
The motor endplate is an excitatory synapse between a motor nerve and a muscle cell; acetylcholine is the neurotransmitter.

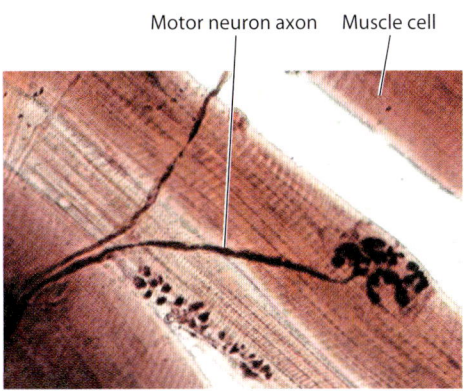

Acetylcholine binds to muscle membrane receptors, causing a depolarization of the muscle cell and contraction.

35.4 NERVOUS SYSTEM ORGANIZATION

The brain is considered the body's command center: It receives sensory information from the eyes, ears, and skin and issues instructions to the rest of the body. Up to this point, we have examined how individual neurons send signals to other nerves and muscles. How are these neurons organized in the body to sense stimuli, process them, and issue an appropriate response? How does the brain coordinate the movement of limbs necessary to walk or run, and adjust heart rate and breathing in response to exercise? In this section, we look at how the nervous system is organized to allow the brain and the body to communicate with each other.

Nervous systems are organized into peripheral and central components.

As animals evolved the ability to sense and coordinate responses to increasingly complex stimuli, their nervous systems became organized into peripheral and central components (**Fig. 35.17**). Not

FIG. 35.17 Human central (yellow) and peripheral (blue) nervous systems.

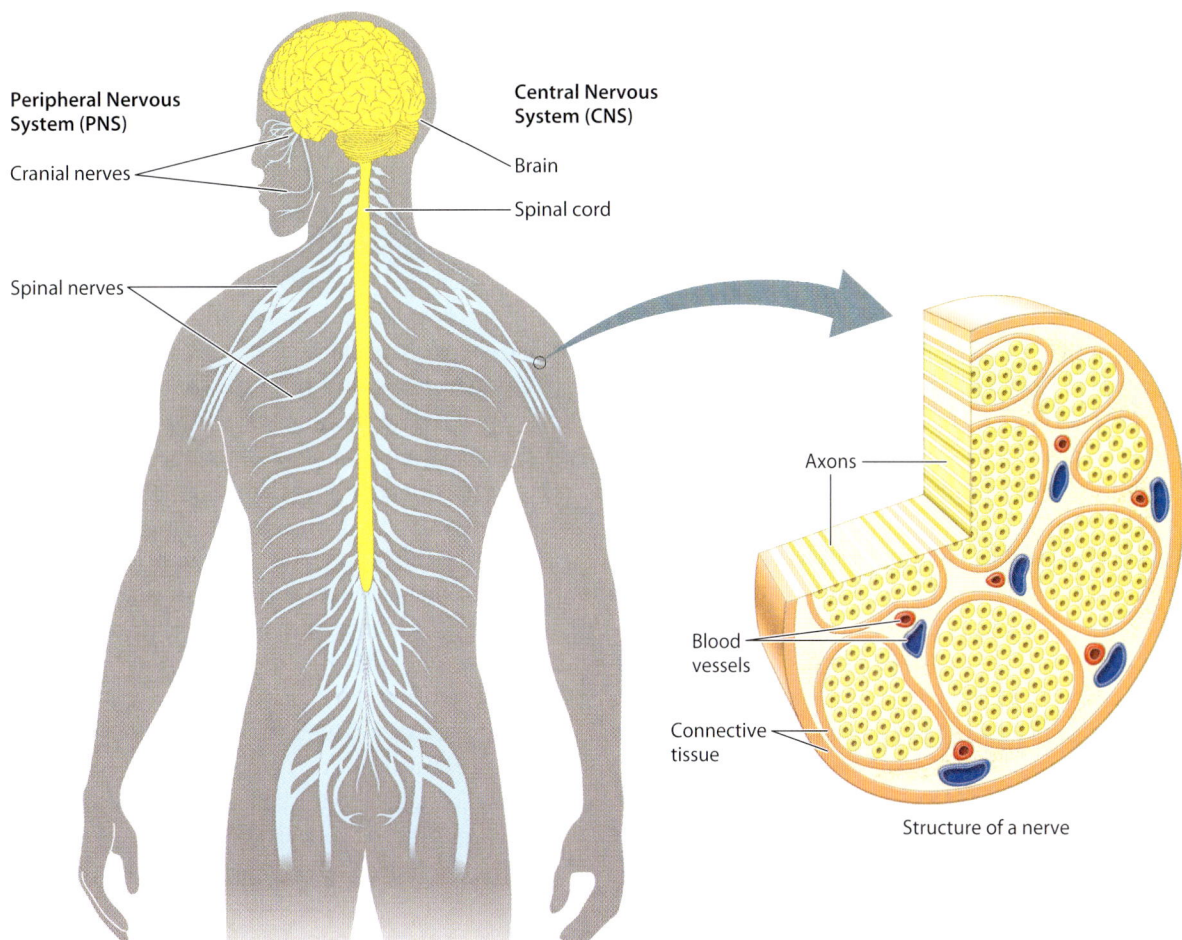

surprisingly, your eyes, sense of touch, and other sensory organs are located on the surface of the body, where they can receive signals from the environment. In contrast, the brain, centrally located ganglia, and a main nerve cord are located in the interior.

How do structures in the interior and near the surface communicate? Nerves form the lines of communication between these nervous system structures. In general, neuron cell bodies are grouped compactly together in sensory organs, ganglia, and a main nerve cord that extends from the brain. In contrast, nerves are composed mainly of axons from many different nerve cells. For example, the optic nerve contains axons that travel from nerve cells in the eye to the brain.

Sensory and motor nerves make up the **peripheral nervous system (PNS).** These nerves communicate with the brain, a main nerve cord, or—in the case of organisms such as flatworms that lack a brain—with centralized information-processing ganglia. The brain, main nerve cord, and centralized ganglia together make up the **central nervous system (CNS).** Information gathered by sensory organs arrives mainly from the animal's periphery, at or near its body surface. After processing the information, the ganglia and brain send commands back to peripheral nerves that coordinate the activity of muscles and glands in different regions of the animal's body. This organization ensures that an animal achieves a coordinated behavioral and physiological response to the stimuli received from its environment. Only the diffuse nerve nets of cnidarians lack defined central and peripheral components.

Specialized sensory organs transmit information from the animal's periphery by **afferent neurons,** which send information toward the CNS. **Efferent neurons** send signals away from the CNS, communicating with muscles and other tissues and organs. The peripheral nervous system also includes interneurons and ganglia that integrate and process information in local regions of the animal's body. For example, vertebrates and invertebrates have ganglia that lie outside each segment of their primary nerve cord. However, the bulk of information processing, particularly in animals that exhibit a greater capacity for learning and memory, occurs in the CNS.

The central nervous system of vertebrates includes both the brain and **spinal cord** (Fig. 35.17). The spinal cord is a central tract of neurons that passes through the vertebrae to transmit information between the brain and the periphery of the body. The vertebrate spinal cord is divided into segments, each controlling body movement in a particular region along the animal's length. Each spinal cord segment contains axons from peripheral sensory neurons, a set of interneurons, and a set of motor neuron cell bodies. These are distinct from, but often associated with, segmental ganglia that lie outside the spinal cord.

In humans and other vertebrate animals, the peripheral nervous system is organized into left and right sets of **cranial nerves** located within the head and **spinal nerves** running from the spinal cord to the periphery. Most of the cranial nerves and all of the spinal nerves contain axons of both sensory and motor neurons. Cranial nerves link specialized sensory organs (eyes, ears, tongue) to the brain. Cranial nerves also control eye movement, facial expression, speech, and feeding. Some cranial nerves, such as the olfactory and optic nerves, contain only sensory axons. Spinal nerves exit from the spinal cord through openings located between adjacent vertebrae to thread through the trunk and limbs of an animal's body. These nerves receive sensory information from receptors in nearby body regions along the length of the body and carry motor signals from the spinal cord back to those regions (Fig. 35.17).

Nervous systems have voluntary and involuntary components.

As bodies with distinct internal organ systems evolved, two separate components of the nervous system emerged. When a gazelle senses a predator, some nerve circuits send a signal to run, an action that is under conscious control, while others signal the heart to beat faster and blood vessels supplying muscles to dilate, actions that occur unconsciously. Conscious reactions are under the control of the **voluntary** component of the nervous system, and unconscious ones are under the control of the **involuntary** component. Voluntary components mainly handle sensing and responding to external stimuli, whereas involuntary components typically regulate internal bodily functions. Both nervous system components are found in invertebrate and vertebrate animals.

Let's look at these components in insects and crustaceans. In these animals, a nerve circuit regulates foregut function together with nerve circuits that regulate other parts of the animal's digestive system. These circuits are an involuntary component of an animal's nervous system. Sensory structures such as the antennae and eyes transmit information to the animal's brain, which allows voluntary responses to stimuli received by these sensory organs. Other invertebrates such as mollusks also have involuntary components that regulate the function of internal organ systems that are distinct from their voluntary sensory–motor system.

In vertebrates, the peripheral nervous system is divided into **somatic** (voluntary) and **autonomic** (involuntary or visceral) components. The somatic nervous system is made up of sensory neurons that respond to external stimuli and motor neurons that synapse with voluntary muscles. This system is considered voluntary because it is under conscious control. However, many reflexes are controlled at lower spinal levels (discussed in the following section) or by the brainstem, independent of conscious control by the central nervous system.

The autonomic nervous system (**Fig. 35.18**) controls internal functions of the body such as heart rate, blood flow, digestion, excretion, and temperature. It includes both sensory and motor components, which usually act without our conscious awareness. The autonomic nervous system, in turn, is divided into two major subdivisions, a **sympathetic division** and a **parasympathetic division.** Both divisions continuously monitor and regulate internal functions of the body. Generally, sympathetic and parasympathetic nerves have opposite effects, but not always. For example, sympathetic neurons stimulate the heart to beat faster, whereas parasympathetic neurons cause the heart to beat slower.

The sympathetic nervous system generally results in arousal and increased activity. It is the pathway activated when animals are exposed to threatening conditions, resulting in what is often referred to as the fight-or-flight response. This response includes an increase in the heart and breathing rates, increase in glucose release by the liver, and inhibition of digestion. The parasympathetic nervous system, in contrast, slows the heart and stimulates digestion as well as metabolic processes that store energy—in other words, it enables the body to "rest and digest."

The sympathetic and the parasympathetic nerves have different anatomical distributions (Fig. 35.18) as well as different functions. Sympathetic nerves leave the CNS from the middle region of the spinal cord, forming ganglia along much of the length (thoracic and lumbar) of the spinal cord. Parasympathetic nerves leave from the brain by cranial nerves, as well as from lower levels (sacral) of the spinal cord.

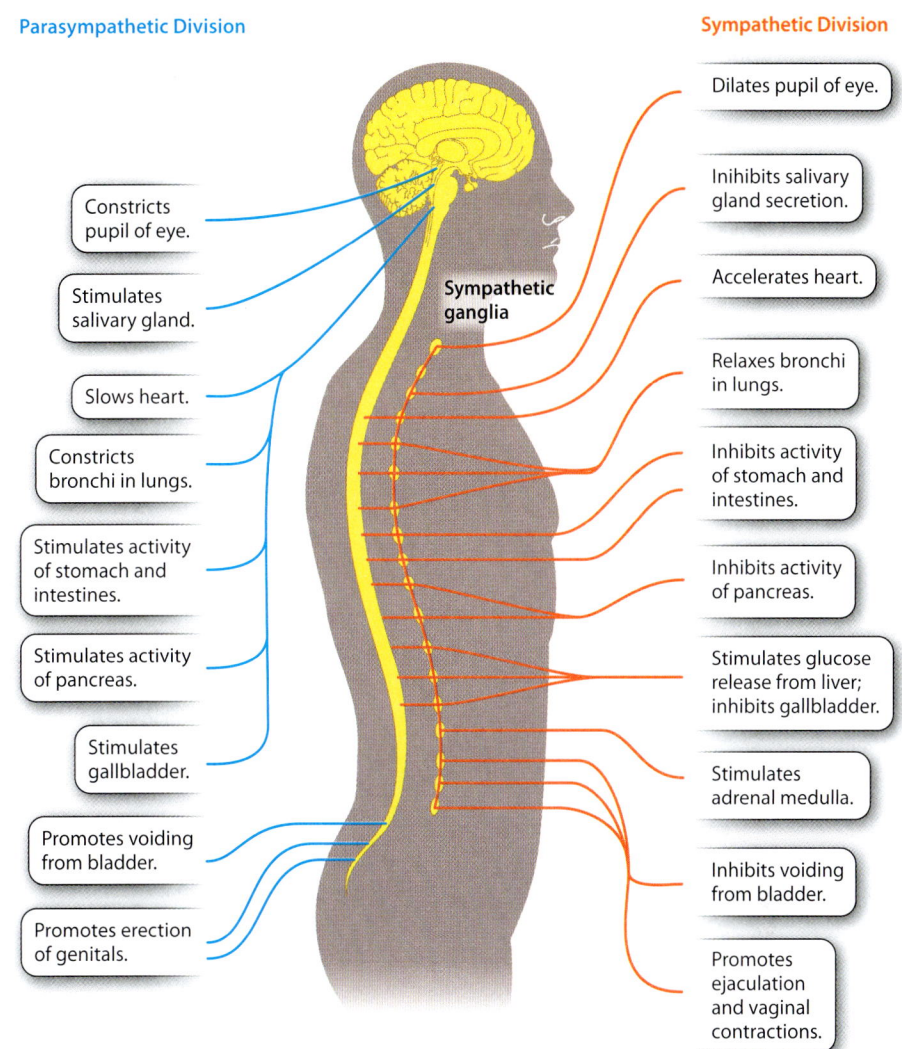

FIG. 35.18 Function and organization of the human involuntary (autonomic) nervous system. The autonomic system, which controls involuntary functions, is divided into sympathetic and parasympathetic systems.

The nervous system helps to maintain homeostasis.

The autonomic nervous system plays a key role in homeostasis, which is the ability to maintain a steady physiological state in the face of changing environmental conditions. Many physiological parameters are maintained in a narrow range of conditions, including pH, temperature, and ion concentrations. Homeostasis is maintained within individual cells as well as within the organism as a whole. Enzymes within a cell, for example, often work effectively only in narrow pH and temperature ranges. Ion concentrations also must be maintained within narrow ranges for normal cell function. As we have seen, the firing of action potentials by neurons requires particular ion concentrations on either side of the membrane. Similarly, the water content of cells and the body as a whole is kept stable through the careful regulation of ions and other solutes, as discussed in Chapter 41.

The concept of homeostasis was first described as regulation of the body's "interior milieu" in the late 1800s by the French physiologist Claude Bernard, who is often credited with bringing the scientific method to the field of medicine. The term "homeostasis" was coined by the American physiologist Walter Cannon, whose book *The Wisdom of the Body* (first published in 1932) popularized the concept.

Maintaining steady and stable conditions takes work in the face of changing environmental conditions. That is, a cell (or organism) actively maintains homeostasis. How does the body, and in particular the nervous system, maintain homeostasis? Homeostatic regulation often depends on **negative feedback**

(**Fig. 35.19**). In negative feedback, a stimulus acts on a sensor that communicates with an effector, which produces a response that opposes the initial stimulus. For example, negative feedback is used to maintain a constant temperature in a house. Cool temperature (the stimulus) is detected by a thermostat (the sensor). The thermostat sends a signal to the heater (the effector), producing heat (the response). In this way, a stable temperature is maintained (Fig. 35.19a).

In a similar fashion, humans and other mammals maintain a steady body temperature even as the temperature outside fluctuates. Nerve cells in the hypothalamus (located in the base of the brain) act as the body's thermostat (Fig. 35.19b). A drop in body temperature signals the hypothalamus to activate the somatic nervous system to induce shivering and metabolic heat, as discussed in Chapter 40. At the same time, the hypothalamus activates the autonomic nervous system, causing peripheral blood vessels to constrict. The reduction in blood flow near the body's surface in turn reduces heat loss to the surrounding air. An increase in temperature signals sweat glands and vasodilation of peripheral blood vessels to aid heat loss from the skin.

Homeostatic regulation, therefore, relies on negative feedback to maintain a set point, which in this case represents an animal's preferred body temperature. The ability to maintain a constant body temperature is known as thermoregulation, and it is just one of many parameters that the body actively maintains, as we discuss in subsequent chapters.

Simple reflex circuits provide rapid responses to stimuli.

An animal that has perceived a predator has an advantage if it can move more quickly than the predator. Fast responses are made possible by simple reflex circuits that bypass the brain by directly connecting sensory neurons (detecting the presence of the predator) with motor neurons (the quick movement necessary to escape the predator). Reflex circuits are common in both the somatic and autonomic components of the nervous system.

Simple spinal reflex circuits connect sensory and motor neurons directly in the spinal cord so that a vertebrate can respond rapidly to sensory stimuli without the need for delay-causing conscious input from the brain. The large Mauthner neurons of fish are an example of such a circuit. Sensory neurons respond to threatening visual, tactile, or vibratory cues from the environment and transmit signals to the hindbrain of a fish, activating a large Mauthner neuron on the side opposite the stimulus. The Mauthner cell in turn rapidly activates motor neurons along the body, causing the fish to bend away from the threatening stimulus to initiate a rapid escape. At the same time, the Mauthner cell and motor nerves on the other side of the fish are inhibited. The giant axons of squid are part of a similar reflex circuit that allows them to swim quickly away from a predator.

The **knee-jerk reflex** in humans is an example of a simple nerve circuit that includes only a single synapse between two neurons—a sensory neuron and a motor neuron (**Fig. 35.20**). Physicians commonly use this reflex to evaluate peripheral nervous and muscular system function. Let's follow this reflex pathway. It starts with specialized "stretch" receptors in the extensor muscles of the leg, including the quadriceps muscle. These stretch receptors sense the stretch of the muscle that occurs during movement or in response to a physician's strike of a reflex hammer on the stretch receptors. The receptors are part of a sensory nerve with dendrites that extend from the stretch receptors to cell bodies in ganglia

FIG. 35.19 Temperature regulation by negative feedback in (a) a house and (b) a mammal. In negative feedback, a response (such as heat) opposes the stimulus (cold), leading to a stable state (a steady temperature).

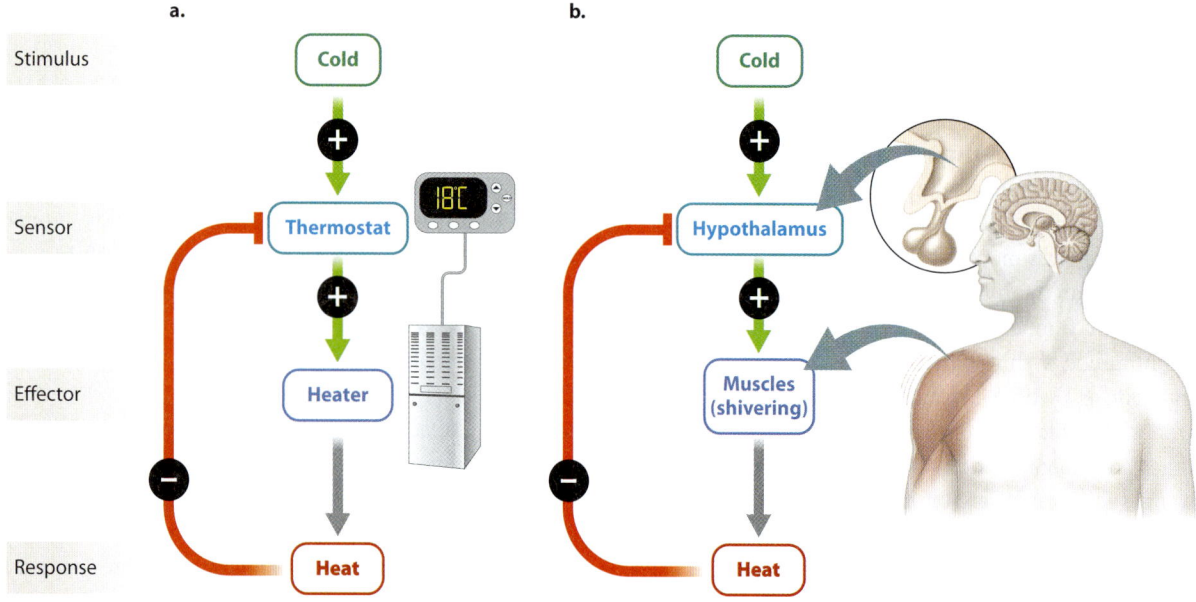

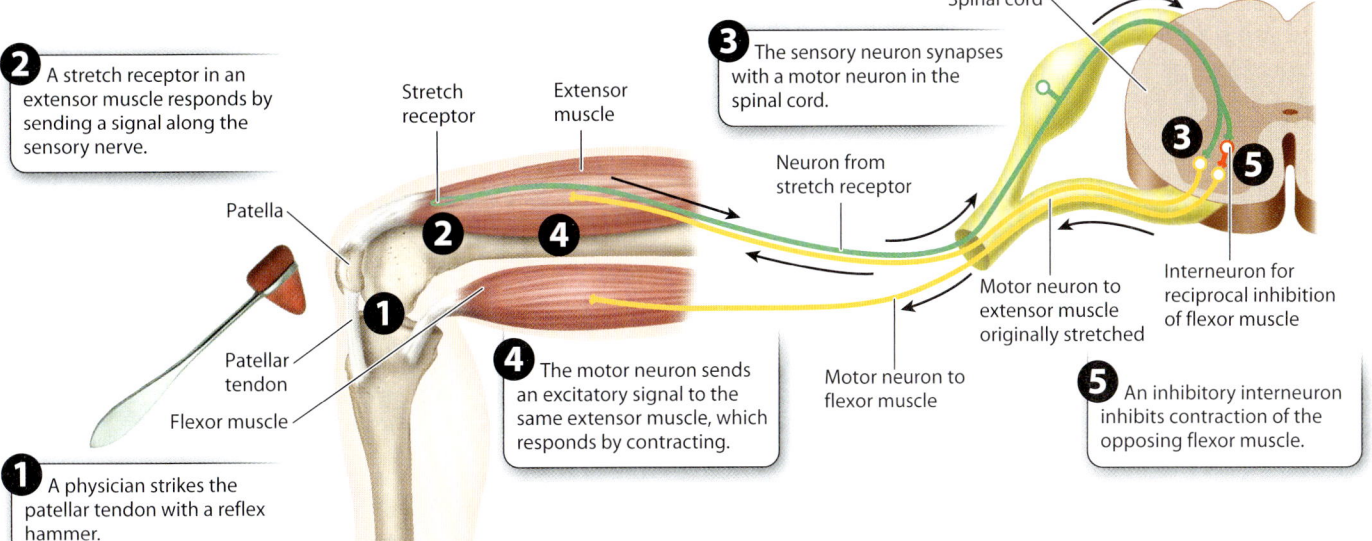

FIG. 35.20 The knee-jerk reflex. The knee-jerk reflex involves a sensory neuron, a single synapse, and a motor neuron.

beside the spinal cord. These cell bodies have axons that extend into the spinal cord. In response to a stretch, a signal is sent from the stretch receptor, through dendrite and cell body, to the axon. In the spinal cord, the axon of the sensory neuron synapses directly on motor neurons that travel from the spinal cord back to the muscle where the stretch originated. The signal from the muscle stretch receptor stimulates the motor neurons to increase the activation of the muscle, causing the muscle to contract and the leg to extend at the knee. This motion is the knee-jerk response.

This reflex arc does not include an interneuron: It is composed of just two neurons and one synapse. It provides a rapid change in muscle contraction because only one synapse is required to relay the sensory information back to the muscle. Neural circuits containing more neurons and synapses take longer to transmit nerve signals because transmission is delayed by communication at the synapses.

As well as being a useful test in examination, the knee-jerk reflex has a normal physiological role. In running or landing from a jump, the knee joint is flexed (that is, bent), stretching the quadriceps muscles that extend (that is, straighten) the knee joint. Knee flexion allows rapid adjustments in muscle force and knee position that help stabilize the body during running and jumping.

Because muscles can only contract, opposing movements at a joint such as flexion and extension require flexor and extensor muscles on either side of the joint (Chapter 37). These are often activated out of phase (that is, at different times) so that, when one set of muscles is activated (contracting), the other is inhibited (relaxed). The alternating movements of the limbs during walking and running provide an example. This pattern of joint and limb movement is achieved by **reciprocal inhibition:**

When stretch receptors of the knee extensor muscles are activated to stimulate these muscles to extend the knee, they also inhibit the activity of opposing muscles that flex the knee (Fig. 35.20).

Reciprocal inhibition of opposing sets of muscles occurs in the spinal cord. Axons of the stretch receptor neurons not only synapse with motor neurons of extensor muscles but also synapse with inhibitory interneurons that inhibit motor neuron stimulation of the opposing flexor muscles. This inhibitory reflex pathway contains two synapses, one between the sensory neuron and interneuron and the second between the interneuron and the motor neuron to the flexor muscle.

Reciprocal inhibition also operates between the right and left sides of the body. For example, reciprocal inhibition helps to control the movement of the right and left limbs. In this case, interneurons cross the spinal cord to control the timing of activity by flexor and extensor muscles of the opposite limb. Reciprocal inhibition is also involved in the Mauthner cell circuit of fish to ensure that only one side of the fish's body bends away from the stimulus to escape. The local spinal circuits that provide reciprocal inhibition are therefore fundamental to the alternating motion of the body and limbs that characterizes the movements of most animals.

In other cases, more complex circuits in the brain may act to coordinate sensory input and motor output. These circuits integrate other sources of sensory information, such as vision and balance, with conscious commands from the brain to voluntarily control motor behavior. In the next chapter, we look more closely at these sensory circuits.

Core Concepts Summary

35.1 ANIMAL NERVOUS SYSTEMS ALLOW ORGANISMS TO SENSE AND RESPOND TO THE ENVIRONMENT, COORDINATE MOVEMENT, AND REGULATE INTERNAL FUNCTIONS OF THE BODY.

Nerve cells, or neurons, receive and send signals and are the functional unit of the nervous system. page 35-1

Animal nervous systems include three types of neuron: sensory neurons that respond to signals, interneurons that integrate and process sensory information, and motor neurons that produce a response. page 35-2

Ganglia are localized collections of nerve cell bodies that integrate and process information. page 35-2

Simply organized animals, such as cnidarians, have a nerve net to coordinate sensory and motor function. page 35-3

Forward locomotion led to the evolution of specialized sense organs in the head, along with concentrated groupings of nerve cells to form ganglia and a brain. page 35-3

35.2 THE NEURON IS THE FUNCTIONAL UNIT OF THE NERVOUS SYSTEM. NEURONS HAVE DENDRITES THAT RECEIVE INFORMATION AND AXONS THAT TRANSMIT INFORMATION.

Neurons share a common organization, with dendrites that receive inputs, a cell body that receives and sums the inputs, and axons that transmit signals to other nerve cells. page 35-4

Glial cells provide nutritional and physical support for neurons. page 35-6

In vertebrates, glial cells also produce the myelin sheath that insulates axons, increasing the speed of nerve impulses. page 35-6

35.3 THE ELECTRICAL PROPERTIES OF NEURONS ALLOW THEM TO COMMUNICATE RAPIDLY WITH ONE ANOTHER.

Neurons have electrically excitable membranes that code information by changes in membrane voltage and transmit information in the form of electrical signals called action potentials. page 35-6

Ion channels open and close in response to changes in membrane voltage, underlying the production of action potentials in nerve cells. page 35-8

Action potentials fire in an all-or-nothing fashion followed by a brief refractory period. page 35-8

Action potentials are conducted in a saltatory fashion in myelinated axons, firing at nodes of Ranvier where the axon membrane is exposed and not insulated by myelin. page 35-11

Most neurons communicate by chemical synapses formed between an axon terminal and a neighboring nerve or muscle cell. page 35-12

Communication across the synapse occurs when an arriving action potential triggers the release of neurotransmitters from vesicles within the axon terminal. page 35-12

Neurotransmitters released from presynaptic vesicles bind to receptors in the postsynaptic membrane, causing either an excitatory or an inhibitory stimulus. page 35-13

Excitatory stimuli depolarize the membrane, producing an excitatory postsynaptic potential (EPSP), whereas inhibitory stimuli hyperpolarize the membrane, producing an inhibitory postsynaptic potential (IPSP). page 35-13

35.4 ANIMAL NERVOUS SYSTEMS CAN BE ORGANIZED INTO CENTRAL AND PERIPHERAL COMPONENTS.

Animal nervous systems are organized into central and peripheral components called the central nervous system (CNS) and peripheral nervous system (PNS). page 35-15

The central nervous system includes the brain and one or more main trunks of nerve cells, such as the spinal cord. The peripheral nervous system is distributed throughout the animal's body and is composed of sensory and motor nerve cells. page 35-16

In many invertebrates and vertebrates, the peripheral nervous system is divided into voluntary and involuntary components. page 35-16

In vertebrates, the voluntary component is somatic, and the involuntary is autonomic. page 35-16

The autonomic system regulates body functions through opposing actions of the sympathetic and parasympathetic divisions. page 35-17

The nervous system helps to regulate physiological functions to actively maintain stable conditions inside a cell or an organism, a property known as homeostasis. page 35-17

Homeostasis is often achieved by negative feedback, in which the response inhibits the stimulus. page 35-17

Simple reflex circuits can involve as few as two neurons: a sensory neuron from the periphery that synapses with a motor neuron in the spinal cord that sends a signal to a muscle. page 35-18

Self-Assessment

1. Diagram a simple nervous system of an animal that lacks cephalization and compare that system with the general organizational features of a nervous system that exhibits cephalization.

2. Name the three basic categories of neuron and describe their functions.

3. Diagram and label the basic features of a neuron, indicating where information is received and where it is sent.

4. Graph an action potential, showing the change in electrical potential on the y-axis and time on the x-axis. Indicate on the graph the phases when voltage-gated Na$^+$ and K$^+$ ion channels are opened and when they are closed.

4. Explain what is meant by saying action potentials are "all-or-nothing."

6. Explain why action potentials propagate along an axon only in a single direction.

7. Briefly describe how myelinated axons increase the speed of signal transmission.

8. Diagram a chemical synapse, labeling the vesicles that contain neurotransmitter molecules and the receptors that bind the neurotransmitter to produce either an inhibitory or excitatory stimulus in the postsynaptic cell.

9. Briefly describe how neurotransmitter binding to receptors on a postsynaptic cell causes inhibition or excitation.

10. Describe how vertebrate nervous systems are organized into voluntary and involuntary components, listing which functions of an animal are controlled by each component.

11. Diagram a simple circuit that includes a sensory neuron that synapses with a motor neuron to produce a reflex. Indicate where in the nervous system this synapse is found.

> Do you understand the chapter's Core Concepts? Log in to check your answers to the Self-Assessment questions, then practice what you've learned and reinforce this chapter's concepts by working through the problems and multimedia tutorials provided there.
>
> www.biologyhowlifeworks.com

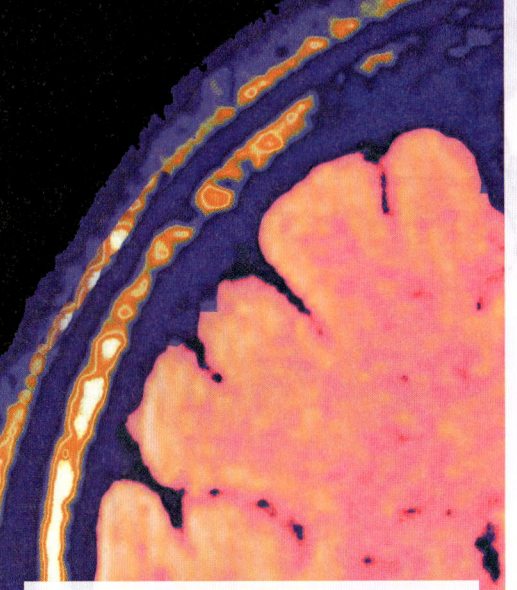

CHAPTER 36

ANIMAL SENSORY SYSTEMS AND BRAIN FUNCTION

Core Concepts

36.1 Animal sensory receptors detect physical and chemical stimuli by changes in membrane potential.

36.2 Specialized chemoreceptors relay information about smell and taste.

36.3 Hair cells convey information about movement and sound.

36.4 The ability to sense light and form images depends on photosensitive cells with light-absorbing proteins.

36.5 The brain processes and integrates information from multiple sensory systems, with tactile, visual, and auditory stimuli mapped topographically in the cerebral cortex.

36.6 Cognition is the ability of the brain to process and integrate complex sources of information, to remember and interpret past events, to solve problems, to reason, and to form ideas.

To see, your eyes detect light, which is a form of electromagnetic radiation; to hear, your ears detect sound waves; to smell, your nose detects odor molecules present in the air. In all these cases, your senses are detecting a physical or chemical stimulus in the environment. The first half of this chapter explores how specialized nerve cells detect these signals and code them as information that can be transmitted and processed by the nervous system. How is a dog able to distinguish thousands of different odors? How is a hawk able to see a small rodent far below? How are humans able to detect small differences in the pitch of a sound?

The second half of the chapter follows nervous system pathways from the senses into the brain, on the way exploring basic principles of brain function. Processing of sensory information in the brain takes place in regions specialized for each sense, and within many of these regions information is represented in the form of topographic maps.

36.1 ANIMAL SENSORY SYSTEMS

Animals are able to sense the physical properties of their environment, including light, chemicals, temperature, pressure, and sound, useful in finding mates and food and avoiding predators and noxious environments. Early in evolutionary history, organisms evolved specialized membrane **receptors** that functioned to detect these critical features in their environment. For example, before the evolution of nerve cells and a nervous system, bacteria evolved receptors located in the cell membrane to sense osmotic pressures that might otherwise rupture the membrane (Chapter 5). Bacteria and sponges also evolved membrane receptors that could detect chemicals—including nutrients—in their environment.

The senses of smell, taste, and sight in multicellular organisms with a nervous system also rely on membrane receptors. However, these receptors are embedded in specialized membranes of sensory neurons, forming a **sensory receptor.** In this case, a sensory receptor refers to the entire neuron, not just a membrane protein. In most multicellular animals, the sensory receptors are organized into specialized **sensory organs** that convert particular physical and chemical stimuli into nerve impulses that are processed by a nervous system and sent to a brain. Cnidarians (which include jellyfish, corals, and anemones) and roundworms evolved sensory receptors to sense physical contact, and cnidarians and flatworms were among the first multicellular animals to evolve simple light-sensing photoreceptive organs.

The conversion of physical or chemical stimuli into nerve impulses is called **sensory transduction.** For example, receptors located in the ear convert the energy of sound waves into nerve impulses that allow an animal to distinguish loud versus soft sounds and high- versus low-pitched sounds. Although the sense organs of different animals

FIG. 36.1 **Diverse sensory receptors.** (a) Many insects can detect ultraviolet light. (b) Rattlesnakes can see at night by detecting infrared radiation. (c) Dogs have a keen sense of smell.

share many similar properties, differences have also evolved. Consequently, animals perceive the world differently from one another. Many insects, for instance, are sensitive to ultraviolet light; nocturnal snakes can see at night by sensing infrared radiation; and dogs can distinguish odor compounds as much as 100 million times lower in concentration than humans can detect (**Fig. 36.1**).

Specialized sensory receptors detect diverse stimuli.

Animals have evolved a diverse array of sensory receptors that respond to different stimuli, among them touch, light, and the oscillations in air pressure we call sound waves. How is a physical phenomenon in the world outside an organism transformed into a nerve impulse? The initial transformation takes place inside the sensory receptor (**Fig. 36.2**). A physical or chemical stimulus is converted into a change in the sensory receptor's membrane potential. Recall from Chapter 35 that a cell's membrane potential is the charge difference between the inside and the outside of the cell membrane, and that a depolarization in membrane potential is the first step in firing an action potential. Sensory receptors either fire action potentials themselves or synapse with other neurons that fire action potentials, which are then transmitted to the brain.

In many cases, the sound wave, touch, or other stimulus causes ion channels in the cell's plasma membrane to open. The influx of ions changes the membrane potential by altering the distribution of charged ions on either side of the membrane. If the influx of ions reduces the charge difference, then the cell is depolarized (Chapter 35).

The most ancient type of sensory detection is chemoreception. All organisms respond to chemical cues in their environment. Even the earliest branching groups of Bacteria and Archaea have protein receptors in their cell membranes that respond to molecules in the environment. In animals, **chemoreceptors** respond to molecules that bind to specific protein receptors on the cell membrane of the sensory receptor (Fig. 36.2a). Many animals detect food in their environment by sensing key molecules such as oxygen (O_2), carbon dioxide (CO_2), glucose, and amino acids. Mosquitoes track CO_2 levels to detect prey for blood meals, and coral polyps respond to simple amino acids in the water, extending their bodies and tentacles toward areas of greater concentration to feed. Other arthropods, such as flies and crabs, have chemosensory hairs on their feet. These animals taste potential food sources by walking on them. Salmon rely on chemoreception to detect chemical traces of the home waters of the river where they hatched, and where they will return to mate and spawn.

Chemoreception also underlies the sense of smell and taste. In most cases, the binding of molecules to a protein receptor on a taste receptor causes the protein receptor to change conformation. That change in conformation in turn triggers the opening of Na^+ channels through G protein signal transduction pathways similar to those discussed in Chapter 9. The influx of Na^+ ions depolarizes the receptor cell membrane. No action potential fires, but the depolarization travels far enough down the receptor's short axon to trigger the release of neurotransmitters.

Mechanoreceptors respond to physical deformations of their membrane produced by touch, stretch, pressure, motion, and sound. Deformation of the receptor membrane opens sodium channels, causing a depolarization of the endings of the cell's dendrites (Fig. 36.2b). Early mechanoreceptors in bacteria sensed local physical forces and internal cell pressure. Other mechanoreceptors in roundworms and anemones are linked to externally projecting cilia at the animal's body surface.

One well-studied example of a mechanoreceptor is a touch receptor in the roundworm *Caenorhabditis elegans*. Deformation of the surface of the worm—its cuticle—exerts pressure on proteins connected to an ion channel. The mechanical force exerted changes the shape of the ion channel, causing it to open and leading to a change in membrane potential.

A mechanoreceptor in humans and other mammals is the sensory receptors found in the skin that sense touch and pressure. The cell bodies of these neurons are located in ganglia near the spinal cord, with extensions going in two directions: to the skin and to the spinal cord. In the skin, the neuron has branched tips, which are the initial sensors of touch and pressure. If the stimulus is strong enough, local depolarization leads to the firing of an action potential that travels all the way to the spinal cord. Thus, in contrast to a chemoreceptor, this type of mechanoreceptor transmits an action potential. The axon is much too long for simple depolarization to spread to its other end.

Mechanoreceptors are also found at the base of whiskers that rodents, cats, dogs, and other mammals use to sense touch with their snouts. Stretch receptors found in muscles are also mechanoreceptors that influence a muscle's motor activation, helping to control its length and force. A very different group of specialized mechanoreceptors called hair cells are the sensory receptors for balance, gravity sensing, and hearing, discussed in section 36.3.

Thermoreceptors in the skin and in specialized regions of the central nervous system (CNS) respond to heat and cold. Thermoreceptors help to control an animal's metabolism and patterns of blood flow, regulating body temperature by affecting rates of heat gain and loss (Chapter 40). As a result, they help to maintain homeostasis. Many invertebrate animals rely on thermoreception when seeking environments that provide favorable temperatures. Thermoreceptors in the skin of vertebrate animals are believed to be simple dendrite endings that respond to changes in skin temperature by altering the firing rate of the cell's action potentials.

Nociceptors, or pain receptors, are another class of nerve cell with dendrites in the skin and connective tissues of the body. When these sensory receptors are exposed to an excessive mechanical, thermal, or chemical stimulus, they send action potentials to the brain or spinal cord, typically stimulating rapid withdrawal from the painful stimulus. Pain is a subjective sensation, and therefore difficult to assess and treat. Although it is clear that many animals experience pain, it is difficult to judge how different species associate the sensation of pain with underlying physical or chemical stimuli.

FIG. 36.2 **Sensory transduction, which converts an external stimulus into a change in membrane potential.** (a) Chemoreceptors are located in the antennae of the luna moth; (b) mechanoreceptors below the cutical of roundworms; and (c) photoreceptors in the eyes of squid.

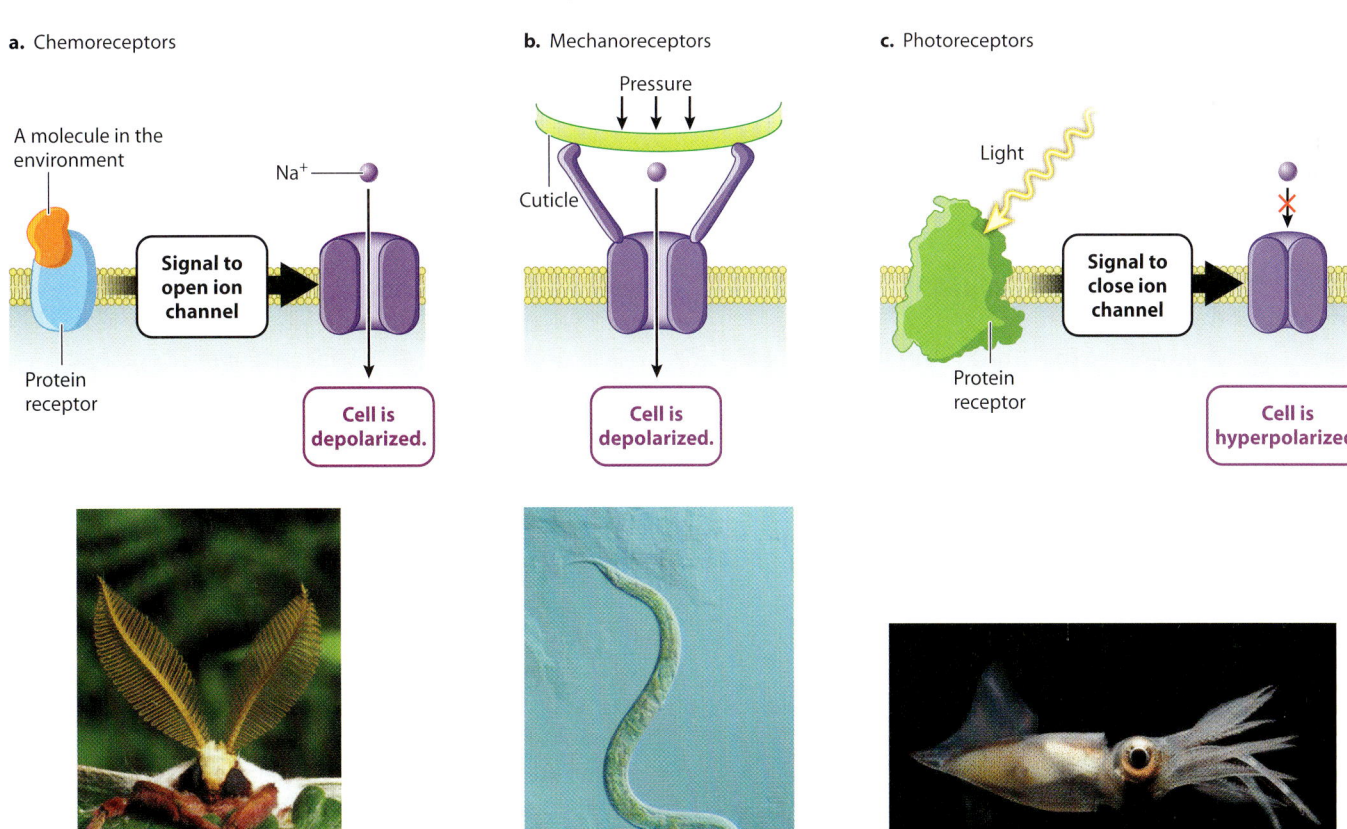

Light-detecting **photoreceptors** are the sensory receptors in eyes (Fig. 36.2c). Photoreceptors respond to individual photons of light energy by closing Na⁺ channels, causing the cell membrane to become hyperpolarized rather than depolarized. Most receptors excite the neurons that they synapse with, but vertebrate photoreceptors are unusual in that they inhibit the firing rate of other neurons within the eye rather than exciting them. Photoreception in the vertebrate eye is discussed in section 36.4.

Some fish, such as catfish, contain specialized **electroreceptors** arranged in a lateral line along their bodies. Electroreceptors enable these fish to detect weak electrical signals emitted by all organisms. They likely evolved as an adaptation for locating prey or potential predators in poorly lit habitats where vision was less useful. Some specialized "weakly electric" fish actually generate an electromagnetic field by emitting pulses from an electric organ located in the tail. Disturbances in the field detected by electroreceptors of the lateral line system signal the location of nearby prey. These fish also inhabit rivers with poor visibility. The bill of the duckbilled platypus also contains electroreceptors that locate prey in dimly lit water.

→ **Quick Check 1** Aspirin and ibuprofen reduce pain by inhibiting the synthesis of chemicals called prostaglandins. What effect would you predict prostaglandins to have on the firing rate of nociceptors?

Stimuli are transmitted by changes in the firing rate of action potentials.

Sensory reception depends on converting the energy of a physical or chemical stimulus into an action potential, either in the sensory receptor itself or in the neuron it synapses with. How do these nerve impulses convey information to the brain? In effect, action potentials can be considered a code that the brain deciphers.

To convert information from the environment into a code, the nerve impulses conveyed by sensory organs carry out the following functions:

1. Convey the strength of the incoming signal—the brightness of light, the loudness of a sound, or the intensity of an odor.

2. Carry information about even weak signals when necessary.

3. Convey the location of a signal's source—for example, where is the body being touched?

4. Filter out unimportant background signals.

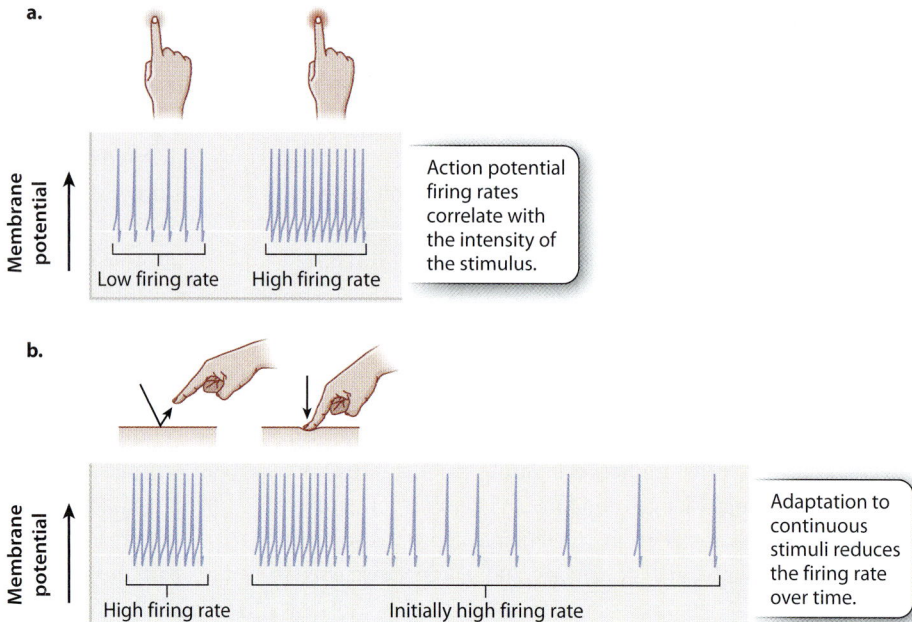

FIG. 36.3 Sensory receptor signals. The firing rate of action potentials increases with (a) the intensity of the stimulus and (b) a novel compared to a continuous stimulus.

Because nerve cells transmit action potentials in an all-or-nothing fashion, information about the strength of the signal is coded by the number of action potentials fired over a period of time, which is the nerve cell's **firing rate.** Most commonly, strong signals produce high firing rates and weak signals produce low or background firing rates (**Fig. 36.3a**). For example, the firing rate of mechanoreceptors increases as the tactile pressure they sense increases.

Several mechanisms increase a nerve cell's sensitivity to even weak signals. Following sensory transduction, the signal typically undergoes initial processing. As discussed in Chapter 35, sensory stimuli may be summed over space or time. Multiple receptors that receive a stimulus often converge onto a neighboring neuron that increases its firing rate proportionally to the number of signals received. This is an example of **spatial summation.** Combining the synaptic input increases the receptor organ's sensitivity to external stimuli. **Temporal summation** is the integration of sensory stimuli that are received repeatedly over time by the same sensory cell. When a sensory cell is stimulated more frequently, the excitatory postsynaptic potentials (EPSPs) sum, making it more likely that the cell is depolarized above threshold and fires an action potential.

Sensory receptors also distinguish between discrete (that is, intermittent) and continuous stimuli, providing a way to ignore background noise. Novel stimuli are generally the most important for an animal to respond to. Consequently, sensory receptors initially respond most strongly when a stimulus is first received (**Fig. 36.3b**). If the stimulus continues over a longer time period, sensory receptors typically reduce their firing rate through a process called **adaptation** (this is *not* the adaptation that results from natural selection). Adaptation of firing rate enables an animal's sensory receptors to focus on novel stimuli, which are those likely to be most important. You can experience adaptation by touching your skin lightly at a particular location. Initially, skin mechanoreceptors fire in response to the touch, but after a while their firing rate diminishes and the sensation of a touch becomes less noticeable. Adaptation also ensures that the clothes touching your skin do not cause an irritating and distracting long-term sensation.

In addition, sensory receptors locate a signal's source by **lateral inhibition** (**Fig. 36.4**). Lateral inhibition enhances the strength of a sensory signal locally but diminish it peripherally. In this process, receptors inhibit interneurons that receive sensory inputs from adjacent regions of the sensory organ, while sending excitatory signals to interneurons within their local region. This increases the contrast in signal strength between adjacent regions. For example, lateral inhibition in the retina of the eye enhances edge detection and sharpens image contrast. It also enables the sense of touch to be localized to a small region of the body surface. The inhibitory and excitatory signals are transmitted by synapses that have different neurotransmitters.

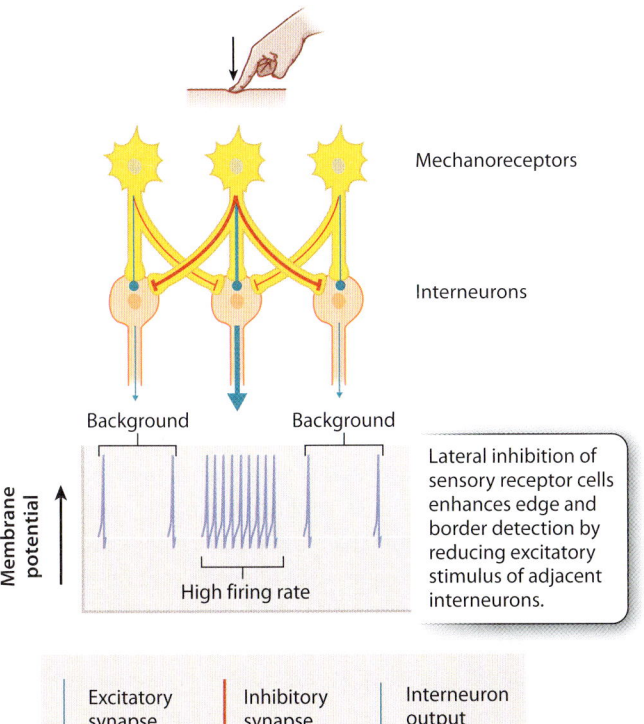

FIG. 36.4 Lateral inhibition. The firing rate of action potentials is higher in the center of the sensory receptive field than in surrounding regions.

We have not yet described all the information in the code. How do action potentials convey pitch or color or a particular odor? This information is conveyed by the identity of the particular sensory receptors that are activated. Sensory organs are so constructed that, for example, tone A activates "tone A" sensory neurons in the ear, red light activates "red" sensory neurons in the eye, and the fragrance of a rose activates "rose" sensory neurons in the nose. The way in which the "correct" sensory neuron is excited varies for each of the senses. In the next three sections, we discuss how this happens for smell and taste, hearing, and vision.

36.2 SMELL AND TASTE

In many animals, specialized sensory organs for smell (**olfaction**) and taste (**gustation**) have evolved that detect a variety of chemical compounds. These organs contain distinct types of protein membrane receptor, and each of them binds to a specific class of molecule of complementary chemical structure. A large number of genes code for different chemical odor receptors, and the presence of so many receptors enables the detection of broad categories of smell and taste.

Smell and taste depend on chemoreception of molecules carried in the environment and in food.

The sense of smell in mammals depends on specialized sensory neurons that extend into the nasal cavity. These neurons communicate directly with the brain by the olfactory nerve, one of the cranial nerves of the head (**Fig. 36.5**). Odor molecules are captured by nasal mucus during inhalation and sensed by the cells' long, thin hairlike extensions, which project into the mucus. Membrane receptors on these extensions bind odor molecules that are complementary in structure to the receptor. When bound, the odor molecules produce excitatory postsynaptic potentials (EPSPs). If enough odor molecules bind to the receptor cell, the EPSPs are summed and transmit an action potential to the brain.

Each sensory receptor has one type of membrane receptor, but there are typically many different sensory receptors for many different molecules. Humans have nearly 1000 genes that express particular odorant receptor proteins. Mice have about 1500 genes that express odorant receptor proteins, constituting about 15% of all their genes. The olfactory system can detect specific odor molecules in quite low concentrations.

Taste, or gustation, is achieved by clusters of chemosensory receptor cells located in specialized **taste buds,** the sensory organs for taste (Fig. 36.5). Some fish have taste buds in their skin that sense amino acids in the water, helping them to localize food. The taste buds of terrestrial vertebrates are in the mouth; most of the taste buds in mammals are on the tongue. In humans, the tongue has about 10,000 taste buds, which are contained within raised structures called papillae, giving the tongue its rough surface. Each taste bud has a pore through which extend fingerlike

FIG. 36.5 **The senses of smell and taste.** Olfactory receptors respond to chemical odors in the nasal passages and communicate directly with the brain. Taste cells are chemosensory receptor cells located in taste buds.

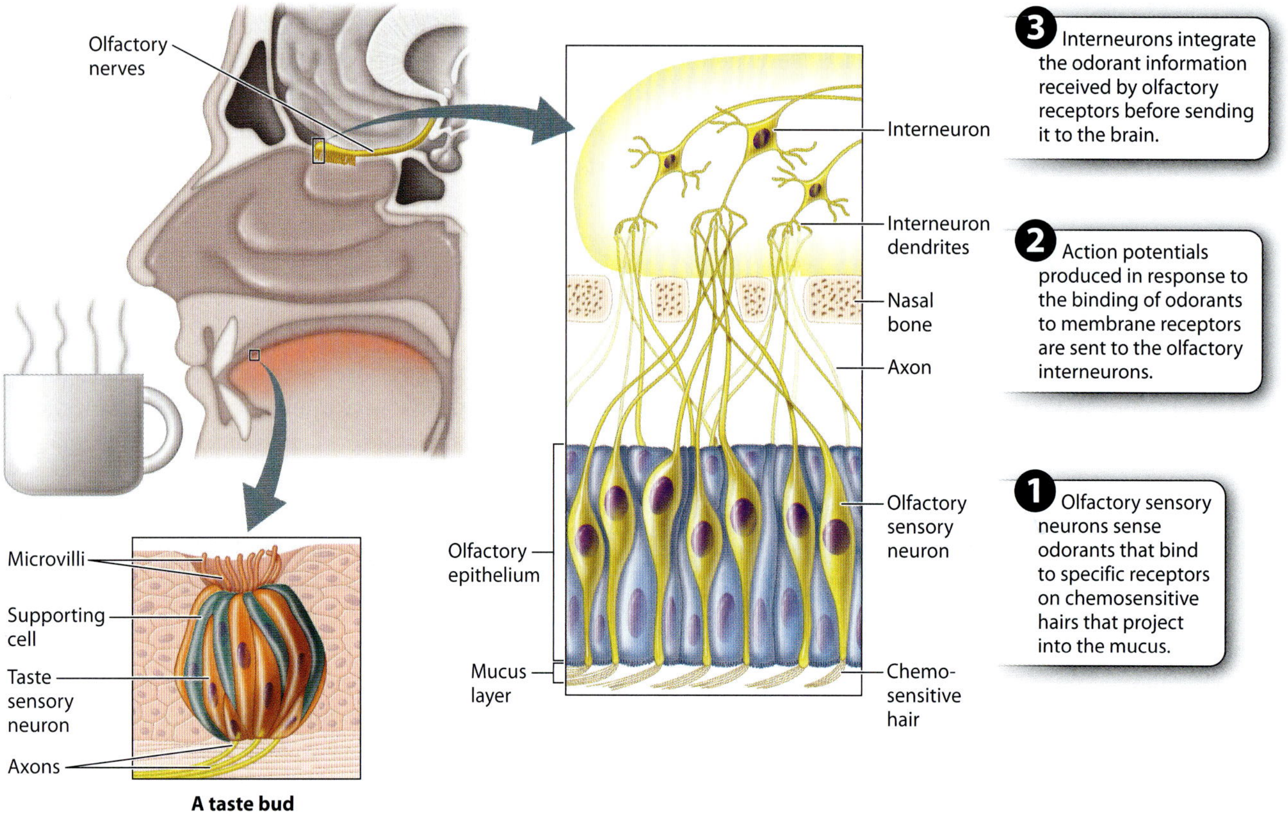

projections of sensory cells called microvilli that contact food items. The microvilli provide a large surface area that is rich in membrane receptors. These receptors bind to specific chemical compounds that give the food its taste. Taste bud sensory cells synapse on interneurons, which fire action potentials when a sufficient number of sensory cells are depolarized by a particular food item. In contrast to olfaction, taste generally requires higher concentrations of food chemicals coming into contact with taste bud receptors to stimulate receptor depolarization.

Human taste can be divided into five categories—sweet, sour, bitter, salty, and savory. The tongue contains only five types of taste receptor, one specialized for each of these categories. Variation in taste is achieved by combining the signals received by receptors in the five categories. Your sense of taste is substantially enhanced by airborne odors that emanate from a food or beverage and are sensed by chemoreceptors in the nose. Signals from taste bud and olfactory receptors are combined in the brain, eliciting the subtle sensation of taste and flavor of a particular food or drink.

→ **Quick Check 2** Why is your sense of smell and taste diminished when you have a cold?

36.3 SENSING GRAVITY, MOVEMENT, AND SOUND

The ability to detect motion, orient with respect to gravity, and hear all depend on specialized mechanoreceptors called **hair cells** that sense movement and vibration. They are found in fishes and amphibians, in which they detect movement of the surrounding water, and in many invertebrates, in which they sense gravity and other forces acting on the animal. They are also found in the ears of terrestrial vertebrates, in which they sense sound, body orientation, and motion.

In all these cases, hair cells sense mechanical vibrations. Movement of small nonmotile cell-surface projections called **stereocilia** causes a depolarization of the cell's membrane by opening or closing ion channels. Despite their name, stereocilia are more similar to microvilli than to cilia, which unlike stereocilia can move on their own (Chapter 10). Stereocilia contain actin filaments but lack the microtubules and dynein motor proteins of motile cilia. Hair cells themselves do not fire action potentials but instead release neurotransmitters that bind to receptors in adjacent sensory neurons, altering their firing rate.

Hair cells sense gravity and motion.

Hair cells detect gravity and motion in many animals. Hair cells contained in the **lateral line system** of fish and sharks sense water vibrations that indicate nearby threats or prey, as well as their own motion (**Fig. 36.6a**). The lateral line is a sensory organ along both sides of the body that uses hair cells to detect movement of the surrounding water.

A sense of gravity helps animals to orient their body within the environment, providing a sense of "up" and "down." Even buoyant aquatic animals depend on a sense of gravity to keep oriented to the water's surface or within the water column. Gravity-sensing organs called **statocysts** are found in most invertebrates (**Fig. 36.6b**). The statocyst is made up of an internal chamber lined by hair cells with stereocilia that project into the chamber. Small granules of sand or other material form a

FIG. 36.6 **Motion and gravity sensation.** (a) Hair cells within the lateral line system of fish sense water vibrations for motion and prey detection. (b) Gravity-sensing organs (statocysts) in lobsters provide cues about body orientation.

a. Lateral line

b. Statocyst

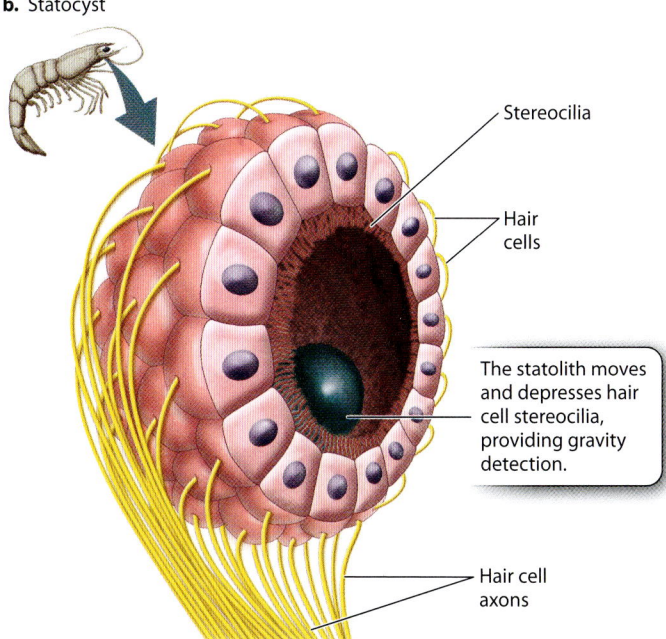

FIG. 36.7 **The vestibular system.** The vestibular system senses angular rotations of the head from movements of fluid within three semicircular canals that deflect hair cells. It detects gravity and posture by means of statoliths in two statocyst chambers.

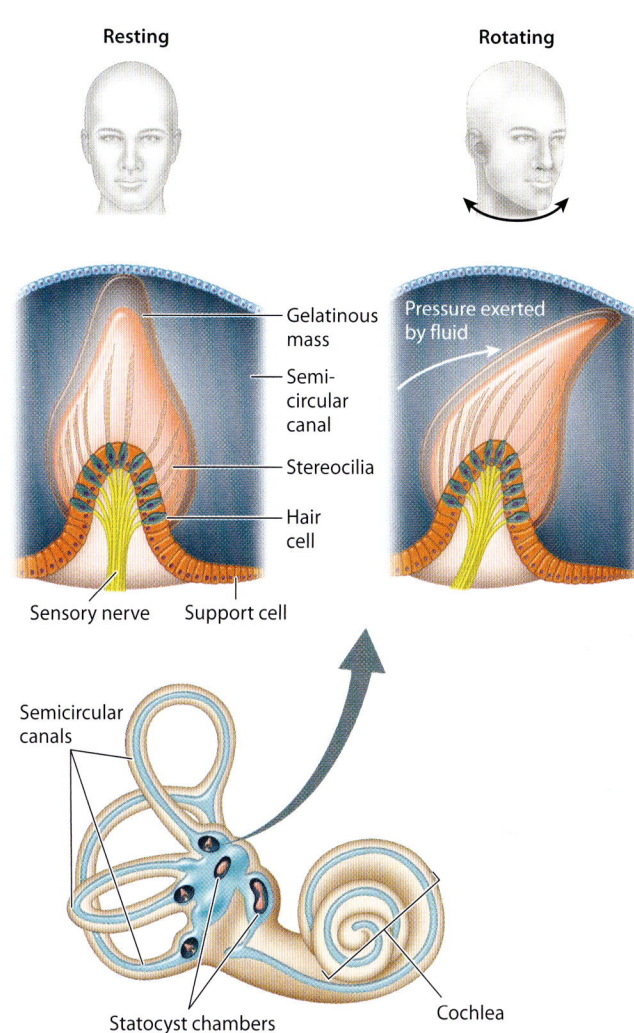

dense particle called a **statolith** that is free to move within the statocyst organ. By pressing down on hair cells at the "bottom" of the chamber, the statolith activates those cells, indicating the direction of gravity. Statoliths help anemones and jellyfish orient themselves in the water and direct their tentacles. When statoliths of lobster and crayfish are replaced by magnetic particles and subjected to an experimental magnetic field, the lobsters and crayfish swim upside down or on their sides in response to the experimental change in the direction of the magnetic field.

The mammalian inner ear contains organs that sense motions of the head and its orientation with respect to gravity. These organs make up the **vestibular system** (**Fig. 36.7**), which consists of two statocyst chambers and three **semicircular canals.** These statocyst chambers are similar to those of invertebrates, providing

FIG. 36.8 **The human ear.** (a) The tympanic membrane and bones of the middle ear transmit external sound vibrations to the oval window of the inner ear. (b) Hair cells in the cochlea respond to the vibrations, sending signals to the brain that distinguish sound amplitude and pitch. (c) A scanning electron micrograph shows the V-shaped arrangement of stereocilia on hair cells.

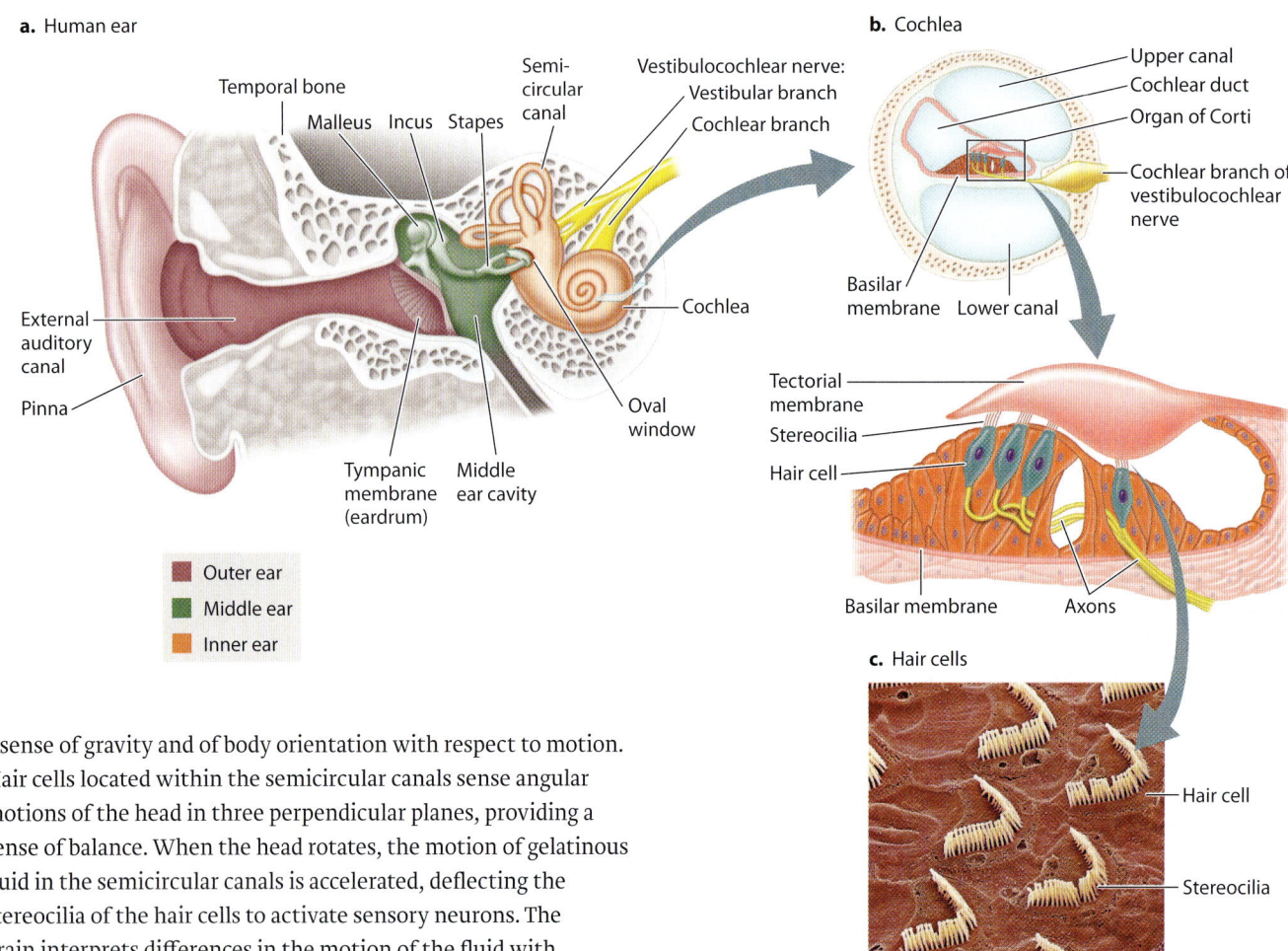

a sense of gravity and of body orientation with respect to motion. Hair cells located within the semicircular canals sense angular motions of the head in three perpendicular planes, providing a sense of balance. When the head rotates, the motion of gelatinous fluid in the semicircular canals is accelerated, deflecting the stereocilia of the hair cells to activate sensory neurons. The brain interprets differences in the motion of the fluid with respect to the hair cells among the three semicircular canals to resolve angular motions of the head in any direction. As a result, mammals and birds react much faster to head motion than to visual cues, and that ability helps them to stabilize their gaze.

→ **Quick Check 3** After you spin in place or take a ride on a merry-go-round, why are you unstable when you try to walk?

Hair cells detect the physical vibrations of sound.

What our ears interpret as sound are alternating waves of pressure traveling through the air. The frequency of the waves—the number of waves per second—determines the pitch of a sound, and the amplitude, or height, of the waves determines its loudness. Sound waves cause the stereocilia of hair cells to bend, and the hair cells become excited, depolarizing and releasing neurotransmitters. The process sounds simple, but how can the bending of stereocilia produce all the variations in pitch and loudness you hear in the world around you?

Insects hear by sensing airborne vibrations with small hairs on their antennae or other regions of their body. Hairs of different length detect different frequencies because the shorter hairs are stiffer than long hairs and vibrate at higher frequencies. Frequency is measured in Hertz (Hz), or cycles per second. Male mosquitoes have small antennal hairs that vibrate in response to the 500-Hz hum of a female mosquito's flight, which attracts them to the female to mate. Many insects also have specialized ears that sense sound by means of a **tympanic membrane.** The tympanic membrane is a thin sheet of tissue at the surface of the ear that vibrates in response to sound waves, amplifying airborne vibrations. The vibration excites hair cells attached to the inside of the tympanic membrane, and these hair cells in turn excite other neurons to fire action potentials. Crickets and other singing insects sense the chirping or buzzing of potential mates by their ears.

Hearing is most widespread and developed in terrestrial vertebrates. The ears of amphibians, reptiles, and birds have a simple external tympanic membrane that vibrates when sound waves strike its surface. This membrane is similar to the tympanic membrane of insects but evolved independently and therefore is a case of convergent evolution. Terrestrial vertebrate ears can detect a broad range of sound frequencies. For humans, the range of audible frequencies is from 20 to 20,000 Hz. For dogs, the audible range is 70 to 44,000 Hz, and some rodents can hear up to 80,000 Hz. Birds generally hear in the same frequency range as mammals—one reason bird songs are pleasing to our ear. Frogs and reptiles generally hear at lower frequencies than mammals and birds. Vertebrate ears can also make fine distinctions between close frequencies and can detect softer sounds because external structures amplify the sounds before they reach the hair cells.

The ears of mammals have an external structure, the **pinna,** that enhances the reception of sound waves contacting the ear. Notice the heightened sensitivity that this structure provides for hearing when you cup your hands behind your ears. It is no coincidence that dogs and rabbits perk up their ears when alerted by a sound: This action orients them to the direction of the sound's source.

Fig. 36.8 shows the structure of the human ear. The pinna is part of the **outer ear,** which also includes the ear canal and tympanic membrane, or mammalian **eardrum**, which transmits airborne sounds into the ear. The **middle ear** contains three small bones or ossicles, called the **malleus**, **incus,** and **stapes**, which amplify the waves that strike the tympanic membrane. The stapes connects to a thin membrane called the **oval window** of the cochlea in the **inner ear.** The **cochlea** (which means "snail" in Latin) is a coiled chamber within the skull that contains hair cells that convert pressure waves into an electrical impulse that is sent to the brain.

The mammalian eardrum evolved from the tympanic membrane of reptiles and earlier amphibians. The three middle-ear bones are also evolutionarily related to three bones in reptiles, but only the stapes functions in reptiles to transmit vibrations of the tympanic membrane to the inner ear. The malleus and incus help support jaw movements during feeding in reptiles. During the evolution of mammals from their reptilian ancestors, these two bones were incorporated together with the stapes to form the middle-ear bones that transmit sounds from the eardrum to the inner ear.

The process of hearing can be separated into three stages:

1. *Amplification.* Sound vibrations received by the outer ear are transmitted by the eardrum and amplified by the three bones in the middle ear (Fig. 36.8a). Through piston-like actions, the stapes transmits the energy of its movements to vibrations of the oval window of the cochlea in the inner ear. Vibrations transmitted from the eardrum through the middle ear to the oval window are amplified more than 30 times because of the larger size of the eardrum compared to the oval window and the lever-like action of the three bones.

2. *Transfer of sound vibration to fluid pressure waves.* The cochlea contains fluid and an upper and a lower canal separated by a **basilar membrane** and **cochlear duct** (Fig. 36.8b). Vibrations of the oval window cause fluid pressure waves in both canals at nearly the same time.

3. *Mechanoreception by hair cells within the cochlea.* The cochlear duct contains the **organ of Corti,** which has specialized hair cells with stereocilia supported by the basilar membrane. The stereocilia of mammalian hair cells form a "V" and project into a rigid **tectorial membrane** that does not move (Fig. 36.8c). The fluid vibrations within the cochlear canals induce motions of the basilar membrane relative to the tectorial membrane, bending the stereocilia of the hair cells back and forth in localized regions of the cochlea (**Fig. 36.9**). Bending of the stereocilia stimulates them to release excitatory neurotransmitters that cause postsynaptic neurons to fire action potentials. These

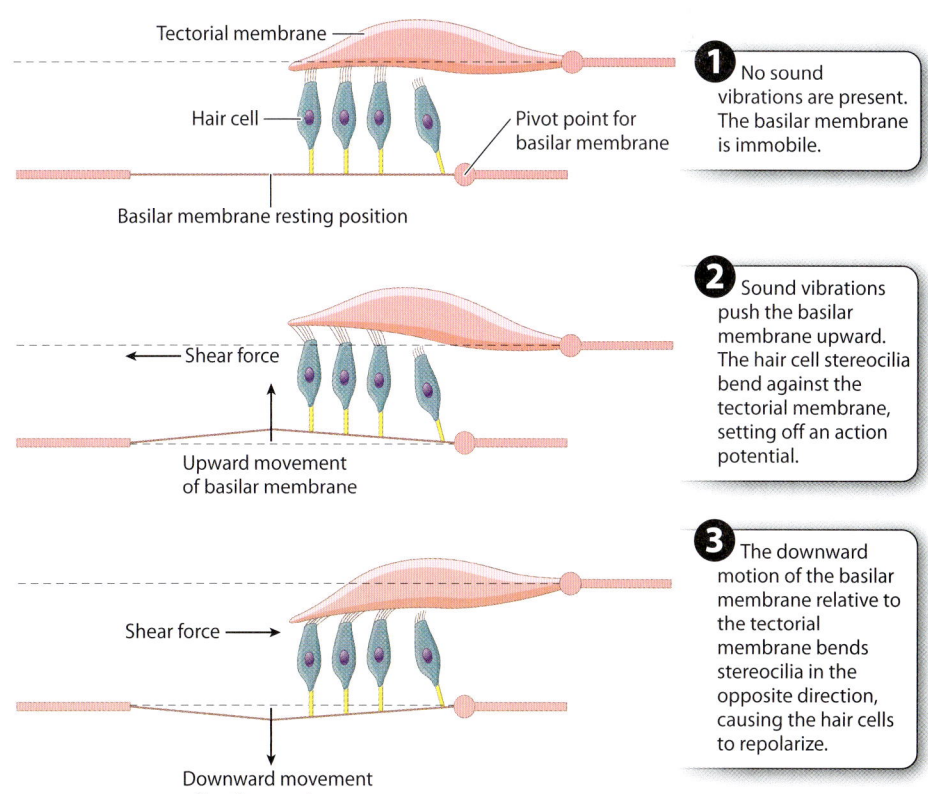

FIG. 36.9 Activation of hair cells. Upward movement of the basilar membrane pushes hair cells into the tectorial membrane, causing bending of the stereocilia.

are sensed as sound by neuronal networks in the **auditory cortex,** the area of the brain that processes sound.

Sound is characterized by amplitude (loudness) and frequency (pitch). Louder sounds produce larger fluid vibrations that cause the hair cell stereocilia to bend more, increasing their release of excitatory neurotransmitters. The release of more neurotransmitters increases the firing rate of the postsynaptic cell. The firing rate indicates intensity, or in the case of hearing, loudness.

Many people can tell pure tones apart that differ by only a fraction of a percentage point in frequency. The ability to discriminate different sound frequencies is largely the result of differences in the mechanical properties of the basilar membrane along its length. At the apex, the basilar membrane is widest but thinnest and most flexible. At the base, the basilar membrane is narrow but thick and stiff. Thus, the basilar membrane is mechanically tuned to respond in different regions to different frequencies. Low-frequency vibrations excite hair cells in the apex of the basilar membrane, midrange-frequency vibrations excite hair cells in the middle region of the basilar membrane, and high-frequency vibrations excite hair cells closest to the base.

Sound amplitude and pitch, together with vestibular sensing of gravity and head motion, are transmitted by the vestibulocochlear nerve (another cranial nerve of the head) to the brain's auditory cortex, discussed later in the chapter.

? CASE 7 Predator–Prey: A Game of Life and Death
How have sensory systems evolved in predators and prey?

Many insect-eating bats rely on **echolocation** to find and apprehend their flying prey (**Fig. 36.10**). As it flies, a bat emits short bursts of high-frequency sound. These sound pulses bounce off surrounding objects and are reflected back to the bat. The echoes are detected by the ears and processed by the bat's brain to locate the prey. Bats typically increase their call rate as they approach the prey to locate it more accurately and judge its flight trajectory and its speed. The sharpness of an image is measured by its resolution, the smallest distance between two features or

FIG. 36.10 A bat about to catch a moth by high-frequency auditory echolocation.

objects that can be perceived. Bats can resolve prey and other objects at less than 1 mm, an ability that exceeds that of the most sophisticated sonar developed by human engineers.

Together with mosquitoes and other flying insects, nocturnal moths are one of the bat's primary food sources. In response to the evolution of echolocation by bats, several moths have evolved the ability to emit sounds that jam the bat's sonar signal. These moths are more likely to escape capture. The evolution of sophisticated sensory systems in both groups of animals, and the brain organization that goes with it, is the result of a kind of evolutionary arms race. It illustrates how animal sensory systems and brains can achieve impressive detection and information processing performance that improves an animal's chance of survival.

36.4 VISION

Electromagnetic receptors respond to electrical, magnetic, and light stimuli. Of these, light-detecting photoreceptors are the most common and diverse. Most animals sense light in their environment. Although animals have evolved different light-sensing organs, all rely on the same light-sensitive photopigment to convert light energy into a nervous signal. Their use of the same photopigment suggests a common evolutionary origin for the considerable diversity of animal eyes that dates from the

Cambrian Period (approximately 500 million years ago), when diverse forms of multicellular animal life first evolved.

All animals use a similar photosensitive protein called opsin to detect light.

Most multicellular animals have cells called photoreceptors that respond to light. Animals with simple forms of photosensitivity can move toward or away from light sources. The photoreceptors of other animals are arranged to form eyes that produce an image. In some animals, detecting the motion of shapes and simple patterns of light is most important. Other animals, such as vertebrates and cephalopod mollusks like squid and octopus, perceive sharper images. Despite the diversity of light-sensitive organs and eyes that have evolved, all of them depend on a similar photosensitive protein known as **opsin** to convert the energy of light photons into electrical signals in the receptor cell.

Genes involved in the development of photoreceptors in flatworms are shared both with other invertebrates and with vertebrates, indicating that an early common ancestor of invertebrates and vertebrates first evolved the underlying biochemical machinery necessary for perceiving light. In these animals, a master regulatory gene, *Pax6*, is now known to function as a transcription factor regulating the development of the eye. In experiments performed by the Swiss developmental biologist Walter Gehring, *Pax6* was recently found to regulate eye development in fruit flies and mice. Gehring and his group were able to insert the mouse *Pax6* gene into fruit flies, leading to the formation of an additional small eye on the fly's antenna (Chapter 20). *Pax 6* has also been found expressed in flatworm photoreceptors. Remarkably, all light-sensing and image-formation systems therefore ultimately derive from the ancestor of the flatworm's original photoreceptor.

Rhodopsin is the specific transmembrane protein found in the photosensitive cells of vertebrates. This protein is covalently bound to **retinal**, a derivative of vitamin A that absorbs light (**Fig. 36.11**). Opsin molecules are arranged in cylindrical groups in the plasma membrane of most photosensitive cells. These cells are modified neurons with leaky Na^+ channels. As a result, their resting membrane potential (in the dark) is less negative (about −35 mV) than that of other nerve cells. When opsin absorbs a light photon, it undergoes a conformational change from a *cis* to a *trans* configuration, and that conformational change causes Na^+ channels to close. Without Na^+ ions entering the cell, the cell membrane becomes hyperpolarized (rather than depolarized) and the cell's neurotransmitter release is reduced. Photoreceptors themselves do not fire action potentials, but when stimulated by light energy they inhibit the firing rate of other neurons within the eye. The decrease in firing rate provides information about the intensity and pattern of light received.

Animals see the world through different types of eyes.

Although all eyes rely on similar opsin photopigments, the way the visual world is perceived—what it looks like to the organism—depends on the structure of the eye in which the receptors are embedded. Here, we look at three main types of

FIG. 36.11 Rhodopsin, the light-sensitive pigment found in vertebrates. Rhodopsin is made up of the transmembrane protein opsin bound to the light-absorbing retinal.

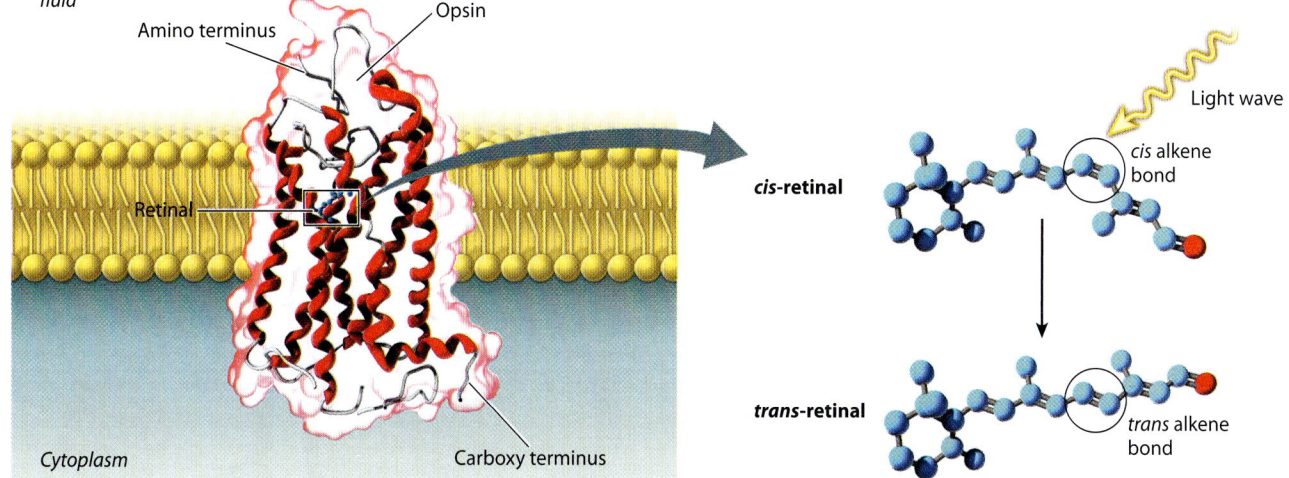

FIG. 36.12 Three types of eye.

a. Eyecup

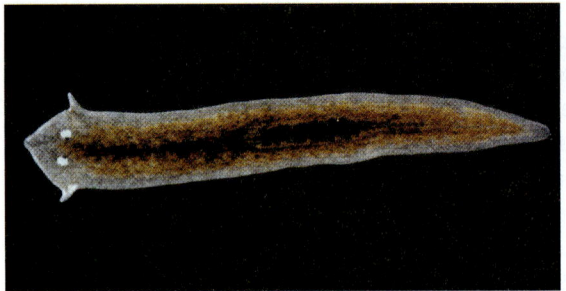

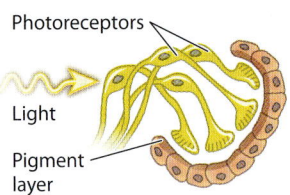

The flatworm *Planaria* uses simple photoreceptors and a pigmented epithelium to sense the direction of light.

b. Compound eye

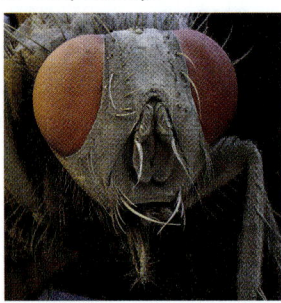

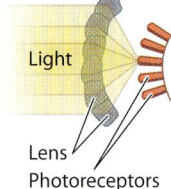

The compound eye of insects, such as the common housefly, are composed of hundreds of ommatidia, each with a lens, that individually sense light.

c. Single-lens eye

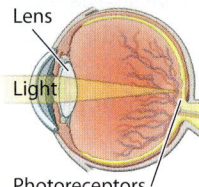

The single-lens eye of a squid focuses light on a retina and allows for a high degree of acuity.

eye structure shown in **Fig. 36.12**: **eyecups**, **compound eyes**, and **single-lens eyes**. Each has very different capabilities.

In flatworms, two simple eyecups on the dorsal, or back, surface of the head detect the direction and intensity of light sources (Fig. 36.12a). Each eyecup contains photoreceptors that point up and to the left or right. A pigmented epithelium blocks light from behind, so the photoreceptors receive light only from above and in front of the animal. In response to light, flatworms turn to move directly away from the light source, seeking a dark region that hides them from potential predators. To find a dark region, the flatworm compares the light intensity received by photoreceptors of the two eyecups, and then moves in the direction of lower light intensity. Many invertebrates use simple light-sensitive photoreceptors in a similar way.

Two major types of image-forming eye evolved in other invertebrates: compound eyes and single-lens eyes. Insects and crustaceans have compound eyes that consist of individual light-focusing elements called **ommatidia** (Fig. 36.12b). The number of ommatidia in the eye determines the resolution, or sharpness, of the image. The compound eyes of ants have only a few ommatidia, and those of fruit flies have about 800. In contrast, the eyes of dragonflies have 10,000 or more. Dragonflies are predators, and their high number of ommatidia is an adaptation for visually tracking their prey.

Within a single ommatidium, light is focused through a lens onto a central region formed from multiple overlapping photoreceptors. Each ommatidium is sensitive to a narrow angle of light (about 1° to 2°) in the animal's visual field. Compound eyes of insects and crustaceans provide a mosaic image because individual light regions are sensed by separate ommatidia. It is likely that the image is sharpened within the brain, but the resolution of images produced by compound eyes is not nearly as good as those produced by single-lens eyes. Nevertheless, compound eyes are extremely good at detecting motion and rapid flashes of light, more than 300 per second compared to 50 per second for a human eye. Insect eyes also have good color vision, and many insects can perceive ultraviolet light. Many pollinating insects sense UV light to locate the flowers they prefer to visit for nectar (**Fig. 36.13**).

The single-lens eyes of vertebrates and cephalopod mollusks like squid and octopus work like a camera to produce a sharply defined image of the animal's visual field (see Fig. 36.12c). In spite of their similar outward appearance, there is clear evidence that single-lens eyes evolved independently in vertebrates and in some mollusks (**Fig. 36.14**). Other invertebrates, such as some spiders and annelid worms, also evolved single-lens eyes independently. An advantage of single-lens eyes compared to compound eyes is that the single

FIG. 36.13 The same flower seen with visual and UV light. Many pollinating insects see ultraviolet light, which can dramatically alter the image that attracts them to a particular flower. (Tormentil is shown here.)

Seen with visual light

Seen with UV light

FIG. 36.14 Convergent evolution of single-lens eyes. Single-lens eyes evolved independently in vertebrates (fish and mammals) and in some mollusks (octopus and squid).

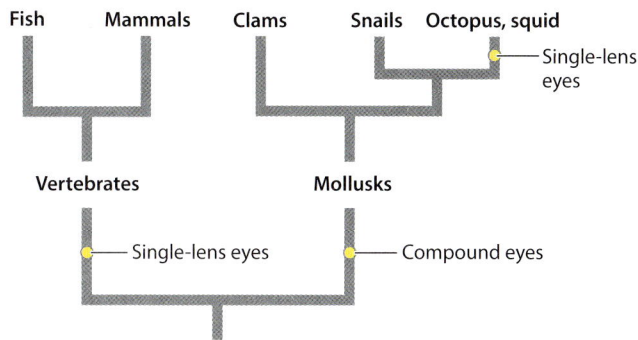

lens can focus light rays on a particular region of photoreceptors, improving both image quality and light sensitivity. As a result, animals with single-lens eyes can detect prey effectively by their motion and shape.

We focus much of our subsequent discussion on the human eye, but many of its organizational features and properties apply to vertebrates more generally.

The structure and function of the vertebrate eye underlie image processing.

Vertebrate eyes are generally protected within the eye socket of the skull. The anatomy of the eye is shown in **Fig. 36.15**.

The eye is surrounded by the **sclera,** a tough, white outer layer. Below the sclera, a thinner layer, called the choroid, carries blood vessels to nourish the eye. Mucus secretions produced over the sclera keep the eye moist. The portion of the sclera in the front of the eye is transparent; this is the **cornea.** Light passes through the transparent cornea, enters the eye through an opening called the **pupil,** and then passes through a convex **lens.** Small muscles called ciliary muscles attach between the sclera and lens. Contraction or relaxation of these muscles adjusts the shape of the lens to focus light images.

At the front of the eye, surrounding the pupil, the choroid forms a donut-shaped **iris** that opens and closes to adjust the amount of light that enters through the pupil, in the manner of the diaphragm of a camera. In bright light, the iris constricts to make the pupil small. In dim light, the iris dilates to allow in more light. The iris also gives the eye its color.

The focusing muscles and lens separate the eye into two regions. The interior region in front of the lens is filled with a clear watery liquid, the **aqueous humor.** This liquid is produced continuously and drained by small ducts at the base of the eye. Blockage of the ducts results in increased pressure within the eye that if sustained can cause glaucoma, which sometimes leads to blindness. The large cavity behind the lens is filled with a gel-like substance, the **vitreous humor,** that makes up most of the eye's volume.

To form an image, the cornea and lens bend incoming light rays, focusing them on the **retina** (Fig. 36.15). In doing so, the light rays cross, inverting the image on the retina and changing up to down and left to right. The retina is a thin tissue located in the posterior of

FIG. 36.15 Anatomy of the human eye. The eyes of other mammals and of birds have a similar organization.

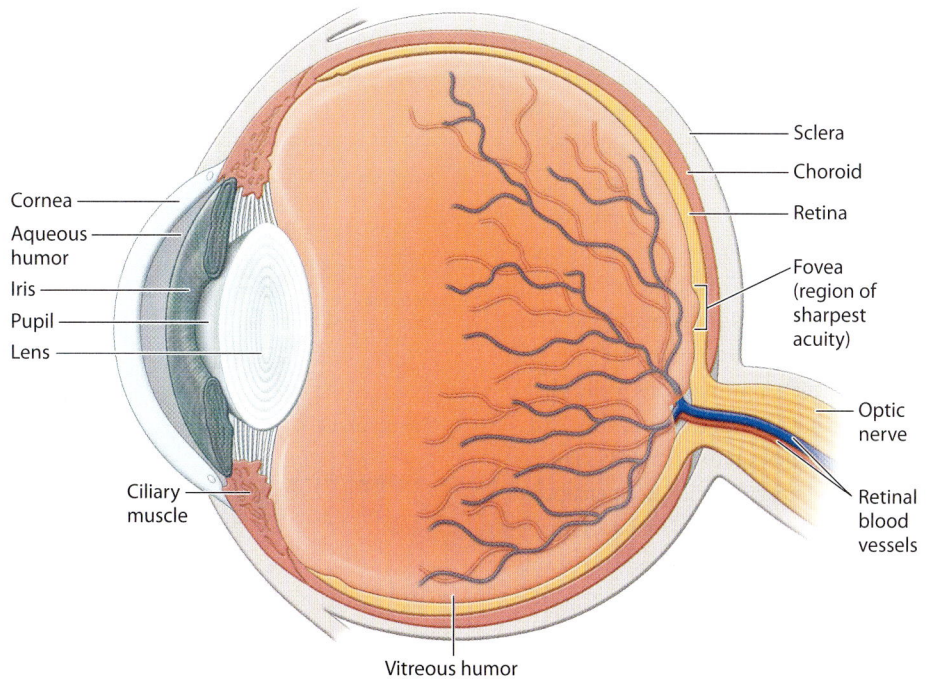

FIG. 36.16 Focusing an image. The eyes of humans and other mammals and of birds focus the light of an image on the retina by changing the shape of the lens.

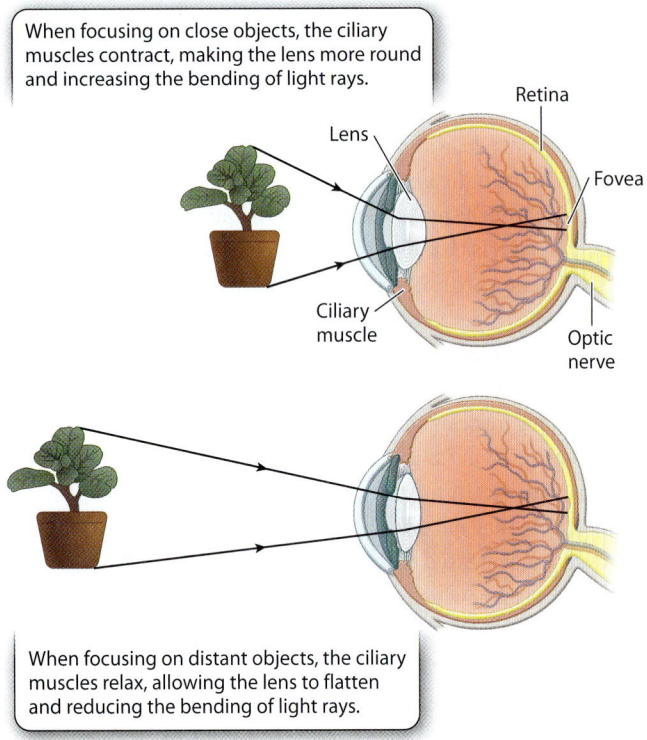

When focusing on close objects, the ciliary muscles contract, making the lens more round and increasing the bending of light rays.

When focusing on distant objects, the ciliary muscles relax, allowing the lens to flatten and reducing the bending of light rays.

the eye. It contains the photoreceptors and other nerve cells that sense and initially process light stimuli. Humans, other mammals, and birds focus light on the retina by changing the shape of the lens (**Fig. 36.16**). When we focus on objects at close range, as when we read a book, the ciliary muscles contract, making the lens more rounded to bend the light rays more. When we focus on distant objects, the ciliary muscles relax, allowing the lens to flatten, reducing the diffraction, or bending, of light. In contrast, the eyes of octopus and squid focus by moving the lens back and forth within the eye, similar to a camera, rather than by changing its shape.

Why do so many animals have two eyes? One reason is that two eyes provide a wider field of vision. Close one eye and you will see the visual field of the other eye. With both eyes open, the visual field is wider than the visual field of either eye alone. You'll also notice considerable overlap in the visual field of both eyes. This overlap exists in humans and other primates, as well as in other mammals, and makes possible binocular vision, the ability to combine images from both eyes to produce a single visual image. Binocular vision allows for depth and distance perception.

Prey animals, such as some birds, antelope, and rabbits, have eyes positioned more to the side, expanding their visual field up to 360°, but limiting or preventing binocular vision. Animals that require greater visual sharpness, such as predatory hawks, cats, and snakes, have eyes pointing forward, increasing their binocular field of view.

Color vision detects different wavelengths of light.

Color vision is crucial for many invertebrate and vertebrate animals. It is achieved by photoreceptor cells that contain visual pigments sensitive to different wavelengths of light (**Fig. 36.17**). Color-sensitive pigments are found within the **cone cells** of vertebrates. Cone cells are distinguished by their shape from the more numerous **rod cells**, which are sensitive to light but not color. Rod cells detect shades ranging from white to shades of gray and black. The human retina contains about 6 million cone cells and about 125 million rod cells. Because of their greater number and sensitivity to light, rod cells enable animals to see in low light. Together, rod and cone cells constitute approximately 70% of all sensory receptor cells in the human body, highlighting the importance of vision to perceive our environment.

Vertebrate cone cells likely evolved from a rod cell precursor. Most vertebrates have cone cells with four photopigments, whereas the cone cells of most mammals have only two. It is likely that two photopigments were lost early in mammal evolution, when mammals were small, nocturnal, and burrowing. Old World primates, apes, and humans re-evolved trichromatic (that is, three-color) vision by means of gene duplication (some New World primates also regained trichromatic vision), likely in response to selection for better ability to locate fruit, a key part of their diet. Each human cone cell therefore has one of three photopigments, which absorb light at blue, green, or red wavelengths (**Fig. 36.18**). Stimulation of cone cells in varying combinations of these three photopigments allows humans and other primates to see a full range of color (violet to red). Fish, amphibians, reptiles, and birds also have good color vision, including for many the ability to see ultraviolet.

The low sensitivity of cone cells to light makes it hard to detect color at night. Cone cells are most concentrated within the **fovea** of the retina, the center of the visual field of most vertebrates (see Fig. 36.15). Cone cells provide the sharpest vision. Animals

FIG. 36.17 Vertebrate photoreceptors: rods and cones.

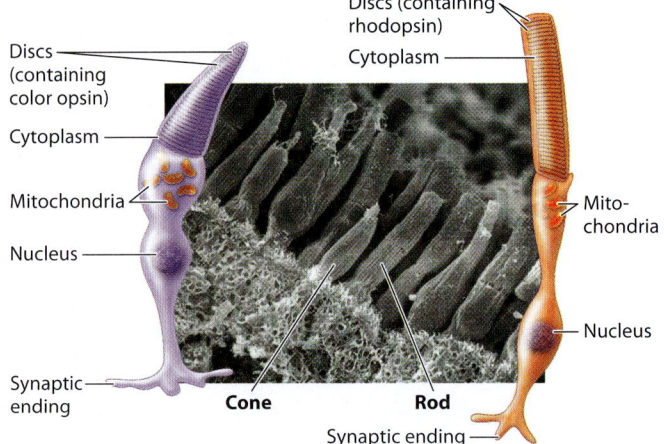

FIG. 36.18 **Absorption spectra of the three opsins in human cone cells.** Each cone cell expresses just one of the three opsins.

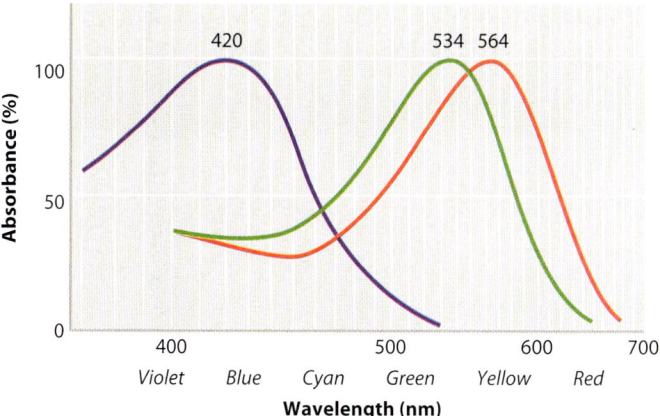

with particularly sharp vision, such as hawks and other birds of prey, have an extremely concentrated number of cone cells, approaching 1 million/mm² (compared to about 150,000/mm² in the human fovea). Birds of prey can see small prey on the ground from high in the air. Many birds also have eyes with two foveae, one projecting forward for binocular vision and one projecting to the side to enhance the sharpness of their side vision.

In contrast, rod cells are absent in the fovea but predominate in the periphery. Their positioning makes it easier to detect motion in the periphery, particularly at night. The eyes of many nocturnal predatory mammals, such as foxes and cats, contain mainly rod cells, enhancing nighttime vision. These animals also have a pigment within the retina that reflects light past the photoreceptors, enhancing the ability to sense light under low-light conditions. The reflection of light by this pigment creates the "eyeshine" that you see when a flashlight or a car's headlights shines on a nocturnal predator's eyes at night. Great white sharks are able to hunt at night because they also have a retinal reflective pigment.

Local sensory processing of light determines basic features of shape and movement.

The rods and cones detect the intensity, color, and pattern of light entering the eye through the lens. However, it is their interaction with a highly ordered array of interneurons in the retina that begins the first steps of visual sensory processing (**Fig. 36.19**). The retina consists of five layers of neurons that form a signal-processing network just in front of a pigmented epithelium. The rods and cones are at the back of the retina, so that light must pass through several layers of neurons before reaching these light-sensitive cells. Because they are photoreceptors, rod and cone cells hyperpolarize in response to light but do not fire action potentials. They synapse on to **bipolar cells**, which also do not fire action potentials, but rather adjust their release of neurotransmitter in response to the input from multiple rod and cone cells. The bipolar cells, in turn, synapse on to **ganglion cells** located on the front of the retina. If activated, ganglion cells transmit action potentials by the **optic nerve** (a cranial nerve) to the **visual cortex** in the brain, the part of the brain that processes images. The optic nerve begins at the front of the retina and exits at the back, creating an area without light-sensitive cells: This is the blind spot. In octopus and squid, photoreceptors are in the front of the retina, not in the back as they are in vertebrates. As a result, octopus and squid don't have a blind spot.

In addition to these three primary layers of neurons, two additional sets of nerve cells combine information laterally across the retina. **Horizontal cells** communicate between neighboring pairs of photoreceptors and bipolar cells, enhancing contrast through lateral inhibition to sharpen the image. **Amacrine cells** communicate between neighboring bipolar cells and ganglion cells, enhancing motion detection and adjusting for changes in illumination of the visual scene.

Photoreceptor cells, bipolar cells, ganglion cells, horizontal cells, and amacrine cells form a neural network that processes visual information before it is sent to the brain for further processing and interpretation. Visual signals from approximately 125 million photoreceptors converge to approximately 1 million

FIG. 36.19 **Cellular organization of the human retina.** The retina is made up of five layers of cells: rods and cones, bipolar cells, ganglion cells, horizontal cells, and amacrine cells.

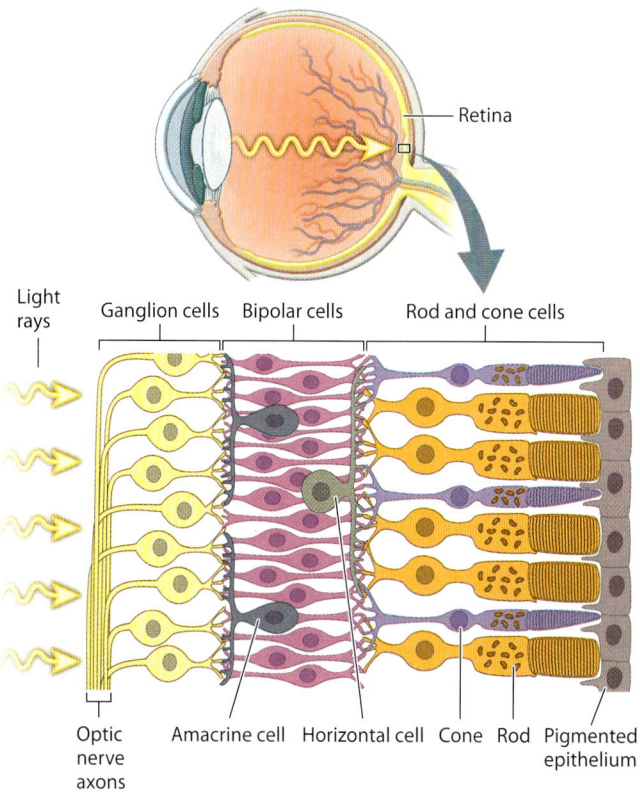

ganglion cell neurons that communicate with the brain. Properties of this network were first studied in the mammalian retina by the American neurophysiologists Stephen Kuffler, David Hubel, and Torsten Wiesel in the 1950s and 1960s (**Fig. 36.20**). Hubel and Wiesel shared the 1981 Nobel Prize in Physiology or Medicine for their later work on how visual information is processed in the brain.

→ **Quick Check 4** What is the primary role of rod cells, and what are the two roles of cone cells in the retina?

HOW DO WE KNOW?

FIG. 36.20

How does the retina process visual information?

BACKGROUND In the 1950s, the American neurophysiologist Stephen Kuffler was interested in understanding how the retina helps to process light information before it is sent to the brain. He focused on the activity of ganglion cells in the retina because they receive input from the photoreceptors and bipolar cells.

EXPERIMENT Kuffler stimulated different regions of the cat's retina with localized points of light while recording the action potentials produced by ganglion cells.

RESULTS Kuffler found that there are two types of ganglion cell: on-center and off-center cells. On-center ganglion cells fire more action potentials when light shines on the center of the cell's receptive field compared to the surrounding region, and off-center cells fire more when light is shown in the periphery and less on the center. These patterns are explained by lateral inhibition of input to the ganglion cells by the photoreceptors and bipolar neurons in the retina.

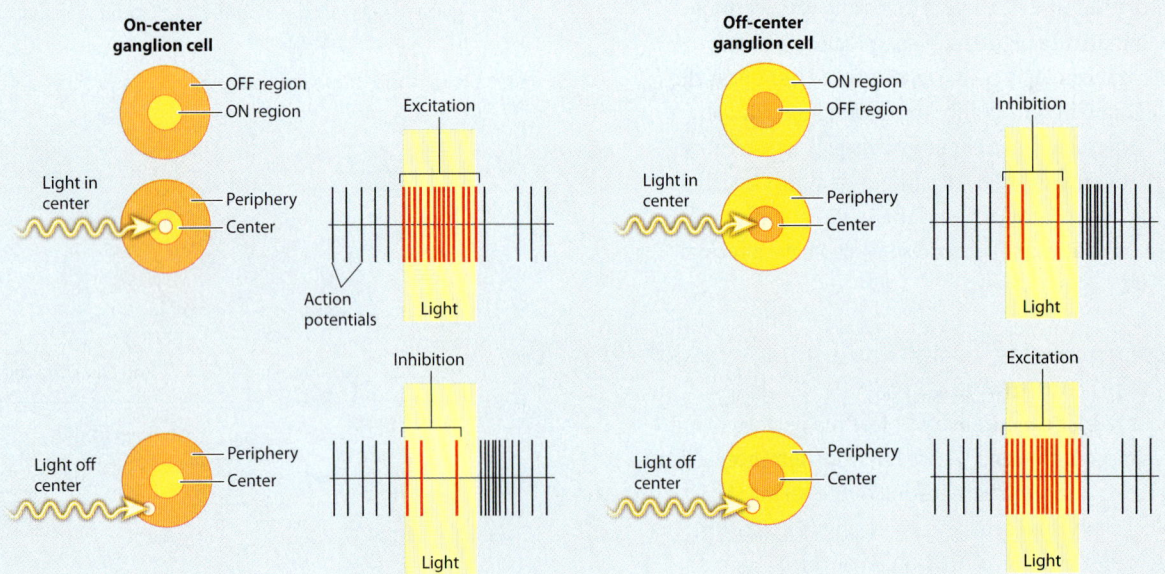

FOLLOW-UP WORK In the 1960s, Hubel and Wiesel found similar center–surround neural receptive fields, though with enhanced opposition, in part of the thalamus and in the visual cortex of the brain. Cells with these fields enable cats and other mammals to detect shapes of a given orientation moving through their visual field. Similar center–surround receptive fields have also been found in the somatosensory and auditory cortex, highlighting the use of lateral inhibition to enhance sensory acuity and edge detection. Other studies have found similar center–surround sensory processing in invertebrates and other vertebrates.

SOURCES Kuffler, S. W. 1953. "Discharge Patterns and Functional Organization of Mammalian Retina." *Journal of Neurophysiology* 16:37–68; Hubel, D. H. 1963. "The Visual Cortex of the Brain." *Scientific American* 209:54–62.

36.5 BRAIN ORGANIZATION AND FUNCTION

Up to this point, we have looked at how sensory information, such as smell, taste, hearing, and vision, is detected by sensory receptors. In each case, the sensory information is sent to the brain, where it is processed and integrated. Much of what we know about how the brain functions has been accumulated over many years by studies of patients who suffered brain injuries. A well-known example is Phineas Gage, a railroad worker who in 1848 survived severe damage to his brain's frontal lobe after a tamping iron he was using to set a dynamite charge caused the dynamite accidentally to explode, sending the iron through his skull and the frontal cortex of his brain (**Fig. 36.21**). Gage had been hard-working, responsible, and well liked; after the accident he was given to fitful arguments, lacked self-control, and showed reduced intellectual capacity. These changes provided some of the first evidence that particular brain regions control particular aspects of a person's behavior and cognitive function. The remainder of this chapter explores the functional organization of the vertebrate brain, highlighting examples that show how the brain integrates, organizes, and maps sensory information, and how memories of sights and sounds, for example, are formed that contribute to learning.

The brain processes and integrates information received from different sensory systems.

With the evolution of greater cognitive ability, animal brains became larger and more complex. Different brain regions became specialized to carry out different functions. The vertebrate brain is organized into a **hindbrain, midbrain,** and **forebrain,** with a

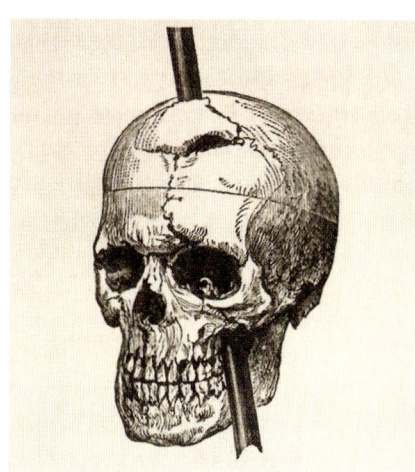

FIG. 36.21 **The skull of Phineas Gage.** Phineas Gage survived an accident in which a tamping iron passed through his skull, but his personality changed and his intellectual capacity diminished.

cerebral cortex formed from a portion of the forebrain (**Fig. 36.22a**). The hindbrain and midbrain control basic body functions and behaviors, and the forebrain, particularly the cerebral cortex, governs more-advanced cognitive functions. The cerebral cortex of mammals—particularly primates and humans—is greatly expanded relative to that of other vertebrates.

In adult vertebrates, the hindbrain develops into the **cerebellum** and a portion of the **brainstem** (Fig. 36.22a). The remainder of the

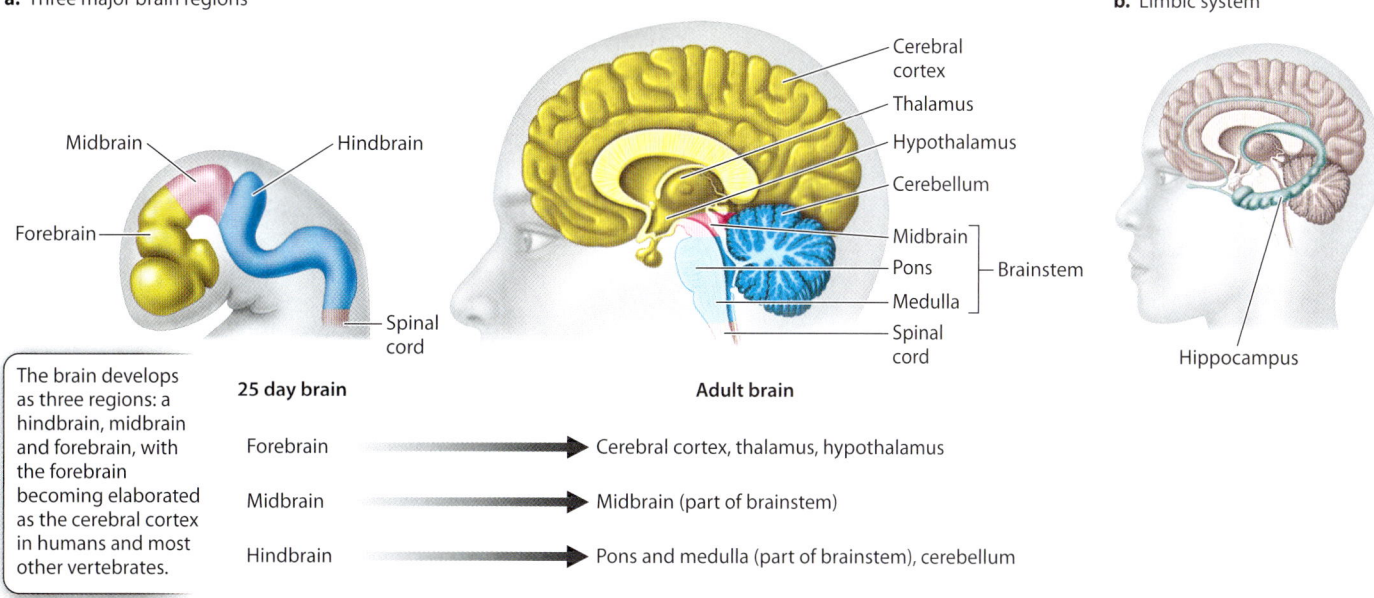

FIG. 36.22 **Human brain development and organization.** (a) The brain develops from the forebrain, midbrain, and hindbrain. (b) Part of the forebrain develops into the limbic system, which controls drives and emotions.

brainstem develops from the midbrain. The cerebellum coordinates complex motor tasks, such as catching a ball or learning to write and talk. It integrates both motor and sensory information. The brainstem, consisting of the **medulla, pons,** and **midbrain,** initiates and regulates motor functions such as walking and controlling posture and coordinates breathing and swallowing. The brainstem activates the forebrain by relaying information from lower spinal levels. High levels of activity within the brainstem maintain a wakeful state; low levels enable sleep. If the midbrain is damaged, loss of consciousness and then coma result.

The forebrain consists of an inner brain region that forms the **thalamus** and the underlying **hypothalamus,** and a more anterior region that develops into the **cerebrum,** the outer left and right hemispheres of the cerebral cortex (Fig. 36.22a). The thalamus is a central relay station for sensory information sent to higher brain centers of the cerebrum. The hypothalamus interacts closely with the autonomic and endocrine systems to regulate the general physiological state of the body. In humans and most other primates, the cerebral cortex is the largest part of the brain, overseeing sensory perception, memory, and learning. Whereas the forebrain is essential to normal behavior in mammals, it appears less important for other vertebrates.

Other inner components of the forebrain constitute the **limbic system,** which controls physiological drives, instincts, emotions, and motivation and, through interactions with midbrain regions, the sense of reward **(Fig. 36.22b).** Stimulation of the limbic system can induce strong sensations of pleasure, pain, or rage, reinforcing learning related to basic physiological drives. A posterior region of the limbic system, the **hippocampus,** is involved in long-term memory formation, discussed in the next section.

Sensory information reaches the cerebral cortex from the cranial nerves and nerves passing through the spinal cord. This information passes through the brainstem, and then through the thalamus, the central relay station for sensory information. From the thalamus, information from each of the senses goes to a different region of the brain specialized to further process that information, in a manner discussed next.

The brain is divided into lobes with specialized functions.

The cerebral hemispheres are the largest structures of the mammalian brain (**Fig. 36.23**). They consist of a highly folded outer layer of **gray matter** about 4 mm thick, called the **cortex,** that is made up of densely packed neuron cell bodies and their dendrites (Fig. 36.23a). The folds greatly increase the surface area of the cortex and are the result of selection for an increased number of cortical neurons within the limited volume of the skull. Deep inside the cortex is the **white matter,** which contains the axons of cortical neurons. It is the fatty myelin produced by glial cells surrounding the axon that makes this region of the brain white. These axons are the means by which neurons in different regions of the gray matter communicate with neurons in other regions of the cortex and in deeper regions of the forebrain, midbrain, and hindbrain.

FIG. 36.23 Human brain organization. (a) The folding of the brain and (b) its lobes.

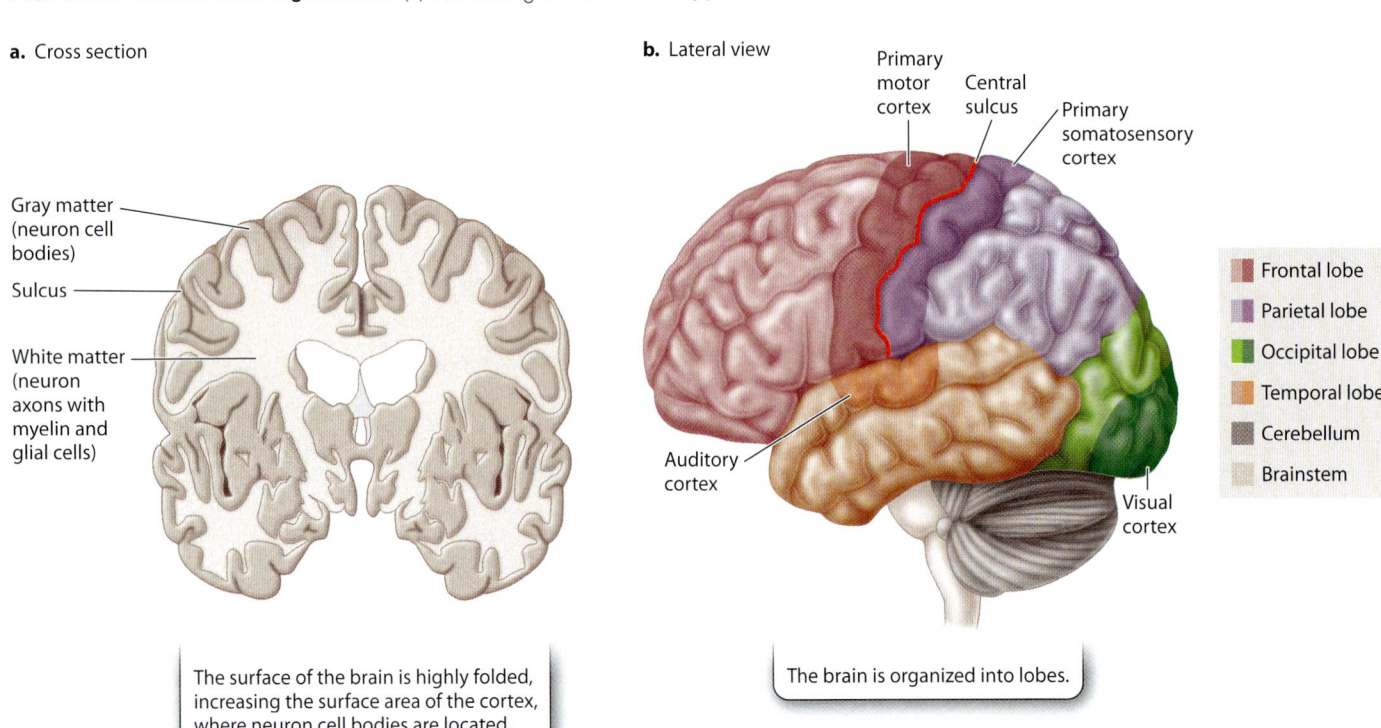

FIG. 36.24 Somatosensory topographic map.

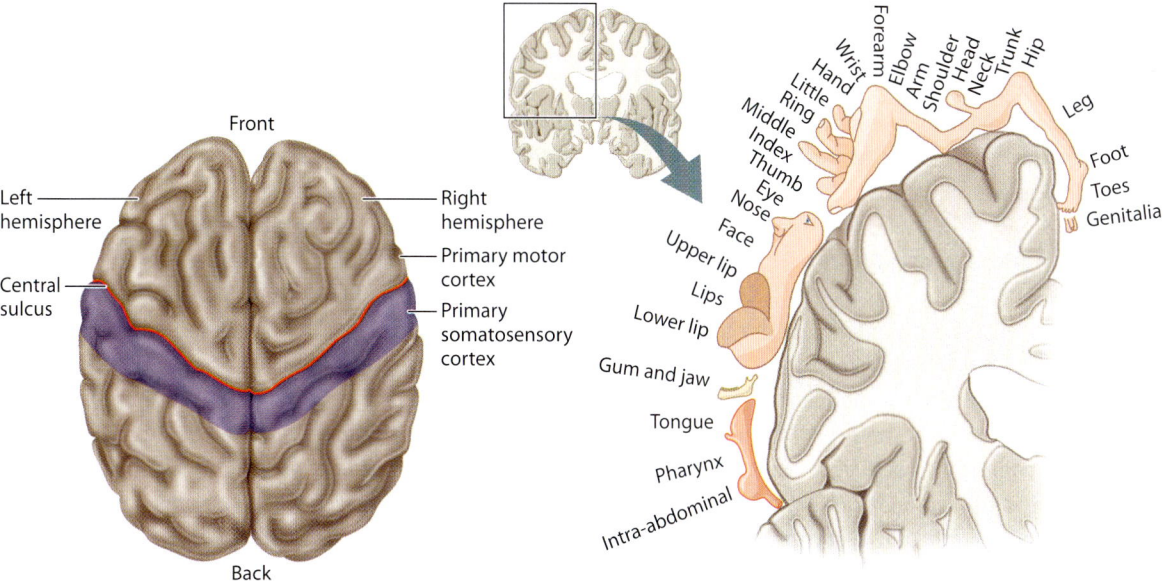

Major regions of the cerebral hemispheres are defined by clearly visible anatomical lobes separated in most cases by deep crevices called **sulci** (singular, sulcus) (Fig. 36.23b). The **frontal lobe** is located in the anterior region of the cerebral cortex (behind your forehead). It is important in decision making and planning. The two **parietal lobes** are located posterior to the frontal lobe. The parietal lobes are separated from the frontal lobe by the central sulcus. The parietal lobe controls body awareness and the ability to perform complex tasks, such as dressing or brushing one's hair. The two **temporal lobes** lie to the side, below the parietal lobes. They are involved in processing sound, as well as performing other functions discussed below. Located behind the parietal lobe, at the back of the brain, is the **occipital lobe,** which processes visual information from the eyes.

The central sulcus separates the **primary motor cortex** of the frontal lobe from the **primary somatosensory cortex** of the parietal lobe (**Fig. 36.24**). "Command" neurons in the primary motor cortex produce complex coordinated behaviors by controlling skeletal muscle movements. The primary somatosensory cortex integrates tactile information from specific body regions and relays this information to the motor cortex. Both cortices make connections to the opposite side of the body by sending axons that cross over in the brainstem and spinal cord. As a result, the right cortex controls the left side of the body and the left cortex controls the right side of the body.

Information is topographically mapped into the vertebrate cerebral cortex.

The primary motor cortex and somatosensory cortex are organized as maps that represent different body regions, as shown in Fig. 36.24. The neurons that control movements of the foot and human lower limb are located near the mid-axis of the motor cortex. Those that control motions of the face, jaws, and lips map to the side and farther down in the primary motor cortex. There is a similar topographic map in the somatosensory cortex for pressure and touch sensation. Regions such as the fingers, hands, and face that are involved in fine motor movements and distinguishing fine sensations are represented by larger areas of the cortex.

The auditory cortex in the temporal lobe is similarly organized as a map. This region processes sound information transmitted from the cochlea of the ear. Neurons in the auditory cortex are organized by pitch: Neurons sensitive to low frequencies are located at one end and neurons sensitive to high frequencies at the other, helping to distinguish fine differences in pitch. With the evolution of language in humans, regions farther back in the temporal lobe developed into language and reading centers. These centers are linked by pathways to a specialized motor center controlling speech located at the base of the primary motor cortex of the frontal lobe. Functions of the temporal lobe include object identification and naming. Damage to the temporal lobe can lead to an individual's loss of facial recognition, even though others may still be recognized by voice or other body features.

The occipital lobe topographically maps visual information received from the optic nerves. The topographic mapping of information from retinal ganglion cells enables the brain to detect patterns and motion that are precisely related to different regions of the visual field. As a result, highly visual animals, such as humans, primates, and birds, are able to detect complex patterns and movements in their environment.

36.6 MEMORY AND COGNITION

The ability of the brain to process and integrate complex sources of information, interpret and remember past events, solve problems, reason, and form ideas is broadly referred to as **cognition.** In this section, we consider memory and learning, as examples of cognitive abilities.

The brain serves an important role in memory and learning.

Memory is the basis for learning. An animal that can remember its encounter with a predator or noxious plant will avoid that danger in the future. A memory is formed by changes to neural circuits within specific regions of the brain—specifically, by changes in the synaptic connections between neurons in a neural circuit. These changes enable an animal, for example, to recognize other individuals in its social group or to recognize the threat of a predator. In humans and other primates, the hippocampus, a brain structure in the limbic region, plays a special role in memory formation. We can remember much of the information we receive for several minutes. However, unless reinforced by repetition or by paying particular attention, the information is typically forgotten. The hippocampus transforms reinforced short-term memories into long-term memories. Long-term memory is essential for learning.

The hippocampus forms long-term memories by repetitively relaying information to regions of the cerebral cortex. The molecular and cellular basis for memory and learning remains uncertain but depends on establishing particular neural circuits in the cortex. These circuits can be activated to recall a memory that is triggered by relevant stimuli. The formation of a memory circuit requires changes in synapses, which establish a particular set of connections between neurons in the circuit. Some synapses are weakened or removed altogether, and others are formed or strengthened to establish and refine the circuit.

The ability to adjust synaptic connections between neurons is called **synaptic plasticity** (**Fig. 36.25**). Long-term potentiation (LTP) is an example of synaptic plasticity that is believed to underlie memory and learning. In response to repeated excitation, neurons in excited circuits within the hippocampus release the neurotransmitter glutamate, opening Na^+ and Ca^{2+} channels. The increased Ca^{2+} stimulates protein synthesis and signaling pathways that make the cells more responsive to subsequent stimulation. The release of glutamate stimulates the placement of new receptors in the postsynaptic cell membrane as well as the formation of new dendrites, strengthening synaptic signaling between the two cells.

Through repeated activation of neurons within the hippocampus and associated regions of the cerebral cortex, neural circuits are established that may be reinforced by LTP to create a memory that can be retrieved over long periods of time. Although

FIG. 36.25 Synaptic plasticity. Memory and learning depend on long-term potentiation (LTP) of synapses within neural circuits.

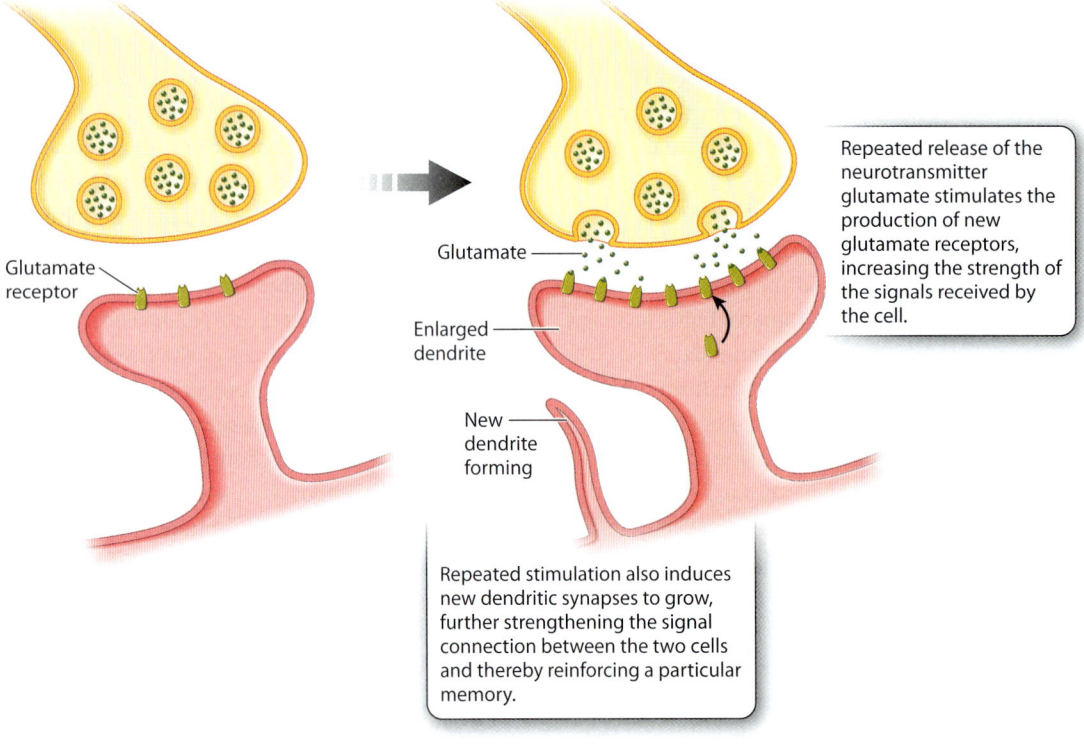

memories may be associated with certain regions of the cortex, they also rely on the interaction of nerve cells and networks within multiple brain regions. Consequently, when a local brain region suffers damage, memories that are lost as a result may sometimes be recovered by reinforcement of the memory stored within other regions.

Although the brain regions involved in memory and learning vary among different animals, the basic features of how memories are constructed by means of synaptic plasticity within neural circuits are likely broadly shared across animals capable of forming memories. For example, the ability of an octopus to learn and remember how to perform motor tasks rivals that of certain vertebrates. Even the roundworm *C. elegans* has been found to be capable of memory and learning: Individual roundworms learn to avoid pathogenic bacteria in favor of bacteria that are nutritious.

→ **Quick Check 5** Which three areas of the human brain are important for remembering and recognizing a face?

Cognition involves brain information processing and decision making.

Cognitive brain function ultimately gives rise to a state of consciousness—an awareness of oneself and one's actions in relation to others. Consciousness allows judgments to be made about events and experiences in one's environment. When you suddenly see someone you recognize from long ago, rapid neural associations within the cortex identify that person by name, retrieve multiple memories of past events you shared with that person, and elicit conscious thoughts, including your opinion of that person. At the same time, extraneous sensory information is filtered and other thoughts recede, allowing you to focus your attention on the person you have just seen and recognized.

Mental processes are studied by relating changes in brain activity to changes in an animal's behavioral state. The study of the brain is made challenging by the complexity of the circuits involved. Nevertheless, considerable progress is being made through non-invasive neuroimaging techniques (**Fig. 36.26**). These techniques allow the interactions of different brain regions within human and other animal subjects to be investigated while the subjects perform a particular mental or motor task. The results of neuroimaging studies can be linked to studies of the molecular and cellular properties of nerve cells, to advance our understanding of how nerve cells form and maintain nerve cell circuits among different brain regions, and of how these circuits are linked to particular behaviors and cognitive processes.

FIG. 36.26 Non-invasive imaging of brain function. PET scan imaging reveals regions of metabolic brain cell activity. The reddish areas show the locations of (upper left) the visual area activated by sight, (upper right) the auditory area activating by hearing, (lower left) the tactile area activated by touching braille, and (lower right) areas of the frontal cortex activated by generating words.

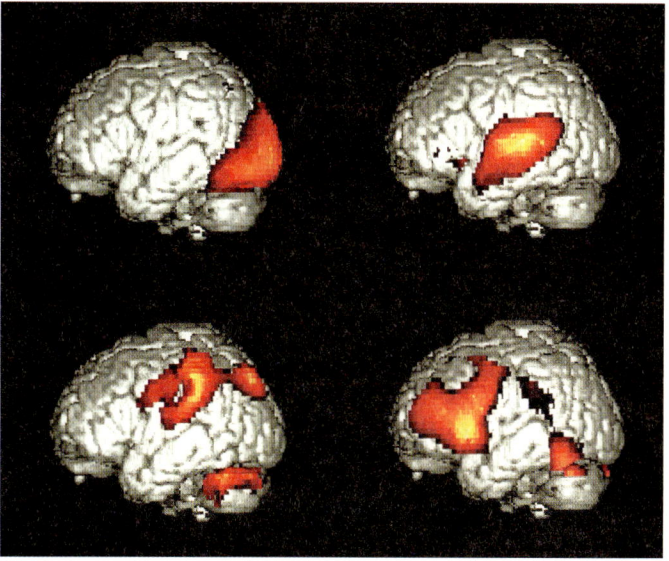

Do animals other than humans have a conscious state of awareness? Although the conscious awareness of other animals is difficult to study, it seems clear that many animals exhibit conscious states of awareness that influence their experience and behavior. Donald Griffin, known for his co-discovery as an undergraduate of bat echolocation, advocated the view not only that animals are conscious, but also that animal consciousness is amenable to scientific study. Behavioral biologist Marion Dawkins summarized her review of studies on animal cognition and consciousness as follows:

Our near-certainty about [human] shared experiences is based, amongst other things, on a mixture of the complexity of their behavior, their ability to "think" intelligently and on their being able to demonstrate to us that they have a point of view in which what happens to them matters to them. We now know that these three attributes—complexity, thinking, and minding about the world—are also present in other species. The conclusion that they, too, are consciously aware is therefore compelling.

Core Concepts Summary

36.1 ANIMAL SENSORY RECEPTORS DETECT PHYSICAL AND CHEMICAL STIMULI BY CHANGES IN MEMBRANE POTENTIAL.

Sensory neurons include chemoreceptors, mechanoreceptors, thermoreceptors, pain receptors, and electromagnetic receptors. page 36-2

Action potential firing rate correlates with the strength of a stimulus. Generally, strong stimuli induce high firing rates and weak stimuli induce low firing rates. page 36-4

Animal sensory receptors increase their sensitivity to stimuli by temporal and spatial summation, enhance their acuity by lateral inhibition, and adapt to continuous stimuli. page 36-4

36.2 SPECIALIZED CHEMORECEPTORS RELAY INFORMATION ABOUT SMELL AND TASTE.

Odorant molecules bind to membrane receptors on the surface of olfactory chemoreceptors. page 36-5

In humans, five taste receptors—for sweet, bitter, sour, salty, and savory—are activated in combination to determine a specific taste. page 36-6

36.3 HAIR CELLS CONVEY INFORMATION ABOUT MOVEMENT AND SOUND.

Hair cells are specialized mechanoreceptors that detect motion and vibration. page 36-6

The statocysts of invertebrates are sensitive to gravity, orienting an organism to "up" and "down." page 36-7

The vestibular system of the vertebrate ear provides a sense of balance and gravity. page 36-7

The tympanic membrane in the outer ear transmits airborne sound waves to the middle ear, where the signal is amplified, and then to the cochlea in the inner ear, where sound waves are converted to fluid pressure waves and then sensed by hair cells that convert the signal to an electrical impulse. page 36-8

36.4 THE ABILITY TO SENSE LIGHT AND FORM IMAGES DEPENDS ON PHOTOSENSITIVE CELLS WITH LIGHT-ABSORBING PROTEINS.

The protein opsin present in photoreceptors converts light energy into chemical signals, altering the firing rate of neurons they inhibit or stimulate. page 36-11

The eyecups of flatworms detect light and dark. page 36-12

The compound eyes of arthropods sense light using individual light-focusing elements called ommatidia, providing low acuity but rapid motion detection. page 36-12

The single-lens eyes of vertebrates and cephalopod mollusks focus images on a retina, which contains photoreceptor cells and interneurons that process the light stimuli. page 36-12

Photoreceptor cells of the retina include rod cells, which detect light but not color, and cone cells, which detect color. page 36-14

Photoreceptor cells, bipolar cells, ganglion cells, horizontal cells, and amacrine cells form a network that processes visual information in the retina. page 36-15

36.5 THE BRAIN PROCESSES AND INTEGRATES INFORMATION FROM MULTIPLE SENSORY SYSTEMS, WITH TACTILE, VISUAL, AND AUDITORY STIMULI MAPPED IN THE CEREBRAL CORTEX.

The vertebrate brain is organized into a hindbrain, midbrain, and forebrain, with the forebrain elaborated into the cerebral cortex in birds and mammals. page 36-17

The brain is divided into frontal, parietal, temporal, and occipital lobes that are specialized for different functions. page 36-19

Somatosensory, motor, auditory, and visual information is topographically mapped to specific areas in the cerebral cortex. page 36-19

36.6 COGNITION IS THE ABILITY OF THE BRAIN TO PROCESS AND INTEGRATE COMPLEX SOURCES OF INFORMATION, TO REMEMBER AND INTERPRET PAST EVENTS, TO SOLVE PROBLEMS, TO REASON, AND TO FORM IDEAS.

The cerebral cortex integrates and processes information from diverse sources, giving rise to problem solving, reasoning, and decision-making. page 36-20

The brain stores memories of past experiences and enables learning through the creation of more permanent neural circuits. page 36-20

Consciousness is an awareness of one's own identity in relation to others. page 36-21

Self-Assessment

1. Give an example of a chemosensory neuron, a mechanosensory neuron, and an electromagnetic sensory neuron.

2. Diagram a generalized receptor neuron, showing how it changes its firing rate in response to detected stimuli.

3. State a hypothesis that explains why all animals have chemoreceptors.

4. Describe the three main stages by which sound is detected and coded by the mammalian ear.

5. Compare and contrast the roles of rod cells and cone cells in the retina.

6. Describe the role of the cornea and lens in vertebrate eyes.

7. Describe three different types of eye in animals.

8. Draw the brain, label the lobes, and describe their primary functions.

9. Describe the importance of topographic mapping of sensory input to the cortex, using the primary somatosensory cortex as an example.

10. Explain how brain function can be understood by studying patients with brain injuries.

Do you understand the chapter's Core Concepts? Log in to check your answers to the Self-Assessment questions, then practice what you've learned and reinforce this chapter's concepts by working through the problems and multimedia tutorials provided there.

www.biologyhowlifeworks.com

CHAPTER 37

ANIMAL MOVEMENT

Muscles and Skeletons

Core Concepts

37.1 Muscles are biological motors composed of actin and myosin that generate force and produce movement for body support, locomotion, and control of the body's internal physiological functions.

37.2 Muscle force depends on muscle size, degree of actin–myosin overlap, contractile velocity, and stimulation rate.

37.3 Hydrostatic skeletons, exoskeletons, or endoskeletons provide animals with mechanical support and protection.

37.4 Vertebrate endoskeletons allow for growth and repair and transmit muscle forces across joints.

The ability to move is a defining feature of animal life, allowing animals to explore new environments and avoid inhospitable ones, escape predators, mate, feed, and play. An animal's motor and nervous systems (Chapters 35 and 36) work together to enable it to sense and respond to its environment. How are movements produced, and how are they controlled? What determines an athlete's performance? What are the mechanical requirements for movement and support? This chapter explores the organization and function of the muscles and skeletons that provide mechanical support and power the movements of larger multicellular animals (**Fig. 37.1**).

37.1 MUSCLES: BIOLOGICAL MOTORS THAT GENERATE FORCE AND PRODUCE MOVEMENT

We rely on our muscles for all kinds of movement. Muscles function as biological motors within the body because they generate force and produce movement. Some of these movements are obvious—running, climbing up stairs, raising your hand, playing the piano. We are less aware of others—moving food through the digestive tract, breathing in and out, pumping blood through the body. A muscle's ability to produce movement depends on a combination of electrically excitable muscle cells and contractile proteins within these cells that can be activated by the nervous system.

The contractile machinery of muscles is ancient. The proteins underlying muscle contraction are found in eukaryotes that lived more than 1 billion years ago, and the first muscle fibers common to all animals are found in cnidarians (jellyfish and sea anemones), which lived at least 600 million years ago. Thus, basic features of muscle organization and function are conserved across the vast diversity of eukaryotic life. As we will discuss, the geometry and organization of these proteins largely determines how muscles contract to produce force and movement.

Muscles can be striated or smooth.

Muscle cells, or **fibers,** produce forces within an animal's body, as well as exert forces on the environment. Despite their diversity, all muscle cells produce force by a similar mechanism. Skeletal muscle fibers produce forces by shortening, thus pulling on bones of the skeleton. Muscles pull on the bones of your leg and foot to swing them forward when you kick a ball. Muscle fibers in the heart contract to pump blood. Still other muscle fibers control blood flow through arteries by contracting to change an artery's cross-sectional shape, or are used for movements involved in digestion and reproduction.

Muscles can shorten very quickly, and they can produce large forces for their weight. The forces they produce can be many times an animal's body weight, allowing

FIG. 37.1 Muscle and bones. Muscle and bones work together to produce movement.

the animal to accelerate quickly. Some small muscles contract extremely quickly: The muscles of many flying insects contract 200 to 500 times per second. Even in vertebrate animals, muscles can contract as many as 50 to 100 times per second. Other muscles, such as the closing muscle of a clam, contract slowly, but do so for prolonged periods without tiring.

All muscles contain contractile proteins that enable them to shorten and produce force, but they are divided into two broad groups on the basis of their function and appearance under a light microscope (**Fig. 37.2**). As their name suggests, **striated muscles** appear striped under a light microscope. They include **skeletal muscles** (Fig. 37.2a), which connect to the body skeleton to move the animal's limbs and torso, and **cardiac muscle** (Fig. 37.2b), which contracts to pump blood. In contrast, **smooth muscles** appear uniform under the light microscope (Fig. 37.2c). They are found in the walls of arteries to regulate blood flow, in the respiratory system to control airflow, and in the digestive and excretory systems to help transport food and waste products. Smooth muscles contract slowly compared to cardiac and skeletal muscles.

Striated and smooth muscles use the same sets of muscle proteins—**actin** and **myosin** (Chapter 10)—to contract and generate force (force is mass times acceleration, and in this case is the ability to accelerate the motion of the body). These proteins are organized into thin threads called **filaments** that interact with one another to cause muscles to shorten.

The actin and myosin filaments in striated muscles are arranged in a regularly repeating pattern, producing the striations that can be observed under the microscope. In contrast, the organization of actin and myosin filaments in smooth muscles is irregular, giving these muscles a smooth appearance when viewed under the light microscope.

Skeletal muscle fibers are organized into repeating contractile units called sarcomeres.

The visible bands in skeletal muscles turn out to be an important clue to the mechanism used to produce muscle contractions.

FIG. 37.2 Light micrographs of (a) skeletal muscle, (b) cardiac muscle, and (c) smooth muscle. Skeletal and cardiac muscles have a striated, or striped, appearance; smooth muscle does not.

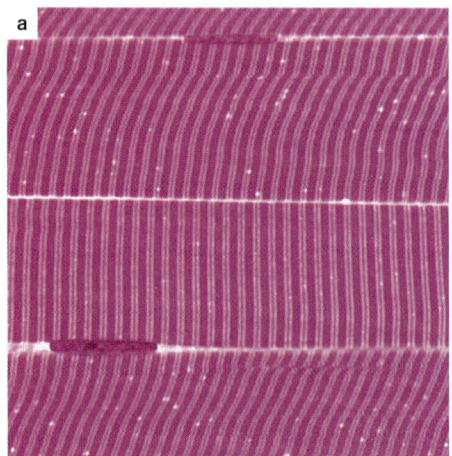
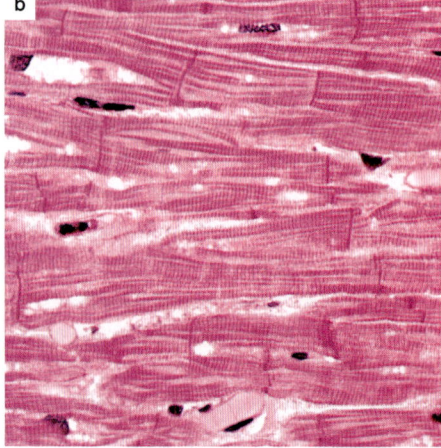
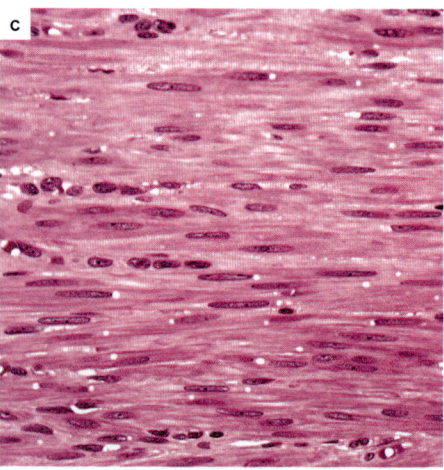

Whole muscles are made up of parallel bundles of individual muscle fibers (**Fig. 37.3**). Each muscle fiber is an elongated single cell that can be up to 20 cm long. It has several nuclei and therefore is described as **multinucleated.** The cell's multinucleated organization means that gene expression must be coordinated among all nuclei in the muscle fiber. Coordinated gene expression is necessary when endurance training increases the cell's rate of ATP production, so that the entire cell contracts uniformly. Each muscle fiber contains hundreds of long rodlike structures called **myofibrils** that contain parallel arrays of the actin and myosin filaments that cause a muscle to contract.

Let's consider the structural organization of myosin and actin filaments in more detail (**Fig. 37.4**). Each myosin molecule consists of two long polypeptide chains coiled together, each ending with a globular head. Consequently, the myosin molecule resembles a double-headed golf club. The myosin molecules are arranged in parallel to form a **thick filament,** with the

FIG. 37.4 Thin and thick filaments. Thin filaments are two actin filaments wound around each other. Thick filaments are made up of numerous myosin molecules arranged in parallel.

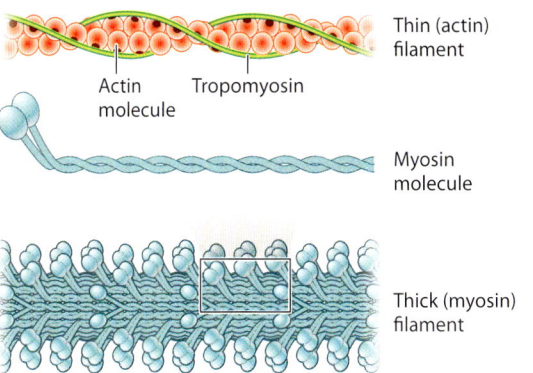

numerous myosin heads extending out from their flexible necks along the myosin filament. Two helically arranged actin filaments are twisted together to form a **thin filament.** The protein **tropomyosin** runs in the grooves formed by the actin helices.

The thin filaments are attached to protein backbones called **Z discs** that are regularly spaced along the length of the myofibril (**Fig. 37.5a**). The region from one Z disc to the next, called the **sarcomere,** is the basic contractile unit of a muscle. The sarcomeres are repeating contractile units arranged in series along the length of the muscle fiber. The shortening of thousands of sarcomeres along a fiber leads to shortening and contraction of a muscle.

Sets of actin thin filaments extend from both sides of the Z discs toward the midline of the sarcomere (Fig. 37.5a). In the middle of the sarcomere, not directly contacting the Z discs, are myosin thick filaments. The thin filaments overlap with the myosin thick filaments, forming two regions of overlap within a sarcomere. Toward either end of the sarcomere, as well as in the middle, are regions where the actin and myosin filaments do not overlap. These regions of non-overlap are lighter in color than the regions that overlap, which appear dark under an electron microscope. A third large protein, called **titin,** links the myosin filaments to the Z discs at the ends of the sarcomere. Titin is believed to help with assembly and to protect the sarcomeres from being overstretched, thus contributing to muscle elasticity.

As we have seen, the regular pattern of actin and myosin filaments within sarcomeres along the length of the fiber gives skeletal muscles their striated appearance (Fig. 37.5a). In cross section, thick and thin filaments are arranged in a hexagonal lattice that allows each thick filament to interact and slide with respect to six adjacent thin filaments (**Fig. 37.5b**). The precise geometry of thin and thick filaments is critical to their interaction, discussed on the following page.

FIG. 37.3 Muscle organization. Muscles are made up of bundles of muscle fibers (cells), each of which contains myofibrils.

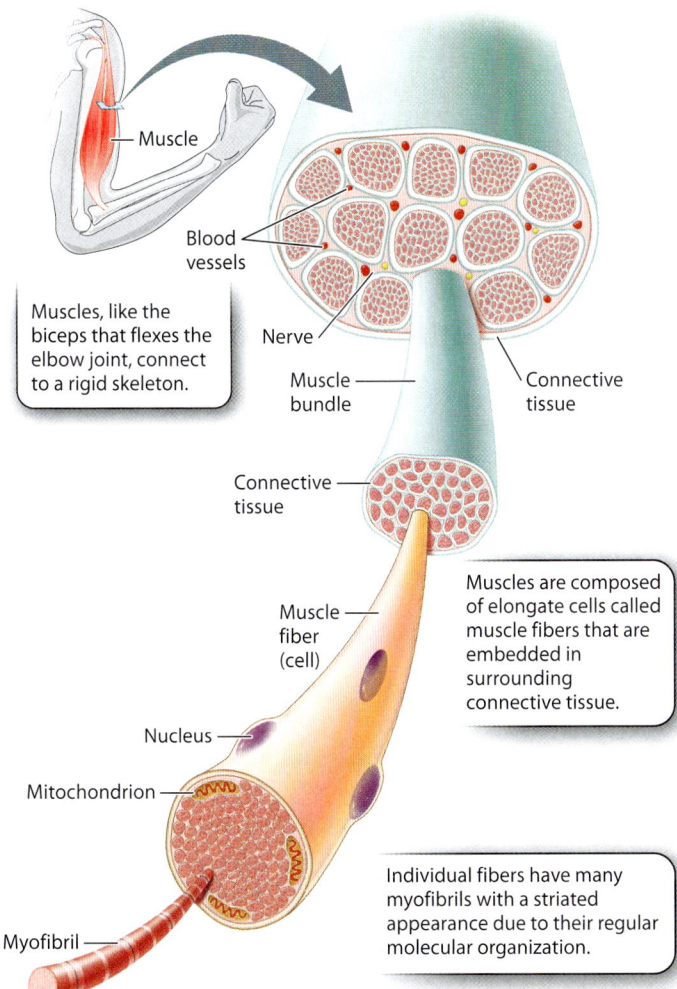

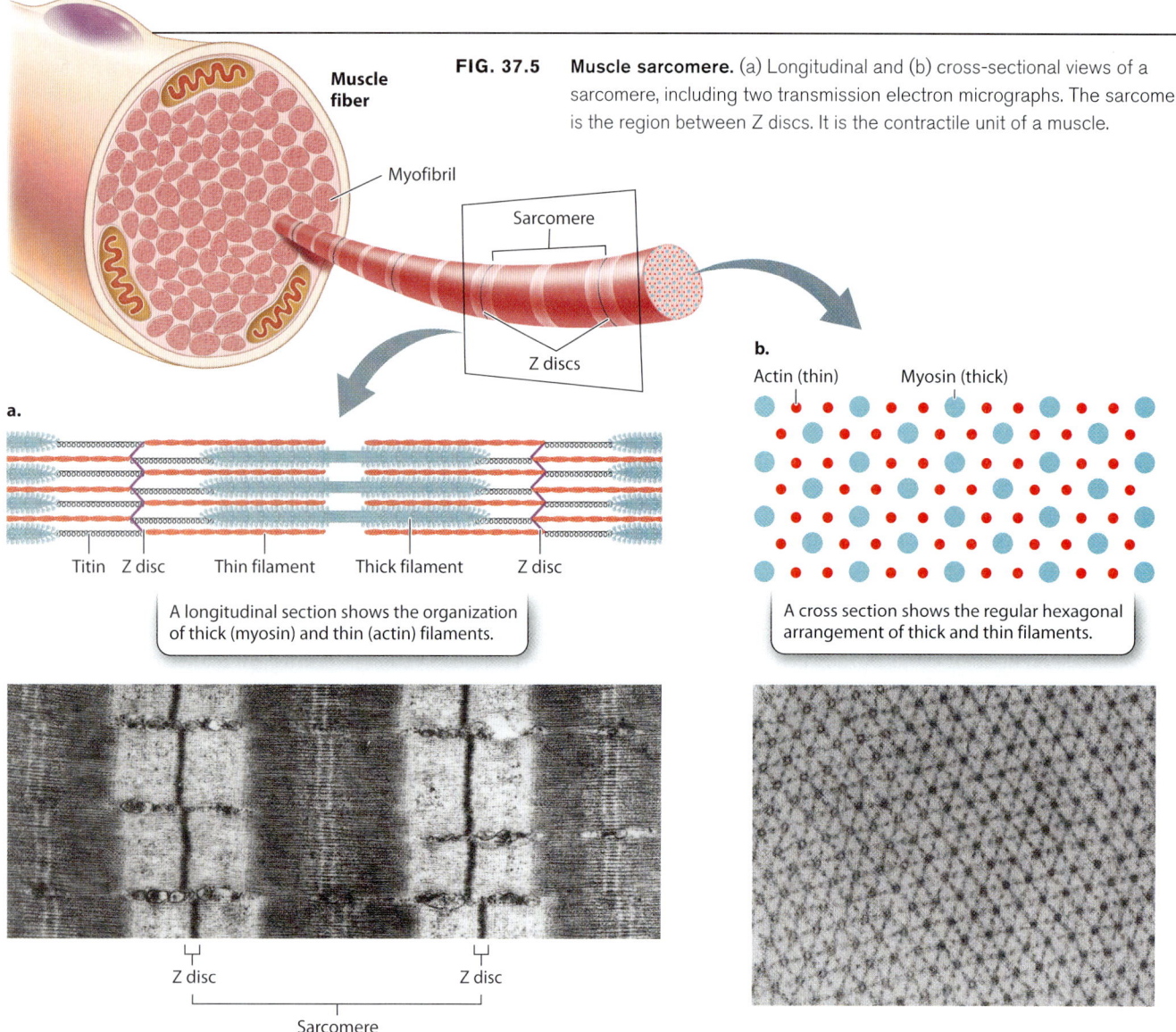

FIG. 37.5 **Muscle sarcomere.** (a) Longitudinal and (b) cross-sectional views of a sarcomere, including two transmission electron micrographs. The sarcomere is the region between Z discs. It is the contractile unit of a muscle.

Muscles contract by the sliding of myosin and actin filaments.

By using high-resolution light microscopy, two independent teams (physiologists Hugh Huxley and Jean Hanson in the United States and Andrew Huxley and Rolf Niedergerke in England) were able to quantify the changing banding pattern of sarcomeres when rabbit or frog myofibrils were stimulated to contract to different lengths. Their results led them to hypothesize that muscles produce force and change length by the sliding of actin filaments relative to myosin filaments. Their theory is known as the **sliding filament model** of muscle contraction.

The banding patterns of sarcomeres reveal the extent of overlap between actin and myosin filaments. When myofibrils contracted to short lengths, the sarcomeres were observed to have increased actin–myosin overlap. When myofibrils were stretched to longer lengths, actin–myosin overlap decreased (**Fig. 37.6**). However, the lengths of the myosin and actin filaments never varied.

Consequently, nearly all of the length change during a muscle contraction results from the sliding of actin filaments with respect to myosin filaments within individual sarcomeres. The length change of the whole muscle fiber is therefore a sum of the fractions by which each sarcomere shortens along the fiber's length.

A muscle's ability to generate force and change length is largely determined by the properties of its sarcomeres. Whereas sarcomere length is quite uniform in vertebrate animals (about 2.3 µm at rest), it is variable in invertebrate animals (ranging from as short as 1.3 µm to as long as 43 µm). Longer sarcomeres allow a greater degree of shortening. Thus, length changes are more limited in vertebrate muscles compared with some invertebrate muscles that have long sarcomeres. For example, when the muscles in an octopus tentacle contract, they can produce large motions of the tentacle. However, because they have shorter sarcomeres, vertebrate muscles can shorten to produce movements more quickly.

FIG. 37.6 Sliding filament model.
Muscle contraction, or shortening, results from the sliding of actin thin filaments relative to myosin thick filaments. Filament sliding changes the sarcomere banding pattern between Z discs.

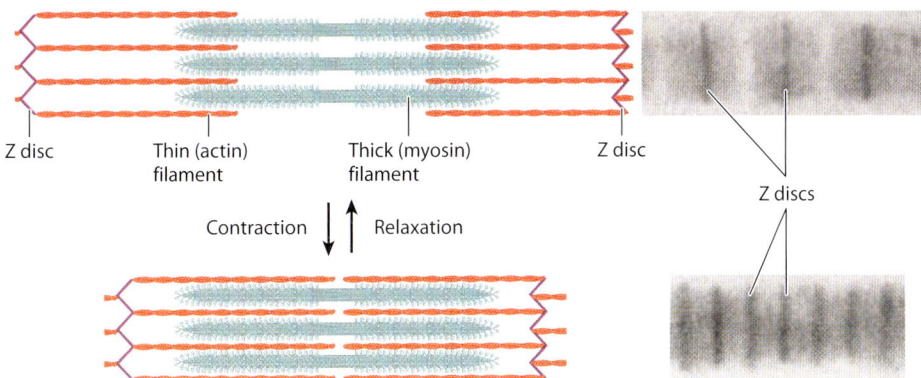

Interactions between the myosin and actin filaments are what cause a muscle fiber to shorten and produce force. Along the myosin filament, the two heads of myosin molecules bind to actin at specific sites to form **cross-bridges** between the myosin and actin filaments. The myosin filaments pull the actin filaments toward each other by means of the cross-bridges they form with actin filaments. The key to making the filaments slide relative to each other is the ability of the myosin head to undergo a conformational change and therefore pivot back and forth. The myosin head attaches to the actin filament and pivots forward, sliding the actin filament toward the middle of the sarcomere. The myosin head then detaches from the actin filament, cocks back, reattaches, and the cycle repeats. The movement of the myosin head is powered by ATP. Together, these events describe the **cross-bridge cycle**.

The cross-bridge cycle is just that—a cycle (**Fig. 37.7**). Let's look at it in more detail:

1. The myosin head binds ATP. Binding of ATP allows the myosin head to detach from actin.

2. The myosin head hydrolyzes ATP to ADP and inorganic phosphate (P_i). Hydrolysis of ATP results in a conformational change in which the myosin head is cocked back. ADP and P_i remain bound to the myosin head. Because ADP and P_i are

FIG. 37.7 Cross-bridge cycle.
The myosin head binds to actin and uses the energy of ATP to pull on the actin filament, causing muscle shortening.

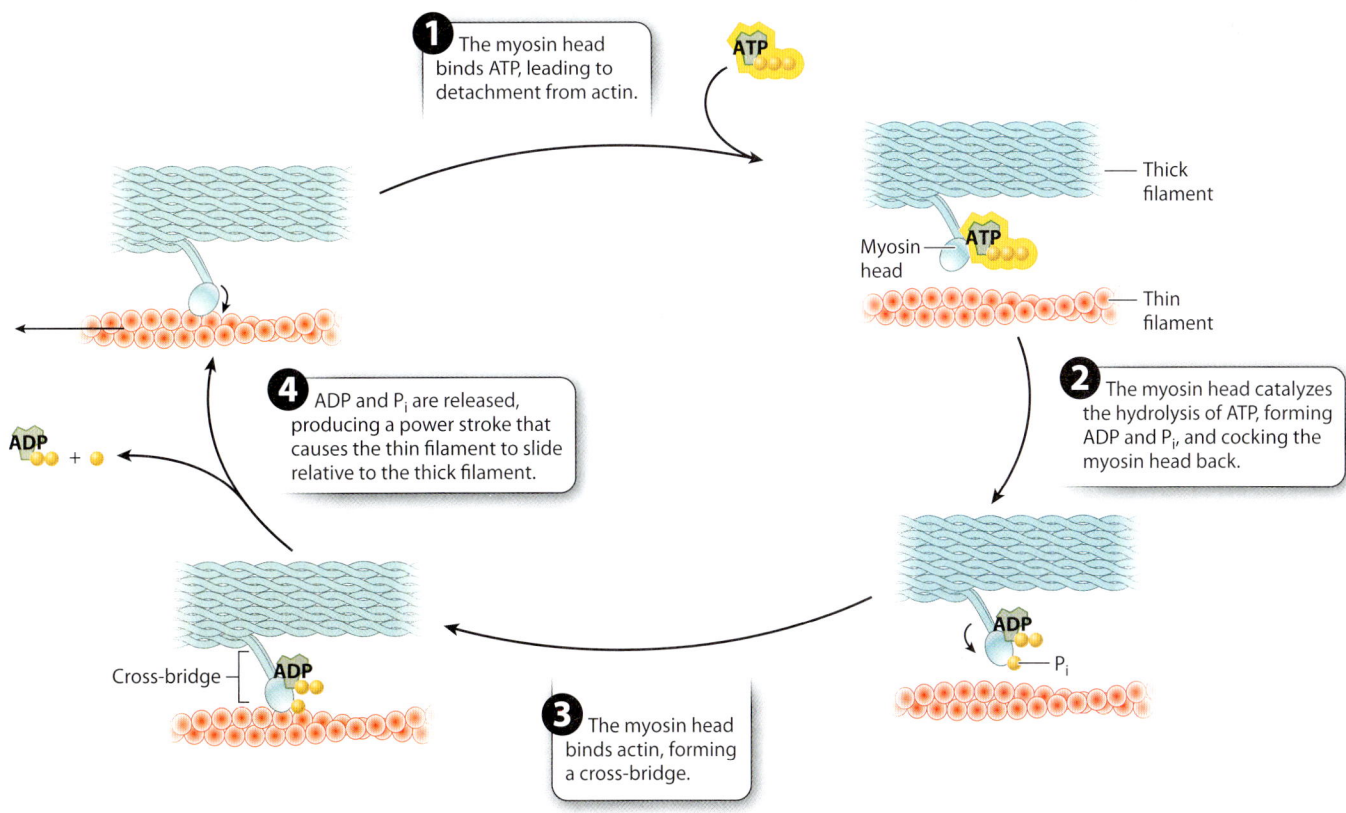

bound rather than released, the myosin head is in a high-energy state.

3. The myosin head then binds actin, forming a cross-bridge.

4. When the myosin head binds actin, the myosin head releases ADP and P_i. The result is another conformational change in the myosin head, called the **power stroke.** During the power stroke, the myosin head pivots forward and generates a force, causing the myosin and actin filaments to slide relative to each other over a distance of approximately 7 nm. The power stroke pulls the actin filaments toward the sarcomere midline.

The myosin head then binds a new ATP molecule (step 1 again), allowing it to detach from actin, and the cycle repeats. After detachment, the myosin head again hydrolyzes ATP, allowing it to return to its original conformation and bind to a new site farther along the actin filament. With the release of ADP and P_i, the myosin head undergoes another step of force generation and movement.

Individual muscle contractions are the result of many successive cycles of cross-bridge formation and detachment, during which the chemical energy released by ATP hydrolysis is converted into mechanical work (the product of force and the distance moved—that is, the amount of shortening). Muscle fibers that contract especially quickly, such as those that power insect wing flapping or produce the sound of cicadas and rattlesnake tails, express myosin molecules that have higher rates of ATP hydrolysis, allowing faster rates of cross-bridge cycling, force development, and shortening. Thus, myosin functions as both a structural protein and an enzyme.

Because each thick filament can interact with as many as six actin filaments, at any one time numerous cross-bridges anchor the lattice of myosin filaments, while other myosin heads are detached to find new binding sites. When summed over millions of cross-bridges, these molecular events generate the force that shortens the whole muscle.

→ **Quick Check 1** When an animal dies, its limbs and body become stiff because its muscles go into rigor mortis (literally, "stiffness of death"). Why would the loss of ATP following death cause this to happen?

Calcium regulates actin–myosin interaction through excitation–contraction coupling.

We have discussed *what* makes muscles contract. We next consider *when* they contract. Striated and smooth muscle fibers are both activated by the nervous system. Whereas vertebrate skeletal muscles are innervated by the somatic nervous system, smooth muscles are innervated by the autonomic nervous system, as are cardiac muscle fibers (Chapter 35).

Like nerve cells, muscle fibers are electrically excitable. Skeletal muscle fibers are activated by impulses transmitted by motor nerves to synaptic junctions (**Fig. 37.8**). Motor neuron axons have branches that allow them to synapse with multiple muscle fibers. When action potentials traveling down a motor neuron arrive at the neuromuscular junction, the neurotransmitter acetylcholine is released into the synaptic cleft. The neurotransmitter binds with receptors on the muscle cell at a region called the **motor endplate,** triggering the opening of Na^+ channels. The resulting influx of Na^+ ions in turn initiates a wave of depolarization that passes along the length of the muscle fiber (Chapter 35).

How does membrane depolarization lead to the cross-bridge cycle? Actin and myosin filaments can form cross-bridges only when the myosin-binding sites on actin are exposed. At rest, these binding sites are blocked by the protein tropomyosin. The wave of depolarization in the muscle cell initiates a chain of events that leads to a change in the conformation of a second protein, called **troponin,** which in turn moves tropomyosin away from these binding sites, allowing cross-bridges between actin and myosin to form and the muscle to contract (Fig. 37.8).

The chain of events that concludes with the exposure of the myosin-binding sites begins at the motor endplate. After the initial membrane depolarization at the motor endplate, the wave of depolarization is conducted inward to deeper regions. You saw in Chapter 5 that eukaryotic cells contain several types of membrane-bound internal organelle. The myofibrils of muscle cells are surrounded by a highly branched membrane-bound organelle called the **sarcoplasmic reticulum (SR),** a modified form of the endoplasmic reticulum (Fig. 37.8). Depolarization initiated in the plasma membrane is conducted to the SR through a specialized transmission system, formed from invaginations of the plasma membrane, known as the **T-tubule system** that is in contact with both the muscle cell's plasma membrane and the SR (Fig. 37.8). When the muscle is at rest, the SR contains a large internal concentration of calcium (Ca^{2+}) ions pumped in by calcium pumps in its membranes.

A muscular contraction is initiated when depolarization of the T-tubules causes the SR to release Ca^{2+}. The Ca^{2+} diffuses into the myofibrils and binds to troponin, causing the troponin molecule to change shape. This conformational change of troponin, in turn, causes tropomyosin to move, exposing myosin-binding sites along the actin filament. Now myosin cross-bridges can form with actin, producing a contraction. Note that at this stage in the cross-bridge cycle, the myosin head has already hydrolyzed ATP, is bound to ADP and P_i, and is in the cocked-back "ready" position. Binding to actin then allows the power stroke to occur. In this way, contraction is initiated immediately following depolarization and release of Ca^{2+}.

FIG. 37.8 Excitation–contraction coupling. Depolarization (excitation) leads to shortening (contraction) of the muscle.

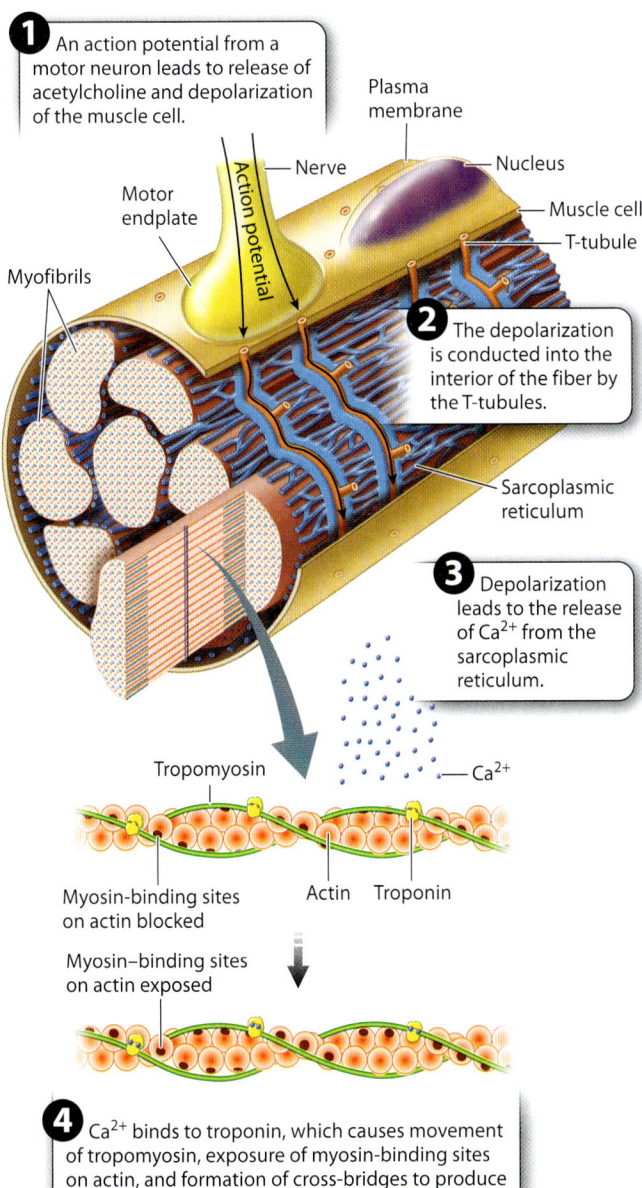

a. Scanning electron micrograph of a motor nerve endplate

b. Muscle excitation–contraction coupling

1. An action potential from a motor neuron leads to release of acetylcholine and depolarization of the muscle cell.

2. The depolarization is conducted into the interior of the fiber by the T-tubules.

3. Depolarization leads to the release of Ca^{2+} from the sarcoplasmic reticulum.

4. Ca^{2+} binds to troponin, which causes movement of tropomyosin, exposure of myosin-binding sites on actin, and formation of cross-bridges to produce shortening of the muscle.

Thus, Ca^{2+} release from the SR and binding to the troponin–tropomyosin complex on the actin filament results in **excitation–contraction coupling** (so called because excitation of the muscle cell is coupled to contraction of the muscle), producing force and movement. Together, these events are the molecular "switch" that causes a muscle to contract. Muscle relaxation follows the end of neural stimulation, when acetylcholine is broken down or reabsorbed and Ca^{2+} is actively transported back into the sarcoplasmic reticulum, causing tropomyosin molecules to block myosin-binding sites along the actin filaments.

Calmodulin regulates Ca^{2+} activation and relaxation of smooth muscle.

How is the contraction of smooth muscle regulated? The smooth muscle that controls many internal organs can be activated by stimulation from the autonomic nervous system but also responds to stretch of the muscle, local hormones, or other local factors such as pH, oxygen, carbon dioxide, or nitric oxide. For example, smooth muscle in the walls of arteries contracts when stimulated by the release of epinephrine to reduce blood flow to a body region (Chapter 39), and smooth muscle within the mammalian uterus contracts when oxytocin is released to stimulate contractions (Chapter 42). As in skeletal muscle, these initiators trigger the release of Ca^{2+} from the SR. However, unique to smooth muscle, Ca^{2+} also enters through voltage-gated and stretch-receptor calcium channels in the cell's plasma membrane.

Smooth muscle lacks the troponin–tropomyosin mechanism for regulating contraction. Instead, the switch that activates smooth muscle cells is the binding of the protein **calmodulin** with Ca^{2+} released from the SR or entering through the cell's membrane. The calmodulin–Ca^{2+} complex activates the enzyme myosin kinase, which phosphorylates the smooth muscle myosin heads, causing them to bind actin and begin the cross-bridge cycle. A second enzyme dephosphorylates the myosin heads, disrupting their ability to bind to actin, and the muscle relaxes.

The SR of smooth muscle cells is less extensive and has many fewer calcium pumps than the SR of skeletal muscle cells. As a result, calcium is returned more gradually from the myofibrils back to the SR, or pumped back out through the cell's membrane, slowing the relaxation of smooth muscle and resulting in longer periods of contraction compared with skeletal muscle fibers.

→ **Quick Check 2** Curare is a paralyzing compound that blocks the action of the neurotransmitter acetylcholine at the muscle fiber's motor endplate. What effect do you think curare has on the release of calcium ions from the sarcoplasmic reticulum of the muscle cell?

37.2 MUSCLE CONTRACTILE PROPERTIES

Actin and myosin are evolutionarily conserved across all animal life, so all muscles have similar force-generating properties. The huge muscles of a weight lifter are evidence that larger muscles can exert more force. Why is that? The summed force produced by adjacent myofibrils of activated muscle fibers determines the force the whole muscle can exert. This means that larger muscles with more fibers produce greater forces than smaller muscles with fewer fibers. However, the force that a muscle produces also depends on its contraction length and the speed of contraction, as we discuss next.

Muscle length affects actin-myosin overlap and generation of force.

Why is it difficult to jump while standing on your toes? A muscle's ability to generate force depends in part on how much it is stretched before contraction begins. The sliding filament model for muscle contraction, like any good model, makes specific predictions. In this case, it predicts that the amount of overlap between actin and myosin filaments within the sarcomere

---

HOW DO WE KNOW?

FIG. 37.9

How does filament overlap affect force generation in muscles?

BACKGROUND From the 1930s through the 1960s, British muscle physiologists studied the properties of striated frog muscle fibers. In the first set of experiments, A. V. Hill examined how a muscle's shortening velocity affects its force production. In the second set of studies, British physiologists Andrew Gordon and Andrew Huxley, together with American physiologist Fred Julian, studied how a muscle's contractile length affects its force-generating ability.

HYPOTHESIS They hypothesized that changes in overlap between myosin and actin affect the number of cross-bridges and thus the force the muscle can produce.

EXPERIMENT Gordon, Huxley, and Julian isolated a single muscle fiber and used optical microscopy to measure sarcomere length. They kept this length constant while the fiber was stimulated to produce force. Measurements of force were obtained at different sarcomere lengths.

RESULTS The top graph shows the data charted by Gordon, Huxley, and Julian, and the bottom graph reinterprets their data in terms of force and percent of muscle length. Their experiments showed that the muscle fiber produced maximal force at an intermediate sarcomere length (approximately 2.3 μm). At this length, the greatest number of cross-bridges were hypothesized to form, generating maximal force. As the sarcomere length was decreased or stretched, and then stimulated to contract, force declined. These results were consistent with a decrease in the number of cross-bridges that could form as a result of decreased actin–myosin overlap.

FOLLOW-UP WORK Other studies confirmed that changes in myosin cross-bridge formation are linked to changes in actin and myosin filament overlap at different sarcomere lengths, supporting the sliding filament model of muscle contraction.

SOURCE Gordon, A. M., A. F. Huxley, and F. J. Julian. 1966. "The Variation in Isometric Tension with Sarcomere Length in Vertebrate Muscle Fibres." *Journal of Physiology* 184:170–192.

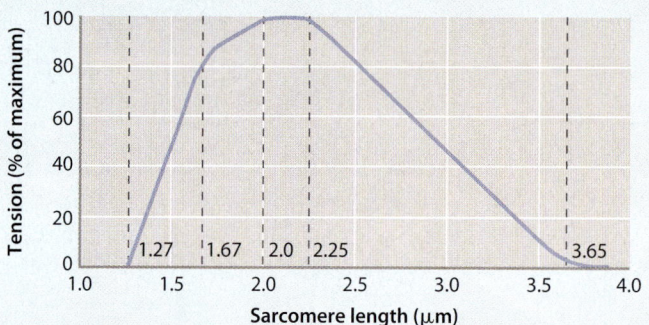

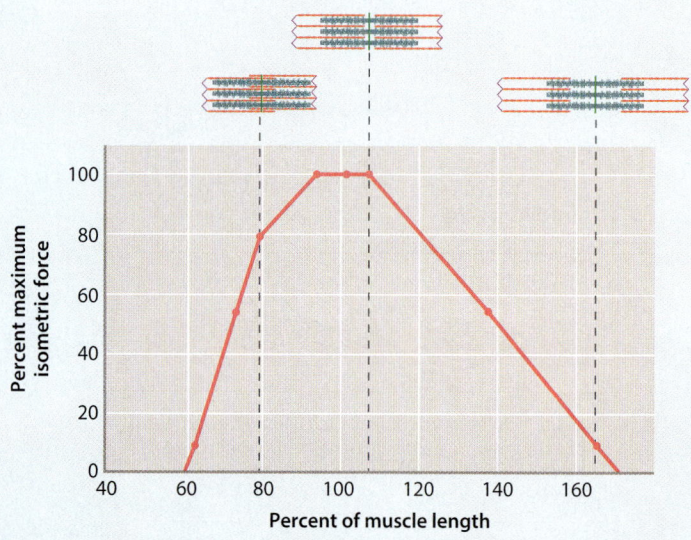

determines the number of cross-bridges that can form. The more overlap, the more force that can be produced. Conversely, when a muscle fiber is pulled to long lengths and then contracts, it generates lower force because the overlap between myosin and actin filaments is reduced. When a muscle fiber contracts at short lengths, it also produces lower force because myosin filaments begin to run into the Z disc at the end of the sarcomere, disrupting the geometry of the myosin filament lattice and hindering cross-bridge formation. At intermediate lengths the muscle fiber generates maximum force because actin–myosin filament overlap is greatest. These predictions were tested by experiments (**Fig. 37.9**).

This influence of muscle fiber length on force generated is the reason that you can't jump well if you are on your toes and your ankles are fully extended: This position shortens your calf muscle fibers. Conversely, if your ankles are overly flexed, your calf muscles are lengthened, and again jumping force is limited. Through training, athletes and dancers learn techniques, such as bending the knees before jumping, so that their leg muscles contract at intermediate lengths to maximize muscle-force output.

The relationship between force and length of striated muscle also explains Starling's Law for the function of the heart as a pump (discussed in Chapter 39), which ensures that the heart contracts more strongly when it is filled by larger amounts of blood returning from the veins.

→ **Quick Check 3** Why can you lift a larger load when your elbow is slightly flexed and your biceps muscle is at an intermediate length than you can when your elbow is fully extended?

Muscle force and shortening velocity are inversely related.

In experiments carried out in the 1930s, A. V. Hill (who had won the 1922 Nobel Prize in Physiology or Medicine for his work on muscle energy use) observed that a muscle shortens fastest when producing low forces (**Fig. 37.10**). To produce larger forces, the muscle must shorten at progressively slower velocities. For example, your arm moves faster when you throw a light ball than when you throw a heavy one.

The term "muscle contraction" suggests that muscles shorten when stimulated to generate force (in a shortening contraction), but in fact muscles may also exert force and remain a uniform length or even be lengthened. Consequently, the term "contraction" needs to be broadened beyond its familiar reference to muscle shortening to include those instances when a muscle exerts force while remaining the same length or being actively stretched.

When a muscle contracts but cannot shorten (for example, when you carry a heavy bucket of water), the muscle's shortening velocity is zero (Fig. 37.10), and it generates an **isometric** force (*iso-* means

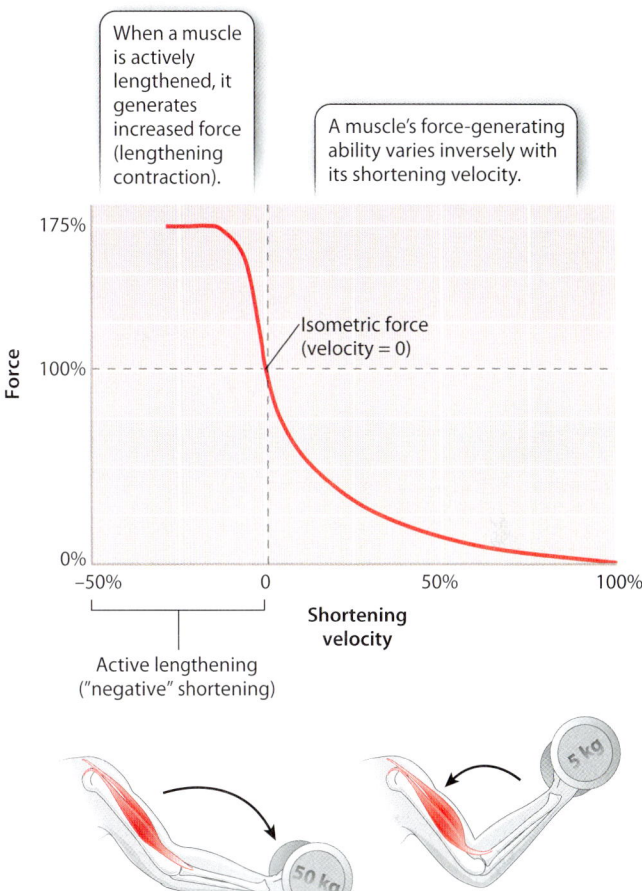

FIG. 37.10 Force-velocity relationship of muscle. Skeletal muscles exert greater force when they shorten more slowly and produce even greater force when they are stretched relative to being isometric.

"same" and *metric* refers to "measure"). That is, the muscle still generates force, even though it stays the same length and does not shorten. Isometric exercises use this type of muscle contraction: Holding a weight in a fixed position or exerting force against an immovable object helps maintain muscle strength and may be beneficial in rehabilitation after injury by avoiding joint motion.

Muscles can also be stretched when the external load against which they contract exceeds their force output. This action is known as a **lengthening contraction.** During a lengthening contraction, the muscle generates much more force (up to 75% more, Fig. 37.10) than when it remains isometric or shortens. This kind of contraction occurs when you lower a box from a shelf or when your leg muscles are stretched to support your body in landing from a jump. It also happens during every stride you take when running. A brief phase of muscle stretch at the onset of contraction is common in animal movement because it allows the muscles to generate more force than if they remained isometric or if they began shortening at the start of

FIG. 37.11 Antagonist muscles. Skeletal muscles can produce force only by pulling on a skeletal attachment, so they are arranged as antagonists to produce opposing motions at a joint.

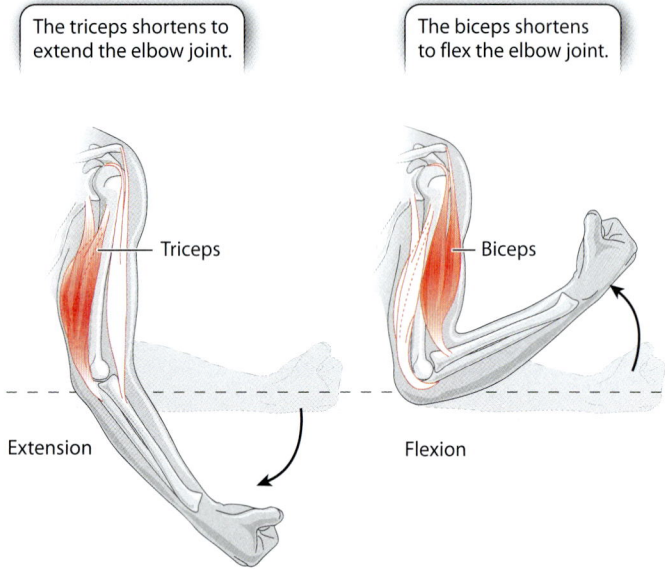

Antagonist pairs of muscles produce reciprocal motions at a joint.

Because muscles can generate force only by pulling, not pushing, on the skeleton, to produce reciprocal joint and limb movements they must be arranged as pairs of **antagonist muscles** that pull in opposite directions. **Flexion** is the joint motion in which bone segments rotate closer together, and **extension** describes the motion when they move apart. For example, muscles at the elbow joint are organized as flexors (the biceps muscle) that contract to bring the arm toward the shoulder, and extensors (the triceps muscle) that contract to extend the arm (**Fig. 37.11**). Additional pairs of antagonist muscles must be present for movements in other directions. Muscles in the body wall of cnidarians (jellyfish and sea anemones) and segmented annelid worms, as well as the smooth muscles in the walls of intestines, are organized in longitudinal (lengthwise) and circumferential (circular) layers, allowing them to function as muscle antagonists that control movement and shape.

In contrast, when muscles combine to produce similar motions they are termed muscle **agonists**. Muscles arranged as agonists increase the strength and improve the control of joint motion. The three heads of the triceps are agonists of each other (Fig. 37.11).

Muscle force is summed by an increase in stimulation frequency and the recruitment of motor units.

How are animals able to vary the amount of force exerted by a muscle? As we have seen, skeletal muscles are stimulated by motor nerves, which conduct action potentials to the neuromuscular junction. The force exerted by a muscle depends on the frequency of stimulation by the motor nerve because more-frequent stimulation increases the amount of calcium

the contraction. Muscle lengthening also reduces the muscle's energy consumption because fewer fibers are recruited to generate a given level of force.

Injury can result if a muscle is too rapidly or repetitively stretched. Rapid and repetitive stretching often occurs during prolonged downhill hiking, causing muscle damage and soreness, particularly in the quadriceps muscles in the front of the thigh. Suitable training and warm-up exercises help muscles to stretch without damage.

FIG. 37.12 (a) Twitch and (b) fused tetanus muscle contractions.

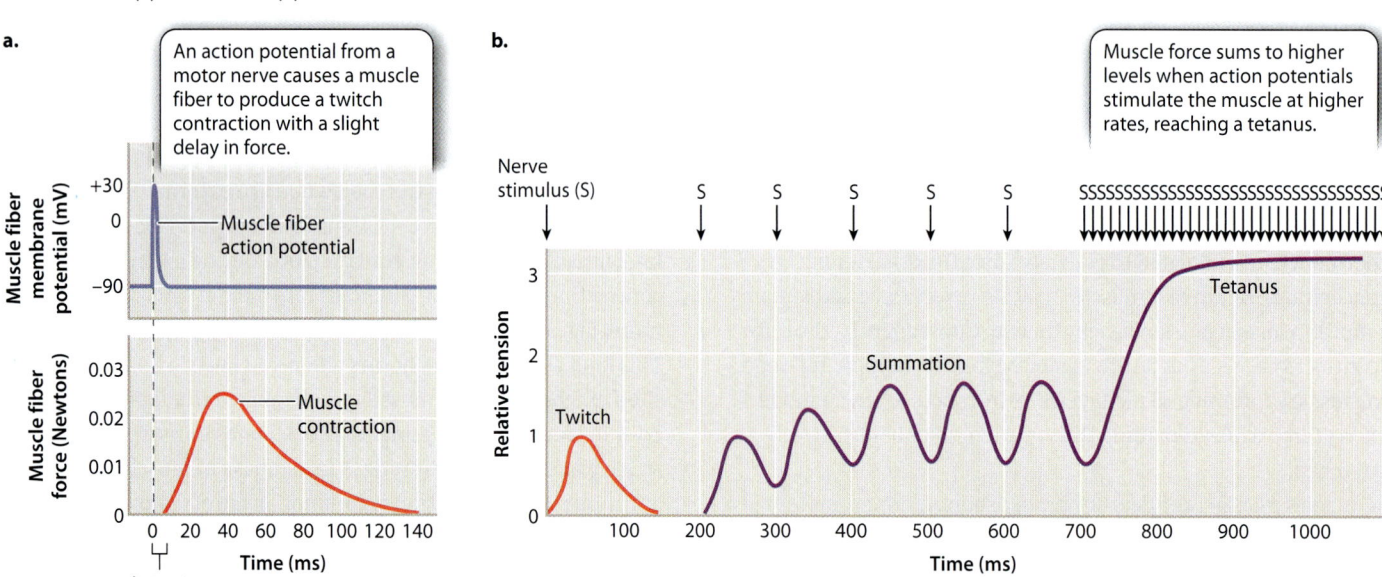

FIG. 37.13 Motor units. A motor unit is the motor neuron and the population of cells (fibers) it innervates.

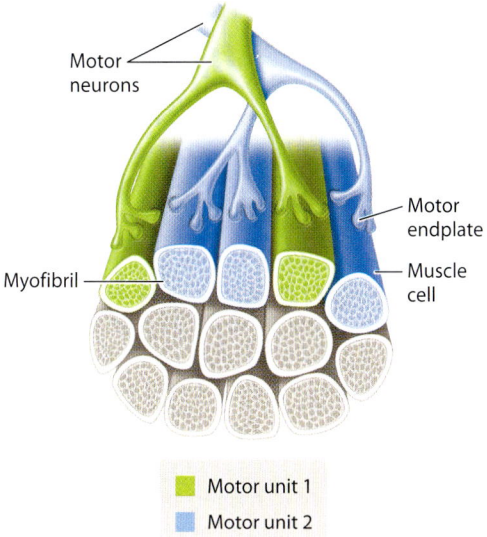

that is released to activate cross-bridge formation in the muscle fibers. A single action potential results in a **twitch** contraction of a certain force (**Fig. 37.12a**). If sufficient time is allowed for the muscle to pump Ca^{2+} back into the sarcoplasmic reticulum, the contraction ends, force falls to zero, and a second stimulus can elicit a twitch contraction of the same force. However, if a second action potential arrives before the muscle has relaxed, a greater force is produced. With increasing frequency of stimulation, muscle force increases (**Fig. 37.12b**). This property of muscle contraction is called force summation.

At a sufficiently high stimulation frequency, muscle force reaches a plateau and remains steady as stimulation continues. This muscle contraction of sustained force is described as a **tetanus.** Measurements in animals indicate that skeletal muscles commonly produce tetanic contractions, which generate steady and large forces. Tetanus is a normal physiological response of a muscle, but it can also be induced by the bacterium *Clostridium tetani*, so named because it causes severe muscle spasms, particularly of the jaw, chest, back, and abdomen.

A motor neuron and the population of muscle fibers (cells) that it innervates is collectively termed a **motor unit** (**Fig. 37.13**). Motor units can include relatively few to several hundred muscle fibers. The number of muscle fibers innervated by a given motor nerve, and hence the size of the motor unit, affects how finely a muscle's force can be controlled. For example, finger muscles have small motor units allowing fine adjustments in muscle force, whereas leg muscles have large motor units providing larger adjustments in force. Together with stimulation frequency, a muscle's force output depends on the number of motor units (and therefore the number of muscle fibers) that are activated. Within each motor unit, muscle fibers generally share similar contractile and metabolic properties. Differences in these properties can give muscles very different capabilities, as discussed in the next section.

Skeletal muscles have slow-twitch and fast-twitch fibers.

The difference between a sprinter and a marathoner is not just training but also the properties of their muscles. There are different types of muscle fiber (**Fig. 37.14a**), allowing

FIG. 37.14 Slow-twitch (darkly stained) and fast-twitch (lightly stained) skeletal muscle fibers. These two types of fiber provide differences in endurance and the speed of contraction. (a) Mammalian limb muscle; (b) muscles of two different fish species.

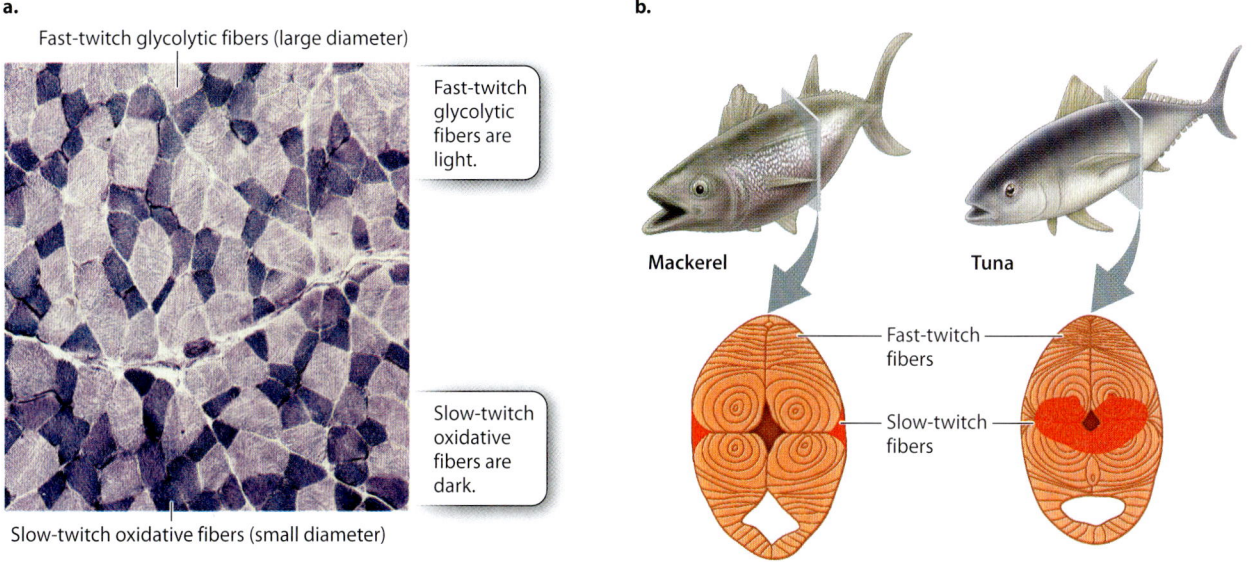

variation in the way muscles contract. In vertebrates, these are most simply classified as red **slow-twitch** fibers and white **fast-twitch** fibers. Slow-twitch fibers are found in muscles that contract slowly and consume less ATP to produce force. An animal uses slow-twitch fibers to control its posture or move slowly and economically. Fast-twitch fibers generate force quickly, producing rapid movements, but they consume more ATP. An animal depends on fast-twitch fibers to sprint or lunge at its prey. A key difference between the two types of fiber is that slow-twitch fibers obtain their energy through oxidative phosphorylation (aerobic respiration), and fast-twitch fibers obtain their energy mainly through glycolysis (Chapter 7).

Although slow-twitch fibers develop force more slowly, they have greater resistance to fatigue in response to repetitive stimulation—that is, their loss of force over time is less. Their greater endurance is explained by the ability of their mitochondria to supply ATP to muscle fibers by means of aerobic respiration. Slow-twitch fibers have many mitochondria and are well supplied by capillaries. They contain an abundance of **myoglobin,** an oxygen-binding protein related to hemoglobin that facilitates oxygen delivery to the mitochondria (Chapter 39). The iron in myoglobin gives muscles with these fibers a red appearance. Slow-twitch muscle fibers express a chemically "slow" form of myosin with a relatively low rate of ATP hydrolysis, limiting their speed of cross-bridge cycling and force development.

White fast-twitch muscle fibers have fewer mitochondria and capillaries, contain little myoglobin, and rely heavily on anaerobic glycolysis (lactic acid fermentation) to produce ATP. Fast-twitch fibers express a chemically "fast" form of myosin with a high rate of ATP hydrolysis, favoring rapid force development and movement. However, fast-twitch fibers fatigue quickly because of the drop in pH caused by the accumulation of lactic acid. Because white fibers are larger than red fibers, they generate more force. In general, most skeletal muscles are composed of mixed populations of red and white fiber types, providing a broad range of contractile function.

Athletes who have more oxidative slow-twitch fibers excel at endurance and long-distance competitions. In contrast, athletes who have larger numbers of fast-twitch fibers perform better in sprint races and weight-lifting competitions. Because the distribution and properties of muscle fiber types are strongly inherited, it is largely true that champions are born and not made. Nevertheless, aerobic training can significantly increase the energy output through oxidative phosphorylation of an individual's muscle fibers, just as weight lifting and speed training can enhance an individual's strength and sprinting ability. Contrary to popular belief that weight lifting builds muscle by straining the muscle to build new fibers, it actually builds muscle by increasing the size of existing muscle fibers through the synthesis of additional myosin and actin filaments. This synthesis increases the cross-sectional area of the fiber (cell) and thus its force-generating capacity.

? CASE 7 Predator–Prey: A Game of Life and Death
How do different types of muscle fiber affect the speed of predators and prey?

We have just seen that slow-twitch muscle fibers provide endurance for sustained activity, whereas larger fast-twitch fibers produce greater speed and strength but fatigue quickly. The relative distribution of these fiber types in the muscles of different animals is influenced mainly by genetic heritage. This affects the locomotive behavior and capacities of both predator and prey animals. Whereas cheetahs have exceptional speed, their muscles tire quickly. As their prey, antelope have muscles that enable endurance long-distance running and contract rapidly for maneuvering.

Highly aerobic animals such as dogs and antelope have large fractions of slow-twitch fibers *and* specialized fast-twitch muscle fibers that also have a high oxidative capacity in addition to using glycolysis for rapid ATP synthesis. These muscle fibers allow the animals to move for extended periods of time at relatively fast speeds. In contrast, cats have mainly fast-twitch fibers, enabling them to sprint and pounce quickly. Nevertheless, cats and other animals use particular muscles with high concentrations of slow-twitch fibers (these are the dark-staining fibers shown in Fig. 37.14a) for slower postural movements and stealth. Animals such as sloths and lorises that move slowly have mostly oxidative slow-twitch fibers and few glycolytic fast-twitch fibers.

More generally, animals with high body temperatures and high metabolic rates (birds and mammals) have larger numbers of oxidative fibers compared with animals that have lower body temperatures and lower metabolic rates (reptiles). As a result, lizards use glycolytic fast-twitch fibers to sprint for brief periods but must then recover from the acid buildup produced by their muscles' anaerobic activity in longer periods of rest (Chapter 40). In contrast, birds and mammals can sustain activity over longer time periods because they have oxidative muscle fibers supported by aerobic ATP supply.

Most fish have a narrow band of slow-twitch red muscle fibers that runs beneath their skin, which they use for slow steady swimming. Tuna and some sharks have evolved deeper regions of red muscle that they can keep warm (**Fig. 37.14b**), enabling their muscles to contract longer and faster to enhance their swimming. When escaping from a predator or attacking prey, fish recruit their larger fast-twitch white trunk musculature for rapid swimming.

The evolutionary arms race between predators and prey has resulted in selection for particular muscle fiber types and other musculoskeletal specializations for rapid movement and maneuvering. Features of the skeleton that contribute to these specializations are discussed next.

37.3 ANIMAL SKELETONS

Most animal skeletons provide a rigid set of elements that articulate, that is, they meet at joints, to transmit muscle forces for movement and body support. They also enable many animals to manipulate their environment by digging burrows, building nests, or constructing webs to catch food. Animals have evolved three types of skeletal system: **hydrostatic skeletons, endoskeletons,** and **exoskeletons.** All three types of skeleton use rigid elements to resist the pull of antagonist sets of muscles.

Hydrostatic skeletons support animals by muscles that act on a fluid-filled cavity.

Hydrostatic skeletons evolved early in multicellular animals with the first cnidarians (jellyfish and anemones) approximately 600 million years ago. They are found in nearly all multicellular animals as well as in many vascular plants. In animals that depend on a hydrostatic skeleton, fluid contained within a body cavity serves as the supportive component of the skeleton. Muscles exert pressure against the fluid to produce movement.

Two sets of muscles surround the fluid, controlling the width and length of the body cavity and, in many cases, of the whole animal (**Fig. 37.15a**). Circular muscles reduce the diameter of the body cavity, and longitudinal muscles reduce its length. A sea anemone can bend its body in various directions to feed or to resist water currents by closing its mouth to keep its fluid volume constant and contracting its circular and longitudinal muscles. Sea anemones extend themselves into the water column by contracting their circular muscles and relaxing their longitudinal muscles. Controlling the amount of fluid in their body cavity also allows sea anemones to adjust their shape. When threatened,

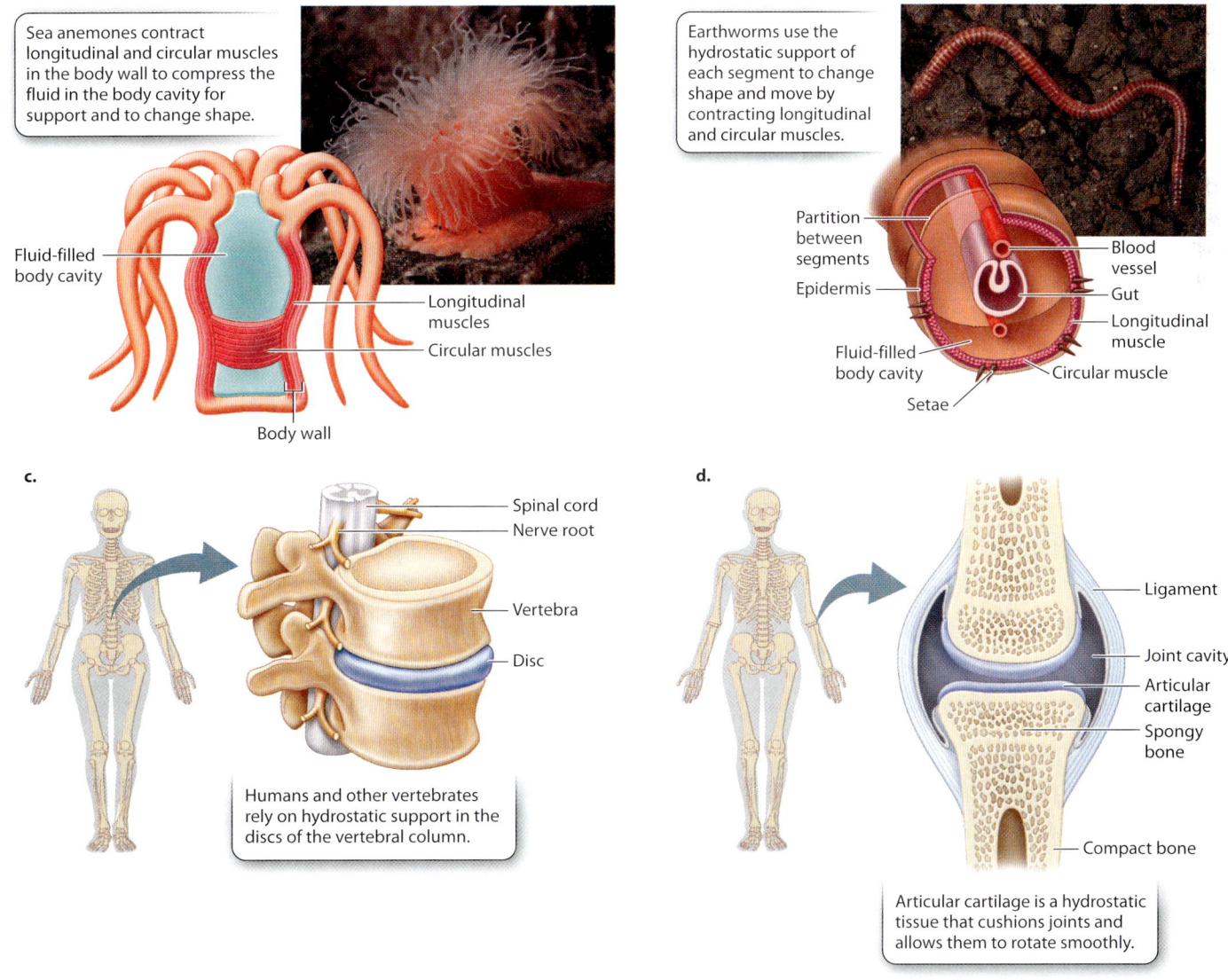

FIG. 37.15 Diverse examples of hydrostatic skeletal support. (a) Sea anemone; (b) earthworm; (c) intervertebral discs; (d) articular cartilage.

a sea anemone rapidly retracts by contracting its longitudinal muscles and allowing water to escape from its body cavity.

Similarly, earthworms burrow through the soil by moving a series of hydrostatic segments along their body (**Fig. 37.15b**). By contracting longitudinal muscles in a few segments, they shorten those segments and also widen them to anchor those segments against the soil. They then extend intervening segments between anchor points by contracting circular muscles, causing the segments to lengthen and move the animal's body through the soil.

Because of their simple but effective organization, hydrostatic skeletons are well adapted for many uses. In groups such as squid and jellyfish, the hydrostatic skeleton has become specialized for jet-propelled locomotion. Large circular muscles contract to eject fluid from the body cavity, propelling the animal in the opposite direction. The muscles of other mollusks also employ a hydrostatic skeleton, allowing a clam, for example, to burrow into the sediment with a muscular "foot" and an octopus to adeptly manipulate its flexible tentacles. (Clams also have shells that form an exoskeleton to protect the body.) These same principles allow an elephant to control its trunk, and our tongue to manipulate food and assist in speech.

Even vertebrate animals with rigid endoskeletons have hydrostatic elements for flexibility and cushioning of loads transmitted by the skeleton. These include **intervertebral discs** and **cartilage.** Intervertebral discs are sandwiched between the bony vertebrae of the backbone (**Fig. 37.15c**). Each disc has a wall reinforced with connective tissue that surrounds a jelly-like fluid. Intervertebral discs enable the backbone to twist and bend. Disc walls damaged by repeated stress can rupture, causing significant lower back pain. Articular cartilage, another fluid-filled tissue, forms the joint surfaces between adjacent bones (**Fig. 37.15d**). The fluid within the cartilage resists the pressure between the ends of articulating bones, cushioning forces transmitted across the joint.

Exoskeletons provide hard external support and protection.

An exoskeleton is a rigid skeleton that lies external to the animal's soft tissues. The first mineralized skeletons arose with sponges about 650 million years ago, but more successful exoskeletons subsequently evolved in other invertebrate groups adapted for life in aquatic and terrestrial environments (**Fig. 37.16**). Arthropod exoskeletons, which first evolved in aquatic crustaceans, protect animals from desiccation and physical insults. Consequently, exoskeletons were key to the success and diversification of insects in terrestrial environments about 450 million years ago. Because they are on the exterior of the animal, exoskeletons provide hard external support and protection, allowing muscles to attach from the inside. However, a main disadvantage is that exoskeletons limit growth.

Different invertebrates have different kinds of exoskeleton. The shells of bivalve mollusks such as clams and mussels form a hard, mineralized calcium carbonate exoskeleton (Fig. 37.16a and b). The calcium carbonate is reinforced by proteins and is therefore an example of a composite material, one that combines substances with different properties. Because of its composite nature, the shell is less brittle and more difficult to break than if it were made of just one substance. The shell expands as the organism grows, as epidermal cells deposit new layers of mineralized protein in regular geometric patterns. You can see the growth rings if you examine a mollusk shell. Although some marine bivalves achieve large size over many years, growth is limited by the rate at which new skeletal material can be deposited on the outside of the shell.

FIG. 37.16 Examples of exoskeletons. The shells of a bivalve mollusk (a) and nautilus (b), made of calcium carbonate, surround and protect internal soft body parts. The exoskeleton of arthropods, such as a grasshopper (c), is made of stiff cuticle that surrounds internal muscles and tendons and other internal body parts.

Arthropods, such as insects, have more complex exoskeletons formed of a **cuticle** that covers their entire body (Fig. 37.16c). The cuticle of insects is composed mainly of **chitin,** a nitrogen-containing polysaccharide. Marine crustaceans, such as crabs and lobsters, also incorporate calcium carbonate in their cuticle. The calcium carbonate makes the exoskeleton hard and stiff, helping to protect them from predators, much like the hardened cuticle of terrestrial insects. When initially formed, the cuticle of arthropods is soft, so the animal can grow before the cuticle is hardened. The cuticle remains flexible at joints, allowing motion between body segments. Mature cuticle consists of two layers. A thin outer, waxy layer minimizes water loss. Water loss is especially dangerous for arthropods, given their small size. The much thicker inner layer, whether flexible or stiff, is tough, making it hard to break.

Because a rigid exoskeleton restricts growth, arthropods shed their cuticle at intervals, a process termed **molting.** Molting allows arthropods to expand and grow before forming a new rigid exoskeleton. While offering several protective benefits, exoskeletons pose risks as well. Animals are vulnerable when the newly formed exoskeleton has not yet hardened. Exoskeletons are also hard to repair. If a skeleton is damaged while growing, the animal must produce an entirely new one. Because the exoskeleton is a thin-walled structure, it is prone to breaking if its surface area is very large. Consequently, the imperviousness of the monstrous insects depicted in science-fiction films is improbable—they would easily break from a blow to their exoskeleton.

The rigid bones of vertebrate endoskeletons are jointed for motion and can be repaired if damaged.

Endoskeletons first evolved in vertebrate animals about 500 million years ago, during the Cambrian explosion. This type of skeleton lies internal to most of an animal's soft tissues. In contrast to exoskeletons, the bony endoskeletons of vertebrate animals can grow extensively and, when broken, can be repaired. Endoskeletons also provide protection for internal organs, such as the brain, lungs, and heart.

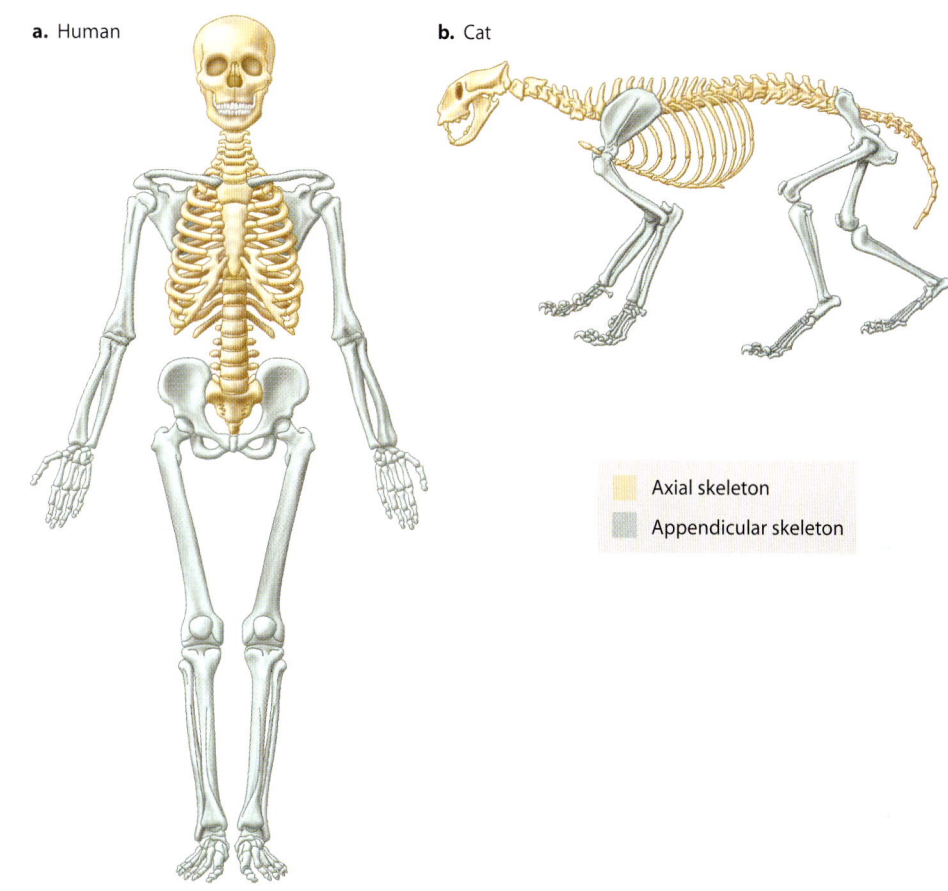

FIG. 37.17 The vertebrate skeleton, showing axial and appendicular regions.

a. Human
b. Cat

Axial skeleton
Appendicular skeleton

The bones of vertebrate skeletons consist of a variety of tubular, rodlike, and platelike elements that form a scaffold to which the muscles attach. Because bones are rigid and hard, there are joints between adjacent bones to enable movement. Muscles attach to the skeleton by connective tissue and specialized **tendons** made of collagen. Tendons transmit muscle forces, allowing the forces to be redirected and transmitted over a wide range of joint motion. Tendons, like a spring, also store and recover elastic energy.

The vertebrate skeleton can be separated into **axial** and **appendicular** regions (**Fig. 37.17**). The axial skeleton consists of the skull and jaws of the head, the vertebrae of the spinal column, and the ribs. Bones of the skull protect the brain and sense organs of the head. The appendicular skeleton consists of the bones of the limbs, including the shoulder and pelvis. The axial and appendicular regions reflect the evolutionary ancestry of vertebrates. The axial skeleton formed first, to protect the head and provide support and movement of the animal's body by bending its body axis. As a result, the axial skeleton is the main skeletal component of fishes. The appendicular skeleton comprises the pectoral and pelvic fins of fishes, which were elaborated into limbs when vertebrate animals first evolved to live on land. In a special group of lobed-fin fishes, the fin bones became adapted for an amphibious lifestyle, evolving into the limb bones that terrestrial vertebrates use for support against gravity and for movement on land.

Skeletal tissues have relatively few cells for their volume, similar to invertebrate exoskeletons. These tissues consist mostly of an extensive **extracellular matrix** that is secreted by specialized cells, forming a connective tissue external to the cells (Chapter 10). Three main extracellular tissues are produced in the formation of the vertebrate skeleton: bone, tooth enamel, and cartilage.

Bone tissue–forming cells called **osteoblasts** synthesize and secrete calcium phosphate as **hydroxyapatite** mineral crystals in close association with the protein **collagen**. Generally, bone consists of two-thirds hydroxyapatite mineral and one-third type I collagen protein. The combination of mineral and protein makes bone a composite material, like exoskeletons. Bone is rigid (because of the properties of the mineral), but also hard to break because it absorbs much energy before fracture (because of the properties of the collagen). The genetic bone disorder osteogenesis imperfecta leads to bones that are brittle and fragile because the osteoblasts produce insufficient collagen in the extracellular matrix. Bone gradually becomes more mineralized and brittle as a normal process of aging. Sharks and other elasmobranchs (the cartilaginous fishes) evolved unusual skeletons of calcified cartilage rather than mineralized bone tissue to achieve rigid mechanical support.

Articular cartilage located at the joint surfaces of a bone forms a gel-like matrix that resists fluid being squeezed out when forces press on the cartilage during movement. The cartilage is reinforced by type II collagen fibers, allowing the cartilage to cushion and distribute loads as the joint moves. Articular cartilage consists of 70% water, 15% large molecules called proteoglycans, and the remaining 15% collagen. Cartilage may degenerate for a variety of reasons, including rheumatoid arthritis and repetitive physical injury, but in general it lasts over an individual's lifetime.

→ **Quick Check 4** Glass is strong and rigid but can be easily broken. In contrast, a bone is strong and rigid but hard to break. How can this be?

37.4 VERTEBRATE SKELETONS

Although both vertebrate skeletons and invertebrate exoskeletons are rigid and have flexible joints, certain features distinguish them. Here, we explore the biology of vertebrate skeletons, focusing on how bones develop and grow and how they are repaired and remodeled.

Vertebrate bones form by intramembranous and endochondral ossification.

Vertebrate bones form and develop by two distinct processes. Bones of the skull and the ribs form when precursor cells differentiate into osteoblasts that immediately begin producing bone as a skeletal sheet or membrane. These bones are often referred to as membranous bones.

Most other bones of the body—the vertebrae, shoulder, pelvis, and limb bones—are formed as cartilage first (**Fig. 37.18**). The precursor cells initially become cartilage-producing cells, known as chondroblasts, that in the embryo grow into cartilage models of most bones of the developing body (Fig. 37.18a). For example, at approximately 9 weeks' development, a human fetal hand consists mainly of this cartilage model (Fig. 37.18b). This embryonic cartilage is the same as the articular cartilage

FIG. 37.18 Endochondral ossification. (a) Most bones other than the skull undergo endochondral ossification, which involves the initial formation of a cartilage model of the bone, followed by vascularization and the transformation into bone tissue by osteoblasts. (b) Cartilage model of a human fetal hand, showing initial sites of ossification at about 9 weeks of embryonic development.

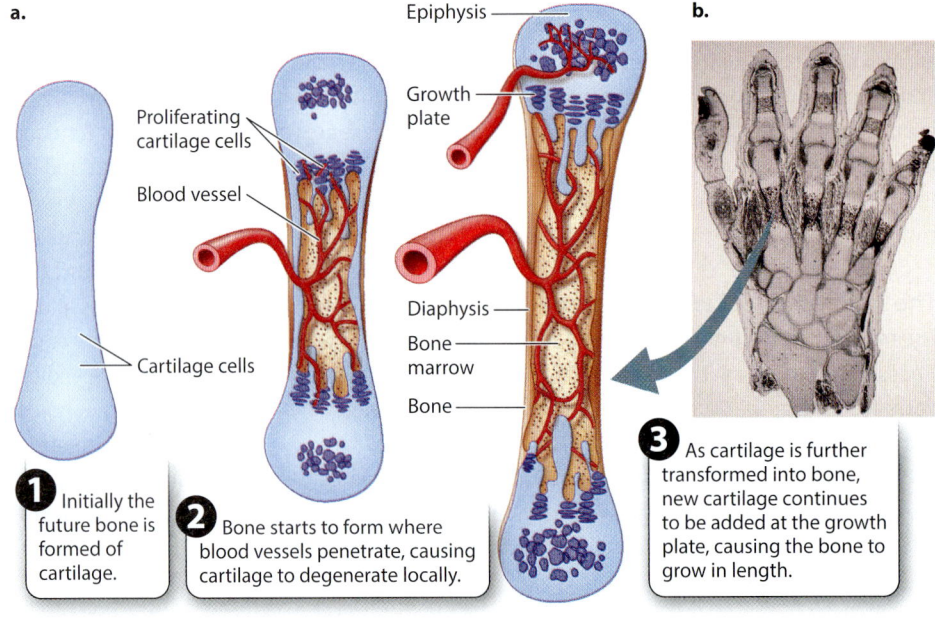

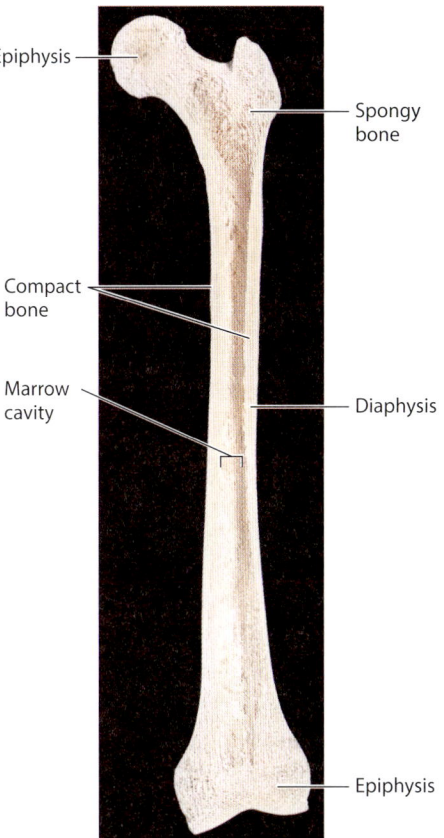

FIG. 37.19 Compact and spongy bone.

that remains at the ends of bones to form the joint surfaces in the growing and mature animal. Because cartilage is a pliable fluid-filled matrix, it grows by expansion within the structure as well as at the surface, enabling the fetal skeleton to grow rapidly. As the fetus grows and its skeleton matures, the cartilage model of the bone is invaded by blood vessels. The presence of blood vessels triggers the transformation of cartilage into hard, mineralized bone.

The cartilage template of a bone is replaced by one of two main types of bone tissue (**Fig. 37.19**). Both types contribute to the overall form of many bones in the body. **Compact bone** forms the walls of the bone's shaft. It consists mainly of dense mineralized bone tissue containing bone cells known as **osteocytes** and the network of blood vessels supplying them with oxygen and nutrients.

A second type of bone, called **spongy bone,** consists of small plates and rods known as **trabeculae** with spaces between them. Spongy bone is found in the ends of limb bones and within the vertebrae. It reduces the weight of bone in these regions, but it is also stiff so that forces are transmitted effectively from the joint to the bone shaft. **Bone marrow,** a fatty tissue found between trabeculae and also within the bone's central cavity, contains many important cell populations, including blood-forming cells, other stem cells (Chapter 20), and immune system cells (Chapter 43), as well as fat cells.

Bones can grow in length as well as in width. Growth in length occurs at a **growth plate** (see Fig. 37.18)**,** a region of cartilage between a middle region, called the **diaphysis,** and the end, called the **epiphysis**. Blood vessels invade the central diaphysis region, as well as each epiphysis, triggering the transformation of cartilage into bone. The growth plate is left in between. The growth plate adds new cartilage toward the bone's diaphysis, enabling bone length to continue to increase after birth (see Fig. 37.18). At maturity, the growth plates fuse as the remaining cartilage is replaced with bone, preventing further growth in length. The fusion of growth plates is typical of most mammals and birds. In amphibians and reptiles, the growth plates often do not fuse and growth may continue at a slower pace over much of their lifetime.

Limb bones grow in diameter when osteoblasts deposit new bone on the bone's external surface. At the same time, bone is removed from the inner surface, expanding the marrow cavity. Bone removal is slower than bone growth, thickening the walls during growth. Bone is removed from the marrow cavity by a group of cells called **osteoclasts** that secrete digestive enzymes and acid to dissolve the calcium mineral and collagen. These dissolved compounds are reabsorbed and recycled for bone formation in other regions of the skeleton. Through this process, the vertebrate skeleton serves as an important store of calcium and phosphate ions. For example, in female birds, calcium removed from the skeleton is regularly used to form the eggshell.

A great advantage of endoskeletons compared with exoskeletons is that a vertebrate's skeleton can be repaired if damaged by osteoblasts and osteoclasts forming and removing mineralized tissue in particular regions of the bone. Generally, physically active younger adults have thicker bones than less-active individuals. However, in older humans, bone tissue is gradually lost as osteoclasts remove more bone than is produced by osteoblasts. Bone loss is particularly severe in women after menopause, in part because of hormone shifts, and can lead to osteoporosis, a condition that significantly increases the risk of bone fracture.

Joint shape determines range of motion and skeletal muscle organization.

The shapes of the bone surfaces that meet at a joint determine the range of motion at that joint. Joints range from simple **hinge joints** that allow one axis of rotation to **ball-and-socket**

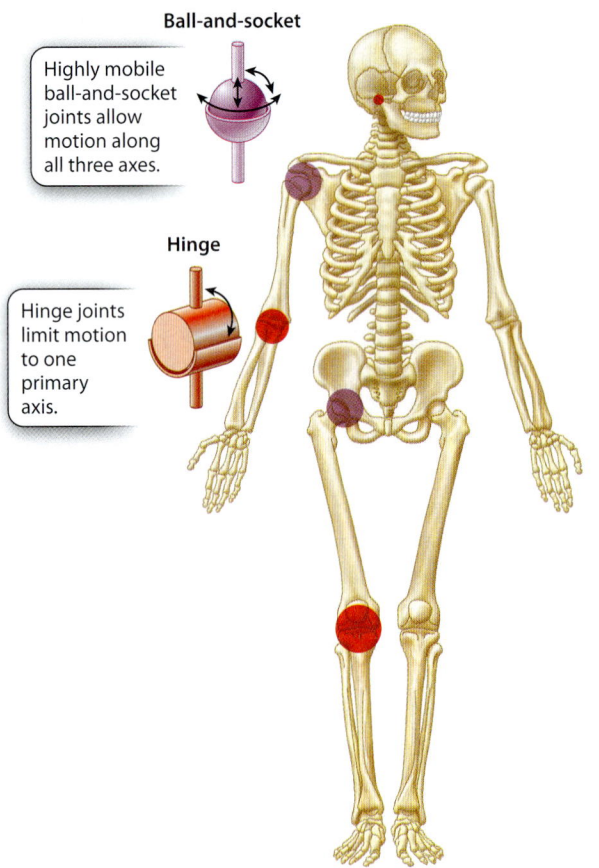

FIG. 37.20 **Hinge and ball-and-socket joints.** The elbow and knee are examples of hinge joints. The shoulder and hip are examples of ball-and-socket joints.

joints that allow rotation in three axes (**Fig. 37.20**). The human elbow joint and the ankle joint of a dog are examples of hinge joints. The shoulder and hip joints are examples of ball-and-socket joints, which allow the widest range of motion, as when you throw or kick a ball. The joints at the base of each finger are intermediate: They allow you to flex and extend your fingers as well as spread them laterally or move them together when making a fist or grasping different-sized objects.

Joints with a broader range of motion are generally less stable. The shoulder is the most mobile joint in the human body, but it is also the most often dislocated or injured. In contrast, the ankle joint of dogs, horses, and other animals is a stable hinge joint that is unlikely to be dislocated, but its range of motion is limited to flexion and extension. Because muscles are arranged as paired sets of antagonists to move a joint in opposing directions, hinge joints are controlled by as few as two antagonist muscles (generally referred to as a flexor and an extensor). In contrast, ball-and-socket joints minimally have three sets of muscle antagonists to control motion in three different planes. As a result, a more complex organization of muscles is needed to control the movements of the arm at the shoulder joint or the leg at the hip joint.

Muscles exert forces by skeletal levers to produce joint motion.

By serving as a rigid set of levers, the skeleton enables muscles to transmit forces that cause joint rotation. Analogous to the gears of a bicycle, muscles that attach farther from a joint's axis of rotation produce slower but stronger movements, whereas muscles that attach closer to a joint produce rapid but weaker movements. This is why you open a door by pulling on a doorknob located opposite from its hinges. The doorknob's location increases the lever distance, making it easier to pull the door open than if it were placed next to the hinge.

The strength of rotational movement, or torque, is determined by the product of a muscle's force (F) and perpendicular lever distance (r) from the axis of joint rotation: $F \times r$. Muscles produce joint torques to counteract torques that are applied by external forces, such as a weight that is carried by your hand. In **Fig. 37.21**, the axis of rotation is the elbow, and contraction of the biceps muscle causes the arm to flex at the elbow. The strength of elbow rotation is determined by the biceps muscle's force (F) and the lever distance between the elbow joint and the site of muscle attachment, where the muscle pulls on the arm bone (r). Muscles produce stronger rotational movements either by exerting a larger force or by having a longer lever distance between the joint

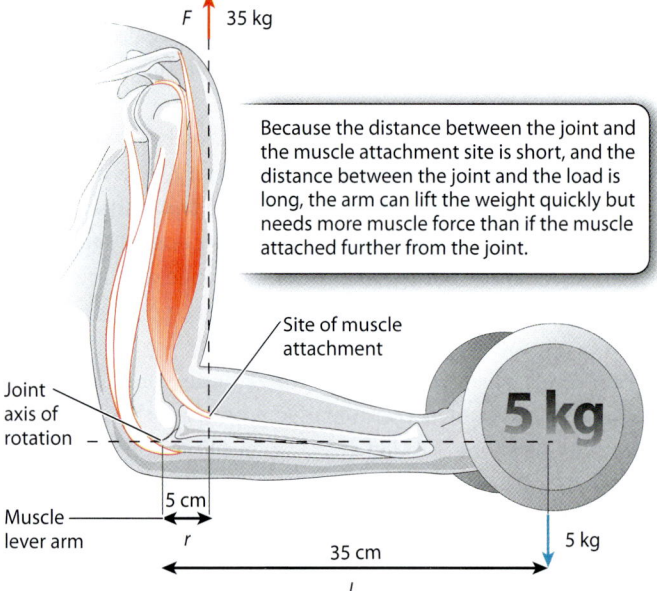

FIG. 37.21 **Torque.** The torque ($F \times r$) depends on the muscle's force (F) and the distance the force acts from the joint's axis of rotation (r), referred to as the muscle's lever arm. This torque must counteract the opposing torque of the weight and its lever arm (L).

FIG. 37.22 A trade-off between force and velocity in animal skeletons. The armadillo forelimb skeleton is adapted for digging, whereas the horse forelimb is adapted for speed and weight support.

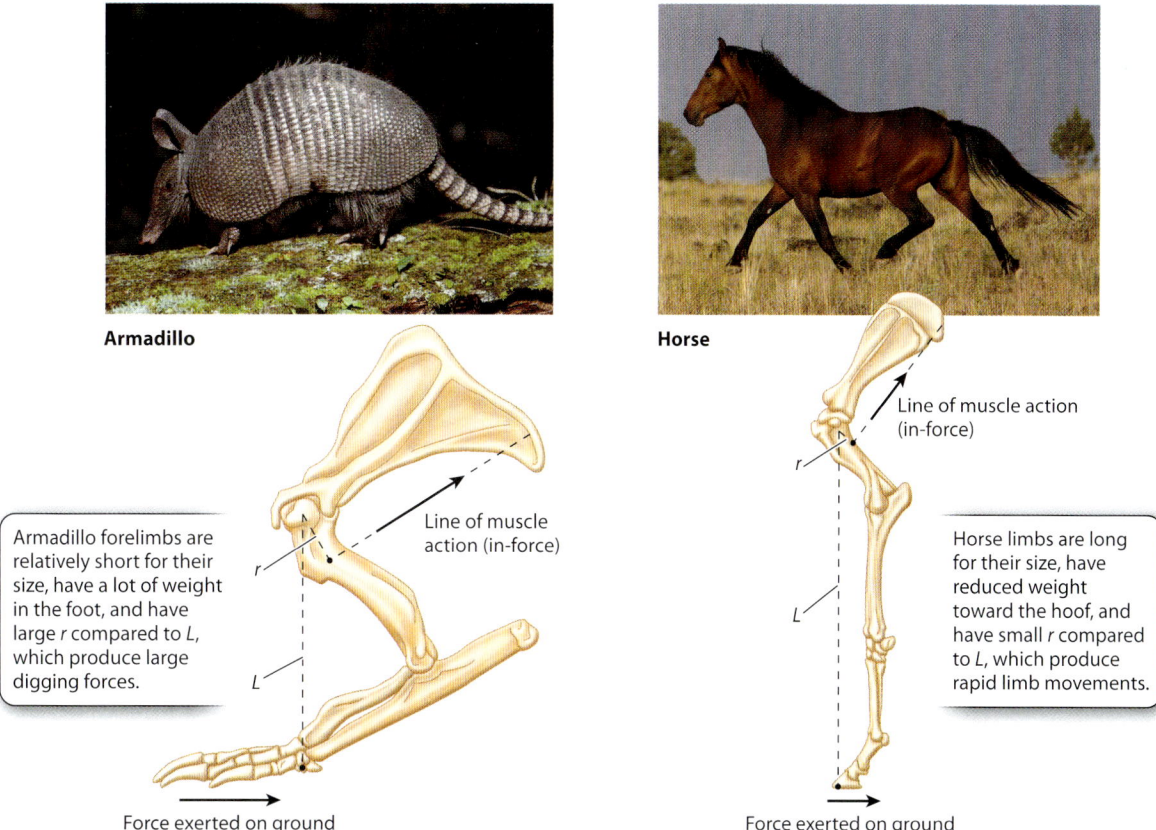

and muscle attachment site. Muscles that transmit force closer to the joint (smaller lever distance) result in less rotational strength but can produce large ranges of movement compared to muscles that attach farther from the joint.

As a result, the trade-off of force versus velocity that we observed earlier as a basic contractile property of a muscle also applies to the way muscles pull on the skeleton to cause joint movement. Muscles that transmit large forces for body movement and weight support, such as the calf muscles that extend your ankle joint, typically have large leverage for strong joint rotation. The heel of your foot gives the Achilles tendon of your calf muscles this greater leverage.

Animal joints located a long distance from the muscle attachment site (high muscle leverage) and a shorter distance from where the output force is exerted produce strong but slow movements. Moles, armadillos, and spade-foot toads are examples of digging animals that have specialized muscle–joint arrangements like this (**Fig. 37.22**). In contrast, joints located a short distance from the muscle attachment site (low muscle leverage) and a longer distance from where the output force is exerted produce faster movements but with less force. These are found in animals adapted for high-speed running, such as antelope, cheetahs, and horses.

Larger, faster animals also increase their economy of movement by reducing the mass of muscles and bones toward the end of their limbs. This lowers the energy cost of swinging the limb, which antelope rely on for fast economical locomotion to escape predators. Predators generally have heavier limbs that require larger muscles to move but which make them stronger.

Core Concepts Summary

37.1 MUSCLES ARE BIOLOGICAL MOTORS COMPOSED OF ACTIN AND MYOSIN THAT GENERATE FORCE AND PRODUCE MOVEMENT FOR BODY SUPPORT, LOCOMOTION, AND CONTROL OF THE BODY'S INTERNAL PHYSIOLOGICAL FUNCTIONS.

There are two main types of muscle: striated (skeletal and cardiac) and smooth. page 37-1

Skeletal and cardiac muscle fibers appear striated when viewed under the microscope because of the regular spacing of sarcomeres along their length. page 37-2

Smooth muscle fibers, which regulate airflow for breathing, blood flow through arteries, and the passage of food through the gut, lack regular sarcomere organization and appear smooth when viewed under the microscope. page 37-2

Skeletal muscle fibers are long thin cells composed of parallel sets of myofibrils built up from smaller parallel arrays of actin and myosin filaments. page 37-2

The sarcomere is the basic contractile unit of a skeletal muscle. Sarcomeres are arranged in series along the length of a myofibril. page 37-2

Muscles change length and produce force by the formation of actin–myosin cross-bridges, causing the sliding of myosin and actin filaments relative to each other. page 37-5

Muscles are stimulated by motor neurons at the fiber's motor endplate, leading to depolarization of the muscle cell that triggers the release of Ca^{2+} ions from the sarcoplasmic reticulum. Ca^{2+} binds troponin, which moves tropomyosin off the myosin-binding sites on actin, allowing myosin heads to form cross-bridges with actin. page 37-6

Excitation–contraction coupling is the process in which depolarization of the muscle cell leads to its shortening. page 37-6

37.2 MUSCLE FORCE DEPENDS ON MUSCLE SIZE, DEGREE OF ACTIN–MYOSIN OVERLAP, CONTRACTILE VELOCITY, AND STIMULATION RATE.

Muscles exert their highest force when actin–myosin filaments have maximal overlap, allowing the most cross-bridges to form. page 37-8

Muscles exert more force when they contract at slow velocities compared to when they contract at high velocities. page 37-9

Because muscles can transmit force only by pulling on the skeleton, they are arranged as antagonist pairs to produce reciprocal motions of a joint or limb. page 37-10

In vertebrates, a motor unit consists of a single motor neuron and the muscle fibers (cells) it innervates. page 37-11

Muscle force is increased by increasing the motor neuron firing rate and, in vertebrates, by the number of motor units that are activated. page 37-11

Vertebrate muscles have two types of fiber: red slow-twitch fibers that contract slowly over longer time periods and white fast-twitch fibers that contract rapidly but fatigue quickly. page 37-11

37.3 HYDROSTATIC SKELETONS, EXOSKELETONS, OR ENDOSKELETONS PROVIDE ANIMALS WITH MECHANICAL SUPPORT AND PROTECTION.

The rigid element of a hydrostatic skeleton is an incompressible fluid within a body cavity, used to support the body and change shape. page 37-13

Invertebrate exoskeletons form an external rigid support system, which protects the animal and limits water loss, but also limits growth and repair. page 37-14

Vertebrates have a bony endoskeleton that provides rigid support and protection of body organs, and that can grow and be repaired. page 37-15

The vertebrate endoskeleton is organized into axial (central) and appendicular (limb) components. page 37-15

Vertebrate endoskeletons consist of bone and cartilage. page 37-16

Bone is a composite tissue that consists of calcium phosphate mineral and type I collagen and that is rigid and hard to break. page 37-16

Cartilage is a fluid-based gel reinforced by type II collagen and proteoglycans, providing cushioning support at joint surfaces. page 37-16

37.4 VERTEBRATE ENDOSKELETONS ALLOW FOR GROWTH AND REPAIR AND TRANSMIT MUSCLE FORCES ACROSS JOINTS.

Vertebrate bones develop by one of two processes: either directly or by way of a cartilage model (endochondral ossification). page 37-16

Cartilage forms much of the embryonic skeleton and remains as growth plates within the bone to provide rapid growth during adolescence, as well as forming the bone's joint (articular) surfaces. page 37-16

Bone has two basic structures: Compact bone forms the solid walls of a bone's shaft, and spongy bone forms a mesh that supports the cartilage at the bone's ends. page 37-17

Bone formation by osteoblasts and removal by osteoclasts is a continuing process of growth, shape change, and repair. page 37-17

The shape of a bone's joint surfaces largely determines the range of motion and stability of a joint. page 37-17

Muscles produce joint movements by transmitting forces by means of rigid skeletal levers. page 37-18

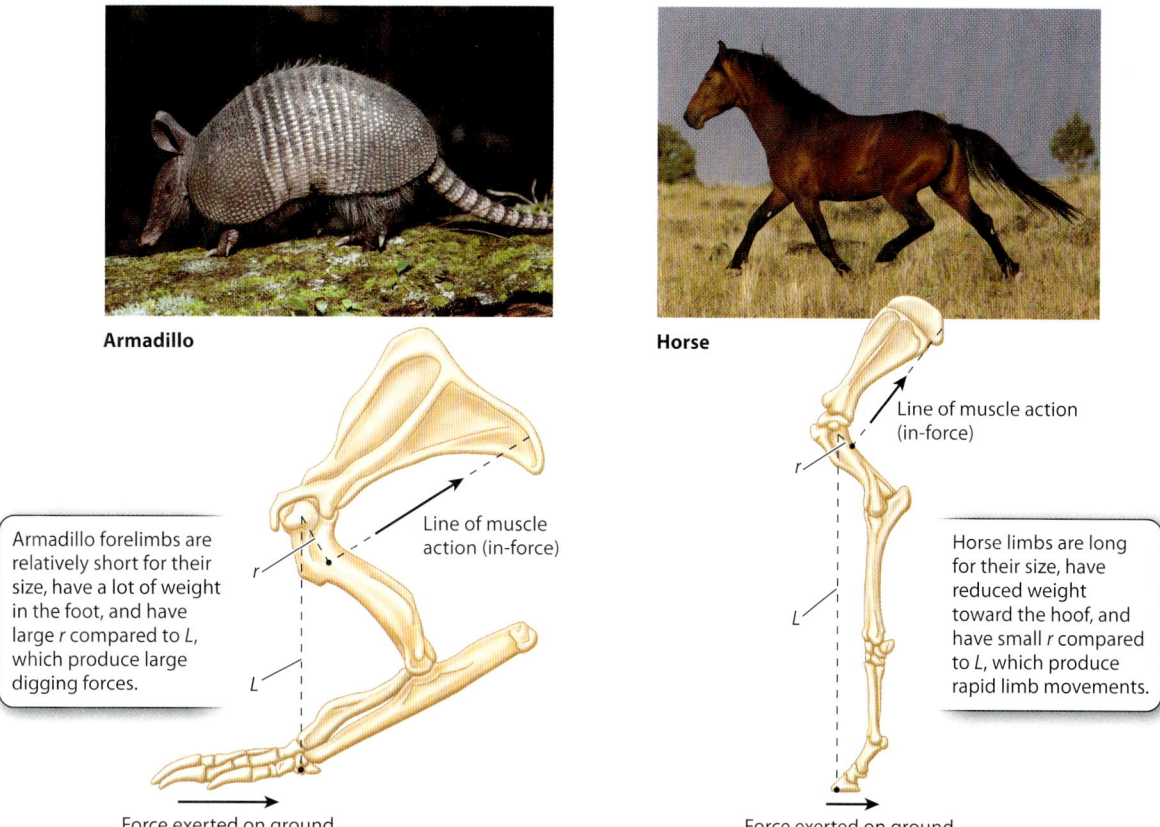

Self-Assessment

1. Diagram a sarcomere, showing the basic organization of thin (actin) and thick (myosin) filaments. Indicate on your diagram the regions where myosin cross-bridges form with actin.

2. Describe the sequence of events that occurs when an action potential arrives at a muscle fiber's motor endplate, causing the muscle to be depolarized.

3. Draw a graph of the isometric force–length relationship of striated muscle, indicating where maximal overlap between actin and myosin filaments occurs.

4. Draw a graph of the force–shortening velocity relationship of striated muscle.

5. Compare the force produced by a muscle over time when it is stimulated by a single twitch stimulus with the force produced by multiple stimuli at low frequency, and then with the force produced when the stimulation frequency is increased.

6. Name the two basic structural elements common to all animal skeletons.

7. Compare and contrast three features of an exoskeleton and an endoskeleton.

8. Identify the two primary components of bone tissue and explain how each of these contributes to a bone's strength, stiffness, and resistance to fracture.

9. Diagram a limb bone, such as the tibia, showing the regions of articular cartilage, spongy trabecular bone, compact bone tissue, and the marrow cavity.

Do you understand the chapter's Core Concepts? Log in to check your answers to the Self-Assessment questions, then practice what you've learned and reinforce this chapter's concepts by working through the problems and multimedia tutorials provided there.

www.biologyhowlifeworks.com

CHAPTER 38

ANIMAL ENDOCRINE SYSTEMS

Core Concepts

38.1 Animal endocrine systems release chemical signals called hormones into the bloodstream. These signals respond to the environment, regulate growth and development, and maintain homeostasis.

38.2 Hormones achieve specificity by binding to receptors on or inside their target cells. Their signals are amplified to exert strong effects on their target cells.

38.3 In vertebrate animals, the hypothalamus and the pituitary gland control and integrate diverse bodily functions and behaviors.

38.4 Chemical communication can also occur locally between neighboring cells or, through the release of pheromones, between individuals, coordinating social interactions.

Animals, like all organisms, respond to their environment, regulating their metabolism to meet changes in energy required by different activities and maintaining stable internal bodily functions. We have seen how the nervous system allows an animal to sense and respond to its environment, coordinating the animal's movement and behavior (Chapters 35–37). The nervous system also works closely with the **endocrine system** to regulate an animal's internal physiological functions. This coordination is accomplished by signals sent from the nervous system to the endocrine system, which distributes chemical signals throughout the body.

The signals communicated by the nervous and endocrine systems differ greatly in their modes of transmission and the times over which they act. The nervous system sends signals rapidly by action potentials running along nerve axons, and communication between adjacent nerve cells occurs by means of neurotransmitters in proximity at synapses. The endocrine system, on the other hand, relies on cells and glands that secrete chemical signals called **hormones,** which are released into the bloodstream and circulate throughout the body. As a result, endocrine communication is generally slower and more prolonged than the rapid and brief signals transmitted by nerve cells. Hormones exert specific effects on particular target cells within the body by binding to receptors on or in target cells. The downstream influence of hormones is amplified in a series of steps that occur along a particular endocrine pathway so that a small amount of hormone can have dramatic effects in the body.

In this chapter, we explore the properties and actions of hormones that help to coordinate and maintain a broad set of physiological functions of the organism. In particular, we examine the mechanisms by which hormones trigger cellular responses in their target organs. Because most hormones are evolutionarily conserved, their chemical organization is similar across a diverse array of animals. However, their functions often evolve rapidly, enabling hormones to serve new and broader roles.

38.1 AN OVERVIEW OF ENDOCRINE FUNCTION

The word "hormone" often popularly connotes teenagers and the changes that occur to their bodies as they grow and mature. Hormones, as we will see, do have key functions in animal growth and development—but that is only one of their roles. Hormones play diverse physiological roles in the body, including regulating an organism's response to the environment and helping to maintain stable physiological conditions within cells or within the animal as a whole. We start by highlighting these functions before addressing the molecular mechanism of how hormones work.

The endocrine system helps to regulate an organism's response to its environment.

Organisms face constant changes in their environment. These changes often present physiological challenges or stresses that the organism must respond to by altering the functional state of its body. Changes in light or temperature, the threat of a predator, or the presence of a potential mate stimulate responses of an animal's nervous system and endocrine system. Endocrine responses are commonly triggered by sensory signals received by the nervous system that are relayed to the endocrine system. The endocrine system, with its slower and more prolonged signaling, reinforces physiological changes in the animal's body that better suit the environmental cues received by its nervous system.

When a gazelle sees or smells a predator, its endocrine system helps to ready its body for rapid escape. When a female cardinal sees a bright red, singing male early in spring, endocrine signals transmitted within the female cardinal's brain initiate changes in its reproductive organs and increase its behavioral responsiveness, encouraging it to select the male as its mate. And when particular smells stimulate a sense of hunger and digestive function in animals, such as dogs and humans, the animals are attracted to food. In each of these cases, the endocrine system helps an organism respond appropriately to environmental cues.

The endocrine system is involved in growth and development.

The endocrine system plays important roles in growth and development, which require broad changes in many different organ systems. The release of circulating hormones accomplishes these changes, regulating how animals develop and grow. For example, we discuss in Chapter 42 the importance of the sex hormones estrogen and testosterone in determining female or male sexual characteristics during embryonic development and in sexual development during puberty.

As another example, consider the well-studied endocrine system of insects, which regulates their growth and development. Insects produce a rigid exoskeleton (Chapter 37), which must be periodically shed and a larger new one made to enable the insect to grow (**Fig. 38.1**). Shedding of the exoskeleton is referred to as **molting**. Each molting produces a new larval stage. In some insects, such as moths, the animal's body also changes dramatically, undergoing **metamorphosis** at key stages in development (Fig. 38.1a). In other insects, such as grasshoppers and crickets, the animal grows larger after each molting, but with little change in body form (Fig. 38.1b).

Molting and metamorphosis are regulated by hormones released from tissues in the insect's head. British physiologist Sir Vincent Wigglesworth studied their effects on the growth and development of the blood-sucking insect *Rhodnius*. Normally, *Rhodnius* goes through five juvenile stages before developing into its final adult body form. Like grasshoppers and crickets, it does not undergo metamorphosis. Each of the developmental stages is triggered by a blood meal. Wigglesworth showed that he could prevent molting if the head of the insect were removed shortly after a blood meal (**Fig. 38.2**). However, if the insect takes a blood meal and the head is removed a week later, the animal's body

FIG. 38.1 Growth and development in (a) moths and caterpillars and (b) grasshoppers.

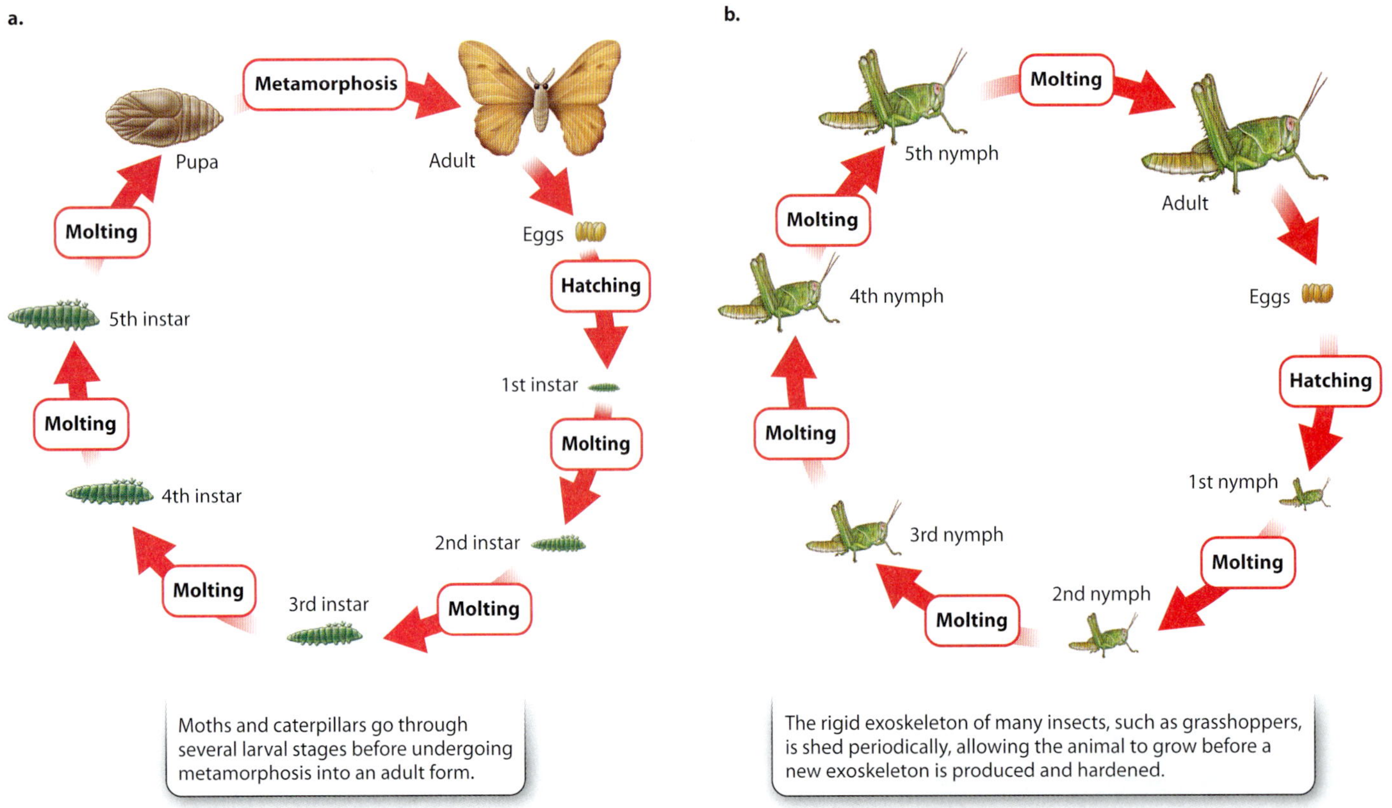

Moths and caterpillars go through several larval stages before undergoing metamorphosis into an adult form.

The rigid exoskeleton of many insects, such as grasshoppers, is shed periodically, allowing the animal to grow before a new exoskeleton is produced and hardened.

HOW DO WE KNOW?

FIG. 38.2

How are growth and development controlled in insects?

BACKGROUND During the 1930s, British physiologist Sir Vincent Wigglesworth studied how a blood meal taken by the bug *Rhodnius* triggered its molting and growth. This work pioneered the discovery of hormonal substances that stimulated its growth. *Rhodnius* goes through five successive larval stages (called nymphs) before becoming a winged adult, as shown in Figure 38.2a. Each developmental step is triggered by a blood meal.

HYPOTHESIS Wigglesworth hypothesized that a substance (specifically, a hormone) that diffuses from the head triggers molt in *Rhodnius*.

EXPERIMENT Wigglesworth decapitated juvenile bugs at different intervals of time after a blood meal and observed whether or not molting occurred. (Decapitation does not kill the insect.)

RESULTS Wigglesworth showed that if a bug is decapitated less than an hour after a blood meal, it fails to molt. If a bug is decapitated 1 week after the blood meal, it molts (Fig. 38.2b, Experiment 1). He also found that if he used a fluid tube to join the body of a bug decapitated immediately after a blood meal with the body of one that wasn't decapitated until a week after feeding, the bodies of both bugs molted (Fig. 38.2b, Experiment 2). This experiment demonstrated that the diffusing hormone could trigger molting in the bug that lacked the hormone because of immediate decapitation.

a.
Development of *Rhodnius*

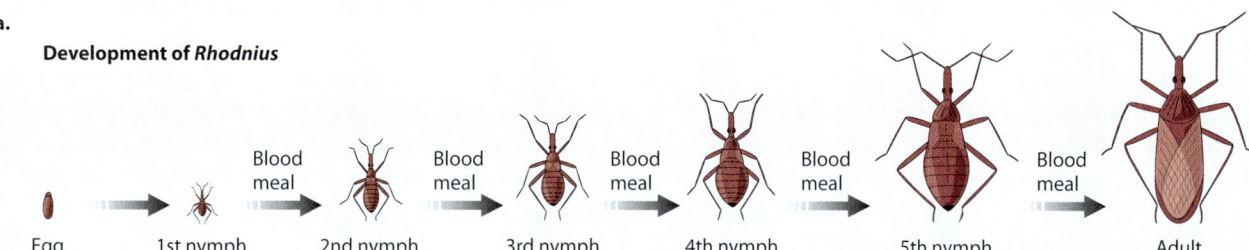

b.

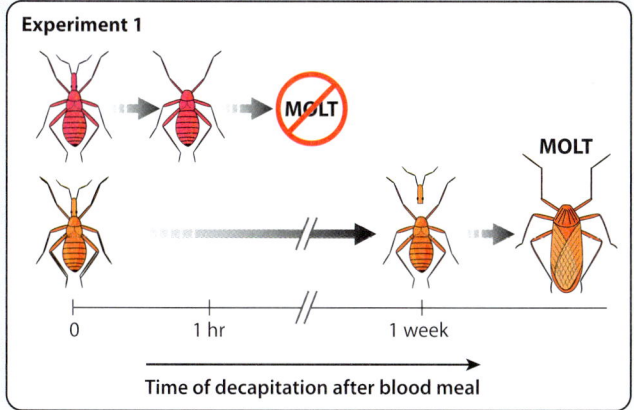

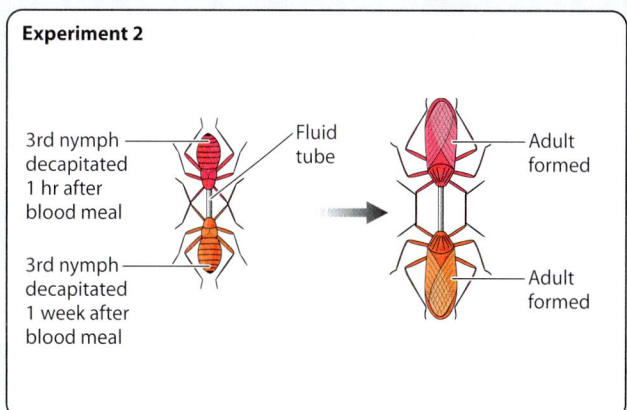

CONCLUSION A substance diffuses from the head in *Rhodnius* and triggers the molting process.

FOLLOW-UP WORK Subsequent studies by the German endocrinologist Alfred Kuhn confirmed the result in experiments performed on moth caterpillars. More recent molecular analyses performed by Japanese molecular endocrinologists Nagasawa and colleagues (1984) and Kawakami and colleagues (1990) isolated and sequenced the structure of the hormone from silkworm moths. It is now known as prothoracicotropic hormone, or PTTH, because it acts on the prothoracic gland of the insect to release the molting hormone, ecdysone.

SOURCES Wigglesworth, V.B. 1934. "The Physiology of Ecdysis in *Rhodnius prolixus* (Hemiptera). II. Factors Controlling Moulting and 'Metamorphosis.'"*Quarterly Journal of Microscopical Sciences* 77:191–223; Nagasawa, H., H. Kataoka, A. Isogai, S. Tamura, A. Suzuki, H. Ishizaki, A. Mizoguchi, Y. Fujiwara, A. Suzuki. 1984. "Amino-Terminal Amino Acid Sequence of the Silkworm Prothoracicotropic Hormone: Homology with Insulin." *Science* 226:1344–1345; Kawakami, A., H. Kataoka, T. Oka, A. Mizoguchi, M. Kimura-Kawakami, T. Adachi, M. Iwami, H. Nagasawa, A. Suzuki, H. Ishizaki. 1990. "Molecular cloning of the *Bombyx mori* Prothoracicotropic Hormone." *Science* 247:1333–1335.

undergoes a molt and grows into its adult form. Wigglesworth hypothesized that a diffusing substance from the animal's head controls its molt. In the first case, this substance does not have time to reach the body to trigger a molt, but in the second it does.

The German endocrinologist Alfred Kuhn found a similar result in experiments performed on moth caterpillars. Tying the caterpillar's head off from the rest of the body prevented the body from molting. Wigglesworth and Kuhn considered the diffusing substance to be a brain hormone. Subsequent work showed that cells in a specialized region of the insect brain secrete the peptide now known as prothoracicotropic hormone (PTTH) (**Table 38.1**). In *Rhodnius*, PTTH is released after a blood meal and acts to trigger the animal's molt.

Two decades later, the German endocrinologist Peter Karlson isolated and purified a second hormone, termed ecdysone. Whereas PTTH triggers molting, ecdysone coordinates the growth and reorganization of body tissues during a molt. Ecdysone is released from a gland when stimulated by PTTH (Table 38.1). Karlson was able to purify a small amount of ecdysone from large numbers of silk moth *Bombyx* caterpillars. He ultimately isolated a mere 25-mg sample from 500 kg of moth larvae! This tiny amount highlights the general principle that relatively few hormone molecules can have a large effect on an organism.

In his studies of *Rhodnius*, Wigglesworth also noted that, whatever the larval stage, the decapitated bug always molted into an adult after its blood meal. By removing just the front brain region of the head, Wigglesworth was able to show that a region in the head close to the brain releases a hormone that normally prevents the earlier larval stages from molting into the adult form (Table 38.1). This hormone, called juvenile hormone, is released in decreasing amounts during each successive larval molt. After the fifth and final stage, the level of juvenile hormone is so low that it no longer blocks maturation, allowing the final molt to undergo metamorphosis into the adult form (**Fig. 38.3**).

These studies reveal how small amounts of hormones released from key glands within the animal's body regulate major stages of growth and changes in body form during metamorphosis. When an animal's body requires coordinated changes in multiple organ systems, hormonal regulation by the endocrine system plays a critical role.

Hormones also regulate growth in humans and other vertebrate animals. Growth hormone controls the growth of the skeleton and many other tissues in the human body. Growth hormone is produced by the **pituitary gland,** which is located beneath the brain. Tumors of the pituitary that cause an overproduction of growth hormone lead to conditions of gigantism, and tumors that result in too little growth hormone cause pituitary dwarfism. The role of the pituitary gland in regulating body function will be described later in the chapter.

TABLE 38.1 Major Invertebrate Hormones

SECRETING TISSUE OR GLAND	HORMONE	TARGET GLAND OR ORGAN	ACTION
Brain	Brain hormone (peptide)	Prothoracic gland	Stimulates release of ecdysone to trigger molt and metamorphosis
Brain	Juvenile hormone (peptide)	Corpora allatum	Inhibits metamorphosis to adult stages; stimulates retention of juvenile characteristics
Prothoracic gland	Ecdysone (steroid)	All body tissues	Stimulates molt and, in absence or low levels of juvenile hormone, stimulates metamorphosis
Brain	Ecdysone (steroid)	All body tissues	Stimulates growth and regeneration in sea anemones, flat worms, nematodes, annelids, snails, and sea stars
Brain	Melanocyte-stimulating hormone (peptide)	Chromatophores	Stimulates pigmentation changes in cephalopods (octopus, squid, and cuttlefish)
Brain and eye stalks	Chromatotropins (peptides)	Chromatophores	Stimulates pigmentation changes in crustaceans
Reproductive gland	Androgen (peptide)	Reproductive tract	Regulates development of testes and male secondary sexual characteristics of crustaceans

FIG. 38.3 Hormonal control of growth and development in *Rhodnius*.

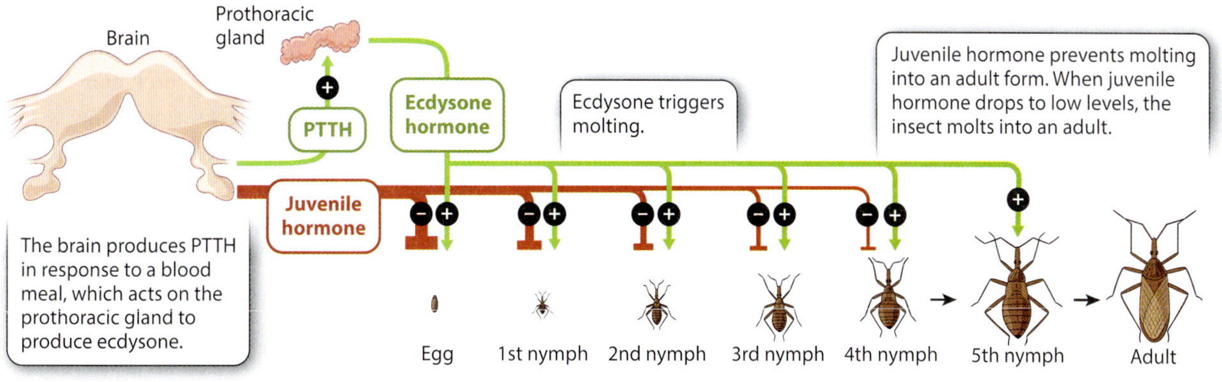

The endocrine system underlies homeostasis.

In addition to helping an animal respond to the environment and regulating growth and development, hormones synchronize and coordinate multiple bodily processes over longer time periods. Consequently, endocrine control of internal body functions is central to **homeostasis,** the maintenance of a steady physiological state within a cell or an organism (Chapters 5 and 35).

Physiological mechanisms that maintain a stable internal environment are necessary because without them, changing environmental conditions would lead to dangerous shifts in an animal's physiological function. For example, an animal's body weight depends on the regulation of its energy intake relative to energy expenditure. Disruption of hormones that regulate appetite and food intake can lead to food disorders that promote obesity or lead to weakness and lethargy. Similarly, hormones that regulate the concentration of key ions in the body, such as Na^+ and K^+, are fundamental to healthy nerve and muscle function (Chapters 35 and 36) and to fluid balance within the body (Chapter 41).

How does the body, and in particular the endocrine system, maintain homeostasis? Hormones are released in response to stimuli and have an effect on distant organs. Homeostatic regulation, or maintenance of homeostasis, depends on **feedback** from the target organ sent back to the endocrine gland that secretes the hormone. In feedback, a stimulus evokes a response, which in turn affects the stimulus. Thus, feedback is the means by which a signal produced by the regulatory organ can modify its own subsequent production, either inhibiting or stimulating it. There are two general types of feedback, **negative feedback** and **positive feedback.**

Homeostasis typically depends on negative feedback for control. In negative feedback, a change in a system causes a response that brings the system back to the starting point. If we consider the components of such a system, we can define a stimulus (the change itself), a sensor (which detects the change), an effector (which leads to a response), and the response (which ultimately opposes the initial change), thereby feeding back information to the sensor that limits a further response (**Fig. 38.4**). In Chapter 35, we saw how both the temperature in a house and the core body temperature of an animal are maintained at a constant level by this type of control.

In fact, many physiological parameters are maintained at steady levels by negative feedback. In the case of the endocrine system, a stimulus leads to release of a hormone, and the hormone causes a response in the body that opposes the initial stimulus so that secretion of the hormone is decreased and there is a limit to further change. For example, blood-glucose and calcium levels are kept within a narrow range by negative feedback. Although the parameter—temperature, glucose levels, calcium levels—does not change, constant feedback between the response and sensor maintains a set point. That is, the maintenance of homeostasis is an active process.

Let's examine an example in more detail. Like core body temperature, the amount of glucose in the blood of animals is

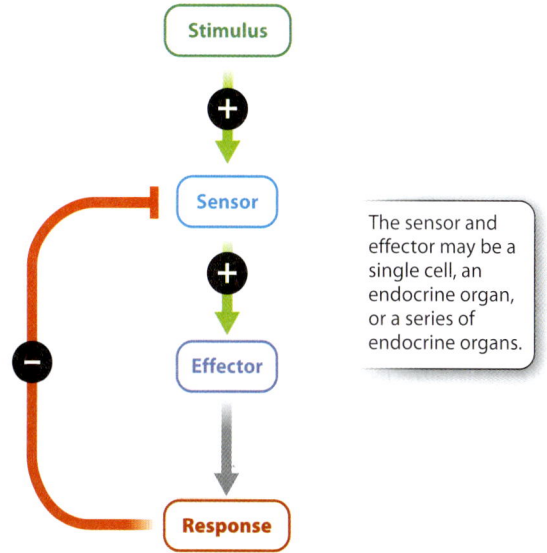

FIG. 38.4 **Negative feedback.** Negative feedback results in steady conditions, or homeostasis.

The sensor and effector may be a single cell, an endocrine organ, or a series of endocrine organs.

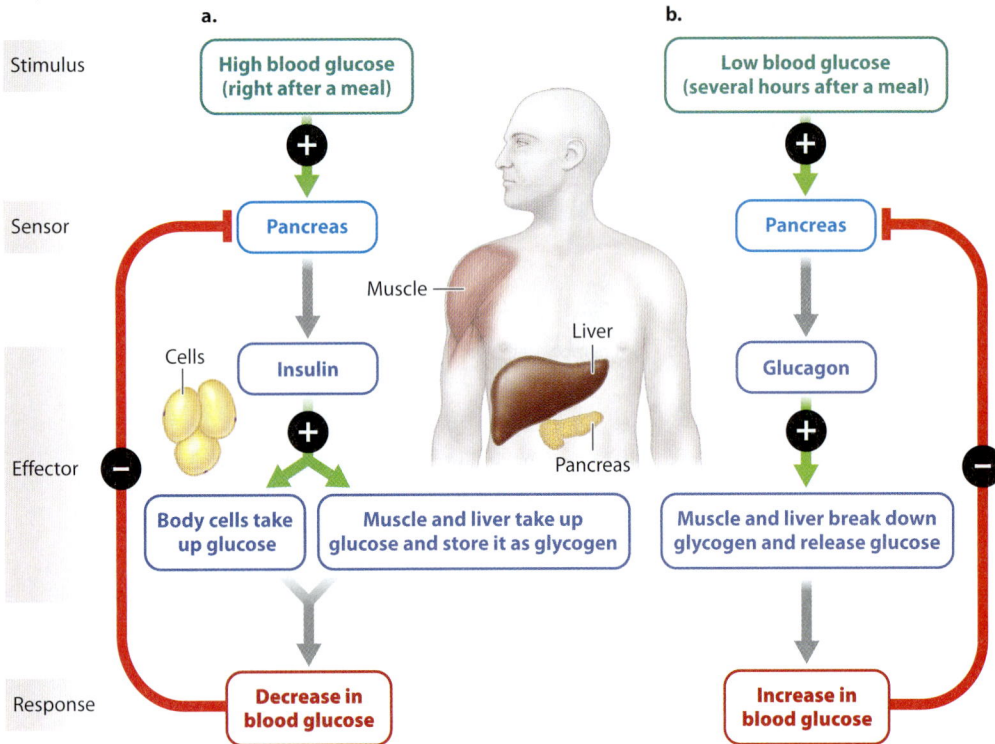

FIG. 38.5 Control of glucose levels in the blood by negative feedback.

maintained at a steady level (**Fig. 38.5**). If glucose levels are too low, cells of the body do not have a ready source of energy. If glucose levels are too high for too long, they can damage organs.

Maintaining steady blood glucose levels is challenging because glucose is commonly absorbed into the bloodstream from the intestine after a meal, and glucose is readily taken up by cells to meet their energy needs. How then are constant levels maintained in the body? After a meal, when blood glucose rises, β (beta) cells of the pancreas secrete the hormone insulin, which circulates in the blood (Fig. 38.5a). In response to insulin, muscle and liver cells take up glucose from the blood and convert it to a storage form called glycogen (Chapter 7). In this way, insulin guards against high levels of glucose in the blood.

What happens if blood glucose levels fall too low several hours after a meal? In this case, a different population of cells in the pancreas, called α (alpha) cells, secrete the hormone glucagon, which has effects opposite to those of insulin (Fig. 38.5b). Glucagon stimulates the breakdown of glycogen into glucose and its release from muscle and liver cells. The result is that blood glucose levels rise. In both cases, the stimulus (either high or low blood glucose levels) is sensed by cells of the pancreas (β or α cells) that trigger a response (secretion of insulin or of glucagon, bringing blood glucose levels back to the set point). Note that in each case the response feeds back to the secreting cells to reduce further hormone secretion.

When the control of blood glucose levels by insulin fails, a disease called **diabetes mellitus** results. When untreated, diabetic individuals excrete excess glucose in their urine because blood glucose levels are too high. Diabetes also causes cardiovascular and neurological problems, including loss of sensation in the extremities, particularly the feet.

In some instances, it is necessary to enhance the production of a hormone, accelerating the response of the target cells for a period of time. Positive feedback provides this enhancement (**Fig. 38.6**). In positive feedback, a stimulus causes a response in the same direction as the initial stimulus, which leads to a further response, and so on. In positive feedback in the endocrine system, a stimulus leads to secretion of a hormone that causes a response, and the response causes the release of more hormone. The result is an escalation of the response. A positive feedback loop reinforces itself until it is interrupted or broken by some sort of external signal.

Positive feedback occurs in mammals during childbirth (Chapter 42). In response to uterine contractions, the hormone oxytocin is released from the pituitary gland. Uterine

FIG. 38.6 **Positive feedback.** In positive feedback, a stimulus causes a response, and that response causes a further response in the same direction.

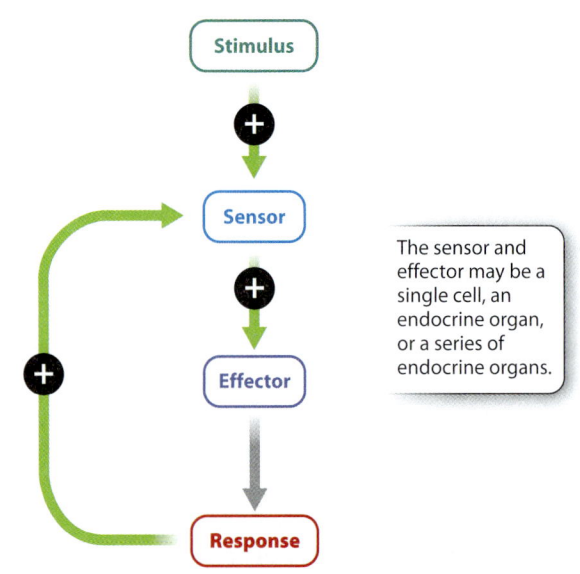

The sensor and effector may be a single cell, an endocrine organ, or a series of endocrine organs.

contractions in turn stimulate the pituitary gland to secrete even more oxytocin, causing the uterine muscles to contract more forcefully and more frequently.

→ **Quick Check 1** Diabetes mellitus is a disease characterized by high blood glucose levels. What two different physiological conditions can produce diabetes?

38.2 PROPERTIES OF HORMONES

Hormones play diverse roles in the body, enabling regulation and coordination of multiple bodily functions in response to environmental cues. Because hormones are released into the bloodstream, they have the potential to affect many organ systems. Before discussing other important hormonal regulatory pathways, we must first consider how hormones target specific tissues and how different hormones exert their effect on different target cell types. Receptors on target organs respond to blood-borne hormones by binding the hormone, allowing the hormone to influence the function of specific cell targets. Certain classes of hormone (amine and peptide hormones) bind to receptors on the cell membrane and influence cellular function by signaling pathways (Chapter 9). The other general class of hormone (steroid hormones) is lipid soluble, so these hormones can diffuse across the cell membrane. Steroid hormones bind to receptors in the cytoplasm or in the nucleus and regulate transcriptional control of protein synthesis of the target cell.

Three main classes of hormone are peptide, amine, and steroid hormones.

Hormones are grouped into three general classes: hydrophilic **peptide hormones,** hydrophilic **amine hormones,** and hydrophobic **steroid hormones,** all of which are illustrated in **Fig. 38.7.** Peptide hormones and amine hormones are both derived from amino acids. Peptide hormones are short chains of amino acids, whereas amine hormones are derived from a single aromatic amino acid, such as tyrosine. All steroid hormones are derived from cholesterol. Whereas peptide hormones (which are sometimes large enough to be considered proteins) can evolve through changes in their amino acid sequence, steroid hormones cannot. Evolutionary changes in steroid hormone function depend instead on changes in the receptors they bind and the cellular responses they trigger.

FIG. 38.7 (a) Peptide, (b) amine, and (c) steroid hormones.

a. Peptide hormones

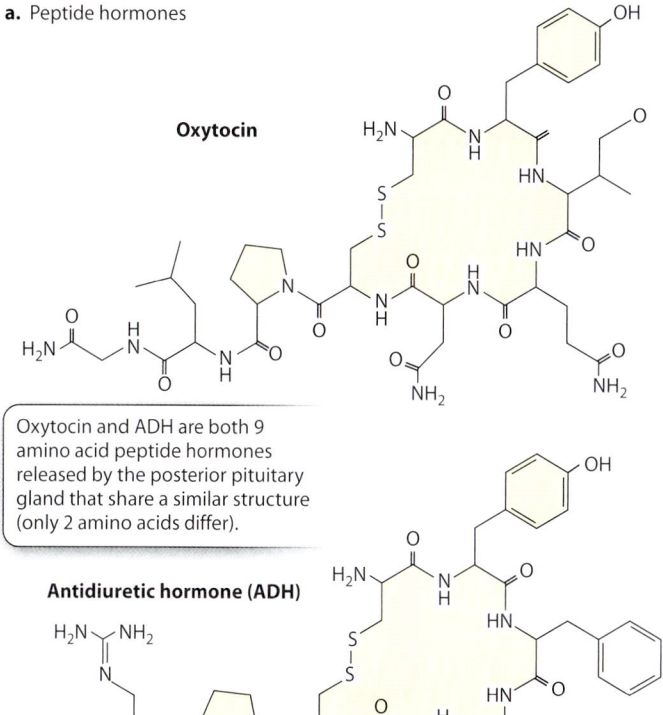

Oxytocin and ADH are both 9 amino acid peptide hormones released by the posterior pituitary gland that share a similar structure (only 2 amino acids differ).

b. Amine hormones

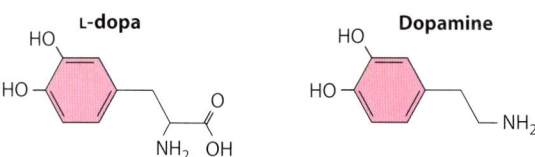

Amine hormones are derived from aromatic amino acids such as tyrosine.

c. Steroid hormones

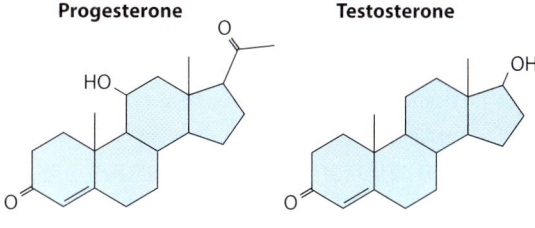
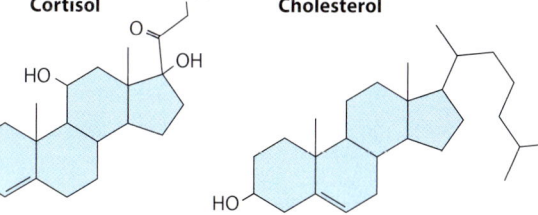

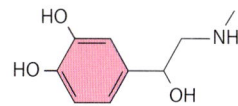

The steroid hormones all share a similar structure and are derived from cholesterol.

Peptide and amine hormones are more abundant than steroid hormones and are more diverse in their actions. Because most peptide and amine hormones are hydrophilic, they cannot diffuse across the plasma membrane. Instead, nearly all peptide and amine hormones bind to membrane receptors on the surface of the cell, triggering the intracellular second messenger pathways discussed in Chapter 9 (**Fig. 38.8a**). Peptide and amine hormones alter the biochemical activity of the target cell by activating or inactivating enzymes within the cell. Typically, they affect protein kinases (which phosphorylate other proteins), initiating signaling cascades within the target cell. These signaling cascades can lead to changes in gene expression or metabolism, trigger a cell to grow, divide, or change shape, or lead to the release of other hormones. Peptide and amine hormones act on timescales of minutes to hours.

There are many examples of peptide hormones. In vertebrates, for instance, growth hormone stimulates protein synthesis and growth of many body tissues, particularly the musculoskeletal system (**Table 38.2**). As we saw, insulin and glucagon are peptide hormones released by the pancreas that regulate glucose metabolism by the liver, muscles, and other tissues. Gastrin and cholecystokinin are peptide hormones that regulate mammalian digestive function (Chapter 40). In addition to serving as a neurotransmitter released by the sympathetic nervous system (Chapter 35), epinephrine and norepinephrine (also known as adrenaline and noradrenaline) act as amine hormones, supporting an animal's fight-or-flight response to stress.

In contrast to peptide and amine hormones, steroid hormones are hydrophobic. Therefore, they diffuse freely across the cell membrane to bind with receptors in the cytoplasm or nucleus (**Fig. 38.8b**), forming a steroid hormone–receptor complex. Hormone–receptor complexes that form in the cytoplasm are transported into the nucleus of the cell. These complexes most commonly act as transcription factors, stimulating or repressing gene expression and, therefore, altering the proteins produced by the target cell. Consequently, steroid hormones typically exert profound and long-lasting effects on the cells and tissues they target, with timescales of days to months. The actions of steroid hormones, therefore, tend to be more diverse than those exerted by amine and peptide hormones.

The sex hormones estrogen, progesterone, and testosterone (Table 38.2), which we discuss in Chapter 42, are steroid hormones that regulate the differentiation, maturation, and functional state of the vertebrate reproductive organs. Cortisol is a steroid hormone that mediates a vertebrate animal's response to stress and inhibits inflammation. Steroid hormones similar to cortisol are used medically to reduce the symptoms of certain inflammation disorders. However, they carry risks because they also depress the immune system, increasing the chance of infection and diminishing wound healing.

FIG. 38.8 Mechanism of action of (a) peptide and (b) steroid hormones.

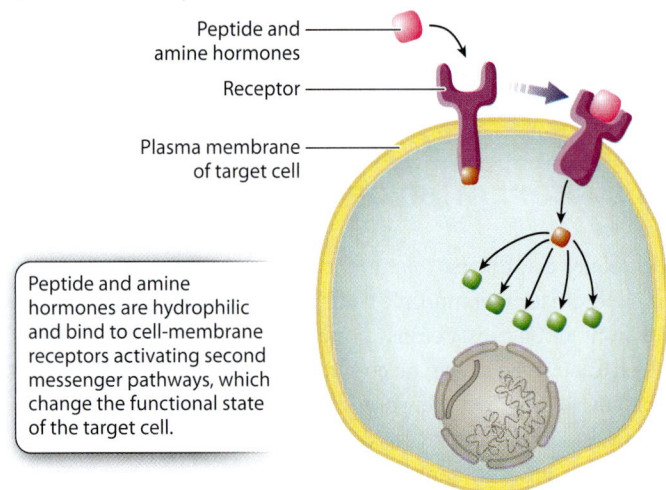

a. Cell-surface receptors

Peptide and amine hormones are hydrophilic and bind to cell-membrane receptors activating second messenger pathways, which change the functional state of the target cell.

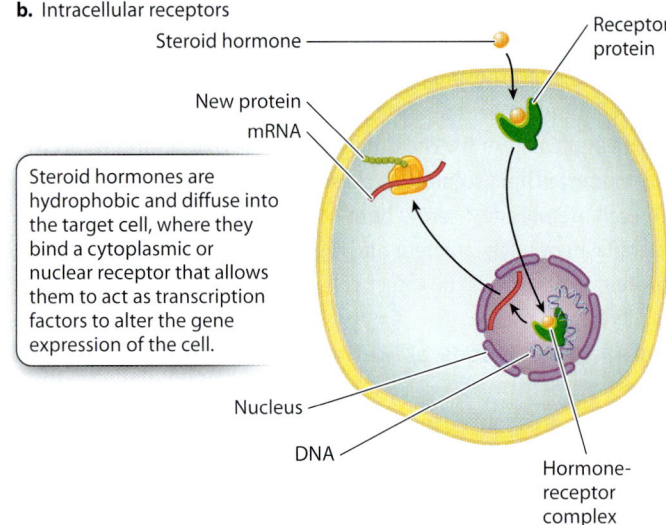

b. Intracellular receptors

Steroid hormones are hydrophobic and diffuse into the target cell, where they bind a cytoplasmic or nuclear receptor that allows them to act as transcription factors to alter the gene expression of the cell.

Hormonal signals are typically amplified.

Hormones are typically released in small amounts. So how is it possible for them to have large effects on the overall physiology of an organism? The answer is that hormone signals are amplified. Amplification occurs through a series of signaling steps between different endocrine glands in a hormonal pathway, as well as signal transduction steps within the target cells after the hormone binds to a cell receptor. Signals are thus amplified at each step of the pathway, resulting in a large effect on the target organ system.

Hormonal signaling pathways between endocrine glands and tissues are often referred to as endocrine axes. For example, in vertebrates, the hypothalamic–pituitary axis involves hormonal signals that are amplified along a pathway from the hypothalamus

CHAPTER 38 ANIMAL ENDOCRINE SYSTEMS

TABLE 38.2   Major Vertebrate Hormones

SECRETING TISSUE OR GLAND	HORMONE	TARGET GLAND OR ORGAN	ACTION
Hypothalamus	Releasing factors (peptides)	Anterior pituitary gland	Stimulate secretion of anterior pituitary hormones
Anterior pituitary gland	Thyroid-stimulating hormone (TSH) (glycoprotein)	Thyroid gland	Stimulates synthesis and secretion of thyroid hormones by the thyroid gland
	Follicle-stimulating hormone (FSH) (glycoprotein)	Gonads	Stimulates maturation of eggs in females; stimulates sperm production in males
	Luteinizing hormone (LH) (glycoprotein)	Gonads	Stimulates production and secretion of sex hormones in ovaries (estrogen and progesterone) and testes (testosterone)
	Adrenocorticotropic hormone (ACTH) (peptide)	Adrenal glands	Stimulates production and release of cortisol
	Growth hormone (GH) (protein)	Bones, muscles, liver	Stimulates protein synthesis and body growth
	Prolactin (protein)	Mammary glands	Stimulates milk production
	Melanocyte-stimulating hormone (peptide)	Melanocytes	Regulates skin (and scale) pigmentation
Posterior pituitary gland	Oxytocin (peptide)	Uterus, breast, brain	Stimulates uterine contraction and milk let-down; influences social behavior
	Antidiuretic hormone (ADH) (vasopressin) (peptide)	Kidneys, brain	Stimulates uptake of water from the kidneys; involved in pair bonding
Thyroid gland	Thyroid hormones (peptides)	Many tissues	Stimulate and maintain metabolism for development and growth
	Calcitonin (peptide)	Bone	Stimulates bone formation by osteoblasts
Ovaries	Estrogen (steroid)	Uterus, breast, other tissues	Stimulates development of female secondary sexual characteristics and regulates reproductive behavior
	Progesterone (steroid)	Uterus	Maintains female secondary sexual characteristics and sustains pregnancy
Testes	Testosterone (steroid)	Various tissues	Stimulates development of male secondary sexual characteristics; regulates male reproductive behavior and stimulates sperm production
Adrenal cortex	Cortisol (steroid)	Liver, muscles, immune system	Regulates response to stress by increasing blood glucose levels and reduces inflammation

(continues on page 38-10)

TABLE 38.2 Major Vertebrate Hormones (continued)

SECRETING TISSUE OR GLAND	HORMONE	TARGET GLAND OR ORGAN	ACTION
Adrenal medulla	(Nor-) Epinephrine (peptide)	Heart, blood vessels, liver	Stimulates heart rate, blood flow to muscles, and elevation of blood glucose level as part of fight-or-flight response
Parathyroid glands	Parathyroid hormone (PTH) (protein)	Bone	Stimulates bone resorption by osteoclasts to increase blood Ca^{2+} levels
Pancreas	Insulin (protein)	Liver, muscles, fat, other tissues	Stimulates uptake of blood glucose and storage as glycogen
	Glucagon (protein)	Liver	Stimulates breakdown of glycogen and glucose release into blood
	Somatostatin (peptide)	Digestive tract	Inhibits insulin and glucagon release; decreases digestive activity (secretion, absorption, and motility)
Stomach	Gastrin (peptide)	Stomach	Stimulates protein digestion by secretion of digestive enzymes and acid; stimulates gastric motility
Small intestine	Cholecystokinin (peptide)	Pancreas, liver, gallbladder	Stimulates secretion of digestive enzymes and products from liver and gallbladder
	Secretin (peptide)	Pancreas	Stimulates bicarbonate secretion from pancreas
Pineal gland	Melatonin (peptide)	Brain, various organs	Regulates circadian rhythms

to the anterior pituitary gland, and then from the anterior pituitary gland to target glands or tissues in the body.

The vertebrate hypothalamic–pituitary axis begins with trace amounts of peptide hormones called **releasing factors** that are initially released by the hypothalamus (**Fig. 38.9**). They signal to the anterior pituitary gland, leading to a much larger release of associated hormones from that organ. For example, corticotropin releasing factor secreted by the hypothalamus stimulates the release of a larger amount of adrenocorticotropic hormone (ACTH) by the anterior pituitary gland.

Hormones released by the anterior pituitary gland in turn bind cell receptors in the target organ. In this case, ACTH acts on cells of the adrenal cortex, stimulating their secretion of the hormone cortisol. Cortisol acts on many different cells and tissues in the body, causing what is known as an acute stress response.

Among its effects, it causes the liver to convert glucose to glycogen (although through effects on other cells it also increases blood glucose levels). This action yields 56,000x more glycogen molecules than the initial number of releasing factors secreted by the hypothalamus.

Similar signaling amplification cascades occur in invertebrates in the regulation of insect molting and metamorphosis (see Fig. 38.3). The amount of brain hormone that is produced is dwarfed by the amount of ecdysone or juvenile hormone that is released to regulate the growth and metamorphosis of the insect's body. As in vertebrates, second messenger signaling cascades within receptor cells of the body respond to the release of these hormones by amplifying the hormone signal in the cell. Thus, amplification by signal transduction pathways applies both to second messenger systems within a cell as well as the chemical

FIG. 38.9 Amplification of a hormonal signal.

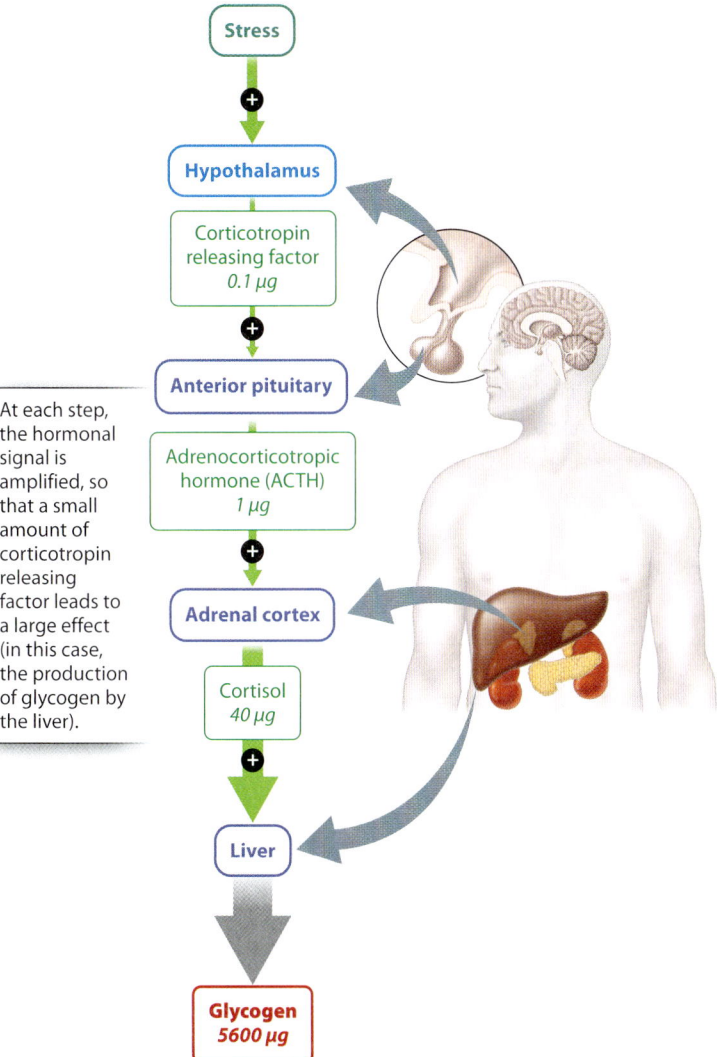

At each step, the hormonal signal is amplified, so that a small amount of corticotropin releasing factor leads to a large effect (in this case, the production of glycogen by the liver).

signals that are transmitted as hormones between glands and tissues in an organism's endocrine system.

→ **Quick Check 2** Why do steroid and peptide hormones bind different kinds of receptor and how does this difference affect the resulting signaling pathways in the target cell?

Hormones act specifically on cells with receptors that bind the hormone.

Hormones are chemical signals that enable communication between different cells in an organism. Because hormones are secreted into the bloodstream, they achieve specificity by binding to receptors on the surface of or inside the target cells. Therefore, it is the presence or absence of a receptor for a given hormone that determines which cells respond and which ones do not. For example, when the hormone oxytocin is released into the bloodstream of mammals, it affects only cells that express a receptor on the cell surface capable of binding oxytocin as it flows by in the bloodstream. These cells are uterine muscle cells and secretory cells in breast tissue. When oxytocin binds these cell-surface receptors, uterine muscle cells are stimulated to contract, and secretory cells in breast tissue release milk during breastfeeding. The effect of oxytocin also depends on cells of other organs *not* expressing the receptor so that the hormone can exert its effect on these specific tissues.

The binding of a hormone to its receptor leads to changes in the target cell, resulting in a cellular response. The specific action of a hormone depends on the kind of response that it triggers in the target cell. For example, the binding of a hormone can lead to changes in ion fluxes across the cell membrane, activation of an intracellular signal transduction cascade, or changes in gene and protein expression.

→ **Quick Check 3** What general features make a chemical compound a hormone, and how do hormones achieve specificity for certain kinds of target cells?

Hormones are evolutionarily conserved molecules with diverse functions.

Most hormones have an ancient evolutionary history. The structures of many hormones are evolutionarily conserved. Some vertebrate hormones can also be found in many invertebrates, but typically they serve different functions. Since the first vertebrate animals diverged from invertebrates over 500 million years ago, this indicates that some vertebrate hormones are even older. And, in many cases, their roles in many invertebrate animals have yet to be discovered.

Thyroid-stimulating hormone (TSH), a hormone released by the anterior pituitary gland that targets the thyroid gland, regulates metabolism in vertebrate animals but triggers metamorphosis in amphibians and feather molt in birds. It has even been found in snails and other invertebrates that lack a thyroid gland. Its function in snails appears to be to stimulate the number of sperm or eggs produced.

Recent genomic analysis has also revealed that the receptors for many hormones evolved well before the hormones with which they now interact. Even though the structure of a hormone or its receptor is often largely unchanged across diverse groups of organisms, the functions of hormones and their receptors can readily be selected to take on new roles. An example is the shift in function of TSH to regulate metabolism in mammals and feather molt in birds from an earlier role stimulating sperm or egg production in snails. Thus, the biochemical evolution of hormones and their target cell receptors can be readily altered as organisms evolve new behaviors and exploit new environments.

Another intriguing finding is that many peptides originally identified as hormones in various tissues have also been found to function as neurotransmitters in the nervous system. For example, oxytocin, which stimulates uterine contractions and the release of milk, also serves as a neurotransmitter in the brain and is believed to influence social behavior, as well as stimulate sexual arousal in mammals. Similarly, another peptide hormone, antidiuretic hormone, which regulates water uptake in the kidneys (Chapter 41), also functions as a neurotransmitter in the brain influencing mammalian mating and pair-bonding behavior. Hence, gene expression of a hormone in one kind of cell may produce the same compound in another cell but for an entirely different function. The roles of hormones as chemical messengers are varied within an organism and easily changed over the course of evolution.

38.3 THE VERTEBRATE ENDOCRINE SYSTEM

The vertebrate endocrine system regulates changes in the animal's physiological and behavioral states in response to sensory cues received by its nervous system both from the environment and from internal body organs. These sensory signals are processed within the brain and transmitted to the endocrine system primarily by the hypothalamus, which is located in the forebrain.

The hypothalamus, in turn, relays the signals to the pituitary gland, which lies just below it. The pituitary gland is a central regulating gland of the vertebrate endocrine system, releasing hormones that coordinate the action of many other endocrine glands and tissues. These glands, illustrated in **Fig. 38.10**, control the growth and maturation of the body, regulate reproductive development functions of the animal, coordinate digestion and metabolism, and control water balance. These effects reflect the broad range of roles that the endocrine and nervous systems perform in maintaining homeostasis.

The pituitary gland communicates with many cells, tissues, and organs of the body. Some of these, like the thyroid and adrenal glands, secrete hormones in response to signals from the pituitary gland and therefore have exclusively endocrine functions. Others, like the lungs, kidneys, and digestive tract, harbor endocrine cells that secrete hormones and also have other physiological roles in the body. Consequently, the vertebrate endocrine system is not localized in one part of the body, but is present throughout.

FIG. 38.10 Human endocrine organs and the hormones they secrete.

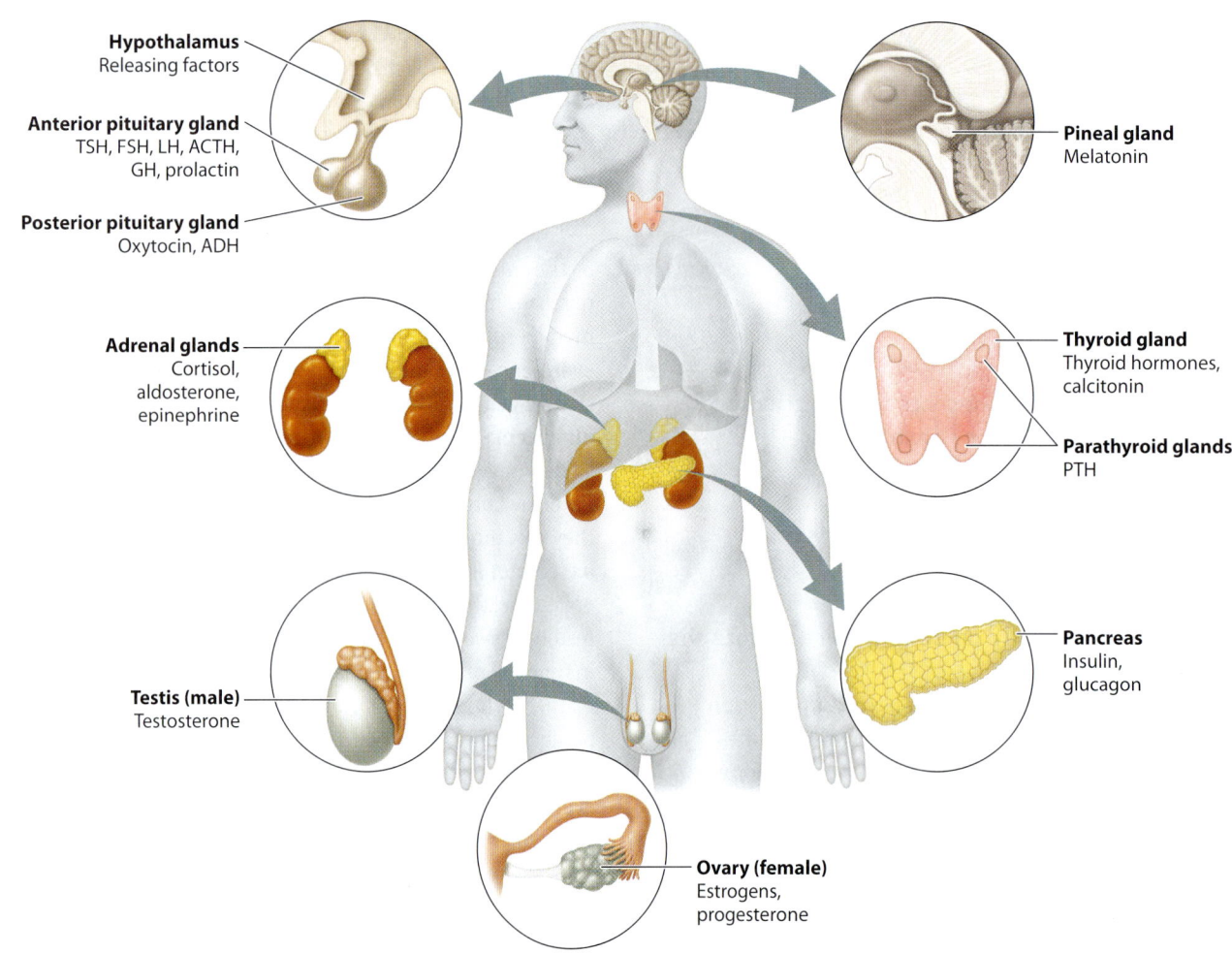

FIG. 38.11 The hypothalamus and the pituitary gland.

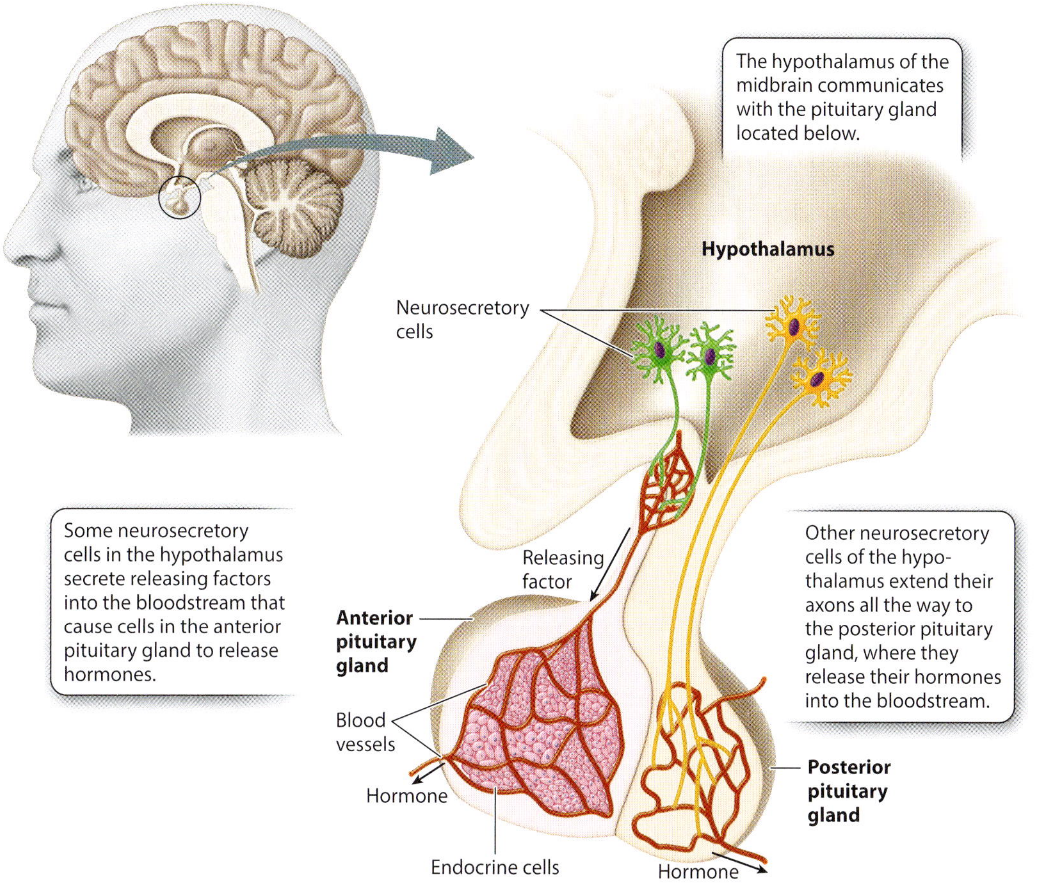

The hypothalamus of the midbrain communicates with the pituitary gland located below.

Hypothalamus

Neurosecretory cells

Some neurosecretory cells in the hypothalamus secrete releasing factors into the bloodstream that cause cells in the anterior pituitary gland to release hormones.

Releasing factor

Anterior pituitary gland

Blood vessels

Hormone

Endocrine cells

Other neurosecretory cells of the hypothalamus extend their axons all the way to the posterior pituitary gland, where they release their hormones into the bloodstream.

Posterior pituitary gland

Hormone

The pituitary gland integrates diverse bodily functions by secreting hormones in response to signals from the hypothalamus.

The hypothalamus is the main route by which nervous system signals are transmitted to the vertebrate endocrine system. The function of the hypothalamus is to transmit these signals to the pituitary gland, the endocrine gland that acts as a control center for most other endocrine glands in the body.

The pituitary gland is divided into anterior and posterior regions (**Fig. 38.11**). This division is not arbitrary—the two regions have distinct functions, organizations, and embryonic origins. The **anterior pituitary gland** forms from epithelial cells that develop and push up from the roof of the mouth, whereas the **posterior pituitary gland** develops from neural tissue at the base of the brain. Both sets of ectodermal cells form pouches that develop into glands, which come to lie adjacent to each other. The anterior and posterior pituitary glands should therefore be thought of as two distinct glands, not one.

Because of these developmental differences, the anterior and posterior pituitary glands receive input from the hypothalamus in different ways. The hypothalamus contains **neurosecretory cells.** These cells are located in the hypothalamus, which is part of the brain, and therefore they are neurons. However, instead of secreting neurotransmitters that bind to another neuron or to muscle, these cells release hormones into the bloodstream. Some of these neurosecretory cells communicate with the anterior pituitary gland. In this case, they secrete hormones called releasing factors into small blood vessels that travel to and supply the anterior pituitary gland (Fig. 38.11). In response, cells of the anterior pituitary gland release hormones into the bloodstream, in which they circulate throughout the body and bind to receptors on target cells, tissues, and organs.

In contrast, communication between the hypothalamus and the posterior pituitary gland does not involve hypothalamic releasing factors. Instead, the posterior pituitary gland contains the axons of neurosecretory cells whose cell bodies are located in the hypothalamus. These axons release hormones directly into the bloodstream, by which they are transported to distant sites (Fig. 38.11). Consequently, the posterior pituitary is part of the nervous system itself.

In response to signals from the hypothalamus, distinct hormones are secreted by the anterior and posterior pituitary glands (Table 38.2). The anterior pituitary gland hormones include thyroid stimulating hormone (TSH), which acts on the thyroid gland, the gonadotropic hormones follicle-stimulating hormone (FSH) and luteinizing hormone (LH), which act on the female and male gonads (the ovaries and testes), and adrenocorticotropic hormone (ACTH), which acts on the adrenal glands (see Fig. 38.10). In each of these cases, the anterior pituitary hormone acts on an endocrine gland to cause release of other hormones. Hormones that control the release of other hormones are called **tropic hormones.** In response to TSH, the thyroid gland releases thyroid hormones that regulate the metabolic state of the body. In response to FSH and LH, the ovaries release estrogen and progesterone and the testes release testosterone. In response to ACTH, the adrenal glands release cortisol, which has diverse effects that include stimulating glucose release into the bloodstream, maintaining blood pressure, and suppressing the immune system.

The anterior pituitary gland also secretes growth hormone (GH), which acts generally on the muscles, bones, and other body tissues to stimulate their growth, and prolactin, which stimulates milk production in the breast of female mammals in response to an infant's suckling at the mother's nipple.

The posterior pituitary gland hormones include **oxytocin** and **antidiuretic hormone** (ADH; also called **vasopressin**). These are two evolutionarily related peptide hormones, each consisting of nine amino acids. As we discussed earlier, oxytocin plays several roles related to female reproduction: It causes uterine contraction during labor and stimulates the release of milk during breastfeeding. Antidiuretic hormone acts on the kidneys (Chapter 41), regulating the concentration of urine that an animal excretes, which is critical to maintaining water and solute balance in the body.

In addition to their roles in reproduction and kidney function, oxytocin and vasopressin are also released by cells in the brain and may play roles in social behaviors. For instance, oxytocin is important in regulating maternal behavior toward infants. Recent research has suggested that oxytocin may also have a role in behaviors such as trust as a prelude to mating in vertebrate animals. These functions reflect the long evolutionary history of oxytocin, as it also plays a role in the social behavior of insects.

Recent evidence suggests that antidiuretic hormone plays a parallel role in regulating male social behavior and parental behavior in mammals. This is not surprising because oxytocin and antidiuretic hormone have very similar chemical structures (see Fig. 38.7). The integration of reproductive function with parental and social behavior for the care of young is likely favored strongly by natural selection. Receptors for these hormones are expressed within the brain. Thus, in addition to acting on reproductive and renal organs of the body, oxytocin and antidiuretic hormone affect mammalian behavior by acting directly on the brain.

Many targets of pituitary hormones are endocrine tissues that also secrete hormones.

As we have seen, some of the hormones released by the anterior pituitary gland act on endocrine organs that themselves then release hormones. These tropic hormones are TSH, FSH and LH, and ACTH. Here, we look at their target organs in more detail.

TSH acts on the **thyroid gland,** which is located in the front of the neck (see Fig. 38.10), and leads to the release of two peptide hormones, thyroxine (T4) and triiodothyronine (T3). These hormones regulate cellular metabolism throughout the body. Overproduction of thyroid hormones (hyperthyroidism) or thyroid hormone deficiency (hypothyroidism) creates symptoms that reflect either an overly active metabolic state (increased appetite and weight loss) or one that is too low (fatigue and sluggishness). Both conditions are diagnosed by blood tests that monitor circulating levels of thyroid hormones, and both can be treated by medication.

Because thyroid hormones require iodine for their formation, individuals who do not acquire enough iodine from their diets produce too little thyroid hormone, resulting in metabolic problems. Additionally, iodine insufficiency stimulates increased production of TSH by the anterior pituitary gland because of the absence of negative feedback. Over time, the thyroid gland enlarges to form a goiter, which is observed as an enlargement of the throat, to compensate and produce more thyroid hormone. The introduction of iodized salt has eliminated goiter formation and hypothyroidism in many countries, but in many underdeveloped areas these remain significant public health problems.

The gonadotropic hormones are FSH and LH. They target the female and male gonads—the **ovary** and **testis** (see Fig. 38.10). In response to FSH and LH, the ovary and testis each secrete sex hormones that regulate their own development as well as the sexual differentiation and maturation of secondary sexual characteristics of the body. Female sex hormones include the steroids estrogen and progesterone. The principal male sex hormone is the androgen testosterone. These sex hormones are common to a vast majority of vertebrates, regulating sexual differentiation, gonadal maturation, and reproductive behavior. The role of these sex hormones in regulating reproductive function is discussed in greater detail in Chapter 42.

Testosterone is a naturally occurring anabolic steroid that stimulates the synthesis in the testes of proteins needed for sperm production and the development of male sexual features and body tissue growth, particularly in muscles. Anabolic steroids promote protein synthesis to build body tissues and anabolic metabolism to store energy within cells. A variety of synthetic anabolic steroids have been developed to treat muscle wasting due to loss of appetite and diseases such as cancer and AIDS. Anabolic steroids are used illegally by both male and female athletes to promote muscle strengthening, undermining fair competition. The risks to health of continued use of anabolic steroids are considerable: liver damage, heart disease, and high blood pressure. Tissue damage occurs because elevated circulating levels of testosterone produced by anabolic steroid use inhibits the normal synthesis of anterior pituitary tropic hormones.

ACTH released by the anterior pituitary gland is also a tropic hormone. It acts on the cortex (the outer portion) of the paired **adrenal glands** (see Fig. 38.10). The adrenal glands are located adjacent to the kidneys ("ad-" meaning "near" and "renal" referring to the kidneys). In humans, each small adrenal gland lies just above each kidney. During times of stress, such as starvation, fear, or intense physical exertion, ACTH stimulates adrenal cortex cells to secrete cortisol.

Other endocrine organs have diverse functions.

Although many endocrine organs receive signals from the hypothalamic–pituitary axis, some respond instead to internal physiological states of the body. One of these, adjacent to the thyroid gland, is the **parathyroid gland** (see Fig. 38.10), which secretes parathyroid hormone (PTH). Calcitonin, which is secreted by the thyroid gland, and PTH together regulate the actions of bone cells (osteoblasts and osteoclasts, Chapter 37) that control bone formation and bone removal and therefore the levels of calcium in the blood.

When circulating levels of calcium fall too low, the parathyroid gland triggers release of PTH, stimulating osteoclasts to reabsorb bone mineral, halting bone formation and raising blood calcium levels. When calcium levels are too high, the production of PTH is inhibited and calcitonin is released to shift bone metabolism toward net bone formation, building bone that stores calcium in the skeleton. Note that, as with blood glucose levels, calcium levels are maintained in a narrow range by negative feedback: The response to the hormone (increasing or decreasing levels of calcium in the blood) is the opposite of the stimulus (low or high levels of calcium in the blood) so that a stable set point is maintained.

Several organs of the digestive system, including the pancreas, stomach, and duodenum, produce hormones that regulate digestion, many of which are discussed in Chapter 40.

The **pineal gland** is located in the thalamic region of the brain (see Fig. 38.10). It responds to autonomic nervous system input by secreting melatonin, a hormone that helps control an animal's state of wakefulness. When melatonin levels rise, animals sleep. In diurnal animals (those that are active during the day), increased levels occur at night; in nocturnal animals (those that are active during the night), increased levels occur during the day. The pineal gland is sensitive to changes in light and darkness. It releases melatonin in response to daily (and seasonal) light cycles, helping to maintain circadian rhythms of the body. Circadian rhythms are cycles of about 24 hours (or longer) in which an animal's biochemical, physiological, and behavioral state shifts in response to changes in daily and seasonal environmental conditions. Exposure to sunlight inhibits the production of melatonin in diurnal animals, whereas darkness stimulates its release. To overcome jet lag, travelers are advised to seek sunlight and often to take melatonin as a medication before sleeping to help reset their 24-hour circadian rhythm. In other vertebrates, such as lampreys, fishes, and reptiles, the pineal gland functions as a "third eye," sensing light through an opening in the skull. Light shining on the pineal gland regulates the animal's general state of activity, influencing its behavioral response to changing light conditions.

In several animals, including ground squirrels, bats, bears, and rattlesnakes, the pineal gland regulates hibernation, controlling the animal's metabolic state over longer time periods. In many animals, release of melatonin also regulates seasonal breeding cycles.

? CASE 7 Predator–Prey: A Game of Life and Death
How does the endocrine system influence predators and prey?

The endocrine system works closely with the nervous system to enable animals to respond to external cues in the environment. In Chapters 35 and 36, we discussed how prey animals, such as warthogs, use visual and olfactory cues processed by the nervous system to detect the distant movement of a predator, such as a lion, which in some instances occurs too late for a successful escape (**Fig. 38.12**). We saw how the sympathetic nervous system,

FIG. 38.12 Predator and prey. Animals rely on rapid integration of sensory information with coordination of bodily functions to escape a predator or to catch their prey.

in the fight-or-flight response, can trigger broad physiological changes, such as increased heart and breathing rates and changes in metabolism that ready the predator to attack or the prey to respond.

These actions are coordinated by the endocrine system. The sympathetic nervous system sends axons to the **adrenal medulla,** the inner part of the adrenal gland. In response to stimulation by the sympathetic nervous system, cells of the adrenal medulla secrete two hormones, epinephrine and norepinephrine (also known as adrenaline and noradrenaline). These are hormones released by the adrenal gland with targets throughout the body, leading to the physiological changes of the fight-or-flight response. Interestingly, epinephrine and norepinephrine also act as neurotransmitters in the brain. In fact, cells of the adrenal medulla are modified nerve cells that have lost their axons and dendrites.

The same set of responses occurs when, walking along a deserted sidewalk, we note a shadowy movement or hear a footstep behind us. The perceived threat sets in motion a coordinated physiological reaction, mediated by the combined action of the nervous and endocrine systems. These changes make us alert and ready for action to avoid or resist a perceived threat. They also inhibit digestive functions and eliminate a sense of appetite.

Earlier, we considered the importance of homeostasis in maintaining a variety of parameters, such as temperature and glucose levels, at a steady level. In this case, there is an adjustment of this set point, which is critical in enabling an organism to respond to external cues and adjust its physiological response appropriately. When the real or perceived threat is gone, the body is able to return to its prior state by reestablishing the earlier set point.

38.4 OTHER FORMS OF CHEMICAL COMMUNICATION

As we have seen, hormones are chemical compounds secreted by glands that enter the circulation and act on distant cells, whose response is determined by the presence of receptors that bind the hormones. However, other chemical compounds act over short distances, and still others are released into the environment to signal other individuals of the same species. Here, we explore additional modes of chemical communication and the compounds that produce them.

Local chemical signals regulate neighboring target cells.

Whereas hormones enter the bloodstream to be transmitted to more distant target cells (**Fig. 38.13a**), other chemical compounds act locally on neighboring cells (**Fig. 38.13b**). In

FIG. 38.13 Modes of chemical signaling. Chemical signaling can occur (a) over long-distances (endocrine), or (b) locally (paracrine and synaptic).

a. Long-distance signaling

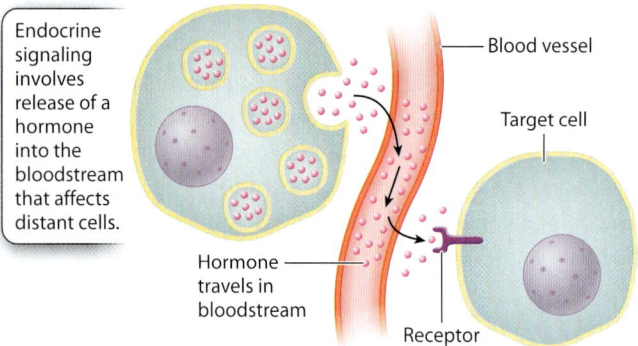

b. Local signaling

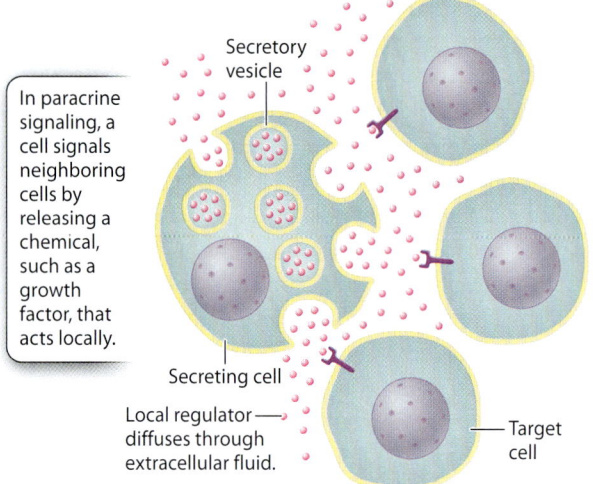

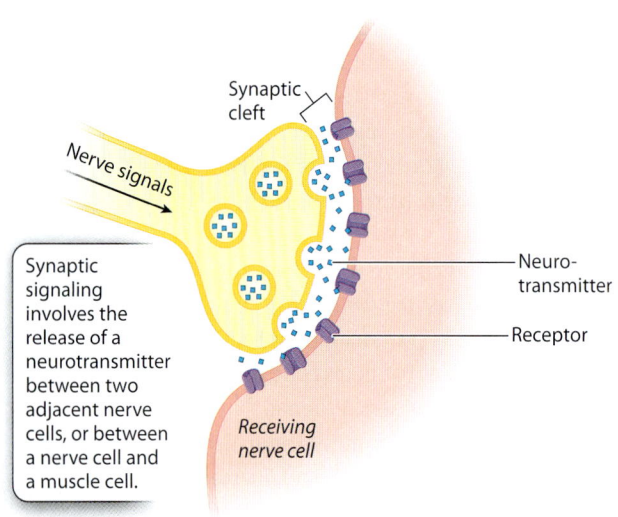

order to take up the chemical signal, these cells must have receptors for the compound to bind to before the signal degrades or diffuses into the bloodstream.

Chemical compounds that act locally are said to have **paracrine** function (Fig. 38.13b). If the compounds act on the secreting cell itself, they are described as having **autocrine** function. That is, the chemical signals may stimulate or inhibit their own secretion. Growth factors (also referred to as cytokines and distinct from the growth hormone released by the anterior pituitary gland) and histamine are examples of paracrine chemical agents released in small amounts that act locally on neighboring cells.

Growth factors (Chapter 9) enhance the differentiation and growth of particular kinds of tissue. For example, bone morphogenetic proteins (BMPs) promote the formation of bone in skeletal growth, and fibroblast growth factors (FGFs) stimulate the formation of connective tissue. Similarly, nerve growth factors (NGFs) promote the survival and growth of nerve cells. When local tissues are damaged, specialized cells release histamine, which dilates blood vessels, allowing blood proteins and white blood cells to move into the region to fight infection and induce repair (Chapter 43). The release of histamine causes the swelling, redness, and warmth that commonly surround a wound. Locally acting signals have immediate, short-term effects because they are quickly degraded.

Another form of local chemical signaling is the synaptic release of neurotransmitters that occurs between directly communicating nerve cells or between nerve cells and muscles at neuromuscular junctions (Fig. 38.13b). Synaptic signaling is extremely rapid and brief compared to the other modes of chemical communication.

Pheromones are chemical compounds released into the environment to signal behavioral cues to other species members.

Animal endocrine systems generally transmit chemical signals that integrate the physiological functions of organ systems within the body. However, many animals release small chemical compounds into the environment to signal and influence the behavior of other members of their species. These water or airborne compounds are called **pheromones.** For example, female bees and wasps release pheromones to attract male mates. Even plants, such as orchids, release pheromones. Some plant pheromones can attract pollinating bees or wasps by mimicking the chemical signals released by the females of these species. There are many different pheromones, which serve very different functions, as shown in **Fig. 38.14**.

Female silk moths use pheromones for long-distance communication to attract males. The sex pheromone bombykol, released from a gland in the female's abdomen, signals her location to males looking for a mate. Sex pheromones play a key role in the mating behavior of many invertebrate and vertebrate species. In most mammals, amphibians, and reptiles, pheromones are detected by a vomeronasal organ with chemosensory neurons in the nasal region of the skull. The release of sex pheromones is one of the most common types of pheromone signaling to attract a mate and trigger ritualistic social and mating behaviors (Chapter 45). Most studies indicate that humans and nonhuman primates lack a vomeronasal organ and do not produce pheromones.

Other pheromones signal territorial boundaries or predatory threats. Dogs and other canids (this family also includes wolves and coyotes), as well as cats, secrete pheromones in their urine that mark their territories. Social seabirds, such as cormorants and boobies, mark their nesting sites by pheromones.

FIG. 38.14 Pheromones. (a) Pheromones are means of attracting mates, as in these ladybugs; (b) wolves secrete pheromones in their urine that mark their territory; (c) honeybees swarm to defend their nest after a bee that stings a threatening animal releases an alarm pheromone.

FIG. 38.15 Trail pheromones in ants. Here, a chemical trail has been laid down by worker ants signaling a food source for the colony.

FIG. 38.16 A brightly colored male cichlid fish. The behavior and markings of this fish are under hormonal control and triggered by social context.

Ants and other social insects commonly deploy alarm pheromones to warn of an attack by an advancing army from another colony. The release of alarm pheromones from a stinging bee causes other bees to swarm and join the attack to defend their nest. In many other species, an animal under attack releases an alarm pheromone, warning its neighbors. Similarly, social mammals such as deer warn of an approaching predator by means of alarm pheromones.

Social insects, such as ants, also deploy trail pheromones that provide chemical cues that other worker ants can follow to a food source (**Fig. 38.15**). An ant returning to the colony from a food source releases trail pheromones from its abdomen onto the soil. Other worker ants use their antennae to detect and track the concentration of the trail pheromone guiding them along the trail. If you have watched ants marching in line across a sidewalk, a trail, or into and out of your house, you have seen worker ants using pheromone cues to signpost the way to food.

Aquatic species also depend on pheromone signaling. Fish, salamanders, and tadpoles release alarm pheromones into the water to warn of a threatening predator. Prey species that share the same habitat may even sense and respond to the release of another species' alarm pheromone.

Fish also release pheromones to coordinate mating and to regulate social interactions. In dense populations of cichlid fish living in Lake Tanganyika in central Africa, the ratio of brightly colored breeding males (**Fig. 38.16**) to females is highly regulated. Only about 10% of males are brightly colored and defend their territory to attract and mate with females. This restriction on the number of dominant breeding males is regulated by behavioral and chemical pheromone cues. When a dominant male dies, subordinate males fight for the vacated territory. When a new male assumes dominance, it becomes brightly colored to attract females and defend its newly won territory. The dominant breeding male reinforces its dominance by releasing pheromones in its urine that signal females and subordinate males in its breeding area. The color change is mediated by hormones released from the male fish's pituitary gland, which act on pigment cells called melanocytes that are located in the epidermis of its scales. Thus, the fish makes use of both endocrine and pheromone signals to change its behavior and the behavior of other members of its species.

→ **Quick Check 4** What distinguishes paracrine and pheromone signals from endocrine signals?

Core Concepts Summary

38.1 ANIMAL ENDOCRINE SYSTEMS RELEASE CHEMICAL SIGNALS CALLED HORMONES INTO THE BLOODSTREAM. THESE SIGNALS RESPOND TO THE ENVIRONMENT, REGULATE GROWTH AND DEVELOPMENT, AND MAINTAIN HOMEOSTASIS.

The nervous system responds rapidly to sensory information, providing immediate physiological responses, whereas the endocrine system acts more slowly and over prolonged periods to affect broad changes in the physiological and behavioral state of an animal. page 38-1

The endocrine system helps an animal respond to its environment. page 38-1

The endocrine system is involved in growth and development. page 38-2

The nervous and endocrine systems regulate physiological functions to maintain homeostasis. page 38-5

Homeostasis is often achieved by negative feedback, in which the response opposes the stimulus. page 38-5

A stimulus can be amplified by positive feedback, in which the response amplifies the initial stimulus. page 38-6

38.2 HORMONES ACHIEVE SPECIFICITY BY BINDING TO RECEPTORS ON OR INSIDE THEIR TARGET CELLS. THEIR SIGNALS ARE AMPLIFIED TO EXERT STRONG EFFECTS ON THEIR TARGET CELLS.

Hormones are chemical compounds that are secreted and transported by the circulatory system, often to distant targets. page 38-7

Hormones bind to specific receptors located on the surface or inside the target cells that receive their signals. page 38-7

The three classes of hormone are peptide, amine, and steroid hormones. page 38-7

Peptide and amine hormones are hydrophilic and bind to cell membrane receptors, stimulating second messenger pathways to alter the biochemical function of target cells. page 38-8

Steroids are hydrophobic and bind to intracellular or nuclear receptors, which function as transcription factors to regulate gene expression and protein synthesis. page 38-8

Hormones communicate by signaling cascades to amplify the strength of their downstream effect on target cells. page 38-8

The hypothalamic–pituitary axis forms the central endocrine pathway by which vertebrates regulate and integrate diverse bodily functions. page 38-10

Hormones are evolutionarily conserved molecules common to diverse groups of animals, but which have evolved novel functions. page 38-11

38.3 IN VERTEBRATE ANIMALS, THE HYPOTHALAMUS AND THE PITUITARY GLAND CONTROL AND INTEGRATE DIVERSE BODILY FUNCTIONS AND BEHAVIORS.

The anterior pituitary gland secretes hormones in response to releasing factors produced and secreted by the hypothalamus. page 38-13

Hormones released by the anterior pituitary gland include thyroid-stimulating hormone (TSH), the gonadotropic hormones follicle-stimulating hormone (FSH) and luteinizing hormone (LH), adrenocorticotropic hormone (ACTH), growth hormone (GH), and prolactin. page 38-14

Tropic hormones are hormones that act on endocrine glands to produce additional hormones. TSH causes the thyroid gland to release thyroid hormones; the gonadotropic hormones FSH and LH act on the female ovaries to produce estrogen and progesterone and the male testes to produce testosterone; and ACTH acts on the adrenal gland to release cortisol. page 38-14

The posterior pituitary gland contains the axons of neurosecretory cells that release the peptide hormones oxytocin and antidiuretic hormone into the bloodstream. page 38-14

Some endocrine glands do not respond to signals from the pituitary gland. These include the parathyroid gland, which releases parathyroid hormone that helps to control calcium levels; the pancreas, which secretes insulin and glucagon to regulate blood glucose levels; and the pineal gland, which secretes melatonin in response to light and dark cycles. page 38-15

38.4 CHEMICAL COMMUNICATION CAN ALSO OCCUR LOCALLY BETWEEN NEIGHBORING CELLS OR, THROUGH THE RELEASE OF PHEROMONES, BETWEEN INDIVIDUALS, COORDINATING SOCIAL INTERACTIONS.

Paracrine signals act locally over short times to affect neighboring cells, compared to the more distant and longer-lasting signals transmitted by blood-borne hormones. page 38-17

Pheromones are chemical signals released into the air or water, signaling behavioral cues between individuals of a species. page 38-17

Sex pheromones attract mates and coordinate ritualized mating behavior; alarm pheromones signal predatory threats; trail pheromones signal the location of food sources; and territorial pheromones signal a male's dominance and breeding area. page 38-17

Self-Assessment

1. Describe three general functions of the endocrine system and provide an example of each.

2. Diagram an endocrine pathway, such as parathyroid regulation of blood calcium levels, that maintains homeostasis of a bodily function by means of negative feedback.

3. Diagram an endocrine pathway, such as oxytocin regulation of female labor during birth, that relies on positive feedback.

4. Identify and provide an example of each of the three general classes of hormone.

5. Explain how hormones act specifically on only some cells or organs despite being transported by the bloodstream throughout the body.

6. Describe how hormones exert substantial and broad effects on their target cells despite being released in small amounts by the initial signaling gland.

7. Explain how peptide hormones affect the function of a target cell.

8. Explain how steroid hormones affect the function of a target cell.

9. Explain the difference between the way in which the hypothalamus controls the anterior pituitary gland and the way it controls the posterior pituitary gland.

10. Explain the difference in action between a hormone, a paracrine signal, a neurotransmitter, and a pheromone.

Do you understand the chapter's Core Concepts? Log in to check your answers to the Self-Assessment questions, then practice what you've learned and reinforce this chapter's concepts by working through the problems and multimedia tutorials provided there.

www.biologyhowlifeworks.com

CHAPTER 39

ANIMAL CARDIOVASCULAR AND RESPIRATORY SYSTEMS

Core Concepts

39.1 Respiration and circulation depend on diffusion over short distances and bulk flow over long distances.

39.2 Respiration provides oxygen and eliminates carbon dioxide in support of cellular metabolism.

39.3 Red blood cells produce hemoglobin, greatly increasing the amount of oxygen transported by the blood.

39.4 Circulatory systems have different-sized vessels that facilitate bulk flow and diffusion.

39.5 The evolution of animal hearts reflects selection for a high metabolic rate, achieved by increasing the delivery of oxygen to metabolically active cells.

As animals evolved to be larger and more complex, they faced the challenge of obtaining an adequate supply of oxygen (O_2) and nutrients and, at the same time, eliminating carbon dioxide (CO_2) and other waste products. In this chapter, we explore how animal respiratory and circulatory systems evolved to meet the metabolic needs of an animal's cells. In particular, we focus on the uptake of O_2 and the release of CO_2, key gases in cellular respiration (Chapter 7).

Gases, nutrients, and wastes are continually being exchanged between an organism and its environment. Because of their small size or thin bodies, single-celled organisms and simple multicellular animals exchange compounds with the environment by diffusion. Because diffusion works effectively only over very short distances, larger and more complex animals cannot rely on diffusion as the sole means of gas, nutrient, and waste exchange (Chapter 28). These animals evolved respiratory structures at the body's surface for taking up O_2 and releasing CO_2, and internal circulatory systems for long-distance transport to different body regions.

Because similar physical laws govern gas and fluid movement, animal respiratory and circulatory systems share many features. However, fish obtain O_2 from water, whereas land animals breathe air. Water and air have very different physical properties. Thus, aquatic and terrestrial environments pose distinct challenges to the animal in obtaining O_2. Because life first began in the water, the evolution of life on land required substantial changes in how animals exchange gases, take up nutrients, and eliminate wastes.

39.1 DELIVERY OF OXYGEN AND ELIMINATION OF CARBON DIOXIDE

Eukaryotic cells use O_2 to burn organic fuels—carbohydrate, lipid, and protein—for ATP production (Chapter 7). The increase in atmospheric O_2 with the evolution of land plants was key to the evolution of O_2-based cellular respiration, and that in turn contributed to the diversification of multicellular animal life during the Cambrian Period about 500 million years ago. In the process of burning fuel by aerobic respiration, CO_2 is produced. Animals have thus evolved mechanisms to acquire O_2 from their environment and, at the same time, to eliminate excess CO_2 from their body to the environment. These mechanisms make up the process known as animal respiration—or "breathing." The transport of O_2 and CO_2 between an animal and its environment is referred to as **gas exchange**, and is fundamental to all eukaryotic animals, as well as to photosynthetic plants.

Diffusion governs gas exchange over short distances.

Single-celled organisms and simple animals and plants exchange gases by diffusion. **Diffusion** is the random movement of individual molecules (Chapter 5). If there is a difference in concentration of a substance between two regions, diffusion results in the net movement of molecules from regions of higher concentration to regions of lower concentration until the substance is evenly distributed.

Given a large enough surface, diffusion is an extremely effective way to exchange gases and other substances over short distances between two compartments, for example between the lung and the blood vessels supplying the lung. The rate of diffusion is directly proportional to the surface area over which exchange occurs and to the concentration difference, and inversely proportional to the distance over which the molecules move (or the thickness of the barrier). Diffusion is generally ineffective over distances exceeding about 100 μm, or 0.1 mm. Consequently, respiratory and circulatory structures have large surface areas and thin membranes.

Because effective transport by diffusion is limited to short distances, some of the earliest and simplest invertebrate animals are composed of thin sheets of cells with a large surface area. For example, sponges and sea anemones have a thin body with a central cavity through which water circulates, and flatworms have a flattened body surface (**Fig. 39.1**).

Net movement of a substance results from diffusion when there is a concentration difference between two regions. When considering the diffusion of a gas, however, we usually refer to its partial pressure rather than its concentration. The **partial pressure (p)** of a gas is defined as its fractional concentration relative to other gases present, multiplied by the overall (that is, the atmospheric) pressure (**Fig. 39.2**). Pressure is measured in units of millimeters of mercury (mmHg). Oxygen makes up approximately 21% of the gases in air (nitrogen contributes about 78%, carbon dioxide about 0.03%, and the remainder is trace gases). At sea level, atmospheric pressure is approximately 760 mmHg, which means that O_2 in the atmosphere exerts a partial pressure (pO_2) of about 160 mmHg (0.21×760). For O_2 to diffuse from the air into cells, the partial pressure of O_2 inside the cells must be lower than the partial pressure of O_2 in the atmosphere.

→ **Quick Check 1** What two structural features favor diffusion?

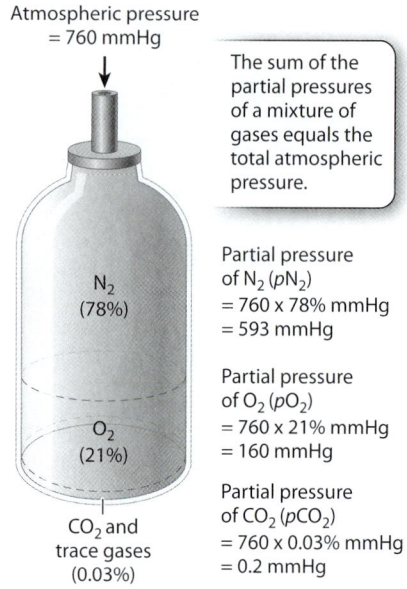

FIG. 39.2 Partial pressure. The partial pressure of a gas is its fractional concentration multiplied by the total atmospheric pressure.

Atmospheric pressure = 760 mmHg

The sum of the partial pressures of a mixture of gases equals the total atmospheric pressure.

N_2 (78%)

Partial pressure of N_2 (pN_2) = 760 x 78% mmHg = 593 mmHg

O_2 (21%)

Partial pressure of O_2 (pO_2) = 760 x 21% mmHg = 160 mmHg

CO_2 and trace gases (0.03%)

Partial pressure of CO_2 (pCO_2) = 760 x 0.03% mmHg = 0.2 mmHg

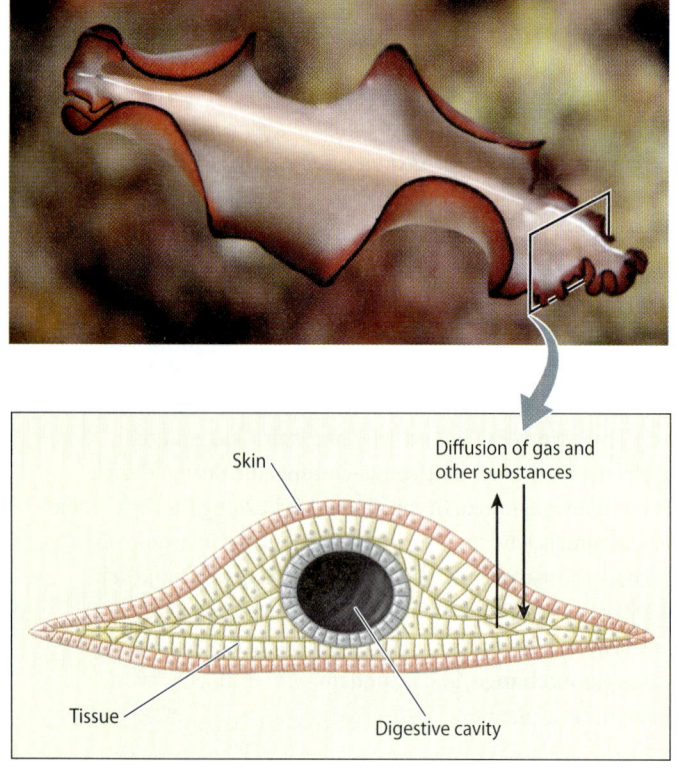

FIG. 39.1 The thin body of a flatworm. Gases can be exchanged by diffusion in flatworms because their bodies are only a few cell layers thick.

Bulk flow moves fluid over long distances.

Larger, more complex animals transport O_2 and CO_2 longer distances to cells within their body. These animals rely on **bulk flow** in addition to diffusion. Bulk flow is the physical movement

FIG. 39.3 The four steps of gas transport and exchange.

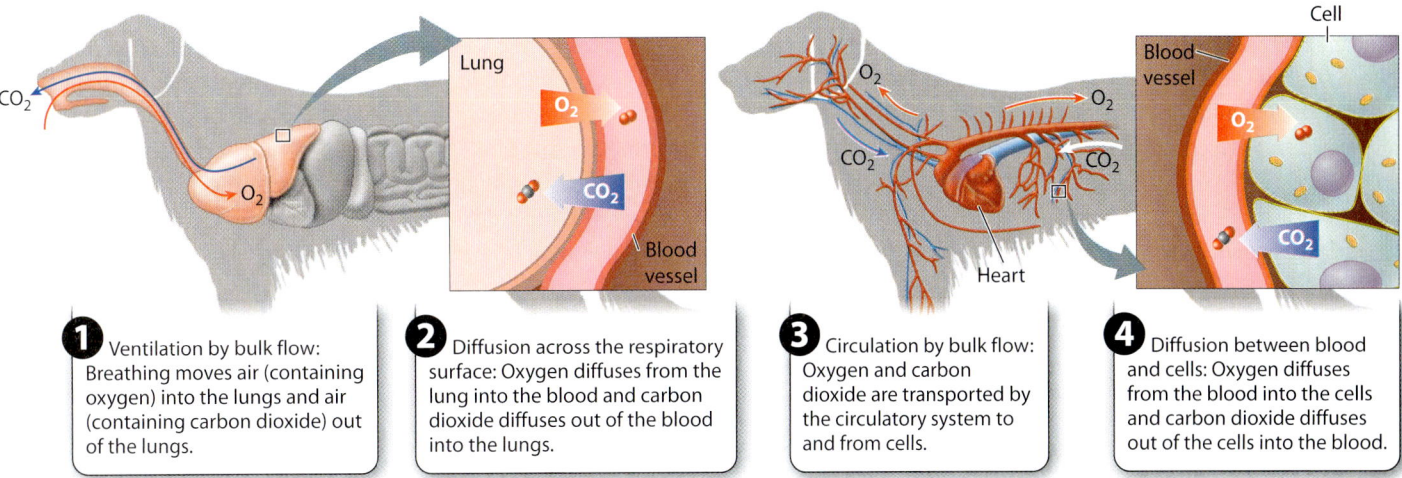

① Ventilation by bulk flow: Breathing moves air (containing oxygen) into the lungs and air (containing carbon dioxide) out of the lungs.

② Diffusion across the respiratory surface: Oxygen diffuses from the lung into the blood and carbon dioxide diffuses out of the blood into the lungs.

③ Circulation by bulk flow: Oxygen and carbon dioxide are transported by the circulatory system to and from cells.

④ Diffusion between blood and cells: Oxygen diffuses from the blood into the cells and carbon dioxide diffuses out of the cells into the blood.

of fluid, and the gases and compounds carried by the fluid, over a given distance. Specialized pumps, like the heart, generate the pressure that is required to move the fluid.

Bulk flow occurs in two steps to meet the gas exchange needs of cells in larger animals. The first is **ventilation.** Ventilation is the movement of the animal's respiratory medium—water or air—past a specialized respiratory surface. The second is **circulation.** Circulation is the movement of a specialized body fluid that carries O_2 and CO_2. The circulatory fluid is called **hemolymph** in invertebrates and **blood** in vertebrates. This fluid delivers O_2 to cells within different regions of the body and carries CO_2 back to the respiratory exchange surface.

Ventilation and circulation each require a pump to produce a pressure (P) that drives flow (Q) against the resistance (R) to flow. The rate of flow is governed by the following simple equation:

$$Q = P/R$$

Resistance to flow measures the difficulty of pumping the fluid through a network of chambers and vessels located within the respiratory and circulatory systems. If the resistance to flow doubles, the flow rate is halved. The longer the network of vessels and the narrower the vessels themselves, the greater the vessels' resistance to the fluid (whether air, water, or blood) moving through them. Consider how much harder it is to blow fluid out of a long thin tube than a short wide one.

In summary, we can view gas exchange in complex multicellular organisms as four steps that are linked in series (**Fig. 39.3**). (1) To deliver O_2 to the mitochondria within their cells, animals move fresh air or water past their respiratory exchange surface in the process of ventilation (bulk flow). Ventilation maximizes the concentration of O_2 in the air or water on the outside of the respiratory surface. (2) The buildup of O_2 favors the diffusion of O_2 into the animal across its respiratory surface. (3) Following diffusion into the blood, O_2 is transported by the circulation (bulk flow) to the tissues. Internal circulation again serves to maximize the concentration of O_2 outside cells. (4) Oxygen then diffuses from the blood across the cell membrane and into the mitochondria, where it burns fuels for ATP production.

The same four steps occur in reverse to remove CO_2 from the body: CO_2 moves from cells to the blood by diffusion, then is carried in the circulation to the respiratory surface by bulk flow, then moves across the respiratory surface by diffusion, and finally moves out of the animal by bulk flow. Because diffusion and bulk flow are coupled, it is important that each step of transport (whether for a gas, nutrient, or waste) has a similar capacity—that is, the same amount of O_2 enters the blood through diffusion as is transported by the circulation. This matching of coupled transport processes reflects an evolutionary principle of minimizing excess capacity, which is costly to the animal.

→ **Quick Check 2** Those who suffer an asthma attack have difficulty breathing. From the discussion of bulk flow, what do you think happens to these individuals' airways that makes it difficult for them to breathe?

39.2 RESPIRATORY GAS EXCHANGE

All animals obtain O_2 from the surrounding air or water. Aquatic animals like crabs, aquatic salamanders, and fish that take in O_2 from water breathe through **gills,** highly folded delicate structures that facilitate gas exchange with the surrounding water

(**Fig. 39.4a**). Some animals, such as frogs and salamanders, also breathe through their skin. In contrast, many terrestrial animals, such as reptiles, birds, and mammals, have internal **lungs** for gas exchange (**Fig. 39.4b**). Instead of lungs, terrestrial insects evolved a system of air tubes called **tracheae** that branch from openings along their abdominal surface into smaller airways. These smaller airways, termed tracheoles, supply air directly to the cells within their body.

For both aquatic and terrestrial animals, ventilation supplies O_2 to gas exchange surfaces, and then O_2 crosses the surface through diffusion. Recall that diffusion is much slower than the bulk flow of ventilation. How can the rate of diffusion through the gas exchange surface in lungs or gills keep up with the rate of O_2 supplied by ventilation? The gas exchange surface must have a large surface area and be extremely thin. The respiratory surfaces of more complex and active animals are highly folded, creating a large surface area within a small space. These surfaces are also only one or two cell layers thick, providing a diffusion distance of as little as 1 to 2 μm. In fact, this short distance is all that separates the air you breathe into your lungs from the bloodstream into which O_2 diffuses.

Many aquatic animals breathe through gills.

Most large aquatic animals (with the notable exception of marine mammals like dolphins and whales) use gills to breathe O_2-containing water. Some sessile invertebrates, including tube worms and sea urchins, as well as nudibranch mollusks and aquatic amphibians, have external gills (Fig. 39.4a). However, because external gills are easily damaged or may be eaten by predators, most mollusks, crustaceans, and fishes have evolved internal gills within protective cavities. Whereas the external gills of invertebrates often rely on the natural motions of the water to move water past the gills, invertebrates with internal gills often have cilia that direct water over the gill's surface.

Fish need sufficient O_2 to meet the energy demands of swimming. As a result, they actively pump water through their mouth and over the gills, which are located in a chamber behind the mouth cavity (**Fig. 39.5**). The effort required to ventilate the gills could be high because of the density and viscosity of water, but fish greatly reduce the energy cost of moving water by maintaining a continuous, unidirectional flow of water past their gills. In bony fishes, a protective flap overlying the gills, called the operculum, expands laterally to draw water over the gills while the mouth is refilling before its next pumping cycle.

The gills of fishes typically consist of a series of gill arches located on either side of the animal behind the mouth cavity and, in bony fishes, beneath the operculum. Each gill arch consists of two stacked rows of flat leaf-shaped structures called gill filaments. Numerous **lamellae**, thin sheetlike structures, are evenly but tightly spaced along the length of each gill filament and extend upward from the filament's surface. A series of blood vessels brings O_2-poor blood from the heart to the lamellar surfaces. The lamellae are composed of flattened epithelial cells and are extremely thin, so the water passing outside and blood passing inside the lamellae are separated by only about 1 to 2 μm. The lamellae give the gills an enormous surface area for their size.

The lamellae are oriented so that the blood flowing through them in a capillary network moves in a direction opposite to the flow of water past the gills (Fig. 39.5). This type of organization, in which fluids with different properties move in opposite directions, is an efficient way to exchange properties between the two fluids. The property can be a chemical property, such as O_2 concentration, or a physical property, such as heat.

FIG. 39.4 Animal gas exchange organs. Gas exchange organs are shaped and folded to provide large surfaces for gas exchange.

a. Aquatic animals

Tube worm: External gills

Salamander: External gills

Fish: Internal gills

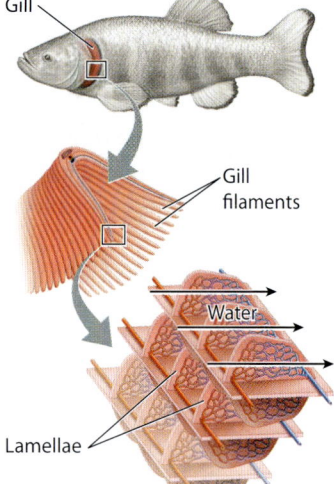

b. Terrestrial animals

Grasshopper: Tracheae

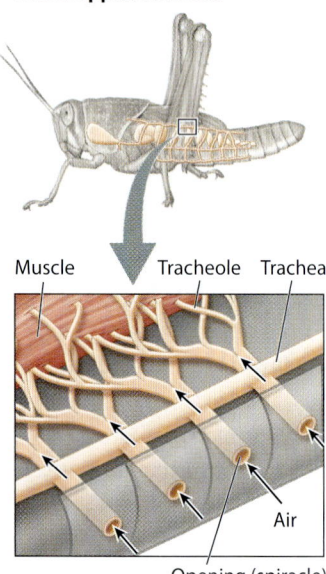

Dog: Internal lungs

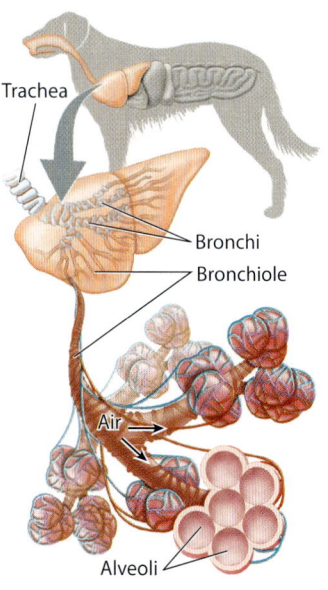

FIG. 39.5 Respiration by fish gills. Fish gills extract O_2 from water efficiently through a unidirectional flow of water and countercurrent flow.

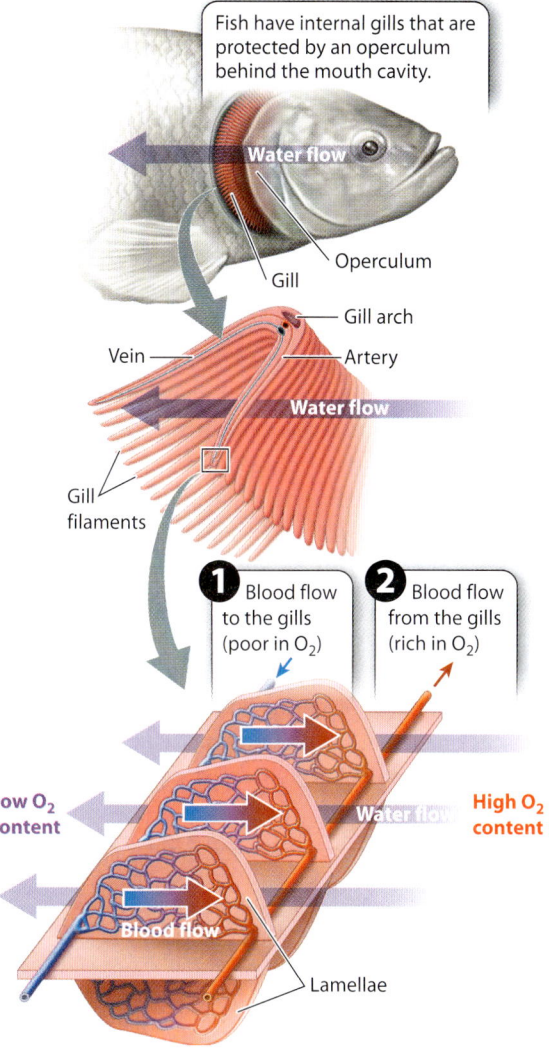

Consider for a moment two tubes right next to each other, one carrying hot water and the other carrying cold water (**Fig. 39.6**). If we want to make the hot water cold and the cold water hot, the efficiency of heat transfer between the two tubes depends on the direction of flow. When the water in the tubes flows in the same direction (concurrent flow), the hot water gets colder and the cold water gets hotter since heat is transferred between the two tubes (Fig. 39.6a). Given enough distance, the water will be at an intermediate (warm) temperature at the end of the tubes, reaching the same average temperature in both tubes.

Now consider what happens when the water in the two tubes flows in opposite directions (countercurrent flow). In this case, the cold water encounters increasingly hot temperatures and, given enough distance, becomes nearly as hot as the hot water entering the other tube (Fig 39.6b). Similarly, the hot water encounters increasingly cold temperatures, and becomes nearly as cold as the cold water entering the other tube. With countercurrent flow, the two essentially exchange properties (heat, in this example).

This mechanism, known as **countercurrent exchange**, is used widely in nature. As a result of countercurrent exchange, fish gills can extract nearly all the O_2 in the water that passes over them (Fig. 39.6c). At the same time, CO_2 readily diffuses out of the blood vessels and into the water that leaves the gill chamber. A set of blood vessels carries the O_2-rich blood away from the gills to supply the fish's body.

Insects breathe air through tracheae.

With the evolution of life on land, animals that breathe air were able to increase their uptake of O_2 and achieve higher rates of metabolism. Oxygen uptake increased for the following reasons:

1. The O_2 content of air is typically 50 times greater than that of a similar volume of water. Fresh air contains approximately

FIG. 39.6 Concurrent and countercurrent exchange. Countercurrent exchange is used widely in nature, such as in fish gills, in cases where two fluids exchange properties (such as O_2, CO_2, or heat) efficiently.

a. Concurrent flow

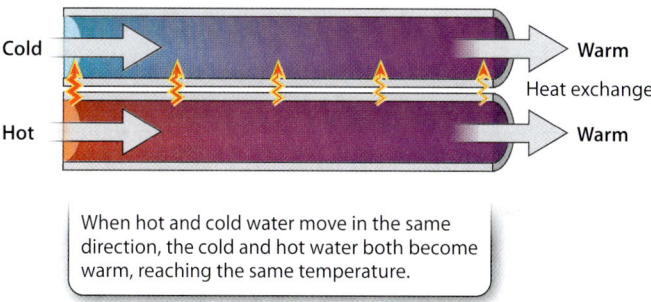

When hot and cold water move in the same direction, the cold and hot water both become warm, reaching the same temperature.

b. Countercurrent flow

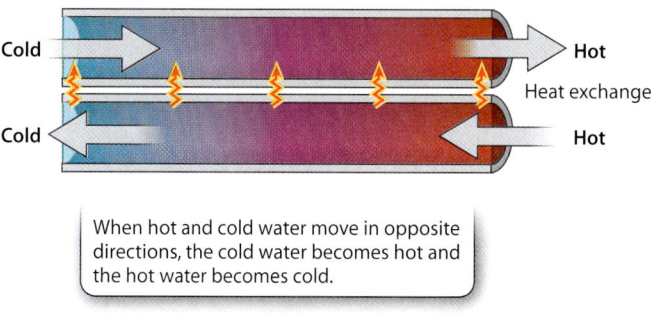

When hot and cold water move in opposite directions, the cold water becomes hot and the hot water becomes cold.

c. Countercurrent flow in fish gills

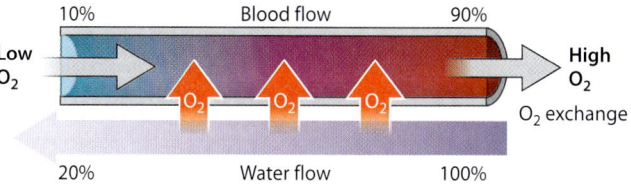

210 ml O_2 per liter of air (21%), whereas the concentration of O_2 in well-mixed fresh or salt water rarely exceeds 4 ml O_2 per liter of water (0.4%).

2. Oxygen diffuses about 8000 times faster in air than in water. For this reason, the O_2 content can be close to zero a few centimeters beneath the surface of a stagnant pond.

3. Water is 800 times denser and 50 times more viscous than air. Thus, it requires more energy to pump water than to pump air past a gas exchange surface.

Because of their small size, insects can employ a direct pathway of air transport that gets air right to their tissues (see Fig. 39.4b). There is no respiratory surface. In contrast to terrestrial vertebrates, insects rely on a two-step process of ventilation and diffusion to supply their cells with O_2 and eliminate CO_2. First, air enters an insect through openings, called **spiracles,** along either side of its abdomen. The spiracles can be opened or closed to limit water loss and regulate O_2 delivery, much like the leaf stomata of plants (Chapter 29). Inside the insect body, air is ventilated through a branching series of air tubes—the tracheae and tracheoles—directly to the cells. Second, diffusion occurs at the cell: O_2 supplied by the fine airways diffuses into the cells, and CO_2 diffuses out and is eliminated through the insect's tracheae.

The mitochondria of the flight muscles and other metabolically active tissues of insects are located within a few micrometers of tracheole airways. Because O_2 and CO_2 diffuse rapidly in air, the tracheal system of insects delivers O_2 to cellular mitochondria at high rates. Insects also have an air sac system connected to their tracheae. The air sacs act like bellows to pump air through the tracheae, speeding the movement of air to tissues so that gas exchange is faster.

Terrestrial vertebrates breathe by tidal ventilation of internal lungs.

Unlike fish gills, the lungs of most land vertebrates inflate and deflate to move fresh air with O_2 into the lungs and expire stale air with CO_2 out of the lungs. The low density and viscosity of air enables these animals to breathe by **tidal ventilation** without expending too much energy. In tidal respiration, air is drawn into the lungs during **inhalation** and then moved out during **exhalation** (**Fig. 39.7a**).

Mammals and reptiles expand their thoracic cavity to draw air inside their lungs on inhalation. The expansion of the lungs causes the air pressure inside lungs to become lower than the air pressure outside the lungs. The resulting negative pressure draws air into the lungs. By contrast, amphibians inflate their lungs by pressure produced by their mouth cavity. In most animals, exhalation is passively driven by elastic recoil of tissues that were previously stretched during inhalation. The contraction of the lungs causes the air pressure inside the lungs to become higher than the air pressure outside the lungs. The resulting positive pressure forces air out of the lungs.

In mammals, inhalation during normal, relaxed breathing is driven by contraction of the **diaphragm,** a domed sheet of muscle located at the base of the lungs that separates the thoracic and

FIG. 39.7 **Tidal ventilation of the lungs.** (a) The movement of the diaphragm causes lung volume to increase and decrease during breathing, drawing in air, then expelling it. (b) The change in lung volume is typically about 0.5 liter.

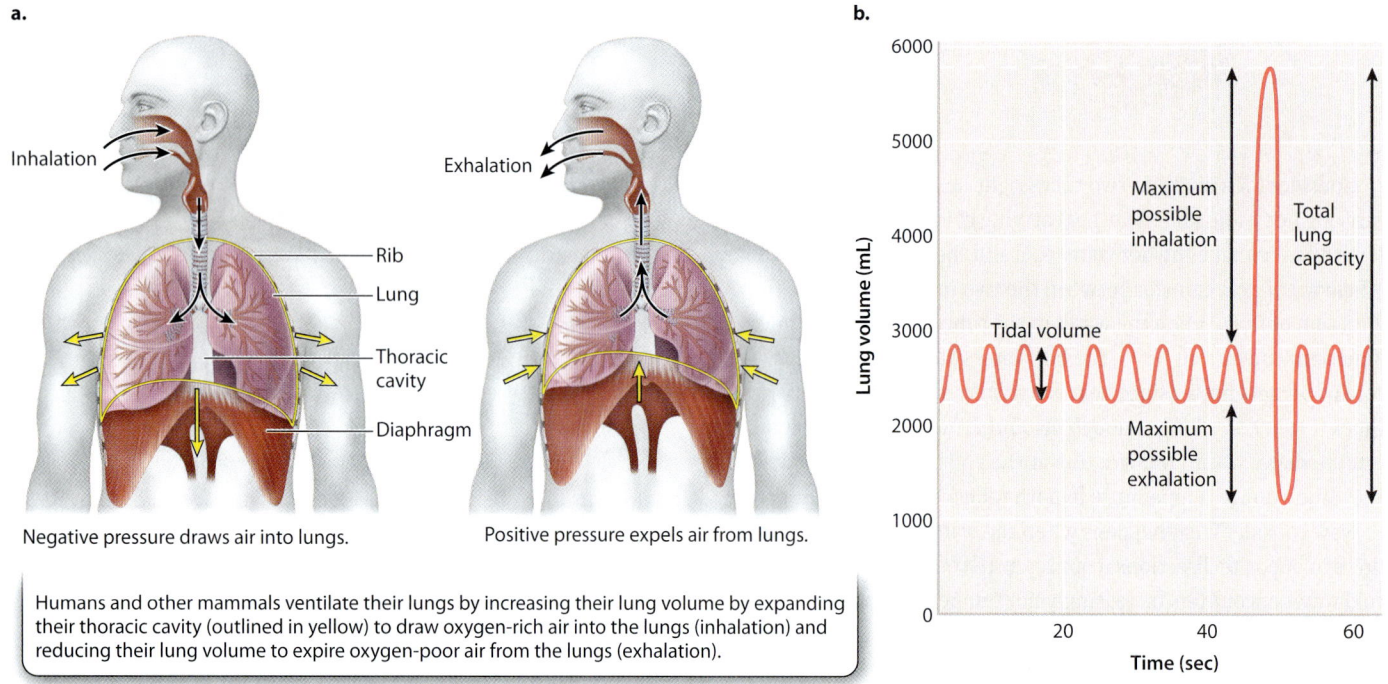

Humans and other mammals ventilate their lungs by increasing their lung volume by expanding their thoracic cavity (outlined in yellow) to draw oxygen-rich air into the lungs (inhalation) and reducing their lung volume to expire oxygen-poor air from the lungs (exhalation).

FIG. 39.8 **Human lung anatomy.** (a) Millions of alveoli provide a vast surface for gas exchange. (b) A scanning electron micrograph of the lung alveolar surface. (c) A transmission electron micrograph showing a cross section of the alveolar wall and lung capillary.

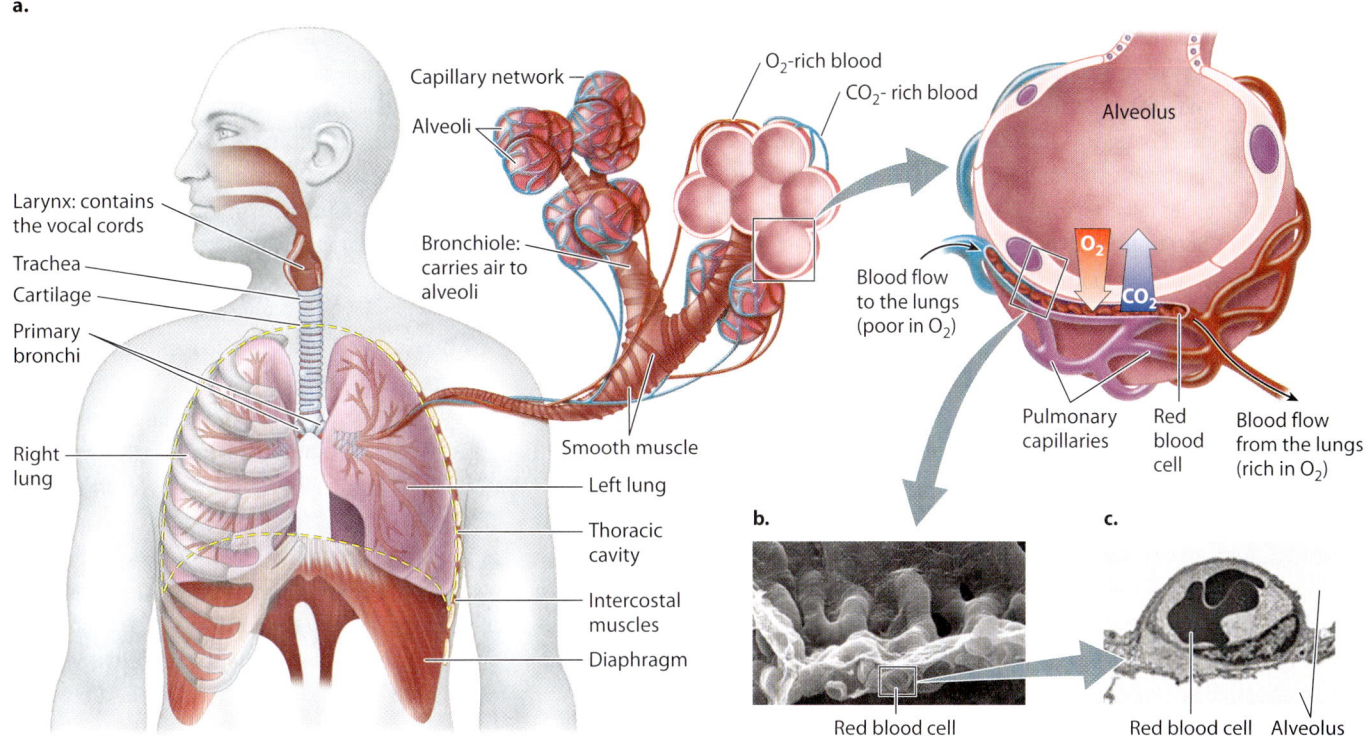

abdominal cavities (Fig. 39.7a). Exhalation occurs passively by elastic recoil of the lungs and chest wall. During exercise, other muscles come into play to assist with inhalation and exhalation. For example, **intercostal muscles**, which are attached to adjacent pairs of ribs, assist the diaphragm by elevating the ribs on inhalation and depressing them during exhalation. The action of the intercostal muscles helps to produce larger changes in the volume of the thoracic cavity, increasing the negative pressure that draws air into the lungs during inhalation and assisting elastic recoil of the lungs and chest wall to pump air out of the lungs during exhalation.

At rest, we inhale and exhale about 0.5 liter of air every cycle (**Fig. 39.7b**). This amount represents the **tidal volume** of the lungs. With a breathing frequency of 12 breaths per minute, the ventilation rate (breathing frequency × tidal volume) is 6 liters per minute. When more O_2 is needed during exercise, both breathing frequency and tidal volume increase to elevate ventilation rate.

The fresh air inhaled during tidal breathing mixes with O_2-depleted stale air that remains in the airways after exhalation. As a result, the pO_2 in the lung (approximately 100 mmHg in humans) is lower than the pO_2 of freshly inhaled air (approximately 160 mmHg at sea level). Consequently, the fraction of O_2 that can be extracted is lower than the fraction that can be extracted by the countercurrent flow of fish gills, which can reach 90% or more.

Typically, mammalian lungs extract less than 25% of the O_2 in the air, and reptile and amphibian lungs extract even less. Nevertheless, the disadvantages of tidal respiration are offset by the ease of ventilating the lung at high rates and the high O_2 content of air.

→ **Quick Check 3** What properties of air make it possible for humans and other mammals to breathe by tidal ventilation?

Mammalian lungs are well adapted for gas exchange.

Despite the limitations of tidal respiration, mammalian lungs are well adapted for breathing air. They have an enormous surface area and a short diffusion distance for gas exchange. Consequently, the lungs of mammals supply O_2 quickly enough to support high metabolic rates.

Except in the case of birds (discussed below), the lungs of air-breathing vertebrates are blind-ended sacs located within the thoracic cavity. Air is taken in through the mouth and nasal passages, and then passes through the **larynx**, within which the **vocal cords** are located. (The vocal cords enable speech, song, and sound production.) Air then enters the **trachea**, the central airway leading to the lungs (**Fig. 39.8**). The trachea divides into two airways, called **primary bronchi**, one of which supplies each lung. The trachea and bronchi are supported by rings of cartilage that prevent them from collapsing during respiration, ensuring that airflow meets with minimal resistance. You can feel some of these cartilage rings at the front of your throat below your larynx (the part of the larynx you can feel is your Adam's apple). The primary bronchi divide into smaller secondary bronchi and again into finer **bronchioles**. This branching continues until the terminal bronchioles have a diameter of less than 1 mm. The very fine bronchioles end with clusters of tiny thin-walled sacs, the **alveoli**, where gas exchange by diffusion takes place. As a result of this branching pattern, each lung consists of several million alveoli, providing a large surface area. Together, human lungs consist of 300 to 500 million alveoli with a combined surface area as large as a tennis court—about 100 m^2!

Small blood vessels, called **pulmonary capillaries**, supply the alveolar wall. Each alveolus is lined with thin epithelial cells in intimate contact with the endothelium of these small blood vessels (Fig. 39.8). As a result, the diffusion distance from the alveolus into the capillary is extremely short (about 2 μm), similar to that of gill lamellae and the terminal airways of bird lungs.

Keeping the surfaces of the alveoli moist is critical because moisture helps move O_2 molecules from the air into solution and thus diffuse across the alveolar wall. Mucus-secreting cells lining the alveoli keep the inside surface of the alveoli coated with a fluid film. Other alveolar epithelial cells produce a compound called **surfactant**, which acts like soap to reduce the surface tension of the fluid film. Surface tension is a cohesive force that holds the molecules of a liquid together, causing the surface to act like an elastic membrane. Because of surface tension, it requires greater pressure to inflate a small balloon than a large one. The surfactant allows the lungs to be inflated easily at low volumes when the alveoli are partially collapsed and small. It also ensures that alveoli of different sizes inflate with similar ease, enabling more nearly uniform ventilation of the lung.

Mucus-secreting cells line all the airways of the lung not only to keep them moist but also to trap and remove foreign particles and microorganisms that an animal may breathe in with air. Beating cilia on the surface of these cells move the mucus and foreign debris out of the lungs and into the throat, where it is swallowed and digested. Smoking even one cigarette can stop the cilia's beating for several hours or even kill them. This is one reason why smokers cough—they must clear the mucus that builds up.

The structure of bird lungs allow unidirectional airflow for increased oxygen uptake.

Birds extract O_2 efficiently to supply the considerable energy needed for flight. An efficient O_2 supply is particularly important for migratory birds that fly at high altitude where the air's partial pressure of O_2 is relatively low. The lungs of birds are unique in that they are rigid and do not inflate and deflate. As a consequence, birds do not use tidal breathing. Instead, they benefit from unidirectional airflow that enhances gas exchange by maintaining larger concentration gradients for diffusion.

A bird's respiratory system consists of two sets of air sacs: one located posterior to the lungs (behind them) and one located anterior to the lungs (in front of them). Air is pumped through these air sacs in a bellows-like action (**Fig. 39.9**). As the set of posterior air sacs expand, air is drawn from the bird's mouth into those sacs. As the sacs are then compressed by surrounding muscles acting on the skeleton, air is pumped through the bird's rigid lungs.

FIG. 39.9 Respiration by bird lungs. Anterior and posterior air sacs move air unidirectionally through a rigid lung, increasing the uptake of O_2 at low pO_2.

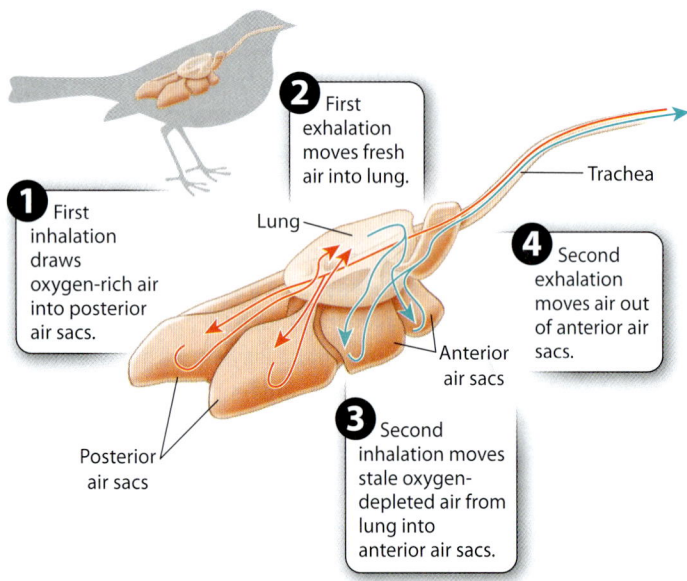
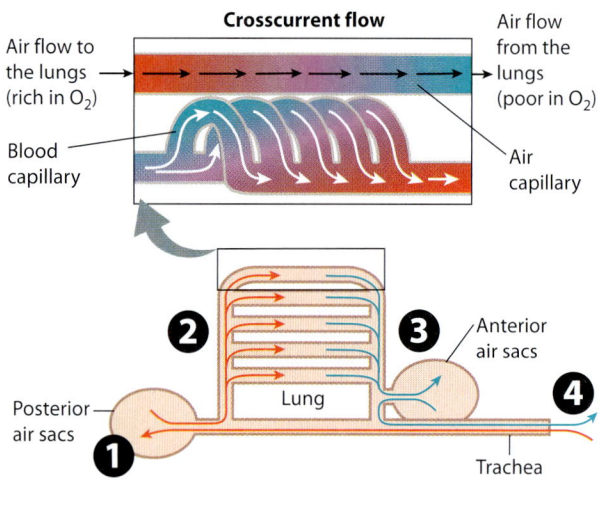

The lungs consist of a series of larger air channels that branch into fine air capillaries having a large surface area. The air capillaries direct the air so that it moves **crosscurrent** (at 90°) to the path of blood flow through the walls of the air capillaries. This crosscurrent flow is not as efficient as the countercurrent flow of fish gills, but it enables birds to extract more O_2 than mammals can.

After losing O_2 and gaining CO_2, the air leaves the lungs and enters the set of anterior air sacs. After fresh air is inhaled into the posterior air sacs, the stale air in the anterior air sacs is exhaled out of the bird's trachea and mouth. As a result, birds achieve a continuous supply of fresh air into the lung during both inhalation *and* exhalation. In contrast to other air-breathing vertebrates, therefore, birds move air through their respiratory system in two cycles of ventilation.

Voluntary and involuntary mechanisms control breathing.

Because an animal's need for O_2 varies with activity level, animals adjust their respiratory rate to meet their cells' changing demand for O_2. Respiration is a unique physiological process in that it is controlled by both the voluntary and involuntary components of the nervous system (Chapter 35). In sleep, breathing is maintained at a resting rate by the involuntary part of the nervous system. Indeed, in most circumstances, breathing is controlled unconsciously.

The regulation of blood O_2 levels is a key example of homeostasis, as we saw in the case of core body temperature (Chapter 35) and blood glucose levels (Chapter 38). Recall that homeostasis often depends on sensors that monitor the levels of the chemical being regulated. In the case of breathing, the sensors are chemoreceptors located within the brainstem and in sensory structures called the **carotid** and **aortic bodies (Fig. 39.10).** The carotid bodies sense O_2 and proton (H^+) concentrations of the blood going to the brain, and the aortic bodies monitor their levels in blood moving to the body. In contrast, chemoreceptors in the brainstem sense CO_2 and H^+ concentrations. The most important factor in the control of breathing is the amount of CO_2 in the blood. If the concentration of CO_2 in the blood is too high, chemoreceptors in the brainstem stimulate motor neurons that activate the respiratory muscles to contract more strongly or more frequently. Stronger or faster breathing rids the blood of excess CO_2 and increases the supply of O_2 to the body.

Breathing can also be controlled voluntarily. It is a simple matter to voluntarily hold your breath. Holding the breath makes it possible for humans and marine mammals to dive under water; it is also critical to the production of speech, song, and sound for communication. Sound is produced by voluntarily adjusting the magnitude and rate of airflow over the vocal cords of mammals, the syrinx of songbirds, or the glottal folds of calling amphibians, such as some toads and frogs.

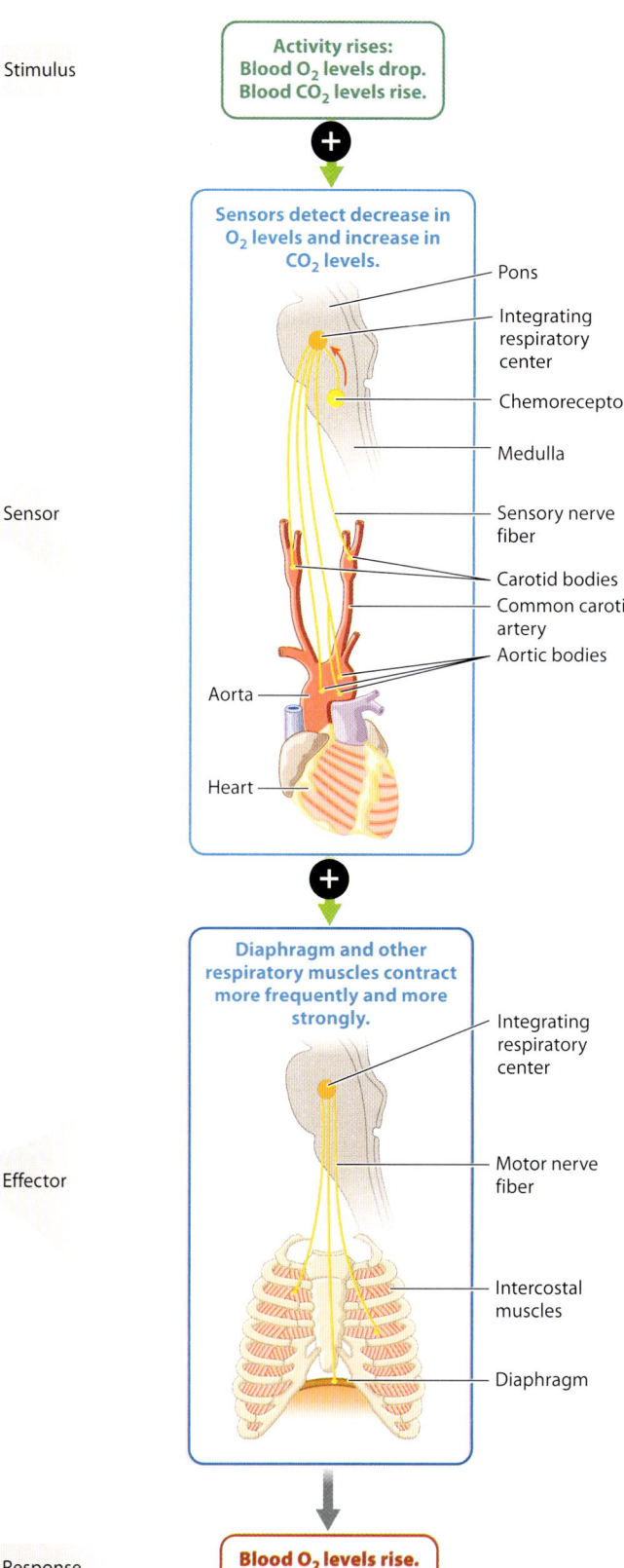

FIG. 39.10 Homeostatic control of breathing. Breathing becomes stronger or faster when O_2 levels drop.

39.3 OXYGEN TRANSPORT BY HEMOGLOBIN

Once O_2 is extracted from the environment and diffuses into the animal, it is transported to respiring cells. Oxygen is transported in the circulation by specialized body fluids—the blood of vertebrates or the hemolymph of invertebrates.

Blood plasma is the fluid portion of blood without the cells. It can hold only as much O_2 or CO_2 as can be dissolved in solution. How much O_2 goes into solution at a given partial pressure is a measure of its **solubility**. Because O_2 is about 30 times less soluble than CO_2, only about 0.2 ml of O_2 can be carried in 100 ml of blood. In both vertebrates and invertebrates, **hemoglobin** evolved as a specialized iron-containing, or heme, molecule for O_2 transport. By binding O_2 and removing it from solution, hemoglobin increases the amount of O_2 in the blood a hundredfold. Because of its greater solubility, CO_2 is transported in solution within the blood. Rather than remaining dissolved in solution as a gas, most of the CO_2 (about 95%) is converted to carbonic acid, which dissociates to form bicarbonate ions (HCO_3^-) and protons (Chapter 6).

Whereas some invertebrate hemoglobins exist in solution in the hemolymph and others exist within cells carried by the hemolymph, all vertebrate hemoglobins are produced by and reside in red blood cells. Hemoglobin gives the cells and the blood their red appearance.

Hemoglobin is an ancient molecule with diverse roles related to oxygen binding and transport.

Hemoglobin is an ancient molecule that has been found in organisms representing all kingdoms of life. In bacteria, archaeons, and fungi, similar forms of hemoglobin evolved as

HOW DO WE KNOW?

FIG. 39.11

What is the molecular structure of hemoglobin and myoglobin?

BACKGROUND In the 1950s, the Austrian Max Perutz worked with John Kendrew in the Cavendish Laboratory of the University of Cambridge to determine the molecular structure of globular proteins. Perutz and Kendrew were interested in understanding how the structures of hemoglobin and myoglobin enabled binding and transport of O_2.

EXPERIMENT Perutz and Kendrew developed and applied the new technique of X-ray crystallography to determine the three-dimensional molecular structures of hemoglobin and myoglobin.

RESULTS Perutz and Kendrew showed that adult hemoglobin consists of four subunits, two α (alpha) and two β (beta) subunits. Each subunit contains a heme group that contains iron, which is the site of O_2 binding. By contrast, myoglobin consists of only a single subunit with one heme group. These differences in molecular structure underlie the O_2-binding and dissociation properties of the two O_2 transport proteins.

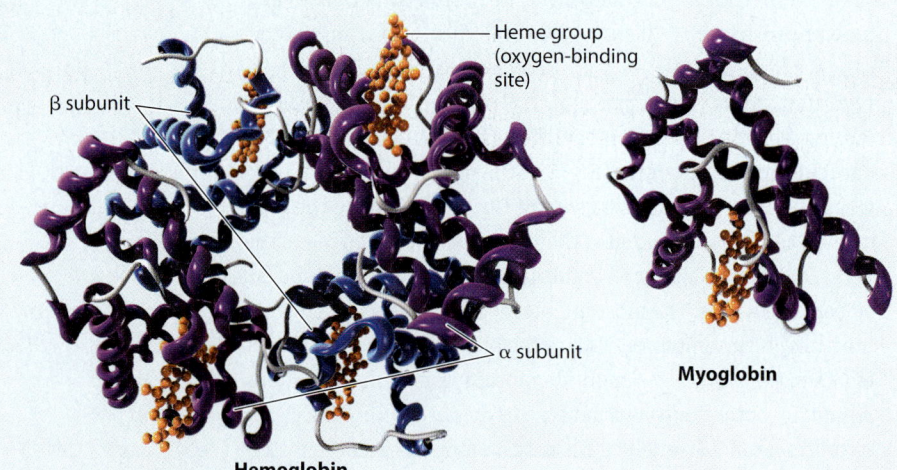

FOLLOW-UP WORK Work by Perutz in the 1960s showed how the molecular structure of hemoglobin changes when it binds O_2 versus when it is in an unbound state. Perutz also supervised Francis Crick as a doctoral student, who later worked with James Watson to resolve the structure of DNA. Perutz and Kendrew received the 1962 Nobel Prize in Chemistry for their work.

SOURCES Kendrew J. C., M. F. Perutz. 1948. "A Comparative X-Ray Study of Foetal and Adult Sheep Haemoglobins." *Proceedings of the Royal Society, London, Series A* 194:375–98; Perutz, M. F. 1964. "The Hemoglobin Molecule." *Scientific American* 211:64–76; Perutz, M. F. 1978. "Hemoglobin Structure and Respiratory Ttransport." *Scientific American* 239:92–125.

FIG. 39.12 (a) Hemoglobin and (b) myoglobin dissociation curves.

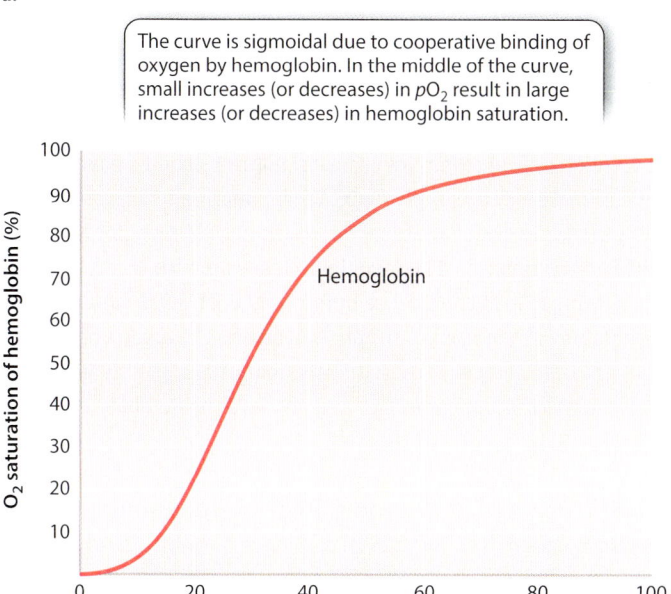
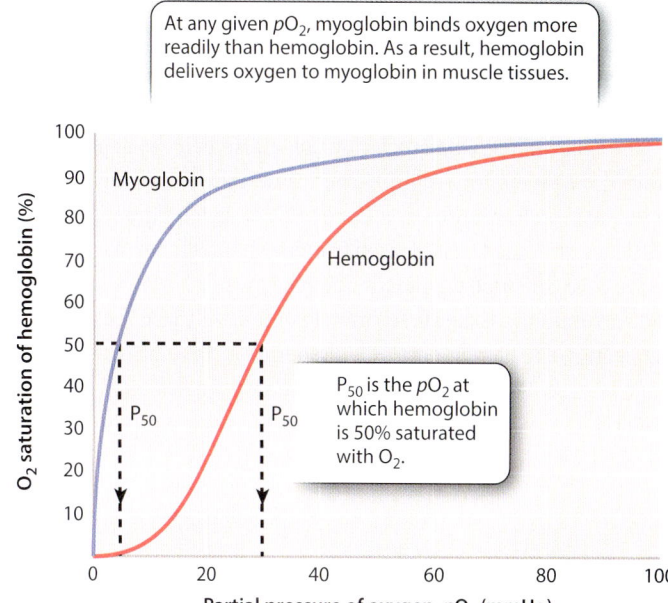

a terminal electron acceptor in cellular respiration and may have also served a role in scavenging O_2 to avoid damage due to oxidative stress.

Plant hemoglobins are known that bind and transport O_2 within the cell. In soybeans, hemoglobin binds O_2 in root nodule cells to keep concentrations low enough to avoid inhibiting nitrogenase, the enzyme used by symbiotic rhizobial bacteria in the root nodules for nitrogen fixation (Chapter 29).

Invertebrate hemoglobins found in roundworms, annelid worms, and arthropods bind O_2 in the circulatory fluid, as discussed above. They share considerable gene and amino acid sequence similarity with vertebrate hemoglobins, indicating that a common ancestor to invertebrate and vertebrate animals evolved a shared form of hemoglobin more than 670 million years ago.

Hemoglobin reversibly binds oxygen.

Hemoglobin exists in large concentrations within red blood cells. It is a globular protein that consists of four polypeptide units (**Fig. 39.11**). Each of these four units surrounds a heme group that contains iron, which reversibly binds one O_2 molecule. After O_2 diffuses into the blood, it diffuses into the red blood cells and binds to the heme groups in hemoglobin. Hemoglobin's binding of O_2 removes O_2 from solution, keeping the pO_2 of the red blood cell below that of the blood plasma so that O_2 continues to diffuse into the cell. The removal of O_2 from the plasma, in turn, keeps the pO_2 of the plasma below that of the lung alveolus, so O_2 continues to diffuse from the lungs into the blood.

When more O_2 is present in blood plasma, we expect more O_2 to become bound to hemoglobin, and that is what happens. But there is an interesting wrinkle that enables hemoglobin to readily bind O_2 leaving the lung. If we plot blood pO_2 against the percentage of O_2 bound to hemoglobin (the relation defined as "hemoglobin saturation"), we get the curve shown in **Fig. 39.12a**. As blood pO_2 increases, hemoglobin saturation rises slowly at first, then more steeply, and then more slowly again until it levels out. At 100% saturation, all hemoglobin molecules bind four O_2 molecules. The curve is called the hemoglobin's **oxygen dissociation curve**, and it has a sigmoidal shape—that is, a shape like an "S."

The shape of this curve can be explained as a result of changes in the ability of hemoglobin to bind O_2 (a property called its binding affinity) at different O_2 partial pressures. At 25% saturation, each hemoglobin molecule binds on average one O_2 molecule. As pO_2 rises, hemoglobin binds O_2 with increasing binding affinity. This increase in affinity results from the interaction of adjacent heme groups in each hemoglobin molecule. After one heme group binds the first O_2 molecule, hemoglobin undergoes a conformational change that increases the binding affinity of the remaining heme groups for additional O_2. The increase in binding affinity with additional binding of O_2 is called **cooperative binding**, and it gives the O_2 dissociation curve for hemoglobin its sigmoidal shape.

The increasing affinity of hemoglobin for O_2 with increasing O_2 concentration has important physiological consequences. In the middle part of the O_2 dissociation curve, small increases in O_2 concentration lead to large increases in hemoglobin saturation. Therefore, under normal circumstances, hemoglobin in the blood leaving the lung is fully saturated with O_2—each hemoglobin molecule is bound to four O_2 molecules.

The shape of the O_2 dissociation curve also helps to explain how O_2 is delivered to respiring cells. When the hemoglobin in red blood cells reaches tissues needing O_2 to supply their mitochondria, the O_2 is released from the hemoglobin. As active cells consume O_2, they reduce the local pO_2 of the cell and surrounding tissues to 40 mmHg or less. At these lower pO_2 values, the hemoglobin's O_2 dissociation curve has a steep slope (Fig. 39.12a). The steepness of the slope indicates that for a relatively small decrease in pO_2, large amounts of O_2 can be released from hemoglobin to diffuse into the cell.

Myoglobin stores oxygen, enhancing delivery to muscle mitochondria.

Myoglobin is a specialized O_2 carrier within the cells of vertebrate muscles. In contrast to hemoglobin, myoglobin is a monomer that contains only a single heme group. Because there are no interacting subunits, the O_2 dissociation curve for myoglobin has a different shape from that of hemoglobin (**Fig. 39.12b**). In fact, over the physiological range of pO_2, myoglobin has a greater affinity for O_2 than hemoglobin does and binds O_2 more tightly. As a result, hemoglobin releases O_2 to exercising muscles. Once the exercising muscles consume the available O_2, the intracellular pO_2 drops, and the myoglobin releases its bound O_2 to the muscle cell's mitochondria.

Red muscle cells that depend mainly on aerobic respiration to produce ATP (Chapter 37) store large amounts of myoglobin. The myoglobin in these cells can release O_2 quickly at the onset of activity, before respiratory and circulatory systems have had time to increase the supply of O_2. Diving marine mammals, such as whales and seals, carry large amounts of myoglobin within their muscles. The myoglobin loads up with O_2 when the animals breathe at the surface before a dive. The O_2 bound to the myoglobin is then used to supply ATP during the dive, when the animal cannot breathe. These marine mammals can stay under water for 30 minutes or longer and dive to considerable depths (**Fig. 39.13**).

Many factors affect hemoglobin–oxygen binding.

Obtaining enough O_2 is a special challenge in some circumstances. How does a mammalian fetus in its mother's uterus take up O_2 from its mother's blood? How do animals that live at high altitude obtain O_2 under conditions of low pO_2? In each case, the species adapts by evolving forms of hemoglobin with higher binding affinity.

In mammals, the mother's respiratory and cardiovascular systems must provide the fetus with O_2 and remove CO_2, as well as supply nutrients and remove waste. Maternal and fetal blood do not mix. Rather, they exchange gases across the placenta. Yet the O_2 concentration gradient at this exchange point does not favor the movement of O_2 from the mother's blood to the fetal blood. Fetal mammals solve the problem of extracting O_2 from their mother's hemoglobin by producing a form of hemoglobin before birth that has a higher affinity for O_2 than their mother's hemoglobin. The O_2 dissociation curve for fetal hemoglobin is shifted to the left of the curve for maternal hemoglobin (**Fig. 39.14a**), allowing the fetal hemoglobin to extract O_2 from its mother's circulation. At birth, when the newborn begins to breathe, its red blood cells rapidly shift to producing the adult form of hemoglobin, maintaining this form throughout life.

Many animals live at high altitude. For example, birds have been observed flying over the highest peaks (taller than 9000 m) despite the fact that air at this altitude has less than one-third the O_2 content of air at

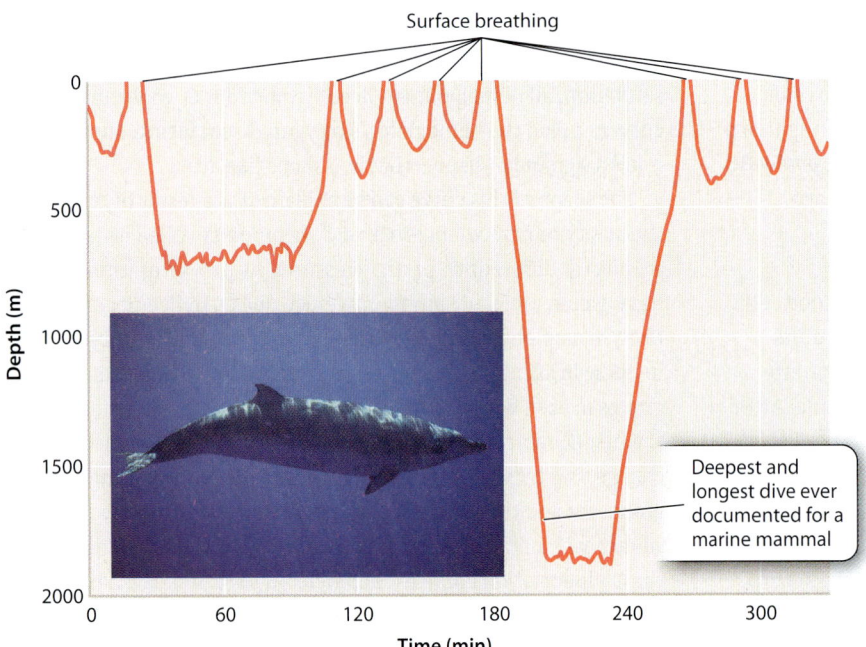

FIG. 39.13 Myoglobin as an O_2 store for long-lasting dives. The graph tracks the dives of a beaked whale (*Ziphius cavirostris*) fitted with a tracking device.

Deepest and longest dive ever documented for a marine mammal

FIG. 39.14 Changes in hemoglobin's O$_2$ dissociation curve.

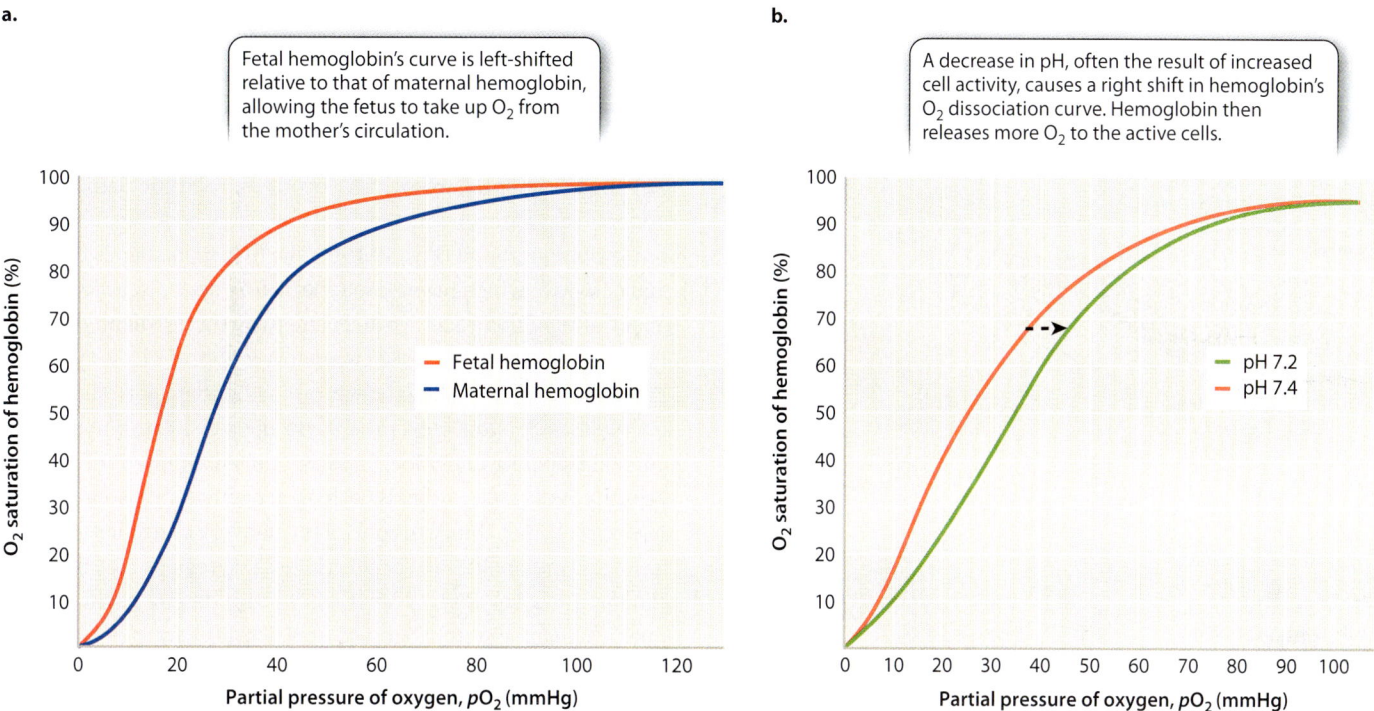

sea level. Similarly, llamas inhabit the Andes at altitudes up to 5000 m, where at a pO_2 of approximately 80 mmHg the air has about one-half the O$_2$ content of air at sea level. Consequently, the lung pO_2 of a llama is only about 50 mmHg. How do high-flying birds and llamas obtain enough O$_2$? Their hemoglobin has a higher affinity for O$_2$ and binds O$_2$ more readily than that of mammals living at sea level. Consequently, their O$_2$ dissociation curve is shifted to the left, favoring the binding of more O$_2$ at a given pO_2.

Tissue and blood pH also has an important physiological effect on the O$_2$ affinity of hemoglobin. When a decrease in pH (that is, an increase in H$^+$ concentration) occurs during exercise, the affinity of hemoglobin for O$_2$ *decreases*, resulting in a rightward shift in the O$_2$ dissociation curve (**Fig. 39.14b**). This phenomenon is called the Bohr effect after Christian Bohr, the Danish physiologist who first described it in 1904. Decreases in pH occur when CO$_2$ is released from metabolizing cells, or when an inadequate supply of O$_2$ leads to the production of lactic acid (Chapter 7). Because hemoglobin's affinity for O$_2$ is reduced, more O$_2$ is released and supplied to the cells for aerobic ATP synthesis.

Carbon dioxide also reacts with the amine (NH$_2$) groups of hemoglobin, reducing hemoglobin's affinity for O$_2$. Thus, when released from respiring tissues, CO$_2$ promotes increased O$_2$ delivery both through its direct effect on hemoglobin and its contribution to a decrease in blood pH by the Bohr effect. While the production of CO$_2$ promotes O$_2$ release at the tissues, its elimination at the lung increases hemoglobin's affinity for O$_2$, enhancing O$_2$ uptake.

→ **Quick Check 4** Sickle-cell anemia is a genetic disorder of individuals homozygous for a mutation of hemoglobin that causes their red blood cells to be sickle shaped and stiff under conditions of low pO_2. Why is this disease life threatening?

39.4 CIRCULATORY SYSTEMS

The O$_2$ carried by hemoglobin in red blood cells is transported to tissues throughout the body by the circulatory system. The **closed circulatory systems** of animals are made up of a set of internal vessels and a pump—the **heart**—to transport the blood to different regions of the body. Circulatory systems have two conflicting requirements: They must produce enough pressure to carry the circulating blood to all the tissues, yet once the blood reaches smaller vessels that supply the cells within the tissues the blood pressure and flow rate must not be too high. High pressure would push fluid through the walls of the smaller vessels and cause the blood to flow so quickly that there would not be enough time to exchange gases.

Generally, smaller animals, such as insects and many mollusks, have **open circulatory systems** that contain few blood vessels: Most of the circulating fluid, the hemolymph, is contained within

FIG. 39.15 (a) Open and (b) closed invertebrate circulatory systems.

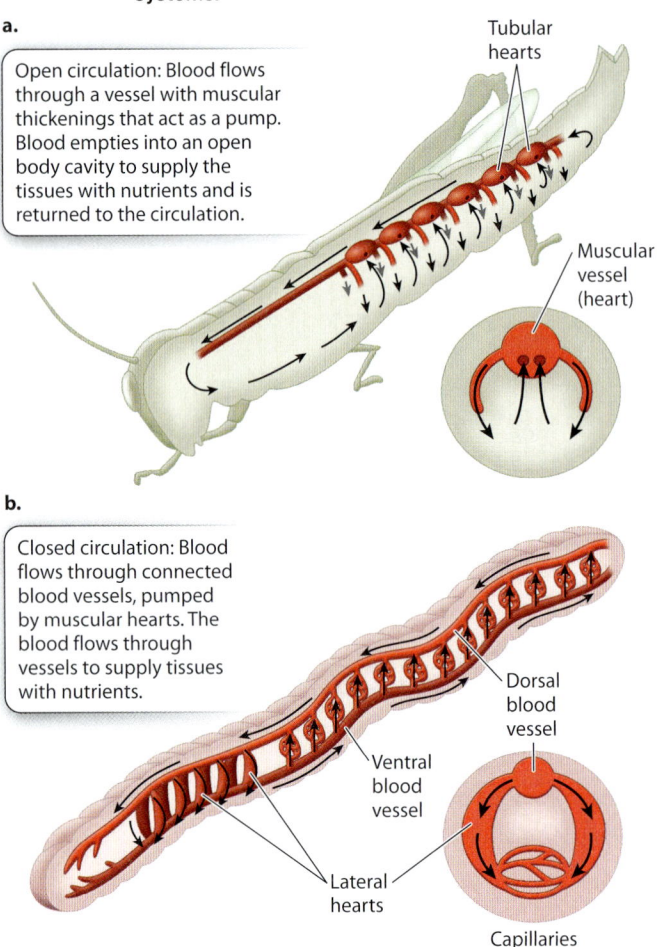

the animal's body cavity. The hemolymph bathes the animal's tissues and organs (**Fig. 39.15a**). Open circulatory systems have limited control of where the fluid moves. Muscles that are active in locomotion can assist circulation of the hemolymph, and some invertebrates have simple hearts with openings that help pump fluid between different regions of the animal's body cavity.

Animals with closed circulatory systems tend to be larger, although segmented worms have evolved closed circulatory systems that allow them to expand individual body segments to burrow underground (**Fig. 39.15b**). The first cephalopods (squid and octopus), which have closed circulatory systems, evolved about 480 million years ago and were early successful predators in Paleozoic oceans. A closed circulatory system delivers O_2 at high rates to exercising tissues, providing them with the energy used to chase prey. In contrast, open circulatory systems generally operate under low pressure and have limited transport capacity. As a result, animals with open circulatory systems are less active. Their ability to control the delivery of respiratory gases and metabolites to specific tissues and regions is also limited. However, insects can be very active despite having an open circulatory system because they obtain O_2 by their tracheal system independently of the circulatory system.

Closed circulatory systems can control blood flow to specific regions of the body by varying the resistance to flow. For example, wading birds reduce blood flow to their legs when they are in cold water to reduce heat loss, and vertebrates can increase blood flow to their muscles to deliver O_2 and nutrients during exercise. In order to pump blood through a set of closed interconnected vessels, a muscular heart is needed to produce sufficient pressure to overcome the flow resistance of the vessels.

FIG. 39.16 Flow and vessel size. (a) Vessel hierarchy of circulatory systems; (b) total vessel cross-sectional area versus flow velocity.

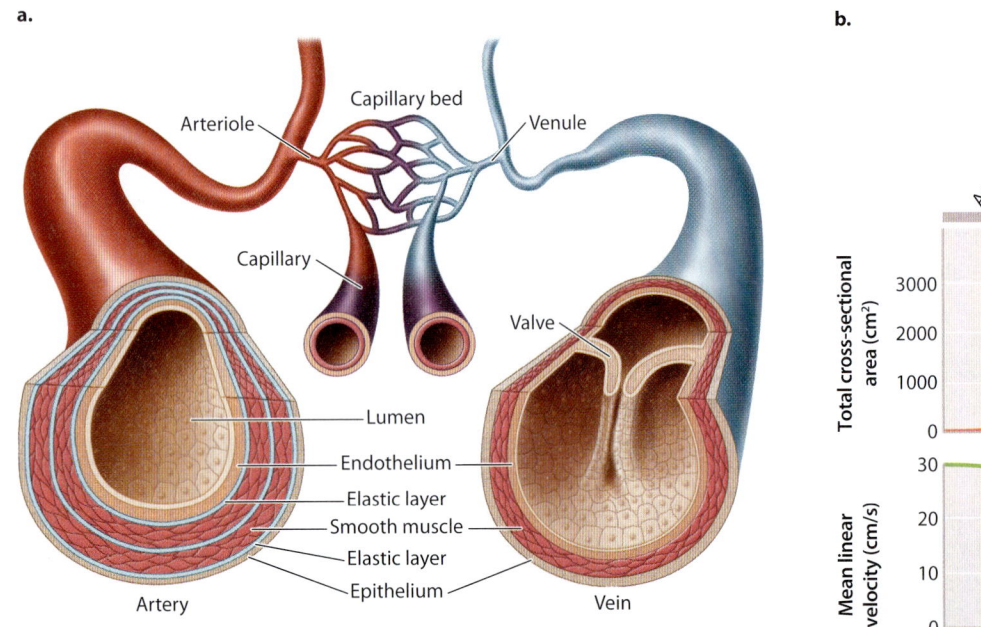
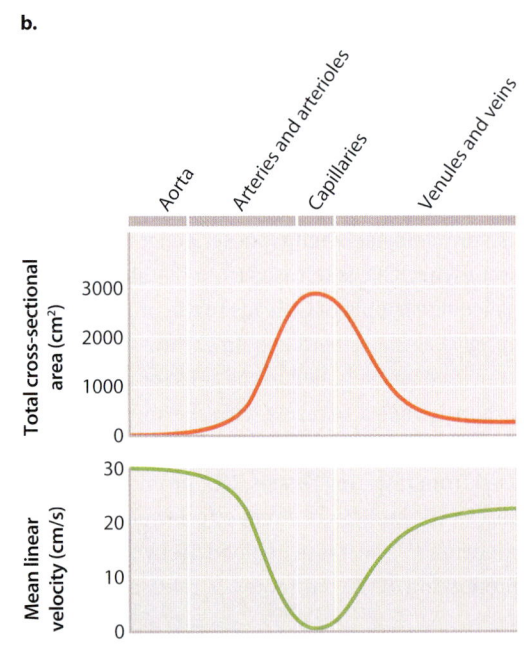

Circulatory systems have vessels of different sizes to facilitate bulk flow and diffusion.

For a fluid like blood to flow through a set of pipes, pressure (P) is required to overcome the resistance (R) to flow. The pressure developed by a heart when it contracts establishes a rate of blood flow that is governed by P/R. That is, the rate of blood flow increases with an increase in pressure and decreases with an increase in resistance. The resistance to flow is determined in part by the fluid's stickiness (its **viscosity**) and the vessel's length. Longer vessels of a given size impose greater resistance. However, the main factor governing resistance to flow is the vessel's radius (r). Resistance is proportional to $1/r^4$, which means that if a vessel's radius is reduced by half, its resistance to flow increases 16 times.

The high resistance of narrow vessels presents a challenge. To keep diffusion distances short, gas exchange between cells and blood requires narrow vessels. How does the circulatory system overcome the dramatic increase in resistance with narrower diameters? Animal circulatory systems are organized to have blood flow over longer distances in a relatively few large-diameter vessels with low resistance to flow (**Fig. 39.16a**).

Arteries are the large, high-pressure vessels that move blood flow away from the heart to the tissues. **Veins** are the large, low-pressure vessels that return blood to the heart. Arteries branch into blood vessels of progressively smaller diameter called **arterioles,** and these arterioles ultimately connect to finely branched networks of very small blood vessels called **capillaries.** It is at the capillaries that gases are exchanged by diffusion with the surrounding tissues. The number of smaller-diameter vessels at each branching point in the circulatory system greatly exceeds the number of larger diameter vessels. The reverse organization is found on the return side of circulation: Numerous capillaries drain into vessels of progressively larger diameter called **venules,** and the venules drain into a few larger veins that return blood to the heart.

This organization has two advantages. First, it maintains the same volume of blood flow at all levels within the circulatory system: The increased resistance to flow in the capillaries is offset by the large increase in the number of capillaries. Second, it enables the blood to flow more slowly in smaller vessels (**Fig. 39.16b**), providing time for gases and metabolites to diffuse into and out of the neighboring cells. Indeed, the diameter of a capillary is so small that red blood cells pass through one at a time, and must be flexible to do even that.

Arteries are muscular, elastic vessels that carry blood away from the heart under high pressure.

When an animal becomes active after a period of rest and feeding, blood flow must be increased to its muscles and reduced to its digestive organs (**Fig. 39.17**). How do animals accomplish this? The answer comes from physics: the strong effect of vessel radius on flow resistance. Arterioles supplying regions of the body that need less blood shrink in diameter by contracting the circular smooth muscle fibers in their wall. The reduction in diameter increases the resistance to flow and thereby reduces the rate of blood flow. For example, when you put your hands or feet into cold water, the arterioles contract, reducing the loss of heat from blood flowing to your hands and feet.

Arterioles supplying regions of the body that need more blood expand in diameter by relaxing their smooth muscles. The larger vessels offer less resistance, so more blood can flow through them. What is the mechanism that causes the

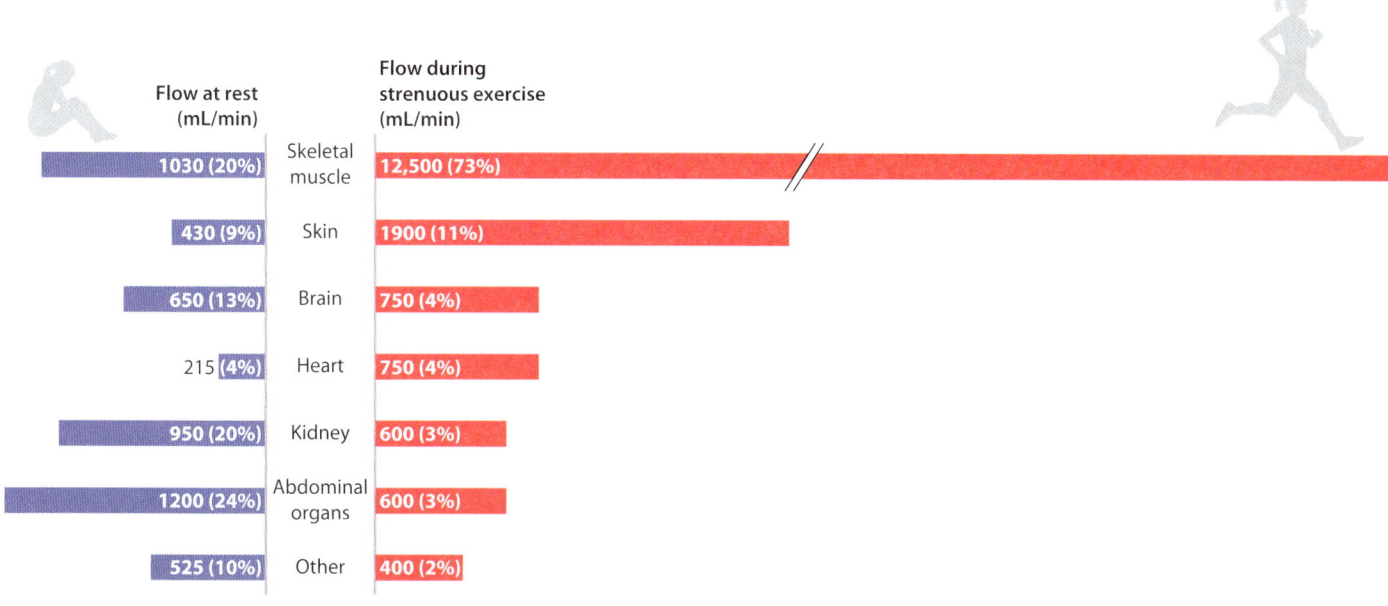

FIG. 39.17 Changes in blood flow from resting to exercise as measured in humans. Numbers in parentheses indicate percent of total blood flow.

smooth muscles to relax when cells in neighboring tissue become more active? As the rate of cellular respiration in these cells increases, they begin releasing more CO_2 as well as other metabolites, such as lactate and hydrogen ions, that are produced in respiration. The increase of these molecules in the bloodstream acts as a signal to the smooth muscle to begin relaxing. In this way, blood flow increases to the capillaries to match O_2 and nutrient delivery to the metabolic need of the cells.

A strong pulse of flow pushes against the arterial walls about every second in a human as the heart beats to pump blood through the arteries. This pulse of flow causes a momentary expansion of the arteries. Artery walls are able to withstand these repeated pressure pulses because they contain multiple elastic layers composed of two proteins: **collagen** and **elastin.** Collagen fibers are strong and resist the expansion of the arterial wall during these pressure pulses. Together with elastin fibers, the collagen provides an elastic rebound of the arterial wall once the pulse has passed, returning energy to help smooth out blood flow. The collagen and elastin resist overexpansion of the arterial wall at local sites. If an artery wall deteriorates, the collagen and elastin can become so thin that the artery bulges outward. The outward bulge, called an aneurysm, can lead to a life-threatening rupture.

→ **Quick Check 5** What change in vessel shape most affects resistance to blood flow?

Veins are thin-walled vessels that return blood to the heart under low pressure.

Blood collected from local capillary networks returns to the heart through progressively larger veins that ultimately drain into the two largest veins, the **venae cavae** (singular, vena cava). The venae cavae drain blood from the head and body into the heart. There is little pressure available to push the blood forward in the veins because pressure has been lost to the resistance of the arterioles and capillaries. Consequently, veins are thin walled and have little smooth muscle or elastic connective tissue. Because of the low pressure, blood tends to accumulate within the veins: As much as 80% of your total blood volume resides in the venous side of your circulation at any one time. When you stand or sit for a long time, blood may pool within the veins of your limbs.

Veins located in the limbs and in the body below the heart have one-way valves that help prevent blood from pooling due to gravity (see Fig. 39.16a). However, the most important aid to returning blood to the heart is the voluntary muscle contraction that occurs during walking and exercise, which exerts pressure on the veins. Above the level of the heart, the return of blood to the heart through the veins is assisted by gravity.

Compounds and fluid move across capillary walls by diffusion, filtration, and osmosis.

We have seen that O_2 and CO_2 move between capillaries and tissues by diffusion. In addition, water, certain ions, and other small molecules, but not proteins, move from capillaries into the surrounding interstitial fluid (the extracellular fluid surrounding vessels) by filtration, forced by blood pressure.

Why doesn't the blood plasma lose all its water and ions over time? As water is filtered out of the capillaries, some ions, proteins, and other large compounds remain inside, increasing in concentration (**Fig. 39.18**). As a result, there is a tendency of water to flow back into the capillary by osmosis (Chapter 5). At the same time, blood pressure decreases because of the flow resistance imposed by the capillaries. When the pressure is no longer high enough to counter osmosis, water moves back into the capillaries.

In vertebrates, excess interstitial fluid is also returned to the bloodstream by means of the **lymphatic system,** a network of vessels distributed throughout the body. The fluid that enters the lymphatic system is called the **lymph.** Small lymphatic vessels merge with progressively larger thin-walled vessels, draining the lymph into lymphatic ducts, which empty into the venous system and then the heart. Lymphatic vessels have one-way valves, similar to those in the veins, that assist the return of lymph to the circulatory system. Local muscle contractions also

FIG. 39.18 Pressures pushing fluid out of blood vessels (blood pressure) and into them (osmotic pressure).

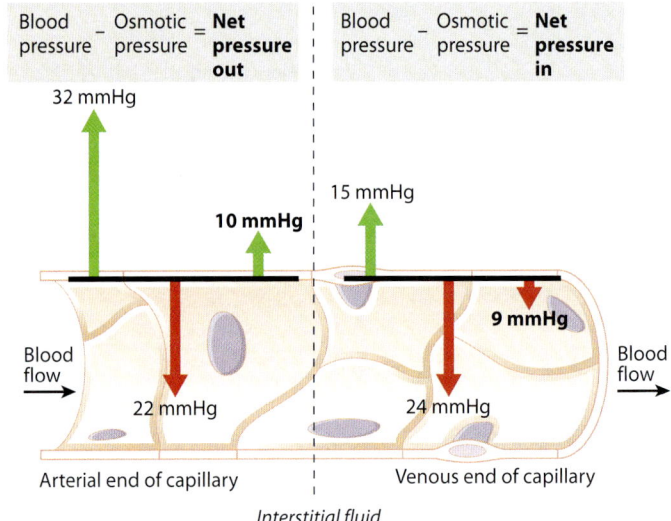

help to pump the lymph toward the thoracic duct. In addition to its role in returning plasma to the bloodstream, the lymphatic system has important functions in the immune system (Chapter 43).

? CASE 7 Predator–Prey: A Game of Life and Death
How do hormones and nerves provide homeostatic regulation of blood flow as well as allow an animal to respond to stress?

An animal that is dehydrated or has lost blood following an injury may experience a prolonged drop in blood pressure. In such cases, the posterior pituitary gland releases antidiuretic hormone (ADH), also known as vasopressin (Chapter 38), into the circulation. ADH causes the arteries to contract, increasing their resistance to blood flow. The higher resistance elevates blood pressure throughout the body.

Even under less stressful conditions, blood pressure is continually being adjusted to remain within normal physiological limits, another example of homeostasis. The autonomic nervous system plays an important role in keeping blood pressure at healthy levels (Chapter 35). If blood pressure in the general circulation drops, animals reduce the supply of blood to the limbs by constricting arteries that supply the limbs. This response helps maintain blood pressure to the heart, brain, and kidneys. In this case, sympathetic neurons synapsing on the smooth muscles of arterioles in the limbs stimulate these muscles to contract. In contrast, when blood pressure is too high, sympathetic neurons that synapse on the smooth muscles are inhibited. The smooth muscles relax, reducing resistance in the arteries and increasing blood flow. These two responses are called **vasoconstriction** and **vasodilation**.

In some circumstances, it may be advantageous for an animal's blood pressure to rise higher than normal levels. One example is during the fight-or-flight response that occurs when animals are faced with a stressful situation, as when prey animals are alerted to the presence of a predator, and when predators locate a potential prey. The sympathetic nervous system stimulates the adrenal gland to release the hormone **adrenaline** (also called **epinephrine**). The surge of adrenaline into the bloodstream—the "adrenaline rush"—causes an animal to become alert and aroused. At the same time, the sympathetic nervous system acts to increase heart rate and respiratory rate and inhibit digestive processes. Arteries supplying limb muscles dilate to enhance blood flow to exercising muscles, while arteries supplying organs within the digestive system constrict. The increase in heart and breathing rates increases the rate of O_2 delivery to the exercising muscles to meet their increased demand for ATP. These physiological changes, along with a heightening of sensory acuity, prepare an animal for the more intensive activity needed to chase or flee.

In contrast, when an animal has recently eaten a meal, its sympathetic nervous system is inhibited and its parasympathetic nervous system is stimulated, increasing digestive secretions and the activity of gut smooth muscle (Chapter 40). Thus, local regulation of blood flow within different regions and organ systems of the body is integrated with more general feedback from the nervous and endocrine systems. The result is homeostatic regulation of blood pressure that monitors digestion and meets the need for O_2 delivery to all metabolizing cells of the animal. These shifts in blood flow are readily controlled by vasoconstriction or dilation of the arteries supplying the different organs.

39.5 THE EVOLUTION, STRUCTURE, AND FUNCTION OF THE HEART

The pressure required to drive blood through blood vessels is generated by the heart. Animal hearts are pumps made of muscle that rhythmically contract, producing this pressure. Early hearts were as simple as a muscular thickening of a small region of a blood vessel. Such hearts are still found in insects (see Fig. 39.15a). More complex hearts have chambers that expand and fill with deoxygenated blood returning from the animal's tissues. After filling, the heart muscle contracts, pumping deoxygenated blood to lungs or gills and newly oxygenated blood onward to the animal's tissues. The closed circulatory system of cephalopod mollusks has three hearts—two that pump deoxygenated blood through the gills and another that pumps oxygenated blood to the body. In contrast, vertebrates have a single heart with at least two chambers, one for receiving blood and the other for pumping blood to the body during each heartbeat cycle. Hearts also have valves to ensure that blood does not flow backward when the heart contracts.

A theme in the evolution of vertebrate circulatory systems is the progressive separation of circulation to the gas exchange organ from the circulation to the rest of the body. In fish, deoxygenated blood is first pumped to the gills to gain O_2 before flowing to the rest of the body and then returning to the heart. This single circulation path limits the rate of blood flow to metabolically active body tissues: The resistance imposed by the small vessels in the gills slows the blood flow considerably before it even reaches the tissues. By contrast, birds and mammals evolved a separate **pulmonary circulation** to the lungs and **systemic circulation** to the rest of the body. The evolution of separate pulmonary and systemic circulations was made possible by the evolution of a four-chambered heart. This organization has two advantages: It increases the supply of oxygenated blood to active tissues, and it increases the uptake of O_2 at the gas exchange surface.

FIG. 39.19 Fish heart and circulatory system. (a) Fish have a two-chambered heart that pumps deoxygenated blood. (b) Blood flows in a single circuit from the heart, through gills, to the tissues for gas exchange, and back to the heart.

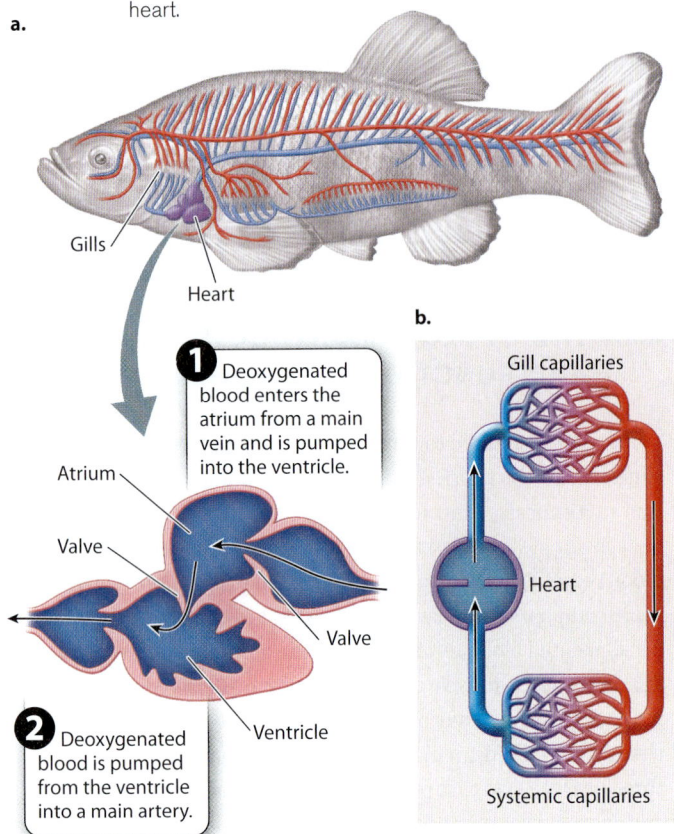

Fish have two-chambered hearts and a single circulatory system.

Fish hearts have two chambers (**Fig. 39.19**), the **atrium** and **ventricle**. Deoxygenated blood returning from the fish's tissues enters the atrium, which fills and then contracts to move the blood into a thicker-walled ventricle. The muscular ventricle pumps the blood through a main artery to the gills for uptake of O_2 and elimination of CO_2. Oxygenated blood collected from the gills travels to the tissues through a large artery called the aorta.

The small gill capillaries impose a large resistance to flow. As a result, much of the blood pressure is lost in moving blood through the gills. This loss of pressure limits the flow of oxygenated blood to body tissues.

Amphibians and reptiles have three-chambered hearts and partially divided circulations.

As animals moved onto land, the transition from breathing water to breathing air had important consequences for the organization of vertebrate circulatory systems. Land vertebrates evolved hearts that separated the circulation of deoxygenated blood pumped to their gas exchange organs from circulation of oxygenated blood delivered to their body tissues. Gas exchange became more efficient and O_2 delivery increased. Metabolic rates rose, and animals became capable of greater activity.

Reflecting their lifestyle, both aquatic and terrestrial, amphibians evolved a variety of ways to breathe. Some retain gills, while others use lungs or their skin, or some combination of these, to breathe. Amphibians evolved an additional atrium, dividing their heart into two separate atria and a single ventricle

FIG. 39.20 Amphibian heart and circulatory system. (a) The frog heart has three chambers: two atria and one ventricle. (b) Mixed blood leaving the common ventricle goes either to the lung and back to the heart, or to the tissues and back to the heart.

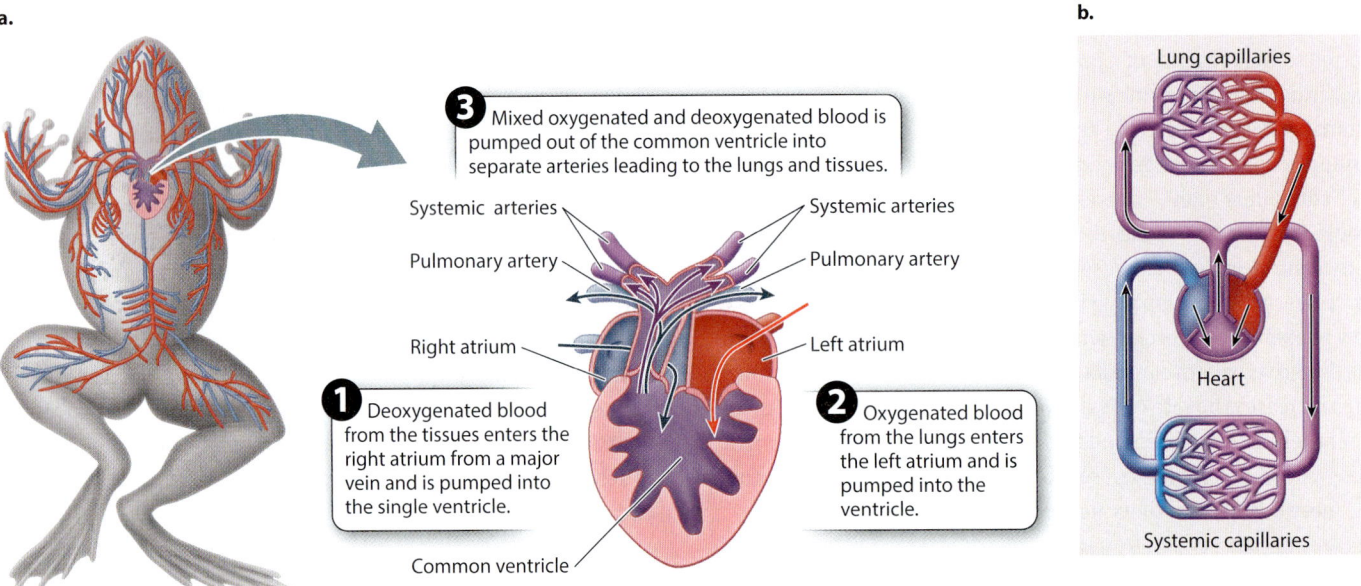

FIG. 39.21 The mammalian heart and circulatory system. (a) Mammals have a fully divided four-chambered heart with two atria and two ventricles. (b) Mammals have separate pulmonary and systemic circulations.

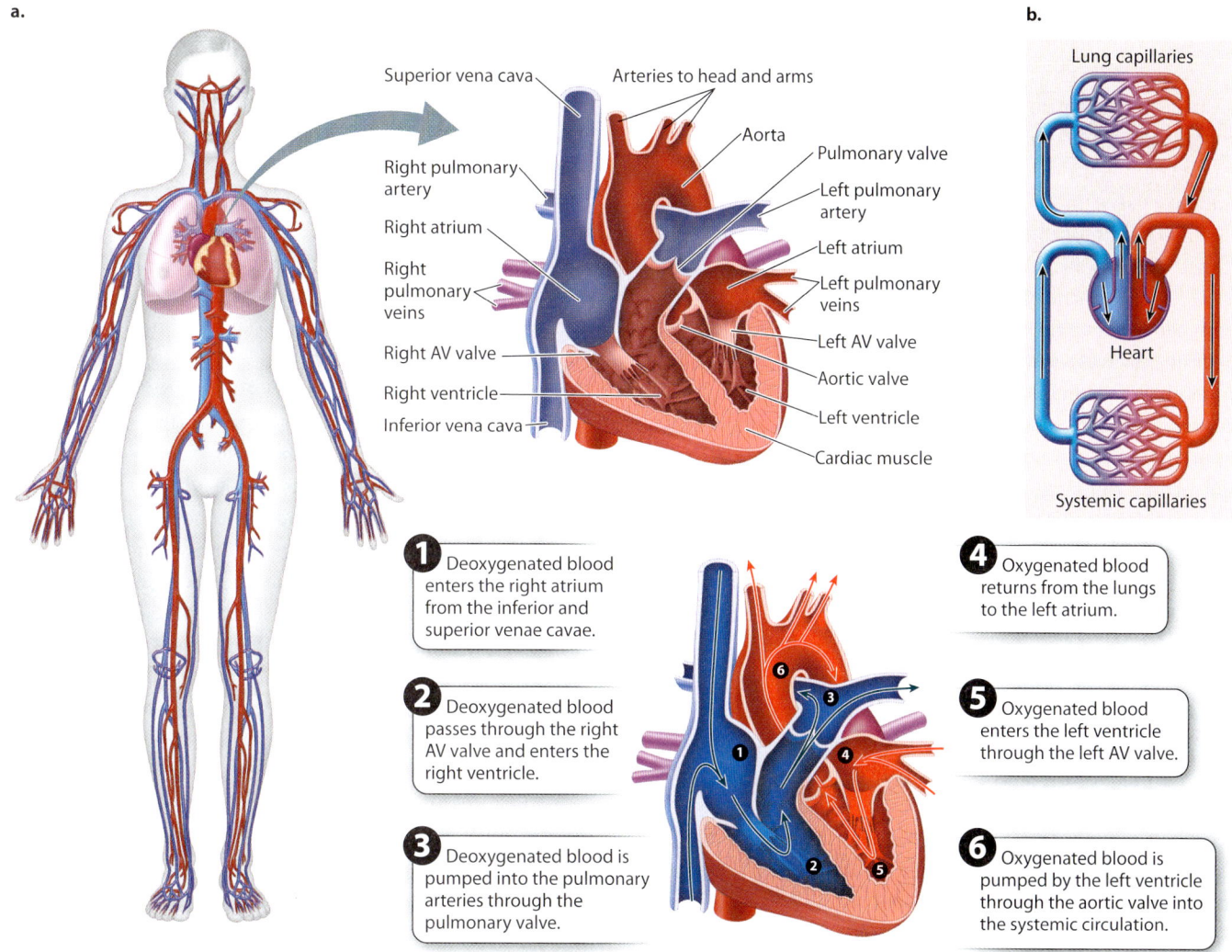

(Fig. 39.20). This arrangement partially separates the pulmonary and systemic circulations, so in amphibians freshly oxygenated blood can be pumped under higher pressure to the body than is the case in fishes. However, because amphibians have a single ventricle, oxygenated blood returning to the heart from the gills or lungs mixes with deoxygenated blood returning from the animal's body before being pumped from the ventricle. Anatomical features of the ventricle help to limit this mixing, but nonetheless mixing still hinders the supply of oxygenated blood to an amphibian's body.

Most reptiles also have a three-chambered heart, but their single ventricle has internal ridges to improve the separation of deoxygenated and oxygenated blood. As a result, reptiles are able to deliver more O_2 to their tissues than amphibians can. At the same time, the incompletely divided ventricle provides reptiles with versatile circulatory responses during exercise and when diving underwater. For example, when they dive, the blood of turtles and sea snakes bypasses their lungs, which they cannot use underwater.

Mammals and birds have four-chambered hearts and fully divided pulmonary and systemic circulations.

Mammals and birds evolved four-chambered hearts with separate atria and separate ventricles (**Fig. 39.21**). This arrangement completely separates blood flow to the lungs from blood flow to the tissues. As a result, these animals can pump blood to their lungs under lower pressure, allowing increased uptake of O_2, while at the same time supplying blood gases and nutrients to their tissues at high pressure. Whereas the systemic blood pressure of fish is only a few mmHg, in a bird or mammal it is 100 mmHg or higher. Moreover, blood flowing to the tissues is fully oxygenated at the outset, delivering more O_2 to body tissues at a higher rate. This change in heart structure and circulatory pattern was linked to an increase in the metabolic rates and the evolution of endothermic lifestyles in birds and mammals (Chapter 40).

Following the pattern established in amphibians and reptiles, deoxygenated blood enters the right atrium from the venae cavae and, when the atrium contracts, moves through an **atrioventricular (AV) valve** into the right ventricle (Fig. 39.21). When the right ventricle contracts, blood is pumped through the **pulmonary valve** into the pulmonary trunk, which divides into the left and right **pulmonary arteries**, and then to the lungs for oxygenation. The walls of the right ventricle are thinner than those of the left ventricle, and its weaker contractions eject blood at a lower pressure. As a result, the blood of the pulmonary circulation moves at a slower rate, allowing greater time for gases to diffuse into and out of the lungs.

The oxygenated blood returns from the lungs through the **pulmonary veins** and enters the left atrium of the heart. When the left atrium contracts, blood is pumped through a second atrioventricular valve into the left ventricle. The thick muscular walls of the left ventricle eject the blood under high pressure to the body. Oxygenated blood leaving the left ventricle passes through the **aortic valve** and then flows to the head and the rest of the body through a large artery called the **aorta**.

The contraction of the two atria followed by contraction of the two ventricles makes up the **cardiac cycle (Fig. 39.22)**. The cardiac cycle is divided into two main phases. **Systole** is the contraction of the ventricles, and **diastole** is the relaxation of the ventricles. Systole, therefore, is the phase of the cardiac cycle in which blood is pumped from the heart into the pulmonary and systemic circulations, and diastole is the phase in which the atria contract and the ventricles fill with blood.

FIG. 39.22 The four-chamber cardiac cycle.

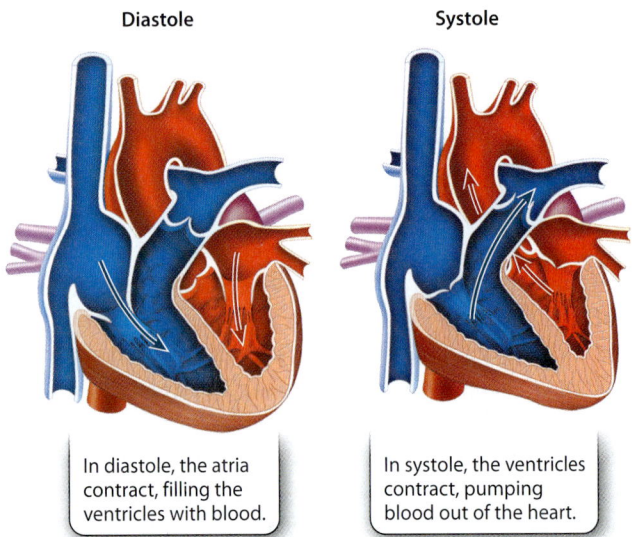

In diastole, the atria contract, filling the ventricles with blood.

In systole, the ventricles contract, pumping blood out of the heart.

→ **Quick Check 6** What pressure difference between the ventricle and the atrium causes the atrioventricular valve to close?

Cardiac muscle cells are electrically connected to contract in synchrony.

In order for the heart to function effectively as a pump, the **cardiac muscle** cells that make up the walls of the atria and ventricles must contract in a coordinated fashion. Atrial muscle cells must be activated to contract in synchrony during diastole to fill the ventricles, and ventricular muscle cells must contract in synchrony during systole to eject the blood from the heart. For cardiac muscle cells to pump in unison, they must be depolarized in unison by an action potential. In vertebrates, this action potential is transmitted by autonomic nerves that control the heart beat cycle.

Cardiac muscle cells have contractile properties similar to those of skeletal muscle cells (Chapter 37). However, cardiac muscle cells are distinct in two ways. First, specialized cardiac muscle cells can generate action potentials on their own, independently of the nervous system. Second, cardiac muscle cells are in electrical continuity with one another, meaning that they can pass their action potentials to adjacent cells. These two features of cardiac muscle cells ensure that all muscle cells in a region surrounding a heart chamber are activated and contract in unison. The specialized cardiac muscle cells capable of generating action potentials independently function as a **pacemaker** that causes the heart to beat with a basic rhythm. These cells stimulate neighboring cells to contract in synchrony. They are found in two specialized regions of the heart—the **sinoatrial (SA)** and **atrioventricular (AV) nodes (Fig. 39.23)**.

Whereas most nerve cells maintain their resting membrane potential until they receive a signal from another cell, the critical feature of pacemaker cells is that their resting membrane potential gradually becomes less negative on its own until it reaches threshold and the cell fires an action potential. The tendency of their membrane potential to become less negative is due to slow leakage of sodium ions into the cell and decreased flow of potassium out of the cell. When the cell reaches threshold, voltage-gated calcium ions open, causing more rapid depolarization. Like nerve cells, after firing an action potential, the pacemaker cells repolarize to reach their resting potential, keeping the heart relaxed between contractions. Also like nerve cells, pacemaker cells repolarize by opening voltage-gated potassium channels to allow potassium ions to leave the cell.

The heartbeat is initiated at the sinoatrial node, located at the junction of the vena cava and right atrium (Fig. 39.23). Because adjacent cardiac muscle cells are in electrical contact, action potentials initiated at the sinoatrial node spread rapidly from one cell to the next. Thus, when the pacemaker cells in the SA node fire an action potential, their depolarization spreads electrically throughout the right and left atria, causing them to contract in unison. Because there is no electrical contact between the atria and ventricles, the ventricles do not contract. Instead, the

FIG. 39.23 Control of heart beat rhythm. The pacemaker cells of the sinoatrial and atrioventricular nodes keep the heart beating with a basic rhythm. This rhythm can be assessed by recording the electrocardiogram (EKG) of the heart.

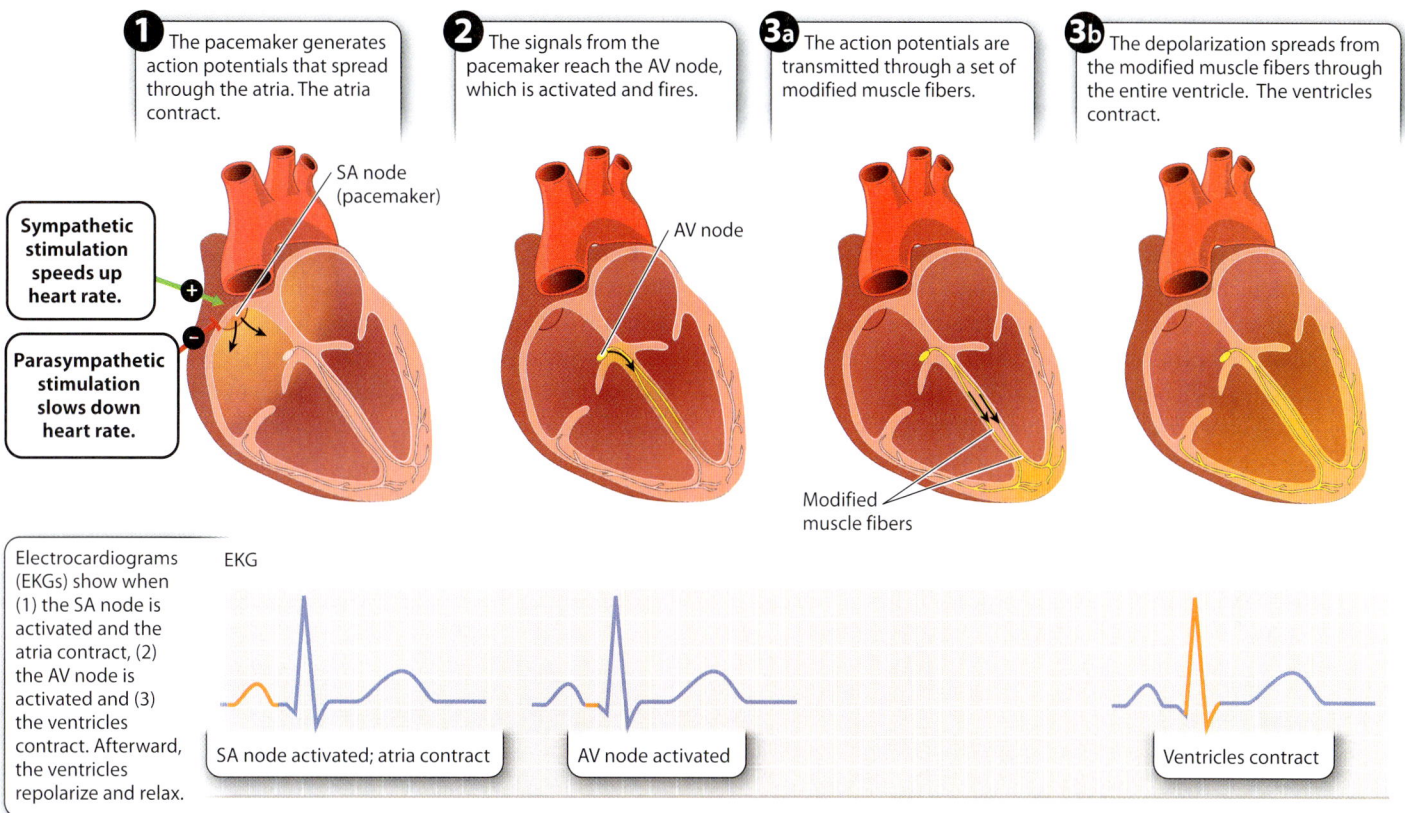

depolarization in the atria reaches a second set of pacemaker cells, located in the atrioventricular (AV) node.

Activation of the AV node transmits the action potential to the ventricles. A modified set of cardiac muscle fibers transmits the action potential from the AV node to the base of the ventricles. From there, the depolarization spreads throughout the ventricle walls, causing the ventricles to contract in unison. The delay in transmission from the AV node and through the conducting fibers ensures that the ventricles do not contract until they are fully filled with blood from the atria.

Electrodes placed over the surface of the chest and other body regions can record the electrical currents produced by the depolarization of the heart while it beats. This type of recording is called an electrocardiogram (EKG, because "kardia" is Greek for heart, although ECG is also used; Figure 39.23). Cardiologists use EKGs to diagnose heart problems or to determine what region of the heart has been damaged following a heart attack.

Cardiac output is regulated by the autonomic nervous system.

The volume of blood pumped by the heart over a given interval of time is its **cardiac output (CO)**. This quantity is the key measure of heart function. Cardiac output is determined by calculating the product of **heart rate (HR)** and the volume of blood pumped during each beat, the **stroke volume (SV)**: $CO = HR \times SV$. For example, if heart rate increases 50% and the heart's stroke volume increases 10%, the heart's cardiac output increases 65% $((1.5 \times 1.1) - 1 \times 100\%)$. Cardiac output rises or falls in response to the metabolic demand for O_2, which in turn depends on an animal's state of activity. When an animal sleeps, cardiac output is minimal, but when it runs, cardiac output increases to meet the demand for O_2 of its active muscles. Animals can increase their cardiac output by increasing heart rate or stroke volume or both. Whereas humans and other mammals rely most heavily on increases in heart rate to increase cardiac output, fish rely more on increases in stroke volume.

The nervous system controls the rate of the heartbeat. Nerve signals influence the rate at which the cardiac pacemaker cells depolarize in between action potentials. When there is no signal from the nervous system, the depolarization rate of the pacemaker cells sets the resting heart rate. Stimulation by sympathetic nerves causes the pacemaker cells of the sinoatrial node to depolarize more rapidly and the heartbeat to speed up. Stimulation by parasympathetic nerves causes these cells to depolarize more slowly and the heartbeat to slow down. By acting as a hormone (Chapter 38), adrenaline released by the adrenal gland into the circulation also elevates the heart rate for more prolonged periods of time. This hormone is a key component of the fight-or-flight response of many vertebrate animals.

Stroke volume is adjusted directly in response to changes in blood flow and heartbeat strength. It depends on how much blood returns to fill the heart and how strongly the heart contracts. These two processes are linked in order to match heart filling (that is, venous return) to heart emptying. During exercise, when an animal's limb muscles regularly contract, their action increases the return of blood to the heart. The influx of additional blood stretches the walls of the atria and ventricles to a greater extent, causing them to contract more forcefully. As a result, more blood is ejected during each heartbeat.

The correspondence of stroke volume to changes in heart filling is described by **Starling's Law.** It provides the means by which venous return and cardiac output are adjusted similarly to meet changing rates of blood supply. Increased sympathetic stimulation to the heart and circulating adrenaline also increase the contractile strength of the heart muscle, further increasing the heart's stroke volume. These changes in heart rate and stroke volume underlie the increase in cardiac output that occurs when an animal changes from resting to activity (see Fig. 39.17).

In addition to the transport of respiratory gases, nutrients, and wastes, circulation is linked to several other bodily functions. The distribution of blood flow, controlled by changes in arterial resistance through vasodilation or vasoconstriction, is an important factor controlling heat loss or gain between an animal's body and its environment. Thus, the circulatory system plays a central role in temperature regulation of many animals (Chapters 35 and 40). The circulatory system also provides the route by which hormones (Chapter 38) and immune cells (Chapter 43) are distributed throughout the body, enabling the endocrine system to maintain homeostatic control of diverse bodily functions and the immune system to defend the body against injurious microbes and toxic compounds. In doing so, the circulatory system serves as a general pathway by which hormonal and immune communication occurs to regulate, integrate, and protect the functional state of the whole organism.

Core Concepts Summary

39.1 RESPIRATION AND CIRCULATION DEPEND ON DIFFUSION OVER SHORT DISTANCES AND BULK FLOW OVER LONG DISTANCES.

Diffusion is effective only over short distances and requires large exchange surface areas with thin barriers. page 39-2

Ventilation and circulation provide bulk transport over long distances. page 39-3

Oxygen is delivered to tissues in four steps: (1) bulk flow of water or air past the respiratory surface (gills and lungs); (2) diffusion of O_2 across the respiratory surface into the circulatory system; (3) bulk flow through the circulatory system; and (4) diffusion of O_2 into tissues and cells. page 39-3

39.2 RESPIRATION PROVIDES OXYGEN AND ELIMINATES CARBON DIOXIDE IN SUPPORT OF CELLULAR METABOLISM.

Many aquatic animals exchange respiratory gases with water through gills. page 39-4

Countercurrent flow of water relative to blood in gills enhances O_2 extraction from water. page 39-5

Terrestrial animals breathe air by means of internal tracheae or lungs. page 39-6

Terrestrial vertebrates inflate and deflate their lungs by bidirectional tidal ventilation driven by changes in pressure. page 39-6

A unidirectional flow of air maximizes O_2 uptake by bird lungs. page 39-8

Both the voluntary and involuntary nervous systems control breathing. page 39-9

39.3 RED BLOOD CELLS PRODUCE HEMOGLOBIN, GREATLY INCREASING THE AMOUNT OF OXYGEN TRANSPORTED BY THE BLOOD.

Hemoglobin greatly increases the transport capacity of O_2 in the blood. page 39-10

Red blood cells contain hemoglobin with iron-containing heme groups that reversibly bind and release O_2. page 39-11

A sigmoidal O_2-dissociation curve describes the change in the binding affinity of hemoglobin for O_2 with changes in partial pressure. The shape of the curve results from cooperative binding of O_2 by hemoglobin. page 39-11

Myoglobin binds and stores O_2 in muscle cells, increasing the delivery of O_2 to muscle mitochondria for activity in general and in marine mammals for diving. page 39-12

Fetal hemoglobin is expressed by the mammalian fetus to allow O_2 uptake from the mother's blood. page 39-12

The affinity of hemoglobin for O_2 is affected by pH and CO_2. page 39-13

39.4 CIRCULATORY SYSTEMS HAVE DIFFERENT-SIZED VESSELS THAT FACILITATE BULK FLOW AND DIFFUSION.

The fluid in a closed circulatory system travels through a set of internal vessels, moved by a pump, the heart. The fluid in an open circulatory system moves through only a few vessels and is mostly contained within the animal's body cavity. page 39-13

Resistance to blood flow is most strongly controlled by vessel diameter. page 39-15

Arteries are thick-walled vessels that contain elastic fibers for support against pressure. page 39-15

Arteries control blood flow within the body by changing their resistance through contraction and relaxation of the smooth muscle in their walls. page 39-15

Capillaries are small-diameter, thin-walled vessels that facilitate diffusion of O_2 and CO_2. page 39-15

Veins are thin-walled vessels that return blood to the heart under low pressure. page 39-16

Fluids, gases, and other compounds move across capillary walls by diffusion, filtration, and osmosis. In vertebrates, some fluid is returned to the bloodstream by the lymphatic system. page 39-16

39.5 THE EVOLUTION OF ANIMAL HEARTS REFLECTS SELECTION FOR A HIGH METABOLIC RATE, ACHIEVED BY INCREASING THE DELIVERY OF OXYGEN TO METABOLICALLY ACTIVE CELLS.

Birds and mammals have a separate pulmonary circulation, from the heart to the lungs, and systemic circulation, from the heart to the rest of the body. page 39-17

Fish have two-chambered hearts and a single circulatory path. page 39-18

Amphibians and reptiles have three-chambered hearts and a partially divided circulatory path. page 39-18

Birds and mammals have four-chambered hearts and fully divided pulmonary and systemic circulations to enhance O_2 uptake and delivery to metabolizing tissues. page 39-19

One-way valves control blood flow through the heart. page 39-20

The cardiac cycle has discrete filling (diastole) and emptying (systole) phases. page 39-20

Cardiac muscle cells are electrically connected, allowing them to contract in synchrony as a unified heart beat. page 39-20

Contraction of the heart is controlled by signals from pacemaker cells of the sinoatrial and atrioventricular nodes, transmitted by electrical conducting fibers to the ventricles. page 39-20

The vertebrate autonomic nervous system regulates the heart's cardiac output by changes in heart rate and stroke volume. page 39-21

Self-Assessment

1. Diagram the four basic steps of O_2 transport from an animal's respiratory medium (air or water) to its cells.

2. For an aquatic animal, describe what features of its gills favor diffusion of O_2 and CO_2.

3. Name two differences in the physical properties of water and air that affect gas exchange in aquatic versus terrestrial animals.

4. Explain how the tracheal respiratory system of insects enables high metabolic rates.

5. Explain how gill respiration is facilitated by unidirectional water flow, whereas lung ventilation can depend on bidirectional airflow.

6. Describe the pressure changes that are needed to draw air into the lungs and to expel air out of the lungs.

7. Explain how cooperative binding by hemoglobin facilitates O_2 uptake into blood.

8. Explain how the branching of larger arteries into many smaller vessels affects the rate of and resistance to blood flow in the smaller vessels.

9. Diagram the path of blood flow through the heart, lungs, and body of a fish and of a mammal or bird.

10. Describe two mechanisms by which the cardiac output of an animal's heart can be adjusted and indicate which is more important for increasing cardiac output in mammals.

Do you understand the chapter's Core Concepts? Log in to check your answers to the Self-Assessment questions, then practice what you've learned and reinforce this chapter's concepts by working through the problems and multimedia tutorials provided there.

www.biologyhowlifeworks.com

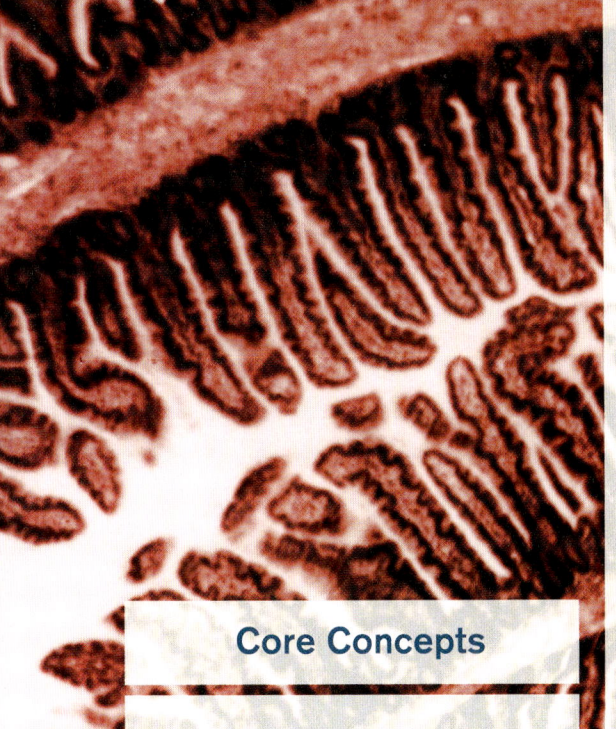
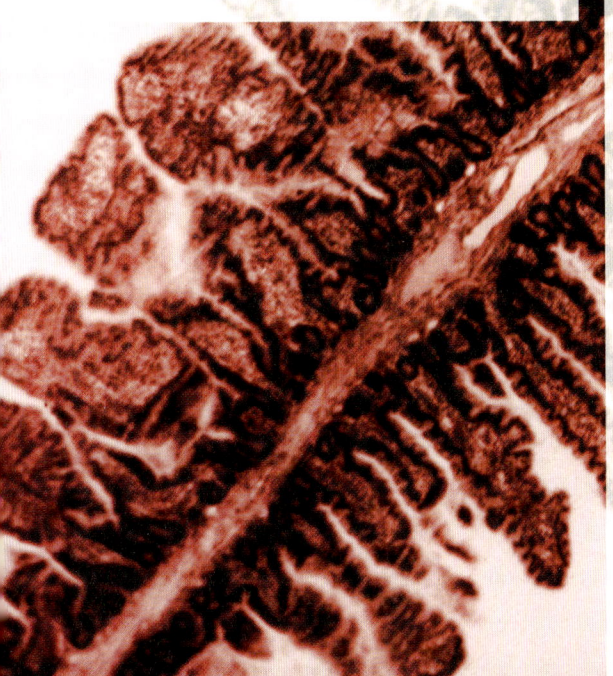

CHAPTER 40

ANIMAL METABOLISM, NUTRITION, AND DIGESTION

Core Concepts

40.1 Metabolic rate depends on level of activity, body size, and body temperature.

40.2 An animal's diet supplies the energy it needs for homeostasis and essential nutrients it cannot synthesize on its own.

40.3 Different animals have different adaptations for feeding.

40.4 The digestive tract is a tubelike structure with regions specialized for different functions.

It is literally true that "you are what you eat." Not only do animals build their bodies and obtain energy from the food they eat, but also their very appearance often reflects what they eat. Consider a cat. We know it's a carnivore—an animal that eats other animals—just from looking at it. It has sharp teeth and powerful jaws specialized for tearing meat; strong legs, sharp claws, and keen eyesight for hunting prey; and a short gut because breaking down animal protein is easier than breaking down plant material. Humans are omnivores, adapted for eating both other animals and plants, and this is reflected in our anatomy. Our teeth include sharp canines for tearing meat and flat molars for grinding plant material. Finally, consider herbivores, such as cows, that eat plants. Cows have evolved a variety of adaptations, including a four-chambered stomach, that allow them to digest cellulose, which is tough but rich in energy.

Animals are heterotrophs, obtaining food from other organisms (Chapter 6). The energy animals obtain from food is essential for building bodies, moving, surviving, and reproducing. Much of an animal's activity therefore centers on obtaining food. In turn, the biology of an animal is in large part shaped by the type of food the animal consumes. Furthermore, an animal's metabolism, the chemical reactions by which it breaks down its food, is closely linked to the energy and nutrients contained in that food.

The food that an animal eats also affects its place in an ecosystem. Primary producers, such as plants and some algae, are food for a great diversity of animals. Animals that feed on plants tend to exist in large numbers. Large animals that feed on other, smaller animals exist in smaller numbers and commonly represent fewer species.

This chapter looks at how animals obtain and break down the food they eat. We begin at the cellular level, building on the biochemistry of metabolism discussed earlier (Chapter 7). Then, we examine the dietary needs and the organization and function of animal digestive systems: how an animal eats, digests, and absorbs food to supply its cells with the energy required for its function. The relationship between an animal's diet and its place in an ecosystem is reserved until Chapter 47.

40.1 PATTERNS OF ANIMAL METABOLISM

As discussed in Chapter 7, the biochemical pathways that make up animal metabolism are remarkably conserved among diverse groups of organisms. These pathways evolved early in the history of life and have been retained throughout the evolution and diversification of life. In Chapter 7, we examined the pathway of glucose breakdown by cellular respiration, which provides ATP to fuel the work of a cell. Here, we put this

40-1

and related pathways in an organismal context. We explore how carbohydrates, proteins, and fats are metabolized for building body structures and supplying energy needs and how metabolic rate varies with an animal's level of activity, internal body temperature, and size.

Animals rely on anaerobic and aerobic metabolism.

Nearly all animals depend on three main molecules as sources of energy and building blocks for growth and development: carbohydrates, fats, and proteins. The conversion of these food sources into biologically useful forms typically involves

FIG. 40.1 Cellular metabolism, which comprises anaerobic and aerobic pathways. Animals obtain energy to fuel their activities and development from the breakdown of carbohydrates, fats, and proteins.

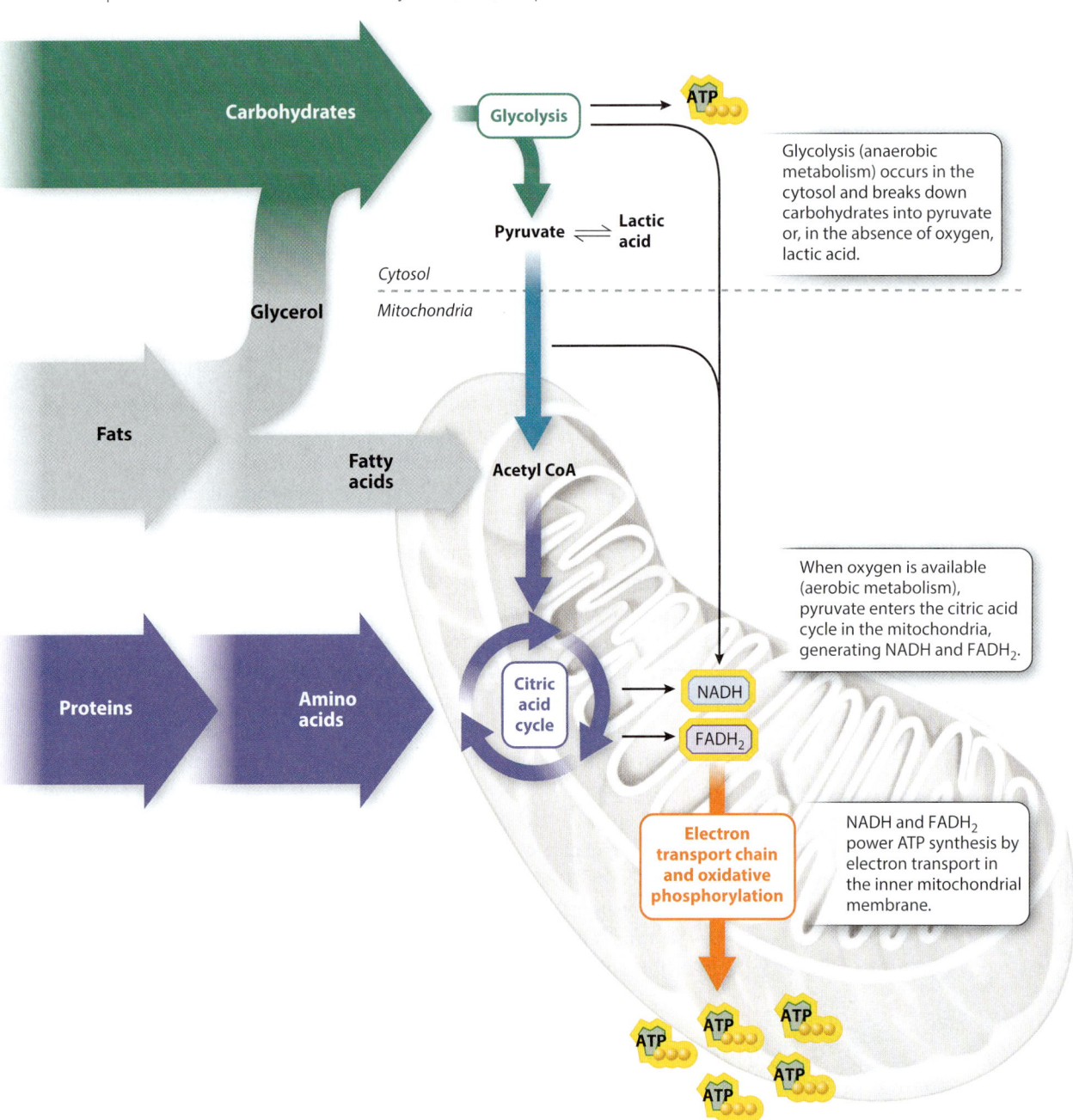

synthesizing **adenosine triphosphate (ATP).** While other nucleic acid triphosphates also serve as usable forms of energy, ATP is by far the most common one used by animals. The breakdown of carbohydrates, fats, and proteins to produce ATP involves a linked set of chemical reactions. Reactions that break down food sources to fuel the energy needs of a cell are referred to as **catabolic.** In contrast, reactions that result in net energy storage within cells and the organism are **anabolic.**

Let's first review the breakdown of carbohydrates, including sugars (such as glucose) and starches, food sources that are chemically similar but clearly different in taste and texture. Glucose can be partially broken down in the absence of oxygen by glycolysis (**Fig. 40.1**). Glycolysis occurs in the cytosol of the cell and results in the production of two molecules of pyruvate and two molecules of ATP from the breakdown of each glucose molecule. If oxygen is not present, or is present only in small amounts, pyruvate is converted by fermentation into lactic acid. The production of ATP by **anaerobic** ("absence of oxygen") **metabolism** provides rapid but short-term energy to the cell and organism.

In contrast, **aerobic** ("of oxygen") **metabolism** (specifically, cellular respiration) carried out within the mitochondria of eukaryotic animals provides a steady supply of ATP for longer-term, sustainable activity. When enough oxygen diffuses into the mitochondria, pyruvate can be processed by the **citric acid cycle** rather than converted to lactic acid (Fig. 40.1). In a series of oxidation–reduction reactions, energy-rich NADH and $FADH_2$ molecules are produced. These molecules then enter the electron transport chain in the inner mitochondrial membrane. The **electron transport chain** couples the transfer of electrons to the pumping of protons across the inner mitochondrial membrane, resulting in a proton electrochemical gradient that enables the synthesis of ATP by oxidative phosphorylation. Because oxygen is reduced to water in the process, oxygen is consumed and water is produced.

Overall, aerobic respiration results in the formation of a total of 32 molecules of ATP from 1 molecule of glucose, much more than the 2 molecules of ATP from 1 molecule of glucose produced anaerobically. The metabolic efficiency, defined as the amount of energy captured in a usable form (approximately 286 kcal of useful ATP energy) per amount of energy in the starting molecule (686 kcal in one mole of glucose), is therefore approximately 42% (or 286/686 x 100). The remaining 58% of this energy is converted to heat and either lost or used to warm the animal. Glucose that is not metabolized is stored in the form of glycogen, primarily in the liver and muscles.

Lipids are another important energy source for most animals. Lipids, such as triacylglycerol, consumed in the diet are broken down to glycerol and free fatty acids, which enter glycolysis or the citric acid cycle to yield ATP by mitochondrial electron transport (Chapter 7). Alternatively, they can be stored as fat (adipose tissue), a particularly efficient storage form of energy. This is because fat can be stored in a form that yields more than twice the energy supply per unit weight (9.4 kcal/g) as glycogen (4.2 kcal/g). Fats are broken down into glycerol and free fatty acids, which enter glycolysis or the citric acid cycle to yield ATP by mitochondrial electron transport.

Proteins consumed in the diet are also a useful energy source for animals. Carnivores such as cats get most of their energy from the protein in the meat that they eat. Proteins, which are needed for building and maintaining the body, constitute the enzymes and structural elements of cells and tissues. Only following prolonged food deprivation, when fat and carbohydrate reserves are depleted, do animals break down protein reserves (mainly from their muscles) to form ATP. Animals preferentially rely on fats for long-term energy supply. Well-nourished humans have enough fat reserves to support their metabolic needs for 5–6 days before they must rely on protein reserves. Because nerve cells depend primarily on glucose to produce ATP, carbohydrates are spared for the brain and nervous system when animals face starvation.

→ **Quick Check 1** Muscle protein, fat, and glycogen are all reservoirs of energy. In what order are they used during a prolonged fast?

Metabolic rate varies with activity level.

An animal's overall rate of energy use defines its **metabolic rate.** Metabolic rate can be measured by the animal's rate of oxygen consumption, which in turn reflects the aerobic production of ATP (**Fig. 40.2**). In 1773, Antoine Lavoisier first demonstrated that animal respiration is the same as combustion, as observed in the burning of a candle. A burning candle placed inside a sealed container is eventually extinguished. Similarly, a bird placed inside a sealed container eventually loses consciousness. When the candle and the bird were placed in the chamber together, both succumbed more quickly. Borrowing from Joseph Priestley's discovery of oxygen as the gas in air that reacts with metals (in the process now called oxidation), Lavoisier recognized that both the bird and candle consumed oxygen, or "pure air" as he called it, converting it into "fixed air" (carbon dioxide), or air of "poorer quality."

Metabolic rate is affected by many factors, one of which is the activity level of an organism. When an animal shifts from rest to activity, its metabolic rate and oxygen consumption

FIG. 40.2 Oxygen consumption in metabolism. The amount of oxygen consumed provides a measure of metabolic rate.

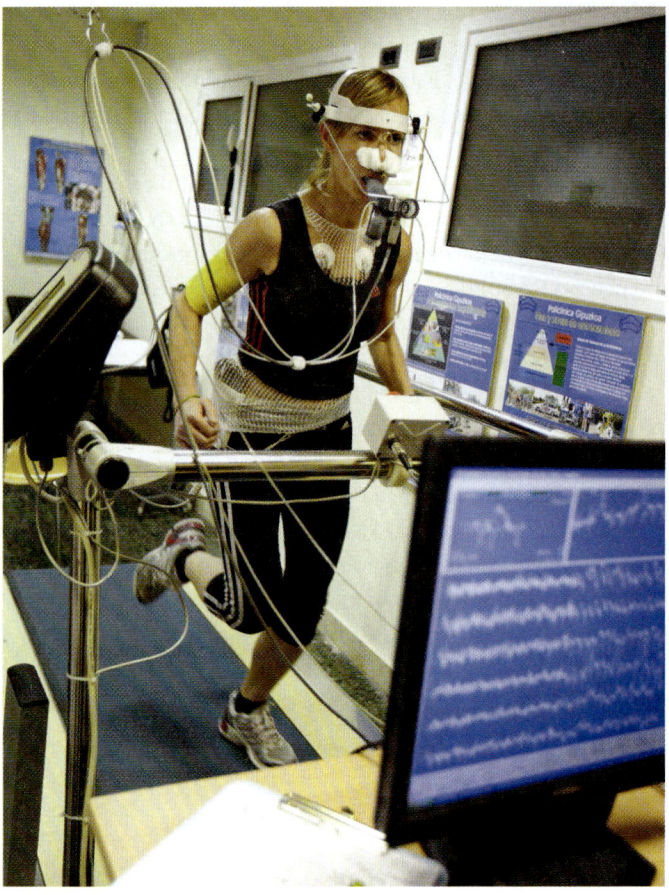

rise to meet its increased demand for ATP. The onset of activity requires immediate energy, which in animals is provided by specialized energy stores in their tissues. In vertebrate muscle cells, for example, phosphocreatine, a ready source of high-energy phosphates, is hydrolyzed to synthesize ATP directly from ADP. This reaction is followed by the relatively rapid production of ATP by anaerobic glycolysis. Although glycolysis produces relatively few ATP molecules per molecule of glucose broken down, the reactions are extremely fast, providing a rapid short-term supply of energy for animals. Animals rely on anaerobic glycolysis for short bursts of intensive activity.

Eventually, enough oxygen diffuses into the mitochondria to allow aerobic respiration to occur at a rate that meets the ATP needs of the animal. Therefore, the rate of oxygen consumption initially increases, then levels off (**Fig. 40.3**). At this point, the animal's need for energy is being met entirely by aerobic respiration.

With even more prolonged or intense activity, more ATP is needed, and it is harder to meet that need by aerobic metabolism alone. Your experience tells you that you can walk at a comfortable pace for a long period of time, but when you run, you tire more quickly. And the faster you run, the more quickly you tire. This is because more intensive physical exercise requires a greater reliance on anaerobic glycolysis to produce ATP. The resulting buildup of lactic acid and decrease in pH inside the muscle cells force an animal to decrease its activity. The production of lactic acid through fermentation limits an animal's performance by lowering the pH of the animal's blood, producing a condition called metabolic acidosis. Metabolic acidosis contributes to fatigue.

Sprinters rely heavily on anaerobic ATP production for intense short-term bouts of activity, but longer races are run at progressively slower speeds for increased endurance as the runner relies more heavily on aerobic metabolism to produce ATP. The world record for the 100-m sprint averages just above 10 m/s, but for the 1500-m race it is about 6.6 m/s.

When activity ends, the animal's oxygen consumption rate declines but does not immediately return to resting levels, as shown in Fig. 40.3. The elevated consumption of oxygen following activity is the animal's **recovery metabolism**. It represents the continued metabolic energy required to reestablish the resting metabolic state of the cells. During recovery metabolism, cells re-synthesize depleted ATP stores and metabolize the end products of fermentation, particularly lactic acid. The difference between an animal's immediate

FIG. 40.3 Changes in oxygen consumption during physical activity.

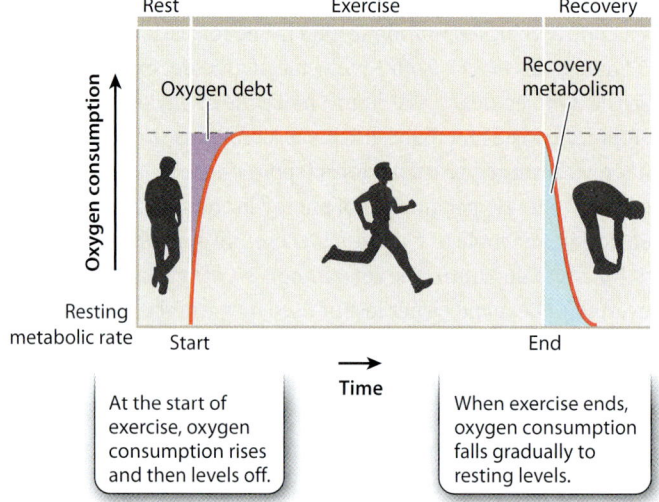

energy need at the onset of activity versus energy supplied by aerobic metabolism is often referred to as the animal's "oxygen debt." This debt is "paid back" following exercise by the animal's recovery metabolism. Recovery metabolism is associated with elevated breathing and heart rates that you (and other animals) experience when resting after moderate to intense exercise.

? CASE 7 Predator–Prey: A Game of Life and Death
Does body temperature limit activity level in predators and prey?

We just saw that metabolic rate increases with activity level. A by-product of metabolism is the generation of heat, which is a necessary consequence of the second law of thermodynamics (Chapters 1 and 6). Animals have several mechanisms for dissipating excess heat, including sweating, changes in blood flow, and (in dogs, for example) panting.

Cheetahs, as top predators, are one of the fastest land animals known today, capable of speeds up to 70 mph (114 km/h). Such extreme speed generates a lot of heat. Interestingly, instead of releasing the excess heat as it builds up, the cheetah stores it, dissipating it only after a chase.

The amount of heat that cheetahs store during a sprint can be considerable, on the order of 60 times the heat produced at rest. Heat storage raises the cheetah's body temperature. Cheetahs can therefore reach remarkable speeds but can sustain them only for short durations and must rest for long periods of time between sprints. Because their bodies can support only so much heat storage, cheetahs do not run when the temperature of their environments reaches approximately 105°F (41°C). In other words, the amount of time a cheetah runs and the distance it covers are limited by its body temperature.

Gazelles, common prey of cheetahs, have similar heat storage mechanisms and are similarly known for their speed and agility. Goats, by contrast, do not store heat as they increase their activity levels, instead dissipating it through evaporative mechanisms. But goats are not known for their speed. Heat storage therefore seems to enable quick bursts of speed, and has evolved as an adaptation among both predators and prey to achieve short, rapid sprints.

Metabolic rate is affected by body size.

In addition to activity level, an animal's size influences its metabolic rate. At rest, larger animals consume more energy and have higher metabolic rates than smaller ones. However, resting, or basal, metabolic rate does not increase linearly with an animal's mass. Instead, measurements of resting metabolic rate in a wide range of organisms show that metabolic rate increases with animal mass raised to the 3/4 power (**Fig. 40.4**). This

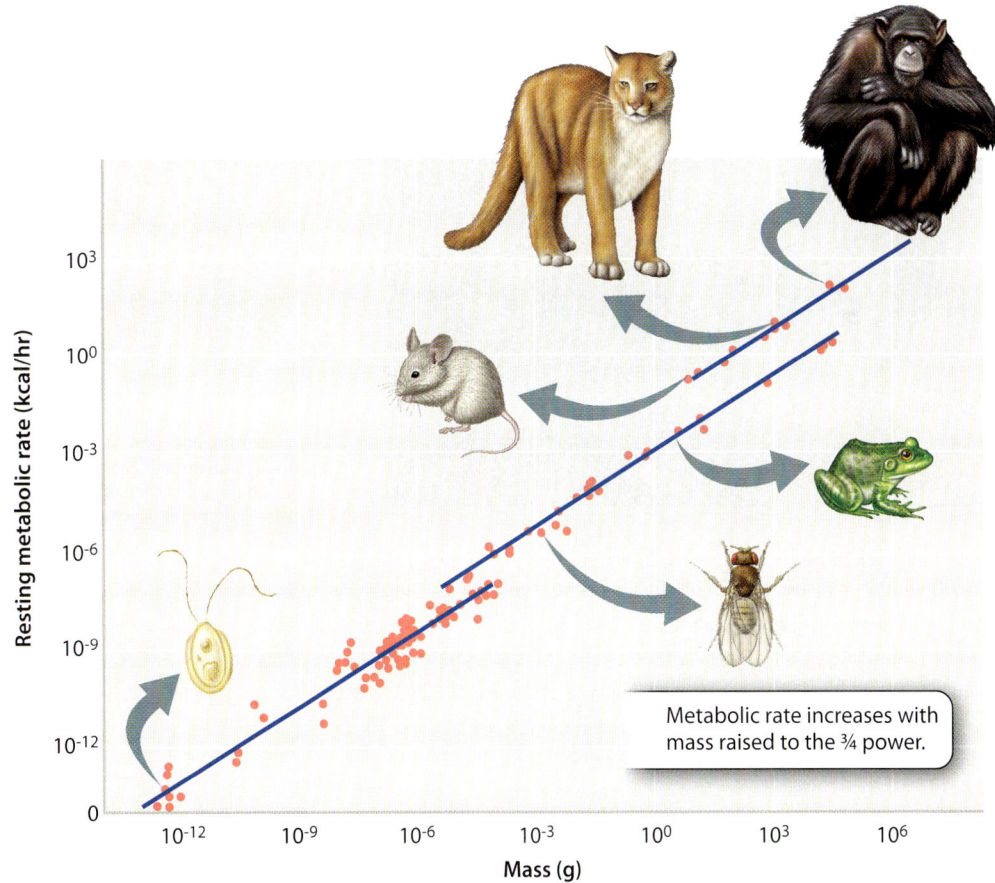

FIG. 40.4 **Resting metabolic rate and body size in diverse organisms.** The relationship plotted here shows that larger animals have lower metabolic rates per gram of tissue than smaller animals.

Metabolic rate increases with mass raised to the ¾ power.

HOW DO WE KNOW?

FIG. 40.5

How is metabolic rate affected by running speed and body size?

BACKGROUND In the 1970s and 1980s, the American physiologist C. Richard Taylor and his colleagues performed studies on the relationship between metabolic rate and running speed in mammals of different sizes. They were interested in understanding the energetic costs of running in different animals.

EXPERIMENT Taylor and colleagues measured oxygen consumption during running for a wide range of organisms. For each, they trained the animals to run on a treadmill at different speeds and measured oxygen consumption using either a face mask or enclosure. Oxygen consumption was used as a measure of metabolic rate.

RESULTS The researchers found that there is a linear increase in metabolic rate with speed in different-sized animals and that larger animals expend less energy per unit body mass to move a given distance compared to smaller ones, as shown on the graph below.

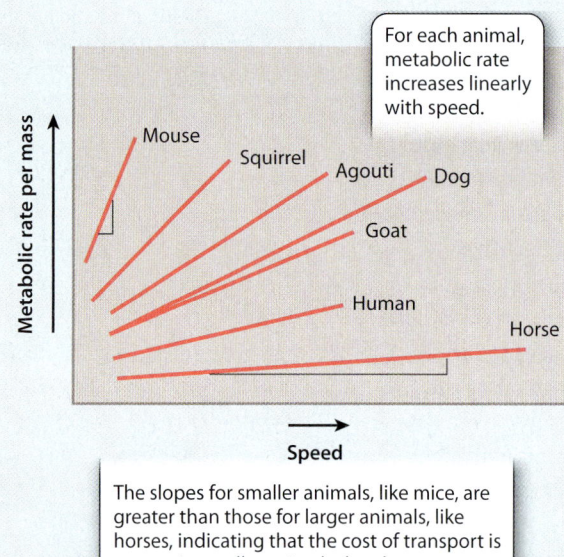

For each animal, metabolic rate increases linearly with speed.

The slopes for smaller animals, like mice, are greater than those for larger animals, like horses, indicating that the cost of transport is greater in smaller animals than larger ones.

FOLLOW-UP WORK Similar studies were performed with kangaroos, which move by hopping rather than running. Interestingly, it was found that it is "cheaper" (that is, it requires less energy) to hop than to run. This finding perhaps explains why kangaroos and other hopping animals survived, while many animals that run on four legs became extinct in Australia 20,000 to 30,000 years ago with the arrival of early humans. Studies have also compared the energetic costs of running on two legs and running on four legs, with the finding that the cost is the same.

SOURCES Dawson, T. J., and C. R. Taylor. 1973. "Energetic Cost of Locomotion in Kangaroos." *Nature* 246:313–314; Taylor, C. R., N. C. Heglund, and G. M. O. Maloiy. 1982. "Energetics and Mechanics of Terrestrial Locomotion." *Journal of Experimental Biology* 97:1–21.

means that the average rate at which each gram of body tissue consumes energy is *less* in larger animals compared to smaller ones. Or, put another way, the larger the organism, the *lower* the metabolic rate per gram of body tissue. This scaling pattern of cellular energy metabolism is remarkable because it holds across a diverse size range of unicellular and multicellular organisms, as can be seen in Fig. 40.4.

In addition to basal metabolic rate, we can also examine how metabolic rate varies with respect to exercise in different-sized animals. In this case, for example, we are looking at how energy is used to allow animals to move at different speeds. The American comparative physiologist C. Richard Taylor has extensively studied how metabolic rate varies with running speed in different-sized terrestrial animals (**Fig. 40.5**). He found that metabolic rate increases linearly with speed, and that larger animals expend less energy per unit mass than smaller ones.

→ **Quick Check 2** Does a dog that is twice as heavy as a cat have twice the resting metabolic rate?

Metabolic rate is linked to body temperature.

Metabolic rate is also affected by an animal's internal body temperature. Temperature affects the rates of chemical reactions, which in turn determine how fast fuel molecules can be mobilized and broken down to supply energy for a cell. Animals can be categorized by the sources of most of their heat. Animals that produce most of their own heat as by-products of metabolic reactions, including the breakdown of food, are **endotherms.** These animals usually, but not always, maintain a constant body temperature that is higher than that of their environment. As we saw earlier (Chapter 35), the maintenance of a constant core body temperature is a form of homeostasis and requires balancing heat production and heat loss. Mammals, including humans, and birds are endotherms.

By contrast, animals that obtain most of their heat from the environment are **ectotherms.** Ectotherms often regulate their body temperatures by behavioral means such as moving into or out of the sun. Think of a turtle or another reptile sunning itself on rock: It is raising its body temperature. The temperatures of these animals usually fluctuate with the outside environment. Most fish, amphibians, reptiles, and invertebrates are ectotherms.

Endotherms are sometimes referred to as "warm-blooded" and ectotherms as "cold-blooded," but these terms are misleading because, depending on environmental conditions, cold-blooded animals can have core body temperatures higher than warm-blooded animals. It should also be kept in mind that these thermoregulatory mechanisms are not completely distinct, but instead represent extremes of a continuum.

Nevertheless, thermoregulatory mechanisms have profound implications for an animal's metabolic rate. Endotherms have a higher metabolic rate than ectotherms. As a result, they are able to be active over a broader range of external temperatures than ectotherms. The activity level and metabolic rate of ectotherms both increase with increasing body temperature. However, ectotherms have metabolic rates that are about 25% of endotherms of similar body mass and at similar body temperatures. Although they can achieve activity levels similar to those of endotherms when their body temperatures are similar, ectotherms cannot sustain prolonged activity. Therefore, ectotherms such as lizards or insects rely on brief bouts of activity, followed by longer periods of inactivity compared to endotherms.

Endotherms benefit from having an active lifestyle. However, their high metabolic rate requires long periods of foraging to acquire enough food to keep warm and be active. By contrast, although ectotherms are limited in their ability to sustain activity, they can survive much longer periods without food.

As we have discussed, the control of core body temperature, or thermoregulation, is a form of homeostasis and requires the coordinated activities of the nervous, muscular, endocrine, circulatory, and digestive systems. We summarize how these systems work together in endotherms and ectotherms in **Fig. 40.6** on the following page.

40.2 ANIMAL NUTRITION AND DIET

Running, jumping, moving, growing, reproducing, and all other activities performed by animals require energy. Autotrophs like plants and certain microbes are able to capture energy from the sun or from inorganic compounds (Chapters 8 and 26), but animals must acquire food in the form of plants or other animals to support their growth, reproduction, and other activities. Food does not consist just of carbohydrates, fats, and proteins, however. To enable various cell processes, animals must also obtain in their diet certain minerals and chemical compounds that they cannot synthesize on their own. In this section, we explore what animals must obtain from their food, and how much of it, in order to sustain their cells and tissues.

Energy balance is a form of homeostasis.

Like core body temperature (Fig. 40.6), blood-glucose levels (Chapter 37), and blood pressure (Chapter 39), the **energy balance** of an organism is often maintained at a constant level.

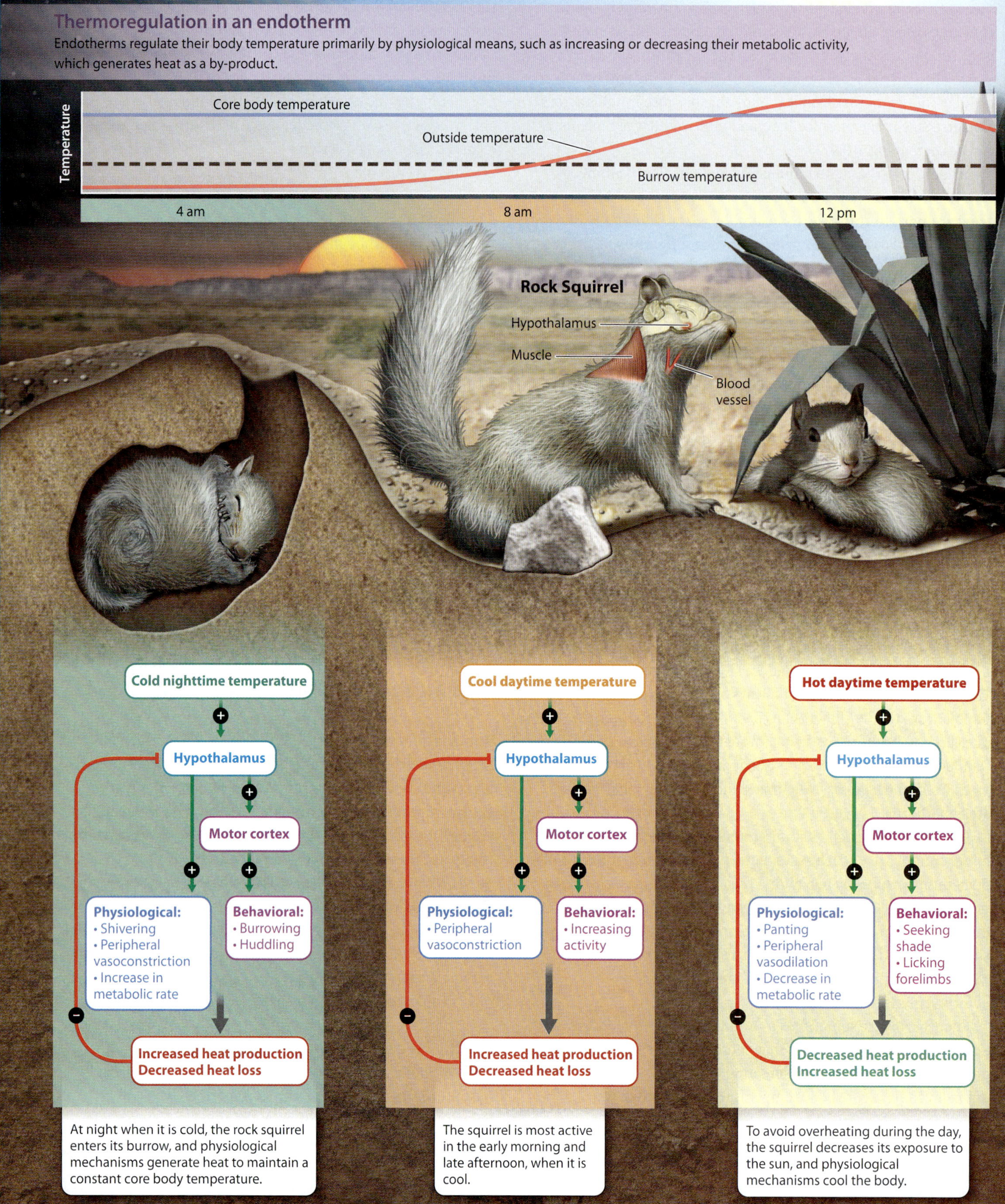

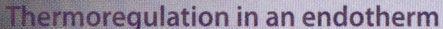

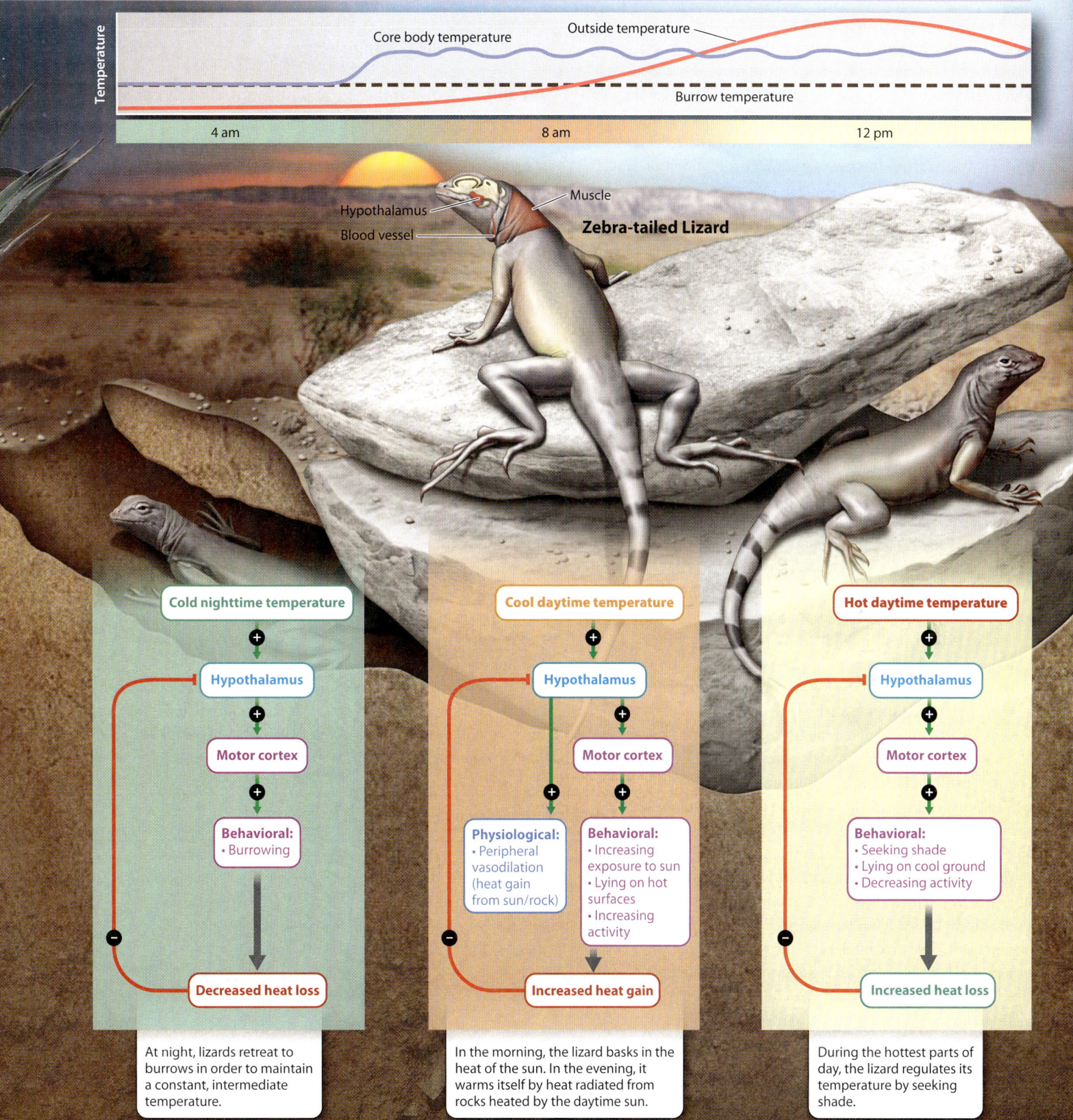

An animal in energy balance takes in the same amount of calories of energy from food that it uses over time to meet its metabolic needs. Energy balance can be thought of as a form of homeostasis. We consider sources of energy, or **energy intake,** and ways in which energy is expended, or **energy use.**

For animals, the source of energy is the diet. In turn, energy is used to do work (maintain tissues, grow, move about, and the like). Energy required for basic life processes accounts for the majority (about 70%) of energy use. The amount of energy used also depends on the level of physical activity, with higher levels of activity using more energy, as we saw. Finally, digestion and absorption of food from the diet itself requires energy.

When energy intake does not equal energy used, there is an energy imbalance. If an animal eats more food than it requires, energy stores such as fat deposits grow over time. The result is that the body shifts its metabolism mostly to anabolic processes that build energy stores. Many animals achieve a net positive energy intake during the late summer and fall when food is plentiful, before it becomes scarce in winter. Other animals maintain a constant energy intake throughout the year by migrating to areas with more abundant food and avoiding colder temperatures that require increased energy expenditure to remain active and warm. Still other animals hibernate, or become less active, to conserve their energy use over the winter when food is scarce or unavailable.

Animals that cannot acquire enough food are in negative energy balance and become undernourished. During prolonged periods of an inadequate food supply or starvation, an animal consumes its own internal fuel reserves. Starvation forces animals to deplete their glycogen and fat reserves first, and then, if no food is found, to resort to protein stores, primarily in muscle tissue. This ultimately leads to muscle wasting. Undernourishment and starvation are particularly serious human health problems in many developing countries, especially those ravaged by war and internal strife.

Humans, like most other animals, store excess food calories as fat. This is because over much of our evolutionary history food was less abundant and was more unpredictable in its availability than it is today. With the development of agriculture and domestication of livestock, food supplies rapidly increased, allowing many human populations to grow and consume increasing amounts of food. With the rise of mechanized agricultural food production and highly processed foods developed by the modern food industry, excessive intake of food calories has led to an increasing and now critical public health problem: obesity.

Obesity is now an epidemic in many industrialized nations, including the United States, where about 36% of the adult population is considered obese. Obesity is a major public health concern because it increases the risk of diabetes, heart disease, and stroke and contributes to a shorter life-span. For most animals, however, acquiring food and storing its products efficiently in the body allow them to have a fuel reserve to meet seasonal energy requirements. This rationing of stored energy remains an essential part of their metabolic and digestive physiology.

An animal's diet must supply nutrients that it cannot synthesize.

Animals' metabolic pathways enable them to obtain energy from the environment as well as to synthesize many of the compounds needed to sustain life. However, many nutrients necessary for life cannot be synthesized by an animal's metabolism and therefore must be acquired in the food that they eat.

Recall that amino acids are the basic units of proteins (Chapter 4). There are 20 different amino acids that are typically found in proteins (see Fig. 4.2). Although all of them are necessary for life, an **essential amino acid** is formally defined as one that cannot be synthesized by cellular biochemical pathways and instead must be ingested. Most animals can synthesize about half of their amino acids. Humans are unable to synthesize 8 of the 20 amino acids (**Table 40.1**). We have to obtain these eight essential amino acids in our diets. Infants require additional amino acids. The most reliable source of all eight essential amino acids is meat. This is not to say that only omnivorous humans can lead healthy lives, though. While most plant proteins lack at least one of the eight amino acids essential to humans, and diets heavy in a particular food, such as corn (which lacks lysine), can lead to protein deficiency, vegetarians can achieve a healthy intake of all essential amino acids by combining plant foods. For example, beans supply the lysine that corn lacks, and corn supplies the methionine that beans lack.

Dietary minerals are chemical elements other than carbon, hydrogen, oxygen, and nitrogen that are required in the diet and must be obtained in the food that an animal eats (**Fig. 40.7**).

TABLE 40.1 Essential Amino Acids for Humans

ESSENTIAL AMINO ACID	COMMON SOURCES
Isoleucine	Meat, dairy, lentils, wheat
Leucine	Meat, cottage cheese, lentils, peanuts
Lysine	Meat, beans, lentils, spinach
Methionine	Meat, dairy, whole grains, corn
Phenylalanine	Meat, dairy, peanuts, seeds
Threonine	Meat, dairy, nuts, seeds
Tryptophan	Meat, dairy, oats
Valine	Meat, dairy, grains

FIG. 40.7 Commonly recognized minerals required by humans. Dietary minerals are chemical elements other than carbon, hydrogen, oxygen, and nitrogen that are required in the diet.

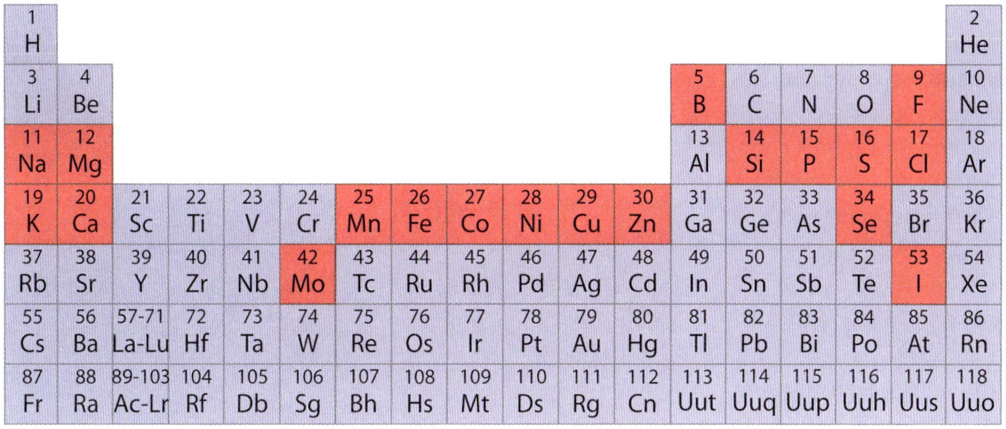

They include such elements as calcium, iron, phosphorus, potassium, zinc, and magnesium. Calcium is required for building skeletons (Chapter 37), and iron in hemoglobin binds oxygen and transports it in the blood (Chapter 39). Humans typically obtain the minerals sodium and chloride, which ionize to form salt crystals, through common table salt and other foods. Many wild animals seek exposed rock that they lick to obtain minerals and salts.

In addition to amino acids and minerals, animals must ingest essential **vitamins,** organic molecules that are required in very small amounts in the diet. Vitamins have diverse roles, some binding to and increasing the activity of particular enzymes, others acting as antioxidants or chemical signals. Different animals have different vitamin requirements, so a molecule that functions as a vitamin for one may not be a vitamin for another. Thirteen essential vitamins have been identified for humans (**Table 40.2**). Knowing which vitamins are required for an animal's health is critical to proper medical and veterinary care. Animals have evolved diets that ensure that their need for vitamins, as well as for essential amino acids and minerals, is met through the combination of foods that they eat.

In humans as well as in other animals, vitamin deficiency can have serious consequences. Whereas most mammals can synthesize ascorbic acid (vitamin C), which is necessary for building connective tissue, primates (including humans) cannot. Without ingesting sufficient vitamin C, humans develop scurvy, a disease characterized by bleeding gums, loss of teeth, and slow wound healing. Until it was discovered that green vegetables and fruit supply vitamin C, scurvy was a prevalent disease that could lead to death for sailors who were deprived of fresh produce for long periods of time while at sea.

Deficiencies of vitamins B_1, B_6, and B_{12} can cause nervous system disorders and various forms of anemia (evidenced by low levels of red blood cells or hemoglobin in the blood). Vitamin D is essential for the absorption of calcium in the diet and thus to skeletal growth and health. With adequate exposure to ultraviolet solar radiation, skin cells synthesize enough vitamin D to sustain a growing body. However, fairer-skinned people inhabiting more northern regions of the world produce less vitamin D and therefore require more of it in their diet. A clear role for vitamin E remains uncertain, but its absence is often linked to anemia.

→ **Quick Check 3** In the context of nutrition, what does "essential" mean?

TABLE 40.2 Essential Vitamins for Humans

VITAMIN DESIGNATION	NAME	COMMON SOURCES
Vitamin A	Retinol	Green and orange vegetables, liver, milk
Vitamin B_1	Thiamine	Pork, grains
Vitamin B_2	Riboflavin	Dairy products, eggs, leafy green vegetables
Vitamin B_3	Niacin	Meat, dairy products, eggs, vegetables
Vitamin B_5	Pantothenic acid	Meat, grains
Vitamin B_6	Pyridoxine	Meat, grains
Vitamin B_7	Biotin	Meat, eggs
Vitamin B_9	Folic acid	Leafy green vegetables, eggs
Vitamin B_{12}	Cobalamin	Liver, eggs
Vitamin C	Ascorbic acid	Fruit, vegetables
Vitamin D	Calciferol	Fish, eggs
Vitamin E	Tocopherol	Fruit, vegetable oils
Vitamin K	Phylloquinone	Leafy green vegetables

40.3 ADAPTATIONS FOR FEEDING

Animals have evolved a variety of adaptations to enhance their ability to acquire energy from the environment. Animals eat other organisms, whether they are other animals, plants, or fungi. Herbivores are adapted to eat plants; carnivores are adapted to eat animals; and omnivores are adapted to eat both. To acquire energy effectively from the food that they eat, animals have evolved a diversity of ways to capture food and mechanisms for breaking down that food before its digestion in the gut.

Suspension filter feeding is common in many aquatic animals.

The most common form of food capture by animals is **suspension filter feeding,** in which water with food suspended in it is passed through a sievelike structure (**Fig. 40.8**). Suspension filter feeding is possible only in aquatic environments, and its ubiquity there speaks to the great diversity of animals found in oceans and freshwater bodies. This form of feeding evolved multiple times independently on many different branches of the tree of life, and so represents a form of convergent evolution (Chapter 23).

Many worms and bivalve mollusks, such as scallops, clams, and oysters, pump water over their gills to trap food particles suspended in the water (Fig. 40.8a). Cells of the gills produce mucus that acts as an adhesive to trap food particles, and the cells' beating cilia sweep food trapped in the mucus to the mouth and gut.

Other aquatic organisms move water with food particles suspended in it through filters in their oral cavity and then convey the food into the gut. Large baleen whales have long comblike blades of baleen (which is made of keratin, the same material that makes up hair) in their mouth to capture small shrimp called krill that are plentiful in ocean waters (Fig. 40.8b).

Large aquatic animals apprehend prey by suction feeding and active swimming.

Instead of relying on suspension filter feeding, many large aquatic animals capture their prey in other ways. Many fish feed by suction. A rapid expansion of the fish's mouth cavity draws water and the desired prey into the mouth (**Fig. 40.9**). After the fish closes its mouth, the water is pumped out of the mouth cavity past the gills. The prey is trapped inside the mouth, moves into the pharynx (part of the throat), and is broken up by specialized pharyngeal jaws (a second set of jaws in the throat) before being

FIG. 40.9   Suction feeding. This form of feeding involves a rapid expansion of the fish's mouth that draws in water and prey.

FIG. 40.8   Suspension filter feeding by (a) a scallop and (b) a baleen whale.

FIG. 40.10 Feeding by active swimming. Large marine predators, like this great white shark, can use their size and speed to capture prey.

vertebrates, the jawed fishes, became dominant in their aquatic environment through the ability to swim and bite forcefully to obtain their food. **Fig. 40.11** shows a phylogenetic tree of vertebrates indicating the appearance of jaws. Jawed fish evolved from jawless ancestors, and jaws can be found in present-day fish, amphibians, reptiles, and mammals. As you can see from the number of jawed groups, jaws are key to the evolutionary success of vertebrates. They are thought to have evolved from cartilage that supported the gills, providing a spectacular example of an organ adapted for one function (gill support) changing over time to become adapted for an entirely different function (predation).

Among vertebrates, mammals evolved a specialized jaw joint, the **temporomandibular joint,** as well as a great diversity of specialized forms of teeth. The temporomandibular joint allowed the teeth of the lower and upper jaws to fit together precisely. This anatomy facilitated the specialization of teeth with

swallowed. Suction feeding is the most generalized form of prey capture among fish, as it allows them to be "sit-and-wait" predators, hiding within a coral reef or under a rock before rapidly striking to capture prey moving in front of them. Suction feeding is effectively used by a wide variety of fish, as well as by aquatic salamanders, and has contributed to their evolutionary diversification and success.

Many insects that bite to obtain a blood meal also rely on suction to draw the blood of their prey into their digestive system. Young mammals also feed by suckling milk from their mother's breast, using their tongue and also generating suction as fish do.

As top predators in marine and freshwater environments, larger fish and marine mammals like sharks, whales, and dolphins actively swim to capture their prey (**Fig. 40.10**). Their larger size and speed enables them to catch smaller or similar-sized prey. As discussed below, the evolution of jaws and teeth considerably enhanced the ability of these and other vertebrate animals to capture greater amounts of food, allowing them to lead a more active and energetically demanding lifestyle.

Jaws and teeth provide specialized food capture and mechanical breakdown of food.

Jaws and teeth were an important evolutionary innovation for active predators. Some of the first

FIG. 40.11 A phylogeny of vertebrates, showing the appearance of jaws. Jaws, which evolved more than 400 million years ago from a group of jawless ancestors, contributed to the success of many groups of animals.

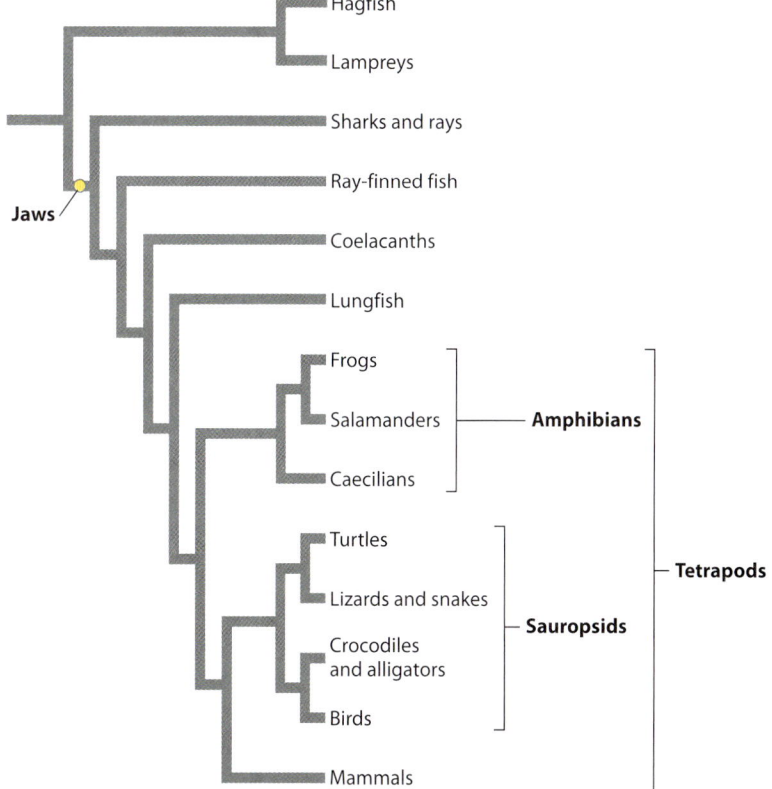

FIG. 40.12 Jaws and teeth of carnivorous, herbivorous, and omnivorous mammals.
Mammals evolved a diverse arrangement of teeth for diverse diets.

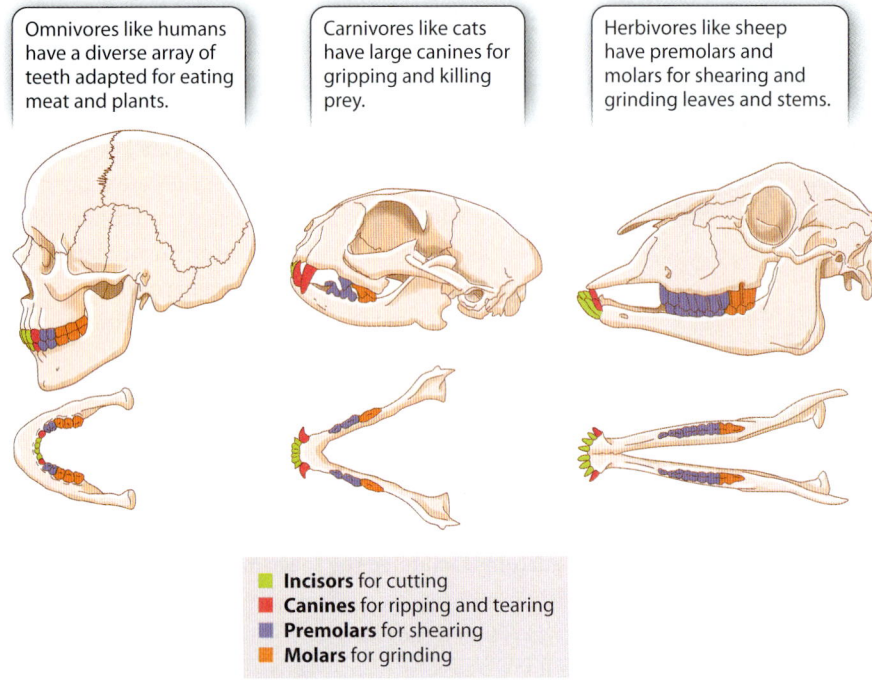

cutting and crushing surfaces, enabling mammals to break down a variety of foods mechanically before swallowing. **Fig. 40.12** illustrates the arrangement of specialized teeth in mammals with different diets. Teeth in the front of the mouth, called **incisors,** are specialized for biting. Others, such as the **canines** of dogs, cats, and other carnivores, are specialized for piercing the body of their prey. Saber-toothed cats, which lived as recently as 10,000 years ago, represent some of the most remarkable examples of animals with long canines (up to 16 inches or 40 cm!) specialized for killing large prey. **Molars** and **premolars** are teeth in the back of the mouth that are specialized for crushing and shredding tougher foods, such as meat and fibrous plant material.

Herbivorous animals, such as cattle, sheep, and horses, have specialized molars and premolars with prominent surface ridges. These teeth enable them to shred the tough plant material that they eat before it is swallowed and digested. Mammalian herbivores use their front incisors and canines to bite grasses and leaves, and then move the food to the rear of the mouth, where it is ground and crushed between their ridged premolars and molars.

40.4 DIGESTION AND ABSORPTION OF FOOD

After animals obtain food, they break it down through the process of digestion. Digestion requires that the food be isolated in a specialized compartment so that it can be broken down chemically without damage to other cellular organelles or structures of the body. Organisms on different branches of the tree of life have evolved very different means of isolating and digesting food. The process in single-celled protists, which obtain food particles by phagocytosis, is **intracellular digestion,** in which food is broken down within cells. Lysosomes containing hydrolytic enzymes fuse with vacuoles containing food particles, so the enzymes are mixed with the food (Chapter 5). Following their chemical breakdown, the food products of intracellular digestion are absorbed into the cell for use. In most animals, however, the process is **extracellular digestion,** in which food is isolated and broken down outside a cell, in a body compartment.

The digestive tract has regional specializations.

Extracellular digestion involves a specialized cavity or body compartment in which food is broken down. Some animals, like sea anemones, have an internal cavity in which digestion occurs. Cells lining this cavity secrete digestive enzymes that break down the food, allowing the smaller chemical compounds to be absorbed by the cells.

Other animals have more elaborate digestive systems that allow the transport of food by a digestive tube that runs from an animal's mouth to its anus. Collectively, the passages that connect the mouth, digestive organs, and anus constitute an animal's **gut** or **digestive tract.** Because the food is moved in a single direction through a tubelike gut, particular regions of the digestive tract can be specialized for different functions. These functions include storage, chemical breakdown of different kinds of food, absorption of released nutrients, and elimination of waste products.

The digestive tracts of animals are commonly divided into a **foregut, midgut,** and **hindgut** (**Fig. 40.13**). Most animals' digestive systems have the same basic plan, though some animals have also evolved specialized structures. The foregut includes the **mouth, esophagus,** and **stomach** or **crop,** which serves as an initial storage and digestive chamber. Next is the midgut, which includes the **small intestine,** where the remainder of digestion and most nutrient absorption takes place. Here, specialized organs secrete enzymes and other chemicals that aid in the breakdown of particular macromolecules, such as fats and carbohydrates. Finally, the digested material reaches the hindgut, which includes the **large intestine** and **rectum.** Within the hindgut, water and minerals are reabsorbed, leaving

FIG. 40.13 Organization of animal digestive tracts: foregut, midgut, and hindgut.

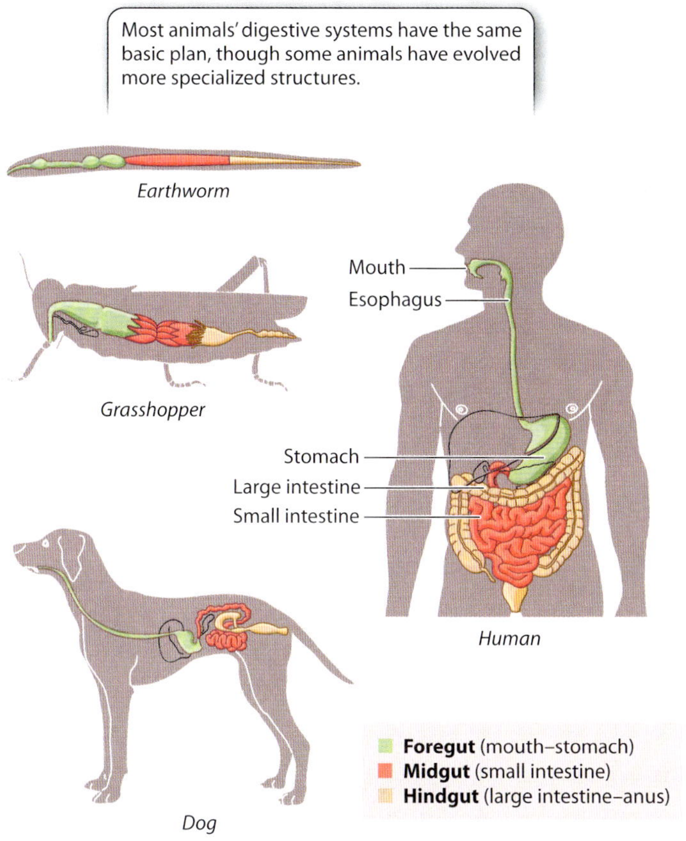

Most animals' digestive systems have the same basic plan, though some animals have evolved more specialized structures.

- **Foregut** (mouth–stomach)
- **Midgut** (small intestine)
- **Hindgut** (large intestine–anus)

the waste products, or **feces,** which are stored in the rectum until being eliminated from the body. Elimination of the feces, or **defecation,** involves relaxation of an anal sphincter combined with muscular contraction of the rectum and body wall.

Digestion begins in the mouth.

Most animals jump-start digestion by breaking down food mechanically. After ingesting food, many animals manipulate the food in the mouth or **buccal cavity.** As we saw in the previous section, many animals break down food with the aid of jaws and teeth. Insects, which lack jaws and teeth, manipulate and break down their food with mouthparts called mandibles before it passes into their gut. In some animals, such as mammals, food entering the mouth is mixed with salivary secretions that contain **amylase,** an enzyme that breaks down carbohydrates, to begin the digestion of sugars and starches.

Mammals and other land vertebrates have a muscular **tongue** that facilitates food manipulation and transport within the mouth cavity. The tongue moves the food into position within the mouth for effective cutting and maceration into smaller pieces. When a bolus of food has been chewed and is ready to be swallowed, the tongue moves it to the rear of the mouth cavity.

Swallowing, which is controlled by the autonomic nervous system, is a complex set of motor reflexes that involves several muscles and structures in the rear of the mouth and the **pharynx,** the region of the throat that connects the nasal and mouth cavities (**Fig. 40.14**). Once initiated, swallowing reflexes

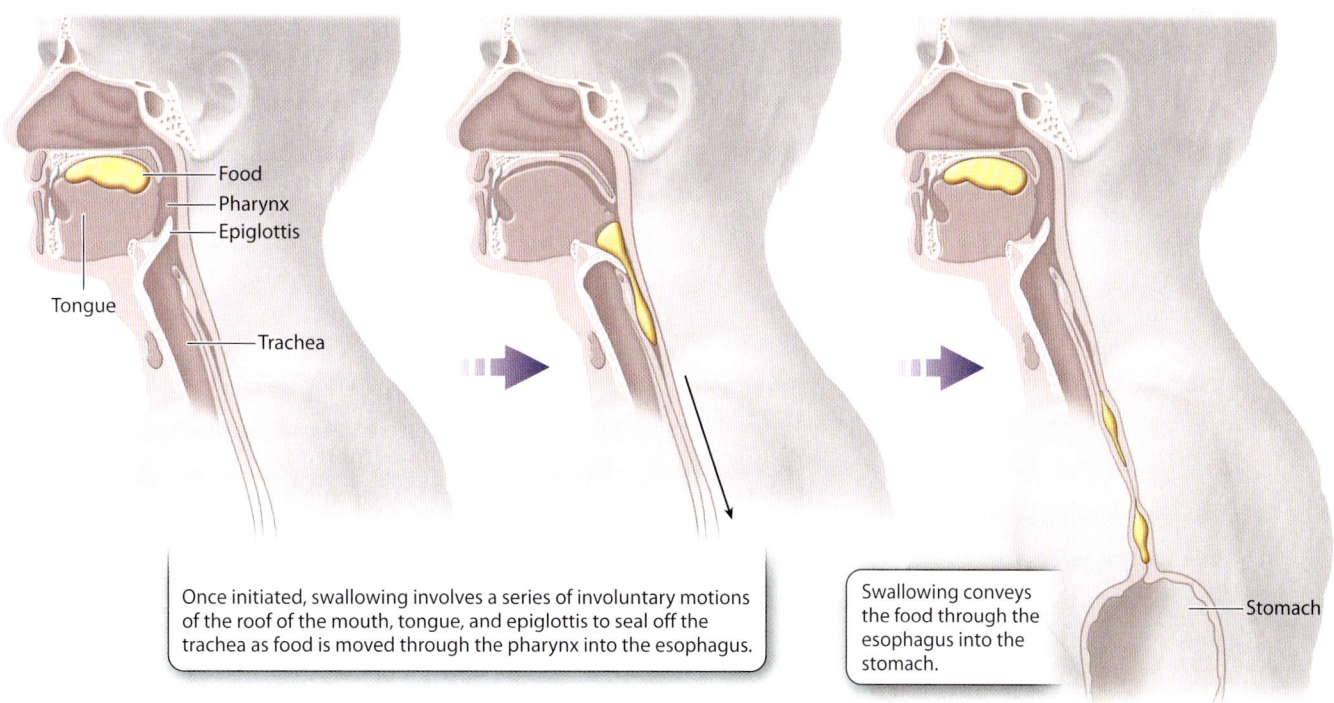

FIG. 40.14 Swallowing. Swallowing moves food from the mouth to the stomach.

Once initiated, swallowing involves a series of involuntary motions of the roof of the mouth, tongue, and epiglottis to seal off the trachea as food is moved through the pharynx into the esophagus.

Swallowing conveys the food through the esophagus into the stomach.

occur without voluntary control. During swallowing, food is pushed through the pharynx, over the **epiglottis,** a flap of tissue that protects food from entering the trachea and lungs, and into the esophagus, the tube that connects the pharynx to the stomach.

Birds, alligators, crocodiles, and earthworms break down food into smaller pieces further along their digestive tracts in the **gizzard,** a compartment with thick muscular walls. Birds and earthworms often ingest small rocks or sediment to help grind the food into smaller pieces within the muscular gizzard. Breaking food into smaller sizes aids the chemical digestion that follows.

→ **Quick Check 4** The rocks and sediments that break apart food mechanically play a role similar to what structures in vertebrates?

The stomach is an initial storage and digestive chamber.

After being ingested and mechanically broken down, the food is transported into a stomach or crop (see Fig. 40.13). Stomachs and crops further fragment and mix the food with digestive enzymes, storing the food in a partially digested form before releasing it for further digestion and nutrient absorption, as needed by the animal.

The stomach is one of the main sites of protein breakdown. The sight, smell, and taste of food send signals to the brain that stimulate a sense of appetite, and to the stomach, stimulating the secretion of hydrochloric acid (HCl) and digestive enzymes that break down proteins into amino acids, which are then absorbed into the bloodstream. Cells secrete these digestive enzymes in an inactive form; otherwise, they would be digested themselves. After secretion, the inactive enzymes are activated by a change in

FIG. 40.15 Control of digestion in the stomach.

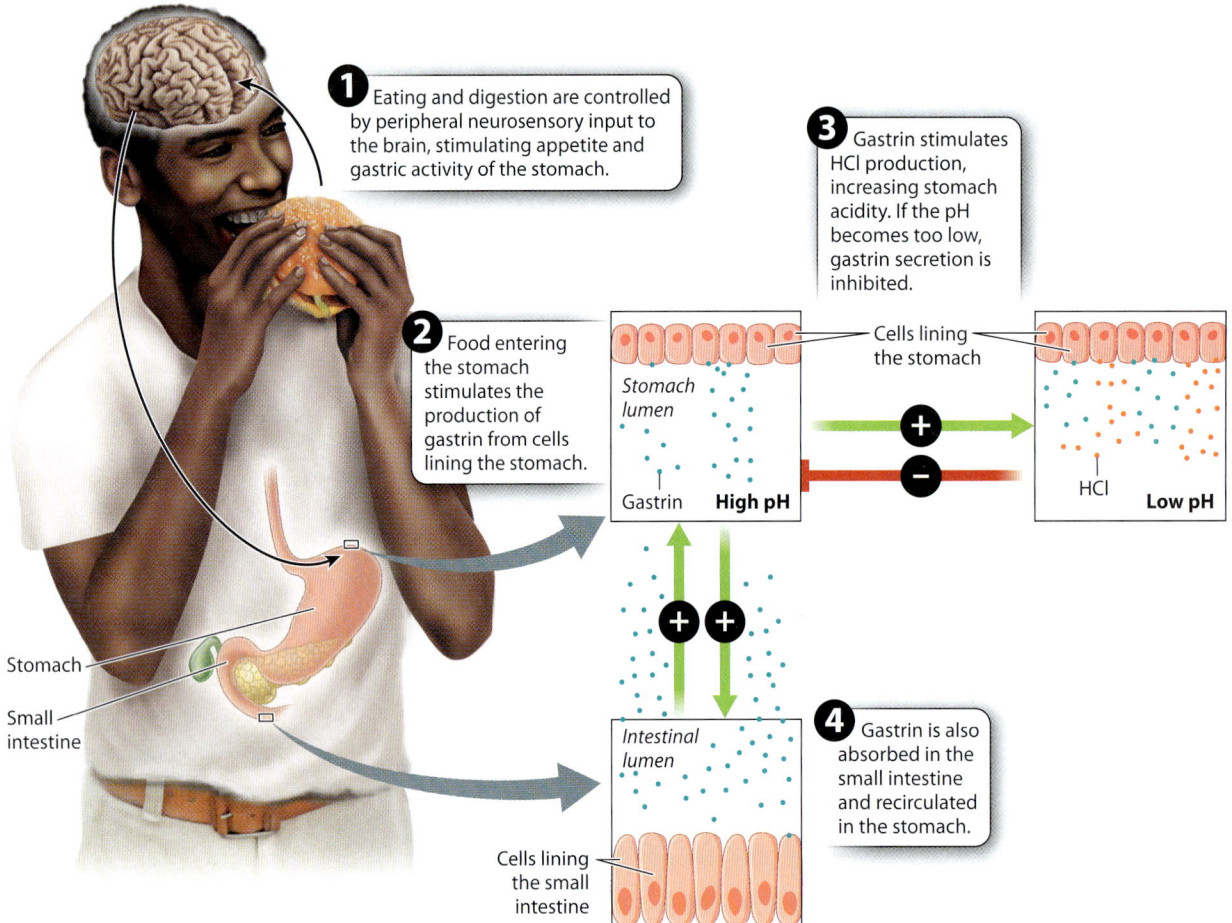

their chemical structure. The primary digestive enzyme produced in the stomach is **pepsin,** an enzyme that breaks down proteins. Cells lining the stomach release pepsin in an inactive form called pepsinogen that is activated by the low pH of the stomach, which is in the range of pH 1–3. The low pH also aids active pepsin in cleaving bonds between amino acids of the peptides. Finally, the acidic environment and digestive enzymes of the stomach kill most pathogens that may be ingested with the food.

In addition to digestive enzymes and HCl, the cells lining the stomach secrete a peptide hormone called **gastrin** in response to food in the stomach (**Fig. 40.15**). Gastrin stimulates the cells lining the stomach to increase their production of HCl. Gastrin secreted by the stomach is absorbed within the small intestine and recirculated to the cells lining the stomach wall. There, the recirculated gastrin stimulates the stomach lining cells to increase their production of gastrin, forming a positive feedback loop (Chapter 38). The feedback loop, starting with the secretion of pepsin to the amplified secretion of gastrin, ensures that protein digestion occurs in response to the start of a meal as food enters the stomach. If the pH of the stomach becomes too low, gastrin secretion is inhibited by negative feedback control.

Glands in the lining of the stomach secrete mucus to protect the stomach wall from the acid its cells secrete to digest food. Sometimes, this defense can break down, resulting in an **ulcer,** a break or erosion in the cells lining the stomach. Although stress and oversecretion of digestive juices can cause an ulcer, Australian pathologists Barry Marshall and Robin Warren made the remarkable discovery that most ulcers result from an infection by the bacterium *Heliobacter pylori*, which can survive in extremely acidic environments. Ulcers produced by *H. pylori* are worsened by the stomach's secretions of HCl and pepsin. The two researchers won the Nobel Prize in Physiology or Medicine in 2005 for their work.

The stomach walls contract to mix the contents of the stomach, aiding their digestion. Waves of muscular contraction, called **peristalsis,** move the food toward the base of the stomach. There, the **pyloric sphincter,** a band of muscle, opens and allows small amounts of digested food to enter the small intestine. Opening and closing of the pyloric sphincter regulates the rate at which the stomach empties, allowing time for digestion and absorption of the food products released into the small intestine.

Animals often eat large amounts of food in a short time, but it takes much longer to digest that food. The stomach serves as a storage compartment as well as a digestion compartment. In humans, it typically takes about four hours for the stomach to empty, allowing digestion and absorption of nutrients to occur between meals. In other animals, stomach emptying can take

FIG. 40.16 Python and prey. Because the prey (here, a duck) is eaten whole, the python's metabolic rate is high during digestion, which takes place over several days.

much longer. Most carnivorous animals consume large and infrequent meals, resulting in long periods of digestion and nutrient absorption compared with humans and other animals that eat more regularly. Snakes such as pythons that engulf large prey whole (**Fig. 40.16**) elevate their metabolic rate to high levels during digestion and spend several days digesting and absorbing the nutrients from their meal. This requires an extensive remodeling of their gut to produce new secretory and absorptive cell surfaces to digest their single large meal. These extra cells are not retained between meals because of the high energy cost of maintaining them.

The small intestine is specialized for nutrient absorption.

After the food enters the small intestine, protein and carbohydrate digestion continue, and fat digestion begins. The small intestine is so named because of its small diameter, but it is a long (about 20 feet or 6 m in an adult human) and therefore large organ. Because of its length, the small intestine is coiled to fit within the abdominal cavity. The small intestine is made up of three sections (**Fig. 40.17**). The initial section is the **duodenum,** into which the food enters from the stomach. Most of the remaining digestion takes place in the duodenum. The next two sections, the **jejunum** and **ileum,** carry out most nutrient absorption.

Two accessory organs, the **liver** and **pancreas,** aid in the digestion of proteins, carbohydrates, and fats in the duodenum of the small intestine. The liver produces **bile,** which is composed of bile salts, bile acids, and bicarbonate ions. Bile

FIG. 40.17 The small intestine and associated organs. Digestion in the small intestine is aided by secretions from the liver, gallbladder, and pancreas. The small intestine is the main site of nutrient absorption.

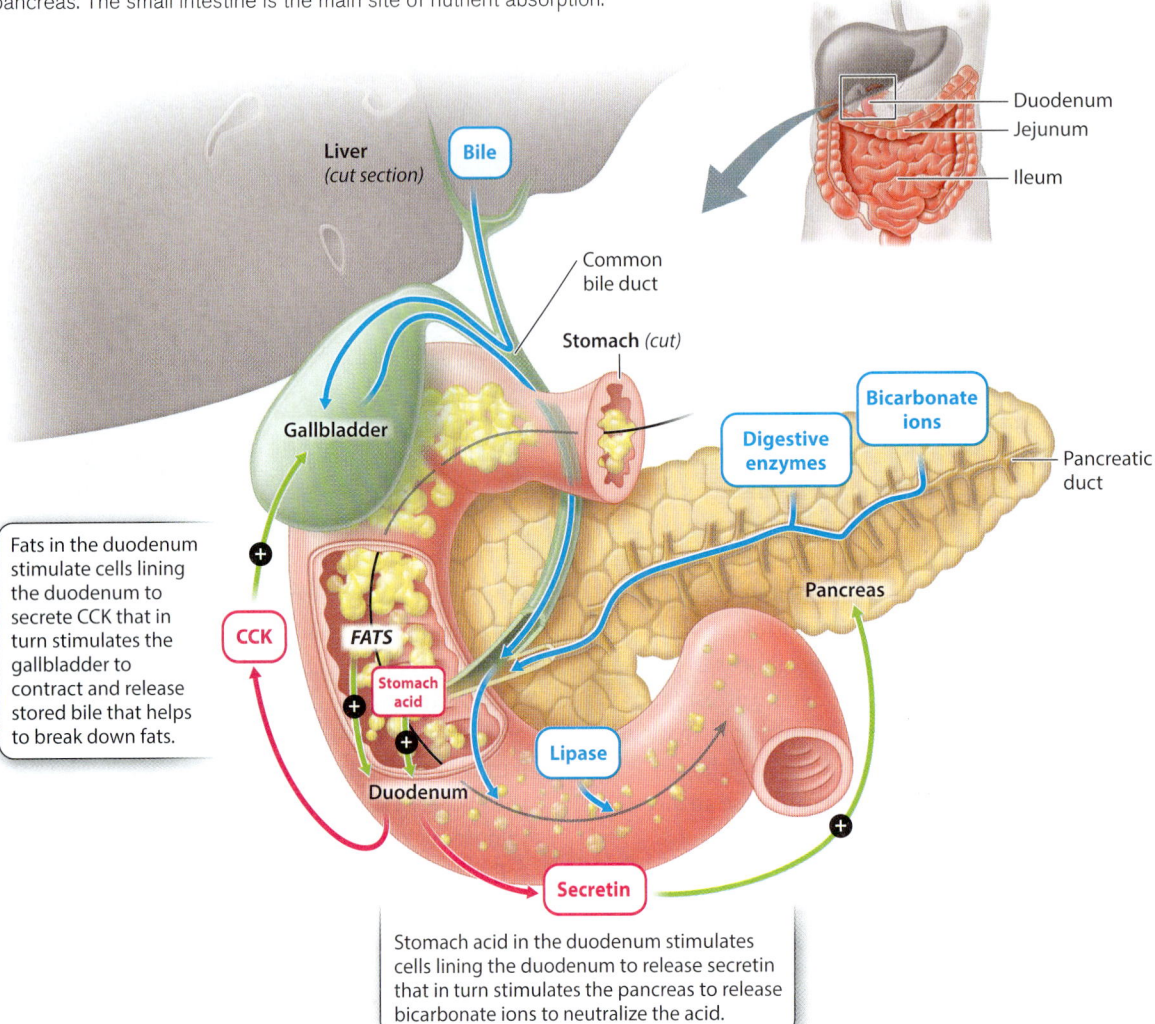

aids in fat digestion by breaking large clusters of fats into smaller lipid droplets, a process called emulsification. Bile produced by the liver is stored in the **gallbladder** (Fig. 40.17). When fats enter the duodenum, cells lining the duodenum release a peptide hormone called **cholecystokinin (CCK)**, which causes the gallbladder to contract, thus releasing the bile into the duodenum. Bile in turn breaks up the fats into smaller droplets, allowing **lipases,** enzymes that break apart lipids such as triacylglycerol, to digest the lipids more effectively, breaking them down into free fatty acids and glycerol that can be readily absorbed across the wall of the small intestine and into the bloodstream.

The pancreas is a secretory gland that lies below the stomach, posterior to the duodenum (Fig. 40.17). It functions as both an endocrine gland, secreting hormones directly into the blood as discussed in Chapter 38, and an exocrine gland, secreting substances into ducts that connect to the duodenum. The pancreas produces a variety of digestive enzymes, including lipase, which breaks down fats, and **trypsin,** which further breaks down proteins. Like pepsin produced by the stomach, the pancreas produces trypsin in an inactive form called trypsinogen to avoid digesting itself.

The pancreas also secretes bicarbonate ions, which neutralize the acid produced by the stomach. Bicarbonate secretion by the pancreas is stimulated by the hormone **secretin** (the first hormone ever to be identified), which is released by cells lining the duodenum in response to the acidic pH of the stomach contents entering the small intestine. Enzymes that

FIG. 40.18 Intestinal villi and microvilli. These structures increase the surface area for absorption of nutrients.

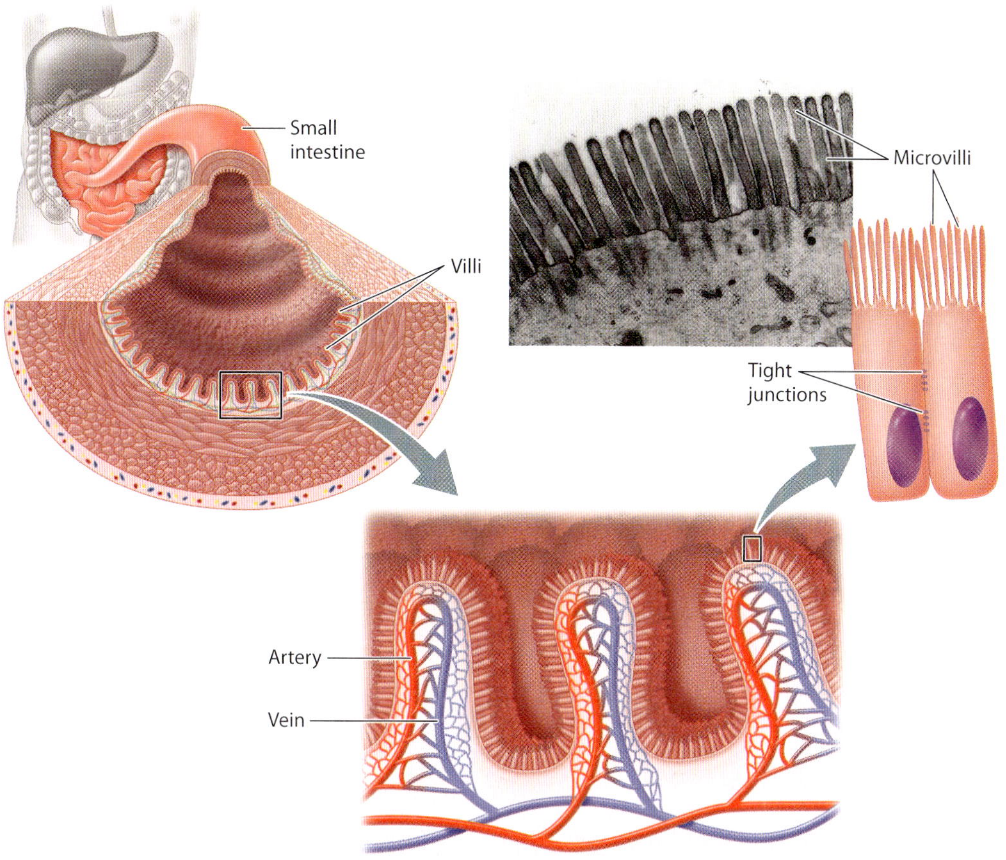

break down proteins work best in acidic environments like that of the stomach, but most proteins, like those that break down carbohydrates and fats, are denatured (that is, unfolded and thus made inactive) in acidic environments. In order to digest carbohydrates and fats, the stomach acid is neutralized by the secretion of bicarbonate ions by the pancreas, which then allows the enzymes that digest carbohydrates and fat to operate.

Final digestion of proteins and carbohydrates occurs in the jejunum and ileum. These two sections of the small intestine have highly folded inner surfaces, called **villi,** and the cells constituting the villi themselves have highly folded surfaces, called **microvilli (Fig. 40.18)**. Together, villi and microvilli greatly increase the surface area for the absorption of nutrients. Maximizing surface area by folding into villi and microvilli is much like the way the lung maximizes surface area for gas exchange with alveoli (Chapter 39). Cells that line the small intestine are connected by **tight junctions** (Fig. 40.17 and Chapter 10). These junctions force the products of digestion to be absorbed across the microvilli surfaces, controlling their movement through the cell and into the bloodstream, rather than leaking between the cells.

The microvilli of the cells lining the small intestine secrete enzymes that cleave peptides into amino acids, which can then be absorbed across the plasma membrane. They also secrete enzymes that break sugars into their subunits, which can also be readily absorbed. Lactose intolerance can arise in humans if they stop producing lactase, an enzyme that normally breaks lactose into its component parts. This can be a critical problem for infants who fail to produce lactase, as milk is their principal source of nutrition. Lactose intolerance is readily treated by lactase supplements to the diet.

Nutrient molecules, such as glucose and amino acids, are often co-transported into cells lining the intestine with sodium ions from gut contents. The sodium ion binds to a transmembrane protein on the surface of cells lining the intestine that also binds the nutrient molecule. Absorption of the nutrient molecule into the cell is driven by the movement of the sodium ion down its

FIG. 40.19 Glucose absorption by cells of the small intestine. Absorption of glucose into the cell is driven by the movement of the sodium ions down a concentration gradient.

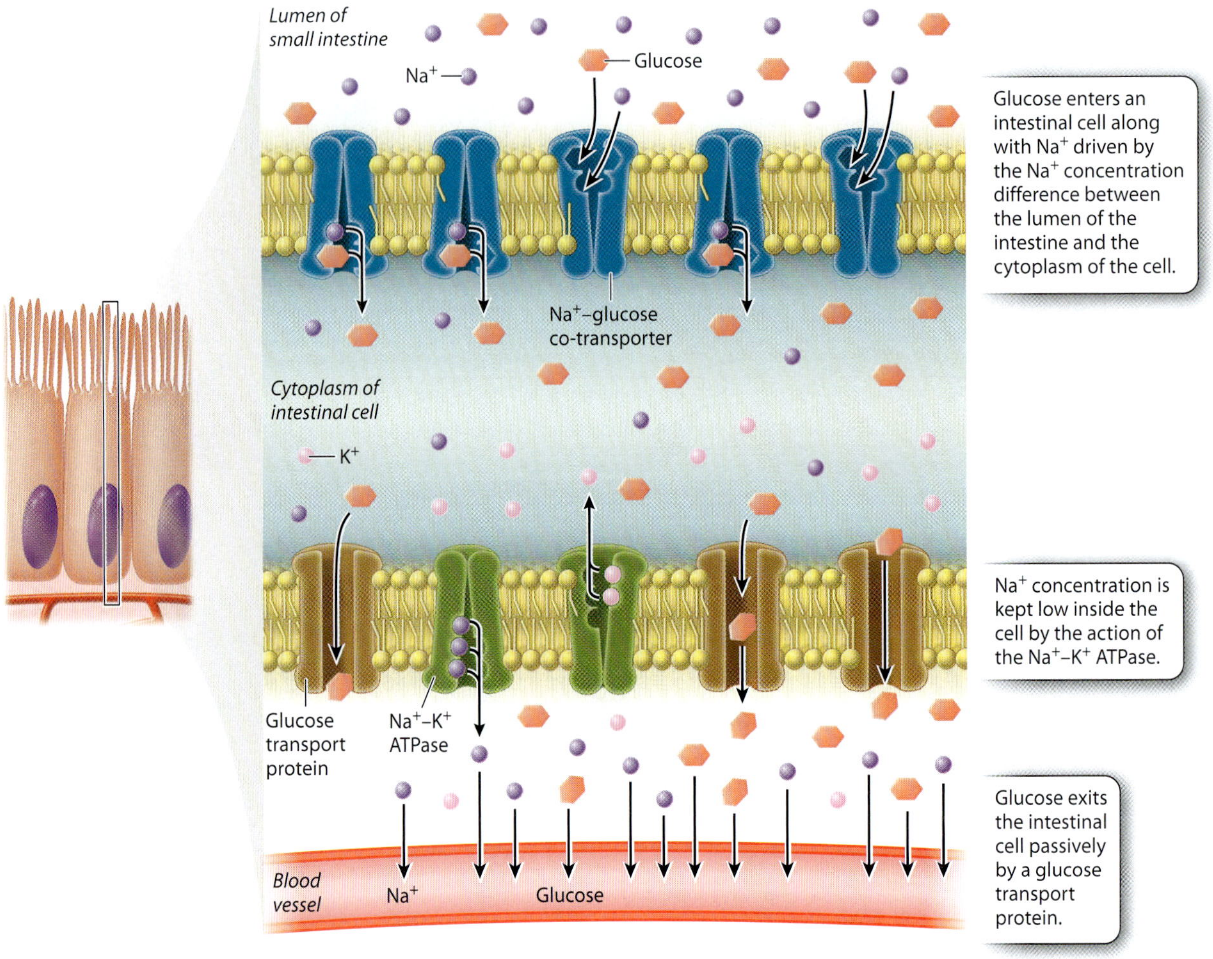

Glucose enters an intestinal cell along with Na⁺ driven by the Na⁺ concentration difference between the lumen of the intestine and the cytoplasm of the cell.

Na⁺ concentration is kept low inside the cell by the action of the Na⁺–K⁺ ATPase.

Glucose exits the intestinal cell passively by a glucose transport protein.

concentration gradient (**Fig. 40.19**). The concentration of sodium ions inside the cell is kept low by the action of the sodium-potassium pump (Chapter 5). Molecules such as glucose are then moved from the intestinal cell into the blood by a pump on the basal side of the cell. Because the products of fat digestion—fatty acids and glycerol—are lipid soluble, they do not require a carrier protein for transport. Instead, they readily diffuse across the lipid plasma membrane, entering the mucosal cell, where they are further broken down and transported to the bloodstream.

The large intestine reabsorbs water and stores waste.

After the absorption of nutrients in the small intestine, water and mineral ions are reabsorbed in the large intestine, or **colon,** of the hindgut (see Fig. 40.13). Peristalsis moves the contents of the small intestine into the colon at a pace that allows time for nutrient absorption in the small intestine. By the time the gut contents reach the large intestine, the nutrients have been absorbed into the body, but water and inorganic ions remain. These are absorbed in the large intestine until the contents form semisolid feces. Excess water absorption can cause constipation, whereas too little water reabsorption results in diarrhea. Diarrhea and constipation also can be caused by toxins released by microorganisms, such as the bacterium *Vibrio cholera*, the cause of cholera, in the food an animal eats. Feces are stored in the final segment of the colon, the **rectum,** until they are periodically eliminated through the anus.

As we saw in Case 5: The Human Microbiome, digestion is not a solo endeavor. Large populations of bacteria reside in the small and large intestine and help extract nutrients that the animal's body cannot extract itself. The principle gut resident is *Escherichia coli*, the bacterium commonly used for research in molecular biology laboratories. In the process of nourishing themselves by aiding in the digestion of the host's gut contents, the bacteria provide nutrients and certain vitamins, such as biotin and vitamin K, that the animal cannot produce itself. Vitamin deficiency therefore can sometimes result from prolonged antibiotic medication that kills large numbers of gut bacteria. Another by-product of bacterial action in the gut is gases such as methane and hydrogen sulfide. These are expelled when foods contain compounds that the bacteria can digest but which cannot be digested and absorbed by the host animal, as when humans eat beans. The mutual benefits to the host animal and the bacteria ensure the success of their symbiotic relationship.

The lining of the digestive tract is composed of distinct layers.

As we have seen, the digestive tract is not a passive tube. It secretes enzymes and other chemical compounds, absorbs nutrients, and actively moves food through the body. How does a series of tubes accomplish all of this? As shown in **Fig. 40.20,** the digestive tract is made up of several layers of tissue, each with a specialized function. The central space, or **lumen,** through which the gut contents travel is surrounded by an inner tissue layer, the **mucosa,** which has secretory and absorptive functions. The cells of the mucosa secrete mucus to protect the gut wall from digestive enzymes, and, in the stomach, hydrochloric acid. Surrounding the mucosa is the **submucosa,** a layer containing blood vessels, lymph vessels, and nerves.

Outside these layers are two smooth muscle layers (Fig. 40.20). An inner **circular muscle** layer contracts to reduce the size of the lumen. An outer **longitudinal muscle** layer contracts to shorten small sections of the gut. These two muscle layers contract alternately to mix gut contents, acting as a traveling wave during peristalsis to move the contents along the digestive tract from compartment to compartment. Between the two muscle layers are autonomic nerves that control the contractions of the two sets of smooth muscle.

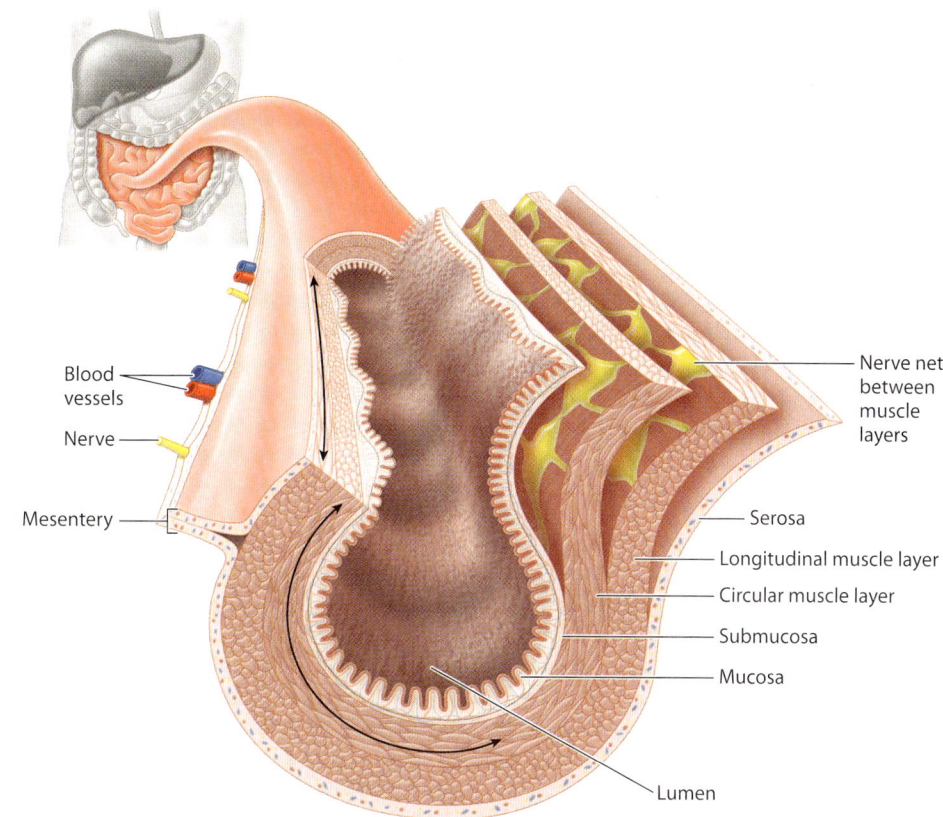

FIG. 40.20 Cross section of the mammalian intestine.

An outer layer of cells and connective tissue called the **serosa** covers and protects the gut (Fig. 40.20). The gut is supported in the abdominal cavity by a membrane called the **mesentery,** through which blood vessels, nerves, and lymph travel to supply the gut.

Plant-eating animals have specialized digestive tracts that reflect their diets.

Herbivorous (that is, plant-eating) animals evolved specialized digestive tracts that enhance their ability to digest cellulose and other plant compounds. Most herbivores, such as cows and termites, lack **cellulase,** the enzyme that breaks down cellulose. Instead, they have specialized compartments in their digestive tract that contain large bacterial populations, which do produce cellulase. The bacteria are able to digest the cellulose, a main constituent of plants. This association of herbivores and their gut microbes is another example of a symbiotic relationship.

Rather than the single stomach of other mammals, ruminants (cattle, sheep, and goats) have a four-chambered stomach that is highly specialized to enhance the ability of their gut bacteria to digest plants (**Fig. 40.21a**). The first two chambers—the **rumen** and the **reticulum**—harbor large populations of anaerobic bacteria that break down cellulose. To facilitate breakdown, the rumen contents are frequently regurgitated (in a process called rumination) for further chewing in the mouth. This repeated chewing decreases the size of the plant fibers, increasing the surface area available for bacterial digestion in the rumen. The bacteria break down cellulose by fermentation, and some of the products are used as nutrients by the host. (The site of fermentation provides the name "foregut fermenters" for these mammals.) Carbon dioxide and methane gas are also produced as a result of bacterial fermentation. The amount of methane produced by domesticated ruminants is second only to the production of industrial methane. Since methane is a greenhouse gas that traps heat from the sun, methane production by cattle contributes to atmospheric increases in methane and global climate change (Chapter 25).

After leaving the reticulum, the mixture of food and bacteria passes into a third chamber, the **omasum,** where water is reabsorbed. It then enters the **abomasum,** the last of the four chambers, where protein digestion begins as it does in other mammals. The acid and protease secretions kill the bacteria, and the nutritional products are absorbed in the small intestine. The bacteria are an important protein source for the ruminant, yielding as much as 100 g of protein each day for a cow. This is important because plants are a poor source of protein but contain considerable nitrogen that the bacteria use to synthesize their own amino acids. The bacteria lost to digestion are balanced by the continual reproduction of bacteria in the rumen.

In contrast to the foregut fermentation of cows and sheep, other mammalian herbivores, such as koalas, rabbits, and horses, digest the plant material they eat by hindgut fermentation (**Fig. 40.21b**). Hindgut fermentation occurs in the colon and in the **cecum,** a chamber that branches off the large intestine. Because the fermentation products released from the cecum of a koala or horse have already passed through the small intestine, the main site of nutrient absorption, hindgut fermentation is less efficient than foregut fermentation in terms of nutrient extraction.

In humans and other animals such as lemurs and squirrels, the cecum is present but small and includes the **appendix,** a narrow,

FIG. 40.21 Digestive systems of (a) a foregut fermenter and (b) a hindgut fermenter.

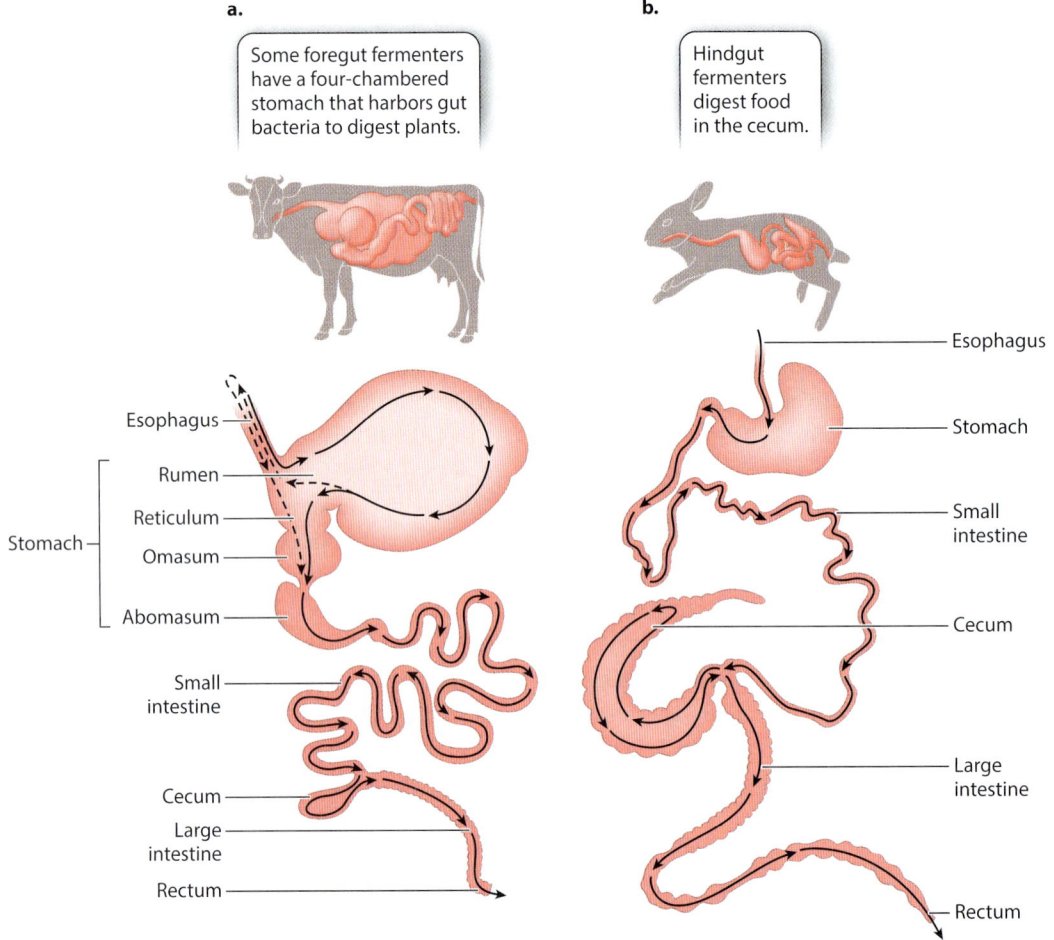

tubelike structure that extends from the cecum. The appendix is an example of a **vestigial structure,** a structure that has lost its original function over time and is much reduced in size. Although the appendix has no clear function in nonherbivorous animals, evidence suggests it may play a role in the immune system. The appendix can cause damage to the organism if it becomes infected by partially digested material falling into it. Unless the infected appendix is removed, it can burst, releasing gut contents into the abdominal cavity, a condition that can be life threatening.

Charles Darwin put forth a hypothesis explaining how the appendix might have evolved to take on its current form in humans. He suggested that in the ancestors of modern humans the cecum once harbored bacteria that aided in the digestion of plant material. However, as their diet changed, they became less reliant on plant material such as cellulose to meet their nutritional needs, so the cecum became smaller and lost its function over time, becoming what we see today as the appendix. Vestigial structures, such as the appendix in humans, remnants of hind limbs in whales, small eyes in cave-dwelling animals, and wings in flightless birds, provide an illuminating window into the past and strong evidence for evolution.

→ **Quick Check 5** Some animals, such as rabbits, whose digestive systems include hindgut fermentation, re-ingest the products of digestion. What function do you think this serves?

Core Concepts Summary

40.1 METABOLIC RATE DEPENDS ON LEVEL OF ACTIVITY, BODY SIZE, AND BODY TEMPERATURE.

Animals acquire carbohydrates, fats, and protein from their diet for energy to fuel cellular processes and activities. page 40-1

Carbohydrates are broken down rapidly in the cytosol by anaerobic glycolysis to produce a small amount of ATP in the absence of oxygen. page 40-2

Carbohydrates and fats produce large amounts of ATP by aerobic respiration through the citric acid cycle and oxidative phosphorylation. page 40-2

An animal's metabolic rate is its overall rate of energy use. page 40-3

Metabolic rate increases with increased physical activity. page 40-3

As the end product of anaerobic glycolysis, lactic acid decreases cellular pH and limits sustainable exercise. page 40-4

Larger animals consume more energy and have higher metabolic rates than smaller ones, but the increase in metabolic rate is not proportional to mass. page 40-5

Endothermic animals maintain elevated body temperature by metabolic heat production and have high metabolic rates. page 40-7

Ectothermic animals depend on external heat sources to warm their bodies. They have lower metabolic rates compared to endothermic animals. page 40-7

40.2 AN ANIMAL'S DIET SUPPLIES THE ENERGY IT NEEDS FOR HOMEOSTASIS AND ESSENTIAL NUTRIENTS IT CANNOT SYNTHESIZE ON ITS OWN.

An animal is in energy balance when the amount of energy that it takes in is equal to the amount it uses to sustain life and perform its functions. page 40-10

Animals must ingest essential amino acids, minerals, and vitamins, those that cannot be synthesized in their cells. page 40-10

40.3 DIFFERENT ANIMALS HAVE DIFFERENT ADAPTATIONS FOR FEEDING.

Aquatic animals take in food by suspension filter feeding, suction feeding, and active swimming. page 40-12

Vertebrates have specialized jaws and teeth for biting and mechanically processing food. page 40-13

Carnivorous animals have enlarged canines and slicing molars for catching prey and eating meat. page 40-13

Herbivorous animals have molars and premolars with surface ridges that enable them to grind plant matter. page 40-13

40.4 THE DIGESTIVE TRACT IS A TUBELIKE STRUCTURE WITH REGIONS SPECIALIZED FOR DIFFERENT FUNCTIONS.

The animal digestive tract consists of a series of compartments linked by a gut tube that allows for spatial and temporal control of digestion and absorption of nutrients. page 40-14

Most animals break down their food into small pieces in the mouth, making it easier to digest. page 40-14

Peristalsis mixes and moves food and digestion products through the gut by means of smooth muscle contractions. page 40-15

The stomach stores food and initiates protein digestion through secretions of hydrochloric acid and digestive enzymes, including pepsin. page 40-16

Carbohydrates and lipids are broken down in the duodenum of the vertebrate midgut, facilitated by the secretion of digestive enzymes, bile salts, and bicarbonate. page 40-17

Small intestine nutrient absorption is enhanced by an enlarged surface area of villi and smaller microvilli on the surface of cells. page 40-17

The colon of the hindgut reabsorbs water before waste products are eliminated as semisolid feces from the rectum through the anus. page 40-20

Herbivorous animals have specialized digestive chambers (rumen and cecum) that enable bacterial fermentation to break down plant cellulose. page 40-21

Self-Assessment

1. Discuss how metabolic rate changes with levels of activity, body temperature regulation, and size.
2. Describe how ATP is produced for short and rapid activities versus long and sustained activities.
3. Explain how you would measure the metabolic rate of an animal.
4. Explain why the world record speed for a 10-m race is faster than that for a marathon.
5. Describe what happens in an endotherm and an ectotherm when outside temperature gets cold, and what happens when it gets hot.
6. Describe the order in which energy reserves are used in negative energy balance, such as starvation.
7. Name four ways in which animals capture prey and, for each, name one organism that uses it.
8. Draw the path of food through the vertebrate digestive tract, naming and describing the major function of each part.
9. Name the principal sites of digestion of proteins, carbohydrates, and fats, and indicate the principal enzymes involved in breaking down each.
10. Compare and contrast foregut and hindgut fermentation.

Do you understand the chapter's Core Concepts? Log in to check your answers to the Self-Assessment questions, then practice what you've learned and reinforce this chapter's concepts by working through the problems and multimedia tutorials provided there.

www.biologyhowlifeworks.com

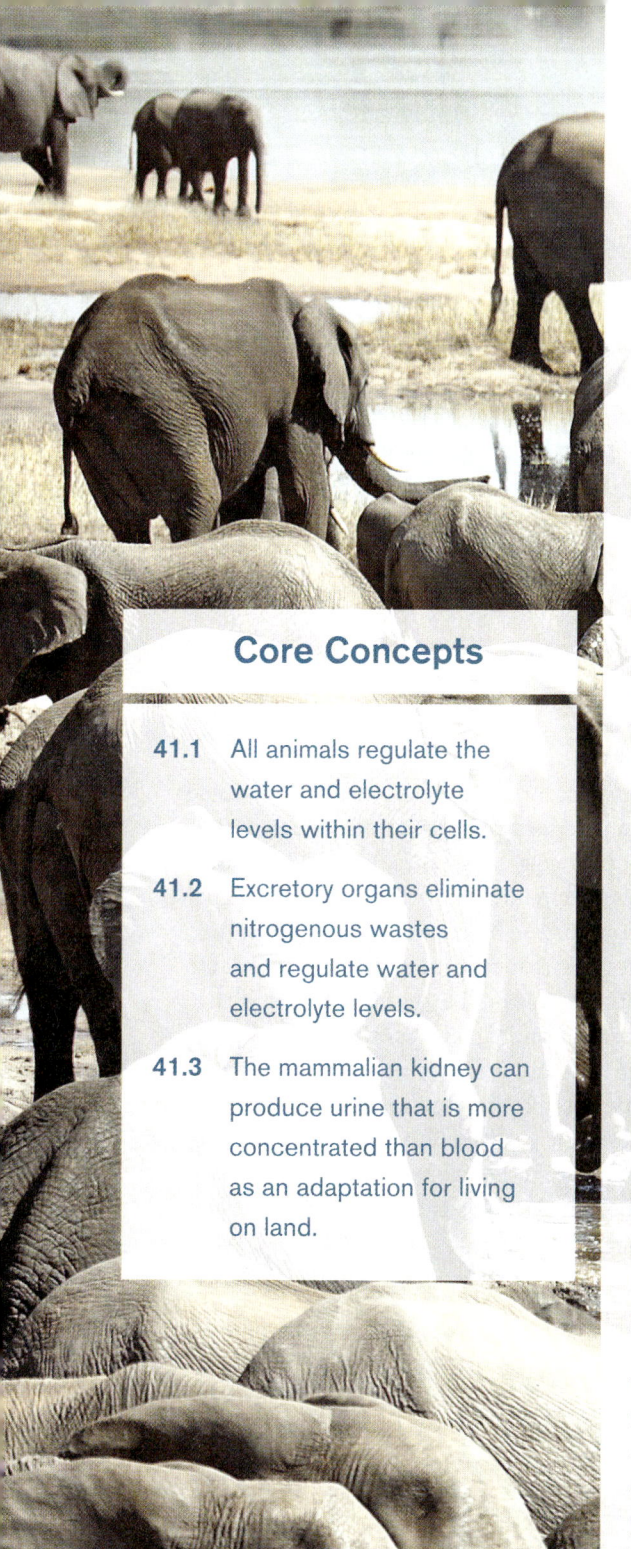

CHAPTER 41

ANIMAL RENAL SYSTEMS

Water and Waste

Core Concepts

41.1 All animals regulate the water and electrolyte levels within their cells.

41.2 Excretory organs eliminate nitrogenous wastes and regulate water and electrolyte levels.

41.3 The mammalian kidney can produce urine that is more concentrated than blood as an adaptation for living on land.

In the 1960s, researchers at the University of Florida developed a new sports drink for players on the football team, the Gators. In addition to water and sugar, the drink contained electrolytes, or ions such as sodium (Na^+) and potassium (K^+) that give the solution electrical properties. The researchers who came up with the formula for the drink reasoned that it was important to replace both water and electrolytes lost in sweat. The drink—known as Gatorade—was a success.

Water and electrolytes are two key substances that allow cells and organisms to maintain their size and shape, as well as to carry out essential functions. Life originated and diversified in a watery environment. Consequently, the chemical functions of cells and organisms crucially depend on the properties of water (Chapter 2) and the relative amounts of water and electrolytes inside a cell. In general, the internal environment of a cell is maintained within a narrow window of conditions. The set of chemical reactions that constitute metabolism, for example, requires enzymes that work best at a particular pH (Chapter 6). **Homeostasis**—the maintenance of a stable internal environment in the face of a changing external environment—is a dynamic process.

Animals must also maintain homeostasis in terms of the molecules produced by normal activity. Metabolism generates wastes that are not needed and, in many cases, are actually toxic to the cell. These must be eliminated, or **excreted.** In most organisms, the excretion of wastes is closely tied to the maintenance of water and electrolyte balance.

This chapter explores the physiological and cellular mechanisms that underlie water and electrolyte balance, as well as waste elimination. It examines how animals meet the demands for water and electrolyte balance in different environments. It also examines how water and electrolyte balance is linked to waste elimination and explores the diversity of excretory organs animals have evolved to regulate these processes, which are critical to maintaining a stable internal body chemistry. In vertebrates, a **renal system** with paired **kidneys** serves these roles. The functional organization of mammalian kidneys is covered in the last section of the chapter.

41.1 WATER AND ELECTROLYTE BALANCE

A healthy person can live several weeks without food but cannot survive much more than a few days without water. A loss of more than 10% of body water threatens survival. Homeostasis of water and electrolyte concentration is critical to normal physiological function, not just in humans, but in all organisms.

41-1

Osmosis governs the movement of water across cell membranes.

Water, like all molecules, moves by diffusion, or the random motion of molecules. When there is a difference in water concentration between two regions, there is a net movement of water molecules from regions of higher *water* concentration to regions of lower *water* concentration, a process called **osmosis** (Chapter 5). Water often contains other dissolved molecules, such as electrolytes, amino acids, and sugars, which are collectively called **solutes.** Because the concentration of a solute is typically known and the concentration of water is not, it is sometimes easier to think about water moving from regions of lower *solute* concentration (higher water concentration) to higher *solute* concentration (lower water concentration). Whether in terms of water concentration or solute concentration, the direction of net water movement is the same.

In organisms, the ability to control osmosis, and therefore to regulate levels of water and electrolyte in cells, depends on the properties of their plasma membranes and embedded proteins. When a barrier allows water or solutes to diffuse freely, it is said to be permeable. When it blocks the diffusion of water or solutes entirely, the membrane is considered impermeable. When it allows the movement of some molecules but not others, it is **semipermeable,** or selectively permeable. The membranes of cells are selectively permeable because of the combination of the lipid bilayer and transmembrane proteins, as we saw in Chapter 5.

Osmosis governs the movement of water across semipermeable membranes (**Fig. 41.1**). Plasma membranes and sheets of cells that are semipermeable allow water and smaller solutes to diffuse freely but prevent larger solutes, like proteins, from passing through them. Water is able to move directly through lipid bilayers by diffusion or through channels called **aquaporins** by facilitated diffusion (Chapter 5).

Consider then a situation in which there is a difference in solute concentration on the two sides of a semipermeable membrane, and the membrane is impermeable to the solute (Fig. 41.1a). Water moves by osmosis from regions of higher *water* concentration to regions of lower *water* concentration, or, from regions of lower *solute* concentration to regions of higher *solute* concentration (Fig. 41.1b).

→ **Quick Check 1** What happens when a cell is placed in a hypotonic solution (one with a lower solute concentration than that of the cell)? What happens when a cell is placed in a hypertonic solution (one with a higher solute concentration than that of a cell)?

The difference in concentration of solutes on the two sides of the membrane creates **osmotic pressure** that drives the movement of water. Osmotic pressure is the tendency of water to move from one solution into another by osmosis. The higher the solute concentration, the higher the osmotic pressure of that solution (that is, there is a greater tendency for water to move *into* that solution). Water diffuses into a region with a higher solute concentration (higher osmotic pressure) until it exerts a hydrostatic pressure (the pressure exerted by water itself due to gravity, not differences in solute concentration) that balances the osmotic pressure. When osmotic pressure equals hydrostatic pressure, the net movement of water stops and equilibrium is reached (Fig. 41.1c). At equilibrium, water

FIG. 41.1 Osmosis.

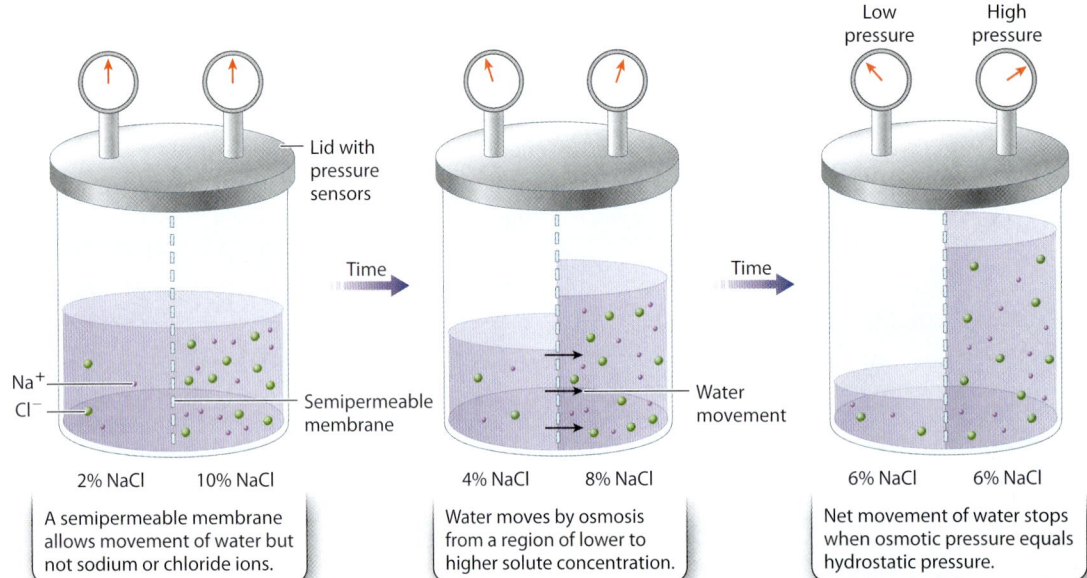

molecules still move across the membrane in both directions by diffusion, but there is no net movement of water molecules.

Whether osmotic pressure drives water into or out of a cell depends on the solute concentration of the inside of the cell relative to the outside. Similarly, whether osmotic pressure drives movements of water into or out of an organism depends on solute concentration inside an animal relative to its environment. Note that osmosis is driven by differences in *total* solute concentration between the two sides of a cell membrane. That is, if a particular electrolyte such as potassium has a high concentration inside the cell and low concentration outside a cell, but another electrolyte, such as sodium, has exactly the opposite distribution, there will be no net water movement because the total concentrations of the solutes on the two sides of the membrane are the same. Also note that water moves across semipermeable membranes in response to concentration differences of *any* solute. That is, the solute can be an electrolyte, such as sodium or potassium, or any other dissolved molecule, such as glucose or an amino acid or even waste, as long as the membrane is impermeable to it.

→ **Quick Check 2** A solution of water and dissolved ions is separated by a semipermeable membrane that allows the passage of water but not small ions. The concentration of sodium is higher on one side of the membrane than on the other side. However, there is no net movement of water across the membrane. How is this possible?

Osmoregulation is the control of osmotic pressure inside cells and organisms.

Animal life on Earth exists in a wide range of environments by adopting different means of **osmoregulation,** or the regulation of water and solute levels to control osmotic pressure. Osmoregulation, like energy balance (Chapter 40), is a form of homeostasis. High osmotic pressures inside a cell can damage the cell, sometimes even causing it to burst and thus disrupt some of the animal's functions. Although cells and tissues can tolerate dehydration, often as high as 50% or more, excessive dehydration impairs a cell's metabolic function, as many chemical reactions, such as the hydrolysis reactions that build proteins and nucleic acids from individual subunits, depend on the presence of water.

How do cells control their internal osmotic pressure? To regulate water and solute levels, and hence the osmotic pressure inside cells, the cell controls the solute concentration of the inside of the cell relative to the outside of the cell because water follows solutes across semipermeable cell membranes. At the level of the cell, then, osmoregulation is achieved by the movement of solutes, particularly electrolytes.

At the level of the organism, osmoregulation is achieved by balancing input and output of water and electrolytes. The water that animals need is gained and lost in different ways. **Fig. 41.2** shows sites of water gain and loss in a saltwater fish, which sites

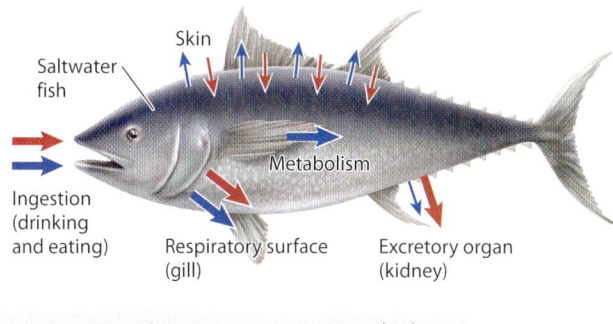

FIG. 41.2 Sites of water and electrolyte gain and loss in a saltwater fish. Osmoregulation is achieved by balancing water and electrolyte input and output.

→ Direction of electrolyte movement (Na⁺, K⁺, Cl⁻)
→ Direction of water movement

vary depending on the organism. In general, animals gain water by drinking water that is less concentrated in solutes (hypotonic) than their body fluids. Humans and many other vertebrates, both freshwater and terrestrial, gain most of their water this way. Water is also acquired through the food that animals eat. Additionally, animals produce water as a product of cellular respiration when two protons (H^+) combine with oxygen (O_2) to form water (Chapter 7). This is the primary source of water for some animals adapted to deserts and oceans, where drinkable water is not readily available. Animals lose water in their urine and feces. Terrestrial animals lose water through evaporation from their lungs, and, in the case of humans, by sweating. Freshwater animals gain water through their gills, whereas most marine animals lose water through their gills.

Similarly, electrolytes are taken in and lost in different ways. Animals gain electrolytes in the food they eat. If the water they drink has a higher solute concentration than their body fluids, animals can gain electrolytes by drinking. Marine aquatic animals gain electrolytes as hypertonic water moves across their gills and body surface. In contrast, freshwater aquatic animals lose electrolytes through diffusion across their gills and skin into the hypotonic watery environment. Humans lose electrolytes as well as water when they sweat. Drinking beverages high in electrolytes is helpful before demanding physical activities—including sports—because these activities can result in substantial water and electrolyte loss through sweating. Electrolytes are also lost by specialized glands (discussed below) and in urine and feces.

The ability of many organisms to regulate their water and electrolyte levels has enabled them to maintain homeostasis and a stable body chemistry in the face of changing environmental conditions, and to live in diverse environments. Most forms of life originated in salty marine environments, but many groups of organisms subsequently moved onto land and into freshwater

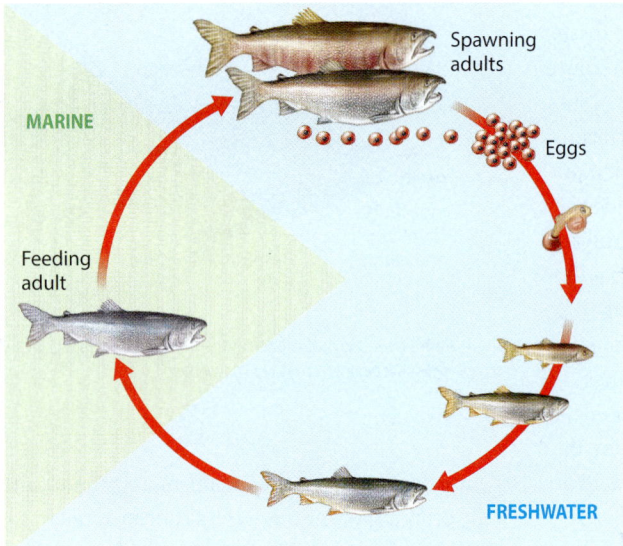

FIG. 41.3 The salmon life cycle. Salmon live in a range of environments at different stages of their life cycle, which require different osmoregulatory strategies.

lakes and rivers that present challenges for water and electrolyte balance. Living in freshwater environments requires animals to avoid excess water uptake into their tissues and control the loss of electrolytes from their body. Living on land requires animals to minimize their water loss in order to avoid desiccation, or drying out. Only insects and terrestrial vertebrates have achieved the ability to restrict water loss.

An extreme example of the ability of an animal to regulate water and electrolyte levels is provided by the life cycle of salmon. Mature salmon swim in open ocean saltwater environments but return upriver to spawn in freshwater environments (**Fig. 41.3**). As they do, they swim from high-salt ocean water through water of intermediate saltiness into freshwater rivers and streams that have little salt. After spawning, the young feed and grow in freshwater over a period of up to 3 years, eventually returning to the open ocean, where they live until they become reproductively mature and return to their home river system to spawn again.

These changes in aquatic habitat demand that the salmon make dramatic adjustments in the way they handle water and solutes. When in the ocean, the salmon must avoid excessive uptake of electrolytes and loss of water. However, when mature salmon migrate upriver to spawn and as young salmon continue to grow in freshwater streams and rivers, the challenge is reversed: They must avoid losing critical electrolytes and gaining too much water across their gills. In salmon, radical changes in osmoregulation are required over the course of an animal's lifetime to enable reproduction in freshwater streams and a return to the sea for further growth and maturation. Next, we explore how gills and kidneys meet these challenges.

Osmoconformers match their internal solute concentration to that of the environment.

Animals regulate their internal osmotic pressure in different ways. Some animals match their internal osmotic pressure to that of their external environment. These animals are **osmoconformers.** They keep their internal fluids at the same osmotic pressure as the surrounding environment, which reduces the movement of water and solutes into or out of their bodies. Because the solute concentrations inside and outside osmoconformers are similar, osmoconformers don't have to spend a lot of energy regulating osmotic pressure. However, they have to adapt to the solute concentration of their external environment, which is typically seawater. Osmoconformers tend to live in environments with stable solute concentrations, which provide an effective way of maintaining stable internal solute concentrations.

Although osmoconformers generally match the *overall* concentration of their tissues with their external environment, they often expend energy to regulate the concentrations of *particular* ions such as sodium, potassium, and chloride (Cl^-), and other solutes such as amino acids and glucose that are important to their function. For example, whereas the intracellular space of nearly all multicellular animals has a relatively high concentration of potassium ions and low concentration of sodium ions, the proportion in the extracellular space of most animals is just the opposite—low in potassium but high in sodium. This means that these cells are actively pumping sodium out and potassium in (Chapter 5).

Most marine invertebrates are osmoconformers, matching their total intracellular solute concentration to the solute concentration of seawater. These organisms maintain high concentrations of sodium and chloride in their cells to achieve an overall solute concentration close to that of seawater. Consequently, they have few specializations for osmoregulation beyond the need to regulate specific internal ion concentrations relative to their environment. Many marine vertebrates are also osmoconformers. Some, like hagfish and lampreys, match their total internal solute concentration to that of seawater but maintain high internal concentrations of particular electrolytes, just as marine invertebrates do.

Other marine vertebrates that are osmoconformers, like sharks, rays and coelocanths, match seawater's solute concentration by maintaining a high internal concentration of a compound called **urea.** Urea is a waste product of protein metabolism that many animals excrete, so it does not build up to high concentrations (section 41.2). By retaining this solute, these marine vertebrates are able to achieve osmotic equilibrium with the surrounding seawater. Even though marine osmoconformers seek to match the osmotic concentration of their environment, these animals also regulate specific ions for particular physiological functions within their body, as discussed in more detail below.

Osmoregulators have internal solute concentrations that differ from that of their environment.

Many animals maintain internal solute concentrations that differ from that of their environment. These animals are **osmoregulators.** Osmoregulators expend considerable energy pumping ions across cell membranes in order to regulate the movement of water into or out of their bodies. Some freshwater and marine fishes that are osmoregulators are estimated to expend 50% of their resting metabolic energy on osmoregulation. However, the ability to regulate osmotic pressure allows osmoregulators to live in diverse environments—saltwater, freshwater, and land.

The external surface (such as the skin) of osmoregulators controls the movement of water and solutes between the environment and the inside of the animal. Although the animal's skin maintains different total concentrations of solutes between the outside and inside of the animal, most of the internal cells do not maintain differences in total solute concentrations on the two sides of the plasma membrane. As a result, in general, there are usually no lasting differences in osmotic pressure between the inside and outside of internal cells of osmoregulators.

Teleosts, or bony fish (the largest group of marine vertebrates), as well as all freshwater and terrestrial animals, are osmoregulators. These animals have evolved a variety of specializations for water and electrolyte balance and waste excretion. Whereas marine teleosts maintain concentrations of electrolytes much lower than the surrounding seawater, freshwater fishes and amphibians maintain concentrations that are higher than those of their aquatic environment.

Aquatic animals use their gills to breathe as well as to regulate electrolyte levels. The thin gill filaments, with their extensive surface area, are well adapted for gas exchange and also facilitate electrolyte transport (**Fig. 41.4**). In marine bony fishes, specialized cells in the gills called **chloride cells** counter the ingestion and diffusion of excess electrolytes into the animal by pumping chloride ions into the surrounding seawater. By pumping negatively charged chloride ions out of the body, chloride cells create an electrical gradient that is balanced by positively charged sodium ions moving out of the body as well (Fig. 41.4a). In contrast, freshwater fishes have gill chloride cells with reversed polarity. Their chloride cells pump chloride ions (with sodium ions following) in the reverse direction, moving the chloride ions into the body to counter their ongoing loss from the gills into the surrounding freshwater (Fig. 41.4b). In both cases, the gills maintain a substantial difference in solute concentration between the fish's internal and external environment.

Sharks and rays excrete excess salt by a rectal gland, which eliminates salt with their digestive wastes. Similarly, many

FIG. 41.4 Osmoregulators. Osmoregulators actively maintain osmotic pressure different from that of their environment.

a. Marine bony fish

Marine fish eliminate excess salt by active transport from gill chloride cells, which pump chloride ions out with sodium ions following.

Drinks seawater

Gill filaments

Gill chloride cells

Active ion elimination by chloride cells and osmotic H_2O loss through gills

b. Freshwater bony fish

Freshwater fish use gill chloride cells to pump chloride ions in, with sodium ions following.

Freshwater fish do not drink water.

Active ion uptake by chloride cells and osmotic H_2O uptake

→ Direction of electrolyte movement (Na^+, K^+, Cl^-)
→ Direction of water movement

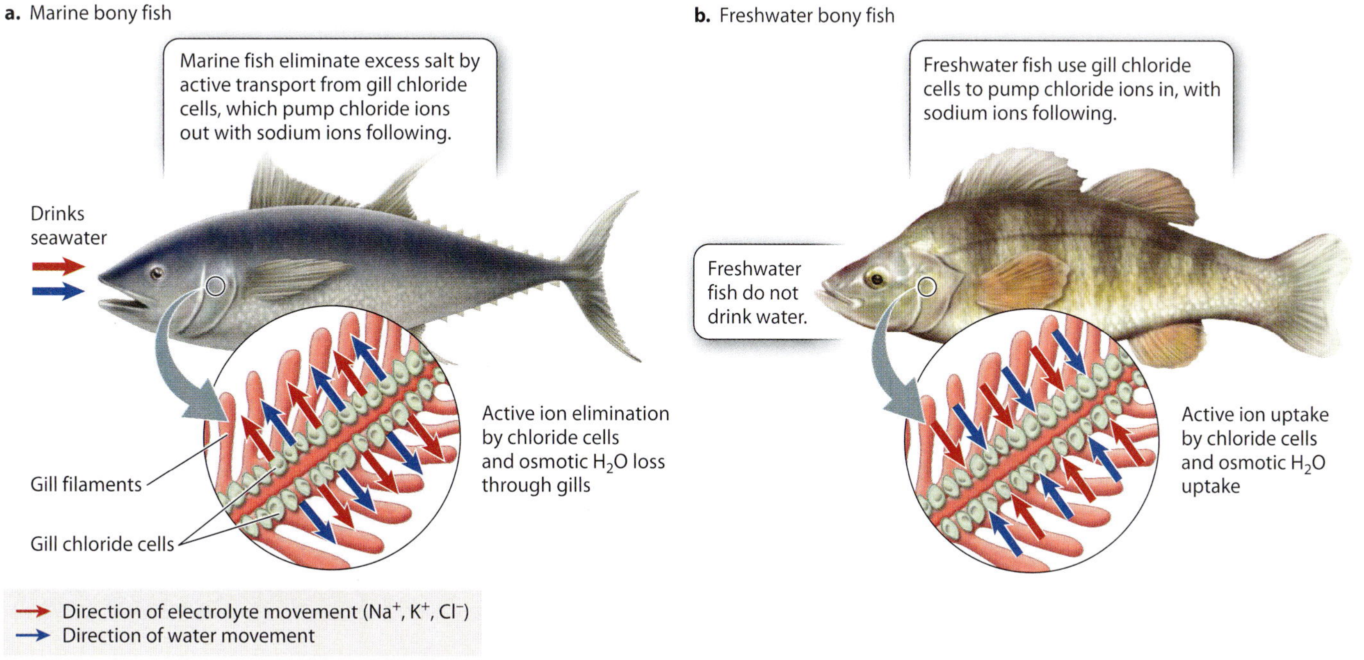

FIG. 41.5 Salt excretion from glands. Many marine birds have glands on the tops of the skull that actively excrete salt.

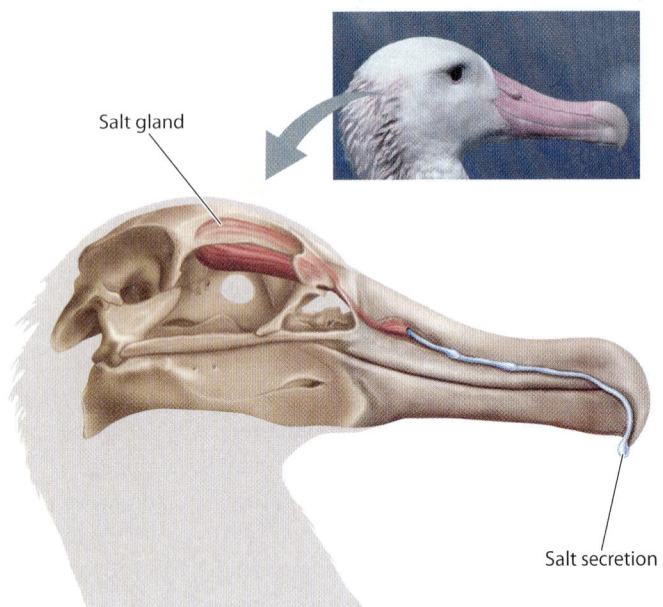

marine birds have evolved specialized nasal salt glands to rid themselves of the high salt content of their diet (**Fig. 41.5**). These salt glands work by actively pumping ions from the blood into cells that make up the gland, followed by excretion of salt from the body through the gland and the nasal cavity. Their salt glands allow these animals to gain net water intake by drinking seawater. In contrast, humans and many other non-marine-adapted animals are incapable of deriving useful net water gain by drinking seawater because they cannot eliminate salt in high enough concentration. The high concentration of magnesium sulfate in seawater results in diarrhea if too much is consumed, exacerbating the net loss of water in the feces.

Some organisms that encounter a range of environments use different means at different times. For example, the brine shrimp *Artemia* usually acts as an osmoconformer but can act as an osmoregulator in extreme environments. *Artemia* lives in salt lakes such as the Great Salt Lake of Utah, where it usually matches its internal solute concentration to that of the outside environment. However, it sometimes encounters exceptionally high salt concentrations. It can survive in this high-salt environment by excreting excess salt ions from its gills to maintain an internal concentration that is hypotonic relative to its environment. Alternatively, in low saltwater environments (but not in freshwater, which they cannot tolerate), *Artemia* reverse their gill ion exchange and maintain a hypertonic internal body fluid concentration relative to the surrounding water.

In addition to respiratory surfaces such as gills and specialized salt glands, the other major regulator of water and electrolyte levels is the kidney, which we discuss in the next two sections.

? CASE 7 Predator–Prey: A Game of Life and Death
Can the loss of water and electrolytes in exercise be exploited as a strategy to hunt prey?

As we have seen, exercise can lead to loss of water and electrolytes, a response that is exploited by African bushmen as they hunt large prey. African bushmen are hunter-gatherers; their tribes live in hot, dry grasslands. Equipped only with wood spears, bows, and poisoned arrows, they have historically hunted kudus, antelope, buffalo, and other large grazing herbivores (**Fig. 41.6**). Bushmen hunters pursue and track down their prey, exploiting effects of the physiological stress that occurs when their prey run during the heat of the day.

Before hunting, bushmen drink as much water as possible, hydrating their bodies. They select a large prey animal, which

FIG. 41.6 Exploiting an animal's physiological stress. (a) A kudu will be forced to run in the heat of the day by (b) African bushmen hunters. The animal will lose water and electrolytes, and will eventually become too dehydrated to escape.

they chase from a shaded refuge into the open sun when temperatures exceed the animal's core body temperature. The bushmen can deal with the high temperatures, cooling their bodies through sweating, but because they hydrated thoroughly before the hunt, their sweating does not dehydrate them. Not so for the prey. The bushmen hunters track and chase their prey until the prey becomes dehydrated and hyperthermic, thus rendered unable to escape.

Even if the bushmen hunters strike the animal successfully with a poisoned arrow, they still must track it for long distances and over long periods of time. Hunts typically last several hours, but sometimes as much as 1 or 2 days, and cover distances of 25 km (about 16 miles) or more. The loss of water in intensive heat disrupts the animal's water and electrolyte balance and thus contributes to its inability to move and defend itself.

41.2 EXCRETION OF WASTES IN RELATION TO ELECTROLYTE BALANCE

In most organisms, the regulation of water and solute levels is closely tied to waste excretion. Excretion is the elimination of waste products and toxic compounds from the body. Some of these wastes are generated as a by-product of metabolism, particularly protein and nucleic acid breakdown. Others were originally ingested, like the undigested remains of food, and need to be eliminated from the organism.

Most multicellular animals have evolved specialized excretory organs, such as the vertebrate kidney, to isolate, store, and eliminate toxic compounds and metabolic waste products. Because these compounds are eliminated as solutes that are dissolved and transported within the blood or other body fluids, their elimination from the body is intimately tied to water and electrolyte balance. The excretory organs of most animals therefore function to maintain water and electrolyte balance in addition to eliminating waste products.

Both the gastrointestinal tract (Chapter 40) and excretory organs like the vertebrate kidneys eliminate wastes from the body. However, the gastrointestinal tract is continuous with the outside world, and therefore its role in the elimination of wastes is quite different from that of excretory organs that separate toxic compounds from essential nutrients, electrolytes, and cells circulating in the blood.

The excretion of nitrogenous wastes is linked to an animal's habitat and evolutionary history.

When proteins and nucleic acids are broken down by metabolism, one of the by-products is ammonia (NH_3). Ammonia contains nitrogen and is toxic to organisms, so it is a form of **nitrogenous waste** (**Fig. 41.7**). By far, most ammonia is produced from the breakdown of proteins. Ammonia can disturb the pH balance of

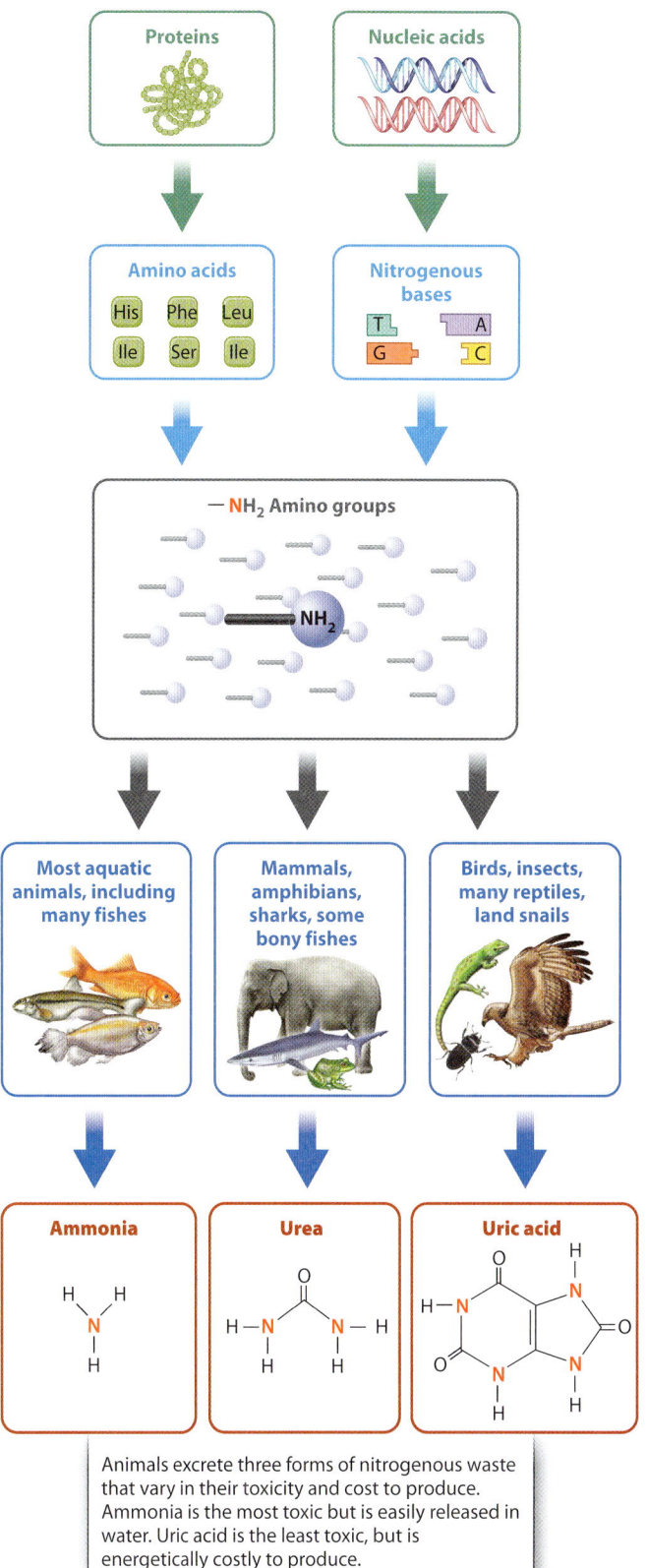

FIG. 41.7 **Nitrogenous waste.** Animals excrete nitrogenous waste in three major forms—ammonia, urea, and uric acid—which vary in their toxicity and cost to produce.

Animals excrete three forms of nitrogenous waste that vary in their toxicity and cost to produce. Ammonia is the most toxic but is easily released in water. Uric acid is the least toxic, but is energetically costly to produce.

cells and cause neuronal cell injury. At high levels, it is lethal to organisms. It is either eliminated from the body or converted into a less toxic form. In either case, ammonia is kept at very low levels in the body.

Some organisms, such as fish, are able to excrete ammonia directly into the surrounding water, primarily through their gills. As a result, ammonia does not accumulate to toxic levels. This process is not energetically expensive, but it requires high volumes of water in the surrounding environment to dilute the highly toxic ammonia.

The movement of animals from water to land posed many challenges, among them the problem of waste elimination. No longer could animals rely on simple diffusion of ammonia into the surrounding water. Instead, organisms evolved biochemical pathways to convert ammonia to less toxic forms. One of these forms is urea (Fig. 41.7). Because urea is less toxic than ammonia, it can be stored in a fairly concentrated form before being excreted. Mammals (including humans), many amphibians, sharks, and some bony fish excrete nitrogenous waste in the form of urea. In mammals, urea is produced in the liver and is then carried by the blood to the kidneys, where it is eliminated. Urea is less toxic than ammonia but requires energy to produce it and water to eliminate it.

Ammonia can also be converted into uric acid (Fig. 41.7). Uric acid is the least toxic of the three forms of nitrogenous waste, so animals can store it at a higher concentration than urea. Birds and many reptiles, arthropods (including insects), and land snails excrete nitrogenous waste in the form of uric acid. These groups evolved the ability to convert ammonia to uric acid independently of one another as an adaptation to living on land, providing a good example of convergent evolution (Chapter 23). Uric acid precipitates from solution and forms a semisolid paste (familiar as white bird droppings). Because uric acid is not dissolved in water, it does not exert osmotic pressure and is eliminated with minimal water loss for the animal. This allows many reptiles and insects to live in extremely hot and dry environments. However, uric acid is energetically expensive to produce.

Many amphibians, like frogs, excrete ammonia early in their development, when they live in water, but then excrete urea when they move onto land after metamorphosis. Alligators and turtles also adjust what they excrete: ammonia when in the water and uric acid when on land.

Ammonia, urea, and uric acid are the three major forms of nitrogenous wastes excreted by animals. However, they are not the only forms. Animals excrete nitrogen in the forms of trimethylamine oxide, creatine, creatinine, and even amino acids. The form of nitrogenous waste that an animal excretes is ultimately linked to its environment and evolutionary history.

→ **Quick Check 3** Why do mammals convert ammonia to urea rather than simply excreting it, as fish do?

Excretory organs work by filtration, reabsorption and secretion.

Excreting wastes does not occur by simply pumping the wastes out of the body. Instead, animals filter or secrete wastes out of the blood along with water, electrolytes, and even essential nutrients, but then reabsorb the nutrients they need, excreting the waste and enough water and electrolytes to maintain appropriate levels of water and electrolytes for hydration and osmotic balance. While this process may seem inefficient, it has several advantages compared to simply pumping wastes out. First, as we will see, it allows organisms to excrete wastes and at the same time balance their water and electrolyte levels. Second, it allows any toxic compound that ends up in the blood to be excreted. If organisms relied on molecule-specific active transport of wastes, novel toxic compounds ingested by organisms could not be excreted.

The first step is to isolate the waste into an extracellular space. In the case of unicellular organisms such as *Paramecium* and simple multicellular animals such as sponges and cnidarians (jellyfish and sea anemones), waste compounds are isolated in a cellular compartment called a **contractile vacuole** (**Fig. 41.8**). Fusion of the contractile vacuole with the cell membrane eliminates the waste contents by exocytosis from the cell (Chapter 5).

Multicellular animals with pressurized circulatory systems (annelid worms, mollusks, and vertebrates) also isolate wastes

FIG. 41.8 Contractile vacuole of a *Paramecium* when (a) full and (b) after emptying.

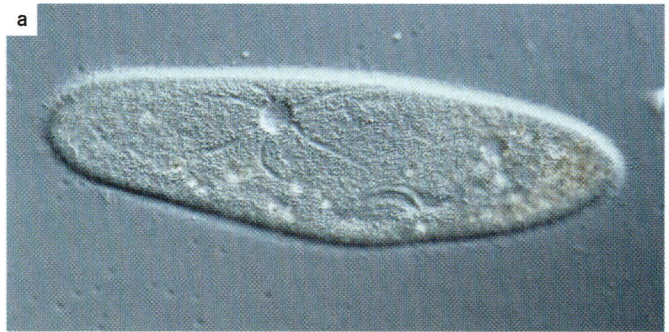

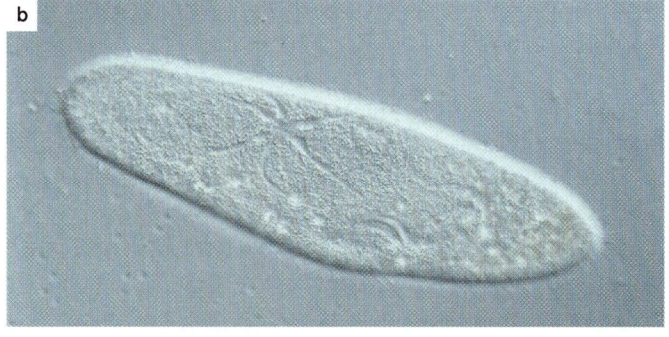

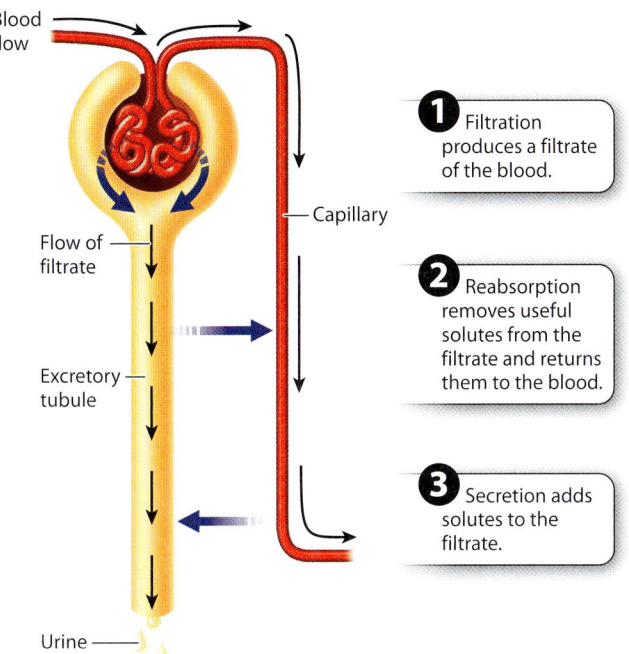

FIG. 41.9 Filtration, reabsorption, and secretion. These three processes eliminate wastes and provide osmoregulation in multicellular animals with pressurized circulatory systems.

1. Filtration produces a filtrate of the blood.
2. Reabsorption removes useful solutes from the filtrate and returns them to the blood.
3. Secretion adds solutes to the filtrate.

Animals have diverse excretory organs.

The processes of filtration, reabsorption, and secretion serve two important and complementary functions: They allow wastes to be isolated and subsequently removed from the body, and they allow the organism to adjust the amounts of water and electrolytes required to meet its osmoregulatory needs. These two functions are carried out by different excretory organs in different animals. For example, in flatworms, excretion involves isolating the fluid from the body cavity in excretory organs called **protonephridia** without filtering it first (**Fig. 41.10**). Cells with beating cilia move the fluid down the lumen of the tubule to an excretory pore. Freshwater flatworms, such as *Planaria,* have an extensive network of tubules that eliminate excess fluid along with waste products. As the fluid passes down the tubule, cells lining the tubule modify the fluid's contents through reabsorption and secretion. The urine leaving a freshwater flatworm contains nitrogenous waste and is less concentrated (that is, more dilute) than its body to balance the water it ingests as it feeds.

In segmented annelid worms such as earthworms, the body fluid is filtered through small capillaries into a pair of excretory organs

FIG. 41.10 Excretory organ of a flatworm.

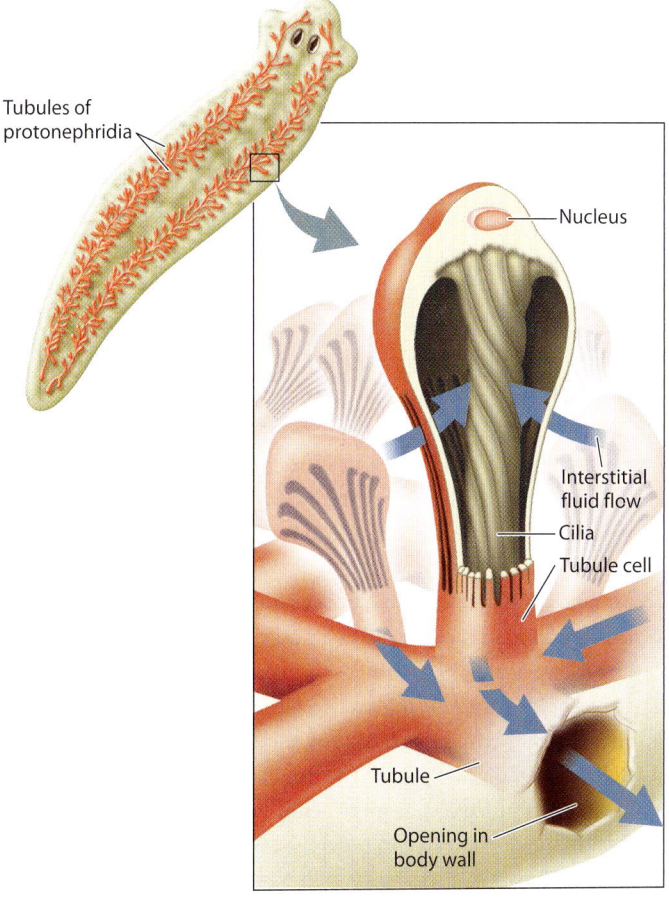

from the blood, but they do so by **filtration** (**Fig. 41.9**). Circulatory pressure pushes fluid containing the wastes through specialized filters into an extracellular space. The filtered fluid, called the filtrate, contains waste products along with water, electrolytes, and other solutes, which drain into **excretory tubules** that connect to the outside. The excretory filter, like a cell's plasma membrane, is selectively permeable. It allows smaller ions and solutes to pass into the tubule lumen along with water but retains larger solutes such as proteins and cells.

Filtration is followed by **reabsorption** of key ions and solutes (Fig. 41.9). Reabsorption is an active process for some molecules and a passive one for others. In both cases, substances that are important for the animal to retain, such as electrolytes, amino acids, vitamins, and simple sugars, are taken up by cells of the excretory tubule and returned to the bloodstream. Water is reabsorbed into the cells by osmosis, driven by the osmotic pressure established by the active transport of solutes into the cell.

The final step in waste removal is the **secretion** of additional toxic compounds and excess ions (Fig. 41.9). Secretion is also an active process and involves eliminating substances that were not filtered from the blood earlier. Filtration, reabsorption, and secretion are the three steps that all animal excretory systems perform to eliminate toxic wastes, retain key electrolytes and solutes, and regulate water volume.

FIG. 41.11 Excretory organ of an earthworm.

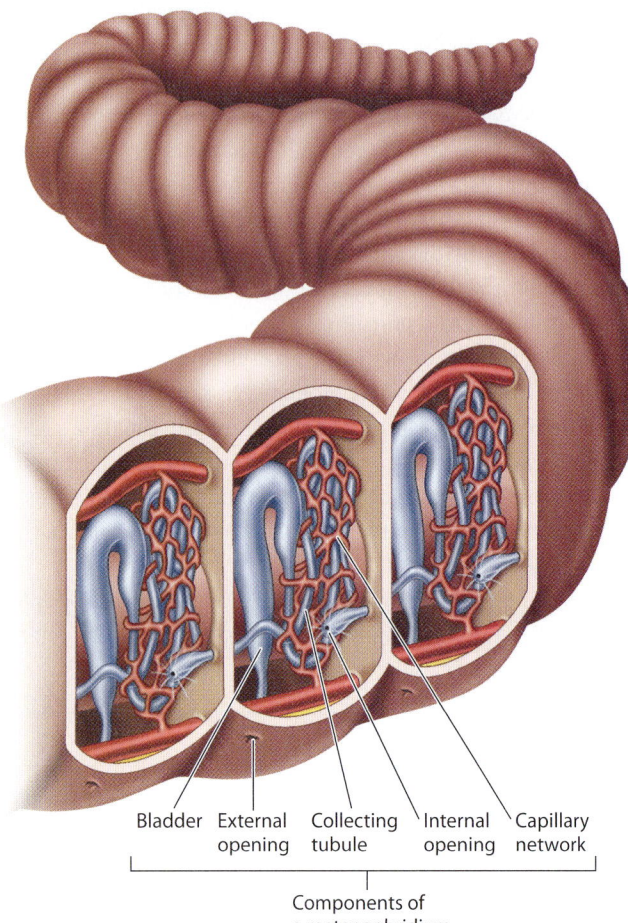

called **metanephridia** within each body segment (**Fig. 41.11**). The fluid enters a metanephridium at a funnel-shaped opening surrounded by cilia that drains body fluid into a series of tubes in the next body segment and then out of the animal through an excretory pore. Blood vessels and cells surrounding the tubules reabsorb and secrete additional compounds as urine is formed. Like freshwater flatworms, earthworms reabsorb electrolytes and eliminate dilute urine containing dissolved nitrogenous wastes.

In insects and other terrestrial arthropods, fluid passes from the main body cavity into a series of tubes called **Malpighian tubules** that empty into the hindgut (**Fig. 41.12**). (These structures are named for Marcello Malpighi, the seventeenth-century Italian physician who first described them.) An insect can have two Malpighian tubules or more than 100. The tubules actively secrete uric acid and excess electrolytes from the animal's tissues while retaining key solutes. Muscle fibers in the walls of the tubules help push the fluid to the hindgut. In the hindgut, the fluid becomes more acidic because of the buildup of uric acid, causing uric acid to precipitate from solution when it enters the rectum. Because uric acid is removed from the fluid, it does not influence the movement of water. Water is therefore free to diffuse into the tissues with the reabsorbed electrolytes that are transported from the rectum before the uric acid and remaining dried digestive wastes are eliminated. Malpighian tubules represent one of the most successful adaptations for excreting nitrogenous wastes with minimal water loss, enabling insects to live in diverse and arid terrestrial environments.

Vertebrates filter blood under pressure through paired kidneys.

The excretory organs of vertebrates are the kidneys. Like all excretory organs, they not only serve to eliminate wastes, but also play an important role in water and electrolyte balance. Vertebrate kidneys also reflect the evolutionary transition of vertebrates from an aquatic to a terrestrial environment, where there is a need to minimize water loss. Mammals and birds, for example, can excrete urine that is more concentrated than their blood and tissues.

The kidneys lie posterior to the abdominal cavity. Whereas the kidneys of humans and other mammals are bean-shaped, the kidneys of birds and other vertebrate animals are elongated and segmented (**Fig. 41.13**). Paired vertebrate kidneys receive blood supply from the heart by renal arteries that branch from the abdominal aorta.

Vertebrates filter their blood through specialized capillaries that have openings in their walls. These porous capillaries form a

FIG. 41.12 Excretory organ of an insect.

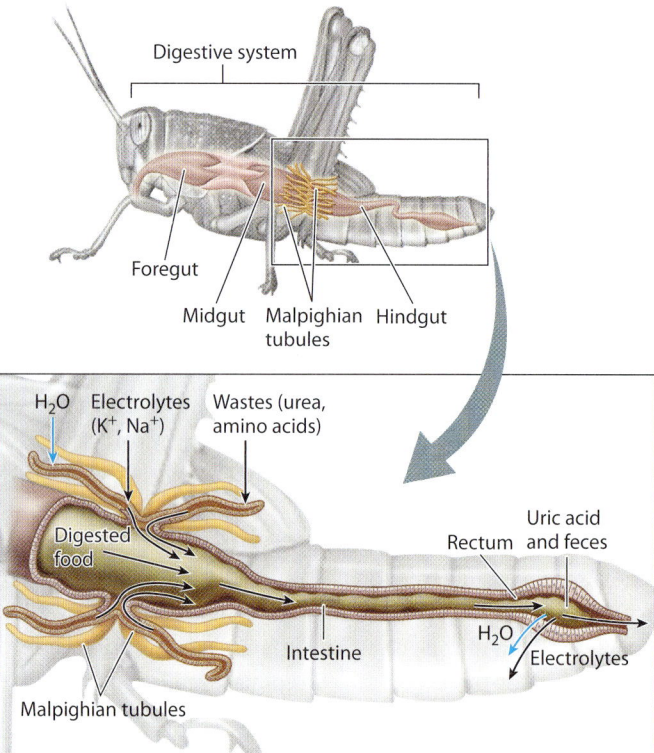

CHAPTER 41 ANIMAL RENAL SYSTEMS: WATER AND WASTE   41-11

FIG. 41.13 Location and shape of kidneys in different vertebrates.

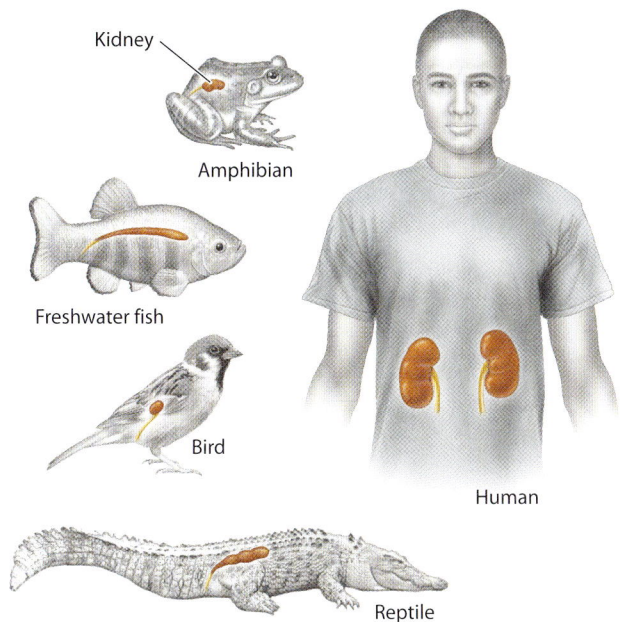

tufted loop called a **glomerulus** (**Fig. 41.14**). There are thousands to millions of glomeruli in each kidney. Driven by circulatory pressure, nitrogenous wastes, electrolytes, other small solutes, and water move through small holes in the capillary walls into an extracellular space surrounded by a capsule. The filtrate then moves through a series of **renal tubules,** which further process the filtrate by reabsorption and secretion before it enters the **collecting ducts** as urine. The collecting ducts converge on a larger tube called the **ureter,** which brings urine from the kidneys to a hollow organ called the **bladder** in mammals and fishes, or the **cloaca** in the case of amphibians, reptiles, and birds, for storage and elimination from the body.

The glomerulus, capsule, renal tubules, and collecting ducts make up a **nephron,** the functional unit of the kidney. Nephrons perform the three basic steps of excretion and osmoregulation: filtration of blood passing through the glomerulus, reabsorption from the renal tubule back to the bloodstream of key electrolytes and solutes, and secretion of additional wastes by the renal tubules.

In freshwater and saltwater fish, the kidneys run along the length of the body, with nephrons arranged segmentally. Because fish excrete nitrogenous waste in the form of ammonia through their gills, the kidney is primarily an organ of water and electrolyte balance rather than one of waste elimination. In saltwater fish, the renal tubules are readily permeable to water, allowing water to diffuse into and out of the tubule passively as the cells of the tubules actively reabsorb electrolytes and other valuable solutes. However, no net water uptake occurs. Consequently, the urine produced has the same concentration as the fish's body fluids. By contrast, the renal tubules in freshwater fish are relatively impermeable to water while allowing reabsorption of valuable electrolytes, sugars, and amino acids. This enables freshwater fish to excrete very dilute urine, eliminating the excess water that continually enters through their gills.

The kidneys of amphibians are similar to those of freshwater fish, reflecting their shared evolutionary history. Like freshwater fish, amphibians take in excess water across their gills and skin as they breathe. Amphibian kidneys therefore produce extremely dilute urine, enabling them to maintain an electrolyte concentration in their body that exceeds that of their freshwater environment.

Reptiles are found in a variety of habitats, from oceans to deserts, and are well adapted to their environments. Marine iguanas, for example, ingest salt water with their food, and so rely on specialized salt glands to eliminate excess electrolytes from their body. Terrestrial reptiles living in desert regions face the challenge of water loss and limited water availability in their environment. Like saltwater fish, these animals produce urine with the same salt content as the rest of their bodies after

FIG. 41.14 Vertebrate kidney and nephron.

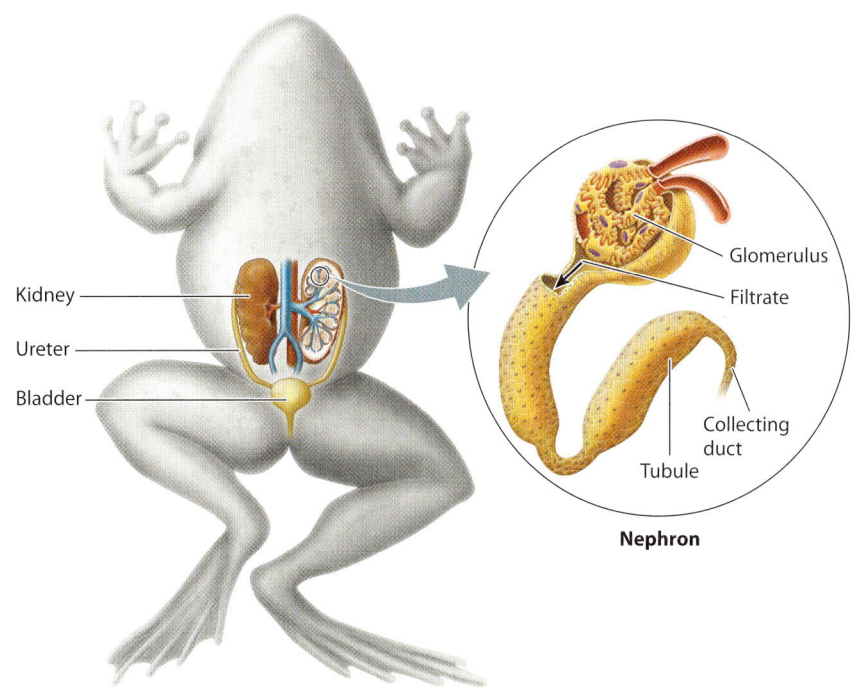

reabsorbing needed compounds in their bodies and secreting toxic wastes not filtered by their kidneys. However, because reptiles excrete nitrogenous waste in the form of uric acid, they are able to reabsorb most of the water from the urine after it drains into their cloaca. This process allows many reptiles to inhabit extremely dry environments.

Mammals also live on land, but excrete most of their nitrogenous waste in the form of urea dissolved in water. Their kidneys are well adapted for excreting waste and balancing their water and electrolyte levels, as we discuss now.

41.3 STRUCTURE AND FUNCTION OF THE MAMMALIAN KIDNEY

Mammals excrete urea dissolved in solution but can concentrate their urine more than their body fluids to minimize the loss of water, an adaptation to living on land. They concentrate their urine by generating concentrated (hypertonic) interstitial fluid deep in the kidney, through which the collecting ducts pass. Then, as the filtrate passes through the collecting ducts, water moves out of collecting ducts and into the interstitial fluid by osmosis, leaving concentrated urine in the collecting ducts. In this way, the urine is concentrated without the kidney ever actively transporting water itself. Instead, water flows passively by osmosis out of the collecting ducts. In addition, as we will see, nowhere along the length of the nephron are there large differences in osmotic pressure between the interstitial space and the renal tubules.

This process is quite efficient, and allows animals to thrive in diverse environments. For example, the giant kangaroo rat shown in **Fig. 41.15,** a desert mammal, reabsorbs so much water from the filtrate in its collecting ducts that it does not need to drink, obtaining enough water from seeds and cellular respiration. The giant kangaroo rat is thought to produce the most concentrated urine of all mammals. Marine mammals, such as whales, dolphins, and seals, also have exceptional abilities to concentrate their urine. They can remain at sea for long periods of time but cannot drink seawater for net water uptake. They, too, rely on metabolic water production in combination with the water in the food they eat to remain in water balance.

In this final section, we examine the mammalian kidney in more detail. The kidneys of birds share many features with those of mammals. However, these features arose independently in birds and mammals and are therefore the result of convergent evolution, presumably as adaptations for efficient processing of wastes in endothermic vertebrates with high metabolic rates (Chapter 40).

The mammalian kidney has an outer cortex and inner medulla.

The paired kidneys of mammals are organized into two layers, an outer layer called the **cortex** and a deeper layer, surrounded by the cortex, called the **medulla (Fig. 41.16)**. Nephrons are organized into wedge-shaped renal pyramids (Fig. 41.16). Each nephron's glomerulus is located within the renal cortex or outer layer of the medulla, and feeds into a renal tubule that starts in the renal cortex before dipping into the medulla at the base of the renal pyramid and then looping back up to the cortex. The collecting ducts of the renal tubules also descend into the kidney's base, emptying into the **renal pelvis** and draining through the ureter into the bladder.

About 20% of the blood circulated through the body with each beat of the heart passes through the kidneys for filtration. The kidneys very finely regulate the blood's water and electrolyte content as they eliminate urea as a nitrogenous waste.

Glomerular filtration isolates wastes carried by the blood along with water and small solutes.

The glomerulus is a tuft of capillaries, with blood entering by an afferent ("toward") arteriole and leaving by an efferent ("away") arteriole (**Fig. 41.17**). The tuft of capillaries is encased in a membranous sac called **Bowman's capsule,** and the space enclosed by the capsule is called Bowman's space. Blood is filtered as it passes through the tuft of capillaries, the filtrate passing out of the capillaries, through Bowman's capsule, and into Bowman's space. Water, wastes, and solutes from the blood pass through the capillary wall into Bowman's space before moving into the renal tubule.

To allow filtration of blood, the capillaries of the glomerulus have a different structure from that of capillaries elsewhere in the body. The cells that line the capillaries, known as endothelial cells, have pores called fenestrae that allow fluid to pass through.

FIG. 41.15 A giant kangaroo rat (*Dipodomys ingens*). Giant kangaroo rats do not drink, instead obtaining water through cellular respiration and from the seeds they eat, which contain small amounts of water.

CHAPTER 41 ANIMAL RENAL SYSTEMS: WATER AND WASTE 41-13

FIG. 41.16 The organization of the mammalian kidney and nephron.

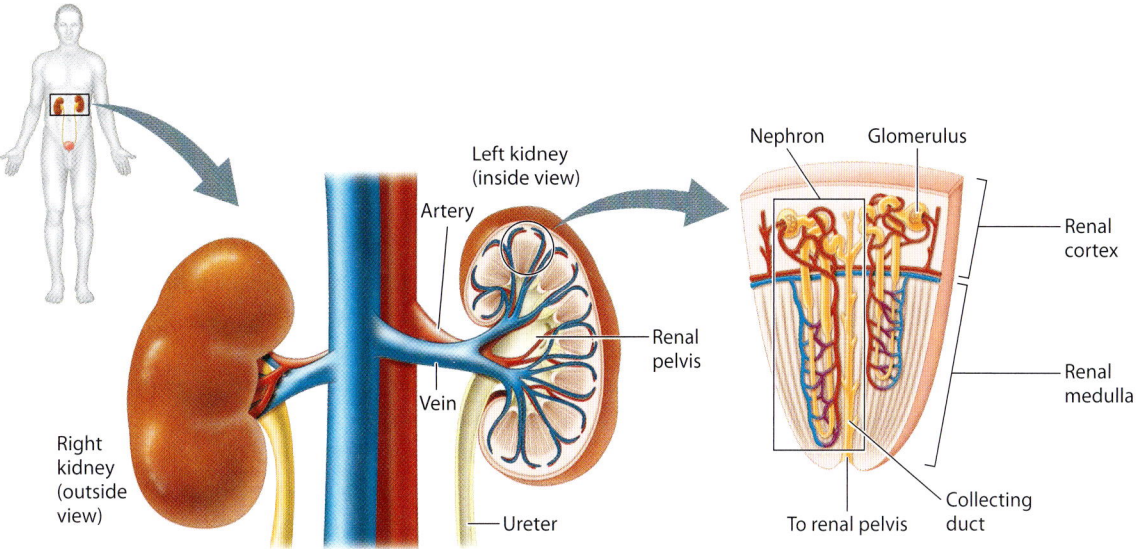

The filtration barrier of the glomerulus is made up of three layers (Fig. 41.17): the endothelial cells; a basal lamina; and a layer of cells with footlike processes called **podocytes,** which loosely interlock to create slits in the cell layer. These three layers together create a filter that allows small molecules to pass through but blocks the passage of large proteins and cells, leaving them in the blood. As a result, wastes such as urea, as well as water and key solutes such as electrolytes, glucose, and amino acids, all end up in the filtrate. After passing into the Bowman's space, the filtrate enters the renal tubule.

FIG. 41.17 The mammalian glomerulus and Bowman's capsule. Three layers create the filter: endothelial cells, basal lamina, and podocytes.

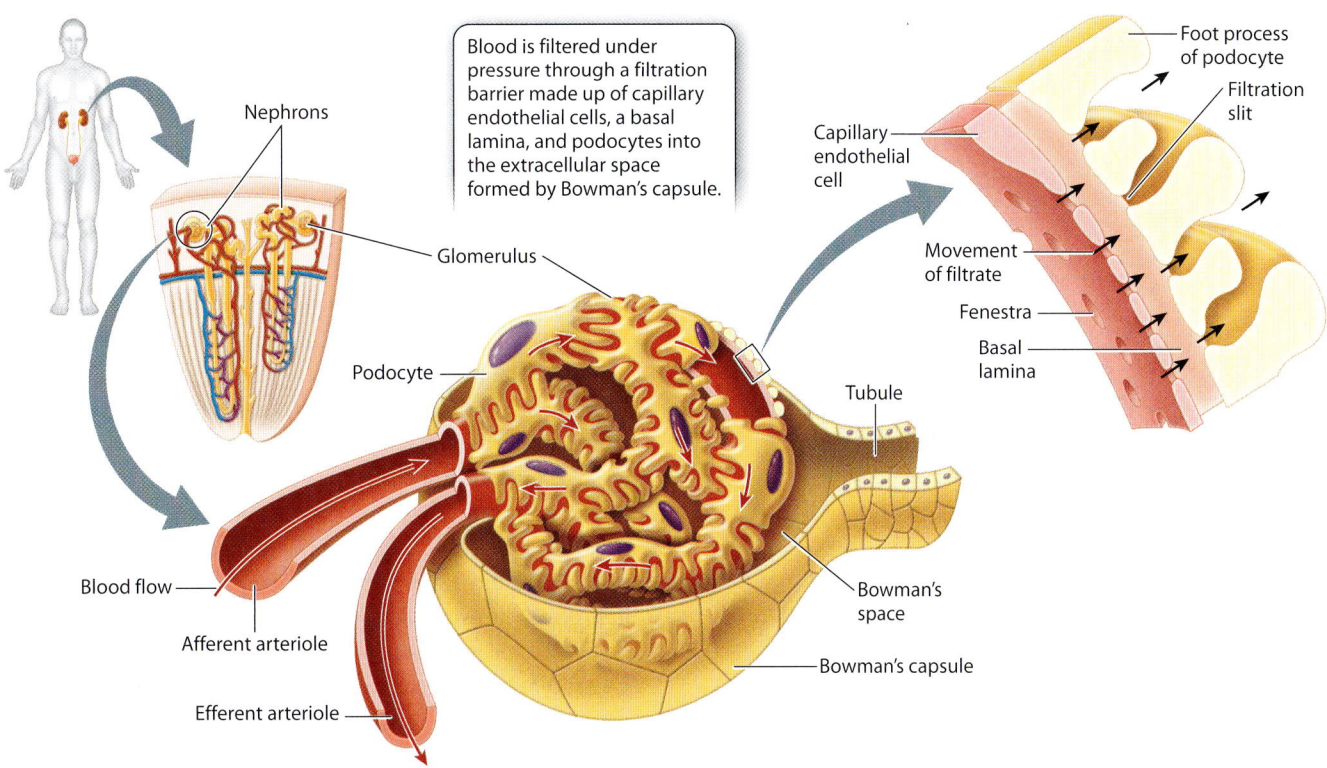

The proximal convoluted tubule reabsorbs solutes by active transport.

The renal tubule of the mammalian kidney is divided into three sections—a **proximal convoluted tubule**, a **loop of Henle**, and a **distal convoluted tubule**—as shown in **Fig. 41.18**. Although the renal tubule is one long tube, these three sections are specialized for different functions due to differences in permeability to different substances.

After passing through the filtration barrier and into Bowman's space, the filtrate moves into the first portion of the renal tubule, the proximal convoluted tubule (Fig. 41.18). Here, electrolytes and other nutrients that the organism requires are reabsorbed into the blood. To maximize absorption, the proximal convoluted tubule has thick walls composed of epithelial cells with fingerlike projections called **microvilli** on their surfaces, similar in structure to those found in the small intestine (Chapter 40). The microvilli form a brush border along the lumen of the tubule, greatly increasing the cells' surface area for solute and water reabsorption. The epithelial cells possess numerous mitochondria that produce the ATP needed to actively reabsorb solutes from the filtrate into the bloodstream.

The proximal tubule reabsorbs all the glucose and amino acids filtered by the glomerulus, as well as most sodium and chloride ions. Because the proximal tubule is permeable to water, water diffuses by osmosis in the same direction as the electrolytes and solutes. By the time the filtrate leaves the proximal convoluted tubule, about 75% of the water, along with most of the electrolytes and solutes, has been reabsorbed into the bloodstream.

The loop of Henle acts as a countercurrent multiplier to create a concentration gradient from the cortex to the medulla.

The proximal convoluted tubule connects to the loop of Henle, the second portion of the renal tubule (Fig. 41.18). (This structure was first described by the nineteenth-century German anatomist F. G. J. Henle.) Although many nephrons contain short loops of Henle that are confined to the renal cortex, some descend all the way into the medulla before looping back up to the cortex. These longer loops of Henle play an important role: They create a concentration gradient in the interstitial fluid from the cortex to the medulla, with the interstitial fluid of the cortex being less concentrated and the interstitial fluid of the medulla being more concentrated. This concentration gradient allows water passing through the downstream collecting duct to be reabsorbed, thus producing concentrated urine.

How is the gradient generated? The key to understanding how the loop of Henle creates a concentration gradient is to recognize that the two limbs of the loop of Henle run in parallel but in opposite directions, and they differ in their permeability to water and ability to actively transport electrolytes. Although the filtrate enters the descending limb before looping around to the ascending limb, it is easier to consider the ascending limb first.

The thick portion of the ascending limb of the loop of Henle is impermeable to water but actively transports electrolytes out of the filtrate (**Fig. 41.19**). As the filtrate moves up the thick ascending limb of the loop of Henle, electrolytes are actively transported out of it into the interstitial fluid surrounding the loop of Henle.

At the same time, the filtrate is moving from the descending limb to the ascending limb. As a result, the interstitial fluid becomes more concentrated, and the filtrate passing through the ascending limb becomes less concentrated. In contrast to the ascending limb, the descending limb is permeable to water. Therefore, water moves passively out of the descending limb, driven by osmosis. As water moves out of the limb, the filtrate becomes more and more concentrated, matching the concentration in the surrounding interstitial space. As

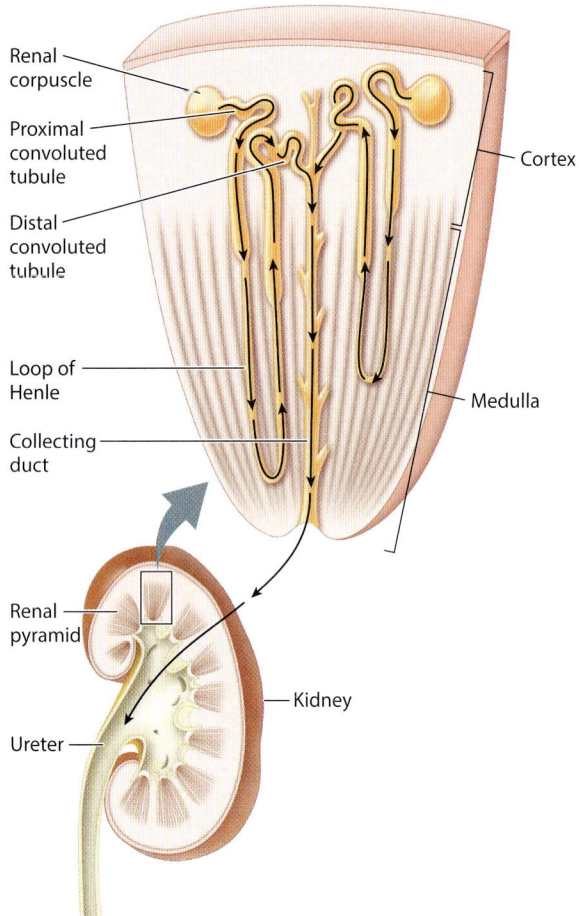

FIG. 41.18 The organization of the mammalian renal tubules and collecting duct.

FIG. 41.19 The mammalian nephron. The loop of Henle creates a concentration gradient, with higher concentration in the medulla and lower concentration in the cortex.

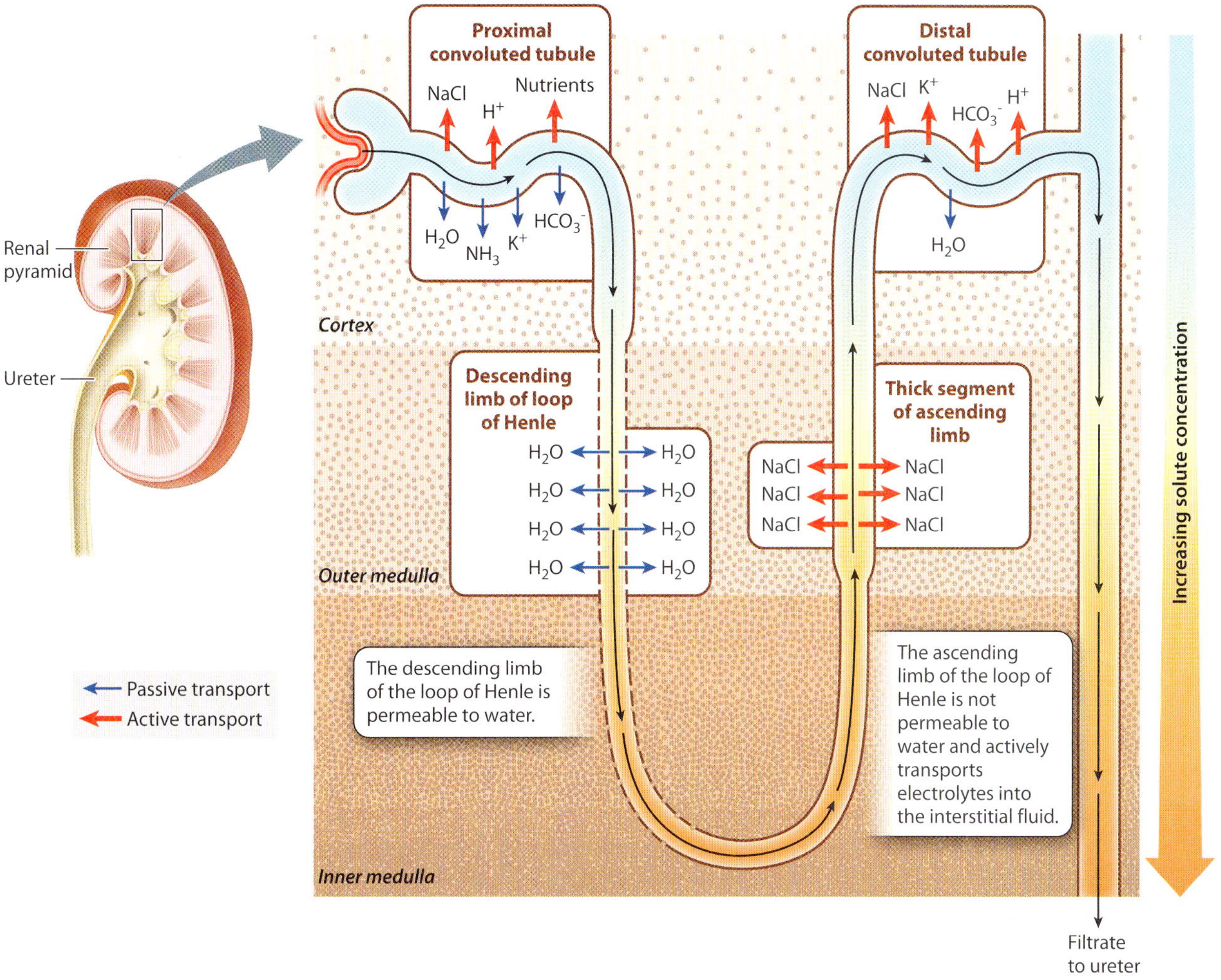

electrolytes are pumped out of the thick ascending limb, the filtrate becomes less and less concentrated. Although the filtrate enters and leaves the loop of Henle at about the same concentration, the trip around the loop creates a concentration gradient in the interstitial fluid surrounding the loop of Henle, with fluid in the inner medulla much more concentrated than that in the outer cortex.

At the base of the loop, the filtrate still contains some electrolytes, but the main solute is urea because electrolytes were actively reabsorbed into the bloodstream from the proximal tubule and because the descending portion of the tubule is impermeable to urea. Then, as additional electrolytes are actively transported out of the thick ascending limb, the concentration of the filtrate (now mainly urea) decreases because the tubule is impermeable to water.

In Chapter 39, we saw how the movement of water and blood in opposite directions in fish gills provides an efficient way to extract oxygen from the water and move it into the blood. This mechanism, called countercurrent exchange, is an efficient mechanism for two fluids to exchange properties, in this case, the amount of oxygen in each of the fluids. In the loop of Henle, active transport of electrolytes creates a concentration gradient, which is then multiplied because the descending and ascending limbs move in parallel but in opposite directions. This system is therefore known as a **countercurrent multiplier.** While countercurrent exchangers *maintain* a concentration gradient, countercurrent multipliers *generate* them.

There is one more important point to consider. Typically, a high concentration of solutes in the interstitial fluid would be quickly lost due to water flowing out of blood vessels by osmosis.

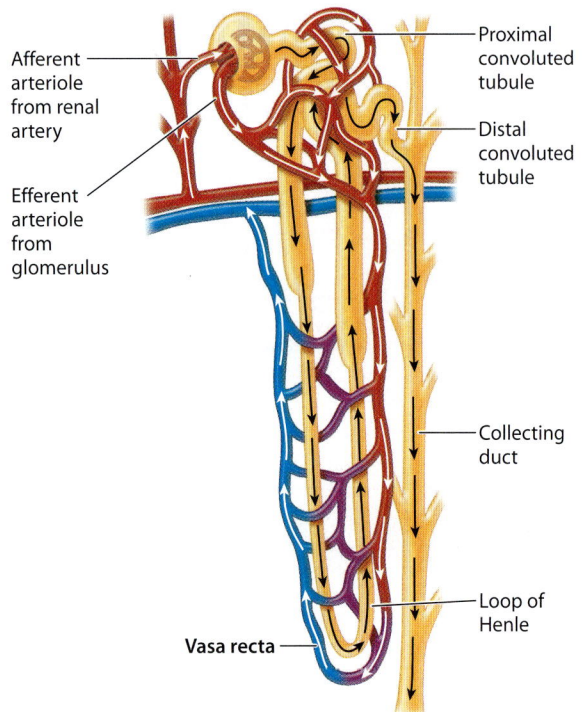

FIG. 41.20 Vasa recta. The blood in the vasa recta moves in a countercurrent fashion, just like the filtrate in the loop of Henle, to maintain the concentration gradient from the cortex to the medulla.

To avoid dissipating the concentration gradient, the blood vessels in the kidneys are also arranged in a particular fashion. The blood vessels, collectively called the **vasa recta,** have descending and ascending vessels arranged in a countercurrent organization, just like the loops of Henle (**Fig. 41.20**). The blood within these vessels flows down into the medulla in one arm and back up through the cortex in the other. This arrangement allows the concentration gradient to be maintained. Blood flowing from the cortex to the base of the kidney encounters increasingly concentrated interstitial fluid, allowing the passive diffusion of water out of the blood. However, the descending vessels pass in proximity to ascending vessels, which reabsorb the water. In effect, water passes straight from one arm of the vasa recta to the other, meaning the interstitial fluid and the bloodstream end up with no net gain or loss of water. As a result, the concentration gradient from the cortex to the medulla is maintained.

The generation of a concentration gradient by a countercurrent multiplier in the loop of Henle is a key step in the production of urine that is more concentrated than the blood. It was first proposed by the Swiss chemist Werner Kuhn in the 1950s. Although initially met with skepticism, subsequent experimental work confirmed predictions of his hypothesis (**Fig. 41.21**).

→ **Quick Check 4** The filtrate is dilute at the start and end of the loop of Henle. What, then, is the function of the loop of Henle?

HOW DO WE KNOW?

FIG. 41.21

How does the mammalian kidney produce concentrated urine?

BACKGROUND The mammalian kidney produces urine that is more concentrated than the blood. How it concentrates urine was one of the most perplexing problems in renal physiology.

HYPOTHESIS One hypothesis to explain how the kidney concentrates urine is that the kidney actively transports water out of the collecting ducts as the filtrate leaves the kidney. Another, first suggested by the Swiss chemist Werner Kuhn in the early 1950s, is that the loop of Henle creates a concentration gradient from the cortex to the medulla by a countercurrent multiplier mechanism. According to this model, the solute concentration is low in the cortex and high in the medulla, so water leaves the collecting ducts passively as it passes from the cortex to the medulla.

PREDICTION A prediction of Kuhn's hypothesis is that the solute concentration of fluid taken at the same level from the medulla, loop of Henle, and collecting ducts should be similar to one another, and higher than that of the blood. If, instead, water is simply pumped out of the collecting ducts, then the fluid in the collecting ducts should be more concentrated than that in the loop of Henle.

EXPERIMENT American physiologists Carl W. Gottschalk and Margaret Mylle tested Kuhn's hypothesis. They used innovative micropuncture techniques in hamsters to measure the solute concentration of fluid taken from the bend of the loop of Henle and the collecting ducts at the same level.

Solute concentration (milliosmoles per kilogram of water)

HAMSTER NO.	LOOP OF HENLE	COLLECTING DUCT	PLASMA
1	1391	1402	308
2	725	720	336
3	1270	1206	325
4	453	453	

FIG. 41.22 Hormonal control of urine concentration by ADH. (a) In the presence of ADH, the collecting duct is more permeable to water than (b) in the absence of ADH.

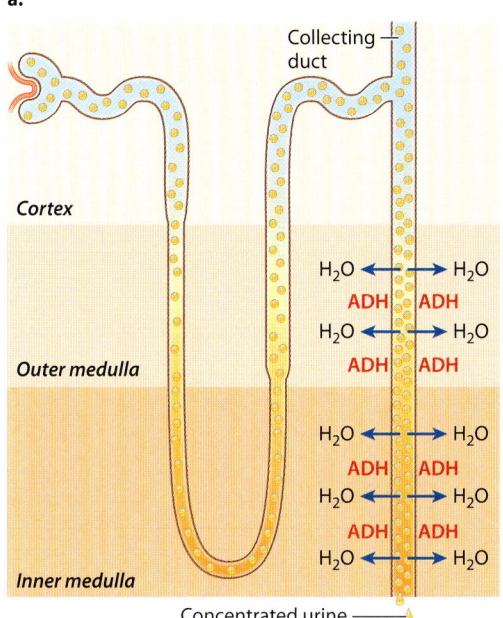
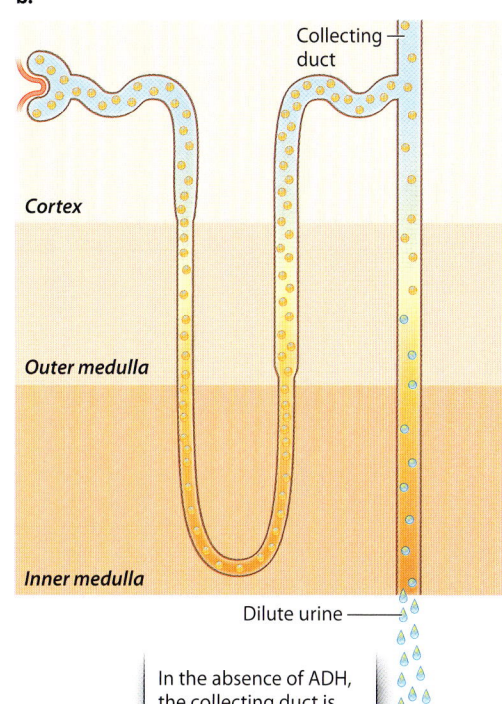

a. ADH released from the posterior pituitary gland increases the collecting ducts' permeability to water, allowing water to diffuse out of the filtrate with the gradient created by the loop of Henle.

b. In the absence of ADH, the collecting duct is less permeable to water, resulting in dilute urine.

RESULTS Their results, shown in the table, are consistent with the predictions of Kuhn's hypothesis. That is, the solute concentrations of the fluid in the loop of Henle and in the collecting ducts are similar to each other and higher than that of the blood.

CONCLUSION These results are consistent with a model in which concentrated urine is produced by first generating a concentration gradient between the cortex and the medulla, and then allowing water to move out of the collecting ducts by osmosis to produce concentrated urine.

FOLLOW-UP WORK Similar measurements in other areas of the kidney and in other mammals and birds supported the model.

SOURCES Gottschalk, C. W., and M. Mylle. 1958. "Evidence That the Mammalian Nephron Functions as a Countercurrent Multiplier System." *Science* 128:594; Gottschalk, C. W., and M. Mylle. 1959. "Micropuncture Study of the mammalian urinary concentrating mechanism: evidence for the countercurrent hypothesis." *American Journal of Physiology* 196 (4):927–936.

The distal convoluted tubule secretes additional wastes.

The filtrate entering the distal convoluted tubule, which is located within the cortex, is hypotonic relative to the interstitial space, having lost nearly all of its electrolytes. As a result, urea remains the principal solute. Other wastes from the bloodstream are also actively secreted into the distal convoluted tubule.

At this point, then, the filtrate is dilute, and the main solutes are urea and other wastes. The dilute filtrate enters the collecting ducts, which drain into the renal pelvis at the base of the kidney, passing from the cortex to the medulla. The concentration gradient established by the loop of Henle from the cortex to the medulla allows water to be reabsorbed as needed depending on the water levels of the organism, as we will see next.

The final concentration of urine is determined in the collecting ducts and is under hormonal control.

The final segment of the nephron is the collecting duct. At this point, the key processes of filtration, reabsorption, and secretion have occurred. The filtrate is dilute but contains wastes and water. The collecting duct is the site where water levels are adjusted to meet the osmoregulatory needs of the organism and therefore maintain homeostasis. The concentration gradient from cortex to medulla established by the loop of Henle allows the collecting ducts to regulate water levels.

The permeability of the collecting ducts to water determines how dilute or concentrated the urine is. The water permeability of the collecting ducts is controlled by the peptide hormone **antidiuretic hormone (ADH),** also called **vasopressin,** which is secreted by the posterior pituitary gland (Chapter 38). In the presence of ADH, the collecting duct walls become permeable to water (**Fig. 41.22**). This change in permeability occurs because aquaporins insert in the membrane of cells of the collecting duct. The aquaporins allow water in the collecting duct to diffuse into the interstitial fluid, thus making the urine in the tubule more concentrated. When a person is dehydrated, the urine leaving the collecting ducts can in the interstitial space at the base of the kidney.

In the absence of ADH, the collecting duct walls are impermeable to water, preventing the water from leaving. This produces dilute urine. Diuretic substances ("diuretic" is from a Greek word meaning "to flow through"), such as caffeine

FIG. 41.23 Control of blood volume and blood pressure by the kidney. Cells of the efferent arteriole of the glomerulus sense blood pressure and start a hormone cascade when blood pressure gets too low.

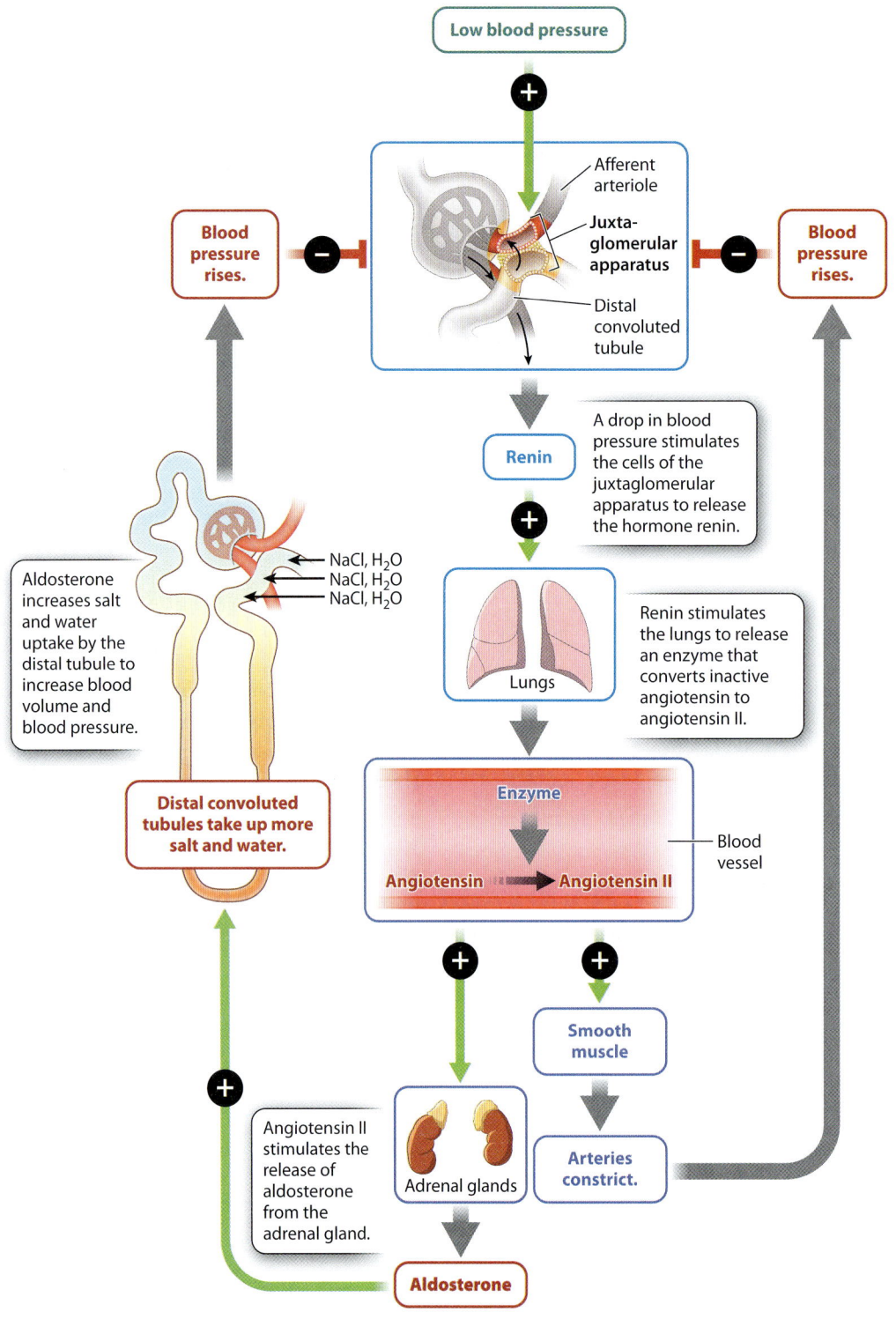

and alcohol, promote the dilution of urine by interfering with ADH receptors and therefore making the collecting duct walls more impermeable to water. If hypotonic urine is produced, there is net water loss. Consequently, diuretic drinks often cause dehydration.

The regulation of water and solute reabsorption by the kidney is remarkably precise. The kidneys of a human filter 180 liters (L) of blood each day (125 ml/min). The total blood volume of a human is about 5 L, which means that the blood is filtered about 60 times per day. Urine production in humans averages 0.5 L per day, which means that under normal conditions, the kidneys reabsorb 99.9% of the water entering the renal tubules through glomerular filtration. In addition, the kidneys reabsorb all the glucose and amino acids that were filtered out of the blood in the glomerulus, together with 99.9% of the electrolytes in the blood. Human kidneys achieve this efficient transport of key ions, solutes, and water because of the renal tubules' extensive microvilli and the large number of mitochondria and membrane transporters in their epithelial cells.

The exceptional ability of the kidneys to filter and reabsorb key ions, solutes, and water also shows why kidney damage can have such serious consequences. People with kidney disease require that their blood be filtered by dialysis treatment as frequently as three times per week, each treatment lasting several hours. Otherwise, toxic wastes would build up to dangerous levels in the body. Renal failure resulting from kidney disease remains a substantial health problem in the world today, costing more than $34 billion in the United States alone.

The kidneys help regulate blood pressure and blood volume.

Because the kidneys receive a large fraction of the blood leaving the heart, they are well suited for monitoring changes in blood pressure and maintaining it at relatively constant levels, another example of homeostasis. Specialized cells of the efferent arteriole leaving the glomerulus of each nephron form the **juxtaglomerular apparatus** (**Fig. 41.23**). In response to a drop in blood pressure, these cells secrete the hormone **renin** into the bloodstream. Renin stimulates epithelial cells of the lung to release an enzyme that converts an inactive hormone (angiotensinogen) in the bloodstream into its active form, **angiotensin II.** Angiotensin II acts on the smooth muscles of arterioles throughout the body, causing them to constrict, thus increasing blood pressure and directing more blood back to the heart.

Angiotensin II also stimulates the release of **aldosterone,** a hormone produced by the adrenal glands that stimulates the distal convoluted tubule and collecting ducts to take up more salt and water. Increased salt and water uptake in turn leads to increased blood volume and therefore blood pressure. Thus, the kidneys participate in the control of blood pressure that is also monitored by baroreceptors (specialized mechanoreceptors that sense blood pressure) in the aorta and carotid bodies. These communicate with the autonomic nervous system to control heart rate and contractile force, as discussed in Chapter 39.

The kidneys, therefore, play many roles. In addition to the elimination of wastes and other toxic compounds, they serve important homeostatic functions, including water and electrolyte balance and the regulation of blood volume and blood pressure.

Core Concepts Summary

41.1 ALL ANIMALS REGULATE THE WATER AND ELECTROLYTE LEVELS WITHIN THEIR CELLS.

Cell membranes and sheets of cells act as semipermeable membranes, allowing the passage of water but restricting or controlling the movement of many solutes. page 41-2

Water moves across semipermeable membranes by osmosis from regions of lower solute concentration to regions of higher solute concentration. page 41-2

Osmotic pressure is the pressure exerted by water as it moves from one solution to another by osmosis. page 41-2

Osmoregulation is the regulation of osmotic pressure inside cells or organisms. page 41-3

Osmoconformers maintain their internal solute concentration similar to that of the environment, whereas osmoregulators have internal solute concentrations different from that of the environment. page 41-4

Both osmoconformers and osmoregulators regulate the levels of particular solutes, especially sodium, potassium, and chloride. page 41-4

Osmoregulators that live in high-salt environments excrete excess electrolytes and minimize water loss. page 41-5

Osmoregulators that inhabit low-salt or freshwater environments excrete excess water and minimize electrolyte loss. page 41-5

Osmoregulators that live on land minimize water loss. page 41-5

41.2 EXCRETORY ORGANS ELIMINATE NITROGENOUS WASTES AND REGULATE WATER AND ELECTROLYTE LEVELS.

Osmoregulation and excretion are closely coordinated processes controlled by the excretory organs of animals. page 41-7

Ammonia, urea, and uric acid are three major forms of nitrogenous waste that animals excrete, depending on their evolutionary history and habitat. page 41-7

Ammonia is toxic but readily diffuses into water; urea is less toxic and can be stored before it is eliminated; uric acid is a semisolid and can be eliminated with minimal water loss as an adaptation for living on land. page 41-7

All animal excretory systems first isolate or filter fluid into an extracellular space. After being isolated or filtered, key electrolytes, solutes, and water are reabsorbed from the filtrate. Additional wastes are then secreted into the remaining filtrate before it is eliminated from the body. page 41-8

Examples of excretory organs are the protonephridia of flatworms, the metanephridia of segmented annelid worms, the Malpighian tubules of insects, and the kidneys of vertebrates. page 41-9

The urine produced by the kidneys and other excretory organs is stored in a bladder or the cloaca until being eliminated from the body. page 41-10

The functional unit of the vertebrate kidney is the nephron, which consists of a glomerulus, renal tubules, and collecting duct. page 41-11

41.3 THE MAMMALIAN KIDNEY CAN PRODUCE URINE THAT IS MORE CONCENTRATED THAN BLOOD AS AN ADAPTATION FOR LIVING ON LAND.

The kidney has an outer cortex and an inner medulla. page 41-12

The renal tubules have specialized regions, including the proximal convoluted tubule, the loop of Henle, and the distal convoluted tubule. page 41-14

The loop of Henle of some nephrons extends all the way into the medulla before looping back to the cortex. These nephrons create a concentration gradient from the outer cortex to the deeper medulla of the kidney. page 41-14

The ability to create a concentration gradient from the cortex to the medulla results from a countercurrent multiplier mechanism of the loops of Henle. page 41-14

The final concentration of urine is regulated by the control of water permeability within the walls of the collecting ducts, which pass from the cortex to the medulla. page 41-17

Release of antidiuretic hormone by the posterior pituitary gland increases the water permeability of the collecting ducts, concentrating the urine that empties from vertebrate kidneys. page 41-17

The kidneys help to regulate blood volume and pressure by secreting the hormone renin, leading to the production of angiotensin II, which constricts blood vessels, and aldosterone, which increases reabsorption of salt and water by the kidneys. page 41-19

Self-Assessment

1. Given a semipermeable membrane that is permeable to water but not to a particular solute with different concentrations on the two sides of the membrane, show the direction of water movement and label the side with the higher osmotic pressure.

2. Describe how animals gain and lose water and electrolytes.

3. Name two animals that are osmoconformers and two that are osmoregulators. Explain the difference between the two types of animal.

4. List three forms of nitrogenous waste and describe how each is an adaptation for the environment in which the animal lives.

5. Describe the three steps in which organisms excrete wastes.

6. Draw a mammalian nephron, label and describe the primary function of each part, and show the direction of water and electrolyte in each part.

7. Explain how the loop of Henle creates a concentration gradient from the cortex to the medulla.

8. Describe the role of ADH in the regulation of urine concentration.

9. Explain how the kidneys help to regulate blood volume and blood pressure.

Do you understand the chapter's Core Concepts? Log in to check your answers to the Self-Assessment questions, then practice what you've learned and reinforce this chapter's concepts by working through the problems and multimedia tutorials provided there.

www.biologyhowlifeworks.com

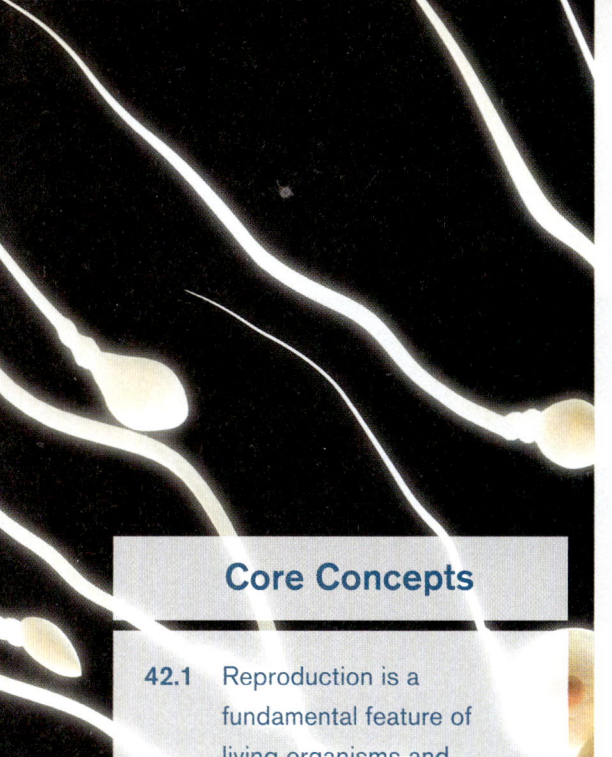

CHAPTER 42

ANIMAL REPRODUCTION AND DEVELOPMENT

Core Concepts

42.1 Reproduction is a fundamental feature of living organisms and occurs asexually and sexually.

42.2 The movement of vertebrates from water to land involved changes in reproductive strategies, including internal fertilization and the amniotic egg.

42.3 The male reproductive system is adapted for the production and delivery of sperm, and the female reproductive system is adapted for the production of eggs and, in many cases, support of the developing fetus.

42.4 Human reproduction involves the formation of gametes, fertilization, and growth and development.

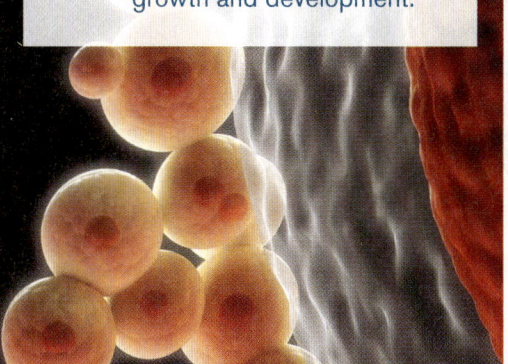

Children are sometimes gently introduced to the subject of human reproduction by referring to "the birds and the bees." Since birds and bees reproduce, it is assumed that they can substitute for humans in discussions about reproduction. But how relevant is bird and bee reproduction to human reproduction? It might come as a surprise that most male birds don't have penises, and that male bees don't even have fathers. Perhaps, then, the lesson of the "birds and the bees" is not so much in what it teaches us about human reproduction, but, as we explore in this chapter, in showing us that reproductive strategies among organisms are spectacularly diverse.

Reproduction is a striking and conspicuous feature of the natural world. Flowers burst into bloom each spring to attract pollinators, male fireflies light up brilliantly on warm summer nights as they signal to females in the grass below, birds sing and display ornate plumage to attract mates. Most living organisms have the ability to reproduce—that is, to make new organisms like themselves. This remarkable ability is not unique to living organisms—after all, computer viruses and self-replicating molecules are capable of reproduction but are not themselves living. However, reproduction is a key attribute of life. All living things came about by reproduction, and most are capable in turn of reproduction. Of course, this statement does not address the question of how life originated in the first place, a topic considered in Case 1: The First Cell.

This chapter focuses on reproduction and development. We begin by considering reproduction broadly across the living world, then turn our attention to animal, and specifically vertebrate, reproduction and the surprise that even among vertebrates there is quite an array of reproductive strategies. In the last two sections, we focus on human reproduction, starting with reproductive anatomy and physiology and then following reproduction from gamete formation to fertilization, pregnancy, and birth.

42.1 THE EVOLUTIONARY HISTORY OF REPRODUCTION

While the ability to reproduce is shared across all domains of life, the particular way in which animals reproduce varies among species, from the familiar to what might strike us as truly bizarre. To take just one of many examples: After birth, a male anglerfish finds a female anglerfish, bites her skin, and fuses with her. His body atrophies, leaving the male reproductive organs physically attached to the female and allowing her eggs to be fertilized when she is ready. To begin to make sense of the diversity of reproductive strategies, we consider how different forms of reproduction evolved, what organisms use them, and their adaptive significance.

FIG. 42.1 Asexually reproducing organisms. (a) Bacteria divide by binary fission. (b) The yeast *Saccharomyces cerevisiae* reproduces by budding. (c) Corals are propagated by fragmentation. (d) Komodo dragons can reproduce by parthenogenesis.

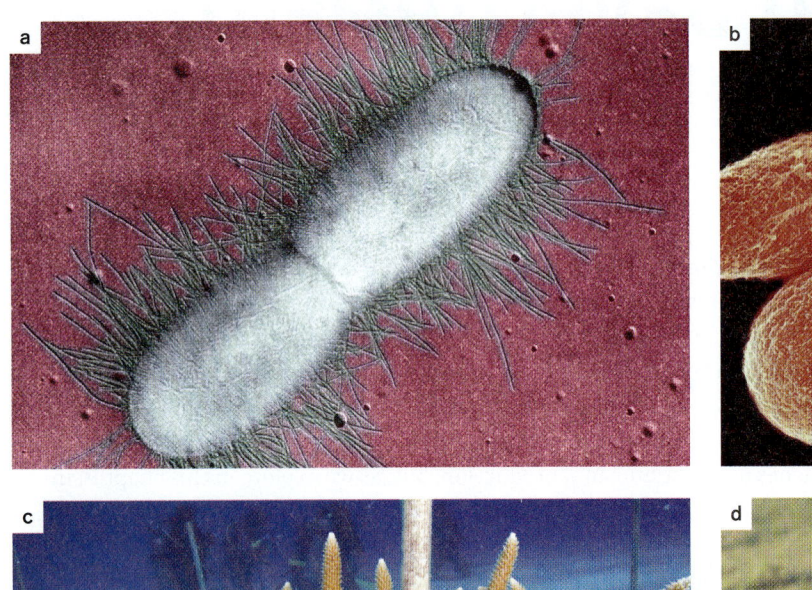
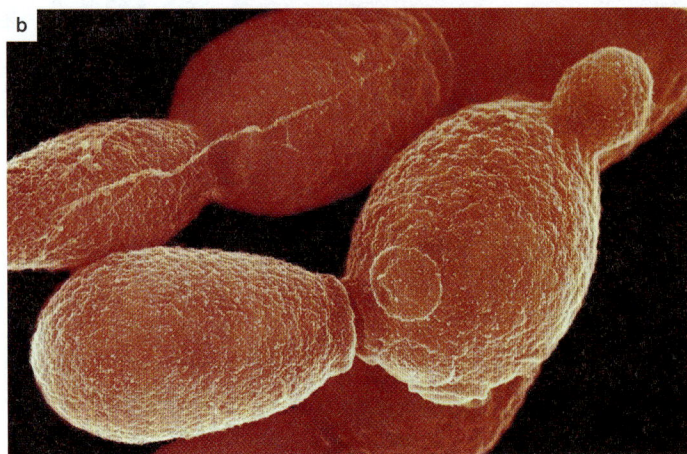

Asexual reproduction produces clones.

Organisms have two potential ways to reproduce. One is the production of genetically identical cells or individuals called **clones.** This form of reproduction is called **asexual reproduction,** and it occurs in a wide variety of organisms (**Fig. 42.1**).

For example, bacteria (Fig. 42.1a) and archaeons reproduce by binary fission, a form of cell division in which the genome replicates and then the cell divides in two (Chapter 11). The result is two cells that are either identical to the parent cell or nearly so, differing only because of chance mutations.

Prokaryotes can nevertheless increase genetic variation in various ways (Chapter 26). For example, they can transfer DNA from one individual to another, usually of the same species, and much more rarely to an individual of another species, during conjugation. They can also obtain DNA directly from the environment and by viral infection, incorporating genes that might have evolved in even distantly related species. These processes allow prokaryotes to acquire novel genetic sequences that can affect fitness.

Eukaryotic cells divide by mitosis (Chapter 11), a process that evolved early and was present in the common ancestor of all living eukaryotes. Mitosis is a form of growth and development in all eukaryotes, and is the basis of asexual reproduction in many eukaryotes.

One form of asexual reproduction in fungi, plants, and some animals is **budding,** in which a bud, or protrusion, forms on an organism and eventually breaks off to form a new organism that is smaller than its parent. The budding yeast, *Saccharomyces cerevisiae*, which is used to make bread and beer, is so named because of its ability to produce new individuals by budding (Fig. 42.1b). Budding can also occur in animals such as the water-dwelling hydra, which can grow and subsequently break off part of itself to form a new, free-living individual. Whether used to describe reproduction in unicellular or multicellular organisms, budding involves an unequal division between a mother and daughter cell or organism. Budding is the result of mitosis and therefore, like binary fission, results in an offspring that is genetically identical to the parent.

Another form of asexual reproduction is **fragmentation,** in which new individuals arise asexually by the splitting of one organism into pieces, each of which develops into a new individual. Certain molds, algae, worms, sea stars, and corals reproduce by fragmentation. While fragmentation occurs naturally, it is also done intentionally to propagate corals in reef restoration projects (Fig. 42.1c). Once again, the new individual is a clone of the parent.

A final example is **parthenogenesis,** literally "virgin birth." In this form of asexual reproduction, females produce eggs that are not fertilized by males but divide by mitosis and develop into new individuals. Parthenogenesis occurs in invertebrates, such as bees, other insects, and crustaceans, as well as in vertebrates, including some fish and reptiles (Fig. 42.1d). The details of parthenogenesis differ depending on the organism. Egg cells are usually the product of meiosis and are haploid, having one set of chromosomes. In some organisms, this haploid egg cell divides by mitosis to become a haploid adult; in others, there is a doubling of the DNA content during development, so the adult is diploid, having two sets of chromosomes; and in still others, the egg cell does not undergo meiosis (it is diploid) and divides by mitosis to become a diploid adult. In all cases, parthenogenesis involves a single maternal parent.

Sexual reproduction involves the formation and fusion of gametes.

The second way that organisms reproduce is by combining a complete set of genetic information from two individuals to make a new genetically unique individual. This form of reproduction is called **sexual reproduction.**

At its core are two basic biological processes—**meiosis** and **fertilization** (**Fig. 42.2**). Meiosis is a form of cell division that halves the number of chromosomes; fertilization involves fusion of specialized cells to restore the original chromosomal content. Meiosis evolved in the ancestors of modern eukaryotes, and so sexual reproduction occurs only in eukaryotes.

As we saw in Chapter 11, meiosis consists of one round of chromosome replication followed by two rounds of cell division, resulting in four cells each with half the number of chromosomes of the parent cell. So, in a diploid species with $2n$ chromosomes, the result of meiosis is four cells with n chromosomes. The resulting cells are haploid and are called **gametes** or **spores.** Humans, for example, have 23 pairs of chromosomes ($2n$ = 46 chromosomes) and human gametes have half this number (n = 23 chromosomes).

FIG. 42.2 Sexual reproduction. This form of reproduction involves the production of gametes by meiosis and their fusion by fertilization.

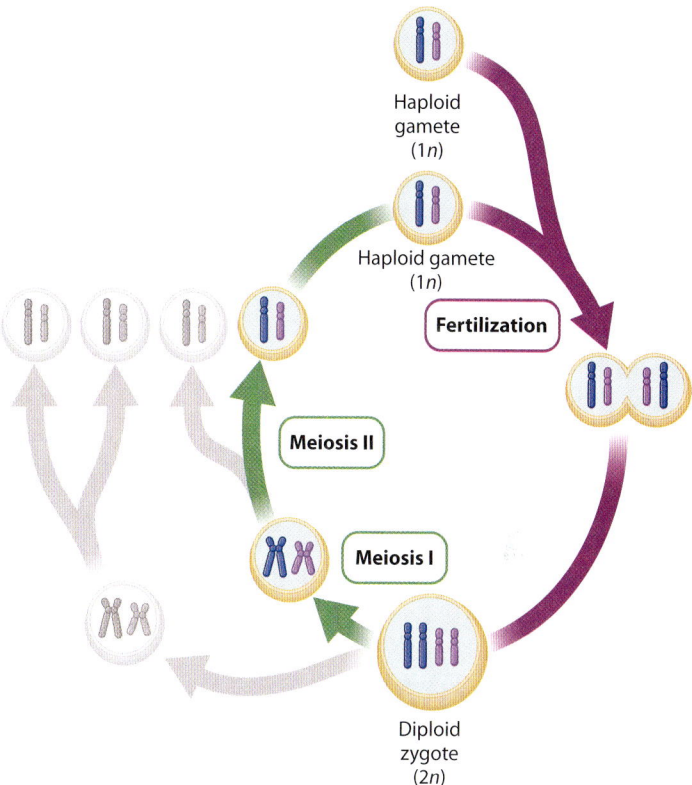

Many species produce two types of gametes that differ in shape and size. The smaller, male gametes are called **spermatozoa,** or **sperm,** and the larger, female gametes are called **ova,** or **eggs.** For other organisms, including green algae, diatoms, slime molds, and most fungi, the two types of gametes are the same size. In spite of their similar size and appearance, however, they come in two (or more) distinct mating types and cannot reproduce sexually with individuals of the same type.

Fertilization involves the fusion of two gametes, one from each parent (Fig. 42.2). The result is a **zygote.** If each gamete has n chromosomes, the resulting zygote has $2n$ chromosomes. Fertilization results in a net reduction in cell number—two haploid gametes fuse to make one diploid zygote. For unicellular organisms with a prominent haploid phase, increase in population size occurs when the zygote undergoes meiosis and the four resulting cells divide by mitosis (Chapter 27). For diploid multicellular organisms, however, fertilization is the direct basis for reproduction. The zygote divides by mitosis and develops into an **embryo,** an early stage of multicellular development.

Meiosis and fertilization have significant genetic consequences. During meiosis, pairs of homologous chromosomes undergo recombination, shuffling the combination of alleles between the chromosomes. In addition, meiosis randomly sorts homologous chromosomes into different gametes. As a result, the products of meiosis, the haploid gametes, are each genetically unique and represent a mixture of the parental genetic makeup. Then, during fertilization, two of these unique haploid gametes fuse, creating new genetic combinations and a unique diploid individual.

→ **Quick Check 1** Sexual reproduction results in offspring that are genetically different from one another and from their parents. Describe four mechanisms that produce this genetic variation. Are any of these mechanisms shared with asexual reproduction?

Many species reproduce both sexually and asexually.

Most organisms that reproduce asexually are also capable of reproducing sexually. For example, hydras reproduce asexually by budding, but this ability to reproduce asexually does not preclude their ability to reproduce sexually. Even budding yeast are capable of sexual reproduction. Most aphids, fish, and reptiles capable of parthenogenesis can also reproduce sexually.

What determines when an organism reproduces sexually and when it reproduces asexually? This question has been examined in the freshwater crustacean *Daphnia*. *Daphnia* reproduce asexually by parthenogenesis in the spring, when food is plentiful, perhaps because asexual reproduction is a rapid form of reproduction that allows the organism to take advantage of abundant resources (**Fig. 42.3**). Later in the season, when conditions are less favorable because of increased crowding and decreased food and temperature, *Daphnia* reproduce sexually, mating to produce a zygote that starts to form an embryo. At an early stage, however, development is arrested and metabolism slows dramatically. The embryo forms a thick-walled structure, called a cyst, which enables it to resist freezing and drying and therefore last through the winter.

This link between environmental conditions and mode of reproduction has been observed in many other species, from algae to slime molds. The life cycle of *Daphnia* hints that asexual and sexual reproduction have costs and benefits, a subject we turn to now.

Exclusive asexuality is often an evolutionary dead end.

Asexual reproduction has some advantages compared to sexual reproduction. It does not involve finding and attracting a mate, which takes time and energy. Certain birds, for example, go to elaborate lengths to attract mates. Male birds of paradise invest a significant amount of energy in growing colorful plumage, and male bowerbirds take time to build elaborate homes for their potential mates (**Fig. 42.4**). For other organisms, finding a mate

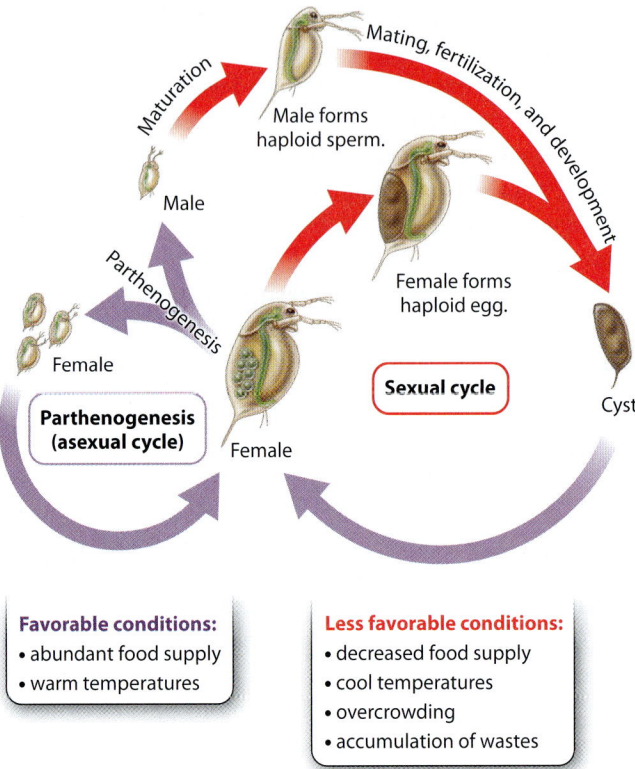

FIG. 42.3 Sexual and asexual reproduction in *Daphnia*. *Daphnia* can reproduce sexually or asexually, depending on environmental conditions.

can be difficult because of very low population density—this is the case for anglerfish, and may help to explain their apparently unusual reproductive strategy, described earlier.

In addition, asexual reproduction is rapid, allowing organisms to increase their numbers quickly. Consider rates of population growth for organisms that reproduce asexually and those that reproduce sexually. In asexual organisms, all offspring can produce more offspring, allowing for exponential population growth if conditions are favorable. By contrast, in sexually reproducing organisms, only females produce new offspring. The consequence of this difference is that an asexual group can have a much greater rate of population growth compared to that of a sexually reproducing one. The British evolutionary biologist John Maynard Smith called this effect the **twofold cost of sex** (**Fig. 42.5**), and it is one of the key disadvantages of sexual reproduction.

In spite of the advantages of asexual reproduction, a survey of the natural world reveals a striking pattern: Most organisms reproduce sexually at least part of the time. And even prokaryotes, as we have seen, have mechanisms for genetic exchange. The observation that the vast majority of organisms reproduce

FIG. 42.4 Time and energy costs of sexual reproduction. (a) A male bird of paradise has colorful plumage to attract a mate. (b) A bowerbird builds an elaborate hut to entice a female.

sexually at some point suggests that sexual reproduction offers advantages over asexual reproduction. Asexual reproduction is quick, but it produces offspring that are genetically identical to one another and to the parent. The only source of genetic variation is chance mutations. By contrast, one of the defining features of sexual reproduction is the production of offspring that are genetically different from one another and from their parents.

Many hypotheses have been developed to explain the advantages of producing genetically distinct offspring in spite of the costs of sexual reproduction. One suggests that sexual reproduction allows organisms to adapt faster than asexual ones because rare beneficial mutations that arise in different organisms can be brought together, increasing the overall fitness of the population. Another set of hypotheses suggests that sexual reproduction allows a population to purge itself of harmful mutations more quickly than could a population of asexual individuals.

A third hypothesis—the Red Queen hypothesis—suggests that sexual reproduction is a mechanism of parasite defense. This hypothesis takes its name from Lewis Carroll's *Through the Looking-Glass*, in which the Red Queen tells Alice that she needs to keep running to stay in the same place. Similarly, in host–parasite coevolution, the host evolves defenses against the parasite, and the parasite evolves new ways to infect the host, so each evolves rapidly, or keeps moving, simply to maintain its same ecological position—that is, to stay in the same place. This kind of rapid

FIG. 42.5 The twofold cost of sex. Population size increases more rapidly in asexually reproducing organisms compared to sexually reproducing organisms.

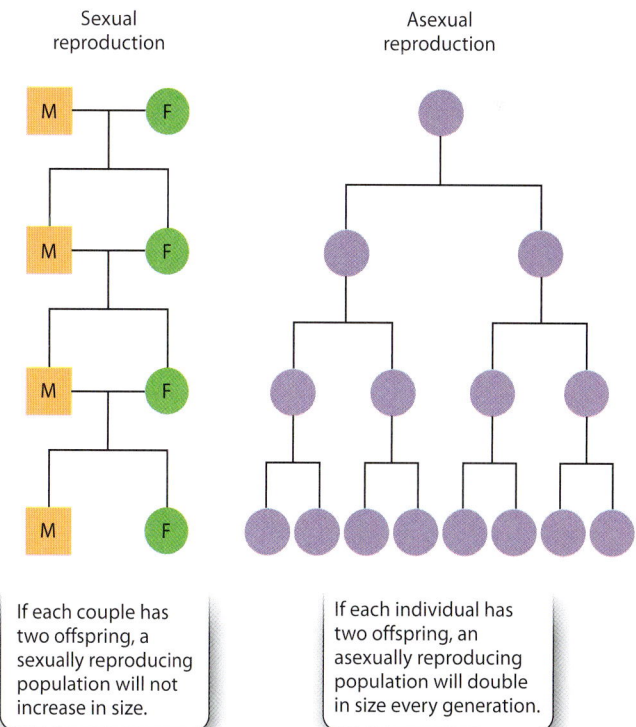

If each couple has two offspring, a sexually reproducing population will not increase in size.

If each individual has two offspring, an asexually reproducing population will double in size every generation.

evolution requires lots of genetic variation on which natural selection can act, and sexual reproduction increases genetic variation among individuals more than asexual reproduction does.

These different explanations for the prevalence of sexual reproduction are not mutually exclusive. It may be that several mechanisms work together, or that different ones are important for different organisms.

It is instructive to consider groups of eukaryotes that have lost the ability to reproduce sexually, and those that have lost the ability to reproduce asexually. While there are organisms such as the New Mexico whiptail lizard (*Cnemidophorus neomexicanus*) that reproduce only asexually, these species are not very old in evolutionary terms. In other words, there are very few, if any, *ancient* asexual groups. It appears that asexual reproduction on its own rarely allows a species to persist for very long in evolutionary terms.

A possible exception to this rule may be bdelloid rotifers, a group of microscopic freshwater invertebrates that are thought to have reproduced asexually for at least 35 to 40 million years (**Fig. 42.6**). The biology of this group is being intensely scrutinized for hints of its pattern of reproduction over time. If indeed bdelloid rotifers have persisted for tens of millions of years without sexual reproduction, then we have an interesting puzzle. We might predict that the rotifers have either an unusual way to escape the long-term

HOW DO WE KNOW?

FIG. 42.6

Do bdelloid rotifers only reproduce asexually?

BACKGROUND Many groups of animals reproduce asexually, but most of these organisms reproduce sexually at least some of the time. The bdelloid rotifer, one of several species of rotifer, may be an exception.

OBSERVATION All bdelloid rotifers found to date are females. No males or hermaphrodites (organisms with both male and female reproductive organs) have ever been observed. Meiosis has never been documented.

HYPOTHESIS Bdelloid rotifers are asexual and reproduce solely by parthenogenesis, in which unfertilized eggs develop into adult females.

PREDICTION In the absence of meiosis, the two copies (alleles) of a gene do not recombine and mutations that accumulate are not reshuffled between them. As a result, two alleles of a gene in an asexually reproducing organism will show a much higher degree of sequence difference compared to two alleles of a gene in a sexually reproducing organism.

RESULTS Following is the sequence from the two copies of the *hsp82* gene in a sexually reproducing rotifer (Fig. 42.6a) and an asexually reproducing rotifer (Fig. 42.6b). As can be seen, the two copies are similar in Fig. 42.6a but very different in Fig. 42.6b.

CONCLUSION All evidence to date suggests that bdelloid rotifers are exclusively asexual.

SOURCE Welch, M. D., and M. S. Meselson, 2000. "Evidence for the Evolution of Bdelloid Rotifers Without Sexual Reproduction or Genetic Exchange." *Science* 288 (5469):1211–1215.

a. The two copies of *hsp82* found in a sexually reproducing rotifer.

The two copies of the *hsp82* gene are very similar in the sexually reproducing rotifer.

b. The two copies of *hsp82* found in an asexually reproducing bdelloid rotifer.

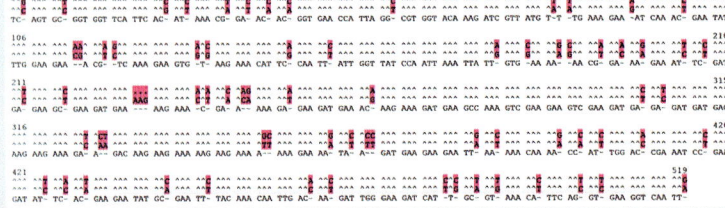

The two copies of the *hsp82* gene are very different in the asexually reproducing rotifer.

Asterisks (*) indicate bases that are identical in the two gene copies, and red shading indicates bases that are different.

evolutionary costs of genetic uniformity, or an as yet unknown mechanism for exchange of genetic material.

Mammals are unusual in the opposite way. Mammals have lost the ability to reproduce asexually. Parthenogenesis, as we have seen, occurs in some vertebrates, but does not occur in mammals. Although this form of reproduction may seem odd from our point of view, scientists have wondered just the opposite—why can't female mammals reproduce without a father? Asked another way, why do mammalian embryos need both a maternal and a paternal genome? Wouldn't two maternal genomes from a single parent do the trick? Evidently not. Scientists have known for some time that mammalian maternal and paternal genomes are not quite the same. There are some genes, called imprinted genes, that are expressed differently depending on their parent of origin. Imprinting ensures that mammals develop from embryos with one maternal and one paternal genome. When researchers manipulated key imprinted genes in a maternal genome, they were able to breed a parthenogenetic mouse—one with two copies of a maternal genome from a single parent.

We can speculate that early mammals gained fitness advantages in the evolution of imprinting of genes. However, a consequence was the loss of parthenogenesis in mammals. Parthenogenesis might offer greater fitness advantages than strictly sexual reproduction in some species of mammals living today, but that option is no longer available.

42.2 MOVEMENT ONTO LAND AND REPRODUCTIVE ADAPTATIONS

We have considered the two modes of reproduction observed across all living organisms. In this section, we narrow our focus to vertebrates: fish, amphibians, reptiles, birds, and mammals. Interestingly, even among vertebrates, there are many different reproductive strategies. Understanding these strategies not only gives us a picture of how these organisms reproduce but also sheds light on how they live and behave.

How vertebrates reproduce is also intimately connected to the transition of vertebrates from water to land 375 to 400 million years ago. Moving from an aquatic environment to a terrestrial one involved many adaptations. Among these were changes in the way that reproduction took place, so it is helpful to consider reproductive strategies in an evolutionary context.

Fertilization can take place externally or internally.

Eggs and sperm require a wet environment to survive. In aquatic organisms like fish and amphibians, eggs and sperm can be released directly into the water in **external fertilization**.

FIG. 42.7 External fertilization. Tree frogs reproduce by external fertilization, with the male frog holding onto the female and fertilizing her eggs as she lays them.

External fertilization, as you might imagine, is a chance affair: Somehow, sperm must meet egg in a large environment. Animals that release gametes externally therefore have developed strategies to increase the probability of fertilization. One strategy involves releasing large numbers of gametes. In addition, many female fish, including salmon and perch, lay eggs on or near a substrate, such as a rock, and males release their sperm onto the eggs to increase the probability of fertilization. Externally fertilizing animals may also come close together physically to improve chances that the sperm will fertilize the eggs. Male frogs grasp females, even though fertilization occurs externally, releasing sperm as the female lays her eggs (**Fig. 42.7**).

A challenge for land-dwelling organisms is how to keep gametes from drying out. **Internal fertilization,** in which fertilization takes place inside the body of the female, is an adaptation for living on land. Reptiles, birds, and mammals all use internal fertilization. However, this mode of fertilization is not exclusive to land dwellers. Marine mammals such as whales and dolphins use internal fertilization, as did the land-dwelling ancestors from which they evolved. Even some fish, such as guppies, reproduce by internal fertilization.

r-strategists and K-strategists differ in number of offspring and parental care.

While the terms "external" and "internal" fertilization focus our attention on where fertilization takes place, the two modes of reproduction are associated with a host of other behavioral and reproductive differences. One key difference is the number of offspring produced.

External fertilization is generally associated with the release of large numbers of gametes and the production of large numbers of offspring, each of which has a low probability of survival in part because the offspring typically receive little, if any, parental care. External fertilization is a game of numbers. In internal fertilization, however, there is much more control over fertilization and subsequent development of the embryo because both fertilization and embryonic development take place inside the female. As a result, animals that use internal fertilization typically produce far fewer offspring and invest considerable time and energy into raising those offspring.

These two strategies represent two ends of a continuum of reproductive strategies that was first described by the American ecologists Robert H. MacArthur and E. O. Wilson in 1967. Organisms that produce large numbers of offspring without a lot of parental investment are called **r-strategists,** and those that produce few offspring but put in a lot of parental investment are **K-strategists** (**Fig. 42.8**). Fish are r-strategists; humans are K-strategists.

According to this model, the environment places selective pressures on organisms to drive them in one or the other direction. In general, r-strategists evolve in unstable, changing, and unpredictable environments. In such an environment, there is an advantage to reproducing quickly and producing many offspring when conditions are favorable. By contrast, K-strategists often evolve in stable, unchanging, and predictable environments, where there tends to be more crowding and larger populations. Here, there is intense competition for limited resources, so traits such as increased parental care and few offspring are favored.

These two strategies represent ends of a spectrum, and there are notable exceptions. An octopus, for example, can lay up to 150,000 eggs, typical of r-strategists, but the female guards and takes care of them for a month or more, typical of K-strategists. However, the two strategies highlight different reproductive patterns across the animal kingdom and suggest important links between number of offspring, parental care, and the environment (Chapter 46).

Oviparous animals lay eggs, and viviparous animals give birth to live young.

When we speak of parental care, it is natural to think of tending the young. However, parental care actually begins much earlier in development, in the provisioning of nutrients for the developing embryo. The eggs of animals with external fertilization have a substance called **yolk** that provides all the nutrients that the developing embryo needs until it hatches.

FIG. 42.8 *r*-strategists and *K*-strategists. *r*-strategists, like fish, produce many offspring with low probability of survival, whereas *K*-strategists, like primates, produce few offspring and put a lot of care into raising their young.

Water and oxygen are obtained directly from the aquatic environment and metabolic wastes easily diffuse out.

Animals that are internally fertilized can either lay eggs or give birth to live young. These two strategies are called **oviparity** ("egg birth") and **viviparity** ("live birth"). Oviparity is used by fish and amphibians, as well as reptiles, birds, and most insects. Even some mammals—the monotremes, which include the spiny anteater and platypus—lay eggs. For oviparous animals there is very little, if any, embryonic development inside the mother, so all the nutrients for the developing embryo come from the yolk.

The movement onto land was associated with important changes in the anatomy of the egg. No longer could the developing embryo depend simply on diffusion of key substances into and out of a watery environment. A key innovation is the **amnion,** a membrane surrounding a fluid-filled cavity that allows the embryo to develop in a watery environment. Reptiles, birds, and mammals are known as amniotes to reflect the presence of this key membrane (**Fig. 42.9**). A second membrane, called the **allantois,** encloses a space where metabolic wastes collect. Finally, a third membrane, called the **chorion,** surrounds the entire embryo along with its yolk and allantoic sac. The embryo is further protected in many species by a hard shell that prevents drying out and facilitates gas exchange. The yolk sac, amnion, allantois, and chorion are **extraembryonic membranes,** sheets of cells that extend out from the developing embryo (**Fig. 42.10**).

FIG. 42.9 **The evolution of the amnion.** The amnion is a fluid-filled cavity surrounding the developing embryo that evolved during the transition from water to land.

FIG. 42.10 **The amniotic egg.** The amniotic egg includes four extraembryonic membranes: the yolk sac, amnion, allantois, and chorion.

Viviparous animals give birth to live young. Most mammals are viviparous, but viviparity is not unique to mammals; it evolved independently in several different groups. For example, some sharks and snakes are capable of giving birth to live young. In viviparous animals, the embryo develops inside the mother. In placental mammals, such as humans, the chorion and allantois fuse to form the **placenta,** an organ that allows nutrients to be obtained directly from the mother.

42.3 HUMAN REPRODUCTIVE ANATOMY AND PHYSIOLOGY

In this and the next section, we continue to narrow our focus in order to consider reproduction in more detail. Here, we cover human reproduction and development. While many aspects of human reproduction are shared with other mammals, others are unique. Our focus on human reproduction allows for a detailed treatment of the topic in one organism and also gives us a chance to explore ways we intervene in the process to promote, prevent, and follow pregnancy. Humans, like all mammals, are amniotes. They reproduce sexually, use internal fertilization, and are viviparous.

The male reproductive system is specialized for the production and delivery of sperm.

Let's start by considering the human male reproductive system. Male gametes, called spermatozoa, or sperm, are produced in the

FIG. 42.11 The male reproductive system. Haploid sperm are produced in the testes, then travel through the epididymis, vas deferens, ejaculatory duct, and urethra.

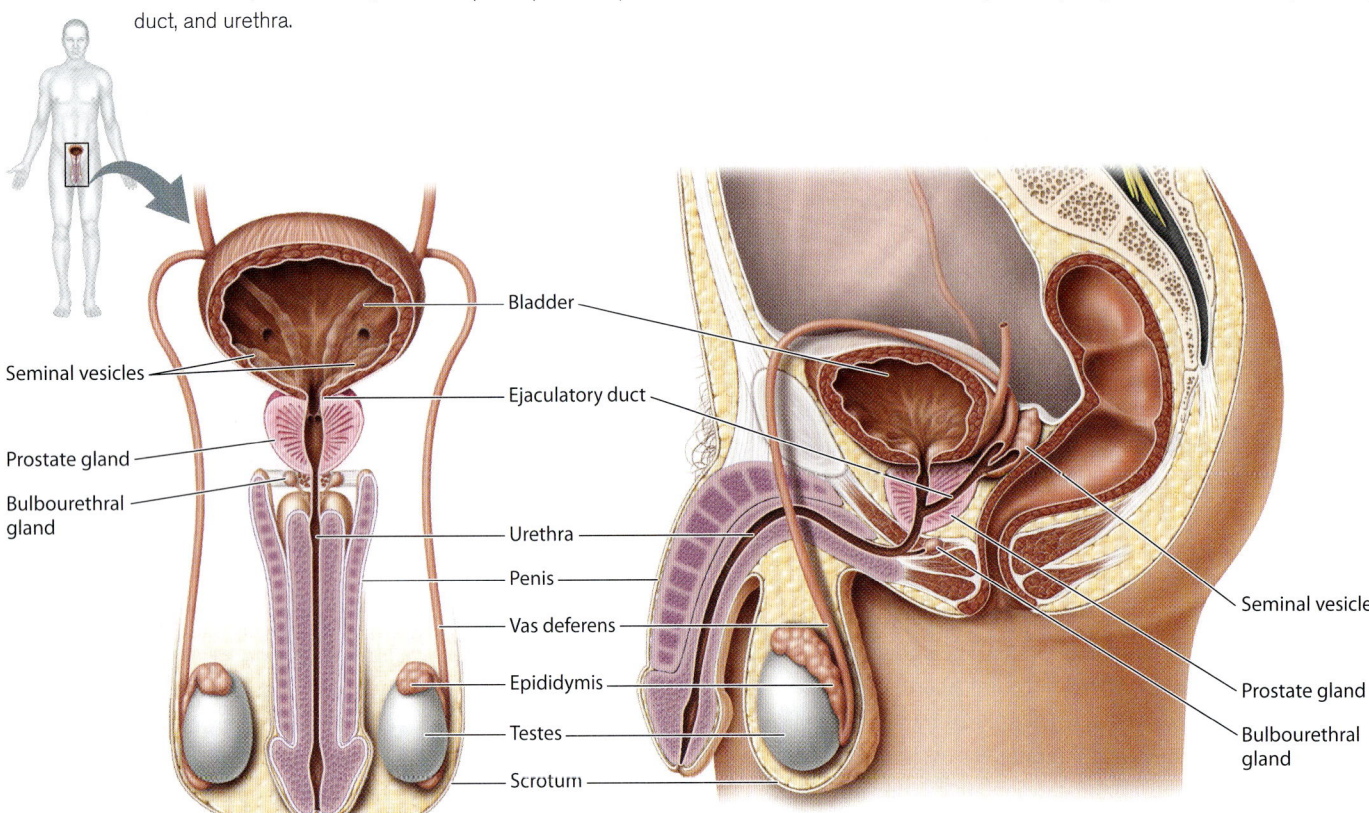

two male **gonads,** or **testes** (singular, testis; **Fig. 42.11**). Over the course of his life, a man is capable of making hundreds of billions of sperm. In fact, a single ejaculate typically contains several hundred million sperm. If it takes just one sperm to fertilize an egg, why do males make so many? A likely explanation involves sperm competition. In species like humans, where females can mate with more than one male, a male's reproductive success is at least partially correlated with the number of sperm—the more sperm, the better chance of fertilizing the egg. A second explanation relates to the female reproductive tract, which is hostile to sperm; many sperm are needed to ensure that at least one makes it to the egg. Of the several hundred million sperm released by a male during a single ejaculation, only several hundred sperm make it to the egg.

Sperm are haploid cells with a unique shape and properties adapted for their function of delivering genetic material to the egg (**Fig. 42.12**). They have a small head containing minimal cytoplasm and a very densely packed nucleus. The head is surrounded by a specialized organelle, the **acrosome,** that contains enzymes that are used by sperm to transverse the outer coating of the egg. Sperm have a long tail, or **flagellum,** which moves the cell by a whipping motion powered by the sperm's mitochondria.

The testes are located outside the abdominal cavity in a sac called the **scrotum** (see Fig. 42.11). This location is critical for the production of sperm, which require cooler temperatures than are found in the abdominal cavity. However, the testes of some mammals, such as elephants, whales, and bats, are located inside the abdominal cavity, indicating that there are species-specific differences in sperm temperature requirements.

Sperm are produced within the testes in a series of tubes called the **seminiferous** ("seed-bearing") **tubules** (**Fig. 42.13**). This is where diploid cells undergo meiosis to produce haploid sperm. When sperm are first produced, they are not fully mature: They are unable to move on their own and incapable of fertilizing an egg. They mature as they move through the male reproductive system.

From the seminiferous tubules, sperm travel to the **epididymis,** where they become motile (Fig. 42.13). This is also where sperm are stored prior to ejaculation. From there, sperm enter the **vas deferens,** a long, muscular tube that follows a circuitous path, starting off in the scrotum, passing into the abdominal cavity, running along the bladder, and then connecting with the **ejaculatory duct** (see Fig. 42.11). The two ejaculatory ducts merge at the **urethra,** just below the bladder. The urethra is used by both the urinary and reproductive systems in males.

FIG. 42.12 Sperm. Sperm are specialized cells with minimal cytoplasm, a densely packed nucleus, an acrosome with digestive enzymes, and a flagellum.

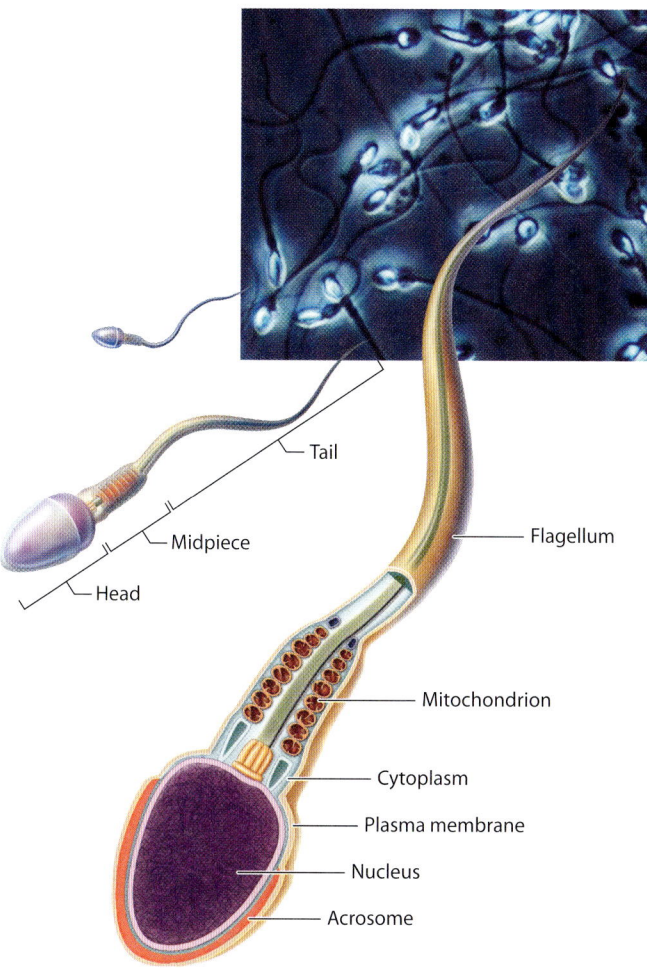

FIG. 42.13 Seminiferous tubules, epididymis, and vas deferens.

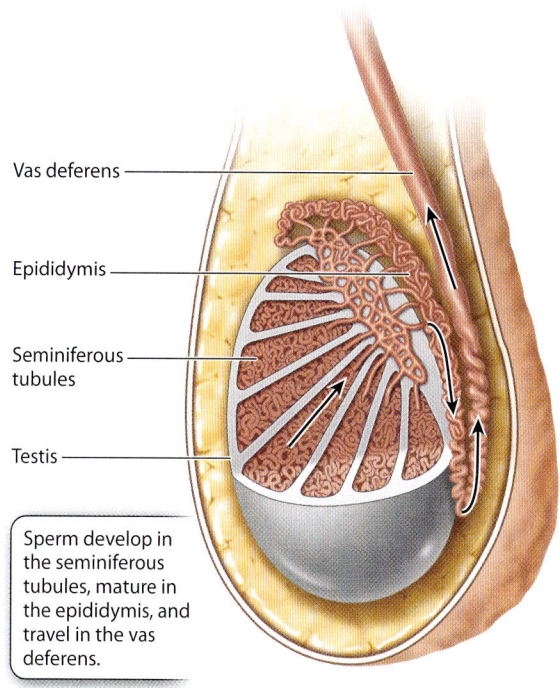

Sperm develop in the seminiferous tubules, mature in the epididymis, and travel in the vas deferens.

Along this path, there are several exocrine glands that produce components of **semen,** the fluid that nourishes and sustains sperm as they travel in the male and then female reproductive tracts. The first is the **prostate gland,** located just beneath the bladder (see Fig. 42.11). The urethra and ejaculatory ducts pass directly through this gland. The prostate gland produces a thin, slightly alkaline fluid that helps maintain sperm motility and counteracts the acidity of the female reproductive tract. Two **seminal vesicles,** located at the junction of the vas deferens and the prostate gland, secrete a protein- and sugar-rich fluid that makes up most of the semen and provides energy for sperm motility. Finally, the **bulbourethral glands,** below the prostate gland, produce a clear fluid that lubricates the urethra for passage of sperm.

The urethra is enclosed in the **penis,** the male copulatory organ (see Fig. 42.11). When a male is sexually aroused, the penis becomes erect as a result of changes in blood flow—the arteries become dilated and the veins compressed. During ejaculation, semen is expelled from the penis.

The male reproductive system is well adapted for the production, storage, and delivery of sperm. Therefore, male contraceptive methods focus on blocking sperm. For example, a condom is a sheath of rubber or latex that covers the penis and blocks sperm from entering the female during sexual intercourse.

→ **Quick Check 2** What effect would cutting the vas deferens and tying off the ends have on male fertility?

The human penis is a relatively simple organ, consisting of a shaft and a head called the **glans penis.** In other animals, the penis can be quite elaborate, with spikes, hooks, and grooves, whose function is to stimulate the female, grip the female, or in some cases even remove sperm from a previous copulation. In marsupial mammals like kangaroos and koalas, males have forked penises, and in snakes and lizards, males have two penises, called hemipenes. In insects, the external genitalia are often distinct enough to be used in species identification. By contrast, most species of birds do not have a penis.

The female reproductive system produces eggs and supports the developing embryo.

The female reproductive system, like the male reproductive system, produces gametes, but it is also specialized to support the developing embryo. Developing female gametes, called **oocytes,**

FIG. 42.14 The female reproductive system. Oocytes are produced in the ovaries and then travel through the fallopian tubes to the uterus, where they implant if they are fertilized; if they are not fertilized, they are shed during menstruation

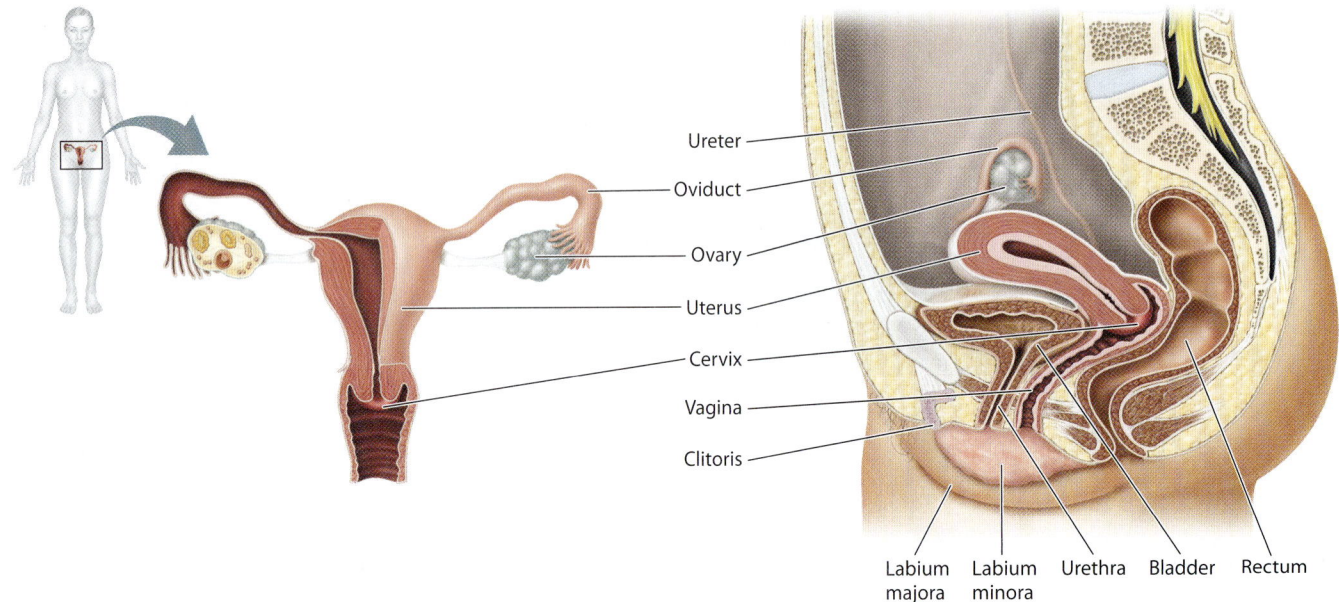

which mature into ova, are produced in the two female gonads, called **ovaries** (**Fig. 42.14**), and are released monthly in response to hormones (discussed in the next section).

The oocyte is a remarkable and unique cell (**Fig. 42.15**). In contrast to sperm, which are small and motile, the oocyte is the largest cell by volume in the human body and is visible to the naked eye; it is about the size of the period at the end of this sentence. Its large size reflects the presence of large amounts of cytoplasm containing molecules that direct the early cell divisions following fertilization and food supply in the form of yolk. While a male produces millions of sperm, a female usually produces just one oocyte each month, as she will nourish and support the developing embryo if the oocyte is fertilized.

Following release from the ovary, the oocyte travels through the **fallopian tube,** or **oviduct** (see Fig. 42.14). There are two fallopian tubes, one on each side. The fallopian tubes have featherlike projections on their ends that help channel the egg into the tube and not into the abdominal cavity.

The oocyte travels from the fallopian tube to the **uterus,** or womb (see Fig. 42.14). The uterus is a hollow organ with thick, muscular walls that is adapted to support the developing embryo if fertilization occurs and to deliver the baby during birth. In humans, the uterus is usually a single pear-shaped structure, but it can sometimes have two horns, which can interfere with pregnancy. Other mammals differ in the shape of the uterus, ranging from pear-shaped (humans and other primates), to two horns (pigs, horses), to more completely divided but still connected at the base (cats, dogs), to completely separate (rodents, rabbits).

The end of the uterus is called the **cervix,** or neck, of the uterus (see Fig. 42.14). Under the influence of hormones, the cervix produces different kinds of mucus capable of either blocking or guiding sperm through the cervix.

The uterus is continuous with the **vagina,** or birth canal (see Fig. 42.14). The vagina is a tubular channel connecting the uterus to the exterior of the body. It is highly elastic, allowing it to stretch during sexual intercourse and birth.

The external genitalia of the female are collectively called the **vulva** (**Fig. 42.16**). The vulva includes two folds of skin, the **labia**

FIG. 42.15 A female gamete. Oocytes are produced in the ovaries, mature into ova (eggs), and are by volume the largest cells in the human body.

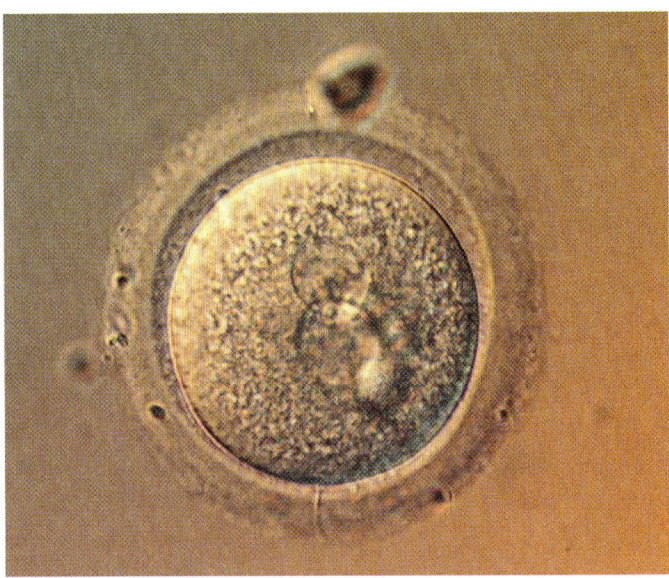

FIG. 42.16 The female external genitalia.

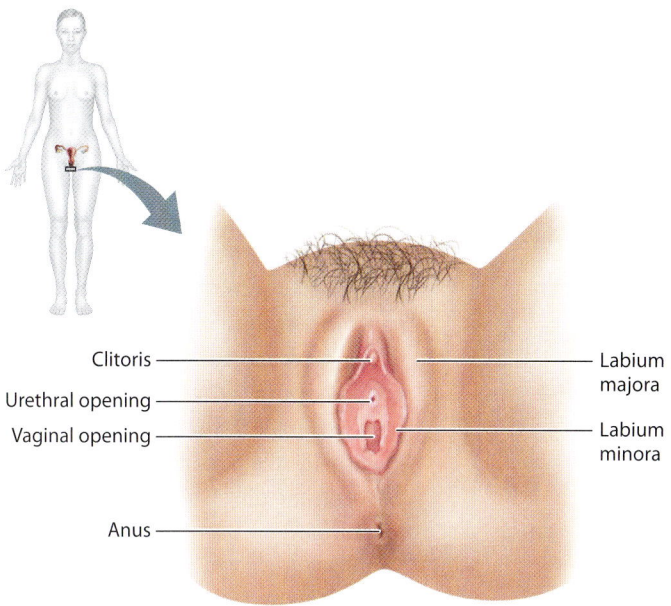

majora and **labia minora**. The labia minora meet at the **clitoris**, the female homolog of the glans penis.

Hormones regulate the human reproductive system.

The male testes and female ovaries not only are the site of gamete production, but also are part of the endocrine system (Chapter 38). That is, they respond to hormones and in turn secrete hormones. These hormones play key roles during embryonic development and puberty, and also in the adult. Hormonal control of reproduction is critically important, as it allows reproduction to take place only when, for example, a woman is able to support the developing embryo.

The master regulator of the endocrine system is the hypothalamus, which is located in the brain (Chapter 38). The hypothalamus secretes hormones that act on the pituitary gland, which in turn releases hormones that act on target organs (**Fig. 42.17**). In the case of the reproductive system, the hypothalamus releases **gonadotropin-releasing hormone (GnRH),** which stimulates the anterior pituitary gland to secrete **luteinizing hormone (LH)** and **follicle-stimulating hormone (FSH).** These hormones then act on the male and female gonads. In males, these hormones are secreted almost continually, whereas in females, they are secreted cyclically. In response, the testes secrete the hormone **testosterone** and the ovaries secrete the hormones **estrogen** and **progesterone.**

In males, LH acts on cells in the testes called **Leydig cells,** which secrete testosterone. Testosterone is a steroid hormone derived from cholesterol that plays key roles in male growth, development, and reproduction. Testosterone is important during embryogenesis to direct development of the male reproductive organs. At puberty, levels of testosterone increase, leading to the development of male **secondary sexual characteristics,** traits that characterize and differentiate the two sexes but that do not relate directly to reproduction. For males, these include a deep voice, facial, body, and pubic hair, and increased muscle mass. Testosterone together with FSH act on another population of cells in the testes called **Sertoli cells** to stimulate sperm production. Sertoli cells, located in the seminiferous tubules, support the process of sperm production. In this way, hormones help to regulate the timing of sexual maturity.

In females, FSH acts on cells surrounding the oocyte called **follicle cells,** which support the developing oocyte and secrete estrogen. At puberty, an increase in the level of estrogen leads to the development of female secondary sexual characteristics, such as enlargement of the breasts, growth of body and pubic hair, and changes in the distribution of muscle and fat. Also at

FIG. 42.17 Hormonal control of the reproductive system.

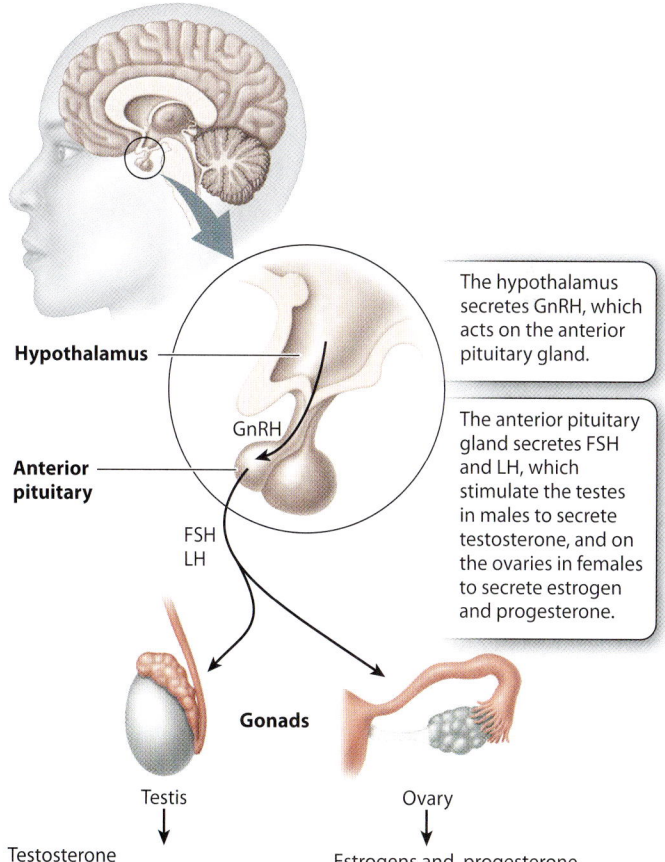

FIG. 42.18 The menstrual cycle. Note the cyclical and coordinated changes in the anterior pituitary hormones, ovary, ovarian hormones, and uterine lining.

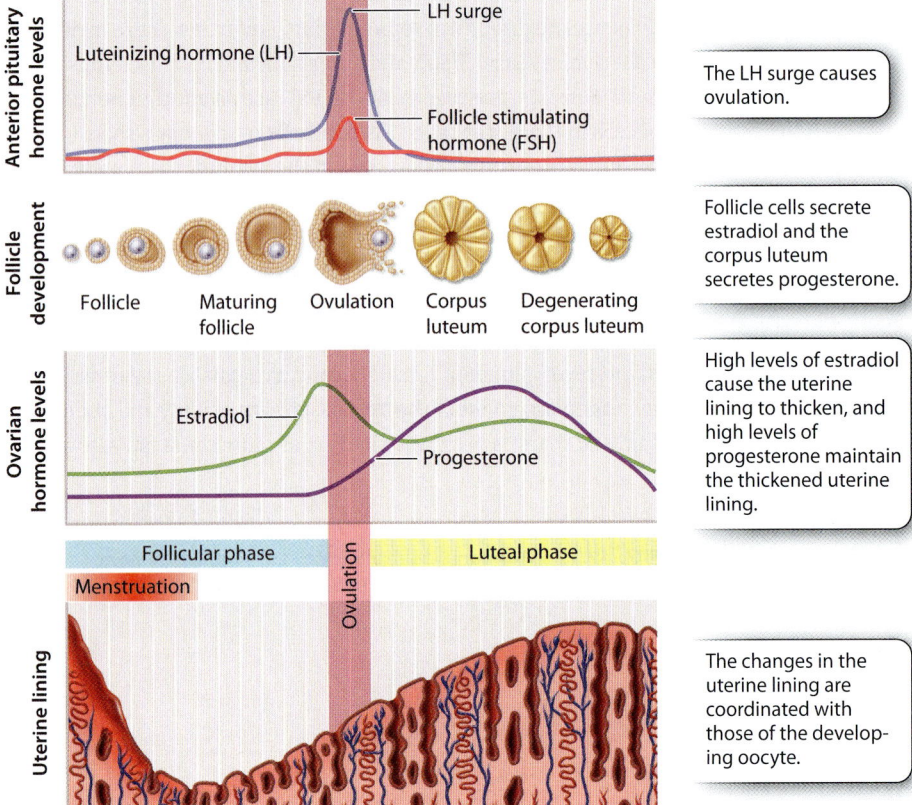

puberty, females begin a monthly cycle, the **menstrual cycle,** in which oocytes mature and are released from the ovary under the influence of hormones (**Fig. 42.18**). The menstrual cycle consists of cyclical and coordinated changes in the ovaries and uterus, timing the release of an oocyte from the ovary with the growth of the uterine lining so it can support the developing embryo if fertilization occurs.

The menstrual cycle has two phases, the **follicular phase** and the **luteal phase.** In the follicular phase, FSH acts on a subset of follicle cells called granulosa cells. As a result, several oocytes begin to mature, but usually only one of these becomes completely mature and the others die off. The granulosa cells secrete a form of estrogen known as estradiol. Estradiol in turn acts on the lining of the uterus, causing it to thicken.

As a result in part of rising estradiol levels, there is a rapid increase followed by a sharp decrease in the level of LH produced by the anterior pituitary gland. This surge causes ovulation, the release of the oocyte from the ovary. Ovulation marks the beginning of the luteal phase. The follicle cells, now devoid of the oocyte, are converted to a structure known as the **corpus luteum,** literally "yellow body," so called because of its

appearance. The corpus luteum is a temporary endocrine structure that secretes the hormone progesterone, which maintains the thickened and vascularized uterine lining.

The oocyte, meanwhile, is swept into the fallopian tube and travels to the uterus. If it is fertilized, the developing embryo implants into the uterine lining. The corpus luteum continues to secrete progesterone, and is first maintained by LH and then by the hormone **human chorionic gonadotropin (hCG)**, which is released by the developing embryo. Eventually, the placenta takes over estrogen and progesterone production to maintain the uterine lining and stimulate growth of the uterus. High levels of estrogen and progesterone during pregnancy also block ovulation because they suppress the release of GnRH, FSH, and LH.

If the oocyte is not fertilized, the corpus luteum degenerates, estrogen and progesterone levels drop, and the uterine lining is shed. The monthly shedding of the uterine lining is known as **menstruation.** Menstrual cycles usually start occurring around age 12 in the United States and continue until approximately age 45 to 55. The cessation of menstrual cycles, called **menopause,** results from decreasing production of estradiol and progesterone by the ovaries.

As we have seen, the brain, anterior pituitary, ovaries, uterus, and placenta pass hormonal messages back and forth to make sure that behavior, gamete release, the uterine lining, and embryo development are coordinated. If hormone levels are low or absent, as occurs normally before puberty or as the result of malnutrition or some diseases, ovulation does not occur. Similarly, too much hormone at the wrong time can interfere with fertility, which is the basis for oral contraceptive pills ("the pill"). Oral contraceptive pills contain various combinations of synthetic estrogens or progesterones or both.

→ **Quick Check 3** From what you know about hormone levels and ovulation during pregnancy, how do you think the pill works?

Humans and chimpanzees both have menstrual cycles. Other placental mammals have an **estrus cycle,** which is characterized by phases in which females are sexually receptive. In these mammals, the uterine lining is reabsorbed instead of shed if fertilization does not occur.

42.4 GAMETE FORMATION TO BIRTH IN HUMANS

With an understanding of male and female reproductive anatomy and endocrinology, we are ready to follow human reproduction and development, from the formation of sperm and eggs, to fertilization itself, to subsequent development of the embryo, and finally to birth. Along the way, we consider how these processes vary in different organisms. Our understanding of these processes is in large part due to studies in model organisms, including the sea urchin (*Strongylocentrotus purpuratus*), roundworm (*Caenorhabditis elegans*), fruit fly (*Drosophila melanogaster*), African clawed frog (*Xenopus laevis*), and house mouse (*Mus musculus*). These organisms have particular reproductive and developmental features that make them amenable to laboratory study.

Male and female gametogenesis have shared and distinct features.

The formation of gametes is called **gametogenesis.** Gametogenesis follows the same steps in males and females, but there are striking differences in timing (**Fig. 42.19**). **Spermatogenesis** is the formation of sperm. It occurs in the seminiferous tubules of the testes, begins at puberty, and continues throughout life. The process starts with spermatogonia, diploid stem cells that divide by mitosis to ensure a continuous population of precursor cells. These cells differentiate into diploid **primary spermatocytes,** which undergo the first meiotic division to produce two **secondary spermatocytes** and the second meiotic division to produce four haploid spermatids joined by cytoplasmic bridges. They subsequently separate from each other to become distinct haploid cells. The maturation process, in which spermatids become spermatozoa or sperm, begins in the testes and continues in the epididymis, where they acquire motility. Spermatogenesis takes about 2 to 3 months.

Oogenesis is the formation of ova or eggs. The diploid precursor cells are called oogonia. These cells form during embryonic development and divide by mitosis. Some oogonia differentiate into diploid **primary oocytes** during fetal development. These enter the first meiotic division but arrest immediately in prophase I, the first stage of meiosis. All these steps take place before the female is even born. At the time

FIG. 42.19 Spermatogenesis in the male and oogenesis in the female.

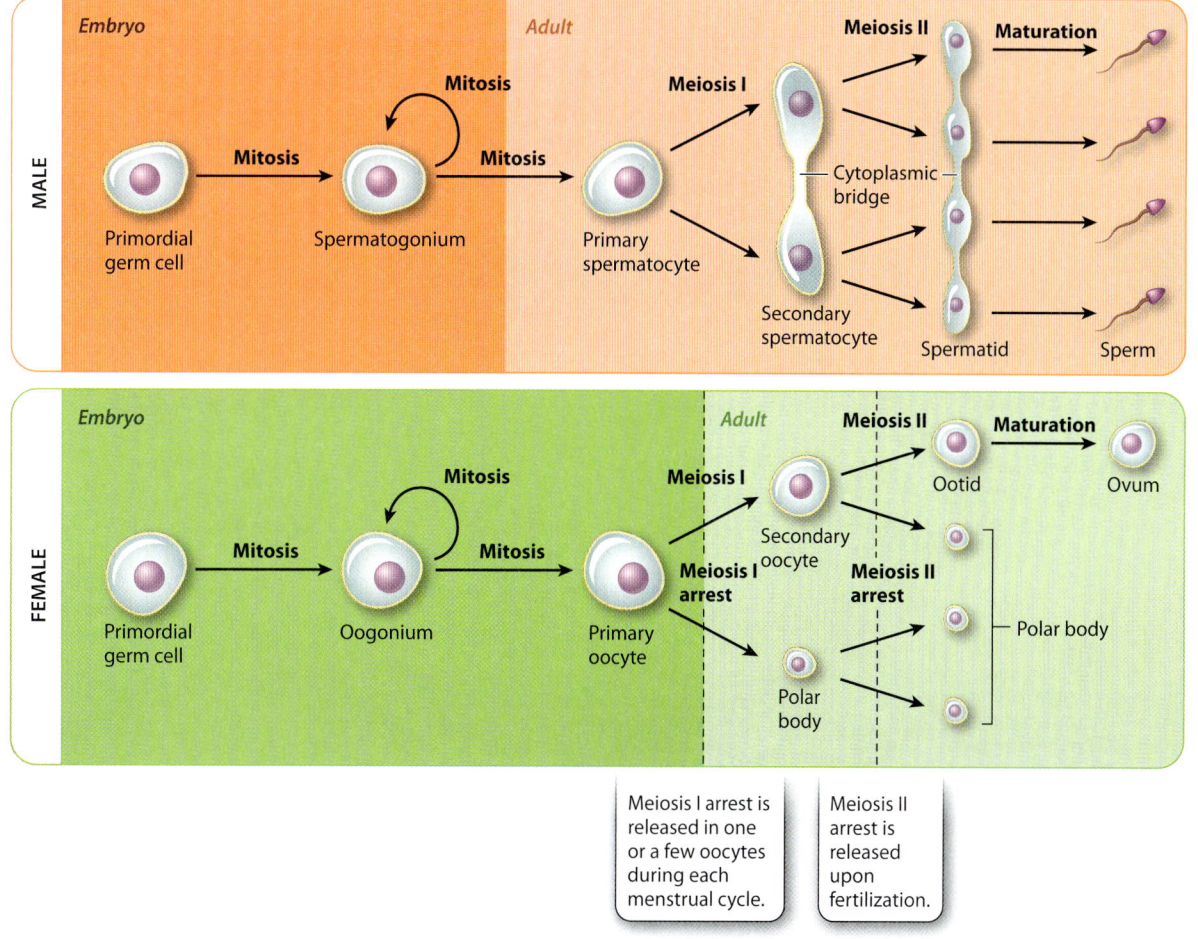

of birth, a female has about 1 to 2 million primary oocytes, all arrested in prophase I of meiosis. They do not resume the first meiotic division until a menstrual cycle. Therefore, remarkably, they all remain arrested for at least 12 years (from birth to the age of menarche) and some as long as 50 years (to the age of menopause).

In the first phase of the menstrual cycle, approximately 10 to 20 follicles begin to mature because of rising FSH levels. One of these follicles becomes dominant, allowing the primary oocyte to complete the first meiotic division. This division is asymmetric, leading to the formation of a large **secondary oocyte** and a much smaller **polar body**. The unequal division allows most of the cytoplasm to be partitioned into the secondary oocyte. The secondary oocyte is released from the ovary and immediately enters the second meiotic division, but again arrests, this time in metaphase II. The secondary oocyte remains arrested in metaphase II until fertilization takes place. Upon fertilization, the second meiotic division continues, producing an ootid and another polar body, again allowing most of the cytoplasm to end up in the developing egg. The ootid quickly develops into a mature ovum or egg.

→ **Quick Check 4** When are primary oocytes and spermatocytes produced, and how many gametes result from one diploid cell in females and males?

Fertilization occurs when a sperm fuses with an oocyte.

Fertilization involves the fusion of gametes, restoring the diploid chromosome content (**Fig. 42.20**). In humans, male and female gametes are usually brought together by sexual intercourse. Sexual arousal leads to a host of physiological changes, including erection of the penis and erection of the clitoris and lubrication of the vagina. The erect penis is inserted into the vagina until orgasm occurs. Orgasm is characterized by feelings of pleasure and contractions of the pelvic muscles. In males, it is usually accompanied by ejaculation, in which semen, containing sperm, is delivered into the vagina.

Sperm travel through the cervix and into the uterus and fallopian tubes. Fertilization usually occurs in one of the fallopian tubes. As a result, many of the details of how sperm travel through the female reproductive tract are poorly understood. Sperm are motile, but from studies of the time it takes sperm to reach the fallopian tubes, it appears that the female reproductive

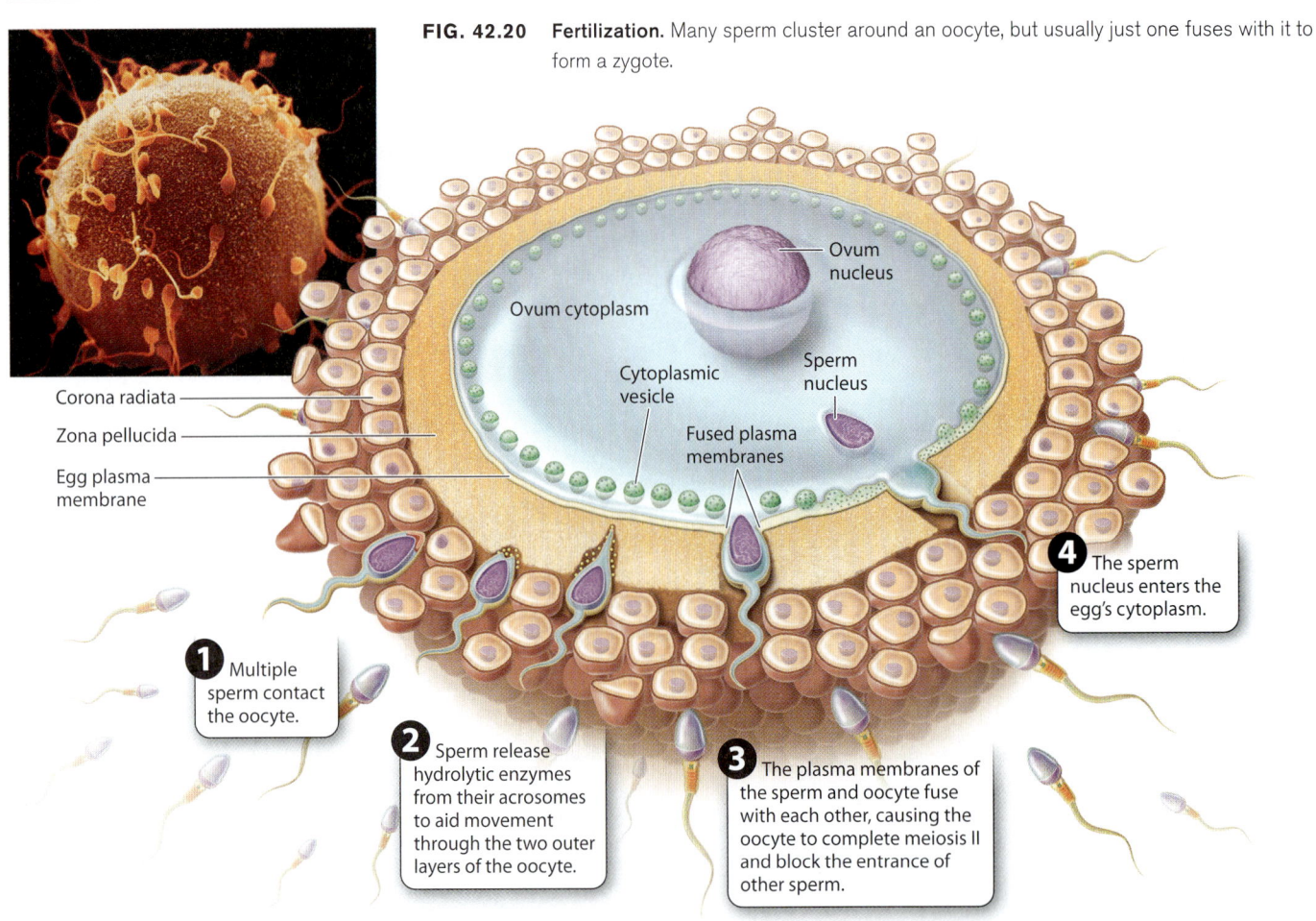

FIG. 42.20 Fertilization. Many sperm cluster around an oocyte, but usually just one fuses with it to form a zygote.

tract plays an active role in moving sperm to the oocyte, for example through muscular contractions.

In the female reproductive tract, sperm go through a key developmental step, called **capacitation,** a process that is unique to mammals. Capacitation is a series of physiological changes that allow the sperm to fertilize the egg. These changes include alterations in the fluidity of the plasma membrane, loss of some surface membrane proteins, and changes in the charge across the membrane, called the membrane potential. As a result, sperm show increased motility.

In many species, the oocyte releases soluble molecules, creating a concentration gradient that sperm use as a guide to find the oocyte. On reaching the oocyte, the sperm passes through two layers—an outer layer of follicle cells called the corona radiata and an inner matrix of glycoproteins called the zona pellucida—before fusing with the plasma membrane of the oocyte (Fig. 42.20). In sea urchins, by contrast, there is a single vitelline membrane surrounding the egg. To facilitate its passage, the head of the sperm releases enzymes from the acrosome. In addition, surface proteins on the gametes allow them to interact in a species-specific manner. For organisms with external fertilization, such as sea urchins, these proteins are especially important.

After passing through the layers surrounding the oocyte, the plasma membranes of the sperm and oocyte fuse (Fig. 42.20). The oocyte then completes meiosis II and changes occur that prevent **polyspermy,** or fertilization by more than one sperm. These mechanisms are classified according to how quickly they act and have been extensively studied in sea urchins. The first line of defense, or fast block to polyspermy, is caused by changes in the membrane potential of the egg, which is usually negative but quickly becomes positive on fertilization. The fast block is quick but temporary and does not occur in mammals.

There is also a second, more stable defense, or slow block to polyspermy. In this process, vesicles in the oocyte fuse with the plasma membrane and release their contents, leading to modification of the structure of the zona pellucida in mammals and the vitelline membrane in sea urchins. The fertilized, diploid egg is a zygote—the first cell in the development of a new organism.

Although sexual intercourse is one way that gametes are brought together, there are now a variety of other methods to achieve fertilization and pregnancy in humans. These are collectively called assisted reproductive technologies and are typically used to treat infertility. For example, **in vitro fertilization (IVF)** is a process in which eggs and sperm are brought together in a petri dish, where fertilization and cell divisions occur. The developing embryos are then inserted into the uterus. The first IVF or "test-tube" baby was born in 1978, and since then several million babies have been conceived in this way. Assisted reproductive technologies such as IVF have opened the door to pre-implantation genetic diagnosis, in which oocytes or embryos are screened for genetic diseases before implantation. This technique raises ethical concerns, as it can be used to determine and select the gender of the embryo and other traits not associated with disease.

The first trimester includes cleavage, gastrulation, and organogenesis.

How does a single cell become a complex, multicellular organism? Development depends on a precisely orchestrated set of cell divisions, movements, and specializations coordinated by both maternal and zygotic genetic instructions (Chapter 20). In this section, we focus on three key processes in early embryonic development: cleavage, gastrulation, and organogenesis.

After fertilization, the single-celled zygote divides by mitosis, a process called **cleavage (Fig. 42.21).** This process divides the single large egg into many smaller cells. In most organisms, these mitotic divisions are rapid, occur at the same time, and are not accompanied by an overall increase in size of the embryo. Furthermore, they occur in the absence of expression of the zygote's genes. Instead, molecules in the egg's cytoplasm, including mRNA and protein encoded by maternal effect genes (Chapter 20) and stored in the egg, direct these early cell divisions.

Cleavage in mammals is unique in several respects. It is slow compared to cleavage in other organisms. In addition, in mammals the early divisions do not occur at the same time, so it is not unusual to find an odd number of cells in the developing embryo. Finally, zygotic transcription begins very early, as early as the first mitotic division, or two-cell stage.

The pattern of cleavage is different for different organisms and is determined in part by the amount of yolk present

FIG. 42.21 Cleavage divisions.

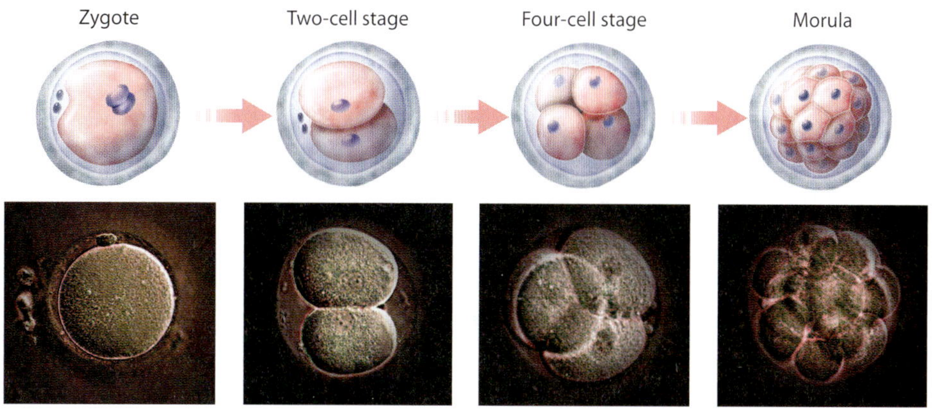

Zygote → Two-cell stage → Four-cell stage → Morula

FIG. 42.22 Cleavage patterns. (a) Partial cleavage, as shown in this zebrafish embryo. (b) Full cleavage, as shown in this sea urchin embryo.

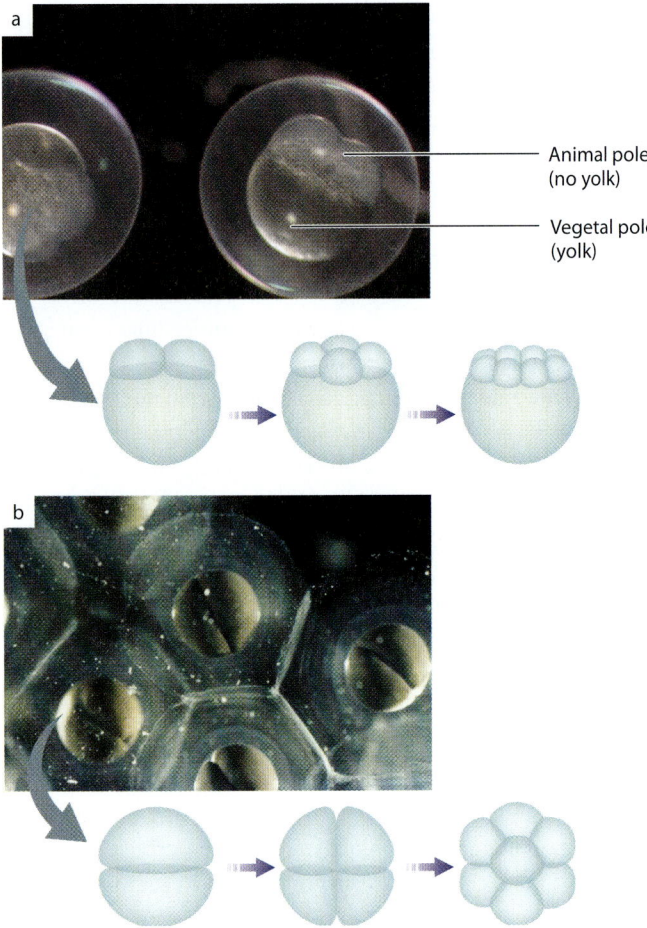

in the egg (**Fig. 42.22**). In animals with lots of yolk, such as insects, fish, and birds, the embryo has two poles—a vegetal pole, where the yolk is concentrated, and an animal pole, where it is less concentrated. The presence of yolk inhibits cell division. Therefore, in these animals, cell division does not extend through the entire egg (Fig. 42.22a). In animals with very little yolk, such as mammals and sea urchins, cell division extends completely through the egg (Fig. 42.22b). In the absence of yolk, the developing embryo needs another source of nutrition, which in humans and other placental mammals is provided through the placenta.

Cleavage results in a solid ball of cells called a **morula** (see Fig. 42.21). Further cell divisions result in a fluid-filled ball of cells called a **blastula.** It is around this time, about 5 days after fertilization, that the developing embryo implants into the uterine lining, which is thickened and highly vascularized under the influence of progesterone from the corpus luteum.

Implantation marks the beginning of **pregnancy,** the carrying of one or more embryos in the uterus. It lasts approximately 38 weeks from fertilization. Because the time of fertilization is often not known, human pregnancy is typically measured as 40 weeks from the last menstrual period. Pregnancy is divided into three periods called **trimesters**, each lasting about 3 months (**Fig. 42.23**). The divisions between the three trimesters are arbitrary, but they provide a convenient way to follow the major changes that occur during pregnancy.

The first trimester is characterized by spectacular changes in the developing embryo under the influence of genes and developmental pathways that are largely conserved across vertebrates (Chapter 20). During this time, the embryo develops all its major organs and begins to be recognizably human. Toward the end of the first trimester, a human embryo is known as a **fetus.**

In mammals, such as humans, the blastula has an inner cell mass and an outer layer of cells, and is known as a blastocyst (**Fig. 42.24a**). The inner cell mass becomes the embryo itself, and the outer cell layer forms part of the placenta. The placenta is composed of both maternal and embryonic cells and is the site of gas exchange, nutrient uptake, and excretion of waste between maternal and fetal blood. The placenta also secretes the hormones estrogen and progesterone, which maintain the uterine lining during pregnancy. Meanwhile, the inner cell mass differentiates into a two-layered (bilaminar) structure consisting of the epiblast and hypoblast (**Fig. 42.24b**).

Gastrulation is the second key developmental process in embryogenesis. It is a highly coordinated set of cell movements that leads to a fundamental reorganization of the embryo. In humans, cells of the epiblast divide and migrate inward (**Fig. 42.24c**). These migrating epiblast cells form three layers or sheets of cells called

FIG. 42.23 The stages of human pregnancy.

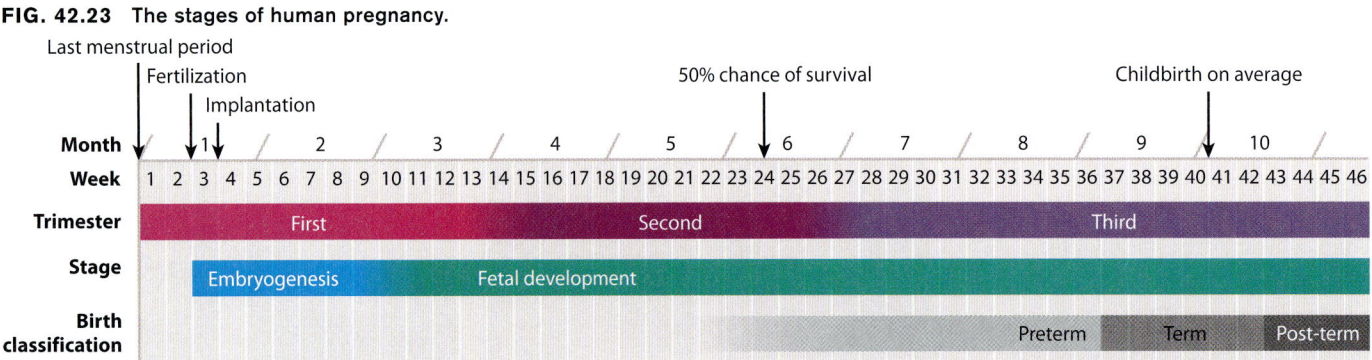

germ layers, which include the outer **ectoderm**, intermediate **mesoderm**, and inner **endoderm** (**Fig. 42.24d**). All of the adult tissues and organs are made from the three germ layers. At this stage, the embryo is known as a trilaminar (three-layered) embryo to reflect the three germ layers.

As with cleavage, the details of gastrulation vary in different organisms. One notable difference is the fate of the opening through which cells migrate during gastrulation. In protostomes ("mouth first"), such as worms, mollusks, and arthropods, it becomes the mouth, whereas in deuterostomes ("mouth second"), such as vertebrates and sea urchins, it becomes the anus and the mouth forms later (Chapter 44).

Organogenesis, the third key developmental process, is the transformation of the three germ layers into all the organ systems of the body. The ectoderm becomes the outer layer of the skin and nervous system, the mesoderm makes up the circulatory system, muscle, and bone, and the endoderm becomes the lining of the digestive tract and lungs (Fig. 42.24d). Because all the basic structures and organ systems develop during the first trimester of pregnancy, it is during this time that the developing fetus is most vulnerable to toxins, drugs, and infections.

It is also during the first trimester that the embryo begins to develop male or female characteristics. In humans, **sex determination** is achieved by the presence or absence of the Y chromosome. The Y chromosome contains a gene, *SRY*, that encodes a protein called testis-determining factor (TDF). TDF in turn acts on the developing gonad to cause it to differentiate into a testis. In the absence of TDF (in females), the immature gonad becomes an ovary. Once sex is determined, sexual differentiation follows under the influence of hormones produced by the gonads. In humans, then, sex determination is genetic, as is the case in other mammals and in birds. For other organisms, such as some reptiles and fish, sex determination depends instead on environmental factors, relying on such cues as temperature and crowding.

The second and third trimesters are characterized by fetal growth.

The second trimester, from 13 to 27 weeks, is characterized by continued development and growth of the fetus. This is the time when fetal movements can first be felt by the mother. Babies born in the last few weeks of the second trimester are often able to survive, and are considered premature. Because of improvements in the intensive care of premature babies, babies born as early as 22 weeks after fertilization can survive.

The third trimester, from 28 to 40 weeks, is characterized by rapid growth, with the fetus doubling in size during the last two months. Weight gain is critical, as birth weight is a major factor in infant mortality. The fetus often moves less frequently during this trimester as it becomes increasingly compressed by limited space. Although 40 weeks (from the last menstrual period) is the average length of pregnancy, a baby born between 37 and 42 weeks is considered term. Human development from fertilization to the end of pregnancy is summarized in **Fig. 42.25**.

FIG. 42.24 Early human embryonic development.

a. Blastocyst

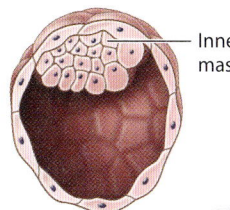

Inner cell mass

The inner cell mass differentiates into two layers of cells – the epiblast and hypoblast.

b. Bilaminar embryo

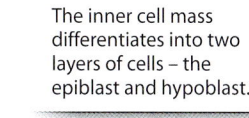

Epiblast
Hypoblast

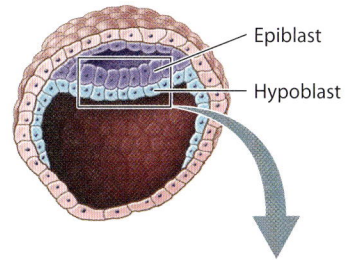

Epiblast
Hypoblast

c. Gastrulation

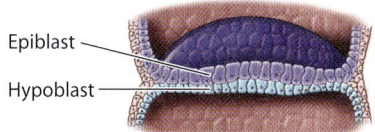

The epiblast cells migrate inward, displace the hypoblast cells, and differentiate into all 3 germ layers – ectoderm, mesoderm, and endoderm.

d. Trilaminar embryo

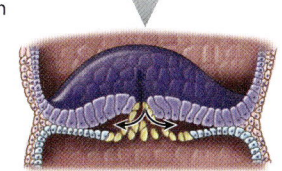

Ectoderm
Mesoderm
Endoderm

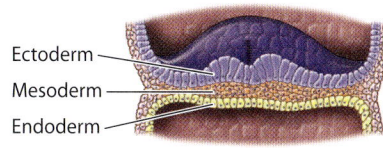

- Outer layer of the skin
- Nervous system
- Posterior pituitary gland
- Cornea and lens of the eye

- Muscle
- Bone
- Connective tissue
- Circulatory system
- Kidneys
- Gonads

- Lining of the respiratory tract
- Lining of the digestive tract
- Liver
- Gallbladder
- Pancreas
- Thyroid gland

VISUAL SYNTHESIS
FIG. 42.25

Reproduction and Development
Integrating concepts from chapters 10, 20, and 42

Early development
The fertilized egg is a totipotent stem cell because it can give rise to a complete organism. During development, cells become increasingly restricted in their developmental potential as a result of changes in gene expression. During gastrulation, cells further differentiate and move relative to one another until the embryo is organized in three germ layers.

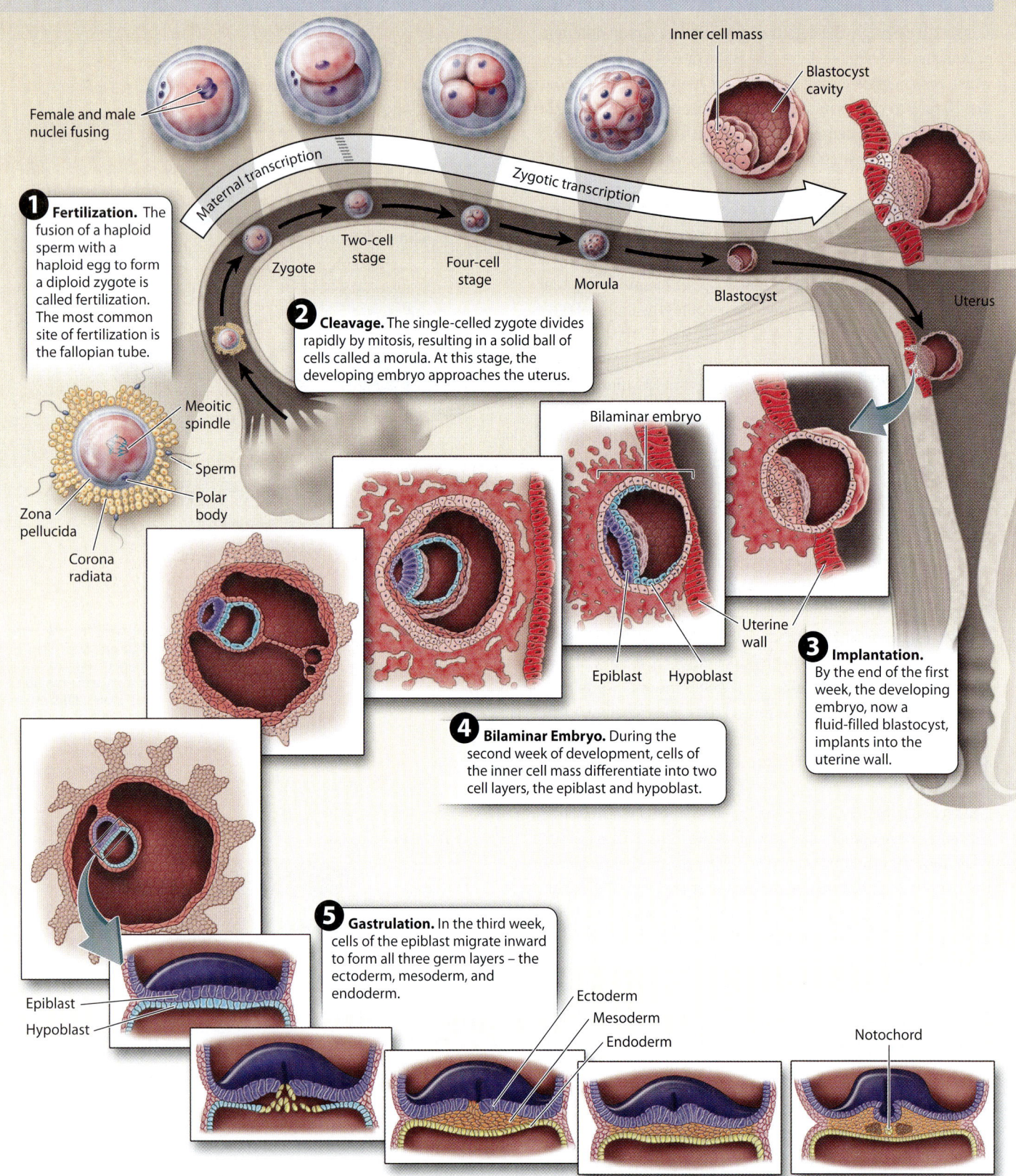

1 Fertilization. The fusion of a haploid sperm with a haploid egg to form a diploid zygote is called fertilization. The most common site of fertilization is the fallopian tube.

2 Cleavage. The single-celled zygote divides rapidly by mitosis, resulting in a solid ball of cells called a morula. At this stage, the developing embryo approaches the uterus.

3 Implantation. By the end of the first week, the developing embryo, now a fluid-filled blastocyst, implants into the uterine wall.

4 Bilaminar Embryo. During the second week of development, cells of the inner cell mass differentiate into two cell layers, the epiblast and hypoblast.

5 Gastrulation. In the third week, cells of the epiblast migrate inward to form all three germ layers – the ectoderm, mesoderm, and endoderm.

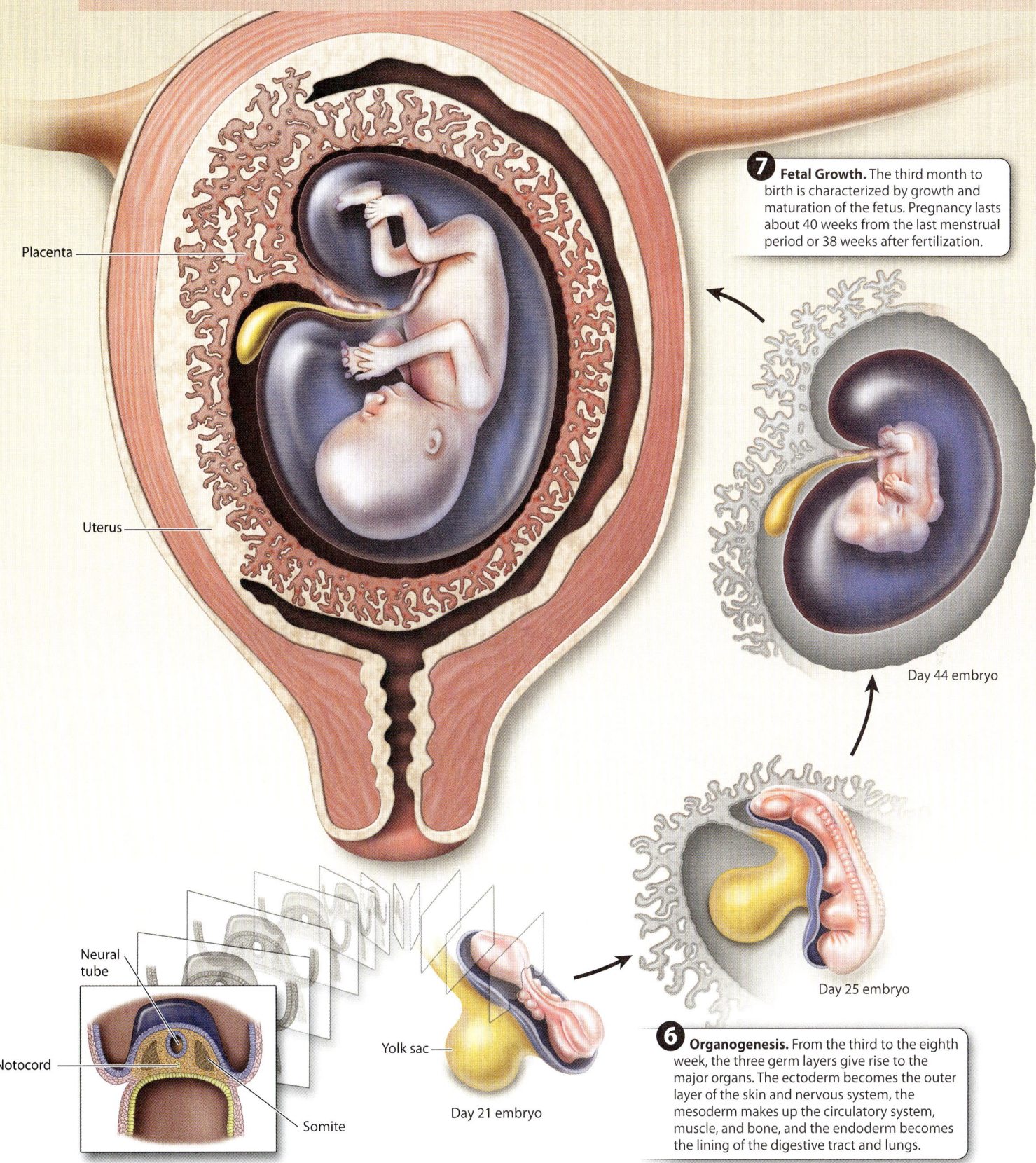

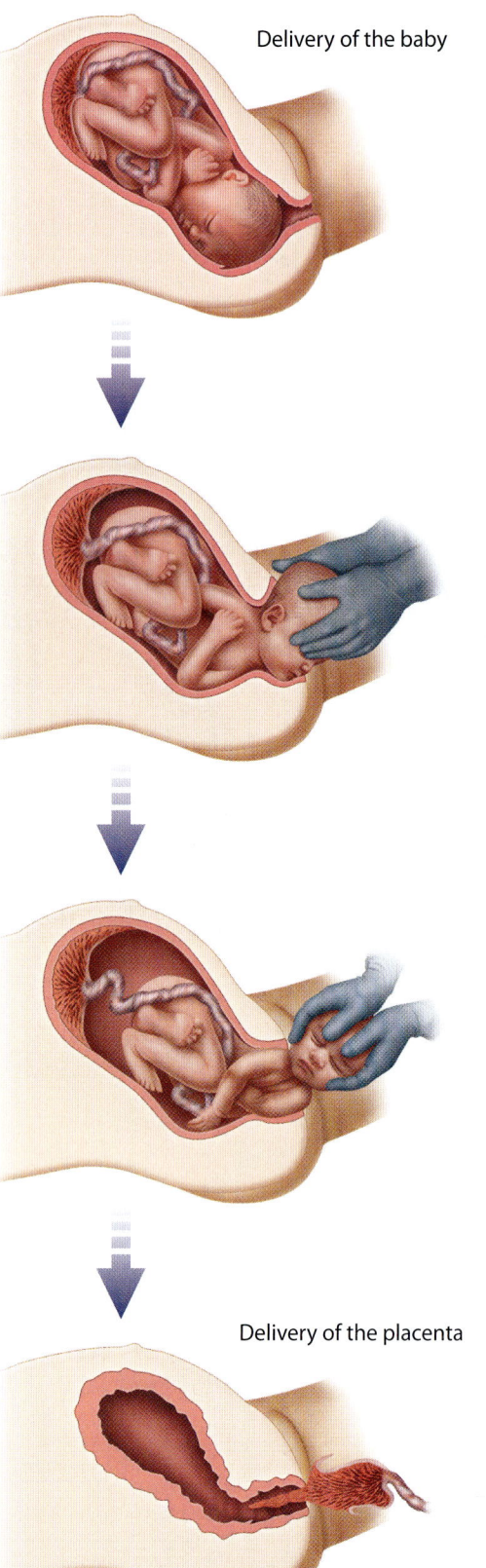

FIG. 42.26 Childbirth. Childbirth involves changes to the cervix, delivery of the baby, and delivery of the placenta.

→ **Quick Check 5** A multicellular organism starts off as a single cell, which divides by mitosis to produce many cells. Therefore, for the most part, all the cells in a multicellular organism are genetically identical—they are clones—yet in time different groups of cells look different and function quite differently from one another. How can this be?

Childbirth is initiated by hormonal changes.

Childbirth marks the end of pregnancy. Although we tend to think of childbirth as the delivery of the baby, it actually consists of three events—changes in the cervix to allow passage of the baby, descent and delivery of the baby, and delivery of the placenta. These three events result from strong muscular contractions of the uterus mediated by the hormone oxytocin from the posterior pituitary gland. Oxytocin causes uterine contractions, which stimulate more oxytocin release, which in turn leads to further contractions, and so on.

During pregnancy, the cervix is a tube about 4 cm long closed with a mucus plug. To allow passage of the infant, the cervix shortens and widens. These changes are not enough to accommodate passage of the baby during delivery. The head and shoulders of the fetus are wide relative to the size of the pelvis. Humans have a narrow pelvis relative to that of other primates as an adaptation for bipedalism. How, then, is the baby born? The baby goes through a series of movements that facilitate its passage through the pelvis (**Fig. 42.26**). Babies are usually born head first with the face pointing down (toward the mother's back) when passing through the birth canal. Other positions make vaginal birth more difficult, sometimes impossible.

The third and final stage of childbirth is the delivery of the placenta, which usually occurs about 15 to 30 minutes after the baby is born.

Although birth normally occurs vaginally, in some cases, such as when the baby is too large or positioned feet first, it is necessary to remove the baby by making an incision in the abdomen and uterus in a surgical procedure known as a Caesarian section. Interventions such as Caesarian sections, the use of antibiotics, and pre- and postnatal care have greatly reduced maternal and fetal mortality associated with childbirth.

Whichever way the baby is delivered, when the umbilical cord is cut and the baby takes its first breath, it begins its life independent of the mother—anatomically independent, that is. Otherwise, human newborns are completely dependent on others. Humans are unusual among animals in having an extended period of helplessness and immaturity while their brains and bodies grow and develop. This time means that parents usually put considerable effort into raising and caring for their offspring, and it allows for extended periods of learning, the development of strong social bonds, and the transmission of culture to the next generation.

Core Concepts Summary

42.1 REPRODUCTION IS A FUNDAMENTAL FEATURE OF LIVING ORGANISMS AND OCCURS ASEXUALLY AND SEXUALLY.

Asexual reproduction involves a single parent. Offspring are produced by binary fission or mitosis and are clones of the parent. page 42-2

Examples of asexual reproduction include budding, fragmentation, and parthenogenesis. page 42-2

Sexual reproduction involves the production of haploid gametes by meiosis and the fusion of gametes to make a diploid zygote. page 42-3

Many organisms reproduce both sexually and asexually. For some organisms, environmental conditions favor one mode or the other. page 42-4

Asexual reproduction allows for rapid population growth but produces clones and results in genetic uniformity. page 42-4

Sexual reproduction produces genetically unique offspring, a feature that is thought to explain why most eukaryotes reproduce sexually at least some of the time. page 42-4

42.2 THE MOVEMENT OF VERTEBRATES FROM WATER TO LAND INVOLVED CHANGES IN REPRODUCTIVE STRATEGIES, INCLUDING INTERNAL FERTILIZATION AND THE AMNIOTIC EGG.

Fertilization can occur outside or inside the body of the female. External fertilization can occur only in an aquatic environment. Internal fertilization is an adaptation to terrestrial living. page 42-7

External fertilization is usually associated with the production of many offspring, little or no parental care, and high mortality. These are characteristics of *r*-strategists. page 42-8

Internal fertilization is often associated with the production of fewer offspring, increased parental care, and low mortality. These are characteristics of *K*-strategists. page 42-8

A watertight sac, called the amnion, is an adaptation for reproduction on land. page 42-8

Oviparous animals lay eggs and development occurs outside the female; viviparous animals give birth to live young and development occurs inside the female. page 42-8

42.3 THE MALE REPRODUCTIVE SYSTEM IS ADAPTED FOR THE PRODUCTION AND DELIVERY OF SPERM, AND THE FEMALE REPRODUCTIVE SYSTEM IS ADAPTED FOR THE PRODUCTION OF EGGS AND, IN MANY CASES, SUPPORT OF THE DEVELOPING FETUS.

Male gametes, called sperm, develop in the seminiferous tubules of the testes. page 42-10

Sperm travel from the seminiferous tubules to the epididymis to the vas deferens to the ejaculatory duct and finally to the urethra before being released during ejaculation. page 42-10

Semen is a mixture of sperm and fluid from the prostate gland, seminal vesicles, and bulbourethral glands. page 42-10

Developing female gametes, called oocytes, are produced in the ovaries. Mature oocytes are called eggs, or ova. page 42-11

An oocyte is released from the ovary, travels through the fallopian tube, and implants in the uterus if it is fertilized. page 42-11

The reproductive system responds to and produces hormones. page 42-13

The hypothalamus releases GnRH, which stimulates the anterior pituitary gland to release FSH and LH, which in turn stimulate the ovaries to release estrogen and progesterone or the testes to secrete testosterone. page 42-13

In males, LH stimulates Leydig cells to secrete testosterone, and FSH and testosterone act on Sertoli cells to support sperm production. page 42-13

In females, monthly menstrual cycles are driven by the interplay of hypothalamic, anterior pituitary, and ovarian hormones, leading to ovulation, which is the maturation and release of an oocyte from the ovary, and to changes to the uterine lining. page 42-13

42.4 HUMAN REPRODUCTION INVOLVES THE FORMATION OF GAMETES, FERTILIZATION, AND GROWTH AND DEVELOPMENT.

Gametogenesis is called spermatogenesis in males and oogenesis in females. page 42-15

In males, spermatogonia differentiate into primary spermatocytes that undergo meiosis to make immature haploid spermatids and mature spermatozoa or sperm. page 42-15

Spermatogenesis begins at puberty and continues throughout life. page 42-15

In females, oogonia undergo differentiation into primary oocytes during fetal development, which begin meiosis but arrest in prophase I. During a menstrual cycle, primary oocytes are released from arrest, finish meiosis I, but then arrest in metaphase II. They remain arrested in metaphase II until they are fertilized. page 42-15

Fertilization is the fusion of the oocyte and sperm plasma membranes. It usually occurs in the fallopian tube and results in the formation of a diploid zygote. page 42-16

Cleavage is the division of the zygote by mitosis into smaller cells. At the blastocyst stage, it implants in the uterus. page 42-17

Gastrulation results in the formation of the three germ layers—ectoderm, mesoderm, and endoderm. page 42-17

Organogenesis is the formation of organs. page 42-17

Pregnancy lasts about 38 weeks from fertilization until birth and is divided into three trimesters. page 42-19

Childbirth, mediated by the posterior pituitary hormone oxytocin, involves changes in the cervix, delivery of the baby, and delivery of the placenta. page 42-22

Self-Assessment

1. List similarities and differences between asexual and sexual reproduction.

2. Explain the roles of meiosis and fertilization in sexual reproduction.

3. Describe the costs and benefits of asexual and sexual reproduction.

4. Provide an explanation for the observation that there are very few, if any, ancient asexual organisms.

5. Name three adaptations that allow reproduction to take place on land.

6. Describe the pathway of sperm from its site of production to the urethra.

7. Describe the pathway of an oocyte (and the egg it develops into) from its site of production to the uterus.

8. Explain the relationships among changes in levels of anterior pituitary hormones, ovarian hormones, oocyte development, and the uterine lining during a menstrual cycle.

9. Describe similarities and differences between male and female gametogenesis.

10. Name and describe three key developmental steps in the development of a single-celled zygote to a multicellular individual.

Do you understand the chapter's Core Concepts? Log in to check your answers to the Self-Assessment questions, then practice what you've learned and reinforce this chapter's concepts by working through the problems and multimedia tutorials provided there.

www.biologyhowlifeworks.com

CHAPTER 43

ANIMAL IMMUNE SYSTEMS

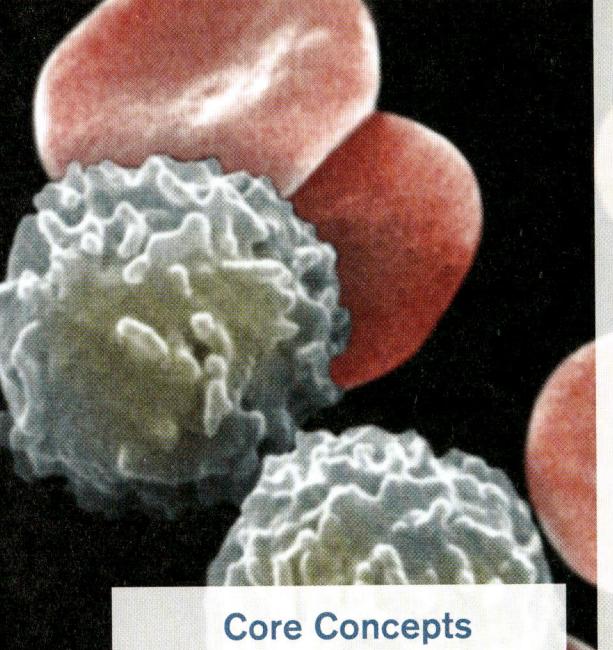

Core Concepts

43.1 The innate immune system acts against a diversity of foreign pathogens.

43.2 The adaptive immune system includes B cells that produce antibodies against specific pathogens.

43.3 The adaptive immune system includes helper and cytotoxic T cells that attack pathogens, foreign cells, and diseased cells.

43.4 Some pathogens have evolved mechanisms that enable them to evade the immune system.

As our bodies' main line of defense, the immune system keeps watch against harmful pathogens. Some pathogens in turn have evolved a means of hiding within our own cells. In this way, they evade easy detection by our immune system, not unlike the Greek soldiers who hid inside a wooden horse during the Trojan War. As the story goes, the Trojans wheeled the magnificent horse within the protective gates of Troy. That evening, the stealthy Greek soldiers emerged and opened the gates, letting in the rest of the Greek army.

As pathogens evolve new strategies, the immune system in turn evolves new responses. The result of this evolutionary back and forth is an elaborate and sophisticated immune system. The immune system is remarkable in its ability in most cases to protect us from harmful pathogens, including viruses like influenza and HIV, bacteria that cause strep throat and tuberculosis, fungi such as ringworm, and single-celled eukaryotes like those that cause malaria. It has evolved defenses that act against a broad range of pathogens and defenses that target specific pathogens.

The activity of the immune system also must be carefully regulated. If it is overactive or acts when it is not supposed to, it may attack the cells of its own body, leading to allergies or debilitating autoimmune diseases. If it is underactive, it won't adequately protect an organism. A weak immune system is most likely to occur at the extremes of age or when infections or disease compromise its function.

We begin with an overview of general defenses that provide a first line of protection against diverse pathogens. We then turn to a powerful part of the immune system that is tailored to attack specific pathogens and that has the ability to learn and remember, attributes we usually associate with the nervous system (Chapters 35 and 36). Finally, we consider three infections that illustrate how pathogens have evolved ways to evade these defenses.

43.1 INNATE IMMUNITY

The immune system consists of two parts that work in different ways and interact with each other to protect against infection. The first, called **innate**, or **natural**, **immunity**, provides protection in a nonspecific manner against all kinds of infection. This form of immunity is present in plants, fungi, and animals and is an evolutionarily early form of immunity. It is considered *innate* because it does not depend on exposure to a pathogen. By contrast, the second form of immunity, called **adaptive**, or **acquired, immunity**, is specific to a given pathogen. This form of immunity "remembers" past infections, and subsequent encounters with the same pathogen generate a stronger response from the host. In other words, the immune system *adapts* over time, and immunity is *acquired* after an initial exposure. This form of immunity evolved later than the innate form

TABLE 43.1 Features of the Innate and Adaptive Immune Systems

	INNATE IMMUNE SYSTEM	ADAPTIVE IMMUNE SYSTEM
Targets	Diverse pathogens	Diverse pathogens
Ability to distinguish self from non-self?	Yes	Yes
Specificity?	No	Yes
Memory?	No	Yes
Organisms	All organisms	Vertebrates

and is unique to vertebrates. **Table 43.1** summarizes the major differences between these two forms of immunity.

We begin by considering forms of defense in the innate immune system. These include tissues, such as the skin, that act as a barrier to infection, cells that act as sentries against pathogens, and proteins and other molecules circulating in the blood that signal the presence of infection. All these defenses are nonspecific, act immediately, and do not depend on past exposure to elicit a response.

The skin and mucous membranes provide the first line of defense against infection.

Organisms have a barrier that surrounds them, like the walls of Troy, which can at times be breached. Ordinarily, the barrier around organisms, among its other functions, protects against infection. Examples include the cell wall of bacteria, the bark of trees, the cuticle of leaves, the exoskeleton of insects, the scales of fish, the shells of eggs, and the skin of mammals (**Fig. 43.1**). These structures act as physical barriers against infection.

To understand the importance of the barrier, consider what happens when it is breached. Burn victims who lose the outer layer of their skin are particularly vulnerable to infection. The bite of a dog or the sting of an insect can seed viruses, bacteria, or parasites directly into the bloodstream, bypassing the protective layer of skin. Even a small puncture wound can provide entry for the bacterium *Clostridium tetani*, the cause of tetanus.

In humans, this physical barrier is the skin (**Fig. 43.2**). It has two layers, the outer epidermis and the inner dermis (Chapter 10). The **epidermis** consists of several layers of cells covered with the protein keratin. The **dermis** consists of connective tissue, hair follicles, blood and lymphatic vessels, and glands. Together, these two layers provide a physical barrier against microorganisms.

Any surface that is in contact with the external environment is a potential entry site for pathogens. Like the skin, the mucous membranes of the respiratory, gastrointestinal, and genitourinary tracts act as physical barriers. However, they have additional protective mechanisms. For example, parts of the respiratory tract have **cilia,** hairlike projections that sweep mucus along with trapped particles and microorganisms out of the body. Saliva and tears contain enzymes that help to protect against bacteria. The stomach, with its low pH and enzymes that break down food, is inhospitable to many microorganisms.

In addition to acting as a physical and chemical barrier, the skin and mucous membranes provide a home to many nonpathogenic microorganisms. By one estimate, of the 100 trillion cells in the human body, only 10% are human—the rest are mostly bacteria. These microorganisms in some cases provide protection against harmful pathogens by competing with them for food and space. They may also maintain health in other ways, such as by aiding digestion. In this way, we live in a mutualistic relationship with these microorganisms (Case 5: The Human Microbiome). The Human Microbiome Project, sponsored by the National Institutes of Health, is an effort to describe and understand the ecology of these microorganisms.

→ **Quick Check 1** Certain diseases, as well as cigarette smoke, can damage cilia lining the respiratory tract. What effects would you expect from damage to respiratory cilia?

FIG. 43.1 Physical barriers against infection. (a) Tree bark; (b) leaf cuticle; (c) insect exoskeleton; (d) egg shell.

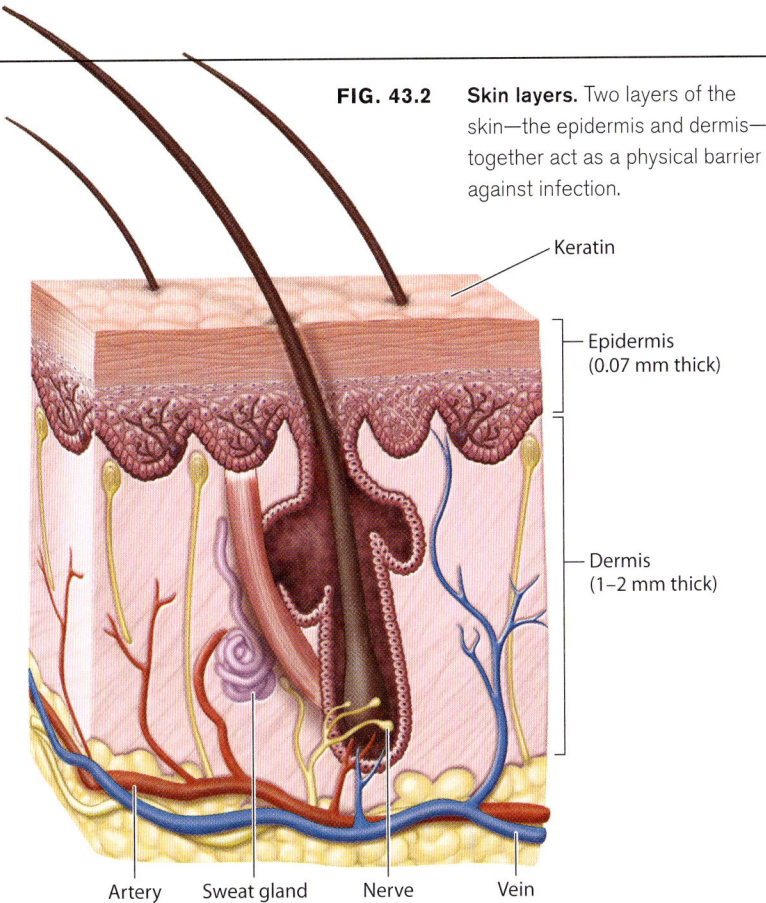

FIG. 43.2 **Skin layers.** Two layers of the skin—the epidermis and dermis—together act as a physical barrier against infection.

Some cells act broadly against diverse pathogens.

The innate immune system has numerous cells in addition to those in the skin and mucous membranes, found in different locations in the body and having different mechanisms of action. These cells are **white blood cells,** or **leukocytes,** and they arise by differentiation from stem cells in the bone marrow (**Fig. 43.3**). Many of these cells have important roles in the adaptive immune response as well. Here, we focus on their role in innate immunity.

Phagocytes (literally, "eating cells") are immune cells that engulf and destroy foreign cells or particles. The engulfing of a cell or particle by another cell is called **phagocytosis.** The Russian microbiologist Ilya Mechnikov first described this remarkable process, for which he shared the Nobel Prize in Physiology or Medicine with the German scientist Paul Ehrlich in 1908.

The process begins when a phagocyte encounters a cell or particle that it recognizes as foreign and binds to it. The phagocyte then extends its plasma membrane completely around the cell or particle until it is within a separate compartment inside

FIG. 43.3 Development of white blood cells important in the immune system.

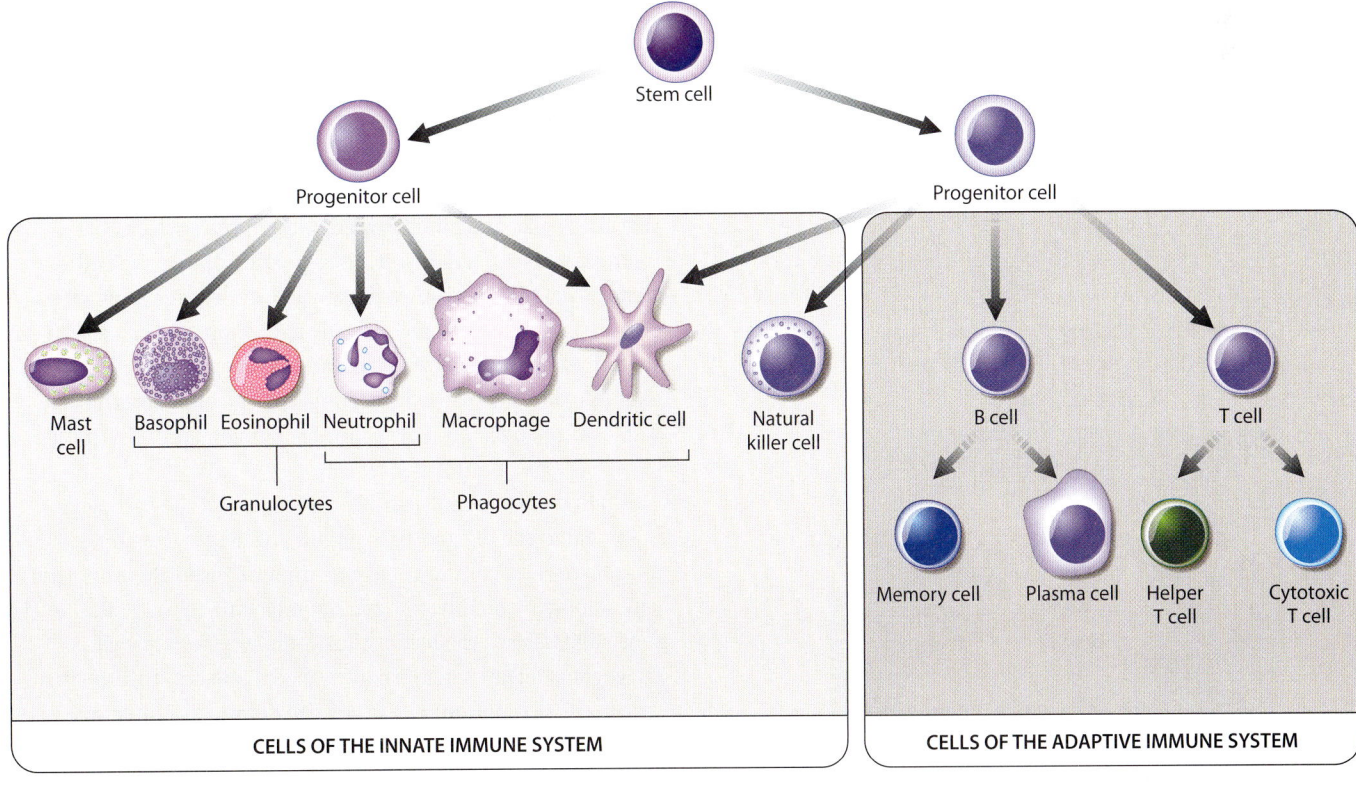

the phagocyte (**Fig. 43.4**). In some cases, the compartment merges with a lysosome, and enzymes within the lysosome digest the foreign cell or particle. Phagocytosis can also trigger a **respiratory burst,** a process that generates reactive oxygen species, such as superoxide radicals and hydrogen peroxide, and reactive nitrogen species, such as nitrogen oxide and nitrogen dioxide. These molecules have antimicrobial properties that directly damage pathogens.

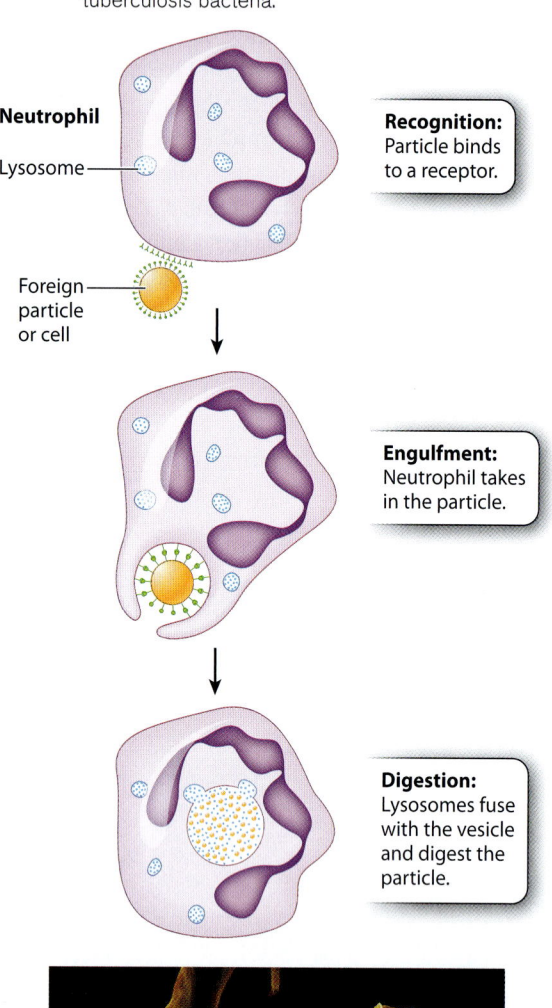

FIG. 43.4 Phagocytosis. A phagocyte recognizes, engulfs, and digests foreign particles and cells. The scanning electron micrograph at the bottom shows a macrophage engulfing tuberculosis bacteria.

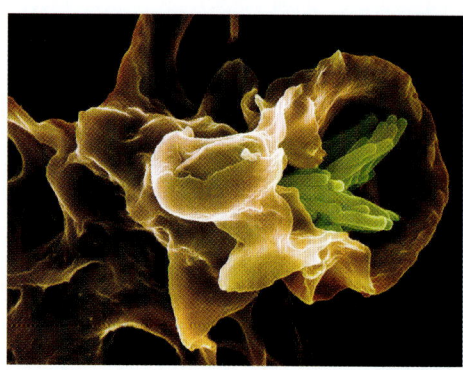

There are three major types of phagocytic cell: **macrophages, dendritic cells,** and **neutrophils** (see Fig. 43.3). In each case, their names give clues to their morphology. Macrophages are large cells that patrol the body (*macro* means "large"). Dendritic cells have long cellular projections reminiscent of the dendrites of neurons (Chapter 35), but they are not part of the nervous system. These cells are typically part of the natural defenses found in the skin and mucous membranes. Neutrophils are members of a group of cells called **granulocytes** because of the presence of granules in their cytoplasm. The granules of neutrophils take up both acidic and basic dyes, so their staining pattern is considered neutral. Neutrophils are very abundant in the blood and are often the first cells to respond to infection.

The innate immune system contains several other cells in addition to phagocytic cells (see Fig. 43.3). Two types of granulocyte, **eosinophils** and **basophils,** defend against parasitic infections but also contribute to allergies. **Mast cells** release **histamine,** an important contributor to allergic reactions and inflammation, discussed below. Finally, **natural killer cells** do not recognize foreign cells, but instead recognize and kill host cells that are infected by a virus or have become cancerous or otherwise abnormal.

Phagocytes recognize foreign molecules and send signals to other cells.

Phagocytes attack foreign cells that they encounter but leave host cells alone. How is a phagocyte able to distinguish a foreign cell from a host cell? All cells have an array of proteins and other molecules on their surfaces. Phagocytes detect specific surface molecules that are present on pathogens but not on the body's own cells.

Toll-like receptors (TLRs) play a critical role here. TLRs are a family of transmembrane receptors present on phagocytes that recognize and bind to molecules on the surface of microorganisms. As a result, they provide one of the earliest signals that an infection is present. TLRs detect a wide range of microorganisms by recognizing evolutionarily conserved surface molecules shared by many microorganisms. For example, one type of TLR recognizes a glycolipid on the cell wall of certain bacteria. Another recognizes a protein that is part of the bacterial flagella.

Binding of the TLR to surface molecules on the pathogen is a signal to the phagocyte to engulf and destroy its target. Phagocytes also contribute to the immune response in another way. Following binding of a foreign molecule to the TLR, they send a message to the rest of the immune system as well. Phagocytes release chemical messengers called **cytokines** that recruit other immune cells to the site of injury or infection. Cytokines, like hormones, provide long-distance communication between cells (Chapters 9 and 38).

FIG. 43.5 Inflammation. Inflammation involves increased blood flow, leaky blood vessels, and activation of the immune system. Thus, a mosquito bite causes redness and swelling (photograph at bottom).

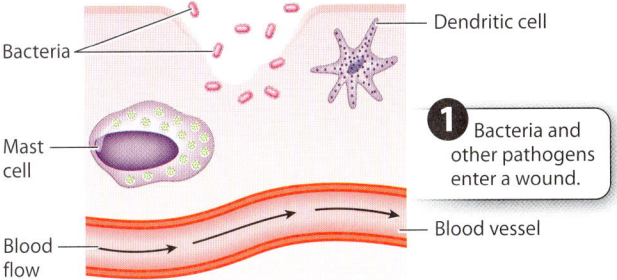

1. Bacteria and other pathogens enter a wound.

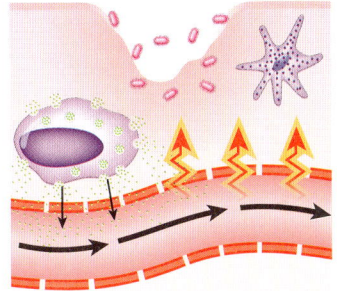

2. Mast cells release histamine that increases blood flow (causing redness and heat) and makes blood vessels leaky (causing swelling).

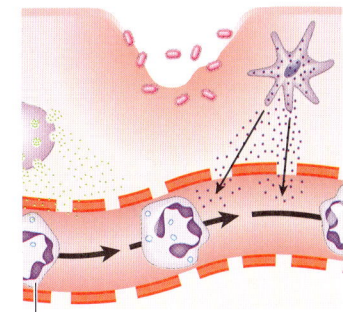

3. Immune system cells in the tissue release chemical messengers called cytokines that bind to and recruit phagocytes in nearby blood vessels.

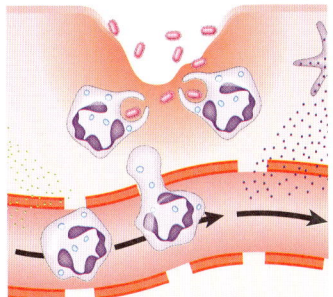

4. Phagocytes enter the infected site from the blood and remove pathogens by phagocytosis.

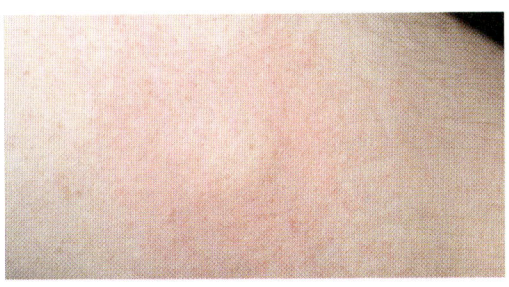

Inflammation is a coordinated response to tissue injury.

While there are generally some immune cells already present at the site of tissue injury or infection, many more must be recruited to fight infection successfully. That recruitment happens as part of a process called **inflammation**. The redness and swelling you see following an injury, a break in the skin, or an infection are signs of inflammation. More precisely, inflammation is a physiological response of the body to injury that removes the inciting agent if present and begins the healing process. It works through the coordinated responses of the local tissue, the vascular system (blood and lymph), and the innate immune system. Inflammation therefore is a useful, beneficial process. However, left unchecked or allowed to continue for a long time, it can be debilitating.

Inflammation is characterized by four classic signs, described by their Latin names: *rubor* (redness), *calor* (heat), *dolor* (pain), and *tumor* (swelling). The process begins following tissue injury or infection (**Fig. 43.5**). Certain cells in the tissue, such as dendritic and mast cells, are activated and release cytokines and other chemical messengers. Some of these chemical messengers recruit additional white blood cells to the site of infection or injury. Others make it easier for white blood cells to reach this site quickly. For example, histamine, released by mast cells and basophils, acts directly on blood vessels to cause vasodilation, increasing blood flow to the site of infection or injury. Vasodilation causes the characteristic redness and heat of inflammation. In addition, histamine increases the permeability of the blood vessel wall. Fluid leaks out of the blood vessel, carrying white blood cells into the damaged tissue. The increased fluid in the tissue surrounding the blood vessels is visible as swelling. Some chemical messengers act directly on nerve fibers, causing pain.

Phagocytes that travel in the blood can move from a blood vessel to the site of infection. This process is called **extravasation** (**Fig. 43.6**). The phagocyte travels along the vessel wall in a rolling motion, grabbing weakly onto the wall as glycoproteins on the phagocyte surface bind transiently to proteins on the endothelial cells. Stronger interactions put a brake on this rolling motion and allow the phagocyte to adhere more firmly to the vessel wall. As it nears the site of infection, the phagocyte encounters and binds to cytokines. The phagocyte then changes shape and moves in between cells lining the blood vessel and into the surrounding tissue.

Several proteins play key roles in inflammation. In the early 1900s, researchers studied patients with pneumonia caused by the bacterium *Streptococcus pneumoniae*. They looked for and found a change in the level of proteins in the blood as a marker for the initial phase of the disease, which became known as the **acute phase response.** One of these proteins, C-reactive protein or CRP, recognizes and binds to molecules on the surface of some bacteria. These CRP-coated bacteria are more easily taken up by

FIG. 43.6 Extravasation. In extravasation, an activated phagocyte moves from a blood vessel to the surrounding tissue.

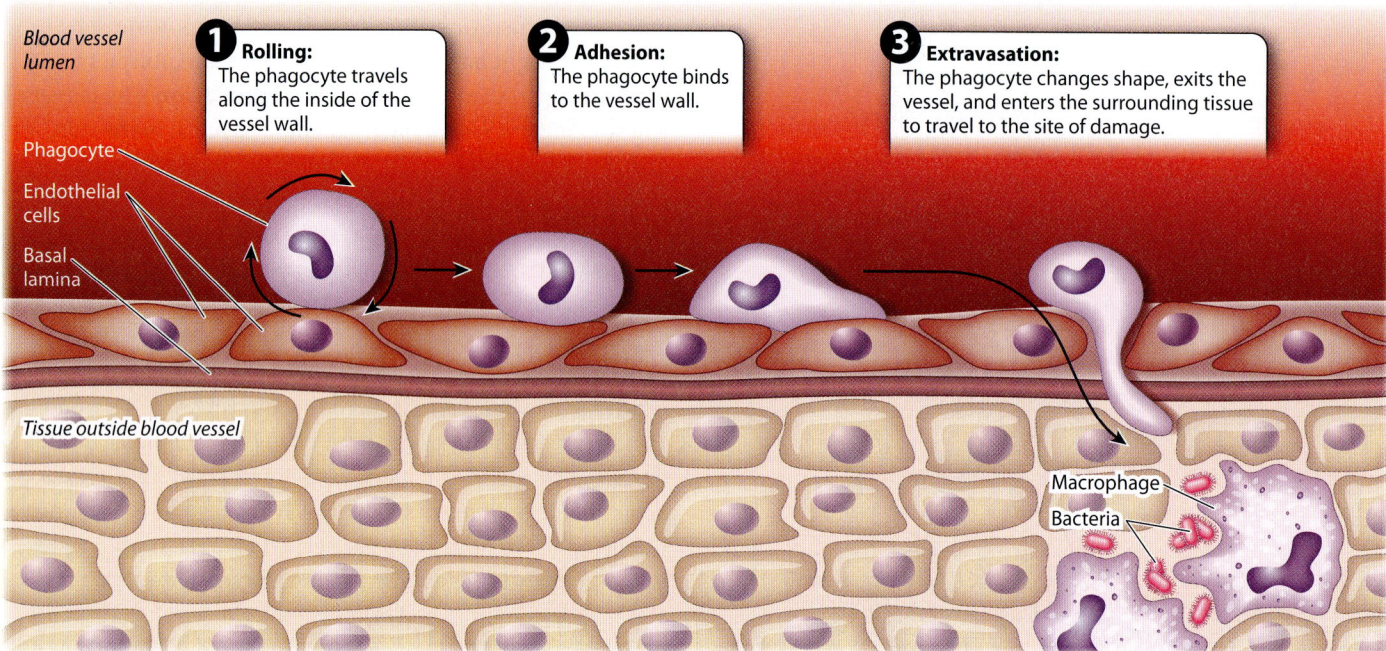

phagocytes. The binding of a molecule like CRP to a pathogen to facilitate uptake by a phagocyte is called **opsonization**.

The complement system participates in the innate and adaptive immune systems.

In addition to individual proteins, such as CRP, there is an entire system of proteins circulating in the blood that participates in innate immune function. Collectively, these proteins make up the **complement system**. First described by Paul Ehrlich in the late 1800s, this system is so named because it supplements or *complements* other parts of the immune system.

The complement system consists of more than 25 proteins that circulate in the blood in an inactive form (**Fig. 43.7**). The system is activated when these proteins bind to molecules specific to microorganisms or to antibodies (discussed below). Activation in turn sets off a biochemical cascade in which the product of one reaction is the enzyme that catalyzes the next. Such sequential activation leads to amplification of the response, as we saw earlier in cell signaling (Chapter 9) and in the endocrine system (Chapter 38), where a small stimulus can have a dramatic response.

Activation of the complement system has three effects. The most dramatic is breaking open cells, or lysis. Lysis results when complement proteins form a **membrane attack complex (MAC)** that makes holes in bacterial cells. Second, the complement system targets pathogens for phagocytosis by coating bacteria with a protein that phagocytes recognize (opsonization). Finally, activation of the complement system leads to the production of

FIG. 43.7 The complement system. The proteins of the complement system contribute to the defense against pathogens by cell lysis, phagocytosis, and inflammation.

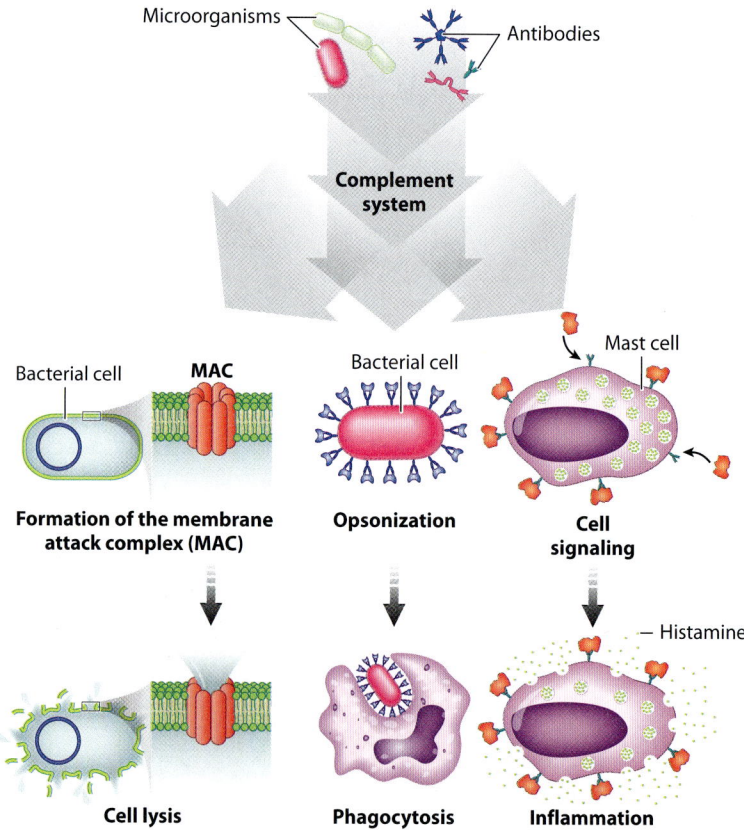

activated proteins that attract other components of the immune system. For example, some of these proteins stimulate mast cells, which release histamine and initiate the inflammatory response.

You can appreciate the importance of the complement system in the immune system by considering what occurs in individuals who lack key components. Complement deficiencies are a rare form of **immunodeficiency,** any disease in which part of the immune system does not function properly. Their effects depend on which component of the complement system is lacking. For example, one type of complement deficiency makes individuals vulnerable to bacterial infections because fewer pathogens undergo cell lysis and opsonization.

Because antibodies are an important signal activating the complement system, the complement system provides an important link between the innate, nonspecific arm of the immune system and the adaptive, specific arm, a topic we consider next.

43.2 ADAPTIVE IMMUNITY: B CELLS, ANTIBODIES, AND HUMORAL IMMUNITY

The innate immune system can react to diverse pathogens and has the capacity to distinguish self (host cells) from non-self (foreign cells). The adaptive immune system shares these features, and has two additional features not present in the innate immune system. The first is specificity: The adaptive immune system targets responses to specific pathogens. As a result, it is often more effective at eliminating a pathogen. The second is memory: The adaptive immune system remembers past infections and mounts a stronger response on re-exposure. How can the adaptive immune system respond to a single pathogen among the diversity of pathogens that it meets? How is immunological memory achieved? We explore answers to these questions in the next two sections.

While many cells and tissues play a role in adaptive immunity, two types of cell are particularly important: **B lymphocytes,** or **B cells,** and **T lymphocytes,** or **T cells** (see Fig. 43.3). B and T cells are named for their site of maturation. In birds, B cells mature in the bursa of Fabricius, an organ of the lymphatic system located off the digestive tract (**Fig. 43.8**). In mammals, B cells mature in the bone marrow. T cells mature in the thymus, an organ of the immune system located just behind the sternum, or breastbone. B and T cells circulate in blood and lymphatic vessels, and also can be found in the spleen, liver, and lymph nodes (Fig. 43.8). We begin with a discussion of the role of B cells in adaptive immunity.

B cells produce antibodies.

One of the hallmarks of adaptive immunity is the ability to target specific pathogens. But there are a tremendous number of different viruses, bacteria, single-celled eukaryotes, fungi, and worms that can lead to infection. How does the adaptive immune system recognize each one specifically? The specificity of the adaptive immune system is in part the result of **antibodies** produced by B cells.

An antibody is a large protein that carries sugar molecules attached to some amino acids (hence it is a glycoprotein). An antibody binds to foreign molecules that occur naturally on or in microorganisms and participate in normal cellular functions. Such a molecule is termed an **antigen,** which is a molecule that binds to and leads to the production of antibodies. (The word "antigen" is a contraction of "antibody" and "generator.") The great diversity of antigens on different microorganisms is matched by a great diversity of antibodies. It is estimated that humans produce approximately 10 billion different antibodies, each with the ability to recognize one or a few antigens on microorganisms.

Antibodies can be found on the surface of B cells or free in the blood or tissues. Collectively, antibodies make up a significant portion of the proteins present in the blood, typically about 20%. Because antibodies are present in the blood, tissue fluid, and secretions, which used to be called humors, this part of

FIG. 43.8 Major organs of the immune system.

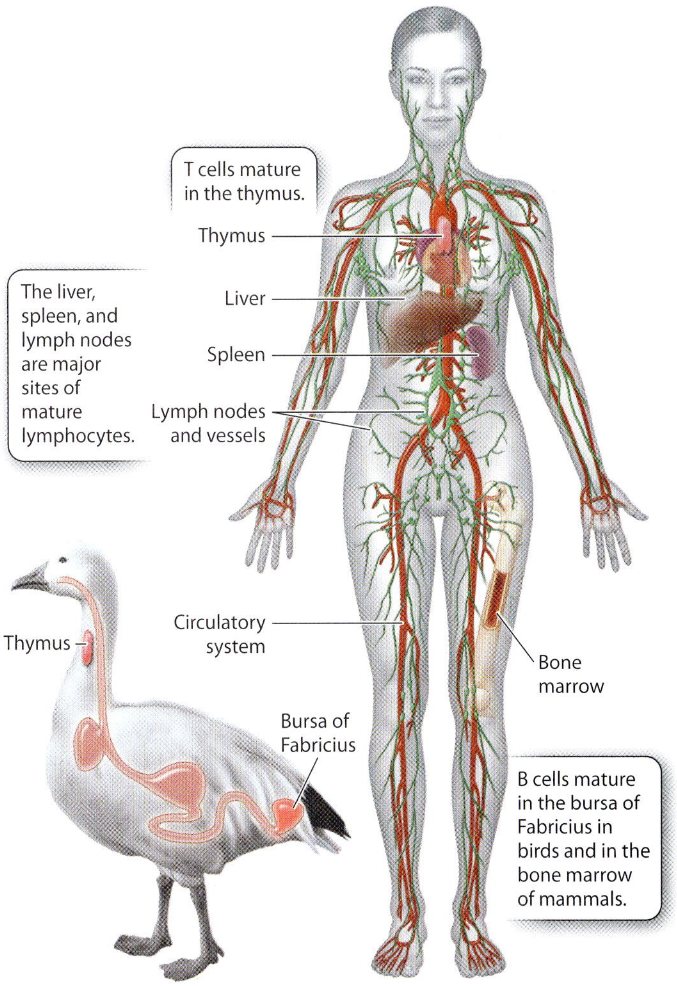

T cells mature in the thymus.

Thymus

The liver, spleen, and lymph nodes are major sites of mature lymphocytes.

Liver

Spleen

Lymph nodes and vessels

Thymus

Circulatory system

Bursa of Fabricius

Bone marrow

B cells mature in the bursa of Fabricius in birds and in the bone marrow of mammals.

FIG. 43.9 Antibody structure. An antibody has variable regions that distinguish it from the billions of other antibodies and allow it to bind a specific antigen.

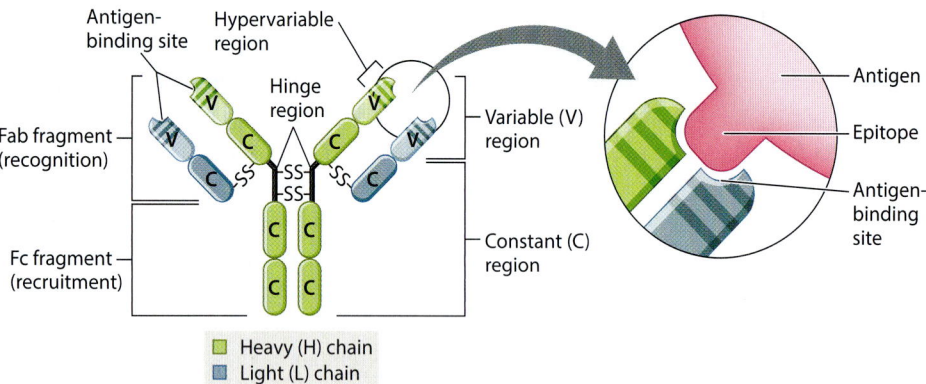

the adaptive immune system is sometimes called **humoral immunity.**

Antibodies have a distinctive structure that reflects their specificity for a given antigen. The simplest antibody molecule, shown in **Fig. 43.9,** is a Y-shaped protein made up of four polypeptide chains—two identical **light (L) chains** and two identical **heavy (H) chains,** so called because of their relative molecular weights. Covalent disulfide bonds hold the four chains together.

The light and heavy chains are further subdivided into **variable (V)** and **constant (C)** regions (Fig. 43.9). Within the variable regions of the L and H chains are regions that are even more variable, which are called **hypervariable regions.** Together, these hypervariable regions interact in a specific manner with a portion of the antigen called the **epitope.** Each of the billions of different antibodies has a distinct set of hypervariable regions that recognizes a unique epitope.

Binding of antibodies to antigens is the first step in recognition and removal of microorganisms. Binding alone is sometimes enough to disable a microorganism. In these cases, binding can lead to precipitation of the antibody–antigen complex (agglutination). More often, the microorganism is destroyed by a different component of the immune system, for example, by a phagocyte or the complement system. In this case, the function of an antibody is to recognize the pathogen, then recruit other cells of the immune system or activate the complement system. These latter functions are handled by a different part of the antibody from recognition. This can be demonstrated by treatment with proteolytic enzymes (enzymes that cut proteins at specific locations). For example, one type of enzyme breaks the molecule at the hinge region, producing two **Fab fragments** and one **Fc fragment** (Fig. 43.9). (The letters "ab" stand for "antigen-binding"; "c" stands for "crystallizing.") Each Fab fragment has an antigen-recognition site, and the Fc fragment activates the complement system and binds to cell-surface receptors of other cells of the immune system.

Mammals produce five classes of antibody with different biological functions.

Antibodies are members of a family of proteins with common structural features. As a group, they are called **immunoglobulins (Ig).** There are five classes of immunoglobulin—**IgG, IgM, IgA, IgD,** and **IgE**—each with a different function (**Fig. 43.10**). These classes are defined by their heavy chains, which differ in the amino acid sequences of their constant regions. There are also two types of light chain. These two types of light chain occur in all five classes of antibody, but any given antibody contains only one type of light chain.

IgG is by far the most abundant of the five antibody classes. It is the Y-shaped antibody depicted in Figs. 43.9 and 43.10. IgG circulates in the blood and is particularly effective against bacteria and viruses. In addition, IgG is the only class of antibody that can cross the placenta because of the presence of receptors on placental cells for the Fc portion of IgG. The ability of IgG to cross the placenta provides protection to the developing fetus.

IgM is a pentamer in mammals and a tetramer in fish. (A pentamer is a molecule made of five monomers; a tetramer is a molecule made of four monomers.) The individual units—the monomers—are linked by a joining chain (Fig. 43.10). IgM also exists as a monomer on the surface of B cells. IgM is particularly important in the early response to infection

FIG. 43.10 Five classes of antibody.

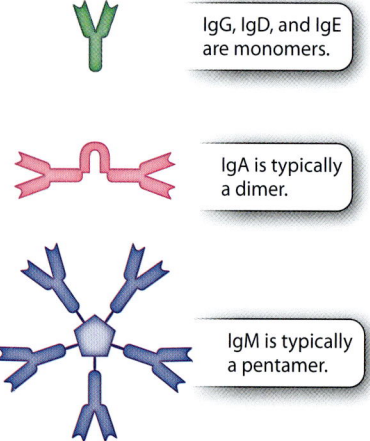

(discussed below) and is very efficient in activating the complement system and stimulating an immune response.

IgA is usually a dimer consisting of two antibody molecules linked by a joining chain (Fig. 43.10). It is the major antibody on mucosal surfaces, such as those of the respiratory, gastrointestinal, and genitourinary tracts. It helps to protect mucous membranes from infection. It is also present in secretions, including tears, saliva, and breast milk.

IgD and IgE, like IgG, are monomers. IgD is typically found on the surface of B cells. This immunoglobulin helps initiate inflammation. IgE plays a central role in allergies, asthma, and other **immediate hypersensitivity reactions,** which are characterized by a heightened or an inappropriate immune response to common antigens. Immediate hypersensitivity reactions result from the binding of the Fc portion of IgE to receptors on the surface of mast cells and basophils. Binding leads to the release of histamine and cytokines, resulting in inflammation that can be life threatening.

The five classes of antibody are also called **isotypes.** B cells can change the class of antibody they make, a process known as **isotype** or **class switching,** discussed below.

Clonal selection is the basis for antibody specificity.

A central question in immunology is how antibody specificity is achieved. There are seemingly limitless types of antigen that an organism might come across. How can B cells generate antibodies that interact only with specific antigens among this tremendous diversity?

Two alternative hypotheses were suggested to explain antibody specificity. In 1900, Paul Ehrlich suggested that there exists a large pool of B cells, each with a different antibody on its surface. The antigen binds to one or a few of these antibodies, stimulating the B cell to divide. In this model, specificity is achieved before any exposure to the antigen. Ehrlich's model drew on concepts from chemistry such as the binding of an enzyme to a substrate in the manner of a lock and key (Chapter 6). Similarly, the antigen and antibody interact in a specific way.

Later, in the early 1900s, a different hypothesis was suggested. This hypothesis proposed that the antigen instructs the antibody to fold in a particular way so that the two interact in a specific manner. Here, the antigen plays a more active role in the process and antibody specificity is determined only after interaction with the antibody.

Which of these two hypotheses is correct? Does the antigen *select* a B cell with a preexisting surface antibody with which it interacts specifically, as in the first hypothesis, or does it *instruct* the B cell to make a specific antibody, as in the second?

Experimental data on the genetic basis for antibody diversity (described below) provided strong evidence for the first hypothesis. Now called **clonal selection,** it is one of the central principles of immunology (**Fig. 43.11**). Every individual has a very large pool of B cells, each with a different antibody on its cell surface. Collectively, these antibodies recognize a great diversity of antigens, but each individual antibody can recognize only one antigen or a few antigens. An antigen interacts with the

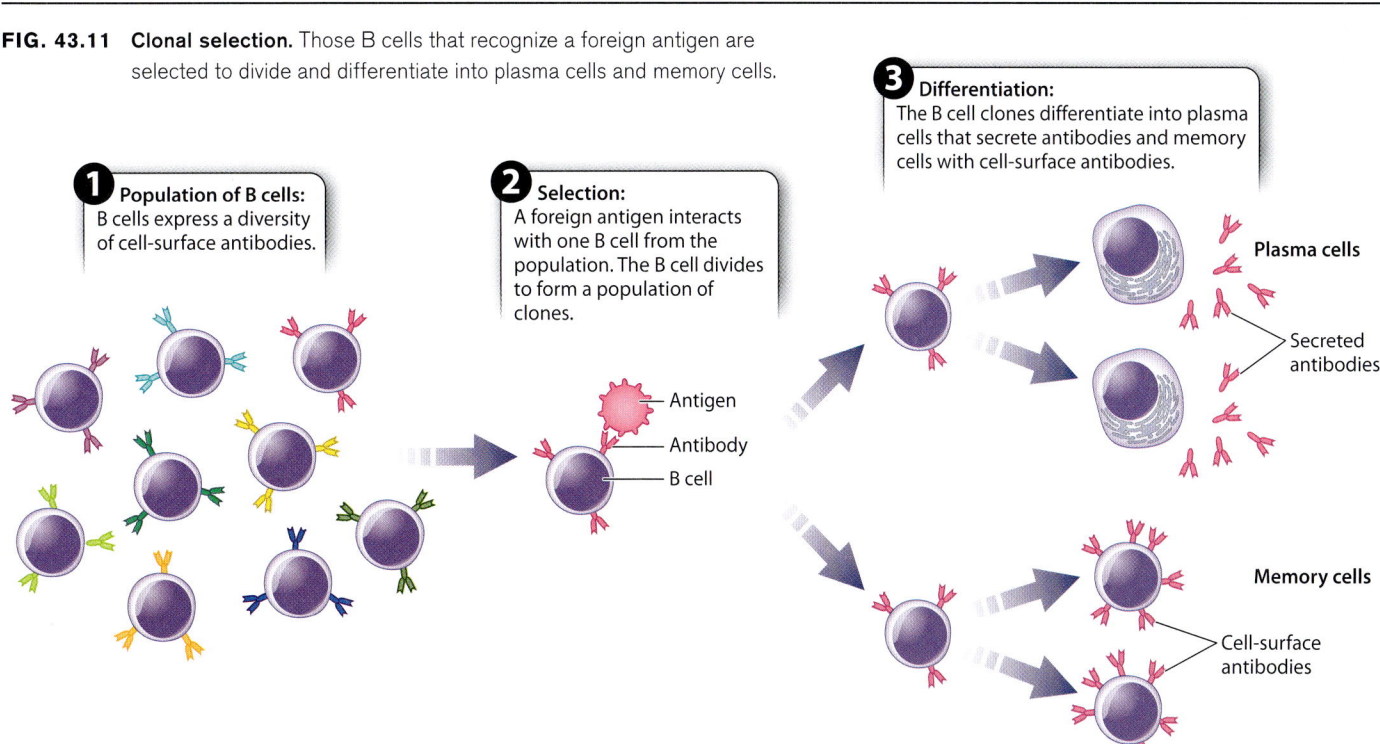

FIG. 43.11 Clonal selection. Those B cells that recognize a foreign antigen are selected to divide and differentiate into plasma cells and memory cells.

B cell, or a small set of B cells, that has the best fit between the antigen and the membrane-bound antibody. Binding leads the B cell to divide and differentiate into two types of cell. **Plasma,** or **effector, cells** secrete antibodies. **Memory cells** are long-lived cells that contain membrane-bound antibody having the same specificity as the parent cell. Since the resulting population of B cells arose from one B cell, they are clones of one another. Almost counterintuitively, antibody specificity is achieved before any exposure to the antigen.

Clonal selection also explains immunological memory.

Clonal selection provides a mechanism for explaining antibody specificity. It also explains the second feature of the adaptive immune system: the ability to remember past infections—that is, to react more vigorously on re-exposure. The first encounter with an antigen leads to a **primary response** (**Fig. 43.12**). In this response, there is a short lag before antibody is produced. The lag is the time required for B cells to divide and form plasma cells. The plasma cells secrete antibodies, typically IgM. The level of IgM in the blood increases, peaks, and then declines.

On re-exposure to the same antigen, even months or years after the first exposure, there is a **secondary response,** which is quicker, stronger, and longer than the primary response (Fig. 43.12). During this response, the antibodies are released by memory B cells. These cells respond more quickly to antigen exposure than naïve B cells (B cells that haven't been previously exposed to antigen). In addition, the pool of memory B cells is larger than the pool of naïve B cells for a given antigen, explaining why the secondary response produces more antibodies for a longer period of time. IgG is the most abundant antibody in the secondary response.

It is the presence of long-lived memory cells produced following the primary response that provides natural, long-lasting immunity after an infection such as chicken pox. We have learned to take advantage of immunological memory in **vaccination.** An antigen from a pathogen is deliberately given to a patient in a vaccine to induce a primary response but not the disease, thereby providing future protection from infection.

Until the early 1700s, it was common practice to inoculate people with smallpox in an effort to induce a mild disease and prevent a more severe, even lethal, infection. However, the English scientist Edward Jenner is usually credited with the first demonstration of the protective effects of vaccination. It was well known that milkmaids, exposed to the relatively benign cowpox virus from milking cows, were immune to the related but more deadly smallpox virus. In 1796, Jenner tested the consequences of exposure to cowpox by deliberately inoculating a young boy with cowpox and demonstrating that he became immune to smallpox. In fact, the modern word "vaccine" is derived from the Latin *vacca*, which means "cow."

Today, vaccines take many forms: They can be a protein or a part of a protein from the pathogen, a live but weakened form of the pathogen, or a killed pathogen. They are among the most effective public health measures ever developed.

→ **Quick Check 2** Look again at Fig. 43.12. Imagine that a different novel antigen is added on day 40 rather than a second injection of the same antigen. Would you predict that there would be a primary or a secondary response to this second antigen?

Genomic rearrangement creates antibody diversity.

We pointed out earlier that antibody specificity is determined before exposure to the antigen. In other words, clonal selection requires a mechanism for producing a great diversity of antibodies. How is this diversity achieved? One possibility is that each antibody is encoded by a separate gene. However, this cannot be the case, as the number of antibodies produced by an organism far exceeds the total number of genes in the genome. For example, the human genome consists of about 25,000 protein-coding genes, but humans can produce about 10 billion different antibodies.

In 1965, American biologists William Dreyer and J. Claude Bennett suggested a novel hypothesis to explain antibody diversity. As they put it, this hypothesis was "radically different from anything found in modern molecular genetics." They proposed that a single antibody is made by separate gene segments that are brought together by recombination. According to this model, many copies of each gene segment are present, each slightly different from the others. Only one copy of each gene segment ends up in the final, recombined antibody gene (and hence in the antibody protein). Diversity is therefore

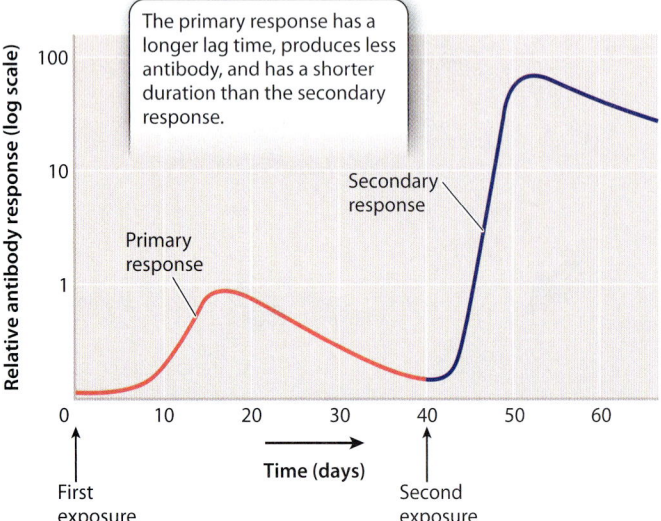

FIG. 43.12 Primary and secondary responses. The immune system responds more strongly to the second exposure to an antigen than to the first exposure because of the presence of memory B cells.

achieved by combining different copies of each segment in different B cells as they mature.

At the time Dreyer and Bennett proposed their model, evidence for it was indirect. More than 10 years later, direct experimental evidence showed that their hypothesis is correct (**Fig. 43.13**). As a B cell differentiates, different gene segments are joined in a process called **genomic rearrangement** that produces a specific antibody.

HOW DO WE KNOW?

FIG. 43.13

How is antibody diversity generated?

BACKGROUND Humans produce an estimated 10 billion different antibodies. Antibodies are proteins, encoded by genes. However, there are only about 25,000 genes in the human genome. How then can humans (and other vertebrates) produce such a diversity of antibodies from a limited set of genes? This was one of the central problems in immunology until the 1960s and 1970s.

HYPOTHESIS In 1965, American biologists William Dreyer and J. Claude Bennett proposed a model in which there are many copies of two gene segments, which they called the V (variable) and C (constant) segments. An antibody gene is assembled by recombining single copies of each of these segments. At the time, there was no direct experimental evidence to support this hypothesis.

EXPERIMENT In 1976, Japanese immunologists Nobumichi Hozumi and Susumu Tonegawa tested the Dreyer and Bennett hypothesis. Using mice as a model system, they isolated DNA from embryonic cells (before recombination was thought to occur) and from adult B cell tumor lines that produce one type of antibody (after recombination was thought to occur). The procedure they followed was slightly different from what is described here but conceptually the same. DNA is cut by restriction enzymes, the fragments are separated on a gel and transferred to a filter paper, and a light chain (containing both V and C regions) is used as a radioactive probe to determine the size and positions of the V and C regions (Southern blot, Chapter 12).

RESULTS Hozumi and Tonegawa found that the V and C regions were far apart in the DNA of embryonic cells and close together in adult cells, suggesting that the gene segments were brought together during B cell differentiation.

CONCLUSION Antibody genes are assembled by recombination of individual gene segments, providing a mechanism for generating antibody diversity.

FOLLOW-UP WORK Further studies showed that there are multiple copies of several different gene segments, now called the V, D (diversity), J (joining), and C regions for heavy chains, and V, J, and C regions for light chains. These are recombined in unique patterns in different B cells, so each B cell produces one antibody and a population of B cells produces a diversity of antibodies.

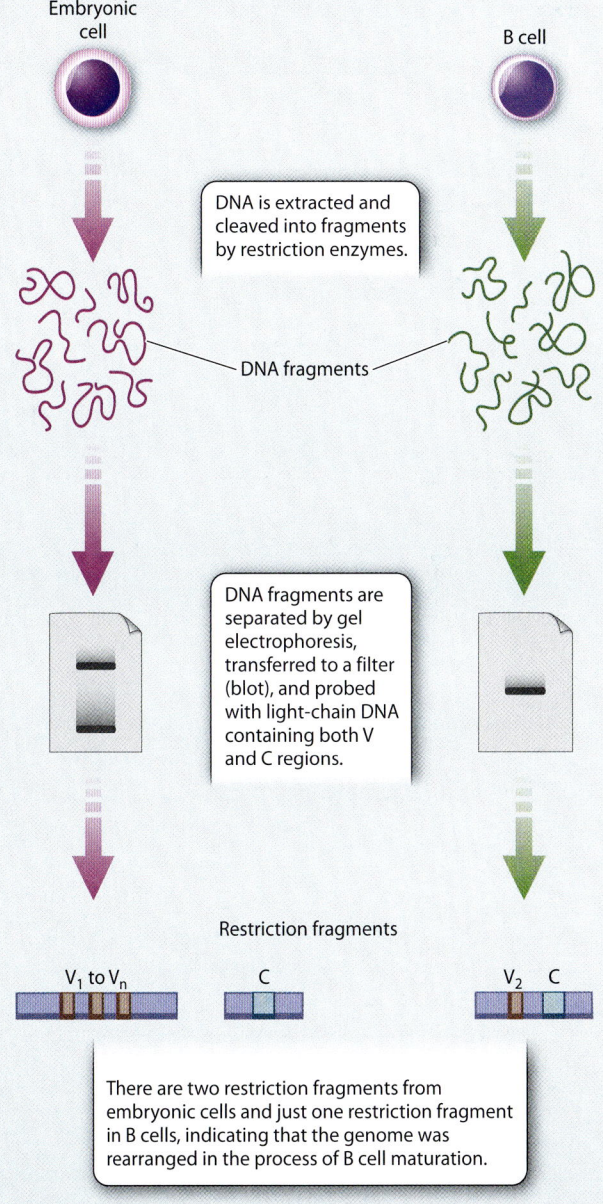

SOURCES Dreyer, W. J., and J. C. Bennett. 1965. "The Molecular Basis of Antibody Formation: A Paradox." *Proceedings of the National Academy of Sciences* 54:864–869; Hozumi, N. and S. Tonegawa. 1976. "Evidence for Somatic Rearrangement of Immunoglobulin Genes Coding for Variable and Constant Regions." *Proceedings of the National Academy of Sciences* 73:3628–3632.

FIG. 43.14 Genomic rearrangement in mice. In genomic rearrangement, different gene segments recombine to generate a functional gene encoding the parts of an antibody.

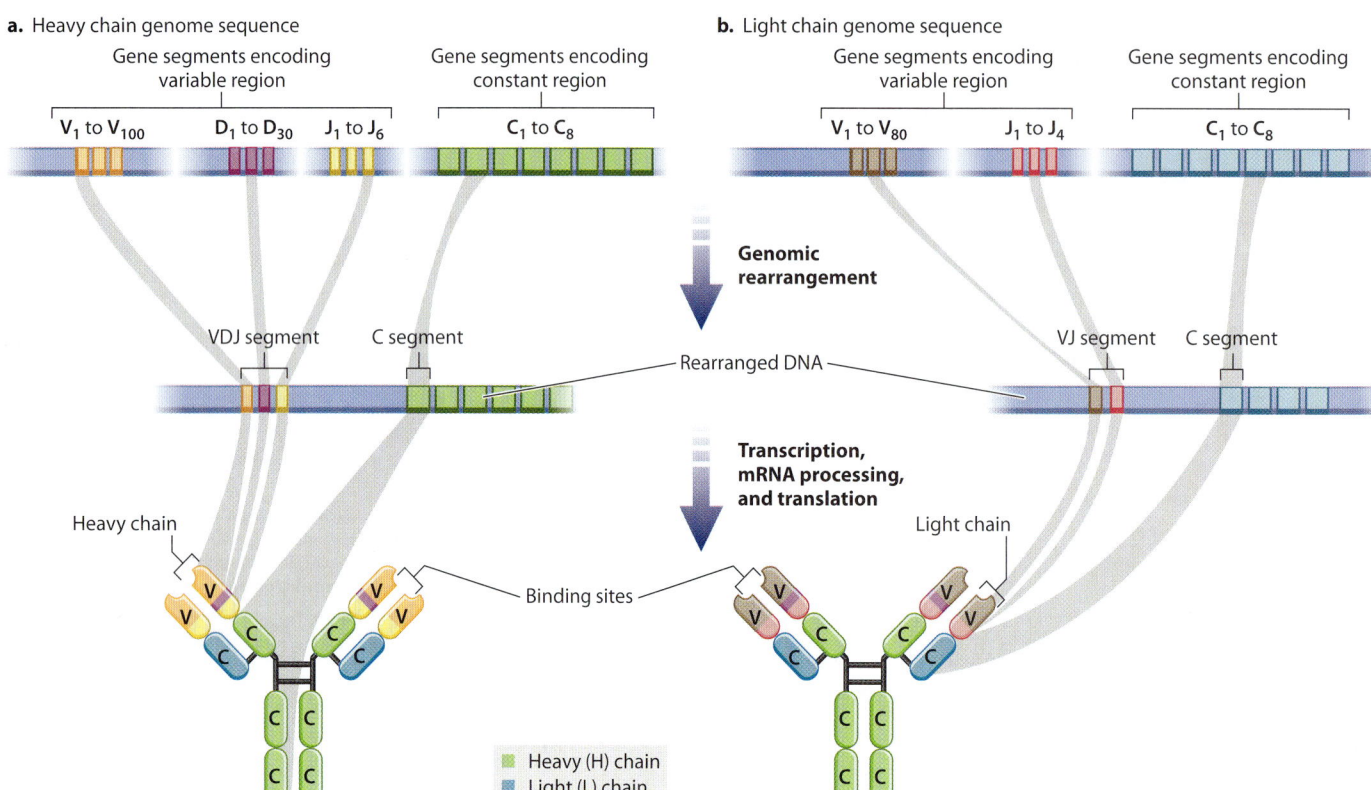

For H chains, there are multiple V (variable), D (diversity), J (joining), and C (constant) gene segments (**Fig. 43.14**). In mice H chains, for example, there are 100 different V segments, 30 D segments, 6 J segments, and 8 C segments. During B cell differentiation, the DNA in this region undergoes recombination so that just one of each of these segments makes part of a functional H chain gene, and the intervening DNA is deleted. The assembled VDJ segment encodes the variable portion, and the C segment encodes the constant region. L chain genes also contain multiple copies of gene segments, except they do not have a D gene segment, so the VJ segment encodes the variable portion and the C segment encodes the constant region. The result of this process is that each B cell encodes a single H chain and a single L chain, and each B cell expresses a unique antibody.

It is worth pausing here to consider the significance of this mechanism. We generally think of DNA as being a stable blueprint present in identical copies in all of the cells in our bodies. B cells are an exception. As a result of genomic rearrangement, the DNA in each mature B cell is different from the DNA in every other mature B cell and different from the DNA in other cells in the body.

Genomic rearrangement is the key mechanism that creates antibody diversity, but other mechanisms contribute as well. One is simply the association of different L and H chains to make a functional antibody. In addition, recombination is sometimes imprecise, creating different sequences at the junctions between the different gene segments. There are also mechanisms for adding nucleotides at the junctions by a process called N-nucleotide addition, and replacing nucleotides within VJ or VDJ segments by a process called somatic hypermutation. Finally, alternative splicing of mRNA (Chapter 3) allows a single B cell to express both membrane-bound and secreted antibodies.

We have seen that B cells as a group produce an enormous diversity of antibodies, whereas a given B cell makes one and only one antibody. But B cells are diploid, containing maternal and paternal copies of the genes that encode for antibodies. Maternal and paternal alleles of the antibody genes are similar, but not identical. As a result, a B cell able to use both homologs would make two different antibodies, yet B cells always make one. Remarkably, once one allele undergoes genomic rearrangement, the other allele is prevented from doing so. As a result, B cells act as if they have a single allele for each antibody gene—they express the maternal or the paternal gene, but not both.

As we have seen, B cells are able to switch between different classes, making IgM early in the response to infection and IgG later, while maintaining specificity to the same antigen. How is isotype switching achieved? Isotype switching results from further genome rearrangement, in which a given V(D)J region can join to different C segments during B cell differentiation.

The mechanisms we have considered pertain to humans and mice. Whereas B cells mature in the bone marrow in humans and mice, they mature in gut-associated lymphoid tissue (GALT) in many other vertebrates. Furthermore, genomic rearrangement is less important in many vertebrates. Instead, somatic hypermutation appears to play a more important role.

→ **Quick Check 3** If a given B cell produces only one type of antibody, how do organisms produce a great diversity of antibodies?

43.3 ADAPTIVE IMMUNITY: T CELLS AND CELL-MEDIATED IMMUNITY

Antibodies are remarkable for their diversity and specificity. They provide protection against many kinds of bacteria, especially those encased in a polysaccharide capsule, such as the bacteria that cause pneumonia and meningitis. The B cells that generate them exhibit immunological memory, as we saw by contrasting the primary and secondary immune responses. However, B cells and their antibodies have limitations. For example, B cells on their own can make antibodies only against some antigens; for other antigens, they require the assistance of other cells. Furthermore, they are sometimes ineffective against pathogens that take up residence inside a cell, as do the bacteria that cause tuberculosis.

To handle these and other kinds of pathogens, there is another part of the adaptive immune system, which depends on the second type of lymphocyte—the T cell. T cells do not secrete antibodies. Instead, they participate in **cell-mediated immunity**, so named because cells, not antibodies, recognize and act against pathogens. The importance of cell-mediated immunity is most dramatically demonstrated by individuals taking immunosuppressive drugs or infected by HIV. Both of these suppress the activity of T cells. The result can be overwhelming infections and even tumors.

T cells include helper and cytotoxic cells.

T cells, like B cells, originate in the bone marrow. However, unlike B cells, T cells mature in the thymus. A mature T cell

TABLE 43.2 Comparison of Helper and Cytotoxic T Cells

	HELPER T CELL	CYTOTOXIC T CELL
Functions	Activation of macrophages Activation of B cells Activation of cytotoxic T cells Secretion of cytokines	Cell killing
Surface molecule	CD4	CD8
MHC molecule recognized by the T cell	Class II	Class I

is characterized by the presence of a **T cell receptor** (TCR) on the plasma membrane. The T cell receptor is a protein receptor that recognizes and binds to the antigen. In this way, the T cell receptor is similar to an antibody. However, whereas antibodies may be either on the cell surface of B cells or secreted, TCRs are always found on the cell surface of T cells.

There are two major subpopulations of T cells with different names, functions, and cell-surface markers—**helper T cells** and **cytotoxic T cells** (Table 43.2; see Fig. 43.3). Helper T cells do just that—they *help* other cells of the immune system by secreting cytokines. One of their key roles is to activate B cells to secrete antibodies. Although B cells can work on their own, particularly against encapsulated bacteria, in most cases they require the participation of helper T cells. Helper T cells also activate macrophages, cytotoxic T cells, and other cells of the immune system.

Cytotoxic T cells, as their name implies, are cytotoxic—they can *kill* other cells. Like B cells, they are activated by cytokines released from helper T cells. They are particularly effective against altered host cells, such as those infected with a virus or that have become cancerous.

These two types of T cell are distinguished by the presence of different glycoproteins on their surface—CD4 on helper T cells and CD8 on cytotoxic T cells. The ratio of these two classes of T cell can be used to assess immune function. In healthy individuals, the CD4:CD8 ratio is usually about 2. A decreased ratio, indicating the presence of fewer helper T cells relative to cytotoxic T cells, is typical of individuals infected with HIV, the virus that causes AIDS. HIV infects and kills helper T cells, disarming a key player in the immune system.

FIG. 43.15 The T cell receptor (TCR). The TCR is a protein on the surface of T cells that binds foreign antigens that are bound to MHC proteins, triggering the division of the T cell into clones.

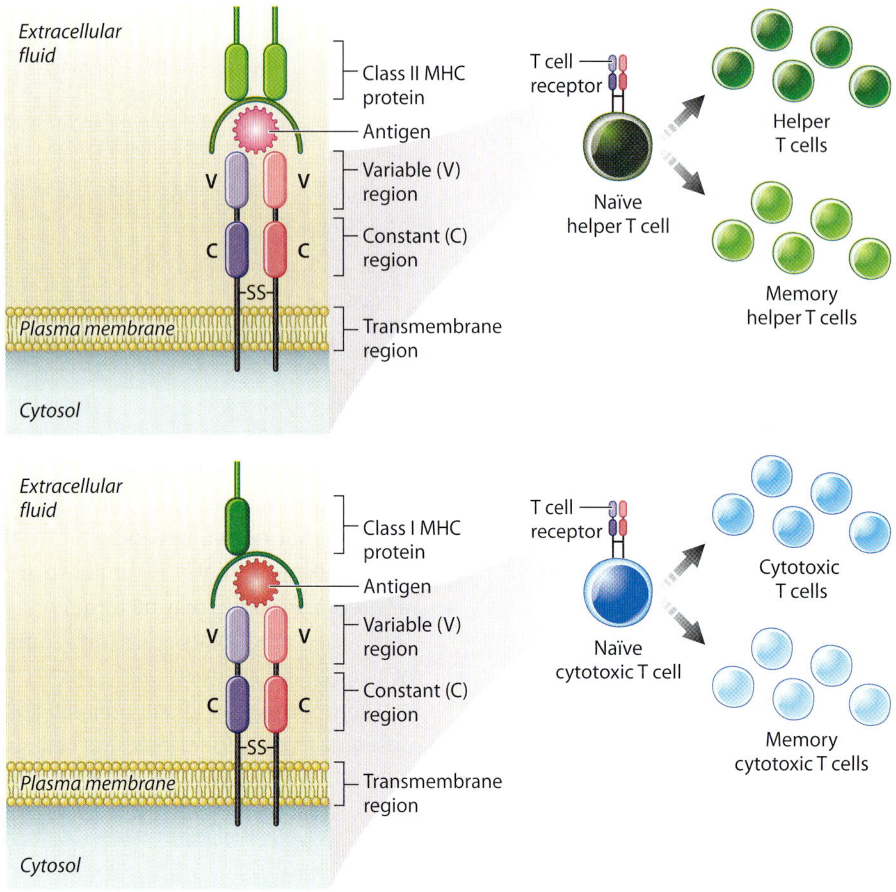

T cells have T cell receptors on their surface.

T cells have TCRs on their surface that recognize and bind antigens (**Fig. 43.15**). In several respects, TCRs are similar to antibodies on the surface of B cells. For example, TCRs recognize antigens with a specific structure. In addition, there is a great diversity of TCRs that differ from one another, but each T cell has just one type of TCR on its surface. Binding of TCR to an antigen triggers the T cell to divide into clones, resulting in a pool of T cells that are each specific for a given antigen. Finally, the diversity of TCRs among different T cells results from genomic rearrangement of V, D, J, and C gene segments.

However, TCRs are different from antibodies in important ways. First, they are composed of two, rather than four, polypeptide chains. Second, they are not secreted like antibodies, but are always membrane bound on the T cell surface. Finally, the TCR does not recognize free antigen. It only recognizes antigen in association with proteins of the **major histocompatibility complex,** or **MHC,** proteins that appear on the surface of most mammalian cells (Fig. 43.15).

Once activated, T cells divide and form helper and cytotoxic T cells. Some cells of each type are memory cells that provide long-lasting immunity following an initial infection, as in the case of B cells (Fig. 43.15). Also like B cells, T cells can sometimes be activated too strongly. Earlier, we saw how IgE bound to mast cells and basophils can lead to immediate hypersensitivity reactions characteristic of allergies and asthma. The counterpart in T cells is a **delayed hypersensitivity reaction,** which, as its name suggests, does not begin right away. For example, if you touch poison ivy, your skin will turn red and start itching only after a delay of several hours or days. Delayed hypersensitivity reactions are initiated by helper T cells, which release cytokines that attract macrophages to the site of exposure, which is typically the skin.

T cell activation requires the presence of antigen in association with MHC proteins.

Because of their central role in activating different components of the immune system, T cells themselves require not one signal but two signals to become activated. One of these signals is an antigen, which indicates the presence of foreign cells or particles. The second signal is MHC proteins. T cells interact only with antigens that are bound to molecules of the MHC on the surface of host cells. The requirement for antigen in association with MHC proteins explains why T cells target infected or cancerous host cells. MHCs were first discovered in transplantation biology because it is their presence that leads to acceptance or rejection of transplanted tissues.

The MHC is a cluster of genes present in all mammals that encode proteins on the surface of cells. The MHC is composed of many genes with a high rate of polymorphism, meaning that there is a lot of variability in the gene sequence (and consequently the protein sequence) among individuals. In humans and mice, the genes are divided into three classes: **class I** genes are expressed on the surface of all nucleated cells; **class II** genes are expressed on the surface of macrophages, dendritic cells, and B cells; and **class III** genes encode several proteins of the complement system and proteins involved in inflammation.

Let's consider the activation of helper T cells (**Fig. 43.16**). When an antigen enters the immune system, it may be recognized by an antibody directly or be taken up by **antigen-presenting cells.** These cells, which include macrophages, dendritic cells, and B cells, take up the antigen and return portions of it to the cell surface bound to MHC class II proteins. Helper T cells recognize processed antigen along with MHC class II molecules by their T cell receptors. As a result of TCR binding to antigen and MHC class II proteins, the helper T cells release cytokines that activate other parts of the immune system, including macrophages, B cells, and cyotoxic T cells.

Cytotoxic T cells also recognize antigen displayed by host cells, but in this case the antigen is presented in association with MHC class I molecules (Fig. 43.16). Because class I molecules are present on virtually all cells, cytotoxic T cells recognize and kill any host cell that becomes abnormal in some way. For example, a virus-infected cell often expresses viral antigens and MHC class I molecules on its surface. Cytotoxic T cells recognize the antigen and MHC class I molecules and kill the cell. Tumor cells express novel antigens along with MHC class I proteins, and may also be eliminated in some cases by cytotoxic T cells.

→ **Quick Check 4** How does T cell activation differ from B cell activation?

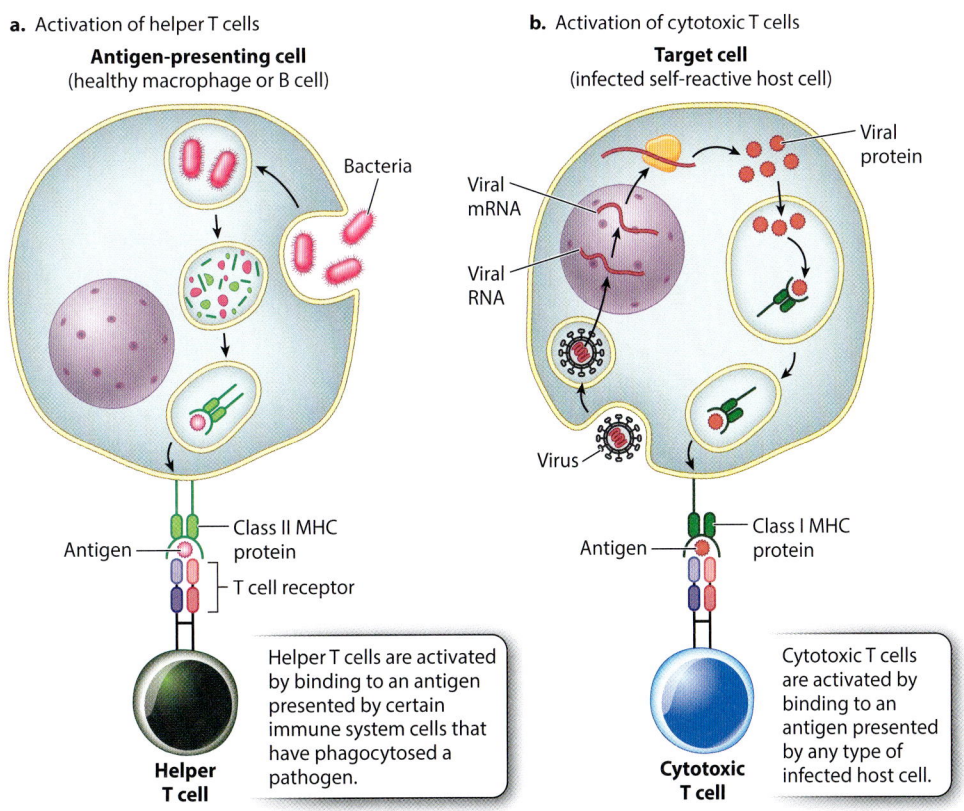

FIG. 43.16 Three-way interactions among the T cell receptor, MHC molecule, and processed antigen. Helper T cells recognize processed antigen in association with class II MHC protein, and cytotoxic cells recognize processed antigen in association with class I MHC protein.

The ability to distinguish between self and non-self is acquired during T cell maturation.

Of the possible T cell receptors that are generated by genomic rearrangement, only some are useful: those that react with the host's own MHC molecules. In addition, useful T cell receptors must not react with molecules normally present in or on cells of the host. In other words, T cells must not be activated in response to an organism's own cells. Molecules normally present on a host cell that can bind antibodies are called self antigens.

A sorting process is therefore necessary so that only some T cells mature and others are eliminated (**Fig. 43.17**). As T cells mature in the thymus, they interact with cells of the epithelium of the thymus. Those that recognize self MHC molecules on epithelium cells are **positively selected** and continue to mature. Those that react too strongly to self antigens in association with MHC are **negatively selected** and eliminated through cell death. Note that, in spite of their names, both processes involve elimination of some T cells. By far, the majority of T cells are eliminated through a combination of positive and negative selection.

The result of this sorting process is twofold. First, T cells become MHC restricted, so helper T cells interact with foreign antigen plus MHC class II proteins and cytotoxic T cells interact with foreign antigen plus MHC class I proteins. Second, T cells exhibit **tolerance**—in other words, they do not respond to self antigens, even though the immune system functions normally otherwise. Those antigens present as T cells mature are labeled as self and do not elicit a response; those not present are non-self and, if encountered, do elicit a response.

B cells are not MHC restricted. However, they, too, exhibit tolerance to self antigens because they go through a similar process of negative selection in the bone marrow. During that process, B cells that react strongly to self antigens are eliminated through cell death. Although most self-reactive T and B cells are eliminated, some T and B cells that react to self antigens escape this developmental check and end up in the circulation. As a result, additional mechanisms are in place to eliminate these self-reactive cells, either through the activity of T cells or through cell death.

The ability to distinguish self from non-self is critical. Failure leads to **autoimmune diseases**, in which tolerance is lost and the immune system becomes active against antigens of the host. Autoimmune diseases can be debilitating as T cells or antibodies attack cells and organs of the host. In rheumatoid arthritis, self-reactive T cells attack the joints, causing inflammation and tissue damage; in type I diabetes, attack is directed against insulin-producing cells of the pancreas; and in multiple sclerosis, the target is the myelin sheath surrounding nerves of the central nervous system. It is not fully understood how these self-reactive T cells evade checks on their development. Perhaps one clue is that some autoimmune diseases are associated with particular MHC alleles.

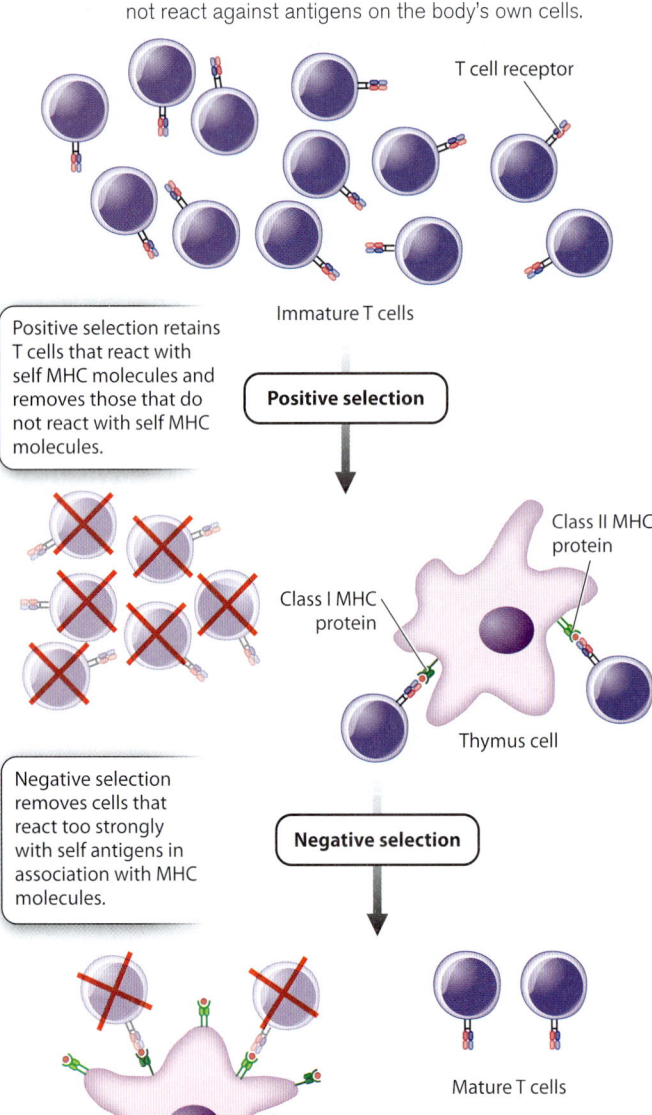

FIG. 43.17 Positive and negative selection of T cells. Selection ensures that T cells react with self MHC molecules but do not react against antigens on the body's own cells.

43.4 THREE INFECTIONS: A VIRUS, BACTERIUM, AND EUKARYOTE

Although we have presented the different components of the immune system one at a time, in life they work together to provide a coordinated response to pathogens. We have seen how B cells often require T cell help, how the complement system can be activated by antibodies produced by B cells, and how macrophages present antigens in association with MHC molecules to T cells. In turn, pathogens have evolved means of evading the immune system. In this section, we consider the flu virus, the bacterium that causes tuberculosis, and the malaria parasite. These three pathogens illustrate how pathogens evolve and adapt in response to the immune system, and how the immune system in turn evolves and adapts in response to pathogens.

The flu virus evades the immune system through antigenic drift and shift.

Everyone knows the symptoms: fever, chills, sore throat, cough, weakness, and fatigue, perhaps a headache and muscle pain. Many viruses cause these common flu-like symptoms. However, the common cold, stomach flu, and 24-hour flu are caused by unrelated viruses, not by the flu virus.

The flu—its full name is influenza—is caused by an RNA virus that infects mammals and birds. It is notable for causing seasonal outbreaks, or epidemics. Four times in the last century—the Spanish flu of 1918, the Asian flu of 1957–58, the Hong Kong flu of 1968–69, and the swine flu of 2009—the virus spread more widely and infected more people, and the disease reached pandemic levels. The flu is more than just a nuisance: It causes hundreds of thousands of deaths each year, and millions of deaths during pandemics.

Part of the success of the virus lies in its ability to spread easily. The flu virus can spread in the air by small droplets produced when an infected person coughs or sneezes. It can also spread by direct contact or by touching a contaminated surface, such as a doorknob. Hand washing, therefore, is one of the most effective means of protection.

Another characteristic of the flu is its ability to evade the immune system. Like the Greek soldiers in their wooden horse, the flu virus can change its outer coat so memory cells, produced from earlier infections, no longer recognize it. There are three major types of flu virus—called A, B, and C—and many different strains. For example, the swine flu pandemic of 2009 was caused by a type A flu virus of the strain H1N1. These strains differ in several respects, including the structure of a cell-surface glycoprotein called hemagglutinin (HA). HA binds to epithelial cells and controls viral entry into these cells. It can bind only to cells that display a complementary cell-surface protein. As a result, hemagglutinin determines in part which organisms the flu virus infects—humans, pigs, or birds—and which part of the body it infects—the nose, throat, or lungs.

Cells infected by viruses secrete cytokines that bring macrophages, T cells, and B cells to the site of infection. These cytokines, produced in abundance, lead to many of the symptoms commonly associated with the flu. For example, the cytokine interleukin-1 produces fever.

Once in the cell, the virus replicates its genome and makes more virus particles. However, replication is prone to error, so

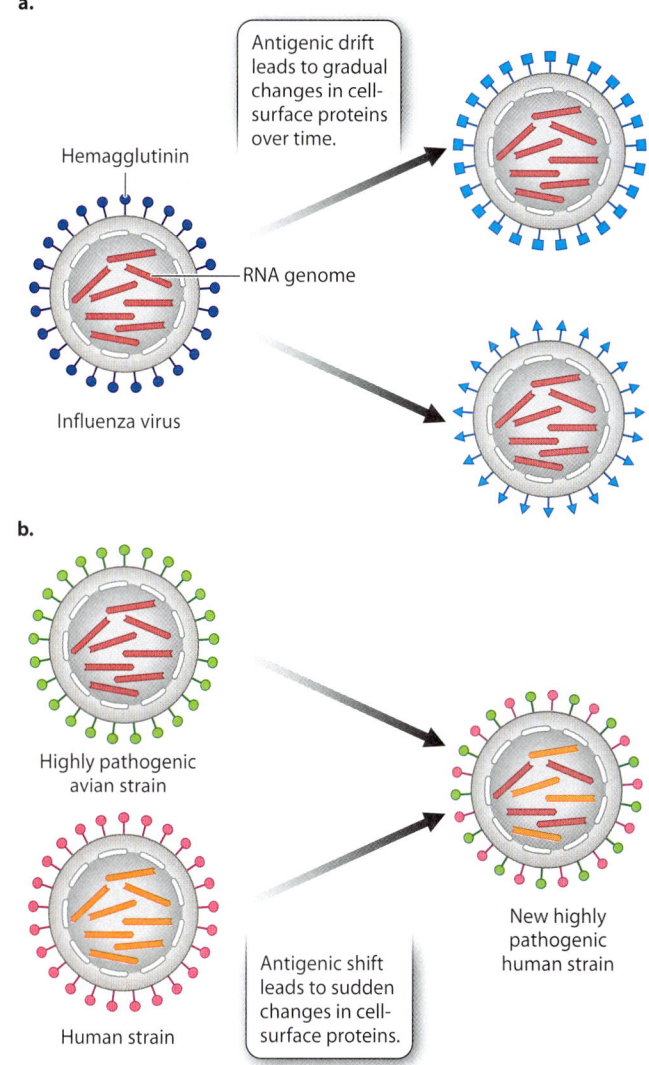

FIG. 43.18 (a) Antigenic drift and (b) antigenic shift. Virus populations evolve over time and evade the immune system's memory cells.

there is a high rate of mutation that leads to changes in the amino acid sequences of antigens present on the viral surface, including HA. This process is called **antigenic drift (Fig. 43.18a).** It allows a population of viruses to evolve over time and evade memory T and B cells that remember past infections.

Antigenic drift leads to a gradual change of the virus over time. The flu virus is also capable of sudden changes by a process known as **antigenic shift (Fig. 43.18b).** The viral genome consists of eight linear RNA strands. If a single cell is co-infected with two or more different flu strains, the RNA strands can reassort to generate a new strain. H1N1, for example, has genetic elements from human, pig, and bird flu viruses. Antigenic drift and shift make it difficult to predict from year to year which strains will be most prevalent and therefore what vaccine will be most effective.

Tuberculosis is caused by a slow-growing, intracellular bacterium.

Our second example is tuberculosis, also known as TB. Like the flu, it is spread through small droplets dispersed in the air when a person with active disease coughs or sneezes. Also like the flu, the TB pathogen replicates inside host cells. But unlike the flu, TB is caused by a bacterium, not a virus. TB results from infection with *Mycobacterium tuberculosis*, a bacterium with several unusual properties. It is very small with a compact genome, lacks a plasma membrane outside of the cell wall, and replicates exceedingly slowly. *M. tuberculosis* takes about 16–20 hours to divide, compared to 20 minutes for the common intestinal bacterium and model organism *E. coli*.

TB is very common. It is estimated that about a third of the world's population has TB. Symptoms are absent in most cases because the bacteria are in a dormant state, kept in check by the immune system. About 10% of cases are active, characterized by chronic cough, fever, night sweats, and the weight loss that gives TB its other name: consumption.

TB primarily affects the lungs. The bacteria enter macrophages in the alveoli of the lungs, where they replicate. The macrophages release cytokines, recruiting T and B cells to the site of infection. These lymphocytes surround the infected macrophages, forming a structure called a **granuloma (Fig. 43.19)**. The granuloma helps to prevent the spread of the infection and aids in killing infected cells.

Many TB cases remain asymptomatic. However, the bacteria can sometimes overcome host defenses, especially if the immune system is compromised or suppressed, as in the case of co-infection with HIV. Treatment for active disease requires a long course of multiple antibiotics. In many areas of the world, however, TB has evolved antibiotic resistance, a growing and alarming problem.

Diagnosis poses challenges, especially in the case of asymptomatic infections. The tuberculin skin test is performed by injecting protein from the bacteria under the skin and observing whether or not a reaction, in the form of a hard, raised area, occurs. A positive skin test is an example of a delayed hypersensitivity reaction caused by the recruitment of memory T cells to the site of inoculation. Unfortunately, the test is not always accurate, and newer tests are being developed.

Developing an effective vaccine is challenging. The only one currently available is BCG (Bacillus Calmette-Guérin, named for Albert Calmette, a French bacteriologist, and Camille Guérin, a French veterinarian, who developed it early in the twentieth century). BCG is a live but weakened strain of a related bacterium that infects cows. In fact, this vaccine was developed following the success of using cowpox to immunize against smallpox. However, BCG is only sometimes effective, and a public health disaster resulting from contaminated vaccine occurred in 1930 in Germany. Consequently, it is not used for mass immunization in the United States.

The malaria parasite uses antigenic variation to change surface molecules.

"Bad air." That's the literal translation of "malaria," a contraction of the Italian *mala* ("bad") and *aria* ("air"). But malaria is not caused by the air. Nor is it caused by a virus or bacterium. Malaria is caused by a single-celled eukaryote of the genus *Plasmodium* (Case 4: Malaria). Several species infect humans, but *P. falciparum* is the most common and most virulent.

Malaria is both devastating and common: There are about 500 million cases and 2 million deaths every year, mostly in sub-Saharan Africa, and mostly of children. Malaria causes cyclical fevers and chills and can lead to coma and even death.

FIG. 43.19 A granuloma in a tubercular lung.

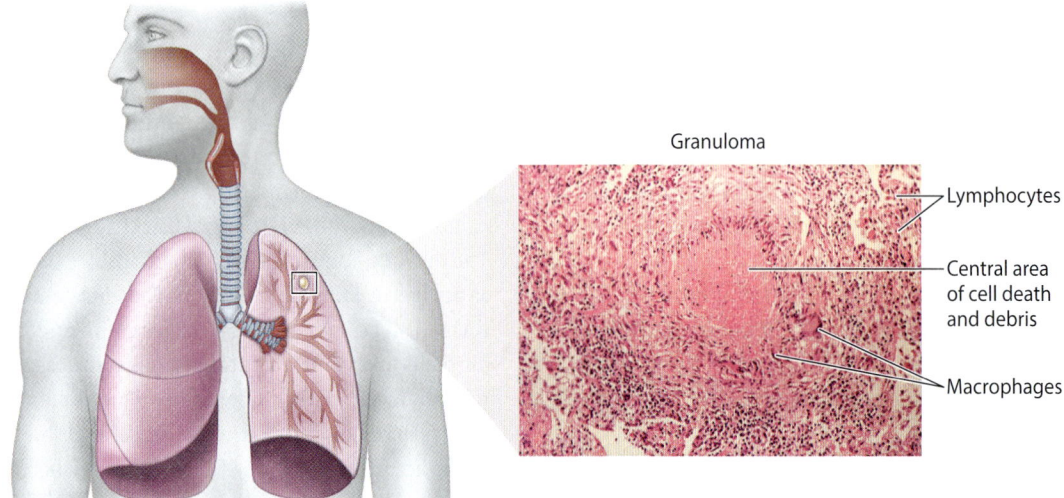

FIG. 43.20 Malaria transmission. (a) The malaria parasite, a single-celled eukaryote, is transmitted by the bite of a mosquito. (b) Following infection, red blood cells adhere to blood vessel walls by interactions with a malaria-encoded cell-surface protein and receptors on the blood vessel wall.

a.

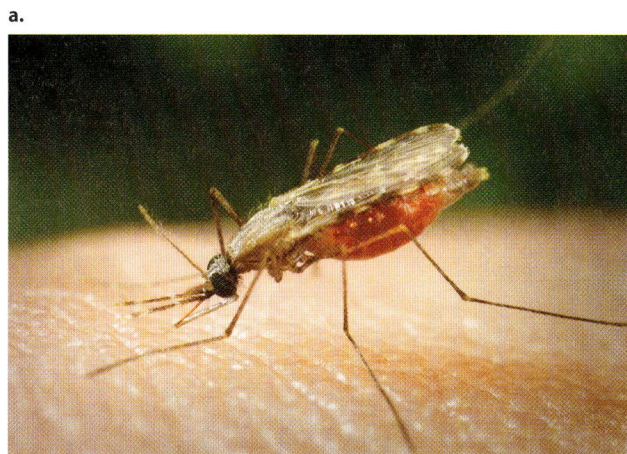

b.

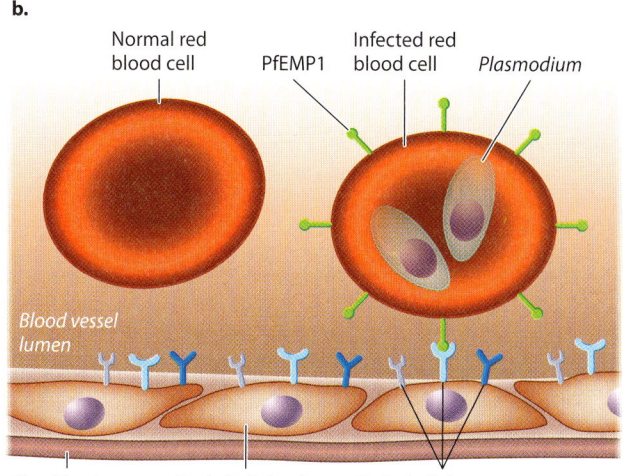

The malaria parasite is transmitted to humans by the bite of an infected mosquito (**Fig. 43.20a**). Using mosquitoes as a vector, the malaria parasite deftly bypasses the natural protective barrier provided by the skin. The malaria parasite spreads to the liver and then to red blood cells, where it completes its life cycle. The progeny can be taken up again by mosquitoes.

Malaria is an ancient parasite that has coevolved with humans (Case 4: Malaria). We have already seen how altered hemoglobin provides some protection against infection (Chapter 21). But the malaria parasite, too, has adapted—it is able to evade our immune system. It resides in liver cells and red blood cells, where it cannot be easily detected by macrophages, B cells, or T cells. Infected red blood cells can be removed by the spleen, but the malaria parasite has evolved a way to avoid this filtering action. It expresses an adhesive protein that inserts itself on the surface of red blood cells. This protein interacts with proteins on the surface of cells lining blood vessels, helping infected cells stick to blood vessel walls and keeping them in the circulatory system and out of the spleen (**Fig. 43.20b**). The protein, PfEMP1 (*Plasmodium falciparum* erythrocyte membrane protein 1), could conceivably be a target for antibodies or TCRs. However, it is encoded by one of about 60 genes, each slightly different from the others. The parasite expresses just one of these genes at a time and can change which one is expressed, in a process called **antigenic variation.** The great and ever-changing diversity of this protein, both in a single parasite and in the population as a whole, makes it a moving target for the immune system.

Despite continuing research, there is no vaccine. Researchers hope that the sequence of the malaria genome, completed in 2002, will provide new targets for vaccine research. In the meantime, efforts to control malaria focus on measures to control mosquitoes, such as nets and insecticides, and on treatment with antibiotics. The evolution of antibiotic resistance, however, is an all too familiar problem.

Core Concepts Summary

43.1 THE INNATE IMMUNE SYSTEM ACTS AGAINST A DIVERSITY OF FOREIGN PATHOGENS.

The skin is a physical barrier and contains chemicals, cells, and microorganisms that provide protection against infection. page 43-2

Phagocytes, including macrophages, dendritic cells, and neutrophils, engulf other cells and debris. page 43-3

Toll-like receptors on the surface of phagocytes play a central role in recognition of foreign molecules. page 43-4

Eosinophils, basophils, mast cells, and natural killer cells are key players in the innate immune system. page 43-4

Inflammation is the response of the body to infection or injury. It entails increases in blood flow and blood vessel permeability and recruitment of phagocytes. page 43-5

The complement system consists of circulating proteins that activate one another. Activation results in the formation of a membrane attack complex that lyses bacteria and other cells, production of proteins that enhance phagocytosis, and generation of cytokines. page 43-6

43.2 THE ADAPTIVE IMMUNE SYSTEM INCLUDES B CELLS THAT PRODUCE ANTIBODIES AGAINST SPECIFIC PATHOGENS.

B cells produce antibodies that bind to foreign antigens. page 43-7

Antibodies have two light chains and two heavy chains joined by covalent disulfide bonds. page 43-8

Light and heavy chains have constant and variable regions. page 43-8

Mammals produce five types of antibody: IgG, IgM, IgA, IgD, and IgE. IgG is the most abundant; IgM and IgD are found on the surface of B cells; IgA is secreted; and IgE is involved in allergic reactions. page 43-8

Antibody specificity is achieved by clonal selection, in which an antigen binds to one or a few B cells on the basis of the fit with its cell-surface antibody receptor, leading to B cell differentiation into plasma and memory cells. page 43-9

The primary immune response is characterized by a long lag period and a short response. The secondary immune response is quicker and more robust and provides the basis for immunity following infection or vaccination. page 43-10

Antibody diversity is generated by genomic rearrangement, where individual gene segments are joined to make functional genes. page 43-10

43.3 THE ADAPTIVE IMMUNE SYSTEM INCLUDES HELPER AND CYTOTOXIC T CELLS THAT ATTACK PATHOGENS, FOREIGN CELLS, AND DISEASED CELLS.

Two types of T cell are helper T cells and cytoxic T cells. page 43-13

Helper T cells activate other cells of the immune system, including macrophages, B cells, and cytotoxic T cells. page 43-13

Cytotoxic T cells kill altered host cells. page 43-13

T cells are activated by the binding of a T cell receptor to an antigen in association with an MHC protein. page 43-14

As T cells mature in the thymus, they undergo positive and negative selection, so only T cells that recognize self MHC, but not self antigens, survive. page 43-15

The ability to distinguish self from non-self antigens is critical for proper immune function; autoimmune diseases result when the immune system attacks cells of the host. page 43-15

43.4 SOME PATHOGENS HAVE EVOLVED MECHANISMS THAT ENABLE THEM TO EVADE THE IMMUNE SYSTEM.

The flu is caused by an RNA virus that can change its surface proteins through antigenic drift and antigenic shift. page 43-16

The bacterium that causes tuberculosis grows slowly and resides inside cells, particularly macrophages. page 43-17

Malaria is caused by a single-celled eukaryote with a complex life cycle having stages spent in mosquitoes and humans. The malaria parasite has evolved several mechanisms, including antigenic variation, that help it reproduce in humans without being eliminated by the immune system. page 43-18

Self-Assessment

1. Compare and contrast features of the innate and adaptive immune systems.
2. List components of the innate immune system.
3. Describe how cells of the innate immune system recognize pathogens.
4. Explain how the four signs of inflammation relate to the changes that occur in the underlying tissues.
5. Draw a picture of an antibody molecule, labeling each of the parts. Include the light and heavy chains, the variable and constant regions, and the antigen-binding site.
6. Describe how B cells produce a great variety of different antibodies, each specific for a particular antigen.
7. Explain how immunological memory is achieved.
8. Compare and contrast the activation of T cells and B cells.
9. Describe how positive and negative selection leads to a population of T cells that react with foreign antigens but not self antigens, and what happens when tolerance to self antigens is lost.
10. Describe three mechanisms that allow pathogens to evade the immune system.

Do you understand the chapter's Core Concepts? Log in to check your answers to the Self-Assessment questions, then practice what you've learned and reinforce this chapter's concepts by working through the problems and multimedia tutorials provided there.

www.biologyhowlifeworks.com

CASE 8

Biodiversity Hotspots

RAIN FORESTS AND CORAL REEFS

When we think about mass extinctions, the demise of the dinosaurs is often the first example to spring to mind. But many conservationists believe we are in the midst of a great mass extinction at this very moment. The International Union for the Conservation of Nature has estimated that 800 plant and animal species have died out in the last 500 years, and more than 16,000 are currently at risk of extinction.

The cause of this mass die-off isn't an asteroid impact or a cataclysmic volcanic event. Instead, human activity is dramatically altering the planet and its biological diversity.

Despite our role in the extinctions, it's clearly not possible to save every single species threatened with extinction. Conservationists must make difficult choices, prioritizing which species to protect. To get the most bang for every conservation buck, ecologists have identified 25 priority areas, or biodiversity hotspots.

These hotspots are areas with disproportionately large concentrations of native species. Scientists estimate that as many as 44% of all vascular plants and 35% of vertebrate species, for example, are confined to these 25 hotspots. Those figures are impressive, considering that the identified hotspots make up just 1.4% of Earth's land surfaces.

The islands of the Caribbean—the West Indies—constitute one such biodiversity hotspot. The Caribbean Sea contains more than 7000 islands rich in biological diversity, including an estimated 6550 native plant species. All the Caribbean islands are flush with flora and fauna, but one island in particular epitomizes the concept of a biodiversity hotspot. Hispaniola—the second largest of the Caribbean islands—was the site of the first New World colonies, founded by Columbus. Today, Hispaniola is shared by two countries, Haiti to the west and the Dominican Republic to the east.

Hispaniola is only about the size of Ireland, yet it contains a striking assortment of habitats including desert, savanna, rain forests, dry forests, pine forests, coastal estuaries, and interior mountains. With such a rich abundance of habitats, Hispaniola is, in some ways, a microcosm of the planet's many ecosystems. And like the Earth's other natural places, Hispaniola faces many of the same challenges.

Of all the habitats on Hispaniola, the forests have the greatest variety of plants, animals, and other organisms. Ferns, orchids, and bromeliads—spiky relatives of the pineapple—grow in the treetops far above the forest floor. These plants, known as epiphytes, absorb water from dew, rainwater, and even the air. Their roots never reach the soil.

The forest is alive with activity. Ants march silently. Crickets and cicadas buzz and whine in the treetops. Vertebrate vocalists such as frogs and birds fill the air with chirps and whistles, making themselves known both to potential mates and rivals. Bats swoop through the darkened skies, using ultrasonic calls to locate insect prey. About 30,000 species of multicellular organisms are found on Hispaniola, and a third of those are endemic—they are found nowhere else in the world. One example is the Hispaniolan solenodon, a long-nosed mammal that resembles an overgrown shrew and secretes toxic saliva. Listed as endangered, the solenodon has flirted with extinction for decades.

> **Conservationists must make difficult choices, prioritizing which species to protect. To get the most bang for every conservation buck, ecologists have identified 25 priority areas, or biodiversity hotspots.**

How did all these species come to live on the island? Over tens of millions of years, organisms traveled to Hispaniola by air and by sea. Geographically separated from their ancestors on the mainland, many of these organisms evolved to become distinct species. As they adapted to fill vacant niches on the island, animals such as *Anolis* lizards, frogs, and crickets evolved an impressive diversity of forms.

In some cases, scientists have been able to explore Hispaniola's evolutionary history directly. They have found remains of insects and other species entombed in fossilized tree resin, or amber, 30–23 million years ago. Many of these preserved specimens resemble species still living on Hispaniola. Others are unlike any creatures found on the island today.

The stunning diversity of Hispaniola doesn't just exist on land. Conservationists consider the warm, shallow reefs of the Caribbean one of the world's marine biodiversity hotspots. Coral reefs in general are among the world's most biologically diverse ecosystems. Species from all recognized phyla of animals are found on reefs.

Around the world, reefs support hundreds of thousands of species, including snails, corals, lobsters, and fish. Many, but not all, of these creatures spend their entire life in this habitat. Reefs also act as vital nursery grounds for animals that live in other parts of the ocean, from sea turtles to sharks. Reef biodiversity is critical to maintaining fisheries that provide food and income for millions of people.

Hispaniola's rain forests and reefs—and indeed all the biodiversity hotspots identified by conservationists—aren't just notable for their biological richness. They have also been given priority status because they are facing serious threats.

Reefs around the world are being damaged by human activities. Overfishing and pollution take their toll on reef ecosystems. Climate change is increasing seawater temperature and altering ocean chemistry. Those changes are threatening the coral organisms that form the backbone of the reefs.

Ecosystems on land, too, face significant challenges. The Caribbean islands retain only about 11% of their original primary vegetation. Humans have cleared great swaths of land for agriculture and chopped trees to use as fuel. Humans have also introduced non-native, or invasive, species to the islands, both intentionally and accidentally. On Hispaniola, Europeans introduced the Javanese mongoose to control rats in sugarcane fields. The mongoose has devastated local populations of reptiles, amphibians, and birds.

If Hispaniola serves as a microcosm of the planet's ecosystems, it also serves as a cautionary tale. Two

Biodiversity hotspots. The areas highlighted in orange have particularly high species diversity.

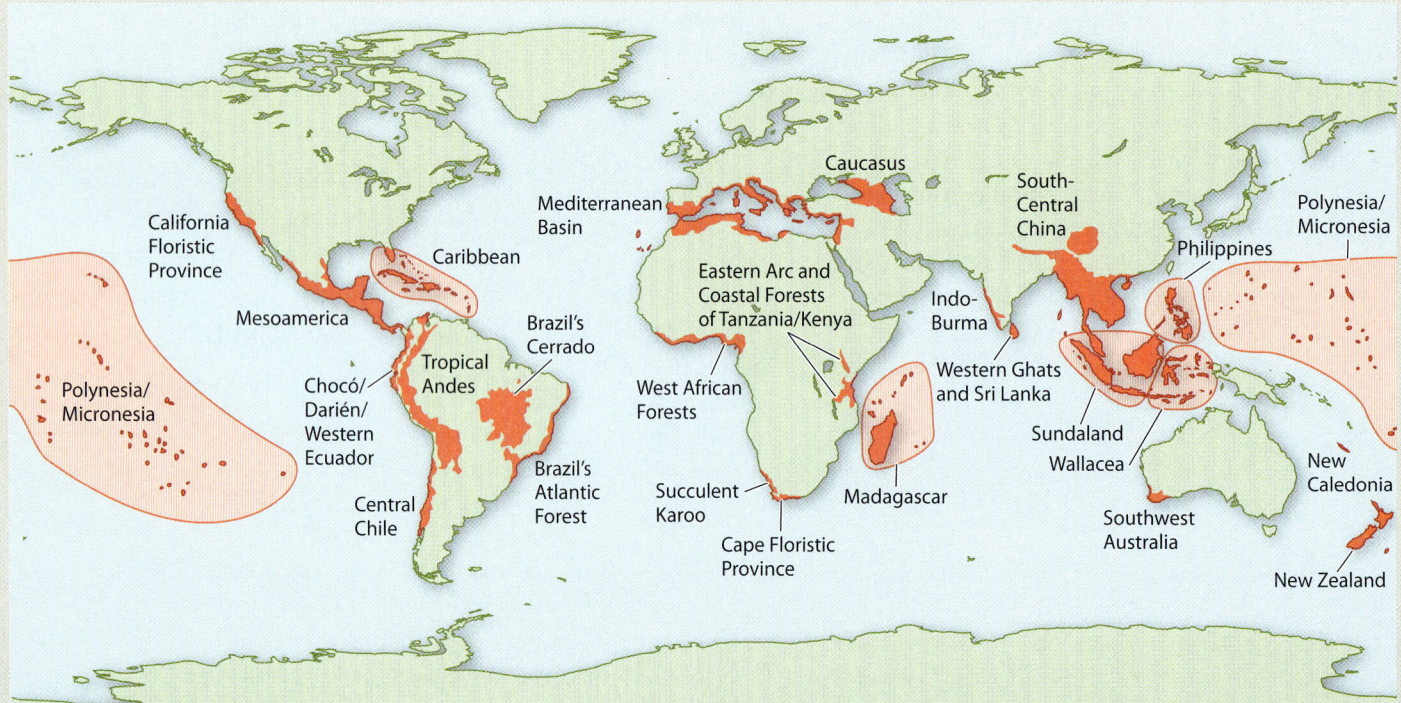

Coral reefs. Coral reefs support many species but are threatened by ocean warming and acidification.

divergent paths have emerged on the single island. Haiti has historically suffered greater poverty than the Dominican Republic, and has much greater population density. Those factors put more pressure on the environment in Haiti, and have contributed to much greater deforestation on the western half of the island.

The case of Hispaniola illustrates the complex challenges facing conservationists. Solutions to environmental challenges must take into consideration the effects of poverty, culture, and other social pressures. Balancing the needs of humans with the needs of natural ecosystems is no easy task. But as a relatively small island, Hispaniola may in fact be in a better position to turn things around than larger, more politically complex countries. The world's biodiversity hotspots might yet lead the charge in finding ways for humans and the rest of the planet's inhabitants to peacefully coexist.

? CASE 8 QUESTIONS

Answers to Case 8 questions can be found in Chapters 44–48.

1. How have coral reefs changed through time? *See page 44-29.*
2. How do islands promote species diversification? *See page 46-15.*
3. Why are tropical species so diverse? *See page 47-21.*
4. How has climate change affected coral reefs around the world? *See page 48-7.*
5. How has human activity affected biological diversity? *See page 48-15.*

CHAPTER 44

ANIMAL DIVERSITY

Core Concepts

44.1 The animal tree of life consists of more than a million species.

44.2 The earliest branches of the animal tree include sponges, cnidarians, ctenophores, and placozoans.

44.3 Bilaterians, including protostomes and deuterostomes, have bilateral symmetry and develop from three germ layers.

44.4 Vertebrates have a bony cranium and vertebral column, and include fish, reptiles, birds, and mammals.

44.5 Animals first evolved more than 600 million years ago in the oceans, and by 500 million years ago, the major structural and functional body plans of animal phyla were in place.

Biologists have described about 1.8 million species of eukaryotic organisms from the world's forests, deserts, grasslands, and oceans. Of these, about 1.3 million species are animals. There is reason to believe that animal diversification began in the oceans, but today most animal species are found on land. The majority of all known animal species are insects.

In previous chapters, we saw how the organ systems of animals enable them to move, feed, and behave in different ways. Clearly, this wide range of functions has permitted animals to diversify to a degree unmatched by plants, fungi, algae, or protozoans. And just as clearly, species diversity is unevenly distributed among major animal groups. How can we bring phylogenetic order to animal diversity, and how, in turn, can we understand the biological traits that have given rise to such great diversity on a few branches of the animal tree of life?

44.1 A TREE OF LIFE FOR MORE THAN A MILLION ANIMAL SPECIES

As discussed in Chapter 23, it is relatively easy to understand how the visible features of humans, chimpanzees, and gorillas show them to be more closely related to one another than any of them is to other animal species. Furthermore, it isn't hard to see that humans, chimpanzees, and gorillas are more closely related to monkeys than they are to lemurs, and that humans, chimpanzees, gorillas, and monkeys are more closely related to lemurs than to horses. Anatomy and morphology reveal key evolutionary relationships among vertebrate animals, but how do we come to understand the place of vertebrates in a broader tree that includes all animals? More generally, how can we construct an animal phylogeny that includes organisms with body plans as different as those of sponges, sea stars, earthworms, and mussels?

Phylogenetic trees propose an evolutionary history of animals.
The anatomy of animals ranges from simple (think of sponges) to remarkably complex (lobsters, for example, or humans). All animals are multicellular heterotrophs, gaining both food and energy from organic molecules (Chapter 6). Most fungi are also multicellular heterotrophs (Chapter 34), but fungi have cell walls and animals do not. Animals are also distinguished by embryos that include a gastrula stage (Chapter 42) and by the presence of collagen, a fiberlike protein that makes up connective tissues such as basal lamina, tendon, and cartilage.

While they share some general features, animals vary widely in many respects. For example, they differ in the ways that they gather food. Tiny animals called placozoans that live on the seafloor absorb organic compounds directly from the environment and take in particulate matter by phagocytosis. In contrast, bilaterally symmetrical animals from insects to eagles have highly adapted nervous and muscular systems that enable them to capture plants or other animals and digest them in a specialized gut.

44-1

FIG. 44.1 A simple phylogenetic tree of animals. The tree shows increasing complexity from the closest protistan relatives of animals, the choanoflagellates, to bilaterian animals such as insects and mammals.

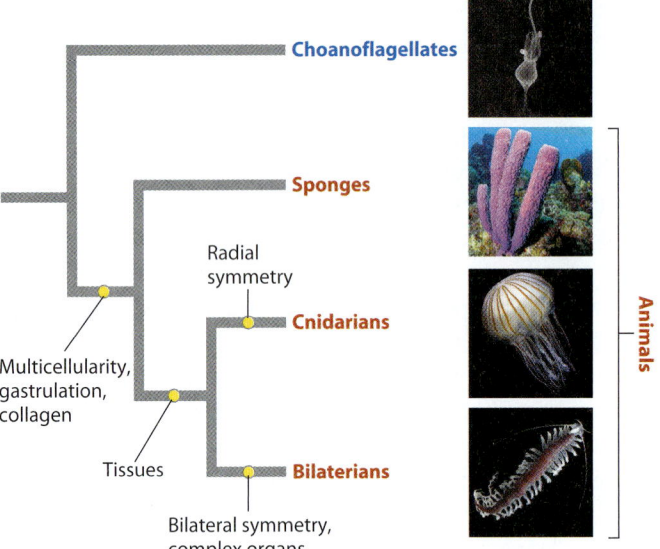

Sponges have no nerve cells, and sea anemones have only simple networks of nerves, but humans have brains containing trillions of cells that permit us to make and communicate discoveries about sponges, sea anemones, and other organisms we encounter. How the diverse morphologies, anatomies, and behaviors that characterize animal species came to be is a question of evolution, and phylogenetic trees provide hypotheses about the evolution of animal diversity.

Early in the history of biology, taxonomists such as Linnaeus recognized that for all their diversity, animals can be sorted into a limited number of distinctive body plans, each group today formally called a **phylum** (plural, phyla). The problem for biologists has long been how to understand the evolutionary relationships among animal phyla, still an active field of research today.

Fig. 44.1 shows a simple phylogenetic tree. Recall the basic logic of phylogenies, presented in Chapter 23. The fundamental mechanism by which biological diversity increases is speciation, the divergence of two populations from a common ancestor. Played out repeatedly through time, speciation gives rise to a treelike pattern of evolutionary relatedness, with more closely related taxa branching from points close to the tips of the tree and more distantly related taxa diverging from branch points nearer its base. Distinctive features of organisms, called characters, evolve throughout evolutionary history, and the time when they arose can be estimated from their shared presence in the descendants of the population in which the character first evolved. But a prominent feature of evolution is change, and in fact many characters change from one form to another and sometimes back again. Thus, mammals are descended from egg-laying ancestors, but most living mammals do not lay eggs—that character has been lost, replaced by live birth.

A character can evolve more than once in separate groups by convergent evolution. Such characters, called analogies, include the single-lens eye in octopus and vertebrates, and wings in birds and bats. These characters are not helpful in reconstructing evolutionary relationships. Close studies of morphological and molecular characters can often distinguish analogies from similarities that result from common ancestry, called homologies, which are useful in understanding evolutionary relationships. A large number of studies have resolved many but not all of the evolutionary relationships among animals.

With these points in mind, Fig. 44.1 proposes that animals are closely related to protists called choanoflagellates (Chapter 27), but differ from them in the presence (among other features) of persistent multicellularity, the formation of a gastrula during early development, and the synthesis of collagen. **Poriferans** (the phylum that is made up of sponges) diverged from other animals early, and **cnidarians** (the phylum that contains jellyfish and sea anemones) diverged later from animals whose descendants have well-defined and complex organs. Note that this diagram does not tell us that sponges are older than more complex animals. Instead, it says that sponges and more complex animals diverged from a common ancestor but doesn't tell us what that common ancestor looked like. Similarly, cnidarians and more complex animals with bilateral symmetry diverged from their last common ancestor at a single point in time and so are equally old, but structurally distinct.

We'll use phylogeny as our framework for discussing the immense diversity of animals, but first, we need to consider how the tree itself has taken shape through more than a century of biological research.

Nineteenth-century biologists grouped animals by anatomical and embryological features.

Since the earliest days of comparative anatomy, it has been clear that animals vary in their structural complexity. Sponges have only a few cell types, and those cells are not organized into tightly coordinated tissues or organs. Jellyfish and sea anemones have simple tissues but not complex organs. Crabs and alligators have complex organs that govern movement, behavior, digestion, and gas exchange.

Organisms also differ in their symmetry. Sponges are often very irregular in form. Jellyfish and sea anemones display **radial symmetry,** meaning that their bodies have an axis that runs from mouth to base with many planes of symmetry through this axis (**Fig. 44.2a**). This organization allows jellyfish to move up and down in the water column by flexing muscles around their bell-like bodies, and permits sea anemones and corals to wave their ring of food-gathering tentacles in all directions at once. Most simple animals with radial symmetry are grouped together on one branch of the animal tree, in the phylum Cnidaria.

Other organisms show **bilateral symmetry:** Their bodies have a distinct head and tail, marking front and back, with a

FIG. 44.2 Symmetry in animal form. (a) Radial or (b) bilateral symmetry defines distinct groups of animals.

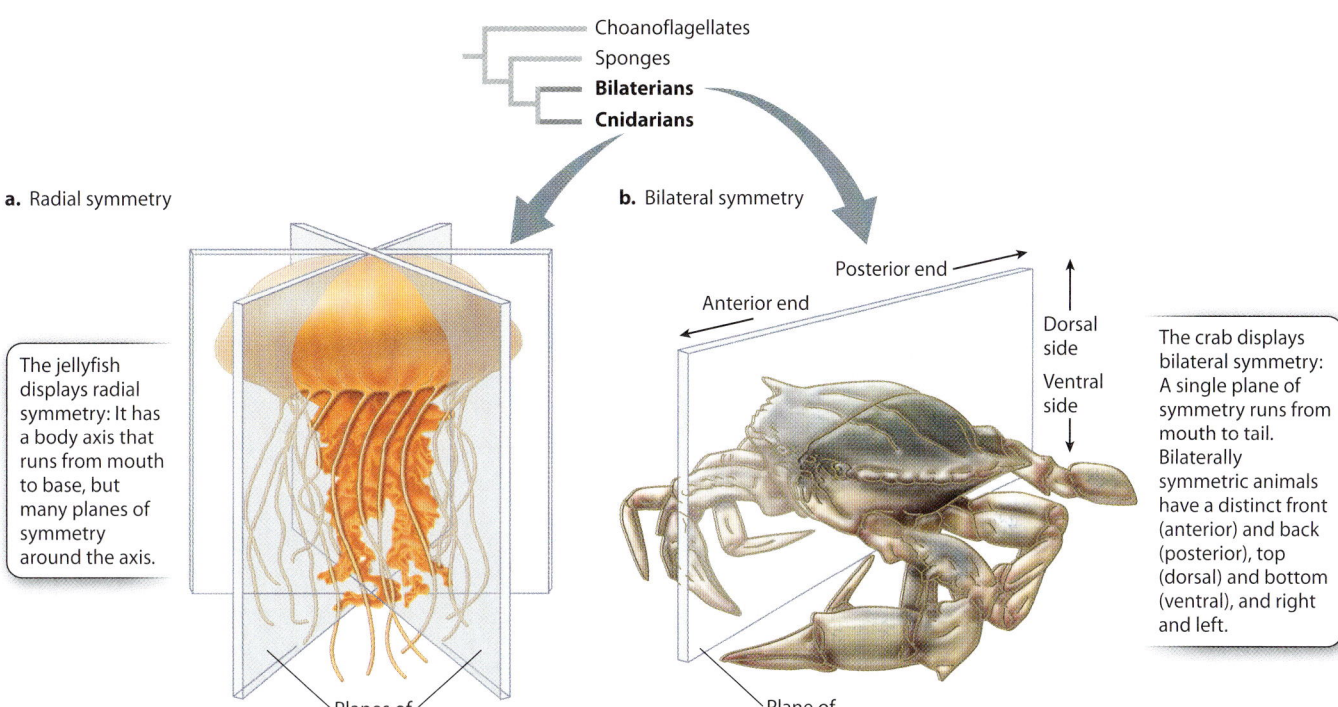

single plane of symmetry running between them at the midline (**Fig. 44.2b**). Bilateral symmetry enables animals to move in one horizontal direction to capture prey, find shelter, or escape from enemies. It also allows the development of specialized sensory organs at the front end for guidance, and specialized appendages along both sides for locomotion, sensing the environment, or defense. Animals with bilateral symmetry cluster together on one branch of the animal tree and are called **Bilateria**.

Biologists understood early that cnidarians are simpler than animals with bilateral symmetry. Bilaterian animals develop, to varying degrees, digestive and circulatory systems, nervous systems, and muscles. From these observations, early biologists concluded that sponges branched near the base of the animal tree, cnidarians in the middle, and bilaterians closer to the top.

Within the bilaterian group, however, understanding evolutionary relationships proved more difficult. Some biologists pointed to the presence or absence of a cavity, or coelom, in tissues surrounding the gut, dividing bilaterians into three groups: **acoelomates** (without a body cavity), **coelomates** (with a body cavity), and **pseudocoelomates** (with a body cavity that does not completely surround the internal organs). These groups are shown in **Fig. 44.3**. A fluid-filled body cavity cushions the internal organs against hard blows to the body and enables the body to turn without twisting these organs. A body cavity also allows internal organs like the stomach to expand, enhancing digestive function.

Early biologists also noted that some animals, notably insects and earthworms, show a striking pattern of body segmentation, but others do not. These variations in anatomy were critically important in defining phyla within bilaterian animals, but in the end they did not provide a way to understand the details of evolutionary relationships among these phyla.

New insights became possible with the advent of microscopes that enabled the direct study of development from its earliest

FIG. 44.3 Major types of anatomical organization of bilaterian animals. Bilaterians vary in the presence, absence, or structure of a body cavity.

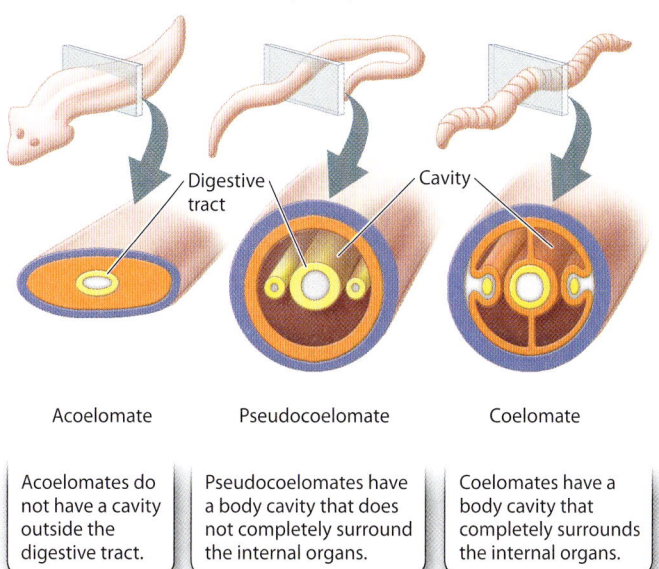

Acoelomates do not have a cavity outside the digestive tract.

Pseudocoelomates have a body cavity that does not completely surround the internal organs.

Coelomates have a body cavity that completely surrounds the internal organs.

FIG. 44.4 Variation in embryonic development. Early development of germ tissues in the embryo separates the diploblastic cnidarians from the triploblastic bilaterians.

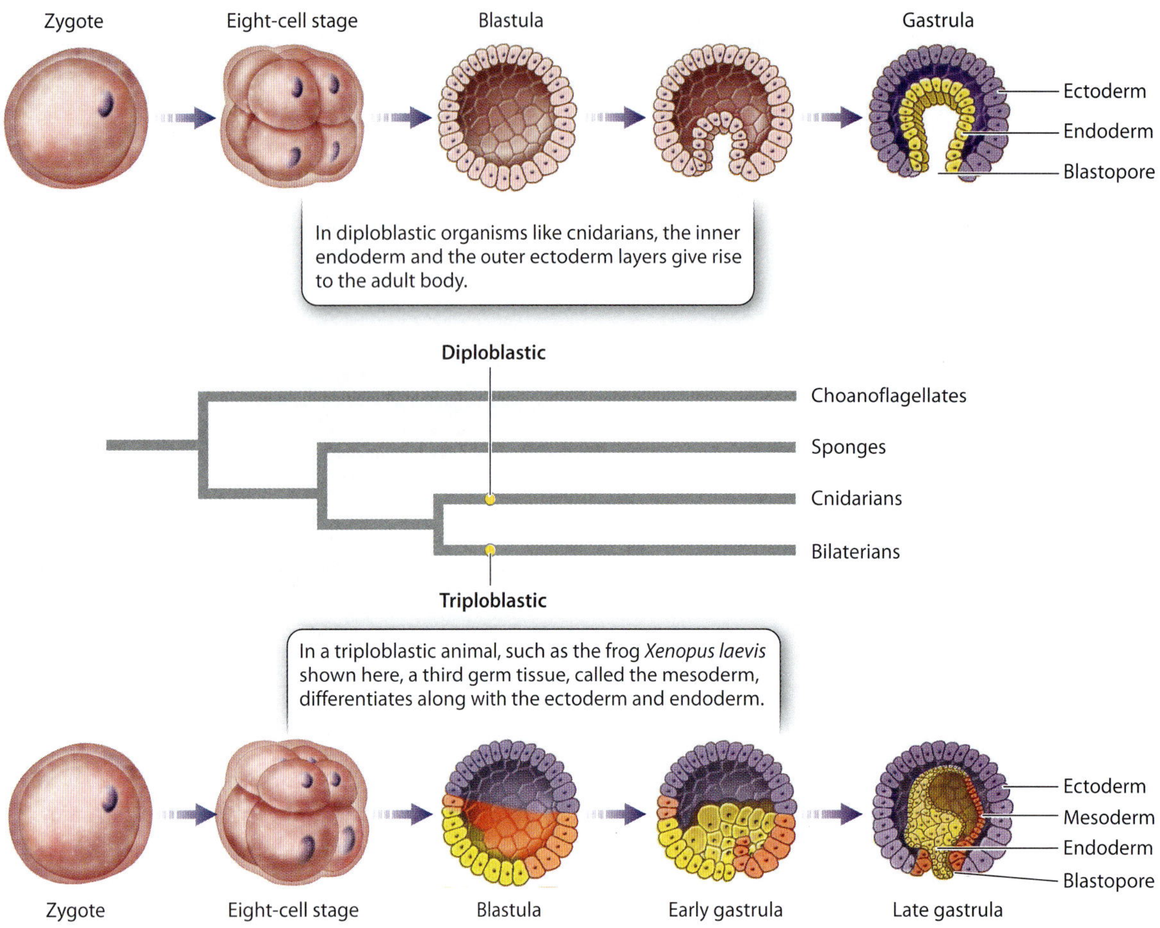

In diploblastic organisms like cnidarians, the inner endoderm and the outer ectoderm layers give rise to the adult body.

In a triploblastic animal, such as the frog *Xenopus laevis* shown here, a third germ tissue, called the mesoderm, differentiates along with the ectoderm and endoderm.

stages. One important observation was that some animals that look very different as adults share patterns of early embryological development. For example, adult sea stars and catfish have few morphological features in common, but their embryos show a number of key similarities, including the number of embryonic tissue layers they develop and the way that the early cell divisions occur. In the Cnidaria, the embryo has two germ layers, termed endoderm and ectoderm, from which the adult tissues develop. These animals are therefore called diploblastic (**Fig. 44.4**). The bilaterian animals are triploblastic, with a third germ layer, the mesoderm, lying between the endoderm and ectoderm, and developing into muscles and connective tissues (Fig. 44.4).

Comparative embryology also enabled biologists to divide bilaterian animals into the two groups shown in **Fig. 44.5,** called **protostomes** (from the Greek for "first mouth") and **deuterostomes** (from the Greek for "second mouth"). In protostomes, the first opening to the internal cavity of the developing embryo, called the blastopore, becomes the mouth. In deuterostomes, the blastopore becomes the anus. However, only in the age of molecular sequence comparisons have the relationships among phyla within each of these two groups become clear.

Molecular sequence comparisons have confirmed some relationships and raised new questions.

Over the past two decades, comparisons among DNA, RNA, and amino acid sequences have greatly improved our understanding of evolutionary relationships among major groups of animals. Molecular comparisons support many of the conclusions reached earlier on the basis of comparative anatomy and embryonic development, including the early divergence of sponges. Other hypotheses have been rejected. For example, the traditional phylogenetic division of bilaterians into acoelomate, coelomate, and pseudocoelomate groups gets little support from molecular studies. Neither do molecular sequence comparisons support the once widespread view that segmented bodies indicate a close relationship between earthworms and lobsters.

Some of the biggest surprises provided by molecular data concern the relationships of phyla within the protostomes, showing that the protostome phyla can be divided into two subgroups, the **lophotrochozoans** and the **ecdysozoans** (Fig. 44.5). As we will see, biologists have identified developmental and anatomical characters that unite the phyla on each branch, but the phylogenetic significance of these features was recognized

CHAPTER 44 ANIMAL DIVERSITY

FIG. 44.5 **A simple phylogenetic tree of bilaterian animals.** Embryological features and similarities in the molecular sequences of genes show that bilaterian animals can be divided into two major groups, the deuterostomes and protostomes. Protostomes can be further subdivided into lophotrochozoans and ecdysozoans.

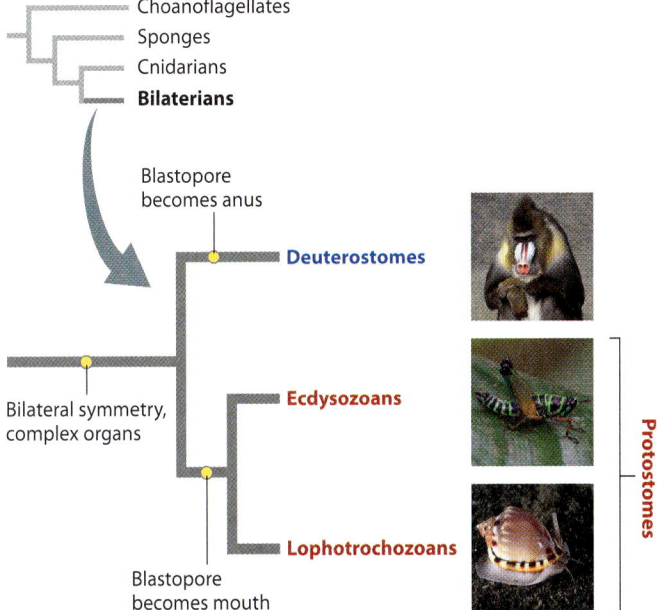

Sponges are simple and widespread in the oceans.

As we discussed in Chapters 27 and 28, the unicellular ancestors of animals carried out all functions from metabolism to reproduction as individual cells. The cells making up sponges retain much of this independence, while also having some benefits of multicellularity. Coordination among cells means that there can be distinct kinds of cell specialized for different functions—cells that form a protective skin, for example, and others that secrete enzymes for digestion. Most dramatically, multicellularity permits sponges to extend above the seafloor, giving them access to food suspended in water currents.

The sponge body plan resembles a flower vase, with many small pores along its sides and a larger opening at the top (**Fig. 44.6**). The cells on the outer surface of the sponge are tough and act as the sponge's skin. The interior surface is lined by cells called **choanocytes** that have flagella and function in nutrition and gas exchange. Choanocyte cells have a collar of small cilia around their flagellum, much like the cells of our closest unicellular relatives, the choanoflagellates (Chapter 27). Between the interior and exterior cell layers lies a gelatinous mass called the **mesohyl** (Fig. 44.6). Mesohyl is mostly noncellular, but it contains some amoeba-like cells that function in skeleton formation and the dispersal of nutrients.

Sponges gain nutrition by intracellular digestion (Chapter 40). The choanocytes that surround the interior chamber of the sponge

only after gene sequences revised the thinking about protostome animals. Molecular sequence comparisons have also overturned traditional views on relationships among the deuterostome phyla, discussed later in this chapter.

→ **Quick Check 1** Which animals are more closely related to sponges, Cnidaria or Bilateria?

44.2 THE SIMPLEST ANIMALS: SPONGES, CNIDARIANS, CTENOPHORES, AND PLACOZOANS

The human body is complex, our diverse functions made possible by sophisticated, interacting organ systems: from muscle pairs attached to skeletons that allow us to run, jump, and dance to lungs that exchange gases with the atmosphere and a brain and nervous system that enable us to sense our environment and respond to it. None of these organ systems is well developed in early-branching animal groups. How then do sponges, cnidarians, ctenophores, and placozoans function, and what are the consequences of their anatomical organization for diversity and ecology? On the other hand, these animals have survived a half billion years of Earth history. What can we learn from the diversity and continuity of their remarkable adaptations?

FIG. 44.6 Anatomy of a sponge.

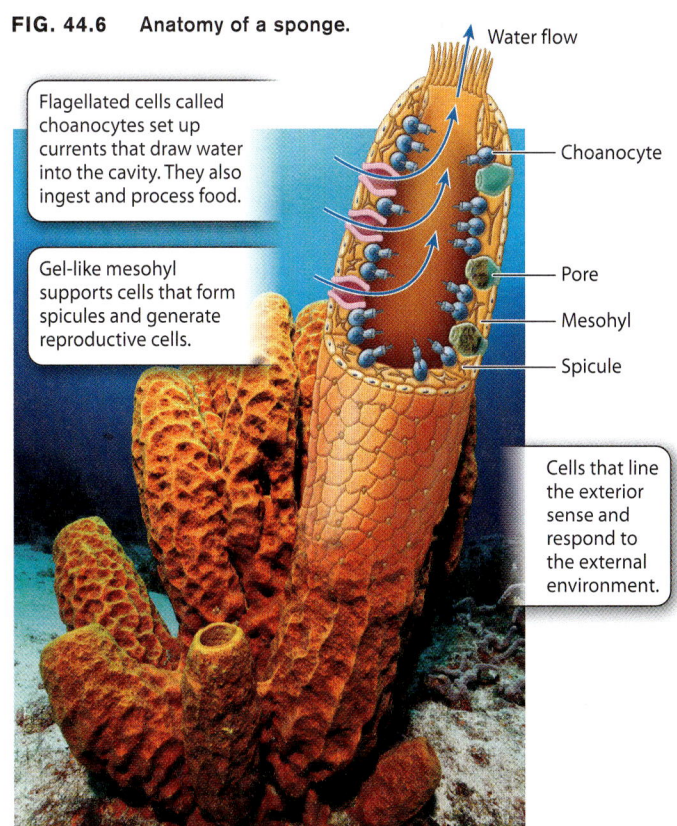

Flagellated cells called choanocytes set up currents that draw water into the cavity. They also ingest and process food.

Gel-like mesohyl supports cells that form spicules and generate reproductive cells.

Cells that line the exterior sense and respond to the external environment.

beat their flagella, creating a current that draws water from outside the body, through the pores in its walls and upward through the central cavity of the sponge, where it exits through the large opening at the top. The circulating water contains food particles and dissolved organic matter, which cells lining the cavity capture by endocytosis. The individual cells then metabolize the food in their interiors, much the way that protozoans feed.

From this observation, we might conclude that sponges are simply a group of uncoordinated cells, each working for itself, but that isn't the case. The choanocyte flagella beat in a coordinated pattern, helping to draw water into the body interior through the pore system and outward again through the vase opening. Moreover, the shape of the sponge body itself directs water movement across feeding cell surfaces. These features effectively move water across the cells, facilitating food uptake. Sponge cells require oxygen for respiration and must get rid of the carbon dioxide respiration generates. Gas exchange occurs by diffusion, aided by the movement of water through the sponge cavity.

Sponges don't have highly developed reproductive organs. Instead, cells recruited from the choanocyte layer migrate into the mesohyl, where they undergo meiosis and differentiate as sperm or eggs. Sperm released into the water fuse with eggs in the mesohyl of other sponges.

Many sponges build skeletons of simple structures called spicules. Some sponges precipitate spicules of glass-like silica (SiO_2). Others, however, make their spicules—and sometimes massive skeletons—of calcium carbonate ($CaCO_3$). Still other sponges have a skeleton made up of proteins (these are often used as bath sponges). So we see that sponges are intermediate in the sophistication of their bodies. Their cells function much as single-celled protozoa do, but they coordinate their activities and so are more efficient in extracting food and oxygen from seawater.

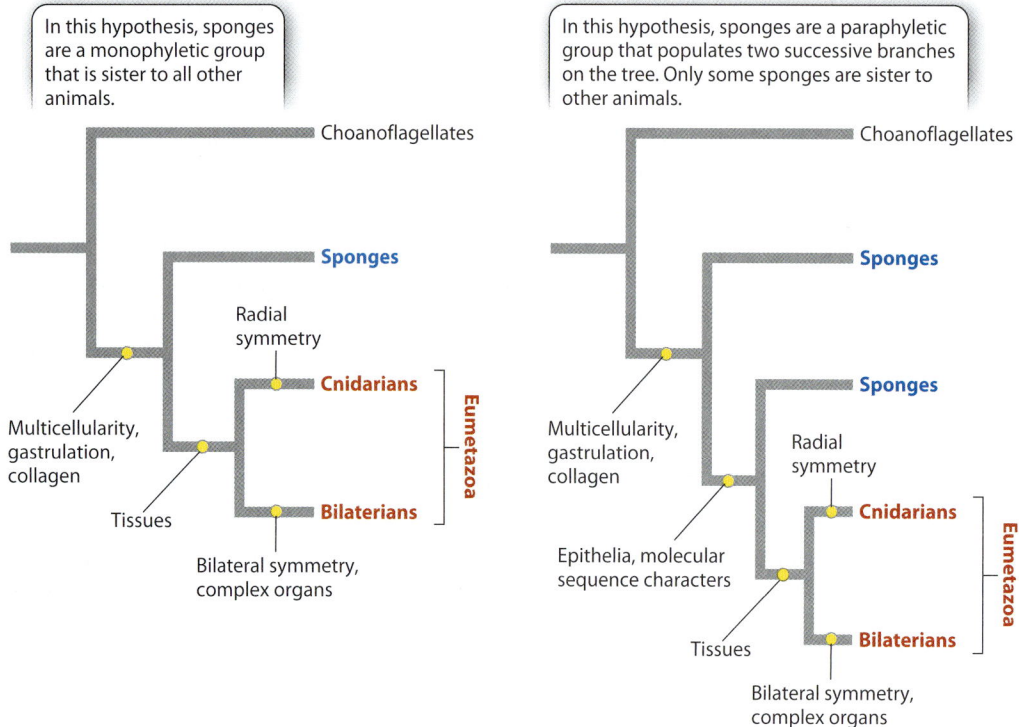

FIG. 44.7 **Two hypotheses for the phylogenetic relationships of sponges to other animals.** One group of sponges, the homoscleromorphs, has epithelium-like cell sheets. In the left-hand hypothesis, this character is interpreted as a convergence with eumetazoans; in the right-hand hypothesis, it is viewed as a shared derived character for a homoscleromorph sponge + eumetazoan group.

Despite their anatomic simplicity, sponges are major contributors to seafloor communities. Approximately 9000 species of sponge have been described, most of them ocean dwelling, but a few that live in freshwater lakes. Many sponges obtain at least part of their nutrition from symbiotic algae living within their bodies. In fact, all sponges are full of bacterial cells. Many of these are probably symbionts, but only a few experiments have demonstrated their function.

Until recently, most biologists regarded sponges as a monophyletic group, one that includes all the descendants of a common ancestor. Some molecular data, however, now suggest that sponges are paraphyletic—that is, that they contain some but not all the descendants of their common ancestor (Chapter 23).

Fig. 44.7 shows these two possibilities. If the tree on the left is correct, the two main limbs of the animal tree contain very different organisms—the simple sponges on one branch and all more complex animals on the other. As noted earlier, this figure indicates that sponges are distinct from other animals but leaves open the question of what ancestral animals were like. In contrast, the hypothesis illustrated by the tree on the right suggests that the last common ancestor of all living animals, in essence, *was* a sponge. At present, biologists are divided on this issue; more research is needed to demonstrate how sponges relate to each other and to other animals, which are grouped together as the **Eumetazoa** (Fig. 44.7).

Cnidarians are the architects of life's largest constructions: coral reefs.

Many of us have encountered at least a few cnidarians. Jellyfish flourish in marine environments from coastlines to the deep sea; sea anemones cling to the seafloor, extending their tentacles upward to gather food and deter predators; corals secrete massive skeletons of calcium carbonate, forming reefs that fringe continents and islands in tropical oceans.

Jellyfish and sea anemones look strikingly different, but they share a body plan common to all cnidarians (**Fig. 44.8**). At one end of the radially symmetrical body is a mouth surrounded by tentacles armed with stinging cells that subdue prey and defend against enemies. An anemone is like a jellyfish stuck upside down onto the seafloor with its tentacles and mouth facing upward. We call a free-floating jellyfish a medusa, and the sessile form of an anemone a polyp. Some cnidarians develop into one form or the other, but many have life cycles that alternate between the two.

The mouth of both medusa and polyp opens into a closed internal **gastric cavity,** the site of extracellular digestion and excretion. Rather than digesting food particles inside individual cells, cnidarians receive both food and digestive enzymes in the gastric cavity. (The digestive enzymes are secreted by cells that line the walls of the cavity.) This arrangement permits cnidarians to digest large food items, such as a whole fish (paralyzed by the stinging cells on the tentacles), whose dissolved nutrients can then be absorbed through the walls of the cavity. Cnidarians, then, can consume many types of food that are unavailable to sponges.

The cnidarian body develops from a diploblastic embryo (one with two germ layers). It has an outer layer, the **epidermis,** which develops from the ectoderm, and an inner lining, the **endodermis,** that derives from the endoderm. These tissues enclose a gelatinous mass called the **mesoglea** (the "jelly" of jellyfish). This organization sounds a bit like the sponge body plan, but there are important differences.

First, in cnidarians, the cells that form the epidermis and endodermis occur as closely packed layers of cells embedded in a protein-rich matrix, forming an epithelium (Chapter 10). Epithelia line compartments (like the gut) within animal bodies, but they do not occur in most sponges. These tissues are formed from an unusually tight layer of specialized cells, each connected to its neighbors with specialized junctions that regulate the passage of ions or other molecules. Epithelial layers often absorb or secrete substances into the compartments they line, making possible organs such as those of the digestive system (Chapter 40).

Second, cnidarians have a wider array of cell types than sponges do, permitting more sophisticated tissue function. Muscle cells allow jellyfish to swim through the ocean. A simple network of nerve cells permits cnidarians to sense their environment and respond to it by directional movement. No such cells occur in sponges. Cnidarians don't have a brain, but at least a few have light-sensitive cells that function as simple eyes.

FIG. 44.8 Two cnidarians: (a) jellyfish and (b) sea anemones. All cnidarians have a similar organization.

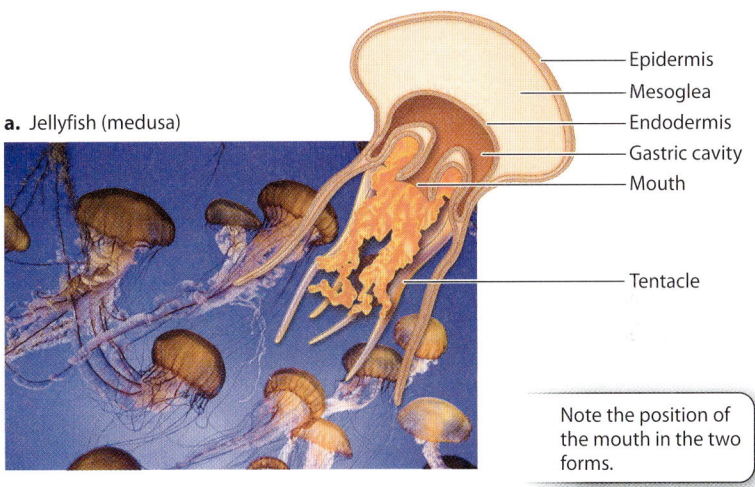

a. Jellyfish (medusa)

- Epidermis
- Mesoglea
- Endodermis
- Gastric cavity
- Mouth
- Tentacle

Note the position of the mouth in the two forms.

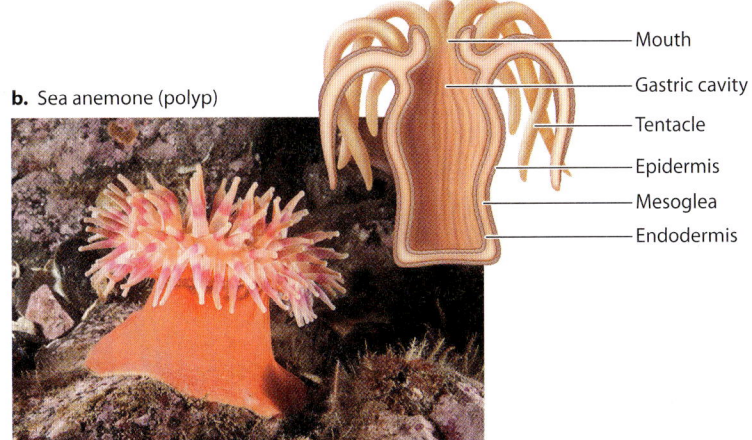

b. Sea anemone (polyp)

- Mouth
- Gastric cavity
- Tentacle
- Epidermis
- Mesoglea
- Endodermis

Moreover, whereas sponges filter water to gain food, cnidarians are predators, capturing prey with their tentacles and digesting it in the gastric cavity. Specialized cells on the tentacles contain a tiny harpoon-like organelle called a nematocyst, often tipped with a powerful neurotoxin that greatly aids prey capture and defense against other predators (**Fig. 44.9**). Other specialized cells lining the gastric cavity secrete digestive enzymes that break down ingested food into molecules that can be taken up by cells lining the cavity by endocytosis. There is no specialized passage for waste removal. Instead, waste is excreted back into the gastric cavity and leaves by way of the mouth. Oxygen uptake and carbon dioxide release occur by diffusion.

Many cnidarians reproduce asexually to form colonies. For example, corals form extensive rounded, fan-shaped, or hornlike colonies by budding, building potentially massive structures on the seafloor (**Fig. 44.10a**). In the Portuguese man-of-war, different individuals in the same colony develop distinct morphologies, some specialized for flotation, others for prey capture, and still others for reproduction (**Fig. 44.10b**). Among animals without complex organs, such colonies represent the height of morphological complexity.

About 9000 species of cnidarians live in the oceans. Reef corals are remarkable not only for the size of their skeletal colonies, but also for their mode of nutrition. Most have lost the ability to capture prey and instead gain nutrition from symbiotic algae in their surface tissues.

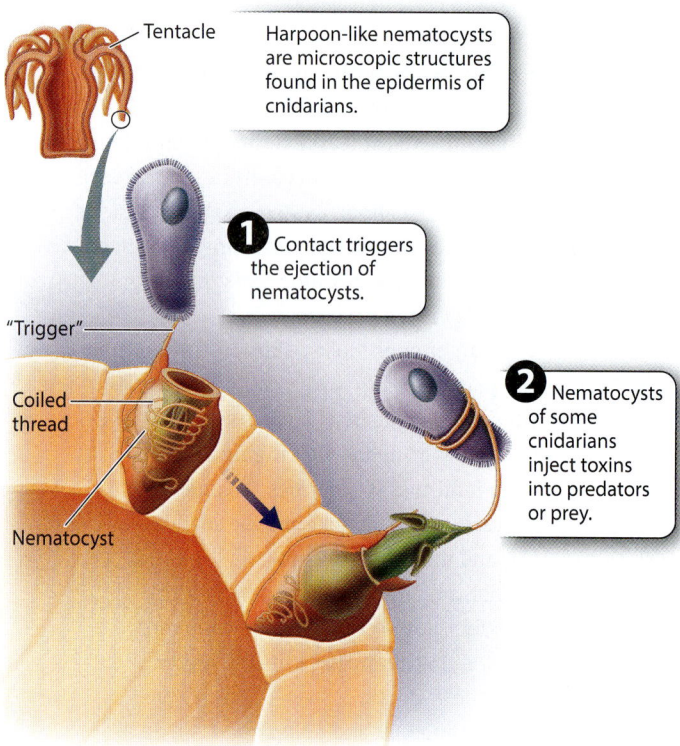

FIG. 44.9 Nematocysts. Nematocysts are harpoon-like organelles that capture prey.

FIG. 44.10 Colony formation in cnidarians. (a) In soft corals, the individual members of the colony are similar in form and function. (b) In the Portuguese man-of-war, the members of the colony differentiate to serve distinct functions, such as floating, feeding, and reproduction.

Ctenophores and placozoans represent the extremes of body organization among early branching animals.

Ctenophores, or comb-jellies, resemble cnidarians in body plan, and for many years most biologists thought that the two were close relatives. Like cnidarians, comb-jellies have radial symmetry, with an outer epithelium and an inner endodermis that enclose a gelatinous interior (**Fig. 44.11**). Comb-jellies also differentiate muscle cells and a simple nerve net, as well as rudimentary gonads. They are predators that feed by ingestion, digesting prey within their gastric cavity by enzymes secreted from the cells lining the gut cavity. Gas exchange occurs by diffusion. However, there are important differences between cnidarians and comb-jellies.

Comb-jellies propel themselves through the oceans by the coordinated beating of cilia that extend from epidermal cells. These cilia are usually arranged in comb-like groups,

FIG. 44.11 Ctenophores, or comb-jellies.

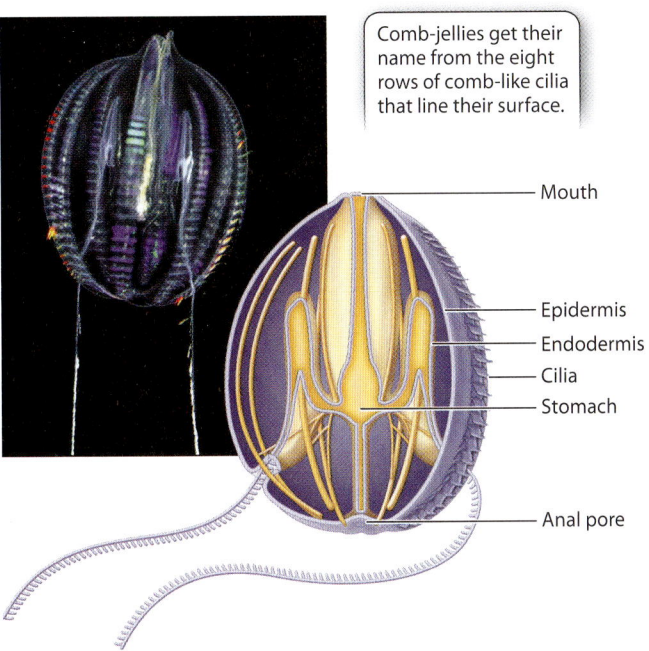

Comb-jellies get their name from the eight rows of comb-like cilia that line their surface.

Mouth
Epidermis
Endodermis
Cilia
Stomach
Anal pore

hence the common name for this phylum. Some but not all comb-jellies also have a pair of long tentacles that aid in feeding. Like cnidarians, comb-jellies have specialized cells on their tentacles, but in comb-jellies, the cells secrete adhesive molecules that entangle prey.

Unlike cnidarians, digestive wastes generated by comb-jellies move through a gut cavity for elimination through an anal pore opposite the mouth. This simple flow-through gut provides comb-jellies with an anterior–posterior axis that cnidarians do not have. This arrangement allows each section of the gut to be specialized for a particular function, such as temporary storage (near the mouth), digestion (in the middle), and absorption and excretion (at the rear), increasing the overall efficiency in processing food. Cnidarians digest and absorb and store food in the same pouch, and so are much less efficient.

Also, the pattern of muscle cell development in comb-jellies suggests that these cells originate from a rudimentary mesoderm situated between the ectoderm and endoderm. Flow-through guts, mesoderm, and axial symmetry are characteristic of bilaterian animals, as is another feature of comb-jellies—the use of acetylcholine as a neurotransmitter (Chapter 35). This unique combination of traits makes placement on the tree of life difficult. About 100–150 species of comb-jelly have been named, and photographs from deep-sea submersibles make it clear that many more remain to be discovered in the oceans.

→ **Quick Check 2** Which would you expect to use available nutrients more effectively and reproduce more rapidly, a jellyfish or a sponge of the same size?

If comb-jellies are the most complex animals discussed so far, **placozoans** are the simplest. These tiny (millimeter-scale) animals each contain only a few thousand cells arranged into upper and lower epithelia that sandwich an interior fluid crisscrossed by a network of multinucleate fiber cells (**Fig. 44.12**). Placozoans can absorb dissolved organic molecules, but commonly feed by surrounding food particles and secreting digestive enzymes to break them down. Individual cells then bring food particles in by endocytosis. Placozoans have no specialized tissues and few differentiated cell types. Cilia on cell surfaces allow movement, and gas exchange occurs by diffusion. Placozoans reproduce asexually but can also form egg and sperm cells for sexual reproduction.

FIG. 44.12 The placozoan, *Trichoplax adhaerens*.

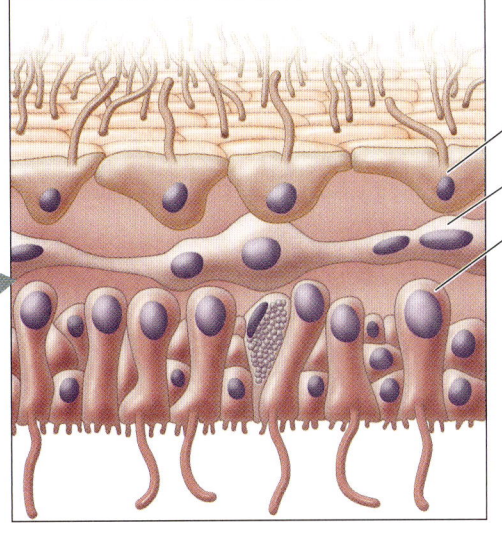

Upper epithelium
Multinucleated fiber cell
Lower epithelium

Placozoans may be the simplest of all animals, consisting of a few thousand cells organized into upper and lower epithelia, with multinucleated fiber cells in between.

Despite their morphological simplicity, placozoans have a genome that contains many of the genes for transcription factors and signaling molecules that are present in cnidarians and bilaterian animals. What roles these genes play in placozoan biology remains unclear. A single species of placozoan, *Trichoplax adhaerens*, is known. First observed adhering to the inner wall of a marine aquarium, *Trichoplax* occurs widely in tropical and subtropical oceans. A second species was described in 1896 but has not been

TABLE 44.1 Animal Phyla.

PHYLUM	COMMON NAME	LIMB OF ANIMAL TREE	APPROXIMATE NUMBER OF DESCRIBED SPECIES
Acanthocephala	Thorny-headed worms	Lophotrochozoa	1000
Acoelomorpha	Acoel flatworms	Basal Bilateria (?)	350
Annelida	Segmented worms	Lophotrochozoa	15,000
Arthropoda	Arthropods	Ecdysozoa	800,000–1,000,000
Brachiopoda	Lamp shells	Lophotrochozoa	350
Bryozoa	Moss animals, sea mats	Lophotrochozoa	5000
Chaetognatha	Arrow worms	Protostomia	100
Chordata	Chordates	Deuterostomia	50,000
Cnidaria	Corals, jelly fish, etc.	Early branching Eumetazoa	9000
Ctenophora	Comb-jellies	Early branching Eumetazoa	100–150
Cycliophora	*Symbion*	Lophotrochozoa	At least 3
Echinodermata	Echinoderms	Deuterostomia	7000
Echiura	Spoon worms	Lophotrochozoa	150
Entoprocta	Goblet worms	Lophotrochozoa	150
Gastrotricha	Hairy backs	Lophotrochozoa	700
Gnathostomulida	Jaw worms	Lophotrochozoa	100
Hemichordata	Acorn worms, pterobranchs	Deuterostomia	90
Kinorhyncha	Mud dragons	Ecdysozoa	150
Loricifera	Brush heads	Ecdysozoa	120
Micrognathozoa	Jaw animals	Lophotrochozoa	1
Mollusca	Mollusks	Lophotrochozoa	80,000
Nematoda	Roundworms	Ecdysozoa	20,000
Nematomorpha	Horsehair worms	Ecdysozoa	320
Nemertea	Ribbon worms	Lophotrochozoa	1200
Onychophora	Velvet worms	Ecdysozoa	75
Phoronida	Horseshoe worms	Lophotrochozoa	20
Placozoa	*Trichoplax adhaerens*	Early branching Eumetazoa	1
Platyhelminthes	Flatworms	Lophotrochozoa	25,000
Porifera	Sponges	Basal animals	9000
Priapulida	Priapulid worms	Ecdysozoa	17
Rhombozoa	—	Lophotrochozoa	75
Rotifera	Rotifers	Lophotrochozoa	2000
Sipuncula	Peanut worms	Lophotrochozoa	350
Tardigrada	Water bears	Ecdysozoa	More than 600
Xenoturbellida	—	Deuterostomia	2

FIG. 44.13 A phylogenetic tree of animals, with placozoans and ctenophores added. The placements of sponge subgroups, placozoans, and ctenophores remain uncertain and should be treated as hypotheses to be tested.

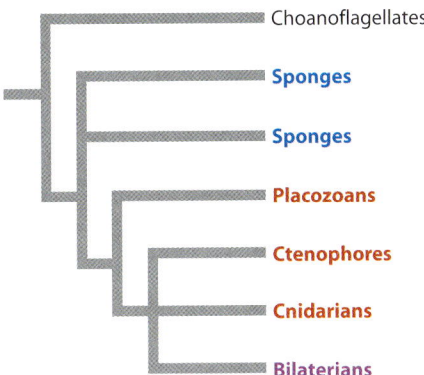

observed since that time. **Fig. 44.13** shows one current view of the phylogenetic relationships among sponges and eumetazoans. It should be regarded as a set of hypotheses to be tested against new data rather than an unambiguously "correct" tree. See **Table 44.1** for a list of all currently recognized animal phyla.

44.3 BILATERIAN ANIMALS

The phyla discussed in the preceding section include about 20,000 species, a tiny fraction of all animal diversity. The final stop in this broad tour is the bilaterian branch, on which most animal species reside. Earlier, we pointed out that basic patterns of embryo formation divide bilaterians into two major groups, the Protostomia and the Deuterostomia (see Fig. 44.5). Molecular sequence data support the protostome–deuterostome groupings of bilaterian animals and show that the protostome animals can be further divided into two groups: the Lophotrochozoa (which include mollusks and annelid worms) and the Ecdysozoa (which include insects and other arthropods).

As noted above, bilaterian animals have bilateral symmetry and complex organs that develop from a triploblastic (three-germ layer) embryo. The anatomical complexity of bilaterian animals makes possible types of locomotion, feeding, gas exchange, behavior, and reproduction that are unknown in earlier branching groups. These capabilities underpin the remarkable diversity of bilaterian animals.

Lophotrochozoans make up nearly half of all animal phyla, including the diverse and ecologically important annelids and mollusks.

The Lophotrochozoa are a group of animals first recognized on the basis of unique molecular characters and now known to include several very different types of animal. On the one hand, there are the brachiopods, Bryozoa, and related phyla that have a tentacle-lined organ for filter feeding called a lophophore. Then, there are the mollusks, annelid worms, and their close relatives, characterized by a type of larva called a trochophore. The odd name "lophotrochozoan" is a contraction of the terms "lophophore" and "trochophore," with "zoan" added to indicate that these organisms are animals (*zoon* is Greek for "animal"). The majority of lophotrochozoan species also have a distinctive form of spiral cleavage early in development.

The Lophotrochozoa contains 17 phyla, mostly small marine animals of limited diversity, but it also includes the diverse and ecologically important **annelid worms** and **mollusks** (**Fig. 44.14**). We'll focus on these two phyla, which are closely

FIG. 44.14 A phylogenetic tree for lophotrochozoan animals, showing all 17 phyla in this group. Phyla discussed in the text are shown in red. Uncertainties are shown by nodes with more than two branches.

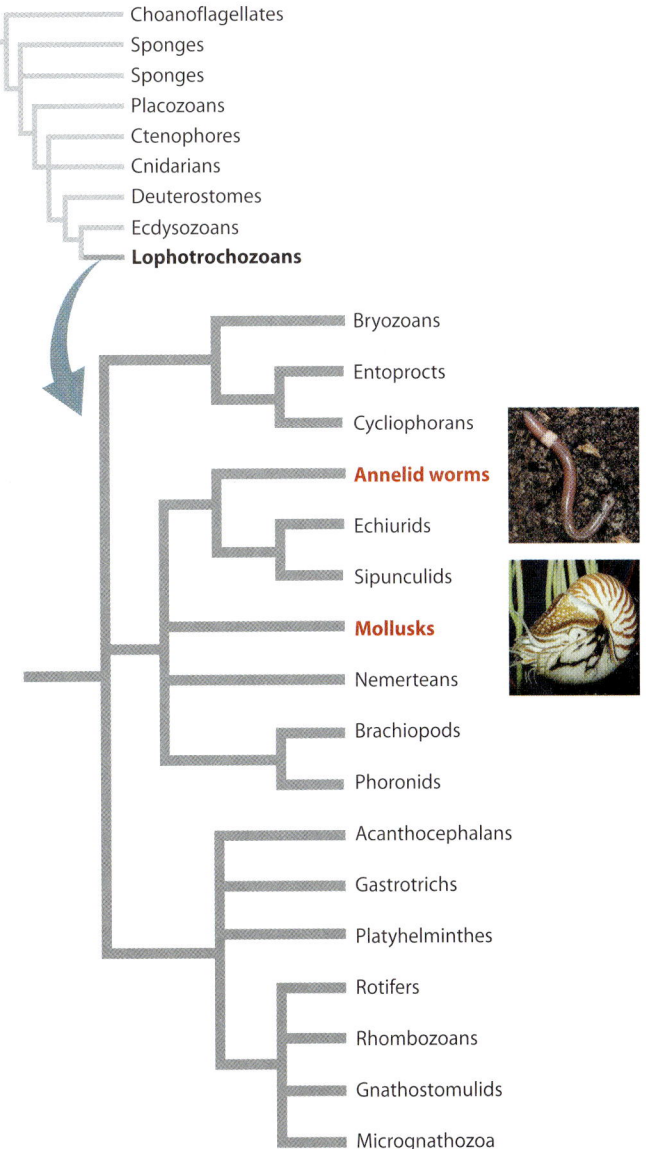

FIG. 44.15 Annelids. (a and b) Earthworms typify the segmented body plan found throughout the Annelida. (c) The most diverse annelids are polychaete worms, which live in the oceans. (d) Leeches are specialized for ingesting blood meals from mammal hosts. (e) Vestimentiferan worms live along hydrothermal rift vents in the oceans, gaining nutrition from chemosynthetic bacteria in specialized organs (red in the image).

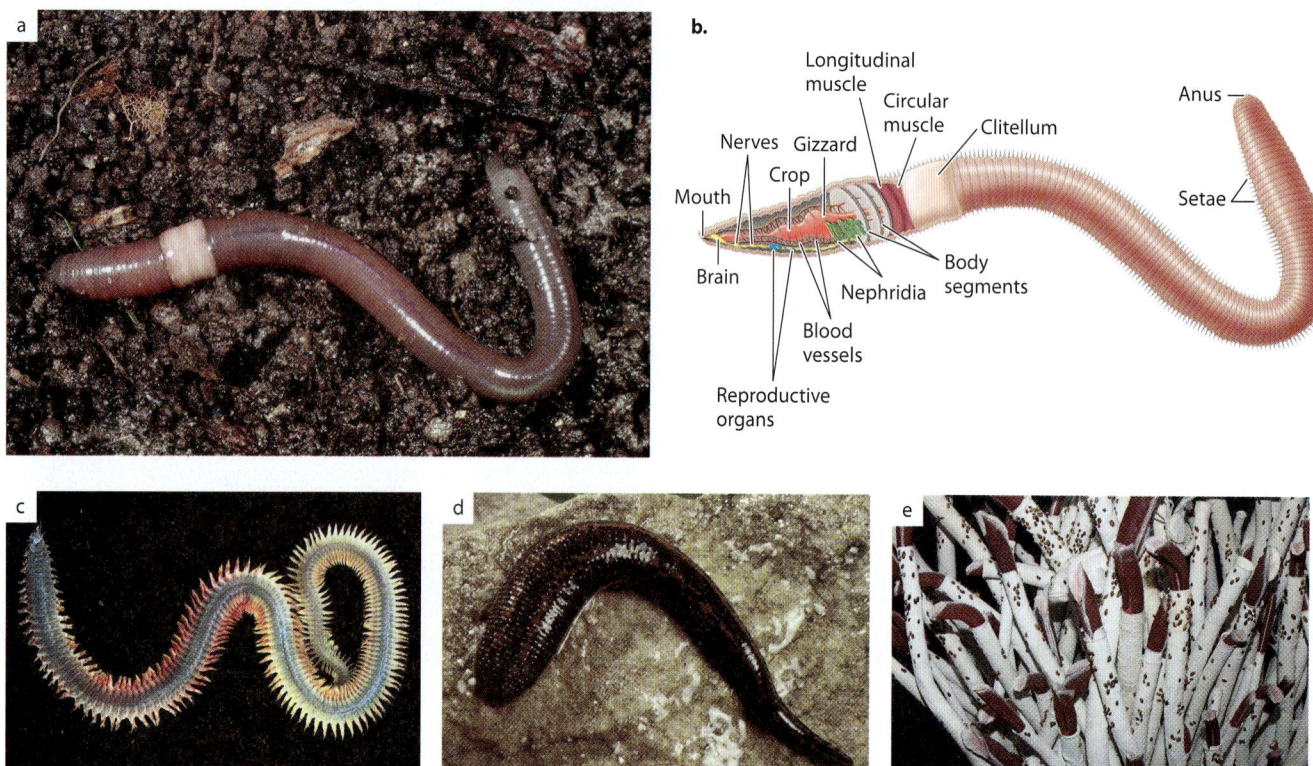

related but have nonetheless evolved along quite different anatomical trajectories in the half billion years since their origins.

The annelids most familiar to us are the earthworms, commonly found in soil (**Figs. 44.15a** and **44.15b**). Most of the 15,000 known annelid species, however, live in the oceans (**Fig. 44.15c**). All annelids have a cylindrical body with distinct segments, and they illustrate the advantages of a bilaterian body plan (see Fig. 44.2b). At one end, the head has a well-developed mouth and, internally, a cerebral ganglion (a collection of nerve cells) that connects to an extensive nervous system. A digestive system extends through the body from the head to an anus, with a sequence of specialized organs for crushing, then digesting, and finally excreting ingested food, much like the digestive system in our own bodies. Working together, these features enable annelids to move through their environment, actively searching for food and digesting it efficiently after ingestion.

Aquatic annelids have gill-like organs for gas exchange, but terrestrial earthworms exchange gases through their skin. In both cases, a closed circulatory system moves dissolved gases through the body. Annelids have waste-filtering organs called nephridia, gonads (repeated in most segments), and a fluid-filled coelom, or body cavity. Fluids in the coelom form a hydrostatic skeleton that works in coordination with paired muscles in each segment to direct movement.

Many annelids are predators that capture and ingest prey, but some, like earthworms, ingest sediment, digesting the organic matter it contains and excreting mineral particles. A few marine annelids have evolved tentacles that enable them to filter food particles from water, while leeches, a specialized group of freshwater annelids, attach themselves to vertebrates to suck out a meal of blood (**Fig. 44.15d**). And, remarkably, the meter-long vestimentiferan worms that live around hydrothermal vents in the oceans have given up ingestion altogether (**Fig. 44.15e**). Without mouths, these enormous worms gain nutrition from chemosynthetic bacteria that live within a collar of specialized tissue. Discovered in the 1970s, vestimentiferan worms were originally placed in their own phylum, but molecular sequence comparisons indicate that they are annelids.

Mollusks are the second major phylum of the Lophotrochozoa. Mollusks develop a distinctive larva called a trocophore that has a tuft of cilia at its top and additional cilia bands around its middle. Marine annelids develop from trochophores as well, and the presence of this unique form of larva in both phyla suggests a close evolutionary relationship between the two. This hypothesis is supported by molecular sequence comparisons. Nonetheless,

FIG. 44.16 A gastropod—here, a snail—showing a body plan typical of the phylum Mollusca.

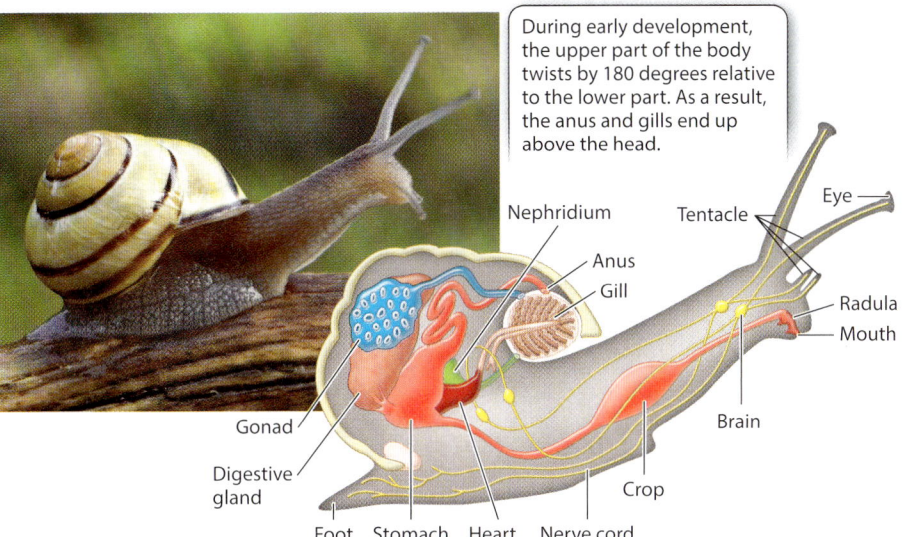

During early development, the upper part of the body twists by 180 degrees relative to the lower part. As a result, the anus and gills end up above the head.

Gastropods—snails and slugs—provide an illuminating example of the mollusk body plan (**Fig. 44.16**). Gastropods have a head with a well-developed mouth that contains a toothlike structure called a radula for feeding. The mouth connects to a gut cavity that extends to an anus. Feather-like gills facilitate gas exchange, and a muscular foot is used for locomotion. A neural ganglion in the head coordinates a nervous system that extends through a body that also contains a well-developed circulatory system, gonads, and nephridia. Gastropods are coelomates, but the body cavity is generally reduced to small pouches that surround the heart and other organs. The outer surface of the body consists of the mantle. In many gastropods the mollusks and annelids share very few features of the adult body plan. Mollusks are distinguished by a unique structure called the mantle that plays a major role in breathing and excretion, and which forms their shells when present. In fact, mollusks by themselves exhibit a remarkable variety of forms and functions. More than 80,000 species of mollusks have been described, most of them gastropods.

mantle tissues secrete external skeletons of calcium carbonate, which form shells.

Some gastropods eat algae, but many are predators. About half occur in the ocean, and the other half are freshwater and terrestrial species that primarily feed on plants. In land snails and slugs, the only terrestrial mollusks, the gills have been lost, and gas exchange occurs in an internal cavity that has been modified to function as a lung.

The second great class of mollusks is the **cephalopods**—about 700 species of squid, cuttlefish, octopus, and the chambered nautilus (**Fig. 44.17**). Cephalopods share a number of features with gastropods, including much of the internal anatomy, feather-like gills, and mantle. But they also have distinctive adaptations for their unique modes of life. Most obviously, cephalopods have muscular tentacles that capture prey and sense the environment. Surprisingly, these leglike appendages are found on the head region rather than on the sides of the body as in other animals.

FIG. 44.17 Cephalopods. *Nautilus* (a and b), octopus (c), and squid are ecologically important predators in the oceans.

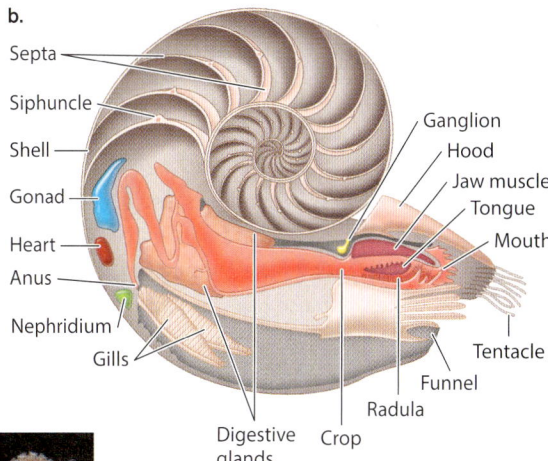

Fig. 44.18 *Mercenaria mercenaria*, a bivalve mollusk commonly harvested from coastal waters of eastern North America.

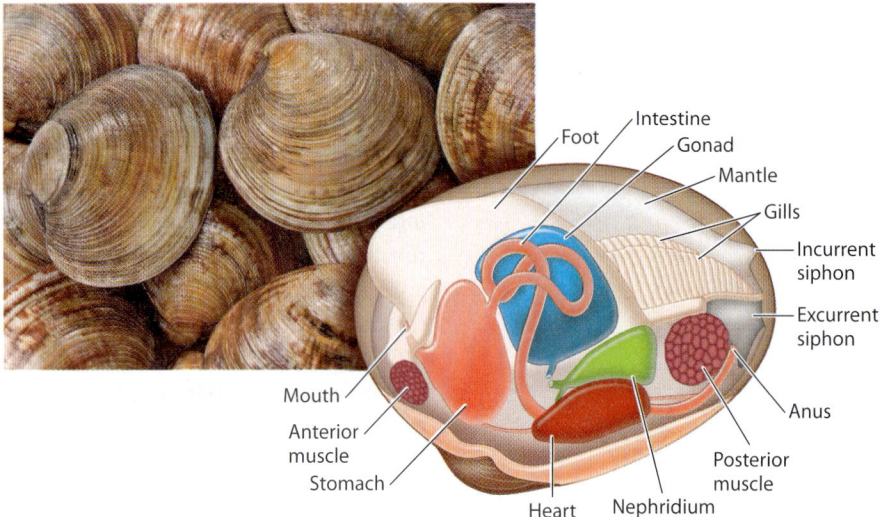

Cephalopods are superb swimmers, able to dart through the water by means of a jet propulsion system that forces water through mantle tissue that is fused to form a siphon. It is this combination of tentacles and rapid locomotion that makes cephalopods important predators in the oceans. Moreover, cephalopods have exceptional eyesight and exhibit the most complex behavior of any invertebrate animals, able to learn visual patterns and solve puzzles to gain food. The chambered nautilus secretes a coiled shell of calcium carbonate (Figs. 44.17a and 44.17b), but in other cephalopods, like squid and octopus, the shell is reduced or absent (Fig. 44.17c). In oceans of the Mesozoic Era (252–65 million years ago), however, shelled cephalopods called ammonites were among the most abundant and diverse predators.

A third major group of mollusks, the **bivalves,** includes the familiar clams, oysters, and mussels. Once again, these animals have anatomical features that point to their close relationship to snails and squids, but also have distinctive features that set them apart (**Fig. 44.18**). Notably, bivalves have (literally) lost their heads. As their name implies, bivalves have evolved a skeleton in which two hard shells called valves are connected by a flexible hinge.

Most bivalves obtain food by filtering particles from seawater. However, many bivalves live within marine sediments, burrowing into sand or mud with their muscular foot. How can bivalves filter seawater within the sediment? Once again, this group of mollusks has a modified mantle that allows a new function. Flaps of mantle tissue have fused to form a pair of siphons that extend upward from the bivalve's body to the surface of the seafloor above it. One siphon draws water containing food and oxygen into the body. The second siphon then returns water and waste materials to the environment. About 2000 species of bivalves have been described, mostly from the oceans, but also many from freshwater.

→ **Quick Check 3** From the shared traits of present-day snails, squids, and clams, what features do you think the common ancestor of all three had?

Ecdysozoans include arthropods, the most diverse animals.

The second major group of protostome animals is the Ecdysozoa (**Fig. 44.19**), which get their name from the process of ecdysis, or molting. All ecdysozoans secrete a cuticle made of protein that covers their bodies. This tough cuticle is like a suit of flexible, lightweight armor, protecting bodies from injury and physical challenges of the environment such as drying. This "armor" has a special property: The hard cuticle can be used to form appendages that function as tools, weapons, or even wings. Like a suit of armor, the cuticle is always a perfect fit for the wearer but does not stretch, and so it must be exchanged episodically during growth for a larger size to fit a larger, growing body. Ecdysozoans, then, are animals that molt their external cuticle during growth, producing a very soft, larger replacement underneath that soon hardens into a new protective covering.

The Ecdysozoa include eight phyla, but one dwarfs the rest in anatomical complexity, diversity, and ecological importance. As noted earlier, the **Arthropoda,** including insects, contains more than half of all known animal species. What features have led to the remarkable evolutionary success of this branch?

Perhaps the most obvious feature of arthropods is the one for which they are named: jointed legs (*arthro* means "jointed," and *pod* means "foot" or "leg"). The jointed leg is the most versatile part of the arthropod body; it has been modified through evolution into structures that function as paddles, spears, stilts, pincers, needles, hammers, and more. The jointed leg allowed arthropods to colonize land by providing a multipoint, independent suspension—much like that of an all-terrain vehicle—able to support a relatively large body and move over the most uneven ground. So well adapted for locomotion is the arthropod leg that NASA engineers are using it as a model for planetary exploration vehicles. Researchers have conducted

FIG. 44.19 A phylogenetic tree for ecdysozoan animals, showing all eight phyla in this group. Phyla discussed in the text are shown in red. Uncertainties are shown by nodes with more than two branches.

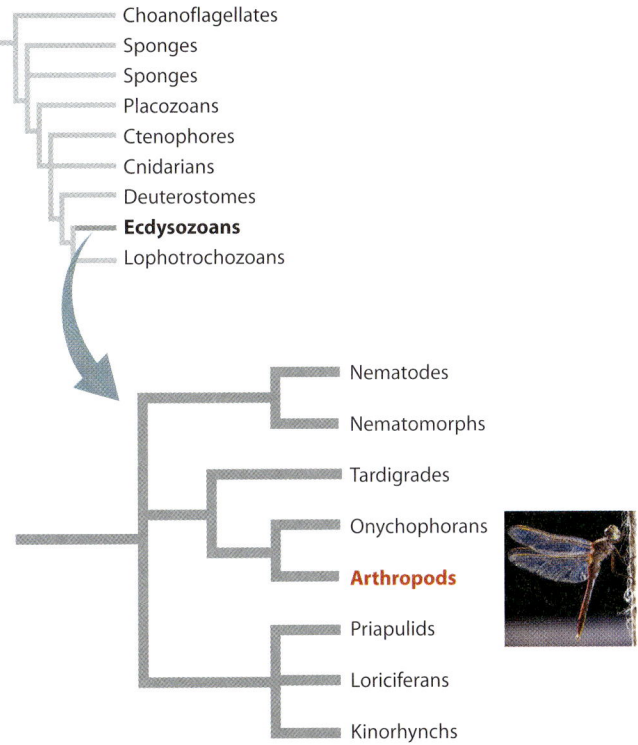

many studies to understand how this one type of limb could give rise to so many diverse structures with different functions (**Fig. 44.20**).

The arthropods' other defining characteristic is the material that forms their hard external skeleton: a strong, lightweight and nearly indestructible polysaccharide called **chitin.** Chitin, you may recall from Chapter 34, also forms the wall of fungal cells. Next to the cellulose that forms the wall of plant cells, chitin may be the most abundant biomolecule on Earth.

There are four main groups of arthropods (**Fig. 44.21**), of which the most diverse by far is the **insects.** The other three are **chelicerates,** including spiders, scorpions, and their relatives; **myriapods**, including centipedes and millipedes; and **crustaceans**, including lobsters, shrimp, crabs, and their relatives.

Chelicerates, named for the pincer-like claws called chelicerae, are the only arthropods that lack antennae. The chelicerates are mostly carnivores, except for plant-eating mites. Scorpions and spiders are known for their venoms, used to subdue prey. In spiders, the venom is injected through chelicerae modified to form fanglike appendages, while in scorpions the venom is injected through the curved tail segment of their abdomen. Other familiar chelicerates include the harvestmen (often called daddy long-legs in the United States), ticks, and horseshoe crabs. Giant sea scorpions called eurypterids, up to 2 m long, were dominant predators in warm shallow seas 450–250 million years ago. The chelicerates were among the very first animals to invade land, about 400 million years ago, aided by their protective cuticle and supported by their arrays of jointed legs.

Myriapods are named for their many pairs of legs, which make them easy to recognize. Centipedes ("hundred legs") may have up to 300 pairs of legs and are fast-moving predators with front legs modified into venomous, fanglike organs. Millipedes ("thousand legs") may have several hundred pairs of legs (though never a thousand) and are slow-moving herbivores and detritivores (eaters of dead organisms). Millipedes often have glands lining their long bodies that produce a substance that contains cyanide, which they secrete to defend against predators.

Crustaceans also have distinctive appendages, notably their branched legs. They navigate their world with the aid of two pairs of highly sensory antennae, the larger of which has two branches. The unique crustacean larva, called a nauplius, swims with the aid of the smaller antenna and has a single eye that typically disappears in the adults. In the sea, these versatile arthropods fill many of the ecological roles that insects play on land, eating plants, other animals, or detritus. The tiny shrimplike plankton known as krill occur in vast numbers in polar seas, supporting ecosystems of fish, birds, and mammals such as whales. One Antarctic krill species is estimated to total over 500 million tons of biomass, more than the global total for humans. Lobsters and crabs are largely detritivores, scouring the ocean bottoms for dead and dying animals. Some crustaceans are even parasites, attaching themselves to the skin or inside the mouths of fish or sea mammals in order to feed.

Insects were the first animals to evolve wings, nearly 350 million years ago, and modern-day mayflies and dragonflies still bear structural similarities to these first fliers. The earliest branching members of this group, the silverfish and rockhoppers, are referred to as primitively wingless because they evolved before the origin of wings. Some other insects, like fleas, have lost wings that were present in their common ancestor, presumably because their parasitic lifestyles are better served by jumping than flying. Like all arthropods, insects repeatedly shed their exoskeleton as they grow into adulthood. No insect can molt wings, however, and so only adults can fly.

Within the insects, the main evolutionary dividing line is between those that undergo **metamorphosis,** a major change in form from one developmental stage to another, and those such as grasshoppers and water bugs that do not. In insects that do not

HOW DO WE KNOW?

FIG. 44.20

How did the diverse feeding appendages of arthropods arise?

BACKGROUND Arthropod species diversity is matched by their diversity of form and function, challenging scientists to recognize and relate the equivalent parts of the very different front ends of, for example, spiders, crabs, and centipedes.

HYPOTHESIS Researchers hypothesized that the diverse front appendages of the different classes of arthropods develop from similar limb buds clearly visible in developing embryos, much as the limb buds leading to bird wings, whale flippers, and human arms are present in vertebrate embryos.

EXPERIMENT The early stages of arthropod embryos have a definite head region and several body segments, each with a pair of appendages. *Hox* genes specify the identity of each body segment (Chapter 20). The expression patterns of different *Hox* genes can be visualized by means of labeled probes on whole embryos. In this way, the development of appendages on the same segment in different types of arthropod can be compared.

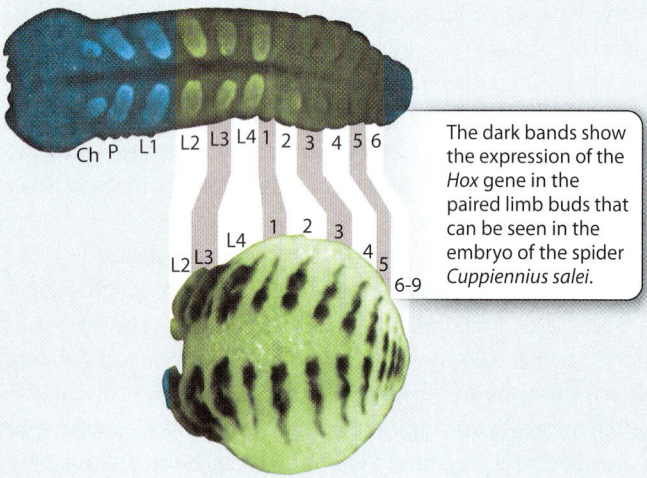

The dark bands show the expression of the *Hox* gene in the paired limb buds that can be seen in the embryo of the spider *Cuppiennius salei*.

Feeding structures in the head regions of arthropods develop from paired appendages in the embryo.

Blue segments are in the head region, and the green regions bear walking legs.

Crustaceans*	Insects	Myriapods	Chelicerates
pre-Ant	pre-Ant	pre-Ant	pre-Ant
Antenna	Antenna	Antenna	x
Antenna	Mandibles	Mandibles	Chelicerae
Mandibles	Maxillae	Maxillae	Secondary maxillae
Maxillae	Maxillae	Maxillae	Walking legs
Maxillae	Bottom of mouth	Walking legs	Walking legs
Walking legs	Walking legs	Walking legs	Walking legs
Walking legs	Walking legs	Walking legs	Walking legs
Walking legs	Walking legs	Walking legs	A1
Walking legs	A1	Walking legs	A2
Walking legs	A2	Walking legs	A3

*Crustaceans are paraphyletic.

RESULTS Study of the embryos revealed that, for example, the jawlike mandibles of insects and crustaceans form from appendages on the same segments of the developing body, whereas this segment on spiders and scorpions bears a pair of leglike feelers. Behind the mandibles are two pairs of feelers that are represented by legs in spiders and scorpions. On the other hand, the fangs of spiders and their counterparts in scorpions arise from a segment closer to the front end than the mandibles.

CONCLUSIONS The diverse feeding structures of arthropods arise from limb buds in the first segments of the developing embryo. The mandibles of crustaceans and insects arise from the same body segment and indicate the common ancestry of these animals, while spiders, scorpions, and their relatives are more distantly related. Evolution has modified simple legs into the remarkable diversity of forms, enabling structures as specialized as a spider's fangs and the grinding mandibles of crabs.

FOLLOW-UP WORK There has been recent controversy over the front appendages of sea spiders and whether or not they are equivalent to the great appendage on some Cambrian arthropod fossils.

SOURCE Damen, W. G. M., M. Hausdorf, E.-A. Seyfarth, and D. Tautz. 1998. "A Conserved Mode of Head Segmentation in Arthropod Revealed by the Expression Pattern of HOX Genes in a Spider." *Proceedings of the National Academy of Sciences USA* 95:10665–10670.

FIG. 44.21 Arthropods. This diverse phylum includes four main groups: chelicerates (spiders, scorpions, and ticks), myriapods (centipedes and millipedes), crustaceans (including lobsters, shrimp, and crabs), and insects. Molecular data indicate that insects are derived from crustacean ancestors.

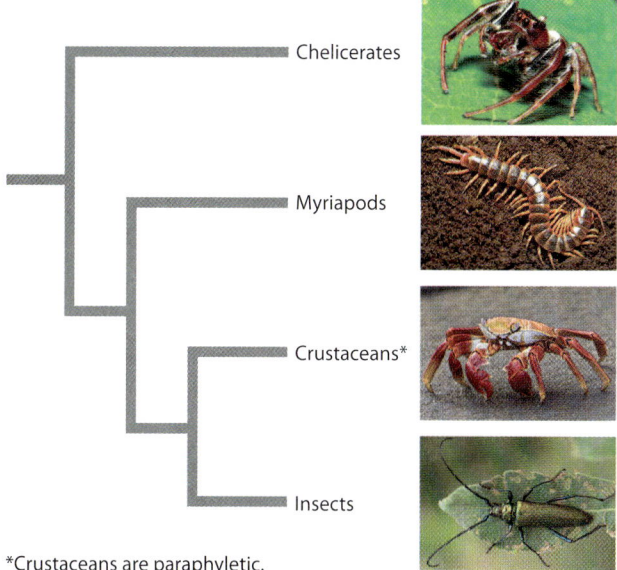

*Crustaceans are paraphyletic.

undergo metamorphosis, the main change in body form involves the appearance of wings, though some, like dragonflies, may also change the form of their legs or eyes as they leave the aquatic environment.

However, for those insects undergoing metamorphosis, the body changes from a wormlike larva specialized for feeding to a stage called a **pupa** (**Fig. 44.22**). During the pupa stage, the body tissues undergo a remarkable transformation from the relatively simple larva to a very different looking adult such as a fly, butterfly, wasp, or beetle, usually specialized for reproduction. Insects undergo a nearly complete dissolution of their muscles and other body parts, and so require the immobile, coffin-like pupa in which to transform their bodies from something as simple as a maggot, which lacks even an obvious head or tail, into an animal as sophisticated as a fly, able to walk on water, hover motionless in front of a flower, or fly at 60 mph. This evolutionary step has little parallel elsewhere in life and is one of the keys to insect diversification.

Unlike other terrestrial arthropods, all insects, whether or not they undergo metamorphosis, have highly specialized eggshells that can withstand desiccation while still allowing gas exchange. This shell allows insect eggs to survive in dry settings such as the surfaces of leaves, and so they can occupy habitats that are difficult for other terrestrial arthropods to access.

Insects have evolved one other critical adaptation for life on land. Aquatic arthropods obtain oxygen through gills, which don't work well in air because their large surface area dries out. Some terrestrial chelicerates have evolved simple lungs, called book lungs: modified appendages filled with hemolymph (blood) that have a large surface area for gas exchange, but are kept moist inside a protective pocket. Some crabs are able to make extended forays on land by keeping their gills moist underneath their carapace. In contrast, insects exchange gases through small pores in their exoskeletons called **spiracles** (see Fig. 39.4b). The spiracles connect to an internal system of tubes, the **tracheae,** that direct oxygen to and remove carbon dioxide from respiring tissues, much like air exchange ducts in a building.

Insects are extraordinarily diverse, with approximately 1 million described species making up 80% or more of all known animal species. Their success may be a function of the three adaptations that allow them to live in diverse habitats: desiccation-resistant eggs, wings, and metamorphosis. The ability of insect eggs to resist drying out means that they can be placed on virtually any surface. Their wings allow them to reach habitats and feed in ways other animals cannot. Metamorphosis allows them to have two bodies in their life cycles: a larval form specialized for feeding and an adult form specialized for mating and reproduction. These three adaptations enabled insects to use flowering plants directly as food. In turn, many insects pollinate plants, and so pollinating insects are in part responsible for flowering plant diversity (Chapter 33).

FIG. 44.22 Development in butterflies, a group of insects that undergoes metamorphosis.

Complete metamorphosis involves a pupal stage (center) in which the tissues of the larva (caterpillar on left) are reorganized into a very different adult form (butterfly on right).

Deuterostomes include humans and other chordates, but also acorn worms and sea stars.

In a classification recognized long ago by shared features of larval development, and strongly supported by molecular sequence data, the Deuterostomia (**Fig. 44.23**) include three major phyla: **Chordata** (vertebrates and closely related invertebrate animals such as sea squirts), **Hemichordata** (acorn worms), and **Echinodermata** (sea urchins and sea stars).

FIG. 44.23 A phylogenetic tree for deuterostome animals, showing all phyla (and the major subdivisions of the chordates) in this group. Phyla discussed in the text are shown in red.

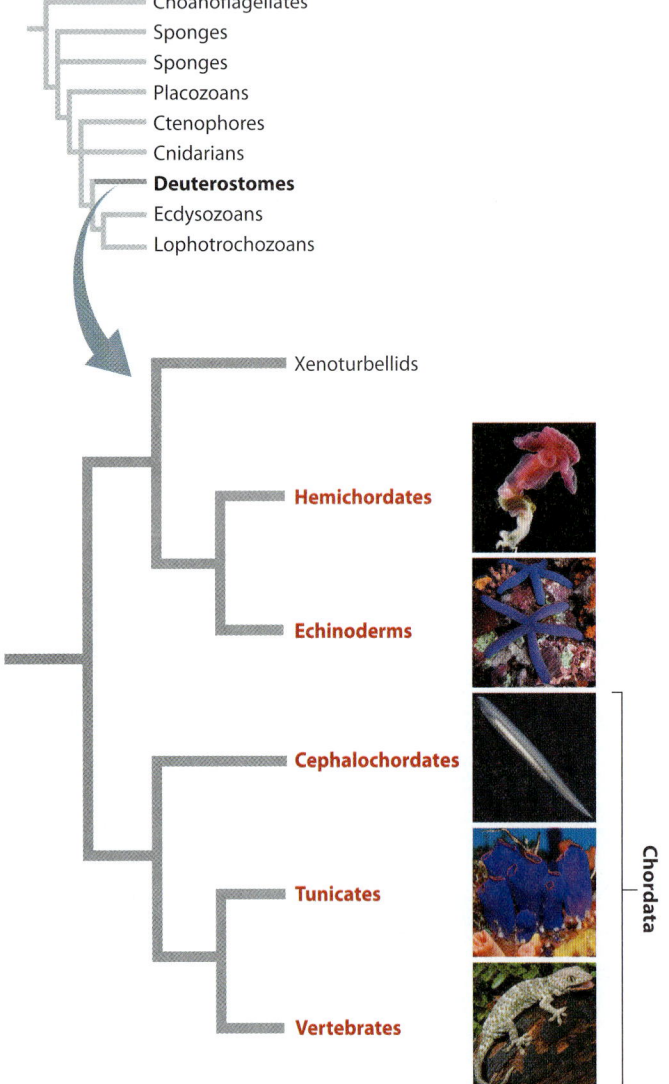

FIG. 44.24 Hemichordates, including (a) acorn worms, with gill slits, and (b) pterobranchs, with their filter-feeding tentacles.

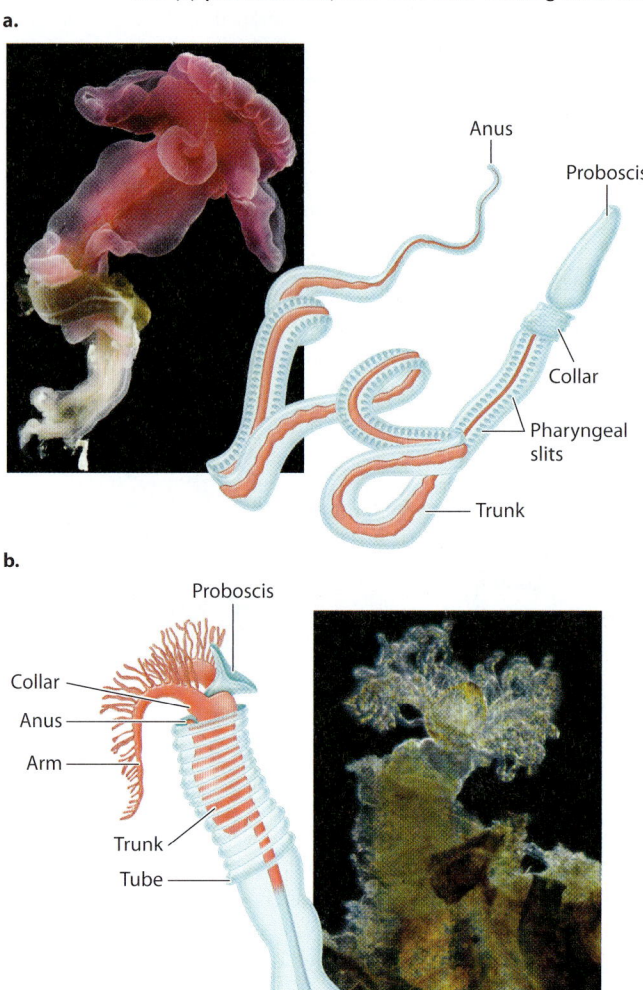

Let's begin with the deuterostomes most distantly related to the vertebrates, the hemichordates and echinoderms. Hemichordates include acorn worms, about 75 species of wormlike animals that move through seafloor sediments in search of food particles. Pterobranchs, also hemichordates, consist of about a dozen species of animals that attach to the seafloor and use tentacles to filter food from seawater (**Fig. 44.24**). Hemichordates all have a mouth on an elongate protuberance called a proboscis that connects to the digestive tract by a tube called the **pharynx**, which contains a number of vertical openings called **pharyngeal slits** separated by stiff rods of protein. They also have a **dorsal nerve cord**. Additional

FIG. 44.25 Echinoderms. (a) Sea stars are familiar sights on the shallow seafloor; (b) sea urchins are motile echinoderms found on most seafloors.

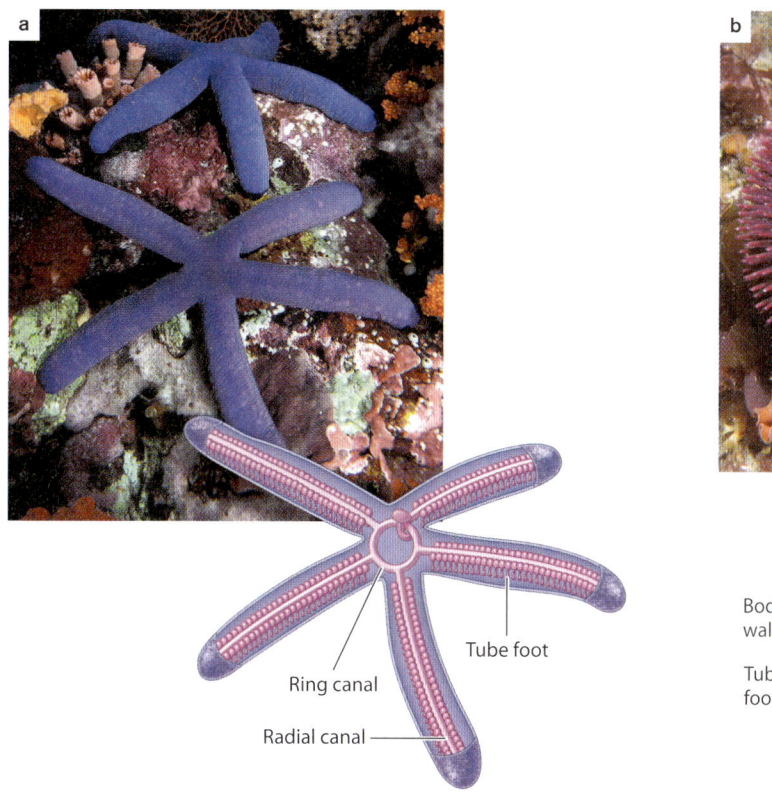

characters that relate hemichordates to other deuterostomes are apparent in embryological and larval development. Nonetheless, living hemichordates have body plans quite distinct from those of chordates and echinoderms.

Another major group of deuterostomes is the echinoderms, among the most distinctive of all animal phyla. Echinoderms include sea stars and sea urchins (including sand dollars), as well as brittle stars, sea cucumbers, and sea lilies—all of which share a unique fivefold symmetry on top of their basic bilaterian organization (**Fig. 44.25**). Echinoderms also form distinctive skeletons made of interlocking plates of porous calcite, a form of calcium carbonate. The third unique feature of echinoderms is the **water vascular system,** a series of fluid-filled canals that permits bulk transport of oxygen and nutrients. **Tube feet,** small projections of the water vascular system that extend outward from the body surface, facilitate locomotion, sensory perception, food capture, and gas exchange. About 7000 echinoderm species reside in present-day oceans, but the fossil record suggests that this phylum was more diverse in the past.

Zoologists have long known that echinoderm and hemichordate larvae share many features. Molecular sequence data now confirm that echinoderms are the closest relatives of hemichordates, with chordates a sister group to the group containing echinoderms and hemichordates (see Fig. 44.23).

Chordates include vertebrates, cephalochordates, and tunicates.

The other great branch of the deuterostome tree is the chordates, the phylum that includes vertebrate animals. Within the phylum Chordata, there are three subphyla: the **cephalochordates,** the **tunicates,** and the **vertebrates** (also called **craniates**). Like hemichordates, chordates all have a pharynx with pharyngeal slits (well illustrated by the invertebrate chordate amphioxus in **Fig. 44.26**). In fish, these pharyngeal slits form the gills, but in terrestrial animals like humans, these slits can be seen only in developing embryos. The **notochord,** a stiff rod of collagen and other proteins, runs along the back, providing support for the axis in some chordates. In vertebrates, the notochord is apparent

FIG. 44.26 Amphioxus, a cephalochordate. Amphioxus clearly has many features in common with vertebrates, but it lacks a well-differentiated head and has no cranium or vertebrae.

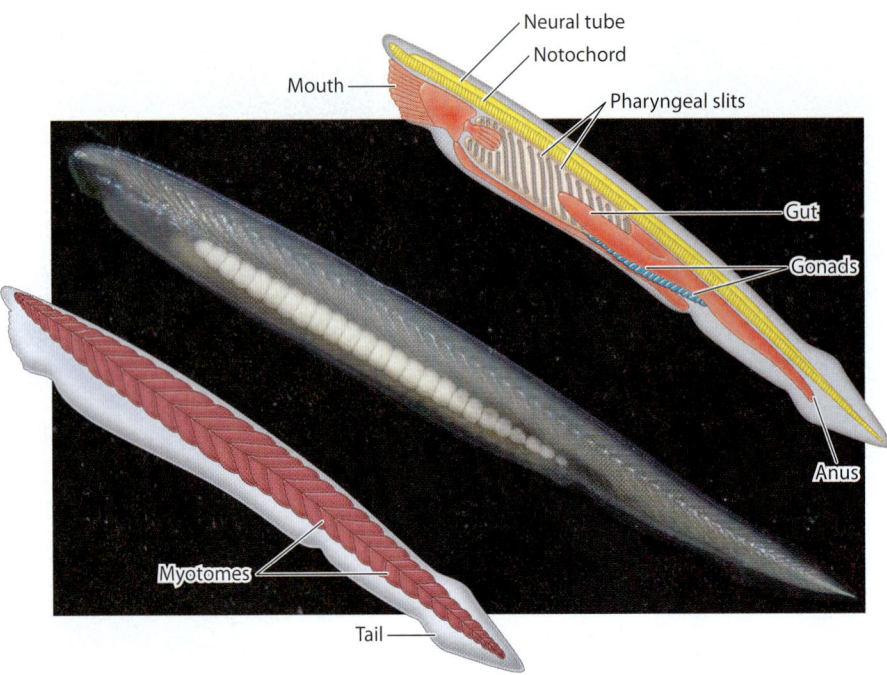

FIG. 44.27 Tunicates. (a and b) Adult form; (c) larval form.

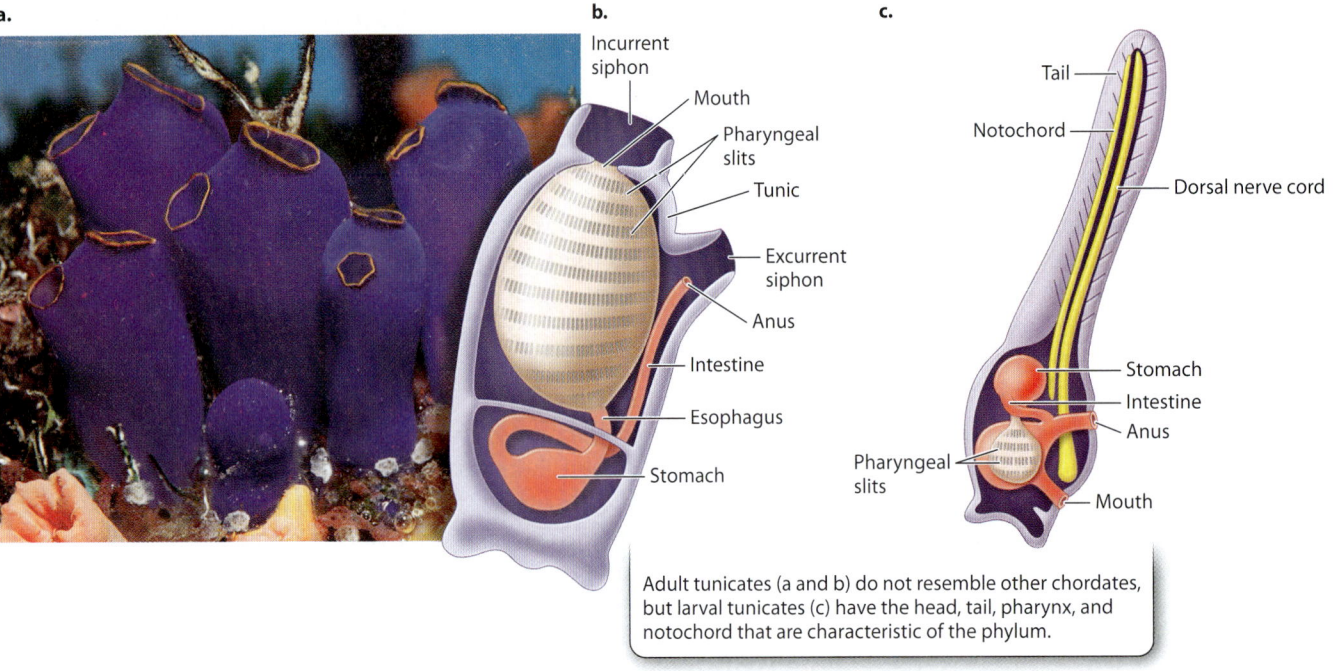

Adult tunicates (a and b) do not resemble other chordates, but larval tunicates (c) have the head, tail, pharynx, and notochord that are characteristic of the phylum.

only during early embryogenesis and is replaced functionally by the development of a **vertebral column.** Also forming during early development is the **neural tube,** a cylinder of embryological tissue that develops into a dorsal nerve cord. Body musculature is organized into a series of segments called **myotomes.** Chordates have a tail (reduced but still internally present in humans and other apes) that extends posterior to the anus and muscularized appendages that include the fins of fish and legs of terrestrial vertebrates.

Amphioxus (Fig. 44.26) belongs to the cephalochordates. These animals share key features of body organization with vertebrates but lack a well-developed brain and eyes, have no lateral appendages, and do not have a mineralized skeleton. Nonetheless, their many similarities to vertebrates suggest they are closely related, and molecular sequence comparisons confirm this hypothesis.

The tunicates also share key features of the chordate body plan during early development, but they have a unique adult form (**Fig. 44.27**). Tunicates include about 3000 species of filter-feeding marine animals, such as sea squirts anchored to the seafloor and salps that float in the sea. The adult tunicate body is a basket-like structure that is highly modified for filter-feeding. It might be mistaken for a sponge, but its complex internal structure shows that it lies far from sponges on the animal tree. The tunicate's body wall, or tunic, has a siphon-like mouth at one end that draws water through an expanded pharynx that captures food particles and exchanges gases. Water and wastes are expelled through an anal siphon. In the adult tunicate, the only obvious similarities to other chordates are the pharynx and its pharyngeal slits. Larval tunicates, however, resemble other chordates like fish or tadpoles with a more typical chordate body plan, including a notochord, neural tube, and a long tail with muscles arranged in myotomes.

For many years, zoologists considered cephalochordates to be the closest relatives of vertebrate animals, but molecular data now suggest that, in fact, the tunicates are our closest invertebrate relatives. This means that the common ancestor of tunicates, amphioxus, and vertebrate animals was a small animal much like amphioxus in organization. After the divergence of the three major groups of chordates, however, the tunicates' adaptation for filter-feeding led to the unique anatomy they have today.

→ **Quick Check 4** You do not have a notochord, a tail, or gill slits. How do we know that you are a chordate?

44.4 VERTEBRATE DIVERSITY

In many ways, the last stop on our tour of animal diversity is the most familiar because it includes our own species, *Homo sapiens*. The animals known as vertebrates are named for their jointed skeleton that runs along the main axis of the body, forming a series of hard segments collectively termed **vertebrae** (singular, vertebra). In addition to features shared with other chordates, vertebrate animals are distinguished by a cranium that protects a well-developed brain, a pair of eyes, a distinctive mouth for food capture and ingestion, and an internal skeleton commonly mineralized by calcium phosphate (**Fig. 44.28**). The vertebrates also have a coelom in which the organs are suspended, and a closed circulatory system. Like the other chordates, vertebrates have pharyngeal slits in at least the early embryonic stages.

Many of the features that separate vertebrates from invertebrate chordates can be found in the head, including a bony cranium that protects the brain, joined in most vertebrates by the mandible (or jawbone) to form a skull. This protective helmet for the delicate neuronal tissues permits the development of a larger brain than would otherwise be possible. The mandible permits eating foods such as other living animals, themselves often protected by hard shells, and plants containing hard lignin. Because the bottom scavenger hagfish have a cranium but not vertebrae (or a mandible), some biologists use the name "Craniata" for this group, rather than the traditional "Vertebrata."

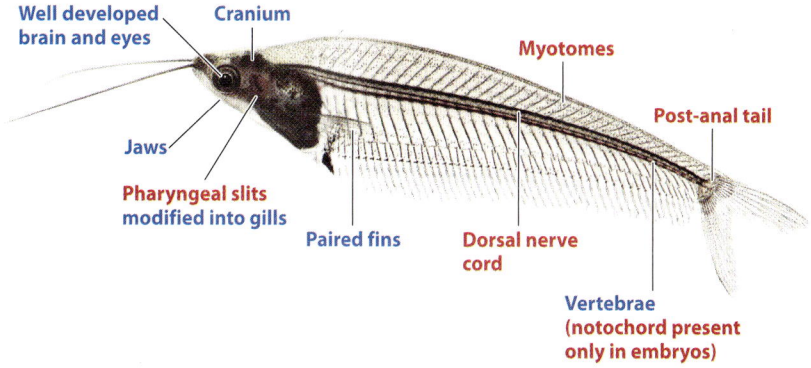

FIG. 44.28 Vertebrate characters. Anatomical features of vertebrates that are characteristic of all chordates are shown in red, and those that are unique to vertebrates are shown in blue.

FIG. 44.29 A phylogeny of the craniates. This phylogeny is favored by recent molecular data, but, like all phylogenies, is a hypothesis. The placements of hagfish relative to other vertebrates and turtles relative to other sauropsids have provoked debate for many years.

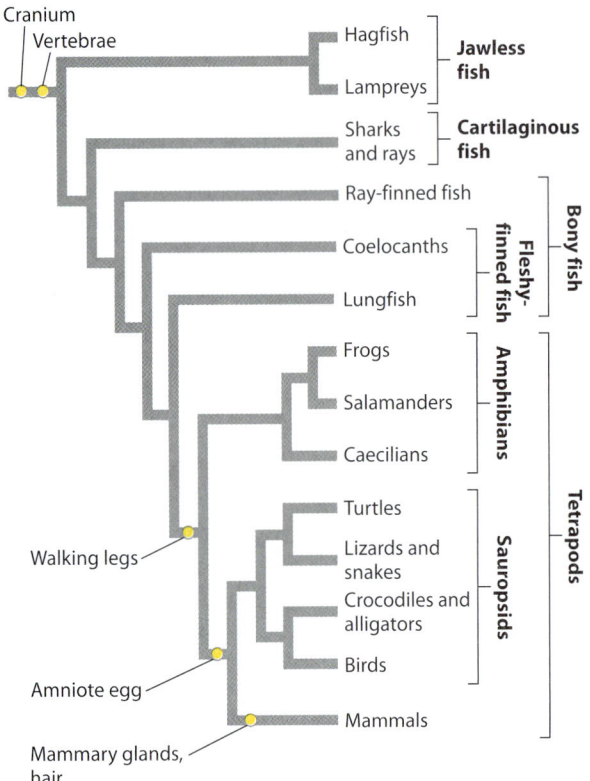

Fish are the earliest-branching and most diverse vertebrate animals.

We sometimes think of fish as one group, but the aquatic animals we commonly call "fish" include four distinct groups of aquatic vertebrates, not all of them monophyletic (**Fig. 44.29**).

The earliest branching craniates are the **hagfish** and **lampreys** (**Fig. 44.30**). These animals have a cranium built of cartilage but lack jaws. Lampreys also have a vertebral column built of cartilage. Hagfish lack vertebrae, though current research suggests that this trait existed in a common ancestor of hagfish and lampreys, but was lost in hagfish. These eel-like organisms feed on soft foods without the aid of jaws because they diverged before jaws evolved from pharyngeal slits. Hagfish feed on marine worms and on dead and dying sea animals, while lampreys live parasitically, sucking body fluids from fish prey. Both hagfish and lampreys have a series of gill slits through which water enters to bring oxygen to the gills.

Biologists have long debated the phylogenetic relationships among hagfish, lampreys, and other vertebrate animals. The presence of a vertebral column in lampreys convinces many biologists that lampreys are the sister group to other vertebrates and hagfish the sister group to all vertebrates, including lampreys. The debate continues, but several lines of molecular data now favor the view shown in Fig. 44.29, that hagfish and lampreys together form the sister group to all other vertebrates. About 65 species of lampreys and hagfish live in freshwater and coastal oceans. Fossils record the existence of many extinct species of jawless fishes. When we include fossils, the jawless

FIG. 44.30 Hagfish and lampreys. Sometimes grouped together as jawless fish, (a) hagfish and (b) lampreys are the earliest branching craniates alive today.

FIG. 44.31 Cartilaginous fish. The skeletons and jaws of sharks are made of cartilage.

fishes as a whole are paraphyletic because one of their groups gave rise to fish with jaws.

Chondrichthyes, or cartilaginous fish, form the next deepest branch on the vertebrate tree. This monophyletic group includes about 800 species of sharks, rays, and chimaeras, all of which have jaws and a skeleton made of cartilage (**Fig. 44.31**). These fish deposit calcium phosphate minerals only in their teeth and in small toothlike structures called denticles embedded in the skin. The best-known cartilaginous fish are the sharks, many of which are predators that put those mineralized teeth to good use. On the other hand, the group also includes whale sharks, filter-feeding animals that, at a length up to 40 feet (12 m), are the gentle giants of the sea. Rays are closely related to sharks but have a flattened body adapted for life on the seafloor, where they feed on clams and crabs. The cartilaginous fish can retain levels of urea in their tissues that would poison other vertebrates, a physiological adaptation that helps them to maintain their salt balance without well-developed kidneys (Chapter 41). Nearly all Chondrichthyes occur in the oceans, but Bull Sharks (*Carcharhinus leucas*) commonly swim into rivers—they've been recorded well up the Mississippi—and there are several freshwater species of rays.

Moving up one more branch in the vertebrate tree, we encounter the **Osteichthyes,** or bony fish (see Fig. 44.29). Bony fish have a cranium, jaws, and bones mineralized by calcium phosphate. Numbering about 20,000–25,000 fresh and seawater species, these are the fish that we most commonly encounter (**Fig. 44.32**). Bony fish are by far the most diverse group of vertebrates, possessing several unique features that facilitate their occupation of diverse

FIG. 44.32 Bony fish. The char, well known to those who fish in freshwater, illustrates the principal anatomical features of bony fish: a cranium, jaws, and mineralized bones.

FIG. 44.33 Fleshy-finned fish. The fish most closely related to tetrapods are the fleshy-finned fish: (a) the coelacanth and, even closer, (b) the lungfish. These fish have paired fins with an anatomical structure similar to that of tetrapod legs.

niches. First, they have a system of movable elements in their jaws that allows them to specialize and diversify their feeding on many different types of food. Second, they possess a unique gas-filled sac called a swim bladder that permits exquisite control over their position in the water column through changes in buoyancy. Third, bony fish have kidneys that allow them to regulate water balance and so occupy waters over a wide range of salinity.

The swim bladder plays a special role in vertebrate evolution. Early vertebrates evolved a gut sac that enabled them to gulp air to obtain additional oxygen. In some fish, this sac evolved into the swim bladder, but in one group the sac was modified to become a lung. Charles Darwin was among the first to recognize that the swim bladder and the vertebrate lung are homologous organs.

The common ancestor of tetrapods had four limbs.

Most animal phyla occur in the oceans, and many are exclusively marine. A subset of animal phyla has successfully colonized freshwater, and a smaller subset has radiated onto the land. Eleven groups on the animal phylogenetic tree contain both aquatic *and* terrestrial species: nematodes, flatworms, annelids (earthworms and leeches), snails, tardigrades (microscopic animals with eight legs, sometimes called water bears), onychophorans (velvet worms), four groups of arthropods (millipedes and centipedes, scorpions and spiders, land crabs, and insects), and vertebrates. In each of these groups, the aquatic species occupy the earliest branches, and the terrestrial species occupy later branches. No two of these groups share a last common ancestor that lived on land; thus, they all made the transition independently. Earlier, we outlined the extraordinary diversity of terrestrial arthropods. Here, we focus on vertebrates as the other great animal colonists of the land.

Most bony fish have fins that are supported by a raylike array of thin bones. About a dozen closely related species, however, have paired pectoral and pelvic fins. These fish are called **fleshy-finned fish** and include the **coelacanth** and **lungfish** (**Fig. 44.33**). Although these animals resemble other fish, the coelacanth and lungfish are the nearest relatives of tetrapods, four-legged

FIG. 44.34 The amphibian life cycle.

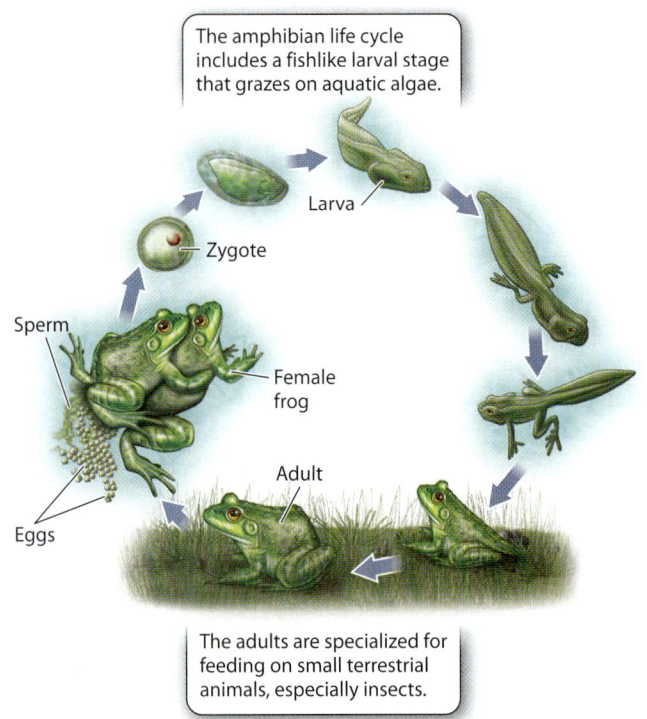

The amphibian life cycle includes a fishlike larval stage that grazes on aquatic algae.

The adults are specialized for feeding on small terrestrial animals, especially insects.

animals. Lungfish can survive periods when their watery habitat dries by burying themselves in moist mud and breathing air. Coelacanths first appeared over 400 million years ago and were thought to have been extinct for over 80 million years until a specimen was discovered in a fish market in South Africa in 1938. Hence, they are sometimes referred to as living fossils. There are two living species of coelacanths and three species of lungfish, one each in South America, Africa, and Australia. Because the fleshy-finned fish are the closest relatives of tetrapods, there is no monophyletic group that corresponds to our popular notion of "fish." To be monophyletic, such a group must include the tetrapods.

Fossils document the anatomical transition in vertebrate animals as they moved from water to land, including changes in the form of the limbs, rib cage, and skull (see Fig. 23.20). Living tetrapods include amphibians, such as frogs and salamanders; lizards, turtles, crocodilians, and birds; and mammals. The last common ancestor of all these animals had four limbs, hence the name **Tetrapoda** ("four legs"). Some, however, like snakes and a few amphibians and lizards, have lost their legs in the course of evolution.

There are over 4500 species of **Amphibia** ("double life"), ranging in size from tiny frogs a few millimeters in length to the Chinese Giant Salamander, which is over 1 m long. Their name reflects their distinctive life cycle (**Fig. 44.34**). Most species have an aquatic larval form with gills that permit breathing under water and an adult form that is terrestrial and usually has lungs for breathing air. Amphibians must reproduce in the water, and so are not completely terrestrial. Whereas many amphibian larvae graze on algae, the adults are predators and often have a muscular tongue for capturing prey. Many are protected by toxins they secrete from glands in their skin, sometimes advertised by their brilliant color patterns. Most amphibian adults require moist skin for breathing and consequently have small lungs, with toads and red efts being notable exceptions.

Amniotes evolved terrestrial eggs.

Like insects, some vertebrates evolved an egg adapted to tolerate dry conditions that accompany life on land. The amniotic egg has a desiccation-resistant shell and four membranes that permit gas exchange and management of waste products produced by the embryo (see Fig. 42.10). These eggs must be fertilized internally before the eggshell is produced by the female because sperm cannot penetrate the shell. Amniotic eggs permit long development times, with the embryo's nutrition supported by a large yolk or the placenta. They keep wastes separated from the embryo so they do not poison it, giving the embryo time to build more complex bodies than otherwise would be possible.

The **amniotic egg** can exchange gases while retaining water, and so permits the group of vertebrates known as **amniotes** to live in dry terrestrial habitats that amphibian eggs cannot tolerate. Amniotes include lizards, snakes, turtles, and crocodilians (about 6000 species of scaly animals commonly referred to as "reptiles"), as well as the birds and mammals (**Fig. 44.35**). Most mammals, many lizards, and some snakes have evolved live birth rather than laying eggs. Nonetheless, they retain the specialized membranes characteristic of all amniotes. Instead of being wrapped around the

FIG. 44.35 Amniotes, including (a) turtles, (b) lizards, (c) mammals, and (d) birds.

embryo inside a tough egg shell, these membranes surround and protect the embryo inside the womb.

As shown in Fig. 44.29, amniotes form a monophyletic group with two major branches. One is made up of mammals, and the other contains turtles, birds, and the amniotes traditionally grouped together as reptiles. Like fish, however, reptiles are not monophyletic—birds are descended from reptiles, although you have to look closely to see the similarities between the two groups. Birds lack teeth, most of their scales have been modified into feathers (except on the legs and toes), and they are generally adapted for flight. Adaptations for flight include hollow bones that are light but remain strong, and a method of breathing that extracts much oxygen from each breath, permitting their high-performance flight (Chapter 39).

One consequence of this streamlined, lightweight body is a lack of the heavy jaws needed for grinding plant leaves or other low-nutrition foods. Instead, birds have their grinding organ, the gizzard (also found in other reptiles), inside their bodies near their center of gravity, where it interferes less with flight. Only one kind of bird, the weak-flying Amazonian Hoatzin, can actually feed on leaves. As discussed in Chapter 23, fossils show that birds diverged from dinosaur ancestors, and molecular sequence comparisons indicate that crocodiles are the closest living relatives of birds.

→ **Quick Check 5** Why do bird feet have scales like those found on snakes and lizards?

All **mammals** are covered with hair and feed their young milk from the mammary glands for which the class Mammalia is named. Like many other groups of animals and plants, the early-branching mammals show intermediate stages in the evolution of the body plans that dominate on Earth today. The first fossils that clearly show hair appeared about 200 million years ago, during the Triassic Period. Like feathers in birds, hair permits retention of body heat. The earliest-branching living mammals, the monotremes, lay eggs, like birds or lizards (**Fig. 44.36**), but their hatched young drink milk secreted from pores in the skin of the mother's belly. Among these are the Australian platypus and the echidna, or spiny anteater.

The first mammals that gave birth to live young appeared in the Jurassic Period, about 125 million years ago. These animals gave rise to the two major groups of living mammals, marsupial and placental mammals. **Marsupials** include kangaroos, koalas, and related groups native to Australia, as well as the opossums found in the Americas. Their young are born at an early stage of development, and the tiny babies must crawl to a pouch where mammary glands equipped with nipples provide them with milk. Only later do fully formed juveniles emerge from the pouch and begin life on the ground.

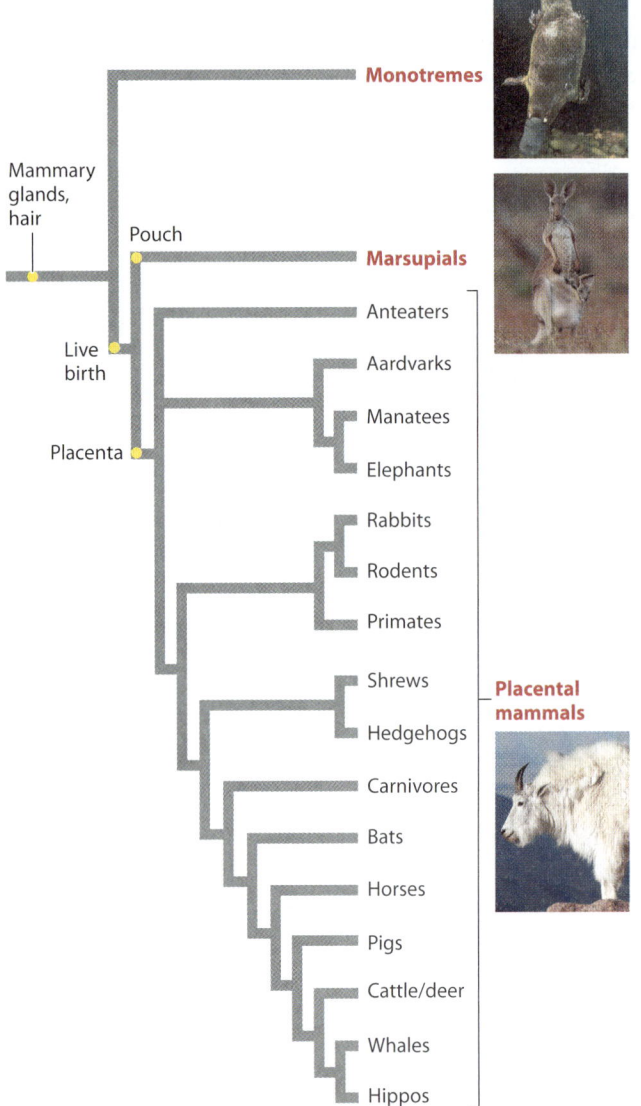

FIG. 44.36 A phylogenetic tree for mammals, illustrating the three major groups: monotremes, marsupials, and placental mammals.

The **placental mammals** are named for a temporary organ called the **placenta** that develops in the uterus along with the embryo, providing nutrition that enables the offspring to be larger and more quickly independent when born (Chapter 42). Most living mammals fall into this group. Placental mammals include the carnivores, such as lions and weasels; the primates, including monkeys, apes, and humans; and the hooved mammals, which include cattle, pigs, deer, and—perhaps surprisingly—their marine off-shoot, the whales.

All together, the familiar mammals mentioned so far number about 1000 species. The most diverse mammals, however, belong to two groups of mostly small animals, the bats and the rodents. Bats, which evolved flight and a form of sonar called echolocation to find food at night, include more than 1000 species. The largely plant-feeding rodents are as diverse as the large mammals and bats combined, numbering more than 2000 species. Beavers, gophers, rats, mice, hamsters, and squirrels are all rodents. Their evolutionary success is often attributed to innovations in tooth development. Their teeth are unusual in growing continually throughout life. This constant growth permits constant wear, and so allows these animals to gnaw through hard protective coverings to obtain nutrition from a remarkably wide variety of otherwise hard-to-access sources like the insides of seeds.

44.5 THE EVOLUTIONARY HISTORY OF ANIMALS

Because many animals form mineralized skeletons, sedimentary rocks deposited during the past 542 million years contain a rich fossil record of shells and bones. Animals also leave a sedimentary calling card in their tracks, trails, and burrows, again widespread in sedimentary rocks formed after 542 million years ago. Not all animal phyla are well represented in the fossil record—fossil annelids, for example, are rare and flatworms are essentially unknown as fossils because they lack mineralized body parts (Chapter 23). In contrast, the fossilized shells and bones of bivalve mollusks, brachiopods, echinoderms, and mammals preserve an excellent record of evolutionary history within these groups. If we step back and look at the big picture, what do fossils tell us about the evolutionary history of the animal kingdom?

Fossils and phylogeny show that animal forms were initially simple but rapidly evolved complexity.

Phylogeny suggests that animals are relative latecomers in evolutionary history, and the fossil record confirms this hypothesis. Life originated more than 3.5 billion years ago, but, as discussed in previous chapters, microorganisms dominated ecosystems for most of our planet's history. As noted in Chapter 28, macroscopic fossils of organisms thought to be animals first appear in rocks deposited only 575 million years ago. Called Ediacaran fossils after the Ediacara Hills of South Australia where they were discovered, these fossils have simple shapes that are not easily classified among living animal groups (**Fig. 44.37**; see also Fig. 28.13).

FIG. 44.37 *Dickinsonia*, an Ediacaran fossil. *Dickinsonia* has been interpreted as an early animal that acquired food by phagocytosis and by absorbing organic molecules, and exchanged gases by diffusion. The structural simplicity of *Dickinsonia* and similar fossils suggests that they record early branches of the animal tree, perhaps as placed in the diagram.

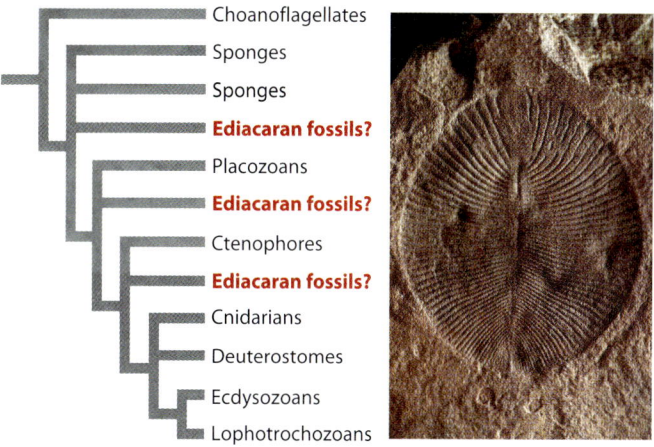

Phylogeny suggests that we should look for sponges, cnidarians, and other diploblastic animals among the oldest animal fossils, but sponges, at least, are rare among Ediacaran fossils. Instead, a majority of Ediacaran fossils show simple, fluid-filled tubes, without identifiable mouths or other organs. Many may have formed colonies, gaining complexity through colonial growth and differentiation, as some living cnidarians do. These early animals had epithelia and are thought to have obtained food by taking in dissolved organic matter while exchanging gases by diffusion, somewhat like living placozoans. Most Ediacaran fossils probably form an early branch or branches on the animal tree, falling among the limbs whose survivors include cnidarians, ctenophores, and placozoans (Fig. 44.37).

Why did the first animal radiation occur so late in the evolutionary day? Scientists continue to debate this question, but part of the answer appears to lie in Earth's environmental history. Geochemical data suggest that only during the Ediacaran Period did the atmosphere and oceans come to contain oxygen in quantities sufficient to support the metabolism of large, active animals.

The animal body plans we see today emerged during the Cambrian Period.

Ediacaran fossils differ markedly from the shapes of living animals, but in the next interval of geologic history, the Cambrian

FIG. 44.38 Fossils of the Cambrian Period. Cambrian rocks preserve early representatives of present-day phyla, including (a) early arthropods called trilobites, (b) annelid worms, (c) early deuterostomes with a pharynx, and even (d) primitive vertebrates.

Period (542–489 million years ago), we begin to see the remains of animals with familiar body plans (**Fig. 44.38**). Cambrian fossils commonly include skeletons made of silica, calcium carbonate, and calcium phosphate minerals, and these record the presence of arthropods, echinoderms, mollusks, brachiopods, and other bilaterian animals in the oceans. Spicules made by sponges are common in Cambrian rocks as well. The rocks also preserve complex tracks and burrows made by organisms with hydrostatic skeletons, muscular feet, and the jointed legs of arthropods. And in a few places, notably at Chengjiang in China and the Burgess Shale in Canada, unusual environmental conditions have preserved a treasure trove of animals that did not form mineralized skeletons (Chapter 23).

These exceptional windows on early animal evolution show that, during the first 40 million years of the Cambrian Period, the body plans characteristic of most bilaterian phyla took shape in a transition sometimes called the **Cambrian explosion.** Sponges and cnidarians radiated as well, producing through time the biomass- and diversity-enhancing habitats of reefs and imparting an ecological structure to life in the sea broadly similar to what we see today.

Scientists sometimes argue about whether the name "Cambrian explosion" is apt. The fossil record makes it clear that bilaterian body plans did not suddenly appear fully formed, so the event was not truly "explosive." Rather, fossils demonstrate a huge accumulation of new characters in a relatively short period of time, during which the key attributes of modern animal phyla emerged. For example, living arthropods combine segmented bodies with a protective cuticle, jointed legs, other appendages specialized for feeding or sensing the environment, and compound eyes with many lenses. Cambrian fossils include the remains of organisms with some but not all of the major features combined today in arthropods. In short, the first 40 million years of Cambrian animal evolution ushered in a world utterly distinct from anything known in the preceding 3 billion years.

Despite the burst of evolution recorded by Cambrian fossils, there were still relatively few species of marine animals at the end of the Cambrian Period. The ensuing Ordovician Period (489–444 million years ago) was a time of renewed animal diversification, especially the evolution of heavily skeletonized animals in the world's oceans. The Ordovician radiation established a marine ecosystem that persisted for more than 200 million years. Interestingly, however, if you had walked along an Ordovician beach, the shells washing about your feet would have been far different from the ones you see today. The dominant shells were those of brachiopods, not clams. Broken corals in the surf were the skeletons of now-extinct cnidarians only distantly related to modern reef-forming corals. And arthropod shells, molted during growth, were those of now-extinct trilobites, not lobsters or crabs. This ecosystem was devastated by a mass extinction 252 million years ago (Chapter 23).

Animals began to colonize the land 420 million years ago.

The colonization of land by animals began only after plants had established themselves there. Land plants evolved about 460 million years ago. Arthropods followed by about 420 million years ago, especially chelicerates that included the ancestors of spiders, mites, and scorpions. Insects, descended from crustacean ancestors that independently gained access to the land surface, appeared at about the same time. The radiation of the major groups of insects, however, began 360 million years ago with a marked diversification of dragonflies and the ancestors of cockroaches and grasshoppers. Some of the dragonfly-like insects that darted among the plants of Carboniferous coal swamps had bodies the size of a lobster and 75-cm wing spans! One might well wonder what prevents insects from attaining this size today. These immense insects flew through an atmosphere that contained more oxygen than is present today, and this higher level of atmospheric oxygen is thought to have been a necessary condition for insect gigantism. After some time, levels of atmospheric oxygen declined to levels closer to those seen today, and the giant insects disappeared. Flies, beetles, bees, wasps, butterflies, and moths radiated later, beginning in the early Mesozoic Era. Their rise in diversity parallels that of the flowering plants and reflects coevolution of these pollinators and flowering plants.

Tetrapod fossils first appear in sedimentary rocks deposited near the end of the Devonian Period, about the same time as the early insect radiations 360 million years ago. As discussed in Chapter 23, the fossil record documents in some detail the shifts in skull, trunk, and limb morphology that allowed vertebrate animals to colonize land. These include the evolution of muscled, articulated legs from fins, together with a set of strong, articulated digits where the limbs meet the ground, internal lungs rather than gills, and an erect, elevated head with eyes oriented for forward vision. The appearance of amphibians and then reptiles follows the predictions of phylogenies based on comparative biology.

The most remarkable animals among the reptiles are surely the dinosaurs. Dinosaurs first evolved about 210 million years ago from small, bipedal ancestors that radiated following the mass extinction 252 million years ago. Once established, the dinosaurs radiated to produce many hundreds of species, dominating terrestrial ecosystems until 65 million years ago, when nearly all dinosaur species disappeared, the most familiar victims of an enormous and Earth-changing asteroid impact (see Fig. 1.3). We say "nearly all" because the fossil record documents the divergence of the ancestors of modern birds from a specific subgroup of dinosaurs, about 150 million years ago (see Fig. 23.20).

Mammals have been the dominant vertebrates in most terrestrial ecosystems since the extinction of dinosaurs, but the group originated much earlier, at least 210 million years ago. During the age of dinosaurs, most mammals were small nocturnal or tree-dwelling animals that stayed out of the way of large dinosaurs, although fossils from China show clearly that the largest mammals of this interval ate small dinosaurs and their eggs.

We summarize the biological diversity of all organisms, including animals, along with a timeline of Earth's history, in **Fig. 44.39**.

? CASE 8 Biodiversity Hotspots: Rain Forests and Coral Reefs

How have coral reefs changed through time?

On land, tropical rain forests are hotspots of diversity, and coral reefs support unusually large numbers of species in the oceans. Reefs have existed through most of Earth's recorded history, and rain forests have existed since the origin of trees more than 370 million years ago. Through time these have been Earth's most diverse communities in the sea and on land. But the taxonomic composition of these communities has changed through time because of both evolutionary innovation and mass extinction.

Bacteria built the first reefs, layered mounds of limestone that rose above the seafloor. With the Cambrian diversification of animals, calcified sponges rose to ecological prominence as participants in reef construction, but these communities were short lived, as mass extinction about 515 million years ago essentially wiped out the reef-building sponges. Subsequent renewed diversification of sponges, bryozoans, and corals with massive calcium carbonate skeletons resulted in new and different reef communities beginning in the Ordovician Period. Eventually these communities were also destroyed by mass extinction during the Devonian Period. Reefs were widespread again by the Permian Period (300–252 million years ago), dominated this time by different groups of sponges and bryozoans, along with skeleton-forming green algae.

The devastating extinction at the end of the Permian Period removed all corals and calcifying sponges, and for several million years following this event, reefs were built only by microorganisms, much as they were before the evolution of animals. Then, the evolution of new skeleton-forming corals (from sea anemone-like ancestors) and a resurgence of sponges capable of massive calcification introduced new reef systems into the ocean. Given the iconic status of coral reefs as examples of marine biodiversity, it is important to note that reefs like those of the present day have developed only over 40 million years of Earth history.

As discussed in Chapter 48, we may be nearing another great change in reef biology as global climate change threatens reef communities in many parts of the world. Today's coral species are at their tolerance limits of ocean temperature and pH, and both may be changing more rapidly than species can adapt. From the perspective of Earth history, the loss of coral reefs would simply be one more in a long series of extinction episodes. For humans, however, the commercial and ecological consequences would be substantial and permanent.

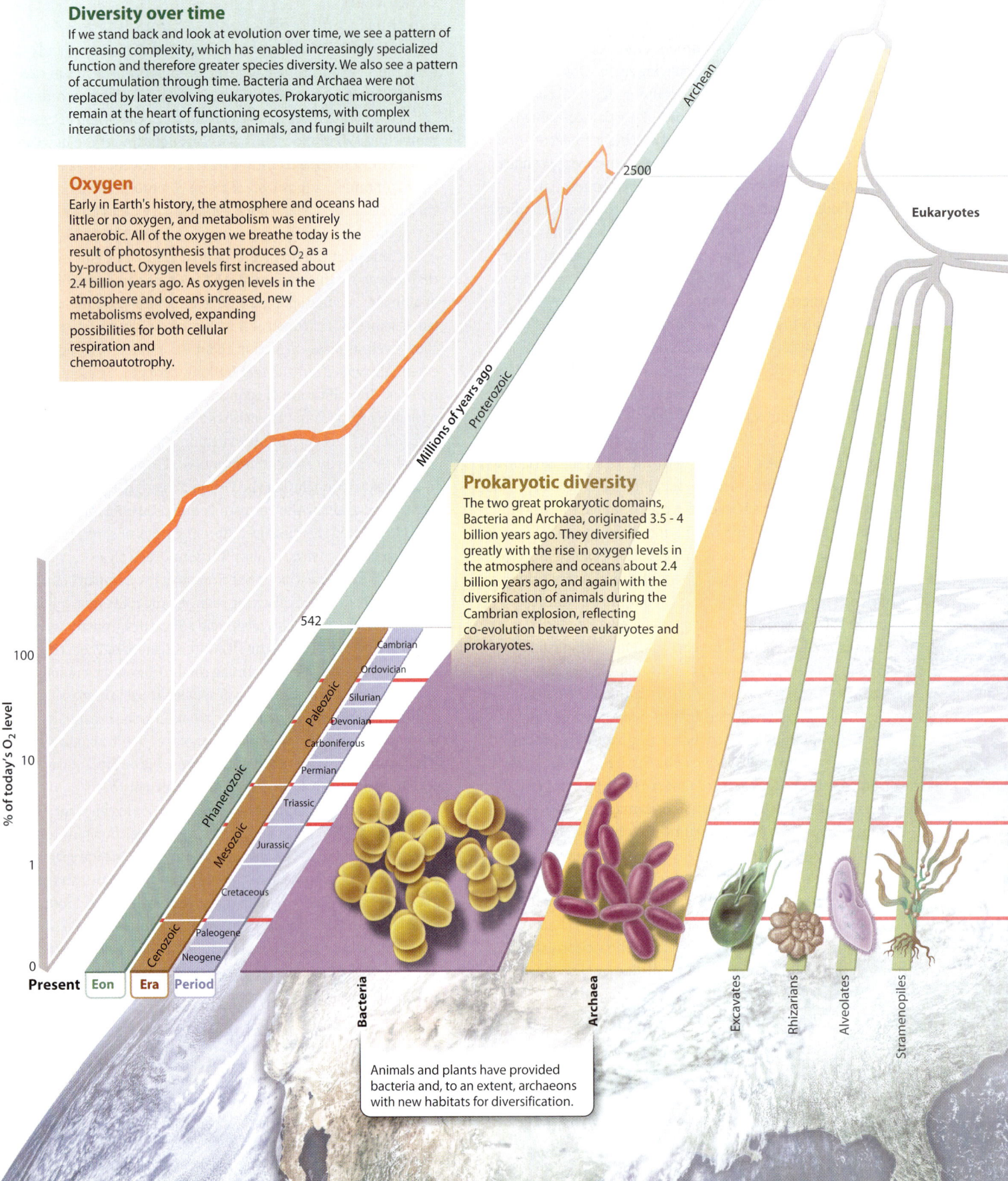

VISUAL SYNTHESIS
FIG. 44.39

Diversity through Time
Integrating concepts from Chapters 26, 27, 33, 34, and 44

Diversity over time
If we stand back and look at evolution over time, we see a pattern of increasing complexity, which has enabled increasingly specialized function and therefore greater species diversity. We also see a pattern of accumulation through time. Bacteria and Archaea were not replaced by later evolving eukaryotes. Prokaryotic microorganisms remain at the heart of functioning ecosystems, with complex interactions of protists, plants, animals, and fungi built around them.

Oxygen
Early in Earth's history, the atmosphere and oceans had little or no oxygen, and metabolism was entirely anaerobic. All of the oxygen we breathe today is the result of photosynthesis that produces O_2 as a by-product. Oxygen levels first increased about 2.4 billion years ago. As oxygen levels in the atmosphere and oceans increased, new metabolisms evolved, expanding possibilities for both cellular respiration and chemoautotrophy.

Prokaryotic diversity
The two great prokaryotic domains, Bacteria and Archaea, originated 3.5 - 4 billion years ago. They diversified greatly with the rise in oxygen levels in the atmosphere and oceans about 2.4 billion years ago, and again with the diversification of animals during the Cambrian explosion, reflecting co-evolution between eukaryotes and prokaryotes.

Animals and plants have provided bacteria and, to an extent, archaeons with new habitats for diversification.

44-30

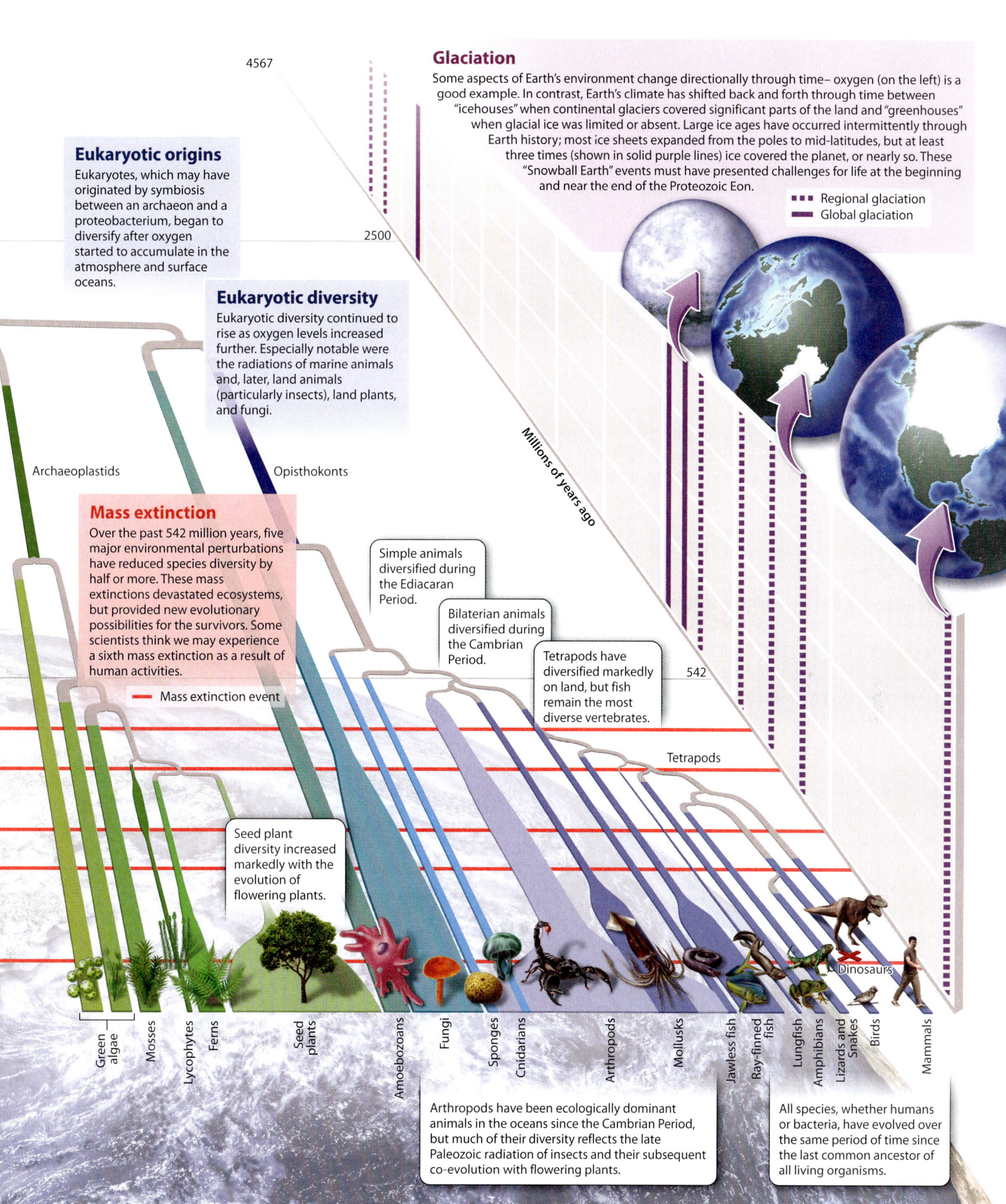

Core Concepts Summary

44.1 THE ANIMAL TREE OF LIFE CONSISTS OF MORE THAN A MILLION SPECIES.

Sponges lie on the deepest branch of the animal tree of life. page 44-1

All animals other than sponges are eumetazoans and fall into four major groups: cnidarians, ctenophores, placozoans, and bilaterians. page 44-1

Early biologists grouped animals on the basis of shared features of adult bodies, such as symmetry and type of body cavity. page 44-2

Bilaterians are characterized by bilateral symmetry, three germ layers, and well-developed organ systems. They include insects, worms, and mammals, and can be grouped as protostomes and deuterostomes. page 44-3

Embryology provided additional information on animal phylogeny, and, in the age of molecular biology, molecular sequence comparisons have provided still more information. page 44-4

44.2 THE EARLY BRANCHES OF THE ANIMAL TREE INCLUDE SPONGES, CNIDARIANS, CTENOPHORES, AND PLACOZOANS.

Sponges have a simple anatomical organization, feed by drawing water containing dissolved food particles into their interiors, and are widespread in the oceans. page 44-5

Cnidarians form true epithelial tissue, differentiate muscle and nerve cells, and act as predators by means of specialized cells called nematocysts. page 44-7

Ctenophores, or comb-jellies, resemble cnidarians but move by the beating of cilia, have an anal pore for waste excretion, and have a rudimentary mesoderm germ layer. They remain a difficult limb to locate on the animal tree of life. page 44-8

Placozoans have a very simple organization, but genomic data support the hypothesis that they are the sister group to all other eumetazoans. page 44-9

44.3 BILATERIANS, INCLUDING PROTOSTOMES AND DEUTEROSTOMES, HAVE BILATERAL SYMMETRY AND DEVELOP FROM THREE GERM LAYERS.

Protostomes are divided into lophotrochozoans and ecdysozoans. page 44-11

Lophotrochozoans consist of 17 phyla, including annelid worms and mollusks. page 44-11

Annelid worms include earthworms and leeches, but most species of annelids live in the oceans. page 44-11

Mollusks include gastropods (snails, slugs), cephalopods (squid, octopus), and bivalve mollusks (clams, oysters). page 44-11

Ecdysozoans include several phyla that molt their cuticle, notably the arthropods. page 44-14

Arthropods can be divided into four main groups—insects, chelicerates (spiders, scorpions), myriapods (centipedes, millipedes), and crustaceans (lobsters, shrimp). page 44-14

Deuterostomes include three main phyla—Chordata, Hemichordata, and Echinodermata. page 44-18

Well-known echinoderms are sea stars and sea lilies, which have outward fivefold symmetry. page 44-18

Chordates include vertebrates, cephalochordates, and tunicates. page 44-19

44.4 VERTEBRATES HAVE A CRANIUM AND VERTEBRAL COLUMN, AND INCLUDE FISH, REPTILES, BIRDS, AND MAMMALS.

The group commonly known as "fish" consists of four distinct groups of aquatic vertebrates: hagfish and lampreys, cartilaginous fish, bony fish, and fleshy-finned fish (lungfish). page 44-22

Lungfish are the closest relatives of tetrapods, which include amphibians, lizards, turtles, crocodilians, birds, and mammals. page 44-24

Amphibians have an aquatic larval form and a terrestrial adult form. page 44-24

Amniotes, such as lizards, snakes, crocodiles, birds, and mammals, have an amniotic egg, permitting movement into dry, terrestrial habitats. page 44-25

Mammals are covered with hair and feed their young milk from mammary glands. page 44-25

44.5 ANIMALS EVOLVED MORE THAN 575 MILLION YEARS AGO IN THE OCEANS, AND BY 500 MILLION YEARS AGO THE MAJOR STRUCTURAL AND FUNCTIONAL BODY PLANS OF ANIMAL PHYLA WERE IN PLACE.

Ediacaran fossils from 575 million years ago provide evidence of early animals. page 44-27

The Cambrian explosion was an interval of rapid diversification beginning 542 million years ago, during which time most of the animal body plans we see today first evolved. page 44-27

Chelicerates and insects were the first animals to colonize the land, sometime after 420 million years ago. page 44-28

Tetrapods first appear in the fossil record about 360 million years ago. page 44-28

Mammals originated at least 210 million years ago, but became dominant only after the extinction of the non-avian dinosaurs 65 million years ago. page 44-28

The fossil record of reef ecosystems reflects both evolutionary innovation and mass extinction. page 44-29

Self-Assessment

1. Draw a simplified animal tree of life, indicating the relationships among sponges and the four other major groups of animals.

2. Describe what kinds of trait are used to classify animals.

3. Compare and contrast the anatomical organization and feeding patterns of sponges, cnidarians, comb-jellies, and placozoans.

4. Describe feeding habits that do not require a head.

5. Name three features that account at least in part for the success of insects.

6. Draw a phylogeny of deuterostomes, indicating the most likely relationships among chordates, hemichordates, and echinoderms, with common examples of each group.

7. List features common to fish, amphibians, amniotes, and mammals.

8. Describe the evolutionary significance of the Cambrian explosion.

Do you understand the chapter's Core Concepts? Log in to check your answers to the Self-Assessment questions, then practice what you've learned and reinforce this chapter's concepts by working through the problems and multimedia tutorials provided there.

www.biologyhowlifeworks.com

CHAPTER 45

ANIMAL BEHAVIOR

Core Concepts

45.1 For any behavior, we can ask what causes it, how it develops, what adaptive function it serves, and how it evolved.

45.2 Animal behavior is shaped in part by genes acting through the nervous and endocrine systems.

45.3 Learning is a change of behavior as a result of experience.

45.4 Orientation, navigation, and biological clocks all require information processing.

45.5 Communication involves an interaction between a sender and a receiver.

45.6 Social behavior is shaped by natural selection.

45.7 Some behaviors are influenced by sexual selection.

There is no Nobel Prize for biology. It is the Prize for Physiology or Medicine that is awarded for biological research. Biologists specializing in other areas, such as behavior and ecology, are not even in the running for science's ultimate prize—except, that is, in 1973, when the Nobel Committee bent its own rules and awarded the Prize for Physiology or Medicine to three scientists with no physiological or medical credentials. The three were Niko Tinbergen, Konrad Lorenz, and Karl von Frisch. The award was made in recognition of their roles in pioneering the study of animal behavior under natural conditions.

Why "under natural conditions"? In the early 1900s, Carl von Hess, a prominent visual physiologist, claimed that bees are color blind, a conclusion based on experiments in laboratory settings. However, by training bees to go to food sources associated with differently colored cards in a natural context, the Austrian behavioral biologist Karl von Frisch demonstrated that, in fact, bees have excellent color vision. He emphasized this point at a scientific meeting where he was showing bees that had been trained to associate blue cards with food. By chance, the audience was wearing blue name tags, which proved highly popular with the blue-fixated bees! Von Frisch showed that behavior cannot be studied in isolation, but can only be fully understood when studied in a natural context—in this case, that of foraging for food.

In this chapter, we explore animal behavior—how animals learn, communicate with one another, form social groups, and choose mates. In each case, animal behavior can be considered at the individual level; individual behavior is determined in part by a complex interplay of the nervous and endocrine systems. Behavior is also influenced by the environment, as we saw with von Frisch's bees. Finally, behaviors, like all traits, are shaped by natural selection.

45.1 TINBERGEN'S QUESTIONS

We begin our exploration of animal behavior by asking a simple question: Why does an animal exhibit a particular behavior? The Dutch behavioral biologist Niko Tinbergen divided this over-arching question into four separate questions, each one focusing on a different aspect of the behavior. Let's consider an example—Why does a bird sing? While apparently straightforward, this question can be broken down into more specific questions and then answered in a number of different ways:

1. *Causation.* What physiological mechanisms cause the behavior? This question can have multiple answers. A bird sings because its hormone levels have changed in response to changes in day length. Or, more immediately, a bird sings because air passing through its specialized singing organ, the syrinx, causes membranes to vibrate rhythmically.

2. *Development.* How did the behavior develop? Here the focus is on the role of genes and the environment in shaping the development of the behavior. In birds, typically it's the male that sings, and he has learned the song from his father.

3. *Adaptive function.* How does the behavior promote the individual's ability to survive and reproduce? In this case, the answer to the question might be that a male bird sings in order to attract a mate and then reproduce.

4. *Evolutionary history.* How did the behavior evolve over time? Complex bird songs may have evolved from vocalizations made by ancestors that were reinforced and became increasingly stereotyped or ritualized over time, so much so that we can often identify a bird species simply by hearing it sing. A behavior may have originated to fulfill a function different from the one it currently serves. For example, the song of a particular species may have first evolved to claim a territory but now is used to attract mates.

Tinbergen's analysis allows us to see that different answers to the question "Why does a bird sing?" can all be correct. The first two, causation and development, are mechanistic explanations for behavior, whereas the second two, adaptive function and evolutionary history, provide evolutionary explanations of how natural selection has shaped a behavior over time. Tinbergen's four questions are complementary ways of looking at the same problem, and are a good starting point for the analysis of behavior.

45.2 GENES AND BEHAVIOR

The answers to Tinbergen's questions rely on an interplay between genes and the environment. The influence of genes is especially clear in **innate** behaviors, those that are instinctive and carried out regardless of earlier experience. **Fig. 45.1** shows that a male silkworm moth of the genus *Bombyx* flies upwind toward the source of a female-produced pheromone, an airborne chemical signal that communicates with members of the same species, in this case to attract the opposite sex. Pheromones released from a female's abdominal gland are sensed by small hairs on the male's antennae. In the presence of pheromones, the antennal sensory hairs fire action potentials (Chapter 35). When about 200 hairs are activated per second, a male flies upwind distances of a kilometer or more, tracking the increasing pheromone concentration until he finds the female. The male does not need to learn this behavior; he performs it spontaneously. His genes encode molecular receptors to which the pheromone binds, and this binding triggers a cascade of events that result in the moth heading up the pheromone's concentration gradient.

A particular behavior may also depend on an individual's experience. In this case, the behavior has been **learned.** As we will see, even neurologically simple organisms have a considerable capacity to learn. Fruit flies, for example, learn to avoid certain locations or substances if they associate them with an unpleasant experience.

We can consider, then, the extent to which a particular behavior is genetically encoded (part of the animal's **nature**) and the extent to which it is conditioned by the environment in which the animal develops (part of the animal's **nurture**). However, as we discussed in Chapter 18, nature and nurture are inextricably linked. For example, most human behaviors are the product of an interaction between nature and nurture. Consider language in the context of Tinbergen's second question, concerning the development of behavior: We have an innate ability for acquiring language, but the specific language that we acquire depends entirely on our nurture. A baby in Italy learns Italian, and one in Finland learns Finnish.

The fixed action pattern is a stereotyped behavior.

Behavior can be complex and therefore influenced by a large number of factors. For that reason, complex behaviors can sometimes defy analysis. Scientists studying behavior tend to focus on simple behaviors because they are more easily studied and understood.

Some of the first behaviors to be carefully analyzed were **displays,** which are patterns of behavior that are species specific and tend to be highly repeatable and similar from one individual to the next. Presumably, natural selection has favored display behaviors that are unmistakable in their function as signals. Consider, for example, courtship displays that are often precursors to mating in birds (**Fig. 45.2**). Experiments have shown that a bird raised in complete isolation from other members of its species will still perform a courtship display with great precision.

Displays are an example of a **fixed action pattern (FAP).** A FAP is a sequence of behaviors that, once triggered, is followed through to completion. A classic example, originally studied by Tinbergen, is the response of a goose to an egg that has fallen from its nest (**Fig. 45.3**). The stimulus that initiates the behavior is the **key stimulus** and, in this case, it is the sight of the misplaced egg. This sight provokes in the goose an egg-retrieval

FIG. 45.1 **An example of innate behavior.** A male moth flies toward the source of a female-produced pheromone.

Female moth

Pheromone

Male moth

FIG. 45.2 Courtship display in the Raggiana Bird-of-paradise. The male (left) is performing a highly repeated set of behaviors to entice the female to mate with him.

FAP, which consists of rolling the egg back to the nest with the underside of its beak (Fig. 45.3a). This response cannot be broken down into smaller subunits and it is always carried out to the end, even if it is interrupted. In fact, the goose will persist in its efforts even if the researcher ties a string around the misplaced egg and removes it while the goose is in mid-action. The goose will still continue the task of rolling the now-absent egg (Fig. 45.3b).

It is possible to understand this behavior by varying attributes of the key stimulus (in this case, the misplaced egg). The researcher can make models that vary in only one attribute to investigate how sensitive the goose is to aspects such as color or shape. A remarkable finding is that many birds respond most strongly to the largest round object provided, even a soccer ball, as illustrated in Fig. 45.3c. A soccer ball in fact not only elicits the egg-retrieval FAP, but it does so even more strongly than does a normal egg. The soccer ball is considered a **supernormal stimulus** because it is larger than any egg the goose would naturally encounter and elicits an exaggerated response. Natural selection has likely favored geese that recognize and respond to large eggs that have rolled outside the nest, but since eggs never grow to the size of soccer balls in normal circumstances, selection has not shaped an appropriate response to unrealistically large egg-shaped objects.

The nervous system processes stimuli and evokes behaviors.

How do animals recognize the stimulus that leads to a particular behavior? As we have seen, a goose attempts to retrieve an egg that is nearby but outside its nest. How does it recognize that the object in question is an egg? At its root, this is an extraordinarily difficult problem. The world is full of stimuli and there are many

FIG. 45.3 A fixed action pattern. A goose displays and completes the behavior to retrieve an egg (a), even if the behavior is interrupted (b) or the stimulus altered (c).

things that look like an egg. Thus, the challenge for the animal is to filter out the correct signal (the egg) from the surrounding noise (everything else). An animal has to process many different kinds of information in three dimensions.

We know that stimulus recognition is often carried out by **feature detectors,** specialized sensory receptors or groups of sensory receptors that respond to important signals in the environment. Frogs provide a good example. Male frogs call to attract females and warn off other males. A single pond may contain many species of frog, each with its specific advertisement vocalization, or call. A heavily populated frog pond is positively cacophonous, but it is critical to an individual frog that it be able to distinguish the call of other members of its own species from those of other species. Further complicating the matter is that different species of frog can have similar calls, often virtually indistinguishable to the human ear.

One approach to the problem of understanding how frogs recognize their species-specific call has been to experiment with recordings of frog calls. By recording frog calls, altering aspects of the call in specific ways, playing them back, and then monitoring the responses of frogs, we can identify the impact of different components—pitch, duration, and pulse frequency, for example—of the call. These kinds of study have demonstrated that the frog's auditory nerves can act as a feature detector, each one "tuned" to a specific component of the call, as shown in **Fig. 45.4**. The combination of different feature detectors stimulated by different components adds up to the recognition by the frog of a specific call. Once the sound has been correctly identified, the appropriate behavior follows.

Hormones can trigger certain behaviors.

As we have seen, stimuli are detected and processed by sensory receptors. In some cases, the response includes the release of a hormone. The power of a hormone lies in its ability to affect multiple cells in target organs simultaneously (Chapter 38). A neuron, on the other hand, is typically much more specific in its action, connecting directly to another neuron or neurons, or to a muscle (Chapter 35).

Hormones can affect behavior. The hormone testosterone, for example, is critical in sex determination in mammals, including humans (Chapter 42). In addition, in many respects males and females behave differently, and testosterone is at least partially responsible for many sex-specific behaviors.

Anolis lizards demonstrate the important role of hormones in sexual behavior, showing both how social stimuli can affect the release of hormones and how hormones can affect behavior (**Fig. 45.5**). Females of *Anolis carolinensis* collected in the spring and housed in the laboratory under a spring light-dark cycle prepare for reproduction, and about 80% of individuals have active egg follicles in their ovaries (Fig. 45.5a). If one male is added to a group of females, however, that figure increases to 100% (Fig. 45.5b). The courting behavior of the male lizard stimulates the females to produce hormones that cause the full development of the ovaries, making the females reproductively active. If a group of males instead of a single male is added to an all-female population, on the other hand, only about 40% of the females undergo ovarian development (Fig. 45.5c). This unexpected result seems to occur because the males interact and fight among themselves rather than court the females, and therefore the courtship stimulus is lacking. Castrated males (which do not produce testosterone) that are added to a group of females have no effect on rates of ovarian development, which remains about 80% (Fig. 45.5d). They fail to court because the behavior is

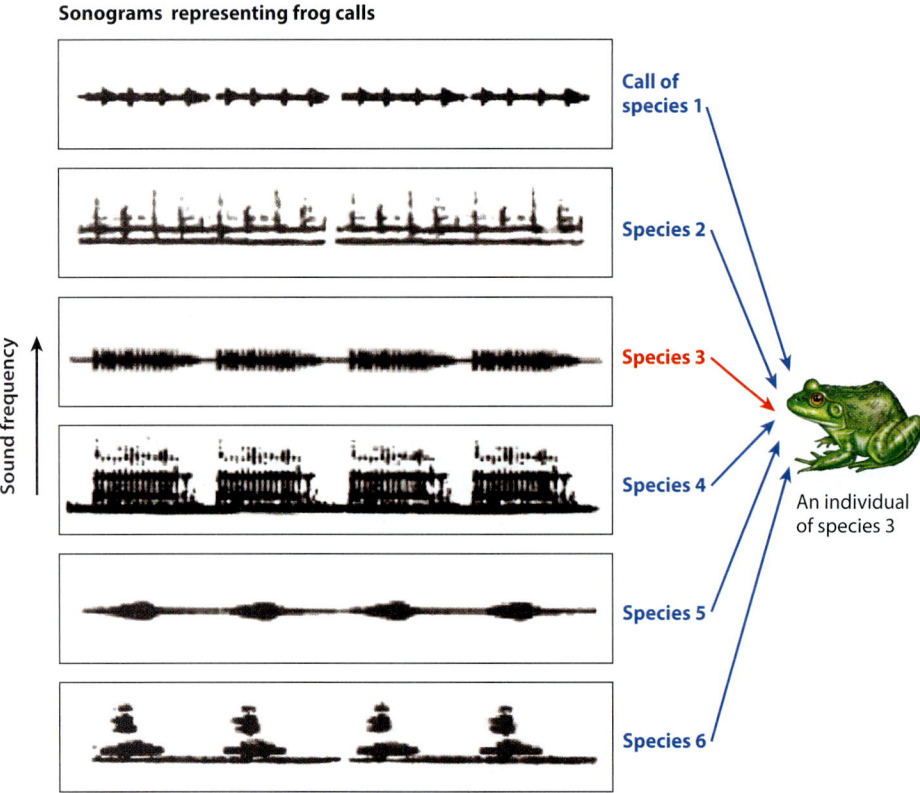

FIG. 45.4 **A feature detector.** The frog has a feature detector tuned to its species-specific call, allowing the frog to distinguish the call of its species (red arrow) from the calls of other frogs (blue arrows).

FIG. 45.5 Role of hormones in behavior, demonstrated by groups of *Anolis* lizards.

a. + 0 males

In a group of all female *Anolis carolinensis* collected in the spring, 80% of the females are prepared for reproduction with egg follicles in their ovaries.

b. + 1 male

A male displaying courtship behavior added to the all-female group causes 100% of females to become reproductively active.

c. + more than one male

More than one male added to a group of females fight with each other instead of courting females. Now only 40% of the females are reproductively active.

d. + castrated male

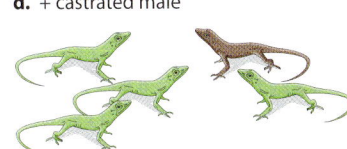

Castrated males do not display courtship behavior, and so do not affect the females' reproductive readiness.

e. + castrated male injected with testosterone

An injection of testosterone causes a castrated male to display courtship behavior, affecting the group of females in the same way as a non-castrated male.

■ Reproductive   ■ Nonreproductive

testosterone-mediated. However, a castrated male that is injected with testosterone does display courting behavior and has the same effect as the presence of a single male, inducing all the females to undergo ovarian development (Fig. 45.5e). These simple laboratory experiments demonstrate the complex interplay between hormones and behavior.

Breeding experiments can help determine the degree to which a behavior is genetic.

The nervous and endocrine systems that in part shape behaviors are the product of genetic instructions. If you take an extreme view, then, all behaviors have a genetic component. After all, it is the instructions in the genome that build the nervous system, muscles, and glands that lead to behavior. On the other hand, all behavior can be considered environmental since, without the appropriate environment, an organism would not develop the ability to perform the behavior. On a very basic level, an organism in an environment without appropriate nutrients would lack the necessary energy to perform a given behavior. How, then, do we study the genetic basis of behavior? Modern approaches harness the power of molecular genetics, but more traditional analyses are also highly informative.

Artificial selection, in which humans breed animals and plants for particular traits, provides strong evidence of the role of genetics in influencing behavior. Dogs were domesticated from wolves about 10,000 years ago. Today, there is great diversity among dogs, although they are all the same species. Obviously, there has been extensive selection on physical traits, with the result that a Dachshund, for example, looks quite different from a German Shepherd or Great Dane. Selection has also been applied to behavior: A Pointer has extraordinary "pointing" behavior that specifies the location of a hunter's prey, and a Border Collie is an excellent herder (**Fig. 45.6**).

FIG. 45.6 **Artificial selection.** Artificial selection can alter behavior, as shown by the remarkable differences in behavior we see among breeds of dogs: (a) a Pointer and (b) a Border Collie.

In the late 1950s, William Dilger studied the nest-building behavior of lovebirds in the genus *Agapornis*. Some species transport their nesting material in their beaks, whereas others tuck pieces into their tail feathers. Is there a genetic basis to the nesting behavior of these birds? To answer this question, Dilger set up crosses between species with different nest-building techniques to produce hybrids. He then observed how the hybrids built their nests. Interestingly, the hybrid offspring showed nest-building behavior intermediate between that of the parents: They tried to tuck material under their feathers, but were not successful. These experiments suggest that this behavior has a genetic basis.

It was not possible to do a true Mendelian dissection of the lovebirds' behavior (Chapter 16). Such an analysis would require additional crosses, which were not possible because the hybrids were sterile. Like most behaviors, the lovebirds' nest-building behavior is probably controlled by a large number of genes, each with a relatively small effect (Chapter 18). Therefore, it is difficult to identify which genes are responsible for the behavior. Even with molecular methods, the task of mapping and identifying the underlying genes when a large number of different genes govern a particular trait is daunting.

Molecular techniques provide new ways of testing the role of genes in behavior.

Molecular biology is changing the way we approach behavioral genetics. Some studies have identified complex behaviors that are strongly influenced by a single gene. One well understood instance of a single gene's effect on behavior comes from the fruit fly *Drosophila*.

In *Drosophila*, there are two different alleles of the *foraging* (*for*) gene—*for^s* and *for^R*—that are common in populations. These two alleles have different effects on the behavior of *Drosophila* larvae. The gene encodes a cGMP-dependent protein kinase, an enzyme expressed in the brain that affects neuronal activity and alters behavior. In the absence of food, both "sitter" (*for^s*) and "rover" (*for^R*) larvae move about in search of food. In the presence of food, however, sitters barely move, feeding on the patch on which they find themselves, whereas rovers move extensively both within a patch of food and between patches (**Fig 45.7a**).

Both forms are present in natural populations (**Fig. 45.7b**); typically, 70% are rovers, 30% are sitters. This proportion suggests that both strategies are adaptive. Furthermore, studies have shown that rovers are selected for in crowded environments in which there is an advantage to seeking new food sources, and sitters are selected for in less-crowded

FIG. 45.7 The effect of genotype at the foraging locus on the feeding behavior of *Drosophila melanogaster* larvae. (a) When food is present, rover larvae travel farther than sitter larvae. (b) Both types of larvae exist in natural populations.

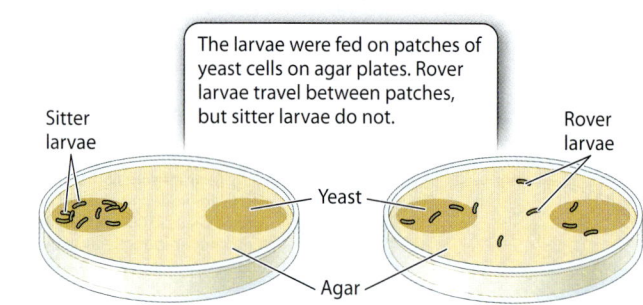

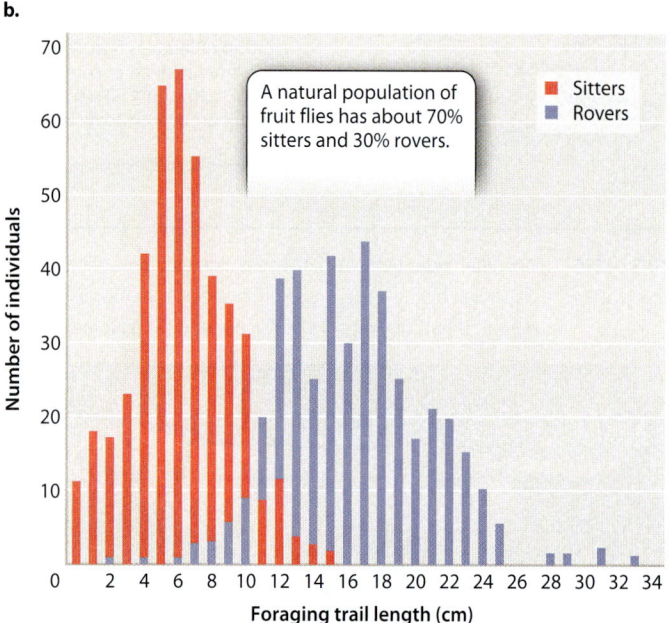

ones because they take maximum advantage of their current food source. In this case, variation at a single gene affects a complex behavior in fruit flies.

The *foraging* gene is present in other insects as well, including honeybees, suggesting that it has been evolutionary conserved. Honeybees with low levels of *for* expression in their brain tend to stay in the hive, whereas those with high levels of *for* expression are likely to be foragers (**Fig. 45.8**).

→ **Quick Check 1** The *for* gene affects foraging behavior in fruit flies and honeybees. In what way does it act differently in the two insects?

HOW DO WE KNOW?

FIG. 45.8

Can genes influence behavior?

BACKGROUND In fruit flies (*Drosophila melanogaster*), variation at the *foraging* (*for*) gene affects feeding behavior of larvae. Two alleles of the *for* gene—*for^s* and *for^R*—exist in natural populations; *for^s* larvae tend to stay on a patch of food ("sitters") and *for^R* larvae tend to move from patch to patch ("rovers"). Honeybees (*Apis mellifera*) also forage for food, but, in contrast to fruit flies, this behavior changes with age. Young honeybees stay at the hive ("nurses"), and older ones forage for nectar ("foragers"). The *for* gene is present in honeybees, raising the possibility that this gene is involved in the developmental change in foraging in honeybees.

HYPOTHESIS Foraging in honeybees is influenced by the expression of the *for* gene, and nurses have lower expression levels than foragers.

EXPERIMENT 1 Levels of *for* mRNA were measured in honeybee nurses and foragers.

RESULTS Levels of *for* mRNA are significantly higher in foragers than in nurses (Fig. 45.8a). However, foragers are older than nurses, so it was unclear whether the increased expression in foragers compared to nurses is associated with differences in age or differences in behavior. Therefore, the experiment was repeated with honeybees that forage at a young age. These young foragers also have higher *for* mRNA levels than do nurses (Fig. 45.8b).

EXPERIMENT 2 The *for* gene encodes a cGMP-dependent kinase that phosphorylates other proteins (Chapter 9). Honeybee nurses were treated with cGMP to activate the kinase and their subsequent behavior was monitored. A related compound, cAMP, with a chemical structure similar to cGMP but which does not affect the kinase, was used as a control, to ensure that any effect observed was specific for the pathway and not the result of the treatment.

RESULTS Treatment with cGMP changed the behavior of nurses, causing them to forage (Fig. 45.8c). Furthermore, cGMP acted in a dose-dependent fashion: The higher the dose, the more foraging behavior was observed. No effect on foraging was seen in the control treatment (Fig. 45.8d).

CONCLUSION The same gene that is involved in two different behavioral phenotypes in fruit flies (sitters vs. rovers) is also involved in a developmental behavioral change in honeybees (nurses vs. foragers). This finding suggests that a gene can have related but different functions in different organisms.

FOLLOW-UP WORK Researchers have looked at the expression profiles of many other genes in the honeybee genome to learn which genes are associated with the change of foraging behavior of honeybees.

SOURCE Ben-Shahar, Y., A. Robichon, M. B. Sokolowski, and G. E. Robinson. 2002. "Influence of Gene Action Across Different Time Scales on Behavior." *Science* 296:741–744.

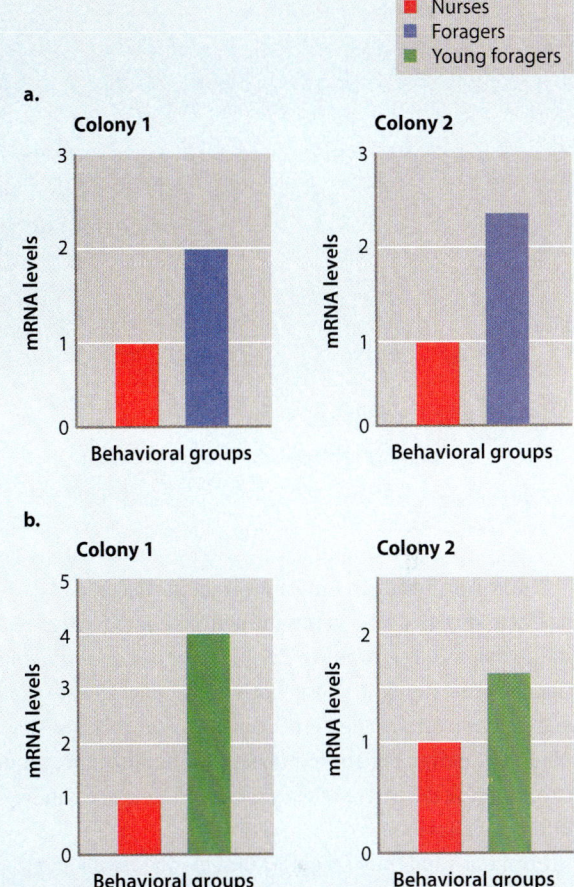

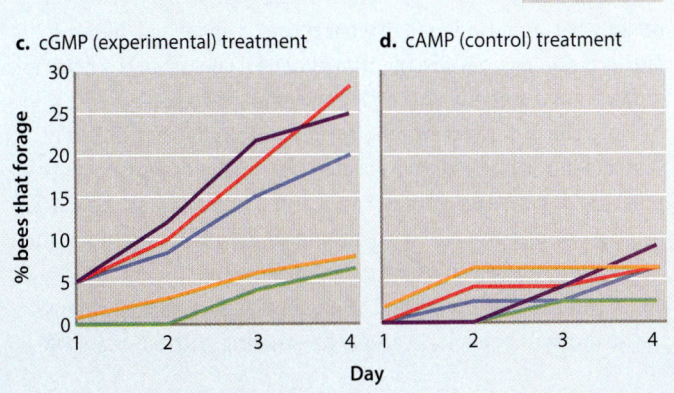

FIG. 45.9 Differing mating behavior in voles. The different behaviors in (a) monogamous prairie voles and (b) promiscuous montane voles results from variation at a single gene.

Studies of the North American vole, in the genus *Microtus*, have also revealed how genes influence behavior (**Fig. 45.9**). The prairie vole is monogamous: Couples pair bond for life, meaning that they mate only with each other. By contrast, the prairie vole's close relative, the montane vole, is promiscuous: A male mates and moves on, and a female, over her lifetime, produces litters by several different males. What differences underlie these different behaviors?

Hormones and their receptors provide an answer. In all mammals, antidiuretic hormone (ADH) controls the concentration of urine (Chapter 41). The receptor for ADH is not only expressed in the kidney, but also in the brain, where it affects behavior. Both the monogamous prairie vole and the promiscuous montane vole produce ADH, but the hormone receptors in the two species are different. The American neuroscientist Thomas Insel analyzed the genes that encode the ADH receptor in the two vole species. He found a difference in the region of the gene that determines under what conditions and in what cells it is expressed. Because of this difference, the distribution of ADH receptors in the brain is different in the prairie vole and in the montane vole.

Does the difference in gene expression explain the different sexual behaviors of the two species? The answer seems to be yes. Insel and his colleague Larry Young inserted the prairie vole ADH receptor gene, complete with its adjacent regulatory region, into a laboratory mouse (a promiscuous species, like the montane vole). Interestingly, Insel and Young observed a marked change in the mouse's behavior. Rather than mating with a female and then moving on to another mate, the transgenic male mouse pair bonded with the female. In short, the addition of a single gene affected a complex behavior such as pair bonding, making the otherwise promiscuous mouse more likely to pair bond.

45.3 LEARNING

Genes are not the only determinants of behavior. We all know that experience often leads to changes in behavior. We call this process **learning**. Humans are extraordinary learners, but even many relatively simple organisms respond to experience. Researchers have identified several types of learning. The categories are not strict, and there is some overlap among them, but they are a useful way to organize our thinking and give us deeper insights into the subject.

Non-associative learning occurs without linking two events.

Non-associative learning is learning that occurs in the absence of any particular outcome, such as a reward or punishment. One type of non-associative learning is **habituation,** which is the reduction or elimination of a behavioral response to a repeatedly presented stimulus. Chicks presented with silhouettes flying overhead provide an example. Initially, any overhead silhouette provokes a defensive, crouching posture, but eventually, chicks habituate to overhead silhouettes that have proved not to be threatening. In the absence of any consequences, chicks no longer crouch in response to a harmless silhouette passing repeatedly overhead.

Sensitization is another form of non-associative learning. Sensitization is the enhancement of a response to a stimulus that is achieved by presenting a strong or novel stimulus *first*. This pre-stimulus makes the animal more alert and responsive to the next stimulus. The sea slug, *Aplysia,* a model organism among scientists interested in the neuronal basis of behavior, exhibits sensitization. *Aplysia* withdraws its gills in response to a touch on the siphon, the tube through which it draws water over its gills. Interestingly, if the animal is given a weak electric shock first, its response to a touch on its siphon is much more rapid, suggesting that the shock has made the slug more sensitive to later stimuli.

Associative learning occurs when two events are linked.

Associative learning (also called **conditioning**) occurs when an animal learns to link (or associate) two events. Perhaps the most famous example of conditioning is Ivan Pavlov's dogs. Pavlov first presented the dogs with meat powder, and they salivated in response. He then presented the dogs with an additional cue, a ringing bell, whenever he presented the dogs with meat powder. After repeated experience of the two stimuli together, the dogs salivated at the sound of the bell alone, in expectation of the meat reward. This form of conditioning, in which two stimuli are paired, is called **classical conditioning.** In this case, a stimulus that leads to a behavior (the meat powder) was paired with a neutral stimulus (the bell) that initially had nothing to do with salivation. Eventually, a novel association is made between the sound of the bell and food, and the sound of the bell alone elicits salivation.

A second form of associative learning involves linking a behavior with a reward or punishment. If the behavior is rewarded (positively reinforced), it is more likely to occur the next time around. It is not just out of the goodness of her heart that the trainer at Sea World gives a sea lion or dolphin a fish reward at the end of each trick that the animal performs. However, if the behavior is punished (negatively reinforced), the response becomes less likely. This form of associative learning is called **operant conditioning.**

In classical conditioning, an association is made between a stimulus and a behavior, whereas in operant conditioning, an association is made between a behavior and a response. In addition, operant conditioning involves a novel, undirected behavior that becomes more or less likely over time. Consider a rat in a cage with some levers in it, each a different color. The rat has no impulse to press any of the levers, but by chance, it presses the red one. A food reward tumbles into the cage. Eventually the rat learns to associate pressing the red lever with the food reward; it has been operantly conditioned. It is thought that play in young animals provides them with the opportunity to explore their environment much in the same way as the rat does when first placed in the cage. Like the rat, they may find that some behaviors are more rewarding or more punishing than others.

→ **Quick Check 2** Both classical and operant conditioning involve learning to associate two events. What then is the difference between the two kinds of conditioning?

Learning takes many forms.

The capacity to learn often has adaptive functions. Much learning, including human learning, is based on **imitation:** One individual copies another. Learning by imitation was famously observed in the days when people in Britain had milk deliveries left outside on the doorstep: A number of birds, including Great Tits and Blue Tits, learned to open foil-topped milk bottles, and this behavior spread rapidly through populations of these birds as naïve birds copied the successful birds' approach. Individuals can also learn by imitating individuals of a different species. An octopus can learn to open a jar with a reward inside by watching a human or another octopus do it.

The capacity to learn a particular task seems in many instances to be innate. Female digger wasps are foragers. Tinbergen showed that a digger wasp learns the landmarks around her nest and then uses this information to find her way back to it from hunting. In his experiment, illustrated in **Fig. 45.10,** Tinbergen placed a ring of pine cones around a wasp's nest and, after she had flown out and returned a few times, moved the circle of pine cones to the side of the nest. The wasp flew past her own nest and went directly to the center of the ring. Such innate responses improve the survival or reproductive success of the individual.

This adaptive aspect of animal learning is also revealed by taste aversion experiments, in which an animal typically learns to avoid certain flavors associated with a negative outcome. Rats learn to avoid flavored water if consuming it is associated with nausea. However, they do *not* learn to avoid flavored water if it is associated with a different kind of negative reinforcement, such as a mild electric shock. Rats, it turns out, can make some associations, but not others. The ones they can learn are the biologically meaningful ones, those that favor survival. In the course of evolution, the rat's ancestors encountered poisoned food that resulted in nausea and the aversion response evolved. Until humans started doing experiments on them, rats had never encountered a bad meal that resulted in an electric shock. It is not surprising, then, that the ability to pair these two phenomena never evolved. These experiments also show that the specific

HOW DO WE KNOW?

FIG. 45.10

To what extent are insects capable of learning?

BACKGROUND European digger wasps, *Philanthus triangulum*, live in the sand. These wasps are sometimes called "bee wolves" because they specialize on hunting for honeybees to feed their developing young. After mating, each female digs a long burrow with a few chambers at the end where she lays her eggs. She then forages for honeybees that she brings back to these chambers for her larvae to eat. The wasp faces a navigational challenge: Having captured her prey, how can she find her way back to and recognize her nest? Niko Tinbergen noticed that wasps lingered briefly in the vicinity of a new nest before heading off to hunt, and thought that they were learning local landmarks associated with the nest.

HYPOTHESIS The wasp learns visual cues around her nest to help her locate it upon her return.

METHOD Tinbergen recognized that a good test of the learning abilities of an insect should take place in its natural environment. He combined his skills as a naturalist and as an experimentalist to devise an elegant demonstration of the way in which female wasps learn landmarks for navigation. Tinbergen's approach was simple. He placed a ring of pine cones around the nest of a wasp, and then, once she had left to hunt, he shifted them to a new location away from the nest entrance. If visual cues are key, the wasp would return to the displaced pine cone ring. On the other hand, if cues are, for example, olfactory, she would return directly to the nest.

RESULTS Females carried out a brief landmark-learning flight on departure from the nest (Fig. 45.10a). When the landmarks were displaced, the females returned to the wrong location (Fig. 45.10b).

CONCLUSION Female digger wasps learn and then use local landmarks, such as a ring of pine cones, as cues to the nest location.

SOURCE Tinbergen, N. 1958. *Curious Naturalists*. New York: Basic Books.

a.

b.

evolutionary history of the species in question matters when analyzing animal behavior.

In addition to adaptive predispositions for *what* can be learned and not learned, many species exhibit predispositions for *when* learning takes place. This is particularly evident in **imprinting**, a form of learning typically seen in young animals in which they acquire a certain behavior in response to key experiences during a critical period of development. Konrad Lorenz made imprinting famous by exploiting the observation that newly hatched goslings and ducklings rapidly learn to treat any animal they happen to see shortly after hatching as their mother. Lorenz found that, if he was the first person the hatchlings saw, they would follow him as though he were their mother (**Fig. 45.11**). This behavior is adaptive because the first being that a hatchling normally sees is a parent. This type of imprinting is called **filial imprinting**, and it is most common in species whose offspring leave the nest and walk around while still young, like chicks and ducklings. Filial imprinting is rare in species of birds whose young stay in the nest until they are able to fly away.

Experiments have shown that filial imprinting typically occurs during a specific, sensitive period in the animal's life and that the results are usually irreversible. After Lorenz's baby ducks had

FIG. 45.11 Imprinting on (a) Konrad Lorenz or (b) an ultralight aircraft. By manipulating newly hatched birds' first sight of a moving being, it is possible to induce powerful attachments in the birds.

imprinted on him, they would not change their minds about who their parent was even when presented with their real mother duck. The timing of the sensitive period varies from species to species. In some cliff-nesting sea birds, it has been shown that imprinting on auditory stimuli (like the call of the parents) begins while the chick is still in the egg.

45.4 ORIENTATION, NAVIGATION, AND BIOLOGICAL CLOCKS

Learning is a form of information processing, in which experience shapes behavior. At a mechanistic level, it typically involves a change in the strength of connections between neurons. The central nervous system of even relatively simple animals is capable of extraordinary feats of information processing. The ways in which animal nervous and endocrine systems and the input of experience are integrated to generate adaptive behaviors are remarkable. Some of these adaptive behaviors include ways of moving, navigating, and keeping time. Here, we look at how organisms integrate environmental stimuli to produce these behaviors.

Orientation involves a directed response to a stimulus.

Even the simplest bacteria and protozoa are capable of moving in response to stimuli. A *Paramecium* that finds itself in an unfavorable environment, such as water that is too warm or too salty, increases its speed and begins to make random turns. When it finds favorable conditions, such as cooler water, it slows and reduces its turning rate. These random, undirected movements are termed **kineses.**

In contrast, **taxes** (singular, taxis) are movements in a specific direction in response to a stimulus. An interesting example of a taxis is movement oriented to a magnetic field, a behavior called magnetotaxis—the term was first used by the American microbiologist Richard Blakemore in 1975. Blakemore had found that anaerobic bacteria in the genus *Aquaspirillum,* which swim by means of flagella, tend to swim toward magnetic north. These bacteria can be attracted to the side of a dish with a bar magnet. Little bits of magnetized iron oxide, arranged in a row inside the bacterial cell, allow the bacteria to sense the magnetic field. Blakemore hypothesized that the bacteria swim north in order to swim deeper. In the northern hemisphere, the north magnetic pole is inclined downward. At Woods Hole, Massachusetts, where these bacteria were found, the magnetic pole is at about a 70-degree incline. Since the bacteria are anaerobic, they must stay buried in sediment, and they remain buried by moving downward along the magnetic gradient.

To test his hypothesis, Blakemore looked for and found bacteria in New Zealand that exhibited similar behavior to the Woods Hole bacteria except that they swam toward the *south* magnetic pole, bringing them downward in the southern hemisphere. Finally, he took New Zealand cultures back to Woods Hole, where the bacteria swam up into the oxygen and died. These experiments demonstrated that bacteria are able to sense a magnetic field and move in a directed fashion relative to that field.

Navigation is illustrated by the remarkable ability of homing in birds.

Many animals use environmental cues to migrate long distances. The navigational achievements of homing pigeons are legendary. They can home—that is, find their way back to the place where they are housed and fed—over extremely long distances, even more than a thousand miles. It seems that they use a wealth of cues when homing. A compass may tell you which way is north, but you must also have **map information**—you must know where you are with respect to your goal—if the compass is to be useful for finding a particular location. So pigeons must have both compass and map senses. That map sense is presumably based on landmarks. Given the long distances the pigeons travel, these landmarks must vary in their type and probably include olfactory as well as visual cues.

The pigeon compass relies on different kinds of information. For example, pigeons can navigate during the day using the sun as a compass, and at night they use the stars. And, like *Aquaspirillum*, they can detect the Earth's magnetic field. The magnetic navigation system is presumably important to pigeons on cloudy days. Researchers have performed experiments with pigeons wearing magnetic helmets, which disrupt the pigeons' ability to detect the Earth's magnetic field. On cloudy days, the pigeons wearing the helmets are unable to home. On sunny days, however, the pigeons' sun compass overrides the erroneous magnetic information of the helmets.

The sun compass requires information about time as well. Every hour, the sun moves 15 degrees through the sky. To determine where north is, a pigeon needs information from the sun as well as some way to keep time. It turns out that pigeons have a clock—a biological one.

Biological clocks provide important time cues for many behaviors.

Like us, other animals live in space *and* time. So far, we have focused on their behavior in space. For many species, time is a life-and-death matter. When to migrate or mate is a critical decision in a seasonal environment. Researchers are beginning to unravel the neural and genetic underpinnings of the clocks in some species. In model organisms like *Drosophila*, for example, researchers have identified a number of genes that, when mutated, cause the clock to run slow or fast.

A biological clock is a set of molecular mechanisms that cycles on its own and therefore keeps a regular rhythm. It helps to control physiological or behavioral aspects of an animal. Different clocks work on different timescales. Daily cycles are governed by a **circadian clock.** Circadian clocks regulate many daily rhythms in animals, such as feeding, sleeping, hormone production, and core body temperature. Some species, especially seacoast species living in habitats where the tides are important, time activities by a **lunar** (moon-based) **clock.** There are also **annual** (yearly) **clocks**. For example, periodical cicadas are insects in the genus *Magicicada* that have a generation time of either 13 or 17 years: In a given location, there will be a cicada outbreak every 13 or 17 years.

Circadian clocks are observed in many organisms, including plants, fungi, and animals. Humans have their own circadian clocks, as jetlag never fails to remind us. The circadian clock is based on a set of clock genes. The protein products of these genes oscillate through a series of feedback loops in a roughly 24-hour cycle. Thus, when animals that, like us, are active during the day and inactive at night are placed in artificial conditions that are always lit, they continue to follow a basic day–night, active–inactive cycle. The clocks are not perfect, though: As the period spent in constant light is prolonged, the circadian clock drifts slowly until the animal is eventually no longer synchronized with the true day–night cycle.

Biological clocks remain synchronized with the day–night cycle because they are often reset, or entrained, by external inputs. For the circadian clock, light is the primary input, so the natural light–dark cycle keeps the clock from drifting. For clocks related to the seasons, day length (known as **photoperiod**) is the critical input because it is a good indicator of the time of the year (Chapter 30). For example, many mammals produce their offspring in the spring so that they can grow over the

HOW DO WE KNOW?

FIG. 45.12

Does a biological clock play a role in birds' ability to orient?

BACKGROUND One suggestion for how pigeons home is that they use a sun compass. If you are in the northern hemisphere and you know the time is 12 noon, then the sun is due south of you. Orienting yourself by this method is possible only if you know the time, so the question arises whether homing birds have the ability to tell the time. One way to answer this question is to "clockshift" the birds. Researchers raise birds in an artificial day–night cycle that is out of sync with the actual one. When released into a sunlit environment, these birds' sense of time is shifted by a set number of hours:

event	dawn	dusk
actual time	6 am	6 pm
clockshifted time	12 noon	12 midnight

HYPOTHESIS If a bird's ability to home is dependent on an internal clock, clockshifting should affect the bird's homing ability in a predictable way. Given that the sun travels 360 degrees in 24 hours, a 6-hour clockshift will result in a 90-degree error in homing direction because $360/(24/6) = 90$.

summer, when resources are most abundant. These mammals have been selected over many generations to synchronize their reproduction at a time when the young have the greatest chance of survival. Photoperiod determines the timing of many kinds of behaviors, including migration, development, and reproduction.

The importance of the sun compass coupled with the biological clock in homing pigeons can be demonstrated by experimentally disturbing the birds' clock, as discussed in **Fig. 45.12**.

→ **Quick Check 3** Why is photoperiod such a widely used time cue?

45.5 COMMUNICATION

Up to this point, we have focused on individual behaviors. Behaviors often depend on environmental cues, as we saw with the examples of navigation and keeping time. In addition, behaviors are often shaped by interactions with other individuals. Learning sometimes depends on communication between a teacher and student, and, as anyone knows who has ever watched a mother duck tending her ducklings, communication is central to filial imprinting.

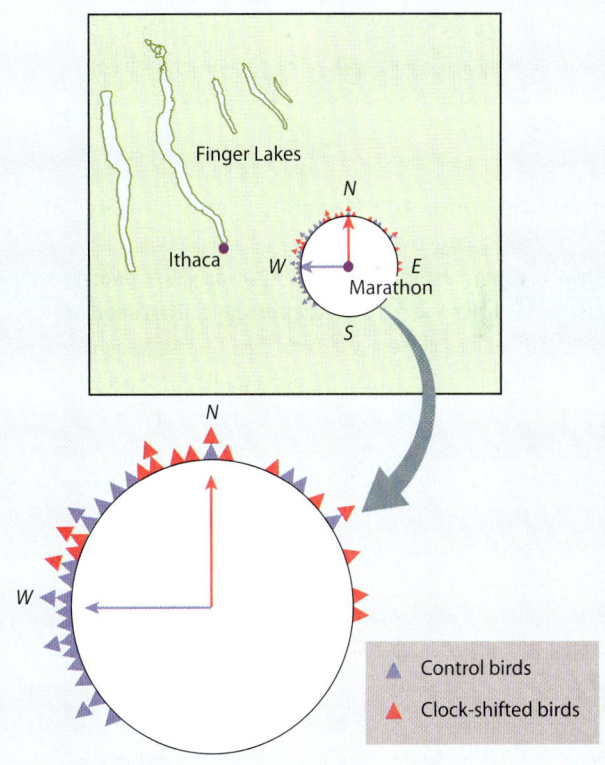

EXPERIMENT Birds were clockshifted by raising them in a chamber under an artificial light. Birds from a home loft in Ithaca, New York, were released on a sunny day at Marathon, New York, about 30 km east of Ithaca. Release on a sunny day made it possible for the birds to use the sun's position to navigate.

RESULTS As expected, the control birds (those that were not clockshifted) were usually good at picking the direction of their home loft, heading approximately westward toward Ithaca. The results for the clockshifted birds were very different: They miscalculated the appropriate direction. These birds headed approximately northward, as shown by the positions of the red triangles on the compass in the figure.

INTERPRETATION Assume the birds are released at 12 noon, when the sun is due south. The control birds know to fly in a direction 90 degrees clockwise from the direction of the sun, but the clockshifted birds "think" it is 6 p.m., so they expect the sun to be in the west. Their 90-degree clockwise correction, then, has them flying due north.

CONCLUSION The clear difference between control and clockshifted birds in the experiment shows that an internal time-based sun compass is an important component of the birds' homing abilities. However, the scatter of points (for both experiment and control) suggests that other factors are also important. This conclusion is reinforced by the observation that birds home well on cloudy days, when they cannot use a sun compass, suggesting that birds use multiple cues and navigational systems when they are homing.

SOURCE Keeton, W. T. 1969. "Orientation by Pigeons: Is the Sun Necessary?" *Science* 165:922–928.

FIG. 45.13 Types of signal in different forms of communications. Signals can be (a) visual, as in the male frigatebird's display, (b) auditory, such as a male frog's call, (c) electrical, as in African electric fish, which generate species-specific electric pulses, (d) chemical, such as the pheromones with which a queen ant controls her colony, and (e) mechanical, as when spiders "twang" webs to interact.

Communication is the transfer of information between a sender and receiver.

The sophistication of animal communication varies enormously, as can be seen by the variety of types of signal illustrated in **Fig. 45.13.** Even our closest relatives, the chimpanzees, cannot rival the human facility with language, but a number of species have evolved forms of communication that convey remarkably complex and specific information. For example, a vervet monkey that perceives a threat to its group will utter an alarm call that not only warns of a threat, but actually specifies the nature of the threat—whether it is a hawk, leopard, snake, or some other type of predator.

The simplest definition of **communication** is the transfer of information between two individuals, the **sender** and the **receiver.** The sender supplies a signal that elicits a response from the receiver. For example, the bright petals of a flower signal to an insect that nectar and pollen are available. This definition, however, has its problems. An owl hears the rustling of a mouse and responds accordingly. Most people would agree that the owl is not really communicating with the mouse (nor the mouse with the owl). For this reason, some biologists prefer to define communication as attempts by the sender to manipulate in some way the behavior of the receiver.

How has communication evolved? It is thought that communication has often evolved through co-opting and modifying behaviors used in another context. This process is called **ritualization,** and it involves (1) increasing the conspicuousness of the behavior; (2) reducing the amount of variation in the behavior so that it can be immediately recognized; and (3) increasing its separation from the original function. The scent markings with which mammals mark their territory provide an example. The original function was simply the elimination of waste, but this function has been modified through evolution: Strategically placed marks communicate with other individuals, indicating, for example, the extent of an individual's territory. The communication advantage is that an intruder detecting a territorial scent may be less likely to invade, thereby avoiding what could be an expensive fight.

Forms of communication have frequently evolved that prevent animals from coming to harm in what has sometimes been called limited-war strategies. Thus, two males, instead of battling it out, may engage in elaborate displays to size each other up, either literally by standing side by side or through displays of physical prowess such as roaring, in order to determine who is dominant. Fighting would be disadvantageous to both since even the winner might be seriously injured. In this case, communication may have evolved as a way for individuals to assess each other.

In some cases, communication in the natural world can be deceitful. A male may attempt to convince other males (or a female) that he is bigger (and stronger) than he really is. Possibly one reason that dogs circling in a fight raise the hair along their back is to inflate their apparent size. A potential prey may attempt to convince a predator that it is not, in fact, prey. This form of deceit is especially evident in some species of butterfly that mimic leaves or bark when their wings are at rest. Alternatively, predators may emit deceitful signals to entice their prey. Females of one species of firefly, for instance, mimic the flashes of the females of another species in order to lure males of the second species, which they then eat.

Some forms of communication are complex and learned during a sensitive period.

Bird song is one of the best known and richest forms of animal communication. Because of the clear connection between sensory perception and motor output involved in this form of communication, songbirds have served as a model system in the study of learning and communication. These songs are complex sequences of sounds, often repeated over and over. Like cricket and frog calls, bird songs are **advertisement displays,** behaviors by which individuals draw attention to their status (for example, that they are sexually available or are holding territory). Bird songs are typically produced in the breeding season and usually only by males, although in some species females also sing and perform "duets" with their mates.

In all species of birds that sing, some or all of the song is learned, often during a specific, sensitive period. Detailed studies pioneered by British-American neurobiologist and ethologist Peter Marler of the White-Crowned Sparrow, shown in **Fig. 45.14,** have yielded the following general picture of the process.

Song in White-Crowned Sparrows is learned by imprinting: The young male hears adult song during a sensitive period, 10–50 days after hatching. Then, shortly afterward, young male White-Crowned Sparrows produce unstructured twittering sounds, known as a subsong, comparable to the babbling of human babies. If deprived of hearing adult song, for example by isolation during the sensitive period, the bird sings for the rest of his life a song not much different from unstructured twittering, even if he hears his father sing both before and after isolation.

Between about 250 and 300 days, the male sings an imperfect copy of his father's song (known as plastic song), and by about 300 to 350 days after hatching, song acquisition is complete (this song is known as structured song). At this point, even if the bird is deafened in the lab, he will still sing correctly. The song he produces is a precise copy of the one he learned, typically by hearing his father's song during the sensitive period, complete with most of its individual as well as species-specific characteristics.

So, White-Crowned Sparrows have a programmed predisposition for *when* song learning takes place. Even more interestingly, *what* can be learned is similarly constrained. If a tape of another species' song is played during the sensitive period, even the song of closely related Swamp Sparrows, the White-Crowned male cannot learn that song; his adult song ends up being not much different from his unstructured twittering. However, if he is provided with a live tutor, such as a live male Swamp Sparrow rather than a tape, his ability to learn the other bird's songs is greatly improved. Furthermore, if he hears tapes of both the song of his own species and that of closely related species played together, he can pick out the correct elements and sing a perfect White-Crowned Sparrow song. Thus, birds preferentially learn their species-specific song, but they cannot sing without learning it.

Other forms of communication convey specific information.

Honeybees have an elaborate means of communication that is quite different from the sound-based communication of birds. To collect food for a honeybee hive, about 10,000 foragers leave the nest each day, making several million foraging trips in the course of a summer, each of which may take them up to 10 km from the nest. The hive can be considered a kind of giant organism, fixed in one spot but sending out foragers in all directions. As social insect expert and evolutionary biologist E. O. Wilson points out, if honeybee workers were as large as people, they would cover an area the size of Texas in their daily search for food. Given the

---

FIG. 45.14 Song acquisition in the White-Crowned Sparrow. Sonograms show differences between the unstructured twittering of early life and the final, structured song.

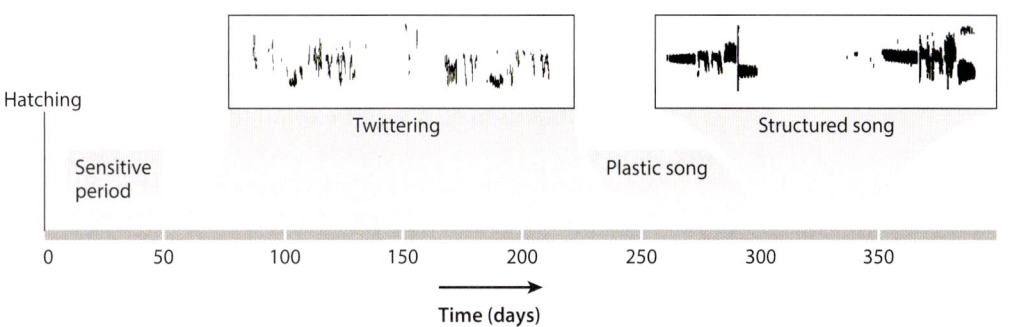

vastness of this task, it is not surprising that returning workers use a variety of cues to inform outgoing foragers about the location and nature of the food resources they have found.

Returning workers pass samples of the collected food to other workers that gather around the forager when she returns to the hive. The returning foragers then begin a series of movements in specific patterns that encode information about the direction and distance of food sources. This is the dance language of honeybees, originally discovered by Karl von Frisch.

If the food source is within about 50 m of the hive, a simple dance called the round dance—rapid movement in little circles, in both clockwise and counterclockwise directions—is sufficient. Outgoing foragers then know that the source is close and they search at random in the vicinity of the hive until they find the source.

If the food source is farther away, the incoming forager does a more complex dance, the waggle dance. The steps of the dance are illustrated in **Fig. 45.15**. The forager moves quickly back and forth ("waggles") as it moves forward, circles back and repeats the waggle, then circles back the other way, and so on. This dance usually goes on for several minutes and may take much longer. The distance to the food source is conveyed by the length of time it takes to waggle up the middle of the circle. The direction, with reference to the position of the sun, is conveyed by the angle of the line of the waggle run: If the food source is in a direct line with the sun, the dancer's waggle line points straight up; if food is at an angle of 45 degrees to the left of the sun, the dancer's line is 45 degrees to the left of vertical. Short of human language, the bees' dance language is among the natural world's most extraordinary and effective methods of communication.

45.6 SOCIAL BEHAVIOR

Returning to Tinbergen's questions, we can consider further the adaptive function and evolutionary history of animal behavior. The evolutionary advantages of most behaviors are readily apparent. The imprinting of Lorenz's goslings on the first moving object they encountered makes sense when you consider that that object is usually their parent (not Konrad Lorenz!) and that sticking close to their parent is an excellent survival strategy in a world filled with predators.

Consider cooperative hunting seen in killer whales. Killer whales participate in a behavior called wave washing, in which a pod of whales collectively creates waves to wash a prey species—typically a seal—off an iceberg (**Fig. 45.16**). This behavior is collaborative, and it is clear why each individual participates. By itself, a single killer whale is incapable of generating a big enough wave, but, through cooperation, an individual can contribute to prey capture and gain some portion of the prey.

Sometimes, however, how a behavior enhances survival and reproduction is less clear. Perhaps the most prominent examples of this kind of behavior are sacrifices made by individuals apparently for the good of others. Formally, we describe an act of self-sacrifice as **altruistic**. Such acts seem to fly in the face of the natural selection, which posits that every individual competes for resources and mates to maximize its genetic contribution to the next generation. This contradiction is particularly stark in the case of species like honeybees: Honeybee workers, which are females, do not reproduce but instead assist the reproduction of the queen. Darwin confessed that such behaviors caused him sleepless nights. How, Darwin wondered, could altruism arise?

Group selection is a weak explanation of altruistic behavior.

Darwin himself suggested a solution to the altruism problem, a solution that today we call **group selection**. The idea is that natural selection operating on individuals is a less powerful force than another form of selection that operates on groups. Consider two groups, one with individuals that are altruistic, helping

FIG. 45.15 The waggle dance of honeybees. This form of communication provides information to other foragers about the distance and direction of a food source.

| The distance to the food source is communicated by the length of time it takes to move up the middle of the circle (the "waggle"). | The direction of the food source relative to the sun is communicated by the angle of the waggle run relative to vertical. |

FIG. 45.16 Mutually beneficial cooperation. Killer whales cooperate to create a wave to wash prey off an iceberg.

one another out, and another with individuals that act entirely selfishly. Darwin argued that, in a conflict between the two groups, the first group would beat the second because the first group works better as a unit. Here, then, we see that selection in favor of altruism that benefits the group has trumped selection in favor of selfish behavior.

Although an attractive idea, group selection is probably not important in the evolution of altruism. Altruism under group selection is typically not an **evolutionarily stable strategy,** meaning that this kind of behavior can be readily driven to extinction by an alternative strategy. It used to be thought that lemmings, a species of Scandinavian rodent, would collectively kill themselves by plunging into the ocean when the population became too large to sustain. It was claimed that the lemmings were altruistic: Their self-sacrificial act would ensure that the population did not exhaust available resources, preventing a population crash.

This is a nice story, but untrue. Lemmings do not kill themselves for the good of the species. An individual that "cheats" by not performing the altruistic act will survive and breed, and its altruistic companions will not. Now consider that this behavior—to kill oneself or not—is genetically encoded. A mutation that arises in a population that causes the individual to "cheat" and not kill itself will spread rapidly through the population by natural selection until, after many generations, it has completely replaced the altruistic allele that causes the individual to kill itself. The lemming behavior, then, is not an evolutionarily stable strategy because natural selection favors an alternative strategy.

This description is oversimplified because there is unlikely to be a simple genetic basis to traits such as these, but the point stands: Group selection is in general not an evolutionarily stable strategy because it can readily be overthrown by "selfish" strategies.

→ **Quick Check 4** It's sometimes said that the reason an animal does something is that it's "for the good of the species." Why is this argument incorrect?

Reciprocal altruism is one way that altruism can evolve.

If group selection does not provide a strong explanation for the evolution of altruism, how can we explain this behavior? **Reciprocal altruism,** whereby individuals exchange favors, is one way that altruism can evolve. This idea is perhaps best summarized as "You scratch my back, and I'll scratch yours." For this kind of behavior to work, individuals have to be able to recognize each other and remember previous interactions. If I did you a favor, I need to be able to recall who you are so that I can be sure that I get the return I expect.

A famous example of reciprocal altruism comes from an unlikely source—vampire bats. Adult female vampire bats live in groups of 8 to 12. Individuals have been recorded as living as long as 18 years, and in one study, two tagged females shared the same roost for 12 years. As is the case for most mammals, young males disperse from their natal group, but females remain. Thus, the female groups are composed partly of kin, although not entirely, since they are occasionally joined by unrelated females. In addition, because they have long-term associations, there is opportunity for reciprocal interactions.

Feeding for these bats is a "boom or bust" activity—either a bat has found and successfully fed on a blood source (boom), or a bat has failed to feed (bust). As a result, bats are often either overfed or underfed. Underfed bats, in turn, are at risk of dying by starvation. Bats returning from a successful feeding expedition often regurgitate blood to unsuccessful ones, on the expectation that the favor will be returned at a future date.

Kin selection is based on the idea that it is possible to contribute genetically to future generations by helping close relatives.

Only in the 1960s did we come to understand that the relatedness of participating individuals has played an important role in the evolution of altruism. This understanding lay in the insight that there are at least two ways to contribute genetically to the next generation: Either an individual can produce its own offspring, or it can produce no offspring and help close relatives reproduce instead. The latter strategy can become established in a population through a process called **kin selection,** a form of natural selection that favors the spread of alleles that promote behaviors that help close relatives, or kin.

What relationships qualify as kin? Consider an animal (gray shading in the figure) that has an allele A at a particular locus (**Fig. 45.17**). If the animal reproduces, the probability of each offspring inheriting an A allele is 0.5. But what if the animal does not reproduce? According to Mendel's laws, there is a 50% chance that the animal's parents passed on the same A allele to the animal's sibling (Chapter 16). If the sibling does, in fact, inherit allele A, the probability that it passes allele A to each of its offspring is also 0.5. Given the two probabilities (that the sibling has an A, and that the allele passed on to an offspring), the probability that the animal's niece or nephew has A is 0.25. In other words, a diploid animal (such as humans) is twice as closely related to its own offspring than to its nephew or niece. Therefore, if the animal does not reproduce in order to help its sibling have two offspring, the animal will make the same net genetic contribution to the next generation as it would having one of its own offspring.

→ **Quick Check 5** How many cousins are "genetically equivalent" to a single offspring?

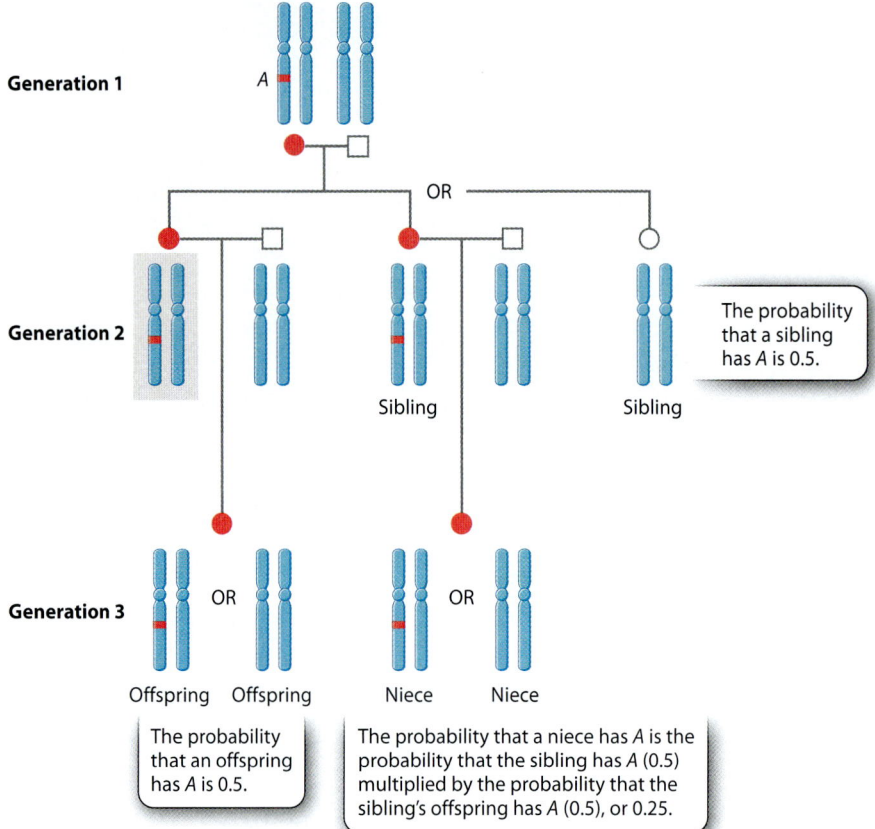

FIG. 45.17 Computing relatedness. One's relatedness (r) to another individual is the probability that one shares, through inheritance, a given gene with that individual. Here, relatedness is calculated with reference to the individual indicated by the gray background.

This logic was formalized in 1964 in a famous paper by British evolutionary biologist William D. Hamilton in terms of three key variables: B, the benefit of the behavior to the recipient (how many individuals does the behavior save?); C, the cost of the behavior to the donor (how many offspring are lost?); and r, the degree of relatedness between the recipient and donor. Hamilton asserted that as long as rB exceeds C, altruism can evolve.

How in practice does kin selection work? Let's start with an extreme case of relatedness: clonal organisms that are genetically identical to each other (meaning that the probability of a sibling having allele A is 1). Many aphid species have a phase of their life cycle in which a "mother" aphid reproduces parthenogenetically to produce a small colony of clones of herself (Chapter 42). Any aphid in the colony is capable of reproducing in this way, but in a case of reproductive altruism, some individuals do not reproduce at all. They are "soldier" aphids, who defend the colony against predators, typically other insects. As specialized defenders, they do not have the ability to reproduce. In this case, the Hamilton calculation is simple. By helping their clone-mates survive, the sterile soldier aphids are ensuring that their own genes survive to be passed on to the next generation.

The most famous cases of reproductive altruism occur in the insect group Hymenoptera, which includes ants, bees, and wasps. Many species of Hymenoptera are **eusocial,** meaning that they have overlapping generations in a nest, cooperative care of the young, and clear and consistent division of labor between reproducers (the queen of a honeybee colony) and nonreproducers (the workers). Often called social insects, these species are one of evolution's most extraordinary success stories.

What is it about Hymenoptera that predisposed them to eusociality? E. O. Wilson argues that natural selection has strongly favored group behavior in these insects. They typically receive group benefits from living in and defending a nest. For example, they can easily communicate with one another about the location of food resources such as nectar.

William Hamilton also noted that most Hymenoptera have an unusual mode of sex determination: Females (the queen and workers) are diploid, whereas males (drones) are haploid, produced from unfertilized eggs. As a result, the degree of relatedness, r, is higher for the sister–sister relationship (0.75) than for the mother–daughter one (0.5), as shown in **Fig. 45.18.** Put another way, a female is more closely related to her sister than she is to her daughter. Therefore, it makes more evolutionary sense, in terms of genetic representation in the next generation, for a worker to help her mother, the queen, to produce more sisters than for her to produce her own offspring.

This form of social organization can result in a huge colony of related individuals all working collectively, much as different cells work collectively within a body. Often, we see specialization into castes: Some ants are soldiers, defending the colony, others are foragers, and so on. If we extend the analogy of individuals acting like cells in a body, the multiple castes are equivalent to organs, all coordinating and working together. The term "superorganism" is sometimes used to describe a single social insect colony.

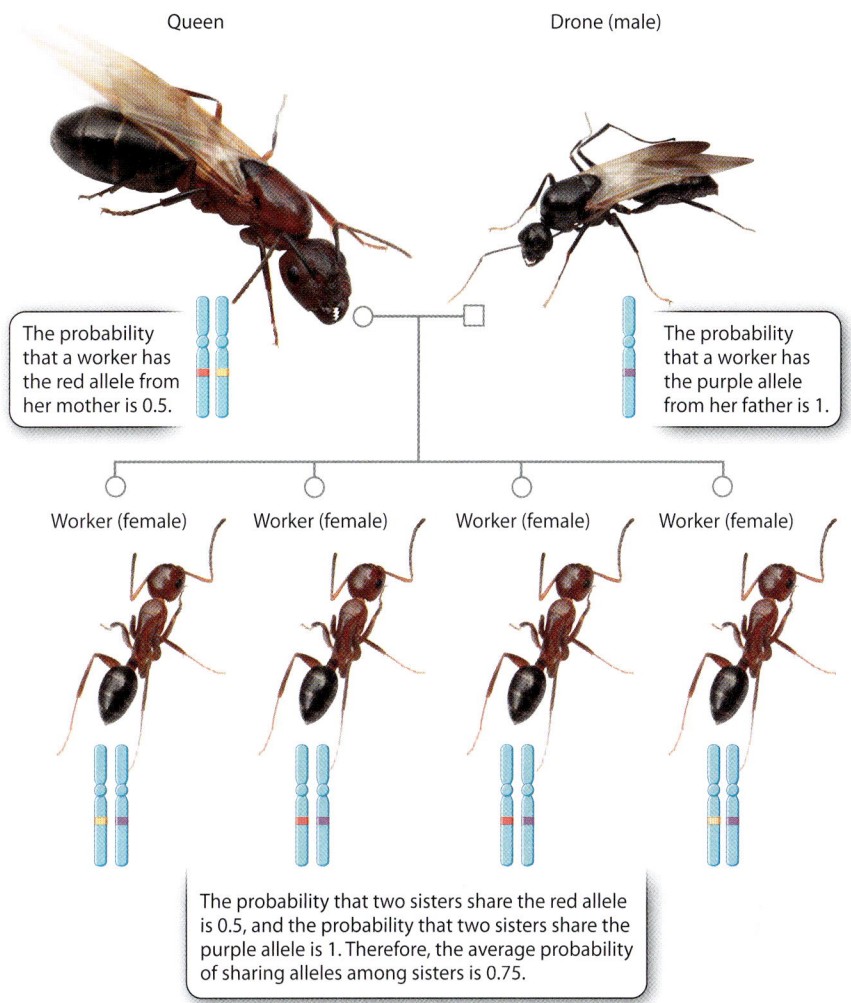

FIG. 45.18 Relatedness in haplodiploid insects, such as ants. Sisters are more related to each other (0.75) than to their offspring (0.5).

The probability that a worker has the red allele from her mother is 0.5.

The probability that a worker has the purple allele from her father is 1.

The probability that two sisters share the red allele is 0.5, and the probability that two sisters share the purple allele is 1. Therefore, the average probability of sharing alleles among sisters is 0.75.

45.7 BEHAVIOR AND SEXUAL SELECTION

Throughout this chapter, we have discussed the development of certain behaviors in terms of natural selection. In many species, behavior is also affected by **sexual selection,** in which traits evolve that increase the probability of finding and attracting mates (Chapter 21). Traits favored by sexual selection sometimes appear to run counter to those favored by natural selection. For example, a peacock's tail is a compromise between sexual selection, which acts to increase the size and color of the display, and natural selection, which acts to minimize the amount a male invests in its cumbersome and metabolically expensive tail.

Patterns of sexual selection are governed by differences between the sexes in their investment in offspring.

In peacocks, it is the male that has the showy tail, not the female. Male birds often have bright colors, complex songs, and elaborate displays, and females usually do not. Differences in color or size between males and females are examples of **sexual dimorphisms,** or phenotypic differences between the sexes (**Fig. 45.19**).

Why do we see such differences between males and females? The key is the level of investment that an individual makes in an offspring. Consider a typical rodent species in which a male mates with a female and then moves on in search of additional mates. The female invests time, energy, and resources in the offspring, both during gestation and, after birth, during lactation. The male, on the other hand, has invested one ejaculate, no more.

In many species, eggs are fertilized externally and neither males nor females care for offspring. But, even in these species, there are differences in parental investment between males and females. Eggs are larger than sperm, and it therefore takes more energy and resources to make eggs than to make sperm. As a result, females still invest more in offspring than males do in these species.

These differences in parental investment lead to different selection pressures on females and on males. The reproductive success of females (that is, her number of offspring) is limited by the number of eggs she can produce or the number of pregnancies she can carry. By contrast, the reproductive success of males is limited by the number of mates. As a result, any trait that increases a male's ability to attract a female is favored by sexual selection and will evolve over time.

Therefore, sexual selection tends to act more strongly on males than on females, leading to the evolution of the bright colors and striking displays we often associate with male birds. Tellingly, in species in which this typical pattern of parental investment is reversed (the male invests more in the offspring than the female), we see a corresponding reversal of sexual selection (females that are more colorful than males). For example, in phalaropes and sandpipers, males incubate the eggs and females are brightly colored.

Sexual selection can be intrasexual or intersexual.

In *On the Origin of Species*, Darwin recognized two forms of sexual selection (**Fig. 45.20**), and later, in *The Descent of Man* and *Selection in Relation to Sex*, he developed his ideas more fully. In one form, members of one sex (usually the males) compete with one another for access to the other sex (usually the females). This form is called **intrasexual selection** since it focuses on interactions between individuals of one sex (Fig. 45.20a). Because competition typically occurs among males, it is in males that we see physical traits such as large size and horns and other elaborate weaponry, as well as behaviors such as fighting. Larger, more powerful males tend to win more fights, hold larger territories, and have access to more females.

Darwin also recognized a second form of sexual selection. Here, males (typically) do not fight with one another, but instead compete for the attention of the female with bright colors or advertisement displays. In this case, females choose their mates. This form of selection is called **intersexual selection,** since it focuses on interactions between females and males (Fig 45.20b). The peacock's tail is thought to be the product of intersexual selection: Its evolution has been driven by a female preference for ever-showier tails.

FIG. 45.19 Sexual dimorphism, or phenotypic differences between the sexes. Males (on the left in each photograph) look different from females, as shown by (a) lions, (b) cardinals, and (c) stag beetles.

FIG. 45.20 Sexual selection. (a) Intrasexual selection often involves competition between males, as shown by this battle between two male Elk. (b) Intersexual selection often involves bright colors and displays by males to attract females, as shown by this male and female Japanese Red-crowned Crane.

Often, these eye-catching ornaments are associated with elaborate behaviors on the part of the male. Male sage grouse, for example, congregate and put on displays in the mating season so that females can select their mates. Male bowerbirds accumulate shiny, bright objects in carefully tended displays to attract females (see Fig. 42.4b).

In general, therefore, females have evolved behaviors to ensure the quality of their mate (they tend to be choosy), whereas males have evolved behaviors to maximize the number of matings they have (they tend to be competitive). However, as we saw earlier, when the male invests more in the offspring than do the females, there is a reversal of these patterns so that males are choosy and females competitive. In seahorses, for example, males brood the offspring in a specialized pouch and behave in the choosy way we see in females of other species. It is also important to keep in mind that other factors can influence and modify these general behavioral patterns. Nevertheless, sexually selected traits remind us of the power of selection in shaping all kinds of behaviors.

Core Concepts Summary

45.1 FOR ANY BEHAVIOR, WE CAN ASK WHAT CAUSES IT, HOW IT DEVELOPS, WHAT ADAPTIVE FUNCTION IT SERVES, AND HOW IT EVOLVED.

Tinbergen pointed out that questions about animal behavior can be asked at four different levels of analysis, focusing on cause, development, function, and evolution. page 45-1

The four questions have different answers, and all may be correct explanations of the behavior. page 45-2

45.2 ANIMAL BEHAVIOR IS SHAPED IN PART BY GENES ACTING THROUGH THE NERVOUS AND ENDOCRINE SYSTEMS.

Behavior can be innate, meaning that it is carried out in the absence of experience, or learned, meaning that it depends on experience. page 45-2

A fixed action pattern is a stereotyped sequence of behaviors that, once initiated by a key stimulus, always goes to completion. page 45-2

Stimuli are recognized by feature detectors, which are specialized sensory receptors of the nervous system. page 45-4

Hormones can affect behavior. page 45-4

The extent to which genes influence a behavior can be determined by crossing closely related species having different behaviors and analyzing the behavior of the offspring. page 45-5

Most behaviors are influenced by many genes, but a few are strongly influenced by a single gene. page 45-6

Molecular studies are providing new ways to understand the role of genes in behavior in laboratory animals. page 45-6

45.3 LEARNING IS A CHANGE OF BEHAVIOR AS A RESULT OF EXPERIENCE.

Non-associative learning is a form of learning that occurs without linking two events. page 45-8

Habituation, a form of non-associative learning, is the reduction or elimination of a response to a repeated stimulus. page 45-8

Sensitization, another form of non-associative learning, is the enhancement of a response to a stimulus that is achieved by first presenting a novel stimulus. page 45-9

Associative learning occurs when an animal links two separate events. page 45-9

Classical conditioning is a form of associative learning that occurs when two stimuli are paired, and a novel association is made between a formerly neutral stimulus and a behavior. page 45-9

Operant conditioning is a form of associative learning that occurs when a behavior is rewarded or punished, making the behavior more or less likely to occur, respectively. page 45-9

Imitation and imprinting are also forms of learning. page 45-9

45.4 ORIENTATION, NAVIGATION, AND BIOLOGICAL CLOCKS ALL REQUIRE INFORMATION PROCESSING.

Kineses are undirected movements, whereas taxes are directed movements. page 45-11

Long-distance navigation may require the processing of many external cues, including the position of the sun, the Earth's magnetic field, and information about time. page 45-11

Biological clocks control some physiological and behavioral aspects of animals. page 45-12

45.5 COMMUNICATION INVOLVES AN INTERACTION BETWEEN A SENDER AND A RECEIVER.

Communication is an interaction between two individuals, the sender and the receiver, and often involves attempts by the sender to manipulate the behavior of the receiver. page 45-14

Communication can evolve by the modification of noncommunicative aspects of an animal's behavior. page 45-14

Bird song is a rich form of communication that is often learned during a sensitive period. Acquisition of bird song proceeds through distinct stages. page 45-15

The dance language of honeybees communicates information about the location and direction of a food source. page 45-15

45.6 SOCIAL BEHAVIOR IS SHAPED BY NATURAL SELECTION.

One of the most perplexing behaviors from the standpoint of natural selection is altruism, in which an organism reduces its own fitness by acts that benefit others. page 45-16

Group selection is the idea that altruistic behaviors can evolve if they benefit the group, not the individual, but this idea is not widely accepted. page 45-16

Reciprocal altruism requires that individuals interact regularly and remember each other so that they can "pay back" past favors. page 45-17

Kin selection is the concept that altruistic behaviors can evolve if they benefit close relatives. page 45-18

45.7 SOME BEHAVIORS ARE INFLUENCED BY SEXUAL SELECTION.

In most animal species, females invest more in offspring than do males. page 45-20

Differences in parental investment lead to different selection pressures on males and females. Typically, the reproductive success of males depends on the number of mates, so sexual selection favors traits in males that increase the probability of finding and attracting females. page 45-20

Intrasexual selection involves competition among individuals of one sex (usually the males) for access to the other. page 45-20

Intersexual selection involves bright colors, ornaments, songs, and other advertisements usually by males to attract females to mate, and accounts for many of the spectacular displays we see among animals. page 45-20

Self-Assessment

1. Choose a behavior described in this chapter and ask the four kinds of questions that Tinbergen might have asked about it.
2. Provide an example of an innate and a learned behavior.
3. Describe two approaches used to determine the extent to which genes influence a particular behavior.
4. Give two examples of non-associative learning.
5. Give two examples of associative learning.
6. Differentiate between a kinesis and a taxis.
7. Explain why many biologists would not consider any transfer of information between a sender and receiver to be a form of communication.
8. Describe how an altruistic behavior might evolve by kin selection.
9. Differentiate between intra- and intersexual selection, and give one example of each.

Do you understand the chapter's Core Concepts? Log in to check your answers to the Self-Assessment questions, then practice what you've learned and reinforce this chapter's concepts by working through the problems and multimedia tutorials provided there.

www.biologyhowlifeworks.com

CHAPTER 46

POPULATION ECOLOGY

Core Concepts

46.1 A population consists of all the individuals of a given species that live and reproduce in a particular place and is characterized by its size, range, and density.

46.2 The age structure of a population helps ecologists understand past and predict future changes in population size.

46.3 The dynamics of populations are influenced by the colonization and extinction of smaller, interconnected populations that make up a metapopulation.

In 2004, undergraduates on a field trip to the Dominican Republic, on the island of Hispaniola, swept up an unfamiliar black-and-white butterfly in their nets. That butterfly was a Lime Swallowtail (*Papilio demoleus*), the very first of its species, native to Asia, to be captured in the New World.

Soon, Lime Swallowtails would spread to all corners of Hispaniola, a serious concern because their caterpillars strip the leaves off young lime and orange trees. Within 3 years, the Lime Swallowtail had also established itself in nearby Puerto Rico and Jamaica, and it has just reached Cuba. From there, it is just a short flight to Florida and the multibillion-dollar citrus industry of the United States. Can we enlist basic principles of biology to help understand this butterfly's Caribbean expansion and predict its future as it continues to spread?

Ecology is the study of the relationships of organisms to one another and to the environment. Ecological relationships sustain a flow of energy and materials from the sun, Earth, and atmosphere through organisms, with carbon and other elements eventually returning to the environment (Chapter 25). And, in the end, it is these interactions that determine the shifting ranges of Lime Swallowtails and other species on Earth.

In previous chapters, we discussed the anatomical, physiological, reproductive, and behavioral characteristics of organisms. These attributes are the products of evolution, and they determine the ways in which individual organisms interact with their physical and biological environment. Abiotic factors that influence evolution include climate and nutrient availability; biotic factors include competitors, predators, parasites, and prey, as well as organisms that provide shelter or food. Ecology and evolution, then, are closely intertwined.

46.1 POPULATIONS AND THEIR PROPERTIES

When the first United States census was completed in 1790, the British scholar Thomas Malthus supposed that the remarkable recent growth of the American population, which had doubled in the previous 25 years, was a consequence of ample food supply and the active encouragement of marriages, neither of which characterized Europe at the time. In his 1798 volume *An Essay on the Principle of Population,* Malthus argued that a similar periodic doubling of population in the British Isles, on the other hand, would outstrip the more slowly increasing food supply in 50 years and lead to starvation. That is, increasing the number of births per year would eventually result in an increased number of deaths because the amount of food the land could produce was limited. This view of the human condition would influence the thinking of Charles Darwin, who as a young, inquisitive student read the most exciting works of the day. Darwin realized that the tension between population growth and resource limitation applies to all species, leading to a struggle for existence that results in adaptations that enhance survival (Chapter 21).

In Chapter 21, we introduced the concept of population as a fundamental unit of evolution—it is populations, and not individuals, that evolve. Here, we return to populations, this time as equally fundamental units of ecology. A **population** consists of all the individuals of a given species that live and reproduce in a particular place. Natural selection enables populations to adapt to the physical and biological components of their environment. It is the traits evolved under natural selection that determine whether a population will succeed or fail under a given set of physical and biological conditions—whether it will grow and expand its range or shrink, perhaps disappearing entirely from a local environment.

What controls changes in population size? We first look at what happens when there are more births than deaths. We then examine what happens when resources such as food start to run out. We consider how individuals of different ages contribute to population growth, and finally, in the next chapter, we introduce interactions with other species and see how they shape the growth of a population. All these factors directly influence the flow of energy and materials through producers and consumers and, ultimately, the flow of genes through a population.

Three key features of a population are its size, range, and density.

Populations are defined by three key features: size, range, and density (**Fig. 46.1**). **Population size** is simply the number of individuals of all ages alive at a particular time in a particular place. For example, the population of krill in the cold waters around Antarctica is estimated to be 800 trillion, making them by mass the most abundant animals on Earth (Fig. 46.1a).

The size of a population may increase, decrease, or stay the same over time, and trends in population size are a key focus of ecological research and its application to conservation biology. A decrease in size through time, for example, might signal declining health in a local population—a matter of practical concern when the population in question is a commercially harvested fish or an endangered species. In the 1990s, catastrophic declines in cod populations led to a ban on commercial cod fishing on the Grand Banks of Newfoundland. The decline affected both cod and cod fishers, and it also led to changes in Grand Banks ecosystems—crab and lobster populations have increased as predation pressure by cod has declined.

A second key feature of a population is its **geographic range:** how widely a population is spread out. Fig. 46.1b, for example, shows that the krill population ranges through an area of 19 million square kilometers around Antarctica. The geographic range of any species determines how much variation in climate a population can tolerate and how many other species the population encounters. The geographic range of a plant population, for example, influences how many kinds of insect eat it, and the size of the range of an insect determines how many kinds of plant it encounters.

Taken together, population size and range contribute to a third key attribute of a population, its density: how crowded or dispersed the individuals are that make up the population. **Population density** is defined as a population's size divided by

its range. For krill, 800 trillion animals are spread over 19 million square kilometers—a density of 42 million animals per square kilometer (Fig. 46.1c). As we will see, population density can affect population size.

FIG. 46.1 (a) Size, (b) range, and (c) density of a population. The places where populations occur can be mapped, and the numbers of individuals within those populations can be estimated. This information helps us understand what factors limit population distributions.

a. Size

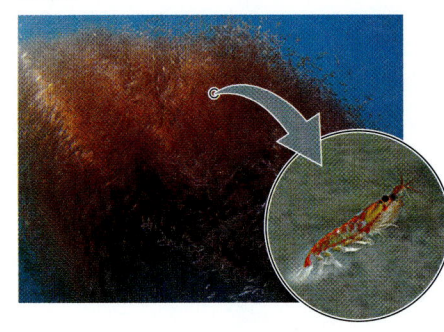

The size of the worldwide population of krill is estimated to be 800 trillion animals.

b. Range

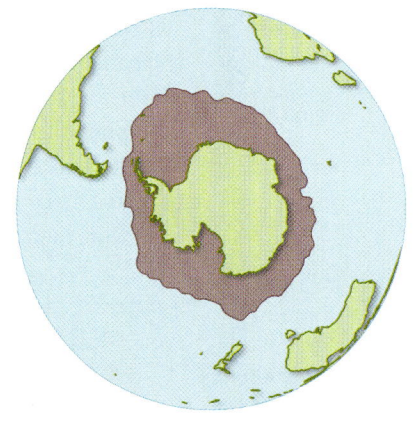

Their range is approximately 19 million square kilometers in the Southern Ocean.

c. Density

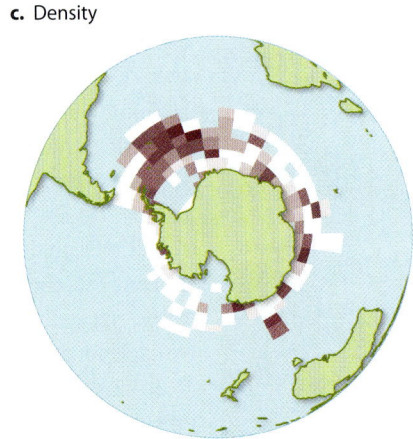

800 trillion krill over 19 million square kilometers of the Southern Ocean means the population density is 42 million animals per square kilometer on average.

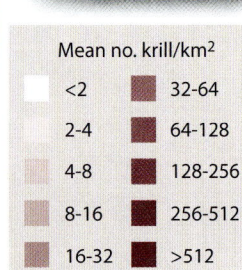

Mean no. krill/km²

<2	32-64
2-4	64-128
4-8	128-256
8-16	256-512
16-32	>512

FIG. 46.2 Distribution of individuals within populations. A population may be (a) randomly distributed, (b) clumped, or (c) overdispersed.

a.

The distributions of individual trees or other organisms can appear to be random, with no clear pattern to where they occur.

b.

If resources are clustered or spatial proximity to other individuals enhances fitness, populations may be clustered.

c.

When resources are limited or predators target a single species, an individual might be better off if it is as far from others as possible, producing an over-dispersed pattern of distribution.

This way of characterizing population density, as average density across the range of a species, is often useful, but it can be misleading if organisms occur unevenly within their range. In nature, the individuals within a population may be distributed randomly—that is, a new individual has an equal chance of occupying any position within the range, and the location of one individual has no influence on where the next will occur (**Fig. 46.2a**). Sometimes, however, resources are distributed patchily within the range, or the chances of a new individual's surviving are enhanced by the presence of other individuals. In these circumstances, the population may be clumped, with individuals clustered together (**Fig. 46.2b**). In some cases, one individual of a population might prevent another from settling nearby, for example if all individuals depend on the same set of resources. Or, predators, having located one prey individual, may search for other members of the same species nearby. Ecologists call such populations overdispersed, meaning that they are distributed more nearly uniformly than would be predicted by chance (**Fig. 46.2c**). Also, habitats suitable for growth may themselves occur in isolated patches. We discuss the implications of habitat patchiness in greater detail later in the chapter.

Population size can increase or decrease over time.

Many factors affect population size. An obvious process that contributes to population increase is birth. Another is immigration: Individuals may arrive in a population from elsewhere. In contrast, mortality (death) and emigration, the departure of individuals from a population, both act to decrease population size (**Fig. 46.3**).

Population size changes continually as individuals are born or immigrate into the population, and as they die or emigrate away from a population. We can describe the changes in a population's size mathematically, defining population size as N and the change in population time over a given time interval as ΔN, where the Greek letter Δ (delta) means "change in." ΔN is the number of individuals at a given time (time 2) minus the number of individuals at an earlier time (time 1), which is symbolized as $N_2 - N_1$. As the processes leading to a change in population size through time include births (B), deaths (D), immigration (I), and emigration (E), we can quantify ΔN as:

$$\Delta N = N_2 - N_1 = (B - D) + (I - E)$$

FIG. 46.3 Factors affecting the size of a population: birth, mortality, immigration, and emigration.

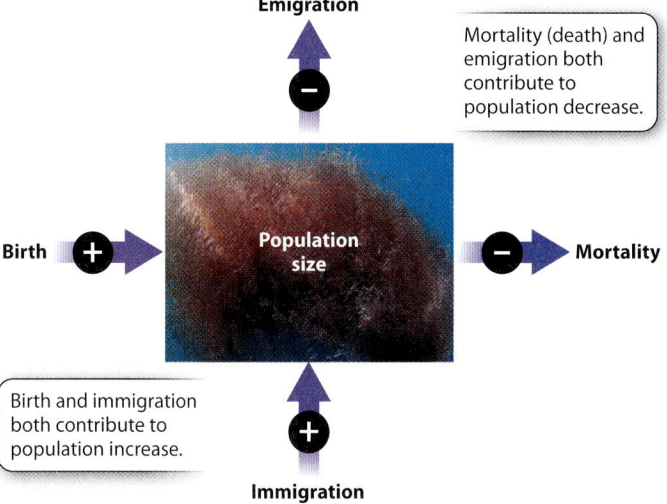

Commonly, ecologists want to know not just whether population size is increasing or decreasing, but also the rate at which population size is changing—that is, the change in population size (ΔN) in a given period of time (Δt), or $\Delta N/\Delta t$. Let's say a population starts with 80 individuals and after 2 years has grown to 120 individuals. The population, then, has gained 40 individuals in 2 years, for a rate of $40/2 = 20$ individuals per year.

The importance we place on such a number strongly depends on the actual population size. An increase of 20 individuals in a year is a big deal if the starting population had 80 individuals, but it is small if the starting population numbered 10,000. Usually, therefore, we are most interested in the proportional increase or decrease over time, especially the growth rate per individual, or the per capita growth rate (*capita* in Latin means "head"). The per capita growth rate is commonly symbolized by *r* and is calculated as the change in population size per unit of time divided by the number of individuals at the start (time 1):

$$r = (\Delta N/\Delta t)/N_1$$

In our example above, *r* equals 20 individuals per year divided by 80 individuals, for a per capita growth rate of 0.25 per year.

It is important to realize that while rates are important in ecology, the data we usually have in hand actually consist of the numbers of individuals that we have counted, and we often do not know exactly how large a natural population really is and so must rely on estimates. We consider how we make such population size estimates later in this chapter, but for the rest of this discussion we focus on the per capita increase (or decrease) in a population over time and assume we know the population sizes for species of interest.

→ **Quick Check 1** If a population triples in size in a year, what is the per capita growth rate?

In the example discussed in the preceding paragraphs, an initial population of 80 individuals increased by 20 per year, numbering 120 after 2 years, for a per capita rate of increase of 0.25 per year. Should this per capita rate of increase continue, how large will the population be after 10 years? We can answer this question using the following formula:

$$N_t = N_1(1 + r)^t$$

The population size after 10 years (N_t, where $t = 10$) will equal the initial population size ($N_1 = 80$) times 1 plus the per capita growth rate (0.25) raised to the power *t* (10 years). Solving the equation for our example shows that in 10 years the population size would be 745. After 15 years, the population would swell to 2273.

This example illustrates **exponential growth,** the pattern of population increase that results when *r* is constant through time. A prominent feature of exponential growth is that the number of individuals added to the population in any time interval is proportional to the size of the population at the start of the interval. That is, while the per capita growth rate remains constant, the actual number of individuals added per time unit increases as the population grows. This is because the new individuals added also reproduce themselves, so through time more and more individuals contribute to the growing population. **Fig. 46.4** shows the characteristic shape of an exponential growth curve.

A second example brings home the remarkable consequences of exponential growth. A chessboard has 8 rows of 8 squares each, a total of 64 squares. Suppose we place a grain of wheat on the first square of the chessboard, two on the second square, four on the third, and keep doubling the number at each successive square. By the time we reach the 64th square, the total number of grains on the board is 18,446,744,073,709,551,615, more than 80 times what the Earth could produce in a year if wheat were planted on every square inch of farmland, and about a million times the number of stars in the Milky Way galaxy.

→ **Quick Check 2** Would you rather have a million dollars today, or start with one cent and double the amount each day for a month? Why?

The per capita growth rate, *r*, is commonly called the **intrinsic growth rate** of a population, the maximum rate of growth when no environmental factors limit population increase. Intrinsic rates of growth in mammals range from 82% per year in muskrats to 50% in snowshoe hares and 10% in moose. Malthus recognized that exponential growth cannot continue for long in nature because the number of individuals will eventually outstrip the resources available to support the population. Darwin, in turn, recognized that when the number of young produced at birth exceeds the number

FIG. 46.4 **Exponential growth.** When per capita growth rate, *r*, is constant, the curve shows ever-larger changes in population size over time.

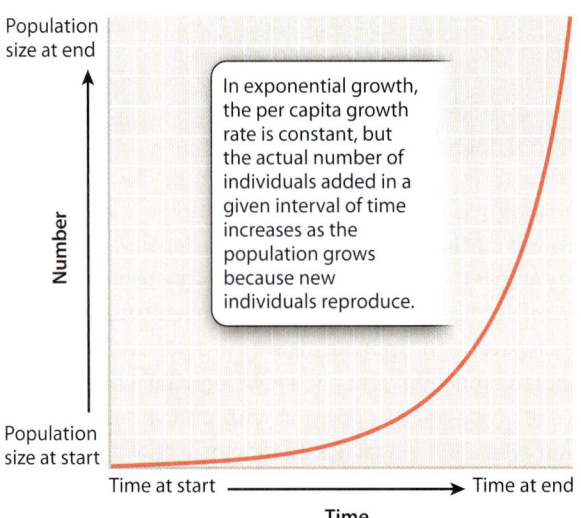

of adults that can be supported by available resources, individuals within the population will compete to obtain the resources needed for growth and reproduction. This **intraspecific** (within-species) **competition** results in natural selection (Chapter 21), linking the ecological and evolutionary aspects of populations. Individuals from different species can also compete for resources. Such **interspecific** (between-species) **competition** can result in an increase or decrease in population size. We focus for the most part on intraspecific competition in this chapter, reserving a detailed discussion of the effects of interspecific competition until Chapter 47.

Carrying capacity is the maximum number of individuals a habitat can support.

As a population grows, crowding commonly results in a decrease in the availability or quality of resources required for continuing growth. As a consequence, growth rate declines. On Hispaniola, for example, Lime Swallowtails have probably reached the limit of available citrus groves. Individual females have to search longer to find suitable plants that do not have eggs already laid by other females. With that delay, and the pressure of eggs that are continuing to mature inside their bodies, females may end up laying too many eggs in one place or on poor-quality food, so there is not enough food for all larvae to survive. All of these factors lead to fewer births than occurred immediately after the Lime Swallowtails first arrived in Hispaniola, when resources were not limiting.

For many organisms, crowding affects the death rate as well as the birth rate. For example, as the fish population of a pond increases, there is less food available for each individual, and oxygen may be used up by microbes that proliferate because of nutrients in the fish waste.

In other words, birth and death rates are themselves affected by population density, which we defined earlier as the number of individuals per unit of area or (for aquatic species) volume of habitat. Increasing density is not always detrimental to a population. For example, eggs of a salmon or sturgeon are more likely to be fertilized in rivers or streams swarming with their species than when there are few reproducing adults. Generally, however, at some point increasing population density causes the environment to deteriorate because of a scarcity of resources. When the density of a growing population reaches this point, the rate of growth slows and eventually levels off as the death rate increases to equal the birth rate.

We call the maximum number of individuals any habitat can support its **carrying capacity** (*K*). In a sense, *K* represents the interplay among the functional requirements of individuals for growth and reproduction and the environmental resources such as food and space available to support these needs.

Many factors can keep a population below *K*. Predation and parasitism are two of them. For example, increasing numbers of Lime Swallowtails on Hispaniola have no doubt attracted predators and parasites that feed on them. Similarly, a high population density of fish in a pond increases the likelihood of infection by viruses, bacteria, and fungi because these agents are transmitted from fish to fish. Predation and parasitism can reduce the population size below *K*, but they do not change *K* itself.

As noted, when a population is small, resources are not limiting and the population can grow at or near its intrinsic growth rate, *r*. As the population increases, however, drawing closer to *K*, the worsening conditions due to crowding slow the rate of growth. For any given population at any given moment, such as deer today in New England forests or striped bass in Chesapeake Bay, we may wish to know what percentage of the carrying capacity is available for further population growth. The answer, expressed mathematically, is $(K - N)/K$.

The effects on population growth as a habitat fills up with individuals can be seen in **Fig. 46.5**. Growth is exponential at first and then slows down as the population size approaches its maximum sustainable size, *K*. The resulting pattern of population growth, with its characteristic S-shaped curve, is termed **logistic growth**. Population size at any time t (N_t) can be calculated using the equation:

$$N_t = \frac{KN_1 e^{rt}}{K + N_1(e^{rt} - 1)}$$

Note that N_t depends on both *K* and *r* (*e* is Euler's number, a constant used in natural logarithms and approximately equal to 2.718).

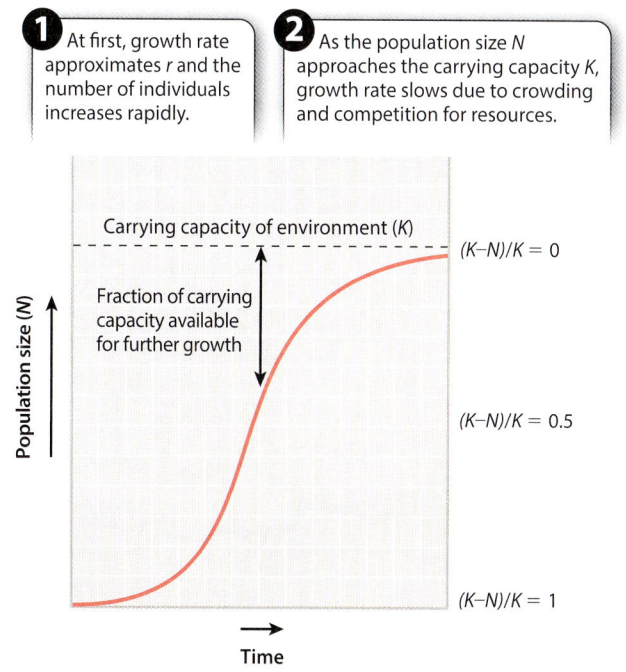

FIG. 46.5 **Logistic growth in a population.** The graph shows the characteristic S-shaped growth curve, which is determined by *r* and *K*.

Logistic growth curves, then, are S shaped because they actually reflect two sets of processes, each related to population size. At first, growth reflects only the rate at which individuals can reproduce themselves, so the curve rises steeply. The second part of the curve, where it starts to level off, reflects the onset of additional factors that become important above a certain population size, such as decreasing availability of food and space.

Virtually all natural populations grow this way. Logistic curves also apply to many other phenomena in biology, as well as in chemistry and such diverse fields as sociology, political science, and economics. Logistic growth therefore is a mathematical concept worth knowing. For example, a logistic curve describes the rate of binding of oxygen to a population of hemoglobin molecules, as seen in Chapter 39. Similarly, the number of fertilized salmon eggs goes up with the amount of sperm (which reflects the number of males in the population) deposited in a stream by males but slows again as the percentage of unfertilized eggs available for the sperm to fertilize drops.

→ **Quick Check 3** Why is the growth curve for most populations S shaped?

Factors that influence population growth can be dependent on or independent of its density.

In the preceding discussion, the factors that limit population size depended on the density of the population. At low density, food and other resources do not limit growth, but as density increases, they exert more and more influence on population growth, spurring competition for available resources. Similarly, at high population density, individuals may be more vulnerable to predation or to infection. Factors such as resources and predation are called **density-dependent** factors.

Not all changes in population growth rates result from population density, however. **Density-independent** factors influence population size without regard for the population's density. They include events like severe drought or a prolonged cold period, either of which can cause widespread mortality independent of population density. A graph of density-independent effects on populations would show instantaneous, rather than gradual, drops in population size, and these changes would occur at different starting densities.

For example, population growth in Song Sparrows is a function of the number of nesting territories established by males each year, and of their young's success in those territories. **Fig. 46.6** shows that the number of territories generally increases for several years until an unusually dry spring occurs, which greatly decreases the amount of vegetation available to support the insects on which the sparrows feed. The number of territories then plummets, regardless of territory numbers the previous year.

Ecologists estimate population size and density by sampling.

Imagine trying to calculate the growth rate of a population of Lime Swallowtails by counting every individual in the Dominican Republic. The task is hopeless. Like that of the Lime Swallowtail, the populations of most species are either too large or too widely dispersed for every individual to be counted. Instead, biologists estimate population size by taking repeated samples of a population and from these samples estimating the total number of individuals and their density. Because of the importance of population size in ecology, biologists have developed a number of sophisticated sampling strategies.

For sessile, or sedentary, organisms like milkweeds in a field or barnacles on a rocky beach, ecologists commonly lay a rope across the area of interest and count every individual that touches it, or they throw a hoop and count the individuals inside it. Each such count gives one sample. In practice, ecologists take a number of samples and use them to estimate the number and distribution of individuals in the population. For example, suppose 1 m² hoops are placed in three different spots in a field, encircling respectively 4, 5, and 3 individual milkweeds. We can multiply the average number of individuals (in this case, 4) per 1 m² sample by the total square meters in the area of interest (in this case, the size of the entire field) to give an estimate of population size.

It is relatively easy to count sedentary organisms along a straight line or within a grid area, but many species move around.

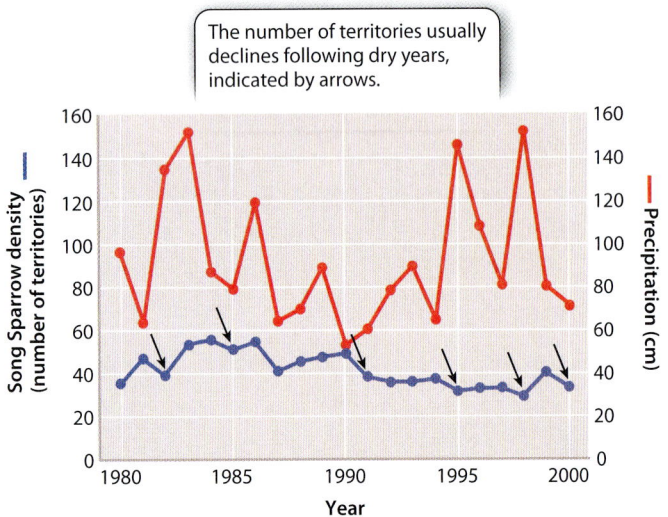

FIG. 46.6 **Density-independent effects on population size.** Independent of population density, extreme weather (such as drought) lowers the quality of habitat for Song Sparrows, reducing the number of territories occupied.

How do ecologists estimate population size for turtles in a pond, or mice in a forest? In these cases, a method called **mark-and-recapture** is useful. Biologists capture individuals, mark them in a way that doesn't affect their function or behavior, release them back into the wild, and then capture a second set of individuals, some of which were previously marked.

In **Fig. 46.7,** for example, an ecologist has marked a butterfly with a painted number, which does not hinder its activity in any way. Later, ecologists return to the area and capture another set of samples from the population. This second sample will most likely contain a few individuals who were captured the first time around, identifiable by the paint mark; these are the recaptures. The rest are newly caught individuals that have no paint mark.

If a large number of butterflies are caught, marked, released, and then recaptured, it is possible to make an estimate of the total population by extrapolating from the proportion of the second sample that is marked.

These methods are simple, and they make a number of assumptions about the population. For example, we assume that the population has not changed in size between our first and second samples. Nonetheless, mark-and-recapture methods have proved effective in estimating the sizes of populations in nature. More sophisticated analyses are available for more complicated situations, for example when more than two samples are caught, or when immigration or emigration occurs during the time between samples.

HOW DO WE KNOW?

FIG. 46.7

How many butterflies are there in a given population?

BACKGROUND Individuals in many populations of butterflies (and other animals) most often live and die within the patch of habitat where they were born. Population sizes of such animals can be estimated by a technique called mark-and-recapture, in which organisms are captured and released on two successive days. We assume that the population size is about the same on the second day as the first day. We also assume that the captured and released animals mixed with the population and do not avoid recapture on the second day.

METHOD Butterflies, such as the Karner Blue Butterfly, are captured in an area, marked with a spot or identification number on the wing as shown here, and then released. The number of marked butterflies is recorded. The next day, butterflies are caught and the number of marked and unmarked butterflies is recorded. Butterflies with marks are the recaptures.

RESULTS To find the population size (N), we take the total number of marked butterflies and unmarked butterflies caught on the second day (C) and divide by the number of recaptures (R). We multiply this number by the number of butterflies marked on the first day (M), as follows:

$$N = C/R \times M$$

CONCLUSION Let's say we capture 100 butterflies on the first day, mark them, and release them. On the second day, we capture 120 butterflies, 30 of which are marked and the rest unmarked. We conclude that the population size is (120/30) × 100, or 400 butterflies total in the population sampled.

RELATED METHODS Mark-and-recapture techniques have proved extremely useful in estimating population sizes of organisms from polar bears to disease victims. There are refinements to the mark-and-recapture approach that take into account the movement of individuals between populations; these methods involve taking multiple samples and adding variables to the equation. Many animals have distinctive color markings, such as the blotches on humpback whale tailfins and frog skin-spotting, and so can be recorded without having to be marked.

SOURCE Lincoln, F.C. 1930. "Calculating waterfowl abundance on the basis of banding returns." *Cir. U.S. Department of Agriculture* 118:1–4.

46.2 AGE-STRUCTURED POPULATION GROWTH

Direct counting and mark-and-recapture methods permit ecologists to assess the size and density of populations in a particular place. However, not all organisms in a population contribute equally to population growth because some reproduce more than others. Therefore, it is also useful to take into account individuals that have differing reproductive capacities, which most often depend on differences in age. Biologists interested in the potential for population growth seek information on the **age structure** of a population, the number of individuals within each age group of the population studied. A population in which most individuals are past their age of greatest reproduction (because for some reason young are not surviving) or a population mostly composed of individuals younger than this age (because older individuals are not surviving) will not increase as rapidly as one dominated by individuals capable of greatest reproductive output.

Birth and death rates vary with age and environment.

To understand past and predict future changes in populations, ecologists use birth rates, expected longevity of individuals, and the proportion of individuals in a population that is able to reproduce. Insights into these parameters come from the age structure of the population.

Commonly, researchers divide the population, or a sample of the population, into age classes, for example individuals born within a single year or within an interval of several years. Age can often be determined by measuring an aspect of the organism's anatomy (for example, counting growth rings in fish scales or a core drilled into a tree trunk) or estimated by overall size (for example, length of a fish, or tree height).

Estimating the number of individuals of differing ages enables ecologists to predict whether the size of a population will increase or decrease. This method applies to human populations as well as those of other species. A growing population usually shows a pyramid-shaped age distribution, with the youngest classes much more abundant than older classes, whereas a stable population shows a more even distribution of age classes. **Fig. 46.8** shows the difference in age structure between France, a country with a stable population, and India, a country with a growing population.

In humans, such differences reflect, in part, socioeconomic variation among countries, but differences in the age structure of other species can reveal threats from various sources. For example, the age structure of fish populations can be affected by the preferential capture of the largest fish, a recurring problem in commercial fisheries. Because the larger fish produce most of the eggs each year, this pattern can eventually lead to a rapid decrease in the population.

Fig. 46.9 shows the effects of overfishing on the age structure of a species of long-lived Pacific rockfish, determined by comparing the ages of fish sampled in unfished waters (Fig. 46.9a) with those in heavily fished waters (Fig. 46.9b). In the unfished population, the curve shows many individuals close to and past the age of 10 years, when reproductive ability starts to climb. In the fished populations, the curve shows a sharp decline of individuals older than 6 years because mostly older and larger fish are caught. Because these older fish produce most of the eggs and young fish, the annual reproductive effort of fished populations is severely affected by their removal.

FIG. 46.8 **Age structure of two populations, with data from 2010.** The data are depicted as histograms, bar graphs in which the length of each bar indicates the number of individuals in the group (here, age group).

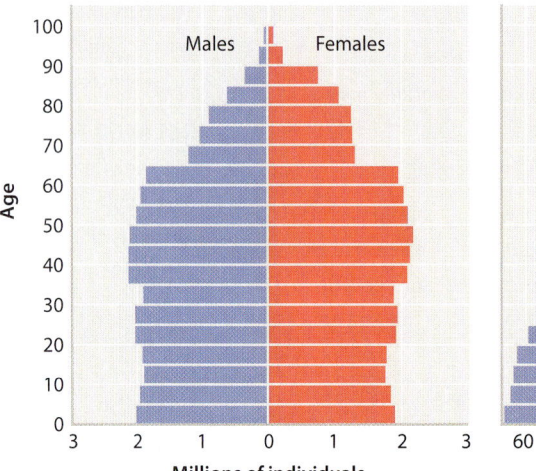
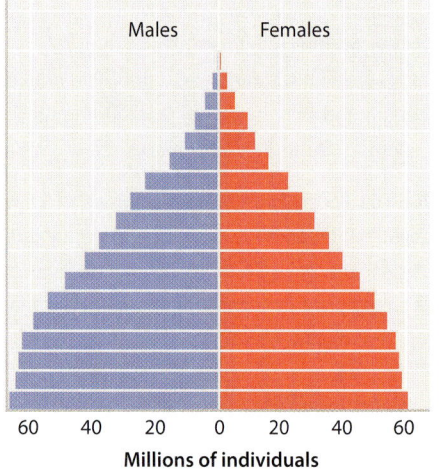

FIG. 46.9 Age structure in Pacific rockfish populations in fished and unfished waters. Harvesting large individuals reduces the population's capacity for growth, jeopardizing commercial stocks.

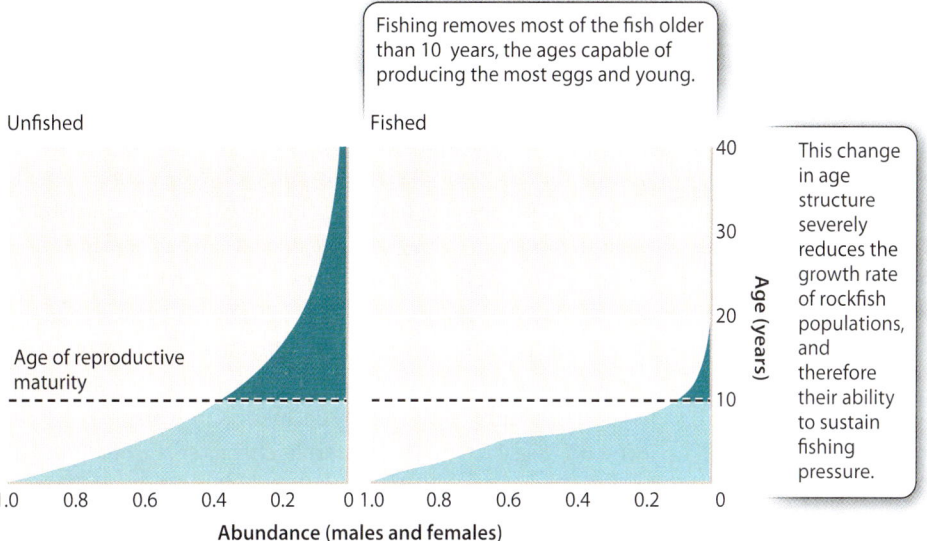

Fishing removes most of the fish older than 10 years, the ages capable of producing the most eggs and young.

This change in age structure severely reduces the growth rate of rockfish populations, and therefore their ability to sustain fishing pressure.

Survivorship curves record changes in survival probability over an organism's life-span.

The age structure of a population reflects death rates as well as births. **Demography** is the study of the size, structure, and distribution of populations over time and includes changes in response to birth, aging, migration, and death. The demography of the invasive Lime Swallowtail butterfly has been studied in Southeast Asia, where it is native, and this investigation may be helpful in predicting the population dynamics of this species in coming years as it spreads from Hispaniola across the Americas. Like many other insects, Lime Swallowtails have a pupal stage in which they metamorphose from a caterpillar larva specialized for feeding on leaves into a winged adult that can search for other plants on which to lay eggs (Chapter 44). Neither the larvae nor the pupae reproduce, but the adults that emerge from the pupae do.

The factors that cause insect mortality commonly differ from one life stage to the next, and knowing their effects can help us understand the physical and biological influences on population growth. **Fig. 46.10** shows the different dangers to each stage of the Lime Swallowtail's life. These butterflies lay 30 or more eggs, but only a fraction of their eggs survive because of parasitism, drying out, falling from leaves, or predation by ants or other insects. The larvae have many defenses against predators, among them camouflage and chemical toxins, but they still fall prey to wasps, as well as to lizards and other animals, so the faster they develop into adults, the more likely they are to escape discovery by their enemies. The pupae, which are large and immobile, are perhaps the most vulnerable stage in the Lime

FIG. 46.10 Lime Swallowtail life cycle and natural enemies. The factors that cause insect mortality commonly differ from one life stage to the next.

In one study, parasites and other factors were observed to remove 32% of the eggs.

Parasites and other factors remove 55% of the pupae. Because pupae are the relatively few survivors of egg and larval mortality, the impact of pupal losses on population growth is magnified.

Swallowtail life cycle, and they are often highly parasitized by a particular kind of wasp.

Fig. 46.11 presents data on the duration of each stage in the Lime Swallowtail life cycle and the numbers killed in that stage. We can consider the individual eggs that begin our study to form a **cohort,** a group defined as the individuals born at a given time, and we can trace their survival as they age. The proportion of individuals from the initial cohort that survive to each successive stage of the life cycle is called **survivorship.** When survivorship curves are plotted through time, they show an ever-declining fraction of the starting number of eggs (Fig. 46.11).

As expected from the preceding discussion, in this example, survival is lowest among eggs (68% survive to the next stage) and pupae (only 45% survive to the next stage), which of course cannot move to escape harm. If we divide the number lost at each stage by the number of days at that stage, we see that the egg stage loses 16 per day, the pupal stage loses 6 per day, but the larval stage loses only 0.6 per day. There is therefore strong selective pressure to decrease the length of each stage, especially those stages most vulnerable to mortality. Nevertheless, if we remember that each adult female can lay 30 or more eggs, only a very few eggs need to survive to the adult egg-laying stage to ensure the growth of the population. In warm climates like the Caribbean, these butterflies may reproduce eight or nine times in a year. Most of their eggs will not survive to become adults; if they did, available resources would soon be depleted.

Patterns of survivorship vary among organisms.

Populations differ in the ages at which they experience the most mortality. In some species, many offspring are produced, but high mortality early in life means that few survive to reproductive age. In other species, mortality is highest in older individuals, while in still other species, mortality occurs with equal frequency at all ages. **Fig. 46.12** illustrates the three principal types of survivorship curve—Type I, Type II, and Type III— which correspond to the pattern of mortality: whether most mortality occurs late in life, is even throughout life, or is concentrated early in life, respectively.

Human populations in affluent countries, which experience most mortality late in life, typically show a Type I survivorship curve (Fig. 46.12a). Other large animals, such as elephants and whales, show similar survivorship curves. Birds and many small mammals, on the other hand, commonly experience consistent levels of mortality throughout life, which can be graphed as a declining straight line. These animals, for which the probability of death does not change much through the life cycle, exhibit a Type II survivorship curve (Fig. 46.12b). Finally, organisms like Turtle Grass reflect high mortality at the earliest stages of their life cycles, resulting in a sharply declining survivorship curve that levels off (Fig. 46.12c). This pattern is characteristic of a Type III survivorship curve, which is typical of small herbaceous plants, small animals such as mice, insects like the Lime Swallowtail, and other organisms that grow fast and reproduce early.

Reproductive patterns reflect the predictability of a species' environment.

In the preceding sections, we discussed population size and growth curves, and we outlined age structure and survivorship. These topics converge when we consider patterns of reproduction. Some species produce large numbers of offspring but provide few resources for their support. In Chapter 42, such species were described as ***r*-strategists.** In contrast, some species produce relatively few young but invest considerable resources into their support. These are termed ***K*-strategists.** How do these concepts relate to r and K in population growth?

In general, species that live in unpredictable environments produce many young. For example, when the environment is full of predators watching for vulnerable prey, such as gulls that watch for young turtles on a beach, or when resources are patchily distributed and abundant for only a short time, the chances of a young organism surviving to adulthood are low. Laying an egg, then, is like buying a lottery ticket—there is a limited chance of success. It is more advantageous to buy many tickets cheaply (that is, to have many offspring and make little parental investment in any of them) than to pay a lot for one. Thus, salmon and other fish produce many thousands of young, each with an extremely small chance of survival. As we saw earlier, Type III strategists like insects and plants also experience high mortality in the first stages of life. Because their reproductive success depends on the small chance that a few of their young will survive all the hazards they encounter, parents produce many offspring and invest relatively little in any of them. We refer to species like most fish as *r*-strategists because they reproduce at rates approaching r, the intrinsic rate of growth.

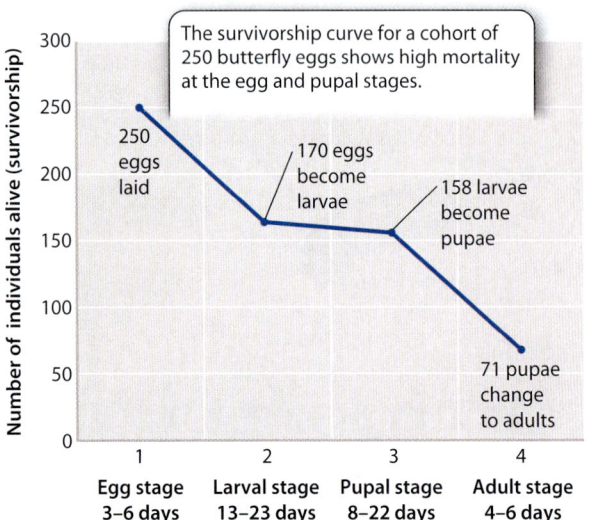

FIG. 46.11 Survivorship curve. A survivorship curve plots the proportion of the original cohort alive at the end of each stage or time period.

FIG. 46.12 Three different types of survivorship curve.

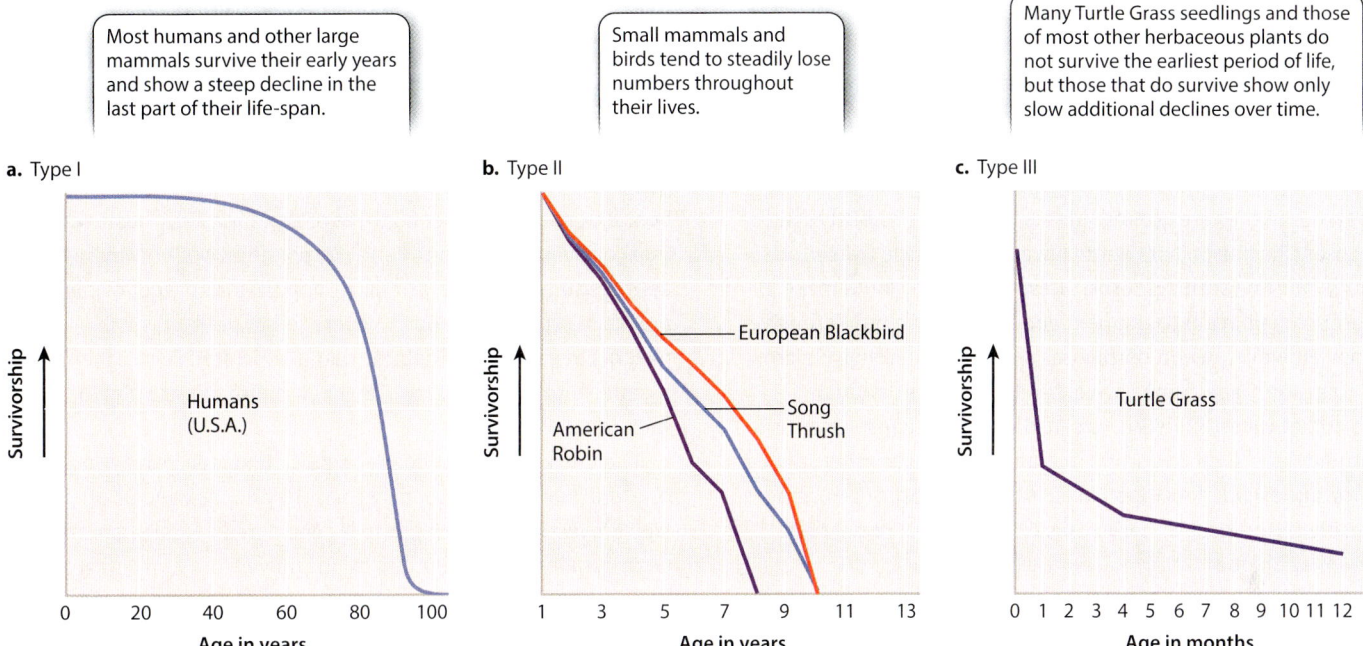

In contrast, species living where resources are predictable often produce a lower number of offspring and invest their reproductive resources in just a few, often larger and better-cared-for, offspring. These offspring are therefore better protected from predators, able to compete with others when they are young, and more likely to survive periods of bad weather. Large animals, such as eagles and bears, produce only two or three young, each of which is much more likely to survive to reproductive age than the offspring of small animals like mice, which have many predators and are easily killed by cold weather. Animals like eagles and bear are called K-strategists because their population densities commonly lie near the carrying capacity K, when resources become scarce.

These two strategies reflect an evolutionary trade-off. A reproductive individual has access to a finite amount of resources and can invest either in many inexpensive young or few well-provisioned offspring. Depending on the combination of physiology and environment, natural selection in a given set of circumstances tends to favor one strategy over the other. In fact, r- and K-strategists represent the opposite ends of a continuum in reproductive pattern. Many species lie between these extremes.

The life history of an organism shows trade-offs among physiological functions.

We've noted that population growth can be limited by resource availability. How an organism makes use of the resources that are available can also affect population growth. Each individual must devote some of its available food and energy intake to growth, some to the maintenance of cells and tissues, and some to reproduction. This allocation is very much like a household budget, in which incoming resources are divided among the day-to-day expenses of acquiring food and paying for housing and transportation, meeting present and future medical and educational expenses, and savings held in reserve for emergencies. When one category goes up in cost—it needs more money, or more resources—another must come down. In nature, individual organisms strike a balance in how they allocate resources to the basic needs of growth, maintenance, and reproduction, a balance that has evolved to minimize waste while retaining the flexibility to use resources in different ways depending on environmental circumstances. These kinds of trade-offs evolve through natural selection and are the rule in ecology, explaining in part why species tend to produce either many small offspring or a few larger ones.

The typical pattern of resource investment in each stage of a given species' lifetime is called its **life history.** The amount of energy available for reproduction varies over an individual's lifetime, and the degree of control an individual has over resource allocation also varies from species to species. Lizards, for example, show considerable flexibility in allocating resources to reproduction. Many lizards show a physiological trade-off from year to year between the number of eggs produced and the average size of each egg. There is a finite amount of resources to put into offspring, and lizards adjust the number and size of eggs depending on conditions in the environment. Larger eggs require more resources to build, leaving less for additional eggs, and larger eggs hatch into larger offspring that are better able to fend for themselves. But if the environment carries a high predation risk, for example, then laying more eggs scattered in different places may be a better strategy than laying a few larger eggs in one place. Does this flexibility make these lizards *K*-strategists or *r*-strategists? The answer is neither. Which strategy is optimal depends on the nature of the risks faced by young.

Plants also show physiological trade-offs in the allocation of resources to seeds. For example, flowers may abort some of their developing seeds to favor the growth of others. Plants also make trade-offs between defense and reproduction. When herbivores are around, plants that invest resources in defenses like poisons or thorns have an advantage over plants that do not invest in defenses. These defensive plants make a greater contribution to the next generation than nondefensive plants. However, when herbivores are absent, defensive plants are commonly at a competitive disadvantage. Individuals that do not invest in defenses can allocate more resources to reproduction. Such trade-offs in plant defense are especially important on nutrient-poor soils, where the cost of replacing valuable tissues lost to herbivores is higher than in nutrient-rich soils (Chapter 32).

The key point is that the various ways in which animals and plants allocate resources to growth and reproduction is rooted in their physiology, and their physiological traits are, themselves, evolved traits. Thus, as stated at the beginning of this chapter, ecology both reflects and determines how organisms function in nature.

→ **Quick Check 4** What ecological conditions require a plant or animal to experience trade-offs among growth, reproduction, defenses, and other adaptive traits?

46.3 METAPOPULATION DYNAMICS

In nature, the populations of many, if not most, species are patchily distributed across the landscape. Deer occur only where there are forest patches of suitable size, milkweeds occur only in open fields, and ducks need to be near water. Local populations are therefore usually separated from one another in space and are connected only occasionally by individuals that migrate between habitat patches. Moreover, some habitat areas that could, in principle, support populations of a given species are unoccupied. How does this kind of distribution influence the dynamics—that is, the change in abundances—of populations in nature?

A metapopulation is a group of populations linked by immigrants.

Groves of trees, separated from other groves by fields, are patches of habitat for the birds, insects, and fungi that rely on them for food and shelter. Similarly, humans are patches of habitat for infectious viruses and certain bacteria, coral reefs are patches for fish, and mountaintops are patches for mountain goats and other high altitude species. For nuthatches or bark beetles, crossing between groves of trees is a risky business, marked by predatory birds and the possibility of not finding another suitable grove. Similarly, populations of reef corals are often separated by open waters full of predatory fish. A **patch,** then, is a bit of habitat that is separated from other bits by inhospitable environments that are difficult or risky for individuals to cross.

The nuthatches that occupy a particular tree grove contact one another more frequently than they do individuals of other tree groves, and so each group is considered a local population. Occasionally, a few migrants might move from grove to grove, enabling gene flow between local populations. We can think of local populations of nuthatches in separate tree groves as each part of a larger, more inclusive group called a metapopulation. A **metapopulation** is a large population made up of smaller populations linked by migration (**Fig. 46.13**).

In order to understand the dynamics of populations, then, we must consider not only the birth and death of individuals, but also

FIG. 46.13 Metapopulations. A metapopulation is a group of independent populations connected by occasional immigrants, illustrated here by populations of nuthatches in groves of trees.

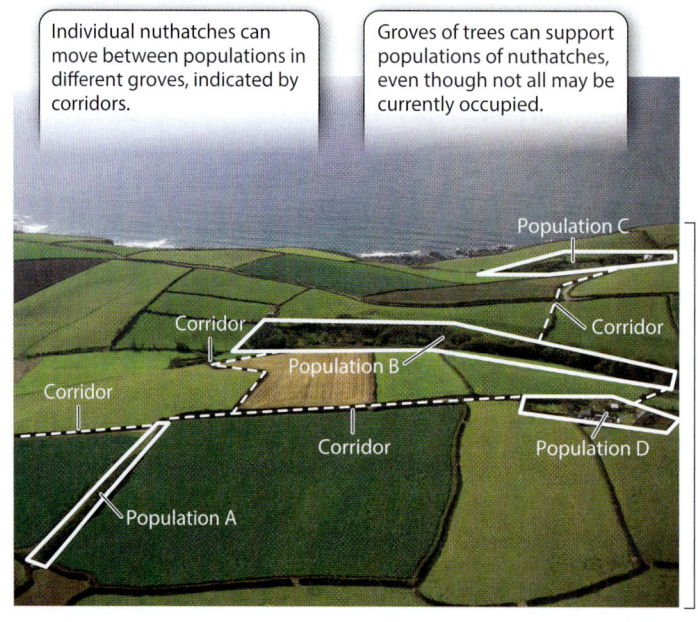

Individual nuthatches can move between populations in different groves, indicated by corridors.

Groves of trees can support populations of nuthatches, even though not all may be currently occupied.

The independent populations, connected by migrants, constitute a metapopulation of nuthatches.

FIG. 46.14 **Experimental patches of habitat.** This study investigated the effects of patch size and connections in moss and the rates of colonization and extinction of the tiny arthropods that live there.

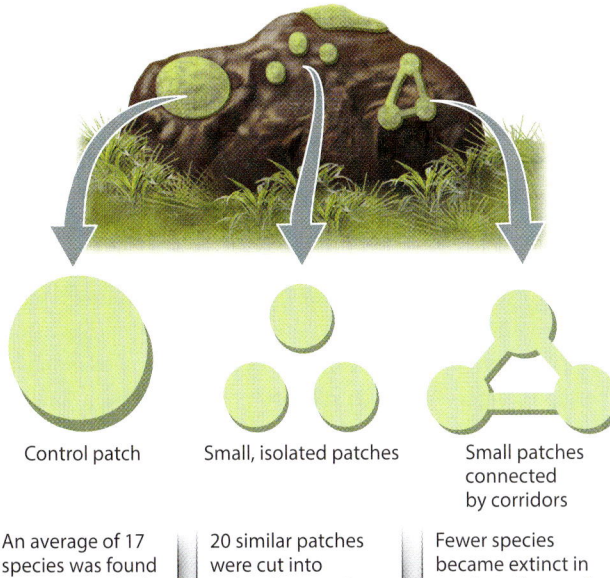

Control patch — An average of 17 species was found on 20 untouched patches of moss.

Small, isolated patches — 20 similar patches were cut into isolated spots of moss. A year later, on these fragmented patches, an average of 7 species had become extinct.

Small patches connected by corridors — Fewer species became extinct in patches of moss that were connected by narrow corridors. An average of only 2 species became extinct.

Experiments with patches of moss growing on boulders showed that the size and connectedness of the patches of moss affected the probability that species of arthropods inhabiting the moss would become extinct (**Fig. 46.14**). Researchers cut large patches of moss into several small fragments and sampled the number of insect species in each patch over time. As expected, isolated patches showed decreases in species diversity, up to 40% in 1 year; some species became extinct. However, when nearly isolated moss patches were left with a narrow corridor of moss connecting them, the rate of extinction was lower. Even narrow corridors of suitable habitat can allow species to move between patches of habitat and prevent extinction, or serve to reintroduce species that become locally extinct.

In a world increasingly marked by habitat fragmentation due to human activities, studies of metapopulations are playing an important role in conservation biology (**Fig. 46.15**). Species survival depends not only on the health of local populations, but also on the ability of individuals to colonize new habitat patches. As the simple experiment with mosses shows, corridors that connect endangered populations can dramatically increase the probability that a species will survive. However, corridors are important only to species that use them to cross barriers between patches. For species like ferns that produce spores able to float between patches on breezes, corridors may not be as important to patch colonization. Dispersal ability differs among organisms, strongly influencing their capacity to colonize distant patches.

the colonization and extinction of local populations. Local populations within a metapopulation have independent fates and commonly become extinct because of their relatively small sizes. In general, small populations become extinct more often than large populations because they are more susceptible to random factors that increase mortality, such as weather, natural disasters, or predation. For example, a nest of predators can eliminate a local prey species; or, if population density falls too low in a small local population because of disease or lack of resources, survivors may not be able to find each other for mating. However, as long as some local populations remain—if they do not all become extinct at the same time—these populations can recolonize old habitats or find new habitat patches, thus ensuring survival of the metapopulation.

FIG. 46.15 **The Biological Dynamics of Forest Fragments Project.** Researchers have monitored nine forest patches near Manaus, Brazil, for more than 30 years and have found that fragmentation strongly affects forest biodiversity, providing clues to the design of effective refuges.

Island biogeography explains species diversity on habitat islands.

Islands represent extremes in habitat patchiness for colonizing species—organisms that live on a mainland must successfully cross a barrier of water to establish a new population on an island. Ecologists commonly expand the use of the term "island" to include any habitat patch that is surrounded by a substantial expanse of inhospitable environment. While this use certainly includes parcels of land surrounded by water, it can also apply to a body of water surrounded by land, an expanse of forest surrounded by grasslands, or a mountaintop alpine field surrounded by forests at lower elevation. Where do colonists of these habitat islands come from, and what determines how many succeed in establishing a new population?

Species diversity reflects both the rate at which new species arrive on the island, and the rate at which species already on the island become extinct (**Fig. 46.16**). As species arrive on an initially uninhabited island, the colonization rate is high because there are no predators yet and the small number of colonists means that competition for resources is limited. Inevitably, as more and more species successfully establish themselves on the island, the rate of colonization goes down. This occurs both because competition and predation increase, making successful colonization more difficult, and because of the diminishing number of species left in the pool of potential colonists. In contrast, extinction rate goes up as newly arrived species compete.

At some point, there is a balance between the arrival of new species and the loss of species due to extinction. At this point, diversity is said to be at equilibrium (Fig. 46.16). This does not mean that no new species will colonize the island, but that each new colonist species will be accompanied by the extinction of one already in place. There is generally turnover in species composition through time.

The **theory of island biogeography,** articulated in 1967 by the American biologists Robert MacArthur and E. O. Wilson, states that the number of species that can occupy a habitat island depends on two factors. The first is the size of the island. Because of their size, larger islands receive more colonists than smaller ones. Furthermore, larger islands can support more species than smaller ones can, so extinction rate goes down. For these reasons, the equilibrium number of species expected for larger islands will be greater than that for smaller islands (**Fig. 46.17a**). The second key factor is the distance of the island from a source of colonists. More distant islands have lower rates of colonization, moving equilibrium species diversity toward lower values (**Fig. 46.17b**).

A number of observations and experiments have provided support for the theory of island biogeography. Wilson and his student Daniel Simberloff tested the theory of island biogeography in a novel way. Working among small mangrove islands off the Florida Keys, they used short-term insecticides to remove all the insects living on individual islands. Over the next few years, new insect populations established themselves on the mangrove islands. The researchers found that the species diversity of islands differing in size and distance from shore corresponded to those predicted by the theory. Larger islands had more successful colonists, whereas smaller and more distant islands had fewer.

At a larger scale, the number of animal species counted on tropical islands generally fits the predictions of island biographic theory. **Fig. 46.18** shows not only that larger islands support more species than smaller ones, but also that there is a quantitative relationship between island size and species number. Similar relationships between island size and equilibrium species diversity have been observed repeatedly. The relationship can be described mathematically as a **species–area relationship:**

$$S = cA^x$$

where S is the number of species at equilibrium, c is a mathematical constant, A is the habitat area, and x is an experimentally determined exponent, relating species number to

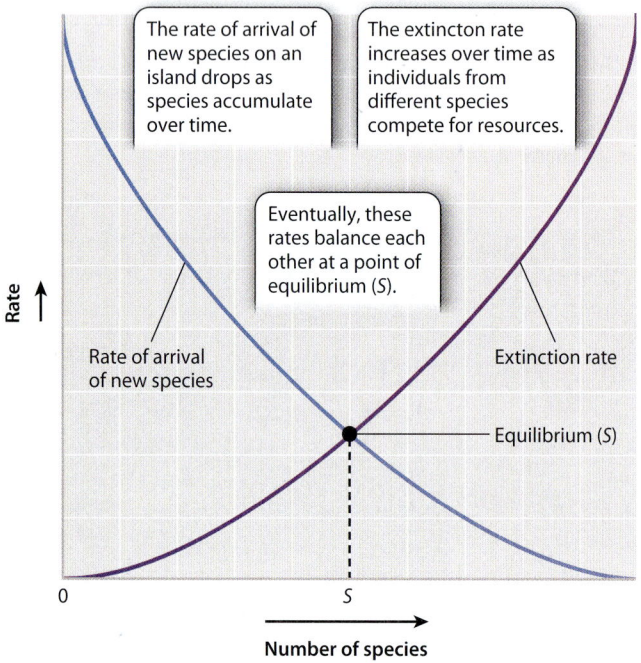

FIG. 46.16 Island species diversity, a balance of colonization and extinction.

FIG. 46.17 Effect of (a) island size and (b) distance from source populations on the equilibrium number of species on an island.

a.
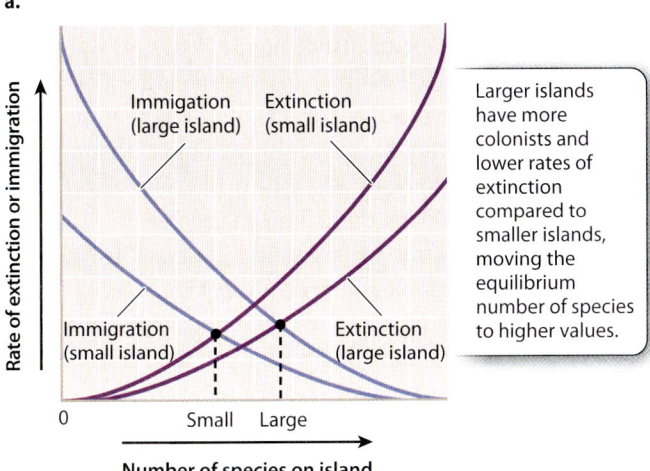

Larger islands have more colonists and lower rates of extinction compared to smaller islands, moving the equilibrium number of species to higher values.

b.
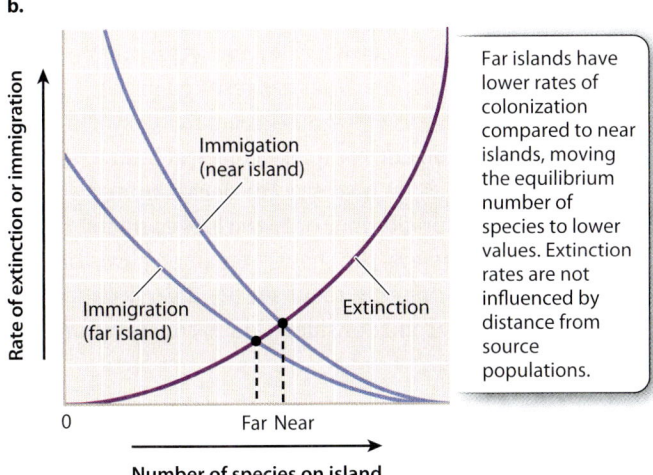

Far islands have lower rates of colonization compared to near islands, moving the equilibrium number of species to lower values. Extinction rates are not influenced by distance from source populations.

habitat area. In many experiments, x has fallen within the narrow range of 0.1 to 0.4.

A useful back-of-the-envelope calculation is that a tenfold increase in island size results in a doubling of species diversity. Perhaps more relevant for issues of conservation, a 90% reduction of habitat area results in a 50% loss of species diversity if no other factors are taken into account. This relationship has many applications in conservation biology, among them the determination of how large a protected area must be to preserve biodiversity and how close reserves must be to one another to ensure the movement of at least some individuals between them.

Island biogeography theory has been widely applied, but it has some limitations in representing island colonizations over long evolutionary time frames. For example, the theory treats all potential colonists as equally able to colonize a new island, when in fact they are not all equally able to arrive or form an association with a species already in place. These abilities are often determined by a species' particular evolutionary history. For example, flying animals like bats and birds more frequently arrive on oceanic islands than nonflying animals like rodents, lizards, and frogs. In contrast, the nonflying animals that colonize successfully more often evolve into multiple new species on islands.

Among the species that have colonized the island of Hispaniola, for example, there are many birds, bats, and strong flying insects like butterflies and dragonflies. Nonflying animals such as frogs, lizards, and fish became established by relatively few colonizations, but successful colonists diversified, evolving a larger number of closely related species now found nowhere else in the world; these are called endemic species.

→ **Quick Check 5** Two islands lie off the coast of Florida. One is larger, the other is closer to the mainland. What determines which one will have more species?

? **CASE 8 Biodiversity Hotspots: Rain Forests and Coral Reefs**
How do islands promote species diversification?
Anolis lizards provide a classic and dramatic example of evolutionary diversification on the island of Hispaniola,

FIG. 46.18 Island size and number of species. The diversity of amphibian and reptile species on Caribbean islands fits the predictions of island biogeography theory.

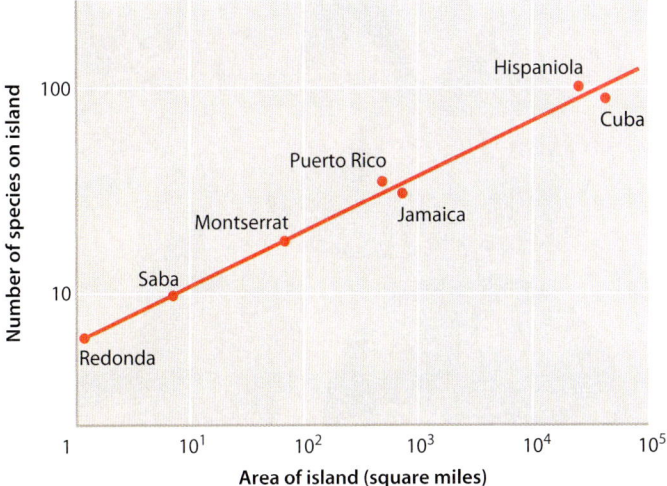

FIG. 46.19 *Anolis* lizards in Hispaniola. These lizards arrived as colonists from elsewhere in the Caribbean and subsequently diversified to produce a variety of species reflecting specialization in their insect-hunting in different parts of local vegetation.

where many dozens of species of closely related species coexist. Phylogenies suggest that *Anolis* lizards colonized Hispaniola in five or six separate events, but the dozens of species present today indicate marked evolutionary diversification after colonization.

All *Anolis* feed on insects and other invertebrates, but species differ in how and where they search for prey. Some with long legs run on the ground, while others with shorter legs and long tails climb on vegetation. Still other forms are large and robust and live in the high canopies of trees. Altogether, *Anolis* have evolved six different feeding strategies with corresponding morphological traits (**Fig. 46.19**).

All six types can coexist in one place because they hunt insects in different ways and in different parts of the vegetation. Careful research is revealing the pattern of population expansion and fragmentation, followed by natural selection within different subpopulations, that has resulted in the diversity observed on Hispaniola today. Through evolution and complex ecological interactions, then, habitat diversity can depart from the simple predictions of island biogeography. Nonetheless, the theory remains a powerful way of thinking about species diversity and an important consideration in conservation biology.

Species coexistence depends on habitat diversity.

Species diversity on islands depends not only on the size of the island and its distance from the mainland, but also on interactions among species. Early experiments by the Russian biologist Georgii Gause showed that a simple system with one predator and one prey population is inherently unstable. The predator overexploits the prey, driving it to extinction, and then becomes extinct itself.

In 1958, the American scientist Carl Huffaker demonstrated that if the prey had refuges where some individuals could escape from predators, they could persist while predator populations declined. Through time, the prey population would recover and rise in population density to a point where predators would again expand and cause prey abundance to decline—and then decline again

HOW DO WE KNOW?

FIG. 46.20

Can predators and prey coexist stably in certain environments?

BACKGROUND In the 1950s, it was not clear whether a simple system of predators and prey could be stable and coexist indefinitely or whether it would become extinct as predators consumed all available prey. The results of experiments up to that time were equivocal, and ecologists knew that predators introduced to islands could hunt their prey to extinction.

HYPOTHESIS Ecologist Carl Huffaker hypothesized that predators and prey could stably coexist if temporary refuges were available for the prey.

EXPERIMENT Huffaker studied two kinds of mites: One feeds on the surface of oranges, and the other preys on the orange-eating mite. Huffaker put two sets of oranges on a table, separating them with barriers of petroleum jelly, essentially making each orange its own habitat patch. He added toothpicks to one set of oranges so that the orange-eating mites could let out silk and float over the petroleum jelly barriers and, at least temporarily, escape predation.

RESULTS In the setup in which prey could not escape, a single cycle of increase and decline was observed: The population size of prey increased but was closely tracked by rising populations of their predators, and eventually both declined to extinction.

In experiments in which toothpicks were provided, the orange-feeding mites climbed the toothpicks and dispersed on silken threads they produced, moving away from predators and toward other oranges without mites. These populations of predators and prey went through three cycles of increase and decline, and cycles would probably have continued if oranges continued to be supplied.

CONCLUSION Predator–prey systems can be stable if there are sufficient areas available where prey can escape predators, at least temporarily.

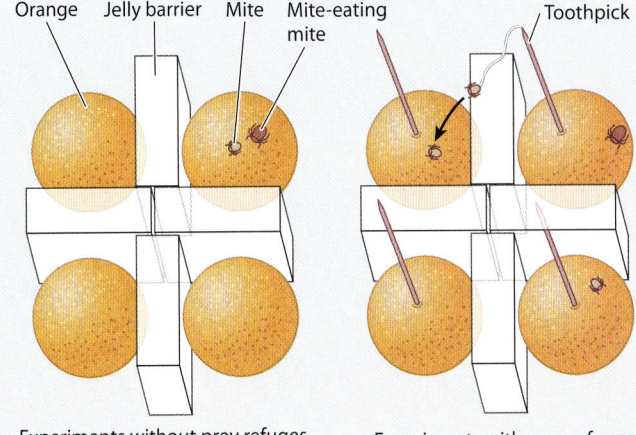

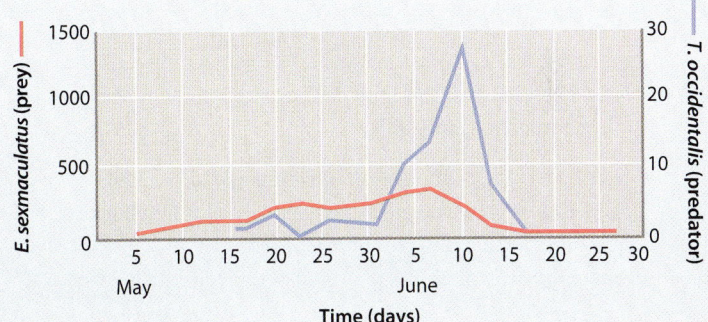

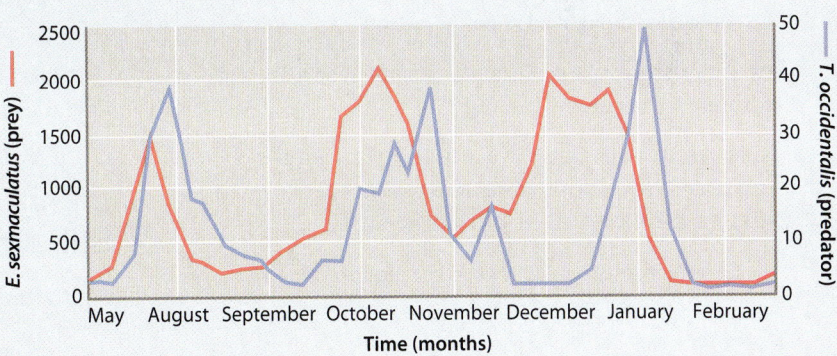

SOURCE Huffaker, C. B. 1958. "Experimental Studies on Predation: Dispersion Factors and Predator–Prey Oscillations." *Hilgardia: A Journal of Agricultural Science* 27:795–834.

themselves (**Fig. 46.20**). Huffaker showed that predators and prey cycle repeatedly through periods of increasing and then decreasing density, as predators track their prey and some prey escape predation. In other words, a long-term, stable oscillation pattern can be achieved that links the population densities of predator and prey. We consider predator–prey interactions in more detail in the next chapter.

Core Concepts Summary

46.1 A POPULATION CONSISTS OF ALL THE INDIVIDUALS OF A GIVEN SPECIES THAT LIVE AND REPRODUCE IN A PARTICULAR PLACE AND IS CHARACTERIZED BY ITS SIZE, RANGE, AND DENSITY.

Population density is calculated by the population size divided by its range. page 46-2

Population size is influenced by births, deaths, immigration, and emigration. page 46-3

The population growth rate is the change in population size, commonly expressed as the per capita rate of growth—the rate of growth (or decline) per individual in the initial population. page 46-3

Initially, growth rates are often exponential, so that the number of individuals added to the population increases over time. page 46-3

Population density often limits population growth because of decreasing resources. This limit defines the carrying capacity of the environment. page 46-5

The growth curve of populations is commonly logistic, an S-shaped curve in which growth is exponential at first but tapers off as population size approaches the carrying capacity of the environment. page 46-5

Population density can be limited by density-dependent factors, such as resources and predation, and by density-independent factors, such as drought and fire. page 46-6

Population size can be estimated by sampling and mark-and-recapture techniques. page 46-6

46.2 THE AGE STRUCTURE OF A POPULATION HELPS ECOLOGISTS UNDERSTAND PAST AND PREDICT FUTURE CHANGES IN POPULATION SIZE.

The age structure of a population is the number of individuals of each age group and is often depicted as a histogram. page 46-8

A growing population shows a pyramid-shaped age distribution, with the youngest classes more abundant than older classes, whereas a stable population shows an even distribution of age classes. page 46-8

Survivorship is the proportion of individuals from an initial cohort that survives to each successive age or stage of the life cycle. page 46-9

Type I, Type II, and Type III survivorship curves correspond to mortality that occurs predominantly late in life, is even throughout life, or is concentrated early in life, respectively. page 46-10

r-strategists are organisms that live in unpredictable environments, producing many young but investing relatively little in each offspring. page 46-10

K-strategists live where resources are predictable and produce a lower number of offspring but invest more in each. page 46-10

r- and K- strategists represent extremes of a continuum of reproductive strategies. page 46-11

Organisms often exhibit trade-offs between reproduction and other physiological functions. page 46-11

The life history of an organism is the evolved pattern of resource investment in each stage of a given species' lifetime. page 46-11

46.3 THE DYNAMICS OF POPULATIONS ARE INFLUENCED BY THE COLONIZATION AND EXTINCTION OF SMALLER, INTERCONNECTED POPULATIONS THAT TOGETHER MAKE UP A METAPOPULATION.

The size of local populations (patches) and the degree of connectedness between patches determines in part whether a given patch will persist or not. page 46-12

An ecological island may be a true island surrounded by water or any habitat surrounded by uninhabitable areas. page 46-12

The theory of island biogeography proposes that the number of species that can occupy an island depends on the size of the island and its distance from the mainland. page 46-14

Species diversity on habitat islands reflects both the rate at which new species arrive and the rate at which colonist species become extinct. page 46-14

Species diversity on islands depends not only on size of the island and distance from the mainland, but also on interactions within and between species. page 46-16

Self-Assessment

1. Name the four factors that affect population size.

2. Draw an exponential and a logistic growth curve, and explain what accounts for their different shapes.

3. Name two density-dependent and two density-independent factors that can limit the size of a population.

4. Draw a graph showing the age structure of a population that is growing rapidly and a graph of one that is not.

5. Plot a survivorship curve for a species with high rates of predation early in life and one for a species with high mortality late in life. Name the type of survivorship this species displays.

6. Explain how r and K strategies relate to the predictability of the environment and in what kinds of environment each strategist would be more successful.

7. Describe what is meant by a trade-off in physiological functions and give an example.

8. Give three examples of a habitat island and explain what makes them islands.

9. Name factors that determine the diversity of species on a habitat island and explain the relevance of these factors in managing the habitat of an endangered species.

Do you understand the chapter's Core Concepts? Log in to check your answers to the Self-Assessment questions, then practice what you've learned and reinforce this chapter's concepts by working through the problems and multimedia tutorials provided there.

www.biologyhowlifeworks.com

CHAPTER 47

SPECIES INTERACTIONS, COMMUNITIES, AND ECOSYSTEMS

Core Concepts

47.1 The niche is the ecological role played by a species.

47.2 Competition, predation, and parasitism are antagonistic interactions in which one species benefits at the expense of another.

47.3 In mutualisms, the benefits to both species outweigh the costs of participation.

47.4 Communities are shaped by the evolved features of organisms, which shape the interactions with other organisms and with the physical environment.

47.5 The cycling of materials and the flow of energy through ecosystems reflect the ways in which organisms interact in communities.

47.6 Biomes are broad, ecologically uniform areas distinguished by their climate, soil, and characteristic species.

The chocolate tree, *Theobroma cacao*, known simply as cacao, is native to tropical America. Cacao grows where the environment is consistently warm and humid. Physical environment alone, however, does not determine where cacao can live. Cacao reproduction, and ultimately distribution, is limited by the requirement that tiny flies called midges pollinate them. The midges' eggs and young need deep shade and a thick layer of moist rotting vegetation, whether from the natural fall of leaves onto the forest floor or from banana plants left to decay in cacao plantations. The midges' needs mean that cacao cannot reproduce in areas with full sun or on clean ground that otherwise would support their growth—at least not without human help. The natural distribution of cacao plants, therefore, depends on interactions with both the physical environment and other species.

The reliance of chocolate trees on midges is not unusual. Every species interacts with other species in ways that affect growth and reproduction. These other species may be the insects that help a plant pollinate its flowers or ones that eat its leaves. They may be a fungus that helps a tree gain nutrients or another that attacks it. They may be other species that compete for resources, predators and pathogens that reduce population size, prey species that provide a specialized food resource, or microbes that provide or take nutrients while living inside their hosts (Case 5: The Human Microbiome). Associations among species, whether positive, negative, or neutral, form a complex web of interactions in which each species influences the population size and distribution of other species, whether they are partners, enemies, or food.

In this chapter, we explore interactions that shape the distributions of populations and, therefore, the composition, diversity, and flow of material and energy in communities. We also discuss how adaptations to the physical and biological aspects of habitats result in the stunning variety of ecosystems seen around the world, from the poles to the tropics.

47.1 THE NICHE

As discussed in Chapter 46, the habitat occupied by a species is sometimes patchy in its distribution, like islands in a sea, and each patch can contain a number of different species. Distinct species can coexist within shared habitat patches because they use different resources or are active at different times. Nearly a century ago, the American ecologist Joseph Grinnell defined the niche as the sum of the habitat requirements needed for a species' survival and reproduction. Later, ecologist Charles Elton redefined the niche as the role a species plays in a community, switching emphasis from the habitat to the population itself. By the middle of the twentieth century, ecologist G. Evelyn Hutchinson had combined these ideas, popularizing the concept of the **niche** as a multidimensional habitat that allows a species to practice its way of

FIG. 47.1 Adaptations of *Anolis* lizards for feeding. (a) *Anolis* species that feed on twigs have short limbs and a prehensile tail that they use to support themselves when they sleep; (b) those that feed on trunks have long limbs.

life. The niche is therefore determined by both physical (abiotic) parameters such as climate and soil chemistry and biological (biotic) factors based on interactions with other species.

The niche is the ecological role played by a species in its community.

The niche, then, reflects the interactions between populations and the physical and biological components of their surroundings. These interactions lead to adaptations, the products of natural selection. Adaptations in turn enable different species to coexist by exploiting different combinations of resources.

Anolis lizards, for example, hunt small insects in humid forests on Hispaniola and other islands in the Caribbean (see Fig. 46.19). The lizards have features of anatomy and physiology that determine, in part, where they can live. They cannot, for example, endure freezing temperatures. Within the forest, different *Anolis* species eat insects found in distinct parts of forest trees. That is, over time, they have evolved different adaptations for feeding as a result of competition for space and resources (**Fig. 47.1**). *Anolis* lizards, then, show that species whose niches overlap may diverge to minimize the overlap, a pattern called **resource partitioning**.

Interactions with predators such as lizard cuckoos and curly-tailed lizards further influence, and limit, the distributions of different species. Moreover, there is reason to believe that the ability (or inability) to disperse contributes to the distributions of individual species. For example, when *Anolis* species are introduced by accident or deliberately to places beyond their dispersal capabilities, they commonly thrive. Therefore, competition, predation, and dispersal are all factors that influence the distribution of species.

FIG. 47.2 Fundamental and realized niches. The fundamental niche includes all the habitat potentially available to a population, whereas the realized niche is the subset of that habitat actually occupied by the population.

The realized niche is smaller than the fundamental niche because of competition and other biotic factors. Tadpoles of small frog species may be excluded from some ponds by larger frog species.

Species are commonly associated with a specific habitat, but "niche" and "habitat" are not interchangeable, as the niche refers to the ways that organisms respond to the resources and other species found within the habitat. American ecologist Eugene Odum summarized the distinction succinctly: The habitat is the organism's "address" and the niche its "profession." This distinction illustrates the dual nature of niches: They reflect where organisms occur and what they do there. The profession of *Anolis* lizards is insect-hunting; their addresses are the trees and shrubs of humid tropical America.

FIG. 47.3 **A mutualistic interaction, in which both participants benefit.** Development of seed-rich cacao fruits (a), whether for the plant's reproduction or to provide raw material for chocolate, depends on pollination by midges (b).

The realized niche of a species is more restricted than its fundamental niche.

The **fundamental niche** comprises the full range of climate conditions and food resources that permits the individuals in a species to live. In nature, however, many species do not occupy all the habitats permitted by their anatomy and physiology. That is because other species compete for available resources, prey on the organisms in question, or influence their growth and reproduction, reducing the range actually occupied. This actual range of habitats occupied by a species is called its **realized niche (Fig. 47.2)**.

For example, tree frogs can reproduce in small ponds near forests throughout eastern North America, their fundamental niche. However, tree frogs are excluded from ponds that also have larger frogs because the larger frogs eat most of the algal food available, limiting the tree frogs' realized niche to ponds without larger frogs. Therefore, competition with other species—that is, **interspecific competition**—can determine the size of the realized niche.

→ **Quick Check 1** Which is larger—the fundamental or the realized niche?

47.2 ANTAGONISTIC INTERACTIONS BETWEEN SPECIES

The realized niche of tree frogs compared with their fundamental niche makes it clear that interactions with other species play key roles in the distribution and abundance of populations. Close interactions between species that have evolved over long periods of time are called **symbioses** (Chapter 26). Many of these interactions enhance the reproduction and population growth of both ecological partners, like chocolate trees and midges (**Fig. 47.3**). These mutually beneficial interactions are called **mutualisms**. Many other interactions are one sided: One organism actually consumes another or competes with it for food or some other resource (**Fig. 47.4**). Such one-sided interactions, in which at least one participant loses more than it gains, are called **antagonisms**.

All interactions combine costs (often in physical resources) and benefits (usually increases in reproduction), and the ratio of costs to benefits for each of the participants reveals whether an overall

FIG. 47.4 An antagonistic interaction, in which one participant benefits and one participant loses.

interaction is mutualistic or antagonistic. Some interactions are direct, where species physically interact (such as plants and their pollinators). Others are indirect, where they influence one another through competition for a shared resource, such as food. Virtually every organism is involved in both mutualisms and antagonisms with other species—think of plants and their pollinators or predators and their prey. These interactions often narrow the geographic distributions of species within the larger boundaries of their physical requirements, and so help shape their realized niches.

Limited resources foster competition.

Whenever there are fewer resources than individuals seeking them, there can be competition. **Competition** is an interaction in which the use of a mutually needed resource by one individual or group of individuals lowers the availability of the resource for another individual or group. Different species of sea anemones compete for space in rocky shallows, swallows compete with one another for nesting holes in tree or cliffs, and many different species of dung beetles compete for animal droppings (some even cling to monkeys, waiting for feces to emerge). Male birds, mammals, insects, and other animals commonly compete for female mates.

In Chapter 46, we saw that in many populations the tension between intrinsic rate of reproduction and environmental carrying capacity results in intraspecific competition—competition within a species. The increase in competition between individuals of a species that accompanies increased population density is a main reason why population growth slows as the species' environment approaches its carrying capacity. As noted earlier in this chapter, competition can also occur between individuals of different species, in which case it is interspecific competition. Whether competition occurs between individuals of one species or two different ones, it is a lose–lose situation: Both sides spend energy they would not in the absence of the other. Indeed, Darwin recognized that the "struggle for existence," as he called it, is a primary driver for natural selection, leading to the evolution of traits that aid survival and reproduction by reducing competition in one way or another.

Competition promotes niche divergence.

In general, when two co-occurring species overlap strongly in resource use, one will either become extinct in that place or change its resource use. For example, coyotes and foxes eat the same small prey, mice and rabbits, but when they meet, a coyote will waste energy by chasing the fox and of course the fox loses energy by running away from the coyote. In regions where foxes and coyotes occur together, the foxes avoid interactions with the larger coyotes by leaving the area, placing their dens in locations where coyotes do not occur. This result of such an antagonistic interaction, in which one species is prevented from occupying a particular habitat or niche, is called **competitive exclusion**. This is another way that a fundamental niche can be reduced to a smaller realized niche. Over time, competitive exclusion leads to resource partitioning, as we saw with the *Anolis* lizards (Chapter 46).

Species can compete with each other directly or indirectly. For example, some trees compete with other trees directly by producing toxins in the soil around them that the seedlings of other species cannot tolerate, and corals may compete by stinging overlapping colonies. Some species compete with others indirectly by interfering with the resources used by a different species. For example, diatom algae make skeletons of silica

---

FIG. 47.5 Competition for space. (a) Herons, egrets, and cormorants compete for nest sites in wetland trees. (b) In the sea, corals compete for space on reef surfaces.

(SiO₂), depleting this substance in the surface waters of the world's oceans. Radiolaria also make skeletons of silica, and for the past 65 million years, radiolarians have evolved silica-saving modifications of their skeletons in response to the increasing sequestration of SiO₂ by diatoms.

Species compete for resources other than food.

Food is the resource we most often think of as the limiting factor for population growth and reproduction, but many species just require a place to stand, so to speak, or suitable conditions for a nest. Barnacle and mussel larvae compete for a place to attach on submerged rocks. Their food, tiny plankton, is almost always abundant and therefore is not as likely to limit population growth as attachment space is.

Similarly, plants compete for space in fields and forests so they can collect sunlight and obtain water and nutrients from the soil. Sunlight cannot itself be exhausted by use, but room to collect sunlight on the ground below can be filled. Nesting sites for birds that lay their eggs in tree hollows or on cliff faces can also be strongly limited by the availability of space, as can the settling of corals on reef surfaces (**Fig. 47.5**).

Predators and parasites can limit prey population size, minimizing competition.

Predation is another form of antagonism, as the predator benefits at the expense of the prey. Predators can limit the population sizes of their prey, preventing prey populations from increasing to the level where competitive exclusion occurs. We have seen that when species overlap in resource use and their populations increase to the point where they compete for resources, they must be separated in space or time or risk local extinction through competitive exclusion when resources become limited.

If, however, populations do not rise to densities at which resources are limiting, competition is reduced and species can overlap in niches without excluding one another. For example, in ponds throughout the southern United States, tadpoles of the large Southern Toad, the Eastern Spadefoot, and the tiny Spring Peeper all use the same food source. In the presence of the two larger species, Peeper tadpoles compete poorly for food (**Fig. 47.6**). As a result, they have low survival rates, and those individuals that do make it to maturity are often undersized

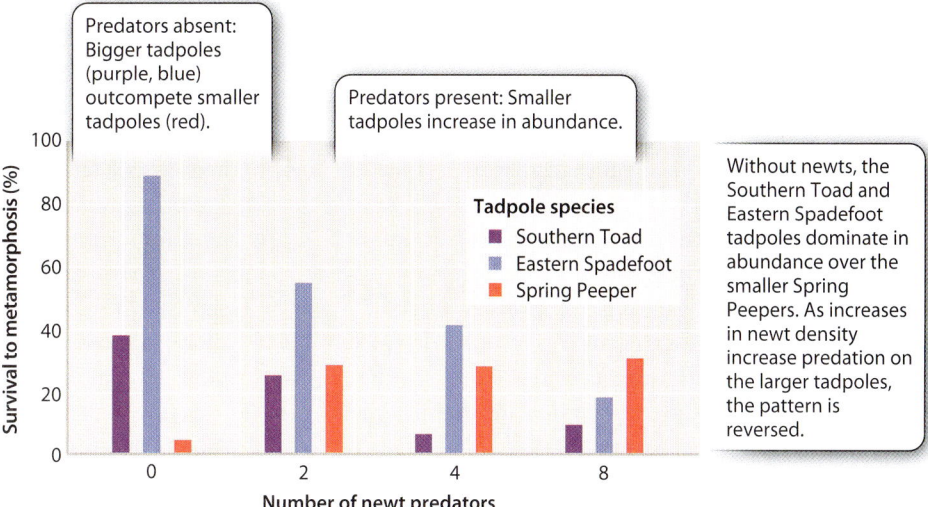

FIG. 47.6 **Effect of predators on the outcome of competition.** Newt predators in ponds prefer to eat bigger tadpoles that compete well for food, enabling smaller tadpoles that are poorer competitors to survive to adulthood.

because as tadpoles they did not have access to adequate nutrition. In the presence of a predatory Red-Spotted Newt, however, the tables are turned. The newt prefers to eat the larger tadpole species, and so prevents them from becoming sufficiently abundant to outcompete the smaller Peeper tadpoles. The Peeper tadpoles, then, have the highest survival rates. With competition for food reduced, these small tadpoles grow to their maximal body size, and become the most abundant tadpole species in the ponds.

Parasites can also limit the population size of their host, keeping numbers well below the carrying capacity of the environment. An extreme example is the fungus that infects the American chestnut, a tree once dominant in eastern North American forests. The fungus attacks the vascular system of young chestnut trees, which usually succumb when they reach 2 to 3 m in height. Today, there are only a few isolated patches of large chestnut trees in Vermont and a handful of other places that have escaped the fungus. As chestnut populations have declined, populations of oak, beech, and other trees that grow in the same forest have increased.

47.3 MUTUALISTIC INTERACTIONS BETWEEN SPECIES

Interactions between species do not necessarily pair a gain for one participant with a loss for the other. As the example of cacao and midges shows, some interactions benefit both participants. Typical benefits include access to nutrients, shelter from enemies or weather, and direct help in reproduction, but all benefits are ultimately measured by natural selection in terms of reproductive output. Some typical costs include the proteins, fats, and carbohydrates that are invested in building structures such as flowers that attract pollinators, or specialized tissues that house bacteria or algae. Costs can also be energy-consuming activities such as the transport of pollen or seeds, or the loss of food resources used by a partner. When the benefits for each participant outweigh their costs, the interaction is a mutualism.

Mutualisms are interactions between species that benefit both participants.

The midges in our opening example obtain the benefit of food from the cacao blossoms at the cost of unwitting pollen transport between flowers; the chocolate tree obtains the benefit of pollination at the cost of producing the sugars and amino acids in the flowers' nectar.

The nitrogen-fixing bacteria that live in nodules on the roots of soybeans (Chapters 26 and 29) provide another illuminating example. The bacteria occur naturally in the soil and are attracted to the plant roots, where they take up residence in nodules produced by the plant. The bacteria provide their host plant with nitrogen in a biologically useful form, and the plant provides the bacteria with food and a stable environment. In response to signals from the bacteria, the plant builds the bacteria's home—the nodules—out of root tissue, a cost outweighed by the benefit of enhanced access to nitrogen, a benefit measured in greater growth and reproductive output. In the interaction between plant and bacteria, both sides win.

Despite the observation that both partners in a mutualism benefit, it is important to remember that each side is acting in its own self-interest and bears costs that are weighed against benefits in terms of growth and reproduction. Associations that, overall, are beneficial to one partner are often also beneficial to the other partner and, therefore, more often represented in the next generation. In other words, mutualisms are subject to natural selection, just like any other adaptation (Chapter 21).

Mutualisms may evolve increasing interdependence.

One of the best-studied mutualisms involves aphids (and their insect relatives) and closely associated bacteria. The insects suck plant sap for food, and the bacteria live in tissues that develop in the insects adjacent to their digestive system. The insects provide a home for the bacteria at some physiological cost but gain an important benefit: Plant sap contains relatively few nutrients, and the bacteria provide their hosts with essential amino acids. The bacteria, in turn, benefit from a stable and favorable environment at the cost of some loss of nutrients supplied to the aphids. The aphids pass the bacteria from mother to daughter in egg cells, guaranteeing that offspring will have bacteria as effective as those associated with earlier aphid generations. When the interaction between species drives reciprocal adaptations in both participants, this opens the possibility for long-term coevolution because the descendants of each side will also be associated with each other, just as were past generations. Any mutations that arise can affect both sides of the partnership because neither aphids nor bacteria are found without the other.

DNA evidence suggests that aphids and their bacteria have been associated for 100 million years or more, and that the bacterial genomes have been gradually losing genes they would need for life outside the insects (**Fig. 47.7**). Some bacterial genes have in fact been acquired by the insects. We saw a similar process of coevolution in Chapter 26. As we discussed there, mitochondria and chloroplasts are probably nature's most dramatic examples of coevolved mutualism.

While some mutualisms involve close interactions between specific species, others are less particular. For example, sweet-tasting fruits evolved in flowering plants, attracting mammals and birds that disperse their seeds, but not in response to any particular bird or mammal species. Similarly, flowers evolved in response to insect pollinators such as bees and flies, whose own adaptations for visiting flowers evolved in response to flower availability.

Mutualisms may be obligate or facultative.

When one or both sides of a mutualism cannot survive without the other, the association is said to be **obligate**. For example, the association of aphids and bacteria is obligatory for both sides: Bacteria cannot live without the shelter the aphids provide, and aphids cannot live without the nutrients provided by the bacteria.

Many associations, however, are not so tightly intertwined, and one or both participants can survive without the other. These interactions are called **facultative**. For the midges that pollinate cacao, the association is facultative because the midges have other sources of food, including other flowers, in the wet forests they inhabit. For cacao, however, the relationship is obligate because bees and other insects do not usually pollinate them, and so their reproduction more or less relies on visits by midges.

Many obligate mutualisms are thought to have begun as facultative relationships that became reinforced over time by natural selection. Thus, the mutation that originally made a cacao flower sprout from the tree trunk close to the ground rather than high out on a branch made the flower easier to pollinate by low-flying midges while decreasing the chances of discovery and pollination by bees. If midges were consistently more reliable pollinators than bees, the mutation would have been selected and then spread throughout the population of cacao (Chapters 21 and 22).

HOW DO WE KNOW?

FIG. 47.7

Have aphids and their symbiotic bacteria coevolved?

BACKGROUND Aphids are small insects that are common pests of garden plants. Specialized cells in the aphids harbor populations of the bacterium *Buchnera* that provide their host with amino acids essential for growth. Most, but not all, aphid reproduction occurs without males, with mothers producing daughters. Mother aphids pass the bacteria through their eggs to their daughters.

HYPOTHESIS Aphids and *Buchnera* bacteria have coevolved for millions of years.

EXPERIMENT Evolutionary ecologist Nancy Moran sequenced the DNA of bacteria and the aphids they come from to establish the phylogenetic relationships of aphid species and their symbionts. She reconstructed the phylogeny of both groups and compared their relationships. If the two phylogenies matched each other, the hypothesis of coevolution would be supported.

RESULTS The phylogenetic trees of aphids and their associated bacteria matched perfectly, just as if the bacteria were a gene of the aphids instead of a separate lineage. This matching of phylogenetic trees was seen both among aphid species in different genera and for aphid populations within a single species. A DNA-based molecular clock further showed that aphids, and the insects related to them, have all coevolved with these bacteria over nearly 200 million years.

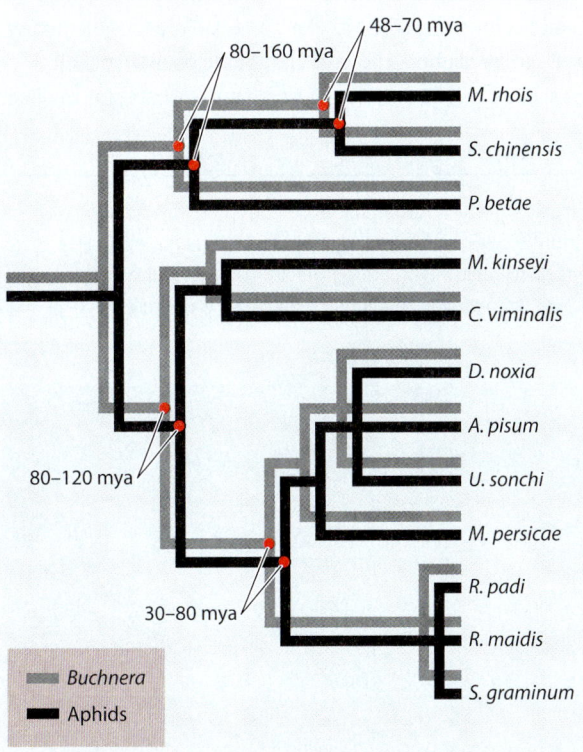

CONCLUSION A single lineage of *Buchnera* has been passed down through the generations and the diversification of its aphid hosts for nearly 200 million years. Host-symbiont systems can be stable over long time intervals and show long-term coevolution.

FOLLOW-UP WORK Researchers are studying the changes in the genomes of bacteria living inside aphids to see how they lose genes necessary for living independently of their aphid hosts.

SOURCE Moran, N. A., M. A. Munson, P. Baumann, and H. Ishikawa. 1993. "A Molecular Clock in Endosymbiotic Bacteria Is Calibrated Using the Insect Hosts." *Proceedings of the Royal Society. London, Series B*, 253:167–171.

TABLE 47.1 Types of Species Interactions

INTERACTION	EFFECT ON EACH SPECIES	EXAMPLE
Mutualism	+/+	Flowers and bees: Flowers gain pollination; bees gain nectar and some pollen.
Antagonism	+/− or −/−	
Competition	−/−	A grass and a wildflower: Each species loses the water, nutrients, and access to sunlight that the other takes.
Predation	+/−	Arctic foxes and lemmings: Foxes benefit from eating lemmings; lemmings lose opportunities to reproduce.
Herbivory	+/−	Bison and grass: Bison benefit from eating grass; grass loses biomass that is eaten.
Parasitism	+/−	Tapeworms and humans: Tapeworms benefit from absorbing nutrients in human intestine; humans lose nutrients.
Commensalism	+/0	Egrets and cattle: Egrets benefit from insects stirred up by cattle; cattle are unaffected by egrets.

The costs and benefits of species interactions can change over time.

In addition to interactions that benefit both partners (mutualism) and those in which at least one partner is harmed (antagonism), there are interactions in which one partner benefits with no apparent effect on the other (**commensalism**). For example, Grey Whales in the Pacific Ocean are commonly festooned with barnacles. The barnacles benefit from the association, obtaining both a substrate for growth and a free ride through waters rich in planktonic food, but the presence or absence of barnacles doesn't seem to affect the whales one way or the other.

Some ecologists also recognize a fourth class of interaction, called amensalism, in which one partner is harmed with no apparent effect on the other. A commonly cited example is the Black Walnut, a forest tree that produces compounds that inhibit the growth of other plant species in its immediate vicinity. A fair criticism of this and many other cited examples of amensalism is that the Black Walnut actually benefits by limiting competition for nutrients in the soil. The spectrum of costs and benefits associated with interactions among organisms is summarized in **Table 47.1**.

Associations are not fixed—they can change over time. A mutualism can in some cases become antagonistic if one of the partners "cheats" by imposing a larger cost than benefit on the other. In fact, mutualisms that are loose associations among changing partners can become one sided rather quickly. For example, many plants have evolved tubular flowers that guide bees past their anthers or stigma on the way to the nectar at the base, taking just

FIG. 47.8 Commensalism and mutualism. Egrets feed on insects stirred up by water buffalo, which are usually not affected by the egrets—a commensalism. If, however, the egrets spot lions and the poor-sighted buffalo take warning, the relationship becomes a mutualism.

a little more time from bees who try to visit as many flowers as possible. Some bees have short-circuited this plant mechanism by nipping the flower base from the outside and then drinking the nectar without pollinating the flower. Most plants have enough successful pollination to ensure seed production, but losses from cheating can still be costly.

Some interactions begin as a benefit to one species, with no benefit or cost to the other. For example, cattle egrets follow water buffalo to pick up the insects stirred up by the buffalo passing. The egrets gain food, but the buffalo are not affected either way—an example of commensalism (**Fig. 47.8**). However, cattle–egret commensalism can become a mutualism over time if the egrets give early warnings of nearby predators such as lions or eat insects like tsetse flies that carry buffalo diseases.

→ **Quick Check 2** What are some of the costs and benefits to an apple tree and to the honeybee that pollinates it?

47.4 ECOLOGICAL COMMUNITIES

We have been discussing how species interact and how these interactions can shape the distributions of populations, limiting distributions in one way or another. Now we look collectively at all the species that occur together in the same place. How, and to what extent, do they interact? And what are the properties of these spatial aggregations of species?

Species that live in the same place make up communities.

A **community** is the set of all populations found in a given place. That definition seems straightforward, but it raises some large questions about the nature of communities. Two different views were proposed nearly a century ago, and ecologists have debated their merits ever since. Frederic Clements, a prominent American plant ecologist of the early twentieth century, likened communities to a "superorganism" in which species interact strongly and predictably, like the organs within a body. In contrast, his contemporary, Henry Gleason viewed communities as simply the products of species acting individually in time and space. Not surprisingly, research inspired by their disagreement suggests that most communities lie somewhere between these extremes.

Populations in a community are tied together by the various interactions that secure their spot in a food web (Chapter 25), as well as by their physical location. But when we look at the details of where particular species occur, it turns out that almost no two species have exactly the same geographic distributions. This is partly a result of competitive exclusion (unless they are obligate associates of each other, like aphids and their bacteria). The Black Spruce and Bog Club Moss that inhabit freshwater bogs in northern New England overlap in their mutual preference for water-saturated, acidic, sandy soils found in cold temperate bogs, but they do not occur in the same places over most of their ranges.

Even within a local habitat like a meadow, species have different distributions that may reflect differences in soil moisture or patterns of sunlight exposure. For example, two kinds of milkweed occur in meadows, but one prefers damper sites and the other drier areas. For milkweeds and other plants, interactions also vary from place to place as the local set of species changes. Because a given species' nearest neighbors differ over its geographic range, natural selection may also differ from place to place as particular competitors, partners, predators, or prey drop out or enter its distribution. For example, in the northern part of their range, Least Weasels prey only on lemmings, and so are under selection to coordinate their reproduction with that of the lemmings. But in the southern part of their range, lemmings are not present and so the weasels hunt mice and rabbits, placing different selection pressures on them.

The activities of different populations in a community also vary in time. For example, flies are active during the day, pollinating flowers that open in the morning and falling prey to dragonflies, birds, and other predators. In contrast, moths are important pollinators at night, when spiders and bats that prey on them are also active. Similarly, in coastal marine communities, oysters, crabs, and many other animals follow the daily rise and fall of tides. And, of course, the activities of many species change seasonally in responses to sunlight, temperature, and rainfall. Protein-rich insects that birds prey on to feed their young are most abundant in the spring, and fruits and seeds rich in the oils that migratory birds need to supply fat reserves are mostly available in late summer and fall.

A single herbivore species can affect other herbivores and their predators.

Predators and prey interact in settings as different as a tropical rain forest or Arctic tundra. In habitats with many different species of predators and prey, identifying the way each species affects others can be challenging. For this reason, much research on how predators and prey influence community structure has been carried out in the Arctic, where communities tend to have relatively few species.

One well-studied community is found on Bylot Island in the Canadian territory of Nunavut. There, two species of lemming, the main herbivores in tundra communities, affect the populations of other herbivores and predators. Snow Geese are herbivores that live alongside the lemmings, and both geese

FIG. 47.9 Predator–prey relationships in the Arctic grassland community on Bylot Island, Canada. Species have either a positive effect (+) or a negative effect (−) on other species.

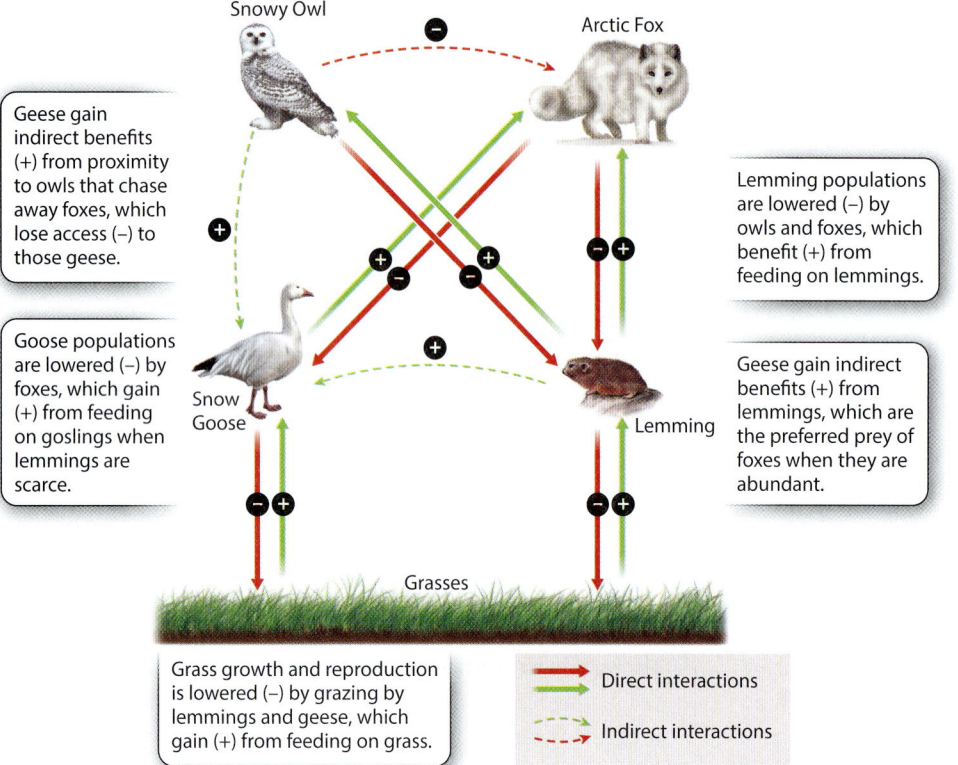

Keystone species have disproportionate effects on communities.

While all or nearly all species in a given place have some effect on one another, the integrity of a community may depend on a single species. This situation can occur when one species influences the transfer of a large proportion of biomass and energy from one trophic level to another or when the species modifies its physical environment, as beavers do when they make ponds. Such pivotal populations are called **keystone species** because they support a community in much the way a keystone supports an arch. Though all species contribute to the structure of a community, a keystone species affects other members of the community in ways that are disproportionate to its abundance or biomass.

A classic experiment by American zoologist Robert Paine gave birth to the concept of keystone species. Along the northwestern coast of the United States, mussels adhere to rocks in the intertidal zone, where they coexist with about two dozen other species of invertebrate animals and seaweeds. Mussel populations are regulated by their principal predator, a sea star called *Pisaster ochraceus*. When Paine removed the sea star from experimental plots, mussel populations exploded, outcompeting other animals and seaweeds for space and so eliminating them from the rocky shoreline. Control plots in which the sea star remained showed no such change. From these results, Paine concluded that the activities of the predatory sea star had a singularly strong influence on the diversity and distribution of other populations in the rocky intertidal community. *Pisaster ochraceus* is a keystone species.

Another experiment, this time unintended, cements the point. Along the west coast of the United States and Canada, giant kelp grow more than 50 m tall, providing a home for a remarkable diversity of fish and invertebrates (**Fig. 47.10**). The kelp community depends strongly on a single predator: the sea otter. Sea otters are the predators of sea urchins, which in turn graze on the rootlike holdfasts that anchor kelp to the seafloor. When the otters were removed by overhunting

and lemmings eat the grasses and other low plants that grow on the island. The main predators of lemmings and Snow Geese are Arctic Foxes and Snowy Owls. Lemmings, foxes, Snowy Owls, geese, and plants, then, are linked by a set of mutualisms and antagonisms, shown in **Fig. 47.9**. Changes in one species can affect all of the others.

Lemming populations rise and fall every 4 years or so. When lemming populations decline, foxes turn to other prey, including the eggs of Snow Geese (while the now harder-to-find lemmings escape and their populations begin to increase again). Snow Geese, therefore, are indirectly influenced by lemming population densities. The population density of foxes also reflects that of lemmings—as lemming populations increase to the point at which they can again become a significant part of fox diets, the foxes have access to enough resources to have more pups, increasing the fox population. In time, the additional predators again cause the lemming population to decline, and fox reproduction drops as a result. The population density of lemmings also influences grass abundance through its indirect effect on whether or not foxes eat Snow Goose eggs; young geese are heavy grazers in years when goose reproduction is high.

FIG. 47.10 Keystone species in Pacific kelp forests. The activities of keystone species strongly influence the communities in which they live.

Sea otter predation is key to the health of a kelp forest. Notice the many fish beneath the kelp canopy.

When otters are absent, their sea urchin prey rise in abundance and consume most of the kelp holdfasts. Without kelp forests, many animals decline in abundance.

decades ago, sea urchin abundance soared and the kelp nearly disappeared, along with many other populations that live in kelp forests. Now, where otters are protected they once again keep urchin populations in check, enabling the kelp to return, and along with it, the diversity of fish and other animals for which kelp provides a habitat.

Keystone species called ecosystem engineers actively shape the physical environment, creating habitat for others. The classic example is beavers, which produce ponds by damming streams in forests, creating habitat that would not otherwise exist (**Fig. 47.11**).

→ **Quick Check 3** Lemmings affect many other species. Why are lemmings not considered a keystone species?

Disturbance can modify community composition.

We have seen how predation, herbivory, and parasitism can reduce the effects of competition. Severe climatic events, such as unusual rainfall or extreme temperatures, can also dramatically lower the abundances of some species, for example when an early frost kills mosquitoes and other insects. Severe physical impacts on a habitat, such as those caused by storms, earthquakes, or road building, are known as **disturbances**, and they have effects on populations of interacting species that are independent of their densities, as we saw in Chapter 46. Disturbances often affect multiple species in the same community and so can exert a strong influence over community composition. We usually think of disturbances as abiotic, but the line is not so clear with road-building by humans, for example, or dam-building by beavers. The main point is that catastrophic changes in a habitat affect populations in ways independent of their densities. Such changes are not uncommon, but are usually unpredictable.

Severe weather in forests demonstrates how disturbance can affect a community. When fire or a hurricane takes down tall trees in a forest, it also takes away the shade that had prevented some species from establishing new seedlings. Fires are frequent sources of disturbance in regions with a seasonally dry climate, such as parts of western North America and southwestern Australia. Like storms, fires remove incumbent plants and animals, opening up habitat for recolonization. Some species have, in fact, adapted to frequent fires. For example, several pine species have evolved thick bark and high branches, traits that help protect them from fire. Others, which have evolved

FIG. 47.11 A beaver dam visible from space. This NASA satellite image shows a beaver dam in northern Alberta, Canada.

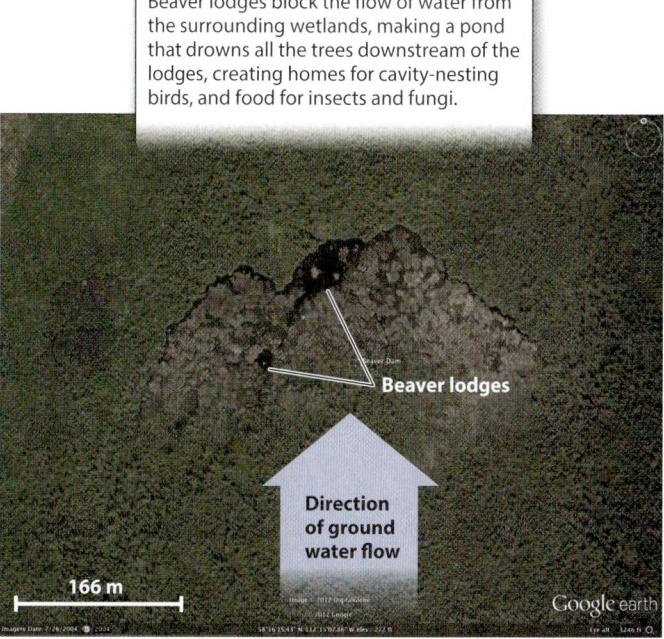

Beaver lodges block the flow of water from the surrounding wetlands, making a pond that drowns all the trees downstream of the lodges, creating homes for cavity-nesting birds, and food for insects and fungi.

cones that open only in the wake of fire, use the habitat space opened up by the fire and the nutrients that dead plant matter releases into the soil.

Disturbances allow species to enter or remain in the community as long as they can establish their young in the newly opened habitat patch. Noting the influence of storms on community composition in forests and coral reefs, American ecologist Joseph Connell proposed that the diversity of species in a given community reflects, at least in part, the frequency and intensity of disturbance. If disturbance is intense, few species can tolerate the physical conditions of the habitat and diversity will be low. On the other hand, if disturbance is rare or weak, competition may take over and only a few, stronger, competitors may remain. Species diversity tends to be highest when disturbance is frequent or intense enough to inhibit competition, but not so strong as to limit the number of species that can tolerate the environment.

Succession describes the community response to new habitats or disturbance.

Following disturbance or the appearance of a new habitat, a community goes through a series of changes. These changes involve a predictable sequence of species that colonize and then transform the community, in what can appear to be a linear process of maturation. For example, following a large storm or fire, newly vacated habitat is commonly colonized by plant species adapted for rapid growth and strong sun. Eventually, these species are replaced by others able to grow in increasing shade. The pace of this transformation may be rapid, occurring over a few decades, or it may be much slower, depending on the environment.

This process of species replacing each other in time is called **succession** (**Fig. 47.12**). Clements viewed the pattern of species replacement in succession as strong evidence in support of his view that communities are highly integrated. Gleason, in contrast, explained succession purely in terms of the habitat preferences of individual populations.

The pin cherry, a resident of New England forests, illustrates how the life histories of individual species can contribute to succession. If you walk through a forest in New Hampshire, you will see oak trees and beeches, pine and hemlock. A mature forest may contain few or no pin cherry trees, but if you examine seeds lying dormant within the forest soil, you will see numerous pin cherry seeds. How did they get there?

Years earlier, disturbance from a storm or fire opened up the forest floor and pin cherries established themselves in the sunlit soils, growing from seeds left in the droppings of birds that had consumed the cherries elsewhere. The seedlings of pin cherries grow well in full sunlight, so the trees grew rapidly to reproductive maturity. A forest canopy was eventually

FIG. 47.12 Succession. In new habitats or following disturbance, species colonize and transform the habitat in a predictable sequence that reflects the adaptations of participating species.

reestablished, but pin cherry seeds do not germinate in the shade. In contrast, oak seedlings prosper in shade, so through time oaks and other species came to dominate this patch of forest. They will continue to dominate it until the next disturbance, when the removal of mature trees will once more expose the forest floor to bright sunlight, literally giving pin cherries another day in the sun.

The succession of species reflects the adaptations of each to the changes in sunlight, soil, water, and space that occur as the first organisms arrive mature and die. These organisms' bodies influence their habitat in both life and death, whether aquatic or terrestrial. Succession over thousands of years can turn bare rock into forest. Lichens are often the first organisms to colonize bare rock newly formed by cooling lava or exposed by landslides. The lichens may start off with only a few mites and other tiny invertebrates living on them. As the stone weathers, and microorganisms feed on the lichen and one another, they form particles of soil that accumulate in cracks and crevices. Eventually, small plants, such as sedges and grasses, take root, and the growing plant community begins to gain height and diversity as other plants establish themselves.

Soon, larger insects, spiders, and other invertebrates join the community, as well as the first small mammals and birds. Many of these early colonizing species are *r*-strategists (Chapter 46), growing and reproducing quickly, and spreading their young far and wide. Birds carry the seeds of shrubs and other plants in their droppings, and the gradually deepening soil eventually permits the establishment of shrubs that in turn provide habitat for larger mammals and birds, and perhaps snakes and salamanders.

The first sun-loving trees, such as birches and pines, appear next, providing shade for the seedlings of later colonizing trees like oaks and maples. In many forests, these stable, long-lived *K*-strategists (Chapter 46) make up a final stage in the succession, forming a mature assembly commonly called a **climax community.** A climax community is one in which there is little further change in species composition.

47.5 ECOSYSTEMS

A community of organisms and the physical environment it occupies together form an **ecosystem (Fig. 47.13).** The physical and biological components of ecosystems are linked by the processes that cycle nutrients and transfer energy through the system. In Chapter 25, we introduced the concepts of the **food web** and the **trophic pyramid** as depictions of carbon cycling and energy flow through ecosystems, respectively. Now we can revisit these concepts, understanding them as reflections of the functional properties of individual species and the mutualisms, antagonisms, and other interactions among species within a community.

Species interactions result in food webs that cycle carbon and other elements through ecosystems.

Of all the interactions observed among species, none are more prominent than predation and herbivory. Heterotrophic organisms, from bacteria to blue whales, obtain carbon and other elements needed for growth from other organisms. Thus, when a sea star preys on a mussel, we can view this interaction both as an antagonism between two species and as a link in the carbon

FIG. 47.13 A savanna ecosystem in Africa. The ecosystem includes the grasses, trees, animals, and microorganisms in the community, as well as the climate, soil chemistry, and other features that define its physical environment.

cycle. Heterotrophic organisms commonly consume other heterotrophs, but, ultimately, ecosystem function depends on autotrophs—photosynthetic or chemosynthetic organisms that can form organic molecules by the reduction of carbon dioxide available in the physical environment (Chapter 6).

In ecology, photosynthetic and chemosynthetic organisms are known as **primary producers**, organisms that take up inorganic carbon, nitrogen, phosphorus, and other compounds from the environment and convert them biochemically into proteins, nucleic acids, lipids, and more (Chapter 25). **Consumers**, heterotrophic organisms of all kinds, depend on primary production, directly consuming primary producers or consuming those that do. Predators hunt and eat prey, herbivores eat plants, and parasites infect their hosts.

The various species present in an ecosystem can therefore be placed in an order that describes how one organism feeds on another, moving carbon through the system. This order is commonly depicted as linear, and it is called a food chain. In nature, however, every species is connected to many others. Many ecologists also stress that most communities contain species that are omnivorous, eating both plants and animals. For these reasons, the movement of carbon through an ecosystem is more realistically thought of as a **food web** (**Fig. 47.14**).

An organism's typical place in a food web is its trophic level (Fig. 47.14). The first trophic level consists of primary producers; consumers occupy the levels above the first trophic level. At the second trophic level are primary consumers; these are the herbivores, organisms that eat plants or other primary producers. Predators, organisms that eat other consumers, are secondary consumers, at the third trophic level, and potentially higher levels as well. Top or apex predators, which have few, if any, natural predators, occupy the highest trophic level. Parasites exploit all levels of the system, obtaining carbon and other nutrients from primary producers and consumers alike. Finally, there are the all-important decomposers and detritivores. Decomposers feed on the dead cells or bodies of other organisms, and detritivores do much the same, consuming partially decomposed plants or animals in soil or sediment, as well as organic compounds in feces. These organisms return carbon dioxide and other inorganic compounds to the environment, completing the cycle of elements through the ecosystem.

FIG. 47.14 A food web. A food web diagrams the cycling of carbon and other elements from the environment through a succession of organisms and back to the environment.

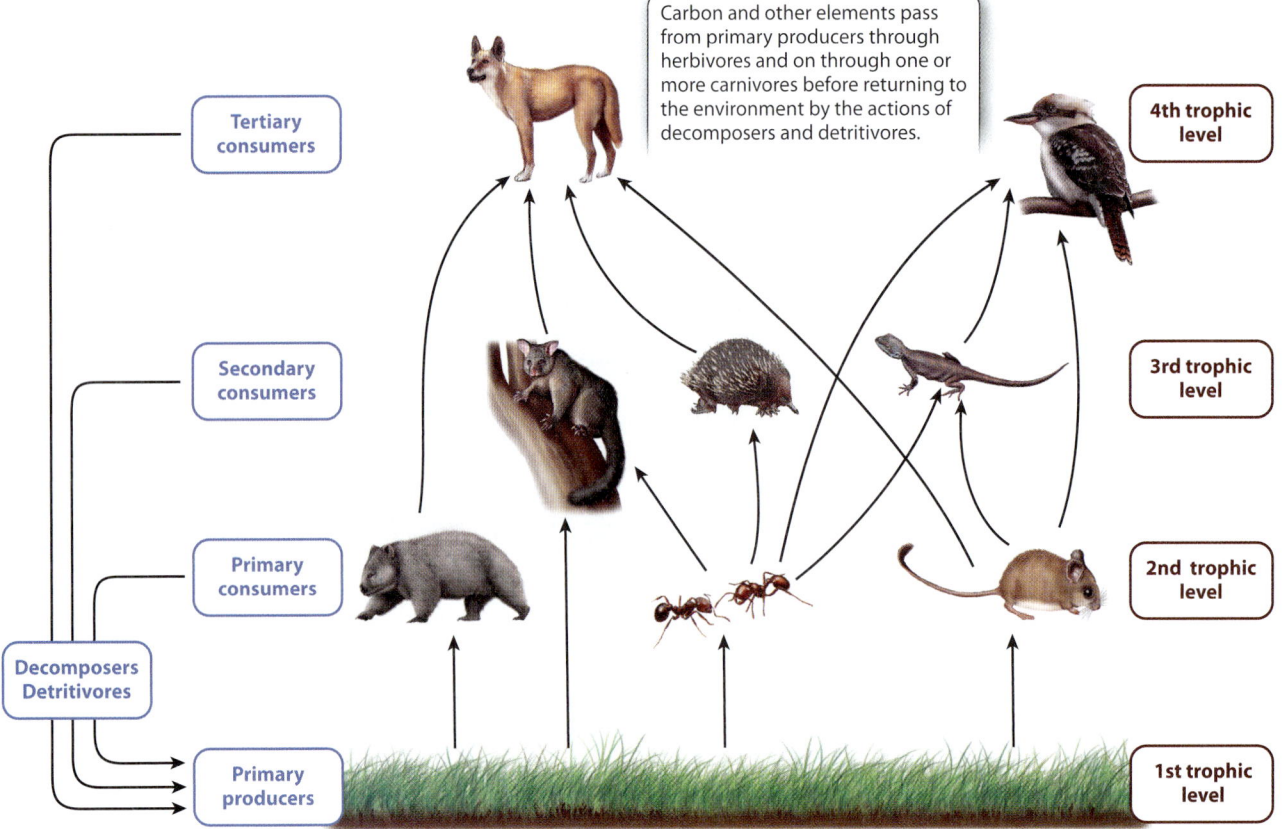

Species interactions form trophic pyramids that transfer energy through ecosystems.

The interactions among species transfer energy as well as carbon and other elements through ecosystems. Unlike carbon, energy does not cycle through an ecosystem, and so new energy must continually be harvested from the environment to sustain the community. In essentially all ecosystems where sunlight is available, photosynthesis lies at the heart of ecosystem function. Plants, algae, and photosynthetic bacteria capture energy from the sun and use it to synthesize organic molecules. Where sunlight is absent, especially in the vast depths of ocean, Bacteria and Archaea fuel primary production by chemical reactions.

Some of the energy harvested by primary producers is stored as chemical bonds in organic molecules such as carbohydrates (Chapter 8) and so will be available to organisms at the next trophic level. Some, however, will be dissipated as heat or used to do work. Because building organic molecules is not 100% energy efficient, and because organisms at one trophic level rarely consume all the resources in the level below them, biomass and the energy it represents generally decrease from one trophic level to the next. The result is a **trophic pyramid**.

The broad base of biomass available in primary producers decreases steadily with each ascending step in trophic level (**Fig. 47.15**). In general, about 10% of the energy and biomass available at one trophic level is passed to the next. So 1000 kg of grass biomass will support about 100 kg of lemmings (about 400 animals), and that, in turn, will support 10 kg of small predators like foxes. At the apex of the pyramid, we find a single kilogram of eagle tissue. In this way, primary production exerts a powerful influence on the rest of the community. To get more foxes and eagles, we would need more photosynthesis.

Light, water, nutrients, and diversity all influence rates of primary production.

If most ecosystems rest on a base of primary production, what controls rates of photosynthesis? Sunlight plays a role. Although all ecosystems, from polar deserts to equatorial forests, receive the same total number of hours of sunlight annually, the intensity of solar radiation declines from the equator to the poles because of Earth's curvature. Seasonality also becomes more pronounced toward the poles, influencing the ability of primary producers to sustain growth throughout the year at higher latitudes.

In addition to sunlight, in most ecosystems water and nutrients govern rates of primary production. In Chapter 29, we saw that plants perform a physiological balancing act, opening their stomata to allow carbon dioxide into leaves but losing water vapor as a result. Because photosynthesis requires both carbon dioxide and water, photosynthetic rate—the amount of CO_2 fixed into organic matter in a given time interval—is closely tied to water

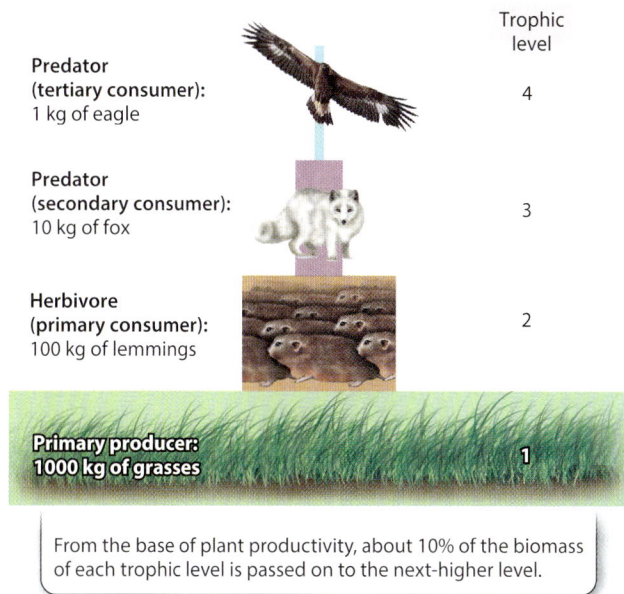

FIG. 47.15 A trophic pyramid. A trophic pyramid graphically shows the transfer of energy through an ecosystem.

From the base of plant productivity, about 10% of the biomass of each trophic level is passed on to the next-higher level.

availability. Lack of regular access to water limits the productivity of photosynthetic organisms in deserts and seasonally arid landscapes, explaining why irrigation increases crop yields in places like southern California, northern Africa, and the Middle East.

Even where water is not limiting, rates of photosynthesis are constrained by the availability of nutrients. This is why fertilizer generally increases crop yields (Chapter 29). Plants, algae, and photosynthetic bacteria all need nitrogen, phosphorus, sulfur, iron, and an array of trace elements to grow, and any one of these can limit primary production. In forests, nitrogen is commonly the soil nutrient that is least abundant relative to the needs of plants. In the open ocean, nitrogen can also be limiting, but commonly it is a scarcity of phosphorus that limits photosynthesis. And in some parts of the ocean, nitrogen and phosphorus are both present in abundance, but primary production is still limited—in this case by iron, needed for many enzymes and electron transport molecules. The idea that primary production is limited by the nutrient that is least available relative to its use by primary producers is called **Liebig's Law of the Minimum**, after Justus von Liebig, a nineteenth-century German scientist who first popularized the concept.

Biological diversity in a given region is, in some ways, a result of high productivity. We have seen that the population sizes of predators depend on prey population size, which depends on primary production, and greater primary production also means a greater variety of plants of different sizes and kinds that can support a greater variety of herbivorous animals. While this

HOW DO WE KNOW?

FIG. 47.16

Does species diversity promote primary productivity?

BACKGROUND The different plant species found in communities have combinations of leaves, stems, and roots that tap different sources of water and nutrients in the soil and capture light at different levels of sun and shade. Do the differences among plant species result in higher levels of primary production than would be possible with fewer species?

HYPOTHESIS Plant diversity promotes primary production within a community.

EXPERIMENT Ecologist David Tilman conducted a long-term experiment in Minnesota, seeding each of 11 plots with 1, 2, 4, or 16 grassland species, and controlling each plot's amount of nutrients, water, and carbon dioxide, as well as manipulating the plots' exposure to herbivory and disturbance by fire. The primary production of each plot was measured annually for up to 23 years.

RESULTS The graph shows that the addition of nitrogen fertilizer [quantified as kg of N per hectare (ha)] increased primary production significantly, and additional water, higher CO_2, herbivory, and disturbance also resulted in at least a modest increase in productivity. The difference in primary production between high-diversity (16 species) and low-diversity (1, 2, or 4 species) plots was at least as great as the largest additions of fertilizer—and much greater than any of the other treatments.

CONCLUSION The hypothesis was supported: In the grassland community under study, species diversity promotes primary production.

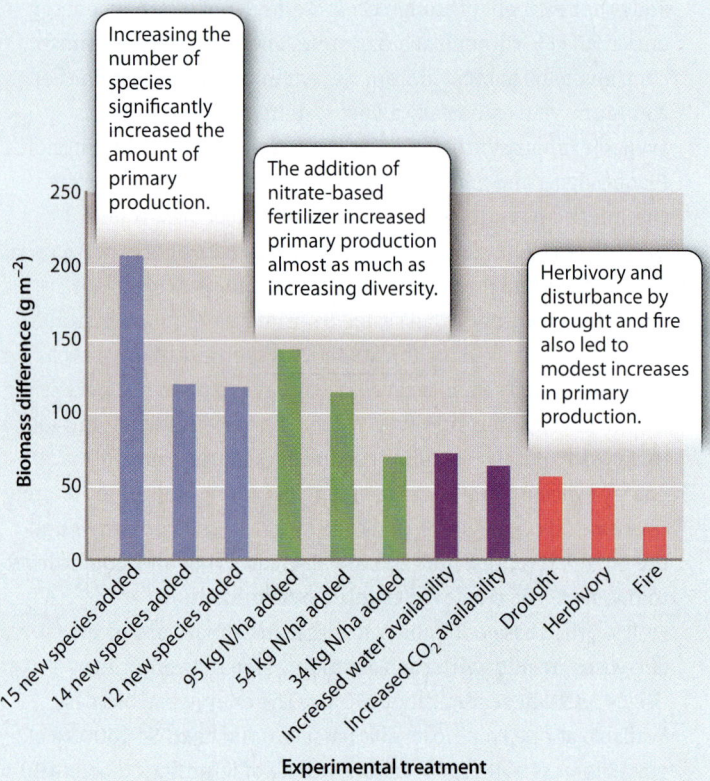

FOLLOW-UP WORK Researchers continue to study the ecosystem effects of species diversity, steadily accumulating data that show the importance of conserving the species diversity found in nature.

SOURCE Tilman, D., P. B. Reich, and F. Isbell. 2012. "Biodiversity Impacts Ecosystem Productivity as Much as Resources, Disturbance, or Herbivory." *Proceedings of the National Academy of Sciences USA*, 109:10394–10397.

---

suggests that high productivity can foster high diversity, recent experiments show that the reverse may also be true: Biological diversity may itself play an important role in ecosystem productivity (**Fig. 47.16**).

→ **Quick Check 4** What is the difference between a community and an ecosystem?

47.6 BIOMES AND DIVERSITY GRADIENTS

Populations are dynamic, continually increasing and decreasing in response to local physical and biological influences. Communities and ecosystems are also dynamic, with species coming and going as a result of disturbance and succession. Nevertheless, the assemblages of species found over broad regions of Earth are both distinctive and stable. Such broad, ecologically uniform areas are called **biomes**. On land, different biomes are recognized by their characteristic vegetation, which reflects the evolutionary adaptation of form and physiology to climate. Terrestrial biomes range from the low-growing tundra found at high latitudes to tropical rain forests, and from grass-dominated steppes and deserts to coastal estuaries (**Fig. 47.17**). Biomes occur in the oceans as well as on land, and the largest and least-understood biome of all is the vast expanse of cold dark water in the deep sea.

FIG. 47.17 Some common land biomes.

Tundra
Tundra occurs close to the North Pole, above 65° N. The South Pole is largely surrounded by the ice and seas of Antarctica, and so there is very little area with plants. Tundra is the coldest biome, and precipitation and evaporation are minimal. Even with little precipitation, the lack of evaporation and drainage means the ground is usually waterlogged, and permanent ice occurs below a few centimeters of soil. The plants are mostly mosses, lichens, herbs, and low shrubs. Grasses and sedges occur in drier places, as do other flowering plants. Plant diversity is low, and most plants are small.

Alpine
The Alpine biome is similar to tundra but lacks permanent ice below the soil, and the temperatures vary more widely. Alpine areas occur throughout the world, often at about 10,000 feet (3000 m) at lower latitudes, but always just below the snow line. Because of their altitude, these are windy, cold places. The thin atmosphere provides only limited protection from UV radiation. Many Alpine plants are therefore low and slow growing.

Taiga
These cool, moist forests occur from 50° to 65° N. The short summer brings rain, and most of the plants are conifers like spruce, fir, larch, and pine, with an understory dominated by shrubs in the blueberry and rose families. The soils are deep with accumulated organic matter because the low temperatures result in slow decomposition, but they are acidic and poor in nutrients.

Temperate coniferous forest
Two broad areas of temperate coniferous forests occur below 50° N in North America, northern Japan, and parts of Europe and continental Asia. Along the Pacific coast of the United States, abundant precipitation permits growth of enormous conifers such as Douglas-fir, Redcedar, Sitka Spruce, and redwoods. Much of the undergrowth is ferns and members of the blueberry family. In the interior of North America, much less precipitation and colder winter temperatures support drought-resistant conifers such as Ponderosa and Lodgepole Pines and Englemann Spruce.

continues on page 47-18

FIG. 47.17 Some common land biomes. *(continued from page 47-17)*

Deciduous forest
A moderate climate and dominance of hardwood deciduous trees occurs across much of North America, Europe, and Asia. Much of this biome has been subjected to human disturbance for agriculture and urban development. There are usually 15 to 25 species of trees, including maples, oaks, poplars, and birches. Springtime sun passes through seasonally leafless trees to reach a diverse understory flora. Soils are rich in nutrients from annual leaf fall, and the moderate temperatures and precipitation promote decomposition, while the cool winters promote accumulation of organic materials.

Temperate grassland
Before settlement, this biome occupied most of the midwestern United States, where it is dominated by blue-stem and buffalo grasses. Fire helps to maintain grass populations in this biome. Lack of precipitation also prevents many species of tree from growing, and those trees that grow are usually in low areas with some moisture. Where there is enough moisture to support decomposition, the soils accumulate nutrients, providing some of the most productive agricultural lands in the world.

Desert
Desert occurs in continental interiors around the world north and south of the equator from 25° to 35°. Wind patterns prevent this biome from receiving more than a few centimeters of precipitation annually. The deep-rooted plants are adapted to store water, like cactus and euphorbia. Primary production is low, and soils are poor in nutrients but may have high surface salt due to evaporation.

Chaparral
Like deserts, the distribution of chaparral reflects a narrow range of climate conditions and occurs on the western edge of continents from 32° to 40° north and south of the equator. Precipitation ranges from 30 to 75 cm per year, usually falling in 2 to 4 months. Typical plants are annual herbs, evergreen shrubs, and small trees. Olives, eucalyptus, acacia, and oaks are typical woody species, always drought resistant and often adapted to withstand fire. Limited precipitation means soils are not rich in organic materials.

Savanna

Tall, perennial grasses dominate this biome, which occurs in eastern Africa, southern South America, and Australia. Rain is seasonal and ranges from 75 to 150 cm per year. Scattered trees and shrubs usually drop their leaves in the dry season to conserve moisture. Animal diversity can be high and includes the large mammals well known in Africa.

Rain forest

This moist, highly diverse forest extends north and south of the equator from 10° N to 10° S. Annual precipitation is commonly more than 250 cm, and tree diversity alone often exceeds 300 species per hectare. Trees grow tall, and many have buttressed roots for support. Lianas and other epiphytic plants are common. Most leaves are evergreen and leathery and many have long pointed tips that facilitate drainage of excess moisture. Because of the high temperatures and heavy rains, decomposition is very rapid, preventing the accumulation of organic materials in clay-rich or sandy soils.

Biomes reflect the interaction of Earth and life.

Globally, the distribution of biomes corresponds closely to climate (**Fig. 47.18**). As noted earlier, this correspondence reflects the need for plants to balance CO_2 intake against water loss, a balance commonly struck by means of adaptations in the shape and physiology of leaves, stems, and roots (Chapter 29). To explore this further, we need to discuss a new concept: **evapotranspiration ratio**.

The intensity of solar radiation varies with latitude and altitude, as does the amount of annual precipitation and wind. These factors determine the amount of evaporation from Earth's surface to the atmosphere, including the amount transpired by plants. The

FIG. 47.18 Climate regions. Distributions of biomes around the world reflect regional climates.

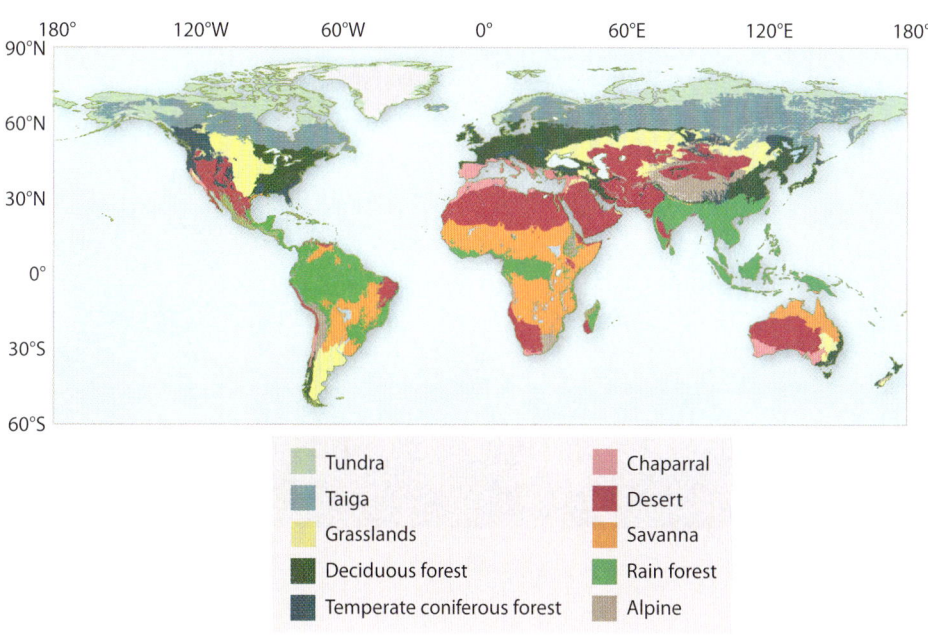

FIG. 47.19 Evaporation and transpiration. The amount of evaporation and transpiration changes with latitude and altitude.

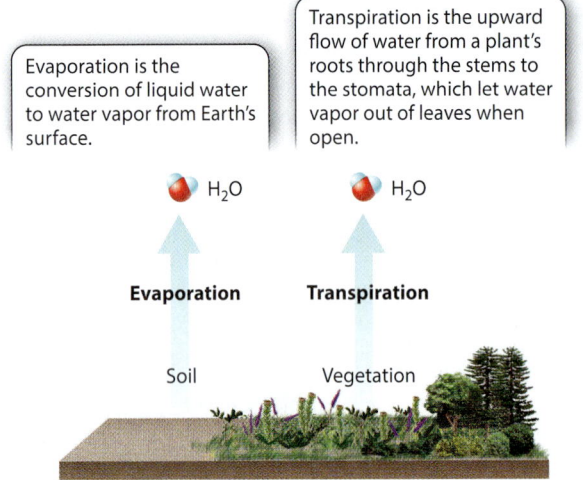

evapotranspiration ratio is the ratio of the amount of water evaporated from the Earth's surface directly from ponds, rivers, and soils, to the amount of water vapor transpired by plants (**Fig. 47.19**).

The ratio of evaporation to transpiration effectively tells us what type of vegetation can be supported in a given region. Deserts, with low annual precipitation, have a high evapotranspiration ratio: Soil moisture evaporates rapidly, and plants must keep transpiration (and therefore photosynthetic rate) to a minimum. At the other extreme, rain forests have low evapotranspiration ratios: High precipitation permits nearly unlimited transpiration (and therefore high photosynthetic rates, leading to luxuriant growth), enabling the development of wet forests with high biomass. Intermediate evapotranspiration ratios occur in the temperate zones that encompass much of the United States, Europe, and parts of eastern Asia, as well as at mid-altitudes, below the highest mountain peaks but well above the coastline. Both latitude and altitude influence the evapotranspiration ratio through their effects on temperature and precipitation, and the combined effects of these factors are reflected in the distribution of biomes around the world.

In similar biomes on different continents, dominant plants generally look similar but are not closely related. That is, plants have independently evolved structures by convergent evolution as adaptations to particular climatic regimes. For example, the cacti that grow in deserts in western North America look very similar to plants in drylands of Africa, but in Africa, these plants belong to a different family entirely (**Fig. 47.20**). Convergent evolution is also seen in rain forest trees that commonly have leaves with long tips that allow excess water to drip away, a reddish color in new leaves that protects them from intense sunlight, and roots with high buttresses that support tall trees atop thin soil.

Tropical biomes usually have more species than temperate biomes.

On a half-hour walk through a temperate forest—for example, in the United States or southern Canada—we can expect to see maybe a dozen different kinds of tree: two or three

FIG. 47.20 Convergence of plant form. Cacti in the deserts of North America (a) and euphorb plants in Africa (b) look similar because they are both adapted to conserve and store water, but they are not closely related.

FIG. 47.21 Latitudinal diversity gradient in mammals. The number of mammal species decreases from the equator to the poles.

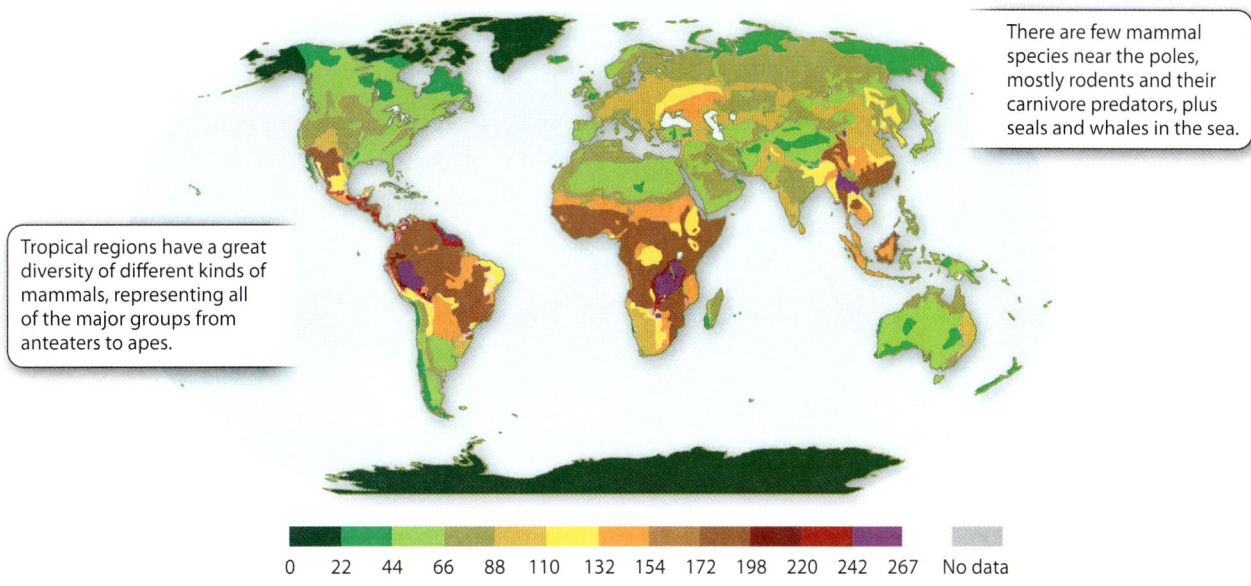

kinds of oak perhaps, a couple of maple species, birch, beech, hornbeam, ironwood, one or two pine species, and possibly a sycamore, poplar, or spruce, depending on just where we are. Standing in front of a tree and turning around, we can almost always see several more trees of the same kind. In contrast, on a similar walk through a forest in the Amazon, or in Borneo, we would pass 300 or 400 different tree species, with almost no near neighbors of the same kind. The great Victorian naturalist Alfred Russel Wallace took note of the low population density of trees in tropical forests and observed, "If the traveller notices a particular species and wishes to find more like it, he may often turn his eyes in vain in every direction."

It isn't just trees. Mammals, birds, insects, and many other kinds of animals show similar patterns of species diversity (**Fig. 47.21**). One survey found more than 700 species of leaf beetles in one small set of trees in the Amazon, a number equal to the total diversity of leaf beetles in all of eastern North America. A single tree in this same forest harbored 43 species of ants, exceeding the total for all of Great Britain.

When we compare biomes across the continents, the pattern just described stands out clearly: Species diversity generally peaks near the equator and declines toward the poles. What are causes of this **latitudinal diversity gradient,** and what are its consequences? Does this global pattern reflect the ecological processes that characterize populations?

? CASE 8 Biodiversity Hotspots: Rain Forests and Coral Reefs
Why are tropical species so diverse?

Theories abound on why the tropics contain so many species. One hypothesis, for which there is broad support, looks to the relative ages of tropical and higher-latitude biomes. Tropical biomes tend to be older, having evolved over tens of millions of years, whereas biomes at higher latitudes have been shaped by climatic change, and ultimately an ice age, within the past few million years. Because tropical habitats have existed much longer than temperate ones, biologists hypothesize that there has been more time for species to evolve and diversify at low latitudes. Several studies of diversification rate, in frogs for example, show comparable rates of diversification in temperate and tropical biomes, supporting the hypothesis that the higher tropical diversity reflects the longer interval through which diversification has taken place.

Time, however, is just one of several factors that influence diversity. Another hypothesis is that temperate communities have fewer species because it is relatively difficult to adapt successfully to cold, dry winters and the range of weather typically experienced at higher latitudes. A tree growing in a temperate area like New England experiences a wide range of temperatures over a year (and sometimes even over a single day) as winds blow and snow falls, or heat waves move through the area. Because species living at high latitudes are adapted to a greater range of conditions on average than those nearer the equator, they tend to have larger geographic ranges than tropical species, which are adapted to the narrow range of environmental variation experienced in any particular place.

It has also been suggested that the many insects and fungi specialized to attack particular kinds of trees are generally more abundant in wet tropical forests, and so trees of a given species tend to be as far apart as they can be, thus reducing the chance of being found by their enemies. This means that there would be more room for different species in any single hectare of land. It has also been noted that there is more land surface area at low latitudes, so the species-area curve (Chapter 46) plays a role in the latitudinal diversity gradient.

Very likely, the latitudinal diversity gradient reflects all these influences and more not yet appreciated. What is clear, however, is that where there are more species of trees, there are also more species of animals and other consumers. Tropical temperatures and high humidity have permitted a wide range of plant species to evolve over the past 100 million years, in turn creating many niches for the animals that depend on them.

The high diversity—and therefore the low population density—of many tropical forests prompts another question. What determines how far apart individual trees can be and still maintain a viable population? One important consideration goes back to mutualisms: Animal pollination ensures that pollen from one tree can reach another individual of the same species if it is within the foraging range of a pollinator carrying its pollen. Of course, if trees are farther apart than any single pollinator can visit within a foraging excursion, they will not be pollinated. Animal pollination, therefore, contributes to the diversity differences between tropical trees and higher-latitude species, which are more likely to be pollinated by wind.

→ **Quick Check 5** From what you know about how species diversity changes with latitude, predict how species diversity changes with altitude.

Evolutionary and ecological history underpins diversity.

Earth's biomes are the historical outcome of natural selection and environmental change through time. For this reason, the fossil record can shed light on how current patterns of diversity came to be. In general, fossils only tell us a little about the origins of tropical forests, but on the island of Hispaniola, there is a

FIG. 47.22 Fossils in 30–23-million-year-old Dominican amber. These document the evolutionary history of (a) orchid-pollinating stingless bees, now extinct on Hispaniola, and (b) *Anolis* lizards that, like their modern descendants, hunted insects in trees.

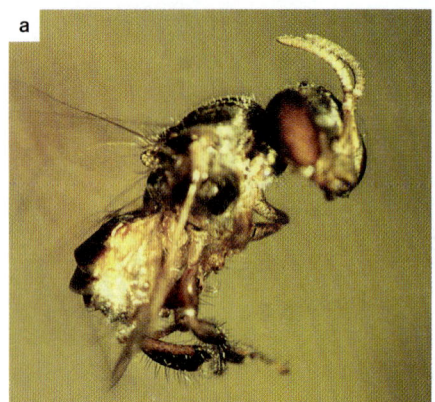 

remarkable exception: nuggets of amber that provide a time capsule of ancient rain forest life. Dominican amber provides an extraordinary window on the diversity of life in the rain forest of Hispaniola 30 to 23 million years ago.

Amber is fossilized tree resin, and in Hispaniola, it comes from the tree *Hymenea*, a member of the pea family. *Hymenea* trees grew in the Hispaniola rain forest 30 to 23 million years ago, much as they do today. Lizards, tree frogs, insects, plants, and fungi were all captured in their sticky sap, which hardened into amber, preserving a record of forest diversity and ecology (**Fig. 47.22**). Studies of these fossils reveal ecological relationships similar to some observed on Hispaniola today. Not surprisingly, some of the most abundant animals in the amber are insects whose modern counterparts have ecological associations with *Hymenea*. Today, ants commonly patrol the bark of *Hymenea* and other tropical trees in search of food, and beetles often feed on trees that secrete resin as a defense mechanism, while stingless bees gather resins for their nests. These insects are preserved intact in the amber as if they were just caught. Indeed, many of the insects seen in this amber, perhaps even the majority, are practically indistinguishable from species that continue to exist today on Hispaniola.

Dominican amber also reveals species that are no longer found on this island. For example, many of the more specialized kinds of ants found in the amber no longer occur on Hispaniola. A kind of beetle that feeds on palm trees in South America is found in amber but not on the island today, even though there are plenty of palms. There are no longer stingless bees on Hispaniola, although the flowers they visit persist, and the bees occur on nearby Jamaica.

Vertebrate fossils are rare in the Dominican amber, but they include *Anolis* lizards closely related to anoles specialized for life in the canopies and high on the trunks of forest trees—resource partitioning began long ago in the Hispaniola rain forest. More vertebrate fossils occur in other settings, and these show that most of the birds and mammals found 25 million years ago are quite different from those in modern forests.

Such fossils allow us to add a deeper dimension of time to our understanding of how ecosystems have come to be. Whether forest, reef, or grassland, every ecosystem contains the present-day survivors of groups representing a broad range of origins in evolutionary time and geographic space. The amber fossils remind us that what G. Evelyn Hutchinson called "the ecological theater and the evolutionary play" have been intertwined throughout our planet's history, producing the diversity of species we see today.

Core Concepts Summary

47.1 THE NICHE IS THE ECOLOGICAL ROLE PLAYED BY A SPECIES.

The niche reflects the biological and nonbiological factors that determine where an organism lives and how it functions. page 47-1

The fundamental niche is the full range of physical conditions and food resources that permit an organism to live. page 47-3

The realized niche is the actual range of habitats where an organism lives, and it is often smaller than the fundamental niche because of competition, predation, and other interactions between species. page 47-3

47.2 COMPETITION, PREDATION, AND PARASITISM ARE ANTAGONISTIC INTERACTIONS IN WHICH ONE SPECIES BENEFITS AT THE EXPENSE OF ANOTHER.

Competition is a form of antagonism that occurs when two individuals of the same or different species use the same limited resource. page 47-4

Species can compete for food, space, mates, or other resources. page 47-4

Competition promotes resource partitioning, which occurs when two co-occurring species overlap in resource use and as a result one changes its resource use. page 47-4

Predators and parasites can reduce competition by limiting the population sizes of successful competitors. page 47-5

47.3 IN MUTUALISMS, THE BENEFITS TO BOTH SPECIES OUTWEIGH THE COSTS OF PARTICIPATION.

Mutualisms can be obligate, meaning that they are required for survival or reproduction, or facultative, meaning that they are not required. page 47-6

Interactions in which one species benefits and the other is unaffected are called commensalisms. page 47-8

Species interactions can change over time. For example, partners can become parasites. page 47-8

47.4 COMMUNITIES ARE SHAPED BY THE EVOLVED FEATURES OF ORGANISMS, WHICH SHAPE THE INTERACTIONS WITH OTHER ORGANISMS AND WITH THE PHYSICAL ENVIRONMENT.

An ecological community consists of all the organisms living in one place. page 47-9

Species that have a large influence on the composition of a community disproportionate to their numbers are known as keystone species. page 47-10

Physical disturbances, such as a storm or drought, can reshape communities. page 47-11

Succession is the predictable order of species colonization and replacement in a new or newly disturbed patch of habitat. page 47-12

47.5 THE CYCLING OF MATERIALS AND THE FLOW OF ENERGY THROUGH ECOSYSTEMS REFLECT WAYS IN WHICH ORGANISMS INTERACT IN COMMUNITIES.

An ecosystem is the community and physical habitat in which the organisms live. page 47-13

Carbon cycles through ecosystems, but energy is used and dissipated. page 47-13

A food web describes how carbon cycles through ecosystems, and a trophic pyramid describes how energy flows through ecosystems. page 47-15

Because of energy lost to heat, work, and the inefficiencies of consumers, each trophic level in an ecosystem generally has about 10% of the biomass found at the next-lower level. page 47-15

In most terrestrial ecosystems, water and nutrients limit rates of primary production. In the sea, primary production is commonly limited by nutrients. page 47-15

47.6 BIOMES ARE BROAD, ECOLOGICALLY UNIFORM AREAS CHARACTERIZED BY THEIR CLIMATE, SOIL, AND PLANT SPECIES.

The distribution of plants reflects in large part abiotic factors such as climate and soil. page 47-19

Annual precipitation and sun intensity affect the amount of evaporation and plant transpiration. page 47-19

Earth's biomes are the historical outcome of environmental change through time and natural selection. page 47-20

Species diversity generally declines from the equator toward the poles, a pattern known as a latitudinal diversity gradient. page 47-20

Self-Assessment

1. Choose an organism, such as an oak tree, and define its niche.
2. Give an example of an antagonism and a mutualism, and in each case, describe the benefits and costs to the participants.
3. Name three factors that help determine the species composition of a community.
4. Describe how a physical disturbance, such as a drought, can affect community composition.
5. Explain what is meant by "ecological succession" and give an example.
6. Describe how herbivores can affect the abundances of organisms at higher and lower trophic levels.
7. Choose five biomes, describe their climate and vegetation, and explain why they differ from one another.
8. Describe the general pattern of diversity from the equator to the poles. Provide two hypotheses to explain this pattern of diversity.

Do you understand the chapter's Core Concepts? Log in to check your answers to the Self-Assessment questions, then practice what you've learned and reinforce this chapter's concepts by working through the problems and multimedia tutorials provided there.

www.biologyhowlifeworks.com

CHAPTER 48

THE ANTHROPOCENE

Humans as a Planetary Force

Core Concepts

48.1 Some scientists call the time in which we are living the Anthropocene Period to reflect our significant impact on the planet.

48.2 Humans have a major impact on the carbon cycle, primarily through the burning of fossil fuels, which returns carbon dioxide to the atmosphere.

48.3 Humans have an important impact on the nitrogen and phosphorus cycles, primarily through the use of fertilizer in agriculture.

48.4 The impact of humans on the environment is changing the stage on which evolution acts.

48.5 In the 21st century, biologists, doctors, engineers, teachers, and informed citizens have vital roles to play in understanding our changing planet and making wise choices for our future.

By almost any measure, you were born into humanity's golden age. Beginning with the invention of the steam engine, humans have learned to harness Earth's store of chemical energy, dramatically extending the power of muscles. You probably don't find it remarkable to drive across a continent in less than a week or fly across an ocean overnight, but your great-great-grandparents would have been astonished. At the same time, advances in antibiotics and public health have drastically reduced infectious disease in many parts of the world. Smallpox, long one of life's deadliest realities, has been eliminated, and polio may eventually follow suit. And the Green Revolution of the last 50 years provides unprecedented amounts of food for a hungry planet (Case 6: Agriculture).

One consequence of this success is an increasing human population: We recently crossed the 7-billion mark. Another is that our actions affect the world around us, directly or indirectly. In this chapter, we look at some of these impacts, and what we can do about them. Can we sustain or control current patterns of population growth and resource use, and if not, will we approach the end of humanity's golden age? As discussed in the following sections, there is reason both for concern and for optimism. Together, scientific insight, technical ingenuity, and a wise citizenship provide our best hope of sustaining hard-won gains in food, energy, and health through the 21st century and beyond.

48.1 THE ANTHROPOCENE PERIOD

Increasingly, scientists refer to the modern era as the **Anthropocene Period**, emphasizing the dominant (and, in part, conscious) impact of humans on the Earth. Our impact is partly a consequence of our sheer numbers, which in turn drive our energy and land use, but it also reflects the individual and societal choices we make.

Humans are a major force on the planet.

There are more humans than ever before, and our numbers are climbing rapidly. A world that had about 500 million inhabitants when Christopher Columbus set sail in 1492 now supports more than 7 billion people (**Fig. 48.1**). By the time you retire, our numbers will probably exceed 9 billion.

→ **Quick Check 1** Take a look at Fig. 48.1. How long did it take our numbers to double, from 1 billion to 2 billion? How about from 2 billion to 4 billion? What does this say about the rate of population growth?

Of course, population size doesn't tell the whole story. Whether driving a car, living in a heated apartment, plugging in a computer or television, or eating fruit

48-1

FIG. 48.1 Growth through time of the human population.

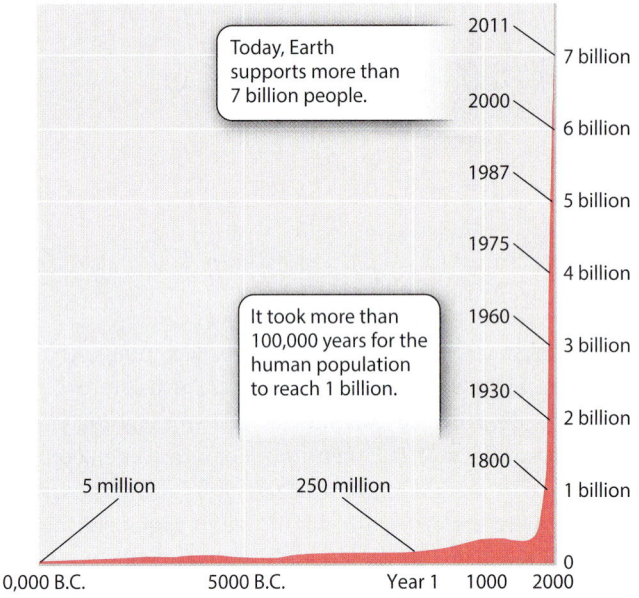

shipped from another hemisphere, you use a great deal more energy than your ancestors did. To give just one example, it has been suggested that the electrical outlets in our homes let us accomplish about the same amount of work performed by 30 servants in a Roman villa. **Fig. 48.2** shows annual per capita energy use in selected countries. Energy use by individuals varies greatly from one part of the world to another. Americans, Canadians, Europeans, and Australians live energy-intensive lives, whereas populations in rural Africa and parts of Asia consume relatively little energy per person.

Our direct use of energy, however, reflects only part of our impact on the planet. Nearly everything we use requires both land and energy. All of us consume food grown in fields, fertilized and harvested through energy, and shipped to a local store. We live in houses built of wood, brick, and stone, and use clothing, furniture, and appliances fashioned from plant materials grown in fields and forests or from minerals extracted from the Earth by mining. It takes land to grow plants, and energy to sow seeds, make and distribute fertilizer, harvest useful materials, and manufacture, ship, and clean the finished products.

The concept of the **ecological footprint** is an attempt to quantify our individual claims on global resources by adding up all the energy, food, materials, and services we use and estimating how much land is required to provide those resources. **Fig. 48.3** plots the average ecological footprint per person against a ranking of the Human Development Index, a measure of standard of living maintained by the United Nations, for many countries. Not surprisingly, developed countries tend to have large ecological footprints—it takes about 7 hectares of land to support an average American, more than 6 to support an average Australian. In contrast, in many parts of Asia and Africa, only a single hectare supports the average citizen.

In many countries, living standards have risen over the past 50 years, nearly always increasing the mean ecological footprint of their citizens. By some estimates, the total ecological footprint of humanity now exceeds the actual surface area of the Earth. Taken together, then, trends in population size and the ecological footprint highlight both the remarkable technological successes of the past century and the challenges we face as citizens, policy makers, scientists, and engineers in the decades to come.

FIG. 48.2 **Energy use by individuals among countries.** Energy use is distributed unevenly among countries.

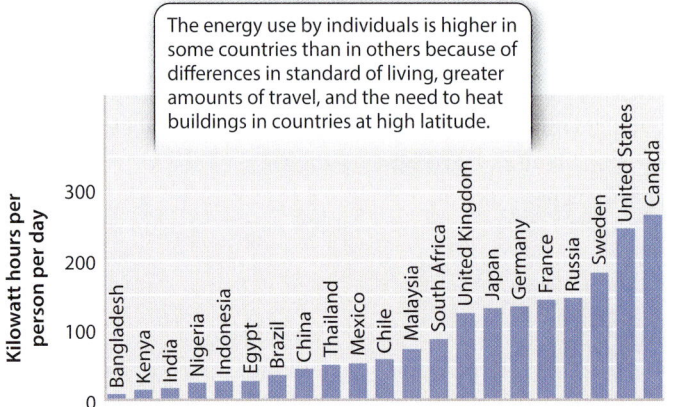

FIG. 48.3 **Our ecological footprint.** The graph shows the number of hectares needed to support an average person compared to a ranking of the Human Development Index, a measure of standard of living.

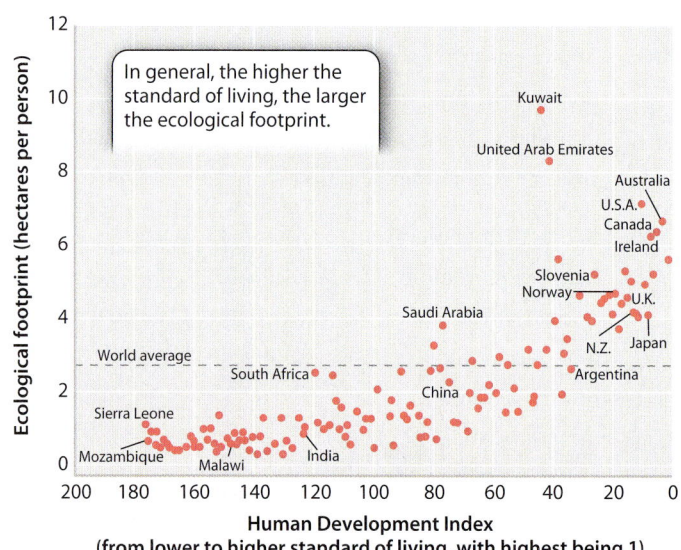

48.2 HUMAN INFLUENCE ON THE CARBON CYCLE

In Chapter 25, we discussed the carbon cycle, explaining how photosynthesis removes carbon dioxide from the atmosphere and respiration replaces it. The remarkable data plotted in the Keeling curve (see Fig. 25.1) show that atmospheric CO_2 levels have increased annually over the past 60 years, and analyses of air bubbles trapped in glacial ice show that this pattern of persistent increase began earlier, with the onset of the Industrial Revolution in the nineteenth century (see Fig. 25.3).

There is no longer any doubt that humans have become major contributors to the carbon cycle, particularly the part of the cycle in which ancient organic matter, or fossil fuel, is oxidized to CO_2, returning CO_2 to the atmosphere (**Fig. 48.4**). As discussed in Chapter 25, burning fossil fuel adds CO_2 to the atmosphere at rates about 100 times greater than all Earth's volcanoes together. As we saw in Chapter 25, measurements over the past several decades of the relative abundances of ^{12}C, ^{13}C, and ^{14}C in atmospheric CO_2 make it clear that the CO_2 now being added to our air comes primarily from the combustion of fossil fuels. No other source of CO_2 has the right isotopic composition to explain the measurements (see Fig. 25.4). Clearing forests for agriculture generates still more carbon dioxide (Fig. 48.4). When land is cleared for agriculture, the existing vegetation is generally burned off, converting much of the organic carbon in biomass and soil organic matter to CO_2. No counteracting process removes CO_2 at comparable rates.

As atmospheric carbon dioxide levels have increased, so has mean surface temperature.

For the past century, scientists, sailors, and interested citizens have monitored temperature at weather stations around the world. More recently, satellites have enabled us to measure temperature in places as remote as the high Arctic and the middle of the ocean. The results are clear: In most parts of the world, mean annual temperature during the decade 1999–2008 was warmer than 1940–1980 averages (**Fig. 48.5**). In some places, the temperature change has been slight, but in others, especially at high latitudes, the increase has been as much as 2°C.

We can measure CO_2 levels in the atmosphere, and they are increasing. We can measure global temperature, and it is increasing. Is increasing CO_2 responsible for observed

FIG. 48.4 **Human-generated sources of carbon dioxide.** Every year, humans convert organic matter to carbon dioxide at rates far higher than those found in nature, predominantly by (a) burning fossil fuels and also through (b) biomass burning, as seen in this Amazon forest.

FIG. 48.5 Global warming.

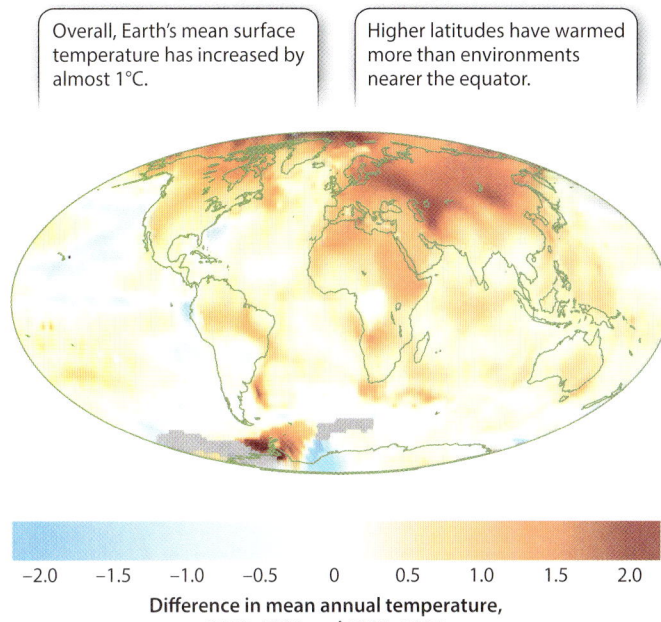

Overall, Earth's mean surface temperature has increased by almost 1°C.

Higher latitudes have warmed more than environments nearer the equator.

−2.0 −1.5 −1.0 −0.5 0 0.5 1.0 1.5 2.0
Difference in mean annual temperature, 1940–1980 and 1999–2008

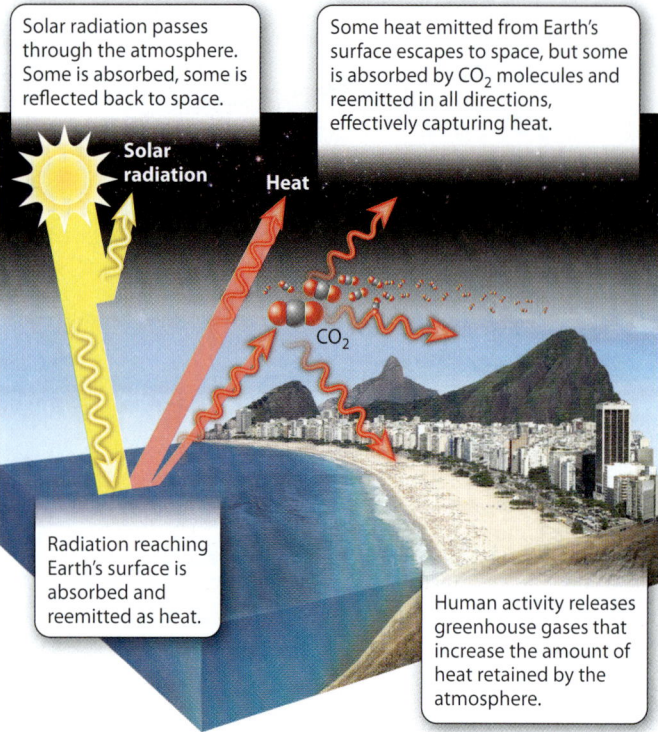

FIG. 48.6 The greenhouse effect. Carbon dioxide is a greenhouse gas: It traps heat from the sun, warming the planet.

temperature changes? To address this question, we must understand that carbon dioxide is a **greenhouse gas**—that is, a gas that absorbs heat energy and then emits it in all directions. Why is that important? As shown in **Fig. 48.6**, solar radiation passes freely through the atmosphere, from top to bottom. Some incoming radiation is reflected from Earth's surface, and the rest is absorbed by the land and sea. And some of the energy absorbed by Earth's surface is radiated back again as infrared radiation, or heat. Greenhouse gases in the atmosphere absorb the infrared radiation reflected up from Earth's surface and emit it in all directions—some will be directed upward, out of the atmosphere, but half will be directed downward, back toward Earth, effectively trapping heat. The net effect is like that of the glass panes of a greenhouse, which allow sunlight into the structure but prevent heat from leaving.

Carbon dioxide is only one of several important greenhouse gases in the atmosphere—water vapor is another, and methane a third—and without these gases absorbing and trapping heat, average surface temperatures would fall below freezing and life would not be possible. However, because CO_2 is increasing rapidly, its greenhouse effect is also increasing. Methane levels are also rising rapidly, in large part because of increasing food production. Most of the methane delivered each year to the atmosphere is generated biologically by methane-producing archaeons (Chapter 26), which thrive in the guts of cattle and in the waterlogged paddies where rice is cultivated. More beef and expanded rice paddies translate into higher rates of methane production. The thawing of permafrost at high latitudes releases additional methane that was trapped in frozen soils when the ice formed long ago.

How much atmospheric warming is due to increases in greenhouse gases is a question of physics. Each molecule of greenhouse gas absorbs and emits a specific amount of heat, and calculations show that the increases in atmospheric greenhouse gases measured over the past 50 years have increased the difference between incoming (solar) and outgoing radiation by about 2.5 watts per square meter. This difference is what adds heat to the oceans and atmosphere. Scientific consensus, reflected in reports from the Intergovernmental Panel on Climate Change, is that this greenhouse effect is the principal cause of observed twentieth-century temperature change. Scientific consensus also holds that global temperature will continue to rise as atmospheric CO_2 continues to increase in this century.

If this view is correct, human activities are changing the world. Can we, however, eliminate the possibility that the observed increases in greenhouse gases and temperature have natural causes? After all, as we saw in Chapter 25, the long-term geologic record indicates that climate and atmospheric composition have changed dramatically and repeatedly throughout our planet's history. Perhaps volcanic emissions have increased, driving the observed changes. Measurements of the isotopic composition of atmospheric CO_2 effectively eliminate volcanic eruptions as a principal source of rising CO_2 (see Fig. 25.4).

Moreover, we can monitor the effects of volcanoes as they occur, and we can gauge the effects of past eruptions because volcanic ash accumulates along with the ice in continental glaciers. Historically, the major effect of large volcanic eruptions has been to *decrease* temperature, because volcanic ash and aerosols reflect incoming solar radiation back into space. What's more, the impact of volcanic eruptions lasts only a few years. It doesn't drive the century-long temperature increase that we are currently observing.

What about another possibility: that the amount of solar radiation entering the atmosphere has varied through time? The sun's output oscillates, and we know from direct measurements how solar radiation has varied during the past 50 years. The effect of this variation can be calculated, and it turns out to be small relative to greenhouse effects. It does not seem that variable solar radiation can account for the amount of temperature change that we have observed.

Up to this point, then, there is overwhelming agreement among scientists that atmospheric CO_2 levels are rising, that temperature is increasing, and that the physics of greenhouse gases relates the two effects. Beyond these facts, however, it is difficult to say with certainty how human-induced global trends in climate will affect a given area.

Why don't we know for certain? Modeling future climatic conditions is difficult because of the many complex interactions that contribute to climate. Climate models are attempts to understand how climate works by fashioning equations that relate a simplified set of variables and interactions. Using mathematical models that accurately approximate current climate, modelers can change inputs into the model, adding CO_2 to the atmosphere, for example, or changing the extent of forest cover. The model then generates a set of results that can be used to predict future climate. All models must be checked against actual observations, and most are sensitive to assumptions made in constructing the model. That said, climate models do a pretty good job of explaining global-scale features of climate and climatic change.

Nevertheless, questions remain. How, in detail, might cloud cover change over the next century, and what effect would this change have on temperature and precipitation? What will be the effect of additional air pollutants, like the black carbon particles released into the air when wood burns incompletely? Will black carbon increase warming by absorbing solar radiation, or decrease it by reflecting radiation back into space? How will oceanic and atmospheric circulation patterns change, and will any changes that may occur enhance or dampen climatic change?

We still seek definitive answers for all these questions. Nonetheless, most climate models suggest that mean global temperature will increase 2°–5°C during the 21st century (**Fig. 48.7**). That increase is not expected to occur uniformly throughout the globe. As has been true during the past 50 years, changing circulation patterns may cool some places and warm others. Likewise, rainfall is likely to increase in some areas, and decline in others.

Climate can be considered average weather over a long time interval, but from one year to the next, weather is tremendously variable—there will be cold years and warm years, wet ones and dry ones. Because weather is so variable, evidence for climate change comes not from individual weather events but from records kept over decades.

Models predict that rainfall patterns should change as Earth warms. To test this, oceanographers analyzed 1.7 million measurements of seawater salinity taken over the past century. The oceans were chosen because they contain 97% of our planet's water and receive 80% of its rainfall. The salinity of surface seawater reflects both the addition of freshwater by rain, which decreases salinity, and evaporation, which increases it. For this reason, changing salinity can indicate whether the balance of rainfall and evaporation is shifting over broad regions of Earth. Consistent with many model predictions, wet areas of the Earth are becoming wetter and dry regions drier.

→ **Quick Check 2** What is the difference between global warming and the greenhouse effect?

Changing environments affect species distribution and community composition.

What will be the consequences of climate change? Without question, some locations will benefit. For example, the increase in temperature in New England will mean a longer growing season. Other regions, however, will suffer. As precipitation patterns change, many places will become drier. Indeed, a number of climate models predict that some of the strongest declines in rainfall will occur in regions that currently produce much of the corn and wheat that feed the world. Already, farmers in southeastern Australia have experienced the worst droughts in a century, and with them unprecedented damage from brush fires. Over much of North America, 2012 was also the driest summer in many decades. Such events are consistent with model predictions, but because of the great natural variability of weather, they do not by themselves confirm climate models. Scientists will be watching carefully to see whether such extreme events increase in frequency in coming years.

To understand how different species will respond to global climate change, we return to the basic principles of ecology and evolution. Natural selection describes the changes in allele frequencies that result from the interactions between individuals of a population and their environments. We tend to think about

FIG. 48.7 Global warming. This map shows predictions for global warming based on the assumption that current trends in atmospheric CO_2 accumulation will continue—the "business as usual" scenario.

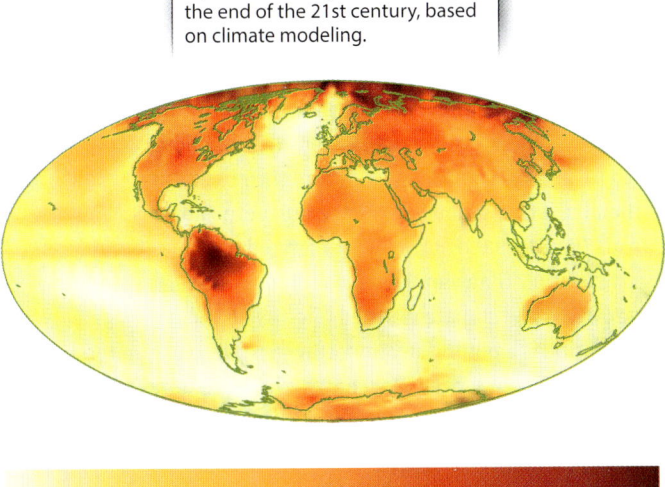

A prediction for global warming by the end of the 21st century, based on climate modeling.

Temperature increase (°C)

natural selection in terms of individuals' adapting to a constant or slowly changing environment. In the 21st century, however, the environment will probably be a rapidly moving target, and so genotypes that conferred high fitness in the past may not be the most advantageous in the future.

Plants illustrate the complicated ways in which organisms respond to environmental change. Meticulous notes taken during the 1840s by Henry David Thoreau record inventories of plant species diversity in woodlands west of Boston, Massachusetts. Climate records show that mean annual temperature in this area has increased by about 2.5°C since Thoreau began his survey, and in response, many but not all species now flower a week earlier than they did in the nineteenth century (**Fig. 48.8**). A number of species have declined in abundance or even disappeared since Thoreau's time, and these tend to be the plants least able to change their flowering time (Chapter 30). Earlier onsets of leaves, flowers, and fruit have been documented in locations distributed widely throughout the temperate zones of North America, Europe, and Asia. The Massachusetts example highlights the simple but important observation that different plant species respond in distinct ways to changing climate, favoring some populations over others.

→ **Quick Check 3** Why should the capacity to shift flowering time influence the distribution of a species during the 21st century?

The example of Thoreau's woods shows that as climate changes, some plants can modify flowering time—a physiological response to environmental cues that now occur earlier in the year. Experimental evidence also shows that many plants also show a direct growth response to elevated CO_2. Plants grown in air with CO_2 levels like those predicted for the end of this century often change the way that they distribute the products of photosynthesis to different organs: The amount of nitrogen in plant tissues commonly declines, as does the proportion of resources devoted to reproduction. Again, this doesn't reflect genetic change, just a physiological response to altered environmental conditions. But if environmental change persists for a long time, as predicted, the varying abilities of different plants to respond physiologically will result in changing allele frequencies, as some variants survive and reproduce better than others. That is, plant populations may evolve by natural selection in response to climate change.

Evolution is one response to global change, but given the pace at which atmospheric composition and climate are changing, many populations may not have time to adapt. In this case, populations will either migrate or become extinct. Fossils deposited as Earth's climate warmed at the end of the last ice age show that many plant species dramatically changed their geographic distributions in response to past climate change. As discussed in Chapter 30, trees don't move, but they can migrate by dispersing seeds to new areas (**Fig. 48.9**). For example, between 18,000 and 8000 years ago, Red Pine and Jack Pine migrated thousands of kilometers to the north from the Gulf of Mexico, where they lived during the ice age, to their current distributions in the northern United States and Canada (Fig. 48.9a). Hickory also migrated northward, expanding its range rather than simply shifting from one place to another as ice retreated and climate warmed (Fig. 48.9b).

Migration is a potentially important response to environmental changes projected for the 21st century, but it requires a continuous route to get from one place to another, and it is a challenge to find these direct paths on continents broken up by agricultural lands and cities. For this reason, some conservation biologists advocate "assisted migration": the deliberate transplantation of plant populations from existing habitats to new ones more favorable to

FIG. 48.8 **Plant responses to warming temperatures.** Observations over a century and a half show that (a) the time of first flowering has become earlier as (b) regional temperatures have become warmer in the woods of Concord, Massachusetts.

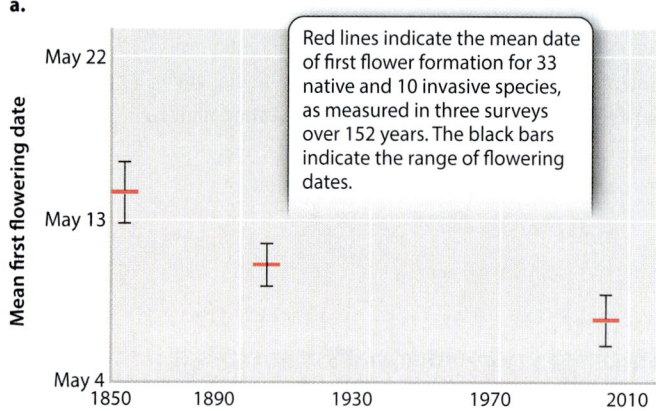

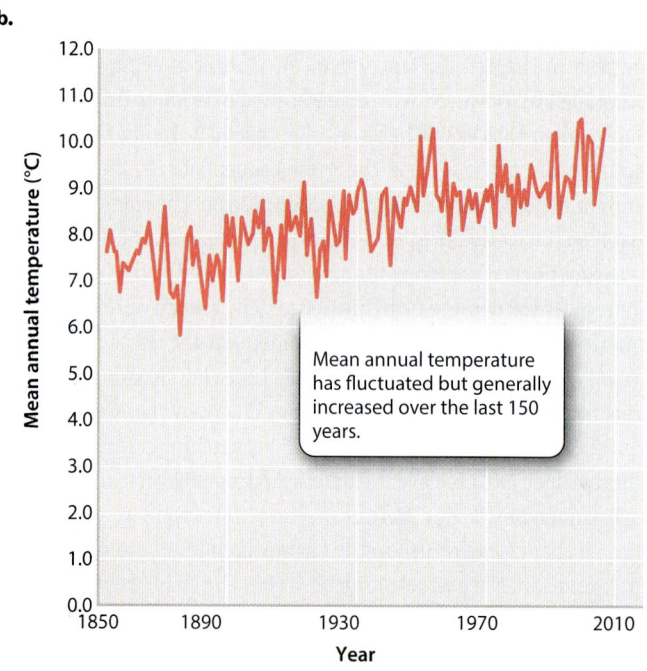

FIG. 48.9 Plant migration in response to climatic change. Records of fossil pollen over the last 13,000 years since the last ice age document (a) the northward migration or (b) expansion of selected tree species.

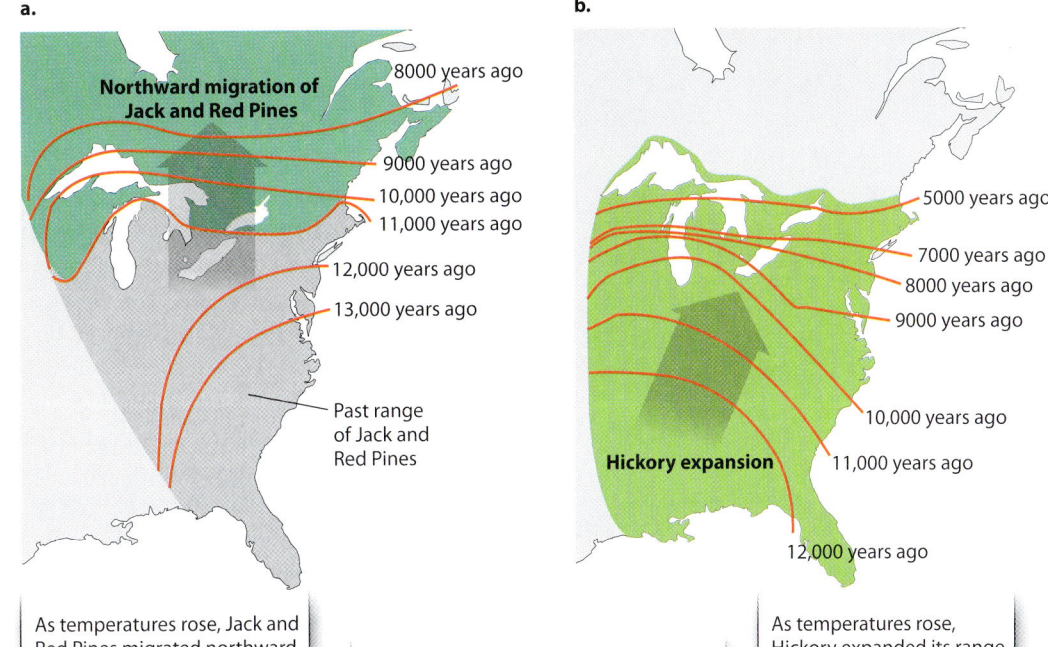

CASE 8 Biodiversity Hotspots: Rain Forests and Coral Reefs
How has climate change affected coral reefs around the world?

growth. To make sure that assisted migration does more good than harm, we need to do more experiments to understand how plants grow and compete in new ecological situations and under different environmental conditions.

Increasing atmospheric CO_2 is not just influencing life on land. It is also having an observable impact on organisms in the oceans, not least on coral reefs that provide hotspots of biological diversity in the marine realm. The Great Barrier Reef is visible from space, its wave-washed necklace of limestone stretching more than 2000 km along the northeastern coast of Australia (**Fig 48.10**). A global diversity hotspot, the reef's 1500 fish species (a tenth of all known fish in the sea), more than 4000 kinds of clams and snails, and myriad other species, prompt basic questions about ecology and diversity, many of which we have addressed in previous chapters. Lately, however, one question has come to dominate scientific discussion of the Great Barrier Reef: Will it exist a century from now?

Corals are the principal architects of the Great Barrier and most other reefs (Chapter 44), and so coral biology is the key to understanding both the reef's unusual diversity and its vulnerability in the face of climate change. Most reef corals gain

FIG. 48.10 The Great Barrier Reef. The world's largest coral reef, lying off the coast of Australia, is visible from space.

FIG. 48.11 Coral bleaching. This coral shows a pattern of whitening that occurs when corals lose their algal symbionts in response to increases in ocean temperature.

In the past 40 years, as the concentration of CO_2 in the atmosphere has grown from about 315 to more than 390 ppm, the pH of surface oceans has correspondingly dropped by 0.1 unit (**Fig. 48.12**). That may not seem like a lot, but remember that the pH scale is logarithmic, and so the 0.1 unit drop represents an increase in hydrogen ions of about 30% since the Beatles recorded their first hit song in the 1960s.

Ocean acidification causes carbonate ions in seawater to decrease, making it more difficult for corals to build their $CaCO_3$ skeletons. In 2000, the German biologist Ulf Riebesell and his colleagues grew coccolithophorids, tiny marine algae that form scales of calcium carbonate, under experimental conditions approximating pre-industrial, present-day, and predicted 21st-century CO_2 levels (**Fig. 48.13**). They found that decreasing pH had little impact on growth rate, but rates of carbonate precipitation declined markedly. Apparently, the algae had difficulty producing nutrition from unicellular algae that live within their tissues. The algal symbionts in reef corals provide food for their host, receiving nutrients, waste disposal, and a stable environment in return. It is the efficient cycling of carbon, nitrogen, and phosphorus between corals and their symbionts that underpins the biological richness of coral reefs. Reef corals accomplish another noteworthy feat: They make skeletons of calcium carbonate ($CaCO_3$) that build the reef's three-dimensional framework through time.

The very processes that facilitate reef growth—biomineralization and microbial symbiosis—are exposing their Achilles' heel in the 21st century. Today, on the Great Barrier Reef, throughout the Caribbean, and elsewhere, corals are dying. Bleaching, indicated by the white skeletons of dead corals, occurs when the symbiotic algae that feed the corals abandon their hosts, thereby sentencing them to death (**Fig. 48.11**). The details are still being researched, but the increase in seawater temperature appears to signal the algal exodus. Increasing seawater temperature also facilitates the spread of infections that can kill coral.

A greater long-term threat, however, may come from increasing CO_2 levels in the ocean. As noted in Chapter 25, not all the CO_2 produced by humans during the past century has accumulated in the atmosphere: About one-third has been absorbed by the ocean. That's good news for slowing global warming because it means there is less CO_2 in the atmosphere to trap heat, but it is potentially bad news for marine habitats because CO_2 is a weak acid (Chapters 6 and 39). Unlike climate change, the effects of increasing CO_2 on ocean pH are straightforward: Increasing the abundance of CO_2 in the oceans causes the pH of seawater to go down, a phenomenon known as **ocean acidification.**

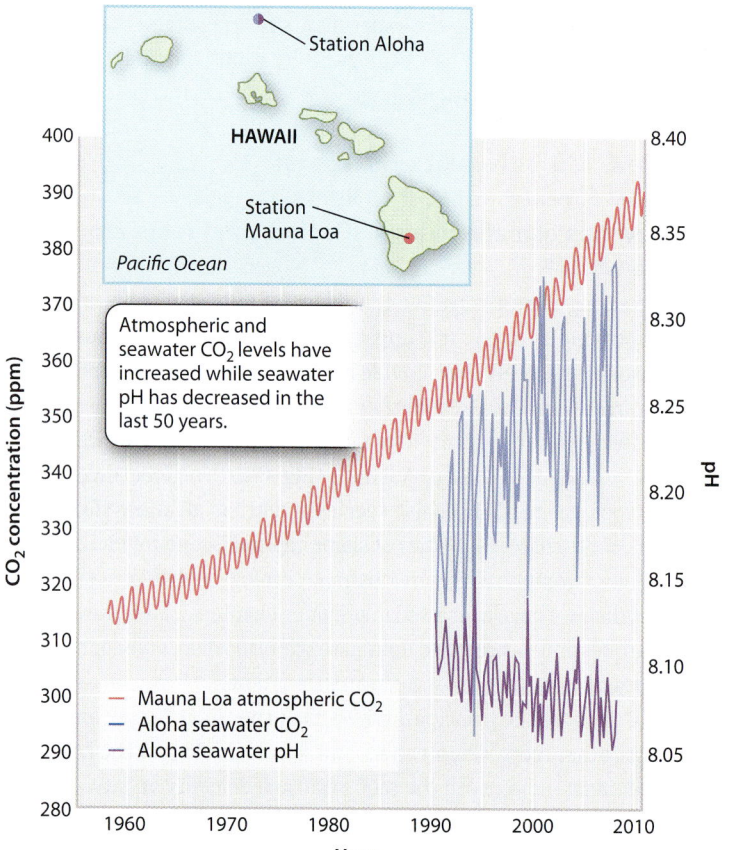

FIG. 48.12 Trends in atmospheric CO_2 levels (red), seawater CO_2 levels (blue), and seawater pH (purple) in the central Pacific Ocean.

Atmospheric and seawater CO_2 levels have increased while seawater pH has decreased in the last 50 years.

skeletons of calcium carbonate in lower pH. Motivated by this result, many laboratories have explored skeleton formation under conditions of ocean acidification. Researchers have found that some marine organisms can regulate the ionic composition of their internal fluids to counter the increased CO_2 in their environment and so precipitate calcium skeletons normally, but other organisms, such as reef-forming corals, cannot.

Essentially all attempts to model our environmental future predict that by the end of this century, atmospheric CO_2 will rise from its current level of nearly 400 ppm to more than 500 ppm, and changing pH in many parts of the ocean will limit the ability of corals to build skeletons. Some corals can survive this change. For example, experiments on two Mediterranean corals showed that at a pH level 0.8 units below the present level, the corals grew naked—essentially as sea anemones—and recovered their skeletons upon return to normal seawater. But if ocean acidification persists over timescales much longer than coral generations, the structural framework for tropical reefs may be irrevocably lost.

HOW DO WE KNOW?

FIG. 48.13

What is the effect of increased atmospheric CO_2 and reduced ocean pH on skeleton formation in marine algae?

BACKGROUND It is well established that atmospheric CO_2 levels are increasing, which in turn decreases the pH of ocean water. In the late 1990s, experiments showed that a decrease in pH affected the ability of some marine organisms to build skeletons made of calcium carbonate ($CaCO_3$). The German biologist Ulf Riebesell and his colleagues carried out experiments to investigate whether ocean acidification affects algae called coccolithophorids, which account for the majority of carbonate precipitation in the ocean.

HYPOTHESIS From the effects of decreased pH on $CaCO_3$ skeleton formation in other marine organisms, Riebesell hypothesized that increasing CO_2 would interfere with skeleton formation in coccolithophorids.

EXPERIMENT Riebesell and his colleagues studied two species of coccolithophorids, *Emiliania huxleyi* and *Gephyrocapsa oceanica*. They grew both species in the laboratory under conditions of increasing CO_2, from pre-industrial levels, to present-day levels, and then to levels predicted for the future. They measured pH and rates of calcification and used scanning electron microscopy to observe skeleton formation.

RESULTS With increasing CO_2 and decreasing pH, they observed decreasing rates of calcification as well as deformed and incomplete coccolithophorids under the scanning electron microscope.

CONCLUSION The results support Riebesell's hypothesis: Ocean acidification interferes with normal skeleton formation in marine plankton.

FOLLOW-UP WORK This work has been extended to investigate the effects of ocean acidification on skeleton formation in corals and other organisms. For example, the Mediterranean coral *Oculina patagonica* grows normally at pH 8.2. When grown at pH 7.4, however, the coral skeletons dissolve. Upon restoration of normal conditions, the corals resume skeleton formation. This experiment shows that corals might survive ocean acidification, although the ecosystem benefits provided by reefs would be lost.

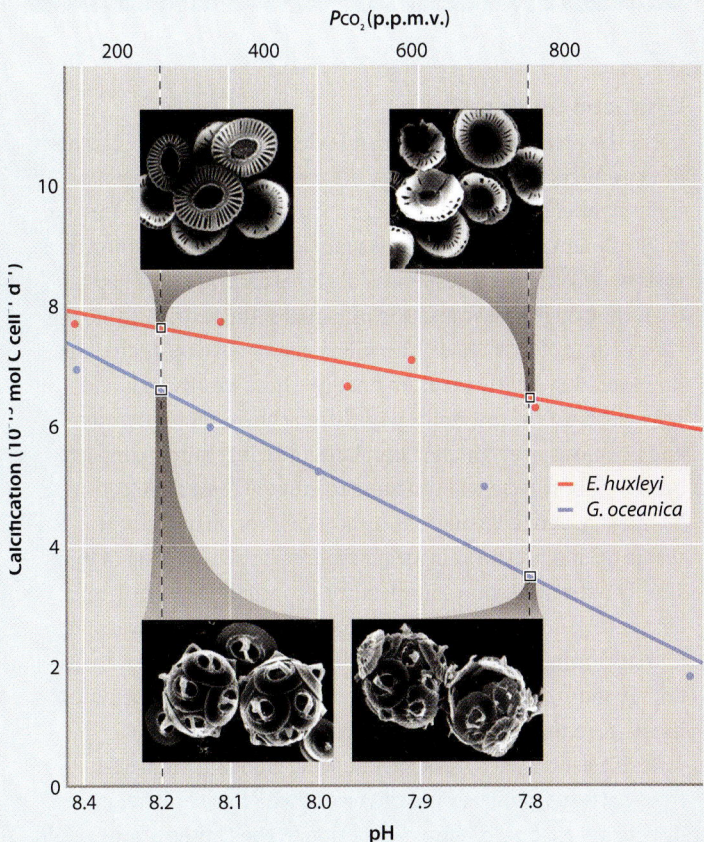

SOURCES Riebesell, U., I Zondervan, B. Rost, P. D. Tortell, R. E. Zeebe, and F. M. M. Morel. 2000. "Reduced Calcification of Marine Plankton in Response to Increased Atmospheric CO_2." *Nature* 407:364–367; Fine, M., and D. Tchernov. 2007. "Scleractinian Coral Species Survive and Recover from Decalcification." *Science* 315: 1811–1811.

FIG. 48.14 Alternative energy sources, including (a) solar and (b) wind energy.

What can be done?

From the preceding discussion, it is clear that rising CO_2 has both direct effects on the physiology of land and sea organisms and indirect effects related to climate change. In order to ensure our environmental future, scientists must understand the climate system and biological responses to global change much better than they do now. Because the principal source of rising CO_2 is the burning of fossil fuels, the solutions we develop as citizens, however, will clearly focus on energy—how we obtain it and how we use it. As the global population continues to increase and people aspire to higher standards of living, human impact on the carbon cycle will continue to grow, at least if we continue business as usual. Business as usual, however, may not be a viable option for the long term because population growth may simply outstrip resource availability, especially for petroleum. Ultimately, our success in both slowing rates of global warming and sustaining energy supplies will depend on our ability to use energy more efficiently and generate new forms of power, most likely from the sun, wind, and nuclear sources.

Half a century ago, geologist M. King Hubbard predicted from patterns of oil use and the rate at which new oil fields are discovered that petroleum production in the United States would peak in the 1970s and decline thereafter. Global production, he suggested, would peak later, in the early decades of the 21st century. King's predictions proved accurate for the United States: Domestic petroleum production peaked in 1972, although petroleum use continues to climb. In addition, global petroleum production also seems on course to fulfill Hubbard's predictions, although the timescale remains a contentious question. Coal, natural gas, and unconventional reserves such as the oil sands found widely in western Canada can quench the world's thirst for fossil fuels for another century or more, and biofuels—fuels generated from plant biomass—may also help conserve petroleum, but the energy produced will be expensive and global change will continue.

Like fossil fuels, alternative energy sources—wind, the sun, tidal energy, and nuclear power (**Fig. 48.14**)—all have important pros and cons, not the least of which, in the case of nuclear power, is long-term storage of radioactive waste. On the other hand, all these sources have the potential to benefit humanity at least three ways: by reducing the amount of CO_2 emitted per kilowatt of energy generated, by conserving finite natural resources, and by helping to contain the cost of energy production. The development of more fuel-efficient cars, airplanes, and buildings will also help us to maintain energy-intensive lifestyles while decreasing their environmental consequences. As individuals, we can choose energy-efficient transportation and appliances and develop habits that decrease energy waste. When one person turns off the light in an empty room, the energy saved is tiny, but when millions of people do it, the benefits add up.

Wise choices by citizens and consumers can slow the growth of atmospheric CO_2, but the CO_2 already there isn't going to go away quickly. As noted earlier, we have dramatically increased the rate at which CO_2 is added to the atmosphere, but we have not changed the rate at which it is removed. The geologic processes that control atmospheric CO_2 work on longer timescales. Therefore, even if we could cut CO_2 emissions by 90% tomorrow, CO_2 would not return to nineteenth-century levels for hundreds or even thousands of years.

Another possibility is to actively remove CO_2 from the air. Reforestation of previously cleared landscapes removes carbon from the atmosphere since plants build biomass from CO_2 during photosynthesis (Chapter 8). It is also possible to capture CO_2 as it rises upward through smokestacks, although it is expensive to do so currently. We do not yet know how to scrub industrial exhaust of CO_2, but if CO_2 could be removed from the air in this way at reasonable costs, it could apply a significant brake to global warming.

Clearly, forestry and 21st-century technology have potentially important roles to fill in undoing the consequences of the past hundred years. For the most part, however, gaining control over our environmental future will entail wise personal and governmental choices about how we obtain and use energy.

48.3 HUMAN INFLUENCE ON THE NITROGEN AND PHOSPHORUS CYCLES

Human activities influence the environment not because we operate outside the carbon cycle, but because we have become an important part of it. Our interest in improved crop production also results in unprecedented human impact on the nitrogen and phosphorus cycles. Once again, environmental changes tied to human nitrogen and phosphorus use are altering the ecological landscape for species in lakes and the sea.

Nitrogen fertilizer transported to lakes and the sea causes eutrophication.

As discussed in Chapter 26, all organisms require nitrogen to synthesize proteins, nucleic acids, and other molecules. Although nitrogen is plentiful in the atmosphere as N_2 gas, this form is not biologically available to animals and plants. As we saw in Chapter 26, bacteria and archaeons are able to convert N_2 to biologically available forms of nitrogen, called fixed nitrogen, such as ammonia (NH_3) and nitrate (NO_3^-). Plants then obtain nitrogen from ammonia and nitrate in the soil, and animals get the nitrogen needed for growth from the food they eat. Crops, harvested year after year in the same fields, deplete available nitrogen, and so high and sustained yields require that farmers add biologically available nitrogen to the soil as fertilizer.

Through the use of fertilizer and other activities, humans add about 150 million tons of fixed nitrogen to the biosphere each year, an amount comparable to that contributed by microbial nitrogen fixation and lightning combined. The majority of this fixed nitrogen (125 million tons per year) is created industrially by reacting nitrogen gas and hydrogen in the presence of a catalyst to produce ammonia. Most commercially produced ammonia is further processed to nitrate that will be spread across fields as fertilizer to increase crop production. Success in feeding the world's people owes much to fertilization with nitrates, but in fact, only about 10% of the nitrogen added to croplands ends up in food. Much of the nitrate fertilizer leaves fields as surface runoff and travels by river to lakes or the sea. Denitrifying bacteria also use some of the added nitrate for respiration, returning N_2 to the atmosphere—and also generating N_2O, yet another greenhouse gas.

In lakes and oceans, the fertilizing effects of agricultural runoff are finally realized, but in unintended ways. The added nutrients lead to a great increase in the populations of algae and cyanobacteria in a process called **eutrophication.** These algal and cyanobacterial masses eventually sink to the bottom, where heterotrophic organisms, mostly bacteria, feed on them, fueling high rates of aerobic respiration. The bacteria's demand for oxygen can completely deplete O_2 in bottom waters, with potentially catastrophic consequences for animal life on the sea or lake floor.

In the Gulf of Mexico, nutrients leached from croplands and transported down the Mississippi River result in an ominously named dead zone of oxygen-depleted bottom water. This dead zone expands seasonally from the mouth of the river west toward Texas, covering an area about the size of New Jersey (**Fig. 48.15**). Within the dead zone, fish and seafloor invertebrates die in masses. Several hundred dead zones have been identified in coastal oceans around the world, and most are expanding year by year. Those in the Gulf of Mexico and the Baltic Sea are among the largest.

Nitrogen applied as fertilizer but spread by runoff and streamflow can impact the full range of ecological interactions discussed in chapters 46 and 47 (**Fig. 48.16**). We benefit tremendously from increased crop production but pay a price in the eutrophication of lakes and oceans. We pay another price as well—as much as 1% to 2% of all fossil fuel consumption goes to make nitrogen fertilizer.

FIG. 48.15 A dead zone, visible in this NASA satellite image as billows of green.

VISUAL SYNTHESIS
Succession: Ecology in Microcosm
FIG. 48.16 Integrating concepts from Chapters 46 through 48

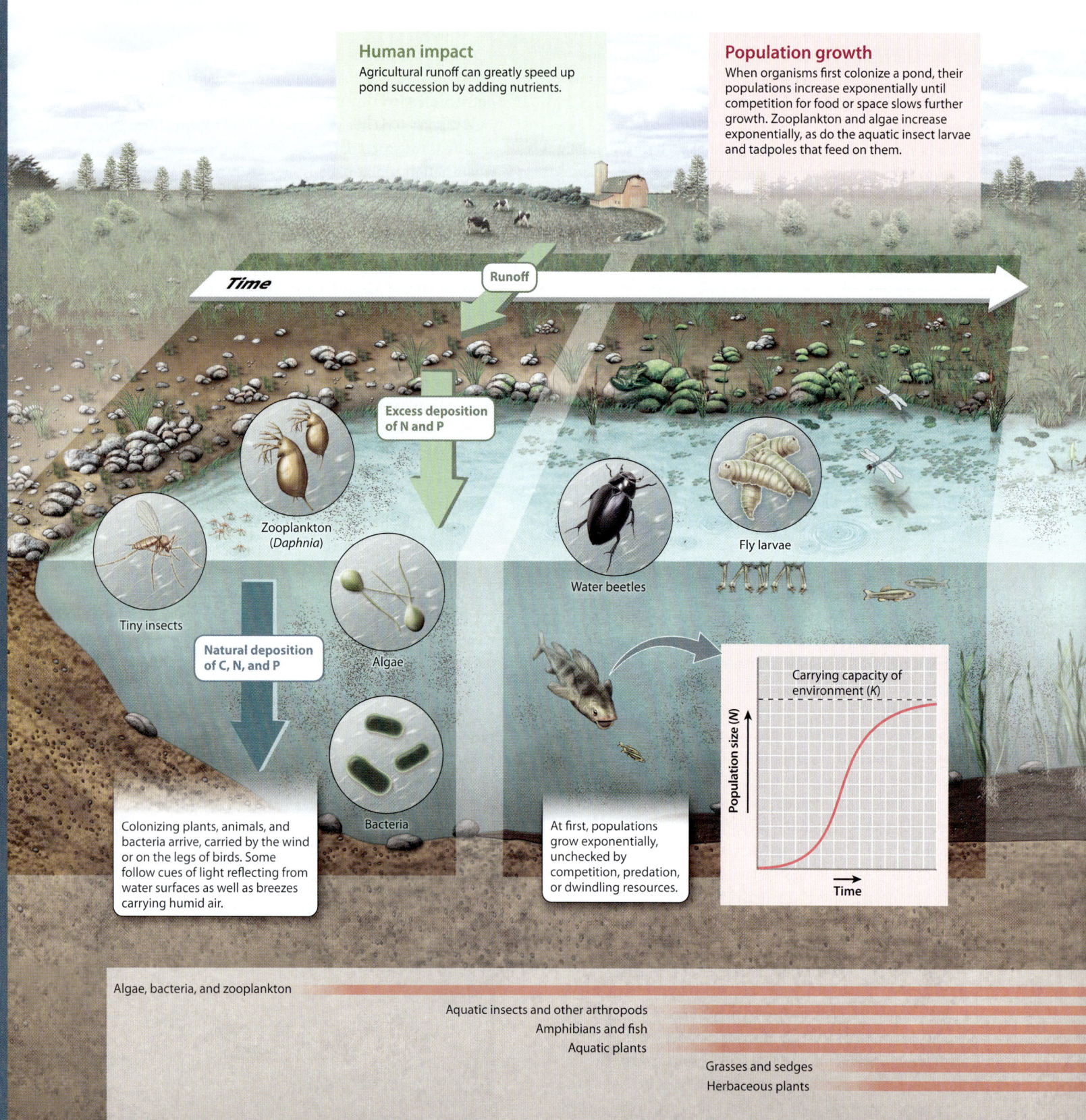

Species interactions
With increasing diversity comes an increase in the numbers of connections between species as predators, prey, or symbionts. Interactions help to control population sizes.

Succession
As forest communities mature and stabilize, shade-tolerant species may only be disrupted by fire or other disturbances. This can set communities back to earlier successional stages. As soils deepen and store the seeds of sun-loving species, they establish themselves in any openings that are created.

Time

Human impact
Logging, starting fires (or suppressing natural fires), and introducing invasive species like Purple Loosestrife can alter the composition of a forest or field community.

Increased accumulation of dead organic material

A succession of plant communities and associated animals increases the biomass dependent on the water source, but eventually sediment runoff from surrounding land fills in the pond, and a terrestrial ecosystem becomes established.

The dominant plants change the physical habitat by shading and spurring organic decay, excluding some animals and plants while permitting others to establish.

Terrestrial animals
Shrubs and trees

48-13

Phosphate fertilizer is also used in agriculture, but has finite sources.

The other major nutrient added to fields as fertilizer is phosphate (PO_4^{3-}), and here the problem is a different one. Like nitrate, much of the phosphate added to fields as fertilizer leaves as runoff, and phosphate runoff can be an important contributor to eutrophication. The Everglades of south Florida provide a good example. This ecosystem is built of plants that have adapted to low phosphate levels. However, phosphate levels in parts of the Everglades have increased up to a hundredfold. Chemical tracers show that these increased phosphate levels stem from phosphate fertilizer used on agricultural lands to the north. As a consequence, phosphate-loving cattails and other introduced plants are expanding at the expense of native plants, decreasing the diversity of this remarkable ecosystem.

There's no question that overuse of phosphates has detrimental effects on certain natural ecosystems, and yet we need to be concerned that we will eventually run out of phosphates to use as fertilizers. Unlike for nitrate, there is no industrial process to produce phosphate. Thus, the only source of phosphate fertilizer is mining. Only a few places in the world have sedimentary rock with phosphates in high enough concentration to be worth mining, and these stores are being depleted.

Globally, phosphate production peaked about 1990, raising a serious question of how the world's need for phosphate fertilizer will be satisfied a century from now. In many ways, the problem of phosphate availability parallels that of petroleum production. As abundant sources become depleted, we will have to mine poorer and poorer resources, raising production costs and ultimately calling into question our capacity to sustain agricultural production.

Human encroachment on the nitrogen cycle and long-term phosphate availability, then, converge on a single question: Can we manage future agricultural production in ways that feed the growing human population in a sustainable fashion?

What can be done?

The Green Revolution in agriculture began in the 1940s with the introduction of high-yield, disease-resistant varieties of wheat to Mexico (Case 6: Agriculture). In subsequent decades, new strains of corn, wheat, and rice, widespread application of industrial fertilizer, and the spread of farm machinery increased global crop production to levels that could scarcely have been imagined at the beginning of the twentieth century. As population growth has accelerated, however, the need to grow ever more food remains strong. And other factors, notably rising meat consumption in many countries and the increasing use of corn to generate biofuel, now place still greater demands on global agriculture. It is estimated that over the next decade alone global corn and wheat production will have to increase by 15% to meet demand.

How can we meet the needs of a growing population? Conventionally, agronomists see two options: We can devote more land to crops, or we can increase the yield of fields already in place. **Fig. 48.17** shows that globally both the amount of land devoted to growing corn and corn yield have increased steadily over the past 50 years. It is possible to increase yield further, as the yield per hectare in the United States, Canada, China, and western Europe far exceeds that of most other countries. Increasing yields, however, will require increasing the use of fertilizer and the use of fossil fuels to power farm machinery, with all the attendant problems of climate change and eutrophication noted earlier. Increasing crop area also comes at a cost, as forests and natural grasslands go under the plow, decreasing the biological storage of carbon and threatening biological diversity, discussed more fully below.

For more than a century, biologists have developed improved varieties of crop plants by breeding, the highly effective practice that Darwin called "variation under domestication" and used to frame his arguments about natural selection. Today, we have another tool available: genetic engineering. Biologists can now introduce genes for desirable traits into crop plants, improving their yield, disease resistance, or nutritional value (Chapter 32). The potential benefits of genetically modified crops are obvious, but some biologists see risks as well, many of which are detailed in Chapter 12.

Conventional breeding and genetic modification may improve crop yields, but in a world of changing climate and evolving pests, these gains may have something of a "Red Queen" flavor to them (Chapter 42). In Lewis Carroll's *Through the Looking-Glass*, the Red Queen tells Alice, "Now, here, you see, it takes all the running you can do, to keep in the same place. If you want to get somewhere else, you must run at least twice as fast as that!" Much of evolution appears to work this way: As the physical or biological environment changes, continued adaptation is needed just to maintain fitness. Strains of wheat, modified to improve drought tolerance, may simply maintain high yields in a changing environment, not increase them. Increases in yields through genetic modification may be possible, but limited. Thus, massive fertilization by nitrate and phosphate fertilizers will continue to be a key component of agriculture.

Another type of biological intervention is possible, however. We are not the only species with a nutritional interest in croplands. Fungi, bacteria, protists, nematodes, insects, birds, and mice all target cornfields, sharply reducing yield. In the United States alone, such unwanted guests cause about $1.5 billion worth of damage per year. Conventional breeding and genetic modification have produced improved resistance to pests, but

FIG. 48.17 Corn yield and area in the United States and globally from 1961 to 2007. Global corn yield and area have increased at about the same rate, whereas U.S. corn yield has risen much more quickly than corn area because of energy-intensive agriculture and use of fertilizer.

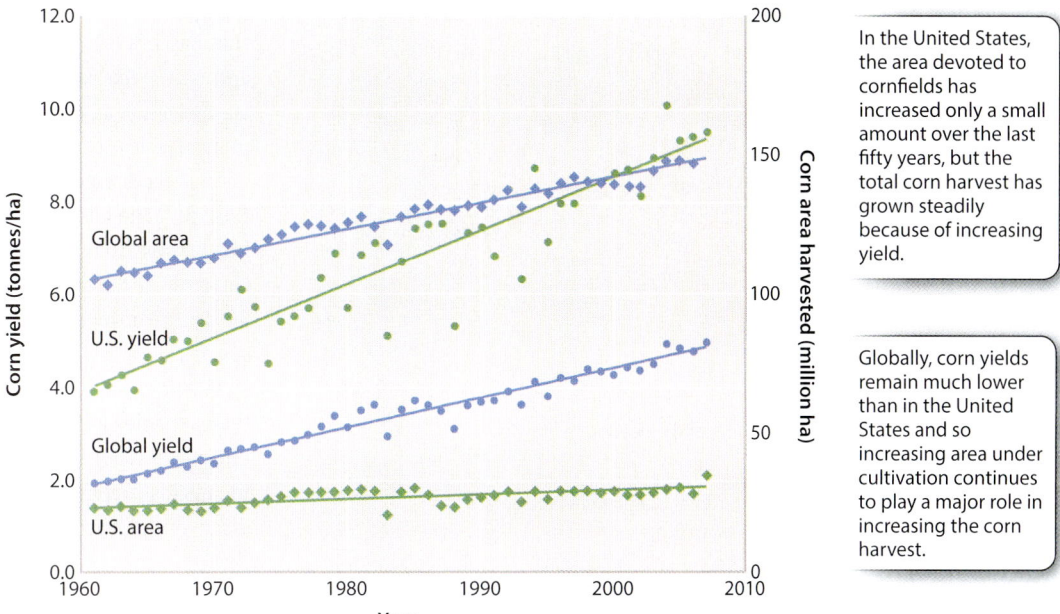

In the United States, the area devoted to cornfields has increased only a small amount over the last fifty years, but the total corn harvest has grown steadily because of increasing yield.

Globally, corn yields remain much lower than in the United States and so increasing area under cultivation continues to play a major role in increasing the corn harvest.

again, the Red Queen looms large. Fungus-resistant plants provide strong selective pressure for fungi that can circumvent the mechanisms of resistance, gaining access to otherwise unavailable food.

Food spoilage after harvest also reduces the effective yield of agricultural lands. It is estimated that one-third of the food grown in the United States is never eaten, and figures for Europe are similar. Fungi and bacteria spoil food during storage and transportation, on supermarket shelves, and in refrigerators. The problem also exists in less-developed countries. The United Nations estimates that in some parts of Africa a quarter of all crops spoil before they ever reach the market. More efficient storage, transportation, and packaging could sharply increase the effective yield of fields and pastures.

As with energy, a sustainable future for agriculture will require creative biologists, ingenious engineers, and wise citizens. Geneticists and physiologists can help to develop plant breeds that improve crop yield while using less fertilizer. Working with microbiologists, engineers can discover new ways of storing and transporting food that increase the likelihood that it will be eaten before it spoils. As individuals, we make daily choices about the food we eat. We can choose to eat more plant products, for example. The basic ecological logic of the trophic pyramid (Chapter 47) tells us that a hectare of grain fed directly to humans feeds more people than the meat of cattle fed on the same crops. We can also choose to eat locally grown products where possible, saving energy while decreasing the likelihood of spoilage.

48.4 HUMAN INFLUENCE ON EVOLUTION

It has sometimes been claimed that as humans exert increasing influence on nature, evolution will grind to a halt. Nothing could be further from the truth. As examples in the preceding sections show, what is changing, and what will continue to change as the human footprint grows, is the selective landscape within which evolution operates. Some species, for example rats and cockroaches, thrive in the urban environments we have constructed for our own benefit. Others will expand or decline as we replace prairies and forests with crop and pasture land. What are our conscious and inadvertent effects on evolution and biodiversity?

? CASE 8 Biodiversity Hotspots: Rain Forests and Coral Reefs
How has human activity affected biological diversity?

As noted in Chapter 1, it has been estimated that humans make use of nearly 25% of the entire photosynthetic output on land. That is, through the crops and livestock we harvest, the biomass we burn, and primary production we eliminate when we clear-cut

FIG. 48.18 Global agriculture land. Through time, land under cultivation has expanded, diminishing natural landscapes and the species diversity they support.

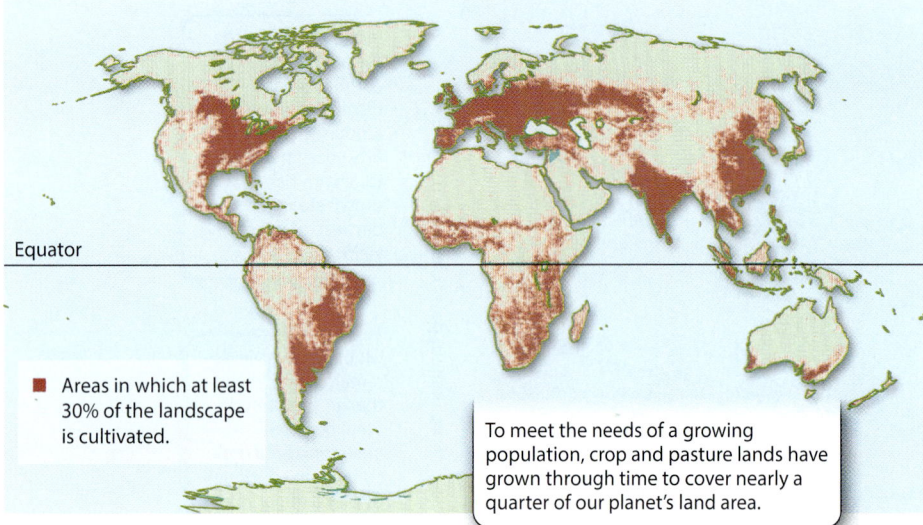

■ Areas in which at least 30% of the landscape is cultivated.

To meet the needs of a growing population, crop and pasture lands have grown through time to cover nearly a quarter of our planet's land area.

forests, we use one in every four grams of organic matter produced by land plants. Ecologists also estimate that crops cover 10% to 15% of Earth's land surface, and pasture land another 6% to 8% (**Fig. 48.18**).

The populations of a few species have expanded dramatically during the Anthropocene Period. For example, 10,000 years ago, corn (*Zea mays*) was an inconspicuous herb in southwestern Mexico; today, corn fields cover nearly 150 million hectares (ha) across the globe. As crop and pasture lands expand, along with urban and suburban environments, indigenous populations become displaced and their natural habitats disrupted.

Habitat loss in tropical rain forests poses one of the most important threats to biological diversity (**Fig. 48.19**). These forests represent the greatest hotspots of biodiversity on land, but clear-cutting for grazing and croplands has destroyed more than half of their pre-industrial area. It has been estimated that deforestation over the next century could eliminate 40% to 50% of all tree species in the Amazon forest.

As the forest disappears, so, too, will untold numbers of insect, mite, and other animal species. The renowned biologist E. O. Wilson has estimated that Earth is losing 0.25% of its species annually—that is, 5000 to 25,000 species per year, depending on how many species actually exist. In the words of another distinguished biologist, Paul Ehrlich, "The fate of biological diversity for the next 10 million years will almost certainly be determined during the next 50–100 years by the activities of a single species."

Without question, there will be fewer species when the 21st century ends than when it began. Lost species means lost opportunities for the discovery of novel compounds for

FIG. 48.19 Forests under threat. (a) The map shows the extent of deforestation worldwide. (b) This desolate scene in Amazonia was once an expanse of tropical rain forest. Nearly 20% of the Amazon rain forest has been cleared for crops, pasture, or mining.

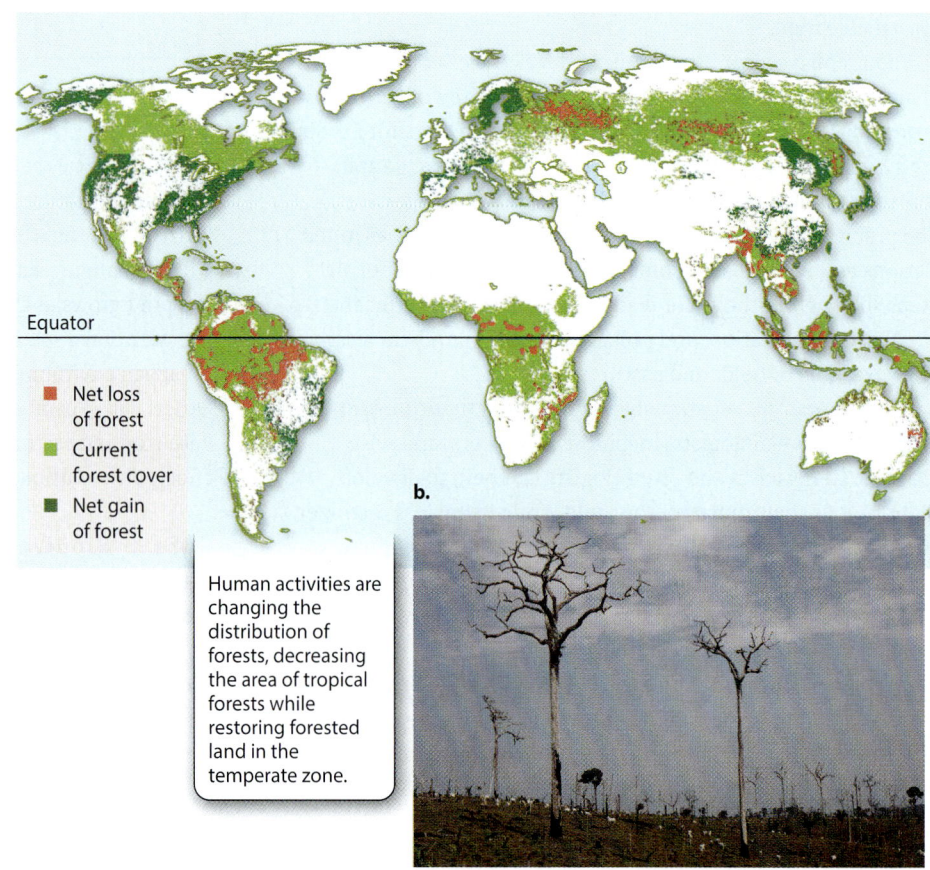

a.

■ Net loss of forest
■ Current forest cover
■ Net gain of forest

Human activities are changing the distribution of forests, decreasing the area of tropical forests while restoring forested land in the temperate zone.

b.

medical research. Diminished diversity may make communities less productive (Chapter 47) and less resilient to fires, hurricanes, or other environmental events. As discussed in the preceding section, there is a pressing need to feed the world, and there are also many good reasons to conserve the biological diversity that has evolved over 4 billion years. Balancing these two imperatives will entail creative approaches to land policy, including thoughtful ways of using land in multiple ways that serve both people and conservation. Reforestation of former crop and pasture lands will help, as will the preservation or establishment of ecological corridors to facilitate migration of populations.

Humans play an important role in the dispersal of species.

Biodiversity loss has many causes. In addition to habitat destruction, one of the most insidious is the introduction of non-native species, often called **invasive species**. Removed from natural constraints on population growth, invasive species can expand dramatically when introduced into new areas, often with negative consequences for native species and ecosystems.

Some species can be transported long distances by natural processes. For example, the Gulf Stream and North Atlantic Drift occasionally carry coconuts from the Caribbean to the coast of Ireland. Migratory birds can transport microorganisms across continents. In the Anthropocene Period, however, dispersal has increased dramatically at the hands of humans. Ships fill their ballast tanks with seawater in Indonesia and empty it into Los Angeles Harbor, carrying organisms across the Pacific Ocean. Insects in fruit from South America wind up on a dock in London. Humans have become major agents of dispersal, adding new complexity to 21st-century ecology.

It has been estimated that 49,000 invasive species have been introduced into the United States (**Fig. 48.20**). Many of these provide food or pleasure (for example, ornamental plants), and only a few have escaped into natural ecosystems. Some of the invasive species that have become established in their new surroundings do little beyond increasing community diversity, but others put strong pressure on native species.

Kudzu, for example, a plant imported from Japan to retard soil erosion, now covers more than 3 million hectares in the southeastern United States, displacing native plants through competition for space and resources (Fig. 48.20a). The Zebra Mussel, a European bivalve that hitchhiked to the Great Lakes in the ballast water of cargo ships, has also multiplied dramatically in its new habitat, displacing native species and clogging intake pipes to power plants (Fig. 48.20b). Nearly 500 introduced plant species have become weeds, contributing disproportionately to crops lost each year to competition from invasive plants.

Invasive species have particularly devastating effects on islands because island species have commonly evolved with relatively few competitors or predators. Bones in Hawaiian lava caves record more than 40 bird species that existed 1000 years ago but are now extinct, eliminated by rats and pigs introduced by Polynesian colonists, as well as by landscape alteration. Introduced predators have had a similar impact on the island of Guam. Brown tree snakes introduced from Australasia since World War II have reduced native bird and reptile diversity by 75% (Fig. 48.20c).

FIG. 48.20 Invasive species. (a) Kudzu (*Pueraria lobata*), a Japanese vine, was introduced to the United States in 1876 and now covers more than 3 million hectares. (b) The Zebra Mussel (*Dreissena polymorpha*), introduced into the Great Lakes from Europe in 1985, disrupts lake ecosystems, threatening native biological diversity. (c) The brown tree snake (*Boiga irregularis*), introduced into Guam from Asia and Australia shortly after World War II, has decimated indigenous bird populations across the island.

Such introductions contribute to the changing diversity and ecology of 21st-century landscapes, but it may be the spread of modern diseases that should impress (and concern) us most. In the world of airplanes and oceangoing ships, bacteria and viruses spread much more rapidly than they did in our pre-industrial past. The H1N1 (swine) flu, a global concern in 2009, is a recent example. The first diagnosed case was identified on April 13 in Mexico, and by mid-May this flu had been reported on five continents.

→ **Quick Check 4** How do invasive species affect species diversity of communities and ecosystems?

Humans have altered the selective landscape for many pathogens.

From the Black Death of the fourteenth century to the flu pandemic of 1918, disease has played a major role in human history. In the Anthropocene Period, we have unprecedented opportunity to control or even eradicate some of the great diseases of the past and present. At the same time, however, the strong selective pressures placed on pathogens by antibiotics pose a key challenge for 21st-century medicine.

Malaria, discussed in Case 4, provides an illustrative case. In any given year, malaria infects an estimated 500 million people, causing a million or more deaths. Infection is particularly severe in parts of Africa, where the direct and indirect effects of malaria account for 40% to 49% of all childhood mortality. Insecticide-soaked bed nets have proven highly effective in affected areas, reducing transmission by as much as 90%. Natural selection, however, has begun to chip away at success in controlling the disease: Mosquito strains have begun to emerge that are resistant to DDT and another family of insecticides called pyrethroids.

Quinine, a chemical extracted from the bark of the cinchona tree, was the first effective treatment for malaria, and remained in widespread use until the 1940s. Since that time, a small arsenal of antibiotics has been developed, including chloroquine, mefloquine, primaquine, and doxycycline. Their usefulness, however, has declined through time as resistance to each of these drugs has evolved in populations of *Plasmodium*, the infectious agent of malaria. More recently, artemisinins, antibiotics based on a traditional Chinese remedy, have proved effective in treatment. Again, however, resistant strains of malaria have been identified.

→ **Quick Check 5** The interplay of pathogens and antibiotics again brings up the problem of the Red Queen. How does the Red Queen hypothesis apply to these kinds of interaction?

The arms race between antibiotics and pathogens is not limited to malaria. It characterizes nearly the whole range of diseases caused by viruses, bacteria, protists, and fungi that infect humans. For example, tuberculosis, a lung infection spread by inhalation of the bacterium *Mycobacterium tuberculosis*, causes 1.8 million deaths annually, mostly in sub-Saharan Africa, where many people are also infected with HIV (Chapter 43). Antibiotic treatment is effective, but many who are infected do not or are unable to comply with the strict regimen of daily treatment for 6 months, hastening the evolution of drug-resistant *M. tuberculosis* strains. This occurs because an incomplete course of antibiotics does not eradicate all the bacteria, selecting for those bacteria that are least susceptible to the antibiotic. In 2007, about half a million cases of antibiotic-resistant tuberculosis were reported. More troubling, a small percentage of these infections have been characterized as "extremely drug resistant," meaning that the pathogens resist both standard treatments and alternative antibiotics developed over the past 20 years (**Fig. 48.21**). In effect, extremely drug-resistant tuberculosis is untreatable.

Because bacteria can pass genes horizontally (Chapter 26), antibiotic resistance evolved in one species can jump to another.

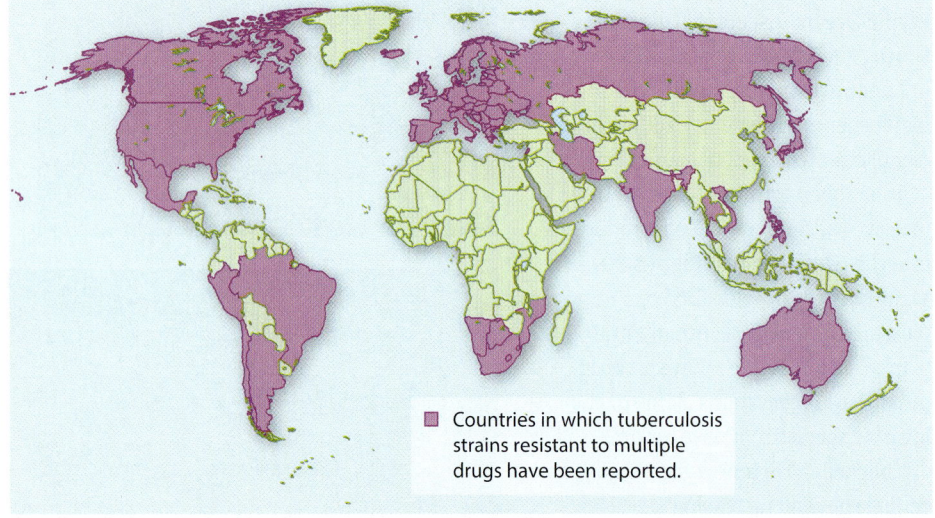

FIG. 48.21 Distribution of extremely drug-resistant tuberculosis. By 2008, confirmed cases had been reported in nearly 50 countries.

■ Countries in which tuberculosis strains resistant to multiple drugs have been reported.

FIG. 48.22 Amphibians under threat. The Harlequin Toad (*Atelopus varius*), from Costa Rica and Panama, has been classified as critically endangered.

Thus, the bacterial incubators for drug resistance can arise anywhere and spread rapidly.

Clearly, despite the remarkable gains of the twentieth century, public health will remain a key issue in the decades ahead. Improved sanitation—clean drinking water and safe food—can do much to reduce transmission of Earth's great diseases, and the search is on for vaccines that can further reduce or even eliminate some historic scourges. But the ongoing evolution of drug resistance will require the continuing efforts of scientists to stay one step ahead of pathogens.

Are amphibians ecology's "canary in the coal mine"?

In the nineteenth and well into the twentieth century, British, American, and Canadian coal miners sometimes took canaries with them down long mine shafts. Canaries are more sensitive than humans to carbon monoxide and other toxic gases that can accumulate in mines. When the canaries passed out, the miners knew it was time to leave. Because many amphibian species have narrow environmental tolerances and exchange gases through their skin, they can be especially sensitive to environmental disruption. For this reason, some biologists see them as "canaries" for the global environment.

If amphibians are indeed our early-warning system, we should be concerned. Fully one-third of all amphibian species are threatened with extinction, and more than 40% have undergone significant population declines over the past decade (**Fig. 48.22**). What accounts for this pattern? Just about all the environmental issues discussed in this chapter come into play. Amphibian habitat is being destroyed in many parts of the world, especially old-growth forests that provide damp, food-rich environments in abundance. Pesticides and other toxins of human manufacture take their toll as well. For example, the common herbicide atrazine has been shown experimentally to impair development in frogs. Studies on placental cells grown in the laboratory suggest that atrazine in our food may also affect human reproduction. When atrazine was applied to human placental cells, it caused overexpression of genes associated with abnormal fetal development. Because of concerns about public health, atrazine has been banned in several European countries.

Recently, infection has also been implicated in amphibian decline. The fungus *Batrachochytrium dendrobatidis*, or *Bd*, infects the skin of frogs and impairs many of the skin's functions. Recent studies of *Bd* show the classic features of epidemics—rapid geographic spread and high mortality—along the expanding front of fungal populations. How *Bd* spreads and whether it acts in concert with other facets of global change remain controversial, but there can be no doubt that disease has joined habitat destruction and pollution as a major cause of amphibian decline.

48.5 SCIENTISTS AND CITIZENS IN THE 21ST CENTURY

The decades ahead will be a time of great opportunity, but they will also present increasing challenges as the human footprint continues to grow around the world. How we will meet those challenges is, in large measure, up to you.

If you become a biologist, you will be able to help improve our planet's health in many different ways (**Fig. 48.23**). Field biologists are needed to document Earth's true biological diversity, even as much of it comes under threat. Physiologists are needed to understand how species respond to environmental change. And ecologists and population geneticists are needed to discover the principles that govern species interactions in communities, so that we can understand how physiological responses to global change in one species will affect the fortunes of others. Only then will we be able to design effective ways to preserve the world's biodiversity. Molecular biologists will help us to combat disease in new ways, and geneticists will develop crops that yield more food, even as temperature and rainfall patterns change. Perhaps most important, biologists will work with economists, lawmakers, and others to ensure that government policies reflect what we know about the natural world on which we all depend.

FIG. 48.23 Human impact. Scientists, doctors, teachers, and informed citizens can make a difference.

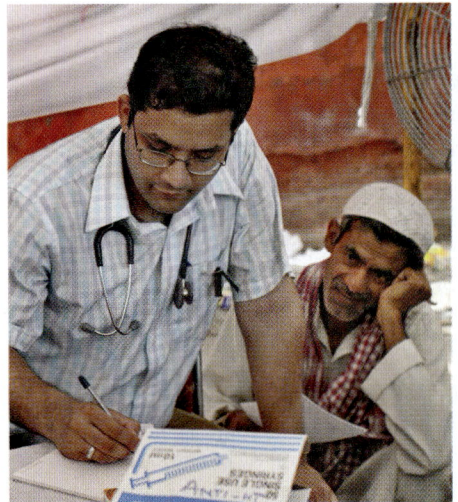

If you become a doctor, you will have many opportunities to improve the health of people in countries around the world (Fig. 48.23). Through laboratory research physicians will discover new ways to treat cancer, malaria, HIV, and a host of other diseases. And by treating patients in new ways, doctors will reduce the threat of antibiotic-resistant pathogens. Working with government and private agencies, doctors will help to ensure clean water and adequate food for people in parts of Africa, Asia, and wherever else the basic needs of life pose an ongoing challenge.

If you become a teacher, you will have a particularly important role in making sure that our children have the knowledge they need to meet the challenges of life in the Anthropocene Period (Fig. 48.23). And, perhaps most important, by choosing to be a responsible citizen, you can help to ensure that your family, your workplace, and your government make wise choices. As individuals, we all have a responsibility to make wise choices in our lives, and in an age of environmental change, knowledge of Earth and life provides a key framework for personal choice. As citizens, our votes shape governments, and biological education can help to ensure wise leadership.

We cannot change the world we've been given, but we can and will influence the one our children inherit. It is easy to despair about the future, but as the novelist Philip Pullman put it, "Hope is not the name of a temperament, or of an emotion, but of a virtue." Hope inspires action, and our collective capacity for positive change is great. Knowledge of biology will not by itself ensure that humanity's golden age can be sustained in coming years, but it will surely be an important part of our solutions.

CHAPTER 48 THE ANTHROPOCENE: HUMANS AS A PLANETARY FORCE

Core Concepts Summary

48.1 SOME SCIENTISTS CALL THE TIME IN WHICH WE ARE LIVING THE ANTHROPOCENE PERIOD TO REFLECT OUR SIGNIFICANT IMPACT ON THE PLANET.

The human population recently passed the 7-billion mark. page 48-1

The ecological footprint is the amount of land required to support an individual at an average standard of living. page 48-1

Our ecological footprint reflects all the activities that support our daily life, from growing and transporting the foods that we eat to supplying the materials and energy for our homes and the vehicles that provide our transportation. page 48-1

48.2 HUMANS HAVE A MAJOR IMPACT ON THE CARBON CYCLE, PRIMARILY THROUGH THE BURNING OF FOSSIL FUELS, WHICH RETURNS CARBON DIOXIDE TO THE ATMOSPHERE.

Carbon dioxide is a greenhouse gas, trapping heat and thereby keeping the surface of the Earth warm. page 48-3

Greenhouse gases play an important role in maintaining temperatures on Earth compatible with life, but increasing levels of greenhouse gases are resulting in higher mean surface temperatures. page 48-3

Changing climate has already affected the distribution of species and the composition of communities, and will continue to do so in the 21st century. page 48-5

Increasing levels of CO_2 are increasing the temperature and decreasing the pH of the oceans, affecting marine organisms, including corals. page 48-7

We can reduce CO_2 emissions, conserve finite natural resources, and contain the cost of energy by developing energy sources other than fossil fuels and improving energy efficiency. page 48-10

48.3 HUMANS HAVE AN IMPORTANT IMPACT ON THE NITROGEN AND PHOSPHORUS CYCLES, PRIMARILY THROUGH THE USE OF FERTILIZER IN AGRICULTURE.

Humans add large amounts of industrially fixed nitrogen to the environment, which enters the biosphere as fertilizer to increase food production. page 48-11

Only a small percentage of fixed nitrogen is taken up by crops; the rest is returned to the atmosphere by soil microorganisms or ends up in lakes and oceans, where it results in eutrophication, a boost in primary production that depletes oxygen and leads to the death of fish and other aquatic organisms. page 48-11

A second major nutrient in fertilizer is phosphorus, which is mined rather than produced industrially. The finite supply of phosphorus raises questions about our ability to maintain agricultural production in the future. page 48-14

We have increased agricultural yields through the use of fertilizer, high-yield and disease-resistant plants, and farm machinery. page 48-14

To feed a growing population, we can devote more land to agriculture or increase productivity further, but each option brings with it costs. page 48-14

48.4 THE IMPACT OF HUMANS ON THE ENVIRONMENT IS CHANGING THE STAGE ON WHICH EVOLUTION ACTS.

Humans have altered the landscape in dramatic ways, leading to the expansion of some species, like corn and wheat, and the decrease and extinction of others. page 48-15

Habitat loss may cause extinction in some of Earth's most diverse ecosystems, such as tropical rain forests. page 48-15

Invasive species, spread through human activity, are major players in biodiversity loss. page 48-17

The use of antibiotics is one of the success stories of twentieth-century medicine, but has also led to the spread of antibiotic resistance, a major challenge for 21st-century biologists. page 48-18

Amphibian populations have declined dramatically because of a combination of habitat loss, pesticide use, and the spread of infection. page 48-19

48.5 IN THE 21ST CENTURY, BIOLOGISTS, DOCTORS, ENGINEERS, TEACHERS, AND INFORMED CITIZENS HAVE VITAL ROLES TO PLAY IN UNDERSTANDING OUR CHANGING PLANET AND MAKING WISE CHOICES FOR OUR FUTURE.

Human impact can be positive as well as negative. page 48-19

Self-Assessment

1. Describe what is meant by the term "ecological footprint."
2. Name three sources of atmospheric CO_2.
3. Explain the relationship between atmospheric CO_2 levels and mean temperature.
4. Provide evidence that indicates that human activities are responsible for increases in atmospheric CO_2 over the last century.
5. Describe two ways in which organisms respond to changes in CO_2 levels and temperatures.

6. Describe several possible solutions to the problem of increased atmospheric CO_2.

7. Describe the causes and consequences of eutrophication.

8. Describe several possible solutions to the problem of feeding a growing human population.

9. Give three examples of species that have benefited from human activity and three examples of species that have been harmed by human activity.

10. Explain why amphibians are sometimes called "canaries in a coal mine."

Do you understand the chapter's Core Concepts? Log in to check your answers to the Self-Assessment questions, then practice what you've learned and reinforce this chapter's concepts by working through the problems and multimedia tutorials provided there.

www.biologyhowlifeworks.com

QUICK CHECK ANSWERS

Chapter 21

1. The allele frequency of *a* was calculated as follows:

   frequency(*a*) = [2 × (number *aa*) + 1 × (number *Aa*)] / [2 × (total number)]

   This equation can be rewritten as

   frequency(*a*) = [(number *aa*) + ½ × (number *Aa*)] / (total number)

   Note that

   number *aa* / total number = freqency(*aa*)

   and

   ½ × (number *Aa*) / total number = ½ frequency(*Aa*)

   Therefore,

   frequency(*a*) = frequency(*aa*) + ½ frequency(*Aa*)

   Stated in words, the frequency of allele *a* equals the frequency of *aa* homozygotes plus half the frequency of *Aa* heterozygotes. Because we were given genotype frequencies (50% *aa*, 25% *Aa*, and 25% *AA*), we can simply substitute into the preceding equation to solve for the frequency of *a*:

   0.50 + ½(0.25) = 0.625, or 62.5%

   By similar logic,

   frequency(*A*) = frequency(*AA*) + ½ frequency(*Aa*)

   which equals

   0.25 + ½(0.25) = 0.375, or 37.5%

   These equations are very useful for determining allele frequencies directly from genotype frequencies.

2. There is no next step. DNA sequence provides *full* genetic resolution.

3. We can conclude that the population is evolving. What we cannot tell is what mechanism—selection, migration, mutation, genetic drift, or non-random mating—is causing it to evolve. To determine what mechanism(s) is (are) driving the process requires more detailed population genetics analysis.

4. Adaptation is the fit between an organism and its environment. Of all the evolutionary mechanisms, only selection causes allele frequencies to change based on how they contribute to the success of an individual in terms of survival and reproduction. This means that allele frequencies in the next generation are ultimately governed by the environment. Because phenotype is in part determined by genotype, organisms become adapted to their environment under the influence of selection over time.

Chapter 22

1. Species change over time, making a single definition that can be applied in all cases difficult. The biological species concept (BSC) works well in many cases and is useful in that it gives us a specific phenomenon—reproductive isolation—to study. In addition, it focuses our attention on the flow of genes between individuals of a single species. However, there are plenty of cases, including asexual organisms and organisms known only from fossils, in which it is not applicable.

2. Volcanic island chains offer many examples of adaptive radiation because they are dependent on the serendipitous process of colonization. This means that only some plants and animals are able to get there. In addition, because of the absence of competitors on these islands, there are often many available ecological opportunities for the colonizers. For example, there may be no insect-eating birds on the islands, providing an ecological niche to be filled.

3. If populations of fish became separated from other populations of fish in different ponds during dry periods, this scenario suggests that speciation was more likely to be allopatric rather than sympatric. The allopatrically evolved species became sympatric only when the lake flooded again, combining all the separate ponds into a single body of water.

Chapter 23

1. No. The two trees are equivalent. The node shared by humans and mice is simply rotated in one tree compared to the other tree. In both trees, the closest relative of humans is mice.

2. A group called "fish" is paraphyletic because it includes some, but not all, of the descendants of a common ancestor. The descendants of the common ancestor of fish also include amphibians, sauropsids, and mammals.

3. The traits are analogous, the result of convergent evolution. Both animals are adapted for swimming in water and converged independently on similar traits, including a streamlined body and fins. However, they are only distantly related, as one is a fish and the other is a mammal.

4. Character traits can be measured for fossils as well as living organisms, so you can examine the features of the skeleton and then construct a phylogenetic tree that takes into account the skeletal traits they share with known vertebrates.

Chapter 24

1. No. Modern humans and modern chimpanzees share a common ancestor. We are not descended from chimpanzees because chimpanzees are living today.

2. The multiregional hypothesis suggests that the most recent common ancestor of all living humans lived about 2 million years ago, when *H. ergaster* first migrated from Africa to colonize Europe and Asia. The out-of-Africa hypothesis suggests that *H. sapiens* evolved much

more recently in Africa and then moved out of Africa, replacing the remnants of populations originally established by *H. ergaster* when it left Africa. The out-of-Africa hypothesis suggests a much more recent common ancestor for all present-day humans, one that lived about 200,000 years ago.

3. Studies of mtDNA, which is maternally transmitted, indicated that there was no input of Neanderthal mtDNA into the modern human population. In contrast, studies of Neanderthal genomic DNA showed that 1% to 4% of non-Africans' genomes are composed of Neanderthal DNA, implying that there was interbreeding between the ancestors of non-Africans and Neanderthals, presumably when the *H. sapiens* population first left Africa. If there is evidence of genetic input but no evidence of a female-based genetic contribution, we can hypothesize that male Neanderthals were the key contributors.

4. It is inaccurate to label *FOXP2* the "language gene" for several reasons. First, the gene is expressed in many tissues, so its effects are not limited to speech and language. Second, many genes are required for language, not just *FOXP2*. Nevertheless, its association with vocal communication in multiple species is striking.

Chapter 25

1. The coincidence in timing between the Industrial Revolution and an increase in CO_2 levels—following a millennium of little change—suggests that human activities played a role in recent changes in atmospheric composition. Alternative hypotheses would focus on other processes that add CO_2 to the atmosphere. Perhaps, for example, increased volcanic activity added more CO_2 to the atmosphere, or perhaps warming induced by changes in solar radiation caused thawing of permafrost, facilitating increased respiration of soil organic matter at high latitudes. These hypotheses can be tested against the historical record of volcanic activity and measurements of solar radiation over the past century.

2. All else being equal, increasing the elevation of mountains should increase rates of chemical weathering and erosion. As chemical weathering of continental rocks consumes CO_2, atmospheric CO_2 levels should decline.

3. Antelopes are primary consumers—that is, animals that eat vegetation. Lions are secondary consumers that eat antelopes. Because the transfer of energy from one level in the food web to the next higher level is inefficient, a given abundance of antelopes can support only a much smaller biomass of lions.

Chapter 26

1. In prokaryotic organisms, horizontal gene transfer transports genes from one organism to another, facilitating the generation of new gene combinations.

2. Archaeal cell and genome organization are similar to those of Bacteria. It was only through the use of molecular-sequence comparisons that Archaea were shown to occupy a distinct branch on the tree of life.

3. Primary production could be accomplished by anoxygenic photosynthesis, which does not generate oxygen. Oxidation of organic matter to carbon dioxide could be accomplished by anaerobic respiration (and fermentation), which does not use oxygen.

4. It wouldn't. Bacteria and archaeons cycle nitrogen between the atmosphere (nitrogen gas) and biologically useful forms. In the absence of these prokaryotes, life as we know it would not be possible.

5. Not necessarily. Today, hyperthermophiles live in hot springs and hydrothermal ridges on the ocean floor. These might have been the environment in which the last common ancestors of these organisms thrived.

Chapter 27

1. In general, eukaryotes employ a subset of the metabolic pathways used by bacteria. Most eukaryotes are capable of aerobic respiration; most can also gain at least some energy by fermentation; and some can also photosynthesize. In many ways, then, carbon cycling by eukaryotes is much like carbon cycling by aerobic prokaryotes. The novel contribution of eukaryotes is the ability to capture and ingest other cells, thus introducing predation into the carbon cycle.

2. No. Plants form one branch of the green algal tree, whose early branches are all occupied by unicellular forms. Similarly, the branch containing animals has choanoflagellates and other unicellular forms in its lower branches. Thus, animals and plants evolved multicellularity independently of each other. (Note that genetic evidence further supports this phylogenetic hypothesis. Plants and animals use different sets of genes to regulate multicellular development; see Chapter 28.)

3. Molecular sequences for genes in red and green algae show the close evolutionary relationship of these organelles to free-living cyanobacteria. This supports the hypothesis that eukaryotes initially gained the ability to photosynthesize by the symbiotic incorporation of cyanobacterial cells. Gene sequences in the chloroplasts of other photosynthetic eukaryotes (brown algae, diatoms, euglenids, and other algae) are very similar to those of red and green algal chloroplasts, indicating that photosynthesis spread throughout the eukaryotic tree by the symbiotic incorporation of green or red algal cells into other eukaryotes.

Chapter 28

1. In simple multicellular organisms, all or nearly all cells are in direct contact with the environment. In complex multicellular organisms, most cells are completely surrounded by other cells.

2. Without mechanisms for bulk transport, the movement of oxygen, nutrients, and molecular signals through organisms is limited by diffusion. Rates of diffusion, in turn, limit the size and shape that an organism can achieve. Bulk transport allows key molecules to be transported over distances much greater than those possible by diffusion alone, making larger organisms possible.

3. All multicellular organisms require molecules that promote adhesion between cells, communication between cells, and differentiation into different cell types within the body. Plants and animals have largely distinct sets of molecules for these functions. Many of the differences between plants and animals reflect the fact that plant cells have cell walls, whereas animal cells do not.

Chapter 29

1. Plants transpire to obtain CO_2 from the atmosphere. Plants cannot completely seal themselves off against water loss because that would prevent them from obtaining the CO_2 needed for photosynthesis. As CO_2 diffuses into a leaf to reach the photosynthetic cells, water vapor evaporates from surfaces within the leaf and diffuses out of the leaf. Thus, transpiration is an unavoidable byproduct of acquiring CO_2 by diffusion from the atmosphere.

2. CAM plants open their stomata at night when rates of evaporation are low and close them during the day to conserve water. During the night, CO_2 that diffuses into the leaf becomes incorporated into 4-carbon organic acids that are stored in the vacuole. During the day, these organic acids are retrieved from the vacuole and broken apart, and CO_2 is released into the cytoplasm. Because the stomata are closed, this CO_2 does not diffuse out of the leaf but instead can be used in the Calvin cycle to synthesize carbohydrates. C_4 plants produce 4-carbon organic acids in mesophyll cells, which then diffuse through plasmodesmata into bundle-sheath cells. In the bundle-sheath cells, each 4-carbon acid is broken up into a 3-carbon compound and CO_2. The CO_2 is used in the Calvin cycle to produce carbohydrates, while the 3-carbon compound diffuses back to the mesophyll cells, completing the C_4 cycle. Because the C_4 cycle operates much faster than the Calvin cycle, the concentration of CO_2 in the bundle-sheath cells is greatly increased. It is this ability to concentrate CO_2 that allows C_4 plants to suppress photorespiration.

3. The statement is correct in that the force that pulls water from the soil results from the evaporation of water from leaf cells, the energy for which comes from the heating of the leaf by the sun. It is incorrect in that there are structures within the plant that make this possible, notably the xylem conduits, and also the roots and the stem. The plant must expend energy to produce all these structures.

4. The movement of carbohydrates in phloem is from source to sink. Sources can be actively photosynthesizing leaves, or any organ with a large amount of stored carbohydrates. Carrots are examples of roots with large amounts of stored carbohydrates. In contrast, sinks are organs that require an import of carbohydrates to meet their growth and respiratory needs. Examples of sinks include young leaves, developing fruits, and most roots. Typically, the shoot is the source and the roots are a sink. At the end of winter or following a severe drought, however, a plant may regrow its shoot by drawing upon carbohydrates that had been stored in the roots. In this case, the movement of carbohydrates is from roots to shoot.

Chapter 30

1. Spores can be dispersed by the wind. In addition, their walls contain sporopollenin, a tough, resistant covering that allows spores to withstand environmental stresses such as ultraviolet radiation and desiccation.

2. Both mosses and ferns, like all plants, exhibit an alternation of a haploid (gametophyte) generation and a diploid (sporophyte) generation. However, in mosses, the sporophyte is dependent on the gametophyte, whereas in ferns the gametophyte and sporophyte are independent of each other.

3. Pollen is the male gametophyte of seed plants that develops within the spore wall. A seed develops from an ovule, which is the female gametophyte enclosed in layers of the sporophyte.

4. Compared to spores, seeds can store more resources, slow down their metabolism, and exhibit dormancy, all of which aid their dispersal.

5. A flower consists of four whorls of organs: From outside in, they are the sepals, petals, stamens, and carpels. The sepals generally protect developing flowers; the petals commonly serve to attract pollinators; the stamens produce pollen; and the carpel protects the ovules inside that develop into seeds following fertilization.

6. A fruit forms from the ovary that encloses the seeds and, in some plants, other parts of the flower, including the petals and sepals.

Chapter 31

1. The shoot apical meristem is the site of cell division and therefore of the growth of stems. It controls the pattern of nodes and internodes and therefore determines the patterns of leaves. Finally, it produces new apical meristems, allowing stems to branch.

2. One of a plant's two lateral meristems produces new vascular tissues, and the other maintains an outer protective layer.

3. Roots and stems both grow from apical meristems, have regions of cell division, elongation, and differentiation, and respond to light and gravity. However, root apical meristems are covered by a root cap, have a single vascular bundle, and initiate branching from the pericycle; all these features are different from what is seen in stems.

Chapter 32

1. No. The specific branch of the plant immune system can defend only against biotrophic pathogens, not necrotrophic ones, because this branch of the immune system is based on receptor molecules within living cells. Necrotrophic pathogens kill cells before colonizing them.

2. In the hypersensitive response, plants block off an infected area. In SAR, plants send a signal to uninfected tissue, enabling it to mount a defense and prevent infection.

3. Preventing jasmonic acid synthesis results in plants that are unable to mount a defense throughout the plant body and as a result suffer more damage from herbivores.

4. The trade-off between growth and defense creates a diversifying force that favors different species that are specialized for each habitat, rather than a single species that dominates across all soil types.

Chapter 33

1. The green algae are paraphyletic because the last common ancestor of all green algae also gave rise to land plants; the bryophytes are paraphyletic because the last common ancestor of mosses, liverworts, and hornworts also gave rise to vascular plants. Spore-bearing vascular plants are paraphyletic because the last common ancestor of lycophytes, ferns, and horsetails also gave rise to seed plants. You can't tell by the figure, but when fossils are considered, gymnosperms are also paraphyletic because the last common ancestor of all gymnosperms also gave rise to the angiopserms.

2. An alternative explanation is that stomata evolved once, in the common ancestor of bryophytes and vascular plants.

3. Xylem vessels evolved convergently in gnetophytes and angiosperms; insect pollination evolved convergently in cycads and angiosperms; fleshy tissues to facilitate seed dispersal evolved convergently in early seed plants (reflected today by ginkgo and cycad seeds), conifers such as juniper and yew, and flowering plants.

4. Monocots produce only a single cotyledon; monocot roots are produced directly from the stem; many monocots produce creeping stems; and the strap-shaped leaves of many monocots elongate from the base and the many parallel veins connect to the stem. Because the vascular bundles of monocot stems are distributed throughout the cross section, as opposed to arranged in a ring, they do not form a vascular cambium.

Chapter 34

1. Hyphae enable fungi to seek out new food resources; their mechanical strength allows them to penetrate large food bodies such as rotting logs or animal carcasses. In addition, hyphae transport nutrients from one part of the fungus to another.

2. A diploid cell has a single nucleus with two complete sets of chromosomes; a dikaryotic cell has two genetically distinct haploid nuclei.

3. Complex multicellular fruiting bodies evolved independently on two branches, the Basidiomycetes and the Ascomycetes.

4. You are eating the multicellular fruiting body built from dikaryotic hyphae.

Chapter 35

1. Nerve signals are transmitted electrically (in the form of an action potential) from one end of a neuron to the other. They are transmitted chemically (by neurotransmitters) across a synapse from one neuron to another.

2. Nerve impulses are slowed or completely stopped in response to loss of myelin surrounding nerve cells.

3. Nerve gases that block acetylcholinesterase, the enzyme that inactivates acetylcholine, cause violent muscle contractions, leading to paralysis of the diaphragm (the main muscle involved in breathing) and death by asphyxiation.

Chapter 36

1. Prostaglandins depolarize nociceptors and increase their firing rate.

2. Excess mucus secretion blocks airborne odor molecules from reaching the chemosensory hairs of olfactory neurons, reducing their ability to bind odorants and therefore reducing the sense of smell. Taste is also affected because much of the sense of taste depends on smell.

3. Spinning causes the hair cells in the semicircular canals to accommodate to the new motion. When you stop spinning, the hair cells detect a change in angular motion of the head (relative to the motion when your head was spinning), which leads to a sense of imbalance when you stand or try to walk.

4. Rod cells are extremely sensitive to light, allowing an animal to see in dimly lit conditions. Cone cells provide color vision through photopigments with different light wavelength absorption peaks, and they provide high spatial acuity, and thus sharp vision.

5. The three areas of the brain that are used to remember and recognize a face are the occipital cortex for processing visual information, the hippocampus for forming long-term memory of a face, and the temporal lobe for facial recognition.

Chapter 37

1. Without newly synthesized ATP, myosin cross-bridges formed with actin cannot detach from their actin binding sites, so they remain in the bound state and make the muscle stiff.

2. Because acetylcholine is blocked and cannot depolarize the muscle cell membrane, calcium is not released from the sarcoplasmic reticulum, and the muscle is unable to contract.

3. By contracting at an intermediate length, actin and myosin filament overlap of the biceps fibers is maximized, allowing the greatest number of cross-bridges to form and produce force to lift a weight.

4. Glass is composed of a single material (silica), which is rigid but can crack easily. Bone is a composite material, built of crystalline mineral reinforced by the protein collagen, enabling bone to absorb much more energy before breaking.

Chapter 38

1. Diabetes can result either from decreased insulin production by the pancreas (type 1 diabetes) or decreased effect of insulin on target cells (type 2 diabetes). Type 1 diabetes is an autoimmune disease in which the insulin-producing cells in the pancreas are attacked by the immune system. The result is inadequate insulin secretion. Daily injections of insulin can maintain stable normal

circulating levels of blood glucose. In individuals with type 2 diabetes, insulin production is unaffected, but cells are not able to respond to normal circulating levels of insulin. Type 2 diabetes is commonly linked to obesity and is most effectively treated by proper diet and exercise.

2. Peptide hormones bind cell membrane receptors because they are hydrophilic and cannot diffuse across the cell membrane, activating intracellular second messengers that stimulate signaling cascades within the target cell. Steroid hormones diffuse across the cell membrane and bind intracellular cytoplasmic or nuclear receptors, leading to changes in gene expression and protein synthesis within the target cell.

3. Hormones are secreted by a cell or gland and transported in the bloodstream to distant sites where they exert their effect. Hormones achieve specificity by binding to receptors, which are present only on their target cells.

4. Paracrine signals are chemicals secreted by a cell that influence the activity of neighboring cells without entering the bloodstream. Pheromones are chemicals secreted into the environment and transmitted to other species members, influencing their behavior. Hormones are released into the bloodstream and have effects on distant cells.

Chapter 39

1. Diffusion is increased by a large surface area for exchange and a short diffusion distance.

2. In individuals with asthma, resistance to airflow increases, thereby decreasing the flow of air. Resistance increases because the diameter of the airways decreases as a result of smooth muscle contraction in the airway walls, as well as mucus secretion.

3. Air has a low density and viscosity, and a high O_2 content relative to water.

4. Sickle-cell anemia is life threatening because transport of the deformed sickle-shaped red blood cells is limited in small blood vessels. Therefore, the supply of O_2 to tissues is reduced and cellular metabolism in the tissues supplied by the blood vessels is disrupted, possibly leading to organ failure.

5. Change in a vessel's diameter (or radius) affects blood flow most. For example, a twofold reduction in radius increases resistance 16 times.

6. The pressure in the ventricle exceeds the pressure in the atrium.

Chapter 40

1. Glycogen stores are used first, then fat, and finally muscle protein.

2. No. Metabolic rate increases with increasing mass, but the relationship between the two is not linear. Instead, metabolic rate increases with mass raised to the ¾ power. This means that a dog that is twice as heavy as a cat, for example, has a metabolic rate that is less than twice as much as the cat.

3. "Essential," when used to describe dietary amino acids, minerals, or vitamins, does not mean that other nutrients are less important or useful. "Essential" in this context means that these nutrients must be obtained in the diet—the organism cannot synthesize them on its own. In other words, it is "essential" that these nutrients are consumed.

4. Rocks and sediments in gizzards break apart food, as teeth do in vertebrates.

5. By eating the products of the cecum, rabbits can pass the partially digested food by the small intestine a second time, where nutrients are then absorbed.

Chapter 41

1. When a cell is placed in a hypotonic solution, water moves into the cell by osmosis and the cell swells or bursts. When a cell is placed in a hypertonic solution, water leaves the cell by osmosis and the cell shrinks.

2. Osmosis depends on the difference in *total* solute concentration on the two sides of a membrane. There is no net movement of water molecules across a semipermeable membrane with a high concentration of a *particular* solute on one side of a membrane if this concentration is matched by a similar concentration of some other solute on the other side of the membrane. For example, the concentration of sodium ions may be high on one side of a semipermeable membrane and low on the other, whereas the distribution of potassium ions may be just the reverse, so the total solute concentration is the same on the two sides of the membrane. The result is no net water movement.

3. Ammonia excreted by fish is rapidly diluted to nontoxic levels in the surrounding water. Mammals convert ammonia to urea, which is much less toxic than ammonia and therefore can be concentrated and stored before being eliminated.

4. Although the filtrate is dilute at the start and end of the loop of Henle, the loop creates a concentration gradient from the cortex to the medulla, which is important for subsequent water reabsorption from the collecting ducts. The loops also leave urea as the main solute.

Chapter 42

1. Sexual reproduction generates genetically unique offspring by (1) chance mutation; (2) recombination between homologous chromosomes during meiosis I; (3) random segregation of homologous chromosomes during meiosis I; and (4) new combinations of chromosomes in the process of fertilization. Asexual reproduction, like sexual reproduction, is also subject to chance mutations.

2. The vas deferens carries sperm from the epididymis to the ejaculatory duct. Cutting the vas deferens on both sides prevents sperm from entering the ejaculatory duct and leads to male sterility. This procedure, called a vasectomy, is a method of male contraception.

3. Estrogens or progesterones (or a combination of these) in the pill suppress GnRH, FSH, and LH and therefore block oocyte development and ovulation.

4. In females, oogonia differentiate into primary oocytes before birth and therefore a female is born with all the primary oocytes she will have. In males, by contrast, spermatogonia do not differentiate into primary spermatocytes until puberty, but sperm can be produced throughout a male's lifetime. In addition, only one of the four products of meiosis becomes a gamete in females, whereas all four products of meiosis become sperm in males.

5. In the process of development, cells differentiate into different cell types. The process of differentiation involves the turning on and turning off of specific genes. So, for example, genes that are expressed in a skin cell are different from those that are expressed in a liver or muscle cell. These changes in gene expression, once established, are for the most part stable through mitotic cell divisions.

Chapter 43

1. Damage to cilia in the respiratory tract increases the probability of lung infection by impairing the ability of the respiratory tract to sweep out microorganisms.

2. A second novel antigen injected on day 40 will produce a primary response, not a secondary response, because it is a novel antigen that has not been encountered before by the immune system.

3. Although a given B cell produces only one antibody, the entire population of B cells produces many different antibodies. Each B cell produces a different antibody as the result of a unique pattern of genomic rearrangement.

4. T cells are activated when their surface T cell receptors bind to an antigen in association with an MHC molecule. By contrast, B cells are activated when their surface antibodies bind to a free antigen.

Chapter 44

1. Neither Cnidaria nor Bilateria is more closely related to sponges. They share a common ancestor that is the closest relative of sponges.

2. A jellyfish would use more available nutrients than a sponge of the same size because it can move toward nutrients and capture and digest larger food items.

3. The shared features among snails, squids, and clams suggest that all mollusks descended from a common ancestor that had a head, radula, foot, mantle, and well-differentiated organ systems.

4. Young humans and other animals often have features that they lose as they age. Our first "baby" teeth are an example. If you could look at your own photograph when you were a one-month-old, inch-long embryo, you would see a notochord, gill slits, and a tail.

5. Scales are found on all of these animals due to their presence in a common ancestor. Even feathers are modified scales.

Chapter 45

1. In fruit flies, there are two alleles of the *for* gene that are expressed in larvae, leading to two different phenotypes in the population. As a result, larvae are either sitters or rovers. By contrast, in honeybees, the *for* gene is up-regulated during development, so a single individual begins by staying in the hive and then becomes a forager.

2. In classical conditioning, one stimulus is exchanged for another (the smell of food is exchanged for the sound of a bell in the case of Pavlov's dogs). As a result, a novel stimulus evokes a behavior. In operant conditioning, a behavior is associated with a reward or punishment, so the behavior becomes more likely (with a reward) or less likely (with a punishment). Therefore, in the case of classical conditioning, an association is made between a stimulus and a behavior, and in operant conditioning, an association is made between a behavior and a response.

3. Photoperiod is a good marker for time of year, and therefore for season. Many species have season-specific behaviors—hibernation in bears, migration in birds—that must be timed appropriately.

4. Organisms do not act "for the good of the species" because natural selection operates on individuals: It is the individual that lives or dies, reproduces or fails to reproduce. Traits that are disadvantageous to the individual are therefore selected against by natural selection even if they are beneficial to the species as a whole.

5. With reference to Fig. 45.17, the probability of each offspring inheriting an A allele is 0.5. The probability of a cousin sharing an A allele is 0.25, so the probability of each of the animal's cousin's offspring having an A is $0.25 \times 0.5 = 0.125$. In this case, the animal would have to substitute four of its cousin's offspring for each of its own.

Chapter 46

1. The per capita growth rate, r, equals $r = (\Delta N/\Delta t)/N_1$. In this case, the change in time Δt is 1 year, so $r = \Delta N/N_1$. If the starting population is x, at triple the size it is $3x$. Therefore, ΔN is $3x - x$, or $2x$, and $r = 2x/x$, or 2. In general, when a population triples in size in a year, it grows by 200%, so the per capita growth rate is 2.

2. One cent per day, doubled each day (a rate of increase of 100%) for 30 days exceeds $10,000,000.

3. Logistic curves are S shaped because exponential increase during early stages of population growth causes the first half of the curve to rise steeply, but then growth becomes limited by resource availability and so slows as population size approaches the carrying capacity of the environment.

4. Every organism invests all available energy in the ways that make most sense for the current environment. Increase in one area means decrease in another. To increase investments in all these areas, more resources in terms of food or energy would have to be acquired.

5. The closer island will initially receive more colonists than the farther one, but the larger island can eventually support more kinds of species.

Chapter 47

1. The fundamental niche is larger than the realized niche because the fundamental niche represents all the habitats where a species could possibly live, but competition among different species usually restricts a species to a smaller area, which is the realized niche.

2. The cost to the apple tree is resources in the form of nectar, which takes energy to produce, and the benefit is pollination. The cost to the bee is the energy required to search for food and transfer pollen, and the benefit is food.

3. Although they can affect many other species in the Arctic, lemmings are not a keystone species because their effects are a function of their great abundance. Keystone species have effects on communities that are disproportionate to their numbers because of the particular roles they play, often as predators.

4. A community is a group of organisms that share a given place. A community together with the physical environment in which the organisms live constitutes an ecosystem.

5. A diversity gradient can be seen with altitude that is similar to the one seen with latitude: Typically, the higher the altitude, the fewer the number of species.

Chapter 48

1. In the 130 years from 1800 to 1930, our numbers doubled from 1 to 2 billion, and they doubled again in just the 45 years from 1930 to 1975. This pattern indicates that the rate of human population growth increased in this time.

2. Global warming is the measured increase in Earth surface temperatures over the past 50 years. The greenhouse effect describes a process by which global warming can occur. The greenhouse effect is the result of the capacity of some molecules—especially carbon dioxide, methane, and water vapor—to absorb heat energy and then emit it in all directions. As a result, the energy incoming from solar radiation exceeds the amount radiated back into space as heat. Without the greenhouse effect, Earth would not be habitable. However, in recent decades, increasing amounts of greenhouse gases in the atmosphere have resulted in global warming.

3. As ecosystems warm, plants able to open flowers earlier appear to have higher fitness, enabling them to maintain or increase their geographic ranges.

4. Invasive species may outcompete native species for available resources, diminishing the population size and, in time, the diversity of native species. Introduced animals may prey on native species, reducing their numbers.

5. The Red Queen hypothesis suggests that organisms need to keep evolving (running) just to stay in the same ecological niche (place). As new antibiotics are developed, pathogens like malaria frequently evolve resistance, so researchers must continually strive to develop new medicines as natural selection erodes the effectiveness of the old.

GLOSSARY

10-nm chromatin fiber A relaxed 30-nm chromatin fiber, the state of the chromatin fiber in regions of the nucleus where transcription is currently taking place.

3′ end The end of a nucleic acid strand that carries a free 3′ hydroxyl.

30-nm chromatin fiber A chromosomal conformation created by the folding of the nucleosome fiber of DNA and histones.

3-phosphoglycerate (3-PGA) A 3-carbon molecule; two molecules of 3-PGA are the first stable products of the Calvin cycle.

5′ cap The modification of the 5′ end of the primary transcript by the addition of a special nucleotide attached in an unusual chemical linkage.

5′ end The end of a nucleic acid strand containing a free 5′ phosphate group.

ABC model A model of floral development that invokes three activities, A, B, and C, each of which represents the function of a protein or proteins hypothesized to be present in the cells of each whorl.

abomasum The fourth chamber in the stomach of ruminants, where protein digestion begins.

abscisic acid A hormone that stimulates root elongation by suppressing ethylene synthesis.

absolute temperature (T) Temperature measured on the Kelvin scale.

accessory pigment A pigment other than chlorophyll in the thylakoid membrane; carotenoids are important accessory pigments.

acidic Describes a solution in which the concentration of protons is higher than that of hydroxide ions (the pH is lower than 7).

acoelomate A bilaterian without a body cavity.

acrosome An organelle that surrounds the head of the sperm containing enzymes that enable sperm to transverse the outer coating of the egg.

actin A protein subunit that makes up microfilaments; used by both striated and smooth muscles to contract and generate force.

action potential A brief electrical signal transmitted from the nerve cell body along one or more axon branches.

activated The state of the receptor after binding the signaling molecule; the activated receptor transmits the information through the cytoplasm of the cell.

activation energy (E_A) The energy input necessary to reach the transition state.

activator A synthesized compound that increases the activity of an enzyme.

active site The portion of the enzyme that binds substrate and converts it to product.

active transport The "uphill" movement of substances against a concentration gradient.

acute phase response A change in the level of proteins in the blood, which is a marker for the initial phase of a disease.

adaptation In sensory reception, the process in which sensory receptors reduce their firing rate when a stimulus continues over a period of time.

adapted Better able to survive and reproduce in a given environment.

adaptive (acquired) immunity The part of the immune system that is specific to given pathogens.

adaptive radiation A bout of unusually rapid evolutionary diversification in which natural selection accelerates the rates of both speciation and adaptation.

addition rule The principle that the probability of either of two mutually exclusive outcomes occurring is given by the sum of their individual probabilities.

adenine (A) A purine base.

adenosine triphosphate (ATP) The molecule that provides energy in a form that all cells can readily use to perform the work of the cell. ATP is the universal energy currency for all cells.

adherens junction A beltlike junctional complex composed of cadherins that attaches a band of actin to the plasma membrane.

adrenal glands Paired glands located adjacent to the kidneys that secrete cortisol in times of stress.

adrenal medulla The inner part of the adrenal gland, which is stimulated by the sympathetic nervous system.

adrenaline (epinephrine) A hormone released by the adrenal gland that causes alertness and arousal.

advantageous Describes mutations that improve their carriers' chances of survival or reproduction.

advertisement display Behavior by which individuals draw attention to their status.

aerobic Utilizing oxygen.

aerobic metabolism Metabolism in the presence of oxygen in the mitochondria of eukaryotic animals that provides a steady supply of ATP for long-term, sustainable activity.

afferent neuron A neuron that sends information toward the central nervous system.

age structure The number of individuals within each age group of the population studied.

agonist muscles Muscle pairs that combine to produce similar motions.

aldose A monosaccharide with an aldehyde group.

aldosterone A hormone produced by the adrenal glands that stimulates the distal convoluted tubule and collecting ducts to take up more salt and water.

alga (plural, algae) A photosynthetic protist.

alkaloid   Any one of a group of nitrogen-bearing compounds that damages the nervous system of animals, produced by some plants as a defensive mechanism.

allantois   In the amniotic egg, a membrane that encloses a space where metabolic wastes collect.

allele frequency   The rate of occurrence of an allele in populations.

alleles   The different forms of a gene, corresponding to different DNA sequences in each different form.

allopatric   Describes populations that are geographically separated from each other.

allosteric effect   A change in the activity or affinity of a protein as the result of binding of a molecule to a site other than the active site.

allosteric enzyme   An enzyme whose activity is affected by binding a molecule at a site other than the active site. Typically, allosteric enzymes change their shape on binding an activator or inhibitor.

alpha (α) carbon   The central carbon atom of each amino acid.

alpha (α) helix   One of the two principal types of secondary structure found in proteins.

alternation of generations   The life cycle in which a haploid gametophyte and a diploid sporophyte follow one after the other.

alternative splicing   A process in which primary transcripts from the same gene can be spliced in different ways to yield different mRNAs and therefore different protein products.

altruistic   Describes an act of self-sacrifice.

Alveolata   A eukaryotic superkingdom, defined by the presence of cortical alveoli, small vesicles that, in some species, store calcium ions.

alveolus (plural, alveoli)   A cluster of tiny thin-walled sacs where gas exchange by diffusion takes place, found at the ends of bronchioles.

amacrine cell   A type of interneuron in the retina that communicates between neighboring bipolar cells and ganglion cells, enhancing motion detection and adjusting for changes in illumination of the visual scene.

amine hormone   A hormones that is derived from a single aromatic amino acid, such as tyrosine.

amino acid replacement   A change in the identity of an amino acid at a particular site in a protein resulting from a mutation in the gene.

amino acid   An organic molecule containing a central carbon atom, a carboxyl group, an amino group, a hydrogen atom, and a side chain. Amino acids are the building blocks of proteins.

amino end   The end of a polypeptide chain that has a free amino group.

amino group NH_2; a nitrogen atom bonded to two hydrogen atoms, covalently linked to the central carbon atom of an amino acid.

aminoacyl (A) site   One of three binding sites for tRNA on the large subunit of a ribosome.

aminoacyl tRNA synthetase   An enzyme that attaches a specific amino acid to a specific tRNA molecule.

amnion   In the amniotic egg, a membrane surrounding a fluid-filled cavity that allows the embryo to develop in a watery environment.

amniotes   The group of vertebrate animals that produces amniotic eggs; this group includes lizards, snakes, turtles, and crocodilians.

amniotic egg   An egg that can exchange gases while retaining water, permitting amniotes to live in dry terrestrial habitats that amphibian eggs cannot tolerate.

Amoebozoa   A superkingdom of eukaryotes with amoeba-like cells that move and gather food by means of pseudopodia.

Amphibia   Species of vertebrates, including frogs and salamanders, with an aquatic larval form with gills and an adult terrestrial form that usually has lungs.

amphipathic   Having both hydrophilic and hydrophobic regions.

amplified   In PCR technology, an alternative term for "replicated."

amylase   An enzyme in salivary secretions that breaks down carbohydrates to begin the digestion of sugars and starches.

anabolism   The set of chemical reactions that build molecules from smaller units utilizing an input of energy, usually in the form of ATP. Anabolic reactions result in net energy storage within cells and the organism.

anaerobic metabolism   Metabolism in the absence of oxygen, which produces ATP that provides rapid but short-term energy to the cell and the organism.

analogous   Describes similar characters that evolved independently in different organisms as a result of adaptation to similar environments.

anammox   Anaerobic ammonia oxidation; the oxidation of ammonium ion to nitrogen gas and the reduction of nitrite to nitrogen gas, with water produced as a by-product.

anaphase   The stage of mitosis in which the sister chromatids separate.

anaphase I   The stage of meiosis I in which the two homologous chromosomes of each bivalent separate as they are pulled in opposite directions, but the sister chromatids remained joined at the centromere.

anaphase II   The stage of meiosis II in which the centromere of each chromosome splits and the separated chromatids are pulled toward opposite poles of the spindle.

anchor   A membrane protein that attaches to other proteins and helps to maintain cell structure and shape.

angiosperms   The flowering plants; angiosperms are one of the two monophyletic groups of vascular plants.

angiotensin II   The active form of the hormone angiotensinogen, which causes the smooth muscles of arterioles throughout the body to constrict, thus increasing blood pressure and directing more blood back to the heart.

annealing   The coming together of complementary strands of single-stranded nucleic acids by base pairing.

annelid worms   A phylum of worms that have a cylindrical body with distinct segments and a bilaterian body plan.

annual clock   A year-based biological clock.

anoxygenic Not producing oxygen; anoxygenic photosynthetic bacteria do not gain electrons from water and so do not generate oxygen gas.

antagonism A one-sided interaction between species in which at least one participant loses more than it gains.

antagonist muscles Muscle pairs that pull in opposite directions.

anterior pituitary gland The region of the pituitary gland that forms from epithelial cells that develop and push up from the roof of the mouth; it receives hormones from the hypothalamus that stimulate it to release hormones in turn.

anther A structure supported by a filament from the stamen that contains several sporangia in which pollen is produced.

Anthropocene Period The modern era, so named to reflect the dominant impact of humans on Earth.

antibody A large protein that carries sugar molecules attached to some amino acids that binds to foreign molecules occurring naturally on or in microorganisms that participate in normal cellular functions.

antidiuretic hormone (ADH) A posterior pituitary gland hormone that acts on the kidneys and controls the water permeability of the collecting ducts, thus regulating the concentration of urine that an animal excretes; also called *vasopressin*.

antigen A foreign molecule that binds to an antibody and leads to the production of more antibodies.

antigenic drift The gradual process in which a high rate of mutation leads to changes in the amino acid sequences of antigens, thus allowing a population of viruses to evolve over time and evade memory T and B cells.

antigenic shift Reassortment of RNA strands in the viral genome, leading to sudden changes in cell-surface proteins, thus making it difficult to predict from year to year which virus strains will be most prevalent and therefore what vaccine will be most effective.

antigenic variation The encoding of a protein at different times by any one of a number of different genes.

antigen-presenting cell A type of cell (including macrophages, dendritic cells, and B cells) that takes up an antigen and returns portions of it to the cell surface bound to MHC class II proteins.

antiparallel Oriented in opposite directions; the strands in a DNA duplex are antiparallel.

aorta A large artery through which oxygenated blood flows from the left ventricle to the head and rest of the body.

aortic body A sensory structure that monitors the levels of oxygen and protons moving to the body.

aortic valve A valve beween the left ventricle and the aorta.

apical dominance The suppression of growth of axillary buds by the shoot apical meristem.

appendicular Describes the part of the vertebrate skeleton that consists of the bones of the limbs, including the shoulder and pelvis.

appendix A narrow, tubelike structure that extends from the cecum. The appendix is a vestigial structure that has no clear function in nonherbivorous animals.

aquaporin A protein channel that allows water to flow through the plasma membrane more readily by facilitated diffusion.

aqueous Watery.

aqueous humor A clear watery liquid that fills the interior region in front of the lens of the vertebrate eye.

Archaea One of the three domains of life, consisting of single-celled organisms without true chromosomes or a nucleus that divide by binary fission, differing from bacteria in many aspects of their cell and molecular biology.

Archaeplastida A eukaryotic superkingdom of photosynthetic organisms; includes the land plants.

Ardi A specimen of *Ardipethecus ramidus*, an important early hominin, dating from about 4.4 million years ago.

arteriole A small branch of an artery.

artery A large, high-pressure vessel that moves blood flow away from the heart to the tissues.

Arthropoda An animal group that includes insects and contains more than half of all known animal species; distinguished by their jointed legs.

artificial selection A form of directional selection analogous to natural selection, but without the competitive element; successful genotypes are determined by the breeder, not by competition.

ascomycetes A monophyletic fungal subgroup of the Dikarya, making up 64% of all fungal species, in which nuclear fusion and meiosis take place in an elongated saclike cell called an ascus; also called *sac fungi*.

asexual reproduction The reproduction of cells or single-celled organisms by cell division; offspring are clones of the parent.

assimilation The process in which plants take up sulfate ions from the soil and reduce them within their cells to hydrogen sulfide that can be incorporated into cysteine and other biomolecules.

associative learning (conditioning) Learning that two events are correlated.

astrocyte A type of star-shaped glial cell that contributes to the blood–brain barrier by surrounding blood vessels in the brain and thus limiting the size of compounds that can diffuse from the blood into the brain.

atom The basic unit of matter.

atomic mass The mass of the atom determined by the number of protons and neutrons.

ATP synthase An enzyme that couples the movement of protons through the enzyme with the synthesis of ATP.

atrioventricular (AV) node A specialized region of the heart containing pacemaker cells that transmit action potentials from the sinoatrial nodes to the ventricles of the heart.

atrioventricular (AV) valve A valve between the right atrium and the right ventricle and between the left ventricle and left atrium.

atrium A heart chamber that receives blood from the lungs or the rest of the body.

auditory cortex The area of the brain that processes sound.

autocrine signaling Signaling between different parts of a cell; the signaling cell and the responding cell are one and the same.

autoimmune disease A disease in which tolerance is lost and the immune system becomes active against antigens of the host.

autonomic nervous system The involuntary component of the peripheral nervous system, which controls internal functions of the body such as heart rate, blood flow, digestion, excretion, and temperature.

autosome Any chromosome other than the sex chromosomes.

autotroph Any organism that is able to convert carbon dioxide into glucose, thus making its own organic source of carbon.

auxin The plant hormone that causes shoots to elongate and guides vascular differentiation.

avirulent Describes pathogens that damage only a small part of the plant because the host plant is able to contain the infection.

axial Describes the part of the vertebrate skeleton that consists of the skull and jaws of the head, the vertebrae of the spinal column, and the ribs.

axillary bud A meristem that forms at the base of each leaf.

axon The fiberlike extension from the cell body of a neuron that transmits signals away from the nerve's cell body; the output end of a nerve cell.

axon hillock The junction of the nerve cell body and its axon.

B lymphocyte (B cell) A cell type that matures in the bone marrow of humans and produces antibodies.

Bacteria One of the three domains of life, consisting of single-celled organisms without true chromosomes or a nucleus that divide by binary fission, differing from archaeons in many aspects of their cell and molecular biology.

bacteriochlorophyll A light-harvesting pigment closely related to the chlorophyll found in plants, algae, and cyanobacteria.

bacteriophage Virus that infects bacterial cells.

balancing selection Natural selection that acts to maintain two or more alleles of a given gene in a population.

ball-and-socket joint A joint that allows rotation in three axes, like the hip and shoulder.

Baltimore system A classification of viruses into seven major groups, I–VII, by type of nucleic acid and method of mRNA synthesis; developed by David Baltimore.

band A crosswise striation in a chromosome, or a horizontal stripe in an electrophoresis gel.

basal lamina A specialized form of extracellular matrix that underlies and supports all epithelial tissues.

base A nitrogen-containing compound that makes up part of a nucleotide.

base excision pair A specialized repair system in which an improper DNA base and its sugar are both removed and the resulting gap is repaired.

base stacking Stabilizing interactions between bases in the same strand of DNA.

basic Describes a solution in which the concentration of protons is lower than that of hydroxide ions (the pH is higher than 7).

basidiomycetes A monophyletic fungal subgroup of the Dikarya, including smuts, rusts, and mushrooms, in which nuclear fusion and meiosis take place in a club-shaped cell called a basidium; also called club fungi.

basilar membrane The membrane that, with the cochlear duct, separates the upper and lower canal of the cochlea.

basophil A type of granulocyte that, along with eosinphils, defends against parasitic infections but also contributes to allergies.

behaviorally isolated Describes individuals that only mate with other individuals on the basis of specific courtship rituals, songs, and other behaviors.

beta (β) sheet One of the two principal types of secondary structure found in proteins.

beta-(β-)oxidation The process of shortening fatty acids by a series of reactions that sequentially remove two carbon units from their ends.

bilateral symmetry Symmetry on both sides of a midline; animals with bilateral symmetry have a distinct head and tail, marking front and back, with a single plane of symmetry running between them at the midline.

bilateria The branch of the animal tree that includes animals with bilateral symmetry.

bilayer A two-layered structure of the cell membrane with hydrophilic "heads" pointing outward toward the aqueous environment and hydrophobic "tails" oriented inward, away from water.

bile A fluid produced by the liver that aids in fat digestion by breaking large clusters of fats into smaller lipid droplets.

binary fission The process by which cells of bacteria or archaeons divide.

binding affinity The tightness of the binding between the receptor and the signaling molecule.

biological species concept (BSC) As described by Ernst Mayr, the concept that "[s]pecies are groups of actually or potentially interbreeding populations that are reproductively isolated from other such groups." The BSC is the most widely used and accepted definition of a species.

biologist A scientist who studies life.

biology The science of life and how it works.

biomass The total mass of organisms in a given area.

biome The distinctive and stable assemblage of species found over a broad region of Earth; terrestrial biomes are recognized by their vegetation.

biomineralization The precipitation of minerals by organisms, as in the formation of skeletons.

biotrophic pathogen A plant pathogen that obtains resources from living cells.

biparental inheritance A type of inheritance in which the organelles in the offspring cells derive from those in both parents.

bipedal Moving by two feet and habitually walking upright.

bipolar cell A type of interneuron in the retina that adjusts its release of neurotransmitter in response to input from rod and cone cells.

bivalent The four-stranded structure consisting of two pairs of sister chromatids aligned along their length and held together by chiasmata.

bivalves A group of mollusks that includes clams, oysters, and mussels; they have evolved a skeleton in which two hard shells are connected by a flexible hinge.

bladder A hollow organ in mammals and fishes for the storage and elimination of urine.

blastocyst A hollow sphere produced by cells in the morula that move in relation to one another, pushing against and expanding the membrane that encloses them. A blastocyst forms from the blastula, has an inner cell mass, and occurs only in mammals.

blastula A fluid-filled ball of undifferentiated cells formed after the fertilized egg has undergone several rounds of mitotic cell division following the morula stage.

blending inheritance The now-discredited model in which heredity factors transmitted by the parents become intermingled in the offspring instead of retaining their individual genetic identities.

blood The circulatory fluid in vertebrates.

bone marrow A fatty tissue between trabeculae and within the central cavity of a bone that contains many important cell populations.

bottleneck An extreme case of genetic drift that occurs when a population falls to just a few individuals.

Bowman's capsule A membranous sac that encases the glomerulus.

brain The centralized concentration of neurons in an organ that processes complex sensory stimuli from the environment or from anywhere in the body.

brainstem Part of the vertebrate brain, formed from the midbrain, which activates the forebrain by relaying information from lower spinal levels.

bronchiole Any one of the fine branches of secondary bronchi.

bryophytes A paraphyletic group of plants that includes the mosses, liverworts, and hornworts.

buccal cavity The mouth.

bud scale One of many like structures formed from leaf primordia that, together, protect shoot apical meristems from desiccation and damage due to cold.

budding A form of asexual reproduction in fungi, plants, and some animals in which a bud forms on the organism and eventually breaks off to form a new organism that is smaller than its parent.

bulbourethral glands Glands below the prostate gland that produce a clear fluid that lubricates the urethra for passage of the sperm.

bulk flow The movement of molecules through organisms in the active circulation of fluids at rates beyond those possible by diffusion across a concentration gradient; also known as *bulk transport*.

bulk transport The movement of molecules through organisms in the active circulation of fluids at rates beyond those possible by diffusion across a concentration gradient; also known as *bulk flow*.

bundle sheath A cylinder of cells that surrounds each vein in C_4 plants in which carbon dioxide is concentrated, suppressing photorespiration.

Burgess Shale A sedimentary rock formation on the seafloor covering what is now British Columbia, Canada, that preserves a remarkable sampling of marine life during the initial diversification of animals.

C_3 plant A plant that does not use 4-carbon organic acids to supply the Calvin cycle with carbon dioxide.

C_4 plant A plant that suppresses photorespiration by increasing the concentration of carbon dioxide in the immediate vicinity of rubisco.

cadherin A calcium-dependent adherence protein, important in the adhesion of cells to other cells.

calmodulin A protein that binds with Ca^{2+} and activates the enzyme myosin kinase.

Calvin cycle The process in which carbon dioxide is reduced to synthesize carbohydrates, with ATP and NADPH as the energy sources.

Cambrian explosion A transition period in geologic time during which the body plans characteristic of most bilaterian phyla developed.

cancer A group of diseases characterized by uncontrolled cell division.

canine One of the teeth in carnivores specialized for piercing the body of prey.

capacitation A series of physiological changes that allow the sperm to fertilize the egg.

capillary A very small blood vessel, arranged in finely branched networks connected to arterioles or venules, where gases are exchanged by diffusion with surrounding tissues.

capsid The protein coat that surrounds the nucleic acid of a virus.

carbohydrate An organic molecule containing C, H, and O atoms that provides a source of energy for metabolism and that makes up the cell wall in bacteria, plants, and algae.

carbon cycle The intricately linked network of biological and physical processes that shuttles carbon among rocks, soil, oceans, air, and organisms.

carboxyl end The end of a polypeptide chain that has a free carboxyl group.

carboxyl group COOH; a carbon atom with a double bond to oxygen and a single bond to a hydroxyl group.

carboxylation The first step of the Calvin cycle, in which carbon dioxide absorbed from the air is added to a 5-carbon molecule.

cardiac cycle The contraction of the two atria of the heart followed by contraction of the two ventricles.

cardiac muscle Muscle cells that make up the walls of the atria and ventricles and contract to pump blood through the heart.

cardiac output (CO) The volume of blood pumped by the heart over a given interval of time, the key measure of heart function.

caroid body A sensory structure that senses oxygen and proton concentrations of the blood moving to the brain.

carpel An ovule-producing floral organ in the center whorl.

carrier A transporter that facilitates movement of molecules.

carrying capacity (K) The maximum number of individuals a habitat can support.

cartilage Fluid-filled tissue that forms the surfaces between adjacent bones, cushioning forces transmitted across the joint.

Casparian strip A thin band of hydrophobic material that encircles each cell of the endodermis of a root, controlling which materials enter the xylem.

catabolism The set of chemical reactions that break down molecules into smaller units and, in the process, produces ATP to meet the energy needs of the cell.

causation A relationship in which one event leads to another.

cavitation The abrupt replacement of the water in a conduit by water vapor, which blocks water flow in xylem.

cecum A chamber that branches off the large intestine; along with the colon, the site of hindgut fermentation.

cell The simplest self-replicating entity that can exist as an independent unit of life.

cell adhesion molecule A cell-surface protein that attaches cells to one another and to the extracellular matrix.

cell cycle The collective name for the steps that make up eukaryotic cell division.

cell division The process by which cells make more cells.

cell plate In dividing plant cells, a new cell wall formed in the middle of the cell from the fusing of vesicles during late anaphase and telophase.

cell theory The theory that the cell is the fundamental unit of life in all organisms and that cells come only from preexisting cells.

cell wall A defining boundary in many organism, external to the cell membrane, that helps maintain the shape and internal composition of the cell.

cell-mediated immunity The ability of T cells, which do not secrete antibodies, to recognize and act against pathogens directly.

cellular blastoderm In *Drosophila* development, the structure formed by the nuclei in the single-cell embryo when they migrate to the periphery of the embryo and each nucleus becomes enclosed in its own cell membrane.

cellular junction A region in the plasma membrane, consisting of cell adhesion molecules and other cytosolic proteins, where a cell makes contact with and adheres to another cell or the extracellular matrix.

cellular respiration A series of chemical reactions that convert the energy stored in nutrients into a chemical form that can be readily used by cells.

cellulase The enzyme that breaks down cellulose.

central dogma The theory that information transfer in a cell usually goes from DNA to RNA to protein.

central nervous system (CNS) In vertebrates, the brain and spinal cord; in invertebrates, centralized information-processing ganglia.

centromere A constriction that physically holds sister chromatids together; the site of the attachment of the spindle fibers that move the chromosome in cell division.

centrosome A compact structure that is the microtubule organizing center for animal cells.

cephalization The concentration of nervous system components at one end of the body.

cephalochordates A subphylum of Chordata that shares key features of body organization with vertebrates but lacks a well-developed brain and eyes, has no lateral appendages, and does not have a mineralized skeleton.

cephalopods A group of mollusks, including squid, cuttlefish, octopus, and chambered nautilus, with distinctive adaptations such as muscular tentacles that capture prey and sense the environment.

cerebellum Part of the vertebrate brain, formed from the hindbrain, which coordinates complex motor tasks by integrating motor and sensory information.

cerebral cortex Part of the vertebrate brain formed from a portion of the forebrain that is greatly expanded in mammals, particularly primates.

cerebrum The outer left and right hemispheres of the cerebral cortex.

cervix The end, or neck, of the uterus.

chain terminator A term for a dideoxynucleotide, which if incorporated into a growing daughter strand stops strand growth because there is no hydroxyl group to attack the incoming nucleotide.

channel A transporter with a passage that allows the movement of molecules through it.

chaperone A protein that helps shield a slow-folding protein until it can attain its proper three-dimensional structure.

character One of the anatomical, physiological, or molecular features that make up organisms' bodies.

character state The observed condition of a character, such as presence or absence of lungs or arrangement of petals.

checkpoint One of multiple regulatory mechanism that coordinate the temporal sequence of events in the cell cycle.

chelicerates One of the four main groups of arthropods, including spiders and scorpions, chelicerates have pincer-like claws and are the only arthropods that lack antennae.

chemical bond Any form of attraction between atoms that holds them together.

chemical reaction The process by which molecules are transformed into different molecules.

chemoautotroph A microorganism that obtains energy from chemical compounds, not from sunlight.

chemoreceptor A receptor that responds to molecules that bind to specific protein receptors on the cell membrane of the sensory receptor.

chemotroph An organism that derives its energy directly from organic molecules such as glucose.

chiasma (plural, chiasmata) A crosslike structure within a bivalent constituting a physical manifestation of crossing over.

chitin A modified polysaccharide containing nitrogen that makes up the cell walls of fungi and the hard exoskeletons of arthropods.

chloride cell A type of specialized cell in the gills of marine bony fishes that counters the ingestion and diffusion of excess electrolytes into the animal by pumping chloride ions into the surrounding seawater; chloride cells in freshwater fishes have opposite polarity.

chlorophyll The major photosynthetic pigment contained in the thylakoid membrane; it plays a key role in the chloroplast's ability to capture energy from sunlight. Chlorophyll appears green because it is poor at absorbing green wavelengths.

chloroplast An organelle that converts energy of sunlight into chemical energy by synthesizing simple sugars.

chloroplast genome In plant eykaryotic cells, the genome of the chloroplast.

choanocyte A type of cell that lines the interior surface of a sponge; choanocytes have flagella and function in nutrition and gas exchange.

choanoflagellate One of a group of mostly unicellular protists characterized by a ring of microvilli around the cell's single flagellum.

cholecystokinin (CCK) A peptide hormone that causes the gallbladder to contract and thus release bile into the duodenum.

cholesterol An amphipathic lipid that is a major component of animal cell membranes.

Chondrichthyes Cartilaginous fish, a monophyletic group that includes about 800 species of sharks, rays, and chimaeras.

Chordata One of the three major phyla of deuterosomes, this group includes vertebrates and closely related invertebrate animals such as sea squirts.

chorion In the amniotic egg, a membrane that surrounds the entire embryo along with its yolk and allantoic sac.

chromatin A complex of DNA, RNA, and proteins that gives chromosomes their structure; chromatin fibers are either 30 nm in diameter or, in a relaxed state, 10 nm.

chromatin remodeling The process in which the nucleosomes are repositioned to expose different stretches of DNA to the nuclear environment.

chromosome In eukaryotes, the physical structure in which DNA in the nucleus is packaged; used more loosely to refer to the DNA in bacterial cells or archaeons.

chromosome condensation The progressive coiling of the chromatin fiber, an active, energy-consuming process requiring the participation of several types of proteins.

chytrid A single-celled aquatic fungus with chitin walls that attaches to decomposing organic matter.

cilium (plural, cilia) A hairlike organelle that propels the movement of cells or of substances within cells or out of the body; shorter than a flagellum.

circadian clock A biological clock, on a daily cycle, that regulates many daily rhythms in animals, such as feeding, sleeping, hormone production, and core body temperature.

circular muscle Smooth muscle that encircles the body or an organ; in the digestive tract, a circular muscle layer contracts to reduce the size of the lumen. A circular muscle layer contracts alternately with longitudinal muscle to move contents through the digestive tract and to enable locomotion in animals with hydrostatic skeletons.

circulation The movement of a specialized body fluid that carries oxygen and carbon dioxide.

cis-regulatory element A short DNA sequence adjacent to a gene, usually at the 5′ end, that interacts with transcription factors.

cisternae The series of flattened membrane sacs that make up the Golgi apparatus.

citric acid cycle The third stage of cellular respiration, in which acetyl-CoA is broken down and more carbon dioxide is released.

cladistics Phylogenetic reconstruction on the basis of synapomorphies.

class A group of closely related orders.

classical conditioning Associative learning in which two stimuli are paired.

cleavage The successive mitotic divisions of the zygote after fertilization, in which the single large egg is divided into many smaller cells.

climax community A mature assembly, a final stage in succession, in which there is little further change in species composition.

clitoris The female homolog of the glans penis.

cloaca A hollow organ in amphibians, reptiles, and birds for the storage and elimination of urine.

clonal selection A hypothesis proposing that the antigen instructs the antibody to fold in a particular way so that the two interact in a specific manner; now a central principle of immunology.

clone An individual that carries an exact copy of the nuclear genome of another individual; clones are genetically identical cells or individuals.

closed circulatory system A circulatory system made up of a set of internal vessels and a heart that functions as a pump to move blood to different regions of the body.

cochlea A coiled chamber within the skull containing hair cells that convert pressure waves into an electrical impulse that is sent to the brain.

cochlear duct A fluid-filled cavity in the cochlea, next to the upper canal, that houses the organ of Corti.

codon A group of three adjacent nucleotides that specifies an amino acid in a protein or that terminates polypeptide synthesis.

coelacanth A species of fleshy-finned fish thought to have been extinct for 80 million years; along with lungfish, the nearest relative of tetrapods.

coelomate A bilaterian with a body cavity.

coenocytic Containing many nuclei within one giant cell; the nucleus divides multiple times, but the nuclei are not partitioned into individual cells.

coenzyme Q The final electron acceptor from both complexes I and II in the electron transport chain.

coevolution The process in which species evolve together, each responding to selective pressures from the other.

cofactor A substance that associates with an enzyme and plays a key role in its function.

cognition The ability of the brain to process and integrate complex sources of information, interpret and remember past events, solve problems, reason, and form ideas.

cohort A group of the individuals born at a given time.

collagen A strong protein fiber found in bone and artery walls.

collecting duct A type of duct in the vertebrate kidney where urine collects.

colon Part of the hindgut and the site of reabsorption of water and minerals; also known as the *large intestine*.

combinatorial control Regulation of gene transcription by means of multiple transcription factors acting together.

commensalism An interaction between species in which one partner benefits with no apparent effect on the other.

communication The transfer of information between two individuals, the sender and the receiver.

community The set of all populations found in a given place.

compact bone Dense, mineralized bone tissue that forms the walls of a bone's shaft.

companion cell A cell adjacent to a sieve element, it carries out cellular functions such as protein synthesis.

comparative genomics The analysis of the similarities and differences in protein-coding genes and other types of sequence in the genomes of different species.

competition An interaction in which the use of a mutually needed resource by one individual or group of individuals lowers the availability of the resource for another individual or group.

competitive exclusion The result of an antagonistic interaction in which one species is prevented from occupying a particular habitat or niche.

competitive inhibitor A reversible inhibitor that reduces the affinity of the substrate for the active site of an enzyme but can be overcome by excess substrate, so the maximum velocity of the reaction is not changed. Competitive inhibitors usually have a structure similar to that of the substrate and therefore bind to the active site of the enzyme.

complement system The collective name for certain proteins circulating in the blood that participate in innate immune function and thus complement other parts of the immune system.

complementary Describes the relationship of purine and pyrimidine bases, in which the base A pairs only with T and G pairs only with C.

complex carbohydrate A long, branched chain of monosaccharides.

complex multicellular organism An organism composed of specialized cells for specific functions, so the organism as a whole can perform a broad range of tasks but individual cells for the most part cannot.

complex trait A trait that is influenced by multiple genes as well as by the environment.

compound eye An eye structure found in insects and crustaceans that consists of a number of ommatidia, individual light-focusing elements.

concordance The percentage of cases in which both members of a pair of twins show the trait when it is known that at least one member shows it.

cone cell A type of photoreceptor cell on the retina that detects color.

conjugation The direct cell-to-cell transfer of DNA, usually in the form of a plasmid.

connective tissue A type of tissue characterized by a few cells and substantial amounts of extracellular matrix; a major component of the dermis.

conserved Describes sequences that are similar in different organisms.

constant (C) Describes an unchanging region of the H and L chains.

constitutive Describes expression of a gene that occurs continuously.

constitutive defense A defense that is always active.

consumer An organism that obtains the carbon it needs for growth and reproduction from the foods it eats and gains energy by respiring food molecules; heterotrophic organisms of all kinds that directly consume primary producers or consume those that do.

continuous Describes variation that occurs across a spectrum.

contractile ring In animal cells, a ring of actin filaments that forms at the equator of the cell perpendicular to the axis of what was the spindle at the beginning of cytokinesis.

contractile vacuole A type of cellular compartment that takes up excess water and waste products from inside the cell and expels them into the external environment.

cooperative binding The increase in binding affinity with additional binding of O_2.

copy-number variation (CNV) Differences among individuals in the number of copies of a region of the genome.

cork cambium Lateral meristem that renews and maintains an outer layer that protects the stem against herbivores, mechanical damage, desiccation, and fire.

cornea The transparent portion of the sclera in the front of the vertebrate eye.

corpus luteum A temporary endocrine structure that secretes progesterone.

correlation The co-occurrence of two events or processes; correlation does not imply causation.

cortex In a stem, the region between the epidermis and the vascular bundles, composed of parenchyma cells. In the mammalian brain, the highly folded outer layer of gray matter, about 4 mm thick, made up of densely packed neuron cell bodies and their dendrites. In the mammalian renal system, the outer layer of the kidney.

co-speciation A process in which two groups of organisms speciate in response to each other and at the same time.

countercurrent exchange A mechanism in which two fluids flow in opposite directions, exchanging properties.

countercurrent multiplier A system that generates a concentration gradient as two fluids move in parallel but in opposite directions.

covalent bond A chemical bond formed by a shared pair of electrons holding two different atoms together.

CpG island A cluster of CpG sites on a DNA strand where cytosine (C) is adjacent to guanosine (G); the "p" represents the phosphate in the backbone.

cranial nerve In vertebrates, a nerve that links specialized sensory organs to the brain; most contain axons of both sensory and motor neurons.

Craniates A subphylum of Chordata, distinguished by a bony cranium that protects the brain.

crassulacean acid metabolism (CAM) The mechanism in plants that helps balance carbon dioxide gain and water loss.

Crenarchaeota One of the three major divisions of Archaea; includes acid-loving microorganisms.

crisscross inheritance A pattern in which an X chromosome present in a male in one generation is transmitted to a female in the next generation, and in the generation after that can be transmitted back to a male.

Cro-Magnon The first population of *Homo sapiens* to arrive in Europe, named for the site in France where specimens were first described.

cross-bridge The binding of the head of a myosin molecule to actin at a specific site between the myosin and actin filaments.

cross-bridge cycle Repeated sequential interactions between myosin and actin filaments at cross-bridges that cause a muscle fiber to contract.

crosscurrent Running across another current at 90 degrees.

crossover The physical breakage, exchange of parts, and reunion between non-sister chromatids.

crustaceans One of the four main groups of arthropods, including lobsters, shrimp, and crabs; their most notable distinction is their branched legs.

cuticle In leaves, a protective layer of a waxy substance secreted by epidermal cells that limits water loss; also, an exoskeleton that covers the bodies of invertebrates such as nematodes and arthropods.

C-value paradox The disconnect between genome size and organismal complexity (the C-value is the amount of DNA in a reproductive cell).

cyanobacteria All bacteria capable of oxygenic photosynthesis.

cyclic electron transport An alternative pathway for electrons during the Calvin cycle that increases the production of ATP.

cyclin-dependent kinase (CDK) A kinase that is always present within the cell but active only when bound to the appropriate cyclin.

cyclin A regulatory protein whose levels rise and fall with each round of the cell cycle.

cytochrome b_6f complex Part of the photosynthetic electron transport chain, through which electrons pass between photosystem II and photosystem I.

cytochrome c The enzyme to which electrons are transferred in complex III of the electron transport chain.

cytokine A chemical messenger released by phagocytes that recruits other immune cells to the site of injury or infection.

cytokinesis In eukaryotic cells, the division of the cytoplasm into two separate cells.

cytokinin A hormone that stimulates the outgrowth of auxiliary buds by increasing the rate of cell division.

cytoplasm The contents of the cell other than the nucleus.

cytosine (C) A pyrimidine base.

cytoskeleton In eukaryotes, an internal protein scaffold that helps cells to maintain their shape and serves as a network of tracks for the movement of substances within cells.

cytosol The region of the cell inside the plasma membrane but outside the organelles; the jelly-like internal environment that surrounds the organelles.

cytotoxic T cell One of a subpopulation of T cells, activated by cytokines released from helper T cells, that can kill other cells.

daughter strand In DNA replication, the strand synthesized from a parental template strand.

day-neutral plant A plant that flowers independently of any change in day length.

decomposer An organism that breaks down dead tissues, feeding on the dead cells or bodies of other organisms.

defecation The elimination of feces.

delayed hypersensitivity reaction Reactions initiated by helper T cells, which release cytokines that attract macrophages to the site of exposure, which is typically the skin.

deleterious Describes mutations that are harmful to an organism.

deletion A missing region of a gene or chromosome.

demography The study of the size, structure, and distribution of populations over time, including changes in response to birth, aging, migration, and death.

denaturation The unfolding of proteins by chemical treatment or high temperature; the separation of paired, complementary strands of nucleid acid.

dendrite A fiberlike extension from the cell body of a neuron that receives signals from other nerve cells or from specialized sensory endings; the input end of a nerve cell.

dendritic cell A type of cell with long cellular projections that is typically part of the natural defenses found in the skin and mucous membranes.

denitrification The process in which some bacteria use nitrate as an electron acceptor in respiration.

density-dependent Describes factors such as resources and predation that depend on the density of the population.

density-independent Describes factors such as severe drought that influence population size without regard for the density of the population.

deoxyribonucleic acid (DNA) A linear polymer of four subunits; the information archive in all organisms.

deoxyribose The sugar in DNA.

depolarization An increase in membrane potential

dermis The layer of skin beneath the epidermis, consisting of connective tissue, hair follicles, blood and lymphatic vessels, and glands. It supports the epidermis both physically and by supplying it with nutrients and provides a cushion surrounding the body.

desiccation Excessive water loss; drying out.

desiccation tolerance A suite of biochemical traits that allows the cells of bryophytes to survive extreme dehydration without damage to membranes or macromolecules.

desmosome A buttonlike point of adhesion that holds the plasma membranes of adjacent cells together.

Deuterostomia The taxonomic name for deuterostomes; this group includes humans and other chordates.

deuterostome A bilaterian in which the blastopore, the first opening to the internal cavity of the developing embryo, becomes the anus. The taxonomic name is Deuterostomia and includes humans and other chordates.

development The process in which a fertilized egg undergoes multiple rounds of cell division to become an embryo with specialized tissues and organs.

diabetes mellitus A disease that results when the control of blood-glucose levels by insulin fails.

diaphragm A domed sheet of muscle at the base of the lungs that separates the thoracic and abdominal cavities and contracts to drive inhalation.

diaphysis The middle region of a bone; blood vessels invading the diaphysis and epiphysis trigger the transformation of cartilage into bone.

diastole The relaxation of the ventricles.

dideoxynucleotide A nucleotide lacking both the 2′ and 3′ hydroxyl groups on the sugar ring.

dietary mineral A chemical element other than carbon, hydrogen, oxygen, or nitrogen that is required in the diet and must be obtained in food.

differentiation The process in which cells become progressively more specialized as a result of gene regulation.

diffusion The random motion of individual molecules, with net movement occurring where there are areas of higher and lower concentration of the molecules.

Dikarya A vast fungal group, in which every mitotic division is accompanied by the formation of a new septum, that includes about 98% of all described fungal species.

dikaryotic (n + n) Describes a stage in the life cycle of some fungi in which mitosis follows plasmogamy and produces two haploid nuclei, one from each parent.

dimerization The mutual attraction of two similar or dissimilar folded polypetide chains that brings them together to form a single molecule.

diploid Describes a cell with two complete sets of chrmosomes.

directional selection A form of selection that selects against one of two extremes and leads over time to a change in a trait.

discrete Describes traits that have clear alternative states.

dispersal The process in which some individuals colonize a distant place far from the main source population.

display A pattern of behavior that is species specific and tends to be highly repeatable and similar from one individual to the next.

disruptive selection A form of selection that operates in favor of extremes and against intermediate forms, selecting against the mean.

distal convoluted tubule The third portion of the renal tubule, in which urea is the principal solute and and into which other wastes from the bloodstream are secreted.

disturbance A severe physical impact on a habitat that has density-independent effects on populations of interacting species.

DNA ligase An enzyme that uses the energy in ATP to close a nick in a DNA strand, joining the 3′ hydroxyl of one end to the 5′ phosphate of the other end.

DNA microarray A waferlike supporting surface to which are attached millions of different oligonucleotides of known sequence; used in genotyping DNA and measuring levels of gene expression.

DNA polymerase An enzyme that is a critical component of a large protein complex that carries out DNA replication.

DNA replication The process of duplicating a DNA molecule, during which the parental strands separate and new partner strands are made.

DNA transposable element (DNA TE) A sequence that replicates and can move from one location to another in the genome.

DNA typing The analysis of a small quantity of DNA to uniquely identify an individual.

domain   One of the three largest limbs of the tree of life: Eukarya, Bacteria, or Archaea.

dominant   The trait that appears in the heterozygous offspring of a cross between homozygous genotypes.

donor   In recombinant DNA technology, the source of the DNA fragment that is inserted into a cell of another organism.

dormancy   A state in which seeds are prevented from germinating.

dorsal nerve cord   A nerve cord that develops in a location dorsal to the notochord; this embryonic feature is unique to chordates.

dosage   The number of copies of each gene in a chromosome.

dosage compensation   The differential regulation of *X*-chromosomal genes in females and in males.

double bond   A covalent bond in which covalently joined atoms share two pairs of electrons.

double fertilization   The process in which two sperm from a single pollen tube fuse with the egg and the diploid cell.

double helix   The structure formed by two strands of complementary nucleotides that coil around each other.

Down syndrome   A condition resulting from the presence of an extra copy of chromosome 21; also known as *trisomy 21*.

downstream gene   A gene that functions later than another in development.

duodenum   The initial section of the small intestine, into which food enters from the stomach.

duplex DNA   Double-stranded DNA.

duplication   A region of a chromosome that is present twice instead of once.

duplication and divergence   The process of creating new genes by duplication followed by progressive change in sequence through evolutionary time.

dynamic instability   Cycles of shrinkage and growth in microtubules.

dynein   A motor protein that carries cargo away from the plasma membrane toward the minus ends of microtubules.

eardrum   In mammals, another name for the *tympanic membrane*, which transmits airborne sounds into the ear.

Ecdysozoa   A group of animals that includes insects and other arthropods.

Echinodermata   One of the three major phyla of deuterosomes, this group includes sea urchins and sea stars.

echolocation   Using sound waves to locate an object; bats find insect prey by emitting short bursts of high-frequency sound that bounce off surrounding objects and are reflected to the bat's ears.

ecological footprint   The quantification of individual human claims on global resources by adding up all the energy, food, materials, and services used and estimating how much land is required to provide those resources.

ecological niche   A complete description of the role a species plays in its environment.

ecological separation   The pre-zygotic isolation of both plants and animals in space.

ecological species concept (ESC)   The concept that there is a one-to-one correspondence between a species and its niche.

ecology   The study of how organisms interact with one another and with their physical environment in nature.

ecosystem   A community of organisms and the physical environment it occupies.

ectoderm   The outer germ layer, which differentiates into epithelial cells, pigment cells in the skin, nerve cells in the brain, and the cornea and lens of the eye.

ectomycorrhizae   One of the two main types of mycorrhizae; ectomycorrhizae produce a thick sheath of fungal cells (hyphae) that surround, but do not penetrate, root cells.

ectotherm   An animal that obtains most of its heat from the environment.

efferent neuron   A neuron that sends signals away from the central nervous system, communicating with muscles and other tissues and organs.

ejaculatory duct   The duct through which sperm travel from the vas deferens to the urethra.

elastin   A protein fiber found in artery walls that provides elasticity.

electrochemical gradient   A gradient that combines the charge gradient and the chemical gradient of protons and other ions.

electromagnetic receptor   A receptor that responds to electrical, magnetic, or light stimuli.

electron   A negatively-charged particle that moves around the atomic nucleus.

electron acceptor   A molecule that gains electrons.

electron carrier   A molecule that stores and transfers energy in the form of "high-energy" or "excited" electrons.

electron donor   A molecule that loses electrons.

electron transport chain   The system that transfers electrons along a series of membrane-associated proteins to a final electron acceptor, releasing the energy of the electrons to produce ATP.

electronegativity   The ability of atoms to attract electrons.

electroreceptor   A sensory receptor found in some fish that enables them to detect weak electrical signals emitted by all organisms.

element   A pure substance, such as oxygen, copper, gold, or sodium, that cannot be further broken down by the methods of chemistry.

elongation   The process in protein translation in which successive amino acids are added one by one to the growing polypeptide chain.

elongation factor   A protein that breaks the high-energy bonds of the molecule GTP to provide energy for ribosome movement and elongation of a growing polypeptide chain.

embryo   An early stage of multicellular development that results from successive mitotic divisions of the zygote.

endergonic Describes reactions with a positive ΔG that are not spontaneous and so require an input of energy.

endocrine signaling Signaling by molecules that travel through the bloodstream.

endocrine system A system of cells and glands that secretes hormones and works with the nervous system to regulate an animal's internal physiological functions.

endocytosis The process in which a vesicle buds off from the plasma membrane, bringing material from outside the cell into that vesicle, which can then fuse with other organelles.

endoderm The germ layer that differentiates into cells of the lining of the digestive tract and lung, liver cells, pancreas cells, and gallbladder cells.

endodermis In plants, a layer of cells surrounding the xylem and phloem at the center of the root that controls the movement of nutrients into the xylem. Also, the inner lining of the cnidarians body.

endomembrane system A cellular system that includes the nuclear envelope, the endoplasmic reticulum, the Golgi apparatus, lysosomes, the plasma membrane, and the vesicles that move between them.

endomycorrhizae One of the two main types of mycorrhizae; endomycorrhizal hyphae penetrate into root cells, where they produce highly branched structures (arbuscules) that provide a large surface area for nutrient exchange.

endophyte A fungus that lives within leaves.

endoplasmic reticulum (ER) The organelle involved in the synthesis of proteins and lipids.

endoskeleton The bony skeletal system of vertebrate animals, which lies internal to most of the animal's soft tissues.

endosperm A tissue formed by many mitotic divisions of a triploid cell, it supplies nutrition to the angiosperm embryo.

endosymbiosis A symbiosis in which one partner lives within the other.

endotherm An animal that produces most of its own heat as by-products of metabolic reactions.

energetic coupling The driving of a non-spontaneous reaction by a spontaneous reaction.

energy balance A form of homeostasis in which the amount of energy calories from food taken in equals the amount of calories used over time to meet metabolic needs.

energy intake Sources of energy.

energy use The ways in which energy is expended.

enhancer A specific DNA sequence necessary for transcription.

enthalpy (H) The total amount of energy in a system.

entropy (S) The degree of disorder in a system.

envelope A lipid structure that surrounds the capsids of some viruses.

environmental risk factor A characteristic in a person's surroundings that increases the likelihood of developing a particular disease.

environmental variation Variation among individuals that is due to differences in the environment.

enzyme A protein that functions as a catalyst to accelerate the rate of a chemical reaction; enzymes are critical in determining which chemical reactions take place in a cell.

eosinophil A type of granulocyte that, along with basophils, defends against parasitic infections but also contributes to allergies.

epidermis In mammals, the outer layer of skin, which serves as a water-resistant, protective barrier. In plants, sheets of cells that line the leaf's upper and lower surfaces and constitute the outer cell layer of roots. Also, the outer layer of the cnidarian body.

epididymis An organ that lies above the testes where sperm become motile and are stored prior to ejaculation.

epigenetic Describes effects on gene expression due to differences in DNA packaging, such as modifications in histones or chromatin structure.

epiglottis A flap of tissue at the bottom of the pharynx that prevents food from entering the trachea and lungs.

epiphysis The end region of a bone; blood vessels invading the epiphysis and diaphysis trigger the transformation of cartilage into bone.

epiphyte A plant that grows high in the canopy of other plants, or on branches or trunks of trees, without contact with the soil.

epistasis Interaction between genes that modifies the phenotypic expression of genotypes.

epithelial tissue A type of animal tissue, made up of epithelial cells, that covers the outside of the body and lines many internal structures.

epitope The portion of an antigen that interacts in a specific manner with a hypervariable region of an antibody and binds to it.

equational division Another name for meiosis II because cells in meiosis II have the same number of chromosomes at the beginning and at the end of the process.

esophagus Part of the foregut; the passage from the mouth to the stomach.

essential amino acid An amino acid that cannot be synthesized by cellular biochemical pathways and instead must be ingested.

estrogen A hormone secreted by the ovaries that stimulates the development of female secondary sexual characteristics.

estrus cycle A cycle in placental mammals other than humans and chimpanzees characterized by phases in which females are sexually receptive.

ethanol fermentation The fermentation pathway in plants and fungi during which pyruvate releases carbon dioxide to form acetaldehyde and electrons from NADH are transferred to acetaldehyde to produce ethanol and NAD^+.

Eukarya One of the three domains of life, consisting of cells with a true nucleus containing chromosomes that divide by mitosis; cells of Eukarya are eukaryotes.

eukaryote A cell with a true nucleus containing chromosomes that divide by mitosis.

Eumetazoa Animal groups other than sponges.

Euryarchaeota One of the three major divisions of Archaea; includes acid-loving, heat-loving, methane-producing, and salt-loving microorganisms.

eusocial Describes behavior of many species of Hymenoptera insects, in which they have overlapping generations in a nest, cooperative care of the young, and clear and consistent division of labor between reproducers (the queen of a honeybee colony) and nonreproducers (the workers).

eutrophication The process in which added nutrients lead to a great increase in the populations of algae and cyanobacteria.

evaporation The amount of water evaporated from the Earth's surface, including ponds, rivers, and soil.

evapotranspiration ratio The amount of water evaporated from the Earth's surface divided by the amount of water transpired by leaves of plants.

evo-devo The short name for evolutionary-developmental biology, a field of study that compares the genetic programs for growth and development in species on different branches of phylogenetic reconstructions.

evolution Changes in the genetic make-up of populations over time, resulting in some cases in adaptation to the environment and the origin of new species.

evolutionarily conserved Little changed through evolution and therefore similar from one organism to the next.

evolutionarily stable strategy A type of behavior that cannot readily be driven to extinction by an alternative strategy.

evolutionary species concept (EvSC) The concept that members of a species all share a common ancestry and a common fate.

excitation–contraction coupling The process that produces muscle force and movement, by excitation of the muscle cell coupled to contraction of the muscle.

excitatory postsynaptic potential (EPSP) A positive change in the postsynaptic membrane potential.

excreted Eliminated (referring to waste generated by metabolism).

excretory tubule In renal systems, a type of tube that drains waste products and connects to the outside of the body.

exergonic Describes reactions with a negative ΔG that release energy and proceed spontaneously.

exhalation The expelling of oxygen-poor air by the elastic recoil of the lungs and chest wall.

exit (E) site One of three binding sites for tRNA on the large subunit of a ribosome.

exocytosis The process in which a vesicle fuses with the plasma membrane and empties its contents into the extracellular space or delivers proteins to the plasma membrane.

exon A sequence that is left intact in mRNA after RNA splicing.

exoskeleton A rigid skeletal system that lies external to the animal's soft tissues.

experimentation A disciplined and controlled way of learning about the world and testing hypotheses in an unbiased manner.

exponential growth The pattern of population increase that results when r (the per capita growth rate) is constant through time.

expressed Turned on or activated, as a gene or protein.

extension (PCR) A step in the polymerase chain reaction (PCR) for producing new DNA fragments in which the reaction mixture is heated to the optimal temperature for DNA polymerase, and each primer is elongated by means of deoxynucleoside triphophosphates.

extension (joint) The joint motion in which bone segments move apart.

external fertilization Fertilization that takes place outside the body of the female; in aquatic organisms, for example, eggs and sperm are released into the water.

extracellular digestion The process in most animals in which food is isolated and broken down outside a cell, in a body compartment.

extracellular matrix A meshwork of proteins and polysaccharides outside the cell; the main constituent of connective tissue.

extraembryonic membrane In the amniotic egg, one of several sheets of cells that extend out from the developing embryo and form the yolk sac, amnion, allantois, and chorion.

extravasation The process in which phagocytes that travel in the blood move from a blood vessel to the site of infection.

eyecup An eye structure found in flatworms that contains photoreceptors that point up and to the left or right.

F_1 generation The first filial, or offspring, generation.

F_2 generation The second filial generation; the offspring of the F_1 generation.

Fab fragment The part of an antigen molecule broken by an enzyme that has an antigen-recognition site.

facilitated diffusion Diffusion through a membrane protein, bypassing the lipid bilayer.

facultative Describes a mutualism in which one or both sides can survive without the other.

fallopian tube (oviduct) A tube from each ovary, through one of which a released oocyte passes.

family A group of closely related genera.

fast-twitch Describes muscle fibers that generate force quickly, producing rapid movements, but consume much more ATP than do slow-twitch fibers.

fatty acid A long chain of carbons attached to a carboxyl group; three fatty acid chains attached to glycerol form a triacylglycerol, a lipid used for energy storage.

Fc fragment The part of an antigen molecule broken by an enzyme that activates the complement system and binds to cell-surface receptors on other cells of the immune system.

feature detector A specialized sensory receptor or group of sensory receptors that respond to important signals in the environment.

feces Waste products that are eliminated from the body.

feedback A response evoked by a stimulus that in turn affects that stimulus; feedback is the means by which a signal produced by a regulatory organ can modify its own subsequent production, either inhibiting or stimulating it.

fermentation A process of breaking down pyruvate through a wide variety of metabolic pathways that extract energy from fuel molecules such as glucose; the partial oxidation of complex carbon molecules to molecules that are less oxidized than carbon dioxide.

ferns and horsetails A monophyletic group of vascular plants.

fertilization The union of gametes to produce a zygote that restores the original chromosomal content of the parental organisms.

fetus In humans, the embryo toward the end of the first trimester.

fiber In angiosperm plants, a narrow cell with thick walls that provides mechanical support in wood, allowing vessels to be specialized for water transport. In animals, a term for a muscle cell, which produces forces within an animal's body and exerts forces on the environment.

filament A thin thread of proteins that interacts with other filaments to cause muscles to shorten.

filial imprinting Imprinting in which newborn offspring rapidly learn to treat any animal they see shortly after birth as their mother.

filtration The separation of solids from fluids, as when circulatory pressure pushes fluid containing wastes through specialized filters into an extracellular space.

firing rate The number of action potentials fired over a given period of time.

first law of thermodynamics The law of conservation of energy: Energy can neither be created nor destroyed—it can only be transformed from one form into another.

first-division nondisjunction Failure of chromosome separation in meiosis I.

fitness A measure of the extent to which an individual's genotype is represented in the next generation

fixation The point at which a given allele has a frequency of 1; it has become the only allele of that gene.

fixed Describes a population that exhibits only one allele at a particular gene.

fixed action pattern (FAP) A sequence of behaviors that, once triggered, is followed through to completion.

flagellum (plural, flagella) An organelle that propels the movement of cells or of substances within cells; longer than a cilium.

fleshy-finned fish Species of fish with paired pectoral and pelvic fins.

flexion The joint motion in which bone segments rotate closer together.

fluid Describes lipids that are able to move in the plane of the cell membrane.

fluid mosaic model A model that proposes that the lipid bilayer is a fluid structure that allows molecules to move laterally within the membrane and is a mosaic of two types of molecules, lipids and proteins.

flux The rate at which a substance, for example carbon, flows from one reservoir to another.

folding domain A region of a protein that folds in a similar way across a protein family relatively independently of the rest of the protein.

follicle cell A type of cell surrounding the oocyte that supports the developing oocyte and secretes estrogen.

follicle-stimulating hormone (FSH) A hormone secreted by the anterior pituitary gland that stimulates the male and female gonads to secrete testosterone in males and estrogen and progesterone in females.

follicular phase The phase of the menstrual cycle during which FSH acts on granulosa cells, resulting in the maturation of several oocytes, of which, usually, only one becomes completely mature.

food chain The linear transfer of carbon from one organism to another.

food web An interaction among organisms within the carbon cycle; the movement of carbon through an ecosystem.

forebrain The region of the vertebrate brain that governs cognitive functions.

foregut The first part of an animal's digestive tract, including the mouth, esophagus, and stomach.

fossil The remains of a once-living organism, preserved through time in sedimentary rocks.

founder event A type of bottleneck that occurs when only a few individuals establish a new population.

fovea The center of the visual field of most vertebrates, where cone cells are most concentrated; the region of greatest acuity.

fragmentation A form of asexual reproduction in which new individuals arise by the splitting of one organism into pieces, each of which develops into a new individual.

frameshift mutation A mutation in which the insertion or deletion of a single nucleotide causes a one-nucleotide shift in the reading frame of the mRNA, changing all following codons.

fraternal (dizygotic) twins Twins that arise when two separate eggs, produced by double ovulation, are fertilized by two different sperm.

frequency of recombination The proportion of recombinant chromosomes among the total number of chromosomes observed.

frontal lobe The region of the brain located in the anterior region of the cerebral cortex, important in decision making and planning.

fruit The structure that develops from the ovary and serves to protect immature seeds and enhance dispersal once the seeds are mature.

fruiting body A multicellular structure in some fungi that facilitates the dispersal of sexually produced spores.

fundamental niche The full range of climate conditions and food resources that permit the individuals in a species to live.

fungi An abundant and diverse group of heterotrophic eukaryotic organisms, principally responsible for the decomposition of plant and animal tissues.

GLOSSARY

G protein-coupled receptor A receptor that couples to G proteins, which bind to the guanine nucleotides GTP and GDP.

G protein A protein that binds to the guanine nucleotides GTP and GDP.

G_0 phase The gap phase in which cells pause in the cell cycle between M phase and S phase; may last for periods ranging from days to more than a year.

G_1 phase The gap phase in which the size and protein content of the cell increase and specific regulatory proteins are made and activated in preparation for S-phase DNA synthesis.

G_2 phase The gap phase in which the size and protein content of the cell increase in preparation for M-phase mitosis and cytokinesis.

gain-of-function mutation Any mutation in which a gene is expressed in the wrong place or at the wrong time.

gallbladder The organ in which bile produced by the liver is stored.

gamete A reproductive haploid cell resulting from meiotic cell division (in some species gametes are called spores). In many species, there are two types of gametes: eggs in females, sperm in males.

gametogenesis The formation of gametes.

gametophyte Describes the haploid multicellular generation that gives rise to gametes.

ganglion (plural, ganglia) A group of nerve cell bodies that processes sensory information received from a local, nearby region, resulting in a signal to motor neurons that control some physiological function of the animal.

ganglion cell A type of interneuron in the retina that synapses with bipolar cells and, if activated, transmits action potentials along the optic nerve to the visual cortex in the brain.

gap junction A type of connection between the plasma membranes of adjacent animal cells that permits materials to pass directly from the cytoplasm of one cell to the cytoplasm of another.

gas exchange The transport of oxygen and carbon dioxide between an animal and its environment.

gastric cavity In cnidarians, a closed internal site where extracellular digestion and excretion take place.

gastrin A peptide hormone produced in the stomach that stimulates cells lining the stomach to increase their production of HCl.

gastropods A group of mollusks consisting of snails and slugs.

gastrula A layered structure formed when the inner cell mass cells of the blastocyst migrate and reorganize.

gastrulation A highly coordinated set of cell movements in which the cells of the blastoderm migrate inward, creating germ layers of cells within the embryo.

gel electrophoresis A procedure to determine the size of a DNA fragment, in which DNA samples are inserted into slots or wells in a gel and a current passed through. Fragments move toward the positive pole according to size.

gene The unit of heredity; the stretch of DNA that affects one or more traits in an organism, usually through an encoded protein or noncoding RNA.

gene expression The production of a functional gene product.

gene family A group of genes with related functions, usually resulting from multiple rounds of duplication and divergence.

gene flow The movement of alleles from one population to another.

gene pool All the alleles present in all individuals in a species.

gene regulation The various ways in which cells control gene expression.

general transcription factors A set of proteins that bind to the promoter of a gene whose combined action is necessary for transcription.

genetic code The correspondence between codons and amino acids, in which 20 amino acids are specified by 64 codons.

genetic drift A change in the frequency of an allele due to the random effects of small population size.

genetic incompatibility Genetic dissimilarity between two organisms, such as different numbers of chromosomes, that is sufficient to act as a pre-zygotic isolating factor.

genetic information Information carried in DNA, organized in the form of genes.

genetic map A diagram showing the relative positions of genes along a chromosome.

genetic risk factor Any mutation that increases the risk of a given disease in an individual.

genetic test A method of identifying the genotype of an individual.

genetic variation Differences in genotype among individuals in a population.

genetically modified organism (GMO) An organism that has been genetically engineered, such as modified viruses and bacteria, laboratory organisms, agricultural crops, and domestic animals. Also known as a *transgenic organism*.

genome The genetic material transmitted from a parental cell or organism to its offspring.

genome annotation The process by which researchers identify the various types of sequence present in genomes.

genomic rearrangement The process of joining different gene segments as a B cell differentiates to produce a specific antibody.

genotype The genetic makeup of a cell or organism; the particular combination of alleles present in an individual.

genotype-by-environment interaction Variation in the effects of the environment on different genotypes, resulting in different phenotypes.

genus A group of closely related species.

geographic range How widely a population is spread and the factors that determine its distribution.

geologic timescale The series of time divisions that mark Earth's long history.

germ cells The reproductive cells that produce sperm or eggs and the cells that give rise to them.

germ layers   Three sheets of cells, the ectoderm, mesoderm, and endoderm, formed by migrating cells of the gastrula that differentiate further into specialized cells.

germ-line mutation   A mutation that occurs in eggs and sperm or in the cells that give rise to these reproductive cells and therefore is passed on to the next generation.

gibberellic acid   A plant hormone that stimulates the elongation of stems.

Gibbs free energy (G)   The amount of energy available to do work.

gills   Highly folded delicate structures in aquatic animals that facilitate gas exchange with the surrounding water.

gizzard   In birds, alligators, crocodiles, and earthworms, a compartment with thick muscular walls in the digestive tract where food mixed with ingested rock or sediment is broken down into smaller pieces.

glans penis   The head of the human penis.

glial cell   A type of cell that surrounds neurons and provides them with nutrition and physical support.

glomeromycetes   A monophyletic fungal group of apparently low diversity but tremendous ecological importance that occurs in association with plant roots.

glomerulus   A tufted loop of porous capillaries in the vertebrate kidney that filters blood.

glycerol   A 3-carbon molecule with OH groups attached to each carbon.

glycogen   The form in which glucose is stored in animals.

glycolysis   The breakdown of glucose to pyruvate; the first stage of cellular respiration. (The second of the four stages is the conversion of pyruvate to acetyl-CoA and the release of CO_2.)

glycosidic bond   A covalent bond that attaches one monosaccharide to another.

Golgi apparatus   The organelle that modifies proteins and lipids produced by the ER and acts as a sorting station as they move to their final destinations.

gonad   The part of the reproductive system where haploid gametes are produced. Male gonads are testes, where sperm are produced. Female gonads are ovaries, where eggs are produced.

gonadotropin-releasing hormone (GnRH)   A hormone released by the hypothalamus that stimulates the anterior pituitary gland to secrete luteinizing hormone and follicle-stimulating hormone.

gram-positive bacteria   Bacteria that retain, in their thick peptidoglycan walls, the diagnostic dye developed by Hans Christian Gram. (Bacteria with thin walls, which do not retain the dye, are said to be gram negative.)

grana   Interlinked structures that form the thylakoid membrane.

granulocyte   A type of phagocytic cell that contains granules in its cytoplasm.

granuloma   A structure formed by lymphocytes surrounding infected macrophages that helps to prevent the spread of an infection and aids in killing infected cells.

gravitropic   Bending in response to gravity. A negative gravitropic response, as in stems, is growth upward against the force of gravity; a positive gravitropic response, as in roots, is with the force of gravity.

gray matter   Densely packed neuron cell bodies and dendrites that make up the cortex, a highly folded outer layer of the mammalian brain about 4 mm thick.

greenhouse gas   A gas that absorbs heat energy from incoming solar radiation and allows it to reach Earth's surface but traps heat that is reemitted from land and sea.

group selection   A form of natural selection that operates on groups rather than on individuals.

growth factor   Any one of a group of small, soluble molecules, usually the signal in paracrine signaling, that affect cell growth, cell division, and changes in gene expression.

growth plate   A region of cartilage near the end of a bone where growth in length occurs.

growth ring   One of the many rings apparent in the cross section of the trunk of a tree, produced by decreases in the size of secondary xylem cells at the end of the growing season, that make it possible to determine the tree's age.

guanine (G)   A purine base.

guard cell   One of two cells surrounding the central pore of a stoma.

gustation   The sense of taste.

gut   Collectively, the passages that connect the mouth, digestive organs, and anus; also known as the *digestive tract*.

gymnosperms   One of the two groups of vascular plants; gymnosperms include pine trees and other conifers.

habituation   The reduction or elimination of a behavioral response to a repeatedly presented stimulus.

hagfish   One of the earliest-branching craniates, with a cranium built of cartilage but no jaws; hagfish feed on marine worms and dead and dying sea animals.

hair cell   A specialized mechanoreceptor that senses movement and vibration.

hairpin (structures)   Stems-and-loops formed in self-complementary, single-stranded nucleic acid molecules, stabilized by base pairing in the stem.

half-life   The time it takes for half of the atoms in a given sample of a substance to decay.

haploid   Describes a cell with one complete set of chromosomes.

haplotype   A haploid genotype, as the particular combination of alleles present in any particular region of a chromosome.

Hardy–Weinberg equilibrium   A state in which allele and genotype frequencies do not change over time, suggesting that evolutionary mechanisms are not acting on the population being studied. It also specifies a mathematical relationship between allele frequencies and genotype frequencies.

heart The pump of the circulatory system, which moves blood to different regions of the body.

Heart rate (HR) The number of heartbeats per unit time.

heartwood The center of the stem in long-lived trees, which does not conduct water.

heavy (H) chains Two of the four polypeptide chains that make up the simplest antibody molecule.

helicase A protein that unwinds the parental double helix at the replication fork.

helper T cell One of a subpopulation of T cells that help other cells of the immune system by secreting cytokines, thus activating B cells to secrete antibodies.

Hemichordata One of the three major phyla of deuterosomes, this group includes acorn worms and pterobranchs.

hemidesmosome A type of desmosome in which integrins are the prominent cell adhesion molecules.

hemoglobin An iron-containing molecule specialized for oxygen transport.

hemolymph The circulatory fluid in invertebrates.

hemophilia A trait characterized by excessive bleeding that results from a recessive mutation in a gene encoding a protein necessary for blood clotting.

heterokaryotic Describes a stage in in the life cycle of some fungi, following plasmogamy, in which cells have two haploid nuclei, one from each parent, and the nuclei have not yet fused in karyogamy to produce a diploid zygote.

heterotroph An organism that obtains its carbon from organic molecules synthesized by other organisms.

heterozygote advantage A form of balancing selection in which the heterozygote's fitness is higher than that of either of the homozygotes, resulting in selection that ensures that both alleles remain in the population at intermediate frequencies.

heterozygous Describes an individual who inherits different types of alleles from the parents, or genotypes in which the two alleles for a given gene are different.

hierarchical Describes gene regulation during development, in which the genes expressed at each stage in the process control the expression of genes that act later.

high-energy phosphate bond In nucleic acid polymerization, the bond connecting the innermost phosphate to the next. The cleaved bond provides the energy to drive the reaction that creates the phosphodiester bond attaching the incoming nucleotide to the 3′ end of the growing chain.

highly repetitive DNA A type of noncoding DNA consting of sequences present in many thousands of copies per genome.

hindbrain Along with the midbrain, the region of the vertebrate brain that controls basic body functions and behaviors.

hindgut The last part of an animal's digestive tract, including the large intestine and rectum.

hinge joint A simple joint that allows one axis of rotation, like the elbow and knee.

hippocampus A posterior region of the limbic system involved in long-term memory formation.

histamine A chemical messenger released by mast cells and basophils; an important contributor to allergic reactions and inflammation.

histone A protein found in all eukaryotes that interacts with DNA to form chromatin.

histone code The pattern of modifications of the histone tails that affects the chromatin structure and gene transcription.

histone tail A string of amino acids that protrudes from a histone protein in the nucleosome.

homeobox A DNA sequence within homeotic genes, which function in development, that specifies the homeodomain.

homeodomain The DNA-binding domain in homeotic proteins, a sequence of 60 amino acids whose sequences are very similar from one homeotic protein to the next.

homeostasis The active regulation and maintenance, in animals, organs, or cells, of a stable internal physiological state in the face of a changing external environment.

homeotic gene A gene that specifies the identity of a body part or segment during embryonic development; also known as a *Hox gene*.

hominins A member of one of the different species in the group leading to humans.

homologous Describes characters that are similar in different species because of descent from a common ancestor.

homologous chromosomes Pairs of chromosomes, matching in size and appearance, that carry the same set of genes; one of each pair was received from the mother, the other from the father.

homozygous Describes an individual who inherits an allele of the same type from each parent, or a genotype in which both alleles for a given gene are the same.

horizontal cell A type of interneuron in the retina that communicates between neighboring pairs of photoreceptors and bipolar cells, enhancing contrast through lateral inhibition to sharpen the image.

horizontal gene transfer The transfer of genetic material between organisms that are not parent and offspring.

hormone A chemical signal that influences physiology and development in both plants and animals; in animals hormones are released into the bloodstream and circulate throughout the body.

host cell A cell in which viral reproduction occurs.

host plant A plant species that can be infected by a given pathogen.

hotspot A site in the genome that is especially mutable.

housekeeping gene A gene that is transcribed continually because its product is needed at all times and in all cells.

***Hox* gene** A gene that species the identity of a body part or segment during embryonic development; also known as a *homeotic gene*.

human chorionic gonadotropin (hCG) A hormone released by the developing embryo that maintains the corpus luteum.

humoral immunity The part of the adaptive immune system consisting of blood, tissue fluid, and secretions, where antibodies are found.

hybridization Interbreeding between two different varieties or species of an organism.

hydrogen bond A weak bond between a hydrogen atom in one molecule and an electronegative atom in another molecule.

hydrophilic "Water loving"; describes a class of molecules with which water can undergo hydrogen bonding.

hydrophobic "Water fearing"; describes a class of molecules poorly able to undergo hydrogen bonding with water.

hydrophobic effect The exclusion of nonpolar molecules by polar molecules, which drives biological processes such as the formation of cell membranes and the folding of proteins.

hydrostatic skeleton A skeletal system in which fluid contained within a body cavity is the supporting element; found in nearly all multicellular animals and many vascular plants.

hydroxyapatite Mineral crystals of calcium phosphate, a major component of bone.

hypersensitive response A type of plant defense against infection in which uninfected cells surrounding the site of infection rapidly produce large numbers of reactive oxygen species, triggering cell wall reinforcement and causing the cells to die, thus creating a barrier of dead tissue.

hyperthermophile An organism that requires an environment with high temperature.

hypervariable region In antibodies, an even more variable region within the variable regions of the heavy and light chains.

hyphae In fungi, highly branched filaments that provide a large surface area for absorbing nutrients.

hypothalamus The underlying brain region of the forebrain, which interacts with the autonomic and endocrine systems to regulate the general physiological state of the body.

hypothesis A tentative explanation for one or more observations that makes predictions that can be tested by experiments or additional observations.

identical (monozygotic) twins Twins that arise from a single fertilized egg, which after several rounds of cell division, separates into two distinct, but genetically identical, embryos.

IgA One of the five antibody classes, IgA is typically a dimer and the major antibody on mucosal surfaces.

IgD One of the five antibody classes, IgD is a monomer and is typically found on the surface of B cells; it helps initiate inflammation.

IgE One of the five antibody classes, IgE is a monomer that plays a central role in allergies, asthma, and other immediate hypersensitivity reactions.

IgG The most abundant of the five antibody classes, IgG is a monomer that circulates in the blood and is particularly effective against bacteria and viruses.

IgM One of the five antibody classes, IgM is a pentamer in mammals and a tetramer in fish and is particularly important in the early response to infection, activating the complement system and stimulating an immune response.

ileum A section of the small intestine that, with the jejunum, carries out most nutrient absorption.

imitation Observing and copying the behavior of another.

immediate hypersensitivity reaction A reaction characterized by a heightened or an inappropriate immune response to common antigens.

immunodeficiency Any disease in which part of the immune system does not function properly.

immunoglobulins (Ig) A family of proteins with common structural features that includes antibodies.

imprinting A form of learning typically seen in young animals in which a certain behavior is acquired in response to key experiences during a critical period of development.

in vitro fertilization (IVF) A process in which eggs and sperm are brought together in a petri dish, where fertilization and early cell divisions occur.

inbred line A true-breeding, homozygous strain.

incisor One of the teeth in the front of the mouth, used for biting.

incomplete dominance Describes inheritance in which the phenotype of the heterozygous genotype is intermediate between those of homozygous genotypes.

incomplete penetrance The phenomenon in which some individuals with a genotype corresponding to a trait do not show the phenotype, either because of environmental effects or because of interactions with other genes.

incus A small bone in the middle ear that helps amplify the waves that strike the tympanic membrane.

induced pluripotent cell (iPS cell) A cell that has been reprogrammed to become pluripotent by activation of certain genes, most of them encoding transcription factors or chromatin proteins.

inducer A small molecule that elicits gene expression.

inducible defense A defense that is mounted only when a threat is sensed.

inflammation A physiological response of the body to injury that removes the inciting agent if present and begins the healing process.

inhalation The drawing of oxygen-rich blood into the lungs by the expansion of the thoracic cavity.

inhibitor A synthesized compound that decreases the activity of an enzyme.

inhibitory postsynaptic potential (IPSP) A negative change in the postsynaptic membrane potential.

initiation The stage of translation in which methionine is established as the first amino acid in a new polypeptide chain.

initiation factor A protein that binds to mRNA to initiate translation.

innate Describes behaviors that are instinctive and carried out regardless of earlier experience.

innate (natural) immunity The part of the immune system that provides protection in a nonspecific manner against all kinds of infection; it does not depend on exposure to a pathogen.

inner cell mass A mass of cells in one region of the inner wall of the blastocyst from which the body of the embryo develops.

inner ear The part of the inner ear that includes the cochlea and semicircular canals.

insects The most diverse of the four main groups of arthropods.

instantaneous speciation Speciation that occurs in a single generation.

integral membrane protein A protein that is permanently associated with the cell membrane and cannot be separated from the membrane experimentally without destroying the membrane itself.

integrin A transmembrane protein, present on the surface of virtually every animal cell, that enables cells to adhere to the extracellular matrix.

intercostal muscles Muscles attached to adjacent pairs of ribs that assist the diaphragm by elevating the ribs on inhalation and depressing them during exhalation.

intermediate filament A polymer of proteins, which vary according to cell type, that combine to form strong, cable-like filaments that provide animal cells with mechanical strength.

intermembrane space The space between the inner and outer mitochondrial membranes.

internal fertilization Fertilization that takes place inside the body of the female.

interneuron A neuron that processes information received by sensory neurons and transmits it to motor neurons in different body regions.

internode The segment between two nodes on a shoot.

interphase The time between two successive M phases.

intersexual selection A form of sexual selection that focuses on interaction between males and females, as when males compete for the attention of the female, who chooses her mate.

interspecific competition Competition between individuals of different species.

intervertebral disc A fluid-filled support structure found between the bony vertebrae of the backbone that enables flexibility and provides cushioning of loads.

intracellular digestion The process in single-celled protists in which food is broken down within cells.

intrasexual selection A form of sexual selection that focuses on interactions between individuals of one sex, as when members of one sex compete with one another for access to the other sex.

intraspecific competition Competition within species.

intrinsic growth rate The per capita growth rate; the maximum rate of growth when no environmental factors limit population increase.

intron A sequence that is excised from the primary transcript and degraded during RNA splicing.

invasive species Non-native species; since they are removed from natural constraints on population growth, invasive species can expand dramatically when introduced into new areas, often with devastating consequences for native species and ecosystems.

inversion The reversal of the normal order of a block of genes.

involuntary Describes the component of the nervous system that regulates internal bodily functions.

ion An electrically charged atom or molecule.

ionic bond The association of two atoms resulting from the attraction of opposite charges.

iris A structure found at the front of the vertebrate eye, surrounding the pupil, that opens and closes to adjust the amount of light that enters the eye.

irreversible inhibitor Any one of a class of inhibitors that usually forms covalent bonds with enzymes and irreversibly inactivates them.

island population A distant, isolated population.

isomers Molecules that have the same chemical formula but different structures.

isometric Describes the generation of force without muscle movement.

isotopes Atoms of the same element that have different numbers of neutrons.

isotype In immunology, any one of the five classes of antibody.

isotype (class) switching The ability of B cells to change the class of antibody they make.

jasmonic acid A chemical signal that is triggered by herbivore damage; exposure to jasmonic acid induces the transcription of defensive genes.

jejunum A section of the small intestine that, with the ileum, carries out most nutrient absorption.

juxtacrine signaling Signaling by direct physical contact of one cell with another, with no chemical signal that diffuses or circulates through an external medium.

juxtaglomerular apparatus The structure formed by specialized cells of the efferent arteriole leaving the glomerulus of each nephron, which secretes rennin into the bloodstream.

karyogamy The fusion of two nuclei following plasmogamy.

karyotype A standard arrangement of chromosomes, showing the number and shapes of the chromosomes representative of a species.

ketose A monosaccharide with a ketone group.

key stimulus A stimulus that initiates a fixed action pattern.

keystone species Pivotal populations that affect other members of the community in ways that are disproportionate to their abundance or biomass.

kidneys In vertebrates, paired organs of the renal system that remove waste products and excess fluid; their action contributes to homeostasis.

kin selection A form of natural selection that favors the spread of alleles that promote behaviors that help close relatives.

kinesin A motor protein, similar in structure to myosin, that transports cargo toward the plus end of microtubules.

kinesis (plural, kineses) A random, undirected movement in response to a stimulus.

kinetic energy The energy of motion.

kinetochore The protein complexes on a chromatid where spindle fibers attach.

kingdom A group of closely related phyla.

Klinefelter syndrome A sex-chromosomal abnormality in which an individual has 47 chromosomes, including two X chromosomes and one Y chromosome.

knee-jerk reflex A reflex commonly tested by physicians to evaluate peripheral nervous and muscular system function.

K-strategist A species that produces relatively few young but invests considerable resources into their support.

labia majora Outer folds of skin in the vulva.

labia minora Inner folds of skin in the vulva that meet at the clitoris.

lactic acid fermentation The fermentation pathway in animals and bacteria during which electrons from NADH are transferred to pyruvate to produce lactic acid and NAD^+.

lagging strand A daughter strand that has its 5′ end pointed toward the replication fork, so as the parental double helix unwinds, a new DNA piece is initiated at intervals, and each new piece is elongated at its 3′ end until it reaches the piece in front of it.

lamella (plural, lamellae) One of many thin, sheetlike structures spread along the length of each gill filament, giving gills an enormous surface area relative to their size.

lampreys One of the earliest-branching craniates, with a cranium and vertebral column built of cartilage but no jaws; lampreys live parasitically, sucking body fluids from fish prey.

large intestine Part of the hindgut and the site of reabsorption of water and minerals; also known as the *colon*.

lariat A loop and tail of RNA formed after RNA splicing.

larynx The structure, above the trachea, that contains the vocal cords.

lateral inhibition Inhibition of a process in cells adjacent to the cell receiving a signal inducing that process, enhancing the strength of a signal locally but diminishing it peripherally.

lateral line system In fish and sharks, a sensory organ along both sides of the body that uses hair cells to detect movement of the surrounding water.

lateral meristem The source of new cells that allows plants to grow in diameter.

latex A white sticky liquid produced in some plants.

latitudinal diversity gradient The increase in species diversity from the poles to the equator.

leading strand A daughter strand that has its 3′ end pointed toward the replication fork, so as the parental double helix unwinds, this daughter strand can be synthesized as one long, continuous polymer.

leaf The principal site of photosynthesis in vascular plants.

leaf primordia Young leaves.

learned Describes a behavior that depends on an individual's experience.

learning The process in which experience leads to changes in behavior.

lengthening contraction The contraction of a muscle against a load greater than the muscle's force output, leading to a lengthening of the muscle.

lens A flexible structure in the vertebrate eye through which light passes after entering through the pupil; it is controlled by ciliary muscles that contract or relax to the adjust the shape of the lens to focus light images.

lenticel A region of less tightly packed cells in the outer bark that allows oxygen to diffuse into the stem.

Leydig cell A type of cell in the testes that secretes testosterone.

Liebig's Law of the Minimum The principle that primary production is limited by the nutrient that is least available relative to its use by primary producers.

life history The typical pattern of resource investment in each stage of a given species' lifetime.

ligand Alternative term for a signaling molecule that binds with a receptor, usually a protein.

ligand-binding site The specific location on the receptor protein where a signaling molecule binds.

ligand-gated ion channel A receptor that alters the flow of ions across the plasma membrane when bound by its ligand.

light (L) chains Two of the four polypeptide chains that make up the simplest antibody molecule.

limbic system Inner components of the forebrain that control physiological drives, instincts, emotions, motivation, and the sense of reward.

LINE Long interspersed nuclear element of about 1000 base pairs present in multiple copies in a genome owing to transposition.

linked Describes genes that are sufficiently close together in the same chromosome that they do not assort independently.

lipase A type of enzyme produced by the pancreas that breaks apart lipids, thus enabling their more effective digestion.

lipid An organic molecule that stores energy, acts as a signaling molecule, and is a component of cell membranes.

lipid raft Lipids assembled in a defined patch in the cell membrane.

liposome An enclosed bilayer structure spontaneously formed by phospholipids in environments with neutral pH, like water.

liver An organ that aids in the digestion of proteins, carbohydrates, and fats in the duodenum by producing bile, which breaks down fat.

lock and key In mating, describes a system that requires both components, whether physical or biochemical, to match for a successful interaction to take place.

logistic growth The pattern of population growth that results as growth potential slows down as the population size approaches *K*, its maximum sustainable size.

long-day plant A plant that flowers only when the light period exceeds a critical value.

longitudinal muscle Smooth muscle that runs lengthwise along a body or organ; in the digestive tract, a longitudinal muscle layer contracts to shorten small sections of the gut. A longitudinal muscle layer contracts alternately with circular muscle to move contents through the digestive tract and to enable locomotion in animals with hydrostatic skeletons.

loop of Henle The middle portion of the renal tubule, which creates a concentration gradient that allows water passing through the collecting duct to be reabsorbed.

Lophotrochozoa A group of animals that contains 17 phyla, mostly small marine animals of limited diversity, but also including the annelid worms and mollusks.

loss-of-function mutation A mutation that inactivates the normal function of a gene.

LTR element A type of transposable element characterized by long repeated sequences, called long terminal repeats, at its ends, and that transposes by means of an RNA intermediate.

Lucy An unusually complete early hominin fossil, found in 1974 in Ethiopia.

lumen In eukaryotes, the continuous interior of the endoplasmic reticulum; in plants, a fluid-filled compartment enclosed by the thylakoid membrane; generally, the interior of any tubelike structure.

lunar clock A moon-based biological clock that times activities in some species, especially those living in habitats where tides are important.

lungfish A species of fleshy-finned fish that can survive periods when their watery habitat dries by burying themselves in moist mud and breathing air; along with coelacanths, the nearest relative of tetrapods.

lungs The internal organs for gas exchange in many terrestrial animals.

luteal phase The phase of the menstrual cycle beginning with ovulation.

luteinizing hormone (LH) A hormone secreted by the anterior pituitary gland that stimulates the male and female gonads to secrete testosterone in males and estrogen and progesterone in females.

lycophytes A major vascular plant group, the spore-dispersing lycophytes are the sister group to all other vascular plants.

lymph The fluid in the lymphatic system in which T and B cells ciculate.

lymphatic system A network of vessels distributed through the body with important functions in the immune system.

lysogenic pathway The alternative to the lytic pathway; in the lysogenic pathway, a virus integrates its DNA into the host cell's DNA, which is then transmitted to offspring cells.

lysosome A vesicle derived from the Golgi apparatus that contains enzymes that break down macromolecules such as proteins, nucleic acids, lipids, and complex carbohydrates.

lytic pathway The alternative to the lysogenic pathway; in the lytic pathway, a virus bursts, or lyses, the cell it infects, releasing new virus particles.

M phase The stage of the cell cycle consisting of mitosis and cytokinesis, in which the parent cell divides into two daughter cells.

macrophage A type of large phagocytic cell that patrols the body.

mainland population The central population of a species.

major groove The larger of two uneven grooves on the outside of a DNA duplex.

major histocompatibility complex (MHC) A group of proteins that appear on the surface of most mammalian cells; only the antigen associated with MHC proteins is recognized by T cell receptors.

malleus A small bone in the middle ear that helps amplify the waves that strike the tympanic membrane.

Malpighian tubule One of the tubes in the main body cavity of insects and other terrestrial arthropods through which fluid passes and which empties into the hindgut.

mammals A class of vertebrates distinguished by body hair and mammary glands from which they feed their young.

map information The knowledge of where an individual is in respect to the goal.

MAP kinase pathway A series of kinases that are triggered by activated GTP-bound Ras; the final kinase enters the nucleus, where it phosphorylates its target proteins.

map unit A unit of distance in a genetic map equal to the distance between genes resulting in 1% recombination.

mark-and-recapture A method in which individuals are captured, marked in way that doesn't affect their function or behavior, and then released.

marsupials A group of mammals that includes kangaroos, koalas, and opposums; their young are born at an early stage of development and must crawl to a pouch where mammary glands equipped with nipples provide them with milk.

mass extinction A catastrophic drop in recorded diversity, which has occurred several times in the past 542 million years.

mast cell A cell that releases histamine.

maternal inheritance A type of inheritance in which the organelles in the offspring cells derive from those in the mother.

maternal-effect gene A gene that is expressed by the mother that affects the phenotype of the offspring, typically through the composition or organization of the oocyte.

mating types Genetically distinct forms of individuals of a fungus species that, by enabling fertilization only between different types, prevent self-fertilization.

mechanoreceptor A sensory receptor that responds to physical deformations of its membrane produced by touch, stretch, pressure, motion, and sound.

mediator complex A complex of proteins that interacts with the Pol II complex and allows transcription to begin.

medulla A part of the brainstem; also, the inner layer of the mammalian kidney.

meiosis I Reductional division, the first stage of meiotic cell division, in which the number of chromosomes is halved.

meiosis II Equational division, the second stage of meotic cell division, in which the number of chromosomes is unchanged.

meiotic cell division A form of cell division that includes only one round of DNA replication but two rounds of nuclear division; meiotic cell division makes sexual reproduction possible.

membrane attack complex (MAC) A complex of complement proteins that makes holes in bacterial cells, leading to cell lysis.

membrane potential A difference in electrical charge across the plasma membrane.

memory cell A type of long-lived cell that contains membrane-bound antibodies having the same specificity as the parent cell.

menopause The cessation of menstrual cycles resulting from decreased production of estradiol and progesterone by the ovaries.

menstrual cycle A monthly cycle in females in which oocytes mature and are released from the ovary under the influence of hormones.

menstruation The monthly shedding of the uterine lining.

meristem A discrete population of actively dividing, totipotent cells near the tip of stems and roots to which cell division is confined.

meristem identity gene A gene that contributes to meristem stability and function.

mesentery A membrane in the abdominal cavity through which blood vessels, nerves, and lymph travel to supply the gut.

mesoderm The intermediate germ layer, which differentiates into cells that make up connective tissue, muscle cells, red blood cells, bone cells, kidney cells, and gonad cells.

mesoglea In cnidarians, a gelatinous mass enclosed in by the epidermis and endodermis.

mesohyl A gelatinous mass that lies between the interior and exterior cell layers of a sponge that contains some amoeba-like cells that function in skeleton formation and the dispersal of nutrients.

mesophyll A leaf tissue of loosely packed photosynthetic cells.

Messel Shale A sedimentary rock formation in a lake in Germany, preserving fossils that document fish, birds, mammals, and reptiles from the beginning of the age of mammals.

messenger RNA (mRNA) The RNA molecule that combines with a ribosome to direct protein synthesis; it carries the genetic "message" from the DNA to the ribosome.

metabolic rate An animal's overall rate of energy use.

metabolism The chemical reactions that convert molecules into other molecules and transfer energy in living organisms.

metamorphosis The process in some animals in which the body changes dramatically at key stages in development.

metanephridia A pair of excretory organs in each body segment of annelid worms that filters the body fluid.

metaphase The stage of mitosis in which the chromosomes are aligned in the middle of the dividing cell.

metaphase I The stage of meiosis I in which the meiotic spindle is completed and the bivalents move to lie on an imaginary plane cutting transverely across the spindle.

metaphase II The stage of meiosis II in which the chromosomes line up so that their centromeres lie on an imaginary plane cutting across the spindle.

metapopulation A large population made up of smaller populations linked by migration.

MHC class I MHC genes and proteins in humans and mice that are expressed on the surface of all nucleated cells.

MHC class II MHC genes and proteins in humans and mice that are expressed on the surface of macrophages, dendritic cells, and B cells.

MHC class III MHC genes and proteins in humans and mice that encode several proteins of the complement system and proteins involved in inflammation.

micelle A spherical structure in which lipids with bulky heads and a single hydrophobic tail are packed.

microfilament A helical polymer of actin monomers, present in various locations in the cytoplasm, that helps make up the cytoskeleton.

microfossil A microscopic fossil up to about 300μm in diameter.

microRNA Small, regulatory RNA molecules that can inhibit translation; also called miRNA.

microtubule A hollow, tubelike polymer of tubulin dimers that helps make up the cytoskeleton.

microvilli Highly folded surfaces of villi, formed by fingerlike projections on the surfaces of epithelial cells.

midbrain Along with the hindbrain, the region of the vertebrate brain that controls basic body functions and behaviors; it is a part of the brainstem.

middle ear The part of the human ear containing three small bones, the malleus, incus, and stapes, which amplify the waves that strike the tympanic membrane.

midgut The middle part of an animal's digestive tract, including the small intestine.

migration The movement of individuals from one population to another.

minor groove The smaller of two unequal grooves on the outside of a DNA duplex.

mismatch repair The correction of a mismatched base in a DNA strand by cleaving one of the strand backbones, degrading the sequence with the mismatch, and resynthesizing from the intact DNA strand.

mitochondria Specialized organelles that harness energy for the cell from chemical compounds like sugars and convert it into ATP.

mitochondrial DNA (mtDNA) A small circle of DNA, about 17,000 base pairs long, found in every mitochondrion.

mitochondrial genome In eukaryotic cells, the DNA in the mitochondria.

mitochondrial matrix The space enclosed by the inner membrane of the mitochondria

mitosis In eukaryotic cells, the division of the nucleus, in which the chromosomes are separated into two nuclei.

mitotic spindle A structure in the cytosol made up predominantly of microtubules that pull the chromosomes into separate daughter cells.

moderately repetitive DNA A type of noncoding DNA consisting of repeated sequences present in hundreds of copies per genome.

Modern Synthesis The current theory of evolution, which combines Darwin's theory of natural selection and Mendelian genetics.

molar One of the teeth in the back of the mouth specialized for crushing and shredding tough foods such as meat and fibrous plant material.

molecular clock The relative constancy of rates of evolutionary change in a DNA nucleotide sequence or a protein amino acid sequence; the correlation between the time two species have been evolutionarily separated and the amount of genetic divergence between them.

molecular evolution Evolution at the level of DNA, which in time results in the genetic divergence of populations.

molecular fossils Sterols, bacterial lipids, and some pigment molecules, which are relatively resistant to decomposition, that accumulate in sedimentary rocks and document organisms that rarely form conventional fossils.

molecular orbital A merged orbital traversed by a pair of shared electrons.

molecular self-assembly The process by which, when conditions and relative amounts are suitable, viral components spontaneously interact and assemble into mature virus particles.

molecule A substance made up of two or more atoms.

mollusks A phylum distinguished by a mantle, which plays a major role in breathing and excretion.

molting Periodic shedding, as of an exoskeleton.

monophyletic Describes groupings in which all members share a single common ancestor not shared with any other species or group of species.

monosaccharide A simple sugar.

morphospecies concept The principle that members of the same species usually look alike.

morula The solid ball of cells resulting from early cell divisions of the fertilized egg.

motor endplate The region on a muscle cell where acetylcholine binds with receptors.

motor neuron A neuron that, on receiving information from interneurons, effects a response in the body.

motor protein A small accessory protein that causes muscle contraction by moving the actin microfilaments inside muscle cells.

motor unit A motor neuron and the population of muscle fibers that it innervates.

mouth The first part of the foregut, which receives food.

mucosa An inner tissue layer with secretory and absorptive functions surrounding the lumen of the digestive tract.

multinucleated Describes cells having several nuclei.

multiple alleles Two or more different alleles of the same gene, occurring in a population of organisms.

multiplication rule The principle that the probability of two independent events occurring together is the product of their respective probabilities.

multipotent Describes cells that can form a limited number of types of specialized cell.

multiregional hypothesis The idea that modern humans derive from the *Homo ergaster* populations that spread around the world starting 2 million years ago.

mutagen An agent that increases the probability of mutation.

mutation Any heritable change in the genetic material, usually a change the nuelcotide sequence of a gene.

mutualism A mutually beneficial interaction between species.

mycelium A network of branching hyphae.

mycorrhizae Symbioses between roots and fungi that enhance nutrient uptake.

myelin Lipid-rich layers or sheaths formed by glial cells that wrap around the axons of neurons and provide electrical insulation.

myofibril A long rodlike structure in muscle fibers that contains parallel arrays of the actin and myosin filaments.

myoglobin An oxygen-binding protein in the cells of vertebrate muscles, related to hemoglobin, that facilitates oxygen delivery to mitochondria.

myosin A motor protein found in muscle cells that carries cargo to the plus ends of microfilaments; used by both striated and smooth muscles to contract and generate force.

myotome In chordates, any one of a series of segments that organizes the body musculature.

myriapods One of the four main groups of arthropods, including centipedes and millipedes; distinguished by their many pairs of legs.

natural killer cell A cell type of the innate immune system that does not recognize foreign cells but instead recognizes and kills host cells that are infected by a virus or have become cancerous or otherwise abnormal.

natural selection The process in which, when there is inherited variation in a population of organisms, the variants best suited for growth and reproduction in a given environment contribute disproportionately to future generations. Of all the evolutionary mechanisms, natural selection is the only one that leads to adaptations.

nature Collectively, the innate qualities of an individual; behavior that is genetically encoded.

Neanderthal *Homo neanderthalensis*, a species similar to humans, but with thicker bones and flatter heads that contained brains about the same size as humans'; appeared in the fossil record 60,000–30,000 years ago.

necrotrophic pathogen A plant pathogen that kills cells before colonizing them.

negative feedback Describes the effect in which the final product of a biochemical pathway inhibits the first step; the process in which a stimulus acts on a sensor that communicates with an effector, producing a response that opposes the initial stimulus. Negative feedback is used to maintain steady conditions, or homeostasis.

negative regulation The process in which a regulatory molecule must bind to the DNA at a site near the gene to prevent transcription.

negative selection Natural selection that reduces the frequency of a deleterious allele.

negatively selected Describes T cells that react too strongly to self antigens in association with MHC and are eliminated through cell death.

neoteny The long-term evolutionary process in which the timing of development is altered so that a sexually mature organism retains the physical characteristics of the juvenile form.

nephron The functional unit of the kidney, consisting of the glomerulus, capsule, renal tubules, and collecting ducts.

nerve A bundle of long fiberlike extensions from multiple nerve cells.

nerve cord A bundle of long fiberlike extensions from multiple nerve cells that serves as the central nervous system of invertebrates such as flatworms and earthworms.

nervous system A network of many interconnected nerve cells.

neural tube In chordates, a cylinder of embryological tissue that develops into a dorsal nerve cord.

neuron Nerve cell; the basic fundamental unit of nervous systems.

neurosecretory cell A neuron in the hypothalamus that secretes hormones into the bloodstream.

neurotransmitter A molecule that conveys a signal from the end of the axon to the postsynaptic target cell.

neutral Describes mutations that have no effect or negligible effects on the organism, or whose effects are not associated with differences in survival or reproduction.

neutron An electrically neutral particle in the atomic nucleus.

neutrophil A type of phagocytic cell that is very abundant in the blood and is often one of the first cells to respond to infection.

niche A multidimensional habitat that allows a species to practice its way of life.

nicotinamide adenine dinucleotide phosphate (NADPH) An important cofactor in many biosynthetic reactions; the reducing agent used in the Calvin cycle.

nitrification The process by which chemoautotrophic bacteria oxidize ammonia (NH_3) to nitrate (NO_3^-).

nitrogen fixation The process in which nitrogen gas (N_2) is converted into ammonia (NH_3), a form biologically useful to primary producers.

nitrogenous waste Waste in the form of ammonia, urea, and uric acid, which are toxic to organisms in varying degrees.

nociceptor A type of nerve cell with dendrites in the skin and connective tissues of the body that responds to excessive mechanical, thermal, or chemical stimuli by withdrawal from the stimulus and by the sensation of pain.

node In phylogenetic trees, the point where a branch splits, representing the common ancestor from which the descendant species diverged. In plants, the point on a shoot where one or more leaves are attached.

nodes of Ranvier Sites on an axon that lie between adjacent myelin-wrapped segments, where the axon membrane is exposed.

non-associative learning Learning that occurs in the absence of any particular outcome, such as a reward or punishment.

non-competitve inhibitor A reversible inhibitor that reduces the maximum velocity of the reaction, but does not affect the affinity of the substrate for the active site of the enzyme. It usually has a structure very different from that of the substrate and binds to the enzyme at a site different from the active site.

nondisjunction The failure of a pair of chromosomes to separate normally during anaphase of cell division.

non-random mating Mate selection in which genotype is a factor.

nonrecombinants Progeny in which the alleles are present in the same combination as that present in a parent.

nonsense mutation A mutation that creates a stop codon, terminating translation.

non-sister chromatids Chromatids of differerent members of a pair of homologous chromosomes; although they carry the same complement of genes, they are not genetically identical.

nonsynonymous (missense) mutation A point mutation (nucleotide substitution) that causes an amino acid replacement.

nontemplate strand The untranscribed partner of the template strand of DNA used in transcription.

norm of reaction A graphical depiction of the change in phenotype across a range of environments.

normal distribution A distribution whose plot is a bell-shaped curve.

notochord In some chordates, a stiff rod of collagen and other proteins that runs along the back and provides support for the axis of the body.

nuclear envelope The cell structure, composed of two membranes, inner and outer, that defines the boundary of the nucleus.

nuclear genome In eukaryotic cells, the DNA in the chromosomes.

nuclear localization signal The signal sequence for the nucleus that enables proteins to move through pores in the nuclear envelope.

nuclear pore One of many protein channels in the nuclear envelope that act as gateways that allow molecules to move into and out of the nucleus and are thus essential for the nucleus to communicate with the rest of the cell.

nuclear transfer A procedure in which a hollow glass needle is used to insert the nucleus of a cell into the cytoplasm of an egg whose own nucleus has been destroyed or removed.

nucleic acid A polymer of nucleotides that encodes and transmits genetic information.

nucleoid In prokaryotes, a cell structure with multiple loops formed from supercoils of DNA.

nucleoside A molecule consisting of a 5-carbon sugar and a base.

nucleosome A beadlike repeating unit of histone proteins wrapped with DNA making up the 10-nm chromatin fiber.

nucleotide The subunit of nucleic acids, consisting of a 5-carbon sugar, a nitrogen-containing base, and one or more phosphate groups.

nucleotide excision repair The repair of multiple mismatched or damaged bases across a region; a process similar to mismatch repair, but over a much longer piece of DNA, sometimes thousands of nucleotides.

nucleotide substitution A mutation in which a base pair is replaced by a different base pair; this is the most frequent type of mutation. Also known as a *point mutation*.

nucleus (of a cell) The compartment of the cell that houses the DNA in chromosomes.

nucleus (of an atom) The dense central part of an atom containing protons and neutrons.

nurture Conditioning of behavior by the environment or personal experience.

obligate Describes a mutualism in which one or both sides cannot survive without the other.

observation The act of viewing the world around us.

occipital lobe The region of the brain that processes visual information from the eyes.

ocean acidification An increase in the abundance of carbon dioxide in the oceans that causes the pH of seawater to go down.

Okazaki fragment Any one of the many short DNA pieces in the lagging strand.

olfaction The sense of smell.

oligodendrocyte A type of glial cell that insulates cells in the brain and spinal cord by forming a myelin sheath.

oligonucleotide A short (typically 20 to 30 nucleotides), single-stranded molecule of known sequence produced by chemical synthesis; oligonucleotides are often used as primer sequences in the polymerase chain reaction.

omasum The third chamber in the stomach of ruminants, into which the mixture of food and bacteria passes and where water is reabsorbed.

ommatidia Individual light-focusing elements that make up the compound eyes of insects and crustaceans; the number of ommatidia determines the resolution of the image.

oncogene A cancer-causing gene.

oocyte The unfertilized egg cell produced by the mother; the developing female gamete.

oogenesis The formation of ova or eggs.

open circulatory system A circulatory system found in many smaller animals that contains few blood vessels and in which most of the circulating fluid is contained within the animal's body cavity.

open reading frame (ORF) A stretch of DNA or RNA consisting of codons for amino acids uninterrupted by a stop codon. In genome annotation, this sequence motif identifies the region as potentially protein coding.

operant conditioning Associative learning in which a novel behavior that was initially undirected has become paired with a particular stimulus through reinforcement.

operator The binding site for a repressor protein.

operon A group of functionally related genes located in tandem along the DNA and transcribed as a single unit from one promoter; the region of DNA consisting of the promoter, the operator, and the coding sequence for the structural genes.

opsin A photosensitive protein that converts the energy of light photons into electrical signals in the receptor cell.

optic nerve A cranial nerve that transmits action potentials from ganglion cells in the retina to the visual cortex of the brain.

orbital A region in space where an electron is present most of the time.

order A group of closely related families.

organ Two or more tissues that combine and function together.

organ of Corti A structure in the cochlear duct, supported by the basilar membrane, with specialized hair cells with stereocilia, that functions to convert mechanical vibrations to electrical impulses.

organelle Any one of several compartments in eukaryotes that divide the cell contents into smaller spaces specialized for different functions.

organic molecule A carbon-containing molecule.

organogenesis The transformation of the three germ layers into all the organ systems of the body.

origin of replication Each point on a DNA molecule at which DNA synthesis is initiated.

osmoconformer An animal that matches its internal osmotic pressure to that of its external environment.

osmoregulation The regulation of water and solute levels to control osmotic pressure.

osmoregulator An animal that maintains internal solute concentrations that differ from that of its environment.

osmosis The net movement of water molecules from regions of higher water concentration to regions of lower water concentration.

osmotic pressure The tendency of water to move from one solution into another by osmosis.

osponization The binding of a molecule to a pathogen to facilitate uptake by a phagocyte.

Osteichthyes Bony fish that have a cranium, jaws, and mineralized bones; there are about 20,000–25,000 species.

osteoblast A type of cell that forms bone tissue.

osteoclast A type of cell that secretes digestive enzymes and acid that dissolves the calcium mineral and collagen in bone.

osteocyte Bone cell.

outer ear The part of the human ear that includes the pinna, the ear canal, and tympanic membrane.

Out-of-Africa hypothesis The idea that modern humans arose from *Homo ergaster* descendants in Africa about 200,000 years ago.

oval window The thin membrane between the stapes of the middle ear and the cochlea of the inner ear.

ovary In plants, a hollow structure at the base of the carpel in which the ovules develop and which protects the ovules from being eaten or damaged by animals; generally, the female gonad where eggs are produced.

oviparity Laying eggs.

ovule The female reproductive structure in plants that contains the sporangium, now filled with a female gametophyte.

ovule cone A type of reproductive structure in pines that occurs on upper branches and produces female gametophytes and female gametes.

ovum (egg) (plural, **ova**) The larger, female gamete.

oxidation reaction A reaction in which a molecule loses electrons and releases energy.

oxidation–reduction reaction A reaction involving the loss and gain of electrons between reactants. In biological systems these reactions are often used to store or release chemical energy.

oxidative phosphorylation The fourth stage of cellular respiration, in which electron carriers generated in stages 1–3 donate their high-energy electrons to an electron transport chain.

oxidizing agent An electron acceptor.

oxygen dissociation curve The curve that results when blood pO_2 is plotted against the percentage of O_2 bound to hemoglobin.

oxygenic Producing oxygen.

oxytocin A posterior pituitary gland hormone that causes uterine contraction during labor and stimulates the release of milk during breastfeeding.

P_1 generation The parental generation in a series of crosses.

pacemaker Describes cardiac muscle cells that function as a regulator of heart rhythm.

palindromic Describes sequence identity in the paired strands of a duplex DNA molecule; a symmetry typical of restriction sites.

pancreas A secretory gland that has both endocrine function, secreting hormones, including insulin, directly into the blood, and exocrine function, aiding the digestion of proteins, carbohydrates, and fats by secreting digestive enzymes into ducts that connect to the duodenum.

paracrine signaling Signaling by a molecule that travels a short distance to the nearest neighboring cell to bind its receptor and deliver its message.

paraphyletic Describes groupings that include some, but not all, the descendants of a common ancestor.

parasexual Describes asexual species that generate genetic diversity by the crossing over of DNA during mitosis.

parasympathetic division The division of the autonomic nervous system that slows the heart and stimulates digestion and metabolic processes that store energy, enabling the body to "rest and digest."

parathyroid gland A gland adjacent to the thyroid gland that secretes parathyroid hormone (PTH), which, with calcitonin, regulates the actions of bone cells.

parenchyma Describes thin-walled, undifferentiated cells in the stem of a vascular plant.

parietal lobe The brain region, posterior to the frontal lobe, that controls body awareness and the ability to perform complex tasks.

parsimony Choosing the simplest hypothesis to account for a given set of observations.

parthenogenesis A form of asexual reproduction in which females produce eggs that are not fertilized by males but divide by mitosis and develop into new individuals.

partial pressure (p) The fractional concentration of a gas relative to other gases present multiplied by the atmospheric pressure exerted on the gases.

partially reproductively isolated Describes populations that have not yet diverged as separate species but whose genetic differences are extensive enough that the hybrid offspring they produce have reduced fertility or viability compared to offspring produced by crosses between individuals within each population.

patch A bit of habitat that is separated from other bits by inhospitable environments that are difficult or risky for individuals to cross.

paternal inheritance A type of inheritance in which the organelles in the offspring cells derive from those in the father.

peat bog Wetland in which dead organic matter accumulates.

pedigree A diagram of family history that summarizes the record of the ancestral relationships among individuals.

penis The male copulatory organ.

pepsin An enzyme produced in the stomach that breaks down proteins.

peptide bond A covalent bond that links the carbon atom in the carboxyl group of one amino acid to the nitrogen atom in the amino group of another amino acid.

peptide hormone A hormone that is a short chain of linked amino acids.

peptidoglycan A complex polymer of sugars and amino acids that makes up the cell wall.

peptidyl (P) site One of three binding sites for tRNA on the large subunit of a ribosome.

pericycle In roots, a single layer of cells just to the inside of the endodermis from which new root apical meristem is formed.

periodic selection The episodic loss of diversity as a successful variant outcompetes others.

periodic table of the elements The arrangement of the chemical elements in tabular form, organized by their chemical properties.

peripatric speciation A specific kind of allopatric speciation in which a few individuals from a mainland population disperse to a new location remote from the original population and evolve separately.

peripheral membrane protein A protein that is temporarily associated with the lipid bilayer or with integral membrane proteins through weak noncovalent interactions.

peripheral nervous system (PNS) Collectively, the sensory and motor nerves, including the cranial and spinal nerves, and interneurons and ganglia.

peristalsis Waves of muscular contraction that move food toward the base of the stomach.

personalized medicine An approach in which the treatment is matched to the patient, not the disease; examination of an individual's genome sequence, by revealing his or her disease susceptibilities and drug sensitivities, allows treatments to be tailored to that individual.

petal A structure, often brightly colored and distinctively shaped, that forms the next-to-outermost whorl of a flower with other petals; petals attract and orient animal pollinators.

phagocyte A type of immune cell that engulfs and destroys foreign cells or particles.

phagocytosis The engulfing of a cell or particle by another cell.

pharyngeal slit A vertical opening separated from other slits by stiff rods of protein in the pharynx of hemichordates.

pharynx The region of the throat that connects the nasal and mouth cavities; in hemichordates, a tube that connects the mouth and the digestive tract.

phenol Any one of a class of compounds, produced by some plants as a defensive mechanism.

phenotype The expression of a physical, behavioral, or biochemical trait; an individual's observable phenotypes include height, weight, eye color, and so forth.

pheromone A water- or airborne chemical compound released by animals into the environment that signals and influences the behavior of other members of their species.

phloem The outer tissue of a stem, which transports carbohydrates from leaves to the rest of the plant body.

phloem sap The sugar-rich solution that flows through both the lumen of the sieve tubes and the sieve plate pores.

phosphatase An enzyme that removes a phosphate group from another molecule.

phosphate group A chemical group consisting of a phosphorus atom bonded to four oxygen atoms.

phosphodiester bond A bond that forms when a phosphate group in one nucleotide is covalently joined to the sugar unit in another nucleotide. Phosphodiester bonds are relatively stable and form the backbone of a DNA strand.

phospholipid A type of lipid and a major component of the cell membrane.

photic zone The surface layer of the ocean through which enough sunlight penetrates to enable photosynthesis.

photoheterotroph An organism that uses the energy from sunlight to make ATP and relies on organic molecules obtained from the environment as the source of carbon for growth and other vital functions.

photoperiod Day length.

photoperiodism The effect of the photoperiod, or day length, on flowering.

photoreceptor A molecule whose chemical properties are altered when it absorbs light; also, photoreceptors are the sensory receptors in the eye.

photorespiration The process in which ATP is used to drive the reactions that convert a portion of the carbon atoms in 2-phosphoglycolate into 3-phosphoglyceric acid, which can reenter the Calvin cycle.

photosynthesis The biochemical process in which carbohydrates are built from carbon dioxide and the energy of sunlight; oxygen is released as a waste product.

photosynthetic electron transport chain A series of redox reactions in which electrons are passed from one compound to another.

photosystem A protein-pigment complex that absorbs light energy to drive redox reactions and thereby sets the photosynthetic electron transport chain in motion.

photosystem I The photosystem that energizes electrons with a second input of light energy so they have enough energy to reduce $NADP^+$.

photosystem II The photosystem that supplies electrons to the beginning of the electron transport chain. When photosystem II loses an electron it can pull electrons from water.

phototroph An organism that captures energy from sunlight.

phototropic Bending in response to light. A positive phototropic response, as in stems, is toward light; a negative phototropic response, as in roots, is away from light.

phragmoplast In dividing plant cells, a structure formed by overlapping microtubules that guide vesicles containing cell wall components to the middle of the cell.

phylogenetic tree A diagrammed hypothesis about the evolutionary history, or phylogeny, of a species.

phylogeny The history of descent with modification and the accumulation of change over time.

phylum A group of closely related classes, defined by having one of a number of distinctive body plans.

phytochrome A photoreceptor that switches back and forth between two stable forms, active and inactive, depending on its exposure to light.

pilus (plural, **pili**) A threadlike, hollow structure through which plasmids are transferred between bacteria.

pineal gland A gland located in the thalamic region of the brain that responds to autonomic nervous system input by secreting melatonin, which controls wakefulness.

pinna (plural, **pinnae**) Small unit of the photosynthetic surface on fern leaves; also, the external structure of mammalian ears that enhance the reception of sound waves contacting the ear.

pit A circular or ovoid region in the walls of xylem cells where the lignified cell wall layer is not produced.

pith In a stem, the region inside the ring of vascular bundles.

pituitary gland A gland beneath the brain that produces a number of hormones, including growth hormone.

placenta In placental mammals, an organ formed by the fusion of the chorion and allantois that allows the embryo to obtain nutrients directly from the mother.

placental mammal A mammal that provides nutrition to the embryo through the placenta, a temporary organ that develops in the uterus; placental mammals include carnivores, primates, hooved mammals, and whales.

plasma (effector) cell A cell that secretes antibodies.

plasma membrane The membrane that defines the space of the cell, separating the living material within the cell from the nonliving environment around it.

plasmid In bacteria, a small circular molecule of DNA carrying a small number of genes that can replicate independently of the bacterial genomic DNA.

plasmodesmata Connections in plant cells between the plasma membranes of adjacent cells that permit materials to pass directly from the cytoplasm of one cell to the cytoplasm of another.

plasmogamy The cytoplasmic union of two cells.

plate tectonics The dynamic movement of Earth's crust, the outer layer of Earth.

pleiotropy The phenomenon in which a single gene has multiple effects on seemingly unrelated traits.

pluripotent Describes embryonic stem cells (cells of the inner mass), which can give rise to any of the three germ layers and therefore to any cell of the body.

podocytes Cells with footlike processes that make up one of the three layers of the filtration barrier of the glomerulus.

point mutation A mutation in which a base pair is replaced by a different base pair; this is the most frequent type of mutation. Also known as *nucleotide substitution*.

Pol II The RNA polymerase complex responsible for transcription of protein-coding genes.

polar (molecule) A molecule that has regions of positive and negative charge.

polar body A small cell produced by the asymmetric first meiotic division of the primary oocyte.

polar covalent bond Bonds that do not share electrons equally.

polar transport The coordinated movement of auxin across many cells.

polarity An asymmetry such that one end of a structure differs from the other.

polarized Having opposite properties in opposite parts; describes a resting membrane potential in which there is a buildup of negatively charged ions on the inside surface of the cell's plasma membrane and positively charged ions on its outer surface.

pollen cone A type of reproductive structure in pines that occurs near the tips of lower branches and produces male gametophytes and male gametes.

pollen tube A structure produced by the male gametophyte that grows outward through an opening in the sporopollenin coat.

pollination The process in which pollen is carried to an ovule.

poly(A) tail The nucleotides added to the 3′ end of the primary transcript by polyadenylation.

polyadenylation The addition of a long string of consecutive A-bearing ribonucleotideas to the 3′ end of the primary transcript.

polycistronic mRNA A single molecule of messenger RNA that is formed by the transcription of a group of functionally related genes located next to one another along the bacterial DNA.

polymer A complex organic molecule made up of repeated simpler units connected by covalent bonds.

polymerase chain reaction (PCR) A selective and highly sensitive method for making copies of a piece of DNA, which allows a targeted region of a DNA molecule to be replicated into as many copies as desired.

polymorphism Any genetic difference among individuals sufficiently common that it is likely to be present in a group of 50 randomly chosen individuals.

polypeptide A polymer of amino acids connected by peptide bonds.

polyphyletic Describes groupings that do not include the last common ancestor of all members.

polyploidy The condition of having more than two complete sets of chromosomes in the genome.

polysaccharide A polymer of simple sugars. Polysaccharides provide long-term energy storage or structural support.

polyspermy Fertilization by more than one sperm.

pons A part of the brainstem.

population All the individuals of a given species that live and reproduce in a particular place; one of several interbreeding groups of organisms of the same species living in the same geographical area.

population density The size of a population divided by its range.

population size The number of individuals of all ages alive at a particular time in a particular place.

positive feedback In the nervous system, the type of feedback in which a stimulus causes a response that leads to an enhancement of the original stimulus that leads to a larger response. In the endocrine system, the type of feedback in which a stimulus causes a response, and that response causes a further response in the same direction. In both cases, the process reinforces itself until interrupted.

positive regulation The process in which a regulatory molecule must bind to the DNA at a site near the gene in order for transcription to take place.

positive selection Natural selection that increases the frequency of a favorable allele.

positively selected Describes T cells that recognize self MHC molecules on epithelium cells and continue to mature.

posterior pituitary gland The region of the pituitary gland that develops from neural tissue at the base of the brain and into which neurosecretory cells of the hypothalamus extend that secrete releasing factors.

posttranslational modification The modification, after translation, of proteins in ways that regulate their structure and function.

post-zygotic Describes factors that cause the failure of the fertilized egg to develop into a fertile individual, thus causing reproductive isolation.

potential energy Stored energy that is released by a change in an object's structure or position.

power stroke The stage in the cross-bridge cycle in which the myosin head pivots forward and generates a force, causing the myosin and actin filaments to slide relative to each other.

prediction An informed guess about the outcome of an experiment or observation based on a hypothesis.

pregnancy The carrying of one or more embryos in the uterus.

premolar One of the teeth between the canines and molars that are specialized for shearing tough foods.

pre-zygotic Describes factors that that prevent the fertilization of an egg, thus causing reproductive isolation.

primary active transport Active transport that uses the energy of ATP directly.

primary bronchi The two divided airways from the trachea, supported by cartilage rings, each airway leading to a lung.

primary growth The increase in plant length made possible by apical meristems.

primary motor cortex The part of the frontal lobe of the brain that produces complex coordinated behaviors by controlling skeletal muscle movements.

primary oocyte A diploid cell formed by mitotic division of oogonia during fetal development.

primary producer An organism that takes up inorganic carbon, nitrogen, phosphorus, and other compounds from the environment and converts them into organic compounds that will provide food for other organisms in the local environment.

primary response The response to the first encounter with an antigen, during which there is a short lag before antibody is produced.

primary somatosensory cortex The part of the parietal lobe that integrates tactile information from specific body regions and relays it to the motor cortex.

primary spermatocyte A diploid cell formed by mitotic division of spermatogonia at the beginning of spermatogenesis.

primary structure The sequence of amino acids in a protein.

primary transcript The initial RNA transcript that comes off the template DNA strand.

primate A member of an order of mammals that share a number of general features that distinguish them from other mammals, including nails rather than claws, front-facing eyes, and an opposable thumb.

primer A short stretch of RNA at the beginning of each new DNA strand that serves as a starter for DNA synthesis; an oligonucleotide that serves as a starter in the polymerase chain reaction.

principle of independent assortment The principle that segregation of one set of alleles of a gene pair is independent of the segregation of another set of alleles of a different gene pair.

principle of segregation The principle by which half the gametes receive one allele of a gene and half receive the other allele.

probability Among a very large number of observations, the expected proportion of observations that are of a specified type.

probe A labeled DNA fragment that can be tracked in a procedure such as a Southern blot.

procambial cell A cell that retains the capacity for cell division and gives rise to both xylem and phloem.

product Any one of the transformed molecules that result from a chemical reaction.

progesterone A hormone secreted by the ovaries that maintains the thickened and vascularized uterine lining.

prokaryote A unicellular organism without a nucleus. Often used to refer collectively to archaeons and bacteria.

prometaphase The mitosis in which the nuclear envelope breaks down and the microtubules of the mitotic spindle attach to chromosomes.

promoter A regulatory region where RNA polymerase and associated proteins bind to the DNA duplex.

proofreading The process in which DNA polymerases correct their own errors by excising and replacing a mismatched base.

prophase The stage of mitosis characterized by the appearance of visible chromosomes.

prophase I The beginning of meiosis I, marked by the visible manifestation of chromosome condensation.

prophase II The stage of meiosis II in which the chromosomes in the now-haploid nuclei recondense to their maximum extent.

prostate gland An exocrine gland that produces a thin, slightly alkaline fluid that helps maintain sperm motility and counteracts the acidity of the female reproductive tract.

protease inhibitor An antidigestive protein that binds to the active site of enzymes that break down proteins in a herbivore's digestive system.

protein family A group of proteins that are structurally and functionally related.

protein sorting The process by which proteins end up where they need to be in the cell to perform their function.

proteins The key structural and functional molecules that do the work of the cell, providing structural support and catalyzing chemical reactions. The term "protein" is often used as a synonym for "polypeptide."

proteobacteria The most diverse bacterial group, defined largely by similarities in rRNA gene sequences; it includes many of the organisms that populate the expanded carbon cycle and other biogeochemical cycles.

protist An organism having a nucleus but lacking other features specific to plants, animals, or fungi.

proton A positively charged particle in the atomic nucleus.

protonephridia Excretory organs in flatworms that isolate waste from the body cavity.

proto-oncogene The normal cellular gene counterpart to an oncogene, which is similar to a viral oncogene but can cause cancer only when mutated.

protostome A bilaterians in which the blastopore, the first opening to the internal cavity of the developing embryo, becomes the mouth.

protozoan (plural, **protozoa**) A heterotrophic protist.

proximal convoluted tubule The first portion of the renal tubule from which electrolytes and other nutrients are reabsorbed into the blood.

pseudocoelomate A bilaterian with a body cavity that does not completely surround the internal organs.

pseudogene A gene that is no longer functional.

pulmonary artery One of two arteries, left and right, that carries deoxygenated blood from the right ventricle to the lungs.

pulmonary capillary A small blood vessel that supplies the alveolar wall.

pulmonary circulation Circulation of the blood to the lungs.

pulmonary valve A valve between the right ventrical and the pulmonary trunk, which divides into the pulmonary arteries.

pulmonary vein A vein that returns oxygenated blood from the lungs to the left ventricle.

Punnett square A worksheet in the form of a checkerboard used to predict the consequences of a random union of gametes.

pupa The quiescent stage of metamorphosis in insects, during which the body tissues undergo a transformation from larva to an adult.

pupil An opening in the vertebrate eye through which light enters.

purine In nucleic acids, the bases adenine and gunanine, which have a double-ring structure.

pyloric sphincter A band of muscle at the base of the stomach that opens to allow small amounts of digested food to enter the small intestine.

pyrimidine In nucleic acids, the bases thymine, cytosine, and uracil, which have a single-ring structure.

quantitative trait A complex trait in which the phenotype is measured along a continuum with only small intervals between similar individuals.

quaternary structure The structure that results from the interactions of several polypeptide chains.

R gene Any one of the group of genes that express the R proteins in plants.

R group A chemical group attached to the central carbon atom of an amino acid, whose structure and composition determine the identity of the amino acid; also known as a *residue* or *side chain*.

R protein Any one of a group of receptors in plant cells, each expressed by a different gene, that function as part of the plant's immune system by each binding to a specific pathogen-derived protein.

radial symmetry Symmetry around an axis that in animals runs from mouth to base with an infinite number of planes of symmetry through this axis.

Ras A cytoplasmic signaling protein, very similar to the a subunit of G proteins.

reabsorption In renal systems, an active or passive process in which substances that are important for an animal to retain are taken up by cells of the excretory tubule and returned to the bloodstream.

reactant Any of the starting molecules in a chemical reaction.

reaction center Two specially configured chlorophyll molecules where light energy is converted into electron transport.

reactive oxygen species Highly reactive forms of oxygen produced when $NADP^+$ is in short supply.

reading frame Following a start codon, a consecutive sequence of codons for amino acids.

realized niche The actual range of habitats occupied by a species.

receiver The individual who, during communication, receives from the sender a signal that elicits a response.

receptor A molecule on cell membranes that detects critical features of the environment. Receptors detecting signals that easily cross the cell membrane are sometimes found in the cytoplasm.

receptor kinase A receptor that is an enzyme that adds a phosphate group to another molecule.

receptor molecule The molecule on the responding cell that binds to the signaling molecule.

recessive The trait that fails to appear in heterozygous genotypes from a cross between the corresponding homozygous genotypes.

reciprocal altruism The exchange of favors between individuals.

reciprocal cross A cross in which the female and male parents are interchanged.

reciprocal inhibition The activation of opposing sets of muscles so that one set is inhibited as the other is activated, allowing the movement of joints such as the knee.

reciprocal translocation Interchange of parts between nonhomologous chromosomes.

recombinant An offspring with a different combination of alleles from that of either parent, resulting from one or more crossovers in prophase I of meiosis.

recombinant DNA The joining of DNA molecules from two (or more) different sources into a single molecule.

recovery metabolism An animal's elevated consumption of oxygen following activity.

rectum The part of the hindgut where feces are stored until elimination.

reducing agent An electron donor.

reduction Gain of electrons by a molecule in a reaction; in the second step of the Calvin cycle, energy and electrons are transferred to the molecules formed from carboxylation.

reduction reaction A reaction in which a molecule acquires electrons and gains energy.

reductional division An alternative name for meiosis I, since this division reduces the number of chromosomes by half.

redundant Describes the genetic code, in which many amino acids are specified by more than one codon.

refractory period The period following an action potential during which the inner membrane voltage falls below and then returns to the resting potential.

regeneration The third step of the Calvin cycle, in which the 5-carbon molecule needed for carboxylation is produced.

regenerative medicine A discipline that aims to use the natural processes of cell growth and development to replace diseased or damaged tissues.

regression toward the mean With regard to complex traits, the principle that offspring exhibit an average phenotype that is intermediate between that of the parents and that of the population as a whole.

regulatory transcription factor A protein that recruits the components of the transcription complex to the gene.

reinforcement of reproductive isolation (reinforcement) The process by which diverging populations undergo natural selection in favor of enhanced pre-zygotic isolation to prevent the production of inferior hybrid offspring.

release factor A protein that causes the bond connecting the polypeptide to the tRNA to break.

releasing factor A peptide hormone that signals to the anterior pituitary gland through blood vessels, leading to a much larger release of associated hormones from that organ.

renal pelvis The area of the mammalian kidney into which the collecting ducts empty.

renal system The system in vertebrates that underlies water and electrolyte balance and waste elimination.

renal tubules Tubes in the vertebrate kidney that process the filtrate from the glomerulus by reabsorption and secretion.

renaturation The base pairing of complementary single-stranded nucleic acids to form a duplex; also known as *hybridization*, it is the opposite of denaturation.

renin A hormone that stimulates epithelial cells of the lung to release an enzyme that converts angiotensinogen in the bloodstream to its active form, angiotensin II.

replication The exact copying of DNA so genetic information can be passed from cell to cell or from an organism to its progeny.

replication bubble A region formed by the opening of a DNA duplex at an origin of replication, which has a replication fork at each end.

replication fork The site where the parental DNA strands separate as the DNA duplex unwinds.

repressor A protein that, when bound with the RNA polymerase complex, can turn off transcription.

reservoir A supply or source of a substance. Reservoirs of carbon, for example, include organisms, the atmosphere, soil, the oceans, and sedimentary rocks.

residue An amino acid that is incorporated into a protein.

resource partitioning A pattern in which species whose niches overlap may diverge to minimize the overlap.

respiratory burst A process triggered by phagocytosis that generates reactive oxygen species and reactive nitrogen species.

responding cell The cell that receives information from the signaling molecule.

response A change in cellular behavior, such as activation of enzymes or genes, following a signal.

resting membrane potential The negative voltage across the membrane at rest.

restriction enzyme Any one of a class of enzymes that recognizes specific, short nucleotide sequences in double-stranded DNA and cleaves DNA at or near these sites.

restriction fragment length polymorphism (RFLP) A polymorphism in which the length of the restriction fragments is different in the two alleles.

restriction site A recognition sequence in DNA cutting, which is typically four or six base pairs long. Most restriction enzymes cleave double-stranded DNA at or near these restriction sites.

reticulum The second chamber in the stomach of ruminants, which, along with the rumen, harbors large populations of anaerobic bacteria that break down cellulose.

retina A thin tissue in the posterior of the vertebrate eye that contains the photoreceptors and other nerve cells that sense and initially process light stimuli.

retinal A derivative of vitamin A that absorbs light and binds to rhodopsin, a transmembrane protein in the photosensitive cells of vertebrates.

reversible inhibitor Any one of the class of inhibitors that form weak bonds with enzymes and easily dissociate from them.

rhizosphere The soil layer that surrounds actively growing roots.

ribonucleic acid (RNA) A molecule closely related to DNA that is synthesized by proteins from a DNA template.

ribose The sugar in RNA.

ribosomal RNA (rRNA) Noncoding RNA found in all ribosomes that aid in translation.

ribosome A complex structure of RNA and protein, bound to the cytosolic face of the RER in the cytoplasm, on which proteins are synthesized.

ribulose bisphosphate carboxylase oxygenase (rubisco) The enzyme that catalyzes a carboxylation reaction.

ribulose-1,5-bisphosphate (RuBP) The 5-carbon sugar to which carbon dioxide is added in carboxylation.

ring species Species that contain populations that are reproductively isolated from each other but can exchange genetic material through other, linking populations.

RISC (RNA-induced silencing complex) A protein complex that is targeted to specific mRNA molecules by base pairing with short regions on the target mRNA, inhibiting translation or degrading the RNA.

ritualization The process of co-opting and modifying begaviors used in another context by increasing the conspicuousness of the behavior, reducing the amount of variation in the behavior so that it can be immediately recognized, and increasing its separation from the original function.

RNA editing The process in which some RNA molecules become a substrate for enzymes that modify particular bases in the RNA, thereby changing its sequence and what it codes for.

RNA polymerase The enzyme that carries out polymerization of ribonucleoside triphosphates from a DNA template to produce an RNA transcript.

RNA polymerase complex Aggregate of proteins that synthesize the RNA transcript complementary to the template strand of DNA.

RNA primase An RNA polymerase that synthesizes a short piece of RNA complementary to the DNA template and does not require a primer.

RNA processing Chemical modification that converts the primary transcript into finished mRNA, enabling the RNA molecule to be transported to the cytoplasm and recognized by the translational machinery.

RNA splicing The process of intron removal from the primary transcript.

RNA transcript The RNA sequence synthesized from a DNA template.

RNA world hypothesis The belief that RNA, not DNA, was the original information-storage molecule in the earliest forms of life on Earth.

rod cell A type of photoreceptor cell on the retina that detects light and shades ranging from white to shades of gray and black, but not color.

root apical meristem A group of totipotent cells near the tip of a root that is the source of new root cells.

root cap A structure that covers and protects the root apical meristem as it grows through the soil.

root hair A slender outgrowth produced by epidermal cells in active areas of the root, which greatly increases the surface area of the root.

root nodule A structure, formed by dividing root cells, into which nitrogen-fixing bacteria enter.

roots A major organ system of vascular plants, generally belowground.

rough endoplasmic reticulum (RER) The part of the endoplasmic reticulum with attached ribosomes.

r-strategist A species that produces large numbers of offspring but provides few resources for their support.

rumen The first chamber in the stomach of ruminants, which, along with the reticulum, harbors large populations of anaerobic bacteria that break down cellulose.

S phase The phase of interphase in which the entire DNA content of the nucleus is replicated.

saccharide The simplest carbohydrate molecule, also called a *sugar*; saccharides store energy in their bonds.

saltatory propagation The movement of an action potential along a myelinated axon, "jumping" from node to node.

Sanger sequencing A procedure in which the terminated daughter strands help in determining the DNA sequence.

sapwood In long-lived trees, the layer adjacent to the vascular cambium that contains the functional xylem.

sarcomere The region from one Z disc to the next, the basic contractile unit of a muscle.

sarcoplasmic reticulum (SR) A modified form of the endoplasmic reticulum surrounding the myofibrils of muscle cells.

saturated Describes fatty acids that do not contain double bonds; the maximum number of hydrogen atoms is attached to each carbon atom, "saturating" the carbons with hydrogen atoms.

scaffold A supporting protein structure in a metaphase chromosome.

Schwann cell A type of glial cell that insulates sensory and motor neurons by forming a myelin sheath.

scientific method A deliberate, careful, and unbiased way of learning about the natural world.

sclera A tough, white outer layer surrounding the vertebrate eye.

scrotum A sac outside the abdominal cavity of the male that holds the testes.

second law of thermodynamics The principle that the transformation of energy is associated with an increase in the degree of disorder in the universe.

second messenger An intermediate cytosolic signaling molecule that transmits signals from a receptor to a target within the cell. (First messengers transmit signals from outside the cell to a receptor.)

secondary active transport   Active transport that uses the energy of an electrochemical gradient to drive the movement of molecules.

secondary growth   The increase in plant diameter, which results from meristem that forms only after elongation below the growing tip is complete.

secondary oocyte   A large cell produced by the asymmetric first meiotic division of the primary oocyte.

secondary phloem   New phloem cells produced by the vascular cambium, which appear to the outside of the vascular cambium.

secondary response   The response to re-exposure to an antigen, which is quicker, stronger, and longer than the primary response.

secondary sexual characteristic   A trait that characterizes and differentiates the two sexes but that does not relate directly to reproduction.

secondary spermatocyte   A diploid cell formed during the first meiotic division of the primary spermatocyte.

secondary structure   The structure formed by interactions between stretches of amino acids in a protein.

secondary xylem   New xylem cells produced by vascular cambium, which appear to the inside of the vascular cambium.

second-division nondisjunction   Disjunction in the second meiotic division.

secretin   A hormone released by cells lining the duodenum in response to the acidic pH of the stomach contents entering the small intestine and that stimulates the pancreas to secrete bicarbonate ions.

secretion   In renal systems, an active process that eliminates substances that were not previously filtered from the blood.

seed   A fertilized ovule; seeds are multicellular structures that allow offspring to disperse away from the parent plant.

seed coat   A protective outer structure surrounding the seed formed from tissues that surround the sporangial wall.

segmentation   The formation of discrete parts or segments in the insect embryo.

segregate   Separate; applies to chromosomes or members of a gene pair moving into different gametes.

selection   The retention or elimination of random mutations in a population of organisms.

selective barrier   Describes the plasma membrane, which lets some molecules in and out freely, lets others in and out only under certain conditions, and prevents other molecules from passing through at all.

self-compatible   Describes species in which pollen and eggs produced by flowers on the same plant can unite and produce viable offspring.

self-incompatible   Describes species in which pollination by the same or a closely related individual does not lead to fertilization.

self-propagating   Continuing without input from an outside source; action potentials are self-propagating in that they move along axons by sequentially opening and closing adjacent ion channels.

semen   A fluid that nourishes and sustains sperm as they travel in the male and then the female reproductive tracts.

semicircular canal   One of three connected fluid-filled tubes in the mammalian inner ear that contains hair cells that sense angular motions of the head in three perpendicular planes.

semiconservative replication   The mechanism of DNA replication in which each strand of a parental DNA duplex serves as a template for the synthesis of a new daughter strand.

seminal vesicles   Two glands at the junction of the vas deferens and the prostate gland that secrete a protein- and sugar-rich fluid that makes up most of the semen and provides energy for sperm motility.

seminiferous tubules   A series of tubes in the testes where sperm are produced.

semipermeable   Selectively permeable; a semipermeable cell membrane allows the movement of some molecules but not others.

sender   The indivdual who, during communication, supplies a signal that elicits a response from the receiver.

sensitization   The enhancement of a response to a stimulus that is achieved by first presenting a strong or novel stimulus.

sensory neuron   A neuron that receives and transmits information about an animal's environment or its internal physiological state.

sensory organ   A group of sensory receptors that converts particular physical and chemical stimuli into nerve impulses that are processed by a nervous system and sent to a brain.

sensory receptor   A sensory neuron with specialized membranes in which receptor proteins are embedded.

sensory transduction   The conversion of physical or chemical stimuli into nerve impulses.

sepal   A structure, often green, that forms the outermost whorl of a flower with other sepals and encases and protects the flower during its development.

septum (plural, septa)   In fungi, a wall that partially divides the cytoplasm into separate cells in hyphae.

sequence assembly   The process in which short nucleotide sequences of a long DNA molecule are arranged in the correct order to generate the complete sequence.

sequence motif   Any of a number of sequences or sequence arrangements that indicate the likely function of a segment of DNA.

serosa   An outer layer of cells and connective tissue that covers and protects the gut.

Sertoli cell   A type of cell in the seminiferous tubules that supports sperm production.

sex chromosome   One of the chromosomes associated with sex, in most animals denoted the X and Y chromosomes.

sex determination   The development of male or female characteristics, resulting from the presence or absence of the Y chromosome.

sexual dimorphism A phenotypic difference between the sexes.

sexual reproduction The process of producing offspring that receive genetic material from two parents; in eukaryotes, the process occurs through meiosis and fertilization.

sexual selection A form of selection that promotes traits that increase an individual's access to reproductive opportunities.

shell (of an atom) An energy level.

shoot The collective name for the leaves, stems, and reproductive organs, the major aboveground organ systems of vascular plants.

shoot apical meristem A group of totipotent cells near the tip of a branch that gives rise to new shoot tissues in plants.

short-day plant A plant that flowers only when the day length is less than a critical value.

shotgun sequencing DNA sequencing method in which the sequenced fragments do not originate from a particular gene or region but from sites scattered randomly across the molecule.

sickle-cell anemia A condition in which hemoglobin molecules tend to crystallize when exposed to lower-than-normal levels of oxygen, causing the red blood cells to collapse and block capillary blood vessels.

side chain A chemical group attached to the central carbon atom of an amino acid, whose structure and composition determine the identity of the amino acid; also known as an *R group*.

sieve element A highly modified cell, lacking most intracellular structure, that is connected end to end with other sieve elements to make up a sieve tube.

sieve plate A modified end wall with large pores that links sieve elements.

sieve tube A multicellular unit composed of sieve elements that are connected end to end, through which phloem transport takes place.

sigma factor A protein that associates with RNA polymerase that facilitates its binding to specific promoters.

signal sequence An amino acid sequence that directs a protein to its proper cellular compartment.

signal transduction The process in which an extracellular molecule acts as a signal to activate a receptor, which transmits information through the cytoplasm.

signaling cell The source of the signaling molecule.

signaling molecule The carrier of information transmitted when the signaling molecule binds to a receptor; also referred to as a *ligand*.

signal-recognition particle (SRP) An RNA–protein complex that binds with part of a polypeptide chain and marks the molecule for incorporation into the endoplasmic reticulum (eukaryotes) or the plasma membrane (prokaryotes).

simple multicellularity Multicellularity, achieved by adhesive molecules, in which there is relatively little communication or transfer of resources between cells and little differentiation of specialized cell types.

SINE Any one of many short interspersed nuclear elements of about 300 base pairs present in multiple copies in a genome owing to transposition.

single-gene trait A trait determined by Mendelian alleles of a single gene without much influence from the environment.

single-lens eye An eye structure found in vertebrates and cephalopod mollusks that works like a camera to produce a sharply defined image of the animal's visual field.

single-nucleotide polymorphism (SNP) A site in the genome where a base pair differs among individuals in a population.

single-stranded binding protein A protein that binds single-stranded nucleic acids.

sink In plants, any portion of the plant that needs carbohydrates to fuel growth and respiration, such as a root, young leaf, or developing fruit.

sinoatrial (SA) node A specialized region of the heart containing pacemaker cells where the heartbeat is initiated.

sister chromatids The two identical copies of chromosomes produced by DNA replication.

sister groups Groups that are more closely related to each other than either of them is to any other group.

skeletal muscle Muscle that connects to the body skeleton to move an animal's limbs and torso.

sliding filament model The hypothesis that muscles produce force and change length by the sliding of actin filaments relative to myosin filaments.

slow-twitch Describes muscle fibers that contract slowly and consume less ATP than do fast-twitch fibers to produce force.

small interfering RNA (siRNA) A type of small double-stranded regulatory RNA that becomes part of a complex able to cleave and destroy single-stranded RNA with a complementary sequence.

small intestine Part of the midgut; the site of the last part of digestion and most nutrient absorption.

small nuclear RNA (snRNA) Noncoding RNA found in eukaryotes and involved in splicing, polyadenylation, and other processes in the nucleus.

small regulatory RNA A short RNA molecule that works primarily by blocking transcription or translation.

smooth endoplasmic reticulum (SER) The portion of the endoplasmic reticulum that lacks ribosomes.

smooth muscle The muscle in the walls of arteries, the respiratory system, and the digestive and excretory systems; smooth muscle appears uniform under the light microscope.

solubility The ability of a substance to dissolve.

solute A dissolved molecule such as the electrolytes, amino acids, and sugars often found in water, a solvent.

solvent A liquid capable of dissolving a substance.

somatic cell A nonreproductive cell, the most common type of cell in body.

somatic mutation A mutation that occurs in somatic cells.

somatic nervous system The voluntary component of the peripheral nervous system, which is made up of sensory neurons that respond to external stimuli and motor neurons that synapse with voluntary muscles.

source In plants, a region that produces or stores carbohydrates.

Southern blot A method for determining the size and number of copies of a DNA sequence of interest by means of a labeled probe.

spatial summation The converging of multiple receptors onto a neighboring neuron, increasing its firing rate proportionally to the number of signals received.

speciation The process that produces new and distinct forms of life.

species A group of individuals that can exchange genetic material through interbreeding to produce fertile offspring.

species–area relationship The relationship between island size and equilibrium species diversity.

spermatogenesis The formation of sperm.

spermatozoa (sperm) The smaller, male gametes.

spinal cord In vertebrates, a central tract of neurons that passes through the vertebrae to transmit information between the brain and the periphery of the body.

spinal nerve In vertebrates, a nerve running from the spinal cord to the periphery containing axons of both sensory and motor neurons.

spindle apparatus The organelle formed by microtubules that separates replicated chromosomes during eukaryotic cell division.

spiracle An opening in the exoskeleton on either side of an insect's abdomen through which gases are exchanged.

spliceosome A complex of RNA and protein that catalyzes RNA splicing.

spongy bone Bone tissue consisting of trabeculae, and thus lighter than compact bone, found in the ends of limb bones and within vertebrae.

spontaneous Occurring in the absence of any assignable cause; most mutations are spontaneous.

sporangium A plant structure in which thousands of cells undergo meiosis, producing large numbers of haploid spores.

spore In plants, the haploid cells resulting from meiotic cell division that disperse and give rise to a new haploid generation. In animals, reproductive haploid cells resulting from meiotic cell division (in some species called gametes).

sporophyte Describes the diploid multicellular generation that gives rise to spores.

sporopollenin A complex mixture of polymers that is remarkably resistant to environmental stresses such as ultraviolet radiation and desiccation; a wall containing sporopollenin protects the plant zygote.

stabilizing selection A form of selection that maintains the status quo and selects against extremes.

stamen A pollen-producing floral organ in the center whorl.

stapes A small bone in the middle ear that helps amplify the waves that strike the tympanic membrane; the stapes connects to the oval window of the cochlea.

starch The form in which glucose is stored in plants.

Starling's Law The correspondence between change in stroke volume and change in the volume of blood filling the heart.

statocyst A type of gravity-sensing organ found in most invertebrates.

statolith In plants, a large starch-filled organelle in the root cap that senses gravity; in animals, a dense particle that moves freely within a statocyst, enabling it to sense gravity.

stem cell An undifferentiated cell that can undergo an unlimited number of mitotic divisions and differentiate into any of a large number of specialized cell types.

stereocilia Nonmotile cell-surface projections on hair cells whose movement causes a depolarization of the cell's membrane.

steroid A type of lipid.

steroid hormone A hormone that is derived from cholesterol.

stigma The surface of the top of the style, to which pollen adheres.

stomach (crop) The last part of the foregut, which serves as an initial storage and digestive chamber.

stomata Pores in the epidermis of a leaf that regulate the diffusion of gases between the interior of the leaf and the atmosphere.

Stramenopila A eukaryotic superkingdom including unicellular organisms, giant kelps, algae, protozoa, free-living cells, and parasites; distinguished by a flagellum with two rows of stiff hairs and, usually, a second, smooth flagellum.

striated muscle Skeletal muscle and cardiac muscle, which appear striped under a light microscope.

strigolactone A hormone, produced in roots and transported upward in the xylem, that inhibits the outgrowth of axillary buds.

stroke volume (SV) The volume of blood pumped during each beat.

stroma The region surrounding the thylakoid, where carbohydrate synthesis takes place.

stromatolite A layered structure that records sediment accumulation by microbial communities.

structural gene A gene that codes for the sequence of amino acids in a polypeptide chain.

style A cylindrical stalk at the top of a carpel.

suberin A waxy compound coating cork cells that protects against mechanical damage, the entry of pathogens, and water loss.

submucosa A tissue layer surrounding the mucosa that contains blood vessels, lymph vessels, and nerves.

subspecies Allopatric populations that have yet to evolve even partial reproductive isolation but which have accumulated a few population-specific traits.

substrate (S) A molecule acted upon by an enzyme.

substrate-level phosphorylation A way of generating ATP in which a phosphate group is transferred to ADP from an organic molecule, which acts as a phosphate donor or substrate.

succession The replacement of species by other species over time.

sugar The simplest carbohydrate molecule; also called a *saccharide*.

sulci Deep crevices in the brain that separate the lobes of the cerebral hemispheres.

supercoil A coil of coils; a circular molecule of DNA can coil upon itself to form a supercoil.

superkingdom One of seven major groups of eukaryotic organisms, classified by molecular sequence comparisons.

supernormal stimulus An exaggerated stimulus that elicits a response more strongly than the normal stimulus.

surfactant A compound that reduces the surface tension of a fluid film.

survivorship The proportion of individuals from an initial cohort that survive to each successive stage of the life cycle.

suspension filter feeding The most common form of food capture by animals, in which water with food suspended in it passes through a sievelike structure.

swallowing A complex set of motor reflexes that conveys food through the pharynx and into the esophagus and stomach.

symbiont An organism that lives in closely evolved association with another species.

symbiosis (plural, symbioses) A close interaction between species that have evolved over long periods of time.

sympathetic division The division of the autonomic nervous system that generally produces arousal and increased activity; active in the fight-or-flight response.

sympatric Describes populations that are in the same geographic location.

synapomorphy A shared derived character; a homology shared by some, but not all, members of a group.

synapse A junction through which the axon terminal communicates with a neighboring cell.

synapsis The gene-for-gene pairing of homologous chromosomes in prophase I of meiosis.

synaptic cleft The space between the axon of the presynaptic cell and the neighboring postsynaptic cell.

synaptic plasticity The ability to adjust synaptic connections between neurons.

synonymous (silent) mutation A mutation in a codon that does not alter the corresponding amino acid in the polypeptide.

systemic acquired resistance (SAR) The ability of a plant to resist future infections, occurring in response to a wide range of pathogens.

systemic circulation Circulation of the blood to the body, excluding the lungs.

systole The contraction of the ventricles.

T cell receptor (TCR) A protein receptor on a mature T cell that recognizes and binds to an antigen.

T lymphocyte (T cell) A cell type that matures in the thymus and includes helper and cytotoxic cells.

tannin Any one of a group of phenols found widely in plant tissues that bind with proteins and reduce their digestibility.

taste bud One of the sensory organs for taste.

TATA box A DNA sequence present in many promoters in eukaryotes and archaeons that serves as a protein-binding site for a key general transcription factor.

taxis (plural, taxes) Movement in a specific direction in response to a stimulus.

taxon (plural, taxa) All the species in a taxonomic entity such as family or genus.

tectorial membrane A rigid membrane in the cochlear duct, against which the stereocilia of hair cells in the organ of Corti bend when stimulated by vibration, setting off an action potential.

telomerase An enzyme that synthesizes telomere repeats.

telomere A repeating sequence at each end of a eukaryotic chromosome.

telophase The stage of mitosis in which the nuclei of the daughter cells are formed and the chromosomes uncoil to their original state.

telophase I The stage of meiosis I in which the chromosomes uncoil slightly, a nuclear envelope briefly reappears, and in many species the cytoplasm divides, producing two separate cells.

telophase II The stage of meiosis II in which the chromosomes uncoil and become diffuse, a nuclear envelope forms around each set of chromosomes, and the cytoplasm divides by cytokinesis.

template A strand of DNA or RNA whose squence of nucleotides is used to sythesis a complementary strand.

template strand In DNA replication, the parental strand whose sequence is used to synthesize a complementary daughter strand.

temporal lobe The region of the brain involved in the processing of sound, language and reading, and object identification and naming.

temporal separation Separation in time, as the pre-zygotic isolation of both plants and animals in time.

temporal summation The frequency of synaptic stimuli; the integration of sensory stimuli that are received repeatedly over time by the same sensory cell.

temporomandibular joint A specialized jaw joint in mammals that allows the teeth of the lower and upper jaws to fit together precisely.

tendon A collagen structure that attaches muscles to the skeleton and transmits muscle forces over a wide range of joint motion.

termination In protein translation, the stage in which the addition of amino acids stops and the completed polypeptide chain is released from the ribosome. In cell communication, the stopping of a signal.

terminator A DNA sequence at which transcription stops and the transcript is released.

terpene Any one of a group of compounds that do not contain nitrogen and produced by some plants as a defensive mechanism.

tertiary structure The overall three-dimensional shape of a protein, formed by interactions between secondary structures.

test (of a hypothesis) An experiment or observation to determine whether a prediction made by the hypothesis holds true.

test (of a protist) A "house" constructed of organic molecules that shelters a protist.

testcross Any cross of an unknown genotype with a homozygous recessive genotype.

testis (plural, **testes**) The male gonad, where sperm are produced.

testosterone A steroid hormone, secreted by the testes, that plays key roles in male growth, development, and reproduction.

tetanus A muscle contraction of sustained force.

tetraploid A cell or organism with four complete sets of chromosomes; a double diploid.

Tetrapoda Animals whose last common ancestor had four limbs; this group includes amphibians, lizards, turtles, crocodilians, birds, and mammals (some tetrapods, like snakes, have lost their legs in the course of evolution).

thalamus The inner brain region of the forebrain, which acts as a relay station for sensory information sent to the cerebrum.

thallus A flattened photosynthetic structure produced by some bryophytes.

Thaumarchaeota One of the three major divisions of Archaea; thaumarchaeota are chemotrophs, deriving energy from the oxidation of ammonia.

theory A general explanation of a natural phenomenon supported by a large body of experiments and observations.

theory of island biogeography A theory that states that the number of species that can occupy a habitat island depends on two factors: the size of the island and the distance of the island from a source of colonists.

thermoreceptor A sensory receptor in the skin and in specialized regions of the central nervous system that responds to heat and cold.

thick filament A parallel grouping of myosin molecules that makes up the myosin filament.

thin filament Two helically arranged actin filaments twisted together that make up the actin filament.

threshold potential The critical depolarization voltage of −50 mV required for an action potential.

thylakoid The internal membrane-bound compartment in the center of chloroplasts, consisting of the highly folded thylakoid membrane, which contains light-collecting pigments and is the site of the photosynthetic electron transport chain and the interior lumen.

thylakoid membrane A highly folded membrane in the center of the chloroplast that contains light-collecting pigments and that is the site of the photosynthetic electron transport chain.

thymine (T) A pyrimidine base.

thyroid gland A gland located in the front of the neck that leads to the release of two peptide hormones, thyroxine and triiodothyorine.

Ti plasmid A small circular DNA molecule in virulent strains of *R. radiobacter* containing genes that can be integrated into the host cell's genome, as well as the genes needed to make this transfer.

tidal ventilation A breathing technique in most land vertebrates in which air is drawn into the lungs during inhalation and moved out during exhalation.

tidal volume The amount of air inhaled and exhaled in a cycle; in humans, tidal volume is 0.5 liter.

tight junction A junctional complex that establishes a seal between cells so that the only way a substance can travel from one side of a sheet of epithelial cells to the other is by moving through the cells by a cellular transport mechanism.

tissue A collection of cells that work together to perform a specific function.

titin A large protein that links myosin filaments to Z discs at the ends of a sarcomere.

tolerance The ability of T cells not to respond to self antigens even though the immune system functions normally otherwise.

toll-like receptors (TLRs) A family of transmembrane receptors on phagocytes that recognize and bind to molecules on the surface of microorganisms, providing an early signal that an infection is present.

tongue A muscle in mammals and other land vertebrates that facilitates food manipulation and transport within the mouth.

topoisomerase Any one of a class of enzymes that regulates the supercoiling of DNA by cleaving one or both strands of the DNA double helix, and later repairing the break.

topoisomerase II An enzyme that breaks a DNA double helix, rotates the ends, and seals the break.

totipotent Describes cells that have the potential to give rise to a complete organism; a fertilized egg is a totipotent cell.

trabeculae Small plates and rods with spaces between them, found in spongy bone.

trace fossil A track or trail, such as a dinosaur track or the feeding trails of snails and trilobites, left by an animal as it moves about or burrows into sediments.

trachea The central airway leading to the lungs, supported by cartilage rings.

tracheae An internal system of tubes in insects that branch from openings along the abdominal surface into smaller airways, directing oxygen to and removing carbon dioxide from respiring tissues.

tracheid A unicellular xylem conduit.

trade-off An exchange in which something is gained at the expense of something lost.

trait A characteristic of an individual.

transcription The synthesis of RNA from a DNA template.

transcriptional activator protein A protein that binds to an enhancer to enable transcription to begin.

transcriptional regulation The mechanisms that collectively regulate whether or not transcription occurs.

transduction Horizontal gene transfer by means of viruses.

transfer RNA (tRNA) Noncoding RNA that carries individual amino acids for use in translation.

transformation The conversion of cells from one state to another, as from nonvirulent to virulent, when DNA released to the environment by cell breakdown is taken up by recipient cells. In recombinant DNA technology, the introduction of recombinant DNA into a recipient cell.

transgenic organisms An alternative term for genetically modified organisms.

transition state The brief time in a chemical reaction in which chemical bonds in the reactants are broken and new bonds in the product are formed.

translation Synthesis of a polypeptide chain corresponding to the coding sequence present in a molecule of messenger RNA.

transmembrane proteins Proteins that span the entire lipid bilayer; most integral membrane proteins are transmembrane proteins

transmission genetics The discipline that deals with the manner in which genetic differences among individuals are passed from generation to generation.

transpiration The loss of water vapor from leaves.

transporters Membrane proteins that move ions or other molecules across the cell membrane.

transposable element A DNA sequence that can replicate and move from one location to another in a DNA molecule.

transposition The movement of a transposable element.

tree of life The full set of evolutionary relationships among all organisms.

triacylglycerol A lipid that stores energy.

trimesters The three periods of pregnancy, each lasting about 3 months.

triose phosphate A 3-carbon carbohydrate molecule; triose phosphates are the true products of the Calvin cycle because they are the molecules exported from the chloroplast.

triploid A cell or organism with three complete sets of chromosomes.

trisomy 21 A condition resulting from the presence of three, rather than two, copies of chromosome 21; also known as *Down syndrome*.

trophic pyramid A diagram that traces the flow of energy through communities, showing the amount of energy available at each level to feed the next. The pyramid shape results because biomass and the energy it represents generally decrease from one trophic level to the next.

tropic hormone A hormone that controls the release of other hormones.

tropism The bending or turning of an organism in response to an external signal such as light or gravity.

tropomyosin A protein that runs in the grooves formed by the actin helices and blocks the myosin-binding sites.

troponin A protein that moves tropomyosin away from myosin-binding sites, allowing cross-bridges between actin and myosin to form and the muscle to contract.

true breeding Describes a trait whose physical appearance in each successive generation is identical to that in the previous one.

trypsin A digestive enzyme produced by the pancreas that breaks down proteins.

T-tubule system A system for the conduction of depolarization of a muscle cell through tubules formed from invaginations of the plasma membrane.

tube feet Small projections of the water vascular system that extend outward from the body surface and facilitate locomotion, sensory perception, food capture, and gas exchange in echinoderms.

tubulin Dimers (composed of an a tubulin and a b tubulin) that assemble into microfilaments.

tumor suppressors A family of genes that encode proteins whose normal activities inhibit cell division.

tunicates A subphylum of Chordata that includes about 3000 species of filter-feeding marine animals, such as sea squirts and salps.

turgor pressure The pressure exerted by water against an object, which provides structural support for many plants, fungi, and bacteria.

Turner syndrome A sex-chromosomal abnormality in which an individual has 45 chromosomes, including only one X chromosome.

twitch A muscle contraction that results from a single action potential.

twofold cost of sex Population size can increase more rapidly in asexually reproducing organisms than in sexually reproducing organisms because only female produce offspring, and females have only half the fitness of asexual parents.

tympanic membrane A thin sheet of tissue at the surface of the ear that vibrates in response to sound waves, amplifying airborne vibrations; in mammals, also known as the *eardrum*.

ulcer A break or erosion in the cells lining the stomach.

unbalanced translocation Tranlocation in which only part of a reciprocal translocation (and one of the nontranslocated chromosomes) is inherited from one of the parents.

unsaturated Describes fatty acids that contain carbon–carbon double bonds.

uracil (U) A pyrimidine base in RNA, where it replaces the thymine found in DNA.

urea A waste product of protein metabolism that many animals excrete.

ureter A large tube in the vertebrate kidney that brings urine from the kidneys to the bladder.

urethra A tube from the bladder that in males carries semen as well as urine from the body.

uterus A hollow organ with thick, muscular walls that is adapted to support the developing embryo if fertilization occurs and to deliver the baby during birth.

vaccination The deliberate delivery in a vaccine of an antigen from a pathogen to induce a primary response but not the disease, thereby providing future protection from infection.

vacuole A cell structure that absorbs water and contributes to turgor pressure.

vagina A tubular channel connecting the uterus to the exterior of the body; also known as the birth canal.

valence electrons The electrons farthest from the nucleus, which are at the highest energy level.

van der Waals forces The binding of temporarily polarized molecules because of the attraction of opposite charges.

variable (in experimentation) The feature of an experiment that is changed by the experimenter from one treatment to the next.

variable (V) region A region of the heavy (H) and light (L) chains of an antibody; a change in the variable region distinguishes a given antibody from all others.

variable expressivity The phenomenon in which a particular phenotype is expressed with a different degree of severity in different individuals.

variable number tandem repeat (VNTR) A genetic difference in which the number of short repeated sequences of DNA differs from one chromosome to the next.

vas deferens A long, muscular tube from the scrotum, through the abdominal cavity, along the bladder, and connecting with the ejaculatory duct.

vasa recta The blood vessels in the kidneys.

vascular bundle A bundle of xylem conduits and phloem; multiple vascular bundles are arranged in a ring near the outside of the stem.

vascular cambium Lateral meristem that is the source of new xylem and phloem.

vascular plant A photosynthetic land plant that can draw water from soil and limit water loss from their leaves.

vasoconstriction The process in the supply of blood to the limbs is reduced by constriction of the arteries that supply the limbs.

vasodilation The process in which resistance in the arteries is decreased and blood flow increased following relaxation of the smooth arterial muscles.

vasopressin See *antidiuretic hormone (ADH)*.

vector In recombinant DNA, a carrier of the donor fragment, usually a plasmid.

vegetative reproduction The production of upright shoots from horizontal stems, permitting new plants to be produced at a distance from the site where the parent plant originally germinated.

veins In plants, the system of vascular conduits that connects the leaf to the rest of the plant; in animals, the large, low-pressure vessels that return blood to the heart.

vena cava (plural **venae cavae**) One of the two largest veins in the body, which drain blood from the head and body into the heart.

ventilation The movement of an animal's respiratory medium—water or air—past a specialized respiratory surface.

ventricle A heart chamber that pumps blood to the lungs or the rest of the body.

venule A blood vessel into which capillaries drain as blood is returned to the heart.

vernalization A prolonged period of exposure to cold temperatures necessary to induce flowering.

vertebrae The series of hard segments making up the jointed skeleton that runs along the main axis of the body in vertebrates.

vertebral column A structure in vertebrates that functionally replaces the embryonic notochord.

vesicle A small membrane-enclosed sac that transports substances within the cell.

vessel A multicellular xylem conduit.

vessel element An individual cell that is part of a vessel for water transport in plants; in contrast to fibers, vessel elements can be quite wide.

vestibular system A system in the mammalian inner ear made up of two statocyst chambers and three semicircular canals.

vestigial structure A structure that has lost its original function over time and is much reduced in size.

vicariance The process in which a geographic barrier arises within a single population, separating it into two or more isolated populations.

villi Highly folded inner surfaces of the jejunum and ileum of the small intestine.

virulent Describes pathogens that are able to overcome the host plant's defenses and lead to disease.

virus A small infectious agent that contains a nucleic acid genome packaged inside a protein coat called a capsid.

viscosity The stickiness of a fluid.

visible light The portion of the electromagnetic spectrum apparent to our eyes.

visual cortex The part of the brain that processes images.

vitamin An organic molecule that is required in very small amounts in the diet.

vitreous humor A gel-like substance filling the large cavity behind the lens that makes up most of the volume of the vertebrate eye.

viviparity Giving birth to live young.

vocal cords Twin organs in the larynx that vibrate as air passes over them, enabling speech, song, and sound production.

voluntary Describes the component of the nervous system that handles sensing and responding to external stimuli.

vulva The external genitalia of the female.

water vascular system A series of fluid-filled canals that permit bulk transport of oxygen and nutrients in echinoderms.

white blood cell (leukocyte) A type of cell in the immune system that arises by differentiation from stem cells in the bone marrow.

white matter Collectively, the axons of cortical neurons in the interior of the brain; it is the fatty myelin produced by glial cells surrounding the axons that makes this region of the brain white.

wild type The most common allele, genotype, or phenotype present in a population; nonmutant.

X chromosome One of the two sex chromosomes; a normal human female has two copies of the X chromosome, a normal male has one X and one Y chromosome.

xanthophyll Any one of several yellow-orange pigments that slow the formation of reactive oxygen species by reducing excess light energy; these pigments accept absorbed light energy directly from chlorophyll and convert this energy to heat.

X-inactivation The process in mammals in which dosage compensation occurs through the inactivation of one X chromosome in each cell in females.

X-linked gene A gene in the X chromosome.

xylem The inner tissue of a stem, which transports water and nutrients from the roots to the leaves.

Y chromosome One of the two sex chromosomes; a normal human male has one X and one Y chromosome.

yeast A single-celled fungus found in moist, nutrient-rich environments.

Y-linked gene A gene that is present in the region of the Y chromosome that shares no homology with the X chromosome.

yolk A substance in the eggs of animals with external fertilization that provides all the nutrients that the developing embryo needs until it hatches.

Z disc A protein backbone found regularly spaced along the length of a myofibril.

Z scheme Another name for the photosynthetic electron transport chain, so called because the overall energy trajectory resembles a "Z."

zygomycetes Fungi groups that produce hyphae undivided by septa and do not form multicellular fruiting bodies; they make up less than 1% of known fungal diversity.

zygote The diploid fertilized egg cell formed by the fusion of two haploid gametes.

CREDITS/SOURCES

Chapter 21

Opening photo Roc Canals Photography/Getty Images. *Fig. 21.1* Tim Davis/Corbis. *Fig. 21.3 (L)* Biopix.dk: G. Drange, *(R)* Gerry Ellis/Getty Images. *Fig. 21.6* akg-images. *Fig. 21.7* A. R. Wallace Memorial Fund & G. W. Beccaloni. *Fig. 21.8 (a)* Dr. Tony Brain/Photo Researchers, Inc., *(b)* Omikron/Photo Researchers, Inc. *Fig. 21.10* Data from L. L. Cavalli-Sforza and W. F. Bodmer, 1971, *The Genetics of Human Populations*, San Francisco: W. H Freeman, p. 613. *Fig. 21.14* After Fig. 20-3, p. 733, in A. J. F. Griffiths, S. R. Wessler, S. B. Carroll, and J. Doebley, 2012, *Introduction to Genetic Analysis*, 10th ed., New York: W. H. Freeman.

Chapter 22

Opening photo Tui De Roy/Getty Images. *Fig. 22.1* Museum of Comparative Zoology, Harvard University. *Fig. 22.2* Louise Morgan/Flickr/Getty Images, mike smith/age fotostock. *Fig. 22.3* With kind permission from Springer Science+Business Media. V. A. Lukhtanov and A. V. Dantchenko, 2002, "Principles of the Highly Ordered Arrangement of Metaphase I Bivalents in Spermatocytes of Agrodiaetus (Insecta, Lepidoptera)," *Chromosome Research* 10(1):5–20, Fig. 1, courtesy of Vladimir Lukhtanov. *Fig. 22.4* Courtesy of Darren E. Irwin. *Fig. 22.5* Haruo Katakura, 1997, "Species of Epilachna Ladybird Beetles," *Zoological Science* 14(6):869–881, Fig. 4, courtesy of Dr. Haruo Katakura. *Fig. 22.7* Dr. Arthur Anker, National University of Singapore, Singapore. *Fig. 22.8* After D. J. Futuyma, 2009, *Evolution*, 2nd ed., Sunderland, MA: Sinauer Associates, Fig. 18.7, p. 484. *Photo*: C. H. Greenewalt/VIREO. *Fig. 22.9* Phylogenetic tree adapted from Fig. 2.1, p.15, in P. R. Grant and B. R. Grant, 2008, *How and Why Species Multiply*, Princeton, NJ: Princeton University Press. Islands/finch species plot data from Fig. 2.4, p. 23, in P. R. Grant and B. R. Grant, 2008, *How and Why Species Multiply*, Princeton, NJ: Princeton University Press. *Fig. 22.10* Data from J. P. Huelsenbeck and B. Rannala, 1997, "Phylogenetic Methods Come of Age: Testing Hypotheses in an Evolutionary Context," *Science* 276:230, doi: 10.1126/science.276.5310.227. *Photos: (L)* Photo by Alex Popinga, courtesy of James Demastes, *(R)* Richard Ditch @ richditch.com. *Fig. 22.13 (TL)* Gary A. Monroe @ USDA-NRCS PLANTS Database, *(TR)* Jason Rick, *(B)* Gerald J. Seiler, USDA-ARS. *Fig. 22.14* After Fig. 20-18, p. 752, in A. J. F. Griffiths, S. R. Wessler, S. B. Carroll, and J. Doebley, 2012, *Introduction to Genetic Analysis*, 10th ed., New York: W. H. Freeman.

Chapter 23

Opening photo: Frans Lanting/www.lanting.com. *Fig. 23.11* Journal *Nature*/Xing Xu/Getty Images. *Fig. 23.12* National Park Service. *Fig. 23.13* José Antonio Hernaiz/age fotostock. *Fig. 23.14* Courtesy of Smithsonian Institution. *Photo* by Chip Clark. *Fig. 23.15* Senckenberg, Messel Research Department, Frankfurt a. M. (Germany). *Fig. 23.16 (T, L–R)* Scott Orr/iStockphoto, Hans Steur, T. Daeschler/VIREO, Illustration by Mark A. Klingler/Carnegie Museum of Natural History, Louie Psihoyos/Corbis, DEA/G. Cigolini/Getty Images, *(B, L–R)* Eye of Science/Photo Researchers, Inc.; Andrew Knoll, Harvard University; Antonio Guillén, Proyecto Agua, Spain; Andrew Knoll, Harvard University. *Fig. 23.18* Ron Blakey and Colorado Plateau Geosystems, Inc. *Fig. 23.19* Data from C. Zimmer, 2009, *The Tangled Bank*, Greenwood Village, CO: Roberts and Company. *Photo*: Jason Edwards/Getty Images. *Fig. 23.20* Shubin Lab/University of Chicago. *Fig. 23.21* Data from Sepkoski's Online Genus Database. http://strata.geology.wisc.edu/jack/. A part of J. J. Sepkoski, Jr., 2002, "A Compendium of Fossil Marine Animal Genera," David Jablonski and Michael Foote (eds.), *Bulletin of American Paleontology* 363:1–560.

Chapter 24

Opening photo Frans Lanting/www.lanting.com. *Fig. 24.1 (L–R)* George Holton/Photo Researchers, Inc., Kevin Schafer/kevinschafer.com, Daily Mail/Rex/Alamy, Mitsuaki Iwago/Minden Pictures, Yellow Dog Productions/Getty Images. *Fig. 24.2 (L–R)* Cyril Ruoso/Minden Pictures, Anup Shah/Minden Pictures, J & C Sohns/age fotostock, Michael Dick/Animals Animals–Earth Scenes, Anup Shah/Animals Animals–Earth Scenes, Yellow Dog Productions /Getty Images. *Fig. 24.4* HO/AFP/Newscom. *Fig. 24.5* The Natural History Museum, London/The Image Works. *Fig. 24.6* Data from R. G. Klein, 2009, *The Human Career*, Chicago: University of Chicago Press, p. 244; *Homo floresiensis* skull from R. D. Martin, A. M. MacLarnon, J. L. Phillips, L. Dussubieux, P. R. Williams and W. B. Dobyns, 2006, Comment on "The Brain of LB1, *Homo floresiensis*," *Science* 312:999. *Fig. 24.7* Data from T. Deacon, "The Human Brain," pp. 115–123, in J. Jones, R. Martin, and D. Pilbeam (eds.), 1992, *The Cambridge Encyclopedia of Human Evolution*, Cambridge, England: Cambridge University Press. *Fig. 24.9* Javier Trueba/MSF/Photo Researchers, Inc. *Fig. 24.12 (L)* Gerry Ellis/Minden Pictures, *(R)* ZSSD/Minden Pictures. *Fig. 24.13* Adapted from Fig. 2.4, p. 44, in H. J. Jerison, 1973, *Evolution of the Brain and Intelligence*, New York: Academic Press. *Fig. 24.15* After Fig. 6.18, p. 149, in D. J. Futuyma, 2009, *Evolution*, 2nd ed., Sunderland, MA: Sinauer Associates. *Fig. 24.16 (a)* Ryan Heffernan/Aurora/Getty Images, *(b)* B&C Alexander/ArcticPhoto. *Fig. 24.17* B&C Alexander/ArcticPhoto. *Fig. 24.18 (a)* Colin F. Sargent, *(b)* Dr. Alex Thornton, University of Cambridge, *(c)* Suzi Eszterhas/Minden Pictures. *Fig. 24.19* Frans Lanting/www.lanting.com.

Part 2 Daniela Dirscherl/Getty Images

Chapter 25

Opening photo: Frans Lanting/www.lanting.com. *Fig. 25.1* Data from R. A. Rohde, "Mauna Loa Carbon Dioxide," from Global Warming Art, PNG image, http://www.globalwarmingart.com/wiki/File:Mauna_Loa_Carbon_Dioxide_png. Based on data from P. Tans and R. Keeling, "Trends in Atmospheric Carbon Dioxide," NOAA/ESRL, ftp://ftp.cmdl.noaa.gov/ccg/co2/trends/co2_mm_mlo.txt. *Fig. 25.3 (L)* Courtesy of Ted Scambos & Rob Bauer, National Snow and Ice Data Center, University of Colorado, Boulder, *(R)* Vin Morgan. *Fig. 25.5* Data from J. G. Canadell, C. Le Quéré, M. R. Raupach, C. B. Field, E. T. Buitenhuis, P. Ciais, T. J. Conway, N. P. Gillett, R. A. Houghton, and G. Marland, 2007, "Contributions to Accelerating Atmospheric CO_2 Growth from Economic Activity, Carbon Intensity, and Efficiency of Natural Sinks," *Proceedings of the National Academy of Sciences USA* 104:18866–18870. *Fig. 25.6* Data from P. Rekacewicz, "The Carbon Cycle," from Vital Climate Graphics, UNEP/GRID–Arendal Maps & Graphics Library, JPG file, 2005, http://maps.grida.no/go/graphic/the_carbon_cycle. Adapted from D. Schimel et al., "Radiative Forcing of Climate Change," in J. T. Houghton et al. (eds.), 1996, *Climate Change 1995: The Science of Climate* Change (Cambridge, England: Cambridge University Press), p. 77. *Photo*: Fotosearch Stock Images. *Fig. 25.7* Dr. David Wachenfeld/AUSCAPE/Minden Pictures. *Fig. 25.8* Steve Gschmeissner/Photo Researchers, Inc. *Fig. 25.9* Data from J. Gaillardet and A. Galy, 2008, "Himalaya—Carbon Sink or Source?" *Science* 320:1728, doi: 10.1126/science.1159279. *Fig. 25.10* Data from P. Rekacewicz, "Temperature and CO_2 Concentration in the Atmosphere over the Past 400,000 Years," from Vital Climate Graphics, UNEP/GRID–Arendal Maps & Graphics Library, JPG file, 2005, http://www.grida.no/publications/vg/climate/page/3057.aspx. Based on J. R. Petit et al., 1999, "Climate and Atmospheric History of the Past 420,000 Years from the Vostok Ice Core, Antarctica," *Nature* 399:429–436. *Fig. 25.11* Plate tectonic maps and continental drift animations by

CS-1

C. R. Scotese, PALEOMAP Project (www.scotese.com). *Fig. 25.12* After Fig 3.4 in K. Wallmann and G. Aloisi, 2012, "The Global Carbon Cycle: Geological Processes," in *Fundamentals of Geobiology,* A. H. Knoll, D. E. Canfield, and K. O. Konhauser (eds.), Chichester, England: Wiley-Blackwell, pp. 20–35. *Fig. 25.15* Albert Aanensen/Nature Picture Library.

Case 5

Seated woman Mimi Haddon/Getty Images. *Pie charts* Data from I. Chung and M. J. Blaser, 2012, "The human microbiome: at the interface of health and disease." *Nature Reviews Genetics* 13:260–270, doi: 10.1038/nrg3182. *Bobtail squid* Gary Bell/OceanwideImages.com.

Chapter 26

Opening photo Steve Gschmeissner/SPL/Getty Images. *Fig. 26.2 (a)* Eye of Science/Photo Researchers, Inc., *(b)* Dennis Kunkel Microscopy, Inc./Phototake, *(c)* Courtesy Mike Dyall-Smith, *(d)* David Scharf/Photo Researchers, Inc., *(e)* ©≈ANIMA RES. *Fig. 26.3* Heide Schulz-Vogt, MPI Bremen. *Fig. 26.4* Dennis Kunkel Microscopy, Inc./Phototake. *Fig. 26.6* lecates via Flickr. *Fig. 26.7 (a)* Andrew Knoll, Harvard University, *(b)* With kind permission from Springer Science+Business Media. Jörg Overmann, 2006, *The Prokaryotes,* Springer eBook, Fig. 7. *Photo* courtesy of Jörg Overmann. *Fig. 26.9* Woods Hole Oceanographic Institution/Visuals Unlimited, Inc. *Fig. 26.12* Wally Eberhart/Visuals Unlimited, Inc. *Fig. 26.16 (a)* Dr. Ralf Wagner, *(b)* Wim van Egmond/Visuals Unlimited, Inc., *(c)* Tom Adams/Visuals Unlimited, Inc., *(d)* Jason Oyadomari. *Fig. 26.14* Adapted from W. F. Doolittle, 2000, "Uprooting the Tree of Life," *Scientific American* 282:90–95. *Fig. 26.17* Adapted from M. Groussin and M. Gouy, 2011, "Adaptation to Environmental Temperature Is a Major Determinant of Molecular Evolutionary Rates in Archaea," *Molecular Biology and Evolution* 28(9):2667, doi: 10.1093/molbev/msr098. © The Author 2011. Published by Oxford University Press on behalf of the Society for Molecular Biology and Evolution. *Fig. 26.18* Andrew Knoll, Harvard University. *Fig. 26.20 (a–c)* Andrew Knoll, Harvard University. *Fig. 26.21 (T)* Andrew Knoll, Harvard University, *(B)* François Gohier/Photo Researchers, Inc. *Fig. 26.22* Image by Zoe Veneti and Kostas Bourtzis. *Fig. 26.23* Data from J. Peterson et al., 2009, "The NIH Human Microbiome Project," Genome Research 19:2317–2323; originally published online October 9, 2009. *Fig. 26.24* Data from C. De Filippo, D. Cavalieri, M. Di Paola, M. Ramazzotti, J. B. Poullet, S. M., S. Collini, G. Pieraccini, and P. Lionetti, 2010, "Impact of Diet in Shaping Gut Microbiota Revealed by a Comparative Study in Children from Europe and Rural Africa," *Proceedings of the National Academy of Sciences USA* 107(33):14693, doi: 10.1073/pnas.1005963107.

Chapter 27

Opening photo Biophoto Associates/Photo Researchers, Inc. *Fig. 27.2* V. Mercanti, S. J. Charette, N. Bennett, J. J. Ryckewaert, F. Letourneur, and P. Cosson, 2006, "Selective Membrane Exclusion in Phagocytic and Macropinocytic Cups, *Journal of Cell Science* 119:4079–4067. *Fig. 27.4 (a, T–B)* Dr. Kari Lounatmaa/Photo Researchers, Inc., Dennis Kunkel Microscopy, Inc., Dr. Jeremy Burgess/Photo Researchers, Inc. *Fig. 27.5* Courtesy of Hwan Su Yoon. *Fig. 27.6* Reprinted by permission from Macmillan Publishers Ltd: B. Boxma, et al., 2005, "An Anaerobic Mitochondrion That Produces Hydrogen, *Nature* 434:74–79. Copyright 2005. *Fig. 27.8* Reprinted by permission from Macmillan Publishers Ltd: V. P. Edgcomb, S. A. Breglia, N. Yubuki, D. Beaudoin, D. J. Patterson, B. S. Leander, and J. M. Bernhard, 2011, "Identity of Epibiotic Bacteria on Symbiontid Euglenozoans in O_2-Depleted Marine Sediments: Evidence for Symbiont and Host Co-evolution, *The ISME Journal* 5:231–243. Copyright 2010. Courtesy of Naoji Yubuki. *Fig. 27.10* Nicole King, Steve Paddock and Sean Carroll, HHMI, University of Wisconsin. *Fig. 27.11 (b)* Wim van Egmond/Visuals Unlimited, Inc., *(c)* Carolina Biological Supply Co./Visuals Unlimited, Inc. *Fig. 27.12 (a)* Matt Meadows/Photo Researchers, Inc., *(b)* Masana Izawa/Minden Pictures. *Fig. 27.13* Carolina Biological Supply Co./Visuals Unlimited, Inc. *Fig. 27.14 (b)* Jerome Pickett-Heaps/Photo Researchers, Inc., *(c)* Wim van Egmond/Visuals Unlimited, Inc., *(d)* David Patterson and Bob Andersen, image used under license to MBL (micro*scope). *Fig. 27.15 (a)* Dr. Peter Siver/Visuals Unlimited, Inc./Corbis, *(b)* Wim van Egmond/Visuals Unlimited, Inc., *(c)* Gerd Guenther/Photo Researchers, Inc., *(d)* Frank Fox/Photo Researchers, Inc., *(e)* Linda Sims/Visuals Unlimited, Inc. *Fig. 27.16* D. J. Patterson, image used under license to MBL (micro*scope). *Fig. 27.17* Mark Conlin/Alamy. *Fig. 27.18 (a & b)* Steve Gschmeissner/Photo Researchers, Inc., *(c)* Andrew Syred/Photo Researchers, Inc., *(d)* Wim van Egmond/Visuals Unlimited, Inc. *Fig. 27.19 (a)* Electron Microscopy Unit, Australian Antarctic Division © Commonwealth of Australia, *(b)* Andrew Syred/Photo Researchers, Inc., *(c)* Gert Hansen–SCCAP, University of Copenhagen. *Fig. 27.23 (a)* Andrew Knoll, Harvard University, *(b)* Nicholas J. Butterfield, University of Cambridge, *(c)* Susannah Porter, University of California, Santa Barbara. *Fig. 27.24 (a)* Astrid & Hanns-Frieder Michler/Photo Researchers, Inc., *(b)* Visuals Unlimited, Inc./Corbis. *Fig. 27.25 (a)* Copyright The Natural History Museum, London, *(b)* Provided by the SeaWiFS Project, NASA/Goddard Space Flight Center, and ORBIMAGE.

Chapter 28

Opening photo irawansubingarphotography/Getty Images. *Fig. 28.1 (a)* Jason Oyadomari, *(b)* D. J. Patterson and Aimlee Ladermann, image used under license to MBL (micro*scope), *(c)* Fabio Rindi. *Fig. 28.2 (a)* Marevision/age fotostock, *(b)* Wolfgang Poelzer/WaterFrame/age fotostock. *Fig. 28.3 (a)* Ingo Arndt/Minden Pictures, *(b)* Richard Herrmann/Visuals Unlimited, Inc., *(c)* Filmfoto/Dreamstime.com, *(d)* Jack Milchanowski/age fotostock, *(e)* David M. Cobb/Visuals Unlimited, Inc. *Fig. 28.5 (a)* Andrew J. Martinez/SeaPics.com, *(b)* D. R. Schrichte/SeaPics.com. *Fig. 28.7* Reprinted from M. J. Dayel, R. Alegado, S. Fairclough, T. Levin, S. A. Nichols, K. L. McDonald, and N. King, 2011, "Cell Differentiation and Morphgenesis in the Colony-Forming Choanoflagellate *Salpingoeca rosetta,*" *Developmental Biology* 357:73–82. Copyright 2011, with permission from Elsevier. *Fig. 28.9* Biophoto Associates/Photo Researchers, Inc. *Fig. 28.10* Biodisc/Visuals Unlimited, Inc. *Fig. 28.13 (a)* Guy Narbonne, *(b)* Mikhahil Fedonkin, *(c)* Andrew Knoll, Harvard University. *Fig. 28.15 (a)* Reprinted from F. Hueber, 2001, "Rotted Wood–Alga–Fungus: The History and Life of Prototaxites Dawson 1959," *Review of Paleobotany and Palynology* 116:123–158, p. 146, Smithsonian Institution, copyright 2001, with permission from Elsevier, *(b)* L. E. Graham, M. E. Cook, D. T. Hanson, K. B. Pigg, and J. M. Graham, 2010, "Structural, Physiological, and Stable Carbon Isotopic Evidence That the Enigmatic Paleozoic Fossil Prototaxites Formed from Rolled Liverwort Mats," *American Journal of Botany* 97(2):268–275, Botanical Society of America; Brooklyn Botanic Garden, copyright 2010. Reproduced with permission of Botanical Society of America, Inc., via Copyright Clearance Center. Image courtesy of Martha E. Cook. *Fig. 28.16* Reprinted by permission from Macmillan Publishers Ltd: P. M. Brakefield, J. Gates, D. Keyes, F. Kesbeke, P. J. Wijngaarden, A. Monteiro, V. French, and S. B. Carroll, 1996, "Development, Plasticity and Evolution of Butterfly Eyespot Patterns," *Nature* 384:236–242. Copyright 1996.

Case 6

Wild wheat James King-Holmes/Photo Researchers, Inc. *Domesticated wheat* Oliver Brandt/Getty Images. *Comparison of wild wheat and domesticated wheat* Biophoto Associates/Photo Researchers, Inc.

Chapter 29

Opening photo Frank Krahmer/Getty Images. *Fig. 29.2* Brent Mishler and Michael Hamilton. *Fig. 29.5* Dr. Jeremy Burgess/Photo Researchers, Inc. *Fig. 29.7* Clément Philippe/age fotostock. *Fig. 29.8* Carolina Biological Supply

Co./Visuals Unlimited, Inc. *Fig. 29.10* Carolina Biological Supply Co./Visuals Unlimited, Inc. *Fig. 29.17 (a)* Topic Photo Agency/age fotostock, *(b)* Photoshot/SuperStock. *Fig. 29.20* Inga Spence/Photo Researchers, Inc.

Chapter 30

Opening photo Frans Lanting/www.lanting.com. *Fig. 30.2 (L)* Dr. Ralf Wagner. email<mikroskopie@dr-ralf-wagner.de>/http://www.dr-ralf-wagner.de, *(R)* Show_ryu/http://commons.wikimedia.org/wiki/File:Chara_braunii_1.JPG. *Fig. 30.4* Malcolm Storey, 2010, www.bioimages.org.uk. *Fig. 30.5 (a)* Keith Burdett/age fotostock, *(b)* Malcolm Storey, 2010, www.bioimages.org.uk. *Fig. 30.6 (L)* Biophoto Associates/Photo Researchers, Inc., *(R)* Masana Izawa/Nature Production/Minden Pictures. *Fig. 30.7* Charles O. Cecil/Alamy. *Fig. 30.9* Jim French. *Fig. 30.10 (T)* Perennou Nuridsany/Photo Researchers, Inc., *(B)* Kevin O´Hara/age fotostock. *Fig. 30.11 (a)* Jim Brandenburg/Minden Pictures, *(b)* J S Sira/Gap Photo/Visuals Unlimited, Inc., *(c)* Rob Whitworth/Gap Photo/Visuals Unlimited, Inc., *(d)* Thomas Marent/Visuals Unlimited, Inc. *Fig. 30.12* Cora Niele/Getty Images. *Fig. 30.13* K. Chae and E. M. Lord, 2011, "Pollen Tube Growth and Guidance: Roles of Small, Secreted Proteins," *Annals of Botany* 108(4):627–636. By permission of Oxford University Press. *Fig. 30.14 (a & b)* Michael & Patricia Fogden/Minden Pictures, *(c)* Charles Melton/Visuals Unlimited, Inc., *(d)* Ch'ien Lee/Minden Pictures. *Fig. 30.15* Hans Christoph Kappel/NPL/Minden Pictures. *Fig. 30.16 (T)* Paul Oomen/Getty Images, *(M)* Kevin Schafer/Getty Images, *(B)* Daniela Duncan/Getty Images. *Fig. 30.18* Greg Lawler/Alamy. *Fig. 30.19* Dwight Kuhn. *Fig. 30.20 (a)* Don Johnston/age fotostock, *(b)* David Norton/Alamy, *(c)* Don Johnston/age fotostock, *(d)* David Cavagnaro/Visuals Unlimited, Inc. *Fig. 30.21* F. Valverde, A. Mouradov, W. Soppe, D. Ravenscroft, A. Samach, and G. Coupland, 2004, "Photoreceptor Regulation of CONSTANS Protein in Photoperiodic Flowering," *Science* 303:1003–1006. Reprinted with permission from AAAS. *Fig. 30.24* Scott Smith/Corbis.

Chapter 31

Opening photo Dietrich Rose/Getty Images. *Fig. 31.1 (a)* Elena Elisseeva/Dreamstime.com, *(b)* LianeM/Alamy. *Fig. 31.2 (a, L–R)* Ian Gowland/Photo Researchers, Inc., M. I. Walker/Photo Researchers, Inc., *(b)* Siobhan Braybrook. *Fig. 31.3* M. I. Walker/Photo Researchers, Inc. *Fig. 31.4 (L–R)* Jonathan Buckley/Gap Photo/Visuals Unlimited, Inc., Jerome Wexler/Visuals Unlimited, Inc., Christian Hütter/imagebroker/age fotostock. *Fig. 31.5* Topham/The Image Works. *Fig. 31.6 (a)* Dr. Keith Wheeler/Photo Researchers, Inc. *(b)* Marga Werner/age fotostock, *(c)* J Garden/Alamy, *(d)* Adisa/Dreamstime.com. *Fig. 31.7* Jacqueline S. Wong/ViewFinder Exis, LLC. *Fig. 31.10* Steve Gschmeissner/Photo Researchers, Inc. *Fig. 31.11* Nigel Cattlin/Visuals Unlimited, Inc. *Fig. 31.13* Wally Eberhart/Visuals Unlimited, Inc. *Fig. 31.14 (a)* Alan Majchrowicz/age fotostock, *(b)* Wally Eberhart/Visuals Unlimited, Inc. *Fig. 31.15* Robert & Jean Pollock/Visuals Unlimited, Inc., Scientifica/Visuals Unlimited, Inc. *Fig. 31.16 (a)* Scanning electron micrograph courtesy of the N. Brown Center for Ultrastructure Studies, State University of New York College of Environmental Science and Forestry, Syracuse, NY, *(b)* Andrew Syred/Photo Researchers, Inc. *Fig. 31.17* Carolina Biological Supply Co./Visuals Unlimited, Inc./Corbis. *Fig. 31.18* Science Source/Photo Researchers, Inc. *Fig. 31.19 (a)* Jack Dermid/Visuals Unlimited, Inc., *(b)* A Jagel/age fotostock, *(c)* Jane Grushow/Grant Heilman Photography, *(d)* Patrick Lynch/Alamy. *Fig. 31.22* C. Jourdan, N. Michaux-Ferrière, and G. Perbal, 2000, "Root System Architecture and Gravitropism in the Oil Palm," *Annals of Botany* 85(6):861–868, Fig. 3E. By permission of Oxford University Press. Courtesy of Dr. Christophe Jordan. *Fig. 31.23* G. R. "Dick" Roberts/NSIL/Visuals Unlimited, Inc.

Chapter 32

Opening photo Don Johnston/Getty Images. *Fig. 32.1* Nigel Cattlin/Visuals Unlimited, Inc. *Fig. 32. 2* Clouds Hill Imaging Ltd/Photo Researchers, Inc. *Fig. 32.3 (a)* Clive Varlack/clivevbugs/AGPix, *(b)* Paul Kennedy/Gettty Images, *(c)* Mark Boulton/NHPA. *Fig. 32.5* Norm Thomas/Photo Researchers, Inc. *Fig. 32.6 (a)* Guy Blomme/Bioversity International, *(b)* Bruce Fleming/Cephas Picture Library. *Fig. 32.11 (a)* Fletcher & Baylis/Photo Researchers, Inc., *(b)* Dr. Keith Wheeler/Photo Researchers, Inc., *(c)* Dr. Stanley Flegler/Visuals Unlimited, Inc. *Fig. 32.12* Alex Wild/alexanderwild.com. *Fig. 32.13* Dan L. Perlman/EcoLibrary.org. *Fig. 32.15 (a)* Danny Kessler, Max Planck Institute for Chemical Ecology, Jena, Germany *(b)* Celia Diezel, Max Planck Institute for Chemical Ecology, Jena, Germany. *Fig. 32.17* Data from P. V. A. Fine, I. Mesones, and P. D. Coley, 2004, "Herbivores Promote Habitat Specialization by Trees in Amazonian Forests," *Science* 305:663. *Photos*: Paul Fine. *Fig. 32.19* John L. Obermeyer, IPM Specialist, Department of Entomology, Purdue University.

Chapter 33

Opening photo Frans Lanting/www.lanting.com. *Fig. 33.2* Adapted from A. H. Knoll and K. J. Niklas, 1987, "Adaptation, Plant Evolution, and the Fossil Record," *Review of Palaeobotany and Palynology*, 50:127–149. *Fig. 33.3 (a)* Laurie Campbell /NHPA, *(b)* Densey Clyne/Minden Pictures, Inc., *(c)* Daniel Vega/age fotostock. *Fig. 33.4 (a)* Biopix: A Neumann. *(b)* Piotr Naskrecki/Minden Pictures. *Fig. 33.5 (T)* Science Photo Library/Alamy, *(M)* All Canada Photos/Alamy, *(B)* Daniel Vega/age fotostock. *Fig. 33.6 (T)* Courtesy of Hans Kerp © Palaeobotanical Research Group, University of Münster, *(M & B)* Hans Steur. *Fig. 33.7 (T)* Photo by William Chaloner, Crown Copyright, *(B)* Photo by John Hall © Botanical Society of America. *Fig. 33.8 (a)* Philippe Clement/Nature Picture Library, *(b)* Thomas Marent/Visuals Unlimited, Inc., *(c)* Biopix: JC Schou. *Fig. 33.9 (a)* John Gurche, *(b)* Biophoto Associates/Photo Researchers, Inc., *(c)* Tom Phillips, University of Illinois. *Fig. 33.10 (T & B)* Andrew Knoll, Harvard University. *Fig. 33.11 (a)* Ron Watts/First Light/Corbis, *(b)* David Sieren/Visuals Unlimited, Inc., *(c)* Jonathan Buckley/age fotostock, *(d)* Courtesy of Gary Higgins, *(e)* Doug Sokell/Visuals Unlimited, Inc., *(f)* Biophoto Associates/Photo Researchers, Inc. *Fig. 33.12* Jean-Paul Ferrero/Auscape/Minden Pictures. *Fig. 33.13 (a)* John Shaw/Auscape/Minden Pictures, *(b)* Gerry Bishop/Visuals Unlimited, Inc. *Fig. 33.14* Clay Perry/Corbis. *Fig. 33.15 (T–B)* Pete Ryan/Getty Images, Michael P. Gadomski/Photo Researchers, Inc., Gerald & Buff Corsi/Visuals Unlimited, Inc., Robert Davis/age fotostock. *Fig. 33.16 (a)* Piotr Naskrecki/Minden Pictures, *(b & c)* Andrew Knoll, Harvard University. *Fig. 33.17 (a)* JTB/Photoshot, *(b)* Kathy Merrifield/Photo Researchers, Inc. *Fig. 33.18* Andrew Knoll, Harvard University. *Fig. 33.19 (a)* M. Philip Kahl/Photo Researchers, Inc., *(b)* Ahmad Fuad Morad [Flickr page] 072112, *(c)* Bob Gibbons/Photo Researchers, Inc. *Fig. 33.20 (T–B)* Sangtae Kim, Frans Lanting/National Geographic Stock, Rob & Ann Simpson/Visuals Unlimited, Inc., Jo Whitworth/age fotostock, Tropicals JR Mau/PhotoResourceHawaii.com, Image Source/Getty Images. *Fig. 33.21* Keith Rushforth/FLPA/Minden Pictures. *Fig. 33.22 (a)* Robert Gustafson/Visuals Unlimited, Inc., *(b)* Kerstin Hinze/naturepl.com/NaturePL, *(c)* Axle71/Dreamstime.com, *(d)* Tropicals JR Mau/PhotoResourceHawaii.com. *Fig. 33.24 (a)* Michael P Gadomski/Getty Images, *(b)* Sergio Hayashi/Dreamstime.com, *(c)* Smileyjoanne/Dreamstime.com, *(d)* Danita Delimont/Getty Images.

Chapter 34

Opening photo Chris Ted/Getty Images. *Fig. 34.1* Dennis Kunkel Microscopy, Inc./Visuals Unlimited, Inc. *Fig. 34.3* Simko/Visuals Unlimited, Inc. *Fig. 34.4 (a)* Science Photo Library/Alamy, *(b)* T. R. Filley, R. A. Blanchette, E. Simpson, and M. L. Fogel, 20001, "Nitrogen Cycling by Wood Decomposing Soft-Rot Fungi in the 'King Midas tomb,' Gordion, Turkey," *Proceedings of the National Academy of Sciences USA* 98(23):13346–13350. Copyright 2001 National Academy of Sciences, U.S.A. *Photos* courtesy of Robert A. Blanchette, University of Minnesota. *Fig. 34.5 (a)* Science Source/Photo Researchers, Inc., *(b)* Biophoto Associates/Photo Researchers, Inc. *Fig. 34.6* Reprinted from J. K. Pirri and M. J. Alkema, 2011, "The Neuroethology of *C. elegans* Escape," *Current Opinion in Neurobiology* 22:187–193, Fig. 3. Copyright 2011, with permission from Elsevier. Reproduced with permission from Mark Alkema. *Fig. 34.7* Dr. Jeremy Burgess/Photo Researchers, Inc. *Fig. 34.8 (L–R)* Wallace Garrison/Getty Images, Stephen Sharnoff/National Geographic Stock, Stephen Sharnoff/National Geographic Stock. *Fig. 34.9* Eye of Science/Photo Researchers, Inc. *Fig. 34.10 (a)* Satoshi Kuribayashi/Minden Pictures, *(b, T)* E. R. Degginger/Photo Researchers, Inc., *(b, B)* Garry DeLong/Getty Images. *Fig. 34.11* Matt Meadow/Getty Images. *Fig. 34.12 (L)* Reprinted from R. Dulymamode, P. F. Cannon, and A. Peerally, 1998, "Fungi from Mauritius *Anthostomella* Species on *Pandanus*," *Mycological Research* 102(11): 1319–1324. Copyright 1998, with permission from Elsevier, *(M)* Reprinted from N. L. Glass, R. L. Metzenberg, and N. B. Raju, 1990, "Homothallic Sordariaceae from Nature: The Absence of Strains Containing Only the a Mating Type Sequence," *Experimenal Mycology* 14(3):274–289. Copyright 1990, with permission from Elsevier, *(R)* I. Schmitt, H. T. Lumbsch, and C. Bratt, 2006, "Two New Brown-Spored Species of *Pertusaria* from South-Western North America," *The Lichenologist* 38(5):411–416, Cambridge University Press. *Fig. 34.16* John Taylor/Visuals Unlimited, Inc. *Fig. 34.17* Carolina Biological Supply Co./Visuals Unlimited, Inc. *Fig. 34.18 (a)* Topic Photo Agency/age fotostock, *(b)* Matt Meadows/Getty Images. *Fig. 34.19* Ed Reschke/Getty Images. *Fig. 34.20* David Hughes. *Fig. 34.21 (a)* Inga Spence/Visuals Unlimited, Inc., *(b)* Bruce Perrault, *(c)* David Clapp/Getty Images. *Fig. 34.22 (a)* Albert Lleal/Minden Pictures, *(b)* Wally Eberhart/Getty Images, *(c)* Matthias Breiter/Minden Pictures. *Fig. 34.24* After "The Spread of Wheat Stem Rust UG99 Lineage," Food and Agriculture Organization of the United Nations, JPEG image, 2010, http://www.fao.org/agriculture/crops/rust/stem/rust-report/stem-ug99racettksk/en/. *Photo and inset photo* Yue Jin, USDA-ARS, Cereal Disease Laboratory.

Case 7

Wolf Samuel R. Maglione/Photo Researchers, Inc. *Moose* Ron Erwin/age fotostock. *Graph* Data from J. A. Vucetich and R. O. Peterson, 2012, "The population biology of Isle Royale wolves and moose: an overview," www.isleroyalewolf.org.

Chapter 35

Opening photo Dr. Michael Delannoy/Getty Images. *Fig. 35.1* Winfried Wisniewski/Corbis. *Fig. 35.2 (T)* Andrew J. Martinez/Photo Researchers, Inc., *(M)* UMI NO KAZE/a.collectionRF /Getty Images, *(B)* Robert Pickett/Visuals Unlimited, Inc. *Fig. 35.6 (L)* Cajal Legacy, Instituto Cajal (CSIC), Madrid, Spain, *(R)* Bob Jacobs, Ph.D., Laboratory of Quantitative Neuromorphology, Department of Psychology, Colorado College. *Fig. 35.7* Science VU/C. Raine/Visuals Unlimited, Inc. *Fig. 35.14* Mary B. Kennedy, Division of Biology, Caltech. *Fig. 35.16* David B. Fankhauser, Ph.D., Professor of Biology and Chemistry, University of Cincinnati, Clermont College.

Chapter 36

Opening photo Custom Medical Stock Photo/Getty Images. *Fig. 36.1 (a)* Gerry Bishop/Visuals Unlimited, Inc., *(b)* Joe McDonald/Visuals Unlimited, Inc., *(c)* Monika Ondruová/Dreamstime.com. *Fig. 36.2 (a)* Gary Meszaros/Visuals Unlimited, Inc., *(b)* Sinclair Stammers/Photo Researchers, Inc., *(c)* Danté Fenolio/Photo Researchers, Inc. *Fig. 36.8* SPL/Photo Researchers, Inc. *Fig. 36.10* Michael Durham/Minden Pictures. *Fig. 36.11* Rendered by Dale Muzzey. *Fig. 36.12 (a)* Tom Adams/Visuals Unlimited, Inc., *(b)* Eye of Science/Photo Researchers, Inc., *(c)* Danté Fenolio/Photo Researchers, Inc. *Fig. 36.13* Bjorn Rorslett/Photo Researchers, Inc. *Fig. 36.17* Science Source/Photo Researchers, Inc. *Fig. 36.21* The National Library of Medicine. *Fig. 36.26* WDCN/University College London/Photo Researchers, Inc.

Chapter 37

Opening photo Gustoimages/Science Photo Library/Photo Researchers, Inc. *Fig. 37.1* Tom Brakefield/Corbis. *Fig. 37.2* Dr. Gladden Willis/Visuals Unlimited, Inc. *Fig. 37.5 (a)* Don W. Fawcett/Photo Researchers, Inc. *(b)* Biophoto Associates/Photo Researchers, Inc. *Fig. 37.6* Reprinted by permission from Macmillan Publishers Ltd: H. Huxley and J. Hanson, "Changes in the Cross-Striations of Muscle During Contraction and Stretch and Their Structural Interpretation," 1954, *Nature* 173:973–976. Copyright 1954. *Fig. 37.8* Don W. Fawcett/Photo Researchers, Inc. *Fig. 37.12* Adapted from E. P. Widmaier, H. Raff, and K. T. Strang, 2011, *Vander's Human Physiology*, New York: McGraw-Hill. *Fig. 37.14* Dr. Gladden Willis/Visuals Unlimited, Inc. *Fig. 37.15 (a)* Jan Van Arkel/Minden Pictures, *(b)* Nigel Cattlin/Visuals Unlimited, Inc. *Fig. 37.16 (a)* Rita van den Broek/age fotostock, *(b)* Reinhard Dirscherl/Visuals Unlimited, Inc., *(c)* Thomas Marent/Visuals Unlimited, Inc. *Fig. 37.18* Science Photo Library/Photo Researchers, Inc. *Fig. 37.19* Ralph Hutchings/Visuals Unlimited, Inc. *Fig. 37.22* Based on M. J. Smith and R. J. G. Savage, 1956, "Some Locomotory Adaptations in Mammals," *Zoological Journal of the Linnean Society* 42:603–622. *Photo*: Norbert Wu/Minden Pictures, J. L. Klein & M. L. Hubert/Photo Researchers, Inc.

Chapter 38

Opening photo Heidi & Hans-Jurgen Koch/Minden Pictures/Getty Images. *Fig. 38.12* Peter Blackwell/Minden Pictures. *Fig. 38.14 (L–R)* Jef Meul/Minden Pictures, Jorg & Petra Werner/Animals Animals–Earth Scenes, Steven Smith/Alamy. *Fig. 38.15* Piotr Naskrecki/Minden Pictures/Corbis. *Fig. 38.16* blickwinkel/Alamy.

Chapter 39

Opening photo Science Photo Library/Photo Researchers, Inc. *Fig. 39.1* Amar & Isabelle Guillen/SeaPics.com. *Fig. 39.4 (T)* José B. Ruiz/Minden Pictures, *(B)* Stephen Dalton/Animals Animals. *Fig. 39.8* Reprinted from E. Weibel, B. Sapoval, and M. Filoche, 2005, "Structure and Function in the Periphery of the Lung," *Respiratory Physiology & Neurobiology* 148:3–21. Copyright 2005, with permission from Elsevier. *Fig. 39.11* Rendered by Dale Muzzey. *Fig. 39.13* Based on Fig. 3A in P. L. Tyack, M. Johnson, N. Aguilar de Soto, A. Sturlese, and P. T. Madsen, 2006, "Extreme Diving Behaviour of Beaked Whale Species Known to Strand in Conjunction with Use of Military Sonars, *Journal of Experimental Biology* 209:4238–4253. *Photo*: Bill Curtsinger/Getty Images. *Fig. 39.17* Data from C. B. Chapman and J. H. Mitchell, 1965, "The Physiology of Exercise," *Scientific American* 212:88–96.

Chapter 40

Opening photo Science Source/Photo Researchers, Inc. *Fig. 40.2* age fotostock/SuperStock. *Fig. 40.4* Adapted from A. M. Hemingsen, 1960, "Energy metabolism as related to body size and respiratory surfaces, and its evolution." *Re. Steno Memorial Hospital Nordisk Insulinlaboratorium.* 9:1–110. *Fig. 40.5* Cary Wolinsky/Getty Images. *Fig. 40.8 (a)* Fred Bavendam/Minden Pictures, *(b)* © 2012 Brandon Cole www.norbertwu.com. *Fig. 40.9* Marevision/Getty Images. *Fig. 40.10* C & M Fallows/SeaPics.com. *Fig. 40.16* Frank Woerle/Minden Pictures. *Fig. 40.18* Biophoto Associates/Photo Researchers, Inc.

Chapter 41

Opening photo Steve Allen/Getty Images. *Fig. 41.5* David Wall/Alamy. *Fig. 41.6 (a)* Peter Betts/iStockphoto, *(b)* Roger De La Harpe; Gallo Images/Corbis. *Fig. 41.8 (a)* Michael Abbey/Photo Researchers, Inc., *(b)* Michael Abbey/Photo Researchers, Inc. *Fig. 41.15* Minden Pictures/SuperStock.

Chapter 42

Opening photo Science Faction/Getty Images. *Fig. 42.1 (a)* CNRI/ Photo Researchers, Inc., *(b)* SPL/Photo Researchers, Inc., *(c)* WaterFrame/Alamy, *(d)* Mike Lane/Alamy. *Fig. 42.4* Tim Laman Photography. *Fig. 42.6* D. B. M. Welch and M. Meselson, 2000, "Evidence for the Evolution of Bdelloid Rotifers Without Sexual Reproduction or Genetic Exchange," *Science* 288:1211–1215.

Reprinted with permission from AAAS. *Fig. 42.7* Michael & Patricia Fogden/Minden Pictures. *Fig. 42.10* Thanunkorn Klypaksi/iStockphoto. *Fig. 42.12* Roland Birke/Getty Images. *Fig. 42.15* Claude Cortier/Photo Researchers, Inc. *Fig. 42.20* David M. Phillips/Photo Researchers, Inc. *Fig. 42.21* Anatomical Travelogue/Photo Researchers, Inc. *Fig. 42.22 (a)* Carolina Biological Supply Co./Phototake, *(b)* © 2002 Steve Baskauf.

Chapter 43

Opening photo Dr. David Phillips/Visuals Unlimited/Getty Images. *Fig. 43.1 (a)* Christian Beier/age fotostock, *(b)* Michael P. Gadomski/Photo Researchers, Inc., *(c)* Charles J. Smith, *(d)* Inga Spence/Getty Images. *Fig. 43.4* SPL/Photo Researchers, Inc. *Fig. 43.5* Scientifica/Visuals Unlimited, Inc. *Fig. 43.19* Biophoto Associates/Photo Researchers, Inc. *Fig. 43.20* James Gathany/CDC/Photo Researchers, Inc.

Case 8

Diversity hotspots map Reprinted by permission from Macmillan Publishers Ltd: N. Myers, R. A. Mittermeier, C. G. Mittermeier, G. A. B. da Fonseca, and J. Kent, 2000, "Biodiversity Hotspots for Conservation Priorities," *Nature* 403:854–858. Copyright 2000. *Underwater life on the Great Barrier Reef* Jeff Hunter/Getty Images.

Chapter 44

Opening photo Brian Farrell. *Fig. 44.1 (T–B)* Mark Dayel and Nicole King, www.dayel.com/choanoflagellates, James D. Watt/SeaPics.com, UMI NO KAZE/a.collectionRF/Getty Images, Sonke Johnsen/Visuals Unlimited, Inc. *Fig. 44.5 (T–B)* Glenn Nagel/iSockphoto, Piotr Naskrecki/Minden Pictures, Franco Banfi/WaterF/age fotostock. *Fig. 44.6* imagebroker/Alamy. *Fig. 44.8 (a)* Image Source/Getty Images, *(b)* Andrew J. Martinez/Photo Researchers, Inc. *Fig. 44.10 (a)* Chris Newbert/Minden Pictures, *(b)* George G. Lower/Photo Researchers, Inc. *Fig. 44.11* David Wrobel/SeaPics.com. *Fig. 44.12* Reprinted by permission from Macmillan Publishers Ltd: M. Srivastava et al., 2008, "The *Trichoplax* Genome and the Nature of Placozoans," *Nature* 454:955–960. Copyright 2008. Image courtesy of Ana Signorovitch. *Fig. 44.14 (T)* E. R. Degginger/Photo Researchers, Inc., *(B)* Phillip Colla/Oceanlight. *Fig. 44.15 (a)* E. R. Degginger/Photo Researchers, Inc., *(c)* Alexander Semenov/Photo Researchers, Inc., *(d)* Susan Dabritz/SeaPics.com, *(e)* doc-stock/Visuals Unlimited, Inc. *Fig. 44.16* ARCO/W Rolfes/age fotostock. *Fig. 44.17 (a)* Phillip Colla/Oceanlight, *(c)* Norbert Wu/Minden Pictures. *Fig. 44.18* Glenn Price/Dreamstime.com. *Fig. 44.19* Cindy Singleton/iStockphoto. *Fig. 44.20 (T)* W. G. M. Damen, M. Hausdorf, E. A. Seyfarth, and D.Tautz, 1998, "A Conserved Mode of Head Segmentation in Arthropods Revealed by the Expression Pattern of Hox Genes in a Spider," *Proceedings of the National Academy of Sciences, USA* 95(18):10665–10670. Copyright 1988 National Academy of Sciences, U.S.A., *(B, all photos)* Piotr Naskrecki. *Fig. 44.21 (T–B)* B. G. Thomson/Photo Researchers, Inc., Tom McHugh/Photo Researchers, Inc., Michael S. Nolan/SeaPics.com, Dietmar Nill/Minden Pictures. *Fig. 44.22* Ralph A. Clevenger/Corbis. *Fig. 44.23 (T–B)* David Shale/Minden Pictures, Azure Computer & Photo Services/Animals Animals, Biosphoto/Christian Kšnig, Reinhard Dirscherl/WaterFrame/Getty Images, Dwight Kuhn. *Fig. 44.24 (a)* David Shale/Minden Pictures, *(b)* D. P. Wilson/FLPA/Photo Researchers, Inc. *Fig. 44.25 (a)* Azure Computer & Photo Services/Animals Animals, *(b)* V & W/Ricardo Fernandez/SeaPics.com. *Fig. 44.26* Biosphoto/Christian Kšnig. *Fig. 44.27* Reinhard Dirscherl/WaterFrame/Getty Images. *Fig. 44.28* Stockbroker xtra/age fotostock. *Fig. 44.30 (a)* Copyright © Brandon Cole, *(b)* Gary Meszaros/Visuals Unlimited, Inc. *Fig. 44.31* Michael Patrick O'Neill/Alamy. *Fig. 44.32* U.S. Fish and Wildlife Service/Joel Sartore–National Geographic Stock with Wade Fredenberg. *Fig. 44.33 (a)* Gerard Lacz/Animals Animals–Earth Scenes, *(b)* D. R. Schrichte/SeaPics.com. *Fig. 44.35 (a)* Gary Meszaros/Photo Researchers, Inc., *(b)* Kim Taylor/Minden Pictures, *(c)* David Pattyn/Minden Pictures, *(d)* NASA/Jim Grossmann. *Fig. 44.36 (T–B)* Dave Watts/NHPA/Photoshot, Martin Willis/Minden Pictures, Tom Walker/Visuals Unlimited, Inc. *Fig. 44.37* O. Louis Mazzatenta/National Geographic/Getty Images. *Fig. 44.38 (a)* Western Trilobite Association, *(b)* Smithsonian Institution–National Museum of Natural History. Image: Jean-Bernard Caron, *(c)* S. Bengtson, 2000, "Teasing Fossils out of Shales with Cameras and Computers." *Palaeontologia Electronica* 3(1), article 4, 14 pp. Copyright Palaeontological Association, 15 April 2000, *(d)* Reprinted from Palaeoworld, 20, J.-Y. Chen, 2011, "The Origins and Key Innovations of Vertebrates and Arthropods," Fig. 5d. Copyright 2011, with permission from Elsevier.

Chapter 45

Opening photo Dan L. Perlman/EcoLibrary.org. *Fig. 45.1* Harry Rogers/Photo Researchers, Inc., Fabio Pupin/Visuals Unlimited, Inc. *Fig. 45.2* Phil Savoie/npl/Minden Pictures. *Fig. 45.6 (a)* Adriano Bacchella/naturepl.com, (b) Fred Lord/Alamy. *Fig. 45.7* M. B. Sokolowski, 2001, "Drosophila: Genetics Meets Behaviour," *Nature Reviews Genetics* 2: 879–890, doi:10.1038/35098592. *Fig. 45.9 (a)* Todd Ahern/Emory University, *(b)* Rick & Nora Bowers/Alamy. *Fig. 45.11 (T)* Nina Leen/Time Life Pictures/Getty Images, *(B)* Jeffery Phelps/Getty Images. *Fig. 45.13 (a)* Dan L. Perlman/EcoLibrary.org, *(b)* Michael & Patricia Fogden/Getty Images, *(c)* © Matt E. Arnegard, *(d)* Dan L. Perlman/EcoLibrary.org, *(e)* Ron Rowan Photography/Shutterstock. *Fig. 45.14* After T. J. Carew, 2000, *Behavioral Neurobiology*, Sunderland, MA: Sinauer Associates, p. 244. *Photo*: Paul Sutherland/Getty Images. *Fig. 45.15* After M. B. Sokolowski, 2001, "*Drosophila*: Genetics Meets Behaviour," *Nature Reviews Genetics* 2:879–890, doi:10.1038/35098592. *Fig. 45.16 (T)* John Durban/NOAA, *(B)* Kathryn Jeffs/NPL/Minden Pictures. *Fig. 45.18* Alex Wild/alexanderwild.com. *Fig. 45.19(a)* Nico Tondini/Robert Harding World Imagery/Corbis, *(b)* Steve Byland/iStockphoto, *(c)* Thomas Marent/Minden Pictures. *Fig. 45.20* Kelly Funk/All Canada Photos/Corbis, Steven Kaufman/Getty Images.

Chapter 46

Opening photo Charles J. Smith. *Fig. 46.1* Data from A. Atkinson, V. Siegel, E. A. Pakhomov, M. J. Jessopp, V. Loeb, 2009, "A Re-appraisal of the Total Biomass and Annual Production of Antarctic Krill," *Deep-Sea Research I* 56:727–740. *Photos*: Richard Herrmann/Visuals Unlimited, Inc., *(inset)* Gerald & Buff Corsi/Visuals Unlimited, Inc./Corbis. *Fig. 46.3* Richard Herrmann/Visuals Unlimited, Inc. *Fig. 46.6* Data from M. K. Chase, N. Nur, and G. R. Geupel, 2005, "Effects of Weather and Population Density on Reproductive Success and Population Dynamics in a Song Sparrow (*Melospiza melodia*) Population: A Long-Term Study," *The Auk* 122(2):571–592. *Fig. 46.7* Scientifica/Visuals Unlimited, Inc. *Fig. 46.8* Data from United Nations Department of Economic and Social Affairs, Population Pyramids (France and India), http://esa.un.org/wpp/population-pyramids/population-pyramids.htm. *Fig. 46.9* Data from S. A. Berkeley, M. A. Hixon, R. J. Larson, and M. S. Love, 2004, "Fisheries Sustainability via Protection of Age Structure and Spatial Distribution of Fish Populations," *Fisheries* 29(8): 23–32. *Fig. 46.11* Data from M. Pathak, and P. Q. Rivsi, 2002, "Age Specific Life Tables of *Papilio demoleus* on Different Hosts," *Annals of Plant Protection Sciences* 10(2):375–376. *Fig. 46.12 (a)* Data from R. Pearl, 1978, *The Biology of Population Growth,* 3rd ed., North Stratford, NH: Ayer Company Publishers; *(b)* Data from E. S. Deevey, Jr., 1947, "Life Tables for Natural Populations of Animals. *Quarterly Review of Biology* 22:283–314. *(c)* Data from J. E. Kaldy and K. H. Dunton. "Ontogenetic Photosynthetic Changes, Dispersal and Survival of *Thalassia testudinum* (Turtlegrass) Seedlings in a Sub-tropical Lagoon," *Journal of Experimental Marine Biology and Ecology* 240(2):193–212. *Fig. 46.13* Charles O'Rear/Corbis. *Fig. 46.15* Mark Moffett/Minden Pictures. *Fig. 46.18* After R. H. MacArthur and E. O. Wilson, 1967, *The Theory of Island Biogeography,* Princeton, NJ: Princeton University Press. *Fig. 46.19 (T–B)* Eladio M. Fernandez, Eladio M. Fernandez, Eladio M. Fernandez, © 2013 Jake Scott, Edward Myles/FLPA/Minden Pictures, Eladio M. Fernandez.

Chapter 47

Opening photo Charles J. Smith. *Fig. 47.1* Eladio M. Fernandez. *Fig. 47.3 (a)* Michael & Patricia Fogden/Minden Pictures, *(b)* Mark Moffett/Minden Pictures. *Fig. 47.4* Charles J. Smith. *Fig. 47.5* Courtesy of Jim Maddox, Galloway Twp., NJ; picture-newsletter.com. *Fig. 47.6* Data from P. J. Morin, 1981, "Predatory Salamanders Reverse Outcome of Competition Among Three Species of Anuran Tadpoles," *Science* 212:1284–1286. *Fig. 47.7* Gerry Bishop/Visuals Unlimited, Inc. *Fig. 47.8* Biosphoto/Gérard Lacz. *Fig. 47.10* Carl Gwinn, blackcormorant.net. *Fig. 47.11* © 2012 DigitalGlobe © 2012 Google. *Fig. 47.13* Dan L. Perlman/EcoLibrary.org. *Fig. 47.17 (Tundra)* Helge Schulz /Picture Press/Getty Images, *(Alpine)* Christian Ziegler/Minden Pictures, *(Taiga)* Gerry Ellis/Minden Pictures, *(Temperate coniferous forest)* Charlie Ott/Photo Researchers, Inc., *(Deciduous forest)* Tim Fitzharris/Minden Pictures, *(Temperate grassland)* Robert and Jean Pollock/Photo Researchers, Inc., *(Desert)* Yva Momatiuk & John Eastcott/Minden Pictures, *(Chaparral)* Richard Herrmann/Visuals Unlimited, Inc., *(Savanna)* Richard Du Toit/Minden Pictures, *(Rain forest)* Charles J. Smith. *Fig. 47.18* After D. M. Olson et al., 2001, "Terrestrial Ecoregions of the World: New Map of Life on Earth," *Bioscience* 51:933–938. *Fig. 47.21* After D. M. Kaufman, 1995, "Diversity of New World Mammals: Universality of the Latitudinal Gradients of Species and Bauplans," *Journal of Mammalogy* 76:322–334. *Fig. 47.22* Courtesy of George Poinar.

Chapter 48

Opening photo Michael Utech/Getty Images. *Fig. 48.1* Data from "Historical Estimates of World Population," United States Census Bureau International Data Base, last modified June 2012, http://www.census.gov/population/international/data/worldpop/table_history.php. *Fig. 48.2* Data from *United Nations* Department of Economic and Social Affairs, 2009, *World Economic and Social Survey 2009: Promoting Development, Saving the Planet*, New York: United Nations, p. 44, http://www.un.org/en/development/desa/news/policy/wess-2009.shtml. *Fig. 48.3* Data from *Global Footprint Network, 2006; United Nations Development Programme, 2006. Fig. 48.4* Tim McCaig/iStockphoto, Stockbyte /Getty Images. *Fig. 48.5* Data from R. A. Rohde, Global Warming Art, http://www.globalwarmingart.com/. Based on data from the Hadley Centre Coupled Model version 3 (HadCM3), http://badc.nerc.ac.uk/view/badc.nerc.ac.uk__ATOM__dpt_1162913571289262. *Fig. 48.6* Fotosearch Stock Images. *Fig. 48.7* Adapted from R. A. Rohde, "Global Warming Predictions," from Global Warming Art, JPEG image, http://www.globalwarmingart.com/wiki/File:Global_Warming_Predictions_Map_jpg. Based on data from the Hadley Centre Coupled Model version 3 (HadCM3), http://badc.nerc.ac.uk/view/badc.nerc.ac.uk__ATOM__dpt_1162913571289262. *Fig. 48.8* Data from A. J. Miller-Rushing and R. B. Primack, 2008, "Global Warming and Flowering Times in Thoreau's Concord: A Community Perspective," *Ecology* 89(2):332–341, doi: 10.1890/07-0068.1. *Fig. 48.9* After M. B. Davis, 1976, "Pleistocene Biogeography of Temperate Deciduous Forests," *Geoscience and Man* 13:13–26. *Fig. 48.11* Fred Bavendam/Minden Pictures. *Fig. 48.13* Reprinted by permission from Macmillan Publishers Ltd: U. Riebesell, I. Zondervan, B. Rost, P. D. Tortell, R. E. Zeebe, and F. M. M. Morel, 2000, "Reduced Calcification of Marine Plankton in Response to Increased Atmospheric CO_2," *Nature* 407:364–367. Copyright 2000. *Fig. 48.14* euroluftbild.de/dpa/Corbis, Travel Pix/Robert Harding. *Fig. 48.15* Image courtesy of Liam Gumley, Space Science and Engineering Center, University of Wisconsin–Madison and the MODIS Science Team. *Fig. 48.17* Data from M. D. Edgerton, 2009, "Increasing Crop Productivity to Meet Global Needs for Feed, Food, and Fuel," *Plant Physiology* 149:9, http://www.plantphysiol.org/content/149/1/7. *Fig. 48.18, Fig. 48.18 map* After Millennium Ecosystem Assessment, 2005, *Ecosystems and Human Well-being: Synthesis*, Washington, DC: Island Press). http://www.millenniumassessment.org/documents/document.356.aspx.pdf. *Fig. 48.19* AP Photo/Andre Penner. *Fig. 48.20* James K. York/Dreamstime.com, AP Photo/U.S. Department of Agriculture, Ch'ien Lee/Minden Pictures. *Fig. 48.22* Michael Fogden/Animals Animals. *Fig. 48.23 (Doctor)* AP Photo/Sebastian John, *(Science teacher)* WA/Dann Tardif/Blend Images/Getty Images, *(Scientist in the field)* Alexis Rosenfeld/Photo Researchers, Inc., *(Scientists testifying before congressional subcommittee)* Matthew Cavanaugh/EPA/Newscom, Chas Metivier/ZUMA Press/Newscom.

INDEX

Bold face indicates a definition, *italics* indicates a figure, 1-1*t* indicates a table, C1-1 indicates a Case.

A

Abomasum, **40-22**, *40-22*
Abscisic acid, 31-7*t*, 31-8, **31-20**
   and stomata, 29-5
Acacia trees [*Acacia cornigera*], 32-11–32-12, *32-12*
Acetabularia [alga], *27-15*
Acetylcholine, 35-14
Acoelomates, **44-3**, *44-3*
Acorn worms, 44-18, *44-18*
Acquired immunity, **43-1**. *See also* Adaptive immune system
Acrosomes, **42-10**, *42-11*
ACTH. *See* Adrenocorticotropic hormone (ACTH)
Actin, **37-2**
   and muscle contraction, 37-4–37-6, *37-5*
   and muscle structure, 37-3, *37-3*, *37-4*
   regulation of, 37-6–37-7, *37-7*
Action potentials, 35-4, **35-5**, **35-8**, *35-9*, 35-11
   in cardiac muscle, 39-20
   propagation of, 35-8, *35-10*, 35-10–35-12
   and sensory receptors, 36-4, *36-4*–*36-5*
Acute phase response, **43-5**–43-6
Adaptation [natural selection], **21-3**
Adaptation [sensory receptors], **36-5**
Adaptive immune system, **43-1**, 43-7–43-13
   and antibodies, 43-8, *43-8*–*43-9*
   and B cells, 43-7–43-8
   cells of, *43-3*
   and clonal selection, 43-9, *43-9*–*43-10*
   features of, 43-2*t*
   and genomic rearrangement, 43-10–43-13, *43-12*
   and immunological memory, 43-10, *43-10*
Adenosine triphosphate (ATP), **40-2**
   and cross-bridge cycles, 37-5, *37-5*–*37-6*
   and metabolism, 40-2
   and nutrient uptake by roots, 29-16
   and slow-twitch fibers, 37-12
ADH. *See* Antidiuretic hormone (ADH)
Adrenal glands, *38-12*, **38-15**
Adrenal medulla, **38-16**

Adrenaline. *See also* Epinephrine
   and fight-or-flight response, 38-16
Adrenocorticotropic hormone (ACTH), 38-9*t*, 38-10
Advantageous mutations, 21-2–**21-3**
Advertisement displays, *45-15*, **45-15**, 45-20, 45-21
Aerobic metabolism, 40-2–**40-3**
Aerobic reactions, **26-6**
Afferent neurons, **35-16**
Agave shawii [monocot], 33-17, *33-17*
Age structures, **46-8**, *46-8*
Agonists, **37-10**, *37-10*
Agriculture [Case 6], C6-2–C6-4, *C6-3*
   by ants, 34-6
   and atmospheric carbon dioxide, 25-6, 48-3
   and biodiversity, 48-13–48-17, *48-16*
   and C₄ plants, 29-8
   and centers of origin, 33-20–33-21, *33-21*
   criticisms of, C6-4
   and directional selection, 21-12, *21-12*
   and double fertilization, 30-15
   and ecology, 48-12–48-13
   and fungal pathogens, 34-4, *34-4*
   and fungi, 34-19, *34-19*
   and genetically modified organisms, 32-17–32-18, *32-18*
   and germination, 30-21–30-22
   and hormones, 31-9–31-10
   and human activities, 48-14–48-15, *48-15*
   and nitrogen fixation, 29-18
   and photoperiodism, 30-17
   and pollination modes, 30-12
   sustainable, 48-15
   and symbiotic relationships, 33-10–33-11
   and vernalization, 30-18, 30-22
Aldosterone, **41-19**
Algae, **27-9**, 27-13–27-15, *27-14*, *27-15*
   brown, 27-16, *27-16*, 28-3
   complex multicellular, 28-3, 28-4
   and eutrophication, 48-11, *48-11*
   green, 27-14, *27-14*–27-15, *27-15*
   life cycle of, 30-2, *30-3*
   and the long-term carbon cycle, 25-14

   red, 27-13, *27-14*, 27-20, *27-21*, 28-3
   simple multicellular, 28-2, 28-2–28-3, *28-3*
   and spread of photosynthesis, 27-16–27-19, *27-19*
Algoriphagus machipongenesis [bacterium], 28-6–28-7
Alkaloids, **32-10**, 32-10*t*
Alland, Henry, 30-17
Allele frequencies, **21-3**
   and Hardy–Weinberg equilibrium, 21-7, *21-7*
   and natural selection, 21-10
Alleles, and mutations, 21-2, *21-3*
Allopatric speciation, 22-6–22-11, **22-7**, *22-8*, *22-9*, *22-10*
Alpine biome, 47-17, *47-17*
Alternation of generations, **30-3**, *30-3*
Alternative energy sources, *48-10*, 48-10–48-11
Alternative RNA splicing, and antibodies, 43-12
Altruism, **45-16**
   reciprocal, 45-17–45-18
Alveolates [Alveolata], **27-16**, *27-17*
   Visual Synthesis of history of, *44-30*
Alveoli, 39-7, **39-8**
Amacrine cells, **36-15**, *36-15*
Amanita [fungus], *34-17*
Amber, fossils in, C8-2, *47-22*, 47-22–47-23
Amborella trichopoda [angiosperm], *33-15*, 33-16
Amine hormones, 38-7, **38-7**–38-8
Amino acids
   essential, 40-10, 40-10*t*
   evolution of, 21-16
Ammonia
   and Archaea, 26-18
   and nitrogen fixation, 26-11–**26-12**, 29-17
   as nitrogenous waste, 41-7, *41-7*
Ammonites, 44-14
   extinction of, 23-18
Amnion, **42-9**, *42-9*
Amniotes, 44-25–44-27
Amniotic eggs, 42-8–42-9, *42-9*, **44-25**–44-26
Amoebas
   endosymbiosis in, 27-6, *27-6*
   movement of, 27-9

Amoebozoa, *27-12*, **27-12**–27-13, *27-13*
   history of, 27-21, *27-21*
   Visual Synthesis of history of, *44-31*
Amphibians [Amphibia], **44-25**
   hearts of, *39-18*, 39-18–39-19
   and human activities, 48-19, *48-19*
   kidneys of, 41-11, *41-11*
   life cycle of, *44-24*, 44-25
   Visual Synthesis of history of, *44-31*
Amphioxus [cephalochordate], 44-19, *44-20*, *44-21*
Amylase, **40-15**
Anabaena [cyanobacterium], *26-16*
Anabolism, **40-3**
Anaerobic metabolism, 40-2–**40-3**
Analogous characters, *23-5*, 23-5–**23-6**, 44-2
Anammox reaction, 26-11–**26-12**
Ancestry, and mitochondrial DNA, 24-14
Androgen, 38-4*t*
Angiosperms, **29-2**, 30-9–30-16. *See also* Gymnosperms
   double fertilization of, 30-15
   fruits of, 30-15–30-16, *30-16*
   life cycle of, *30-14*
   and photoperiodism, 30-17–30-22
   phylogenetic tree for, *33-15*
   self-compatible versus self-incompatible, 30-12
   Visual Synthesis of, 33-22–33-23
   wood of, 31-12–31-13, *31-13*
Angiotensin II, **41-19**
Anglerfish, 42-1, 42-4
Animal behavior, 45-1–45-2
   and communication, 45-13–45-16
   and genes, 45-2–45-8
   hormonal control of, 38-14
   and information processing, 45-11–45-13
   and learning, 45-8–45-11
   and sexual selection, 45-19–45-21
   social, 45-16–45-19
Animal circulatory systems, 39-1, 39-13–39-17
   closed versus open, 39-13–39-14, *39-14*
   and gas exchange, 39-1–39-3
   and the heart, 39-17–39-22

I-1

Animal endocrine systems, **38-1**–38-7
   functioning of, 38-1–38-4, *38-2*, *38-5*
   and homeostasis, 38-5–38-7, *38-6*, *38-16*
   and hormones, 38-7–38-12
   and pheromones, 38-16–38-18
   and reproductive systems, 42-13
   vertebrate, 38-12–38-16
Animal immune systems, 43-1
   adaptive, 43-7–43-13
   and appendix, 40-23
   and cell-mediated immunity, 43-13–43-16
   and infections, 43-16–43-19
   innate, 43-1–43-7
   organs of, *43-7*
Animal metabolism, 40-1–40-7. *See also* Cellular respiration; Digestion
   and digestion, 40-14–40-23
   and feeding, 40-12–40-14
   and nutrition, 40-7–40-11
Animal movement, 37-1
   and muscle contraction, 37-8–37-12
   and muscle structure, 37-1–37-7
   and skeletons, 37-13–37-19
Animal nervous systems, **35-1**–35-2, *35-3*. *See also* Brains; Neurons
   and behaviors, 45-3–45-4, *45-4*
   evolution of, 35-1–35-2
   function of, 35-2–35-4, *35-3*
   and homeostasis, 35-2, 35-17–35-18, *35-18*
   organization of, *35-15*, 35-15–35-19, *35-17*, *35-18*
Animal renal systems, **41-1**
   and excretion, 41-7–41-12
   and kidneys, 41-12–41-19
   and osmoregulation, 41-1–41-7
Animal reproduction
   and colonization of land, *42-7*, 42-7–42-9
   and female reproductive systems, 42-11–42-14, *42-12*, *42-13*, *42-14*
   history of, 42-1–42-7
   and male reproductive systems, 42-9–42-11, *42-10*, *42-11*
Animal respiratory systems, 39-1
   and breathing, 39-4–39-9
   and gas exchange, 39-1–39-3
   and oxygen transport, 39-10–39-13
Animal sensory systems, 36-1–36-5. *See also* Senses
   and the brain, 36-17–36-19
   and cognition, *36-20*, 36-20–36-21, *36-21*
   and sense of sight, 36-10–36-16
   and sense of smell, 36-5–36-6, *36-6*
   and sense of sound, *36-8*, 36-8–36-10, *36-9*, *36-10*
Animals. *See also* Bilaterians [Bilateria]; Cnidarians [Cnidaria]; Sponges [Porifera]; Vertebrates [Vertebrata]
   bulk transport in, *28-5*, 28-5–28-6
   cognition in, 36-21
   as complex multicellular organisms, 28-3, *28-4*
   diploblastic versus triploblastic, *44-4*, 44-4
   early evolution of, 28-12–28-13, *28-13*
   evolutionary history of, 44-27–44-29, *44-30*–*44-31*
   life cycle of, *27-4*, 27-4
   phylogenetic tree for, *28-12*, 44-1–44-2, *44-2*
   versus plants, 28-10, *28-11*, 28-11–28-12
   Visual Synthesis of history of, *44-31*
Annelid worms [Annelida], **44-11**–44-12, *44-12*
   Cambrian, *44-28*
Annual clocks, **45-12**–45-13
Anolis [lizard]
   and biodiversity hotspots, C8-2
   fossils of, in amber, *47-22*, 47-22
   hormonal control of behavior in, 45-4–45-5, *45-5*
   and island biogeography, 46-15–46-16, *46-16*
   niches of, *47-2*, 47-2
Anoxygenic bacteria, **26-7**, *26-7*, *26-16*
Antagonisms, *47-3*, **47-3**–47-5, *47-8t*
   and competition, *47-4*, 47-4
   and niche divergence, 47-4–47-5
   and predation, *47-5*, 47-5
Antagonist muscles, **37-10**, *37-10*
Antelope, 37-12, 37-19
Anterior pituitary gland, *38-12*, *38-13*, **38-13**
Anthers, *30-9*, **30-10**
Anthoceros [hornwort], 33-3
Anthropocene, **48-1**–48-2, *48-2*. *See also* Human activities
Antibiotics. *See* Drugs
Antibodies, **43-7**, *43-8*
   classes of, *43-8*, 43-8–43-9
   diversity of, 43-10–43-13, *43-12*
   production of, 43-7–43-8
Antidiuretic hormone (ADH), *38-9t*, **38-14**, **41-17**, *41-17*. *See also* Vasopressin
   and behavior, 45-8
Antigen-presenting cells, **43-15**, *43-15*
Antigenic drift, **43-17**, *43-17*
Antigenic shift, **43-17**, *43-17*
Antigenic variation, **43-19**
Antigens, **43-7**
   and T cell activation, 43-14–43-15, *43-15*
Ants
   agriculture of, 34-6
   communication by, 45-14
   and kin selection, 45-19
   and pheromones, 38-18, *38-18*
   and plant defenses, *32-11*, 32-11–32-12, *32-12*
   zombie, 34-16, *34-16*
Aortas, *39-19*, **39-20**
Aortic bodies, **39-9**, *39-9*
Aortic valves, *39-19*, **39-20**
Apes, 24-1–24-5, *24-2*
Aphanothece [cyanobacterium], *26-16*
Aphids, C5-4, 47-6–47-7, *47-7*
Apical dominance, **31-10**, *31-10*
Appendicular skeletons, **37-15**, *37-15*
Appendix, **40-22**–40-23
Apples, 30-12, *30-15*
Aquaporins, **41-2**
Aquaspirillum [bacterium], 45-11
Arabidopsis thaliana [mouse-ear cress]
   cyanobacterial genes in, 27-6
   and hormones, 31-20
   pollen tubes in, 30-10
   and vernalization, 30-18–30-19, *30-19*
Archaea, **26-1**, **26-5**–26-6, *26-6*. *See also* Bacteria; Eukarya; *specific* archaeon
   diversity of, 26-16–26-19, *26-18*
   evolutionary history of, *26-20*, 26-20–26-23
   and the nitrogen cycle, 26-10–26-12, *26-11*
   phylogenetic tree for, *26-14*, *26-17*, 26-17
   and the sulfur cycle, 26-10, *26-10*
   Visual Synthesis of history of, *44-30*
Archaeopteryx lithographica [dinosaur-bird], 23-16, *23-17*
Archaeplastida, **27-13**–27-15, *27-14*. *See also* Algae; Plants
   history of, 27-20–27-21
   Visual Synthesis of history of, *44-31*
Ardi, 24-4–**24-5**, *24-5*
Ardipithecus ramidus [hominin], 24-4–24-5, *24-5*
Aristotle, 34-12
Armadillos, 37-19, *37-19*
Artemisinin, C4-4
Arteries, *39-14*, **39-15**–39-16
   pulmonary, 39-19, 39-20
Arterioles, *39-14*, **39-15**
Arthrobotrys [fungus], 34-5, *34-5*
Arthropods [Arthropoda], **44-14**–44-17
   Cambrian, *44-28*
   exoskeletons of, *37-13*, 37-13–37-14
   phylogenetic tree for, 44-17
   Visual Synthesis of history of, *44-31*
Artificial selection, *21-12*, **21-12**–21-13
   and behaviors, *45-5*, 45-5–45-6
   and genetic variation, 33-21
   limits of, 21-12, *21-12*
Ascomycetes, *34-14*, **34-14**–34-16
   life cycle of, 34-15, *34-15*
Asexual reproduction, 42-2, **42-2**–42-3
   and biological species concept, 22-3–22-4
   evolutionary problems with, 42-4–42-7
   vegetative, 30-22, 30-22–30-23
Aspen trees, *30-22*, 30-23
Assimilation, **26-10**, *26-10*
Associative learning, **45-9**
Astrocytes, **35-6**
Atelopus varius [toad], 48-19
Athyrium felix-femina [lady-fern], 33-9
Atmosphere
   history of oxygen in, *25-14*, 25-14–25-15, 26-20, 44-27, 44-29
   and the long-term carbon cycle, 25-9–25-12, *25-10*, *25-11*
   and the short-term carbon cycle, 25-1–25-6, *25-2*
ATP. *See* Adenosine triphosphate (ATP)
Atrioventricular (AV) nodes, **39-20**, *39-21*
Atrioventricular (AV) valves, *39-19*, **39-20**
Atrium, **39-18**, *39-18*
Auditory cortex, **36-10**
Australopithecus afarensis [hominin], 24-4, 24-5, *24-5*, 24-6

Autocrine signaling, **38-17**
Autoimmune diseases, **43-16**
Autonomic nervous system, **35-16**–35-17, *35-17*
Auxin, *31-7t*, **31-8**
 and branching, 31-10
 and phototropism, 31-17–31-18
 and root development, 31-14–31-15
 and stem development, 31-8, *31-8*–*31-9*
Avirulent pathogens, **32-3**
AVR proteins, 32-4, *32-6*
Axial skeletons, **37-15**, *37-15*
Axillary buds, **31-5**–31-6, *31-6*
 and branching, 31-10, *31-10*
Axon hillocks, *35-4*, **35-5**
Axons, *34-5*, *35-4*, 35-5
Azolla [fern], 33-10, *33-10*–*33-11*

B

B cells (B lymphocytes), *43-3*, **43-7**
 antibody production in, 43-7–43-9, *43-8*
 and genomic rearrangement, 43-10–43-13
Bacillus thuringiensis (Bt) [bacterium], C6-4, 32-18, *32-18*
Bacteria, **26-1**. *See also* Archaea; Eukarya; *specific* bacterium
 anoxygenic, 26-7, *26-7*
 asexual reproduction in, 42-2, *42-2*
 cells of, 26-1–26-2, *26-2*
 as consumers, 25-12, *25-12*
 cyano-, 26-15–26-16, *26-16*
 and digestion, 40-20–40-21
 diversity of, 26-12–26-16, *26-16*
 evolutionary history of, *26-20*, 26-20–26-23
 and fermentation, 26-8
 gram-positive, 26-15, *26-15*
 infections by, 43-18, *43-18*
 and the long-term carbon cycle, 25-7
 and the nitrogen cycle, 26-10–26-12, *26-11*
 and nitrogen fixation, 29-17, *29-17*–*29-18*
 pathogenic, C5-2
 photosynthetic, 26-7, *26-7*–*26-8*, *26-8*, 26-15–26-16
 phylogenetic tree for, *26-12*, *26-14*, 26-14, *26-15*
 proteo-, 26-15, *26-15*
 sizes of, 26-2–26-3, *26-3*
 and the sulfur cycle, 26-10, *26-10*
 symbiotic, C5-2–C5-4, *C5-3*, 27-9, *27-9*
 Visual Synthesis of history of, *44-30*

Bacteriochlorophyll, 26-7, *26-8*
Balance, 36-6–36-8, *36-7*
Balancing selection, **21-10**
Ball-and-socket joints, **37-17**–**37-18**, *37-18*
Bamboo, 33-17
Bananas
 agriculture of, C6-2
 diseases of, *32-5*
Banksia [eudicot], *33-20*
Bark. *See* Cork cambium
Barnacles, 47-8
Baroreceptors, 41-19
Basidiomycetes, **34-14**, *34-14*, *34-17*, 34-17–34-19
 life cycle of, 34-18, *34-18*
Basilar membranes, *36-8*, **36-9**
Basophils, *43-3*, **43-4**
Batrachochytrium dendrobatidis [fungus], 48-19
Bats
 echolocation in, 36-10, *36-10*
 as pollinators, 30-11
 reciprocal altruism in, 45-17–45-18
Bdelloid rotifers, 42-6–42-7
Beaver dams, 47-11, *47-11*
Bees
 and behavior, 45-6–45-7, *45-7*
 communication by, 45-15–45-16, *45-16*
 and evolution of flowers, 30-10–30-11
 fossils of, in amber, 47-22
 and kin selection, 45-19
 and pheromones, *38-17*, 38-18
 as pollinators, 30-11–30-13
 vision of, 45-1
Beetles, 44-17, 45-20
Behavior. *See* Animal behavior
Behavioral isolation, **22-5**
Bennett, J. Claude, 43-10–43-11
Bernard, Claude, 35-17
β-globin proteins. *See* Hemoglobin
Bilateral symmetry, **44-2**–44-3, *44-3*
Bilaterians [Bilateria], **44-3**
 chordates, 44-19–44-21
 deuterostomes, 44-18–44-19
 ecdysozoans, 44-14–44-17
 lophotrochozoans, 44-11–44-14
 phylogenetic tree for, *44-5*
 protostomes, **44-4**
Bile, **40-17**–40-18, *40-18*
Binary fission, 42-2, *42-2*
Biodiversity hotspots [Case 8], C8-1–C8-3, *C8-2*, *C8-3*
 and biomes, 47-21–47-22
 and climate change, 48-7, 48-7–48-10, *48-8*

and evolution of coral reefs, 44-29
and habitat loss, 48-13–48-17, *48-16*
and island biogeography, 46-15–46-16, *46-16*
Biofuels, C5-3–C5-4
Biological clocks, 45-12–45-13
Biological species concept (BSC), **22-2**–22-3
 application of, 22-3–22-4
 complications for, 22-4, 26-14
 extensions of, 22-4–22-5
Biologists, 48-19–48-20, *48-20*
Bioluminescence, C5-4, *C5-4*, 26-21
Biomass, **25-6**
 and trophic pyramids, 25-13, 47-15
Biomes, **47-16**, 47-17–47-19
 distribution of, *47-19*
 nature of, 47-19, 47-19–47-20, *47-20*
 number of species in, 47-20–47-22, *47-21*
Biomineralization, **25-8**, *25-8*
Biotrophic pathogens, **32-2**–32-3
Bipedalism, **24-5**, 24-9–24-10, *24-10*
Bipolar cells, **36-15**, *36-15*
Bird song, 45-15, *45-15*
Birds, **44-25**–44-26. *See also specific* bird
 communication by, 45-14, 45-15, *45-15*
 evolution of, 23-16, *23-17*, 44-26, 44-29
 hearts of, 39-19–39-20
 intersexual selection in, 45-20–45-21, *45-21*
 learning by, 45-9
 navigation by, 45-11–45-13, *45-13*
 as pollinators, 30-11
 and seed dispersal, 30-16, *30-16*
 sexual dimorphism in, 45-20, *45-20*
 Visual Synthesis of history of, *44-31*
Birds of paradise, 42-4, *42-5*, 45-3
Births
 and population size, 46-3, *46-3*
 rates of, 46-8, *46-8*, 46-9
Bivalves, **44-14**, *44-14*
Bladders, **41-11**, *41-11*
Blakemore, Richard, 45-11
Blastula, **28-11**, *28-11*, **42-18**, *42-19*
Blood, **39-3**. *See also* Red blood cells; White blood cells
 flow of, 39-15, *39-15*
Blood pressure, regulation of, by kidneys, 41-18, 41-19

Blood types, *21-4t*
 and genetic variation, 21-4
Bohr, Christian, 39-13
Bombyx [moth], 45-2, *45-2*
Bone marrow, 37-16, *37-17*, **37-17**
 and B cells, 43-7
 and white blood cells, 43-3
Bones, 37-17. *See also* Joints; Skeletons
 formation of, 37-16, 37-16–37-17
 and movement, 37-1, 37-2
Bonobos, *24-2*, 24-3
Bony fish, **44-23**, 44-23–44-24
Book lungs, 44-17
Borlaug, Norman, 30-21–30-22, 32-18
Borthwick, H. A., 30-18, 30-20
Bosch, Karl, 29-18
Bottlenecks, **21-13**
Bowerbirds, 42-4, *42-5*, 45-21
Bowman's capsule, **41-12**, *41-13*
Bracket fungi, 34-17, *34-17*
Brain hormone, 38-4, *38-4t*, 38-10
Brains, **35-2**, 36-17–36-19. *See also* Animal nervous systems
 anatomy of, *36-17*, *36-18*, 36-18–36-19
 and blood–brain barrier, 35-6
 versus body size, 24-11, *24-11*
 and central nervous system, 35-15
 and cognition, 36-20, 36-20–36-21, *36-21*
 evolution of human, 24-5, *24-6*, 24-11–24-12
 information processing in, *36-17*, 36-17–36-18
 somatosensory map of, 36-19, *36-19*
Brainstem, *36-17*, **36-17**–36-18
Brakefield, Paul, 28-15
Branching, 31-5–31-6, *31-6*
 and hormones, 31-10, *31-10*, 31-15
Breathing
 and smoking, 39-8
 through gills, 39-4, 39-4–39-5, *39-5*
 through lungs, 39-4, 39-6, 39-6–39-7
 through tracheae, 39-4, 39-5–39-6
Breeding. *See also* Hybridization
 and behaviors, 45-5, 45-5–45-6
Bronchi, primary, **39-7**, *39-7*
Bronchioles, *39-7*, **39-8**
Bryophytes, **29-1**, *29-2*, **33-2**
 convergent evolution in, 33-4
 diversity of, 33-2–33-5
 fertilization in, 30-4
 life cycle of, 30-2–30-4, *30-3*

BSC. *See* Biological species concept (BSC)
Bt. See Bacillus thuringiensis (*Bt*) [bacterium]
Buccal cavity, **40-15**
Buchnera aphidicola [bacterium], C5-4, 47-7, *47-7*
Bud scales, **31-4**–31-5, *31-5*
 and winterizing, 31-20
Budding, **42-2**, **42-2**–42-3
Bulbourethral gland, 42-10, **42-11**
Bulk flow, **39-2**–39-3
Bulk transport, 28-5, **28-5**–28-6, 28-10
Bundle sheaths, 29-7, **29-7**–29-8
Burgess Shale, **23-13**, *23-13*, 44-28
Bursa of Fabricius, **43-7**, *43-7*
Bushbabies, 24-1, *24-2*
Butler, Joyce, 24-17
Butterflies
 and biological species concept, 22-2, 22-3
 and color patterns in, 28-15, *28-15*
 estimating population sizes of, 46-7, *46-7*
 and evolution of flowers, 30-10–30-11
 life cycle of, 46-9, *46-9*
 metamorphosis of, 44-17, *44-17*
 plant defenses against, *32-8*, 32-8–32-9
 as pollinators, 30-11

C

C_3 plants, **29-7**
C_4 plants, *29-7*, **29-7**–29-8, *29-8*
Cactus, beavertail [*Opuntia basilaris*], 33-20
Cadherens, and multicellularity, 28-6–28-7
Caenorhabditis elegans [nematode], fungal infections in, 34-5, *34-5*
Caesarian sections, 42-22
Calcitonin, 38-9t, 38-15
Calcium
 and heartbeat, 39-20
 and muscle contraction, 37-6–37-7, *37-7*
Calcium carbonate
 and the long-term carbon cycle, 25-7, 25-7–25-8, 25-9
 and ocean acidification, 48-8–48-9, *48-9*
 sources of, 27-22
Calmette, Albert, 43-18
Calmodulin, **37-7**
Calvatia gigantea [fungus], 34-17
Calvin cycle
 in C_4 plants, 29-7
 in CAM plants, 29-6

CAM. *See* Crassulacean acid metabolism (CAM)
Cambium. *See* Cork cambium; Vascular cambium
Cambrian explosion, 44-27–**44-28**, *44-28*
Camponotus leonardi [ant], 34-16
Candida albicans [yeast], 34-3, 34-4
Canines [teeth], **40-14**, *40-14*
Cann, Rebecca, 24-6–24-8, 24-13
Cannon, Walter, 35-17
Capacitation, **42-17**
Capillaries, 39-14, **39-15**
 pulmonary, 39-7, *39-8*
Carbon, and radiometric dating, 23-15, *23-15*
Carbon cycle, **25-1**
 and ecology, 25-12, 25-12–25-14, *25-13*, *25-14*
 and evolution, 25-13–25-15
 and fungi, 34-3
 history of, 26-20–26-21
 and human activities, 25-1, 25-3–25-4, 48-3–48-11
 long-term, 25-6–25-7, 25-6–25-12, *25-9*
 and prokaryotes, 26-6–26-9
 short-term, 25-1–25-6, *25-2*
Carbon dioxide. *See also* Atmosphere; Carbon cycle
 and gas exchange, 39-1–39-3, *39-3*
 as greenhouse gas, 48-4, *48-4*
 history of atmospheric, 25-3, *25-3*, 25-9–25-12, *25-10*, *25-11*, *48-8*
 sources and sinks of, 25-5, 25-5–25-6
Carbonates. *See* Calcium carbonate
Cardiac cycle, **39-20**, *39-20*
Cardiac muscle, **37-2**, *37-2*, **39-20**
Cardiac output (CO), 39-15, **39-21**–39-22
Cardinals, 45-20
Carotid bodies, **39-9**, *39-9*
Carpels, 30-9, **30-10**
Carroll, Lewis, 42-5, 48-14
Carrying capacity (K), 46-5, **46-5**–46-6
 Visual Synthesis of, *48-12–48-13*
Cartilage, 37-13, **37-14**
 transformation of, into bone, 37-17
Cartilaginous fish, 44-23, *44-23*
Case 4. *See* Malaria [Case 4]
Case 5. *See* Microbiomes [Case 5]
Case 6. *See* Agriculture [Case 6]
Case 7. *See* Predation [Case 7]
Case 8. *See* Biodiversity hotspots [Case 8]
Casparian strips, *29-15*, **29-16**, 31-14, *31-14*

Catabolism, **40-3**
Caterpillars. *See* Butterflies; Moths
Cats, skeletons of, 37-15
Cattle, microbiomes in, C5-3
Caulerpa [alga], 28-2, *28-2*
Causation, **25-4**
Cavitation, **29-12**, *29-12*
 in conifers, 33-14
CCK. *See* Cholecystokinin (CCK)
Cecum, **40-22**, *40-22*
Cell-mediated immunity, **43-13**–43-16
Cell walls, **26-2**
 bacterial, 26-2, *26-2*
Cells. *See also* B cells (B lymphocytes); Egg cells; Sperm cells; T cells (T lymphocytes)
 amacrine, 36-15, *36-15*
 antigen-presenting, 43-15, *43-15*
 bipolar, 36-15, *36-15*
 chloride, 41-5, *41-5*
 coenocytic, 27-12, 28-2, *28-2*
 companion, 29-13, *29-13*
 cone, 36-14, *36-14*, *36-15*
 dendritic, 43-3, 43-4–43-5, *43-5*
 effector, **43-9**, 43-10
 eukaryotic, 27-1–27-4, *27-2*
 follicle, 42-13, *42-14*
 ganglion, 36-15, *36-15*
 glial, 35-6
 guard, 29-5, *29-5*
 hair, 36-6, *36-8*, 36-8–36-10, *36-9*
 horizontal, 36-15, *36-15*
 Leydig, 42-13
 mast, 43-3, 43-4–43-5, *43-5*
 memory, 43-3, **43-9**, 43-10
 multinucleated, 37-3, *37-3*
 natural killer, 43-3, 43-4
 neurosecretory, 38-13, *38-13*
 origin of, 27-7–27-9, *27-8*
 pacemaker, 39-20, *39-21*
 parenchyma, 29-8, *29-9*
 plasma, 43-3, **43-9**, 43-10
 procambial, 31-8, 31-9
 prokaryotic, 26-1–26-2, *26-2*
 pyramidal, 35-5, 35-5–35-6
 red blood, 21-10, *21-11*
 rod, 36-14, *36-14*, *36-15*
 Schwann, 35-6, *35-6*
 Sertoli, 42-13
 stem, 43-3, *43-3*
 totipotent, 31-1
 white blood, 43-3, *43-3*
Cellular junctions, tight, 40-19
Cellular respiration, and the carbon cycle, 25-2, *25-2*, 25-14
Cellulase, **40-21**
Cellulose, and biofuels, C5-3–C5-4
Centipedes, 44-15, *44-16*

Central nervous system (CNS), *35-15*, 35-15–**35-16**
Cephalization, **35-4**
Cephalochordates, **44-19**, 44-20, **44-21**
Cephalopods, 44-13, **44-13**–44-14
Cerebellum, 36-17, **36-17**–36-18
Cerebral cortex, **36-17**, *36-17*
Cerebrum, *36-17*, **36-18**
Cervix, **42-12**, *42-12*
 and childbirth, 42-22
Chaparral biome, 47-18, *47-18*
Chara [alga], 30-2, *30-2*
Character states, **23-5**
Characters, 23-5, **23-5**–23-6, 44-2
Cheetahs
 muscles of, 37-2, 37-12, 37-19
 as predators, 35-1, *35-2*
Chelicerates, **44-15**, 44-17
Chemautotrophs, **26-9**, *26-9*
 and nitrogen fixation, 26-11–26-12
Chemoreceptors, **36-2**, *36-3*
Chengjiang fossils, 44-28
Childbirth, 42-22, *42-22*
 and positive feedback, 38-6–38-7
Chimpanzees, 24-2, *24-3*
 culture in, 24-16, 24-17
 and human origins, 24-6, 24-6–24-7, *24-7*, 24-12, *24-12*
 metabolism of, 40-6
Chitin, **34-2**, **37-15**, **44-15**
 in cell walls, 34-2
 and pathogens, 32-4
Chlamydomonas reinhardtii [alga], *27-15*
 life cycle of, 27-3, 28-9
 and photosynthesis, 27-14
Chloride cells, **41-5**, *41-5*
Chlorophyll, versus bacteriochlorophyll, 26-7, *26-8*
Chloroplasts
 and cellular metabolism, 27-2, *27-2*
 and endosymbiosis, 27-4–27-6, *27-5*, *27-6*
 origin of, C5-4
Choanocytes, 44-5, **44-5**–44-6
Choanoflagellates, **27-10**–27-11, *27-11*
 and cell adhesion, 28-6–28-7, *28-7*
Chocolate trees, 47-1, 47-3, *47-3*, 47-6
Cholecystokinin (CCK), 38-8, 38-10t, **40-18**, *40-18*
Cholesterol, 38-7
 and steroid hormones, 38-7
Chomsky, Noam, 24-17
Chondrichthyes, **44-23**, *44-23*

Chondroblasts, 37-16
Chordates [Chordata], **44-18**, 44-19–44-21
 and vertebrates, 44-21, *44-21*
Chorion, **42-9**, *42-9*
Chromatin remodeling, and vernalization, 30-18
Chromatotropins, 38-4*t*
Chroococcus [cyanobacterium], *26-16*
Chrysanthemums, polyploidy in, 22-14
Chytrids, **34-12**–34-13, *34-13*
Chytriomyces hyalinus [fungus], *34-13*
Cichlid fish, 38-18, *38-18*
Cilia, **43-2**
Circadian clocks, **45-12**–45-13
 in angiosperms, 30-18
Circular muscles, **40-21**, *40-21*
Circulation, **39-3**
Circulatory systems. *See* Animal circulatory systems; Vascular systems
Citric acid cycle, **40-3**
Cladistics, **23-6**
Clams, 44-14, *44-14*
 exoskeleton of, 37-14, *37-14*
Class I genes [MHC], **43-15**
Class II genes [MHC], **43-15**
Class III genes [MHC], **43-15**
Class switching, **43-9**
Classes, **23-4**, *23-5*
Classical conditioning, **45-9**
Classification. *See* Taxonomy
Cleavage, **42-17**, **42-17**–42-18, *42-18*
 Visual Synthesis of, *42-20*
Clements, Frederic, 47-9, 47-12
Climate change
 and climate models, 48-5
 and communities, 48-5–48-10, *48-6*, *48-7*, *48-8*, *48-9*
 mitigating, 48-10, 48-10–48-11
Climax communities, 47-12, **47-13**
Clitoris, **42-13**, *42-13*
Cloacae, **41-11**, *41-11*
Clonal selection, 43-9, **43-9**–43-10
 and immunological memory, 43-10, *43-10*
Clones, **42-2**
Closed circulatory systems, **39-13**–39-14, *39-14*
Clostridium difficile [bacterium], C5-2
Clostridium tetani [bacterium], 37-11, 43-2
Cnidarians [Cnidaria], **44-7**, 44-7–44-8, *44-8*
 hydrostatic skeletons of, 37-13, *37-13*–37-14
 phylogenetic tree for, *44-6*
 Visual Synthesis of history of, *44-31*

CNS. *See* Central nervous system (CNS)
CO. *See* Cardiac output (CO)
Co-speciation, **22-11**, *22-11*
Coccolithophorids, 27-22, *27-22*
 and ocean acidification, 48-8–48-9, *48-9*
Cochlea, 36-7, 36-8, **36-9**
Cochlear ducts, 36-8, **36-9**
Codium [alga], 28-2, *28-2*
Coelacanths, 44-24, **44-24**–44-25
Coelomates, **44-3**, *44-3*
Coelum, 44-3, *44-3*
Coenocytic cells, **27-12**, **28-2**, *28-2*
Coevolution, **26-21**
 of humans and parasites, C4-2–C4-4, *C4-3*, 22-11–22-12
 and mutualisms, 47-6
 and plant defenses, 32-16–32-17
 of prokaryotes and eukaryotes, 26-21–26-23
Cognition, 36-20, **36-20**–36-21, *36-21*
Cohorts, **46-10**
Coleochaete [alga], 30-2, *30-2*
Collagen, 37-15, **37-16**, **39-16**
Collecting ducts, **41-11**, *41-11*
Colon, **40-20**
Colonization of land, 44-28–44-29
 and convergent evolution, 44-24
 and reproduction, 42-7, 42-7–42-9
Columbines [*Aquilegia*], 30-12, *30-13*
Comb-jellies, 44-8–44-9, *44-9*
Commensalism, 47-8, **47-8**–47-9, 47-8*t*
Communication, 45-13–45-16, **45-14**
 sensitive periods for learning, 45-15, *45-15*
Communities, **47-9**–47-13
 and climate change, 48-5–48-10, *48-6*, *48-7*, *48-8*, *48-9*
 climax, 47-12, 47-13
 and disturbances, 47-11–47-12
 keystone species of, 47-10–47-11, *47-11*
 and predator–prey interactions, 47-9–47-10, *47-10*
 succession in, 47-12, 47-12–47-13
 and trophic pyramids, 25-13
 Visual Synthesis of, *48-12*–48-13
Compact bone, **37-17**, *37-17*
Companion cells, **29-13**, *29-13*
Compasses, sun, 45-12–45-13, 45-16
Competition, **47-4**, 47-8*t*
 intraspecific versus interspecific, 46-5
 and niche size, 47-3

 and predation, 47-5, *47-5*
 for space, 47-5
Competitive exclusion, **47-4**–47-5
 and communities, 47-9
Complement system, 43-6, **43-6**–43-7
Complex multicellularity, **28-1**, 28-3, 28-3–28-4
 construction of, 28-6–28-10
 evolution of, 28-12, 28-12–28-16, *28-13*, *28-14*
 plants versus animals, 28-10–28-12
Compound eyes, **36-12**, *36-12*
Computers
 and climate models, 48-5
 and phylogenetic trees, 23-8
Conditioning, **45-9**
Cone cells, **36-14**, *36-14*, *36-15*
Cones, 33-13, *33-13*
 ovule versus pollen, 30-6, 30-7
Conifers, 33-13, 33-13–33-14
Conjugation [DNA transfer], **26-4**, *26-4*
Connective tissue, and endoskeletons, 37-15
Connell, Joseph, 32-16, 47-12
Consciousness, non-human, 24-18
Constant (C) regions, **43-8**, *43-8*
Constitutive defenses, **32-13**
Consumers, **25-12**, *25-12*, 47-14, **47-14**
Continuous traits, **21-9**
Contraception
 female, 42-14
 male, 42-11
Convergent evolution
 and analogous characters, 23-6
 and biomes, 47-20, *47-20*
 in bryophytes, 33-4
 and colonization of land, 44-24
 in fungi, 34-2
 of kidneys, 41-12
 of single-lens eyes, 36-12–36-13, *36-13*
 of suspension filter feeding, 40-12
 of uric-acid production, 41-8
Coral reefs. *See* Biodiversity hotspots [Case 8]
Corals, 44-7–44-8, *44-8*
 asexual reproduction of, 42-2, *42-3*
 bleaching of, 48-8, *48-8*
 and competition for space, 47-4
 symbiosis in, 27-5, *27-6*
Cork, 31-12, *31-12*
Cork cambium, **31-11**, *31-11*
 and bark, 31-12, *31-12*
 and innate immunity, 43-2, *43-2*

Cormorants, 47-4
Corn. *See Zea mays* [corn]
Corona radiata, *42-16*, *42-17*
Corpus luteum, **42-14**, *42-14*, *42-18*
Correlation, **25-4**
Cortex [brain], **36-18**, *36-18*
Cortex [kidney], **41-12**, *41-13*
Cortex [root], **29-15**, *29-15*, 31-14, *31-14*
Cortex [stem], **31-9**, *31-9*, *31-11*
Cortisol, 38-7, 38-8, 38-9*t*, 38-10, 38-14
Costus [monocot], *33-17*
Cotyledons, 30-15
 and monocots, 33-17
Countercurrent exchange, **39-5**, *39-5*, 41-15
Countercurrent multiplier, *41-15*, **41-15**–41-16
Courtship displays, 45-2, *45-3*
Crabs, 44-15
 bilateral symmetry of, 44-3
 mouthparts of, *44-16*
Cranes, 44-21
Cranial nerves, 35-15, **35-16**
Craniates, **44-19**. *See also* Vertebrates [Vertebrata]
Crassulacean acid metabolism (CAM), **29-6**, **29-6**–29-7
Crenarchaeota, *26-17*, **26-17**, 26-18, *26-18*
Cretaceous extinction, 23-18, *23-19*
Cro-Magnons, 24-8, 24-8–24-9. *See also* Humans
Crocodilians, 44-25
Crop, **40-14**
 and digestion, 40-16
Crops. *See* Agriculture [Case 6]
Cross-bridge cycles, **37-5**, *37-5*
Cross-bridges, **37-5**
 and muscle force, 37-8–37-8
Crosscurrent flow, 39-8, **39-9**
Crossing. *See* Hybridization
Crustaceans, **44-15**, *44-17*
Ctenophores [Ctenophora], 44-8–44-9, *44-9*
 phylogenetic tree for, *44-11*
Culture, 24-15–24-17, *24-16*
 non-human, 24-16–24-17, *24-17*
Cuticle [arthropod], 37-14, **37-15**
Cuticle [leaf], 29-3, **29-4**–29-5
 and innate immunity, 43-2, *43-2*
Cuttlefish, 44-13
Cyanobacteria, **26-15**–26-16, *26-16*
 and eutrophication, 48-11
 and origin of chloroplasts, 27-4–27-6, 27-5, *27-6*
Cyathea dealbata [fern], *33-10*
Cycads, 33-11–33-12, *33-12*

Cycas circinalis [cycad], 33-12
Cysts, 42-4, *42-4*
Cytokines, **43-4**
 and inflammation, 43-5, *43-5*
Cytokinins, 31-7*t*, 31-8, **31-10**
Cytoskeleton, **27-1**, *27-2*
Cytotoxic T cells, 43-3, **43-13**, 43-13*t*, 43-14
 activation of, 43-14–43-15, *43-15*

D

Daeschler, Edward, 23-20
Daphnia [crustacean], 42-4, *42-4*
Darwin, Charles, 21-8
 and altruism, 45-16–45-17
 and artificial selection, 48-14
 and evolution, 21-1, 23-1
 and fossils, 23-14
 and human evolution, 24-1, 24-4, 24-14, 24-18
 and natural selection, 21-8–21-9
 and phototropism, 31-17
 and plants, 33-1
 and sexual selection, 21-13, 24-14, 45-20
 and species, 22-1, 22-15
 and struggle for existence, 46-1, 46-4, 47-4
 and swim bladders, 44-24
 and vestigial structures, 40-23
Darwin, Francis, 31-17
Dating, radiometric, 23-15, *23-15*
Dawkins, Marion, 36-21
Dawsonia superba [moss], *33-4*
Day-neutral plants, **30-17**
Dead zones, 48-11, 48-11–48-14
Deaths
 and population size, 46-3, *46-3*
 rates of, 46-8, *46-8*, 46-9
Deceit, 45-14–45-15
Deciduous forest biome, 47-18, *47-18*
Decomposers, **25-12**, *25-12*
 fungi as, 34-1, 34-3–34-4, *34-4*
Defecation, **40-15**
Defenses. *See also* Animal immune systems; Plant defenses
 constitutive versus inducible, 32-13
Delayed hypersensitivity reactions, **43-14**
Deleterious mutations, 21-2–**21-3**
Demography, **46-9**
Dendrites, **34-5**, *35-4*, *35-5*
Dendritic cells, 43-3, **43-4**
 and inflammation, 43-5, *43-5*
Dendrochronology, 23-15
Denitrification, **26-11**, *26-11*
Density, population, **46-2**, *46-2*
 distribution of, 46-3, *46-3*

Density-dependent factors, **46-6**, *46-6*
Density-dependent mortality, 32-16–32-17, *32-17*
Density-independent factors, **46-6**, *46-6*
Deoxyribonucleic acid (DNA)
 "junk," 27-3
Depolarization, **35-8**
Dermis [skin], **43-2**, *43-3*
Descartes, René, 24-18
Descent of Man, and Selection in Relation to Sex [Darwin], *The*, 21-13, 24-1, 45-20, *45-20*
Desert biome, 47-18, *47-18*, 47-20
Desiccation, **29-1**
Desiccation tolerance, **29-2**, *29-2*
 and leaves, 29-4, 29-4–29-6, *29-5*
 and lichens, 34-7
Deuterostomes [Deuterostomia], **44-4**, *44-18*, **44-18**–44-19, *44-19*
 phylogenetic tree for, *44-18*
Development, 42-15–42-22
 of bones, 37-16, 37-16–37-17
 of the brain, 36-17, *36-17*
 conservation of genes involved with, 24-12, *24-12*
 and endocrine system, *38-2*, 38-2–38-4, *38-3*, *38-5*
 in humans, 42-17–42-22, *42-19*, *42-20*–*42-21*
 and multicellularity, 28-8–28-10, 28-15–28-16
 and neoteny, 24-10–24-11
 and taxonomy, 44-4, *44-4*
 Visual Synthesis of, *42-20*–*42-21*
Diabetes mellitus, **38-6**. *See also* Diseases and abnormalities
Diaphragm, 39-6, **39-6**–39-7, *39-7*
Diaphysis, **37-17**, *37-17*
Diastole, **39-20**, *39-20*
Diatoms, 25-8, *25-8*, 27-16, *27-17*
 history of, 27-22
 life cycle of, 27-3–27-4
Dickinsonia [Ediacaran fossil], 44-27, *44-27*
Dictyophora indusiata [fungus], *34-17*
Dictyostelium [slime mold], 27-13
Diet, 40-10–40-11
Dietary minerals, **40-10**–40-11, *40-11*
Differentiation, and multicellularity, 28-8–28-10
Diffusion, **26-2**, **28-4**, **39-2**
 and bacterial size limits, 26-2–26-3, *26-3*
 versus bulk transport, 28-4–28-6, *28-5*

 in capillaries, 39-16
 and gas exchange, 39-1–39-3
Digestion, 40-14–40-15, *40-15*
 intracellular versus extracellular, 40-14
Digestive tract, **40-14**–40-15, *40-15*
 large intestine, 40-20–40-21, *40-21*
 lining of, 40-21, *40-21*
 mouth, *40-15*, 40-15–40-16
 of plant-eaters, 40-21–40-23, *40-22*
 small intestine, 40-17–40-20, *40-18*, *40-19*
 stomach, *40-16*, 40-16–40-17
Dikarya, **34-14**, *34-14*
Dikaryotic stage, **34-10**, *34-10*
Dilger, William, 45-6
Dinoflagellates, 27-16, *27-17*, 27-19
 history of, 27-22
Dinosaurs
 evolution of, 23-16, *23-17*, 44-26, 44-29
 extinction of, 23-18, 44-29
 feathered, 23-11
 tracks of, *23-12*
 Visual Synthesis of history of, *44-31*
Diploblastic animals, 44-4, *44-4*
Diploidy, **27-3**
Directional selection, **21-12**, *21-12*
Discrete traits, **21-9**
Diseases and abnormalities. *See also* Drugs; Malaria [Case 4]
 allergies, 43-9
 amoebic dysentery, 27-12, *27-12*
 arthritis, 37-16, 43-16
 asthma, 43-9
 autoimmune, 43-16
 caused by prokaryotes, 26-22
 cholera, 40-20
 crown gall, *32-7*, 32-7–32-8
 diabetes, 38-6, 43-16
 and dialysis, 41-19
 and drug resistance, 48-18
 Dutch elm disease, 32-5
 hay-fever, 30-12
 immunodeficiency, 43-7
 influenza, 23-10, 43-17, *43-17*
 and invasive species, 48-18
 lactose intolerance, 40-19
 malnutrition, 40-11
 multiple sclerosis, 43-16
 obesity, 40-10
 osteogenesis imperfecta, 37-16
 osteoporosis, 37-17
 plague, 48-18
 sickle-cell anemia, 21-10, 21-11, 24-15
 smallpox, 43-10

 tuberculosis, 43-18, *43-18*, 48-18, *48-18*
 ulcers, 40-17
 yeast infections, 34-3, 34-5
Dispersal, **22-7**
 and speciation, 22-9, *22-13*
Displays, **45-2**
 advertisement, 45-15, *45-15*, 45-20, *45-21*
 and communication, 45-14
Disruptive selection, *21-11*, **21-13**, 22-12, *22-13*
Distal convoluted tubules, **41-14**, *41-14*, **41-17**
Disturbances, **47-11**–47-12
Diuretics, 41-18–41-19
Diversity. *See also* Genetic variation
 of Archaea, 26-16–26-19, *26-18*
 of Bacteria, 26-12–26-16, *26-16*
 of bryophytes, 33-2–33-5
 of chordates, 44-18, 44-19–44-21
 of cnidarians, 44-7, 44-7–44-8, *44-8*
 of ctenophores, 44-8–44-9, *44-9*
 of deuterostomes, 44-4, *44-18*, 44-18–44-19, *44-19*
 of ecdysozoans, 44-11, 44-14–44-17
 of Eukarya, 27-9–27-20
 and evolutionary history, 47-22–47-23
 of ferns, 33-9, 33-9–33-10
 of flowers, 30-9, 30-10–30-12, *30-11*
 of fungi, 34-12–34-19
 of gymnosperms, 33-11, 33-11–33-14
 of horsetails, 33-9, 33-9–33-10
 of lophotrochozoans, 44-11–44-14
 of lycophytes, 33-6–33-7, *33-7*
 of placozoans, 44-9, 44-9–44-11
 of plants, 33-1–33-2, *33-2*
 of roots, 31-15–31-16, *31-16*
 of sponges, 44-5, 44-5–44-7
 of vertebrates, 44-21–44-27
 Visual Synthesis of history of, *44-30*–*44-31*
Diversity gradient, latitudinal, *47-21*, **47-21**–47-22
DNA. *See* Deoxyribonucleic acid (DNA)
DNA sequencing
 and human origins, 24-3–24-4
 measuring genetic variation using, 21-3, 21-4, 21-6
 and phylogenetic trees, 23-8–23-9, *23-9*

Dobzhansky, Theodosius, 24-18
Dodder, *32-3*
Dogs
 breathing of, *39-4*
 breeding of, 45-5, *45-5*
 digestive tracts of, *40-15*
 learning by, *45-9*
 sensory receptors in, *36-2*
Domains, **23-4**, *23-5*. *See also*
  Archaea; Bacteria; Eukarya
 differences among, *26-5t*
 phylogenetic tree for, *26-5*
Domestication. *See also* Agriculture
  [Case 6]
 and centers of origin, 33-20–33-21, *33-21*
Dopamine, *38-7*
Dormancy, **30-8**
 germination after, 30-18, 30-20–30-21, *30-21*
Dorsal nerve cords, **44-18**
Double fertilization, **30-15**
Dragonflies, 44-15
Dreyer, William, 43-10–43-11
Drosophila melanogaster [fruit fly]
 and Bacteria, 26-21, *26-21*
 and behavior, 45-6, *45-6*
 and biological clocks, 45-12
 genetic variation in, 21-1, 21-5, *21-5*, 24-12
Drugs. *See also* Diseases and abnormalities
 and directional selection, 21-12
 from plants, 32-1, 32-10–32-11, *32-10t*
 resistance to, C4-3–C4-4, 26-5–26-6, 43-18, *48-18*, 48-18–48-19
 from trees, 33-13
Duodenum, **40-17**, *40-18*

E

Eardrums, 36-8, **36-9**
Ears, 36-7–36-8, *36-8*. *See also*
  Sound, sense of
Earthworms, 44-12, *44-12*
 digestive tracts of, *40-15*
 excretory organs of, 41-9–41-10, *41-10*
 hydrostatic skeletons of, 37-13, *37-14*
Eccles, John, 35-12
Ecdysone, 38-4, *38-4t*, 38-10
Ecdysozoans [Ecdysozoa], **44-11**, 44-14–44-17
 phylogenetic tree for, *44-15*
Echinoderms [Echinodermata], **44-18**, 44-19, *44-19*
Echolocation, **36-10**, *36-10*
Ecological footprint, **48-2**, *48-2*

Ecological niches, **22-4**
Ecological separation, **22-5**–22-6, *22-6*
Ecological species concept (ESC), **22-5**
Ecology, **46-1**. *See also* Populations
 and the carbon cycle, 25-12, 25-12–25-14, *25-13*, *25-14*
 Visual Synthesis of, *48-12–48-13*
Ecosystems, 47-13, **47-13**–47-16
 and food webs, 47-13–47-14, *47-14*
 and primary production, 47-15–47-16, *47-16*
 and trophic pyramids, 47-15, *47-15*
Ectoderm, *42-19*, **42-19**
Ectomycorrhizae, 29-16, **29-17**, **34-5**, *34-5*
Ectotherms, **40-7**
 Visual Synthesis of, *40-9*
Ediacaran fossils, 44-27, *44-27*
Effector cells, 43-9, **43-10**
Efferent neurons, **35-16**
Egg cells, as gametes, 27-4
Eggs, amniotic, 42-8–42-9, *42-9*, 44-25–44-26
Eggshells, 43-2, *43-2*
Egrets, *47-4*, *47-8*, *47-9*
Ehrlich, Paul, 32-17, 43-3, 43-6, 43-9, 48-16
Ejaculation, 42-16
Ejaculatory duct, **42-10**, *42-10*
EKG. *See* Electrocardiogram (EKG)
Elastin, **39-16**
Electrocardiogram (EKG), 39-21, *39-21*
Electrolytes
 and excretion, 41-7–41-12
 osmoregulation of, 41-1–41-7
Electromagnetic receptors, **36-10**
Electron transport chain, **40-3**
Electroreceptors, **36-4**
Elk, *44-21*
Elm trees, *32-5*
Elongation zones [of stems], 31-3, 31-3–31-4
Elton, Charles, 47-1
Embryonic development. *See* Development
Emigration, 46-3, *46-3*
Emiliania huxleyi [coccolithophorid], *48-9*
Encephalartos transvenosus [cycad], *33-12*
Endangered or threatened species
 cycads, 33-12
 frogs, 48-19, *48-19*
Endemic species, 46-15
Endocrine signaling, 38-16, 38-16–38-17

Endocrine systems. *See* Animal endocrine systems
Endocytosis, 27-1, *27-2*
Endoderm, *42-19*, **42-19**
Endodermis, **29-15**, *29-15*, 31-14, *31-14*, **44-7**
Endomembrane system, 27-2, *27-2*
Endomycorrhizae, 29-16, **29-17**, **34-5**
 Visual Synthesis of, *33-23*
Endophytes, **34-6**
Endoplasmic reticulum (ER), 27-2, *27-2*
Endoskeletons, **37-13**, *37-15*, 37-15–37-16
Endosperm, **30-15**, *30-15*
Endosymbiosis, **27-5**
 and chloroplasts, 27-4–27-6, *27-5*, *27-6*
 and mitochondria, 27-6–27-7, *27-7*
 and spread of photosynthesis, 27-16–27-19, *27-19*
Endotherms, **40-7**
 Visual Synthesis of, *40-8*
Energy, alternative, *48-10*, 48-10–48-11
Energy balance, **40-7**, 40-10
Energy intake, **40-10**
Energy use [metabolic], **40-10**
Enterotypes, C5-2
Eosinophils, 43-3, **43-4**
Ephedra [gnetophyte], 33-14, *33-14*
Epidermis [leaf], 29-3, **29-3**–29-6
Epidermis [root], **29-15**, *29-15*, 31-14, *31-14*
Epidermis [skin], 43-2, *43-3*, **44-7**
Epididymis, **42-10**, *42-10*, 42-11
Epiglottis, **40-15**, *40-16*
Epinephrine, 38-7, 38-8, *38-10t*, 38-16. *See also* Adrenaline
Epiphysis, 37-16, 37-17, **37-17**
Epiphytes, **29-6**
 bryophytes, 29-6
 ferns, 33-10, *33-11*
Epistylis [protozoan], 28-2, *28-3*
Epithelial tissue, 28-6
Epitopes, **43-8**, *43-8*
Equisetum [horsetail], 33-9, *33-9*
ER. *See* Endoplasmic reticulum (ER)
ESC. *See* Ecological species concept (ESC)
Escherichia coli [bacterium], *26-3*
 and digestion, 40-21
 dividing time of, 43-18
Esophagus, **40-14**, *40-15*
ESPS. *See* Excitatory postsynaptic potentials (EPSP)
Essay on the Principle of Population [Malthus], 21-8–21-9

Essential amino acids, **40-10**, *40-10t*
Estradiol, 42-14, *42-14*
Estrogen, 38-8, *38-9t*, 38-14, **42-13**, *42-13*
 and puberty, 42-13
Estrus cycle, **42-14**
Ethics. *See* Public policy
Ethnicity. *See* Ancestry
Ethylene, *31-7t*, 31-8
 and fruit ripening, 30-16
 and phototropism, 30-18, 31-20
Eudicots, 33-19–33-21, *33-20*
Eukarya, **27-9**. *See also* Eukaryotes
 diversity of, 27-9–27-20
 phylogenetic tree for, *26-14*, *27-10*
 spread of photosynthesis in, 27-16–27-19, *27-19*
Eukaryotes. *See also* Animals; Fungi; Plants; Prokaryotes; Protists
 infections by, 43-18–43-19, *43-19*
 life cycle of, 27-3–27-4, *27-4*
 Visual Synthesis of history of, *44-31*
Eumetazoa, 44-6, **44-7**
Euryarchaeota, **26-17**, *26-17*, 26-18
Eurypterids, 44-15
Eusocial species, **45-19**
Eutrophication, 48-11, **48-11**–48-14
Evaporative pumps, 29-10–29-11, *29-11*
Evapotranspiration, **47-19**–47-20, *47-20*
Evo-devo, **28-16**
Evolution. *See also* Coevolution; Convergent evolution; Phylogenetic trees
 of animal nervous systems, 35-1–35-2
 and the carbon cycle, 25-13–25-15
 of complex multicellularity, *28-12*, 28-12–28-16, *28-13*, *28-14*
 definition of, 21-6
 and developmental genes, 28-15–28-16
 of ears, 36-9
 evidence for, in fossils and phylogeny, 23-19–23-20, *23-20*
 and fossils, 23-19–23-20, *23-20*
 and Hardy–Weinberg equilibrium, 21-6–21-8
 and human activities, 48-15–48-19
 molecular, 21-14–21-16
 of plants, 29-1–29-2, 30-1–30-5
 of prokaryotes, 26-20, 26-20–26-23
 of reproduction, 42-1–42-7
 and vestigial structures, 40-23

Evolutionarily stable strategies, **45-17**
Evolutionary species concept (EvSC), **22-5**
Evolutionary trees. *See also* Genetic variation; Phylogenetic trees
    for Bacteria, 26-12, 26-14, *26-14, 26-15*
EvSC. *See* Evolutionary species concept (EvSC)
Excavates [Excavata], 27-17, 27-20
    Visual Synthesis of history of, *44-30*
Excitation–contraction coupling, 37-6–**37-7**, *37-7*
Excitatory postsynaptic potentials (EPSP), **35-13**, *35-14*
Excretion, **41-1**
    of nitrogenous wastes, 41-7, *41-7–41-8*
Excretory organs, **41-8**, 41-8–41-9, *41-9*
    in animals, 41-9, 41-9–41-10, *41-10*
Excretory tubules, **41-9**, *41-9*
Exercise, and metabolic rate, 40-3–40-5, *40-4*
Exhalation, **39-6**, *39-6*
Exocytosis, **27-1**, *27-2*
    and excretion, 41-8
Exoskeletons, **37-13**, 37-14, *37-14–37-15*
    and innate immunity, 43-2, *43-2*
Exponential growth, **46-4**, *46-4*
Extension [muscles], **37-10**, *37-10*
External fertilization, **42-7**, *42-7*
Extinctions. *See also* Mass extinctions
    and biological species concept, 22-3–22-4
    and climate change, 48-6
    and evolutionarily stable strategies, 45-17
    of hominins, 24-5
    and patch habitats, 46-13
    Visual Synthesis of, *22-16–22-17*
Extracellular digestion, **40-14**
Extracellular matrix, 28-6, **37-16**
    and endoskeletons, 37-16
Extraembryonic membranes, **42-9**, *42-9*
Extravasation, **43-5**–43-6, *43-6*
Eyecups, **36-12**, *36-12*
Eyes. *See also* Sight, sense of
    and color vision, 36-14, *36-14–36-15*, 36-15
    compound, 36-12, *36-12*
    function of, 36-13–36-14, *36-14*
    sensory processing by, 36-15, *36-15–36-16*
    types of, 36-11–36-13, *36-12*

F
Fab fragments, **43-8**, *43-8*
Facultative mutualisms, **47-6**
Fallopian tubes, **42-12**, *42-12*
Families, **23-4**, *23-5*
FAP. *See* Fixed action pattern (FAP)
Fast-twitch fibers, **37-11**, 37-11–**37-12**
Fc fragments, **43-8**, *43-8*
Feathers, evolution of, 23-16
Feature detectors, **45-4**, *45-4*
Feces, **40-15**, 40-20
Feedback, **38-5**
    negative, 35-17–35-18, *35-18*, 38-5, *38-5–38-6*
    positive, 35-8, 38-5, 38-6, *38-6*
Feeding, **40-12**, *40-12–40-14*, *40-13, 40-14*
Fermentation, **26-8**
    hindgut versus foregut, 40-22, *40-22*
Ferns, **29-2**
    diversity of, 33-9, *33-9–33-10*
    phylogenetic tree for, *33-9*
    and rice agriculture, 33-10–33-11
    Visual Synthesis of history of, *44-31*
Fertilization, *42-16*, 42-16–42-17
    external versus internal, 42-7, *42-7–42-8*
    in plants, 30-2
    Visual Synthesis of, *42-20*
    in vitro, 42-17
Fertilizers
    eutrophication, 48-11, *48-11–48-14*
    and nitrogen fixation, 26-12, 29-18
Fetuses, **42-18**
    development of, 42-18–42-19, *42-20–42-21*
    Visual Synthesis of, *42-21*
Fibers [muscle], **37-1**, *37-2*. *See also* Muscles
    slow-twitch versus fast-twitch, 37-11, *37-11–37-12*
Fibers [wood], **31-12**, *31-13*
Fiddleheads, 33-9, *33-9*
Fight-or-flight response, 35-17
    and adrenaline, 38-16
Filaments, **37-2**
    thick versus thin, 37-3, *37-3, 37-4*
Filial imprinting, **45-10**–45-11, *45-11*
Filtration, **41-9**, *41-9*
Finches, Galápagos, 22-9, *22-10*, 22-11
Firing rates, **36-4**, *36-4*
Fish, 44-22, *44-22–44-24*, 44-23
    age structure in, 46-8, *46-9*
    breathing of, 39-4, *39-4–39-5*, *39-5*
    cichlid, 38-18, *38-18*
    communication by, *45-14*
    feeding by, 40-12, *40-12*
    hearts of, 39-18, *39-18*
    jawless, 44-21–44-23, *44-22*
    kidneys of, 41-11, *41-11*
    life cycle of, 41-4, *41-4*
    osmoregulation in, *41-3*, 41-3–41-4
    and pheromones, 38-18, *38-18*
Fisher, Ronald, 21-9–21-10
Fitness, **21-9**
Fixation [of alleles], **21-10**
Fixed action pattern (FAP), **45-2**–45-3, *45-3*
Fixed populations, **21-6**
    Visual Synthesis of, *22-16–22-17*
Flagella, **42-10**
    in opisthokonts, 27-10–27-11, *27-11*
    of sperm cells, 42-10, *42-11*
    in sponges, 44-6
Flatworms, **39-2**
    excretory organs of, 41-9, *41-9*
    eyecups in, 36-12, *36-12*
    gas exchange in, 39-2
Fleas, 44-15
Fleshy-finned fish, **44-24**, *44-24*
Flexion, **37-10**, *37-10*
Flies, fruit. *See Drosophila melanogaster* [fruit fly]
Flight, adaptations for, 44-26
Florigen, 30-18, 31-8
Flowering plants. *See* Angiosperms
Flowers
    and climate change, 48-6, *48-6*
    development of, 31-6, *31-6*
    diversity of, *30-9*, 30-10–30-12, *30-11*
    and fruits, 30-15, *30-15–30-16, 30-16*
    and pollen, 30-9–30-10, *30-10*
    structure of, *30-9*, 30-9–30-10
    in ultraviolet light, 36-12, *36-12*
Fluxes, 25-6–**25-7**, *25-6–25-7*
Follicle cells, **42-13**, *42-14*
Follicle-stimulating hormone (FSH), 38-9t, 38-14, **42-13**, *42-13*
    and menstrual cycle, 42-13–42-14, *42-14*
Follicular phase, **42-14**, *42-14*
Food chains, **25-12**
Food webs, **25-12**, *25-12*, 47-9, **47-13**–**47-14**, *47-14*
Foraminifera, 27-21, *27-21*
Forebrain, **36-17**, *36-17*
Foregut, **40-14**, *40-15*
Fossil fuels, and atmospheric carbon dioxide, 25-4, *25-4–25-6, 25-5*, 48-3, *48-3*

Fossils, **23-11**–23-19
    in amber, C8-2, 47-22, *47-22–47-23*
    of angiosperms, 33-15
    conditions for formation of, 23-12
    Ediacaran, 44-27, *44-27*
    and evolution, 44-27
    and history of atmospheric carbon dioxide, 25-11
    and history of life, *23-11*, 23-11–23-13, *23-12, 23-13*, 47-22–47-23
    and history of multicellularity, 28-12–28-16, *28-13, 28-14*
    and human evolution, 24-4, *24-4–24-5, 24-5*, 24-6
    of lycophytes, 33-5–33-6, *33-6*
    micro-, 26-20, *26-20*, 27-20–27-22, *27-21, 27-22*
    molecular, 23-13
    and phylogeny, 23-19–23-20, *23-20*
    of plants, 31-1, *31-5*, 33-1
    of protists, 27-20–27-22, *27-21, 27-22*
    trace, 23-12, *23-12*, 28-13
    and transitional forms, 23-16, *23-17*
Founder effects, **21-13**
Fovea, 36-13, **36-15**
Fox, George, 26-5
Foxes, 47-9–47-10, *47-10*
Fragmentation, 42-2, **42-3**
Frisch, Karl von, 45-1, 45-16
Frogs, 44-25
    communication by, *45-14*
    external fertilization of, 42-7, *42-7*
Frontal lobes, 36-18, *36-19*, **36-19**
Fruit flies. *See Drosophila melanogaster* [fruit fly]
Fruiting bodies, **34-8**–34-9, *34-9*
Fruits, 30-15, **30-15**–30-16, *30-16*
    and mutualisms, 47-6
    Visual Synthesis of, *33-22*
FSH. *See* Follicle-stimulating hormone (FSH)
Fungi, **34-1**. *See also specific* fungus
    bulk transport in, 28-6
    as complex multicellular organisms, *28-3*, 28-4
    as consumers, 25-12, *25-12*
    as decomposers, 34-1, 34-3–34-4, *34-4*
    diversity of, 34-12–34-19
    early evolution of, 28-14, *28-15*
    hyphae of, 34-2, *34-2*
    in lichen, 34-6, *34-6–34-7, 34-7*
    life cycle of, 34-10, *34-10–34-11, 34-15, 34-15*, 34-18, *34-18*

and nutrient uptake by roots, *29-16*, 29-17
as pathogens, 34-4, 34-4–34-5, 34-5, 47-5
phylogenetic tree for, 34-12, *34-12*
reproduction of, 34-7–34-12, 34-8, 34-9, *34-11*
symbiosis in, 34-5, 34-5–34-7
yeasts, 34-2–34-3, *34-3*

G

Gage, Phineas, 36-17, *36-17*
Galápagos Islands. *See* Finches, Galápagos
Gallbladder, **40-18**, *40-18*
Gametes, **27-3**
Gametogenesis, 42-15, **42-15**–42-16
Gametophytes, **30-3**, *30-3*, *30-7*
Ganglia, **35-2**, 35-16
Ganglion cells, **36-15**, *36-15*
Gap junctions, **28-8**, *28-8*
Garner, Wightman, 30-17
Gas exchange, 39-1–39-3
  respiratory, 39-4–39-9
Gastric cavities, **44-7**, *44-7*
Gastrin, 38-8, 38-10t, 40-16, **40-17**
Gastropods, **44-13**, *44-13*
Gastrulas, **28-11**, *28-11*
Gastrulation, 28-11, *28-11*, **42-18**–42-19, *42-19*
  Visual Synthesis of, *42-20*
Gause, Georgii, 46-16
Gazelles, 35-1, 35-2
Geese
  and communities, 47-9–47-10, *47-10*
  fixed action pattern of, 45-2–45-3, *45-3*
  imprinting by, 45-10–45-11, *45-11*
Gehring, Walter, 36-11
Gel electrophoresis
  measuring genetic variation using, 21-3, 21-5, *21-5*
Gene flow, **21-13**
Gene pool, **21-2**
Gene regulation, and multicellularity, 28-8–28-10, 28-15–28-16
Genera [Genus], **23-4**, *23-5*
Genes, and behavior, 45-2–45-8
Genetic drift, **21-13**–21-14, *21-14*
  versus natural selection, 21-14, 22-15
  Visual Synthesis of, *22-16–22-17*
Genetic engineering
  and agriculture, C6-4, 32-8
  and horizontal gene transfer, 26-4

Genetic incompatibility, **22-6**
Genetic variation, 21-1–21-3. *See also* Diversity; Phylogenetic trees
  in Archaea, 26-16–26-19, *26-18*
  and artificial selection, 33-21
  in Bacteria, 26-12–26-16, *26-16*
  in humans, 24-12–24-15, *24-15*
  and the long-term carbon cycle, 25-13–25-15
  measurement of, 21-3–21-6, *21-5*
  and migration, 21-13
  and mutations, 21-13
  and sexual reproduction, 27-3–27-4
Genetically modified organisms (GMOs), *See also* Genetic engineering, and malaria, C4-4
Genetics, transmission. *See* Transmission genetics
Genomes, organization of, 27-2–27-3
Genomic rearrangement, 43-10–43-13, **43-11**, *43-12*
Genotypes, and Hardy–Weinberg equilibrium, 21-7, *21-7*
Geographic range, **46-2**, *46-2*
Geologic timescale, **23-14**, *23-14*
  Visual Synthesis of, *44-30*
Geology, 23-13–23-17. *See also* Fossils; Mass extinctions; Plate tectonics
  Burgess Shale, 23-13, *23-13*, 44-28
  and the Cambrian explosion, 44-27–44-28, *44-28*
  and dendrochronology, 23-15
  Ediacaran beds, 44-27, *44-27*
  and long-term carbon cycle, 25-6–25-12
  Messel Shale, 23-13, *23-13*
  and meteorites, 23-18
  and Pangaea, 23-16, *23-16*, 23-18
  and radiometric dating, 23-15, *23-15*
  and volcanoes, 23-18, 25-7, *25-9*
  and weathering, 25-7–25-8, *25-9*
Gephyrocapsa oceanica [coccolithophorid], *48-9*
Germ layers, *42-19*, **42-19**
Germ-line mutations, **21-2**
Germination, 30-18, 30-20–30-21, *30-21*
GH. *See* Growth hormone (GH)
Gibberellic acid, 31-7t, 31-8, **31-9**–31-10
Gibbons, 24-2, *24-2*
Gills, **39-3**, *39-4*
  breathing through, 39-4–39-5, *39-5*
Ginkgos, 33-12–33-13, *33-13*

Gizzards, **40-16**
Glaciations, 23-16, 25-10–25-11, *25-11*
  Visual Synthesis of history of, *44-31*
Glans penis, **42-11**
Glaucocystophytes, 27-13, *27-14*
Gleason, Henry, 47-9, 47-12
Glial cells, **35-6**
Global warming, 48-3, 48-3–48-5, *48-5*. *See also* Climate change
Glomeromycetes, **34-14**
Glomerulus, **41-11**, *41-11*
  filtration in, 41-12–41-13, *41-13*
Glucagon, 38-6, 38-8, 38-10t
Glucose
  absorption of, 40-19–40-20, *40-20*
  control of, 38-5–38-6, *38-6*
Gnetophytes, 33-14, *33-14*
Gnetum [gnetophyte], 33-14, *33-14*
GnRH. *See* Gonadotropin-releasing hormone (GnRH)
Golgi, Camillo, 35-6
Golgi apparatus, 27-2, *27-2*
Gonadotropin-releasing hormone (GnRH), **42-13**, *42-13*
Gonads, **42-10**
Gordon, Andrew, 37-8
Gorillas, 24-2, 24-3
Gottschalk, Carl W., 41-16
Gram, Hans Christian, 26-15
Gram-positive bacteria, **26-15**, *26-15*
Grand Canyon, 23-12
Granulocytes, 43-3, **43-4**
Granulomas, **43-18**, *43-18*
Grapes, wine [*Vitis vinifera*], diseases of, *32-6*
Grasses
  C$_4$ photosynthesis in, 29-8
  and communities, 47-10, *47-10*
  evolution of, 33-18, 33-18–33-19, *33-19*
  and grazing, 32-12, 32-12–32-13
  as monocots, 33-17
  pollination of, 30-12
Grasshoppers
  breathing of, 39-4
  digestive tracts of, *40-15*
  exoskeleton of, 37-14, *37-14*
  growth and development of, 38-2, *38-2*, 44-15
Gravitropism, **31-17**, 31-18, *31-18*
Gray matter, **36-18**, *36-18*
Green Revolution. *See* Agriculture [Case 6]
Greenhouse gases, **25-10**, **48-4**, *48-4*
Griffin, Donald, 36-21

Grinnell, Joseph, 47-1
Group selection, **45-16**–45-17, *45-17*
Growth
  exponential, 46-4, *46-4*
  intrinsic rate of, 46-4
  logistic, 46-5, 46-5–46-6
Growth hormone (GH), 38-4, 38-9t, 38-14, 38-17
Growth plates, 37-16, **37-17**
Growth [populations], 46-3–46-5
Growth rings, **31-11**–31-12, *31-12*
  and dendrochronology, 23-15
Guard cells, **29-5**, *29-5*
Guérin, Camille, 43-18
Gustation, **36-5**. *See also* Taste, sense of
Gymnosperms, **29-2**. *See also* Angiosperms
  diversity of, 33-11, 33-11–33-14
  life cycle of, 30-6, *30-7*
  phylogenetic tree for, *33-11*
  wood of, 31-12, *31-13*

H

Haber, Fritz, 29-18
Haber–Bosch process, 29-18
Habitat
  and coexistence of species, 46-16–46-17, *46-17*
  and island biogeography, 46-14, 46-14–46-16, *46-15*, *46-16*
  loss of, 48-16, *48-16*
  versus niche, 47-3
  patches of, 46-12, 46-12–46-13, *46-13*
Habituation, **45-8**
Hagfish, **44-22**, *44-22*
Hair cells, **36-6**
  and sense of sound, 36-8, 36-8–36-10, *36-9*
Half-life, **23-15**, *23-15*
Haloquadratum walsbyi [bacterium], *26-3*
Hamilton, William D., 45-18–45-19
Hanson, Jean, 37-4
Haploidy, **27-3**
Hardy, Godfrey H., 21-6
Hardy–Weinberg equilibrium, **21-6**–21-8, *21-7*
Hawks, 47-3
hCG. *See* Human chorionic gonadotropin (hCG)
Health. *See also* Diseases and abnormalities
  and intestinal bacteria, 26-22, 26-22–26-23
Hearing. *See* Ears; Sound, sense of
Heart rate (HR), **39-21**–39-22

Hearts, **39-13**, 39-17–39-22
   in amphibians, *39-18*, 39-18–39-19
   beating of, 39-20–39-21, *39-21*
   in fish, *39-18*, 39-18
   in mammals and birds, *39-19*, 39-19–39-20
   output of, 39-21–39-22
   in reptiles, 39-19
Heartwood, **31-11**
Heavy (H) chains, **43-8**, *43-8*
Height, human, and natural selection, 21-11
Heliobacter pylori [bacterium], 26-22, 40-17
Helper T cells, 43-3, **43-13**, 43-13*t*, 43-14
   activation of, 43-14–43-15, *43-15*
Hemichordates [Hemichordata], 44-18, **44-18**–44-19
Hemipenes, 42-11
Hemoglobin, *39-10*, **39-10**–39-11
   binding of oxygen by, *39-11*, 39-11–39-13, *39-13*
Hemolymph, **39-3**
   in open circulatory systems, 39-13–39-14
   oxygen transport by, 39-10
Henle, F. G. J., 41-14
Herbivory, 40-21–40-22, 47-8*t*
Herons, 47-4
Hess, Carl von, 45-1
Heterokaryotic stage, **34-10**, *34-10*
Heterotrophs, fungi as, 34-1
Heterozygote advantage, **21-10**
Heyne, Benjamin, 29-6
Hibernation, 40-10
Hill, A. V., 37-8, 37-9
Hindbrain, **36-17**, *36-17*
Hindgut, **40-14**, *40-15*
Hinge joints, **37-17**–37-18, *37-18*
Hippocampus, 36-17, **36-18**
Hispaniola, as biodiversity hotspot, C8-1–C8-3
Histamine, **43-4**
   and inflammation, 43-5, *43-5*
Histones
   evolution of, 21-16
HIV. *See* Human immunodeficiency virus (HIV)
"Hobbits," 24-5, *24-5*
Hodgkin, Alan, 35-11
Homeostasis, **35-2**, **38-5**, **41-1**
   and endocrine systems, 38-5–38-7, 38-6, 38-16
   and energy balance, 40-7, 40-10
   and nervous systems, 35-2, 35-17–35-18, *35-18*

   and osmoregulation, 41-1–41-7
   and respiratory systems, 39-9, *39-9*
   and thermoregulation, 40-7, 40-8–40-9
Hominins, 24-4, **24-4**–24-5, *24-5*, *24-6*
Homo erectus [hominin], 24-5, *24-5*, *24-6*
Homo ergaster [hominin], 24-5, *24-5*, *24-6*
Homo floresiensis [hominin], 24-5, *24-5*
Homo habilis [hominin], 24-5, *24-6*
Homo heidelbergensis [hominin], 24-5, *24-6*
Homo neanderthalensis [hominin], 24-5, *24-5*. *See also* Neanderthals
Homo sapiens [hominin], 24-5, *24-5*, *24-6*. *See also* Humans
Homologous characters, 23-5, 23-5–**23-6**, 44-2
Honeybees. *See* Bees
Hooker, J. D., 33-1
Horizontal cells, **36-15**, *36-15*
Horizontal gene transfer, 26-4, **26-4**–26-5
   and bacterial phylogeny, 26-12, 26-14, *26-14*
   and origin of eukaryotic cells, 27-7
Hormones, **31-7**, **38-1**
   amplification of signals from, 38-8, 38-10, 38-11
   and behaviors, 45-4–45-5, *45-5*
   classes of, 38-7, 38-7–38-8
   and development, 42-19
   evolutionarily conserved, 38-11–38-12
   invertebrate, 38-4*t*
   in plants, 31-6–31-10, 31-7*t*, 31-16
   receptors for, 38-11
   and reproductive systems, *42-13*, 42-13–42-14, *42-14*
   and urine concentration, *41-17*, 41-17–41-19
   vertebrate, 38-9*t*–38-10*t*
Hornworts, 29-1, 33-2, 33-3
Horses, 37-19, *37-19*
Horseshoe crabs, 44-15
Horsetails, 29-2
   diversity of, *33-9*, 33-9–33-10
   phylogenetic tree for, *33-9*
Host plants, **32-3**
Hot springs, and Archaea, 26-17–26-18
Hotspots [biodiversity]. *See* Biodiversity hotspots [Case 8]

"How Do We Know?"
   antibody diversity, 43-11, *43-11*
   Archaea abundance, 26-19, *26-19*
   arthropod mouthparts, 44-16, *44-16*
   artificial selection, 21-12, *21-12*
   asexual reproduction, 42-6, *42-6*
   atmospheric carbon dioxide, anthropogenic, 25-4, 25-4–25-5, *25-5*
   atmospheric carbon dioxide, historical, 25-3, *25-3*
   bacterial diversity, 26-13, *26-13*
   behaviors, genetic basis of, 45-7, *45-7*
   biological clocks, 45-12–45-13, *45-13*
   butterfly wing patterns, 28-15, *28-15*
   C_4 photosynthesis, 29-8, *29-8*
   cell adhesion, 28-6–28-7, *28-7*
   chloroplast origin, 27-5, *27-5*
   coevolution, 47-7, *47-7*
   coexistence of predators and prey, 46-17, *46-17*
   diversity and primary productivity, 47-16, *47-16*
   evolution of woody plants, 33-8, *33-8*
   fungal influence of insect behavior, 34-16, *34-16*
   fungal spore shape, 34-9, *34-9*
   hemoglobin and myoglobin structure, 39-10, *39-10*
   human origins, 24-6, 24-6–24-7, *24-7*
   insect growth and development, 38-3, *38-3*
   insect learning, 45-10, *45-10*
   membrane and action potentials, 35-11, *35-11*
   metabolic rate and body size, 40-6, *40-6*
   muscle contraction, 37-8, *37-8*
   ocean acidification, 48-9, *48-9*
   orientation, 45-12–45-13, *45-13*
   phototropism, 31-17, *31-17*
   phylogenetic trees, 23-10, *23-10*
   plant communication, 32-14–32-15
   pollinator shifts, 30-13, *30-13*
   population sizes, 46-7, *46-7*
   relatedness of humans and chimpanzees, 24-3, *24-3*
   resistance of plant to pathogens, 32-6, *32-6*
   seed germination, 30-20, *30-20*
   sensory processing in the retina, 36-16, *36-16*
   spread of photosynthesis in Eukarya, 27-18, *27-18*

   transitional forms, 23-18, *23-18*
   urine concentration by kidneys, 41-16–41-17
   vernalization, 30-19, *30-19*
   vicariance, 22-8, *22-8*
   water transport in plants, 29-10, *29-10*
Hox genes, 44-16
Hozumi, Nobumichi, 43-11
HR. *See* Heart rate (HR)
Hubbard, M. King, 48-10
Hubel, David, 36-16
Huffaker, Carl, 46-16–46-17
Human activities
   and the carbon cycle, 25-1, 25-3–25-4, 48-3–48-11
   and ecology, 48-12–48-13
   and evolution, 48-15–48-19
   to mitigate human impact, 48-10–48-11, 48-14–48-15, 48-19–48-20
   and the nitrogen cycle, *48-11*, 48-11–48-14
   and the phosphorus cycle, 48-14
   red tides, 27-16
Human chorionic gonadotropin (hCG), **42-14**
Human Development Index, 48-2
Human immunodeficiency virus (HIV), and phylogenetic trees, 23-10, *23-10*
Human microbiome. *See* Microbiomes [Case 5]
Humans
   anatomical features of, 24-9–24-12, *24-10*
   colonization by, *24-13*, 24-13–24-14
   evolution of, 24-4, 24-4–24-5, *24-5*, *24-6*
   female reproductive system in, 42-11–42-14, *42-12*, *42-13*, *42-14*
   fossil record of, 24-4, 24-4–24-5, *24-5*, *24-6*
   genetic variation in, 24-12–24-15, *24-15*
   male reproductive system in, 42-9–42-11, *42-10*, *42-11*
   origins of, 24-6–24-9, *24-9*
   phylogenetic tree for, 24-2, 24-2–24-4
   phylogeny of, 24-2, 24-2–24-4
   sense of sight in, 36-13, 36-13–36-14, *36-14*
   sense of sound in, 36-8
   sense of taste in, 36-6
   skeleton of, *37-17*
   as top predators, C7-4
Humoral immunity, **43-8**. *See also* Adaptive immune system

Humors, 43-7
Hutchinson, G. Evelyn, 47-1, 47-23
Huxley, Andrew, 35-11, 37-4, 37-8
Huxley, Hugh, 37-4
Hybridization, **22-4**. *See also* Breeding
    and biological species concept, 22-4
    and human origins, 24-9, *24-9*
    and instantaneous speciation, 22-13–22-15, *22-14*
Hydrochloric acid, 40-16
Hydrogenosomes, 27-7, *27-7*, 27-9
Hydrostatic pressure, *41-2*, *41-2–41-3*
Hydrostatic skeletons, *37-13*, **37-13**–37-14
Hydrothermal vents, 26-9, *26-9*
Hydroxyapatite, **37-16**
Hygrophorus miniatus [fungus], 34-9
Hypersensitive response [plants], **32-4**–32-5, *32-5*
Hypersensitivity reactions [animals]
    delayed, 43-14
    immediate, 43-9
Hyperthermophiles, **26-17**
Hypervariable regions, **43-8**, *43-8*
Hyphae, **34-2**, *34-2*
Hypothalamus, *36-17*, **36-18**, *38-12*, 38-12–38-14, *38-13*
    and hormonal signals, 38-8, 38-10, *38-11*
    and reproductive system, 42-13, *42-13*
    and thermoregulation, 35-18, *35-18*

I

Ice ages. *See* Glaciations
Ice cores
    and atmospheric carbon dioxide, 25-3, *25-3*
    and oceanic oxygen, 25-10, *25-10*
Ig. *See* Immunoglobulins (Ig)
Ileum, **40-17**, *40-18*, 40-19
Imitation, **45-9**
Immediate hypersensitivity reactions, **43-9**
Immigration, 46-3, *46-3*
Immune systems. *See also* Animal immune systems
    in plants, 32-3–32-4, *32-4*
Immunodeficiency, **43-7**
Immunoglobulins (Ig), 43-8, **43-8**–43-9
Immunological memory, 43-10, *43-10*

Implantation, 42-17
    Visual Synthesis of, *42-20*
Imprinting, **45-10**–45-11, *45-11*
In vitro fertilization (IVF), **42-17**
Incisors, **40-14**, *40-14*
Incus, *36-8*, **36-9**
Inducible defenses, **32-13**
Industrial Revolution
    and agriculture, 29-18
    and atmospheric carbon dioxide, 25-3, 25-4
Infections, 43-16–43-17. *See also* Animal immune systems; Diseases and abnormalities
    by bacteria, 43-18, *43-18*
    by eukaryotes, 43-18–43-19, *43-19*
    by viruses, 43-17, *43-17*
Inflammation, *43-5*, **43-5**–43-6
Influenza virus
    infection by, 43-17, *43-17*
    and phylogenetic trees, 23-10
Information processing, by neurons, 35-13–35-14, *35-14*, *35-15*
Inhalation, **39-6**, *39-6*
Inhibitory postsynaptic potentials (IPSPs), **35-13**, *35-14*
Innate behaviors, **45-2**, *45-2*
Innate immune system, **43-1**–43-7
    cells of, 43-3, *43-3*–43-4
    and the complement system, 43-6, 43-6–43-7
    features of, *43-2t*
    and inflammation, 43-5, *43-5*–43-6, *43-6*
    and pathogen recognition, 43-4, *43-4*
    and physical barriers, 43-2, *43-2*, *43-3*
Inner ear, *36-8*, **36-9**
Insects, **44-15**–44-17
    breathing of, 39-4, *39-5*–39-6
    excretory organs of, 41-10, *41-10*
    exoskeletons of, *37-13*, *37-13*–37-14
    giant, 44-29
    learning by, 45-9–45-10, *45-10*
    sense of sight in, 36-12, *36-12*
    sensory receptors in, *36-2*
Insel, Thomas, 45-8
Instantaneous speciation, **22-13**–22-15, *22-14*
Insulin, 38-6, 38-8, *38-10t*
Integrins, and multicellularity, 28-6–28-7
Interbreeding. *See* Hybridization
Intercostal muscles, **39-7**, *39-7*
Internal fertilization, **42-8**
Interneurons, **35-2**, 35-5

Internodes, **31-2**, *31-2*
Intersexual selection, **45-20**–45-21, *45-21*
Interspecific competition, **46-5**, **47-3**
Intervertebral discs, *37-13*, **37-14**
Intracellular digestion, **40-14**
Intrasexual selection, **45-20**–45-21, *45-21*
Intraspecific competition, **46-5**
Intrinsic growth rate (*r*), **46-4**
Invasive species, *48-17*, **48-17**–48-18
    and biodiversity hotspots, C8-2–C8-3
    Visual Synthesis of, *48-12–48-13*
Involuntary nervous system, **35-16**–35-17
IPSPs. *See* Inhibitory postsynaptic potentials (IPSPs)
Island biogeography, **46-14**, **46-14**–46-16, *46-15*, *46-16*
Island populations, **22-7**, 22-9
Isoetes lacustris [lycophyte], 33-7, *33-7*
Isometric forces, **37-9**, *37-9*
Isotopes
    and atmospheric carbon dioxide, 25-4–25-6, *25-5*
    and C$_4$-grass expansion, 33-18, *33-19*
    and geologic timescale, 23-14–23-15, *23-15*
    and paleotemperatures, 25-10, *25-10*
Isotype switching, **43-9**
Isotypes, **43-9**
IVF. *See* In vitro fertilization (IVF)

J

Janzen, Daniel, 32-16
Jasmonic acid, **32-13**
Jawless fish, **44-21**–44-23, *44-22*
Jaws, 40-13, *40-14*, 44-22
Jejunum, **40-17**, *40-18*, 40-19
Jellyfish, **44-7**, *44-7*–44-8, *44-8*
    radial symmetry of, 44-2, *44-3*
    size of, 28-4–28-5, *28-5*
Jenkins, Farish, 23-20
Jenner, Edward, 43-10
Joints
    motion of, 37-18–37-19, *37-19*
    temporomandibular, 40-13
    types of, 37-17–37-18, *37-18*
Julian, Fred, 37-8
Junipers, 33-13
"Junk DNA," 27-3
Juvenile hormone, 38-4, *38-4t*, 38-10
Juxtaglomerular apparatus, *41-18*, **41-19**

K

K-strategists, *42-8*, **42-8**, **46-10**–46-11, *46-11*
Kacelnik, Alex, 24-18
Kangaroo rats, 41-12, *41-12*
Karlson, Peter, 38-4
Karyogamy, **34-10**, *34-10*
Kawakami, A., 38-3
Keeling, Charles, 25-1, 25-3
Keeling curve, 25-1–25-2, *25-2*, 48-3
Kelp, 27-15–27-16, *27-16*
    as complex multicellular organisms, 28-4
    forests of, 47-10–47-11, *47-11*
Key stimuli, **45-2**, *45-3*
Keystone species, **47-10**–47-11, *47-11*
Kidneys, **41-1**
    blood filtering in, 41-10–41-12, *41-11*
    function of, 41-14–41-17, *41-15*, *41-16*
    regulation of blood pressure by, *41-18*, 41-19
    structure of, 41-12–41-13, *41-13*
    and urine production, *41-17*, 41-17–41-19
Kin selection, **45-18**–45-19
Kineses, **45-11**
King, Mary-Claire, 24-3–24-4, 24-10–24-11
Kingdoms, **23-4**, *23-5*
    of Eukarya, 27-9
Kingfishers, 22-9, *22-9*
Knee-jerk reflex, **35-18**–35-19, *35-19*
Komodo dragons, 42-2
Korarchaeota, 26-17, *26-17*
Krebs cycle. *See* Citric acid cycle
Krill, 44-15, 46-2, *46-2*
Krings, Matthias, 24-8
Kudus, 41-6, 41-6–41-7
Kudzu, *48-17*, 48-17
Kuffler, Stephen, 36-16
Kuhn, Alfred, 38-3, 38-4
Kuhn, Werner, 41-16–41-17

L

Labia majora, **42-12**–**42-13**, *42-13*
Labia minora, **42-13**, *42-13*
Laboratory techniques. *See also* "How Do We Know?"
    PET scans, 36-21
    radioactive labeling, 43-13
    sampling strategies, 46-6–46-7, *46-7*
    scanning electron microscopy, 27-20
    transmission electron microscopy, 27-5, 27-20

Ladybugs
   and ecological separation, 22-6, 22-6
   exoskeleton of, 43-2
   genotype versus phenotype in, 21-4, 21-4
   and pheromones, 38-17
Laetiporus sulphureus [fungus], 34-17
Lamellae, **39-4**, 39-5
Lampreys, **44-22**, 44-22
Language
   acquisition of, 45-2
   and anatomy, 24-10, 24-11
   and brain evolution, 36-19
   and communication, 45-14
   non-human, 24-17, 24-17–24-18
Large intestine, **40-14**, 40-15
   and digestion, 40-20–40-21, 40-21
Larynx, **39-7**, 39-7
Lateral buds. *See* Axillary buds
Lateral inhibition, **36-5**, 36-5
   and sight, 36-15, 36-16
Lateral line system, **36-7**, 36-7
Lateral meristems, **31-10**–31-11, 31-11
Latex, 32-8, **32-9**
Latitudinal diversity gradient, 47-21, **47-21**–47-22
Lavender [*Lavandula stoechas*], 30-9
Lavoisier, Antoine, 40-3
Leaf primordia, **31-2**, 31-2, 31-3, 31-6
Learned behaviors, **45-2**, 45-2
Learning, **45-8**–45-11
   associative, 45-9
   modes of, 45-9–45-11
   non-associative, 45-8–45-9
   sensitive periods for, 45-15, 45-15
Leaves, **29-3**, 29-3
   growth and development of, 31-4, 31-4–31-6, 31-5
   non-photosynthetic functions of, 31-4–31-5, 31-5
   transpiration in, 29-3–29-4, 29-4
Leeches, 44-12, 44-12
Legumes
   as eudicots, 33-20
   and nitrogen fixation, 29-18
Lemmings, 45-17, 47-9–47-10, 47-10
Lemurs, 24-1–24-2, 24-2
Lengthening contractions, 37-9, **37-9**–37-10
Lenticels, **31-12**, 31-13
Leptosporangia, 33-9, 33-10
Leucojum vernum [monocot], 33-17

Leukocytes, **43-3**, 43-3
Leydig cells, **42-13**
LH. *See* Luteinizing hormone (LH)
Lichens, 34-6, 34-6–34-7, 34-7
Liebig, Justus von, 47-15
Liebig's Law of the Minimum, **47-15**
Life, origin of [Case 1], 26-20–26-21
Life cycles
   of algae, 30-2, 30-3
   of amphibians, 44-24, 44-25
   of angiosperms, 30-14
   of animals, 27-4, 27-4
   of bryophytes, 30-2–30-4, 30-3
   of butterflies, 46-9, 46-9
   of cells, 27-3, 27-4
   of diatoms, 27-3–27-4
   of eukaryotes, 27-3–27-4, 27-4
   of fish, 41-4, 41-4
   of fungi, 34-10, 34-10–34-11, 34-15, 34-15, 34-18, 34-18
   of gymnosperms, 30-6, 30-7
   of insects, 44-17
   of plants, 27-4, 27-4, 30-1–30-5
   of vascular plants, 30-4–30-5, 30-5
Life histories, 46-11–46-12
Light (L) chains, **43-8**, 43-8
Lignin
   in cell walls, 29-9, 29-12
   and wood, 31-12
Limbic system, 36-17, **36-18**
Lime Swallowtail [butterfly], 46-1, 46-5, 46-6, 46-9, 46-9, 46-10
Limestone. *See* Calcium carbonate
Linnaeus, Carolus, 23-1, 44-2
Lions, 38-15, 38-15, 45-20
Lipases, **40-18**, 40-18
Liver, **40-17**, 40-18
Liverworts, 29-1, 33-2, 33-3
Lizards, 44-25, 44-25
   asexual reproduction of, 42-6
   hormonal triggers of behavior in, 45-4–45-5, 45-5
   and island biogeography, 46-15–46-16, 46-16
   and life history, 46-11–46-12
   niches of, 47-2, 47-2
   thermoregulation in, 40-9
   Visual Synthesis of history of, 44-31
Llamas, 39-13
Loblolly pine trees [*Pinus taeda*], 30-6, 30-7
Lobsters, 44-15
Lock and key systems, **22-5**
   and antibody specificity, 43-9
Logistic growth, 46-5, **46-5**–46-6
Loligo forbesi [squid], 35-11
Long-day plants, **30-17**

Long-term potentiation (LTP), 36-20
Longitudinal muscles, **40-21**, 40-21
Loops of Henle, 41-14, **41-14**–41-16, 41-15
Lophotrochozoans [Lophotrochozoa], **44-11**–44-14
   phylogenetic tree for, 44-11
Lorenz, Konrad, 45-1, 45-10–45-11, 45-11, 45-16
Lovebirds, 45-6
LTP. *See* Long-term potentiation (LTP)
Lucy, 24-4, **24-5**
Lumen, **40-21**, 40-21
Lunar clocks, **45-12**
Lungfish, **44-24**, **44-24**–44-25
   Visual Synthesis of history of, 44-31
Lungs, **39-4**, 39-4
   avian, 39-8, 39-8–39-9
   book, 44-17
   breathing through, 39-6, 39-6–39-7
   evolution of, 44-24
   mammalian, 39-7, 39-7–39-8
Luteal phase, **42-14**, 42-14
Luteinizing hormone (LH), 38-9t, 38-14, **42-13**, 42-13
   and menstrual cycle, 42-14, 42-14
Lycoperdon perlatum [fungus], 34-8
Lycophytes, **29-2**
   ancient giant, 33-7, 33-7–33-9
   diversity of, 33-6–33-7, 33-7
   fertilization in, 30-4
   fossils of, 33-5–33-6, 33-6
   Visual Synthesis of history of, 44-31
Lycopodium annotinum [lycophyte], 33-6, 33-7
Lymph, **39-16**–39-17
Lymphatic system, **39-16**–39-17

M

MAC. *See* Membrane attack complex (MAC)
MacArthur, Robert H., 42-8, 46-14
Mackerel, 37-11
Macrophages, 43-3, **43-4**
Magnetotaxis, 45-11, 45-12
Magnolias [*Magnolia grandiflora*], 30-9
Magnoliids, 33-15, 33-16
Mainland populations, **22-7**
Maize. *See Zea mays* [corn]
Major histocompatibility complex (MHC), **43-14**, 43-14
   and T cell activation, 43-14–43-15, 43-15

Malaria [Case 4], C4-2–C4-4
   and human activities, 48-18
   and human evolution, 24-15
   as infection, 43-18–43-19, 43-19, 48-18
   and natural selection, 21-10
Malleus, 36-8, **36-9**
Malpighi, Marcello, 41-10
Malpighian tubules, **41-10**, 41-10
Malthus, Thomas, 21-8, 46-1, 46-4
Mammals, **44-26**–44-27
   hearts of, 39-19, 39-19–39-20
   intrasexual selection in, 45-20–45-21, 45-21
   kidneys of, 41-12–41-19
   phylogenetic tree for, 44-26
   sexual dimorphism in, 44-20, 44-20
   Visual Synthesis of history of, 44-31
Map information, **45-11**
Marchantia berteroana [liverwort], 33-3, 33-3
Margulis, Lynn, 27-5–27-6
Mark-and-recapture, **46-7**, 46-7
Marler, Peter, 45-15
Marshall, Barry, 40-17
Marsupials, **44-26**, 44-26
Mass extinctions, **23-18**–23-19, 23-19. *See also* Extinctions
   and evolution, 44-29
   and human activities, C8-1
   Visual Synthesis of history of, 44-30–44-31
Mast cells, 43-3, **43-4**
   and inflammation, 43-5, 43-5
Mating, assortative, **21-7**
Mating types, **34-11**
Mayr, Ernst, 22-2, 26-14
Mechanoreceptors, **36-2**–36-3, 36-3
Mechnikov, Ilya, 43-3
Medulla [brain], 36-17, **36-18**
Medulla [kidney], **41-12**, 41-13
Meerkats, culture in, 24-16, 24-17
Meiosis. *See* Meiotic cell division
Meiotic cell division, **42-3**
   and sexual reproduction, 42-2, 42-3–42-4
Melanocyte-stimulating hormone, 38-4t, 38-9t, 38-14
Melanocytes, 38-18
Melatonin, 38-10t, 38-15
Membrane attack complex (MAC), **43-6**, 43-6
Membrane potentials, **35-7**. *See also* Action potentials
   and capacitation, 42-17
   postsynaptic, 35-13, 35-14
   resting, 35-7, 35-7, 35-11

and sensory receptors, 36-4, 36-4–36-5
threshold, 35-8, 35-9
Membranes. *See* Plasma membranes
Memory. *See* Cognition
Memory cells, 43-3, 43-9, **43-10**
Menarche, 42-16
Mendel, Gregor, 21-9–21-10
Menopause, **42-14**, 42-16
Menstrual cycle, **42-14**, *42-14*
Menstruation, **42-14**
Merezhkovsky, Konstantin Sergeevich, 27-5–27-6
Meristem identity genes, **31-3**
Meristems, 28-10, **28-10**, **31-1**. *See also* Shoot apical meristems
lateral, 31-10–31-11, *31-11*
root apical, 31-14, 31-14–31-16, *31-15*
Mesentery, **40-21**, *40-21*
Mesoderm, *42-19*, **42-19**
Mesoglea, 28-5, *28-5*, **44-7**, *44-7*
Mesohyl, *44-5*, **44-5**–44-6
Mesophyll, *29-3*, **29-3**–29-4
Metabolic rate, **40-3**–40-5, *40-4*
and body size, *40-5*, 40-5–40-7, *40-6*
and thermoregulation, 40-7
Metabolism. *See also* Animal metabolism; Cellular respiration; Digestion
aerobic, *40-2*, 40-2–40-3
anaerobic, *40-2*, 40-2–40-3
in animals, 40-1–40-2
recovery, *40-4*, 40-4–40-5
Metamorphosis, **38-2**, *38-2*, **44-15**, 44-17, *44-17*
Metanephridia, **41-10**, *41-10*
Metapopulations, **46-12**–46-13. *See also* Populations
and coexistence of species, 46-16–46-17, *46-17*
and island biogeography, *46-14*, 46-14–46-16, *46-15*, *46-16*
Meteorites, and extinction of the dinosaurs, 23-18
Methanopyrus kandleri [archaeon], 26-17
MHC. *See* Major histocompatibility complex (MHC)
Microbial mats, 26-7, *26-7*
Microbiomes [Case 5], C5-2–C5-4, *C5-3*, 26-22, 26-22–26-23, 43-2
Microfossils, 26-20, *26-20*, **27-20**–27-22, *27-21*, *27-22*
Microsporidia, 27-11
Microvilli, **40-19**, *40-19*, **41-14**
in opisthokonts, 27-10, *27-11*
Midbrain, **36-17**, *36-17*, **36-18**

Middle ear, *36-8*, **36-9**
Midges, 47-1, 47-3, *47-3*, 47-6
Migration, **21-13**
and climate change, 48-6–48-7, *48-7*
and metapopulations, 46-12
Milkweed, *32-8*, 32-8–32-9
Millipedes, 44-15
Minerals, dietary, **40-10**–40-11, *40-11*
Mistletoe, 32-3, *32-3*
Mites, 44-15
Mitochondria
and cellular metabolism, 27-2, *27-2*
and endosymbiosis, 27-6–27-7, *27-7*
origin of, C5-4
Mitochondrial DNA (mtDNA), **24-8**
and human origins, 24-6, 24-6–24-9, *24-7*, *24-9*, 24-13, *24-13*
Model organisms. *See Arabidopsis thaliana* [mouse-ear cress]; *Caenorhabditis elegans* [nematode]; *Dictyostelium* [slime mold]; *Drosophila melanogaster* [fruit fly]; *Escherichia coli* [bacterium]; *Pisum sativum* [garden pea]; *Zea mays* [corn]
Modern Synthesis, 21-9–**21-10**
Molars, **40-14**, *40-14*
Molecular biology
and behavior, 45-6, 45-6–45-8, *45-7*, *45-8*
and human origins, 24-3
and phylogenetic trees, 23-8–23-9, *23-9*, 44-4–44-5, *44-5*
Molecular clocks, **21-15**–21-16, *21-16*
and phylogenetic trees, 23-8–23-9
Molecular evolution, 21-14–21-16, **21-15**
Molecular fossils, **23-13**
Mollusks [Mollusca], **44-11**, 44-12–44-14, *44-13*, *44-14*
exoskeletons of, 37-13, *37-13*
feeding by, 40-12, *40-12*
locomotion of, 37-14
Visual Synthesis of history of, *44-31*
Molting, **37-15**, 38-2, **38-2**, 44-14
Monkeys, 24-1–24-2, *24-2*
and molecular clocks, 21-15
Monocots, 33-16–33-19, *33-17*, *33-18*, *33-19*
Monophyletic groups, **23-4**, *23-4*, **44-6**

Monosiga brevicollis [choanoflagellate], 28-7, 28-10
Monotremes, 44-26, *44-26*
Moose, *C7-2*, *C7-3*
Moran, Nancy, 47-7
Morels, 34-14, *34-14*
Morphospecies concept, **22-3**
Mortality, 46-3, *46-3*
Morula, *42-17*, **42-18**
Mosquitoes, and malaria, C4-2–C4-4, *C4-3*, 43-19, *43-19*
Mosses, 29-1, *29-2*, 33-2, *33-3*
Moths
chemoreceptors in, 36-3
and echolocation, 36-10, *36-10*
growth and development of, 38-2, *38-2*, 38-4
and pheromones, 45-2, *45-2*
Motor cortex, primary, *36-18*, 36-19, **36-19**
Motor endplates, 35-14, *35-15*, **37-6**, *37-7*
Motor neurons, **35-2**, 35-5
Motor units, **37-11**, *37-11*
Mouth, **40-14**, *40-15*
and digestion, *40-15*, 40-15–40-16
mtDNA. *See* Mitochondrial DNA (mtDNA)
Mucosa, **40-21**, *40-21*
Mucus, 43-2
Multicellularity
and cell adhesion, 28-6–28-7
and cell communication, 28-7–28-8, *28-8*, 28-9
and cell growth and differentiation, 28-8–28-10
complex, 28-1, 28-3, 28-3–28-4
evolution of, *28-12*, 28-12–28-16, *28-13*, 28-14
simple, 28-1–28-3, *28-2*
Multinucleated cells, **37-3**, *37-3*
Multiregional hypothesis [of human origins], **24-6**, 24-8
Muscle contraction
force of, **37-9**, 37-9–37-11, *37-10*, *37-11*
and joint motion, 37-10, *37-10*
mechanism of, 37-4–37-7, *37-5*, *37-7*
and muscle length, *37-8*, 37-8–37-9
Muscles
building of, 37-12
organization of, 37-2–37-3, *37-3*, *37-4*
and predation, 37-12
and slow-twitch versus fast-twitch fibers, 37-11, 37-11–37-12
types of, 37-1–37-2, *37-2*, 37-10, *37-10*

Mushrooms. *See also* Fungi
edible, 34-10
Mussels, 44-14, 47-10
as invasive species, 48-17, *48-17*
Mutations
advantageous, 21-3
deleterious, 21-3
and genetic variation, 21-2
neutral, 21-3
Visual Synthesis of, *22-16–22-17*
Mutualisms, **47-3**, 47-5–47-9, 47-8*t*. *See also* Symbiosis
and coevolution, 47-6
obligate versus facultative, 47-6
Mycelium, **34-2**, 34-9
Mycobacterium tuberculosis [bacterium], 43-18, 48-18
Mycorrhizae, 29-16, 29-17
Myelin, **35-6**, *35-6*
Mylle, Margaret, 41-16
Myofibrils, **37-3**, *37-3*
Myoglobin, **37-12**, **39-12**
storage of oxygen by, 39-12, *39-12*
Myosin, **37-2**
and muscle contraction, 37-4–37-6, *37-5*
and muscle structure, 37-3, *37-3*, *37-4*
regulation of, 37-6–37-7, *37-7*
Myotomes, **44-21**
Myriapods, **44-15**, *44-17*
Myxobacteria, 26-3, *26-3*

N

NAD+/NADH. *See* Nicotinamide adenine dinucleotide (NAD+/NADH)
Nagawasa, H., 38-3
Nanoarchaeota, 26-17, *26-17*
Natural immunity, **43-1**. *See also* Innate immune system
Natural killer cells, 43-3, **43-4**
Natural selection, **21-1**, 21-8–21-13
and adaptations, 21-8–21-9
versus genetic drift, 21-14
in humans, 24-14–24-15
and the Modern Synthesis, 21-9–21-10
and mutations, 21-10
and sexual selection, 21-13
types of, 21-10–21-13, *21-11*, *21-12*
Visual Synthesis of, *22-16–22-17*
Nature [behavior], **45-2**
Nauplius, 44-15
Nautilus, 44-13, 44-13–44-14
exoskeleton of, 37-14, *37-14*
Navigation
by bees, 45-15–45-16, *45-16*
by birds, 45-11–45-13, *45-13*

Neanderthals, **24-5**, *24-5*
 and human origins, *24-8*,
  24-8–24-9
Necrotrophic pathogens, **32-2**–32-3
Nectar, 30-11, 32-11
Negative feedback, **35-17**–35-18,
 35-18, 38-5, **38-5**–38-6
Negative selection [evolution],
 21-10
Negative selection [of T cells],
 43-16, *43-16*
Nematocysts, 44-8, *44-8*
Nematodes [Nematoda]. See
 Caenorhabditis elegans
 [nematode]
Neoteny, **24-10**–24-11, *24-11*
Nephrons, **41-11**, *41-11*
 organization of, 41-12, *41-13*
Nerve cords, **35-3**
Nerves, **35-3**
 cranial, 35-15, 35-16
 spinal, 35-15, 35-16
Nervous systems. *See* Animal
 nervous systems
Neural networks
 and cognition, 36-20,
  36-20–36-21
 and sensory processing, 36-15,
  36-15–36-16
Neural tubes, **44-21**
Neurons, **35-1**. *See also* Action
 potentials; Animal nervous
 systems
 afferent versus efferent, 35-16
 communication between, 35-12,
  35-12–35-14, *35-13*, *35-14*,
  35-15
 electrical activity of, 35-6–35-12,
  35-7, *35-9*, *35-10*, *35-11*
 shape of, 35-5, *35-5–35-6*
 structure of, 35-4, *35-4–35-5*
 types of, 35-2, *35-5*
Neurosecretory cells, **38-13**, *38-13*
Neurotransmitters, 35-4, **35-5**,
 35-12, *35-12–35-14*
Neutral mutations, 21-2–**21-3**
Neutrophils, 43-3, **43-4**
Niches, **47-1**–47-3, *47-2*
 and competition, 47-4,
  47-4–47-5
 versus habitat, 47-3
 realized, 47-2, 47-3
 and resource partitioning,
  47-2–47-3
Nicotinamide adenine
 dinucleotide (NAD⁺/NADH)
 and metabolism, 40-3
Niedergerke, Rolf, 37-4
Nim Chimpsky, *24-17*
Nitrification, **26-11**, *26-11*

Nitrogen cycle, 26-10–26-12, *26-11*
 and agriculture, 29-18
 eutrophication, 48-11,
  48-11–48-14
Nitrogen fixation, **26-11**, *26-11*
 and agriculture, 29-18
 Haber–Bosch process for, 29-18
 in lichens, 34-6
 and plant roots, 29-17, *29-17–29-18*
Nitrogenous waste, 41-7, **41-7**–41-8
Nociceptors, **36-3**
Nodes [phylogenetic], **23-2**
Nodes [plant], **31-2**, *31-2*
Non-associative learning,
 45-8–45-9
Non-random mating, **21-7**
Norepinephrine, 38-7, 38-8,
 38-10t, 38-16
Notochords, **44-19**, *44-20*, 44-21
Nuclear envelopes, 27-2, *27-2*
Nurture [behavior], **45-2**
Nymphs, 38-2, 38-3, *38-3*

O

Oak trees, red [*Quercus rubra*], *33-20*
Oats [*Avena sativa*], *31-17*
Obligate mutualisms, **47-6**
Occipital lobes, *36-18*, *36-19*, **36-19**
Ocean acidification, 48-8, **48-8**–48-10, *48-9*
Octopus, *43-13*, 44-13–44-14
 eyes of, 36-12
 learning by, 45-9
Odum, Eugene, 47-3
Olfaction, **36-5**. *See also* Smell,
 sense of
Oligodendrites, **35-6**
Omasum, **40-22**, *40-22*
Ommatidia, **36-12**, *36-12*
Oocytes, 42-12, *42-12*
 development of, 42-14
 primary versus secondary, *42-15*,
  42-15–42-16
Oogenesis, 42-15, **42-15**–42-16
Oomycetes, 27-16
 and phylogenetic trees,
  23-9–23-10
 and potato blight, 32-1–32-2
Open circulatory systems,
 39-13–39-14, *39-14*
Operant conditioning, **45-9**
Opercula, 39-4, *39-5*
Ophiocordyceps [fungus], *34-16*
Ophioglossum [fern], *33-9*
Opisthokonts, **27-10**–27-11,
 27-11. *See also* Animals;
 Choanoflagellates; Fungi
 Visual Synthesis of history of,
  44-31
Opsin, **36-11**

Opsonization, **43-6**, *43-6*
Optic nerves, **36-15**, *36-15*
Orangutans, 22-2, *22-3*, 24-2, *24-3*
Orchids
 Female-bee-mimicking [*Ophrys
  ciliata*], 30-11–30-12, *30-12*
 lady's-slipper [*Cypripedium
  reginae*], *30-9*
Orders, **23-4**, *23-5*
Ordovician radiation, 44-28
Organ of Corti, *36-8*, **36-9**
Organogenesis, **42-19**
 Visual Synthesis of, *42-21*
Organs, sensory, 36-1
Orgasms, 42-16
Orientation, 45-11
Origin of Species [Darwin], *On the*,
 21-1, 21-8, 22-1, 24-1, 45-20
Oscillatoria [cyanobacterium],
 26-16
Osmoconformers, **41-4**–41-5
Osmoregulation, 41-3, **41-3**–41-5
Osmoregulators, 41-5, **41-5**–41-6,
 41-6
Osmosis, 41-2, **41-2**–41-3
 in capillaries, 39-16, *39-16*
 and guard cells, 29-5
 in loops of Henle, 41-14
 and osmoregulation, 41-3,
  41-3–41-5
Osmotic pressure, 41-2, **41-2**–41-3
Osteichthyes, 44-23, **44-23**–44-24
Osteoblasts, *37-16*, **37-16**–37-17
Osteoclasts, **37-17**
Osteocytes, **37-17**
Out-of-Africa hypothesis [of
 human origins], 24-6,
 24-6–24-9, *24-7*, *24-9*
Outer ear, *36-8*, **36-9**
Oval window, *36-8*, **36-9**
Ovaries [animal], *38-12*, **38-14**,
 42-12, *42-12*
Ovaries [plant], *30-9*, **30-10**
Oviducts, **42-12**, *42-12*
Oviparity, 42-8–**42-9**
Ovule cones, *30-6*, **30-7**
Ovules, *30-6*, **30-7**
Owls, 47-9–47-10, *47-10*
Oxygen
 discovery of, 40-3
 dissociation curves for, 39-11,
  39-11, *39-13*
 and gas exchange, 39-1–39-3, *39-3*
 history of terrestrial, 25-14,
  25-14–25-15, 26-20, 44-27,
  44-29
 and metabolic rate, 40-3–40-5,
  40-4
 and multicellularity, 28-13–28-14,
  28-14

 transport of, by hemoglobin,
  39-10–39-13
 Visual Synthesis of history of,
  44-30
Oxygenic reactions, **26-6**
Oxytocin, 38-7, *38-9t*, **38-14**
 and childbirth, 38-6–38-7, 42-22
 receptors for, 38-11
Oysters, 44-14

P

Pääbo, Svante, 24-8
Pacemaker cells, **39-20**, *39-21*
Pain, receptors for, 36-3
Paine, Robert, 47-10
Pancreas, *38-12*, 38-15, **40-17**,
 40-18, 40-18–40-19
Pangaea, 23-16, *23-16*, 23-18
Paracrine signaling, 38-16,
 38-16–38-17
Paramecium [protozoan], 41-8,
 41-8
Paraphyletic groups, **23-4**, *23-4*,
 44-6
Parasexual species, **34-11**, *34-11*
Parasites
 and co-speciation, 22-11, *22-11*
 and competition, 47-5
 microsporidia as, 27-11
Parasitism, *47-8t*
Parasympathetic division, **35-17**,
 35-17
Parathyroid glands, *38-12*, **38-15**
Parathyroid hormones (PTH),
 38-10t, 38-15
Parenchyma cells, **29-8**, *29-9*
Parental investment, 45-20
Parietal lobes, *36-18*, *36-19*, **36-19**
Parsimony, 23-6–**23-8**
Parthenogenesis, 42-2, **42-3**, *42-4*
Partial pressure (*p*), **39-2**, *39-2*
Partial reproductive isolation,
 22-6, *22-7*
Passionflower [*Passiflora caerulea*],
 33-20
Patches, **46-12**, *46-12–46-13*
 experimental, *46-13*, *46-13*
Pathogens. *See also* Animal
 immune systems
 biotrophic versus necrotrophic,
  32-2–32-3
 and human activities, 48-18,
  48-18–48-19
 plant defenses against,
  32-1–32-8
 virulent versus avirulent, 32-3
Paulinella chromatophora
 [amoeba], 27-6, *27-6*, *27-19*
Pavlov, Ivan, 45-9
Peacocks, 45-19–45-20

Peas, garden. *See Pisum sativum* [garden pea]
Peat bogs, **33-5**, *33-5*
   and fungal inhibition, 34-4
Pectins, in cell walls, 28-6
Penguins, Adelie, 21-2, *21-2*, 24-12
Penis, 42-10, **42-11**
Pepsin, **40-17**
Peptide hormones, 38-7, **38-7**–38-8, *38-8*
Peptidoglycan, **26-2**
Pericycles, **31-15**, *31-15*
Periodic selection, **26-14**–26-15
Peripatric speciation, **22-7**, 22-9, *22-9*
Peripheral nervous system (PNS), 35-15, 35-15–**35-16**
Peristalsis, **40-17**, 40-20
Permian extinction, 23-18–23-19, *23-19*
Pesticides, and GMOs, 32-17–32-18
Petals, 30-9, **30-10**
Peziza [fungus], 34-15, *34-15*
Phagocytes, 43-3, **43-3**–43-4
   and extravasation, 43-5–43-6, *43-6*
   and pathogen recognition, 43-4
Phagocytosis, 27-2, 27-3, **43-3**–43-4, *43-4*
Pharyngeal slits, **44-18**, *44-18*, 44-19, 44-20, 44-21
Pharynx, **40-15**, *40-15*, **44-18**
Phenols, 32-10t, **32-11**
Pheromones, 38-17, **38-17**–38-18, *38-18*
   and communication, 45-14
   and pollination, 30-11–30-12
Phloem, 29-8–**29-9**, *29-9*
   carbohydrate transport through, 29-12–29-14, *29-13*
   secondary, 31-11, *31-11*, 31-12
   Visual Synthesis of, 33-23
Phloem sap, **29-13**
Phosphorus, and human activities, 48-14
Photoheterotrophs, 26-8–**26-9**
Photoperiodism, **30-17**, **45-12**
Photoreceptors, **30-17**–30-18, *30-19*, 36-3, **36-4**
   and plant growth, 31-19
Photosynthesis
   in Bacteria, 26-7, 26-7–26-8, *26-8*, 26-15–26-16
   in C_4 plants, 29-7, 29-7–29-8, *29-8*
   in CAM plants, 29-6, *29-6*
   and the carbon cycle, 25-2, 25-2–25-3, 25-13–25-14
   in Eukarya, 27-16–27-19, *27-19*
Photosystems, in Bacteria, 26-7–26-8, *26-8*

Phototropism, *31-17*, **31-17**–31-18, *31-18*
Phyla [Phylum], **23-4**, *23-5*, **44-2**
   of animals, 44-10t
Phylogenetic trees, 23-1–23-5, *23-2*, **23-2**, *23-3*. *See also* Evolutionary trees; Genetic variation
   for angiosperms, 33-15
   for animals, 28-12, 44-1–44-2, *44-2*
   for Archaea, 26-17, *26-17*
   for arthropods, 44-17
   for bilaterians, 44-5
   for chloroplasts, 27-5
   for cnidarians, 44-6
   and coevolution, 47-7, *47-7*
   and computers, 23-8
   construction of, 23-5, 23-5–23-9, *23-6*, *23-7*, *23-9*
   for ctenophores, 44-11
   for deuterostomes, 44-18
   for domains, 26-5
   for ecdysozoans, 44-15
   for Eukarya, 27-10, 28-4
   for ferns, 33-9
   for fungi, 34-12, *34-12*
   for gymnosperms, 33-11
   for horsetails, 33-9
   for humans, 24-2, 24-2–24-4
   for lophotrochozoans, 44-11
   for lycophytes, 33-7
   for mammals, 44-26
   and molecular biology, 23-8–23-9, *23-9*
   for placozoans, 44-11
   for plants, 28-12, 29-2, 30-2, 33-2
   problem-solving with, 23-9–23-11, *23-10*
   for sponges, 44-6
   and spread of photosynthesis, 27-18, *27-18*
   for vertebrates, *23-3*, 40-13, 44-22
Phylogeny, **23-1**
   and fossils, 23-19–23-20, *23-20*
Phytochrome, **30-20**–30-21, *30-21*, 31-19
Phytoliths, 33-18
Phytophthora infestans [protist], 32-1–32-2, *32-2*, 32-4, 32-18
Picrophilus torridus [archaeon], 26-18
Pigeons, homing, 45-11–45-13
Pigmented epithelium, 36-15, *36-15*
Pigments. *See also* Chlorophyll
   and germination, 30-20
   skin, 24-14

Pilobilus [fungus], 34-13, 34-13–34-14
PIN transport proteins, *31-8*, 31-8–31-9
Pine trees, 33-12, 33-13, *33-13*. *See also* Gymnosperms; Trees
   and cork cambium, 31-12
   and growth rings, 31-12
   wood of, 31-13
Pineal glands, *38-12*, **38-15**
Pinna [pinnae] (in plants), 33-10, *33-10*
Pinna [pinnae] (in mammalian ear), 36-8, **36-9**
Pisum sativum [garden pea]
   and allele frequencies, 21-3
Pith, **31-9**, *31-9*, *31-11*
Pits, **29-9**, *29-9*
Pituitary gland, **38-4**, *38-12*, **38-12**–38-14, *38-13*
   anterior versus posterior, *38-13*, 38-13
   and hormonal signals, 38-8, 38-10, 38-11
Placenta, **42-9**, **44-26**
   and childbirth, 42-22, *42-22*
   formation of, 42-18
   hormone production by, 42-14
Placental mammals, **44-26**, **44-26**–44-27
Placozoans [Placozoa], **44-9**, 44-9–44-11
   phylogenetic tree for, 44-11
Plant defenses
   and communication, 32-13–32-14
   against herbivores, 32-8–32-13
   and life histories, 46-12
   against pathogens, 32-1–32-8
   and plant diversity, 31-16–32-18
   resources for, 31-13–31-16
Plant growth
   and environmental conditions, 31-16–31-20, *31-17*, *31-18*, *31-19*
   and hormones, 31-6–31-10
   lateral, 31-10–31-13
   roots, 31-13–31-16
   upward, 31-2–31-6
Plants. *See also* Angiosperms; Bryophytes; Gymnosperms; Lycophytes
   versus animals, 28-10, *28-10*, 28-10–28-11
   bulk transport in, 28-5, 28-5–28-6
   as complex multicellular organisms, 28-3, 28-4
   diversity of, 33-1–33-2, *33-2*
   early evolution of, 28-14–28-15
   evolution of, 29-1–29-2, 30-1–30-5
   life cycle of, 27-4, *27-4*, 30-1–30-5

   phylogenetic tree for, *28-12*, 29-2, 30-2, 33-2
   structure and function of, 29-1–29-2, *29-2*
   vascular, 29-1
   Visual Synthesis of history of, 44-31
Plasma cells, 43-3, 43-9, **43-10**
Plasma membranes, 27-2, *27-2*
Plasmids, **26-2**, *26-2*
Plasmodesmata, **28-8**, *28-9*
   and evolution of plants, 28-14
   and winterization, 31-20
Plasmodia, 27-12, *27-12*
Plasmodium falciparum [protist], 27-16
   and coevolution, 22-11–22-12
   drug resistance in, 48-18
   infection by, 43-18–43-19, *43-19*
   and malaria, C4-2–C4-4, *C4-3*
Plasmogamy, **34-10**, *34-10*
Plate tectonics, **25-8**
   and the geologic timescale, 23-15–23-16, *23-16*
   and the long-term carbon cycle, 25-8–25-9, 25-11, 25-15
Platypus, sense of electricity in, 36-4
Pneumococcus. *See Streptococcus pneumoniae* [bacterium]
PNS. *See* Peripheral nervous system (PNS)
Polar bodies, 42-14, **42-15**
Polar transport, *31-8*, **31-8**–31-9, 31-17
Polarization, **35-7**
Pollen, 30-6, 30-6–30-8, **30-7**
Pollen cones, 30-6, **30-7**
Pollen tubes, **30-7**, *30-10*
Pollination, **30-7**
   and angiosperm evolution, 33-16
   modes of, 30-10–30-12, *30-11*, *30-12*
   Visual Synthesis of, 33-22
Polychaete worms, 44-12
Polyphyletic groups, **23-4**, *23-4*
Polyploidy, **22-14**
   and instantaneous speciation, 22-14, 22-14–22-15
Polysaccharides, 28-6
Polyspermy, **42-17**
Polytrichum commune [moss], 30-2–30-4, *30-3*, *30-4*, 33-3
Pons, *36-17*, **36-18**
Population [Malthus], *Essay on the Principle of*, 21-8–21-9
Population density, **46-2**, *46-2*
   distribution of, 46-3, *46-3*
   and diversity, 47-22
   measuring, 46-6–46-7, *46-7*

Population size, **46-2**, 46-2
  and carrying capacity, 46-5, 46-5–46-6
  changes in, 46-3, 46-3–46-5, 46-4
  factors affecting, 46-6, 46-6
  human, 48-1, 48-2
  measuring, 46-6–46-7, 46-7
  Visual Synthesis of, 48-12–48-13
Populations, 21-1–21-2, 46-1–**46-2**. *See also* Metapopulations
  age structure of, 46-8, 46-8–46-9, 46-9
  features of, 46-2–46-3, 46-3
  fixed, 21-6
  island, 22-7, 22-9
  and life histories, 46-11–46-12
  mainland, 22-7
  and reproductive patterns, 46-10–46-11
  and survivorship curves, 46-9–46-10, 46-10, 46-11
Positive feedback, **35-8**, **38-5**, 38-6, 38-6
Positive selection [evolution], **21-10**
Positive selection [of T cells], **43-16**, 43-16
Post-zygotic isolating factors, **22-5**, 22-6
Posterior pituitary gland, 38-12, 38-13, **38-13**
Potatoes, blight of, 27-16, 32-1–32-2, 32-2, 32-4, 32-18
Potentials. *See* Action potentials; Membrane potentials
Power strokes, **37-6**
Prasiola [alga], 28-2
Pre-zygotic isolating factors, **22-5**–22-6
Predation [Case 7], C7-2–C7-4, C7-3, 47-8t
  and cellular metabolism, 27-2
  and communities, 47-9–47-10, 47-10
  and competition, 47-5, 47-5
  and echolocation, 36-10, 36-10
  and endocrine systems, 38-15, 38-15–38-16
  and jaws and teeth, 40-13–40-14, 40-14
  and multicellularity, 28-7, 28-7, 28-14
  and nervous systems, 35-1, 35-2
  and osmoregulation, 41-6, 41-6–41-7
  stability of, 46-16–46-17
  and thermoregulation, 40-5
  and vision, 36-14
Pregnancy, **42-18**

Premolars, **40-14**, 40-14
Pressure
  osmotic versus hydrostatic, 41-2, 41-2–41-3
  partial (*p*), 39-2, 39-2
  turgor, 29-14
Pressure, sense of
  and communication, 45-14
  receptors for, 36-2–36-3
  and somatosensory cortex, 36-19, 36-19
Prey. *See* Predation [Case 7]
Priestly, Joseph, 40-3
Primary bronchi, **39-7**, 39-7
Primary growth, **31-3**
  Visual Synthesis of, 33-22
Primary motor cortex, 36-18, 36-19, **36-19**
Primary oocytes, 42-15, **42-15**–42-16
Primary producers, 25-12, 25-12, **26-10**, **47-14**, 47-14
Primary production, 47-15–47-16, 47-16
Primary responses, **43-10**, 43-10
Primary somatosensory cortex, 36-18, 36-19, **36-19**
Primary spermatocytes, **42-15**, 42-15
Primates, 21-2, **24-1**–24-2
Procambial cells, 31-8, **31-9**
Progesterone, 38-7, 38-8, 38-9t, 38-14, **42-13**, 42-13
  and menstrual cycle, 42-14, 42-14
Prokaryotes. *See also* Archaea; Bacteria; Eukarya
  evolutionary history of, 26-20, 26-20–26-23
  Visual Synthesis of history of, 44-30
Prolactin, 38-9t, 38-14
Prosimians, 24-1–24-2, 24-2
Prostate gland, 42-10, **42-11**
Protease inhibitors, **32-11**
Proteobacteria, **26-15**, 26-15
Prothoracicotropic hormone (PTTH), **38-4**, 38-5
Protists, **27-9**
  as consumers, 25-12, 25-12
  history of, 27-20–27-22, 27-21, 27-22
Protonephridia, **41-9**, 41-9
Protostomes [Protostomia], **44-4**
Protozoa, **27-9**
Proximal convoluted tubules, **41-14**, 41-14
Pseudocoelomates, **44-3**, 44-3
Pseudogenes, **21-16**
Pseudomonas aeruginosa [bacterium], 26-4

Pseudopodia, 27-9, 27-12, **27-12**
Psilotum [fern], 33-9
Pteridium aquilinum [fern], 30-5, 30-5
Pterobranchs, 44-18, 44-18
PTH. *See* Parathyroid hormones (PTH)
PTTH. *See* Prothoracicotropic hormone (PTTH)
Puberty, 42-13–42-14
Public policy
  and citizenship, 48-2, 48-19–48-20, 48-20
  and vaccines, 43-10, 43-18
Puccinia monoica [fungus], 34-17, 34-19
Puffballs, 34-17, *34-17*, 34-18
Pullman, Philip, 48-20
Pulmonary arteries, 39-19, **39-20**
Pulmonary capillaries, 39-7, **39-8**
Pulmonary valves, 39-19, **39-20**
Pulmonary veins, 39-19, **39-20**
Pupae, **44-17**, 44-17
Pyloric sphincters, **40-17**
Pyramidal cells, 35-5, 35-5–35-6
Pythons, 40-17

Q
Quinine, C4-3, 48-18

R
R genes, **32-4**, 32-4
R proteins, **32-4**
r-strategists, 42-8, **42-8**, **46-10**–46-11, 46-11
Race. *See* Ancestry
Radial symmetry, **44-2**, 44-3
Radioactive decay, and geologic timescale, 23-14–23-15, 23-15
Radiolaria, 27-21, 27-21
Radiometric dating, 23-15, 23-15
Rafflesia, 32-3, 32-3
Rain forest biome, 47-19, 47-19, 47-20
Rain forests. *See* Biodiversity hotspots [Case 8]
Ramón y Cajal, Santiago, 35-5–35-6
Range, geographic, **46-2**, 46-2
Rats
  kangaroo, 41-12, 41-12
  learning by, 45-9
Raven, Peter, 32-17
Ray-finned fish, 44-22, 44-23–44-24
  Visual Synthesis of history of, 44-31
Rays, 44-23
Reabsorption, **41-9**, 41-9
Realized niches, 47-2, **47-3**

Receivers, **45-14**
Receptors, **36-1**
  sensory, 36-1–36-5, 36-2
Reciprocal altruism, **45-17**–45-18
Reciprocal inhibition, **35-19**, 35-19
Recovery metabolism, 40-4, **40-4**–40-5
Rectum, **40-14**, 40-15, **40-20**
Red blood cells, sickled, 21-10, 21-11
Red Queen hypothesis, 42-5, 48-14–48-15
Red tides, 27-16
Redwoods, 33-13
Reefs. *See* Biodiversity hotspots [Case 8]
Reflex, knee-jerk, 35-18–35-19, 35-19
Refractory period, **35-8**, 35-9
Reinforcement of reproductive isolation, **22-15**
Relatedness, 45-18, 45-18–45-19, 45-19
Releasing factors [hormonal signaling], 38-9t, **38-10**, **38-13**
Renal pelvis, **41-12**, 41-13
Renal systems. *See* Animal renal systems
Renal tubules, **41-11**, 41-11
Renin, **41-19**
Renner, Otto, 29-10
Reproduction. *See also* Animal reproduction; Asexual reproduction; Sexual reproduction
  and colonization of land, 42-7, 42-7–42-9
  in fungi, 34-7–34-12
  history of, 42-1–42-7
  in plants, 30-7–30-8
  vegetative, 30-22, 30-22–30-23
  Visual Synthesis of, 42-20–42-21
Reproductive isolation, 22-5–22-6
  and biological species concept, 22-2
  partial, 22-6, 22-7
  reinforcement of, 22-15
Reptiles, 44-25
  hearts of, 39-19
  kidneys of, 41-11, 41-11–41-12
Research. *See* "How Do We Know?"; Laboratory techniques
Reservoirs, **25-6**–25-7, 25-6–25-7
Resource investment, and life histories, 46-11–46-12
Resource partitioning, **47-2**–47-3
Respiration. *See* Animal respiratory systems; Cellular respiration
Respiratory bursts, **43-4**

Respiratory chain. *See* Electron transport chain
Resting membrane potentials, **35-7**, *35-7*, *35-11*
Reticulum, **40-22**, *40-22*
Retinal, **36-11**, *36-11*
Retinas, *36-14*, **36-14**–*36-15*
   sensory processing in, *36-15*, *36-16*, *36-18*–*36-16*
Rhizanthes lowii [flower], *30-11*
Rhizarians, *27-21*, *27-21*, *28-2*
   Visual Synthesis of history of, *44-30*
Rhizobium radiobacter [bacterium], *32-7*, *32-7*–*32-8*, *32-18*
Rhizopus [fungus], *34-8*, *34-13*
Rhizospheres, **29-14**
Rhodnius [insect], *38-2*–*38-4*, *38-3*, *38-5*
Rhodopsin, **36-11**, *36-11*
Ribulose bisphosphate carboxylase oxygenase (Rubisco), *29-7*, *29-7*–*29-8*
Rice
   and hormones, *31-9*
   and symbiosis, *33-10*–*33-11*
Riebesell, Ulf, *48-8*–*48-9*
Ring species, **22-4**, *22-4*
Ritualization, **45-14**
Rod cells, **36-14**, *36-14*, *36-15*
Root apical meristems, *31-14*, **31-14**–*31-16*, *31-15*
Root caps, **31-14**, *31-14*
Root hairs, *29-15*, **29-15**, *31-14*
Root nodules, *29-17*, **29-18**
Roots, **29-2**–*29-3*, *29-3*, *29-10*
   diversity of, *31-15*–*31-16*, *31-16*
   evolution of, *31-13*–*31-14*
   growth and development of, *31-14*–*31-16*, *31-15*, *31-19*–*31-20*
   and mycorrhizae, *29-16*, *29-17*–*29-18*, *34-5*, *34-5*–*34-6*
   nutrient uptake through, *29-15*, *29-15*–*29-17*
   structure of, *31-14*
   Visual Synthesis of, *33-23*
Ross, A. F., *32-5*–*32-6*
Roundworms, *36-3*
Rubisco. *See* Ribulose bisphosphate carboxylase oxygenase (Rubisco)
Rumen, **40-22**, *40-22*

S

Saccharomyces cerevisiae [yeast], *34-3*, *34-3*, *42-2*, *42-2*
Sahelanthropus tchadensis [hominin], *24-4*, *24-4*, *24-5*
Saint-Hilaire, Étienne Geoffroy, *24-10*

Salamanders, *44-25*
   breathing of, *39-4*
Salicylic acid, *32-10t*
Saliva
   and digestion, *40-15*
   and innate immunity, *43-2*
Salmon, life cycle of, *41-4*, *41-4*
Salpingoeca rosetta [choanoflagellate], *28-6*–*28-7*, *28-7*
Salps, *44-21*
Salvinia [fern], *33-9*, *33-10*
Sand dollars, *44-19*
Sapwood, **31-11**
SAR. *See* Systemic acquired resistance (SAR)
Sarcomeres, **37-3**, *37-4*
Sarcoplasmic reticulum (SR), **37-6**–*37-7*, *37-7*
Sarich, Vince, *21-15*
Savanna biome, *47-19*, *47-19*
Scallops, *40-12*
Scholander, Per Fredrick, *29-10*
Schwann cells, **35-6**, *35-6*
Schwendener, Simon, *34-6*
Scientific method. *See* "How Do We Know?"
Scorpions, *44-15*
Scrotum, **42-10**, *42-10*, *42-11*
Sea anemones, *44-7*, *44-7*–*44-8*
   gas exchange in, *39-2*
   hydrostatic skeletons of, *37-13*, *37-13*–*37-14*
Sea cucumbers, *44-19*
Sea lilies, *44-19*
Sea otters, *47-10*–*47-11*
Sea slugs, *45-9*
Sea squirts, *44-18*, *44-21*
Sea stars, *44-18*, *44-19*, *44-19*
   as keystones species, *47-10*
Sea urchins, *44-18*, *44-19*, *44-19*, *47-11*
Seahorses, *45-21*
Seaweeds, *27-13*–*27-14*, *27-15*
   history of, *27-21*
   simple multicellular, *28-3*
Secondary growth, **31-3**
   Visual Synthesis of, *33-22*
Secondary oocytes, *42-15*, **42-16**
Secondary phloem, **31-11**, *31-11*, *31-12*
Secondary responses, **43-10**, *43-10*
Secondary sexual characteristics, **42-13**
Secondary spermatocytes, **42-15**, *42-15*
Secondary xylem, **31-11**, *31-11*
Secretin, *38-10t*, **40-18**
Secretion, **41-9**, *41-9*
Seed banks, *33-21*

Seed coats, **30-8**, *30-8*
Seeds, **30-8**, *30-8*
   and fruits, *30-15*, *30-15*–*30-16*, *30-16*
Selaginella willdenowii [lycophyte], *33-6*–*33-7*, *33-7*
Selection, **21-6**. *See also* Artificial selection; Natural selection
   balancing, *21-10*
   clonal, *43-9*, *43-9*–*43-10*, *43-10*
   directional, *21-12*, *21-12*
   disruptive, *22-12*, *22-13*
   group, *45-16*–*45-17*, *45-17*
   intrasexual versus intersexual, *45-20*–*45-21*, *45-21*
   kin, *45-18*–*45-19*
   periodic, *26-14*–*26-15*
   positive versus negative, *21-10*
   sexual, *21-13*, *21-14*, *45-19*–*45-21*, *45-21*
   and speciation, *22-15*
   types of, *21-10*–*21-13*, *21-11*, *21-12*
Self-compatible angiosperms, *30-12*
Self-incompatible angiosperms, **30-12**
Semen, **42-11**
Semicircular canals, *36-7*, **36-7**–*36-8*
Seminal vesicles, *42-10*, **42-11**
Seminiferous tubules, *42-10*, *42-11*
Semipermeable membranes, **41-2**, *41-2*
Senders, **45-14**
Senses. *See also* Pressure, sense of; Sight, sense of; Smell, sense of; Sound, sense of; Taste, sense of
   of balance, *36-6*–*36-8*, *36-7*
   of electric field, *36-4*, *45-14*
   of gravity, *36-6*–*36-8*, *36-7*
   of magnetic field, *45-11*, *45-12*
   of motion, *36-6*–*36-8*, *36-7*
   of pain, *36-3*
   of temperature, *36-3*
Sensitization, **45-8**
Sensory neurons, **35-2**, *35-5*
Sensory organs, **36-1**
Sensory receptors, **36-1**–*36-2*, *36-2*
   for gravity and movement, *36-7*, *36-7*–*36-8*
   and potentials, *36-4*, *36-4*–*36-5*
   for sight, *36-10*–*36-16*, *36-11*, *36-12*, *36-13*, *36-14*, *36-15*
   for smell and taste, *36-5*–*36-6*, *36-6*
   for sound, *36-8*, *36-8*–*36-10*, *36-9*, *36-10*
   specialized, *36-2*–*36-4*
Sensory transduction, **36-1**–*36-2*, *36-3*

Sepals, *30-9*, **30-10**
Septa, **34-2**, *34-3*
Sequoia trees, giant [*Sequoiadendron giganteum*], *33-13*
Serosa, **40-21**, *40-21*
Sertoli cells, **42-13**
Sex determination, **42-19**
Sex hormones. *See* Estrogen; Progesterone; Testosterone
Sex pheromones, *38-17*, *38-17*
Sexual dimorphism, **45-20**, *45-20*
Sexual intercourse, *42-16*
Sexual reproduction, *42-3*, **42-3**–*42-4*. *See also* Meiotic cell division
   and genetic diversity, *27-3*–*27-4*
   and two-fold cost of sex, *42-4*–*42-5*, *42-5*
Sexual selection, **21-13**, *24-14*, **45-19**–*45-21*, *45-21*
Sharks, *44-23*, *44-23*
   feeding by, *40-13*, *40-13*
   skeletons of, *37-16*
Shiitakes, *34-14*, *34-14*
Shoot apical meristems, *31-2*, **31-2**–*31-3*
   and branching, *31-5*–*31-6*, *31-6*
   and flower development, *31-6*, *31-6*
   and leaf arrangements, *31-4*, *31-4*–*31-5*, *31-5*
Shoots, **29-2**, *29-3*
   growth and development of, *31-2*, *31-2*–*31-3*, *31-3*
   and leaf arrangements, *31-4*, *31-4*–*31-5*, *31-5*
Short-day plants, **30-17**
Shrimp, *22-8*, *22-8*
Shubin, Neil, *23-20*
Sieve elements, **29-13**, *29-13*
Sieve plates, **29-13**
Sieve tubes, *29-13*, **29-13**–*29-14*
Sight, sense of, *36-10*–*36-16*. *See also* Eyes
   and communication, *45-14*
   in humans, *36-13*, *36-13*–*36-14*, *36-14*
   and occipital lobes, *36-19*, *36-19*
   receptors for, *36-4*
Signaling
   and communication, *45-14*
   paracrine versus autocrine, *38-16*, *38-16*–*38-17*
Silica, *25-8*, *25-8*
   sources of, *27-22*
Silverfish, *44-15*
Simberloff, Daniel, *46-14*
Simple multicellularity, **28-1**–*28-3*, *28-2*

Single-lens eyes, **36-12**, *36-12*
  convergent evolution of, *36-12*–*36-13*, *36-13*
Sinks, in plants, 29-13–29-14
Sinoatrial (SA) nodes, **39-20**, *39-21*
siRNA. *See* Small interfering RNA (siRNA)
Sister groups, **23-3**, *23-3*
  of Eukarya, 27-19–27-20, *27-20*
  of humans, 24-3
  of plants, 30-2
  of vascular plants, 33-6
Skeletal muscles, **37-2**, *37-2*. *See also* Muscle contraction; Muscles
Skeletons. *See also* Bones; Joints
  endo-, *37-15*, 37-15–37-16
  exo-, *37-14*, 37-14–37-15
  hydrostatic, 37-13, *37-13*–37-14
  vertebrate, 37-16–37-19, *37-17*, *37-18*, *37-19*
Skin
  and innate immunity, 43-2
  structure of, *43-3*
Sliding filament model, **37-4**–37-6, *37-5*
Slime molds, 27-12, 27-12–27-13, *27-13*
  simple multicellular, 28-2, *28-2*
Slow-twitch fibers, 37-11, *37-11*–**37-12**
Slugs, 44-13
Small interfering RNA (siRNA), **32-6**
  and plant defenses, 32-6–32-7, *32-7*
Small intestine, **40-14**, *40-15*
  and digestion, 40-17–40-20, *40-18*, *40-19*
Smell, sense of, 36-5–36-6, *36-6*
  and communication, 45-14
  receptors for, 36-2
Smith, John Maynard, 42-4
Smooth muscles, **37-2**, *37-2*, **39-15**
  regulation of, 37-7
Snails, 44-13, *44-13*
Snakes
  digestion in, 40-17, *40-17*
  as invasive species, 48-17, *48-17*
  sensory receptors in, 36-2
  Visual Synthesis of history of, 44-31
Snapdragons [*Antirrhinum majus*], 30-12
Social behavior, 45-16–45-19
  and group selection, 45-16–45-17, *45-17*
  and kin selection, 45-18–45-19
  and reciprocal altruism, 45-17–45-18

Sodium-potassium pumps
  and action potentials, 35-8, *35-9*, *35-10*, 35-11
  and digestion, 40-20
  and membrane potentials, 35-7, *35-7*
Solar energy, 48-10, *48-10*. *See also* Photosynthesis
Solubility, **39-10**
Solutes, **41-2**
  in urine, 41-16
Somatic mutations, **21-2**
Somatic nervous system, **35-16**
Somatosensory cortex, primary, *36-18*, *36-19*, **36-19**
Somatostatin, 38-10t
Sonar. *See* Echolocation
Sound, production of, 39-9
Sound, sense of, 36-8, 36-8–36-10, *36-9*, *36-10*. *See also* Ears
  and communication, 45-14
  and temporal lobes, 36-19, *36-19*
Sources, **29-13**
  in plants, 29-13–29-14
Soybeans, and nitrogen fixation, 26-11, *26-11*
Sparrows, bird song of White-Crowned, 45-15, *45-15*
Spatial summation, **35-13**, *35-14*, **36-4**
Speciation, **22-1**, 22-6–22-15
  allopatric, 22-6–22-11, *22-8*, *22-9*, *22-10*
  and genetic divergence, 22-6, *22-7*, 22-15
  instantaneous, 22-13–22-15, *22-14*
  modes of, *22-13*
  peripatric, 22-7, 22-9, *22-9*
  and selection, 22-15
  sympatric, 22-12, 22-12–22-13
  Visual Synthesis of, 22-16–22-17
  without natural selection, 22-15
Species, **21-2**, **22-1**. *See also* Biological species concept (BSC)
  alternative definitions of, 22-3, 22-4–22-5
  bacterial, 26-14–26-15
  endangered or threatened, 33-12
  endemic, 46-15
  eusocial, 45-19
  invasive, 48-17, 48-17–48-18
  keystone, 47-10–47-11, *47-11*
  parasexual, 34-11, *34-11*
  ring, 22-4, *22-4*
  sub-, 22-7
  types of interactions among, 47-8t
Species–area relationship, **46-14**, *46-14*, *46-15*

Sperm cells, 42-10, *42-11*
  flagella of, 42-10, *42-11*
  as gametes, 27-4
  and sperm competition, 42-10
Spermatocytes, 42-15, *42-15*
Spermatogenesis, **42-15**, *42-15*
Sphagnum [moss], 33-5, *33-5*
Spicules, 44-5, *44-6*
Spiders, 44-15, *44-16*
  communication by, 45-14
Spinal cords, 35-15, **35-16**
Spinal nerves, 35-15, **35-16**
Spiracles, 39-4, **39-6**, **44-17**
Spirogyra [alga], 27-15
Splachnum luteum [moss], 33-4
Sponges, 44-5, 44-5–44-7
  exoskeletons of, 37-13
  gas exchange in, 39-2
  phylogenetic tree for, *44-6*
  size of, 28-4, *28-5*
  Visual Synthesis of history of, 44-31
Spongy bone, **37-17**, *37-17*
Sporangia, **30-4**, *30-4*
  of ferns, 33-10
  of fungi, 34-8, *34-8*, 34-13, *34-13*
  of horsetails, 33-10
  of lycophytes, 33-4, 33-6, *33-7*
  of slime molds, 27-12, *27-12*, *27-13*
Spores, **30-3**, *30-3*
  of fungi, 34-7–34-9, *34-8*
Sporophytes, **30-3**, *30-3*, *30-7*
Sporopollenin, **30-4**
Squid, 44-13–44-14
  electrical activity in, 35-11, *35-11*
  eyes of, 36-3, 36-12, *36-12*
  symbiotic bacteria in, C5-4, *C5-4*
Squirrels
  and seed dispersal, 30-16, *30-16*
  thermoregulation in, *40-8*
SR. *See* Sarcoplasmic reticulum (SR)
Stabilizing selection, *21-11*, **21-11**–21-12, *21-12*
  in mainland populations, 22-11
Staghorn ferns, 33-10, *33-11*
Stamens, 30-9, **30-10**
Stapes, 36-8, **36-9**
Star anise [*Illicium verum*], 33-16, *33-16*
Starling's Law, **39-22**
  and muscle contraction, 37-9
Starvation, 40-3, 40-10
Statistics, and phylogenetic trees, 23-8
Statocysts, **36-7**, *36-7*
Statoliths, **31-18**, *31-18*, **36-7**, *36-7*

Stem cells, differentiation of, in bone marrow, 43-3, *43-3*
Stems, 29-2–29-3, *29-3*. *See also* Phloem; Xylem
  carbohydrate transport through, 29-12–29-14
  growth and development of, 31-3, 31-3–31-4
  water transport through, 29-8–29-12
Stereocilia, **36-6**, *36-7*, *36-8*
  mechanoreception by, 36-9, *36-9*
Steroid hormones, 38-7, **38-7**–38-8, *38-8*
Stigmata, 30-9, **30-10**
Stimuli, 45-2–45-5
  key, 45-2, *45-3*
  supernormal, 45-3, *45-3*
Stinkhorns, 34-17, *34-17*, 34-19
Stomach, **40-14**, *40-15*
  and digestion, 40-16, 40-16–40-17
Stomata, 29-4, **29-4**–29-6, *29-5*
Stramenopiles [Stramenopila], 27-15, **27-15**–27-16, *27-16*, *27-17*
  Visual Synthesis of history of, 44-30
Streptococcus pneumoniae [bacterium], 26-3
  and inflammation studies, 43-5
Streptomyces [bacterium], 26-3
Streptophytes, 27-15
Striated muscles, **37-2**, *37-2*
Strigolactone, 31-8, **31-10**, 32-3
Stroke volume (SV), **39-21**–39-22
Stromatolites, **26-20**, *26-20*
Styles, 30-9, **30-10**
Suberin, **31-12**
Submucosa, **40-21**, *40-21*
Subspecies, **22-7**
Succession, *47-12*, **47-12**–47-13
  Visual Synthesis of, 48-12–48-13
Suction feeding, **40-12**, 40-12–40-13
Suess, Hans, 25-4
Sulci, *36-18*, *36-19*, **36-19**
Sulfur cycle, 26-10, *26-10*
Sun compasses, 45-12–45-13, 45-16
Sunflowers, hybridization in, 22-13–22-14, *22-14*
Superkingdoms, **27-10**
  of Eukarya, 28-4
Supernormal stimuli, **45-3**, *45-3*
Superorganisms, 45-19
Surfactants, **39-8**
Survivorship, **46-10**
Survivorship curves, 46-9–46-11, *46-10*, *46-11*
Suspension filter feeding, **40-12**, *40-12*

SV. *See* Stroke volume (SV)
Swallowing, *40-15*, **40-15**–40-16
Swim bladders, 44-24
Switchgrass, and biofuels, C5-3
Symbionts, **27-5**
Symbiosis, **27-5**–27-6, **47-3**
  and agriculture, 33-10–33-11
  and digestion, 40-20–40-22
  and energy uptake by roots, *29-16*, 29-16–29-18, *29-17*
  and fungi, 34-5, 34-5–34-6
  and lichens, *34-6*, 34-6–34-7, *34-7*
  and microbiomes, C5-2–C5-4
  and plant defenses, *32-11*, 32-11–32-12, *32-12*
Symmetry, radial versus bilateral, 44-2–44-3, *44-3*
Sympathetic division, **35-17**, *35-17*
Sympatric speciation, *22-12*, **22-12**–22-13, *22-13*
Synapomorphies, **23-6**, *23-6*, 27-20
Synapses, 35-4, **35-5**
  neuron communication at, *35-12*, 35-12–35-13
Synaptic clefts, 35-4, **35-5**
Synaptic plasticity, *36-20*, **36-20**–36-21
Synaptic signaling, 38-16, 38-17
Systemic acquired resistance (SAR), **32-5**–32-6, *32-6*
Systole, **39-20**, *39-20*

T

T cell receptors (TCRs), **43-10**–43-11, *43-11*
T cells (T lymphocytes), *43-3*, **43-7**
  activation of, 43-14–43-15, *43-15*
  positive and negative selection of, 43-16, *43-16*
  types of, 43-13–43-14, *43-13t*
T-tubule system, **37-6**, *37-7*
Taiga biome, 47-17, *47-17*
Tannins, **32-11**
Taste, sense of, 36-5–36-6, *36-6*. *See also* Tongues
  receptors for, 36-2
Taste buds, **36-5**–36-6, *36-6*
Taxa, **23-4**
Taxes [taxis], **45-11**
Taxonomy, 23-2, 23-4–23-5, *23-5*. *See also* Phylogenetic trees
  history of, 44-2–44-4
Taylor, C. Richard, 40-6, 40-7
TCA cycle. *See* Citric acid cycle
TCRs. *See* T cell receptors (TCRs)
Tears, 43-2
Technologies. *See* Laboratory techniques

Tectorial membranes, *36-8*, **36-9**
Teeth, 40-13–40-14, *40-14*
Temperate coniferous forest biome, 47-17, *47-17*
Temperate grassland biome, 47-18, *47-18*
Temporal lobes, *36-18*, *36-19*, **36-19**
Temporal separation, **22-5**–22-6
Temporal summation, **35-13**, **35-14**, **36-4**
Temporomandibular joints, **40-13**
Tendons, **37-15**
Tennyson, Alfred, Lord, C7-2, 32-1
Tentacles, 44-13
Terpenes, **32-10**–32-11, *32-10t*
"Test-tube" babies, 42-17
Testes, *38-12*, **38-14**, **42-10**, *42-10*, *42-11*
Testosterone, 38-7, 38-8, *38-9t*, 38-14, **42-13**, *42-13*
  and puberty, 42-13
Tests [cellular], **27-9**
Tests [laboratory]. *See* Laboratory techniques
Tetanus, *37-11*, **37-11**, 43-2
Tetraploidy, **22-14**
Tetrapods [Tetrapoda], 44-24–**44-25**
  Visual Synthesis of history of, 44-31
Thalamus, *36-17*, **36-18**
Thallus, **33-3**
Thaumarchaeota, **26-17**, *26-17*, *26-18*
Thermoreceptors, **36-3**
Thermoregulation
  and circulation, 39-22
  and feedback, 38-5
  and the hypothalamus, 35-18, *35-18*
  and metabolism, 40-5, 40-7
  Visual Synthesis of, 40-8–40-9
Thick filaments, **37-3**, *37-3*, 37-4
Thin filaments, **37-3**, *37-3*, 37-4
Thiomargarita namibiensis [bacterium], 26-3, *26-3*
Thomas, Eugen, 34-6
Thoreau, Henry David, 48-6
Threshold potentials, **35-8**, *35-9*
Through the Looking-Glass [Carroll], 42-5, 48-14
Thyroid glands, *38-12*, **38-14**
Thyroid hormones, *38-9t*, 38-14
Thyroid-stimulating hormone (TSH), *38-9t*, 38-11
Ti plasmids, **32-8**
Ticks, 44-15
Tidal ventilation, **39-6**, *39-6*
Tidal volume, *39-6*, **39-7**
Tight junctions, **40-19**

Tiktaalik roseae [fish-amphibian], 23-17–23-18, *23-18*
Tilman, David, 47-16
Tinbergen, Niko, 45-1–45-2, 45-9–45-10, 45-16
Titin, **37-3**, *37-4*
Tits, blue, culture in, 24-16, *24-17*
TLRs. *See* Toll-like receptors (TLRs)
TMV. *See* Tobacco mosaic virus (TMV)
Toads, *48-19*
Toadstools, **34-17**, *34-17*
Tobacco, coyote [*Nicotiana attenuata*], *32-13*, 32-13–32-14, 32-16
Tobacco mosaic virus (TMV), *32-5*, 32-5–32-6, *32-6*, 32-16
Tolerance, **43-16**
Toll-like receptors (TLRs), **43-4**
Tonegawa, Susumu, 43-11
Tongues, 37-14, **40-15**, *40-15*
Torque, *37-18*, 37-18–37-19
Totipotent cells, 31-1
Touch. *See* Pressure, sense of
Trabeculae, **37-17**
Trace fossils, **23-12**, *23-12*, 28-13
Tracheae [insect], **39-4**, *39-4*, **44-17**
  breathing through, 39-5–39-6
Tracheae [mammalian], **39-7**, *39-7*
  protection of, during swallowing, *40-15*, 40-16
Tracheids, **29-9**, *29-9*, 31-12, *31-13*
Tracheoles, 39-4, *39-6*
Trade-offs, *32-15*, **32-15**–35-16
  and life histories, 46-11–46-12
  in skeletons, 37-19, *37-19*
Traits
  discrete versus continuous, 21-9
  measuring genetic variation using, 21-3–21-4, *21-4*
Transduction [DNA transfer], **26-4**, **26-5**
Transduction [sensory], **36-1**–36-2, *36-3*
Transformation, **26-4**, **26-5**
Translocation [bulk transport], 29-14
Transmembrane proteins, and cell adhesion, 28-6
Transmission genetics, and behavior, 45-6
Transpiration, *29-3*, 29-3–**29-4**
Tree of life, 23-2. *See also* Phylogenetic trees
  on-line version, 23-9

Trees. *See also* Conifers; Eudicots
  acacia [*Acacia cornigera*], *32-11*–*32-12*, *32-12*
  aspen, *30-22*, 30-23
  chocolate, 47-1, *47-3*, *47-3*, 47-6
  drugs from, 33-13
  elm, 32-5
  giant sequoia [*Sequoiadendron giganteum*], 33-13
  loblolly pine [*Pinus taeda*], *30-6*, *30-7*
  pine, *33-12*, 33-13, *33-13*
  red oak [*Quercus rubra*], 33-20
Tricarboxylic acid (TCA) cycle. *See* Citric acid cycle
Trichophyton rubrum [fungus], *34-2*
Trichoplax adhaerens [placozoan], *44-9*, 44-9–44-11
Trilobites, 44-28
Trimesters, **42-18**
Triploblastic animals, **44-4**, *44-4*
Trophic pyramids, **25-13**, *25-13*, *47-13*, **47-13**, *47-15*, **47-15**
Tropic hormones, **38-14**
Tropism, **31-17**, *31-17*
Tropomyosin, **37-3**, *37-3*, **37-6**–37-7, *37-7*
Troponin, **37-6**–37-7, *37-7*
Truffles, **34-15**–34-16
Trypsin, **40-18**
TSH. *See* Thyroid-stimulating hormone (TSH)
Tube feet, **44-19**
Tube worms, 39-4
Tuna, 37-11
Tundra biome, 47-17, *47-17*
Tunicates, **44-19**, *44-20*, 44-21
Turgor pressure, 29-14
Turtles, **44-25**, *44-25*
Twitch contractions, *37-10*, 37-10–**37-11**
Two-fold cost of sex, **42-4**–42-5, *42-5*
Tympanic membranes, **36-8**, *36-8*

U

Ug99 [fungus], 34-19, *34-19*
Ulcers, **40-17**
Urea, **41-5**
  as nitrogenous waste, *41-7*, 41-8
Ureters, **41-11**, *41-11*
Urethra, **42-10**, *42-10*
Uric acid, *41-7*, 41-8
Urine, 41-16–41-19, *41-17*
Uroglena [alga], 28-2
Ustilago maydis [fungus], **34-17**, *34-17*
Uterus, **42-12**, *42-12*
  and childbirth, 42-22

V

Vaccination, **43-10**
Vaccines, C4-4, 43-10, 43-18
Vagina, **42-12**, *42-12*
Variable (V) regions, **43-8**, *43-8*
Vas deferens, **42-10**, *42-10*, *42-11*
Vasa recta, **41-16**, *41-16*
Vascular bundles, **31-9**, *31-9*
Vascular cambium, **31-10**–31-11, *31-11*
   and secondary xylem and phloem, 31-11–31-12, *31-12*
Vascular plants, **29-1**–29-2. *See also* Angiosperms; Gymnosperms
   life cycle of, 30-4–30-5, *30-5*
Vascular systems
   and bulk transport, 28-5, 28-5–28-6
   water, 44-19, *44-19*
Vascular wilt diseases, 32-5, *32-5*
   and fungi, 34-4, *34-4*
Vascularization, 37-16, 37-17
Vasilov, Nicolai, 33-21
Vasopressin, 38-7, 38-9*t*, 38-12, **38-14**, **41-17**
Vegetarianism, 40-10
Vegetative reproduction, 30-22, **30-22**–30-23. *See also* Asexual reproduction
Veins [animals], *39-14*, **39-15**, *39-16*
   pulmonary, 39-19, 39-20
Veins [plants], **29-3**, *29-3*
Venae cavae, **39-16**
Ventilation, **39-3**. *See also* Breathing
   tidal, 39-6, *39-6*
Ventricles, **39-18**, *39-18*
Venules, *39-14*, **39-15**
Vernalization, **30-18**
   and agriculture, 30-22
Vertebrae, **44-21**
Vertebral columns, **44-21**
Vertebrates [Vertebrata], 44-21–44-27
   breathing of, 39-4, 39-6, 39-6–39-7
   versus chordates, 44-21, *44-21*

endocrine systems of, *38-12*, 38-12–38-16
phylogenetic tree for, 23-3, 40-13, 44-22
renal systems of, 41-10–41-12, *41-11*
skeletons of, 37-16–37-19, *37-17*, *37-18*, *37-19*
Vesicles, 27-1–27-2, *27-2*
Vessel elements, **29-9**, *29-9*
Vestibular system, **36-7**, *36-7*–36-8
Vestigial structures, **40-23**
Vestimentiferan worms, 44-12, *44-12*
Vibrio cholera [bacterium], 40-20
Vicariance, **22-7**, 22-8, *22-8*
Villi, **40-19**, *40-19*
Viridoplantae, 27-14. *See also* Algae; Plants
Virulent pathogens, **32-3**
Viruses. *See also* Human immunodeficiency virus (HIV); Influenza virus
   and horizontal gene transfer, 26-5
   infections by, 43-17, *43-17*
   as plant pathogens, 32-2
Viscosity, **39-15**
Vision. *See* Eyes; Sight, sense of
Visual cortex, **36-15**, *36-15*
Vitamins, **40-11**, *40-11t*
Viviparity, **42-8**–**42-9**
Vocal cords, **39-7**, *39-7*
Volcanoes
   and the long-term carbon cycle, 25-7, *25-9*
   and mass extinctions, 23-18
Voles, 45-8, *45-8*
Voluntary nervous system, **35-16**–35-17
Volvox [alga], 27-14, *27-15*, 28-9
Vulva, **42-12**, *42-13*

W

Wallace, Alfred Russel, 21-9
   and diversity, 47-21
   and natural selection, 21-8
   and species, 22-2
Warren, Robin, 40-17

Warthogs, 38-15, *38-15*
Wasps, learning by, 45-9–45-10, *45-10*
Water
   and excretion, 41-7–41-12
   osmoregulation of, 41-1–41-7
Water buffalo, 47-8, 47-9
Water bugs, 44-15
Water lilies, 33-15, *33-15*–33-16
Water vascular systems, **44-19**, *44-19*
Weathering, and the long-term carbon cycle, 25-7–25-8, *25-9*
Weinberg, Wilhelm, 21-6
Welwitschia mirabilis [gnetophyte], 33-14, *33-14*
Went, Fritz, 31-17
Whales
   and commensalism, 47-8
   cooperative hunting by, 45-16–45-17, *45-17*
   diving by, 39-12, *39-12*
   feeding by, 40-12, *40-12*
Wheat
   agriculture of, C6-2–C6-3, *C6-3*, 30-21–30-22, 32-18
   and fungi, 34-19, *34-19*
   and hormones, 31-9
White blood cells, **43-3**, *43-3*
White matter, **36-18**, *36-18*
Wiesel, Torsten, 36-16
Wigglesworth, Vincent, 38-2, 38-3
Wilson, Allan, 21-15, 24-3–24-4, 24-6, 24-10–24-11
Wilson, E. O., 45-15, 45-19, 46-14, 48-16
Wilson, K. O., 42-8
Wind energy, 48-10, *48-10*
Wisdom of the Body [Cannon], *The*, 35-17
Woese, Carl, 26-5
Wolbachia [bacterium], 26-21, *26-21*
Wolves
   and pheromones, 38-17, *38-17*
   as predators, C7-2, *C7-3*
Wood, 31-12–31-13, *31-13*
   decomposition of, 34-3–34-4, *34-4*
   evolution of, 33-16

Worms. *See also* Annelid worms [Annelida]; Earthworms; Flatworms
   acorn, 44-18, *44-18*
   polychaete, 44-12
   round-, *36-3*
   tube, 39-4
   vestimentiferan, 44-12, *44-12*

X

Xanthoria flammea [lichen], 34-7
Xylem, 29-8–**29-9**, *29-9*
   and cavitation, 29-11–29-12, *29-12*
   evaporative pump in, 29-10–29-11, *29-11*
   secondary, 31-11, *31-11*
   Visual Synthesis of, 33-23
   water transport through, 29-9–29-10

Y

Yeasts, **34-2**–34-3
   asexual reproduction of, 42-2, 42-2–42-3
   and fermentation, 26-8, 34-1
   and infections, 34-1
Yolks, **42-8**–42-9, *42-9*
Young, Larry, 45-8

Z

Z discs, **37-3**, *37-4*
Zea mays [corn]
   directional selection of, 21-12, *21-12*
   fungal infections of, 34-17
   genetically engineered, C6-4, 32-18, *32-18*
   increasing yields of, 48-14, 48-15
Ziphius cavirostris [whale], 39-12
Zona pellucida, 42-16, 42-17
Zygomycetes, 34-13, **34-13**–34-14
Zygotes, **27-3**
   formation of, at fertilization, 42-16, 42-17